D0150357

De Gruyter Expositions in Mathematics 9

Editors
Victor P. Maslov, Moscow, Russia
Walter D. Neumann, New York City, New York, USA
Markus J. Pflaum, Boulder, Colorado, USA
Dierk Schleicher, Bremen, Germany
Raymond O. Wells, Bremen, Germany

Marek Jarnicki, Peter Pflug

Invariant Distances and Metrics in Complex Analysis

2nd Extended Edition

De Gruyter

Mathematics Subject Classification 2010: Primary: 32-02; Secondary: 32Hxx, 32Exx.

Authors
Marek Jarnicki
Jagiellonian University
Faculty of Mathematics and Computer Science
Institute of Mathematics
ul. prof. Stanisława Łojasiewicza 6
30-348 Kraków
Poland
marek.jarnicki@im.uj.edu.pl

Peter Pflug
Carl von Ossietzky Universität Oldenburg
Faculty of Mathematics and Science
Institute of Mathematics
Ammerländer Heerstraße 114–118
26129 Oldenburg
Germany
peter.pflug@uni-oldenburg.de

ISBN 978-3-11-025043-5
e-ISBN 978-3-11-025386-3
Set-ISBN 978-3-11-218830-9
ISSN 0938-6572

Library of Congress Cataloging-in-Publication Data

A CIP catalog record for this book has been applied for at the Library of Congress.

Bibliographic information published by the Deutsche Nationalbibliothek

The Deutsche Nationalbibliothek lists this publication in the Deutsche Nationalbibliografie; detailed bibliographic data are available in the Internet at http://dnb.dnb.de.

© 2013 Walter de Gruyter GmbH, Berlin/Boston

Typesetting: Da-TeX Gerd Blumenstein, Leipzig, www.da-tex.de
Printing and binding: Hubert & Co. GmbH & Co. KG, Göttingen
∞ Printed on acid-free paper
Printed in Germany

www.degruyter.com

To Mariola and Rosel

Preface to the second edition

The first edition of this book appeared in 1993, i.e., about twenty years ago. In the meantime, activities in the area of complex analysis have led to answers to many questions we posed at that time. Moreover, many new "invariant" objects have appeared and have so far been successfully used in Several Complex Variables. Therefore, we were happy when De Gruyter asked us to update the first edition. Our initial thought was that we could fix the old structure and then add the new results. But looking back this was an overly optimistic way of approaching the new edition. It turned out that there was a huge amount of new material we had to cover. As a compromise we have tried to keep as much as possible of the old book untouched. Nevertheless, the structure has changed. We hope that the reader will appreciate this new *extended version* as it is now. It covers more than twice the material in the old version. We have to confess that, because of the limited number of pages, we could not include all the topics we wanted in this volume. With a small number of exceptions we restrict our discussion to domains in $\mathbb{C}^n$. Moreover, the discussion of strongly linearly convex domains is not as complete as we wanted. Only recently a complete and detailed proof of the main result appeared (see [322]), i.e., equality of all invariant functions on such domains. We encourage the reader to turn to the original paper and study it carefully. Moreover, many details on the symmetrized polydisc and the spectral ball, as well as a lot of results on estimates for invariant metrics on $\mathbb{C}$-convex domains (not covered by our book), may be found in the interesting booklet "Invariant functions and metrics in complex analysis" by N. Nikolov (see [379]).

Each chapter starts with a brief summary of its contents and continues with a short introduction. It ends with a "List of problems" section that collects all the problems from the chapter. Likewise, problems from all chapters are collected in a "List of problems" appendix. We encourage the reader to work on these problems. We hope there will be a similar progress as there has been after the first version.

Moreover, there are many points in the proofs that we have marked with ■. This means that the reader is encouraged to write out the argument in more detail.

We also have to confess that some part of this edition is based on the article "Invariant distances and metrics in complex analysis – revisited" [267], which may be thought of as a step between the first edition and the current one.

Furthermore, we should point out that, in general, we did not check the details of the results presented in the Miscellanea. This is left for the reader interested in such results.

We like to thank all our colleagues who reported to us about the gaps in this book during its writing. In particular, we thank Dr. P. Zapałowski for all the corrections he made. It would not be possible to reach the current presentation layer of this book without his precise and detailed observations. Nevertheless, according to our experiences with our former books, we are sure that many errors have remained, and we are responsible for not detecting them.

We will be pleased if readers inform us about comments and/or remarks they may have while studying this text – please use one of the following e-mail addresses:

- Marek.Jarnicki@im.uj.edu.pl
- Peter.Pflug@uni-oldenburg.de

Finally, it is our great pleasure to thank the following institutions for their support during the writing of this book:
Jagiellonian University in Kraków,
Carl von Ossietzky Universität Oldenburg,
Polish National Science Center (NCN) – grant UMO-2011/03/B/ST1/04758,
Deutsche Forschungsgemeinschaft – grant 436POL113/103/0-2.
We are deeply indebted to the De Gruyter publishing company for giving us the chance to write this extended second version.

Marek Jarnicki
Kraków – Oldenburg, January 2013 Peter Pflug

Preface to the first edition

One of the most beautiful results in classical complex analysis is the Riemann mapping theorem, which says that, except for the whole complex plane, every simply connected plane domain is biholomorphically equivalent to the unit disc. Thus, the topological property "simply connected" is already sufficient to describe, up to biholomorphisms, a large class of plane domains. On the other hand, the Euclidean ball and the bidisc in $\mathbb{C}^2$ are topologically equivalent simply connected domains, but they are not biholomorphic. This observation, which was made by H. Poincaré as early as the end of the last century, shows that even inside the class of bounded simply connected domains there is no single model (up to biholomorphisms) as is the case in the plane. Therefore, it seems to be important to associate with domains in $\mathbb{C}^n$ tractable objects that are invariant under biholomorphic mappings. Provided that these objects are sufficiently concrete, one can hope to be able to decide, at least in principle, whether two given domains are biholomorphically distinct.

An object of this kind was introduced by, for example, by C. Carathéodory in the thirties. His main idea was to use the set of bounded holomorphic functions as an invariant. More precisely, he defined pseudodistances on domains via a "generalized" Schwarz Lemma. A specific property of these pseudodistances is that holomorphic mappings act as contractions. Thus, in particular, biholomorphic mappings operate as isometries. For such objects the name "invariant pseudodistances" has become very popular. This is where the title of our book comes from, although in the text we prefer to talk about holomorphically contractible pseudodistances. Apart from the class of bounded holomorphic functions, other classes of functions are used to obtain, via extremal problems, new objects that are contractible with respect to certain families of holomorphic mappings. For example, the class of square integrable holomorphic functions was used by S. Bergman. Moreover, all these objects admit infinitesimal versions associating with any "tangent vector" a specific length that is contractible under holomorphic mappings. Besides using families of functions to associate (via an extremal problem) tractable objects with domains in $\mathbb{C}^n$, one can consider sets of analytic discs as new biholomorphic invariants. This idea is due to S. Kobayashi.

The main goal of our book is to present a systematic study of invariant pseudodistances and their infinitesimal counterparts, the invariant pseudometrics. To illustrate various aspects of the theory, we add a lot of concrete examples and applications. Although we have tried to make the book as complete as possible, the choice of topics we present obviously reflects our personal preferences.

Our interest in this area started in the middle of the eighties when we, somehow accidentally, came across the "Schwarz Lemma on Cartesian products" (in the terminology of the book, the "product property of the Carathéodory distance"). This result was stated in the 1976 survey article *Intrinsic distances, measures and geometric function theory* of S. Kobayashi but no proof was given there (in fact, as it turned out later, no proof did exist at that time). In our attempts to find a proof of this theorem, we have gone deeper into the field of invariant distances. For instance, we have learned that a lot of seemingly simple questions were still waiting for solutions. We were able to solve some of them but most remain still without answer. We have put many of these problems into the text (marking them by ?). The reader is encouraged to work on some of them.

According to our experience over the last ten years, we feel that we should refrain from discussing manifolds and complex spaces. So we only deal with domains in $\mathbb{C}^n$. Even here, of course, plenty of results are beyond the scope of our book. For the convenience of the reader who would like to go further, we collect (without proofs) part of this material in a supplementary chapter (Miscellanea). We mention that, although many of the results in the book are stated in the domain case, they can be almost literally transformed to the manifold case; see, for instance, [3, 137, 178, 312, 335, 398].

During the preparation of this book, we had to decide what kind of knowledge the reader is supposed to have. We have assumed that he is familiar with standard complex analysis of several variables. Of course, what we mean by "standard" reflects our academic education. As a form of a compromise, we have added an appendix in which we collect results we assume to be known (or which are not easy to find in the literature). Moreover, chapters conclude with rough notes and some exercises.

In the text, we often use certain standard symbols and notation without explicit definitions and the reader is referred to the "list of symbols" at the end of the book. Moreover, abbreviations HF, PSH, PSC, AUT, GR, MA, and H refer to the sections of the Appendix.[1]

It is our deep pleasure to be able to state a debt of gratitude to our teachers: Professors Hans Grauert and Józef Siciak, who have led our first steps in complex analysis. Next, we would like to thank our colleagues for stimulating discussions and help during writing this book. We especially want to thank M. Capiński, H.-J. Reiffen, R. Zeinstra, and W. Zwonek, who also helped us with corrections of the text. We express our gratitude to Mrs. H. Böske, who spent a lot of time typing and retyping our notes. We thank both our universities for support before and during the preparation of the book. Finally, we are deeply indebted to the Walter de Gruyter Publishing Company, especially to Dr. M. Karbe, for having encouraged us to write this book.

Marek Jarnicki
Kraków – Vechta, December 1992 Peter Pflug

[1] Notice that in the second edition of the book the references to the Appendix are organized in a different way.

Contents

Chapter 1

Hyperbolic geometry of the unit disc

Summary. We present the hyperbolic geometry of the unit disc $\mathbb{D}$ (§ 1.1) and give some applications, e.g., Picard's theorem (§ 1.2).

Introduction. The main concept of the theory that will be developed in this book is to describe the holomorphic structure of a given domain $G \subset \mathbb{C}^n$ in terms of geometric properties of the space (G, d_G), where d_G is a suitable pseudodistance on G. We will see that there are various systems $(G, d_G)_G$ that may be useful for this purpose. We will always assume that any such system is holomorphically coherent in the sense that, for two arbitrary domains $G \subset \mathbb{C}^n$, $D \subset \mathbb{C}^m$, any holomorphic mapping $F : G \longrightarrow D$ is a *contraction* of (G, d_G) into (D, d_D), i.e.,

$$d_D(F(z'), F(z'')) \le d_G(z', z''), \quad z', z'' \in G. \tag{$*$}$$

In particular, any biholomorphic mapping $F : G \longrightarrow D$ has to be an *isometry* between (G, d_G) and (D, d_D).

Obviously, if the system $(G, d_G)_G$ is too poor we cannot expect any essential influence of geometric properties of (G, d_G) on holomorphic properties of G. Therefore, one must exclude these trivial systems. We can reach this goal by different methods; the simplest way seems to be the following: assume that $d_\mathbb{D}$ is "very good" (from holomorphic and geometric points of view), where $\mathbb{D}$ denotes the open unit disc in $\mathbb{C}$. Then, we can hope that, in view of $(*)$, at least some of the "very good" properties of $d_\mathbb{D}$ propagate via holomorphic mappings to almost all (G, d_G). The aim of this chapter is to explain what we mean by "very good" properties of $d_\mathbb{D}$.

1.1 Hyperbolic geometry of the unit disc

Let

$$\boldsymbol{m}(\lambda', \lambda'') := \left| \frac{\lambda' - \lambda''}{1 - \lambda' \bar{\lambda}''} \right|, \quad \lambda', \lambda'' \in \mathbb{D},$$

$$\gamma(\lambda) := \frac{1}{1 - |\lambda|^2}, \quad \lambda \in \mathbb{D}.$$

Note that the definition of $\boldsymbol{m}$ may be extended to $\mathbb{C} \times \mathbb{C} \setminus \{(\lambda', \lambda'') : \lambda' \bar{\lambda}'' = 1\}$.

Using the above notation, the Schwarz–Pick lemma may be formulated as follows (cf. [461]):

Lemma 1.1.1 (Schwarz–Pick lemma). *Let* $f \in \mathcal{O}(\mathbb{D}, \mathbb{D})$. *Then:*

(a) $\boldsymbol{m}(f(\lambda'), f(\lambda'')) \le \boldsymbol{m}(\lambda', \lambda'')$, $\lambda', \lambda'' \in \mathbb{D}$.

(b) $\boldsymbol{\gamma}(f(\lambda))|f'(\lambda)| \le \boldsymbol{\gamma}(\lambda)$, $\lambda \in \mathbb{D}$.

(c) *The following statements are equivalent:*

   (i) $f \in \operatorname{Aut}(\mathbb{D})$;

   (ii) $\boldsymbol{m}(f(\lambda'), f(\lambda'')) = \boldsymbol{m}(\lambda', \lambda'')$, $\lambda', \lambda'' \in \mathbb{D}$;

   (iii) $\boldsymbol{m}(f(\lambda_0'), f(\lambda_0'')) = \boldsymbol{m}(\lambda_0', \lambda_0'')$ *for some* $\lambda_0', \lambda_0'' \in \mathbb{D}$ *with* $\lambda_0' \ne \lambda_0''$;

   (iv) $\boldsymbol{\gamma}(f(\lambda))|f'(\lambda)| = \boldsymbol{\gamma}(\lambda)$, $\lambda \in \mathbb{D}$;

   (v) $\boldsymbol{\gamma}(f(\lambda_0))|f'(\lambda_0)| = \boldsymbol{\gamma}(\lambda_0)$ *for some* $\lambda_0 \in \mathbb{D}$.

Roughly speaking, any holomorphic function $f : \mathbb{D} \longrightarrow \mathbb{D}$ is an $\boldsymbol{m}$- and a $\boldsymbol{\gamma}$-contraction. Moreover, the only holomorphic $\boldsymbol{m}$- or $\boldsymbol{\gamma}$-isometries are the automorphisms of $\mathbb{D}$.

Before we continue to discuss the above objects, we mention a form of a Schwarz–Pick lemma for higher order derivatives (see [355, 118]).

Proposition 1.1.2 (Higher order Schwarz–Pick lemma). *Let* $f \in \mathcal{O}(\mathbb{D}, \mathbb{D})$ *and* $k \in \mathbb{N}$. *Then*

$$\frac{|f^{(k)}(\lambda)|}{1 - |f(\lambda)|^2} \le k!\,(1 + |\lambda|)^{k-1} \frac{1}{(1 - |\lambda|^2)^k}, \quad \lambda \in \mathbb{D}.$$

The proof relies on the following simple lemma.

Lemma 1.1.3. *Let* $\varphi \in \mathcal{O}(\mathbb{D}, \mathbb{D})$ *and let* $\varphi(z) = \sum_{j=0}^{\infty} a_j z^j$ *be its power series expansion. Then* $|a_k| \le 1 - |a_0|^2$, $k \in \mathbb{N}$.

Proof. Fix a $k \in \mathbb{N}$ and put $\omega_j := e^{\frac{2\pi i}{k} j}$, $j = 1, \dots, k$. Recall that $\sum_{j=1}^{k} \omega_j^s = 0$, $1 \le s < k$.

Put $\widetilde{\varphi}(z) := \frac{1}{k} \sum_{j=1}^{k} \varphi(\omega_j z)$, $z \in \mathbb{D}$. Obviously, $\widetilde{\varphi} \in \mathcal{O}(\mathbb{D}, \mathbb{D})$, and its power series expansion is given by

$$\widetilde{\varphi}(z) = a_0 + a_k z^k + a_{2k} z^{2k} + \dots, \quad z \in \mathbb{D}.$$

Set $g := \frac{\widetilde{\varphi} - a_0}{1 - \bar{a}_0 \widetilde{\varphi}}$. Then, $g \in \mathcal{O}(\mathbb{D}, \mathbb{D})$ and its power series expansion is given by $g(z) = b_k z^k + \dots$ with $b_k = \frac{a_k}{1 - |a_0|^2}$. Using the Cauchy inequality for the coefficient b_k finally gives the desired inequality. □

Proof of Proposition 1.1.2. Fix a point $\lambda \in \mathbb{D}$ and put

$$\varphi_\lambda(z) := f\left(\frac{z+\lambda}{1+\bar{\lambda}z}\right) = \sum_{j=0}^{\infty} c_j(\lambda)z^j, \quad z \in \mathbb{D}.$$

Then,

$$f(z) = \varphi_\lambda\left(\frac{z-\lambda}{1-\bar{\lambda}z}\right) = \sum_{j=1}^{\infty} c_j(\lambda)\left(\frac{z-\lambda}{1-\bar{\lambda}z}\right)^j, \quad z \in \mathbb{D}.$$

Taking the k-th derivative of f in the point λ we get, after some simple calculations,

$$f^{(k)}(\lambda) = \sum_{j=1}^{k} c_j(\lambda)\frac{\lambda^{k-j}}{(1-|\lambda|^2)^k}\frac{k!(k-1)!}{(k-j)!(j-1)!}.$$

Recall that $c_0(\lambda) = f(\lambda)$ and $|c_j(\lambda)| \le 1-|c_0(\lambda)|^2 = 1-|f(\lambda)|^2$ if $j \in \mathbb{N}$. Hence,

$$\begin{aligned}|f^{(k)}(\lambda)| &\le \frac{k!(1-|f(\lambda)|^2)}{(1-|\lambda|^2)^k}\sum_{j=1}^{k}\frac{(k-1)!}{(k-j)!(j-1)!}|\lambda|^{k-j} \\ &= k!\frac{1-|f(\lambda)|^2}{(1-|\lambda|^2)^k}\sum_{m=0}^{k-1}\frac{(k-1)!}{m!(k-m-1)!}|\lambda|^m \\ &= k!\frac{1-|f(\lambda)|^2}{(1-|\lambda|^2)^k}(1+|\lambda|)^{k-1}.\end{aligned}$$

□

It seems it is still an open problem to determine the best estimate in Proposition 1.1.2.

Now, we list the elementary properties of the functions $\boldsymbol{m}$ and $\boldsymbol{\gamma}$ that will be useful in the following:

1.1.4. $\boldsymbol{m} \in \mathcal{C}^\infty(\mathbb{D}\times\mathbb{D}\setminus\{(\lambda,\lambda) : \lambda \in \mathbb{D}\})$, $\boldsymbol{m}^2 \in \mathcal{C}^\infty(\mathbb{D}\times\mathbb{D})$, $\boldsymbol{\gamma} \in \mathcal{C}^\infty(\mathbb{D})$.

1.1.5. *For any* $a \in \mathbb{D}$, $\boldsymbol{m}(\cdot,a) = |\boldsymbol{h}_a|$, *where*

$$\boldsymbol{h}_a(\lambda) := \frac{\lambda-a}{1-\bar{a}\lambda}, \quad \lambda \in \mathbb{C},\ \lambda \ne 1/\bar{a};$$

cf. Appendix B.3.1. *In particular,* $\boldsymbol{m}(\cdot,a) = 1$ *on* $\mathbb{T}$, $\log \boldsymbol{m}(\cdot,a)$ *is subharmonic on* $\mathbb{D}$ *and harmonic in* $\mathbb{D}\setminus\{a\}$. *Since* $\boldsymbol{m}$ *is symmetric, the same is true for* $\boldsymbol{m}(a,\cdot)$*; see also Remark* 1.1.22.

The function $\log\boldsymbol{\gamma}$ *is subharmonic on* $\mathbb{D}$.

1.1.6. $\displaystyle\lim_{\substack{\lambda',\lambda''\to a\\ \lambda'\neq\lambda''}} \frac{\boldsymbol{m}(\lambda',\lambda'')}{|\lambda'-\lambda''|} = \boldsymbol{\gamma}(a), \quad a \in \mathbb{D}.$

1.1.7. *The function* $-\log \boldsymbol{m}(a,\cdot)$ *is the classical Green function for* $\mathbb{D}$ *with pole at* a*; cf. Appendix* B.5. *If we put* $u := \boldsymbol{m}^2(a,\cdot)$*, then*

$$\boldsymbol{\gamma}^2(a) = \frac{1}{4}(\Delta u)(a) = (\mathcal{L}u)(a;1),$$

where $\mathcal{L}u$ *is the* Levi form *of* u*; cf. Appendix* B.4.

Lemma 1.1.8. *For any* $a, b, c \in \mathbb{D}$, $a \neq b \neq c \neq a$, *we have*

$$\boldsymbol{m}(a,b) < \boldsymbol{m}(a,c) + \boldsymbol{m}(c,b). \tag{1.1.1}$$

In particular, $\boldsymbol{m} : \mathbb{D} \times \mathbb{D} \longrightarrow [0,1)$ *is a distance.*

The function $\boldsymbol{m}$ is called the *Möbius distance*.

Proof. Observe that for any $a, b \in \mathbb{D}$, $a \neq b$, there exists a unique automorphism $h = \boldsymbol{h}_{a,b} \in \operatorname{Aut}(\mathbb{D})$ such that $h(a) = 0$ and $h(b) \in (0,1)$. The function $\boldsymbol{m}$ is invariant under $\operatorname{Aut}(\mathbb{D})$, and therefore, without loss of generality, we may assume that $a = 0$, $b \in (0,1)$. Then, the inequality (1.1.1) reduces to the following one:

$$b < |c| + \left|\frac{c-b}{1-cb}\right|, \quad c \in \mathbb{D} \setminus \{0,b\}. \qquad \square$$

Remark 1.1.9. Since $\boldsymbol{m}$ is invariant under $\operatorname{Aut}(\mathbb{D})$, $B_{\boldsymbol{m}}(a,r) = \boldsymbol{h}_{-a}(\mathbb{D}(r))$, $a \in \mathbb{D}$, $0 < r < 1$, where $B_{\boldsymbol{m}}$ denotes the ball with respect to the distance $\boldsymbol{m}$; cf. Exercise 1.3.2. In particular:

- the topology generated by $\boldsymbol{m}$ coincides with the Euclidean topology of $\mathbb{D}$,
- the space $(\mathbb{D}, \boldsymbol{m})$ is complete.

Remark 1.1.10. The strict triangle inequality (1.1.1) says that the $\boldsymbol{m}$*-segment*

$$[a,b]_{\boldsymbol{m}} := \{\lambda \in \mathbb{D} : \boldsymbol{m}(a,\lambda) + \boldsymbol{m}(\lambda,b) = \boldsymbol{m}(a,b)\}$$

consists only of the ends. Thus, from the geometric point of view, the space $(\mathbb{D}, \boldsymbol{m})$ is trivial.

Therefore, we have to look for a new candidate to be a "good" distance. Property 1.1.6 suggests the following way, which has its roots in differential geometry.

Let $\alpha : [0,1] \longrightarrow \mathbb{D}$ be a piecewise $\mathcal{C}^1$-curve. We define its $\boldsymbol{\gamma}$*-length* by the formula

$$L_{\boldsymbol{\gamma}}(\alpha) := \int_0^1 \boldsymbol{\gamma}(\alpha(t))|\alpha'(t)|dt.$$

Remark 1.1.11. In each case where we assign an object (like $L_{\boldsymbol{\gamma}}(\alpha)$ above) to a (continuous) curve $\alpha : [a,b] \longrightarrow \mathbb{C}^n$, the reader should verify whether the definition of this object is independent of the following standard identifications:

- change of parametrization: $\alpha \simeq \alpha \circ \varphi$, where $\varphi : [a',b'] \longrightarrow [a,b]$ is an increasing bijection which is assumed to be of the same class as α, e.g. piecewise $\mathcal{C}^1$ if α is piecewise $\mathcal{C}^1$,
- cancellation of constant parts: if $\alpha = \text{const}$ on $[t_1,t_2] \subsetneq [a,b]$, then $\alpha \simeq \alpha|_{[a,t_1]} \cup \alpha|_{[t_2,b]}$.

Notice that in most cases the objects associated with curves will also be independent of the orientation of α.

Remark 1.1.12. For any $f \in \mathcal{O}(\mathbb{D},\mathbb{D})$ we have $L_{\boldsymbol{\gamma}}(f \circ \alpha) \le L_{\boldsymbol{\gamma}}(\alpha)$. In particular, the $\boldsymbol{\gamma}$-length is invariant under $\operatorname{Aut}(\mathbb{D})$.

Define

$$\begin{aligned}\boldsymbol{p}(\lambda',\lambda'') := \inf\{L_{\boldsymbol{\gamma}}(\alpha) : \alpha : [0,1] \longrightarrow \mathbb{D},\\ \alpha \text{ is a piecewise } \mathcal{C}^1\text{-curve, } \lambda' = \alpha(0), \lambda'' = \alpha(1)\}, \quad \lambda',\lambda'' \in \mathbb{D}.\end{aligned}$$

Remark 1.1.13. It is clear that $\boldsymbol{p} : \mathbb{D} \times \mathbb{D} \longrightarrow \mathbb{R}_+$ is a pseudodistance dominating the Euclidean distance and that for any holomorphic function $f : \mathbb{D} \longrightarrow \mathbb{D}$ we have

$$\boldsymbol{p}(f(\lambda'), f(\lambda'')) \le \boldsymbol{p}(\lambda',\lambda''), \quad \lambda',\lambda'' \in \mathbb{D}.$$

In particular, $\boldsymbol{p}$ is invariant under $\operatorname{Aut}(\mathbb{D})$.

It is natural to ask whether for given $a, b \in \mathbb{D}$, $a \ne b$, there exists a $\mathcal{C}^1$-curve α joining a and b in $\mathbb{D}$ for which $\boldsymbol{p}(a,b) = L_{\boldsymbol{\gamma}}(\alpha)$; in differential geometry such a curve is called a geodesic. If the answer is positive, then the next problem is to decide whether geodesics are uniquely determined up to the above identifications.

For $0 < s < 1$ let $\alpha_s(t) := ts$, $0 \le t \le 1$, i.e., α_s denotes the interval $[0,s]$ regarded as a curve. For $a, b \in \mathbb{D}$, $a \ne b$, let $\alpha_{a,b} := h^{-1} \circ \alpha_{h(b)}$, where $h = \boldsymbol{h}_{a,b}$ is the automorphism defined in the proof of Lemma 1.1.8. Note that the image $I_{a,b}$ of the curve $\alpha_{a,b}$ lies on the unique circle $C_{a,b}$ that passes through a and b and is orthogonal to $\mathbb{T}$.

Lemma 1.1.14. *For any $a, b \in \mathbb{D}$, $a \ne b$, we have*

$$\boldsymbol{p}(a,b) = L_{\boldsymbol{\gamma}}(\alpha_{a,b}) = \tanh^{-1}(\boldsymbol{m}(a,b)). \tag{1.1.2}$$

Moreover, $\alpha_{a,b}$ is the unique geodesic joining a and b.

Recall that $\tanh^{-1}(t) = \frac{1}{2}\log\frac{1+t}{1-t}$ and $\left(\tanh^{-1}\right)'(t) = \frac{1}{1-t^2}$, $0 \le t < 1$.

Proof. All objects involved in (1.1.2) are invariant under $\mathrm{Aut}(\mathbb{D})$ so we may assume that $a = 0$, $b \in (0,1)$, and consequently $\alpha_{a,b} = \alpha_b$. First, observe that

$$\boldsymbol{p}(0,b) \le L_{\boldsymbol{\gamma}}(\alpha_b) = \int_0^b \frac{dt}{1-t^2} = \frac{1}{2}\log\frac{1+b}{1-b} = \tanh^{-1}(\boldsymbol{m}(0,b)).$$

On the other hand, if $\alpha = u + iv : [0,1] \longrightarrow \mathbb{D}$ is a piecewise $\mathcal{C}^1$-curve joining 0 and b, then

$$L_{\boldsymbol{\gamma}}(\alpha) \ge \int_0^1 \frac{u'(t)}{1-u^2(t)}dt = \frac{1}{2}\log\frac{1+b}{1-b}. \tag{1.1.3}$$

Thus, (1.1.2) is satisfied and, moreover, if $\boldsymbol{p}(0,b) = L_{\boldsymbol{\gamma}}(\alpha)$, then we have equality in (1.1.3). This implies that $v \equiv 0$, $u : [0,1] \longrightarrow [0,b]$, and u is increasing (remember the identifications of Remark 1.1.11). Finally, $\alpha \simeq \alpha_b$. □

Corollary 1.1.15.

(a) *$\boldsymbol{p}$ is a distance with $\boldsymbol{m} \le \boldsymbol{p}$.*

(b) *For any $f \in \mathcal{O}(\mathbb{D},\mathbb{D})$ if $\boldsymbol{p}(f(\lambda_0'), f(\lambda_0'')) = \boldsymbol{p}(\lambda_0', \lambda_0'')$ for some $\lambda_0', \lambda_0'' \in \mathbb{D}$, $\lambda_0' \ne \lambda_0''$, then $f \in \mathrm{Aut}(\mathbb{D})$.*

(c) *$B_{\boldsymbol{p}}(a,r) = B_{\boldsymbol{m}}(a, \tanh(r))$, $a \in \mathbb{D}$, $r > 0$. In particular,*

- *the topology generated by $\boldsymbol{p}$ coincides with the standard topology of $\mathbb{D}$,*
- *$(\mathbb{D}, \boldsymbol{p})$ is complete.*

(d) $$\lim_{\substack{\lambda',\lambda'' \to a \\ \lambda' \ne \lambda''}} \frac{\boldsymbol{p}(\lambda',\lambda'')}{|\lambda' - \lambda''|} = \boldsymbol{\gamma}(a), \quad a \in \mathbb{D}.$$

(e) *$[a,b]_{\boldsymbol{p}} = I_{a,b}$, i.e., the $\boldsymbol{p}$-segments coincide with the images of geodesics. In particular, $\boldsymbol{p}(0,s) = \boldsymbol{p}(0,t) + \boldsymbol{p}(t,s)$, $0 \le t \le s < 1$.*

The distance $\boldsymbol{p}$ is called the *Poincaré (hyperbolic) distance*. Note that, in view of (e), $(\mathbb{D}, \boldsymbol{p})$ is a model of a non-Euclidean geometry (the Poincaré model).

We are going to justify that $\boldsymbol{p}$ is just an example of a "very good" distance we are looking for. We have already seen that from holomorphic, topological, and geometrical points of view the distance $\boldsymbol{p}$ behaves very regularly. We now like to point out other useful properties of $\boldsymbol{p}$.

Let $\alpha : [0,1] \longrightarrow \mathbb{D}$ be a (continuous) curve. Put

$$L_{\boldsymbol{p}}(\alpha) := \sup\left\{\sum_{j=1}^{N} \boldsymbol{p}(\alpha(t_{j-1}), \alpha(t_j)) : N \in \mathbb{N}, 0 = t_0 < \dots < t_N = 1\right\}.$$

The number $L_{\boldsymbol{p}}(\alpha) \in [0, +\infty]$ is called the *$\boldsymbol{p}$-length* of α. If $L_{\boldsymbol{p}}(\alpha) < +\infty$, then we say that α is *$\boldsymbol{p}$-rectifiable*. Note that $L_{\boldsymbol{p}}(\alpha) \geq \boldsymbol{p}(\alpha(0), \alpha(1))$.

Remark 1.1.16.

(a) For any $f \in \mathcal{O}(\mathbb{D}, \mathbb{D})$ we have $L_{\boldsymbol{p}}(f \circ \alpha) \leq L_{\boldsymbol{p}}(\alpha)$. In particular, $L_{\boldsymbol{p}}$ is invariant under $\operatorname{Aut}(\mathbb{D})$.

(b) By Corollary 1.1.15(e) we get $L_{\boldsymbol{p}}(\alpha_{a,b}) = \boldsymbol{p}(a, b)$.

Corollary 1.1.17. *We have $\boldsymbol{p} = \boldsymbol{p}^i$, where*

$$\boldsymbol{p}^i(a, b) := \inf\{L_{\boldsymbol{p}}(\alpha) : \alpha : [0, 1] \longrightarrow \mathbb{D},\ \alpha \textit{ is a curve joining } a \textit{ and } b\}, \quad a, b \in \mathbb{D}.$$

The above corollary shows that $\boldsymbol{p}$ is an *inner* distance.

It is clear that we can repeat the same procedure for the distance $\boldsymbol{m}$: first, we define $L_{\boldsymbol{m}}(\alpha)$ (observe that Remark 1.1.16(a) remains true). Next, we put

$$\boldsymbol{m}^i(a, b) := \inf\{L_{\boldsymbol{m}}(\alpha) : \alpha : [0, 1] \longrightarrow \mathbb{D}, \alpha \text{ is a curve joining } a \text{ and } b\}, \quad a, b \in \mathbb{D}.$$

Lemma 1.1.18.

(a) *For any curve $\alpha : [0, 1] \longrightarrow \mathbb{D}$ we have $L_{\boldsymbol{m}}(\alpha) = L_{\boldsymbol{p}}(\alpha)$. In particular,*

$$\boldsymbol{m}^i = \boldsymbol{p}. \tag{1.1.4}$$

Moreover, α is $\boldsymbol{m}$- or $\boldsymbol{p}$-rectifiable iff α is rectifiable in the Euclidean sense.

(b) *For any piecewise $\mathcal{C}^1$-curve $\alpha : [0, 1] \longrightarrow \mathbb{D}$ we have*

$$L_{\boldsymbol{p}}(\alpha) = L_{\boldsymbol{\gamma}}(\alpha). \tag{1.1.5}$$

Notice that (1.1.4) may be used as an alternative way to define $\boldsymbol{p}$. Moreover, condition (1.1.4) shows that $\boldsymbol{m}$ is not an inner distance.

Proof. (a) First, observe that for any compact $K \subset \mathbb{D}$ there exists an $M > 0$ such that

$$\frac{1}{M}|\lambda' - \lambda''| \leq \boldsymbol{m}(\lambda', \lambda'') \leq \boldsymbol{p}(\lambda', \lambda'') \leq M|\lambda' - \lambda''|, \quad \lambda', \lambda'' \in K.$$

Hence, for any curve $\alpha : [0, 1] \longrightarrow K$ one gets

$$\frac{1}{M} L_{\|\,\|}(\alpha) \leq L_{\boldsymbol{m}}(\alpha) \leq L_{\boldsymbol{p}}(\alpha) \leq M L_{\|\,\|}(\alpha),$$

where $L_{\| \|}(\alpha)$ denotes the length of α in the Euclidean sense. Consequently, all the notions of rectifiability coincide.

By Property 1.1.6 and Corollary 1.1.15(d), for any compact $K \subset \mathbb{D}$ and for any $\varepsilon > 0$ there exists a $\delta > 0$ such that

$$0 \leq \boldsymbol{p}(\lambda', \lambda'') - \boldsymbol{m}(\lambda', \lambda'') \leq \varepsilon|\lambda' - \lambda''|, \quad \lambda', \lambda'' \in K, \ |\lambda' - \lambda''| \leq \delta,$$

which directly implies that $L_{\boldsymbol{m}}(\alpha) = L_{\boldsymbol{p}}(\alpha)$; see also Lemma 2.7.3(a).

(b) Without loss of generality, we may assume that α is of class $\mathcal{C}^1$. By the same argument as above (Corollary 1.1.15(d)), for any $\varepsilon > 0$ there exists an $\eta > 0$ such that

$$\left| \frac{\boldsymbol{p}(\alpha(t'), \alpha(t''))}{|t' - t''|} - \boldsymbol{\gamma}(\alpha(t'))|\alpha'(t')| \right| \leq \varepsilon, \quad 0 \leq t', t'' \leq 1, \ |t' - t''| \leq \eta,$$

which gives (1.1.5); see also Lemma 2.7.3(c). □

The Poincaré (and Möbius) distance also has the following geometric property: first recall that for $\mu \in \mathbb{R}_*$ we have

$$a^\mu := \begin{cases} \{e^{\mu w} : w \in \mathbb{C}, \ e^w = a\}, & \text{if } a \neq 0 \\ \{0\}, & \text{if } a = 0 \end{cases}.$$

Proposition 1.1.19. *Let $a, b \in \mathbb{D}$, $\mu \geq 1$, and let $\widetilde{a}_0 \in a^\mu$ be fixed. Then*

$$\inf\{\boldsymbol{p}(\widetilde{a}_0, \widetilde{b}) : \widetilde{b} \in b^\mu\} \leq \boldsymbol{p}(a, b).$$

Moreover,

$$\boldsymbol{p}(a^\mu, b^\mu) \leq \boldsymbol{p}(a, b), \quad a, b \in [0, 1), \ \mu \geq 1,$$

where $a^\mu, b^\mu \in [0, 1)$ are taken in the standard sense.

Proof. We may assume that $a, b \in \mathbb{D}_*$.

Observe that

$$f_t(\mu) := \frac{\mu t^{\mu - 1}}{1 - t^{2\mu}} \leq \frac{1}{1 - t^2} = f_t(1), \quad t \in (0, 1), \ \mu \geq 1. \tag{1.1.6}$$

Indeed, it suffices to show that $f_t'(\mu) \leq 0$. We have

$$f_t'(\mu) = \frac{t^{\mu - 1}}{(1 - t^{2\mu})^2} \left(1 - t^{2\mu} + (1 + t^{2\mu}) \log t^\mu\right).$$

Thus, it suffices to show that

$$g(x) := \log x + \frac{1 - x^2}{1 + x^2} \leq 0 = g(1), \quad x \in (0, 1],$$

which follows immediately from the fact that

$$g'(x) = \frac{\left(1 - x^2\right)^2}{x\left(1 + x^2\right)^2} \geq 0, \quad x \in (0, 1].$$

Let $\sigma = \alpha_{a,b} : [0, 1] \longrightarrow \mathbb{D}$ be the $\boldsymbol{p}$-geodesic joining a and b (cf. Lemma 1.1.14). First consider the case where $0 \notin \alpha_{a,b}([0, 1])$. Let $U \subset \mathbb{D}_*$ be an open, simply connected neighborhood of $\sigma([0, 1])$ and let $\ell : U \longrightarrow \mathbb{C}$ be the holomorphic branch of the logarithm with $e^{\mu\ell(a)} = \widetilde{a}_0$. Define $\widetilde{\sigma}(t) := \exp(\mu\ell(\sigma(t)))$, $t \in [0, 1]$. Then, $\widetilde{\sigma} : [0, 1] \longrightarrow \mathbb{D}_*$, $\widetilde{\sigma}(0) = \widetilde{a}_0$, and $\widetilde{b}_0 := \widetilde{\sigma}(1) \in b^\mu$. Consequently, using (1.1.6), we get

$$\begin{aligned}
\boldsymbol{p}\left(\widetilde{a}_0, \widetilde{b}_0\right) &\leq \int_0^1 \boldsymbol{\gamma}\left(\widetilde{\sigma}(t)\right) |\widetilde{\sigma}'(t)| dt \\
&= \int_0^1 \boldsymbol{\gamma}\left(|\sigma(t)|^\mu\right) \mu |\sigma(t)|^{\mu-1} |\sigma'(t)| dt \\
&\leq \int_0^1 \boldsymbol{\gamma}\left(\sigma(t)\right) |\sigma'(t)| dt = \boldsymbol{p}(a, b).
\end{aligned}$$

Observe that if $a, b \in (0, 1)$ and $\widetilde{a}_0 = a^\mu$ in the standard sense, then we may take $\ell = \mathrm{Log} =$ the principal branch of the logarithm, which gives $\widetilde{b}_0 = b^\mu$ in the standard sense.

It remains to consider the case where $0 \in \alpha_{a,b}([0, 1])$. Take a point $c \in \mathbb{D}_*$ near b such that $0 \notin \alpha_{a,c}([0, 1])$. The above part of the proof shows that

$$\inf\{\boldsymbol{p}(\widetilde{a}_0, \widetilde{c}) : \widetilde{c} \in c^\mu\} \leq \boldsymbol{p}(a, c), \quad \inf\{\boldsymbol{p}(\widetilde{c}, \widetilde{b}) : \widetilde{b} \in b^\mu\} \leq \boldsymbol{p}(c, b), \quad \widetilde{c} \in c^\mu.$$

Finally,

$$\begin{aligned}
\inf\{\boldsymbol{p}(\widetilde{a}_0, \widetilde{b}) : \widetilde{b} \in b^\mu\} &\leq \inf\{\inf\{\boldsymbol{p}(\widetilde{a}_0, \widetilde{c}) + \boldsymbol{p}(\widetilde{c}, \widetilde{b}) : \widetilde{b} \in b^\mu\} : \widetilde{c} \in c^\mu\} \\
&\leq \inf\{\boldsymbol{p}(\widetilde{a}_0, \widetilde{c}) + \boldsymbol{p}(c, b) : \widetilde{c} \in c^\mu\} \\
&\leq \boldsymbol{p}(a, c) + \boldsymbol{p}(c, b).
\end{aligned}$$

Letting $c \longrightarrow b$ gives the required result. □

One may also ask how close the Poincaré geometry is to the holomorphic one, that is, what are the relations between the set $\mathrm{Isom}(\boldsymbol{p})$ of all $\boldsymbol{p}$-*isometries* of $\mathbb{D}$ and the group $\mathrm{Aut}(\mathbb{D})$. Observe that $\mathrm{Isom}(\boldsymbol{p}) = \mathrm{Isom}(\boldsymbol{m})$. We can also study the set $\mathrm{Isom}(\boldsymbol{\gamma})$ of all $\boldsymbol{\gamma}$-*isometries* of $\mathbb{D}$, i.e., the set of all $\mathcal{C}^1$-mappings $f : \mathbb{D} \longrightarrow \mathbb{D}$ such that

$$\boldsymbol{\gamma}(f(\lambda))|(d_\lambda f)(X)| = \boldsymbol{\gamma}(\lambda)|X|, \quad \lambda \in \mathbb{D}, X \in \mathbb{C},$$

where $d_\lambda f : \mathbb{C} \longrightarrow \mathbb{C}$ denotes the $\mathbb{R}$-differential of f at λ.

The full answer to this question is given by the following:

Proposition 1.1.20. *For any mapping $f : \mathbb{D} \longrightarrow \mathbb{D}$ the following conditions are equivalent:*

(i) $f \in \operatorname{Isom}(\boldsymbol{p})$,

(ii) $f \in \mathcal{C}^1$ *and* $f \in \operatorname{Isom}(\boldsymbol{\gamma})$,

(iii) *either* $f \in \operatorname{Aut}(\mathbb{D})$ *or* $\overline{f} \in \operatorname{Aut}(\mathbb{D})$.

In other words, $\operatorname{Isom}(\boldsymbol{p}) = \operatorname{Isom}(\boldsymbol{\gamma}) = \operatorname{Aut}(\mathbb{D}) \cup \overline{\operatorname{Aut}(\mathbb{D})}$.

Proof. It is clear that (iii) $\Longrightarrow$ (i) and (iii) $\Longrightarrow$ (ii).

(i) $\Longrightarrow$ (iii). Taking $e^{i\theta}\boldsymbol{h}_{f(0)} \circ f$ in place of f, we may assume that $f(0) = 0$ and that $f(x_0) = x_0$ for some $0 < x_0 < 1$. Then, we have

$$|f(\lambda)| = |\lambda| \text{ and } \left|\frac{f(\lambda) - x_0}{1 - f(\lambda)x_0}\right| = \left|\frac{\lambda - x_0}{1 - \lambda x_0}\right|, \quad \lambda \in \mathbb{D}.$$

Hence, $\operatorname{Re} f(\lambda) = \operatorname{Re}\lambda$, $\lambda \in \mathbb{D}$, and consequently either $f(\lambda) \equiv \lambda$ or $f(\lambda) \equiv \overline{\lambda}$.

(ii) $\Longrightarrow$ (iii). Since f is a $\boldsymbol{\gamma}$-isometry, we have

$$|f'_x(\lambda)\alpha + f'_y(\lambda)\beta| = C(\lambda)|\alpha + i\beta|, \quad \lambda \in \mathbb{D},\ \alpha, \beta \in \mathbb{R},$$

where $C(\lambda) := \frac{\boldsymbol{\gamma}(\lambda)}{\boldsymbol{\gamma}(f(\lambda))} > 0$. Hence, for each $\lambda \in \mathbb{D}$ there exists an $\varepsilon(\lambda) \in \{-1, 1\}$ such that $f'_x(\lambda) = \varepsilon(\lambda) i f'_y(\lambda) \neq 0$. Since the partial derivatives are continuous, the function ε has to be constant, and consequently f is either holomorphic or antiholomorphic. Hence, by the Schwarz–Pick lemma, $f \in \operatorname{Aut}(\mathbb{D}) \cup \overline{\operatorname{Aut}(\mathbb{D})}$. □

We add two more properties of $\boldsymbol{p}$.

Proposition 1.1.21 (cf. [162, 511]). *The function* $\log \boldsymbol{p}$ *is strictly plurisubharmonic on* $\mathbb{D} \times \mathbb{D} \setminus \{(\lambda, \lambda) : \lambda \in \mathbb{D}\}$. *In particular,* $\log \boldsymbol{p} \in \mathcal{PSH}(\mathbb{D} \times \mathbb{D})$.

Proof. Put $u := \log \boldsymbol{p}$, fix $a, b \in \mathbb{D}$, $a \neq b$, and let $h := \boldsymbol{h}_{a,b}$. Then,

$$(\mathcal{L}u)((a,b); (\alpha, \beta)) = (\mathcal{L}u)((0, h(b)); (h'(a)\alpha, h'(b)\beta)), \quad \alpha, \beta \in \mathbb{C}.$$

Thus, it suffices to prove that

$$(\mathcal{L}u)((0,t); (\alpha, \beta)) > 0, \quad 0 < t < 1,\ (\alpha, \beta) \in \left(\mathbb{C}^2\right)_*.$$

Elementary (but tedious) calculations give

$$\begin{aligned}(\mathcal{L}u)((0,t);(\alpha,\beta)) = &\left[4t\left(1-t^2\right)^2 T^2(t)\right]^{-1} \times \left[t^2 T(t)\left|\left(1-t^2\right)\alpha + \beta\right|^2\right.\\ &\left.+ (T(t) - t)\left|\left(1-t^2\right)\alpha - \beta\right|^2\right], \quad 0 < t < 1,\ \alpha, \beta \in \mathbb{C},\end{aligned}$$

where $T := \tanh^{-1}$; cf. Exercise 1.3.4. It remains to observe that $T(t) > t$ for $0 < t < 1$. □

Remark 1.1.22 (cf. [511]). We have $\boldsymbol{m} \notin \mathcal{PSH}(\mathbb{D} \times \mathbb{D})$. In fact,

$$(\mathcal{L}\boldsymbol{m})\left((0,t);(\alpha,\beta)\right) = \frac{1}{4t} \cdot \left[\left(1-t^2\right)^2 |\alpha|^2 + 2\left(3t^2-1\right)\operatorname{Re}\left(\alpha\overline{\beta}\right) + |\beta|^2\right],$$
$$0 < t < 1,\ \alpha,\beta \in \mathbb{C};$$

cf. Exercise 1.3.4. In particular, $(\mathcal{L}\boldsymbol{m})((0,t);(1,-1)) < 0$ as $t \longrightarrow 1^-$.

The Poincaré distance may also be introduced axiomatically: namely, we have

Proposition 1.1.23. *Let* $d : \mathbb{D} \times \mathbb{D} \longrightarrow \mathbb{R}$ *be a function such that*

(i) d *is invariant under* $\operatorname{Aut}(\mathbb{D})$,

(ii) $d(0,s) = d(0,t) + d(t,s)$, $0 \le t \le s < 1$,

(iii) $\lim_{t\to 0+} \frac{d(0,t)}{t} = 1$.

Then, $d = \boldsymbol{p}$.

Proof. Let $\varphi(t) := d(0,t)$, $0 \le t < 1$. In view of (ii) and (iii), $\varphi(0) = 0$ and $\varphi'(0) = 1$. We will show that

$$\varphi'(t) = \frac{1}{1-t^2} = \boldsymbol{\gamma}(t), \quad 0 \le t < 1. \tag{1.1.7}$$

Suppose for a moment that (1.1.7) is true. Then,

$$\varphi(s) = \int_0^s \varphi'(t)dt = \int_0^s \frac{dt}{1-t^2} = \frac{1}{2}\log\frac{1+s}{1-s} = \boldsymbol{p}(0,s), \quad 0 \le s < 1,$$

and hence by (i), $d \equiv \boldsymbol{p}$.

To prove (1.1.7), fix $0 < t_0 < 1$ and let $t > 0$ be such that $t_0 + t < 1$. Because of (ii), we get

$$\varphi(t_0+t) - \varphi(t_0) = d(t_0, t_0+t).$$

On the other hand, by (i) we have

$$d\left(t_0, t_0+t\right) = d\left(h_{t_0}(t_0), h_{t_0}(t_0+t)\right) = d\left(0, \frac{t}{1-(t_0+t)\,t_0}\right).$$

Finally,

$$\lim_{t\to 0+} \frac{\varphi(t_0+t) - \varphi(t_0)}{t} = \frac{1}{1-t_0^2}.$$

The proof for the left-hand derivative is analogous. □

For another axiomatic description of $\boldsymbol{p}$, see [159].

1.2 Some applications

At the end of this introductory chapter, we show how to establish the theorems of Picard in an elementary fashion, i.e., without using elliptic modular functions (cf. [365]). The argument here is based on the following characterization of the function $\boldsymbol{\gamma}$ (cf. [16, 364]).

Proposition 1.2.1.

(a) $\Delta(\log \boldsymbol{\gamma}) = 4\boldsymbol{\gamma}^2$ *on* $\mathbb{D}$.

(b) *If* $\beta : \mathbb{D} \longrightarrow [0, +\infty)$ *is a continuous function that is* $\mathcal{C}^2$ *on* $\{\lambda \in \mathbb{D} : \beta(\lambda) > 0\}$ *and which satisfies*

$$\Delta(\log \beta)(\lambda) \geq 4\beta^2(\lambda)$$

for $\lambda \in \mathbb{D}$ *with* $\beta(\lambda) > 0$, *then either* $\beta(\lambda) < \boldsymbol{\gamma}(\lambda)$ *for every* $\lambda \in \mathbb{D}$ *or* $\beta = \boldsymbol{\gamma}$ *on* $\mathbb{D}$.

Proof. (a) Compute!

(b) First we will prove that $\beta \leq \boldsymbol{\gamma}$.

For $0 < r < 1$, we define on $\mathbb{D}(r)$

$$g_r(\lambda) := \beta(\lambda)/\gamma_r(\lambda) \quad \text{with} \quad \gamma_r(\lambda) := \frac{r}{r^2 - |\lambda|^2}, \qquad \lambda \in \mathbb{D}(r).$$

Since β is bounded on $\mathbb{D}(r)$, it follows that $\lim_{|\lambda| \nearrow r} g_r(\lambda) = 0$. Thus, if g_r is not identically zero, then g_r takes on its positive maximum on $\mathbb{D}(r)$ at a point $\lambda_r \in \mathbb{D}(r)$. Then, by assumption, g_r is a $\mathcal{C}^2$-function near λ_r and so we obtain

$$\Delta(\log g_r)(\lambda_r) \leq 0, \ \text{ i.e., } 4(\beta^2(\lambda_r) - \gamma_r^2(\lambda_r)) \leq \Delta(\log \beta)(\lambda_r) - \Delta(\log \gamma_r)(\lambda_r) \leq 0.$$

Thus, $g_r(\lambda) \leq 1$ for every $\lambda \in \mathbb{D}(r)$ which finally gives

$$g(\lambda) = \lim_{r \nearrow 1} g_r(\lambda) \leq 1, \quad \lambda \in \mathbb{D},$$

i.e., $\beta \leq \boldsymbol{\gamma}$ on $\mathbb{D}$.

Now, put $A := \{\lambda \in \mathbb{D} : \beta(\lambda) < \boldsymbol{\gamma}(\lambda)\}$. Suppose that $\varnothing \neq A \neq \mathbb{D}$ and choose $\lambda_0 \in \partial A \cap \mathbb{D}$. Then, there exist $\lambda_1 \in A$ and positive numbers $R_1 < R_2, R_3 < 1$ such that

$$\mathbb{D}(\lambda_1, R_1) \subset\subset A \quad \text{and} \quad \lambda_0 \in \mathbb{D}(\lambda_1, R_2) \subset\subset \mathbb{D}(R_3).$$

On $\mathbb{D}(R_3)$ we define

$$u(\lambda) := \log(\beta(\lambda)/\boldsymbol{\gamma}(\lambda)), \quad \lambda \in \mathbb{D}(R_3).$$

Observe that u is upper semicontinuous, $u \le 0$, and u is $\mathcal{C}^2$ where $\beta > 0$. Fix $\lambda \in \mathbb{D}(R_3)$ with $\beta(\lambda) > 0$. Then, a calculation gives

$$\begin{aligned}\Delta u(\lambda) &= \Delta(\log\beta)(\lambda) - \Delta(\log\boldsymbol{\gamma})(\lambda)\\ &\ge 4(\beta(\lambda)+\boldsymbol{\gamma}(\lambda))(\beta(\lambda)-\boldsymbol{\gamma}(\lambda)) \ge 8M(\beta(\lambda)-\boldsymbol{\gamma}(\lambda)),\end{aligned}$$

where $M := (1-R_3^2)^{-1}$.

The elementary inequality $M\log(t/s) \ge t-s$, $0 < s \le t \le M$, finally leads to $\Delta u(\lambda) \ge 8M^2 u(\lambda)$ whenever $|\lambda| < R_3$ and $\beta(\lambda) > 0$.

Now, we introduce the following auxiliary function $w : \overline{\mathbb{D}(\lambda_1, R_2)} \longrightarrow \mathbb{R}_{-\infty}$ with

$$w(\lambda) := u(\lambda) + \varepsilon v(\lambda), \quad \text{where } v(\lambda) := \exp(-\alpha|\lambda-\lambda_1|^2) - \exp(-\alpha R_2^2).$$

The positive numbers α and ε are chosen such that the following two conditions are satisfied:

(i) $\Delta v(\lambda) - 8M^2 v(\lambda) \ge (4\alpha^2|\lambda-\lambda_1|^2 - 4\alpha - 8M^2)\exp(-\alpha|\lambda-\lambda_1|^2) \ge 0$ whenever $R_1 \le |\lambda-\lambda_1| \le R_2$;

(ii) $w(\lambda) \le 0$ on $\overline{\mathbb{D}(\lambda_1, R_1)}$.

Since $u(\lambda_0) = 0 < v(\lambda_0)$ and $v|_{\partial B(\lambda_1,R_2)} = 0$, the function w attains its positive maximum inside the annulus $\mathbb{D}(\lambda_1, R_2) \setminus \overline{\mathbb{D}(\lambda_1, R_1)}$ at a point λ_2. Then we have $0 \ge \Delta w(\lambda_2) \ge 8M^2 w(\lambda_2)$, which contradicts $w(\lambda_2) > 0$. □

Corollary 1.2.2 (Ahlfors–Schwarz lemma). *Let $G \subset \mathbb{C}$ be any domain and let $\beta : G \longrightarrow [0,+\infty)$ be a continuous function that is $\mathcal{C}^2$ on $\{\lambda \in G : \beta(\lambda) > 0\}$. Moreover, assume that $\Delta(\log\beta)(\lambda) \ge C\beta^2(\lambda)$ for every $\lambda \in G$ with $\beta(\lambda) > 0$; here, C denotes a fixed positive number. Then, for every $f \in \mathcal{O}(\mathbb{D}, G)$ either*

$$\beta(f(\lambda))|f'(\lambda)| < \sqrt{4/C}\,\boldsymbol{\gamma}(\lambda) \quad \textit{for all } \lambda \in \mathbb{D} \quad \textit{or} \tag{1.2.1}$$

$$\beta \circ f \cdot |f'| \equiv \sqrt{4/C}\,\boldsymbol{\gamma} \quad \textit{on } \mathbb{D}. \tag{1.2.2}$$

Proof. Set $\widetilde{\beta}(\lambda) := \sqrt{C/4}\,\beta(f(\lambda))|f'(\lambda)|$, $\lambda \in \mathbb{D}$, and apply Proposition 1.2.1. □

Remark 1.2.3. If $G = \mathbb{D}$ and $\beta = \boldsymbol{\gamma}$, then Corollary 1.2.2 is nothing other than the classical Schwarz–Pick Lemma.

Corollary 1.2.4 (Liouville theorem). *Under the conditions of Corollary 1.2.2, with $\beta(\lambda) > 0$ for every $\lambda \in G$, any $f \in \mathcal{O}(\mathbb{C}, G)$ is necessarily constant.*

Proof. Suppose that $f \in \mathcal{O}(\mathbb{C}, G)$ with $f'(\lambda_0) \ne 0$. Put $g(\lambda) := f(\lambda_0 + \lambda)$, $\lambda \in \mathbb{C}$. Then, (1.2.1), (1.2.2) with $g_R(\lambda) := g(R\lambda)$, $\lambda \in \mathbb{D}$, imply that

$$\beta(g_R(0))|g_R'(0)| \le (4/C)^{1/2}, \quad R > 0.$$

Thus $g'(0) = f'(\lambda_0) = 0$; a contradiction. □

Observe that with $G = \mathbb{D}$ and $\beta = \gamma$, Corollary 1.2.4 is just the classical Liouville theorem. Moreover, it also contains

Theorem 1.2.5 (Little Picard theorem). *Except for the constant function, there is no entire holomorphic function $f \in \mathcal{O}(\mathbb{C})$ omitting two different complex numbers as values.*

Proof. Suppose that $f \in \mathcal{O}(\mathbb{C}, \mathbb{C} \setminus \{w_1, w_2\})$, $w_1 \neq w_2$. Taking $\frac{f-w_1}{w_2-w_1}$ instead of f, we reduce the proof to the case $w_1 = 0$, $w_2 = 1$. Thus, it suffices to find a suitable function $\beta > 0$ on $G := \mathbb{C} \setminus \{0, 1\}$, which satisfies the assumptions of Corollary 1.2.2.

For $\lambda \in \mathbb{C} \setminus \{0, 1\} =: G$, we define (cf. [365])

$$\beta(\lambda) := \frac{\left(1 + |\lambda|^{1/3}\right)^{1/2}}{|\lambda|^{5/6}} \cdot \frac{\left(1 + |1-\lambda|^{1/3}\right)^{1/2}}{|1-\lambda|^{5/6}}. \tag{1.2.3}$$

An easy calculation then gives

$$\Delta(\log\beta)(\lambda) = \frac{1}{18}\left(\frac{1}{|\lambda|^{5/3}\left(1 + |\lambda|^{1/3}\right)^2} + \frac{1}{|1-\lambda|^{5/3}\left(1 + |1-\lambda|^{1/3}\right)^2}\right)$$

and

$$\lim_{|\lambda| \to a} \frac{\Delta(\log\beta)(\lambda)}{\beta^2(\lambda)} = \begin{cases} +\infty & \text{if } a = \infty \\ 1/36 & \text{if } a = 0 \text{ or } a = 1 \end{cases}.$$

Thus, there exists a positive C with

$$\Delta(\log\beta(\lambda)) \geq C\beta^2(\lambda), \quad \lambda \in G, \tag{1.2.4}$$

and Corollary 1.2.4 can be applied. □

To prove the next theorem (the big Picard Theorem 1.2.7) we need the following auxiliary result, which may be also interesting in itself.

Theorem 1.2.6. *The following three theorems are equivalent and true.*

(i) (Schottky theorem) *For any $\alpha > 0$, $\theta \in (0, 1)$ there exists a constant $M(\alpha, \theta) > 0$ such that for any $r > 0$ and $f \in \mathcal{O}(\mathbb{D}(r), \mathbb{C} \setminus \{0, 1\})$, with $|f(0)| \leq \alpha$, we have $|f(z)| \leq M(\alpha, \theta)$, $z \in \mathbb{D}(\theta r)$.*

(ii) ("Big" Montel theorem) *Let $D \subset \mathbb{C}$ be a domain and let $\mathcal{F} \subset \mathcal{O}(D, \mathbb{C} \setminus \{w_1, w_2\})$ with $w_1, w_2 \in \mathbb{C}$, $w_1 \neq w_2$. Then, $\mathcal{F}$ is normal.*[1]

[1] Recall that a family $\mathcal{F} \subset \mathcal{O}(D)$ is *normal* if every sequence $(f_k)_{k=1}^{\infty} \subset \mathcal{F}$ contains either a subsequence $(f_{k_s})_{s=1}^{\infty}$ such that $f_{k_s} \longrightarrow f \in \mathcal{O}(D)$ locally uniformly in D, or a subsequence $(f_{k_s})_{s=1}^{\infty}$ such that $f_{k_s} \longrightarrow \infty$ locally uniformly in D.

(iii) *For any* $\alpha > 0$ *the family* $\mathcal{F}_\alpha := \{f \in \mathcal{O}(\mathbb{D}, \mathbb{C} \setminus \{0,1\}) : |f(0)| \le \alpha\}$ *is normal.*

Proof. (i) $\Longrightarrow$ (ii): Fix D, w_1, w_2, and $\mathcal{F}$. We may assume that $w_1 = 0$, $w_2 = 1$. Fix a sequence $(f_n)_{n=1}^{\infty} \subset \mathcal{F}$. It suffices to prove that every point $a \in D$ has a neighborhood $U \subset D$ such that the family $\{f_n|_U : n \in \mathbb{N}\}$ is normal.

Fix an $a \in D$. If the sequence $(f_n(a))_{n=1}^{\infty}$ is bounded, then, using (i), we conclude that it is uniformly bounded on $\mathbb{D}(a, \varrho) \subset D$ for certain ϱ. Thus, we may apply the standard Montel argument.

If the sequence $(f_n(a))_{n=1}^{\infty}$ is unbounded, then we may assume that the sequence $(1/f_n(a))_{n=1}^{\infty}$ is bounded. Put $g_n := 1/f_n \in \mathcal{O}(D, \mathbb{C} \setminus \{0,1\})$. The first part of the proof shows that $g_{n_k} \longrightarrow g$ locally uniformly in $\mathbb{D}(a, \varrho)$ for certain ϱ. Observe that $g(0) = 0$. Hence, by the Hurwitz theorem, $g \equiv 0$. Thus, $f_{n_k} \longrightarrow \infty$ locally uniformly in $\mathbb{D}(a, \varrho)$.

The implication (ii) $\Longrightarrow$ (iii) is trivial.

(iii) $\Longrightarrow$ (i): Suppose that there exist $\alpha > 0$, $0 < \theta < 1$, and sequences $r_n \in \mathbb{R}_{>0}$, $f_n \in \mathcal{O}(\mathbb{D}(r_n), \mathbb{C} \setminus \{0,1\})$, $a_n \in \mathbb{D}(\theta r_n)$, $n \in \mathbb{N}$, such that $|f_n(0)| \le \alpha$, but $f_n(a_n) \longrightarrow \infty$.

Define $g_n(z) := f_n(r_n z)$, $z \in \mathbb{D}$, $n \in \mathbb{N}$. Observe that $(g_n)_{n=1}^{\infty} \subset \mathcal{F}_\alpha$. Thus, by (iii), we may assume that $g_n \longrightarrow g_0$ locally uniformly in $\mathbb{D}$, where $g_0 \in \mathcal{O}(\mathbb{D})$. Let $b_n := a_n/r_n$. Then, $|b_n| \le \theta$. We may assume that $b_n \longrightarrow b_0 \in \overline{K}(\theta)$. Finally, $f_n(a_n) = g_n(b_n) \longrightarrow g_0(b_0) \in \mathbb{C}$; a contradiction.

Now, we will prove (iii) (cf. [109], Ch. VII, proof of Theorem 3.8). Fix an $\alpha > 0$. We consider $\mathcal{F}_\alpha$ as a family of functions $\mathbb{D} \longrightarrow \overline{\mathbb{C}}$. Due to the Ascoli theorem, the normality of $\mathcal{F}_\alpha$ is equivalent to its equicontinuity in the spherical distance d on $\overline{\mathbb{C}}$. Recall that

$$d(w_1, w_2) = \frac{|w_1 - w_2|}{\sqrt{1+|w_1|^2}\sqrt{1+|w_2|^2}}, \quad w_1, w_2 \in \mathbb{C}.$$

Let β and C be as in (1.2.3) and (1.2.4). By the Ahlfors–Schwarz lemma (Corollary 1.2.2), we get

$$\beta(f)|f'| \le \frac{2}{\sqrt{C}}\gamma, \quad f \in \mathcal{F}_\alpha.$$

On the other hand, formula (1.2.3) implies that

$$\frac{1}{1+|w|^2} \le C_1\beta(w), \quad w \in \mathbb{C} \setminus \{0,1\},$$

for some $C_1 > 0$. Hence,

$$\frac{|f'(z)|}{1+|f(z)|^2} \le C_1\beta(f(z))|f'(z)|$$
$$\le C_1\frac{2}{\sqrt{C}}\frac{1}{1-|z|^2} = C_2\frac{1}{1-|z|^2}, \quad f \in \mathcal{F}_\alpha,\ z \in \mathbb{D}.$$

Fix an arbitrary closed disc $\Delta = \overline{\mathbb{D}}(a,r) \subset\subset \mathbb{D}$. The above inequality implies that there exists a constant $M > 0$ such that

$$\frac{|f'(z)|}{1+|f(z)|^2} \le M, \quad f \in \mathcal{F}_\alpha,\ z \in \Delta.$$

To prove the equicontinuity it suffices to show that

$$d(f(z_1), f(z_2)) \le M|z_1 - z_2|, \quad f \in \mathcal{F}_\alpha,\ z_1, z_2 \in \Delta.$$

Fix $f \in \mathcal{F}_\alpha$, $z_1, z_2 \in \Delta$, and $\varepsilon > 0$. Let $\xi_k := z_1 + (k/n)(z_2 - z_1)$, $k = 0, \dots, n$, where n is so big that

$$\frac{1+|f(\xi_{k-1})|^2}{\sqrt{1+|f(\xi_k)|^2}\sqrt{1+|f(\xi_{k-1})|^2}} < 1+\varepsilon,$$
$$\left|\frac{f(\xi_k)-f(\xi_{k-1})}{\xi_k-\xi_{k-1}} - f'(\xi_{k-1})\right| < \varepsilon, \quad k = 1, \dots, n.$$

Put $c_k := \sqrt{1+|f(\xi_k)|^2}\sqrt{1+|f(\xi_{k-1})|^2}$. We have

$$d(f(z_1), f(z_2)) \le \sum_{k=1}^{n} d(f(\xi_{k-1}), f(\xi_k)) = \sum_{k=1}^{n} \frac{1}{c_k}|f(\xi_k) - f(\xi_{k-1})|$$
$$\le \sum_{k=1}^{n} \frac{1}{c_k}\left|\frac{f(\xi_k)-f(\xi_{k-1})}{\xi_k-\xi_{k-1}} - f'(\xi_{k-1})\right| |\xi_k - \xi_{k-1}|$$
$$+ \sum_{k=1}^{n} \frac{1}{c_k}|f'(\xi_{k-1})||\xi_k - \xi_{k-1}|$$
$$\le \sum_{k=1}^{n} \frac{\varepsilon}{c_k}|\xi_k - \xi_{k-1}| + \sum_{k=1}^{n} \frac{M}{c_k}\left(1+|f(\xi_{k-1})|^2\right)|\xi_k - \xi_{k-1}|$$
$$\le \varepsilon|z_1 - z_2| + M\sum_{k=1}^{n}(1+\varepsilon)|\xi_k - \xi_{k-1}|$$
$$= (\varepsilon + M(1+\varepsilon))|z_1 - z_2|.$$

Letting $\varepsilon \longrightarrow 0$, we get the required estimate. □

Finally, we are going to present the big Picard theorem.

Theorem 1.2.7 (Big Picard Theorem). *Let $f : \mathbb{D}_* \longrightarrow \mathbb{C}$ be a holomorphic function that has an essential singularity at* 0. *Then,*

$$\#\{w \in \mathbb{C} : \#f^{-1}(w) < +\infty\} \leq 1,$$

i.e., the function f takes on all possible complex values, with at most a single exception, infinitely often.

Proof. Suppose that for some $w_1, w_2 \in \mathbb{C}$, $w_1 \neq w_2$, we have $\#f^{-1}(w_j) < +\infty$, $j = 1, 2$. We may assume that $w_1 = 0$, $w_2 = 1$. Consider the sequence of holomorphic functions $f_k(z) := f(z/2^k)$, $z \in \mathbb{D}_*$, $k \in \mathbb{N}$. Observe that $f_k(\mathbb{D}_*) = f(\mathbb{D}_*(1/2^k))$. Consequently, $0, 1 \notin f_k(\mathbb{D}_*)$ for $k \gg 1$. Hence, using the "big" Montel theorem (Theorem 1.2.6(ii)), we find a subsequence $(f_{k_s})_{s=1}^{\infty}$ such that $f_{k_s} \longrightarrow g$ locally uniformly in $\mathbb{D}_*$, where either $g \in \mathcal{O}(\mathbb{D}_*)$ or $g \equiv \infty$.

If $g \in \mathcal{O}(\mathbb{D}_*)$, then $|f_{k_s}| \leq M$ on $L_0 := \overline{\mathbb{A}}(1/3, 2/3)$ for some $M > 0$. Hence, $|f| \leq M$ on $L_s := \overline{\mathbb{A}}(1/(3 \cdot 2^{k_s}), 2/(3 \cdot 2^{k_s}))$, $s \in \mathbb{N}$. Thus, f has a removable singularity at 0; a contradiction.

If $g \equiv \infty$, then for any $M > 0$ we get $|f_{k_s}| \geq M$ on L_0, $s \gg 1$. Hence, $|f| \geq M$ on L_s, $s \gg 1$. Thus, f has a pole at 0; a contradiction. □

Observe that Theorem 1.2.7 implies that any holomorphic function $f : \mathbb{D}_* \longrightarrow \mathbb{C}$ that omits at least two complex values extends either holomorphically or meromorphically to $\mathbb{D}$.

Definition 1.2.8. Let $D, G \subset \mathbb{C}$ be domains. We say that the space $\mathcal{O}(D, G)$ is *normal*, if every sequence $(f_k)_{k=1}^{\infty} \subset \mathcal{O}(D, G)$ contains

- either a subsequence $(f_{k_s})_{s=1}^{\infty}$ such that $f_{k_s} \longrightarrow f \in \mathcal{O}(D, G)$ locally uniformly in D, or
- a subsequence $(f_{k_s})_{s=1}^{\infty}$ that *diverges locally uniformly in D*, i.e., for any $K \subset\subset D$, $L \subset\subset G$ we have $f_{k_s}(K) \cap L = \varnothing$, $s \gg 1$.

Corollary 1.2.9. *Let $D, G \subset \mathbb{C}$ be domains. Then, $\mathcal{O}(D, G)$ is normal iff $\#(\mathbb{C} \setminus G) \geq 2$.*

Proof. First, observe that if $G \in \{\mathbb{C}, \mathbb{C}_*\}$, then $\mathcal{O}(D, G)$ is not normal – one can take for instance $f_k(\lambda) := \exp(k(\lambda - a))$, where $a \in D$ is fixed.

Now, suppose that $\#(\mathbb{C} \setminus G) \geq 2$. We may assume that $G \subset \mathbb{C} \setminus \{0, 1\}$. Fix a sequence $(f_k)_{k=1}^{\infty} \subset \mathcal{O}(D, G)$. By Theorem 1.2.6(ii), it contains

• either a subsequence $(f_{k_s})_{s=1}^{\infty}$ such that $f_{k_s} \longrightarrow f \in \mathcal{O}(D, \mathbb{C})$ locally uniformly in D, or

• a subsequence $(f_{k_s})_{s=1}^{\infty}$ such that $f_{k_s} \longrightarrow \infty$ locally uniformly in D.

In the first case, if $f(D) \subset G$, then we are done. Suppose that $f(a) \in \partial G$ for some $a \in D$. If $f \not\equiv f(a)$, then, by the Hurwitz theorem, $f(a) \in f_{k_s}(D)$, $s \gg 1$, which gives a contradiction. Thus, $f \equiv f(a)$ and then $(f_{k_s})_{s=1}^{\infty}$ diverges locally uniformly in D.

Obviously, in the second case $(f_{k_s})_{s=1}^{\infty}$ diverges locally uniformly in D. □

Remark 1.2.10. The theorem of Landau can be also derived along the lines of the above ideas (see [365], also [208]); cf. Exercise 1.3.5.

1.3 Exercises

Exercise 1.3.1. Using the Schwarz–Pick lemma (Lemma 1.1.1), prove for $\lambda \in \mathbb{D}$ that

$$\left\{\left(f'(\lambda), f(\lambda)\right) : f \in \mathcal{O}\left(\mathbb{D}, \mathbb{D}\right)\right\} = \left\{(A, B) \in \mathbb{C}^2 : |A|\left(1 - |\lambda|^2\right) \le 1 - |B|^2\right\}.$$

Exercise 1.3.2. Prove that

$$B_{\boldsymbol{m}}(a, r) = \mathbb{D}\left(\frac{a\left(1 - r^2\right)}{1 - r^2|a|^2}, \frac{r\left(1 - |a|^2\right)}{1 - r^2|a|^2}\right), \quad a \in \mathbb{D},\ 0 < r < 1.$$

Exercise 1.3.3. Let μ denote a measure on $\mathbb{D}$ defined by

$$\mu(A) = \int_A \gamma^2(\lambda) d\mathcal{L}^2(\lambda),$$

where $\mathcal{L}^2$ denotes the two dimensional Lebesgue measure on $\mathbb{C}$, and $A \subset \mathbb{D}$ is any Lebesgue measurable set. Prove that μ is invariant under $\mathrm{Aut}(\mathbb{D})$.
Let T denote the $\boldsymbol{p}$-triangle with vertices at $a, b, c \in \mathbb{D}$, i.e., T is determined by $C_{a,b}$, $C_{b,c}$ and $C_{c,a}$. Prove that $\mu(T) = \frac{1}{4}[\pi - (\alpha + \beta + \gamma)]$, where α, β, γ denote the angles of T.

Exercise 1.3.4.

(a) Verify the formula for $(\mathcal{L}u)((0, t); (\alpha, \beta))$ in the proof of Proposition 1.1.21.

(b) Verify the formula for $(\mathcal{L}\boldsymbol{m})((0, t); (\alpha, \beta))$ in Remark 1.1.22.

Exercise 1.3.5. Prove the following theorem of Landau. For any $(a_0, a_1) \in \mathbb{C} \times \mathbb{C}_*$ there exists a positive number $R = R(a_0, a_1)$ such that, whenever $f \in \mathcal{O}(\mathbb{D}(r), \mathbb{C} \setminus \{0, 1\})$ with $f(0) = a_0$ and $f'(0) = a_1$, then $r \le R$.

1.4 List of problems

Chapter 2

The Carathéodory pseudodistance and the Carathéodory–Reiffen pseudometric

Summary. In the twenties of the previous century, C. Carathéodory was the first to use the pseudodistance $\boldsymbol{c}_G$ (now bearing his name) as a tool in complex analysis of several variables. The infinitesimal version of this pseudodistance, i.e., $\boldsymbol{\gamma}_G$, was intensively studied no earlier than in the 1960s by H. J. Reiffen [447, 448]. Most of the results we present in this chapter have become part of standard knowledge and can be found in many textbooks (e.g. [137, 317, 434, 178, 316]). Observe that the problems related to the $\boldsymbol{c}$-, $\boldsymbol{c}^i$-, and $\boldsymbol{\gamma}$-hyperbolicity (§§ 2.5, 2.8) and to the Carathéodory topology (§ 2.6) do not occur as long as we study bounded domains (cf. Propositions 2.5.2 and 2.6.1). The first counterexample was found by J.-P. Vigué [514], who constructed a complex $\boldsymbol{c}$-hyperbolic space M with $\operatorname{top} M \neq \operatorname{top} \boldsymbol{c}_M$. Theorem 2.6.3 and Lemma 2.6.6 give counterexamples in the domain and the manifold case, respectively. Nevertheless, it seems worthwhile to look for more explicit examples. The first examples showing that $\boldsymbol{c}_G$ is not inner in general are due to T. J. Barth [34]. Next, J.-P. Vigué [513] constructed a domain of holomorphy $G \subset \mathbb{C}^2$ such that $(G, \boldsymbol{c}_G)$ is complete but $\boldsymbol{c}_G$ is not inner. It is clear that the counterexample given in Example 2.7.9 is the simplest possible one. Example 2.7.10 shows that there is a big difference between the topological and geometrical structure of the space $(G, \boldsymbol{c}_G)$.

Section 2.11 illustrates problems that may appear when we move from the category of complex manifolds to the category of analytic sets.

Introduction. In Chapter 1 we introduced two continuous distances $\boldsymbol{m} : \mathbb{D} \times \mathbb{D} \longrightarrow [0,1)$, $\boldsymbol{p} : \mathbb{D} \times \mathbb{D} \to \mathbb{R}_+$ ($\boldsymbol{p} = \tanh^{-1}(\boldsymbol{m})$) and the $\mathcal{C}^\infty$-function $\boldsymbol{\gamma} : \mathbb{D} \longrightarrow [1, +\infty)$ such that:

1° $\boldsymbol{m}$, $\boldsymbol{p}$, and $\boldsymbol{\gamma}$ satisfy the Schwarz–Pick lemma, cf. Lemma 1.1.1, Remark 1.1.13, Corollary 1.1.15(b);

2° $\operatorname{top} \boldsymbol{m} = \operatorname{top} \boldsymbol{p} = \operatorname{top} \mathbb{D}$ and the spaces $(\mathbb{D}, \boldsymbol{m})$, $(\mathbb{D}, \boldsymbol{p})$ are complete, cf. Remark 1.1.9, Corollary 1.1.15(c);

3° $\displaystyle\lim_{\substack{\lambda',\lambda'' \to a \\ \lambda' \neq \lambda''}} \frac{\boldsymbol{m}(\lambda', \lambda'')}{|\lambda' - \lambda''|} = \lim_{\substack{\lambda',\lambda'' \to a \\ \lambda' \neq \lambda''}} \frac{\boldsymbol{p}(\lambda', \lambda'')}{|\lambda' - \lambda''|} = \boldsymbol{\gamma}(a)$, $a \in \mathbb{D}$, cf. Property 1.1.6, Corollary 1.1.15(d);

4° the notions of $\boldsymbol{m}$- and $\boldsymbol{p}$-rectifiability of curves coincide with the Euclidean rectifiability and, moreover,

$$L_{\boldsymbol{m}} = L_{\boldsymbol{p}}, \text{ cf. Lemma 1.1.18(a),}$$
$$L_{\boldsymbol{m}} = L_{\boldsymbol{p}} = L_{\boldsymbol{\gamma}} \text{ for piecewise } \mathcal{C}^1\text{-curves, cf. Lemma 1.1.18,}$$
$$\boldsymbol{m}^i = \boldsymbol{p}^i = \boldsymbol{p} = \inf\{L_{\boldsymbol{\gamma}}(\alpha) : \dots\}, \text{ cf. Corollary 1.1.17, Lemma 1.1.18(a);}$$

5° $\mathrm{Isom}(\boldsymbol{m}) = \mathrm{Isom}(\boldsymbol{p}) = \mathrm{Isom}(\boldsymbol{\gamma}) = \mathrm{Aut}(\mathbb{D}) \cup \overline{\mathrm{Aut}(\mathbb{D})}$, cf. Proposition 1.1.20;

6° $\log \boldsymbol{p} \in \mathcal{PSH}(\mathbb{D} \times \mathbb{D})$, $\log \boldsymbol{m}$ is separately subharmonic in $\mathbb{D} \times \mathbb{D}$, cf. Proposition 1.1.21, Property 1.1.5.

In what follows, we will discuss generalizations of the functions $\boldsymbol{m}$, $\boldsymbol{p}$, and $\boldsymbol{\gamma}$ for arbitrary domains $G \subset \mathbb{C}^n$. There are several ways to proceed. In this chapter, we present the simplest one dealing with bounded holomorphic functions (this way was the first also from the historical point of view; cf. [82]).

The results on the Carathéodory isometries (§ 2.4 and Exercise 2.12.9) have the following general aspect. Assume that for any domain $G \subset \mathbb{C}^n$ with arbitrary n we are given a pseudodistance $d_G : G \times G \longrightarrow \mathbb{R}_+$ such that $d_{\mathbb{D}} = \boldsymbol{p}$, and suppose that the system $(d_G)_G$ is holomorphically contractible, that is, $d_D(F(z'), F(z'')) \le d_G(z', z'')$ for any domains $G \subset \mathbb{C}^n$, $D \subset \mathbb{C}^m$ and for any holomorphic mapping $F : G \longrightarrow D$; cf. Remark 2.1.2. Suppose that we want to use the spaces (G, d_G) to characterize properties of holomorphic mappings. The question is how to distinguish between "good" and "bad" systems $(d_G)_G$. One possible criterion is that "good" systems have only few isometries $F : (G, d_G) \longrightarrow (D, d_D)$ that are neither holomorphic nor antiholomorphic (at least in the case where G and D are subdomains of the same space $\mathbb{C}^n$).

2.1 Definitions. General Schwarz–Pick lemma

For any domain $G \subset \mathbb{C}^n$, $n \ge 1$, put

$$\begin{aligned}
\boldsymbol{m}_G(z', z') = \boldsymbol{c}^*_G(z', z'') &:= \sup\{\boldsymbol{m}(f(z'), f(z'')) : f \in \mathcal{O}(G, \mathbb{D})\}, \quad z', z'' \in G,\\
\boldsymbol{c}_G(z', z'') &:= \sup\{\boldsymbol{p}(f(z'), f(z'')) : f \in \mathcal{O}(G, \mathbb{D})\}, \quad z', z'' \in G,\\
\boldsymbol{\gamma}_G(z; X) &:= \sup\{\boldsymbol{\gamma}(f(z))\,|f'(z)\,X| : f \in \mathcal{O}(G, \mathbb{D})\}, \ z \in G,\ X \in \mathbb{C}^n,
\end{aligned}$$

where $f'(z) : \mathbb{C}^n \longrightarrow \mathbb{C}$ denotes the $\mathbb{C}$-differential of f at z, i.e.,

$$f'(z)X = \sum_{j=1}^{n} \frac{\partial f}{\partial z_j}(z) X_j.$$

It is clear that

$$c_G = \tanh^{-1}\left(c_G^*\right) \geq c_G^*, \tag{2.1.1}$$

and (by the Schwarz–Pick lemma)

$$m_{\mathbb{D}} = c_{\mathbb{D}}^* = m, \; c_{\mathbb{D}} = p, \; \gamma_{\mathbb{D}}(\cdot; 1) = \gamma.$$

Observe that $c_{\mathbb{C}^n}^* \equiv 0$, $c_{\mathbb{C}^n} \equiv 0$, and $\gamma_{\mathbb{C}^n} \equiv 0$.

In view of (2.1.1), we can always pass from c_G^* to c_G or vice versa. Nevertheless, in what follows, we will use both c_G^* and c_G; c_G^* is less regular but more handy in calculations (as in the case of the unit disc). We write $c_G^{(*)}$ in all the cases where one can take c_G^* as well as c_G.

Since m, p, and γ are invariant under $\operatorname{Aut}(\mathbb{D})$, we get

$$\begin{aligned}
c_G^*\left(z', z''\right) &= \sup\left\{\left|f\left(z''\right)\right| : f \in \mathcal{O}\left(G, \mathbb{D}\right), f\left(z'\right) = 0\right\}, \quad z', z'' \in G, \\
c_G\left(z', z''\right) &= \sup\left\{p\left(0, f\left(z''\right)\right) : f \in \mathcal{O}\left(G, \mathbb{D}\right), f\left(z'\right) = 0\right\}, \quad z', z'' \in G, \\
\gamma_G\left(z; X\right) &= \sup\left\{\left|f'\left(z\right) X\right| : f \in \mathcal{O}\left(G, \mathbb{D}\right), f\left(z\right) = 0\right\}, \quad z \in G, \; X \in \mathbb{C}^n.
\end{aligned}$$

Now, applying Montel's theorem, we find that for any z', $z'' \in G$ (resp. $z \in G$, $X \in \mathbb{C}^n$) there exists an $f \in \mathcal{O}(G, \mathbb{D})$ such that $f(z') = 0$, $|f(z'')| = c_G^*(z', z'')$ (resp. $f(z) = 0$, $|f'(z)X| = \gamma_G(z; X)$). Any such a function f will be called an *extremal function for* $c_G^{(*)}(z', z'')$ (resp. $\gamma_G(z; X)$). In particular,

$$c_G^* : G \times G \longrightarrow [0, 1), \quad c_G : G \times G \longrightarrow \mathbb{R}_+, \quad \gamma_G : G \times \mathbb{C}^n \longrightarrow \mathbb{R}_+.$$

Since m and p are distances, the function $c_G^{(*)}$ is a pseudodistance; c_G^* is called the *Möbius pseudodistance for* G; c_G is the *Carathéodory pseudodistance for* G. The function γ_G is called the *Carathéodory–Reiffen pseudometric for* G. Note that for any $a \in G$, the function $\gamma_G(a; \cdot) : \mathbb{C}^n \longrightarrow \mathbb{R}_+$ is a complex seminorm.

The definitions of $c_G^{(*)}$ and γ_G may be formally extended to the case where $G = \Omega$ is an arbitrary open set in $\mathbb{C}^n$ (not necessarily connected), but then the Montel argument does not work and, for instance, $c_\Omega(z', z'') = +\infty$ if z' and z'' lie in different connected components of Ω. Therefore, *we will only consider the connected case.*

On the other hand, there are no difficulties in generalizing $c_G^{(*)}$ and γ_G to the case where $G = M$ is a connected complex manifold. Even more, the notions of $c_G^{(*)}$ and γ_G may be extended to the case of connected complex spaces. We do not intend to develop the theory into these directions. We recommend the interested reader to consult for instance [447, 448, 449]. *Except for §§ 2.6 and 2.11, we deal only with domains in* $\mathbb{C}^n$.

As a direct consequence of the definitions, we get

Theorem 2.1.1 (General Schwarz–Pick lemma). *For arbitrary domains $G \subset \mathbb{C}^n$, $D \subset \mathbb{C}^m$ and for any holomorphic mapping $F : G \longrightarrow D$ we have*

$$\boldsymbol{c}_D^{(*)}\left(F\left(z'\right), F\left(z''\right)\right) \le \boldsymbol{c}_G^{(*)}\left(z', z''\right), \quad z', z'' \in G,$$
$$\boldsymbol{\gamma}_D\left(F(z); F'(z)X\right) \le \boldsymbol{\gamma}_G(z; X), \quad z \in G,\ X \in \mathbb{C}^n.$$

In particular, if F is biholomorphic, then the equalities hold.

In other words, the systems $(\boldsymbol{c}_G^{(*)})_G$ and $(\boldsymbol{\gamma}_G)_G$ are *holomorphically contractible*.

Remark 2.1.2 (cf. § 4.1). Observe that from the point of view of the general Schwarz–Pick lemma the Carathéodory pseudodistance and the Carathéodory–Reiffen pseudometric are minimal in the following sense:

if $(d_G)_G$ is any system of functions $d_G : G \times G \longrightarrow \mathbb{R}$ (resp. if $(\delta_G)_G$ is any system of functions $\delta_G : G \times \mathbb{C}^n \longrightarrow \mathbb{R}$), where G runs on all domains in all $\mathbb{C}^n$'s, such that

$$d_{\mathbb{D}}\left(F\left(z'\right), F\left(z''\right)\right) \le d_G\left(z', z''\right), \quad z', z'' \in G,\ F \in \mathcal{O}(G, \mathbb{D})$$
$$(\text{resp. } \delta_{\mathbb{D}}\left(F(z); F'(z)X\right) \le \delta_G(z; X), \quad z \in G,\ X \in \mathbb{C}^n,\ F \in \mathcal{O}(G, \mathbb{D}))$$

and $d_{\mathbb{D}} = \boldsymbol{p}$ (resp. $\delta_{\mathbb{D}} = \boldsymbol{\gamma}_{\mathbb{D}}$), then $\boldsymbol{c}_G \le d_G$ (resp. $\boldsymbol{\gamma}_G \le \delta_G$).

Note that in Theorem 2.1.1 we do not claim that $\boldsymbol{c}_D^{(*)}(F(z_0'), F(z_0'')) = \boldsymbol{c}_G^{(*)}(z_0', z_0'')$ for some $z_0', z_0'' \in G, z_0' \neq z_0''$, (resp. $\boldsymbol{\gamma}_D(F(z_0); F'(z_0)X_0) = \boldsymbol{\gamma}_G(z_0; X_0)$ for some $z_0 \in G, X_0 \in \mathbb{C}^n, X_0 \neq 0$) implies that F is biholomorphic (cf. Lemma 1.1.1(c); as a counterexample we have $\mathbb{C}^2 \ni (z_1, z_2) \overset{F}{\longmapsto} (z_1, 0) \in \mathbb{C}^2$). This is not true even for $D = G \subsetneq \mathbb{C}^n$ and even under more restrictive assumptions on z_0', z_0'' (resp. z_0, X_0) – take for instance $D = G = \mathbb{D}^2$; then, using Proposition 2.3.1(c), we easily conclude that $|z_1| = \boldsymbol{m}_{\mathbb{D}^2}((0,0), (z_1, z_2)) = \boldsymbol{m}_{\mathbb{D}^2}(F(0,0), F(z_1, z_2))$, provided that $|z_1| \ge |z_2|$ (resp. $|X_1| = \boldsymbol{\gamma}_{\mathbb{D}^2}((0,0); (X_1, X_2)) = \boldsymbol{\gamma}_{\mathbb{D}^2}(F(0,0); F'(0,0)(X_1, X_2))$, provided that $|X_1| \ge |X_2|$).

Remark 2.1.3. Let G_0, G, $\widetilde{G}$ be domains in $\mathbb{C}^n$ such that $\varnothing \neq G_0 \subset G \cap \widetilde{G}$ and $\widetilde{G} \not\subset G$. Assume that for any $f \in \mathcal{H}^\infty(G)$ there exists an $\widetilde{f} \in \mathcal{O}(\widetilde{G})$ with $\widetilde{f} = f$ on G_0. (Note that $\widetilde{f}$ is uniquely determined by f.) Then, for any $f \in \mathcal{H}^\infty(G)$, the function $\widetilde{f}$ belongs to $\mathcal{H}^\infty(\widetilde{G})$, and $\|\widetilde{f}\|_{\widetilde{G}} \le \|f\|_G$ (cf. Appendix B.1.12). In particular, if $G_0 = G \subset \widetilde{G}$, then

$\widetilde{G}$ is an $\mathcal{H}^\infty$-extension of G iff $\mathcal{H}^\infty(G) \subset \mathcal{O}(\widetilde{G})|_G$ iff $\mathcal{O}(G, \mathbb{D}) = \mathcal{O}(\widetilde{G}, \mathbb{D})|_G$.

Directly from the definitions and Remark 2.1.3 we get

Remark 2.1.4. If $G_0, G, \widetilde{G}$ are as in Remark 2.1.3, then

$$c_G^{(*)} \le c_{\widetilde{G}}^{(*)} \le c_{G_0}^{(*)} \text{ on } G_0 \times G_0, \quad \gamma_G \le \gamma_{\widetilde{G}} \le \gamma_{G_0} \text{ on } G_0 \times \mathbb{C}^n.$$

In particular, if the space $(G, c_G^{(*)})$ is complete, then G is an $\mathcal{H}^\infty$-domain of holomorphy (cf. Appendix B.7). If moreover $G \subset \widetilde{G}$, then

$$c_G^{(*)} = c_{\widetilde{G}}^{(*)}|_{G\times G}, \quad \gamma_G = \gamma_{\widetilde{G}}|_{G\times\mathbb{C}^n}.$$

For example, $c_{\mathbb{D}_*}^{(*)} = c_{\mathbb{D}}^{(*)}|_{\mathbb{D}_*\times\mathbb{D}_*}$, and hence the space $(\mathbb{D}_*, c_{\mathbb{D}_*}^{(*)})$ is not complete.

Remark 2.1.5. Let $D \subset \mathbb{C}^m$ be a *Liouville domain*, i.e., $\mathcal{H}^\infty(D) \simeq \mathbb{C}$. Then, for every domain $G \subset \mathbb{C}^n$, we have:

- $c_{G\times D}^{(*)}((a,b),(z,w)) = c_G^{(*)}(a,z), \quad (a,b),(z,w) \in G \times D,$
- $\gamma_{G\times D}((a,b);(X,Y)) = \gamma_G(a;X), \quad (a,b) \in G \times D, (X,Y) \in \mathbb{C}^n \times \mathbb{C}^m.$

A more general situation will be discussed in Remark 4.2.9.

2.2 Balanced domains

This is a preparatory section for discussing the Carathéodory pseudodistance and pseudometric (and other objects) for domains of special type. Notice that §§ 2.2.2, 2.2.3, 2.2.4 will be used only in §§ 4.3.5, 8.3, 7.1, respectively.

Let us recall the following standard notion. A domain $G \subset \mathbb{C}^n$ is said to be *balanced* if $\overline{\mathbb{D}} \cdot G = G$.

Let $D \subset \mathbb{C}^n$ be a balanced domain. Define its *Minkowski function* $\mathfrak{h}_D : \mathbb{C}^n \longrightarrow \mathbb{R}_+$,

$$\mathfrak{h}_D(X) := \inf\{t > 0 : X/t \in D\}, \quad X \in \mathbb{C}^n.$$

Remark 2.2.1 (The reader is asked to complete the details).

(a) $\mathfrak{h}_D(\lambda X) = |\lambda|\mathfrak{h}_D(X)$, $X \in \mathbb{C}^n, \lambda \in \mathbb{C}$.

(b) $D = \{X \in \mathbb{C}^n : \mathfrak{h}_D(X) < 1\}$.

(c) $\mathfrak{h}_D \equiv 0$ iff $D = \mathbb{C}^n$.

(d) $\mathfrak{h}_D$ is upper semicontinuous.

(e) If $h : \mathbb{C}^n \longrightarrow \mathbb{R}_+$ is an upper semicontinuous function such that $h(\lambda X) = |\lambda|h(X)$, $X \in \mathbb{C}^n$, $\lambda \in \mathbb{C}$, then the set $D := \{X \in \mathbb{C}^n : h(X) < 1\}$ is a balanced domain and $h \equiv \mathfrak{h}_D$.

(f) There exists a $c > 0$ such that $\mathfrak{h}_D \le c\|\ \|$.

(g) Let $\varphi : \partial\mathbb{B}_n \longrightarrow \mathbb{R}_+$ be upper semicontinuous. Then, the function

$$\mathbb{C}^n \ni X \longmapsto \begin{cases} \varphi(\frac{X}{\|X\|})\|X\|, & \text{if } X \neq 0 \\ 0, & \text{if } X = 0 \end{cases}$$

is the Minkowski function of some balanced domain (cf. (e)). Observe that, in fact, every Minkowski function is of the above form.

(h) If $\mathfrak{h}$ is continuous, then $\partial D = \{X \in \mathbb{C}^n : \mathfrak{h}_D(X) = 1\}$ (cf. Exercise 2.12.5).

(i) D is convex iff $\mathfrak{h}_D$ is a seminorm, i.e., $\mathfrak{h}_D(X+Y) \le \mathfrak{h}_D(X) + \mathfrak{h}_D(Y)$, $X, Y \in \mathbb{C}^n$.

(j) If $\mathfrak{h}_D$ is a seminorm, then $\mathfrak{h}_D$ is continuous.

(k) If $\mathfrak{h}_D$ is a seminorm, then $\mathfrak{h}_D^{-1}(0)$ is a linear subspace of $\mathbb{C}^n$ (cf. Exercise 2.12.5).

(l) If D is bounded, then $\mathfrak{h}_D^{-1}(0) = \{0\}$ (cf. Exercise 2.12.5).

(m) If $\mathfrak{h}_D$ is continuous, then D is bounded iff $\mathfrak{h}_D^{-1}(0) = \{0\}$ iff there exists a $c > 0$ such that $\mathfrak{h}_D \ge c\|\ \|$.

(n) If $(D_s)_{s=1}^{\infty} \subset \mathbb{C}^n$ is a sequence of balanced domains with $D_s \nearrow D$, then $\mathfrak{h}_{D_s} \searrow \mathfrak{h}_D$.

(o) If $D_j \subset \mathbb{C}^{n_j}$ is a balanced domain, $j = 1, 2$, then

$$\mathfrak{h}_{D_1 \times D_2}(X) = \max\{\mathfrak{h}_{D_1}(X_1), \mathfrak{h}_{D_2}(X_2)\}, \quad X = (X_1, X_2) \in \mathbb{C}^{n_1} \times \mathbb{C}^{n_2}.$$

(p) If $L : \mathbb{C}^n \longrightarrow \mathbb{C}^n$ is a $\mathbb{C}$-linear isomorphism such that $L(D) = D$, then $\mathfrak{h}_D \circ L \equiv \mathfrak{h}_D$.

(q) Additionally assume, that D is a *Reinhardt domain*, i.e., D is *invariant under n-rotations*

$$\mathbb{C}^n \ni (z_1, \dots, z_n) \overset{R_\theta}{\longmapsto} (e^{i\theta_1} z_1, \dots, e^{i\theta_n} z_n) \in \mathbb{C}^n, \quad \theta = (\theta_1, \dots, \theta_n) \in \mathbb{R}^n;$$

see § 2.10. Then, $\mathfrak{h}_D(z_1, \dots, z_n) = \mathfrak{h}_D(|z_1|, \dots, |z_n|)$, $(z_1, \dots, z_n) \in \mathbb{C}^n$.

(r) Assume that D is a *complete Reinhardt domain*, i.e., for any $z = (z_1, \dots, z_n) \in D$ and $\lambda = (\lambda_1, \dots, \lambda_n) \in \overline{\mathbb{D}}^n$, the point $\lambda \cdot z := (\lambda_1 z_1, \dots, \lambda_n z_n)$ belongs to D (in particular, D must be a balanced Reinhardt domain). Then, $\mathfrak{h}_D(\lambda \cdot z) \le \mathfrak{h}_D(z)$, $z \in \mathbb{C}^n$, $\lambda \in \overline{\mathbb{D}}^n$. Moreover, $\mathfrak{h}_D$ is continuous (cf. [269], Lemma 1.8.3).

Lemma 2.2.2. *Let $D \subset \mathbb{C}^n$ be balanced. Then, there exists a sequence $(D_s)_{s=1}^{\infty}$ of bounded balanced domains with continuous Minkowski functions such that $D_s \nearrow D$.*

Proof. Take an arbitrary sequence of continuous functions $f_s : \partial\mathbb{B}_n \longrightarrow \mathbb{R}_+$ such that $f_s \searrow \mathfrak{h}_D$ pointwise on $\partial\mathbb{B}_n$. Define

$$g_s(X) := \frac{1}{2\pi}\int_0^{2\pi} f_s(e^{i\theta}X)d\theta, \quad X \in \partial\mathbb{B}_n.$$

Then, g_s is continuous, $g_s(\zeta X) = g_s(X)$, $X \in \partial\mathbb{B}_n$, $\zeta \in \mathbb{T}$, and $g_s \searrow \mathfrak{h}_D$ pointwise on $\partial\mathbb{B}_n$. Let $p_s : \mathbb{C}^n \longrightarrow \mathbb{R}_+$

$$p_s(x) := \begin{cases} g_s(\frac{X}{\|X\|})\|X\|, & \text{if } X \neq 0 \\ 0, & \text{if } X = 0 \end{cases}, \quad q_s(X) := \max\{p_s(X), \|X\|/s\}.$$

Observe that q_s is continuous, $q_s(\lambda X) = |\lambda| q_s(X)$, $q_s(x) \geq \|X\|/s$, $X \in \mathbb{C}^n$, $\lambda \in \mathbb{C}$, and $q_s \searrow \mathfrak{h}_D$ pointwise on $\mathbb{C}^n$. Thus, $D_s := \{X \in \mathbb{C}^n : q_s(X) < 1\}$, $s \in \mathbb{N}$, is the required sequence. □

2.2.1 Operator $\mathfrak{h} \longmapsto \widehat{\mathfrak{h}}$

This subsection is based on [315, 308], and [382].

Let $D \subset \mathbb{C}^n$ be a balanced domain and let $\widehat{D} \subset \mathbb{C}^n$ be the convex hull of D.

Remark 2.2.3 (The reader is asked to complete the details).

(a)
$$\widehat{D} = \left\{\sum_{j=1}^m t_j X_j : m \in \mathbb{N},\ t_1,\dots,t_m > 0, X_1,\dots,X_m \in D,\ \sum_{j=1}^m t_j \leq 1\right\}.$$

(b) $\widehat{D}$ is a balanced domain.

(c) $\widehat{D} = \operatorname{int}\bigcap_{U\in\mathfrak{U}(D)} U = \operatorname{int}\bigcap_{U\in\mathfrak{U}_0(D)} U$, where

$$\mathfrak{U}(D) := \{U : U \text{ is a convex domain in } \mathbb{C}^n \text{ with } D \subset U\},$$
$$\mathfrak{U}_0(D) := \{U : U \text{ is a convex balanced domain in } \mathbb{C}^n \text{ with } D \subset U\}.$$

(d) $\mathfrak{h}_{\widehat{D}} = \sup\{q : q \text{ is a } \mathbb{C}\text{-seminorm},\ q \leq \mathfrak{h}_D\}$; in particular, $\mathfrak{h}_{\widehat{D}}$ is a $\mathbb{C}$-seminorm.

(e) If $D_s \nearrow D$, then $\widehat{D}_s \nearrow \widehat{D}$.

(f) $\widehat{D_1 \times D_2} = \widehat{D}_1 \times \widehat{D}_2$.

(g) If $L : \mathbb{C}^n \longrightarrow \mathbb{C}^n$ is a $\mathbb{C}$-linear isomorphism such that $L(D) = D$, then $L(\widehat{D}) = \widehat{D}$. In particular, if D is Reinhardt, then so is $\widehat{D}$.

For $m \in \mathbb{N}$ define $\mathfrak{h}_D^{(m)} : \mathbb{C}^n \longrightarrow \mathbb{R}_+$,

$$\mathfrak{h}_D^{(m)}(X) := \inf\left\{\sum_{j=1}^{m} \mathfrak{h}_D(X_j) : X_1, \dots, X_m \in \mathbb{C}^n,\ X = \sum_{j=1}^{m} X_j\right\}, \quad X \in \mathbb{C}^n.$$

Remark 2.2.4 (The reader is asked to complete the details).

(a) $\mathfrak{h}_D^{(m)}(\lambda X) = |\lambda|\mathfrak{h}_D^{(m)}(X)$, $X \in \mathbb{C}^n$, $\lambda \in \mathbb{C}$.

(b) $\mathfrak{h}_D^{(m+1)} \le \mathfrak{h}_D^{(m)} \le \mathfrak{h}_D^{(1)} = \mathfrak{h}_D$, $\mathfrak{h}_D^{(p+q)}(X+Y) \le \mathfrak{h}_D^{(p)}(X) + \mathfrak{h}_D^{(q)}(Y)$.

(c) If $\mathfrak{h}_D^{(m+1)} \equiv \mathfrak{h}_D^{(m)}$, then $\mathfrak{h}_D^{(k)} \equiv \mathfrak{h}_D^{(m)}$ for all $k \ge m+1$.

(d) $\widehat{\mathfrak{h}}_D := \lim_{m\to+\infty} \mathfrak{h}_D^{(m)}$ is a $\mathbb{C}$-seminorm.

(e) $\widehat{D} = \{X \in \mathbb{C}^n : \widehat{\mathfrak{h}}_D(X) < 1\}$. In particular, $\mathfrak{h}_{\widehat{D}} = \widehat{\mathfrak{h}}_D$.

In fact, the inclusion "$\subset$" is obvious. If $\widehat{\mathfrak{h}}_D(X) < 1$, then there exist m and $X_1, \dots, X_m \in \mathbb{C}^n$ such that $X = \sum_{j=1}^m X_j$ and $\sum_{j=1}^m \mathfrak{h}_D(X_j) < 1$. Take an $\varepsilon > 0$ so small that $\sum_{j=1}^m t_j < 1$, where $t_j := \mathfrak{h}_D(X_j) + \varepsilon$. Put $X_j' := \frac{X_j}{t_j}$ Note that $X_1', \dots, X_m' \in D$ and $X = \sum_{j=1}^m t_j X_j'$. Thus, $X \in \widehat{D}$.

(f) Using the Hahn–Banach theorem, we get

$$\widehat{\mathfrak{h}}_D(X) = \sup\{|Y \bullet X| : Y \in \varGamma_D\},$$

where $\varGamma_D := \{Y \in \mathbb{C}^n : \forall_{Z\in D} : \ |Y \bullet Z| \le 1\}$.

(g) $\mathfrak{h}_{\widehat{D}} = \mathfrak{h}_D^{(2n+1)}$.

In fact, since both functions are absolutely homogeneous, we only need to show that $\widehat{D} = \{X \in \mathbb{C}^n : \mathfrak{h}_D^{(2n+1)}(X) < 1\}$. We already know that the inclusion "$\supset$" is true. Take an $X \in \widehat{D}$. Then, by the Carathéodory theorem,[1] there exist $m \le 2n+1$, $X_1, \dots X_m \in D$, $t_1, \dots, t_m > 0$, $\sum_{j=1}^m t_j = 1$ such that $X = \sum_{j=1}^m t_j X_j$. Hence, $\mathfrak{h}_D^{(2n+1)}(X) \le \sum_{j=1}^m \mathfrak{h}_D(t_j X_j) = \sum_{j=1}^m t_j \mathfrak{h}_D(X_j) < \sum_{j=1}^m t_j = 1$.

[1] **Theorem** (Carathéodory theorem for convex hulls). *Let $A \subset \mathbb{R}^N$. Then,*

$$\operatorname{conv}(A) = \left\{\sum_{j=1}^{m} t_j X_j : m \le N+1,\ X_1, \dots, X_m \in A,\ t_1, \dots, t_m > 0,\ \sum_{j=1}^{m} t_j = 1\right\}.$$

(h) If $\mathfrak{h}_D$ is continuous and $\mathfrak{h}_D(X) > 0$ for $X \neq 0$, then for any $m \in \mathbb{N}$ and $X_0 \in \mathbb{C}^n$ there exist $X_1, \dots, X_m \in \mathbb{C}^n$ such that $X_0 = \sum_{j=1}^m X_j$ and $\mathfrak{h}_D^{(m)}(X_0) = \sum_{j=1}^m \mathfrak{h}_D(X_j)$.
In fact, fix a $c > 0$ such that $\mathfrak{h}_D(X) \geq c\|X\|$, $X \in \mathbb{C}^n$. Let $X_{k,1}, \dots, X_{k,m} \in \mathbb{C}^n$ be such that $X_0 = \sum_{j=1}^m X_{k,j}$ and $\sum_{j=1}^m \mathfrak{h}_D(X_{k,j}) \searrow \widehat{\mathfrak{h}}_D(X_0)$. We have $c\|X_{k,j}\| \leq \mathfrak{h}_D(X_{k,j}) \leq C = \text{const}$, $k \in \mathbb{N}$, $j = 1, \dots, m$. Thus, we may assume that $X_{k,j} \longrightarrow X_j$ when $k \longrightarrow +\infty$, $j = 1, \dots, m$. Obviously, $X_0 = \sum_{j=1}^m X_j$. Moreover, the continuity of $\mathfrak{h}_D$ implies that $\sum_{j=1}^m \mathfrak{h}_D(X_j) = \mathfrak{h}_D^{(m)}(X_0)$.

Lemma 2.2.5.

(a) (cf. [315])

$$\mathfrak{h}_{\widehat{D}}(X) = \inf\Bigg\{\sum_{j=1}^m \mathfrak{h}_D(X_j) : m \in \mathbb{N},\ X_1, \dots, X_m \in \mathbb{C}^n \text{ are } \mathbb{R}\text{-linearly independent},\ X = X_1 + \dots + X_m\Bigg\}, \quad X \in \mathbb{C}^n.$$

In particular, $\mathfrak{h}_{\widehat{D}} = \mathfrak{h}_D^{(2n)}$.

(b) (cf. [382]) $\mathfrak{h}_{\widehat{D}} = \mathfrak{h}_D^{(2n-1)}$.

Proof. (a) In view of Remark 2.2.4(g), we only need to show that if $X = \sum_{j=1}^m X_j$, $m \leq 2n+1$, and $X_1, \dots, X_m \in (\mathbb{C}^n)_*$ are $\mathbb{R}$-linearly dependent, then there exist $X_1', \dots, X_{m-1}'$ such that $X = \sum_{j=1}^{m-1} X_j'$ and $\sum_{j=1}^{m-1} \mathfrak{h}_D(X_j') \leq \sum_{j=1}^m \mathfrak{h}_D(X_j)$. Suppose that $\sum_{j=1}^m s_j X_j = 0$ with $\sum_{j=1}^m |s_j| > 0$. We may assume that

$$\sum_{j=1}^m s_j \mathfrak{h}_D(X_j) \geq 0 \text{ and } \exists_{j_0 \in \{1,\dots,m\}} : s_{j_0} > 0.$$

We take the smallest $t_0 > 0$ such that $1 - t_0 s_j = 0$ for a $j \in \{1, \dots, m\}$ and then we take the minimal index $j_{\min}$ with this property. We may assume $j_{\min} = m$. Then,

$$X = \sum_{j=1}^m X_j - t_0 \sum_{j=1}^m s_j X_j = \sum_{j=1}^{m-1} (1 - t_0 s_j) X_j, \quad 1 - t_0 s_j > 0,\ j = 1, \dots, m-1,$$

$$\begin{aligned}\sum_{j=1}^{m-1} \mathfrak{h}_D((1 - t_0 s_j) X_j) &= \sum_{j=1}^{m-1} (1 - t_0 s_j) \mathfrak{h}_D(X_j) \\ &= \sum_{j=1}^m \mathfrak{h}_D(X_j) - t_0 \sum_{j=1}^m s_j \mathfrak{h}_D(X_j) \leq \sum_{j=1}^m \mathfrak{h}_D(X_j).\end{aligned}$$

(b) Step 1°. *If $D_s \nearrow D$ is a sequence of balanced domains with $\mathfrak{h}_{\widehat{D}_s} = \mathfrak{h}_{D_s}^{(2n-1)}$, $s \in \mathbb{N}$, then $\mathfrak{h}_{\widehat{D}} = \mathfrak{h}_D^{(2n-1)}$.*

We have $\widehat{D}_s \nearrow \widehat{D}$ and hence $\mathfrak{h}_{\widehat{D}_s} \searrow \mathfrak{h}_{\widehat{D}}$. Thus,

$$\mathfrak{h}_{\widehat{D}} \le \mathfrak{h}_D^{(2n-1)} \le \mathfrak{h}_{D_s}^{(2n-1)} = \mathfrak{h}_{\widehat{D}_s} \searrow \mathfrak{h}_{\widehat{D}}.$$

Step 2°. *The result is true if D is a bounded balanced domain with continuous Minkowski function.*

Fix an X_0 with $\mathfrak{h}_{\widehat{D}}(X_0) = 1$. By (a) and Remark 2.2.4(h), there exist vectors $X_1, \dots, X_{2n} \in \mathbb{C}^n$ such that $X_0 = \sum_{j=1}^{2n} X_j$ and $\sum_{j=1}^{2n} \mathfrak{h}_D(X_j) = 1$. Similar to (a), if $X_1, \dots, X_{2n}$ are $\mathbb{R}$-linearly dependent, then we are done. Hence, we suppose that $X_1, \dots, X_{2n}$ are $\mathbb{R}$-linearly independent. In particular, $0 < \mathfrak{h}_D(X_j) < 1$. Let $X_j' := \frac{X_j}{\mathfrak{h}_D(X_j)}$; note that $X_j' \in \partial D$. Define

$$S := \left\{ \sum_{j=1}^{2n} t_j X_j' : t_1, \dots, t_{2n} \ge 0,\ \sum_{j=1}^{2n} t_j = 1 \right\} \subset \overline{\widehat{D}}.$$

Observe that

$$X_0 = \sum_{j=1}^{2n} \mathfrak{h}_D(X_j) X_j' \in \operatorname{int} S \subset \widehat{D};$$

a contradiction.

Step 3°. It remains to use Steps 1°, 2° and Lemma 2.2.2 □

Example 2.2.6 (cf. [382]). For $n \ge 2$ there exists a bounded balanced pseudoconvex domain $D \subset \mathbb{C}^n$ with continuous Minkowski function such that $\mathfrak{h}_{\widehat{D}} \not\equiv \mathfrak{h}_D^{(2n-2)}$.

Put

$$D := \left\{ (z_1, \dots, z_n) \in \mathbb{C}^n : \sum_{j=2}^{n} \left(2|z_1^3 - z_j^3| + |z_1^3 + z_j^3| \right) < 2(n-1) \right\} \qquad (\dagger)$$

and suppose that $\mathfrak{h}_{\widehat{D}} \equiv \mathfrak{h}_D^{(2n-2)}$. Observe that $D \subset \mathbb{D} \times \mathbb{C}^{n-1}$. In particular, $\widehat{D} \subset \mathbb{D} \times \mathbb{C}^{n-1}$. Let $L := \{1\} \times \mathbb{C}^{n-1}$, $S = \sqrt[3]{1} \subset \overline{\mathbb{D}}$, $\widehat{S} := \operatorname{conv}(S)$. One can easily check that $L \cap \partial D = \{1\} \times S^{n-1}$. Consequently, $L \cap \partial \widehat{D} = \operatorname{conv}(L \cap \partial D) = \{1\} \times \widehat{S}^{n-1}$.

In fact, if $X_0 \in L \cap \partial \widehat{D}$, then $X_0 = \sum_{j=1}^{m} t_j X_j$ with $X_1, \dots, X_m \in \overline{D} \subset \overline{\mathbb{D}} \times \mathbb{C}^{n-1}$, $t_1, \dots, t_m > 0$, $\sum_{j=1}^{m} t_j = 1$. In particular, $|X_{j,1}| \le 1$, $j = 1, \dots, m$. Since $1 = X_{0,1} = \sum_{j=1}^{m} t_j X_{j,1}$, we easily conclude that $X_1, \dots, X_m \in L$.

Thus, $L \cap \partial\widehat{D}$ is a convex set of real dimension $(2n-2)$. On the other hand,

$$L \cap \partial\widehat{D} = \left\{X \in L : \mathfrak{h}_{\widehat{D}}(X) = 1\right\} = \left\{X \in L : \mathfrak{h}_D^{(2n-2)}(X) = 1\right\}.$$

Therefore, every $X \in L \cap \partial\widehat{D}$ may be written as $X = \sum_{j=1}^m X_j$ with $m \le 2n-2$, $\sum_{j=1}^m \mathfrak{h}_D(X_j) = 1$, and $\mathfrak{h}_D(X_j) > 0$, $j = 1, \dots, m$ (recall that D is bounded and $\mathfrak{h}_D$ is continuous). Hence, $X = \sum_{j=1}^m t_j X'_j$, where $t_j := \mathfrak{h}_D(X_j)$, $X'_j := X_j/t_j \in \partial D \subset \overline{\mathbb{D}} \times \mathbb{C}^{n-1}$. In particular,

$$1 = \sum_{j=1}^m t_j X'_{j,1} = \sum_{j=1}^m t_j \operatorname{Re} X'_{j,1} \le \sum_{j=1}^m t_j |X'_{j,1}| \le \sum_{j=1}^m t_j = 1.$$

Thus, $X'_1, \dots, X'_m \in L$ and therefore,

$$L \cap \partial\widehat{D} \subset \bigcup_{\substack{Y_1,\dots,Y_m \in \{1\} \times S^{n-1} \\ m \le 2n-2}} \operatorname{conv}\{Y_1, \dots, Y_m\}.$$

It remains to observe that the right hand side is the finite union of convex sets of at most $(2n-3)$ dimensions – a contradiction.

? Find an effective description of $\widehat{D}$, where D is as in (†).

2.2.2 Operator $\mathfrak{h} \longmapsto \widetilde{\mathfrak{h}}$

Let $D \subset \mathbb{C}^n$ be balanced. Define

$$\widetilde{\mathfrak{h}}_D := \sup\{h : h \in \mathcal{PSH}(\mathbb{C}^n),\ h \le \mathfrak{h}_D\}, \quad \widetilde{D} := \{X \in \mathbb{C}^n : \widetilde{\mathfrak{h}}_D(X) < 1\}.$$

Remark 2.2.7 (The reader is asked to complete the details).

(a) $0 \le \widetilde{\mathfrak{h}}_D \le \mathfrak{h}_D$.

(b) $\widetilde{\mathfrak{h}}_D \in \mathcal{PSH}(\mathbb{C}^n)$.

In fact, by Appendix B.4.16, the function $(\widetilde{\mathfrak{h}}_D)^*$ is psh, and $\widetilde{\mathfrak{h}}_D \le (\widetilde{\mathfrak{h}}_D)^* \le \mathfrak{h}_D$. Thus, $\widetilde{\mathfrak{h}}_D = (\widetilde{\mathfrak{h}}_D)^* \in \mathcal{PSH}(\mathbb{C}^n)$.

(c) $\widetilde{\mathfrak{h}}_D(\lambda X) = |\lambda|\widetilde{\mathfrak{h}}_D(X)$, $X \in \mathbb{C}^n$, $\lambda \in \mathbb{C}$; in particular, $\widetilde{D}$ is a balanced domain and $\widetilde{\mathfrak{h}}_D = \mathfrak{h}_{\widetilde{D}}$.

In fact, take an $\lambda \in \mathbb{C}_*$ and consider the function $\mathbb{C}^n \ni X \overset{u}{\longmapsto} \frac{1}{|\lambda|}\widetilde{\mathfrak{h}}_D(\lambda X)$. Then, $u \in \mathcal{PSH}(\mathbb{C}^n)$ and $u \le \mathfrak{h}_D$. Consequently, $u \le \widetilde{\mathfrak{h}}_D$, i.e., $\widetilde{\mathfrak{h}}_D(\lambda X) \le |\lambda|\widetilde{\mathfrak{h}}_D(X)$, $X \in \mathbb{C}^n$, $\lambda \in \mathbb{C}$. To prove the equality, observe that $|\lambda|\widetilde{\mathfrak{h}}_D(X) = |\lambda|\widetilde{\mathfrak{h}}_D(\frac{1}{\lambda}\lambda X) \le |\lambda||\frac{1}{\lambda}|\widetilde{\mathfrak{h}}_D(\lambda X) = \widetilde{\mathfrak{h}}_D(\lambda X)$.

(d) $\widetilde{\mathfrak{h}}_D = \sup\{h : h \in \mathcal{PSH}(\mathbb{C}^n),\ h \le \mathfrak{h}_D,\ h \text{ is a Minkowski function}\}$; in particular, $\widetilde{D} = \operatorname{int} \bigcap_{U \in \mathfrak{U}} U$, where

$$\mathfrak{U} := \{U : U \text{ is a balanced domain in } \mathbb{C}^n,\ D \subset U,\ \mathfrak{h}_U \in \mathcal{PSH}(\mathbb{C}^n)\}.$$

(e) Note that $\widetilde{D}$ is the smallest pseudoconvex balanced domain that contains D. Recall that the envelope of holomorphy of a balanced domain is again a balanced domain in $\mathbb{C}^n$ (cf. [265], Remark 1.9.6(e)). Therefore, $\widetilde{D}$ is nothing other than the envelope of holomorphy of D.

(f) If $(D_s)_{s=1}^{\infty}$ is a sequence of balanced domains such that $D_s \nearrow D$, then $\widetilde{\mathfrak{h}}_{D_s} \searrow \widetilde{\mathfrak{h}}_D$.

(g) If $D_j \subset \mathbb{C}^{n_j}$ is a balanced domain, $j = 1, 2$, then

$$\widetilde{\mathfrak{h}}_{D_1 \times D_2}(X) = \max\{\widetilde{\mathfrak{h}}_{D_1}(X_1), \widetilde{\mathfrak{h}}_{D_2}(X_2)\}, \quad X = (X_1, X_2) \in \mathbb{C}^{n_1} \times \mathbb{C}^{n_2}$$

(use (e) and [265], Proposition 1.8.15(b); another proof can be given via properties of the Kobayashi–Royden pseudometric, see Chapter 3).

(h) If $L : \mathbb{C}^n \longrightarrow \mathbb{C}^n$ is a $\mathbb{C}$-linear isomorphism such that $L(D) = D$, then $L(\widetilde{D}) = \widetilde{D}$. In particular, if D is Reinhardt, then so is $\widetilde{D}$.

Notice that, in fact, the following general result is true (cf. [265], Theorem 2.12.1). If $L : D \longrightarrow D$ is biholomorphic, then there exists a biholomorphic mapping $\widetilde{L} : \widetilde{D} \longrightarrow \widetilde{D}$ such that $\widetilde{L}|_D = L$.

2.2.3 Operator $\mathfrak{h} \longmapsto \mathbb{W}\mathfrak{h}$

Let $h : \mathbb{C}^n \longrightarrow \mathbb{R}_+$ be a $\mathbb{C}$-seminorm. Put:

$I = I(h) := \{X \in \mathbb{C}^n : h(X) < 1\}$ (I is convex),

$V = V(h) := \{X \in \mathbb{C}^n : h(X) = 0\} \subset I$ (V is a vector subspace of $\mathbb{C}^n$),

$U = U(h) :=$ the orthogonal complement of V with respect to the standard Hermitian scalar product $\langle z, w \rangle := \sum_{j=1}^{n} z_j \overline{w}_j$ in $\mathbb{C}^n$,

$I_0 := I \cap U$, $h_0 := h|_U$ (h_0 is a norm on U, $I = I_0 + V$).

For any *pseudo-Hermitian scalar product* $s : \mathbb{C}^n \times \mathbb{C}^n \longrightarrow \mathbb{C}$,[2] let

$$q_s(X) := \sqrt{s(X, X)},\ X \in \mathbb{C}^n, \quad \mathbb{E}(s) := \{X \in \mathbb{C}^n : q_s(X) < 1\}.$$

---

[2] That is,

- $s(\cdot, w) : \mathbb{C}^n \longrightarrow \mathbb{C}$ is $\mathbb{C}$-linear for any $w \in \mathbb{C}^n$,
- $s(z, w) = \overline{s(w, z)}$ for any $z, w \in \mathbb{C}^n$,
- $s(z, z) \ge 0$ for any $z \in \mathbb{C}^n$ (if $s(z, z) > 0$ for any $z \in (\mathbb{C}^n)_*$, then s is a *Hermitian scalar product*).

Consider the family $\mathcal{F}$ of all pseudo-Hermitian scalar products $s : \mathbb{C}^n \times \mathbb{C}^n \longrightarrow \mathbb{C}$ such that $I \subset \mathbb{E}(s)$, equivalently, $q_s \leq h$. In particular,

$$V \subset I = I_0 + V \subset \mathbb{E}(s) = \mathbb{E}(s_0) + V,$$

where $s_0 := s|_{U \times U}$ (note that $\mathbb{E}(s_0) = \mathbb{E}(s) \cap U$). Let $\mathrm{Vol}(s_0)$ denote the volume of $\mathbb{E}(s_0)$ with respect to the Lebesgue measure of U. Since I_0 is bounded, there exists an $s \in \mathcal{F}$ with $\mathrm{Vol}(s_0) < +\infty$. Observe that for any basis $e = (e_1, \dots, e_m)$ of U ($m := \dim_{\mathbb{C}} U$) we have

$$\mathrm{Vol}(s_0) = \frac{C(e)}{\det S},$$

where $C(e) > 0$ is a constant (independent of s) and $S = S(s_0)$ denotes the matrix representation of s_0 in the basis e, i.e., $S_{j,k} := s(e_j, e_k)$, $j,k = 1, \dots, m$. In particular, if $U = \mathbb{C}^m \times \{0\}^{n-m}$ and $e = (e_1, \dots, e_m)$ is the canonical basis, then $C(e) = \mathcal{L}^{2m}(\mathbb{B}_m)$, where $\mathcal{L}^{2m}$ denotes the Lebesgue measure in $\mathbb{C}^m$. We are interested in finding an $s \in \mathcal{F}$ for which $\mathrm{Vol}(s_0)$ is minimal or, equivalently, $\det S(s_0)$ is maximal.

Observe that, if s has the above property with respect to h (i.e., the volume of $\mathbb{E}(s_0)$ is minimal), then, for any $\mathbb{C}$-linear isomorphism $L : \mathbb{C}^n \longrightarrow \mathbb{C}^n$, the scalar product

$$\mathbb{C}^n \times \mathbb{C}^n \ni (X, Y) \overset{L(s)}{\longmapsto} s(L(X), L(Y)) \in \mathbb{C}$$

has the analogous property with respect to $h \circ L$. In particular, this permits us to reduce the situation to the case where $U = \mathbb{C}^m \times \{0\}^{n-m}$ and then to assume that $m = n$ (by restricting all the above objects to $\mathbb{C}^m \simeq \mathbb{C}^m \times \{0\}^{n-m}$).

Lemma 2.2.8. *There exists exactly one element $s^h \in \mathcal{F}$ such that*

$$\mathrm{Vol}(s_0^h) = \min\{\mathrm{Vol}(s_0) : s \in \mathcal{F}\} < +\infty.$$

The ellipsoid $\mathbb{E}(s^h)$ is called the *John ellipsoid (for I)* (cf. [282]).

Proof. ([536, 537]) We may assume $U(h) = \mathbb{C}^n$. First, we prove that the set $\mathcal{F}$ is compact. It is clear that $\mathcal{F}$ is closed. To prove that $\mathcal{F}$ is bounded, observe that

$$|s(\boldsymbol{e}_j, \boldsymbol{e}_k)| \leq \sqrt{s(\boldsymbol{e}_j, \boldsymbol{e}_j) s(\boldsymbol{e}_k, \boldsymbol{e}_k)} = q_s(\boldsymbol{e}_j) q_s(\boldsymbol{e}_k) \leq h(\boldsymbol{e}_j) h(\boldsymbol{e}_k),$$
$$s \in \mathcal{F},\ j,k = 1, \dots, n,$$

where $\boldsymbol{e}_1, \dots, \boldsymbol{e}_n$ is the canonical basis in $\mathbb{C}^n$. Consequently, the entries of the matrix $S(s)$ are bounded (by a constant independent of s).

Recall that

$$\mathrm{Vol}(s) = \frac{\mathcal{L}^{2n}(\mathbb{B}_n)}{\det S(s)},$$

Now, using compactness of $\mathcal{F}$, we see that there exists an $s^h \in \mathcal{F}$ such that

$$\operatorname{Vol}(s^h) = \min\{\operatorname{Vol}(s) : s \in \mathcal{F}\} < +\infty.$$

It remains to show that s^h is uniquely determined. Suppose that $s', s'' \in \mathcal{F}$, $s' \neq s''$, are both minimal, and let S', S'' denote the matrix representation of s', s'', respectively. We know that $\mu := \det S' = \det S''$ is maximal (with respect to any basis $(e_1, \dots, e_n)$) in the class $\mathcal{F}$. Take a basis $e_1, \dots, e_n$ such that the matrix $A := S''(S')^{-1}$ is diagonal and let $d_1, \dots, d_n$ be the diagonal elements. Note that $1 = \det A = d_1 \cdots d_n$ and that for at least one $j \in \{1, \dots, n\}$ we have $d_j \neq 1$. Put $s := \frac{1}{2}(s' + s'')$. Then, $s \in \mathcal{F}$. Let $S = S(s)$ be the matrix representation of s. We have

$$\begin{aligned}\det S = \frac{1}{2^n}\det(S' + S'') &= \frac{1}{2^n}\det(\mathbb{I}_n + A)\det S' \\ &= \frac{1+d_1}{2}\cdots\frac{1+d_n}{2}\,\mu > \sqrt{d_1 \cdots d_n}\,\mu = \mu;\end{aligned}$$

contradiction ($\mathbb{I}_n$ denotes the unit matrix). □

Put $\widehat{s}^h := m \cdot s^h$ ($m := \dim U(h)$), $\mathbb{W}h := q_{\widehat{s}^h}$ ($\mathbb{W}h(X) = \sqrt{m s^h(X, X)}$, $X \in \mathbb{C}^n$). Obviously, $\mathbb{W}h \leq \sqrt{m}h$ and $\mathbb{W}h \equiv \sqrt{m}h$ iff $h = q_s$ for some pseudo-Hermitian scalar product s. For instance, $\mathbb{W}\|\ \| = \sqrt{n}\|\ \|$, where $\|\ \|$ is the Euclidean norm in $\mathbb{C}^n$. Moreover, $\mathbb{W}(\mathbb{W}h) \equiv \sqrt{m}\mathbb{W}h$.

If $G \subset \mathbb{C}^n$ is a balanced domain, then we put $\mathbb{W}\mathfrak{h}_G := \mathbb{W}\widehat{\mathfrak{h}}_G$.

Remark 2.2.9 (The reader is asked to complete the details).

(a) Let $L : \mathbb{C}^n \longrightarrow \mathbb{C}^n$ be a $\mathbb{C}$-linear isomorphism such that $|\det L| = 1$ and $h \circ L \equiv h$. Then, $\operatorname{Vol}(s_0^h) = \operatorname{Vol}((L(s^h))_0)$ and hence $s^h = L(s^h)$, i.e., $s^h(X, Y) = s^h(L(X),$
$L(Y))$, $X, Y \in \mathbb{C}^n$. In particular, $(\mathbb{W}h) \circ L \equiv \mathbb{W}h$.

(b) Assume additionally that D is Reinhardt (cf. Remark 2.2.1(q)). Then, by (a), we get $s^h(X, Y) = s^h(R_\theta(X), R_\theta(Y))$, $X, Y \in \mathbb{C}^n$, $\theta \in \mathbb{R}^n$, which implies that $s^h(X, Y) = \sum_{j=1}^n a_j X_j \overline{Y}_j$ with $a_1, \dots, a_n \in \mathbb{R}_+$. In particular, $(\mathbb{W}h)(X) = \sqrt{m \sum_{j=1}^n a_j |X_j|^2}$, $X \in \mathbb{C}^n$.

(c) (Cf. [286]) Assume additionally that $D \subset\subset \mathbb{C}^2$ is Reinhardt. Define

$$\begin{aligned}&\mathbb{C}^2 \ni (z, w) \overset{\Psi}{\longmapsto} (|z|^2, |w|^2) \in \mathbb{R}_+^2, \\ &\mathcal{E}_{a,b} := \left\{(X, Y) \in \mathbb{C}^2 : \frac{|X|^2}{a} + \frac{|Y|^2}{b} < 1\right\}, \\ &T_{a,b} := \left\{(t, u) \in \mathbb{R}_+^2 : \frac{t}{a} + \frac{u}{b} < 1\right\}, \quad a, b > 0.\end{aligned}$$

Note that $D \subset \mathcal{E}_{a,b}$ iff $\Psi(D) \subset \Psi(\mathcal{E}_{a,b}) = T_{a,b}$. Observe that $\mathcal{E}_{a_0,b_0}$ has the minimal volume in the category of all ellipsoids $\mathcal{E}_{a,b}$ with $D \subset \mathcal{E}_{a,b}$ iff T_{a_0,b_0} has the minimal area in the category of all triangles $T_{a,b}$ with $\Psi(D) \subset T_{a,b}$. Thus, in view of (b), we have an effective geometric tool for finding $\mathbb{W}h$ in the case where D is a bounded balanced Reinhardt domain (cf. Examples 2.2.12, 2.2.13).

Proposition 2.2.10.

(a) $h \le \mathbb{W}h \le \sqrt{m}h$.

(b) *If* $h(X) := \max\{h_1(X_1), h_2(X_2)\}$, $X = (X_1, X_2) \in \mathbb{C}^{n_1} \times \mathbb{C}^{n_2}$, *then*

$$\widehat{s}^h(X,Y) = \widehat{s}^{h_1}(X_1,Y_1) + \widehat{s}^{h_2}(X_2,Y_2), \quad X = (X_1,X_2), \quad Y = (Y_1,Y_2) \in \mathbb{C}^{n_1} \times \mathbb{C}^{n_2}.$$

In particular,

$$\mathbb{W}h(X) = \left((\mathbb{W}h_1(X_1))^2 + (\mathbb{W}h_2(X_2))^2\right)^{1/2}, \quad X = (X_1,X_2) \in \mathbb{C}^{n_1} \times \mathbb{C}^{n_2}.$$

Proof. ([536, 537]) (a) Using a suitable $\mathbb{C}$-linear isomorphism, we may reduce the situation to the case where:

- $U = \mathbb{C}^n$,
- $s^h(X,Y) = \langle X, Y \rangle$, $X, Y \in \mathbb{C}^n$,
- $\min\{\|X\| : h(X) = 1\} = \|X_*\| = a > 0$, $X_* = (0,\dots,0,a) \in \partial I$; in particular, since I is a balanced convex domain, $I \subset \{(X', X_m) \in \mathbb{C}^{m-1} \times \mathbb{C} : |X_m| < a\}$.

We only need to show that $a \ge 1/\sqrt{n}$. Suppose that $a < 1/\sqrt{n}$ and let $0 < b < 1$ be such that $a^2 + b^2 = 1$. Put $c := a/b$. Note that $(n-1)c^2 < 1$. Let $L : \mathbb{C}^n \longrightarrow \mathbb{C}^n$ be the $\mathbb{C}$-linear isomorphism

$$L(X) := (c\sqrt{n-1}X', X_n), \quad X = (X', X_n) \in \mathbb{C}^{n-1} \times \mathbb{C}.$$

Obviously, $s^{h \circ L^{-1}} = L^{-1}(s^h)$, so

$$\mathrm{Vol}(s^{h \circ L^{-1}}) = \mathcal{L}^{2n}(\mathbb{B}_n)|\det L|^2 = \mathcal{L}^{2n}(\mathbb{B}_n)(c\sqrt{n-1})^{2(n-1)}.$$

On the other hand, $L(I) \subset \mathbb{B}(a\sqrt{n}) \subset \mathbb{C}^n$. Indeed, for $X = (X', X_n)$ we have

$$\begin{aligned}\|L(X)\|^2 &= (n-1)\,c^2\|X'\|^2 + |X_n|^2 \\ &= (n-1)\,c^2\|X\|^2 + \left(1-(n-1)\,c^2\right)|X_n|^2 \\ &< (n-1)\,c^2 + \left(1-(n-1)\,c^2\right)a^2 \\ &= a^2 + \left(1-a^2\right)(n-1)\left(a^2/b^2\right) = na^2 < 1.\end{aligned}$$

Consequently, $\mathrm{Vol}(s^{h\circ L^{-1}}) \leq \mathcal{L}^{2n}(\mathbb{B}_n)\left(a\sqrt{n}\right)^{2n}$. Thus, using the above inequality, we get

$$\left(a\sqrt{n-1}/b\right)^{2(n-1)} \leq \left(a\sqrt{n}\right)^{2n}.$$

Put $f(t) := t(1-t)^{n-1}, 0 \leq t \leq 1$. Then,

$$f\left(a^2\right) = a^2\left(1-a^2\right)^{n-1} \geq \frac{1}{n}\left(1-\frac{1}{n}\right)^{n-1} = f(1/n);$$

a contradiction (because f is strictly increasing in the interval $[0, 1/n]$ and $a^2 < 1/n$).

(b) We may assume that $U(h_j) = \mathbb{C}^{n_j}$, $j = 1, 2$. Put

$$s_*(X,Y) := \frac{n_1}{n_1+n_2}s^{h_1}(X_1,Y_1) + \frac{n_2}{n_1+n_2}s^{h_2}(X_2,Y_2),$$
$$X = (X_1, X_2),\ Y = (Y_1, Y_2) \in \mathbb{C}^{n_1} \times \mathbb{C}^{n_2}.$$

We only need to prove that $\det S(s^h) = \det S(s_*)$ (all matrix representations are taken in the canonical bases of $\mathbb{C}^{n_1}$ and $\mathbb{C}^{n_2}$, respectively). Let $s := s^h$. Since

$$I(h) = I(h_1) \times I(h_2) \subset \mathbb{E}(s_*),$$

we get $\det S(s) \geq \det S(s_*)$.

Let $L : \mathbb{C}^{n_1} \times \mathbb{C}^{n_2} \longrightarrow \mathbb{C}^{n_1} \times \mathbb{C}^{n_2}$ be the isomorphism of the form $L(X_1, X_2) := (X_1, -X_2)$. Then, $h \circ L = h$ and, consequently, $s = L(s)$ (Remark 2.2.9), i.e.,

$$s(X,Y) = s(L(X), L(Y)), \quad X, Y \in \mathbb{C}^{n_1} \times \mathbb{C}^{n_2}.$$

Hence, $s((X_1, X_2), (Y_1, Y_2)) = 0$ if ($X_2 = 0$ and $Y_1 = 0$) or ($X_1 = 0$ and $Y_2 = 0$). Indeed,

$$\begin{aligned} s((X_1,0),(0,Y_2)) &= s(L(X_1,0), L(0,Y_2)) = s((X_1,0),(0,-Y_2)) \\ &= s((X_1,0), -(0,Y_2)) = -s((X_1,0),(0,Y_2)). \end{aligned}$$

Consequently,

$$s(X,Y) = s_1(X_1,Y_1) + s_2(X_2,Y_2), \quad X = (X_1,X_2),\ Y = (Y_1,Y_2) \in \mathbb{C}^{n_1} \times \mathbb{C}^{n_2},$$

where s_j is a Hermitian scalar product in $\mathbb{C}^{n_j}$, $j = 1, 2$. It is clear that $I(h_j) \subset \mathbb{E}(s_j)$, $j = 1, 2$. Let $c_j \leq 1$ be the minimal number such that $I(h_j) \subset \mathbb{E}(c_j^{-2}s_j)$, $j = 1, 2$. Assume that $X_j^0 \in \partial I(h_j)$ is such that $s_j(X_j^0, X_j^0) = c_j^2$, $j = 1, 2$. In particular, $q_s(X_1^0, X_2^0) \leq 1$, so $c_1^2 + c_2^2 \leq 1$. We have

$$\det S(s^{h_j}) \geq c_j^{-2n_j} \det S(s_j), \quad j = 1, 2,$$

and, therefore,

$$\begin{aligned}\det S(s) &= \det S(s_1)\det S(s_2) \le c_1^{2n_1}c_2^{2n_2}\det S(s^{h_1})\det S(s^{h_2})\\ &\le c_1^{2n_1}(1-c_1^2)^{n_2}\det S(s^{h_1})\det S(s^{h_2})\\ &\le \Big(\frac{n_1}{n_1+n_2}\Big)^{n_1}\Big(\frac{n_2}{n_1+n_2}\Big)^{n_2}\det S(s^{h_1})\det S(s^{h_2}) = \det S(s_*),\end{aligned}$$

since the maximum of the function $f(t) = t^{n_1}(1-t)^{n_2}$, $0 \le t \le 1$, is attained at $t = n_1/(n_1+n_2)$. □

Remark 2.2.11.

(a) Let $h := \max\{\mathfrak{h}_{\mathbb{D}(r_1)}, \mathfrak{h}_{\mathbb{D}(r_2)}\}$. Then, the former proposition immediately gives $s^h(X,Y) = \frac{X_1\overline{Y}_1}{2r_1^2} + \frac{X_2\overline{Y}_2}{2r_2^2}$, $X, Y \in \mathbb{C}^2$. Therefore, the triangle with minimal area containing $\Psi(\mathbb{D}(r_1)\times\mathbb{D}(r_2))$ (see Remark 2.2.9(c)) is given by $T_{2r_1^2,2r_2^2}$.

(b) Let $h_1 := \max\{\mathfrak{h}_{\mathbb{D}(\sqrt{3}/2)}, \mathfrak{h}_{\mathbb{D}(1/2)}\}$ and $h_2 := \mathfrak{h}_{\mathbb{B}_2}$. Obviously, $\mathbb{D}(\sqrt{3}/2)\times\mathbb{D}(1/2) \subset \mathbb{B}_2$, but $\mathbb{E}_1 := \mathbb{E}(s^{h_1}) \not\subset \mathbb{E}(s^{h_2}) =: \mathbb{E}_2$. Indeed, looking at the corresponding triangles leads to $\Psi(\mathbb{E}_1) = T_{3/2,1/2} \not\subset \Psi(\mathbb{E}_2) = T_{1,1}$. *This example shows that building the John ellipsoids is not a monotonic operation.*

The following two examples show how to use Remark 2.2.9(c). Both will play an important role during the discussion of the regularity of the Wu metric (see section 8.3).

Example 2.2.12 (cf. [286]). For $R > 1$ put

$$\Omega = \Omega_R := (\mathbb{D}\times\mathbb{D}(1/(2R))) \cup (\mathbb{D}(1/2)\times\mathbb{D}(R)).$$

Ω is a Reinhardt domain and its envelope of holomorphy is given by

$$\widetilde{\Omega} = \widetilde{\Omega}_R = \left\{ z \in \mathbb{D}\times\mathbb{D}(R) : 2R|z_2|\cdot|z_1|^{\frac{\log(2R^2)}{\log 2}} < 1 \right\}$$

(see [269]). Moreover, set $C = C_R := 1/(2eR)$, $A = A_R := \frac{32e^2R^2-1}{32e^2R^2-2} > 1$. Then, the following is true:

If $I = I_R \subset \mathbb{C}^2$ is a Reinhardt balanced domain satisfying

- $I \subset (\mathbb{D}\times\mathbb{D}(2)) \cap \widetilde{\Omega}$,
- $(1, C) \in \partial I$,

then $(\sqrt{A}, 0) \in \mathbb{E}(s^{\widehat{h}}) =: \mathbb{E}$, where $h := \mathfrak{h}_I$.

Proof. Assume the contrary, i.e., $(\sqrt{A}, 0) \notin \mathbb{E}$. Let $T_{a,b} = \Psi(\mathbb{E})$ be the "open" triangle of minimal area containing $\Psi(\hat{I})$. Then, $a \leq A$. Observe that the points $(0, b_0), (1/2, 4), (1, C^2), (A, 0)$ are collinear, where $b_0 := 8 - C^2$. Then, there exists a point $(u_0, v_0) \in (4/3, 1] \times (0, 4)$ such that

$$\overline{\Psi(\widetilde{\Omega})} \cap \big([0,1] \times [0,4]\big) \cap T'_{A,b_0} = \big[(u_0, v_0), \big(1, C^2\big)\big],$$

where $T'_{a,b}$ denotes the hypotenuse of the triangle $T_{a,b}$. Indeed, using the formula of $\Psi(\widetilde{\Omega})$ gives

- $(1/2, 4) \in T'_{A,b_0} \setminus \overline{\Psi(\widetilde{\Omega})}$,
- $(2/3, 1) \in T_{A,b_0} \setminus \overline{\Psi(\widetilde{\Omega})}$.

Hence, $[(1/2, 4), (2/3, y)] \subset T'_{A,b_0} \setminus \Psi(\widetilde{\Omega})$ with $y := \frac{4(A-2/3)}{A-1/2}$. What remains is to note that $(1, C^2)$ is a boundary point of $\Psi(I)$. Hence, u_0, v_0 exist, as was claimed.

Since $a \leq A$ and $(1, C) \in \partial I$, the slope of $T'_{a,b}$ is larger than the one of T_{A,b_0}. Therefore, there are numbers $v_1, v_2 > 0$ and $u_2 > 2/3$ such that

$$\overline{\Psi(\widetilde{\Omega})} \cap \big([0,1] \times [0,4]\big) \cap T'_{a,b} = [(u_2, v_2), (1, v_1)].$$

Put $\mathcal{T} = \{T_{c,d(c)} : [0, u_2] \times [0, v_2] \subset \overline{T}_{c,d(c)}, \ (u_2, v_2) \in T'_{c,d(c)}\}$. Note that $(u_2, v_2) \in T'_{c,d(c)}$ for all $T_{c,d(c)} \in \mathcal{T}$ and so $d(c)$ is well defined. Obviously, $T_{a,b} \in \mathcal{T}$. Applying Remark 2.2.11, it follows that $T_{2u_2,2v_2}$ is a triangle of minimal area in $\mathcal{T}$. Moreover, the area function $(a, 2u_2) \ni c \longmapsto \operatorname{Area}(T_{c,d(c)})$ is strictly decreasing. Finally, observe that

$$\Psi(I) \subset \big([0,1] \times [0,4]\big) \cap \overline{\Psi(\widetilde{\Omega})} \cap T_{a,b} \subset \big([0,1] \times [0,4]\big) \cap \overline{\Psi(\widetilde{\Omega})} \cap T_{a+\varepsilon,d(a+\varepsilon)}$$

which contradicts the minimality of $T_{a,b}$. □

Example 2.2.13 (cf. [286]). Fix an $R \geq 9$ and let $\Omega = \Omega_R$, $\widetilde{\Omega} = \widetilde{\Omega}_R$, and $A = A_R$ be as in Example 2.2.12. Then, the following is true:

If $I \subset \mathbb{C}^2$ is a Reinhardt balanced domain satisfying

- $I \subset \widetilde{\Omega}$,
- $(0, R) \in \partial I$, then $(\sqrt{A}, 0) \notin \mathbb{E}(s^{\hat{h}}) =: \mathbb{E}$, where $h := \mathfrak{h}_I$.

Proof. Assume the contrary, i.e., $(\sqrt{A}, 0) \in \mathbb{E}$ and let $T_{a,b} := \Psi(\mathbb{E})$. Then, $A < a$.

Since, by assumption, $(0, R) \in \partial I$, we have $(0, R^2) \in \overline{T}_{a,b}$ and therefore, $T_{A,R^2} \subset T_{a,b}$, i.e., $b \geq R^2$. Then,

$$\big[\big(1/3, R^2/2\big), \big(1, 1/\big(4R^2\big)\big)\big] \subset T_{A,R^2} \setminus \overline{\Psi(\widetilde{\Omega})} \subset T_{a,b} \setminus \overline{\Psi(\widetilde{\Omega})}.$$

So, one finds a point $(u_0, v_0) \in (0, 1/3) \times (0, +\infty)$ such that

$$(u_0, v_0) \in T'_{a,b} \cap \overline{\Psi(\widetilde{\Omega})} \subset [(0, b), (u_0, v_0)].$$

Similar to the former example we put

$$\mathcal{T} := \{T_{c,d(c)} : [0, u_0] \times [0, v_0],\ (u_0, v_0) \in T'_{c,d(c)}\} \ni T_{a,b}.$$

Observe that $T_{2u_0,2v_0}$ is the triangle in $\mathcal{T}$ with minimal area and that the function $(2u_0, a) \ni c \longmapsto \operatorname{Area}(T_{c,d(c)})$ (note that $2u_0 < a$)) is strictly increasing. Note that there is a small $\varepsilon > 0$, $2u_0 < a - \varepsilon$, such that $\Psi(I) \subset \Psi(\widetilde{\Omega}) \cap T_{a,b} \subset \psi(\widetilde{\Omega}) \cap T_{a-\varepsilon,d(a-\varepsilon)}$ and $T_{a-\varepsilon,d(a-\varepsilon)} \in \mathcal{T}$, which contradicts the minimality of $T_{a,b}$. □

2.2.4 d-balanced domains

The notion of a balanced domain may be generalized in the following way (cf. [377]): Let $d = (d_1, \dots, d_n) \in \mathbb{N}^n$. We say that a domain $D \subset \mathbb{C}^n$ is *d-balanced* if for any $a = (a_1, \dots, a_n) \in D$ and $\lambda \in \overline{\mathbb{D}}$, the point $(\lambda^{d_1} a_1, \dots, \lambda^{d_n} a_n)$ belongs to D.

Observe that balanced domains are simply $(1, \dots, 1)$-balanced domains.

For a d-balanced domain D, we define its *d-Minkowski function* $\mathfrak{h}_D : \mathbb{C}^n \longrightarrow \mathbb{R}_+$,

$$\mathfrak{h}_D(X) := \inf\left\{t > 0 : \left(\frac{X_1}{t^{d_1}}, \dots, \frac{X_n}{t^{d_n}}\right) \in D\right\}, \quad X = (X_1, \dots, X_n) \in \mathbb{C}^n.$$

Remark 2.2.14 (The reader is asked to complete the details).

(a)
$$D = \{X \in \mathbb{C}^n : \mathfrak{h}_D(X) < 1\}.$$

(b) $\mathfrak{h}_D(\lambda^{d_1} X_1, \dots, \lambda^{d_n} X_n) = |\lambda| \mathfrak{h}_D(X)$, $X = (X_1, \dots, X_n) \in \mathbb{C}^n$, $\lambda \in \mathbb{C}$.

(c) $\mathfrak{h}_D$ is upper semicontinuous.

(d) If $(D_s)_{s=1}^{\infty} \subset \mathbb{C}^n$ is a sequence of d-balanced domains with $D_s \nearrow D$, then $\mathfrak{h}_{D_s} \searrow \mathfrak{h}_D$.

(e) If $D_j \subset \mathbb{C}^{n_j}$ is d^j-balanced, $d^j \in \mathbb{N}^{n_j}$, $j = 1, 2$, then $D_1 \times D_2 \subset \mathbb{C}^{n_1} \times \mathbb{C}^{n_2}$ is (d^1, d^2)-balanced and $\mathfrak{h}_{D_1 \times D_2}(X_1, X_2) = \max\{\mathfrak{h}_{D_1}(X_1), \mathfrak{h}_{D_2}(X_2)\}$, $X = (X_1, X_2) \in \mathbb{C}^{n_1} \times \mathbb{C}^{n_2}$.

Any upper semicontinuous function $h : \mathbb{C}^n \longrightarrow \mathbb{R}_+$ with $h(\lambda^{d_1} X_1, \dots, \lambda^{d_n} X_n) = |\lambda| h(X)$, $X = (X_1, \dots, X_n) \in \mathbb{C}^n$, $\lambda \in \mathbb{C}$, will be called *d-Minkowski function.* Clearly, $h = \mathfrak{h}_D$, where $D := \{X \in \mathbb{C}^n : h(X) < 1\}$.

Proposition 2.2.15 (cf. [377]). *Let $D \subset \mathbb{C}^n$ be a d-balanced domain. Then, D is pseudoconvex iff* $\log \mathfrak{h}_D \in \mathcal{PSH}(\mathbb{C}^n)$ *iff* $\mathfrak{h}_D \in \mathcal{PSH}(\mathbb{C}^n)$.

Proof. Clearly, if $\log \mathfrak{h}_D \in \mathcal{PSH}(\mathbb{C}^n)$, then D is pseudoconvex (cf. [265], Corollary 2.2.15). Conversely, assume that D is pseudoconvex and define $\mathbb{C}^n \ni z \overset{F}{\longmapsto} (z_1^{d_1}, \dots, z_n^{d_n}) \in \mathbb{C}^n$. Put $D_0 := F^{-1}(D)$. Then, D_0 is pseudoconvex (cf. [265], Corollary 2.2.20). Observe that $D_0 = \{X \in \mathbb{C}^n : h_0(X) < 1\}$, where $h_0 := \mathfrak{h}_D \circ F$. Since $h_0(\lambda X) = |\lambda| h_0(X)$, $X \in \mathbb{C}^n$, $\lambda \in \mathbb{C}$, we conclude that D_0 is balanced and $h_0 = \mathfrak{h}_{D_0}$. Thus, $\log h_0 \in \mathcal{PSH}(\mathbb{C}^n)$ (cf. Appendix B.7.6). Since $\mathfrak{h}_D(z) = h_0(\sqrt[k_1]{z_1}, \dots, \sqrt[k_n]{z_n})$, we conclude that $\log \mathfrak{h}_D \in \mathcal{PSH}(\mathbb{C}^n_*)$. Hence, since $\mathbb{C}^n \setminus \mathbb{C}^n_*$ is pluripolar, we get $\log \mathfrak{h}_D \in \mathcal{PSH}(\mathbb{C}^n)$ (cf. Appendix B.4.1). □

Let $D \subset \mathbb{C}^n$ be d-balanced. Similar to § 2.2.2, we define

$$\widetilde{\mathfrak{h}}_D := \sup\{h : h \in \mathcal{PSH}(\mathbb{C}^n),\ h \le \mathfrak{h}_D\}, \quad \widetilde{D} := \{X \in \mathbb{C}^n : \widetilde{\mathfrak{h}}_D(X) < 1\}.$$

Remark 2.2.16 (The reader is asked to complete the details).

(a) $0 \le \widetilde{\mathfrak{h}}_D \le \mathfrak{h}_D$.

(b) $\widetilde{\mathfrak{h}}_D \in \mathcal{PSH}(\mathbb{C}^n)$.

(c) $\widetilde{\mathfrak{h}}_D(\lambda^{d_1} X_1, \dots, \lambda^{d_n} X_n) = |\lambda| \widetilde{\mathfrak{h}}_D(X)$, $X = (X_1, \dots, X_n) \in \mathbb{C}^n$, $\lambda \in \mathbb{C}$; in particular, $\widetilde{D}$ is a d-balanced domain and $\widetilde{\mathfrak{h}}_D = \mathfrak{h}_{\widetilde{D}}$.

(d) $\widetilde{\mathfrak{h}}_D = \sup\{h : h \in \mathcal{PSH}(\mathbb{C}^n),\ h \le \mathfrak{h}_D,\ h$ is a d-Minkowski function$\}$; in particular, $\widetilde{D} = \operatorname{int}_{U \in \mathfrak{U}} U$, where

$$\mathfrak{U} := \{U : U \text{ is a } d\text{-balanced domain},\ D \subset U,\ \mathfrak{h}_U \in \mathcal{PSH}(\mathbb{C}^n)\}.$$

(e) If $(D_s)_{s=1}^\infty$ is a sequence of d-balanced domains such that $D_s \nearrow D$, then $\widetilde{\mathfrak{h}}_{D_s} \searrow \widetilde{\mathfrak{h}}_D$.

2.3 Carathéodory pseudodistance and pseudometric in balanced domains

Let $G = G_{\mathfrak{h}} = \{z \in \mathbb{C}^n : \mathfrak{h}(z) < 1\}$ be a balanced domain in $\mathbb{C}^n$ ($\mathfrak{h} = \mathfrak{h}_G$ is the Minkowski function of G).

Proposition 2.3.1.

(a) $\boldsymbol{c}_G^*(0,\cdot) \le \mathfrak{h}$ *in* G*;* $\boldsymbol{\gamma}_G(0;\cdot) \le \mathfrak{h}$ *in* $\mathbb{C}^n$.

(b) *For* $a \in G$ *the following statements are equivalent:*

(i) $\boldsymbol{c}_G^*(0,a) = \mathfrak{h}(a)$*;*

(ii) $\boldsymbol{c}_G^*(0,\cdot) = \mathfrak{h}$ *on* $G \cap (\mathbb{C}a)$*;*

(iii) $\boldsymbol{\gamma}_G(0;a) = \mathfrak{h}(a)$*;*

(iv) $\gamma_G(0;\cdot) = \mathfrak{h}$ *on* $\mathbb{C}a$;

(v) *there exists a* $\mathbb{C}$*-linear functional* $L : \mathbb{C}^n \longrightarrow \mathbb{C}$ *with* $|L| \le \mathfrak{h}$ *and* $|L(a)| = \mathfrak{h}(a)$.

(c) *The following conditions are equivalent:*

(i) $c_G^*(0,\cdot) = \mathfrak{h}$ *in* G;

(ii) $\gamma_G(0;\cdot) = \mathfrak{h}$ *in* $\mathbb{C}^n$;

(iii) $\mathfrak{h}$ *is a seminorm;*

(iv) G *is convex.*

(d) $\gamma_G(0;\cdot) = \widehat{\mathfrak{h}} := \sup\{q : \mathbb{C}^n \longrightarrow \mathbb{R}_+ : q$ *is a seminorm,* $q \le \mathfrak{h}\}$; $\widehat{\mathfrak{h}}$ *is the Minkowski function of the convex hull* $\widehat{G}$ *of* G.

Proof. (a) Fix an $a \in G$ (resp. $a \in \mathbb{C}^n$). We have the following two possibilities:

1° $\mathfrak{h}(a) = 0$; then, the mapping $\mathbb{C} \ni \lambda \longmapsto \lambda a \in G$ is well-defined, and hence, by Theorem 2.1.1, we get

$$c_G^*(0,a) \le c_{\mathbb{C}}^*(0;1) = 0 = \mathfrak{h}(a) \qquad (\text{resp. } \gamma_G(0;a) \le \gamma_{\mathbb{C}}(0;1) = 0 = \mathfrak{h}(a)).$$

2° $\mathfrak{h}(a) > 0$; take $\mathbb{D} \ni \lambda \longmapsto \frac{\lambda}{\mathfrak{h}(a)}a \in G$. Then, the holomorphic contractibility leads to

$$c_G^*(0,a) \le c_{\mathbb{D}}^*(0,\mathfrak{h}(a)) = \mathfrak{h}(a) \qquad (\text{resp. } \gamma_G(0;a) \le \gamma_{\mathbb{D}}(0;\mathfrak{h}(a)) = \mathfrak{h}(a)).$$

(b) Observe that if (v) is satisfied for a point a, then it is satisfied for each point from the line $\mathbb{C}a$. Moreover, if (v) is satisfied, then by the definitions:

$$c_G^*(0,a) \ge |L(a)| = \mathfrak{h}(a), \qquad \gamma_G(0;a) \ge |L'(0)a| = |L(a)| = \mathfrak{h}(a).$$

Thus, in view of (a) and the Hahn–Banach theorem, we have: (v) $\Longrightarrow$ (iv) $\Longrightarrow$ (iii) $\Longrightarrow$ (v) $\Longrightarrow$ (ii) $\Longrightarrow$ (i). Suppose that (i) is fulfilled and let $f \in \mathcal{O}(G,\mathbb{D})$, $f(0) = 0$, be an extremal function for $c_G^*(0,a)$, i.e., $|f(a)| = \mathfrak{h}(a)$. Note that in the case $\mathfrak{h}(a) = 0$ the implication (i) $\Longrightarrow$ (v) is trivial. So, assume that $\mathfrak{h}(a) > 0$. Define

$$\mathbb{D} \ni \lambda \overset{\varphi}{\longmapsto} f\Big(\frac{\lambda}{\mathfrak{h}(a)}a\Big) \in \mathbb{D}.$$

Then, $\varphi \in \mathcal{O}(\mathbb{D},\mathbb{D})$ and $|\varphi(\mathfrak{h}(a))| = \mathfrak{h}(a)$. By the classical Schwarz lemma we get $\varphi(\lambda) = e^{i\theta}\lambda$, $\lambda \in \mathbb{D}$, for some $\theta \in \mathbb{R}$. Let $L := f'(0)$. By (a) we have

$$|L(X)| = |f'(0)X| \le \gamma_G(0;X) \le \mathfrak{h}(X), \quad X \in \mathbb{C}^n.$$

Moreover,

$$|L(a)| = |f'(0)a| = |\varphi'(0)\mathfrak{h}(a)| = \mathfrak{h}(a).$$

(c) In view of (b) we have: (i) $\Longleftrightarrow$ (ii). Since $\gamma_G(0;\cdot)$ is a seminorm, we get (ii) $\Longrightarrow$ (iii). If (iii) is satisfied, then, by the Hahn–Banach theorem, condition (v) from (b) is fulfilled for any $a \in \mathbb{C}^n$. Hence, (iii) $\Longrightarrow$ (i). Obviously, (iii) and (iv) are equivalent.

(d) Since $\gamma_G(0;\cdot)$ is a seminorm, (a) implies that $\gamma_G(0;\cdot) \le \widehat{\mathfrak{h}}$. On the other hand, by (c) we obtain $\widehat{\mathfrak{h}} = \gamma_{\widehat{G}}(0;\cdot) \le \gamma_G(0;\cdot)$. □

For d-balanced domains with $d_1 \le \cdots \le d_n$ there is a chance to calculate the Carathéodory–Reiffen pseudometric at the origin for certain directions (which will be used in § 7.2). Fix a j in $\{1,\dots,n\}$ and denote by $\mathcal{L}_j$ the span of the vectors $\boldsymbol{e}_j,\dots,\boldsymbol{e}_k$, where $k = \max\{s \in \{j,\dots,n\} : d_j = d_s\}$. Then,

Proposition 2.3.2 (cf. [388]). *If G is a d-balanced domain in $\mathbb{C}^n$ with $d_1 \le \cdots \le d_n$ and $j \in \{1,\dots,n\}$, then*

$$\gamma_G(0;X) = \sup\{|P'(0)X| : P \in \mathcal{P}_j\}, \quad X \in \mathcal{L}_j,$$

where

$$\mathcal{P}_j := \{P \in \mathcal{P}(\mathbb{C}^n) : \|P\|_G \le 1, \\ P(\lambda^{d_1}z_1,\dots,\lambda^{d_n}z_n) = \lambda^{d_j}P(z),\ z \in \mathbb{C}^n,\ \lambda \in \mathbb{C}\}.$$

Proof. Let us assume that $d_s = d_j = d_k =: m$ with $d_k < d_{k+1}$ if $k < n$ and $d_{s-1} < d_s$ if $s > 1$. Fix a vector $X = (0,\dots,X_s,\dots,X_k,0,\dots,0)$ and choose an extremal function $f \in \mathcal{O}(G,\mathbb{D})$ for $\gamma_G(0;X)$, i.e., $f(0) = 0$ and $|f'(0)X| = \gamma_G(0;X)$. Moreover, let $\sqrt[m]{1} = \{\xi_1,\dots,\xi_m\}$.

Now, we define a new function $g : G \longrightarrow \mathbb{D}$ as

$$g(z) := \frac{1}{m}\sum_{\mu=1}^{m} f\left(\xi_\mu^{d_1}z_1,\dots,\xi_\mu^{d_n}z_n\right).$$

Note that $g \in \mathcal{O}(G,\mathbb{D})$, $g(0) = 0$, and $|g'(0)X| = |f'(0)X|$, i.e., g is also an extremal function for $\gamma_G(0;X)$.

Now, assume that $f(z) = \sum_{|\alpha|>0} a_\alpha z^\alpha$, $z \in \mathbb{P}_n(r)$, is the power series expansion of f. Then,

$$g(z) = \sum_{|\alpha|>0}\left(\frac{1}{m}\sum_{\mu=1}^{m}\xi_\mu^{\alpha_1d_1+\cdots+\alpha_nd_n}\right)a_\alpha z^\alpha = \sum_{\alpha_1d_1+\cdots+\alpha_nd_n \in m\mathbb{N}} a_\alpha z^\alpha.$$

Here, we have used that $\sum_{\mu=1}^{m}\xi_\mu^\sigma = 0$ for $0 < \sigma < m$.

Put $P(z) := \sum_{\alpha_1d_1+\cdots+\alpha_nd_n=m} a_\alpha z^\alpha$. Obviously, if $\lambda \in \mathbb{C}$, then the polynomial P satisfies $\frac{\partial P}{\partial z_\mu}(0) = \frac{\partial f}{\partial z_\mu}(0)$, $s \le \mu \le k$ and $P(\lambda^{d_1}z_1,\dots,\lambda^{d_n}z_n) = \lambda^m P(z)$.

It remains to check that $|P| < 1$ on G. To prove this, fix an arbitrary $z \in G$ and look at the following holomorphic function:

$$\mathbb{D} \ni \lambda \overset{h}{\longmapsto} g(\lambda^{d_1} z_1, \dots, \lambda^{d_n} z_n) = \sum_{\mu=1}^{\infty} \Big(\sum_{\alpha_1 d_1 + \dots + \alpha_n d_n = \mu m} a_\alpha z^\alpha \Big) \lambda^{\mu m} \in \mathbb{D}.$$

Put $\widetilde{h}(\lambda) := h(\lambda)/\lambda^m$, $\lambda \in \mathbb{D}_*$. Then, $\widetilde{h}$ extends holomorphically to 0 by setting $\widetilde{h}(0) = P(z)$. Note that $|h| < 1$ even on $\overline{\mathbb{D}}$. Therefore, the maximum principle leads to $|P(z)| = |\widetilde{h}(0)| < 1$. Since z was arbitrary, we have $|P| < 1$ on G. □

Corollary 2.3.3 (cf. Corollary 3.5.8).

(a) *Let $G_j = G_{q_j} \subsetneq \mathbb{C}^n$ be balanced convex domains with Minkowski functions q_j, $j = 1, 2$. Then, the following conditions are equivalent:*

   (i) *there exists a biholomorphic mapping $F : G_1 \longrightarrow G_2$ with $F(0) = 0$;*

   (ii) *there exists a $\mathbb{C}$-linear isomorphism $L : \mathbb{C}^n \longrightarrow \mathbb{C}^n$ such that $q_2 \circ L = q_1$, i.e., G_1 and G_2 are* linearly equivalent.

(b) (*cf.* [411]) *Let $G = G_{\mathfrak{h}}$ be a bounded balanced domain in $\mathbb{C}^n$. Then, the following conditions are equivalent:*

   (i) *$G_{\mathfrak{h}}$ and $\mathbb{B}_n$ are linearly equivalent ($\mathbb{B}_n$ is the unit Euclidean ball in $\mathbb{C}^n$);*

   (ii) *$\mathfrak{h}^2 \in \mathcal{C}^2(\mathbb{C}^n)$.*

Note that $\mathfrak{h}^2 \in \mathcal{C}^2(\mathbb{C}^n)$ iff $\mathfrak{h}^2$ is $\mathcal{C}^2$ near $0 \in \mathbb{C}^n$.

Proof. (a) Obviously, (ii) $\Longrightarrow$ (i). If F is as in (i), then, by Proposition 2.3.1(c), we have

$$q_2(F'(0)X) = \boldsymbol{\gamma}_{G_2}(0; F'(0)X) = \boldsymbol{\gamma}_{G_1}(0; X) = q_1(X), \quad X \in \mathbb{C}^n.$$

(b) The only problem is to prove that (ii) $\Longrightarrow$ (i). Since $\mathfrak{h}^2(\lambda z) = \lambda\overline{\lambda}\mathfrak{h}^2(z)$, $\lambda \in \mathbb{C}$, $z \in \mathbb{C}^n$, we get

$$\mathfrak{h}^2(z) = \frac{\partial^2}{\partial\lambda\partial\overline{\lambda}}\mathfrak{h}^2(\lambda z) = \sum_{j,k=1}^{n} \frac{\partial^2 \mathfrak{h}^2}{\partial z_j \partial \overline{z}_k}(\lambda z) z_j \overline{z}_k.$$

In particular,

$$0 < \mathfrak{h}^2(z) = \sum_{j,k=1}^{n} \frac{\partial^2 \mathfrak{h}^2}{\partial z_j \partial \overline{z}_k}(0) z_j \overline{z}_k, \quad z \in (\mathbb{C}^n)_*.$$

Consequently, $\mathfrak{h}^2$ is a Hermitian form, and therefore we get (i). □

Corollary 2.3.4. *If $G = G_q$ is a balanced convex domain in $\mathbb{C}^n$, $a \in G$, and if $h_a \in \operatorname{Aut}(G)$ is such that $h_a(a) = 0$, then*

$$\boldsymbol{c}_G^*(a,z) = q(h_a(z)), \quad z \in G, \qquad \boldsymbol{\gamma}_G(a;X) = q(h_a'(a)X), \quad X \in \mathbb{C}^n.$$

Recall that in the case of the unit polydisc $\mathbb{D}^n \subset \mathbb{C}^n$ we can take

$$h_a(z_1,\dots,z_n) = \Big(\frac{z_1 - a_1}{1 - z_1\bar{a}_1}, \dots, \frac{z_n - a_n}{1 - z_n\bar{a}_n}\Big).$$

In the case of the unit Euclidean ball $\mathbb{B}_n \subset \mathbb{C}^n$ we can use

$$h_a(z) = \frac{\sqrt{1 - \|a\|^2}\Big(z - \frac{\langle z,a\rangle}{\|a\|^2}a\Big) - a + \frac{\langle z,a\rangle}{\|a\|^2}a}{1 - \langle z,a\rangle}, \tag{2.3.1}$$

where $\langle z,a\rangle := \sum_{j=1}^n z_j\bar{a}_j$ is the complex scalar product in $\mathbb{C}^n$ and $\| \ \|$ is the Euclidean norm; cf. Appendix B.3.3. Hence, in view of Corollary 2.3.4, we get

Corollary 2.3.5.

$$\begin{aligned} \boldsymbol{c}_{\mathbb{D}^n}^*(a,z) &= \max\{\boldsymbol{m}(a_j,z_j) : j = 1,\dots,n\}, \\ \boldsymbol{\gamma}_{\mathbb{D}^n}(a;X) &= \max\{\boldsymbol{\gamma}(a_j)|X_j| : j = 1,\dots,n\}, \\ \boldsymbol{c}_{\mathbb{B}_n}^*(a,z) &= \left[1 - \frac{(1 - \|a\|^2)(1 - \|z\|^2)}{|1 - \langle z,a\rangle|^2}\right]^{\frac{1}{2}}, \\ \boldsymbol{\gamma}_{\mathbb{B}_n}(a;X) &= \left[\frac{\|X\|^2}{1 - \|a\|^2} + \frac{|\langle a,X\rangle|^2}{(1 - \|a\|^2)^2}\right]^{\frac{1}{2}}. \end{aligned} \tag{2.3.2}$$

Note that the last equation implies in particular that

$$\|f'(a)\| \le \frac{1 - |f(a)|^2}{1 - \|a\|^2}, \qquad f \in \mathcal{O}(\mathbb{B}_n,\mathbb{D}),\ a \in \mathbb{B}_n,\ j = 1,\dots,n.$$

Similar estimates for higher derivatives, even for mappings $f \in \mathcal{O}(\mathbb{B}_n,\mathbb{B}_m)$, may be found in [117] (see also [101]).

Theorem* 2.3.6. *Let $f \in \mathcal{O}(\mathbb{B}_n,\mathbb{B}_m)$, $X \in (\mathbb{C}^n)_*$, and $k \in \mathbb{N}$. Then,*

$$\begin{aligned} \boldsymbol{\gamma}_{\mathbb{B}_m}(f(z); f^{(k)}(z)X) \le k! &\left(1 + \frac{|\langle z,X\rangle|}{\big((1 + \|z\|^2)\|X\|^2 + |\langle z,X\rangle|^2\big)^{1/2}}\right)^{2(k-1)} \\ &\cdot (\boldsymbol{\gamma}_{\mathbb{B}_n}(z;X))^k, \quad z \in \mathbb{B}_n, \end{aligned}$$

where $f^{(k)}(z)X := \sum_{\alpha\in\mathbb{Z}_+^n,\ |\alpha|=k} \frac{k!}{\alpha!} D^\alpha f(z)X^\alpha \in \mathbb{C}^m$.

As a consequence, one deduces from this result a Schwarz–Pick estimate for mappings $f \in \mathcal{O}(\mathbb{B}_n, \mathbb{B}_m)$.

Corollary 2.3.7. *Let $f \in \mathcal{O}(\mathbb{B}_n, \mathbb{B}_m)$ and $\alpha \in (\mathbb{Z}_+^n)_*$. Put $k = |\alpha|$. Then,*

$$|\langle D^\alpha f(z), f(z)\rangle|^2 + (1 - \|f(z)\|^2)\|D^\alpha f(z)\|^2 \\ \leq \frac{k^k}{\alpha^\alpha}\left(\alpha!(1+\|z\|)^{k-1}\frac{1-\|f(z)\|^2}{(1-\|z\|^2)^k}\right)^2, \quad z \in \mathbb{B}_n.$$

Note that for $f \in \mathcal{O}(\mathbb{D}, \mathbb{D})$, the above estimate is the one we know from Proposition 1.1.2.

Remark 2.3.8. In the context of Corollary 2.3.4, let us mention that the situation in which, for *every* point a of a bounded domain $G \subset \mathbb{C}^n$ with $0 \in G$, there exists an $h_a \in \operatorname{Aut}(G)$ with $h_a(a) = 0$, is very exceptional. We already know that this is true for Euclidean balls and polydiscs. First, observe that the existence of the automorphism h_a (for every $a \in G$) is equivalent to the condition saying that for two arbitrary points $a, b \in G$ there exists an $h_{a,b} \in \operatorname{Aut}(G)$ with $h_{a,b}(a) = b$. Domains $G \subset \mathbb{C}^n$ with this property are called *homogeneous*. One also defines the category of *symmetric* domains, i.e., the category of those domains $G \subset \mathbb{C}^n$ for which for every $a \in G$ there exists an automorphism $g_a \in \operatorname{Aut}(G)$ such that $g_a(a) = a$, $g_a \circ g_a \equiv \mathrm{id}$, and a is an isolated point of the set $\{z \in G : g_a(z) = z\}$. Observe that any balanced domain is symmetric at 0 – we simply take $g_0(z) := -z$, $z \in G$. One can easily prove that the category of homogeneous (resp. symmetric) domains is invariant under biholomorphic mappings and under Cartesian products. Moreover, it is clear that a homogeneous domain is symmetric iff it is symmetric at one point.

It is known that for $n \leq 3$ any bounded homogeneous domain is symmetric (cf. [85]) and the result is not true for $n \geq 4$ (cf. [430]). Moreover, any bounded symmetric domain is homogeneous (cf. [85]). In particular, a bounded balanced domain is symmetric iff it is homogeneous.

One can prove (cf. [85]) that each bounded homogeneous balanced domain $G \subset \mathbb{C}^n$ is biholomorphic to a Cartesian product of domains belonging to the following four *Cartan types*:

- $\boldsymbol{I}_{p,q} := \{Z \in \mathbb{M}(p \times q, \mathbb{C}) : \mathbb{I}_p - ZZ^* > 0\}$, $1 \leq p \leq q$, where $\mathbb{M}(p \times q, \mathbb{C})$ stands for the space of all complex $(p \times q)$-matrices, $\mathbb{I}_p$ denotes the unit $(p \times p)$-matrix, and $\mathbb{I}_p - ZZ^* > 0$ means that the matrix $\mathbb{I}_p - ZZ^*$ is positive definite; *observe that $\boldsymbol{I}_{1,n} = \mathbb{B}_n$*;
- $\boldsymbol{II}_p := \{Z \in \mathbb{M}(p \times p, \mathbb{C}) : Z^t = -Z,\ \mathbb{I}_p - ZZ^* > 0\}$, $p \geq 2$;

- $\boldsymbol{III_p} := \{Z \in \mathbb{M}(p \times p, \mathbb{C}) : Z^t = Z,\ \mathbb{I}_p - ZZ^* > 0\},\ p \geq 1$;
- $\boldsymbol{IV_n} = \mathbb{L}_n := \{z \in \mathbb{B}_n : 2\|z\|^2 - |\langle z, \bar{z}\rangle|^2 < 1\}$ = the *Lie ball* (cf. Example 16.1.3; see also [269], Example 2.1.12(c)); notice that the mapping $\mathbb{L}_2 \ni (z_1, z_2) \longmapsto (z_1 + iz_1, z_1 - iz_2) \in \mathbb{D}^2$ is biholomorphic.

A detailed discussion of the automorphism groups of the above domains may be found in [237].

In particular, up to biholomorphisms, the only bounded homogeneous balanced domains in $\mathbb{C}^3$ are: $\mathbb{B}_3$, $\mathbb{D}^3$, $\mathbb{D} \times \mathbb{B}_2$, and $\mathbb{L}_3$.

Corollary 2.3.9 (Poincaré theorem, cf. [450]). *For $n > 1$, there is no biholomorphic mapping of $\mathbb{D}^n$ onto $\mathbb{B}_n$.*

Proof. Use Corollary 2.3.3 and the fact that $\operatorname{Aut}(\mathbb{D}^n)$ acts transitively on $\mathbb{D}^n$. □

The above Poincaré theorem may be generalized in the following way.

Theorem* 2.3.10 (cf. [269], Theorem 2.1.17). *Let $2 \leq n = n_1 + \dots + n_k = m_1 + \dots + m_\ell$, $B_\mu \in \{\mathbb{B}_{n_\mu}, \mathbb{L}_{n_\mu}\}$, $\mu = 1, \dots, k$, $B'_\nu \in \{\mathbb{B}_{m_\nu}, \mathbb{L}_{m_\nu}\}$, $\nu = 1, \dots, \ell$. Assume that if $B_\mu = \mathbb{L}_{n_\mu}$ (resp. $B'_\nu = \mathbb{L}_{m_\nu}$), then $n_\mu \geq 3$ (resp. $m_\nu \geq 3$). Then, the following conditions are equivalent:*

(i) *there exists a biholomorphism $B_1 \times \dots \times B_k \longrightarrow B'_1 \times \dots \times B'_\ell$;*

(ii) *$\ell = k$, and there exists a permutation σ such that $m_{\sigma(\mu)} = n_\mu$ and $B'_{\sigma(\mu)} = B_\mu$, $\mu = 1, \dots, k$.*

Moreover, every biholomorphic mapping $F : B_1 \times \dots \times B_k \longrightarrow B'_1 \times \dots \times B'_k$ is, up to a permutation of $B'_1, \dots, B'_k$, of the form

$$F(z) = (F_1(z_1), \dots, F_k(z_k)), \quad z = (z_1, \dots, z_k) \in B_1 \times \dots \times B_k,$$

where $F_\mu \in \operatorname{Aut}(B_\mu)$, $\mu = 1, \dots, k$.

Remark 2.3.11.

(a) In the case where $B_\mu = \mathbb{B}_{n_\mu}$, $\mu = 1, \dots, k$, $B'_\nu = \mathbb{B}_{m_\nu}$, $\nu = 1, \dots, \ell$, the above theorem states that the following conditions are equivalent:

  (i) there exists a biholomorphism $\mathbb{B}_{n_1} \times \dots \times \mathbb{B}_{n_k} \longrightarrow \mathbb{B}_{m_1} \times \dots \times \mathbb{B}_{m_\ell}$;

  (ii) $\ell = k$ and there exists a permutation σ with $m_{\sigma(\mu)} = n_\mu$, $\mu = 1, \dots, k$.

(b) In the case where $k = 1$, $B_1 = \mathbb{B}_n$, $\ell = n \geq 2$, the result reduces to the Poincaré theorem (Corollary 2.3.9).

(c) In the case where $k = \ell = n$, the result gives a description of $\operatorname{Aut}(\mathbb{D}^n)$.

(d) In the case $k = 1$, $B_1 = \mathbb{L}_n$, $B'_\nu = \mathbb{B}_{m_\nu}$, $\nu = 1, \dots, \ell$, the result shows that for $n \geq 3$ there is no biholomorphic mapping $\mathbb{L}_n \longrightarrow \mathbb{B}_{m_1} \times \dots \times \mathbb{B}_{m_\ell}$.

2.4 Carathéodory isometries

The Poincaré theorem may be generalized to the case of Carathéodory isometries.

A mapping $F : G \longrightarrow D$ is said to be a $\boldsymbol{c}$-*isometry* if

$$\boldsymbol{c}_D^{(*)}(F(z'), F(z'')) = \boldsymbol{c}_G^{(*)}(z', z''), \quad z', z'' \in G.$$

Recall that any biholomorphic mapping is a $\boldsymbol{c}$-isometry.

Let $k \in \mathbb{N}$, $\alpha := (\alpha_1, \dots, \alpha_k) \in \mathbb{N}^k$, and $n := \alpha_1 + \dots + \alpha_k$. Define

$$\mathbb{B}_\alpha := \mathbb{B}_{\alpha_1} \times \dots \times \mathbb{B}_{\alpha_k}$$

and observe that

$$\mathbb{B}_\alpha = \{z \in \mathbb{C}^n : q_\alpha(z) < 1\},$$

where $q_\alpha : \mathbb{C}^n \longrightarrow \mathbb{R}_+$ is the norm given by the formula

$$q_\alpha(z) = \max\{\|z_1\|_{\alpha_1}, \dots, \|z_k\|_{\alpha_k}\}, \quad z = (z_1, \dots, z_k) \in \mathbb{C}^n = \mathbb{C}^{\alpha_1} \times \dots \times \mathbb{C}^{\alpha_k}$$

($\| \ \|_s$ denotes the Euclidean norm in $\mathbb{C}^s$).

For $a \in \mathbb{B}_s$, let $h_a^{(s)}$ denote the automorphism of $\mathbb{B}_s$ defined by (2.3.1). Then, for any $a = (a_1, \dots, a_k) \in \mathbb{B}_\alpha$ the mapping $h_a^{(\alpha)}$, defined by the formula

$$h_a^{(\alpha)}(z) := (h_{a_1}^{(\alpha_1)}(z_1), \dots, h_{a_k}^{(\alpha_k)}(z_k)), \quad z = (z_1, \dots, z_k) \in \mathbb{C}^n,$$

is an automorphism of $\mathbb{B}_\alpha$ such that $h_a^{(\alpha)}(a) = 0$. Hence, by Corollary 2.3.4, we have

$$\boldsymbol{c}_{\mathbb{B}_\alpha}^{(*)}(a, z) = \max\{\boldsymbol{c}_{\mathbb{B}_{\alpha_j}}^{(*)}(a_j, z_j) : j = 1, \dots, k\},$$
$$a = (a_1, \dots, a_k),\ z = (z_1, \dots, z_k) \in \mathbb{C}^n. \quad (2.4.1)$$

We will study $\boldsymbol{c}$-isometries $F : \mathbb{B}_\alpha \longrightarrow \mathbb{B}_\beta$, where $\alpha = (\alpha_1, \dots, \alpha_k) \in \mathbb{N}^k$, $\beta = (\beta_1, \dots, \beta_\ell) \in \mathbb{N}^\ell$, $n := \alpha_1 + \dots + \alpha_k$, $m := \beta_1 + \dots + \beta_\ell$.

Recall that the case $k = \ell = 1$, $\alpha = \beta = 1$ has been already solved in Proposition 1.1.20.

Remark 2.4.1.

(a) Let $F : \mathbb{B}_\alpha \longrightarrow \mathbb{B}_\beta$ be a $\boldsymbol{c}$-isometry. By (2.3.2) and (2.4.1), the mapping F is injective, continuous, and proper. In particular, $n \leq m$; if $n = m$, then F is a homeomorphism and hence F^{-1} is also a $\boldsymbol{c}$-isometry (use the Brouwer theorem on the invariance of domain).

(b) Since the group $\operatorname{Aut}(\mathbb{B}_\beta)$ acts transitively, to characterize all $\boldsymbol{c}$-isometries $F : \mathbb{B}_\alpha \longrightarrow \mathbb{B}_\beta$ it suffices to consider only those for which $F(0) = 0$.

(c) Suppose that $k = \ell$ and let $F_j : \mathbb{C}^{\alpha_j} \longrightarrow \mathbb{C}^{\beta_j}$ be a unitary or antiunitary operator (in particular, we necessarily have $\alpha_j \le \beta_j$), $j = 1, \dots, k$. Set $F = (F_1, \dots, F_k)$. Then, by (2.3.2) and (2.4.1), F is a $\boldsymbol{c}$-isometry of $\mathbb{B}_\alpha$ into $\mathbb{B}_\beta$.

(d) Let $\varphi : \mathbb{D} \longrightarrow \mathbb{D}$ be given by the formula

$$\varphi(\lambda) = \begin{cases} \lambda & \text{if } \operatorname{Im}\lambda \ge 0 \\ \overline{\lambda} & \text{if } \operatorname{Im}\lambda \le 0 \end{cases}.$$

One can easily prove that

$$\boldsymbol{m}(\varphi(\lambda'), \varphi(\lambda'')) \le \boldsymbol{m}(\lambda', \lambda''), \quad \lambda', \lambda'' \in \mathbb{D},$$

with equality if $(\operatorname{Im}\lambda')(\operatorname{Im}\lambda'') \ge 0$. Let $h_1, h_2 \in \operatorname{Aut}(\mathbb{D})$ be such that $h_j(C_j \cap \mathbb{D}) = (-1, 1)$, $j = 1, 2$, where

$$C_1 := \partial\mathbb{D}(\sqrt{1+\varepsilon^2}e^{i\pi/4}, \varepsilon), \ C_2 := \partial\mathbb{D}(\sqrt{1+\varepsilon^2}e^{3i\pi/4}, \varepsilon)$$

are "small" circles, orthogonal to $\mathbb{T}$. Put $F_j = \varphi \circ h_j$, $j = 1, 2$, $F_3 := \varphi$. Obviously,

$$\boldsymbol{m}(F_j(\lambda'), F_j(\lambda'')) \le \boldsymbol{m}(\lambda', \lambda''), \quad \lambda', \lambda'' \in \mathbb{D}, \ j = 1, 2, 3.$$

One can easily check that for any λ_0', $\lambda_0'' \in \mathbb{D}$ there exists a $j \in \{1, 2, 3\}$ such that

$$\boldsymbol{m}(F_j(\lambda_0'), F_j(\lambda_0'')) = \boldsymbol{m}(\lambda_0', \lambda_0'').$$

Hence, the mapping $F = (F_1, F_2, F_3) : \mathbb{D} = \mathbb{B}_{(1)} \longrightarrow \mathbb{D}^3 = \mathbb{B}_{(1,1,1)}$ is a $\boldsymbol{c}$-isometry such that F_1, F_2, and F_3 are neither holomorphic nor antiholomorphic. Moreover, F is not differentiable; cf. Theorem 2.4.6.

Proposition 2.4.2. *If $k = \ell = 1$ (put $\alpha = \alpha_1$ and $\beta = \beta_1$), then any $\boldsymbol{c}$-isometry $F : \mathbb{B}_\alpha \longrightarrow \mathbb{B}_\beta$ with $F(0) = 0$ is either unitary or antiunitary.*

Proof. (See [331] for $\alpha = \beta$.) It suffices to prove that

$$\text{either } \langle F(z'), F(z'')\rangle_\beta = \langle z', z''\rangle_\alpha, \quad z', z'' \in \mathbb{B}_\alpha,$$
$$\text{or } \langle F(z'), F(z'')\rangle_\beta = \overline{\langle z', z''\rangle_\alpha}, \quad z', z'' \in \mathbb{B}_\alpha$$

($\langle\ ,\ \rangle_s$ denotes the complex scalar product in $\mathbb{C}^s$).

By (2.3.2) we get

$$\|F(z)\|_\beta = \|z\|_\alpha, \quad z \in \mathbb{B}_\alpha,$$

and

$$|1 - \langle F(z'), F(z'') \rangle_\beta| = |1 - \langle z', z'' \rangle_\alpha|, \quad z', z'' \in \mathbb{B}_\alpha. \tag{2.4.2}$$

Hence,

$$F(-z) = -F(z), \quad z \in \mathbb{B}_\alpha. \tag{2.4.3}$$

Relations (2.4.2) and (2.4.3) imply that

$$|\langle F(z'), F(z'') \rangle_\beta| = |\langle z', z'' \rangle_\alpha|,$$
$$\operatorname{Re}\langle F(z'), F(z'') \rangle_\beta = \operatorname{Re}\langle z', z'' \rangle_\alpha, \quad z', z'' \in \mathbb{B}_\alpha.$$

Thus, for any pair $(z', z'') \in \mathbb{B}_\alpha \times \mathbb{B}_\alpha \setminus \{\operatorname{Im}\langle z', z'' \rangle_\alpha = 0\} =: \Omega$ there exists an $\varepsilon(z', z'') \in \{-1, 1\}$ such that

$$\operatorname{Im}\langle F(z'), F(z'') \rangle_\beta = \varepsilon(z', z'') \operatorname{Im}\langle z', z'' \rangle_\alpha.$$

It remains to prove that the function ε is constant. The set Ω has two connected components Ω_-, Ω_+ – see Lemma 2.4.3 below. Since ε is continuous, it is constant on each of them: $\varepsilon = \varepsilon_-$ on Ω_- and $\varepsilon = \varepsilon_+$ on Ω_+. Note that

$$(z', z'') \in \Omega_- \quad \text{iff} \quad (-z', z'') \in \Omega_+.$$

Hence, by (2.4.3) we get $\varepsilon_- = \varepsilon_+$. □

Lemma 2.4.3. *Let* $\Omega_\pm := \{(z', z'') \in \mathbb{B}_n \times \mathbb{B}_n : \pm \operatorname{Im}\langle z', z'' \rangle > 0\}$. *Then,* $\Omega_\pm$ *is connected.*

Proof. We will prove that Ω_+ is connected. Fix two points $a = (a', a'')$, $b = (b', b'') \in \Omega_+$. We are going to connect them in Ω_+ by a curve. We may assume that $\operatorname{Im} a'_j \overline{a''_j} > 0$, $j = 1, \dots, s$, and $\operatorname{Im} a'_j \overline{a''_j} \le 0$, $j = s+1, \dots, n$, where $1 \le s \le n$. If $s \le n-1$, then first we connect a in Ω_+ with the point $\widetilde{a} := (a'_1, \dots, a'_s, 0, \dots, 0, a''_1, \dots, a''_s, 0, \dots, 0)$ by the curve

$$[0, 1] \ni t \longmapsto (a'_1, \dots, a'_s, ta'_{s+1}, \dots, ta'_n, a''_1, \dots, a''_s, ta''_{s+1}, \dots, ta''_n).$$

If $s = n$, then we put $\widetilde{a} := a$. Now, we connect $\widetilde{a}$ in Ω with the point $\approx{a} :=$ $(a'_1, 0, \dots, 0, a''_1, 0, \dots, 0)$ by the curve

$$[0, 1] \ni t \longmapsto (a'_1, ta'_2, \dots, ta'_s, 0, \dots, 0, a''_1, ta''_2, \dots, ta''_s, 0, \dots, 0).$$

An analogous procedure connects b to a point $\overset{\approx}{b} = (0, \dots, 0, b'_\nu, 0, \dots, 0, b''_\nu, 0, \dots, 0)$ for some $\nu \in \{1, \dots, n\}$.

Thus, the proof reduces to the case $n = 1$ (if $\nu = 1$) or $n = 2$ (if $\nu > 1$). Moreover, in the case $n = 2$ we may assume that $a = (a'_1, 0, a''_1, 0)$, $b = (0, b'_2, 0, b''_2)$,

$\operatorname{Im} a_1' \overline{a}_1'' > 0$, $\operatorname{Im} b_2' \overline{b}_2'' > 0$. Observe that in this case the segment $[a, b]$ is contained in Ω_+.

In the case $n = 1$ [3] let $a = (a', a''), b = (b', b'') \in \Omega_+$ be fixed. Let $k = k_a \in \mathbb{D}$ be such that $a'\overline{a}'' = k^2$, $\operatorname{Re} k \geq 0$, $\operatorname{Im} k > 0$. Observe that there exists a curve $\alpha : [0, 1] \longrightarrow \mathbb{D}$ such that $\alpha(0) = a'$, $\alpha(1) = k$, and $|\alpha(t)| > |k^2|$, $t \in [0, 1]$. Then, the curve

$$[0, 1] \ni t \longmapsto \left(\alpha(t), \overline{k}^2 / \overline{\alpha}(t)\right) \in \Omega_+$$

connects a with the point $(k, \overline{k})$. Consequently, it remains to connect the points $(k_a, \overline{k}_a)$, $(k_b, \overline{k}_b)$ in Ω_+. Since $\operatorname{Im}(tk_a + (1-t)k_b)^2 > 0$, $t \in [0, 1]$, we may connect them by the segment. □

In the general case the following result holds.

Proposition 2.4.4. *Let $F : \mathbb{B}_\alpha \longrightarrow \mathbb{B}_\beta$ be a $\boldsymbol{c}$-isometry with $F(0) = 0$. Then, $k \leq \ell$. Moreover, if $k = \ell$, then the mapping F has the form described in Remark* 2.4.1(c), *up to permutations of $(\alpha_1, \dots, \alpha_k)$ and $(\beta_1, \dots, \beta_\ell)$ (i.e., F is component-wise unitary or antiunitary). In particular, we get*

Generalized Poincaré theorem. *If $k > \ell$, then there is no $\boldsymbol{c}$-isometry of $\mathbb{B}_\alpha$ into $\mathbb{B}_\beta$ (cf. Corollary* 2.3.9, *where $k = n \geq 2$, $\alpha_1 = \dots = \alpha_n = 1$, $\ell = 1$, $\beta_1 = n$).*

Proof. (Cf. [331] for $k = \ell$, $\alpha_1 = \dots = \alpha_k = \beta_1 = \dots = \beta_k$.) First, recall that for arbitrary $N \in \mathbb{N}$ we have

$$\max\{\boldsymbol{c}^*_{\mathbb{B}_N}(a, b),\ \boldsymbol{c}^*_{\mathbb{B}_N}(a, -b)\} > \|b\|_N, \quad a, b \in \mathbb{B}_N,\ a \neq 0. \tag{2.4.4}$$

Let $e_s : \mathbb{B}_{\alpha_s} \longrightarrow \mathbb{B}_\alpha$ be given by the formula

$$e_s(x_s) := (0, \dots, 0, \underset{\uparrow\, s\text{-th place}}{x_s}, 0, \dots, 0), \quad x_s \in \mathbb{B}_{\alpha_s},\ s = 1, \dots k.$$

Set $G_s = (G_{s,1}, \dots, G_{s,l}) := F \circ e_s : \mathbb{B}_{\alpha_s} \longrightarrow \mathbb{B}_\beta$, $s = 1, \dots, k$. Fix $s \in \{1, \dots, k\}$ and $x_s \in (\mathbb{B}_{\alpha_s})_*$. Since F is a $\boldsymbol{c}$-isometry, we get

$$\begin{aligned} \boldsymbol{c}^*_{\mathbb{B}_{\alpha_s}}(-x_s, x_s) &= \boldsymbol{c}^*_{\mathbb{B}_\alpha}(e_s(-x_s), e_s(x_s)) = \boldsymbol{c}^*_{\mathbb{B}_\beta}(F(e_s(-x_s)),\ F(e_s(x_s))) \\ &= \max\{\boldsymbol{c}^*_{\mathbb{B}_{\beta_t}}(G_{s,t}(-x_s),\ G_{s,t}(x_s)) : t = 1, \dots, \ell\}. \end{aligned}$$

Hence, there exists a $t = t(s, x_s) \in \{1, \dots, \ell\}$ such that

$$\boldsymbol{c}^*_{\mathbb{B}_{\beta_t}}(G_{s,t}(-x_s),\ G_{s,t}(x_s)) = \boldsymbol{c}^*_{\mathbb{B}_{\alpha_s}}(-x_s, x_s).$$

[3] The idea of the proof is due to W. Jarnicki.

Since $\|G_{s,t}(\pm x_s)\|_{\beta_t} \le \|x_s\|_{\alpha_s}$, this implies that

$$\begin{aligned} G_{s,t}(-x_s) &= -G_{s,t}(x_s), &(2.4.5)\\ \|G_{s,t}(x_s)\|_{\beta_t} &= \|x_s\|_{\alpha_s} \quad \text{(cf. the proof of Proposition 2.4.2).} &(2.4.6)\end{aligned}$$

Take $s' \neq s$ and $x_{s'} \in \mathbb{B}_{\alpha_{s'}}$ with $\|x_{s'}\|_{\alpha_{s'}} \le \|x_s\|_{\alpha_s}$. Suppose that $G_{s',t}(x_{s'}) \neq 0$. Then, (2.4.4), (2.4.5), and (2.4.6) give

$$\begin{aligned}\|x_s\|_{\alpha_s} &= \boldsymbol{c}^*_{\mathbb{B}_\alpha}(e_{s'}(x_{s'}),\ e_s(\pm x_s))\\ &\ge \max\{\boldsymbol{c}^*_{\mathbb{B}_{\beta_t}}(G_{s',t}(x_{s'}), G_{s,t}(x_s)), \boldsymbol{c}^*_{\mathbb{B}_{\beta_t}}(G_{s',t}(x_{s'}), G_{s,t}(-x_s))\}\\ &> \|G_{s,t}(x_s)\|_{\beta_t} = \|x_s\|_{\alpha_s};\end{aligned}$$

a contradiction. This shows that

$$G_{s',t}(x_{s'}) = 0,\ s' \neq s,\ \|x_{s'}\|_{\alpha_{s'}} \le \|x_s\|_{\alpha_s}\ (t = t(s, x_s)). \tag{2.4.7}$$

Using a suitable permutation, we may assume that $\alpha_1 \le \cdots \le \alpha_k$. This permits us to identify $\mathbb{B}_{\alpha_s}$ with $\mathbb{B}_{\alpha_s} \times \{0\} \subset \mathbb{B}_{\alpha_{s'}}$ for $1 \le s < s' \le k$. By virtue of (2.4.6) and (2.4.7), one can easily prove that for any $x_1 \in (\mathbb{B}_{\alpha_1})_*$ the numbers $t(1, x_1), \dots, t(k, x_1)$ are different. Thus, $k \le \ell$.

Now, assume that $k = l$. In particular, $\{t(1, x_1), \dots, t(k, x_1)\} = \{1, \dots, k\}$. Fix an $\hat{x}_1 \in (\mathbb{B}_{\alpha_1})_*$ and let

$$t(s) := t(s, \hat{x}_1), \quad s = 1, \dots, k.$$

It is easily seen that

$$t(1) = t(1, x_1) \quad \text{for any} \quad x_1 \in (\mathbb{B}_{\alpha_1})_*$$

and

$$G_{1,j}(x_1) = 0 \quad \text{for any} \quad x_1 \in \mathbb{B}_{\alpha_1} \quad \text{and} \quad j \neq t(1).$$

Now, it suffices to verify the following property, which when applied inductively then gives the required result.

If for some $s \in \{1, \dots, k\}$ and $t \in \{1, \dots, \ell\}$

$$G_{s,j}(x_s) = 0, \quad x_s \in \mathbb{B}_{\alpha_s},\ j \neq t, \tag{2.4.8}$$

then the mapping $G_{s,t} : \mathbb{B}_{\alpha_s} \longrightarrow \mathbb{B}_{\beta_t}$ is a $\boldsymbol{c}$-isometry and

$$F_t(x_1, \dots, x_k) = G_{s,t}(x_s), \quad (x_1, \dots, x_k) \in \mathbb{B}_\alpha.$$

In particular, by Proposition 2.4.2, $G_{s,t}$ is unitary or antiunitary.

For the proof of the above property, observe that for $x'_s, x''_s \in \mathbb{B}_{\alpha_s}$ we have

$$\begin{aligned}\boldsymbol{c}^*_{\mathbb{B}_{\alpha_s}}(x'_s, x''_s) &= \boldsymbol{c}^*_{\mathbb{B}_\alpha}(e_s(x'_s), e_s(x''_s)) = \boldsymbol{c}^*_{\mathbb{B}_\beta}(F(e_s(x'_s)), F(e_s(x''_s))) \\ &= \boldsymbol{c}^*_{\mathbb{B}_\beta}(G_s(x'_s), G_s(x''_s)) = \boldsymbol{c}^*_{\mathbb{B}_{\beta_t}}(G_{s,t}(x'_s), G_{s,t}(x''_s)) \quad \text{(use (2.4.8))},\end{aligned}$$

and therefore $G_{s,t}$ is a $\boldsymbol{c}$-isometry.

Now, suppose that $F_t(x) \neq G_{s,t}(x_s)$ for some $x = (x_1, \dots, x_k) \in \mathbb{B}_\alpha$. Let $h \in \operatorname{Aut}(\mathbb{B}_{\alpha_s})$, $g \in \operatorname{Aut}(\mathbb{B}_{\beta_t})$ be such that $h(0) = x_s$ and $g(G_{s,t}(x_s)) = 0$. Put $\varphi := g \circ G_{s,t} \circ h : \mathbb{B}_{\alpha_s} \longrightarrow \mathbb{B}_{\beta_t}$. Obviously, φ is a $\boldsymbol{c}$-isometry and $\varphi(0) = 0$. Hence, by Proposition 2.4.2, φ is at least $\mathbb{R}$-linear. Take a $z_s \in \mathbb{B}_{\alpha_s}$ such that

$$\|z_s\|_{\alpha_s} = \boldsymbol{c}^*_{\mathbb{B}_\alpha}(x, e_s(x_s)).$$

Then,

$$\begin{aligned}\|z_s\|_{\alpha_s} &= \boldsymbol{c}^*_{\mathbb{B}_\alpha}(x, e_s(h(\pm z_s))) \\ &\geq \max\{\boldsymbol{c}^*_{\mathbb{B}_{\beta_t}}(F_t(x), G_{s,t}(h(z_s))), \boldsymbol{c}^*_{\mathbb{B}_{\beta_t}}(F_t(x), G_{s,t}(h(-z_s)))\} \\ &= \max\{\boldsymbol{c}^*_{\mathbb{B}_{\beta_t}}(g(F_t(x)), \varphi(z_s)), \boldsymbol{c}^*_{\mathbb{B}_{\beta_t}}(g(F_t(x)), -\varphi(z_s))\} \\ &> \|\varphi(z_s)\|_{\beta_t} = \boldsymbol{c}^*_{\mathbb{B}_{\beta_t}}(G_{s,t}(x_s), G_{s,t}(h(z_s))) \\ &= \boldsymbol{c}^*_{\mathbb{B}_{\alpha_s}}(x_s, h(z_s)) = \|z_s\|_{\alpha_s} \quad \text{(use (2.4.4))};\end{aligned}$$

a contradiction. The proof is complete. □

In the case $\alpha_1 + \cdots + \alpha_k = \beta_1 + \cdots + \beta_\ell$, the following result partially generalizes Proposition 2.4.4.

Theorem* 2.4.5 (cf. [558]). *Let $D_j \subset \mathbb{C}^{\alpha_j}$, $j = 1, \dots, k$, $G_j \subset \mathbb{C}^{\beta_j}$, $j = 1, \dots, \ell$, be bounded strictly convex domains. Assume that $\alpha_1 + \cdots + \alpha_k = \beta_1 + \cdots + \beta_\ell$. Suppose that $F = (F_1, \dots, F_\ell) : D_1 \times \cdots \times D_k \longrightarrow G_1 \times \cdots \times G_\ell$ is a bijective $\boldsymbol{c}$-isometry. Then, $k = \ell$ and, after suitable permutations of domains,*

$$F_j(z) = \varphi_j(z_j), \quad z = (z_1, \dots, z_k) \in D_1 \times \cdots \times D_k,$$

$\varphi_j : D_j \longrightarrow G_j$ is a $\boldsymbol{c}$-isometry, $j = 1, \dots, k$.

In the case of $k < \ell$, Proposition 2.4.4 may be extended to the following deep result.

Theorem* 2.4.6 (cf. [558, 557]). *Let $F = (F_1, \dots, F_\ell) : \mathbb{B}_\alpha \longrightarrow \mathbb{B}_\beta$ be a $\boldsymbol{c}$-isometry (in particular, $k \leq \ell$ – cf. Proposition 2.4.4) with $\alpha_1 \leq \cdots \leq \alpha_k$, $\beta_1 \leq \cdots \leq \beta_\ell$. Then, $\alpha \preceq \beta$, i.e.,*

$$\alpha_k \leq \beta_\ell, \ \alpha_{k-1} \leq \beta_{\ell-1}, \dots, \alpha_1 \leq \beta_{\ell-k+1}.$$

Moreover, if the number $s = \Lambda(\alpha, \beta)$, defined inductively by the formula

$$\Lambda(\alpha,\beta) := \begin{cases} \begin{cases} 0, & \text{if } \ell \geq 3,\ \alpha_1 \leq \beta_{\ell-2} \\ 1, & \text{otherwise} \end{cases}, & \text{if } k = 1 \\ \begin{cases} \Lambda((\alpha_1,\dots,\alpha_{k-1}),(\beta_1,\dots,\beta_{\ell-3})), \text{ if } \ell \geq m+2, \\ \qquad \alpha_k \leq \beta_{\ell-2},\ (\alpha_1,\dots,\alpha_{k-1}) \preceq (\beta_1,\dots,\beta_{\ell-3}) \\ 1 + \Lambda((\alpha_1,\dots,\alpha_{k-1}),(\beta_1,\dots,\beta_{\ell-1})), \text{ otherwise} \end{cases}, & \text{if } k \geq 2 \end{cases},$$

is positive, then, after suitable permutations of balls, we have

$$F_j(z) = \varphi_j(z_j), \quad z = (z_1,\dots,z_k) \in \mathbb{B}_\alpha,$$

where $\varphi_j : \mathbb{B}_{\alpha_j} \longrightarrow \mathbb{B}_{\beta_j}$ is a $\boldsymbol{c}$-isometry, $j = 1,\dots,s$.

Remark 2.4.7 (cf. [558, 557]).

(a) If $k = 1$, then

$$\Lambda(\alpha,\beta) = \begin{cases} 1, & \text{if } \#\{j : \beta_j \geq \alpha_1\} \leq 2 \\ 0, & \text{if } \#\{j : \beta_j \geq \alpha_1\} \geq 3 \end{cases}.$$

cf. Remark 2.4.1(d) and Exercise 2.12.9.

(b) If $\ell \in \{k, k+1\}$, then $\Lambda(\alpha,\beta) = k$ (cf. Proposition 2.4.4).

(c) If $\Lambda(\alpha,\beta) = 0$, then $\ell \geq 3k$.

(d) $\Lambda((1,2,3),\ (1,1,1,1,1,1,2,3,3)) = 2,$

$\Lambda((1,2,3),\ (1,1,1,1,1,2,3,3,3)) = 1,$

$\Lambda((1,2,3),\ (1,1,1,2,2,2,3,3,3)) = 0,$

$\Lambda((1,2),\ (1,1,1,1,2,2)) = 1,$

$\Lambda((1,2),\ (1,1,1,2,2,2)) = 0.$

(e) If $\alpha_k \leq \beta_1$, then

$$\Lambda(\alpha,\beta) = \begin{cases} 0, & \text{if } \ell \geq 3k \\ k - \lfloor \frac{\ell-k}{2} \rfloor, & \text{if } k \leq \ell \leq 3k \end{cases}.$$

Remark 2.4.8 (cf. [558, 557]). The numbers $\Lambda(\alpha,\beta)$ are in some sense optimal.

Indeed, let $k \leq \ell \leq 3k$ and let $\beta = (\beta_1,\dots,\beta_\ell)$ be arbitrary. Let $r := \lfloor \frac{\ell-k}{k} \rfloor$. By Remark 2.4.4(d), there exist $\boldsymbol{c}$-isometries $\varphi_j : \mathbb{D} \longrightarrow \mathbb{B}_{(\beta_{3j-2},\beta_{3j-1},\beta_{3j})}$, $j = 1,\dots,r$, whose components are not $\boldsymbol{c}$-isometries.

If k is even, then we define $F : \mathbb{B}_{(1,\dots,1)} \longrightarrow \mathbb{B}_\beta$,

$$F(z_1,\dots,z_k) := (\varphi_1(z_1),\dots,\varphi_r(z_r), z_{r+1},\dots,z_k).$$

It is clear that F has exactly $k-r$ components that are c-isometries.

If k is odd, then we define $F : \mathbb{B}_{(1,\dots,1)} \longrightarrow \mathbb{B}_\beta$,

$$F(z_1,\dots,z_k) := (\varphi_1(z_1),\dots,\varphi_r(z_r), z_{r+1},\dots,z_k, \psi(z_1)),$$

where $\psi : \mathbb{D} \longrightarrow \mathbb{B}_{\beta_\ell}$ is an arbitrary c-contraction that is not a c-isometry (e.g. $\lambda \longmapsto (\lambda^2, 0,\dots,0)$). As above, F has exactly $k-r$ components that are c-isometries.

It is an open problem whether $\Lambda(\alpha,\beta)$ is optimal for fixed α and β. ?

Remark 2.4.9.

(a) In the context of Theorem 2.3.10, one can ask what are characterizations of Carathéodory isometries $B_1 \times\cdots\times B_k \longrightarrow B'_1 \times\cdots\times B'_\ell$, where $B_\mu \in \{\mathbb{B}_{n_\mu}, \mathbb{L}_{n_\mu}\}$, $\mu = 1,\dots,k$, $B'_\nu \in \{\mathbb{B}_{m_\nu}, \mathbb{L}_{m_\nu}\}$, $\nu = 1,\dots,\ell$, and if $B_\mu = \mathbb{L}_{n_\mu}$ (resp. $B'_\nu = \mathbb{L}_{m_\nu}$), then $n_\mu \geq 3$ (resp. $m_\nu \geq 3$).

Notice that even the case where $k = \ell = 1$, $B_1 = B'_1 = \mathbb{L}_n$ is not solved. ?

(b) More results on various isometries may be found in Miscellanea § A.7.

2.5 Carathéodory hyperbolicity

In general, the pseudodistance $\boldsymbol{c}_G^{(*)}$ need not be a distance, e.g. $\boldsymbol{c}_{\mathbb{C}^n}^{(*)} \equiv 0$. Note that

$$\boldsymbol{c}_G^{(*)} \equiv 0 \quad\text{iff}\quad \boldsymbol{\gamma}_G \equiv 0 \quad\text{iff}\quad \mathcal{H}^\infty(G) \simeq \mathbb{C},$$

where "$\mathcal{H}^\infty(G) \simeq \mathbb{C}$" means that all bounded holomorphic functions on G are constant, i.e., G is a Liouville domain. On the other hand,

$$\boldsymbol{c}_G^{(*)} \text{ is a distance iff } \mathcal{H}^\infty(G) \text{ separates points in } G. \tag{2.5.1}$$

If $\boldsymbol{c}_G^{(*)}$ is a distance, then we say that G is *$\boldsymbol{c}$-hyperbolic*. If for any $a \in G$ the seminorm $\boldsymbol{\gamma}_G(a;\cdot)$ is a norm, then we say that G is *$\boldsymbol{\gamma}$-hyperbolic*. Observe that

$$\boldsymbol{\gamma}_G(a;\cdot) \text{ is a norm iff } \forall_{X\in(\mathbb{C}^n)_*}\ \exists_{f\in\mathcal{H}^\infty(G)} : f'(a)X \neq 0. \tag{2.5.2}$$

Proposition 2.5.1. *If $G \subset \mathbb{C}^1$, then the following conditions are equivalent:*

(i) *G is $\boldsymbol{c}$-hyperbolic;*

(ii) *G is $\boldsymbol{\gamma}$-hyperbolic;*

(iii) $\mathcal{H}^\infty(G) \not\simeq \mathbb{C}$.

In other words, if a domain $G \subset \mathbb{C}^1$ is not a Liouville domain, then it is both $\boldsymbol{c}$*- and* $\boldsymbol{\gamma}$*-hyperbolic.*

Proof. (cf. [471]). Obviously, (i) $\Longrightarrow$ (iii) and (ii) $\Longrightarrow$ (iii). Suppose that (iii) is satisfied and let $f_0 \in \mathcal{H}^\infty(G)$, $a_1, a_2 \in G$, be such that $f_0(a_1) \neq f_0(a_2)$. Put

$$f_j(z) := \begin{cases} \frac{f_0(z)-f_0(a_j)}{z-a_j}, & z \in G \setminus \{a_j\} \\ f_0'(a_j), & z = a_j \end{cases}.$$

Then, $f_j \in \mathcal{H}^\infty(G)$, $j = 1, 2$. One can easily verify that the functions f_0, f_1, f_2 separate points in G and that $\operatorname{rank}(f_0', f_1', f_2') = 1$ on G. □

If $n > 1$, then there are domains such that $\boldsymbol{c}_G^{(*)} \not\equiv 0$ (resp. $\boldsymbol{\gamma}_G \not\equiv 0$) but G is not $\boldsymbol{c}$-hyperbolic (resp. $\boldsymbol{\gamma}$-hyperbolic). For example, take any balanced convex domain $G = G_q \subsetneq \mathbb{C}^n$ with $q^{-1}(0) \neq \{0\}$.

? Relations between $\boldsymbol{c}$- and $\boldsymbol{\gamma}$-hyperbolicity are still not understood (see Exercise 2.12.8).

Proposition 2.5.2. *If $G \subset \mathbb{C}^n$ is a domain biholomorphic to a bounded domain, then G is both* $\boldsymbol{c}$*- and* $\boldsymbol{\gamma}$*-hyperbolic.*

Proof. We may assume that G is bounded. Then, $z_1, \dots, z_n \in \mathcal{H}^\infty(G)$ and so the result follows from (2.5.1) and (2.5.2). □

Note that if G is a bounded domain, $R := \operatorname{diam}(G)$ (in the Euclidean sense), then, by Theorem 2.1.1 and Proposition 2.3.1(c), we have

$$\begin{aligned} \boldsymbol{c}_G^*(z', z'') &\geq \boldsymbol{c}^*_{\mathbb{B}(z',R)}(z', z'') = \frac{\|z' - z''\|}{R}, \quad z', z'' \in G, \\ \boldsymbol{\gamma}_G(z; X) &\geq \boldsymbol{\gamma}_{\mathbb{B}(z,R)}(z; X) = \frac{\|X\|}{R}, \quad z \in G,\ X \in \mathbb{C}^n. \end{aligned} \tag{2.5.3}$$

This is an alternative proof of Proposition 2.5.2.

2.6 The Carathéodory topology

Let $\operatorname{top} \boldsymbol{c}_G^{(*)}$ (resp. $\operatorname{top} G$) denote the topology generated by $\boldsymbol{c}_G^{(*)}$ (resp. the Euclidean topology of G). By virtue of (2.1.1) we have

$$B_{\boldsymbol{c}_G}(a, r) = B_{\boldsymbol{c}_G^*}(a, \tanh(r)), \quad a \in G,\ r > 0.$$

Hence, $\operatorname{top} \boldsymbol{c}_G^* = \operatorname{top} \boldsymbol{c}_G$.

Proposition 2.6.1. *$\boldsymbol{c}_G^{(*)}$ is continuous. In particular,* $\operatorname{top} \boldsymbol{c}_G^{(*)} \subset \operatorname{top} G$.

If G is biholomorphic to a bounded domain, then $\operatorname{top} \boldsymbol{c}_G^{(*)} = \operatorname{top} G$.

Proof. (a) Since $\boldsymbol{c}_G^{(*)}$ is a pseudodistance, it suffices to show that $\lim_{z \to a} \boldsymbol{c}_G^{(*)}(a, z) = 0$. In view of (2.1.1), it is enough to consider only the case of $\boldsymbol{c}_G^*$. Recall that

$$\boldsymbol{c}_G^*(a, \cdot) = \sup\{|f| : f \in \mathcal{O}(G, \mathbb{D}),\, f(a) = 0\}. \tag{2.6.1}$$

The family in (2.6.1) is equicontinuous, and therefore $\boldsymbol{c}_G^*(a, \cdot)$ is continuous.

(b) follows from (2.5.3). □

Note that if $\mathbb{B}(a, 3r) \subset G$, then by Proposition 2.3.1(c)

$$\boldsymbol{c}_G^*(z', z'') \le \boldsymbol{c}_{\mathbb{B}(z', 2r)}^*(z', z'') = \frac{\|z' - z''\|}{2r}, \quad z', z'' \in \mathbb{B}(a, r), \tag{2.6.2}$$

$$\boldsymbol{\gamma}_G(z; X) \le \boldsymbol{\gamma}_{\mathbb{B}(z, r)}(z; X) = \frac{\|X\|}{r}, \quad z \in \mathbb{B}(a, r),\ X \in \mathbb{C}^n. \tag{2.6.3}$$

Relation (2.6.2) gives an alternative proof of Proposition 2.6.1(a).

In $\mathbb{C}^1$ the situation is extremely simple, namely we have

Proposition 2.6.2. *If $G \subset \mathbb{C}^1$ is $\boldsymbol{c}$-hyperbolic, then* $\operatorname{top} \boldsymbol{c}_G^{(*)} = \operatorname{top} G$.

Proof. Let $G \ni a_\nu \longrightarrow a_0 \in G$ in $\operatorname{top} \boldsymbol{c}_G^{(*)}$. Observe that

$$|f - f(a_0)| \le \|f - f(a_0)\|_G \cdot \boldsymbol{c}_G^*(a_0, \cdot), \quad f \in \mathcal{H}^\infty(G).$$

Consequently, $f(a_\nu) \longrightarrow f(a_0)$ for any $f \in \mathcal{H}^\infty(G)$. Since $\mathcal{H}^\infty(G) \not\simeq \mathbb{C}$ (cf. Proposition 2.5.1), there exists an $f_0 \in \mathcal{H}^\infty(G)$, $f_0 \not\equiv 0$, with $f_0(a_0) = 0$. Let

$$f_0(z) = (z - a_0)^k g(z), \quad z \in G, \tag{2.6.4}$$

where $g(a_0) \neq 0$. Clearly, $g \in \mathcal{H}^\infty(G)$. Since $f_0(a_\nu) \longrightarrow 0$ and $g(a_\nu) \to g(a_0) \neq 0$, condition (2.6.4) implies that $a_\nu \longrightarrow a_0$ in $\operatorname{top} G$. □

Unfortunately, for $n \ge 3$ there exist $\boldsymbol{c}$-hyperbolic domains with $\operatorname{top} \boldsymbol{c}_G^{(*)} \neq \operatorname{top} G$. For $n = 2$ we do not know whether such a domain exists. ?

Theorem 2.6.3 (cf. [271]). *For any $n \ge 3$ there exists a $\boldsymbol{c}$- and $\boldsymbol{\gamma}$-hyperbolic domain $G \subset \mathbb{C}^n$ such that* $\operatorname{top} \boldsymbol{c}_G^{(*)} \subsetneq \operatorname{top} G$.

This theorem will be proved by a sequence of lemmas. The proof requires the use of manifolds. Observe that the notions introduced so far may be literally extended to the case of manifolds.

Lemma 2.6.4. *Let V be a $\boldsymbol{c}$- and $\boldsymbol{\gamma}$-hyperbolic connected complex submanifold of $\mathbb{C}^n$, $n \geq 2$, $\dim V \geq 1$, such that* $\operatorname{top} \boldsymbol{c}_V^{(*)} \neq \operatorname{top} V$. *Then, there exists a $\boldsymbol{c}$- and $\boldsymbol{\gamma}$-hyperbolic domain of holomorphy $G \subset \mathbb{C}^n$ with $V \subset G$ such that* $\operatorname{top} \boldsymbol{c}_G^{(*)} \neq \operatorname{top} G$.

Proof. It is known that there exist an open neighborhood U_0 of V and a holomorphic retraction $r : U_0 \longrightarrow V$; cf. [140] (Appendix B.7.18). Put

$$U := \{z \in U_0 : z - r(z) \in \mathbb{D}^n\}.$$

Note that U is an open neighborhood of V. Choose a domain of holomorphy G with $V \subset G \subset U$; cf. [480] (Appendix B.7.19). Then,

$$\begin{aligned} \boldsymbol{c}_V^{(*)}(r(z'), r(z'')) &\leq \boldsymbol{c}_G^{(*)}(z', z''), \quad z', z'' \in G, & (2.6.5)\\ \boldsymbol{c}_G^{(*)}(z', z'') &\leq \boldsymbol{c}_V^{(*)}(z', z''), \quad z', z'' \in V, \\ \boldsymbol{\gamma}_V(r(z); r'(z)X) &\leq \boldsymbol{\gamma}_G(z; X), \quad z \in G,\ X \in \mathbb{C}^n, & (2.6.6)\\ \boldsymbol{\gamma}_G(z; X) &\leq \boldsymbol{\gamma}_V(z; X), \quad z \in V,\ X \in T_z V, \end{aligned}$$

where $T_z V$ denotes the tangent space to V at z. In particular,

$$\boldsymbol{c}_G^{(*)}|_{V \times V} = \boldsymbol{c}_V^{(*)}, \qquad \boldsymbol{\gamma}_G(z; X) = \boldsymbol{\gamma}_V(z; X), \quad z \in V,\ X \in T_z V.$$

Put $f_j(z) := z_j - r_j(z)$, $z \in G$. Then, $f_j : G \longrightarrow \mathbb{D}$, $j = 1, \dots, n$.

Take $z', z'' \in G$, $z' \neq z''$. If $r(z') = r(z'')$, then $f_j(z') \neq f_j(z'')$ for at least one $j \in \{1, \dots, m\}$. If $r(z') \neq r(z'')$, then we can use (2.6.5) to conclude that G is $\boldsymbol{c}$-hyperbolic.

For $z \in G$ and $X \in \mathbb{C}^n$, $X \neq 0$, we have the following two possibilities:

- $r'(z)X \neq 0$: then $\boldsymbol{\gamma}_G(z; X) > 0$ by (2.6.6);
- $r'(z)X = 0$: then $f_j'(z)X = X_j$, $j = 1, \dots, n$, and $\boldsymbol{\gamma}_G(z; X) > 0$ by (2.5.2).

Finally, (2.6.5) implies that $\operatorname{top} \boldsymbol{c}_G^{(*)} \neq \operatorname{top} G$. □

Lemma 2.6.5. *If M is a $\boldsymbol{c}$- and $\boldsymbol{\gamma}$-hyperbolic connected Stein manifold such that* $\operatorname{top} \boldsymbol{c}_M^{(*)} \neq \operatorname{top} M$, *then there exists a $\boldsymbol{c}$- and $\boldsymbol{\gamma}$-hyperbolic domain of $G \subset \mathbb{C}^{2 \dim M + 1}$ with* $\operatorname{top} \boldsymbol{c}_G^{(*)} \neq \operatorname{top} G$.

Proof. Use the embedding theorem (cf. Appendix B.7.17) and Lemma 2.6.4. □

Lemma 2.6.6 (cf. [218]). *There exists a $\boldsymbol{c}$- and $\boldsymbol{\gamma}$-hyperbolic connected Riemann domain M spread over $\mathbb{C}$ with* $\operatorname{top} \boldsymbol{c}_M^{(*)} \neq \operatorname{top} M$.

Proof. In the description of the desired Riemann surface M, we omit the details of how to construct a Riemann surface by glueing together local pieces.

Let $a_k, b_k \in \mathbb{D}$ $(k \in \mathbb{N})$ be sequences of real numbers in $(0, 1)$ without accumulation points in $(0, 1)$, such that $0 < a_k < b_k < a_{k+1}$ $(k \geq 1)$, and put $I_k := [a_k, b_k]$.

Moreover, to any $k \in \mathbb{N}$ we assign n_k pairwise disjoint subintervals $J_{k,j}$ $(1 \leq j \leq n_k)$ of I_k and we put

$$D_0 := \mathbb{D} \setminus \Big(\bigcup_{k=1}^{\infty} \bigcup_{j=1}^{n_k} J_{k,j} \Big), \quad D_k := \mathbb{D} \setminus \Big(\bigcup_{l \neq k} I_l \cup \bigcup_{j=1}^{n_k} J_{k,j} \Big), \quad k \geq 1;$$

the value of n_k will be found later.

Then, the Riemann surface M we need is given by the following glueing process:

D_0 and D_k are glued together along the cuts $J_{k,j}$, $1 \leq j \leq n_k$,
by crosswise identification, $k \in \mathbb{N}$.

If we put $D_k^* := \mathbb{D} \setminus \bigcup_{l \neq k} I_l$, then the subdomain $D_k \cup D_0^k$ of M, where D_0^k denotes the part of D_0 corresponding to D_k, is a two-sheeted covering of D_k^*, whose branch points are given by the ends of the subintervals $J_{k,j}$, $j = 1, \ldots, n_k$.

For any holomorphic function $f : M \longrightarrow \mathbb{D}$, we can define a new holomorphic function $f_k : D_k^* \longrightarrow \mathbb{D}$ by

$$f_k(z) = \frac{1}{4}[f(z_k^+) - f(z_k^-)]^2$$

if z is not a branch point, and where $z_k^{\pm}$ are the "two" points in $D_k \cup D_0^k$ over z. Here, we have used the classical Riemann extension theorem for bounded holomorphic functions. Observe that f_k vanishes at $2n_k$ branch points.

Let $z^* := i/2$; then by the Montel theorem we obtain $|f_k(z^*)| < (1/2)^k$ for n_k sufficiently large. Let z_k denote the point "over" z^* in M lying in the k-th "sheet" D_k. Then, the above observation yields

$$\boldsymbol{c}_M^*(z^*, z_k) \underset{k \to \infty}{\longrightarrow} 0,$$

whereas $(z_k)_k$ does not converge in M. Hence, $\operatorname{top} M \neq \operatorname{top} \boldsymbol{c}_M^*$.

What remains to be verified is that M is $\boldsymbol{c}$-hyperbolic, i.e., $\mathcal{H}^\infty(M)$ separates points. We denote by $\pi : M \longrightarrow \mathbb{D}$ the branched covering map of M, where π is given by "identification". It suffices to separate different points $z^1 \neq z^2$ with $\pi(z^1) = \pi(z^2)$. Assume that z^j belongs to the k_j-th "sheet"of M. Without loss of generality, we may assume that $k_1 > k_2 \geq 0$. Now, we construct a new Riemann surface R as follows:

R is obtained from two copies of $\mathbb{D} \setminus \bigcup_{j=1}^{n_{k_1}} J_{k_1,j}$
by glueing them together along the cuts $J_{k_1,j}$ crosswise.

Observe that there is a branched covering map $\hat{\pi} : R \longrightarrow \mathbb{D}$ (as above). Moreover, there is a holomorphic map $\psi : M \longrightarrow R$ such that $\pi = \hat{\pi} \circ \psi$; here ψ maps the k_1-th "sheet"of M into the "upper sheet" of R and the rest of M into the "lower sheet" of R, again by "identification". Since R is obtained by glueing together a finite number of cut unit discs, $\mathcal{H}^\infty(R)$ separates the points of R. Therefore, since $\psi(z^1) \neq \psi(z^2)$, we obtain a function $f \in \mathcal{H}^\infty(M)$ with $f(z^1) \neq f(z^2)$, i.e., M is $\boldsymbol{c}$-hyperbolic. The $\boldsymbol{\gamma}$-hyperbolicity is obtained in an analogous way. □

Observe that, if the Riemann surface M from Lemma 2.6.6 could be embedded in $\mathbb{C}^2$, then Theorem 2.6.3 would also be true for $n = 2$.

Remark 2.6.7. For a better understanding of $\operatorname{top} \boldsymbol{c}_G^{(*)}$, observe that if a $\boldsymbol{c}$-hyperbolic domain $G \subset \mathbb{C}^n$ satisfies one of the following conditions, then $\operatorname{top} \boldsymbol{c}_G^{(*)} = \operatorname{top} G$.

(a) For any $a \in G$ there exists a basis $\mathcal{B}(a)$ of neighborhoods of a in $\operatorname{top} \boldsymbol{c}_G^{(*)}$ such that any $U \in \mathcal{B}(a)$ is connected in $\operatorname{top} G$.

(b) For any $a \in G$ there exists a neighborhood U_a of a in $\operatorname{top} \boldsymbol{c}_G^{(*)}$ such that U_a is relatively compact in $\operatorname{top} G$.

For suppose that (a) is satisfied. Fix $a \in G$ and $r > 0$ with $\mathbb{B}(a, r) \subset\subset G$. Put

$$\varepsilon := \min\{\boldsymbol{c}_G^{(*)}(a, z) : z \in \partial\mathbb{B}(a, r)\}.$$

Then, $\varepsilon > 0$ (hyperbolicity). Let $U \in \mathcal{B}(a)$ be such that $U \subset B_{\boldsymbol{c}_G^{(*)}}(a, \varepsilon)$. Obviously, $U \cap \partial\mathbb{B}(a, r) = \varnothing$. Since U is connected in $\operatorname{top} G$, we get $U \subset \mathbb{B}(a, r)$.

If (b) is satisfied, then we put

$$\varepsilon := \inf\{\boldsymbol{c}_G^{(*)}(a, z) : z \notin \mathbb{B}(a, r)\}.$$

It suffices to prove that $\varepsilon > 0$ (then $B_{\boldsymbol{c}_G^{(*)}}(a, \varepsilon) \subset \mathbb{B}(a, r)$). Suppose that $\varepsilon = 0$ and let $\boldsymbol{c}_G^{(*)}(a, z_\nu) \longrightarrow 0$, $z_\nu \notin \mathbb{B}(a, r)$, $\nu \geq 1$. We may assume that $z_\nu \in U_a$, $\nu \geq 1$, where U_a is as in (b). Since U_a is relatively compact in $\operatorname{top} G$, we may also assume that $z_\nu \longrightarrow z_0 \in G$ in $\operatorname{top} G$. Clearly, $a \neq z_0$ and by the continuity of $\boldsymbol{c}_G^{(*)}$ we have $\boldsymbol{c}_G^{(*)}(a, z_\nu) \to \boldsymbol{c}_G^{(*)}(a, z_0) > 0$, which gives a contradiction.

2.7 Properties of $\boldsymbol{c}^{(*)}$ and $\boldsymbol{\gamma}$. Length of curve. Inner Carathéodory pseudodistance

First, we complete the list of basic properties of $\boldsymbol{c}^{(*)}$ and $\boldsymbol{\gamma}$.

Proposition 2.7.1. *Let $G \subset \mathbb{C}^n$ be a domain.*

(a) *If $(G_\nu)_{\nu=1}^\infty$ is a sequence of subdomains of G such that $G_\nu \nearrow G$ (i.e., $G_\nu \subset G_{\nu+1}$, $\nu \geq 1$, $\bigcup_{\nu=1}^\infty G_\nu = G$), then $\boldsymbol{c}_{G_\nu}^{(*)} \searrow \boldsymbol{c}_G^{(*)}$ and $\boldsymbol{\gamma}_{G_\nu} \searrow \boldsymbol{\gamma}_G$.*

(b) $\log c_G \in \mathcal{PSH}(G \times G)$, $\log c^*_G$ *is separately plurisubharmonic, i.e., for each* $a \in G$ *the function* $\log c^*_G(a, \cdot)$ *is psh.*

(c) *The function* γ_G *is locally Lipschitz and log-psh on* $G \times \mathbb{C}^n$.

(d) $$\lim_{\substack{z',z'' \to a,\, z' \neq z'' \\ \frac{z'-z''}{\|z'-z''\|} \to X}} \frac{c^{(*)}_G(z', z'')}{\|z' - z''\|} = \gamma_G(a; X), \quad a \in G,\ X \in \mathbb{C}^n,\ \|X\| = 1.$$

Note that, in view of the continuity of γ_G, condition (d) is equivalent to the following one.

(d') *For any compact* $K \subset G$ *and for any* $\varepsilon > 0$ *there exists a* $\delta > 0$ *such that*

$$|c^{(*)}_G(z', z'') - \gamma_G(a; z' - z'')| \le \varepsilon \|z' - z''\|, \quad a \in K,\ z', z'' \in \mathbb{B}(a, \delta) \subset G.$$

Proof. (a) Use the Montel argument.

(b) Observe that

$$\log c_G = \sup\{(\log p) \circ \Phi_f : f \in \mathcal{O}(G, \mathbb{D})\},$$

where $G \times G \ni (z', z'') \overset{\Phi_f}{\longmapsto} (f(z'), f(z'')) \in \mathbb{D} \times \mathbb{D}$, and then apply Proposition 1.1.21 and Appendix B.4.16.

(c) Let $\mathbb{B}(a, 2r) \subset G$. We will prove that for $z', z'' \in \mathbb{B}(a, r)$, $X', X'' \in \mathbb{C}^n$ we have

$$|\gamma_G(z'; X') - \gamma_G(z''; X'')| \le \frac{2}{r^2}\|z' - z''\|\|X'\| + \frac{1}{r}\|X' - X''\|. \tag{2.7.1}$$

First, note that

$$|\gamma_G(z'; X') - \gamma_G(z''; X'')| \le |\gamma_G(z'; X') - \gamma_G(z''; X')| + \gamma_G(z''; X' - X'').$$

By Proposition 2.3.1(c) we get

$$\gamma_G(z''; X' - X'') \le \frac{\|X' - X''\|}{r}, \quad z'' \in \mathbb{B}(a, r),\ X', X'' \in \mathbb{C}^n.$$

It remains to estimate $|\gamma_G(z'; X) - \gamma_G(z''; X)|$. Fix $z', z'' \in \mathbb{B}(a, r)$ and $X \in \mathbb{C}^n$. Suppose that $\gamma_G(z'; X) \ge \gamma_G(z''; X)$. Let $f \in \mathcal{O}(G, \mathbb{D})$ be an extremal function for $\gamma_G(z'; X)$, i.e., $f(z') = 0$ and $|f'(z')X| = \gamma_G(z'; X)$. Then,

$$\begin{aligned}
|\gamma_G(z'; X) - \gamma_G(z''; X)| &= |f'(z')X| - \gamma_G(z''; X) \\
&\le |f'(z')X| - \gamma(f(z''))|f'(z'')X| \\
&\le |f'(z')X| - |f'(z'')X| \\
&\le |f'(z')X - f'(z'')X| \\
&\le \|f'(z') - f'(z'')\|\|X\| \\
&\le \max\{\|f''(z)\| : z \in [z', z'']\} \cdot \|z' - z''\|\|X\| \\
&\le \frac{2}{r^2}\|z' - z''\|\|X\|.
\end{aligned}$$

The last inequality follows from the Cauchy inequalities:

$$\frac{1}{k!}\|f^{(k)}(z)\| \le \frac{\|f\|_G}{[\operatorname{dist}(z,\partial G)]^k}, \quad f \in \mathcal{H}^\infty(G),\ k \in \mathbb{N},\ z \in G.$$

Then, the plurisubharmonicity of $\log \boldsymbol{\gamma}_G$ is a direct consequence of Property 1.1.5 and Appendix B.4.9, B.4.16.

(d) In view of the relation $\boldsymbol{c}_G = \tanh^{-1}(\boldsymbol{c}_G^*)$ it suffices to consider only the case of $\boldsymbol{c}_G^*$. Let $\mathbb{B}(a,4r) \subset G$. We will prove that

$$|\boldsymbol{c}_G^*(z',z'') - \boldsymbol{\gamma}_G(z_0; z'-z'')| \le \frac{1}{(3r)^2}(3\|z'-z''\| + 2\|z'-z_0\|)\cdot\|z'-z''\|,$$
$$z_0, z', z'' \in \mathbb{B}(a,r).$$

The proof of (2.7.1) gives

$$|\boldsymbol{\gamma}_G(z_0; z'-z'') - \boldsymbol{\gamma}_G(z'; z'-z'')| \le \frac{2}{(3r)^2}\|z'-z_0\|\|z'-z''\|, \quad z_0, z', z'' \in \mathbb{B}(a,r).$$

It remains to show that

$$|\boldsymbol{c}_G^*(z',z'') - \boldsymbol{\gamma}_G(z'; z'-z'')| \le \frac{3}{(3r)^2}\|z'-z''\|^2, \quad z', z'' \in \mathbb{B}(a,r).$$

Fix $z', z'' \in \mathbb{B}(a,r)$. Put $X := z'' - z'$. Then, by the Cauchy inequalities for $f \in \mathcal{O}(G,\mathbb{D})$ with $f(z') = 0$ we get

$$|f(z'+X) - f'(z')X| \le \sum_{k=2}^{\infty}\frac{1}{k!}\|f^{(k)}(z')\|\|X\|^k \le \sum_{k=2}^{\infty}\left(\frac{\|X\|}{3r}\right)^k \le \frac{3}{(3r)^2}\|X\|^2.$$

□

Definition 2.7.2. For a curve $\alpha : [0,1] \longrightarrow G$ put

$$L_{\boldsymbol{c}_G^{(*)}}(\alpha) := \sup\left\{\sum_{j=1}^{N}\boldsymbol{c}_G^{(*)}(\alpha(t_{j-1}),\alpha(t_j)) : N \in \mathbb{N},\ 0 = t_0 < \cdots < t_N = 1\right\}.$$

The number $L_{\boldsymbol{c}_G^{(*)}}(\alpha)$ is called the $\boldsymbol{c}_G^{(*)}$*-length* of α; if $L_{\boldsymbol{c}_G^{(*)}}(\alpha) < +\infty$, then we say that α is $\boldsymbol{c}_G^{(*)}$*-rectifiable*.

Observe that

$$L_{\boldsymbol{c}_G^*} \le L_{\boldsymbol{c}_G}$$

and for any holomorphic mapping $F : G \longrightarrow D$ we have

$$L_{\boldsymbol{c}_D^{(*)}}(F\circ\alpha) \le L_{\boldsymbol{c}_G^{(*)}}(\alpha)$$

with equality for biholomorphic mappings.

If $\alpha : [0,1] \longrightarrow G$ is a piecewise $\mathcal{C}^1$-curve, then we can also define its *γ_G-length* by the formula

$$L_{\gamma_G}(\alpha) := \int_0^1 \gamma_G(\alpha(t); \alpha'(t))dt.$$

Note that $L_{\gamma_G}(\alpha) < +\infty$. Moreover, for $F \in \mathcal{O}(G, D)$ we get

$$L_{\gamma_D}(F \circ \alpha) \le L_{\gamma_G}(\alpha)$$

with equality for biholomorphic mappings.

Lemma 2.7.3.

(a) $L_{c_G^*}(\alpha) = L_{c_G}(\alpha)$ *and, moreover, if* $\alpha : [0,1] \longrightarrow G$ *is a* $\| \;\|$*-rectifiable curve (i.e., if* α *is rectifiable in the Euclidean sense), then*

$$L_{c_G^{(*)}}(\alpha) < +\infty.$$

(b) *If* G *is* γ*-hyperbolic, then any* $c_G^{(*)}$*-rectifiable curve is* $\| \;\|$*-rectifiable.*

(c) *If* α *is piecewise* $\mathcal{C}^1$*, then*

$$L_{c_G^{(*)}}(\alpha) = L_{\gamma_G}(\alpha).$$

(d) *If* $\alpha : [0,1] \longrightarrow G$ *is a* $\| \;\|$*-rectifiable curve, then for any* $\varepsilon > 0$ *there exists a piecewise* $\mathcal{C}^1$*-curve* $\widetilde{\alpha} : [0,1] \longrightarrow G$ *such that* $\widetilde{\alpha}(0) = \alpha(0)$, $\widetilde{\alpha}(1) = \alpha(1)$, *and* $|L_{c_G^{(*)}}(\alpha) - L_{c_G^{(*)}}(\widetilde{\alpha})| \le \varepsilon$.

Proof. (a) Since $c_G^* = \tanh c_G$, for any $\varepsilon > 0$ there exists an $\eta > 0$ such that

$$c_G^*(\alpha(t'), \alpha(t'')) \le c_G(\alpha(t'), \alpha(t'')) \le (1+\varepsilon)c_G^*(\alpha(t'), \alpha(t'')),$$
$$0 \le t', t'' \le 1, |t' - t''| \le \eta.$$

Consequently,

$$L_{c_G^*}(\alpha) \le L_{c_G}(\alpha) \le (1+\varepsilon)L_{c_G^*}(\alpha).$$

According to (2.6.2), there exist $M > 0$, $\eta > 0$ such that

$$c_G^*(\alpha(t'), \alpha(t'')) \le M\|\alpha(t') - \alpha(t'')\|, \quad 0 \le t', t'' \le 1, \; |t' - t''| \le \eta.$$

Hence,

$$L_{c_G^*}(\alpha) \le M L_{\| \;\|}(\alpha) \; (< +\infty \text{ if } \alpha \text{ is } \| \;\|\text{-rectifiable}).$$

(b) Since $\boldsymbol{\gamma}_G$ is continuous and G is $\boldsymbol{\gamma}$-hyperbolic, there exists an $\varepsilon > 0$ such that

$$\boldsymbol{\gamma}_G(\alpha(t); X) \geq 2\varepsilon\|X\|, \quad 0 \leq t \leq 1, \ X \in \mathbb{C}^n.$$

By Proposition 2.7.1(d') there exists an $\eta > 0$ such that

$$|\boldsymbol{c}_G^{(*)}(\alpha(t'), \alpha(t'')) - \boldsymbol{\gamma}_G(\alpha(t'); \alpha(t') - \alpha(t''))| \leq \varepsilon\|\alpha(t') - \alpha(t'')\|, \\ 0 \leq t', t'' \leq 1, \ |t' - t''| \leq \eta. \quad (2.7.2)$$

Hence,

$$\boldsymbol{c}_G^{(*)}(\alpha(t'), \alpha(t'')) \geq \varepsilon\|\alpha(t') - \alpha(t'')\|, \quad 0 \leq t', t'' \leq 1, \ |t' - t''| \leq \eta,$$

and finally,

$$L_{\boldsymbol{c}_G^{(*)}}(\alpha) \geq \varepsilon L_{\|\ \|}(\alpha).$$

(c) Without loss of generality, we may assume that α is of class $\mathcal{C}^1$. Then, in view of the continuity of $\boldsymbol{\gamma}$, for any $\varepsilon > 0$ there exists an $\eta > 0$ such that

$$|\boldsymbol{\gamma}_G(\alpha(t'); \alpha(t') - \alpha(t'')) - \boldsymbol{\gamma}_G(\alpha(t'); (t' - t'')\alpha'(t'))| \leq \varepsilon|t' - t''|, \\ 0 \leq t', t'' \leq 1, \ |t' - t''| \leq \eta.$$

We may assume that η is such that (2.7.2) is satisfied. Take $0 = t_0 < \cdots < t_N = 1$ with $t_j - t_{j-1} \leq \eta$, $j = 1, \dots, N$. Then,

$$\left|\sum_{j=1}^{N} \boldsymbol{c}_G^{(*)}(\alpha(t_{j-1}), \alpha(t_j)) - \sum_{j=1}^{N} \boldsymbol{\gamma}_G(\alpha(t_{j-1}); \alpha'(t_{j-1}))(t_j - t_{j-1})\right| \\ \leq \varepsilon(L_{\|\ \|}(\alpha) + 1).$$

Thus,

$$|L_{\boldsymbol{c}_G^{(*)}}(\alpha) - L_{\boldsymbol{\gamma}_G}(\alpha)| \leq \varepsilon(L_{\|\ \|}(\alpha) + 1).$$

(d) By Proposition 2.7.1(d') there exists an $\eta > 0$ such that

$$|\boldsymbol{c}_G^{(*)}(\alpha(t'), \alpha(t'')) - \boldsymbol{\gamma}_G(z; \alpha(t') - \alpha(t''))| \leq \varepsilon\|\alpha(t') - \alpha(t'')\|, \\ 0 \leq t', t'' \leq 1, \ |t' - t''| \leq \eta, \ z \in [\alpha(t'), \alpha(t'')] \subset G.$$

Fix $0 \leq t_0 < \cdots < t_N = 1$ with $t_j - t_{j-1} \leq \eta$, $j = 1, \dots, N$, and

$$L_{\boldsymbol{c}_G^{(*)}}(\alpha) - \sum_{j=1}^{N} \boldsymbol{c}_G^{(*)}(\alpha(t_{j-1}), \alpha(t_j)) \leq \varepsilon.$$

Define $\widetilde{\alpha}_j(\tau) := \alpha(t_{j-1}) + \tau(\alpha(t_j) - \alpha(t_{j-1}))$, $0 \le \tau \le 1$, $j = 1, \dots N$, and put $\widetilde{\alpha} := \widetilde{\alpha}_1 \cup \dots \cup \widetilde{\alpha}_N$. Note that there exist $z_j \in [\alpha(t_{j-1}), \alpha(t_j)]$ such that

$$L_{\boldsymbol{\gamma}_G}(\widetilde{\alpha}_j) = \boldsymbol{\gamma}_G(z_j; \alpha(t_{j-1}) - \alpha(t_j)), \quad j = 1, \dots N.$$

Hence, by (c) and in view of the above estimates, we get

$$\begin{aligned}|L_{\boldsymbol{c}_G^{(*)}}(\alpha) - L_{\boldsymbol{c}_G^{(*)}}(\widetilde{\alpha})| &= |L_{\boldsymbol{c}_G^{(*)}}(\alpha) - L_{\boldsymbol{\gamma}_G}(\widetilde{\alpha})| \\ &\le \varepsilon + \sum_{j=1}^{N} |\boldsymbol{c}_G^{(*)}(\alpha(t_{j-1}), \alpha(t_j)) - \boldsymbol{\gamma}_G(z_j; \alpha(t_{j-1}) - \alpha(t_j))| \\ &\le \varepsilon(L_{\|\ \|}(\alpha) + 1).\end{aligned}$$

□

Definition 2.7.4. Define (cf. Corollary 1.1.17 and Remark 2.7.5(e))

$$\begin{aligned}\boldsymbol{c}_G^{(*)i}(z', z'') = \inf\{&L_{\boldsymbol{c}_G^{(*)}}(\alpha) : \alpha : [0,1] \longrightarrow G, \\ &\alpha \text{ is a } \|\ \|\text{-rectifiable curve joining } z' \text{ and } z''\}, \quad z', z'' \in G.\end{aligned}$$

Remark 2.7.5.

(a) By virtue of Lemma 2.7.3(a) we have

$$(\boldsymbol{c}_G^*)^i = \boldsymbol{c}_G^i.$$

Obviously, $\boldsymbol{c}_G \le \boldsymbol{c}_G^i$.

(b) $\boldsymbol{c}_G^i$ is a pseudodistance, and for any $F \in \mathcal{O}(G, D)$ we have

$$\boldsymbol{c}_D^i(F(z'), F(z'')) \le \boldsymbol{c}_G^i(z', z''), \quad z', z'' \in G,$$

with equality for biholomorphic mappings; the pseudodistance $\boldsymbol{c}_G^i$ is called the *inner Carathéodory pseudodistance for* G. The pseudodistance $\boldsymbol{c}_G$ is said to be *inner* if $\boldsymbol{c}_G = \boldsymbol{c}_G^i$.

(c) For any curve $\alpha : [0,1] \longrightarrow G$ we have

$$\sum_{j=1}^{N} \boldsymbol{c}_G^i(\alpha(t_{j-1}), \alpha(t_j)) \le \sum_{j=1}^{N} L_{\boldsymbol{c}_G^{(*)}}(\alpha|_{[t_{j-1}, t_j]}) = L_{\boldsymbol{c}_G^{(*)}}(\alpha).$$

Consequently,

$$L_{\boldsymbol{c}_G^i} = L_{\boldsymbol{c}_G^{(*)}}$$

and so

$$(\boldsymbol{c}_G^i)^i = \boldsymbol{c}_G^i.$$

(d) By Lemma 2.7.3(c)(d) we see that

$$c^i_G(z',z'') = (\textstyle\int \gamma_G)(z',z'') := \inf\{L_{\gamma_G}(\alpha) : \alpha : [0,1] \longrightarrow G,$$
$$\alpha \text{ is a piecewise } \mathcal{C}^1\text{-curve joining } z' \text{ and } z''\}, \quad z',z'' \in G.$$

(e) By Lemma 2.7.3(b), if G is γ-hyperbolic, then in the definition of c^i_G the $\|\ \|$-rectifiability of α may be omitted; in particular, the definitions of p^i and m^i given in Chapter 1 agree with the definition of $c^{(*)i}_G$. We do not know whether this is true in general.

?

Proposition 2.7.6. *For any $a \in G$, $X \in \mathbb{C}^n$, $\|X\| = 1$, we have*

$$\lim_{\substack{z',z''\to a,\ z'\neq z''\\ \frac{z'-z''}{\|z'-z''\|}\to X}} \frac{c^i_G(z',z'')}{\|z'-z''\|} = \gamma_G(a;X);$$

cf. Proposition 2.7.1(d).

Proof. Let $\mathbb{B}(a,r) \subset G$, $z',z'' \in \mathbb{B}(a,r)$. Then,

$$c_G(z',z'') \le c^i_G(z',z'') = (\textstyle\int \gamma_G)(z',z'')$$
$$\le \int_0^1 \gamma_G(z' + t(z''-z'); z'-z'')dt = \gamma_G(\xi; z'-z''),$$

where $\xi = \xi(z',z'') \in [z',z'']$. Thus,

$$\frac{c_G(z',z'')}{\|z'-z''\|} \le \frac{c^i_G(z',z'')}{\|z'-z''\|} \le \gamma_G\Big(\xi; \frac{z'-z''}{\|z'-z''\|}\Big).$$

Now, we can use Proposition 2.7.1(d) and the continuity of γ_G. □

Remark 2.7.7. By Proposition 2.7.6, c^i_G is continuous. Observe that the c^i_G-balls $B_{c^i_G}(a,r)$ are arcwise connected in top G. In particular, by Remark 2.6.7 we get the following implication:
if G is c^i*-hyperbolic*, i.e., c^i_G is a distance (e.g., if G is c-hyperbolic), then top c^i_G = top G.

Consequently, if G is a domain as in Theorem 2.6.3, then $c_G \neq c^i_G$, i.e., c_G is not inner.

To give more concrete examples of c-hyperbolic domains with $c_G \neq c^i_G$, we will need the following

Lemma 2.7.8 (cf. [513]). *Let $a, b \in G$, $a \neq b$. Suppose that there exists an $f \in \mathcal{O}(G, \mathbb{D})$ that is extremal for $\boldsymbol{c}_G^*(a, b)$ and such that*

$$|f'(a)X| < \boldsymbol{\gamma}_G(a; X), \quad X \in \mathbb{C}^n, \ X \neq 0.$$

Then,

$$\boldsymbol{c}_G(a, b) < \boldsymbol{c}_G^i(a, b).$$

Proof. By continuity there exist $\varepsilon, \delta > 0$ such that

$$\boldsymbol{\gamma}_{\mathbb{D}}(f(z); f'(z)X) + \varepsilon\|X\| \leq \boldsymbol{\gamma}_G(z; X), \quad z \in \mathbb{B}(a, \delta) \subset\subset G, \ X \in \mathbb{C}^n.$$

Recall that we always have

$$\boldsymbol{\gamma}_{\mathbb{D}}(f(z); f'(z)X) \leq \boldsymbol{\gamma}_G(z; X), \quad z \in G, \ X \in \mathbb{C}^n.$$

Let $\alpha : [0, 1] \longrightarrow G$ be a piecewise $\mathcal{C}^1$-curve joining a and b. Denote by t_0 the maximal $t \in [0, 1]$ such that $\alpha([0, t]) \subset \overline{\mathbb{B}(a, \delta)}$. Then,

$$\begin{aligned} L_{\boldsymbol{\gamma}_G}(\alpha) &= \int_0^{t_0} \boldsymbol{\gamma}_G(\alpha(t); \alpha'(t))dt + \int_{t_0}^1 \boldsymbol{\gamma}_G(\alpha(t); \alpha'(t))dt \\ &\geq \int_0^1 \boldsymbol{\gamma}_{\mathbb{D}}(f(\alpha(t)); f'(\alpha(t))\alpha'(t))dt + \varepsilon \int_0^{t_0} \|\alpha'(t)\|dt \\ &\geq L_{\boldsymbol{\gamma}_{\mathbb{D}}}(f \circ \alpha) + \varepsilon\delta \geq p(f(a), f(b)) + \varepsilon\delta = \boldsymbol{c}_G(a, b) + \varepsilon\delta. \end{aligned}$$

Hence, by Remark 2.7.5(d), $\boldsymbol{c}_G^i(a, b) \geq \boldsymbol{c}_G(a, b) + \varepsilon\delta$. □

Example 2.7.9 (cf. [254]). Let

$$P := \{z \in \mathbb{C} : 1/R < |z| < R\} \ (R > 1), \quad a := 1, \ b := -1.$$

If $g \in \mathcal{O}(P, \mathbb{D})$ is an extremal function for $\boldsymbol{c}_P^*(1, -1)$, then

$$f(z) = \frac{1}{2}(g(z) + g(1/z)), \quad z \in P,$$

is also extremal and $f'(1) = 0$. Hence, by Lemma 2.7.8, we get

$$\boldsymbol{c}_P(1, -1) < \boldsymbol{c}_P^i(1, -1).$$

The full description of $\boldsymbol{c}_P^i$ will be given in Proposition 9.1.13.

We conclude this section with a discussion of a surprising structure of Carathéodory balls.

Example 2.7.10 (cf. [272]). The aim of this example is to show that, in general,

$$\overline{B_{c_G}(a,r)} \subsetneq \overline{B}_{c_G}(a,r),$$

where $B_{c_G}(a,r)$ denotes the *open Carathéodory ball*

$$B_{c_G}(a,r) := \{z \in G : c_G(a,z) < r\}$$

and $\overline{B}_{c_G}(a,r)$ is the *closed Carathéodory ball*

$$\overline{B}_{c_G}(a,r) := \{z \in G : c_G(a,z) \leq r\}$$

(the closure $\overline{B_{c_G}(a,r)}$ is taken in sense of the standard topology of G).

Observe that, since c_G is continuous, we always have

$$\overline{B_{c_G}(a,r)} \subset \overline{B}_{c_G}(a,r).$$

One can easily prove that the condition

$$\overline{B_{c_G}(a,r)} = \overline{B}_{c_G}(a,r) \quad \text{for any} \quad a \in G,\ r > 0 \tag{2.7.3}$$

is equivalent to the following "minimum principle":

For any $a \in G$ the function $G \ni z \longmapsto c_G(a,z)$ has no local strictly positive minima. (2.7.4)

? We do not know of any example of a domain $G \subset \mathbb{C}^1$ for which (2.7.3) is not satisfied – cf. Proposition 2.7.12.

On the other hand, we will prove that for any $n \geq 2$ there exists a bounded strongly pseudoconvex domain $G \subset \mathbb{C}^n$ with real analytic boundary such that for some $a \in G$ and $0 < r_2 < r_1$ we have:

(a) the ball $B_{c_G}(a,r_1)$ is relatively compact in G and disconnected,

(b) $\overline{B_{c_G}(a,r_2)} \subsetneq \overline{B}_{c_G}(a,r_2)$.

Moreover, we will construct a connected c- and γ-hyperbolic Riemann surface M such that, for some $a \in M, 0 < r_2 < r_1$, conditions (a), (b) are fulfilled with $G = M$; recall that the answer is not known for $G \subset \mathbb{C}^1$. Note that if (a) is satisfied, then c_G is not inner; cf. Remark 2.7.7.

We now begin the construction.

1° *For any c-hyperbolic domain $G \subset \mathbb{C}^n$, condition* (a) *implies* (b).

Proof of 1°. Let S be a connected component of $B_{c_G}(a,r_1)$ such that $a \notin S$. Put

$$r_2 := \min\{c_G(a,z) : z \in \overline{S}\}$$

and let $b \in \overline{S}$ be such that $c_G(a,b) = r_2$. Observe that $b \in S$, $0 < r_2 < r_1$, $b \in \overline{B}_{c_G}(a,r_2)$, and $S \cap B_{c_G}(a,r_2) = \varnothing$. Hence, $b \in \overline{B}_{c_G}(a,r_2) \setminus \overline{B_{c_G}(a,r_2)}$. □

Note that 1° remains true for c-hyperbolic manifolds.

2° *For any domain* $G \subset \mathbb{C}^n$, *we have*

$$c^{(*)}_{G\times\mathbb{D}}((a,0),(z,\lambda)) = \max\left\{c^{(*)}_G(a,z), c^{(*)}_{\mathbb{D}}(0,\lambda)\right\}, \quad a,z \in G, \lambda \in \mathbb{D}.$$

Note that the above product formula is a particular case of the product property in Theorem 18.2.1.

Proof of 2°. Obviously, it suffices to consider only the case of c^*. The inequality "$\geq$" follows from the holomorphic contractibility (with respect to the projections).

Fix a, $z_0 \in G$ and let $C := c^*_G(a, z_0)$. First, consider the case $C = 0$. Then, for every function $f \in \mathcal{O}(G \times \mathbb{D}, \mathbb{D})$ with $f(a,0) = 0$ we have $f(z_0, 0) = 0$. Hence, $|f(z_0,\lambda)| \leq |\lambda|$ for all $\lambda \in \mathbb{D}$, and consequently $c^*_{G\times\mathbb{D}}((a,0),(z_0,\lambda)) = |\lambda|$ for $\lambda \in \mathbb{D}$.

In the case, where $C > 0$ it suffices to prove the formula for all $|\lambda| = C$. For, if 2° is true on the circle $\{|\lambda| = C\}$, then, by the maximum principle, the formula follows for all $|\lambda| \leq C$ (the function $\mathbb{D} \ni \lambda \longmapsto c^*_{G\times\mathbb{D}}((a,0),(z_0,\lambda))$ is subharmonic; cf. Proposition 2.7.1(b)). To prove the inequality "$\leq$" for $C < |\lambda| < 1$, we use the maximum principle for the subharmonic function

$$\mathbb{D}_* \ni \lambda \longmapsto \frac{1}{|\lambda|} c_{G\times\mathbb{D}}((a,0),(z_0,\lambda)).$$

Now fix a λ_0 with $|\lambda_0| = C$ and let f be an extremal function for $c^*_G(a, z_0)$ with $f(a) = 0$, $f(z_0) = \lambda_0$. Consider the holomorphic map

$$G \ni z \overset{F}{\longmapsto} (z, f(z)) \in G \times \mathbb{D}.$$

Then, the holomorphic contractibility gives

$$c^*_{G\times\mathbb{D}}((a,0),(z_0,\lambda_0)) = c^*_{G\times\mathbb{D}}(F(a), F(z_0)) \leq c^*_G(a,z_0) = C. \qquad \square$$

3° *If* $B_{c_G}(a,r)$ *is relatively compact in* G *and disconnected, then* $B_{c_{G\times\mathbb{D}}}((a,0),r)$ *has the same properties with respect to* $G \times \mathbb{D}$.

Proof of 3°. By virtue of 2° we get $B_{c_{G\times\mathbb{D}}}((a,0),r) = B_{c_G}(a,r) \times B_p(0,r)$. $\square$

4° *If* G_0 *is a domain of holomorphy such that a ball* $B_{c_{G_0}}(a,r)$ *is disconnected, then there exists a bounded strongly pseudoconvex domain* $G \subset G_0$ *with a real analytic boundary such that* $B_{c_G}(a,r)$ *is relatively compact in* G *and disconnected (cf. (a)).*

Proof of 4°. Let $G_\nu \nearrow G_0$ be an exhaustion of G_0 by bounded strongly pseudoconvex domains with real analytic boundaries, cf. Appendix B.7.11. Then, by Proposition 2.7.1(a), $B_{c_{G_\nu}}(a,r) \nearrow B_{c_{G_0}}(a,r)$, and therefore, for sufficiently big ν, the ball $B_{c_{G_\nu}}(a,r)$ is disconnected.

On the other hand, if G is a bounded strongly pseudoconvex domain, then for any $z_0 \in \partial G$ there exists a peak function $f \in \mathcal{O}(\overline{G})$ with $|f| < 1$ in G and $|f(z_0)| = 1$ (cf. [327], see also Appendix B.7.13). This shows that for any $a \in G$, $r > 0$, the ball $B_{\boldsymbol{c}_G}(a, r)$ is relatively compact in G; cf. Example 14.2.4. □

5° *Properties* 1° – 4° *show that it is enough to construct a domain of holomorphy* $G \subset \mathbb{C}^2$ *such that at least one ball* $B_{\boldsymbol{c}_G}(a, r)$ *is disconnected.*

6° (Sibony domains; see [471]) *Let* $(a_k)_{k=1}^{\infty} \subset \mathbb{D}_*$ *be a sequence without accumulation points in* $\mathbb{D}$ *and such that any point* $\zeta \in \mathbb{T}$ *is a non-tangential limit of a subsequence of* $(a_k)_{k=1}^{\infty}$. *Choose* $(\lambda_k)_{k=1}^{\infty} \subset \mathbb{R}_{>0}$ *in such a way that*

$$\sum_{k=1}^{\infty} \lambda_k \log \left| \frac{a_k}{2} \right| > -\infty$$

and define

$$\begin{aligned} \varphi(z) &:= \sum_{k=1}^{\infty} \lambda_k \log \left| \frac{z - a_k}{2} \right|, \quad z \in \mathbb{D}, \\ \psi &:= \exp \varphi, \\ G &:= \{(z, w) \in \mathbb{D} \times \mathbb{C} : |w| \exp \psi(z) < 1\}. \end{aligned}$$

Observe that $G \subsetneq \mathbb{D}^2$ is a fat Hartogs domain of holomorphy (cf. Appendix B.7.7). *Then,* $\mathcal{H}^\infty(G) = \mathcal{H}^\infty(\mathbb{D}^2)|_G$. *In particular,*

$$\begin{aligned} \boldsymbol{c}_G^*((z', w'), (z'', w'')) &= \boldsymbol{c}_{\mathbb{D}\times\mathbb{D}}^*((z', w'), (z'', w'')) \\ &= \max\{\boldsymbol{m}(z', z''), \boldsymbol{m}(w', w'')\}, (z', w'), (z'', w'') \in G; \end{aligned} \tag{2.7.5}$$

cf. Remark 2.1.4.

7° *There exists a Sibony domain with disconnected Carathéodory balls; cf.* 5°.

Proof of 7°. Obviously, it suffices to consider $\boldsymbol{c}^*$-balls. We keep the notation of 6° and we assume that $3/4 < |a_k| < 1$, $k \geq 1$. Take $0 < \varepsilon < 1/2$ and define

$$\begin{aligned} \varphi_\varepsilon(z) &:= \varphi(z) + \log \left| \frac{z^2 - \varepsilon^2}{4} \right|, \quad z \in \mathbb{D}, \\ \psi_\varepsilon &:= \exp \varphi_\varepsilon, \\ G_\varepsilon &:= \{(z, w) \in \mathbb{D} \times \mathbb{C} : |w| \exp \psi_\varepsilon(z) < 1\}. \end{aligned}$$

Clearly, the domain G_ε has the same properties as G; cf. 6°. Put

$$A := \left(\log \frac{1}{8}\right) \sum_{k=1}^{\infty} \lambda_k, \quad B := \frac{1}{4} \exp A,$$

and let $0 < C < B$. Note that

$$\psi_\varepsilon(z) \geq B|z^2 - \varepsilon^2| \quad \text{for } |z| \leq \frac{1}{2}, \tag{2.7.6}$$
$$(\pm\,\varepsilon, w(\varepsilon)) := (\pm\varepsilon, \exp(-C\varepsilon^2)) \in G_\varepsilon.$$

Our aim is to prove that for sufficiently small $\varepsilon > 0$ the ball

$$B(\varepsilon) := B_{\boldsymbol{c}^*_G}((-\varepsilon, w(\varepsilon)), 2\varepsilon)$$

is disconnected. Recall that for $-1 < t < 1$ and $0 < r < 1$ we have

$$B_{\boldsymbol{m}}(t, r) = \mathbb{D}\left(\frac{t(1-r^2)}{1-t^2r^2}, \frac{r(1-t^2)}{1-t^2r^2}\right);$$

cf. Exercise 1.3.2. Consequently, if ε is small, then by (2.7.5) we get

$$(\varepsilon, w(\varepsilon)) \in B(\varepsilon) \subset \mathbb{D}(1/2) \times \mathbb{A}\left(\frac{w(\varepsilon) - 2\varepsilon}{1 - 2\varepsilon w(\varepsilon)}, 1\right). \tag{2.7.7}$$

Suppose that $B(\varepsilon)$ is connected (ε small). Then, there is a continuous curve $\alpha : [0, 1] \longrightarrow B(\varepsilon)$ joining the points $(\pm\varepsilon, w(\varepsilon))$. Consequently, for some $y \in \mathbb{R}$ and $w^* \in \mathbb{C}$ we get $(iy, w^*) \in B(\varepsilon)$. By (2.7.7), we have

$$|y| < \frac{1}{2} \quad \text{and} \quad |w^*| > \frac{w(\varepsilon) - 2\varepsilon}{1 - 2\varepsilon w(\varepsilon)}.$$

On the other hand, by (2.7.6), we get

$$|w^*| < \exp\left(-\psi_\varepsilon\,(iy)\right) \leq \exp\left(-B|\,(iy)^2 - \varepsilon^2|\right) \leq \exp\left(-B\varepsilon^2\right).$$

Finally,

$$\frac{w(\varepsilon) - 2\varepsilon}{1 - 2\varepsilon w(\varepsilon)} < \exp(-B\varepsilon^2),$$

which is impossible for arbitrarily small $\varepsilon > 0$; a contradiction. □

The construction of a "regular" domain $G \subset \mathbb{C}^n$, $n \geq 2$, with properties (a) and (b) is complete.

We go to the construction of a connected $\boldsymbol{c}$- and $\boldsymbol{\gamma}$-hyperbolic Riemann surface M with $\operatorname{top} \boldsymbol{c}_M = \operatorname{top} M$ such that (a) and (b) are fulfilled. In view of 1°, the only problem is to realize (a).

Fix $0 < \eta_0 < \eta_1 < \eta_2 < \eta_3 < \eta_4 < 1$ with $2\sqrt{\eta_1\eta_2\eta_3\eta_4} < \eta_0$ and then define $\Delta := \mathbb{D} \setminus \{\eta_1, \eta_2, \eta_3, \eta_4\}$. Let M denote the Riemann surface obtained by joining two copies of Δ along the two sides of each of the intervals (η_1, η_2) and (η_3, η_4) crosswise. Denote by $\pi : M \to \Delta$ the standard covering projection. Obviously, M

is connected, $\boldsymbol{c}$- and $\boldsymbol{\gamma}$-hyperbolic, and such that $\operatorname{top} \boldsymbol{c}_M = \operatorname{top} M$. For $z \in \Delta$ let $\{z_+, z_-\} := \pi^{-1}(z)$. We will prove that

$$0_- \in B_{\boldsymbol{c}_M^*}(0_+, \eta_0). \tag{2.7.8}$$

Since

$$B_{\boldsymbol{c}_M^*}(0_+, \eta_0) \subset \pi^{-1}(\{z \in \mathbb{D} : |z| < \eta_0\}) \subset\subset M,$$

the ball $B_{\boldsymbol{c}_M^*}(0_+, \eta_0)$ is relatively compact and the points $0_+, 0_-$ lie in different connected components.

In order to verify (2.7.8) take an $f \in \mathcal{O}(M, \mathbb{D})$ with $f(0_+) = 0$ and set

$$\widetilde{f}(z) := \frac{1}{4}(f(z_+) - f(z_-))^2, \quad z \in \Delta,$$
$$\widetilde{f}(\eta_j) := 0, \quad j = 1, \dots, 4.$$

One can easily prove that $\widetilde{f} \in \mathcal{O}(\mathbb{D}, \mathbb{D})$. Hence, by the Schwarz lemma,

$$|\widetilde{f}(0)| \le \eta_1\eta_2\eta_3\eta_4 < \frac{1}{4}\eta_0^2.$$

Finally, $|f(0_-)| < \eta_0$.

Remark 2.7.11. If G is a balanced domain, then

$$\boldsymbol{c}_G^*(0, \lambda a) \le |\lambda| \boldsymbol{c}_G^*(0, a), \quad \lambda \in \overline{\mathbb{D}},\ a \in G.$$

For we observe that $\boldsymbol{c}_G^*(0, e^{i\theta}a) = \boldsymbol{c}_G^*(0, a)$ and we use the Schwarz lemma for log-psh functions; cf. Appendix B.4.24 and Proposition 2.3.1(a). In particular, $B_{\boldsymbol{c}_G}(0, r)$ is connected and $\overline{B_{\boldsymbol{c}_G}(0, r)} = \overline{B}_{\boldsymbol{c}_G}(0, r)$, $r > 0$.

Proposition 2.7.12 (cf. [179]). *Let $G \subset \overline{\mathbb{C}}$ be a doubly connected domain such that no boundary component of G reduces to a point. Let $a, b \in G$, $a \neq b$. Then, there exists a curve $\alpha : [0, 1] \longrightarrow G$ such that $\alpha(0) = a$, $\alpha(1) = b$, and the function*

$$[0, 1] \ni t \longmapsto \boldsymbol{c}_G(a, \alpha(t))$$

is strictly increasing. Consequently, for any $a \in G$ and $r > 0$

- *the Carathéodory ball $B_{\boldsymbol{c}_G}(a, r)$ is connected,*
- *$\overline{B_{\boldsymbol{c}_G}(a, r)} = \overline{B}_{\boldsymbol{c}_G}(a, r)$ (cf. (2.7.4)).*

Proof. Fix $a, b \in G$, $a \neq b$. We may assume that $a = 0$, $b = \infty$, and

$$\partial G \subset \{z \in \mathbb{C} : \operatorname{Re} z \le 0\} =: H$$

(cf. [504], Theorem IX.24). Since $\boldsymbol{c}_G^*(a,\cdot)$ is continuous, it suffices to prove that the function

$$(0,+\infty) \ni t \longmapsto \boldsymbol{c}_G^*(0,t)$$

is strictly increasing. Fix $0 < t' < t''$. Observe that the homography $\varphi(z) = \frac{z-t''}{z-t'}$ maps H onto the closed disc symmetric with respect to the real axis and passing through the points $1 = \varphi(\infty)$, $t''/t' = \varphi(0)$. Consequently, $1 < |\varphi(z)| < t''/t'$, $z \in \partial G$. Let $m := \max_{\partial G} |\varphi| \in (1, t''/t')$. Put $\psi := \varphi/m$. Then, $\psi(0) > 1$. Let $f \in \mathcal{O}(G,\mathbb{D})$ be such that $f(t') = 0$ and $\boldsymbol{c}_G^*(t',0) = \boldsymbol{c}_G^*(0,t') = f(0)$. Define $g := \psi f$. Then, $g \in \mathcal{O}(G,\mathbb{D})$ and $g(t'') = 0$. Consequently, $\boldsymbol{c}_G^*(0,t') = f(0) < g(0) \le \boldsymbol{c}_G^*(t'',0) = \boldsymbol{c}_G^*(0,t'')$. □

The phenomenon described in Example 2.7.10 does not occur if we look at small Carathéodory balls. Moreover, such balls are even starlike with respect to sufficiently small neighborhoods of their centers.

Proposition 2.7.13 (cf. [357]). *Let G be a bounded domain in $\mathbb{C}^n$ and let $a \in G$ be such that $\mathbb{B}(a,r) \subset\subset G \subset \mathbb{B}(a,R)$ for fixed $r,\ R > 0$. For $0 < t \le 1$, we put $s = s_t := \frac{t-t^2}{1+4t}$. Then, for any $z' \in \overline{B}_{\boldsymbol{c}_G^*}(a, s(\frac{r}{R})^2)$, $z'' \in \overline{B}_{\boldsymbol{c}_G^*}(a, t(\frac{r}{R})^2)$, the half-open segment $[z',z'') := \{z' + \tau(z''-z') : 0 \le \tau < 1\}$ is contained in $B_{\boldsymbol{c}_G^*}(a, t(\frac{r}{R})^2)$.*

Proof. First observe (cf. (2.6.2)) that if $z \in \overline{B}_{\boldsymbol{c}_G^*}(a, \sigma(\frac{r}{R})^2)$ with $\sigma \in (0,1]$, then $\|z-a\| \le R\boldsymbol{c}_G^*(a,z) \le \sigma(\frac{r}{R})r < r$, i.e., $\overline{B}_{\boldsymbol{c}_G^*}(a,\sigma(\frac{r}{R})^2) \subset \mathbb{B}(a,r)$. In particular, $[z',z'') \subset \mathbb{B}(a,r) \subset G$.

Now, it suffices to prove that for any $f \in \mathcal{O}(G,\mathbb{D})$, $f(a) = 0$, and for any $\hat{z} \in [z',z'')$ the following inequality is true: $|f(\hat{z})| < t(\frac{r}{R})^2$.

Fix f and $\hat{z}$ and assume, without loss of generality, that $f(\hat{z}) = |f(\hat{z})|$. Set $\varrho := \frac{1-s}{t+s}\frac{R}{r}$. We easily see that $\varrho > 1$. Moreover, for $\lambda \in \mathbb{D}(\varrho)$ we obtain

$$\begin{aligned}\|\lambda z'' + (1-\lambda)z' - a\| &\le |\lambda|\|z''-a\| + |1-\lambda|\|z'-a\| \\ &< \varrho t\frac{r}{R}r + (1+\varrho)s\frac{r}{R}r \le r.\end{aligned}$$

This permits us to define the following function $\varphi \in \mathcal{O}(\mathbb{D}(\varrho),\mathbb{D})$:

$$\varphi(\lambda) := \frac{1}{2}\big(f(\lambda z'' + (1-\lambda)z') + \overline{f(\overline{\lambda} z'' + (1-\overline{\lambda})z')}\big), \quad \lambda \in \mathbb{D}(\varrho).$$

Directly from the definition, we see that:

(a) φ has real values on $(-\varrho, \varrho)$,

(b) $\varphi(0) = \operatorname{Re} f(z') \le s(\frac{r}{R})^2$,

(c) $\varphi(1) = \operatorname{Re} f(z'') \le t(\frac{r}{R})^2$.

What remains to show is that $\varphi(\sigma) < t(\frac{r}{R})^2$ for all $\sigma \in [0,1)$. Otherwise, there would exist $\sigma' \in (0,1)$ such that $\varphi(\sigma') = \sup_{0\le\sigma\le 1}\varphi(\sigma) \ge t(\frac{r}{R})^2$. In particular, $\varphi'(\sigma') = 0$. Now put

$$g(\lambda) := \frac{\varphi(\lambda)-\varphi(\sigma')}{1-\varphi(\sigma')\varphi(\lambda)}, \quad \lambda \in \mathbb{D}(\varrho).$$

Then, $g \in \mathcal{O}(\mathbb{D}(\varrho), \mathbb{D})$ and, in addition, we have $g(\sigma') = g'(\sigma') = 0$. Let h denote the following automorphism of $\mathbb{D}(\varrho)$

$$h(\lambda) := \varrho^2 \frac{\lambda+\sigma'}{\varrho^2+\sigma'\lambda}, \quad \lambda \in \mathbb{D}(\varrho).$$

It follows that $g \circ h \in \mathcal{O}(\mathbb{D}(\varrho), \mathbb{D})$ satisfies $g \circ h(0) = (g \circ h)'(0) = 0$. Hence, from the Schwarz lemma we find that

$$|g \circ h(\lambda)| \le (|\lambda|/\varrho)^2.$$

In particular, for $\lambda = -\sigma'$ we get the following inequalities:

$$\frac{\varphi(\sigma')-\varphi(0)}{1-\varphi(\sigma')\varphi(0)} \le |g(0)| \le \left(\frac{\sigma'}{\varrho}\right)^2 < \frac{1}{\varrho^2}.$$

Because of $\varphi(0) \le s(\frac{r}{R})^2 < t(\frac{r}{R})^2 \le \varphi(\sigma') < 1$ we conclude that

$$(t-s)\left(\frac{r}{R}\right)^2 \le \frac{t(\frac{r}{R})^2 - s(\frac{r}{R})^2}{1-st(\frac{r}{R})^4} < \frac{1}{\varrho^2},$$

from which we derive the following inequality

$$0 < s^3 - (1+t)s^2 + (4t+1)s + t^2 - t = s^3 - (1+t)s^2 \text{ ; a contradiction.} \qquad \square$$

Corollary 2.7.14. *Under the assumptions of Proposition* 2.7.13 *we have:* $\mathbb{B}(a,\varrho) \subset B_{c_G^*}(a, (\frac{r}{R})^2) \subset\subset G$, *where* $\varrho := r(\frac{r}{R})^2$.

Moreover, for any $z'' \in \partial\mathbb{B}(a,\varrho)$ *the function* $[0,1] \ni t \longmapsto \boldsymbol{c}_G^*(a, a+t(z''-a))$ *is strictly increasing.*

Remark 2.7.15.

(a) Proposition 2.7.13 shows that small Carathéodory balls in bounded domains are always starlike.

(b) The claim of Corollary 2.7.14 is also proved in [87], but by a slightly different method.

Finally, the following corollary shows that small Carathéodory balls behave nicely, i.e., differently from those in Example 2.7.10.

Corollary 2.7.16. *Under the assumption of Proposition* 2.7.13, *for any* $t \in (0,1]$ *we have* $\overline{B}_{c_G^*}(a, t(\frac{r}{R})^2) = \overline{B_{c_G^*}(a, t(\frac{r}{R})^2)}$.

For more properties of Carathéodory balls, the reader may consult Chapter A, § A.1.

2.8 c^i-hyperbolicity versus c-hyperbolicity

Recall that a domain $G \subset \mathbb{C}^n$ is $\boldsymbol{c}^i$-hyperbolic if $\boldsymbol{c}_G^i$ is a true distance on G. By virtue of the inequality $\boldsymbol{c}_G \le \boldsymbol{c}_G^i$, if G is $\boldsymbol{c}$-hyperbolic, then it is $\boldsymbol{c}^i$-hyperbolic. Recall (Proposition 2.5.2) that if G is biholomorphic to a bounded domain, then G is $\boldsymbol{c}$-hyperbolic. In the general case, the following result due to J.-P. Vigué (cf. [519]) gives a characterization of $\boldsymbol{c}^i$-hyperbolicity.

Theorem 2.8.1. *Let $G \subset \mathbb{C}^n$ be a domain. Then, the following properties are equivalent:*

(i) *G is $\boldsymbol{c}^i$-hyperbolic;*

(ii) *there is no non-constant $\mathcal{C}^1$-curve $\alpha : [0,1] \longrightarrow G$ such that $\boldsymbol{\gamma}_G(\alpha;\alpha') \equiv 0$;*

(iii) *each point $a \in G$ has a neighborhood $U_a \subset G$ such that $\boldsymbol{c}_G(a,z) > 0$, $z \in U_a \setminus \{a\}$.*

Proof. (i) $\Longrightarrow$ (ii): Suppose the contrary, namely, that there exists a $\mathcal{C}^1$-curve $\alpha : [0,1] \longrightarrow G$ such that $\boldsymbol{\gamma}_G(\alpha;\alpha') \equiv 0$ and $\alpha'(t_0) \ne 0$ for a $t_0 \in [0,1]$. Then, for any $0 \le t' < t'' \le 1$ we have $\boldsymbol{c}_G^i(\alpha(t'),\alpha(t'')) = 0$. By virtue of $\alpha'(t_0) \ne 0$ there are two different points $\alpha(t'), \alpha(t'')$ showing that G is not $\boldsymbol{c}^i$-hyperbolic; a contradiction.

(ii) $\Longrightarrow$ (iii): We proceed by assuming the contrary. So, let $a \in G$ be a point such that there exists a sequence of points $(z^j)_{j\in\mathbb{N}} \subset G \setminus \{a\}$, $z^j \longrightarrow a$, with $\boldsymbol{c}_G(a,z^j) = 0$, $j \in \mathbb{N}$. We have to find a $\mathcal{C}^1$-curve which does fulfill the property stated in (ii). Observe that

$$A := \{z \in G : \boldsymbol{c}_G(a,z) = 0\} = \{z \in G : \forall_{f \in \mathcal{O}(G,\mathbb{D})} : f(a) = f(z)\}$$

is an analytic subset of G (cf. Appendix B.1.1) and $a \in A$. Since $A \setminus \{a\} \ni z^j \longrightarrow a$, we conclude that $\dim_a A \geq 1$. Therefore, there is a $\mathcal{C}^1$-curve $\alpha : [0,1] \longrightarrow \operatorname{Reg} A$ such that $\alpha' \not\equiv 0$. Since $\boldsymbol{c}_G(z', z'') = 0$, $z', z'' \in A$, using Proposition 2.7.1(d), we obtain

$$\boldsymbol{\gamma}_G(\alpha(t); \alpha'(t)) = \lim_{0<h\to 0} \frac{\boldsymbol{c}_G(\alpha(t), \alpha(t+h))}{h} = 0, \quad t \in [0,1];$$

a contradiction.

(iii) $\Longrightarrow$ (i): Fix $a, b \in G$, $a \neq b$, and choose a neighborhood U_a according to (iii). Moreover, let $V_a \subset\subset U_a$ be a neighborhood of a such that $b \notin V_a$. Applying the continuity of $\boldsymbol{c}_G$, there is a $C > 0$ such that $\boldsymbol{c}_G(a, z) \geq C$, $z \in \partial V_a$. Thus, for any $\mathcal{C}^1$-curve $\alpha : [0,1] \longrightarrow G$ with $\alpha(0) = a$, $\alpha(1) = b$, there is a $t_0 \in (0,1)$ with $\alpha(t_0) \in \partial V_a$. Consequently, $L_{\boldsymbol{c}_G}(\alpha) \geq \boldsymbol{c}_G(a, \alpha(t_0)) + \boldsymbol{c}_G(\alpha(t_0), b) \geq C$. Hence, $\boldsymbol{c}^i_G(a,b) \geq C > 0$. □

Moreover, there is the following general relation between $\boldsymbol{\gamma}$-hyperbolicity and local $\boldsymbol{c}$-hyperbolicity.

Proposition 2.8.2. *Any domain $G \subset \mathbb{C}^n$ that is $\boldsymbol{\gamma}$-hyperbolic is* locally $\boldsymbol{c}$-hyperbolic *(i.e., each $a \in G$ has a neighborhood $U_a \subset G$ such that $\boldsymbol{c}_G|_{U_a \times U_a}$ is a distance) and, consequently, $\boldsymbol{c}^i$-hyperbolic.*

Proof. Fix an $a \in G$ and suppose that $z^j, w^j \longrightarrow a$, $z^j \neq w^j$, $\boldsymbol{c}_G(z^j, w^j) = 0$, $j = 1, 2, \dots$. We may assume that $\frac{z^j - w^j}{\|z^j - w^j\|} \longrightarrow X_0 \in \partial\mathbb{B}_n$. Then, by Proposition 2.7.1(d), we have

$$\boldsymbol{\gamma}_G(a; X_0) = \lim_{j\to+\infty} \frac{\boldsymbol{c}_G(z^j, w^j)}{\|z^j - w^j\|} = 0;$$

a contradiction. □

Example 2.8.3. There is a domain $G \subset \mathbb{C}^3$ which is not $\boldsymbol{c}_G$-hyperbolic and not $\boldsymbol{\gamma}_G$-hyperbolic, but nevertheless $\boldsymbol{c}^i_G$-hyperbolic (see [519]). This G is constructed via an example of a 1-dimensional complex space and by then applying the Remmert embedding theorem. We here omit the details.

Remark 2.8.4. Notice that Example 2.8.3 is not given explicitly. So it is interesting to find an effective example of that type; moreover, the question as to whether such an example is possible in $\mathbb{C}^2$ is still an open one.

?

2.9 Two applications

In this section, we present two interesting results, the proofs of which are based on the properties of the Carathéodory pseudodistance. Namely, we are going to describe the automorphisms of product domains and to discuss the existence of fixed points.

First, we recall a few facts about the group $\mathrm{Aut}(G)$ of automorphisms of a bounded domain $G \subset \mathbb{C}^n$. We always assume that $\mathrm{Aut}(G)$ is endowed with the compact open topology or, equivalently, with the topology of locally uniform convergence.

Fix a $\Phi_0 \in \mathrm{Aut}(G)$. Then, the following mappings

(a) $\mathrm{Aut}(G) \ni \Phi \longmapsto \Phi^{-1} \in \mathrm{Aut}(G)$,

(b) $\mathrm{Aut}(G) \ni \Phi \longmapsto \Phi_0 \circ \Phi \in \mathrm{Aut}(G)$,

(c) $\mathrm{Aut}(G) \ni \Phi \longmapsto \Phi \circ \Phi_0 \in \mathrm{Aut}(G)$

are homeomorphisms of $\mathrm{Aut}(G)$. In what follows we are interested in the connected component of $\mathrm{Aut}(G)$ that contains the identity. This component is denoted by $\mathrm{Aut}_{\mathrm{id}}(G)$.

Theorem 2.9.1 (cf. [87]). *Let $G_j \subset \mathbb{C}^{n_j}$, $j = 1, 2$, be bounded domains. Then, for every $\Phi \in \mathrm{Aut}_{\mathrm{id}}(G_1 \times G_2)$, there exist $\Phi_j \in \mathrm{Aut}(G_j)$, $j = 1, 2$, such that $\Phi(z, w) = (\Phi_1(z), \Phi_2(w))$, $z \in G_1$, $w \in G_2$.*

The proof of Theorem 2.9.1 will be given via several lemmas which partially deal with a slightly more general situation that may be of independent interest; see also Theorem 18.2.3.

Lemma 2.9.2. *Let $G_j \subset \mathbb{C}^{n_j}$ be bounded domains and let $D \subset \mathbb{C}^{n_1+n_2}$ be a domain satisfying $G_1 \times G_2 \subset D \subset G_1 \times \mathbb{C}^{n_2}$. Fix $(a_1, a_2) \in G_1 \times G_2$ and let $r_1, r_2 > 0$ be such that $\mathbb{B}(a_j, 2r_j) \subset G_j$, $j = 1, 2$. Moreover, assume that a positive constant K is given with K* diam *$G_1 < r_2$.*

Then, for any points $z', z'' \in G_1$, $z' \neq z''$, $w' \in \mathbb{B}(a_2, r_2)$, $w'' \in \mathbb{C}^{n_2}$, satisfying $(z'', w'') \in D$ and $\frac{\|w'-w''\|}{\|z'-z''\|} < K$, the following equality holds:

$$\boldsymbol{c}_D^*((z', w'), (z'', w'')) = \boldsymbol{c}_{G_1}^*(z', z'').$$

Proof. The inequality "$\geq$" is obvious from the contractibility property of $\boldsymbol{c}^*$. To verify the remaining inequality, fix an $f \in \mathcal{O}(D, \mathbb{D})$ with $f(z', w') = 0$ and $f(z'', w'') = \boldsymbol{c}_D^*((z', w'), (z'', w''))$. Then, it is easily seen that the mapping

$$G_1 \ni z \overset{F}{\longmapsto} w' + (w'' - w')\frac{\langle z - z', z'' - z'\rangle}{\|z' - z''\|^2} \in \mathbb{C}^{n_2}$$

belongs to $\mathcal{O}(G_1, \mathbb{B}(w', r_2)) \subset \mathcal{O}(G_1, G_2)$. Therefore, the function g defined by $g(z) := f(z, F(z))$, $z \in G_1$, has the following properties: $g \in \mathcal{O}(G_1, \mathbb{D})$, $g(z') = 0$, $g(z'') = \boldsymbol{c}_D^*((z', w'), (z'', w''))$. □

Lemma 2.9.3. *Let G_1, G_2, a_1, a_2, r_1, and r_2 be as in Lemma 2.9.2. Then, there exist positive numbers α, β with $\alpha < r_1$ and $\beta < r_2$ such that the following statement is true. Let $\Phi = (\Phi_1, \Phi_2) \in \operatorname{Aut}(D)$ and denote by $\psi := (\psi_1, \psi_2) := \Phi^{-1}$. If the following inequalities:*

$$\begin{aligned} &\|\Phi_1(z,w) - z\| < \alpha, \quad \|\psi_1(z,w) - z\| < \alpha, \\ &\|\Phi_2(z,w) - w\| < \beta, \quad \|\psi_2(z,w) - w\| < \beta, \end{aligned} \tag{2.9.1}$$

hold for all $(z,w) \in \mathbb{C}^{n_1+n_2}$ with $\|z - a_1\| < r_1$, $\|w - a_2\| < r_2$, then $\Phi_1(z,w) = \Phi_1(z,a_2)$ for all $(z,w) \in \mathbb{B}(a_1,\alpha) \times \mathbb{B}(a_2,\beta)$.

There is an immediate consequence from Lemma 2.9.3, namely:

Corollary 2.9.4. *Under the conditions of Lemma 2.9.3 it follows that there exists an $F \in \mathcal{O}(G_1, G_1)$ such that $\Phi_1(z,w) = F(z)$ for all $(z,w) \in D$.*

Proof of Lemma 2.9.3. We start this proof by introducing some positive constants. First, we choose $R > 0$ such that $\bigcap_{\|z-a_1\|<r_1} \mathbb{B}(z, R/2) \supset G_1$; in particular $2r_1 < R/2$. Moreover, we fix ϱ, α, β, and K in such a way that the following inequalities hold:

$$\varrho < r_1\Big(\frac{r_1}{R}\Big)^2 < \frac{r_1}{2}, \quad \alpha < \frac{\varrho}{2}, \quad K \operatorname{diam} G_1 < r_2, \quad \beta < \min\Big\{r_2, K\Big(\frac{\varrho}{2} - \alpha\Big)\Big\}.$$

Recall from Corollary 2.7.14 that for any $z \in \mathbb{B}(a_1, r_1)$ and $X \in \mathbb{C}^{n_1}$, $\|X\| = 1$, the function $[0,\varrho) \ni t \longmapsto \boldsymbol{c}^*_{G_1}(z, z+tX)$ is strictly increasing.

Now, we fix a $\Phi = (\Phi_1, \Phi_2) \in \operatorname{Aut}(D)$, $\psi := (\psi_1, \psi_2) := \Phi^{-1}$, such that Φ and ψ satisfy conditions (2.9.1). We then look at an arbitrary but fixed point $(z', w') \in \mathbb{C}^{n_1} \times \mathbb{C}^{n_2}$ with $\|z' - a_1\| < \alpha$, $\|w' - a_2\| < \beta$. Because of (2.9.1) it follows that

$$\|\Phi_1(z',w') - \Phi_1(z',a_2)\| \le \|\Phi_1(z',w') - z'\| + \|z' - \Phi_1(z',a_2)\| < 2\alpha < \varrho.$$

We choose $\zeta \in \mathbb{C}^{n_1}$ with $\|\zeta - \Phi_1(z',a_2)\| = \varrho$ in such a way that $\Phi_1(z',w') = \Phi_1(z',a_2) + \sigma(\zeta - \Phi_1(z',a_2))$ for a suitable $\sigma > 0$. Then,

$$\|\Phi_1(z',w') - \zeta\| \ge \varrho - 2\alpha > 0.$$

Moreover, observe that

$$\|\zeta - a_1\| \le \|\zeta - \Phi_1(z',a_2)\| + \|\Phi_1(z',a_2) - z'\| + \|z' - a_1\| < 2\varrho < r_1.$$

In particular, the function

$$[0,1] \ni t \longmapsto \boldsymbol{c}^*_{G_1}(\zeta, \zeta + t(\Phi_1(z',a_2) - \zeta))$$

is strictly increasing. Let $(z'', w'') := \psi(\zeta, a_2) \in D$. Then, using (2.9.1) we get

$$\|\zeta - z''\| = \|\zeta - \psi_1(\zeta, a_2)\| < \alpha \text{ and } \|w'' - a_2\| = \|\psi_2(\zeta, a_2) - a_2\| < \beta.$$

To be able to apply Lemma 2.9.2 we establish a sequence of simple inequalities; the proofs here are mainly based on (2.9.1) and the choice of the constants:

(i) $\dfrac{\|\Phi_2(z',a_2)-a_2\|}{\|\Phi_1(z',a_2)-\zeta\|} \le \dfrac{\beta}{\varrho} < \dfrac{K(\varrho/2-\alpha)}{\varrho} < \dfrac{K}{2}$;

(ii)
$$\|z'-z''\| \ge \|\zeta-\Phi_1(z',a_2)\| - \|\Phi_1(z',a_2)-z'\| - \|\zeta-z''\| \ge \varrho-2\alpha > 0,$$
$$\frac{\|w''-a_2\|}{\|z''-z'\|} \le \frac{\beta}{\varrho-2\alpha} < \frac{K(\varrho/2-\alpha)}{\varrho-2\alpha} = \frac{K}{2};$$

(iii) $\dfrac{\|w''-w'\|}{\|z'-z''\|} \le \dfrac{\|w'-a_2\|+\|a_2-\Phi_2(\zeta,a_2)\|}{\varrho-2\alpha} < \dfrac{2\beta}{\varrho-2\alpha} < K$;

(iv) $\dfrac{\|\Phi_2(z',w')-a_2\|}{\|\Phi_1(z',w')-\zeta\|} \le \dfrac{\|\Phi_2(z',w')-w'\|+\|w'-a_2\|}{\varrho-2\alpha} < \dfrac{2\beta}{\varrho} < K.$

Then, Lemma 2.9.2 implies the following chain of equalities:

$$\begin{aligned}
&\boldsymbol{c}^*_{G_1}(\Phi_1(z',a_2),\zeta) \overset{\text{(i)}}{=} \boldsymbol{c}^*_D(\Phi(z',a_2),(\zeta,a_2)) = \boldsymbol{c}^*_D(\Phi(z',a_2),\Phi(z'',w''))\\
&=\boldsymbol{c}^*_D((z',a_2),(z'',w'')) \overset{\text{(ii)}}{=} \boldsymbol{c}^*_{G_1}(z',z'') \overset{\text{(iii)}}{=} \boldsymbol{c}^*_D((z',w'),(z'',w''))\\
&=\boldsymbol{c}^*_D(\Phi(z',w'),\Phi(z'',w'')) = \boldsymbol{c}^*_D(\Phi(z',w'),(\zeta,a_2)) \overset{\text{(iv)}}{=} \boldsymbol{c}^*_{G_1}(\Phi_1(z',w'),\zeta).
\end{aligned}$$

Finally, the strict monotonicity of $t \longmapsto \boldsymbol{c}^*_{G_1}(\zeta,\zeta+t(\Phi_1(z',a_2)-\zeta))$ leads to $\Phi_1(z',w') = \Phi_1(z',a_2)$. Since z', w' were arbitrary, the lemma is established. □

After all these preparations, we turn to the proof of Theorem 2.9.1.

Proof of Theorem 2.9.1. First, we show that the set

$$\begin{aligned}
M := \{\Phi \in \operatorname{Aut}_{\mathrm{id}}(G_1\times G_2) :\\
\exists_{F_j\in\mathcal{O}(G_j,G_j)}\ \forall_{(z,w)\in G_1\times G_2} : \Phi(z,w) = (F_1(z),F_2(w))\}
\end{aligned}$$

is a non-empty open and closed subset of $\operatorname{Aut}_{\mathrm{id}}(G_1\times G_2)$, and, therefore, $M = \operatorname{Aut}_{\mathrm{id}}(G_1\times G_2)$.

Obviously, $\mathrm{id}_{G_1\times G_2} \in M$. The closeness of M simply follows from the fact that if $(f_\nu)_{\nu=1}^\infty \subset \mathcal{O}(G_j)$, $f_\nu \overset{K}{\underset{\nu\to\infty}{\Longrightarrow}} f \in \mathcal{O}(G_j)$, then $D^\alpha f_\nu \overset{K}{\underset{\nu\to\infty}{\Longrightarrow}} D^\alpha f$. The details are left to the reader. So it remains to show that M is open.

Fix a $\Phi \in \operatorname{Aut}_{\mathrm{id}}(G) \cap \overline{M}$. Because of the symmetry, Lemma 2.9.3 can be simultaneously applied for the two components of Φ. Take a $\psi \in M$ such that $\psi^{-1}\circ\Phi$ satisfies the conditions of Lemma 2.9.3. Then, we can find functions $F_j \in \mathcal{O}(G_j,G_j)$ such that $\psi^{-1}\circ\Phi(z,w) = (F_1(z),F_2(w))$, $(z,w) \in G_1\times G_2$. Hence, $\Phi(z,w) = \psi(F_1(z),F_2(w))$. On the other hand, ψ belongs to M, i.e., there are $H_j \in \mathcal{O}(G_j,G_j)$ describing ψ. So, we get that $\Phi(z,w) = (H_1\circ F_1(z), H_2\circ F_2(w))$, $(z,w) \in G_1\times G_2$.

To conclude the proof, it suffices to mention that the F_j's in the definition of M are obviously bijective. □

Remark 2.9.5.

(a) Observe that $\mathbb{D}\times\mathbb{C} \ni (z_1, z_2) \longmapsto (z_1, z_1+z_2) \in \mathbb{D}\times\mathbb{C}$ belongs to $\operatorname{Aut}_{\mathrm{id}}(\mathbb{D}\times\mathbb{C})$ but is not of the form stated in Theorem 2.9.1.

(b) We mention that Theorem 2.9.1 remains true also in the manifold case (cf. [412]). Of course, the condition "bounded" then has to be substituted by another property, namely, by the hyperbolicity.

Theorem 2.9.6 (cf. [448]). *Let G be an arbitrary domain in $\mathbb{C}^n$. Then, a holomorphic mapping $f : G \longrightarrow G$ with $f(G) \subset\subset G$ has a unique fixed point.*

Proof. First, we assume that G is bounded. Write $R := \operatorname{diam}(G)$ and let f satisfy the conditions of the theorem. Then, because of $f(G) \subset\subset G$, there exists a positive number r with $f(G) + \mathbb{B}(r) \subset G$. Now, set $\mu := r/(2R)$.

If $z \in G$ is fixed, we define a holomorphic map $g_z \in \mathcal{O}(G, \mathbb{C}^n)$ by setting $g_z(w) := f(w) + \mu(f(w) - f(z))$, $w \in G$. In view of the choice of μ, it is clear that $g_z \in \mathcal{O}(G, G)$ and, moreover, we have $g_z(z) = f(z)$, $g_z'(z) = (1+\mu)f'(z)$. Therefore, we obtain

$$\boldsymbol{\gamma}_G(f(z); f'(z)X) = (1+\mu)^{-1}\boldsymbol{\gamma}_G(g_z(z); g_z'(z)X) \le (1+\mu)^{-1}\boldsymbol{\gamma}_G(z; X), \quad z \in G,\ X \in \mathbb{C}^n.$$

Then, applying Remark 2.7.5, it follows that

$$\boldsymbol{c}_G^i(f(z'), f(z'')) \le (1+\mu)^{-1}\boldsymbol{c}_G^i(z', z''), \quad z', z'' \in G.$$

From this inequality, we conclude that

$$\boldsymbol{c}_G(f^{(k)}(z), f^{(k+1)}(z)) \le (1+\mu)^{-k}\boldsymbol{c}_G^i(z, f(z)), \quad z \in G,\ k \in \mathbb{N}, \tag{2.9.2}$$

where $f^{(k)}$ denotes the k-th iterate of f, i.e., $f^{(k)} := f \circ \cdots \circ f$ (k times). Finally, observing that $G \subset \mathbb{B}(z, R)$ for any $z \in G$ we have

$$\begin{aligned}
\frac{\|f^{(k+\ell)}(z) - f^{(k)}(z)\|}{R} &= \boldsymbol{c}^*_{\mathbb{B}(f^{(k)}(z),R)}(f^{(k)}(z), f^{(k+\ell)}(z)) \\
&\le \boldsymbol{c}_G(f^{(k)}(z), f^{(k+\ell)}(z)) \\
&\le \sum_{j=1}^{\ell} \boldsymbol{c}_G(f^{(k+j-1)}(z), f^{(k+j)}(z)) \\
&\le \sum_{j=1}^{\ell} (1+\mu)^{-(k+j-1)}\boldsymbol{c}_G^i(z, f(z)), \quad z \in G,\ k, \ell \in \mathbb{N}.
\end{aligned}$$

So, we can now see that $(f^{(k)}(z))_{k=1}^{\infty}$ is a Cauchy sequence in $\mathbb{C}^n$ for an arbitrary point $z \in G$.

Fix a $z' \in G$ and set $z'' := \lim_{k\to\infty} f^{(k)}(z')$. By hypothesis, we have $z'' \in \overline{f(G)} \subset G$ and $f(z'') = \lim_{k\to\infty} f(f^{(k)}(z')) = z''$; i.e., z'' is a fixed point of f.

On the other hand, (2.9.2) shows that there is at most one fixed point of f.

The unbounded case is an immediate consequence. For, let f be as in the theorem. Then, choose a bounded domain D with $f(G) \subset\subset D \subset G$ and apply to $f|_D$ what was said above. □

2.10 A class of n-circled domains

Up to now, the only domains for which we have explicit formulas are balanced convex domains (cf. Proposition 2.3.1(c)). We would like to find another class of domains $G \subset \mathbb{C}^n$ for which $c_G^{(*)}$ may be effectively calculated (at least for some points). A natural candidate is the class of n-circled domains (cf. [250, 253], see also [28]).

Throughout this section, G denotes an n-circled domain in $\mathbb{C}^n$. Recall (cf. [269]) that a set $A \subset \mathbb{C}^n$ is said to be *n-circled* if for $\theta = (\theta_1, \dots, \theta_n) \in \mathbb{R}^n$ we have

$$R_\theta(A) = A,$$

where

$$\mathbb{C}^n \ni (z_1, \dots, z_n) \overset{R_\theta}{\longmapsto} (e^{i\theta_1} z_1, \dots, e^{i\theta_n} z_n) \in \mathbb{C}^n;$$

n-circled domains are also called *Reinhardt domains.*

It is well known that any function $f \in \mathcal{O}(G)$ has the Laurent series representation

$$f(z) = \sum_{\alpha\in\Sigma(G)} a_\alpha z^\alpha, \quad z \in G,$$

where

$$\begin{aligned}\Sigma(G) :&= \{\alpha \in \mathbb{Z}^n : \text{ the function } G \ni z \longmapsto z^\alpha \text{ is well defined}\}\\ &= \{\alpha \in \mathbb{Z}^n : \forall_{(z_1,\dots,z_n)\in G}\ \forall_{j\in\{1,\dots,n\}} : (\alpha_j < 0) \Longrightarrow (z_j \neq 0)\} \supset \mathbb{Z}_+^n.\end{aligned}$$

Note that if $0 \in G$, then $\Sigma(G) = \mathbb{Z}_+^n$. Observe that

$$|a_\alpha z^\alpha| \le \max\{|f(R_\theta(z))| : \theta \in \mathbb{R}^n\}, \quad z \in G,\ \alpha \in \Sigma(G).$$

Consequently, if $f \in \mathcal{H}^\infty(G)$, then

$$|a_\alpha| \cdot \|z^\alpha\|_G \le \|f\|_G, \quad \alpha \in \Sigma(G),$$

and so we get

Lemma 2.10.1. *If $f \in \mathcal{H}^\infty(G)$, then*

$$f(z) = a_0 + \sum_{\alpha\in S(G)} a_\alpha z^\alpha, \quad z \in G,$$

where

$$S(G) := \{\alpha \in \Sigma(G)_* : z^\alpha \in \mathcal{H}^\infty(G)\} \text{ is the set of admissible exponents for } G.$$

In particular,

$$\mathcal{H}^\infty(G) \simeq \mathbb{C} \quad \textit{iff} \quad S(G) = \varnothing.$$

Elementary n-circled (*Reinhardt*) domains are defined as follows:

$$\boldsymbol{D}_{\alpha,C} := \{(z_1, \dots, z_n) \in \mathbb{C}^n(\alpha) : |z_1|^{\alpha_1} \dots |z_n|^{\alpha_n} < e^C\},$$

where $\alpha = (\alpha_1, \dots, \alpha_n) \in (\mathbb{R}^n)_*$, $C \in \mathbb{R}$,

$$\mathbb{C}^n(\alpha) := \{(z_1, \dots, z_n) \in \mathbb{C}^n : \forall_{j \in \{1,\dots,n\}} : (\alpha_j < 0) \Longrightarrow (z_j \neq 0)\}.$$

To simplify notation put $\boldsymbol{D}_\alpha := \boldsymbol{D}_{\alpha,0}$.

There are two possibilities:
either $\boldsymbol{D}_{\alpha,C}$ is of *irrational type*, i.e., $\alpha \notin \mathbb{R} \cdot \mathbb{Z}^n$,
or $\boldsymbol{D}_{\alpha,C}$ is of *rational type*, i.e., $\alpha \in \mathbb{R} \cdot \mathbb{Z}^n$.

In the second case we may assume that $\alpha = (\alpha_1, \dots, \alpha_n) \in (\mathbb{Z}^n)_*$ and that $\alpha_1, \dots, \alpha_n$ are relatively prime. Put

$$\boldsymbol{H}_{\alpha,C} := \{x \in \mathbb{R}^n : \langle x, \alpha \rangle < C\}$$

($\langle\ ,\ \rangle$ denotes the Euclidean scalar product in $\mathbb{R}^n$).

For $A \subset \mathbb{C}^n$ let $\log A$ be the *logarithmic image of* A, that is,

$$\log A := \{x = (x_1, \dots, x_n) \in \mathbb{R}^n : e^x = (e^{x_1}, \dots, e^{x_n}) \in A\}.$$

Clearly, $\log \boldsymbol{D}_{\alpha,C} = \boldsymbol{H}_{\alpha,C}$. Note that the domain $\boldsymbol{D}_{\alpha,C}$ is pseudoconvex (cf. Appendix B.7.5). Observe that

$$S(G) = \{\alpha \in (\mathbb{Z}^n)_* : \exists_C : G \subset \boldsymbol{D}_{\alpha,C}\} = \{\alpha \in (\mathbb{Z}^n)_* : \exists_C : \log G \subset \boldsymbol{H}_{\alpha,C}\}.$$

In particular,

$$\begin{aligned} &S(\boldsymbol{D}_{\alpha,C}) = \varnothing \text{ iff } \boldsymbol{D}_{\alpha,C} \text{ is of irrational type,} \\ &S(\boldsymbol{D}_{\alpha,C}) = \mathbb{N}\alpha \text{ iff } \alpha \in (\mathbb{Z}^n)_*, \alpha_1, \dots, \alpha_n \text{ are relatively prime.} \end{aligned} \tag{2.10.1}$$

Using Lemma 2.10.1, one can easily conclude that in the latter case the mapping

$$\mathcal{H}^\infty(\mathbb{D}) \ni \varphi \longmapsto \varphi \circ \Phi \in \mathcal{H}^\infty(\boldsymbol{D}_{\alpha,C}),$$

where $\Phi(z) = e^{-C} z^\alpha$, is an isometry and an algebraic isomorphism. Thus, we have proved the following

Proposition 2.10.2. *If $G = \boldsymbol{D}_{\alpha,C}$, then*

(a) $\boldsymbol{c}_G^{(*)} \equiv 0$ ($\boldsymbol{\gamma}_G \equiv 0$) *iff G is of irrational type;*

(b) *if $\alpha = (\alpha_1, \dots, \alpha_n) \in (\mathbb{Z}^n)_*$ and $\alpha_1, \dots, \alpha_n$ are relatively prime, then*

$$\boldsymbol{c}_G^{(*)}(w,z) = \boldsymbol{c}_{\mathbb{D}}^{(*)}(\Phi(w), \Phi(z)), \quad w, z \in G,$$
$$\boldsymbol{\gamma}_G(z;X) = \boldsymbol{\gamma}_{\mathbb{D}}(\Phi(z); \Phi'(z)X), \quad z \in G,\ X \in \mathbb{C}^n.$$

Now, assume that $G \subsetneq \mathbb{C}^n$, $n \geq 2$, is a Reinhardt domain with $0 \in G$.
For $\alpha \in S(G)$, put $n_\alpha = n_\alpha(G) := 1/\|z^\alpha\|_G$. If $T \subset S(G)$, then we define

$$M_G^T(z) := \sup\{n_\alpha |z^\alpha| : \alpha \in T\}, \quad z \in G,$$

where $M_G^\varnothing := 0$. Let $M_G := M_G^{S(G)}$ and observe that

$$\boldsymbol{c}_G^*(0,\cdot) \geq M_G \geq M_G^T.$$

Our aim is to characterize some classes of n-circled domains $G \subset \mathbb{C}^n$ (with $0 \in G$) such that $\boldsymbol{c}_G^*(0,\cdot) = M_G$, or even more, $\boldsymbol{c}_G^*(0,\cdot) = M_G^T$ for some $T \subset S(G)$. We are interested in characterizing the minimal set T with this property, and in situations where the set T is finite (which is most important from the point of view of applications).

Note that if α is as in Proposition 2.10.2(b), then by (2.10.1)

$$\boldsymbol{c}_{\boldsymbol{D}_{\alpha,C}}^*(0,\cdot) = M_{\boldsymbol{D}_{\alpha,C}}^{\{\alpha\}}.$$

Lemma 2.10.3.

(a) *The function M_G^T is continuous and invariant under the rotations R_θ, $\theta \in \mathbb{R}^n$.*

(b) *If $T \neq \varnothing$, then, for any $a \in G$, there exists an $\alpha \in T$ such that*

$$M_G^T(a) = n_\alpha |a^\alpha|.$$

Proof. (a) The continuity of M_G^T holds, since the family $\{n_\alpha z^\alpha : \alpha \in T\}$ is equicontinuous.

(b) If $M_G^T(a) = 0$, then the result is trivial. Suppose that $M_G^T(a) > 0$. Let $(\alpha^j)_{j=1}^\infty \subset T$ be such that

$$n_{\alpha^j} |a^{\alpha^j}| \longrightarrow M_G^T(a).$$

Put $f_j(z) := n_{\alpha^j} z^{\alpha^j}$, $j \geq 1$. In view of the Montel theorem, we may assume that $f_j \underset{j\to\infty}{\overset{K}{\Longrightarrow}} f$ in G. Clearly, $f(0) = 0$, $f(a) \neq 0$, and $f \in \mathcal{O}(G, \mathbb{D})$. To prove the required result, it is enough to show that $\lim_{j\to\infty} |\alpha^j|$ exists and is finite. Observe that

$$\sum_{k=1}^{n} z_k \frac{\partial f_j}{\partial z_k}(z) = |\alpha^j| f_j(z), \quad j \geq 1.$$

Hence,

$$\lim_{j\to+\infty} |\alpha^j| = \frac{1}{f(a)} \sum_{k=1}^{n} a_k \frac{\partial f}{\partial z_k}(a). \qquad \square$$

Define

$$B^* = B^*(G) := \{\alpha \in S(G) : \exists_{a\in G}\ \forall_{\beta\in S(G)\setminus\{\alpha\}} : n_\alpha |a^\alpha| > n_\beta |a^\beta|\}.$$

Lemma 2.10.4. *For any $T \subset S = S(G)$, we have*

$$M_G = M_G^T \quad \textit{iff} \quad B^*(G) \subset T,$$

that is, $B^*(G)$ is the minimal set that determines the function M_G.

Proof. The case $S = \varnothing$ is trivial. So suppose that $S \neq \varnothing$. For $\alpha, \beta \in S, \alpha \neq \beta$, put

$$V_{\alpha,\beta} := \{z \in G : n_\alpha |z^\alpha| = n_\beta |z^\beta|\}.$$

It is clear that this set is closed and nowhere dense. Hence, the set

$$G_0 := G \setminus \bigcup_{\alpha,\beta\in S,\ \alpha\neq\beta} V_{\alpha,\beta}$$

is dense in G. Take an $a \in G_0$ and let $\alpha \in S$ be such that $M_G(a) = n_\alpha |a^\alpha|$ (Lemma 2.10.3(b)). By the definition of G_0 we see that $n_\alpha |a^\alpha| \neq n_\beta |a^\beta|$ for all $\beta \in S \setminus \{\alpha\}$. Hence, $\alpha \in B^*(G)$, which shows that

$$M_G = M_G^{B^*(G)} \text{ on } G_0.$$

Finally, the continuity (Lemma 2.10.3(a)) gives $M_G = M_G^{B^*(G)}$ on G.

If $M_G = M_G^T$, then we get the inclusion $B^*(G) \subset T$ directly from the definition of $B^*(G)$ and from Lemma 2.10.3(b). $\square$

Let $B = B(G) := S \setminus (S + S)$, where $S = S(G)$; $B(G)$ is called the set of all *irreducible elements of* S. Observe that for any $\alpha \in S$ there exist $k \in \mathbb{N}$ and $\beta^1, \dots, \beta^k \in B$ such that $\alpha = \beta^1 + \cdots + \beta^k$ (in general, k and $\beta^1, \dots, \beta^k$ are not uniquely determined).

We say that G satisfies the *cone condition* if $\log G$ is a cone with its vertex at a point $x^0 \in \mathbb{R}^n$, i.e., $x^0 + t(x - x^0) \in \log G$ for all $t > 0,\ x \in \log G$.

In this case,

$$n_\alpha = e^{-\langle x^0, \alpha\rangle}, \quad \alpha \in S(G),$$

and hence,

$$B^*(G) \subset B(G).$$

Moreover, we may assume that $x^0 = 0$ (use the mapping $\mathbb{C}^n \ni (z_1, \dots, z_n) \longmapsto (e^{-x_1^0} z_1, \dots, e^{-x_n^0} z_n) \in \mathbb{C}^n$).

Proposition 2.10.5. *If G is an n-circled domain, $0 \in G$, satisfying the cone condition, then*

$$\boldsymbol{c}_G^*(0, \cdot) = M_G \; (= M_G^{B(G)} = M_G^{B^*(G)}).$$

Proof. We may assume that $S = S(G) \neq \varnothing$ and that $\log G$ is a cone with its vertex at $0 \in \mathbb{R}^n$. Since both functions $\boldsymbol{c}_G^*(0, \cdot)$ and M_G are continuous and invariant under rotations, we only need to prove that

$$\boldsymbol{c}_G^*(0, e^x) \le M_G(e^x)$$

for x in some dense subset of $\log G$. Note that

$$M_G(e^x) = \exp(\sup\{\langle x, \alpha\rangle : \alpha \in S\}), \quad x \in \log G.$$

Put

$$G_0 := \{(\lambda^{v_1}, \dots, \lambda^{v_n}) : \lambda \in \mathbb{D}_*, \; v = (v_1, \dots, v_n) \in (-\log G) \cap \mathbb{Z}^n\}.$$

Then, the set

$$\log G_0 = \{tv : t < 0, \; v \in (-\log G) \cap \mathbb{Z}^n\}$$

is dense in $\log G$ and

$$M_G(e^{tv}) = \exp(tQ(v)),$$

where $Q(v) := \inf\{\langle v, \alpha\rangle : \alpha \in S\}$. Observe that $Q(v) \in \mathbb{N}$. On the other hand, if $f \in \mathcal{O}(G, \mathbb{D})$, $f(0) = 0$, then the function

$$\mathbb{D}_* \ni \lambda \longmapsto f(\lambda^{v_1}, \dots, \lambda^{v_n}) = \sum_{\alpha \in S} a_\alpha \lambda^{\langle v, \alpha\rangle}$$

extends to the whole $\mathbb{D}$ and the extension has a zero of order $Q(v)$ at the origin.

Hence,

$$\boldsymbol{c}_G^*(0, (\lambda^{v_1}, \dots, \lambda^{v_n})) \le |\lambda|^{Q(v)}, \quad \lambda \in \mathbb{D}_*. \qquad \square$$

In applications, the most interesting case is when

$$G = \boldsymbol{D}_{\alpha^1,C_1} \cap \cdots \cap \boldsymbol{D}_{\alpha^N,C_N}, \tag{2.10.2}$$

$\alpha^1,\dots,\alpha^N \in (\mathbb{R}^n_+)_*$, $C_1,\dots,C_N \in \mathbb{R}$. We will always assume that the intersection is minimal, i.e.,

$$G \neq \bigcap_{j\neq k} \boldsymbol{D}_{\alpha^j,C_j}, \quad k = 1,\dots,N. \tag{2.10.3}$$

The case $N = 1$ is covered by Proposition 2.10.2, so let $N \geq 2$. Put $r := \operatorname{rank}(\alpha^1, \dots, \alpha^N)$.

Lemma 2.10.6.

(a) *The domain G (as in (2.10.2), (2.10.3)) satisfies the cone condition iff*

$$r = \operatorname{rank}\begin{bmatrix} \alpha^1 & C_1 \\ \vdots & \vdots \\ \alpha^N & C_N \end{bmatrix}.$$

(b) $S(G) = (\mathbb{Z}^n_+)_* \cap \Big(\bigcup_{\substack{1\leq i_1<\dots<i_r\leq N \\ \operatorname{rank}(\alpha^{i_1},\dots,\alpha^{i_r})=r}} (\mathbb{R}_+\alpha^{i_1} + \dots + \mathbb{R}_+\alpha^{i_r})\Big)$.

(c) *Let $\alpha^1,\dots,\alpha^N \in (\mathbb{Z}^n_+)_*$, $\alpha^j = (\alpha^j_1,\dots,\alpha^j_n)$, and suppose that $\alpha^j_1,\dots,\alpha^j_n$ are relatively prime, $j = 1,\dots,N$. Then,*

$$B(G) \subset \{\alpha^1,\dots,\alpha^N\} \cup \Big((\mathbb{Z}^n_+)_* \cap \Big(\bigcup_{\substack{1\leq i_1<\dots<i_r\leq N \\ \operatorname{rank}(\alpha^{i_1},\dots,\alpha^{i_r})=r}} ([0,1)\alpha^{i_1} + \dots + [0,1)\alpha^{i_r})\Big)\Big).$$

In particular, the set $B(G)$ is finite.

(d) *If G (as in (2.10.2) and (2.10.3)) is an $\mathcal{H}^\infty$-domain of holomorphy satisfying the cone condition, and if the set $B(G)$ is finite, then*

$$\alpha^1,\dots,\alpha^N \in \mathbb{R}\cdot B(G).$$

In other words, if G satisfies the cone condition and if G is an $\mathcal{H}^\infty$-domain of holomorphy, then the situation described in (c) is essentially the only case where $B(G)$ is finite.

Proof. (a) The proof is left to the reader.

(b) Let $\widetilde{S}$ denote the right hand side of the formula in question. Clearly, $\widetilde{S} \subset S$. Fix an $\alpha \in S$. Then, there exists a $C \in \mathbb{R}$ such that

$$\boldsymbol{H}_{\alpha^1,C_1} \cap \cdots \cap \boldsymbol{H}_{\alpha^N,C_N} \subset \boldsymbol{H}_{\alpha,C}.$$

One can prove (see Appendix in [249] or [269], Lemma 3.3.9) that there exist $1 \le i_1 < \dots < i_r \le N$ with $\operatorname{rank}(\alpha^{i_1}, \dots \alpha^{i_r}) = r$ and such that

$$\boldsymbol{H}_{\alpha^{i_1}, C_{i_1}} \cap \dots \cap \boldsymbol{H}_{\alpha^{i_r}, C_{i_r}} \subset \boldsymbol{H}_{\alpha, C}.$$

To simplify the notation, put $i_\mu = \mu$, $\mu = 1, \dots, r$. By the above inclusion, $\alpha \in \mathbb{R}\alpha^1 + \dots + \mathbb{R}\alpha^r$, i.e., $\alpha = t_1\alpha^1 + \dots + t_r\alpha^r$ for some $t_1, \dots, t_r \in \mathbb{R}$. It remains to show that $t_1, \dots, t_r \ge 0$.

Let

$$\mathbb{R}^n \ni x \overset{L}{\longmapsto} (\langle x, \alpha^1 \rangle, \dots, \langle x, \alpha^r \rangle) \in \mathbb{R}^r.$$

Note that L is surjective. Put $t := (t_1, \dots, t_r)$. Then,

$$\begin{aligned}\{(\xi_1, \dots, \xi_r) \in \mathbb{R}^r : \xi_j < C_j,\ j = 1, \dots, r\} &= L\Big(\bigcap_{j=1}^{r} \boldsymbol{H}_{\alpha^j, C_j}\Big) \\ &\subset L(\boldsymbol{H}_{\alpha, C}) = \{\xi \in \mathbb{R}^r : \langle \xi, t \rangle < C\}\end{aligned}$$

which implies that $t_1, \dots, t_r \ge 0$.

(c) Take an $\alpha \in B(G)$. By (b) we have $\alpha = t_1\alpha^{i_1} + \dots + t_r\alpha^{i_r}$, where $t_1, \dots, t_r \ge 0$ and $\operatorname{rank}(\alpha^{i_1}, \dots, \alpha^{i_r}) = r$. Suppose that $i_\mu = \mu$, $\mu = 1, \dots, r$. Observe that

$$\alpha = \Big(\sum_{j=1}^{r} \lfloor t_j \rfloor \alpha^j\Big) + \Big(\sum_{j=1}^{r} (t_j - \lfloor t_j \rfloor)\alpha^j\Big) =: \beta^1 + \beta^2,$$

where $\lfloor\ \rfloor$ denotes the integer part. It is easily seen that $\beta^1, \beta^2 \in S \cup \{0\}$. In view of the definition of $B(G)$, either $\alpha = \beta^2$ or $\alpha = \alpha^{i_0}$ for some $i_0 \in \{1, \dots, r\}$.

(d) We may assume that $\log G$ is a cone with its vertex at 0, or equivalently $C_1 = \dots = C_N = 0$. In particular, $n_\alpha = 1$ for any $\alpha \in S$. Put

$$\widetilde{G} := \bigcap_{\beta \in B(G)} \boldsymbol{D}_\beta.$$

Since $B(G)$ is finite, it suffices to prove that $\widetilde{G} = G$. Since $G \subset \widetilde{G}$ is an $\mathcal{H}^\infty$-domain of holomorphy, we only need to prove that $\mathcal{H}^\infty(G) \subset \mathcal{O}(\widetilde{G})|_G$; cf. Remark 2.1.3. Take an $f \in \mathcal{H}^\infty(G)$, $f(z) = f(0) + \sum_{\alpha \in S} a_\alpha z^\alpha$, $z \in G$. To prove that the series is convergent on $\widetilde{G}$, it is enough to show that the sequence $a_\alpha z^\alpha$, $\alpha \in S$, is pointwise bounded on $\widetilde{G}$ ($\widetilde{G}$ is a Reinhardt domain!). Fix $b \in \widetilde{G}$ and $\alpha = \beta^1 + \dots + \beta^k \in S$, $\beta^1, \dots, \beta^k \in B$. Then, by the definition of $\widetilde{G}$ and by the Cauchy inequalities, we have

$$|a_\alpha b^\alpha| = |a_\alpha||b^{\beta^1}| \dots |b^{\beta^k}| \le |a_\alpha| \le \frac{\|f\|_G}{\|z^\alpha\|_G} = \|f\|_G, \quad \alpha \in S. \qquad \square$$

Corollary 2.10.7. *If $G = \boldsymbol{D}_{\alpha^1} \cap \dots \cap \boldsymbol{D}_{\alpha^N}$, $\alpha^j = (\alpha_1^j, \dots, \alpha_n^j) \in (\mathbb{Z}_+^n)_*$, $j = 1, \dots, N$, then the set $B^*(G)$ is finite and*

$$\boldsymbol{c}_G^*(0, z) = \max\{|z^\alpha| : \alpha \in B^*(G)\}, \quad z \in G. \qquad \square$$

? Whereas $\boldsymbol{c}_G^*(0, \cdot)$ is given by monomials if G satisfies the cone condition, the general situation is not yet completely understood.

Remark 2.10.8. If G does not satisfy the cone condition, then, in general, $\boldsymbol{c}_G^*(0, \cdot) \neq M_G$. For example, let

$$G := \{(z_1, z_2) \in \mathbb{C}^2 : |z_1| < 1, \ |z_2| < 1, \ |z_1 z_2| < 1/2\}. \tag{2.10.4}$$

Then,

$$M_G(z) = \max\{|z_1|, \ |z_2|, \ 2|z_1 z_2|\} \quad \text{(use Lemma 2.10.6(b))}$$

but

$$\boldsymbol{c}_G^*(0, z) \geq \max\{|z_1|, \ |z_2|, \ (2/3)(|z_1| + |z_2|), \ 2|z_1 z_2|\} \gneqq M_G(z).$$

In fact, one can prove that

$$\boldsymbol{c}_G^*(0, z) \gneqq \sup\{|Q(z)| : Q \text{ is a homogeneous polynomial}, |Q| \leq 1 \text{ on } G\}$$

(cf. Exercise 2.12.13, see also [250]) and that

$$\boldsymbol{c}_G^*(0, z) = M_G(z)$$

if z belongs to the shadowed part in Figure 2.1 (cf. Exercise 9.2.6).

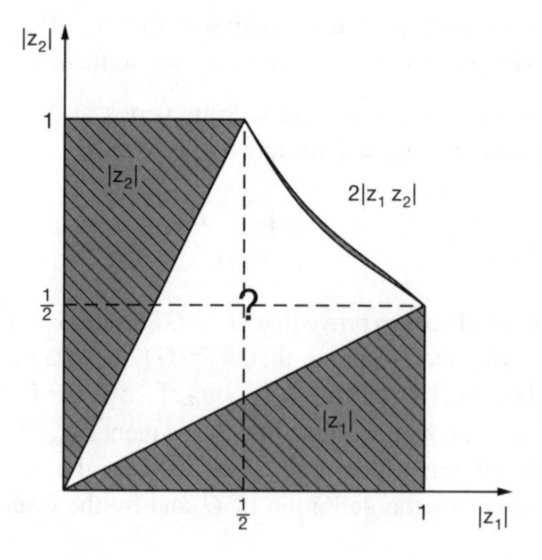

Figure 2.1

The complete formula for $\boldsymbol{c}_G^*(0,\cdot)$ is not known.

Using the same argument, one can easily prove that for the domain

$$G := \{(z_1,z_2,z_3) \in \mathbb{C}^3 : |z_1 z_3| < 1,\ |z_2 z_3| < 1,\ |z_1 z_2 z_3^2| < 1/2\} \tag{2.10.5}$$

we also have

$$\boldsymbol{c}_G^*(0,\cdot) \not\equiv M_G.$$

On the other hand, there are domains without the cone condition for which $\boldsymbol{c}_G^*(0,\cdot) \equiv M_G$. For instance, we have

Proposition 2.10.9. *If* $G = \boldsymbol{D}_{\alpha^1,C_1} \cap \cdots \cap \boldsymbol{D}_{\alpha^N,C_N}$, *where* $\alpha^1,\dots,\alpha^N \in \mathbb{Z}_+^{n-1} \times \{1\}$, *then*

$$\boldsymbol{c}_G^*(0,z) = \max\{e^{-C_j}|z^{\alpha^j}| : j = 1,\dots,N\}\ (= M_G(z)), \quad z \in G.$$

Proof. The inequality "$\geq$" is obvious. Let $\alpha^j = (\beta^j,1)$, $j = 1,\dots,N$. Observe that $\mathbb{C}^{n-1} \times \{0\} \subset G$ and note that G may be written as the Hartogs domain:

$$G = \{(z',z_n) \in \mathbb{C}^{n-1} \times \mathbb{C} : u(z')|z_n| < 1\},$$

where $u(z') := \max\{e^{-C_j}|(z')^{\beta^j}| : j = 1,\dots,N\}$, $z' \in \mathbb{C}^{n-1}$.

If $f \in \mathcal{O}(G,\mathbb{D})$ with $f(0) = 0$, then the Liouville theorem implies $f(z',0) = 0$ for all $z' \in \mathbb{C}^{n-1}$. Hence, by the classical Schwarz lemma (applied to the function $f(z',\cdot)$), we get

$$|f(z',z_n)| \leq u(z')|z_n| = \max\{e^{-C_j}|z^{\alpha^j}| : j = 1,\dots,N\}, \quad (z',z_n) \in G. \qquad \square$$

Example 2.10.10. If

$$G := \{(z_1,z_2,z_3) \in \mathbb{C}^3 : |z_1 z_3| < 1,\ |z_2 z_3| < 1,\ |z_1 z_2 z_3| < 1/2\},$$

then

$$\boldsymbol{c}_G^*(0,z) = \max\{|z_1 z_3|,\ |z_2 z_3|,\ 2|z_1 z_2 z_3|\}\ (= M_G(z)).$$

Note that G seems to be similar to the domain in (2.10.5), but the form of $\boldsymbol{c}_G^*(0,\cdot)$ is totally different.

We conclude this section with some examples for $n = 2$. First, observe that if $G \subset \mathbb{C}^2$ is a 2-circled domain satisfying the cone condition, then

- either $G = \boldsymbol{D}_{\alpha,C}$,
- or $G = \boldsymbol{D}_{\alpha^1,C_1} \cap \boldsymbol{D}_{\alpha^2,C_2}$ with $\operatorname{rank}(\alpha^1,\alpha^2) = 2$.

In the latter case, we may assume without loss of generality that $C_1 = C_2 = 0$. We will consider only the most elementary case where $\alpha^1 = (1,0)$, $\alpha^2 = (p,q) \in \mathbb{N}^2$, and p, q are relatively prime.

Lemma 2.10.11. *If*

$$G := \{(z_1, z_2) \in \mathbb{C}^2 : |z_1| < 1, |z_1^p z_2^q| < 1\} \text{ and } p, q \in \mathbb{N} \text{ are relatively prime,}$$

then

$$B^*(G) \subset B(G) \subset \{(1,0), (p,q)\} \cup \Big\{\Big(\Big\lfloor \frac{\nu p}{q} \Big\rfloor + 1, \nu\Big) : \nu = 1, \dots, q-1\Big\}.$$

In particular, $\# B^*(G) \leq q + 1$.

Proof. Use Lemma 2.10.6(c). □

We will show that the above lemma gives an effective method of finding $B^*(G)$ for

$$G = G_{p,q} := \{(z_1, z_2) \in \mathbb{C}^2 : |z_1| < 1, |z_1^p z_2^q| < 1\}$$

($p, q \in \mathbb{N}$, p, q relatively prime).

Let

$$s(\nu) := \begin{cases} \lfloor \frac{\nu p}{q} \rfloor + 1, & \nu = 0, \dots q-1 \\ p, & \nu = q \end{cases} = \begin{cases} 1, & \nu = 0 \\ \lceil \frac{\nu p}{q} \rceil, & \nu = 1, \dots q \end{cases},$$
$$L_{p,q} := \{(s(\nu), \nu) : \nu = 0, \dots, q\}.$$

By the definition of $B^*(G)$, we have:

$$\begin{aligned} &(s(\nu), \nu) \in B^*(G) \quad \text{iff} \\ &\quad \text{there exists } (z_1, z_2) \in G \text{ such that } |z_1^{s(\nu)} z_2^{\nu}| > |z_1^{s(\mu)} z_2^{\mu}| \text{ for all } \mu \neq \nu. \end{aligned} \tag{2.10.6}$$

In particular, $(1,0)$, $(p,q) \in B^*(G)$, and therefore we are done if $q = 1$. If $q > 1$, then by the definition of $B(G)$ (recall that $B^*(G) \subset B(G)$), we must exclude all multi-indices $(s(\nu_0), \nu_0)$ of the sequence $(s(\nu), \nu)$, $\nu = 0, \dots, q$, for which

$$s(\nu_0) = s(\mu) + s(\nu_0 - \mu) \text{ for some } \mu \in \{0, \dots, q\}$$

(note that this is possible only for $q \geq 3$, $2 \leq \nu_0 \leq q-1$, and $1 \leq \mu \leq \nu_0 - 1$).

Let $(\alpha_0(\nu), \beta_0(\nu))$, $\nu = 0, \dots, q_0$, denote the remaining multi-indices with $\beta_0(\mu) < \beta_0(\nu)$ for $0 \leq \mu < \nu \leq q_0$ (if $q = 2$, then this sequence coincides with the initial one). Obviously,

$$B(G) = \{(\alpha_0(\nu), \beta_0(\nu)) : \nu = 0, \dots, q_0\}.$$

Now, we begin the main reduction procedure.

If $q_0 = 1$, then there is nothing to do and $B^*(G) = \{(1,0),(p,q)\}$. If $q_0 > 1$, then, by (2.10.6), we exclude from the sequence $(\alpha_0(\nu), \beta_0(\nu))$, $\nu = 0,\dots,q_0$, the first multi-index $(\alpha_0(\nu_0), \beta_0(\nu_0))$ such that

$$\max\left\{\frac{\alpha_0(\nu_0)-\alpha_0(\mu)}{\beta_0(\nu_0)-\beta_0(\mu)} : \mu = 0,\dots,\nu_0-1\right\}$$
$$\geq \min\left\{\frac{p}{q}, \frac{\alpha_0(\nu_0)-\alpha_0(\mu)}{\beta_0(\nu_0)-\beta_0(\mu)} : \mu = \nu_0+1,\dots,q_0\right\}$$

(note that this is only possible for $1 \leq \nu_0 \leq q_0 - 1$). If such a multi-index does not exist, then

$$B^*(G) = \{(\alpha_0(\nu), \beta_0(\nu)) : \nu = 0,\dots,q_0\}$$

and we are done. Otherwise, we get a new shorter sequence $(\alpha_1(\nu), \beta_1(\nu))$, $\nu = 0,\dots,q_1 = q_0 - 1$, and we can repeat our reduction procedure again.

Example 2.10.12. Let

$$G := \{(z_1,z_2) \in \mathbb{C}^2 : |z_1| < 1, |z_1^{34} z_2^{55}| < 1\}.$$

Then,

$$c_G^*(0,z) = \max\{|z_1|, |z_1 z_2|, |z_1^2 z_2^3|, |z_1^5 z_2^8|, |z_1^{13} z_2^{21}|, |z_1^{34} z_2^{55}|\},$$

see also Exercise 2.12.14.

The results below (2.10.13–2.10.20) are due to W. Jarnicki.

Proposition 2.10.13. *Assume that $p,q \in \mathbb{N}$ be relatively prime and let*

$$A = \{(\alpha(0),\beta(0)),\dots,(\alpha(q_0),\beta(q_0))\}$$

be such that $B^*(G_{p,q}) \subset A \subset L_{p,q}$ and $(\beta(j))_{j=0}^{q_0}$ is strictly increasing. Then, the following conditions are equivalent:

(i) $\left(\frac{\alpha(j)-\alpha(j-1)}{\beta(j)-\beta(j-1)}\right)_{j=1}^{q_0}$ *is strictly increasing and bounded from above by p/q;*

(ii) $A = B^*(G_{p,q})$.

Proof. (i) $\Longrightarrow$ (ii): Following the effective method outlined after the proof of Lemma 2.10.11, it suffices to show that the reduction procedure mentioned there is impossible for any $1 \le \nu_0 \le q_0 - 1$. Observe that the following holds:

$$\max\left\{\frac{\alpha(\nu_0)-\alpha(\mu)}{\beta(\nu_0)-\beta(\mu)} : \mu = 0,\dots,\nu_0-1\right\} \le \frac{\alpha(\nu_0)-\alpha(\nu_0-1)}{\beta(\nu_0)-\beta(\nu-1)}$$
$$< \frac{\alpha(\nu_0+1)-\alpha(\nu_0)}{\beta(\nu_0+1)-\beta(\nu)} \le \min\left\{\frac{p}{q}, \frac{\alpha(\nu_0)-\alpha(\mu)}{\beta(\nu_0)-\beta(\mu)} : \mu = \nu_0+1,\dots,q_0\right\}.$$

(ii) $\Longrightarrow$ (i): Assuming that the reduction procedure is impossible easily implies (i). □

Let $p, q \in \mathbb{N}$ be relatively prime with $p > 1$. Define

$$C_{p,q} := \begin{cases} \{(\alpha+\beta,\beta) : (\alpha,\beta) \in B^*(G_{p-q,q})\}, & \text{if } p > q \\ \{(1,0)\} \cup \{(\alpha,\beta+k\alpha) : (\alpha,\beta) \in B^*(G_{p,r})\}, & \text{if } p < q,\ q = kp + r,\ 0<r<p \end{cases}.$$

Lemma 2.10.14. *Let $p, q \in \mathbb{N}$ be relatively prime with $p > 1$. Then, $C_{p,q} \subset L_{p,q}$.*

Proof. For $p > q$, observe that $\lceil \nu(p-q)/q\rceil + \nu = \lceil \nu p/q\rceil$, $\nu = 1,\dots,q$.
For $p < q$, observe that $\lceil \nu p/r\rceil = \lceil(\nu + k\lceil \nu p/r\rceil)p/q\rceil$, $\nu = 1,\dots,r$. □

Lemma 2.10.15. *Let $p, q \in \mathbb{N}$ be relatively prime with $p > 1$. Then, $B^*(G_{p,q}) \subset C_{p,q}$.*

Proof. Take $(\alpha,\beta) \in B^*(G_{p,q})$ and $(z_1, z_2) \in G_{p,q}$. We want to prove that $|z_1^\alpha z_2^\beta| \le |z_1^{\alpha_0} z_2^{\beta_0}|$ for some $(\alpha_0,\beta_0) \in C_{p,q}$.

If $(\alpha,\beta) \in C_{p,q}$, then we are done. Otherwise, by Lemma 2.10.11, we have $0 < \beta < q$ and $\alpha = \lceil \beta p/q\rceil$. Consider two cases.

- For $p > q$, put $(\gamma,\delta) := (\alpha-\beta,\beta)$, $(w_1, w_2) = (z_1, z_1 z_2) \in G_{p-q,q}$. Observe that $(\gamma,\delta) \in L_{p-q,q}$. This means that there exists a $(\gamma_0,\delta_0) \in B^*_{p-q,q}$ with $|w_1^\gamma w_2^\delta| \le |w_1^{\gamma_0} w_2^{\delta_0}|$. Putting $(\alpha_0,\beta_0) := (\gamma_0+\delta_0,\delta_0) \in C_{p,q}$ completes the proof.

- For $p < q$, put $(\gamma,\delta) := (\alpha,\beta-k\alpha)$, $(w_1, w_2) = (z_1 z_2^k, z_2) \in G_{p,r}$. Observe that $(\gamma,\delta) \in L_{p,r}$. This means that there exists a $(\gamma_0,\delta_0) \in B^*_{p,r}$ with $|w_1^\gamma w_2^\delta| \le |w_1^{\gamma_0} w_2^{\delta_0}|$. Putting $(\alpha_0,\beta_0) := (\gamma_0, \delta_0 + k\gamma_0) \in C_{p,q}$ completes the proof. □

Theorem 2.10.16. *Let $p, q \in \mathbb{N}$ be relatively prime with $p > 1$. Then, $B^*(G_{p,q}) = C_{p,q}$.*

Proof. Following Lemmas 2.10.14 and 2.10.15, $A := C_{p,q}$ satisfies the assumptions of Proposition 2.10.13. To prove that $A = B^*(G_{p,q})$, consider two cases:

• For $p > q$, let $B^*(G_{p-q,q}) = \{(\alpha'(0), \beta'(0)), \dots, (\alpha'(q_0), \beta'(q_0))\}$ be as in Proposition 2.10.13. Observe that $\frac{\alpha(j)-\alpha(j-1)}{\beta(j)-\beta(j-1)} = 1 + \frac{\alpha'(j)-\alpha'(j-1)}{\beta'(j)-\beta'(j-1)}$, $j = 1, \dots, q_0$, and $\frac{p}{q} = 1 + \frac{p-q}{q}$. This means that condition (i) is satisfied, which completes the proof.

• For $p < q$, let $B^*(G_{p,r}) = \{(\alpha'(0), \beta'(0), \dots, (\alpha'(q_0 - 1), \beta'(q_0 - 1)\}$ be as in Proposition 2.10.13. Observe that $(\alpha(j))_{j=1}^{q_0}$ is strictly increasing and hence $0 = \frac{\alpha(1)-\alpha(0)}{\beta(1)-\beta(0)} < \frac{\alpha(2)-\alpha(1)}{\beta(2)-\beta(1)}$. Moreover, $\frac{\beta(j)-\beta(j-1)}{\alpha(j)-\alpha(j-1)} = k + \frac{\beta'(j-1)-\beta'(j-2)}{\alpha'(j-1)-\alpha'(j-2)}$, $j = 2, \dots, q_0$, and $\frac{q}{p} = k + \frac{r}{p}$. This means that condition (i) is satisfied, which completes the proof. □

Corollary 2.10.17. *Let $p, q \in \mathbb{N}$ be relatively prime. Fix a $k \in \mathbb{N}$. Then, the following holds:*

$$B^*(G_{p+kq,q}) = \{(\alpha + k\beta, \beta) : (\alpha, \beta) \in B^*_{p,q}\}.$$

Proof. Repeatedly apply Theorem 2.10.16 (k times total). □

Remark 2.10.18. Bearing in mind that $B(G_{1,q}) = \{(1,0), (1,q)\}$, Theorem 2.10.16 and Corollary 2.10.17 give an $O((\log q)^2)$-algorithm for computing $B^*(G_{p,q})$ and an $O(\log q)$-algorithm for computing $\#B^*(G_{p,q})$.

Corollary 2.10.19. *Let $f_0 := 0$, $f_1 := 1$, $f_{n+2} := f_{n+1} + f_n$, $n \geq 0$. Fix a $k \in \mathbb{N}$. Then, the following hold:*

(a) $\#B^*(G_{f_{k+2}, f_{k+1}}) = \#B^*(G_{f_k, f_{k+1}}) = 1 + \lceil k/2 \rceil$.

(b) *Let $p, q \in \mathbb{N}$ be relatively prime.*

- *If $p < q$ and $\#B^*(G_{p,q}) > k$, then $p \geq f_{2k-1}$ and $q \geq f_{2k}$.*
- *If $p > q$ and $\#B^*(G_{p,q}) > k$, then $p \geq f_{2k+1}$ and $q \geq f_{2k}$.*

Proof. (a) Check the cases $k = 1, 2$ manually and then use Theorem 2.10.16 to conduct the inductive step:

$$\#B^*(G_{f_{k+4}, f_{k+3}}) = \#B^*(G_{f_{k+2}, f_{k+3}}) = 1 + \#B^*(G_{f_{k+2}, f_{k+1}}).$$

(b) Check the cases $k = 1, 2$ manually and then use Theorem 2.10.16 to conduct the inductive step. □

Remark 2.10.20. Corollary 2.10.19 allows efficient searching for (p, q) with desired $\#B^*(G_{p,q})$ within very large sets of candidate pairs – cf. Exercise 2.12.15.

2.11 Neile parabola

So far, we have discussed domains in $\mathbb{C}^n$ and the corresponding Carathéodory pseudodistance and Carathéodory–Reiffen pseudometric. But we know that there are more complicated sets (e.g., complex manifolds, connected analytic sets, or even so called complex spaces) on which we can study bounded holomorphic functions. In this section, we mostly concentrate on an elementary example, the so called Neile (Neil) parabola $\boldsymbol{N}$, and we try to find a formula of the Carathéodory–Reiffen pseudometric for $\boldsymbol{N}$. In the survey [267] we wrote: "It is a little bit surprising that, despite the elementary description of $\boldsymbol{N}$, an effective formula for $\boldsymbol{c}_{\boldsymbol{N}}$ is not known", and we asked to find such a formula. This was finally done by G. Knese in [307]. This result is the main source for this section.

Let $m < n$ be natural numbers that are relatively prime and let $1 = kn + lm$ with $k, l \in \mathbb{Z}$. Put

$$\boldsymbol{N}_{m,n} := \{(z,w) \in \mathbb{D}^2 : z_1^m = z_2^n\}.$$

Note that $\boldsymbol{N}_{m,n}$ is a connected one-dimensional analytic subset of $\mathbb{D}^2$ with the only singularity at the origin. $\boldsymbol{N}_{m,n}$ is called the *(m,n)-parabola*. Note that $\boldsymbol{N} := \boldsymbol{N}_{2,3}$ is the so-called *Neile parabola*.

Recall (cf. Appendix B.1.1) that a function $f : \boldsymbol{N}_{m,n} \longrightarrow \mathbb{C}$ is *holomorphic on* $\boldsymbol{N}_{m,n}$ if, for any point $(a,b) \in \boldsymbol{N}_{m,n}$, there are an open neighborhood U of (a,b) in $\mathbb{C}^2$ and a function $F \in \mathcal{O}(U)$ such that $F|_{U \cap \boldsymbol{N}_{m,n}} = f|_{U \cap \boldsymbol{N}_{m,n}}$. As usual, $\mathcal{O}(\boldsymbol{N}_{m,n})$ will denote the set of all functions holomorphic on $\boldsymbol{N}_{m,n}$.

The set $\boldsymbol{N}_{m,n}$ has a global bijective holomorphic parametrization

$$\mathbb{D} \ni \lambda \overset{p_{m,n}}{\longmapsto} (\lambda^n, \lambda^m) \in \boldsymbol{N}_{m,n}$$

satisfying the following properties:

- The mapping $q_{m,n} := p_{m,n}^{-1}$ is holomorphic on $\boldsymbol{N}_{m,n}^* := \boldsymbol{N}_{m,n} \setminus \{(0,0)\}$ and continuous on $\boldsymbol{N}_{m,n}$. Note that $q_{m,n}(z,w) = z^k w^l$, $(z,w) \in \boldsymbol{N}_{m,n}^*$, $q_{m,n}(0,0) = 0$.

- The mapping $q_{m,n}|_{\boldsymbol{N}_{m,n}^*} : \boldsymbol{N}_{m,n}^* \longrightarrow \mathbb{D}_*$ is biholomorphic.

  Thus,

$$\begin{aligned}\boldsymbol{c}_{\boldsymbol{N}_{m,n}^*}^*((a,b),(z,w)) &= \boldsymbol{c}_{\mathbb{D}_*}^*(q_{m,n}(a,b), q_{m,n}(z,w)) \\ &= \boldsymbol{m}(q_{m,n}(a,b), q_{m,n}(z,w)), \quad (a,b),(z,w) \in \boldsymbol{N}_{m,n}^*,\end{aligned}$$

  i.e., the Carathéodory distance for $\boldsymbol{N}_{m,n}^*$ is known at least in principle.

- For any $f \in \mathcal{O}(\boldsymbol{N}_{m,n}, \mathbb{D})$, the holomorphic function $h := f \circ p_{m,n} : \mathbb{D} \longrightarrow \mathbb{D}$ satisfies $h^{(s)}(0) = 0$ if $s \in S_{m,n} := \{s \in \mathbb{N} : s \notin \mathbb{Z}_+ m + \mathbb{Z}_+ n\}$. Conversely, if

$h \in \mathcal{O}(\mathbb{D}, \mathbb{D})$ with $h^{(s)}(0) = 0$ for all $s \in S_{m,n}$, then the function $f := h \circ q_{m,n}$ is holomorphic on $\boldsymbol{N}_{m,n}$. Indeed, it is obvious that f is holomorphic on $\boldsymbol{N}^*_{m,n}$. To see that f is also holomorphic at the origin, one has to show that f allows a holomorphic extension in a full-dimensional neighborhood of the origin. Let

$$h(\lambda) = \sum_{j \notin S_{m,n}} a_j \lambda^j, \quad \lambda \in \mathbb{D},$$

be the power series expansion of h around 0. If $j \notin S_{m,n}$, then there are $b_{1,j}, b_{2,j} \in \mathbb{Z}_+$ such that $j = b_{1,j}m + b_{2,j}n$. Thus,

$$\begin{aligned} q^j_{m,n}(z, w) &= z^{kj} w^{lj} = z^{kb_{1,j}m + kb_{2,j}n} w^{lb_{1,j}m + lb_{2,j}n} \\ &= z^{b_{2,j}} w^{b_{1,j}}, \quad (z, w) \in \boldsymbol{N}_{m,n}, \end{aligned}$$

where the equation $z^m = w^n$ has been used. Therefore,

$$h \circ q_{m,n}(z, w) = \sum_{j \notin S_{m,n}} a_j z^{b_{2,j}} w^{b_{1,j}}, \quad (z, w) \in \boldsymbol{N}_{m,n}.$$

In particular, this series converges for all pairs (λ^n, λ^m), $\lambda \in \mathbb{D}$, and therefore on the bidisc $\mathbb{D}^2$; so f is the restriction of a holomorphic function on the bidisc given by this series. Hence, $f \in \mathcal{O}(\boldsymbol{N}_{m,n}, \mathbb{D})$. Put

$$\mathcal{O}_{m,n}(\mathbb{D}) := \{h \in \mathcal{O}(\mathbb{D}, \mathbb{D}) : h^{(s)}(0) = 0, s \in S_{m,n}\}.$$

Thus, we have the bijection $\mathcal{O}(\boldsymbol{N}_{m,n}, \mathbb{D}) \ni f \longmapsto f \circ p_{m,n} \in \mathcal{O}_{m,n}(\mathbb{D})$. Hence,

$$\begin{aligned} &\boldsymbol{c}^*_{\boldsymbol{N}_{m,n}}((a, b), (z, w)) \\ &\quad = \sup\{\boldsymbol{m}(f(a, b), f(z, w)) : f \in \mathcal{O}(\boldsymbol{N}_{m,n}, \mathbb{D})\}, \\ &\quad = \sup\{\boldsymbol{m}(h(q_{m,n}(a, b)), h(q_{m,n}(z, w)) : h \in \mathcal{O}_{m,n}(\mathbb{D})\}, \ (z, w) \in \boldsymbol{N}_{m,n}, \end{aligned}$$

or

$$\begin{aligned} &\boldsymbol{c}^*_{\boldsymbol{N}_{m,n}}(p_{m,n}(\mu), p_{m,n}(\lambda)) \\ &\quad = \sup\{\boldsymbol{m}(h(\mu), h(\lambda)) : h \in \mathcal{O}_{m,n}(\mathbb{D})\}, \\ &\quad = \sup\{\boldsymbol{m}(h(\mu), h(\lambda)) : h \in \mathcal{O}_{m,n}(\mathbb{D}),\ h(0) = 0\}, \quad \mu, \lambda \in \mathbb{D}. \end{aligned}$$

It seems difficult to get a simple formula for $\boldsymbol{c}^*_{\boldsymbol{N}_{m,n}}$ for general (m, n). Therefore, from now on, we restrict our discussion to the Neile parabola for which such a simple formula has been found.

Recall that, so far, we have

$$\begin{aligned} \boldsymbol{c}^*_{\boldsymbol{N}}(p_{2,3}(\mu), p_{2,3}(\lambda)) = \sup\{&\boldsymbol{m}(h(\mu), h(\lambda)) : h \in \mathcal{O}(\mathbb{D}, \mathbb{D}), \\ &h(0) = h'(0) = 0\}, \mu, \lambda \in \mathbb{D}. \end{aligned} \tag{2.11.1}$$

An effective formula will be given in the next theorem.

Theorem 2.11.1 (cf. [307]). *Given $\mu, \lambda \in \mathbb{D}_*$, then*

$$c^*_{\boldsymbol{N}}(p_{2,3}(\mu), p_{2,3}(\lambda)) = \begin{cases} \boldsymbol{m}(\mu^2, \lambda^2), & \textit{if } |\alpha_0| \geq 1 \\ \boldsymbol{m}(\mu^2 \boldsymbol{h}_{\alpha_0}(\mu), \lambda^2 \boldsymbol{h}_{\alpha_0}(\lambda)), & \textit{if } |\alpha_0| < 1 \end{cases},$$

where $\alpha_0 = \alpha_0(\mu, \lambda) := \frac{1}{2}(1/\overline{\mu} + \mu + 1/\overline{\lambda} + \lambda)$. *Moreover,* $c^*_{\boldsymbol{N}}(p_{2,3}(0), p_{2,3}(\lambda)) = |\lambda|^2$, $\lambda \in \mathbb{D}$.

Proof. For simplicity, we will write $p = p_{2,3}$ and $q = q_{2,3}$. Fix $\mu, \lambda \in \mathbb{D}$. Then, equation (2.11.1) may be read as

$$c^*_{\boldsymbol{N}}(p(\mu), p(\lambda)) = \sup\{\boldsymbol{m}(\mu^2 \widetilde{h}(\mu), \lambda^2 \widetilde{h}(\lambda)) : \widetilde{h} \in \mathcal{O}(\mathbb{D}, \mathbb{D}) \text{ or } \widetilde{h} \equiv e^{i\theta}, \theta \in \mathbb{R}\}. \tag{2.11.2}$$

In case of $\mu = 0$ this immediately gives the formula stated in the theorem.

From now on we assume that $\mu \neq 0 \neq \lambda$.

Step 1°. *Reduction to an extremal problem on* $\overline{\mathbb{D}}$:

Recall that for points $a, b \in \mathbb{D}$ the following statement holds: $\boldsymbol{m}(a, b) \leq \boldsymbol{m}(\mu, \lambda)$ if and only if there is an $\widetilde{h} \in \mathcal{O}(\mathbb{D}, \mathbb{D})$ with $a = \widetilde{h}(\mu)$, $b = \widetilde{h}(\lambda)$. Thus, equation (2.11.2) is reduced to the following one:

$$c^*_{\boldsymbol{N}}(p(\mu), p(\lambda)) = \sup\{\boldsymbol{m}(\mu^2 a, \lambda^2 b) : a, b \in \mathbb{D}, \boldsymbol{m}(a, b) \leq \boldsymbol{m}(\mu, \lambda) \text{ or } a = b \in \mathbb{T}\}. \tag{2.11.3}$$

Note that $\mathbb{D} \ni a \longmapsto \frac{\mu^2 a - \lambda^2 b}{1 - \overline{\lambda}^2 \overline{b} \mu^2 a}$ is holomorphic; thus, by the maximum principle, one gets

$$c^*_{\boldsymbol{N}}(p(\mu), p(\lambda)) = \sup\{\boldsymbol{m}(\mu^2 a, \lambda^2 b) : a, b \in \mathbb{D}, \boldsymbol{m}(a, b) = \boldsymbol{m}(\mu, \lambda) \text{ or } a = b \in \mathbb{T}\}. \tag{2.11.4}$$

But then $a = \boldsymbol{h}_\alpha(\mu)$ and $b = \boldsymbol{h}_\alpha(\lambda)$ for some $\alpha \in \mathbb{D}$, implying that

$$c^*_{\boldsymbol{N}}(p(\mu), p(\lambda)) = \sup\{\boldsymbol{m}(\mu^2 \boldsymbol{h}_\alpha(\mu), \lambda^2 \boldsymbol{h}_\alpha(\lambda)) : \alpha \in \overline{\mathbb{D}}\}. \tag{2.11.5}$$

Now put

$$F(\alpha) := \boldsymbol{m}(\mu^2 \boldsymbol{h}_\alpha(\mu), \lambda^2 \boldsymbol{h}_\alpha(\lambda)) \in [0, 1), \quad \alpha \in \overline{\mathbb{D}}.$$

Obviously, F is continuous on $\overline{\mathbb{D}}$ and smooth outside of its zeros.

Step 2°. *Maximalize the function* F:

It suffices to prove that

(a) F has no local maximum on $\mathbb{D}$ except, possibly, at α_0;

(b) if $\alpha_0 \in \mathbb{D}$, then $F(\alpha) \leq F(\alpha_0)$ for every $\alpha \in \mathbb{T}$.

Indeed, if $\alpha_0 \notin \mathbb{D}$, then the supremum is taken on the boundary, meaning it is given by $\boldsymbol{m}(\mu^2, \lambda^2)$. If $\alpha_0 \in \mathbb{D}$, then applying (b) implies that the supremum is taken at an interior point and by (a) this has to be α_0. So it remains to verify (a) and (b).

To (a): A longer calculation leads to the following description of F:

$$F(\alpha) = \boldsymbol{m}(\mu, \lambda) \left| \frac{(\mu + \lambda)(\alpha + \mu\lambda\overline{\alpha} - \mu - \lambda) + \mu\lambda(1 - |\alpha|^2)}{(1 + \mu\overline{\lambda})(1 + \mu\overline{\lambda} - \overline{\alpha}\mu - \alpha\overline{\lambda}) - \mu\overline{\lambda}(1 - |\alpha|^2)} \right| \tag{2.11.6}$$

$$= \boldsymbol{m}(\mu, \lambda) \left| \frac{1 - (\overline{\alpha} - \overline{\alpha}_0 - \overline{\beta}_2)(\alpha - \alpha_0 + \beta_2)}{1 - (\overline{\alpha} - \overline{\alpha}_0 - \overline{\beta}_1)(\alpha - \alpha_0 + \beta_1)} \right|, \tag{2.11.7}$$

where

$$\beta_1 := \frac{1}{2}\Big(1/\overline{\mu} - \mu - 1/\overline{\lambda} + \lambda\Big), \quad \beta_2 := \frac{1}{2}\Big(1/\overline{\mu} - \mu + 1/\overline{\lambda} - \lambda\Big).$$

Using (2.11.7), it suffices to verify that the function

$$G(z) = \left(\frac{F(z + \alpha_0)}{\boldsymbol{m}(\mu, \lambda)} \right)^2 = \left| \frac{1 - (\overline{z} - \overline{\beta}_2)(z + \beta_2)}{1 - (\overline{z} - \overline{\beta}_1)(z + \beta_1)} \right|^2, \quad z \in \mathbb{D} - \alpha_0 =: D',$$

has no local maximum in D' except possibly at $z = 0$. Simple calculations show that G may be written as $G = G_2/G_1$, where

$$G_k(z) := 1 + 2|\beta_k|^2 - |z|^2 + |z^2 - \beta_k^2|^2, \quad z \in D', \ k = 1, 2.$$

Now assume that G has a local maximum at a point $z_0 = x_0 + iy_0 \in D'$, $z_0 \neq 0$. Then,

- $G(z_0) \in (0, 1)$:

  use $F(\alpha) < \boldsymbol{m}(\mu, \lambda)$ for $\alpha \in \overline{\mathbb{D}}$.

- z_0 is a critical point of G: thus,

$$G_1(z_0)\partial_z G_2(z_0) = G_2(z_0)\partial_z G_1(z_0). \tag{*}$$

  Note that $\partial_z G_j(z_0) = -2\overline{z}_0 + 2\overline{z}_0(\overline{z}_0^2 - \overline{\beta}_j^2) = (1/2)(\partial_x - i\partial_y)G_j(z_0)$, $j = 1, 2$.

  Then, we get from (*):

$$\left(\frac{\overline{\beta}_1^2}{G_1(z_0)} - \frac{\overline{\beta}_2^2}{G_2(z_0)} \right) z_0^2 = |z_0|^2(1 - |z_0|^2) \left(\frac{1}{G_2(z_0)} - \frac{1}{G_1(z_0)} \right),$$

which also shows that the left side is real. Now apply (*) to the real derivatives. The last equation then leads to

$$\begin{aligned}\partial^2_{xx}\log G(z_0) =& -4\left(\left(1-|z_0|^2\right)\left(1-\operatorname{Re}\left(z_0^2/|z_0|^2\right)\right)-2x_0^2\right)\\ &\cdot\left(\frac{1}{G_2(z_0)}-\frac{1}{G_1(z_0)}\right),\\ \partial^2_{yy}\log G(z_0) =& -4\left(\left(1-|z_0|^2\right)\left(1+\operatorname{Re}\left(z_0^2/|z_0|^2\right)\right)-2y_0^2\right)\\ &\cdot\left(\frac{1}{G_2(z_0)}-\frac{1}{G_1(z_0)}\right),\\ \partial^2_{xy}\log G(z_0) =& 4\left(2x_0y_0+\left(1-|z_0|^2\right)\operatorname{Im}\left(z_0^2/|z_0|^2\right)\right)\\ &\cdot\left(\frac{1}{G_2(z_0)}-\frac{1}{G_1(z_0)}\right).\end{aligned}$$

- $\Delta(\log G)(z_0) \leq 0$:

  Summing up the previous items, one gets:

$$\Delta\log G(z_0) = -8\left(1-3|z_0|^2\right)\left(\frac{1}{G_2(z_0)}-\frac{1}{G_1(z_0)}\right);$$

  hence, $|z_0|^2 \leq 1/3$ and so $(1-|z_0|^2) > 0$.

- $\widehat{H} := \det H(\log G)(z_0) \geq 0$, where $H(\log G)$ may denote the Hessian-matrix:

  Calculations then give $0 \leq \widehat{H} = 16\left(\frac{1}{G_2(z_0)}-\frac{1}{G_1(z_0)}\right)^2(1-|z_0|^2)(-4|z_0|^2) < 0$;

a contradiction.

To (b): Recall that $F|_{\mathbb{T}} \equiv \boldsymbol{m}(\mu^2,\lambda^2)$. We may assume that $\boldsymbol{m}(\mu^2,\lambda^2) \neq 0$. Then, using (2.11.6), it suffices to verify the following inequality:

$$\left|\frac{\mu+\lambda}{1+\overline{\mu}\lambda}\right|^2 \leq \left|\frac{(\mu+\lambda)(\alpha_0+\mu\lambda\overline{\alpha}_0-\mu-\lambda)+\mu\lambda(1-|\alpha_0|^2)}{(1+\mu\overline{\lambda})(1+\mu\overline{\lambda}-\overline{\alpha}_0\mu-\alpha_0\overline{\lambda})-\mu\overline{\lambda}(1-|\alpha_0|^2)}\right|^2,$$

or, equivalently,

$$\begin{aligned}&\left|(\alpha_0+\mu\lambda\overline{\alpha}_0-\mu-\lambda)+\mu\lambda\frac{(1-|\alpha_0|^2)}{\mu+\lambda}\right|^2-\left|(1+\mu\overline{\lambda}-\overline{\alpha}_0\mu-\alpha_0\overline{\lambda})-\mu\overline{\lambda}\frac{(1-|\alpha_0|^2)}{1+\mu\overline{\lambda}}\right|^2\\ &=|A+B|^2-|C+D|^2=|A|^2-|C|^2+2\operatorname{Re}(A\overline{B}-C\overline{D})+|B|^2-|D|^2 \geq 0.\end{aligned} \tag{2.11.8}$$

Separate calculations then lead to:

(a)
$$|A|^2 - |C|^2 = |\alpha_0 + \mu\lambda\overline{\alpha}_0 - \mu - \lambda|^2 - |1 + \mu\overline{\lambda} - \overline{\alpha}_0\mu - \alpha_0\overline{\lambda}|^2$$
$$= -(1 - |\alpha_0|^2)(1 - |\mu|^2)(1 - |\lambda|^2);$$

(b) $2\,\mathrm{Re}(A\overline{B} - C\overline{D}) = (1 - |\alpha_0|^2)(1 - |\mu|^2)(1 - |\lambda|^2)$;

(c) $|B|^2 - |D|^2 = |\mu\lambda|^2(1 - |\alpha_0|^2)^2 \frac{(1-|\mu|^2)(1-|\lambda|^2)}{|\mu+\lambda|^2|1+\mu\overline{\lambda}|^2}$.

Here, the following identities were used:

- $\alpha_0 + \mu\lambda\overline{\alpha}_0 - (\mu + \lambda) = \frac{\overline{\mu}+\overline{\lambda}}{2\overline{\mu}\overline{\lambda}}(1 + |\mu\lambda|^2)$ and
- $1 + \mu\overline{\lambda} - \overline{\alpha}_0\mu - \alpha_0\overline{\lambda} = -\frac{1+\overline{\mu}\lambda}{2\overline{\mu}\lambda}(|\mu|^2 + |\lambda|^2))$ for (b),
- $|1 + a\overline{b}|^2 - |a + b|^2 = (1 - |a|^2)(1 - |b|^2)$ for (c).

Summing up (a), (b), and (c), we finally get

$$|\mu\lambda|^2(1 - |\alpha_0|^2)^2 \frac{(1 - |\mu|^2)(1 - |\lambda|^2)}{|\mu + \lambda|^2|1 + \overline{\mu}\lambda|^1},$$

which is obviously non negative. Thus, (2.11.8) is true. □

A first application of the former result allows to reformulate a sort of Nevanlinna–Pick interpolation property in terms of the Carathéodory distance for the Neile parabola.

Corollary 2.11.2. *Given pairwise distinct points $z_1, z_2, z_3 \in \mathbb{D}$ and values w_1, $w_2 \in \mathbb{D}$, then the following properties are equivalent:*

(i) *there exists an $f \in \mathcal{O}(\mathbb{D}, \mathbb{D})$ with $f(z_j) = w_j$, $j = 1, 2$, and $f'(z_3) = 0$;*

(ii) $\boldsymbol{m}(w_1, w_2) \le \boldsymbol{c}_N^*(p(\boldsymbol{h}_{z_3}(z_1)), p(\boldsymbol{h}_{z_3}(z_2)))$.

Moreover, if there is equality in (ii), then the solution in (i) is unique and is a Blaschke product of order 2 or 3.

Proof. (i) $\Longrightarrow$ (ii): Take an f as in (i). Thus, $h := f \circ \boldsymbol{h}_{-z_3} \in \mathcal{O}(\mathbb{D}, \mathbb{D})$ with $h'(0) = 0$. Then,

$$\boldsymbol{c}_N^*\Big(p(\boldsymbol{h}_{z_3}(z_1)), p(\boldsymbol{h}_{z_3}(z_2))\Big) \ge \boldsymbol{m}\Big(h(\boldsymbol{h}_{z_3}(z_1)), h(\boldsymbol{h}_{z_3}(z_2))\Big)$$
$$= \boldsymbol{m}(f(z_1), f(z_2)) = \boldsymbol{m}(w_1, w_2). \quad (2.11.9)$$

(ii) $\Longrightarrow$ (i): By Montel there is a $h \in \mathcal{O}(\mathbb{D}, \mathbb{D})$ with $h'(0) = 0$ such that

$$\boldsymbol{c}_N^*\Big(p(\boldsymbol{h}_{z_3}(z_1)), p(\boldsymbol{h}_{z_3}(z_2))\Big) = \boldsymbol{m}\Big(h(\boldsymbol{h}_{z_3}(z_1)), h(\boldsymbol{h}_{z_3}(z_2))\Big) \ge \boldsymbol{m}(w_1, w_2).$$

Therefore, one may choose $g \in \mathcal{O}(\mathbb{D}, \mathbb{D})$ such that $g \circ h \circ \boldsymbol{h}_{z_3}(z_j) = w_j$, $j = 1, 2$. It remains to set $f := g \circ h \circ \boldsymbol{h}_{z_3}$.

Finally assume that f is a solution in (i) satisfying equality in (ii). Put $\alpha_0 = \alpha_0(\boldsymbol{h}_{z_3}(z_1), \boldsymbol{h}_{z_3}(z_2))$. If $\alpha_0 \in \mathbb{D}$, then

$$\boldsymbol{m}\Big(\boldsymbol{h}_{z_3}^2(z_1)\boldsymbol{h}_{\alpha_0}(\boldsymbol{h}_{z_3}(z_1)), \boldsymbol{h}_{z_3}^2(z_2)\boldsymbol{h}_{\alpha_0}(\boldsymbol{h}_{z_3}(z_2))\Big) = \boldsymbol{m}(f(z_1), f(z_2)).$$

Thus, there exists a point $a \in \mathbb{D}$ such that $f = \boldsymbol{h}_a(\boldsymbol{h}_{z_3}^2 \cdot \boldsymbol{h}_{\alpha_0} \circ \boldsymbol{h}_{z_3})$, meaning that f is a Blaschke product of order three.

If $|\alpha_0| \geq 1$, then

$$\boldsymbol{m}(\boldsymbol{h}_{z_3}^2(z_1), \boldsymbol{h}_{z_3}^2(z_2)) = \boldsymbol{m}(f(z_1), f(z_2)).$$

Thus, one may find a $b \in \mathbb{D}$ such that $f = \boldsymbol{h}_b \circ \boldsymbol{h}_{z_3}^2$, saying that f is a Blaschke product, now of second order. □

In a next step the infinitesimal version of the Carathéodory distance will be discussed for the Neile parabola. Fix a point $a \in \boldsymbol{N}$. Recall that a vector $X \in \mathbb{C}^2$ is called a complex tangent vector of $\boldsymbol{N}$ at a, if for every open neighborhood $U = U(0) \subset \mathbb{C}^2$ and every function $F \in \mathcal{O}(U, \mathbb{C})$ satisfying $F|_{\boldsymbol{N} \cap U} = 0$ one has $\sum_{j=1}^2 \frac{\partial f}{\partial z_j}(a)X_j = 0$. We write $X \in T_a^{\mathbb{C}}(\boldsymbol{N})$. Then, it is easy to see that

$$T_a^{\mathbb{C}}(\boldsymbol{N}) = \begin{cases} (3a_1, 2a_2)\mathbb{C}, & \text{if } a \neq 0 \\ \mathbb{C}^2, & \text{if } a = 0 \end{cases}.$$

After this short preparation, we can formulate the effective formula for the Carathéodory–Reiffen pseudometric for $\boldsymbol{N}$.

Proposition 2.11.3. *For a point $a \in \boldsymbol{N}$ one has*

$$\gamma_{\boldsymbol{N}}(a; X) = \begin{cases} |X_2|, & \text{if } |X_2| \geq 2|X_1|,\ a = 0 \\ \frac{4|X_1|^2 + |X_2|^2}{4|X_1|}, & \text{if } |X_2| < 2|X_1|,\ a = 0 \\ |\lambda| \frac{2|a_2|}{1 - |a_2|^2}, & \text{if } X = \lambda(3a_1, 2a_2),\ a \neq 0 \end{cases}.$$

Proof. Let $a = 0$ and $X \in T_a^{\mathbb{C}}(\boldsymbol{N})$. Take an $f \in \mathcal{O}(\boldsymbol{N}, \mathbb{D})$ with $f(0) = 0$ and recall the correspondence between f and h via $f \circ p(\lambda) = \lambda^2 h(\lambda)$, where $h \in \mathcal{O}(\mathbb{D}, \overline{\mathbb{D}})$. Comparing power series expansion on both sides gives

$$\sum_{j=1}^{2} \frac{\partial f}{\partial z_j}(0)X_j = h'(0)X_1 + h_2(0)X_2.$$

Therefore,

$$\gamma_N(0;X) = \sup\left\{\left|\sum_{j=1}^{2} h^{(2-j)}(0)X_j\right| : h \in \mathcal{O}(\mathbb{D}, \overline{\mathbb{D}})\right\}.$$

Hence, by Exercise 1.3.1, we get

$$\sup\{|AX_1 + BX_2| : A, B \in \mathbb{C}, |A| + |B|^2 \le 1\}$$
$$= \sup\{s|X_1| + t|X_2| : s, t \in [0,1], s + t^2 \le 1\}.$$

Thus, maximizing the function $g(t) := (1-t^2)|X_1| + t|X_2|$ over $[0,1]$ immediately gives this part of the formula.

Now, assume that $a \neq 0$ and let $X = (3a_1, 2a_2)$. Take an $f \in \mathcal{O}(N, \mathbb{D})$ with $f(a) = 0$ and write, as above, $f \circ p(\lambda) = h(\lambda)$, $\lambda \in \mathbb{D}$, where $h \in \mathcal{O}(\mathbb{D}, \mathbb{D})$ with $h'(0) = 0$. Put $\theta := a_1/a_2$.Therefore,

$$\sum_{j=1}^{2} \frac{\partial f}{\partial z_j}(a)X_j = \theta h'(\theta).$$

So we end up with

$$\gamma_N(a;X) = |\theta| \sup\{|h'(\theta)| : h \in \mathcal{O}(\mathbb{D}, \mathbb{D}),\ h(\theta) = 0 = h'(0)\}.$$

Now, applying the automorphism $\boldsymbol{h}_{h(0)}$, we may even assume that $h(0) = h'(0) = 0$. Therefore,

$$\gamma_N(a;X) = |\theta| \sup\{\gamma(h(\theta); h'(\theta)) : h \in \mathcal{O}(\mathbb{D}, \mathbb{D}),\ h(0) = h'(0) = 0\}$$

or (recall that now $h(\lambda) = \lambda^2 g(\lambda)$ with $g \in \mathcal{O}(\mathbb{D}, \overline{\mathbb{D}})$)

$$\gamma_N(a;X) = |\theta| \sup\left\{\frac{|\theta^2 g'(\theta) + 2\theta g(\theta)|}{1 - |\theta|^4 |g(\theta)|^2} : g \in \mathcal{O}(\mathbb{D}, \overline{\mathbb{D}})\right\}.$$

As before, the pairs $(g'(\theta), g(\theta))$ vary over all $(A, B) \in \mathbb{C}^2$ satisfying $|A|(1-|\theta|^2) \le 1 - |B|^2$ (see Exercise 1.3.1). So it remains to find

$$\sup\left\{\frac{|\theta|^2 s + 2|\theta| t}{1 - |\theta|^4 t^2} : s, t \in \mathbb{R}_+,\ t^2 + s(1 - |\theta|^2) \le 1\right\}.$$

One may check that the maximum occurs only in case $t = 1$ or $s = 0$. Taking into account that $\theta^2 = a_2$ gives the final formula. □

Remark 2.11.4. There are also other results for (m,n)-parabolas and higher dimensional analogues (see [383] and [547]). For example, one may there find formulas for the inner Carathéodory distance and for the Carathéodory–Reiffen pseudometric. Nevertheless, most of the formulas are still not known.

?

2.12 Exercises

Exercise 2.12.1 (cf. [142, 344]). Prove that for arbitrary $z', z'' \in G \subset \mathbb{C}^n$ we have

$$c_G(z', z'') = 2p(0, (1/2)d_G(z', z'')),$$

or equivalently,

$$d_G(z', z'') = 2\tanh\frac{c_G(z', z'')}{2},$$

where $d_G(z', z'') := \sup\{|f(z') - f(z'')| : f \in \mathcal{O}(G, \mathbb{D})\}$.

Hint. If $\varphi \in \mathcal{O}(\mathbb{D}, \mathbb{D})$, then $|\varphi(-\lambda) - \varphi(\lambda)| \le 2|\lambda|, \quad \lambda \in \mathbb{D}$.

Exercise 2.12.2 (cf. [143]). Let $B = B_q(0, 1) = \{z \in \mathbb{C}^n : q(z) < 1\}$ be the unit ball with respect to a $\mathbb{C}$-norm $q : \mathbb{C}^n \longrightarrow \mathbb{R}_+$ and let d_B be as in Exercise 2.12.1. Observe that, by the Hahn–Banach theorem, we have $d_B(z', z'') \ge q(z' - z'')$, $z', z'' \in B$. Thus,

$$q(z' - z'') \le 2\tanh\frac{c_G(z', z'')}{2}, \quad z', z'' \in B. \tag{2.12.1}$$

Prove that for $z' \ne z''$ the equality holds in (2.12.1) iff there exists a $\mathbb{C}$-linear functional $L : \mathbb{C}^n \longrightarrow \mathbb{C}$ such that:

- $|L(z)| \le q(z)$, $z \in \mathbb{C}^n$,
- $L(z'') = -L(z')$,
- $m(L(z'), L(z'')) = c_B^*(z', z'')$.

Exercise 2.12.3. Let $\mathbb{H}_+ := \{\lambda \in \mathbb{C} : \operatorname{Re}\lambda > 0\}$. Prove that

$$c_{\mathbb{H}_+}^*(\lambda', \lambda'') = \left|\frac{\lambda' - \lambda''}{\lambda' + \overline{\lambda}''}\right|, \quad \lambda', \lambda'' \in \mathbb{H}_+.$$

Exercise 2.12.4 (Borel–Carathéodory lemma). Let $\varphi : \mathbb{D} \longrightarrow \mathbb{C}$ be a holomorphic mapping such that $\varphi(0) = 0$ and $\operatorname{Re}\varphi(\lambda) \le 1$, $\lambda \in \mathbb{D}$. Then,

$$|\varphi(\lambda)| \le \frac{2|\lambda|}{1 - |\lambda|}, \quad \lambda \in \mathbb{D}.$$

Exercise 2.12.5.

(a) Find a bounded balanced domain $G \subset \mathbb{C}^n$ such that $\{X \in \mathbb{C}^n : \mathfrak{h}_G(X) = 1\} \subsetneq \partial G$.

(b) Find a balanced domain $G \subset \mathbb{C}^n$ for which $\mathfrak{h}_G^{-1}(0)$ is not a linear subspace of $\mathbb{C}^n$.

(c) Find a balanced unbounded domain $G \subset \mathbb{C}^n$ with $\mathfrak{h}_G^{-1}(0) = \{0\}$.

Exercise 2.12.6. Generalize the results presented in Remarks 2.2.1, 2.2.3, 2.2.4, and Lemma 2.2.5(a) to the case where $D \subset \mathbb{R}^N$ is a starlike symmetric domain, i.e., $[-1, 1] \cdot D \subset D$.

Exercise 2.12.7. Let $h(z) := \max\{\|z\|, |z_1|/\varepsilon\}$. Calculate $\mathbb{W}h$.

Exercise 2.12.8. Let M be a $\boldsymbol{\gamma}$-hyperbolic Riemann surface such that $\mathcal{H}^\infty(M)$ does not separate points; cf. [216]. Prove via § 2.6 that there exists a $\boldsymbol{\gamma}$-hyperbolic domain of holomorphy $G \subset \mathbb{C}^3$ that is not $\boldsymbol{c}$-hyperbolic.

Exercise 2.12.9. Complete the details of the proof of the following theorem (cf. [555]):
Let $F = (F_1, F_2) : \mathbb{B}_n \longrightarrow \mathbb{B}_{m_1} \times \mathbb{B}_{m_2}$ be a $\boldsymbol{c}$-isometry, $n \le m_1, m_2$. Then, F_1 or F_2 is a $\boldsymbol{c}$-isometry; cf. Remark 2.4.1(d) *and Proposition* 2.4.2. *In particular, up to a permutation of the components of the mapping F, we have $F_1 = \Phi \circ U$, where $U : \mathbb{C}^n \longrightarrow \mathbb{C}^{m_1}$ is unitary or antiunitary and $\Phi \in \operatorname{Aut}(\mathbb{B}_{m_1})$.*

Proof. Consider the following two cases:

$$\forall_{a\in\mathbb{B}_n}\ \exists_{b\in\mathbb{B}_n} : \min\{\boldsymbol{c}_{\mathbb{B}_{m_1}}(F_1(a)), F_1(b)), \boldsymbol{c}_{\mathbb{B}_{m_2}}(F_2(a), F_2(b))\} < \boldsymbol{c}_{\mathbb{B}_n}(a, b), \quad \text{(I)}$$
$$\exists_{a\in\mathbb{B}_n}\ \forall_{b\in\mathbb{B}_n} : \boldsymbol{c}_{\mathbb{B}_{m_1}}(F_1(a)), F_1(b)) = \boldsymbol{c}_{\mathbb{B}_{m_2}}(F_2(a), F_2(b)) = \boldsymbol{c}_{\mathbb{B}_n}(a, b). \quad \text{(II)}$$

Case I. For any $a \in \mathbb{B}_n$ there exist $j \in \{1, 2\}$ and open sets $U_1, U_2 \subset \mathbb{B}_n$, $a \in U_1$, such that

$$\boldsymbol{c}_{\mathbb{B}_{m_j}}(F_j(w), F_j(z)) = \boldsymbol{c}_{\mathbb{B}_n}(w, z), \quad (w, z) \in U_1 \times U_2. \quad (*)$$

First, we prove that

$$\forall_{a\in\mathbb{B}_n} : F_1 \text{ or } F_2 \text{ is holomorphic or antiholomorphic near } a. \quad (**)$$

Fix an $a \in \mathbb{B}_n$ and let j, U_1, U_2 be as in $(*)$. Suppose that $j = 1$. Taking $\Phi \circ F_1 \circ \Psi$ instead of F_1, where $\Psi \in \operatorname{Aut}(\mathbb{B}_n)$ and $\Phi \in \operatorname{Aut}(\mathbb{B}_{m_1})$ are suitable automorphisms, we may assume that $a = 0$, $F_1(0) = 0$, and $F_1((r, 0, \dots, 0)) = (r, 0, \dots, 0)$ for some $0 < r < 1$ with $(r, 0, \dots, 0) \in U_2$. Moreover, we can also assume that U_1, U_2 are disjoint balls. Define

$$u_1 := re_1,$$
$$u_k := re_1 + \varepsilon e_k, \quad k = 2, \dots, n,$$
$$v_k := re_1 + i\varepsilon e_k, \quad k = 1, \dots, n,$$

where ε is so small that $u_k, v_k \in U_2$, $k = 1, \dots, n$ (and where $e_1, \dots, e_n$ denote the vectors of the canonical basis in $\mathbb{C}^n$). Then,

$$F_1(tu_k) = tF_1(u_k) \text{ for all } t \in \mathbb{R} \text{ such that } tu_k \in U_1 \cup U_2, \quad k = 1, \dots, n.$$

Consequently, using induction over k, one can prove that there exists a unitary transformation $T : \mathbb{C}^{m_1} \longrightarrow \mathbb{C}^{m_1}$ such that

$$T \circ F_1(u_k) = u_k, \quad k = 1, \dots, n;$$

as usual, we identify $\mathbb{C}^n$ with $\mathbb{C}^n \times \{0\} \subset \mathbb{C}^{m_1}$. Clearly, formula $(*)$ remains true for $T \circ F_1$. Applying $(*)$ to $(w, z) = (tu_\mu, v_\nu) \in U_1 \times U_2$, we get

$$T \circ F_1(v_k) \in \{v_k, \overline{v}_k\}, \quad k = 1, \dots, n.$$

Now, taking $(w, z) = (tu_k, z) \in U_1 \times U_2$ and $(w, z) = (tv_k, z) \in U_1 \times U_2$, we prove that

$$T \circ F_1(z) \in \{z, \overline{z}\}, \quad z \in U_2.$$

Hence, by the $\boldsymbol{c}$-contractibility of $T \circ F_1$ we get

$$\text{either } T \circ F_1(z) = z,\ z \in U_2, \quad \text{or} \quad T \circ F_1(z) = \overline{z},\ z \in U_2, \tag{$***$}$$

which completes the proof of $(**)$.

Now, we use the $\boldsymbol{c}$-isometricity of F to prove that F_1 or F_2 is holomorphic or antiholomorphic on the whole $\mathbb{B}_n$. Finally, by $(***)$, we get the required result.

Case II. We can assume that $a = 0$ and $F(0) = 0$. Consider the following two subcases:

$$\exists_{b \in \mathbb{B}_n} : \min\{\boldsymbol{c}_{\mathbb{B}_{m_1}}(F_1(b), F_1(-b)), \boldsymbol{c}_{\mathbb{B}_{m_2}}(F_2(b), F_2(-b))\} < \boldsymbol{c}_{\mathbb{B}_n}(b, -b), \tag{a}$$
$$\forall_{b \in \mathbb{B}_n} : \boldsymbol{c}_{\mathbb{B}_{m_1}}(F_1(b), F_1(-b)) = \boldsymbol{c}_{\mathbb{B}_{m_2}}(F_2(b), F_2(-b)) = \boldsymbol{c}_{\mathbb{B}_n}(b, -b). \tag{b}$$

In case (a) we may assume that

$$\boldsymbol{c}_{\mathbb{B}_{m_1}}(F_1(w), F_1(z)) = \boldsymbol{c}_{\mathbb{B}_n}(w, z), \quad (w, z) \in U_b \times U_{-b},$$

where $U_b \subset \mathbb{B}_n$ is an open ball with the center at b, $0 \notin U_b$, and $U_{-b} = -U_b$. Next, we proceed similarly as in Case I.

In case (b) we put $u_k := re_k$, $k = 1, \dots, n$, for some $r \in (0, 1)$. Similar to Case I, we have

$$T_j \circ F_j(u_k) = u_k, \quad j = 1, 2, \ k = 1, \dots, n.$$

The $\boldsymbol{c}$-contractibility of F_j gives

$$\operatorname{Re}(T_j \circ F_j(z))_k = \operatorname{Re} z_k \text{ and } |\operatorname{Im}(T_j \circ F_j(z))_k| \le |\operatorname{Im} z_k|, \quad z \in \mathbb{B}_n, \ j = 1, 2, \ k = 1, \dots, n.$$

Consequently, since the mapping $(T_1 \circ F_1, T_2 \circ F_2)$ is a $\boldsymbol{c}$-isometry, we have $T_j \circ F_j(z) \in \{z, \overline{z}\}$ for z in a certain nonempty open subset of the set

$$D := \{(x_1 + iy_1, \dots, x_n + iy_n) \in \mathbb{B}_n : x_k, y_k > 0, \ k = 1, \dots, n\},$$

and, therefore, either $T_j \circ F_j(z) = z$ or $T_j \circ F_j(z) = \overline{z}$ for z in a nonempty open subset of D, $j = 1, 2$. Proceeding similar to Case I, we complete the proof. □

Exercise 2.12.10. Prove that the Hayashi surface (cf. Lemma 2.6.6) gives another example of a connected $\boldsymbol{c}$- and $\boldsymbol{\gamma}$-hyperbolic Riemann surface with disconnected Carathéodory balls.

Exercise 2.12.11 (cf. [465, 466]). Let

$$G := \{z = (z_1, z_2) \in \mathbb{C}^2 : \|z\|_1 := |z_1| + |z_2| < 1\}$$

be the ℓ_1-unit ball in $\mathbb{C}^2$. Fix $z' = (z_1', z_2') \in G \setminus \{0\}$ and $\lambda', z_1, w_1 \in \mathbb{C}$ such that $\lambda' z' \in G$, $(z_1, z_2') \in G$, and $(w_1, z_2') \in G$.

Prove that

$$\boldsymbol{c}_G^*(z', \lambda' z') = \boldsymbol{m}(\|z'\|_1, \lambda'\|z'\|_1) \quad \text{and } \boldsymbol{c}_G^*((z_1, z_2'), (w_1, z_2')) = \boldsymbol{m}(z_1/(1 - |z_2'|), w_1/(1 - |z_2'|)).$$

Hint. Use the mappings $g \in \mathcal{O}(\mathbb{D}, G)$, $g(\lambda) := \lambda z'/\|z'\|_1$ for $\lambda \in \mathbb{D}$, and $f \in \mathcal{O}(G, \mathbb{D})$, $f(z) := z_1 e^{-i \operatorname{Arg} z_1'} + z_2 e^{-i \operatorname{Arg} z_2'}$ for $z \in G$, to obtain the first equality. Proceed with the second claim in a similar way.

Exercise 2.12.12 (cf. [466]). Let G be as in Exercise 2.12.11 and fix a $z^0 = (z_1^0, z_2^0) \in G$. Assume that there are positive numbers r and ϱ with $0 < r < 1$ and a point $z' \in G$ such that $B_{\boldsymbol{c}_G^*}(z^0, r) = B_{\| \|_1}(z', \varrho)$.

Prove that $z^0 = (0, 0)$ (i.e., the only Carathéodory balls in G that are $\| \|_1$-balls are those with the centers at the origin).

Hint. First, observe the following elementary geometrical fact:
if $|a + r_1 e^{i(\theta_1 + \varphi)}| + |b + r_2 e^{i(\theta_2 + \varphi)}| = \text{constant}$ for every $\varphi \in \mathbb{R}$ (here, $a, b \in \mathbb{C}$, $r_1, r_2 > 0$, and $\theta_1, \theta_2 \in \mathbb{R}$), then, necessarily, $a = b = 0$.

Suppose that $z_1^0 z_2^0 \neq 0$. Then, apply Exercise 2.12.11 to show that

$$\frac{\lambda(\varphi)}{\|z^0\|_1} z^0 \in \partial B_{\boldsymbol{c}_G^*}(z^0, r) = \partial B_{\| \ \|_1}(z', \varrho),$$

where

$$\lambda(\varphi) := \|z^0\|_1 \frac{1 - r^2}{1 - \|z^0\|_1^2 r^2} + r \frac{1 - \|z^0\|_1^2}{1 - \|z^0\|_1^2 r^2} e^{i\varphi}, \quad \varphi \in \mathbb{R}.$$

Use the geometrical fact mentioned above to conclude that

$$z'_j = z_j^0 \frac{1 - r^2}{1 - \|z^0\|_1^2 r^2} \quad \text{for } j = 1, 2 \text{ and } \varrho = r \frac{1 - \|z^0\|_1^2}{1 - \|z^0\|_1^2 r^2}.$$

Finally, employ Exercise 2.12.11 again to show that

$$(w_1(\varphi), z_2^0) \in \partial B_{\boldsymbol{c}_G^*}(z^0, r) = \partial B_{\| \ \|_1}(z', \varrho),$$

where

$$w_1(\varphi) := (1 - |z_2^0|) \frac{z_1^0 (1 - |z_2^0|)(1 - r^2)}{(1 - |z_2^0|)^2 - |z_1^0|^2 r^2} + (1 - |z_2^0|) r \frac{(1 - |z_2^0|)^2 - |z_1^0|^2}{(1 - |z_2^0|)^2 - |z_1^0|^2 r^2} e^{i\varphi}.$$

Then, by the geometrical fact we again have

$$(1 - |z_2^0|) \frac{z_1^0 (1 - |z_2^0|)(1 - r^2)}{(1 - |z_2^0|)^2 - |z_1^0|^2 r^2} = z_1^0 \frac{1 - r^2}{1 - \|z^0\|_1^2 r^2},$$

which is a contradiction.

The remaining case $z^0 \neq 0$, but $z_1^0 z_2^0 = 0$, can be treated in a similar way.

Exercise 2.12.13. Let G be as in (2.10.4). Prove that

$$\begin{aligned} &\sup\{|Q(z)| : Q \text{ is a homogeneous polynomial}, \|Q\|_G \le 1\} \\ &\qquad = \max\{|z_1|, |z_2|, (2/3)(|z_1| + |z_2|), 2|z_1 z_2|\} =: H_G(z), \quad z \in G. \end{aligned}$$

Let $P(z_1, z_2) := a(z_1 + z_2) - b(z_1^2 + z_2^2) + c z_1 z_2$, where $b, c > 0$, $5b < 2c < 4(b+1)$ and $a := (4 + 5b - 2c)/6$. Verify that $\|P\|_G = 1$ and $P(2/3, 2/3) > H_G(2/3, 2/3)$.

Exercise 2.12.14. Using the procedure described after Lemma 2.10.11, write a computer program to calculate $B^*(G_{p,q})$ (p, q relatively prime). Next, verify Example 2.10.12 and the following list of examples:

$$\begin{aligned} B^*(G_{89,144}) &= \{(1,0), (1,1), (2,3), (5,8), (13,21), (34,55), (89,144)\}, \\ B^*(G_{123,199}) &= \{(1,0), (1,1), (2,3), (5,8), (13,21), (34,55), (123,199)\}, \\ B^*(G_{144,199}) &= \{(1,0), (1,1), (3,4), (8,11), (21,29), (55,76), (144,199)\}. \end{aligned}$$

Note that

$$\max\{\#B^*(G_{p,q}) : 1 \le p,q \le 100,\ p,q \text{ are relatively prime}\} = 6,$$
$$\max\{\#B^*(G_{p,q}) : 1 \le p,q \le 200,\ p,q \text{ are relatively prime}\} = 7,$$
$$\#\{(p,q) : 1 \le p,q \le 200,\ p,q \text{ are relatively prime},\ \#B^*(G_{p,q}) = 7\} = 3.$$

Exercise 2.12.15. Using Corollary 2.10.19, check that

$$\max\{\#B^*(G_{p,q}) : 1 \le p,q \le 200\,000\,000,\ p,q \text{ are relatively prime }\} = 21,$$
$$\#\{(p,q) : 1 \le p,q \le 200\,000\,000,\ p,q \text{ are relatively prime},\ \#B^*(G_{p,q}) = 21\} = 79.$$

2.13 List of problems

Chapter 3

The Kobayashi pseudodistance and the Kobayashi–Royden pseudometric

Summary. Around 1966, S. Kobayashi initiated the study of the pseudodistance $\boldsymbol{k}$, which now is called the Kobayashi pseudodistance; cf. [311, 317] (see §§ 3.1, 3.3). An extended study of taut domains and their relation to the topics discussed here can be found in § 3.2. H. L. Royden published the infinitesimal form $\boldsymbol{\varkappa}$ in 1971 [459] (see § 3.5). Most of the material collected in this chapter is due to these two authors and to T. J. Barth (cf. [33, 36]). A fairly complete survey about the results up to 1976 can be found in the report of S. Kobayashi (see [314]). In 1990 S. Kobayashi introduced the pseudometric $\widehat{\boldsymbol{\varkappa}}$, which is here called the Kobayashi–Buseman pseudometric (see [315]) (see § 3.6). The chapter concludes with a discussion of higher-order Lempert functions and Kobayashi–Royden pseudometrics (see § 3.8).

Introduction. In the previous chapter, we discussed the Carathéodory pseudodistance and we observed that if $d_G : G \times G \longrightarrow \mathbb{R}_+$ is a function with $\boldsymbol{p}(f(z'), f(z'')) \le d_G(z', z'')$, $z', z'' \in G$, $f \in \mathcal{O}(G, \mathbb{D})$, then $\boldsymbol{c}_G \le d_G$ (cf. Remark 2.1.2). In this chapter, we will study the opposite case, namely, a family of pseudodistances $(\boldsymbol{k}_G)_G$ such that if d_G is a pseudodistance on G satisfying $d_G(f(\lambda'), f(\lambda'')) \le \boldsymbol{p}(\lambda', \lambda'')$, $\lambda', \lambda'' \in \mathbb{D}$, $f \in \mathcal{O}(\mathbb{D}, G)$, then $d_G \le \boldsymbol{k}_G$. Similar investigations will be presented on the level of pseudometrics. The main tool will be the space $\mathcal{O}(\mathbb{D}, G)$ of "analytic discs" in G "antipodal" to the space $\mathcal{O}(G, \mathbb{D})$ that was the basis for our studies in Chapter 2.

3.1 The Lempert function and the Kobayashi pseudodistance

First, we introduce a family $(\boldsymbol{\ell}_G)_G$ of functions $\boldsymbol{\ell}_G$ from which $\boldsymbol{k}_G$ will be derived as the largest pseudodistance below $\boldsymbol{\ell}_G$.

Before presenting the formal definitions, we make the following observation:

Remark 3.1.1.

(a) Let G be any domain in $\mathbb{C}^n$ and fix two points z_0', z_0'' in G. Then, there exists a curve $\alpha : [0, 1] \longrightarrow G$ connecting the points z_0', z_0''. Using the Weierstrass approximation theorem, we find a polynomial map $P : [0, 1] \longrightarrow G$ with $P(0) = z_0'$ and $P(1) = z_0''$. Then, it is easy to choose a simply connected domain $D \subset \mathbb{C}$, $[0, 1] \subset D$, such that $P(\lambda) \in G$ for $\lambda \in D$. By the Riemann mapping theorem, we can conclude that z_0', z_0'' lie on an analytic disc $\varphi : \mathbb{D} \longrightarrow G$ with $\varphi(0) = z_0'$ and $\varphi(\sigma) = z_0''$ $(0 \le \sigma < 1)$.

(b) The same result remains true in case of complex manifolds – cf. [531] (according to [137], p. 49, this fact was known by J. Globevnik even earlier).

Let $z', z'' \in G$. We put

$$\begin{aligned}\boldsymbol{\ell}_G(z',z'') &:= \inf\{\boldsymbol{p}(\lambda',\lambda'') : \lambda',\lambda'' \in \mathbb{D} : \exists_{\varphi\in\mathcal{O}(\mathbb{D},G)} : \varphi(\lambda') = z',\ \varphi(\lambda'') = z''\}\\ &= \inf\{\boldsymbol{p}(0,\lambda'') : \lambda'' \in \mathbb{D} : \exists_{\varphi\in\mathcal{O}(\overline{\mathbb{D}},G)} : \varphi(0) = z',\ \varphi(\lambda'') = z''\},\\ \boldsymbol{\ell}_G^* &:= \tanh \boldsymbol{\ell}_G,\end{aligned}$$

and we call $\boldsymbol{\ell}_G$ the *Lempert function for* G.

Remark 3.1.2. Observe that

(a) $\boldsymbol{\ell}_G : G \times G \longrightarrow [0, +\infty)$ is a symmetric function;

(b) $(\boldsymbol{\ell}_G)_G$ is a contractible family of functions with respect to holomorphic mappings, i.e., if $F \in \mathcal{O}(G, D)$, then $\boldsymbol{\ell}_D(F(z'), F(z'')) \le \boldsymbol{\ell}_G(z', z'')$, $z', z'' \in G$;

(c) in particular, we have $\boldsymbol{\ell}_D(F(z'), F(z'')) = \boldsymbol{\ell}_G(z', z'')$ whenever $F : G \longrightarrow D$ is a biholomorphic map;

(d) $\boldsymbol{\ell}_{\mathbb{D}} = \boldsymbol{p}$ (use the Schwarz–Pick lemma);

(e) $\boldsymbol{c}_G \le \boldsymbol{\ell}_G$.

Let us summarize what we have found so far: if there is a family $(d_G)_G$ of functions $d_G : G \times G \longrightarrow [0, +\infty)$ contractible under holomorphic mappings with $d_{\mathbb{D}} = \boldsymbol{p}$, then

$$d_G \le \boldsymbol{\ell}_G.$$

But here, in contrast to the situation of the previous chapter, it turns out that, in general, the Lempert function is not a pseudodistance.

Example 3.1.3 (cf. [340]). For $\nu \in \mathbb{N}$, we define domains $G_\nu \subset \mathbb{C}^2$ by

$$G_\nu := \{(z_1, z_2) \in \mathbb{C}^2 : |z_j| < 1,\ j = 1, 2,\ |z_1 z_2| < 1/\nu\}.$$

Moreover, fix $z' := (1/2, 0)$ and $z'' := (0, 1/2)$ and observe that these points are contained in all of the G_ν's. By the holomorphic contractibility of the system $(\boldsymbol{\ell}_G)_G$, we easily obtain

$$\boldsymbol{\ell}_{G_\nu}(z', 0) + \boldsymbol{\ell}_{G_\nu}(0, z'') \le 2\boldsymbol{p}(0, 1/2) =: A.$$

If we assumed the triangle inequality

$$\boldsymbol{\ell}_{G_\nu}(z', z'') \le \boldsymbol{\ell}_{G_\nu}(z', 0) + \boldsymbol{\ell}_{G_\nu}(0, z''),$$

then it is possible to find holomorphic maps $\varphi_\nu \in \mathcal{O}(\mathbb{D}, G_\nu)$ such that

$$\varphi_\nu(0) = z', \ \varphi_\nu(\sigma_\nu) = z'' \text{ with } 0 < \sigma_\nu < 1, \text{ and } \boldsymbol{p}(0, \sigma_\nu) < A + 1.$$

Applying Montel's theorem, we may assume that

$$\varphi_\nu \underset{\nu\to\infty}{\overset{K}{\Longrightarrow}} \varphi \in \mathcal{O}(\mathbb{D}, \mathbb{C}^2), \ \varphi(0) = z', \ \varphi(\sigma) = z'' \text{ with } \sigma = \lim_{\nu\to\infty} \sigma_\nu \in [0, 1).$$

Write $\varphi = (g, h)$, and since $\varphi_\nu \in \mathcal{O}(\mathbb{D}, G_\nu)$, it follows that

$$g \cdot h \equiv 0, \quad g(0) = 1/2, \quad \text{and} \quad h(\sigma) = 1/2,$$

which obviously contradicts the identity theorem.

Hence, for sufficiently large ν the triangle inequality does not hold for the Lempert function $\boldsymbol{\ell}_{G_\nu}$.

Remark 3.1.4. Recall that we still do not know any complete formula of the Carathéodory distance for the domains G_ν (see Remark 2.10.8 for the case $\nu = 2$). Notice that recently some effective formulas for $\boldsymbol{\ell}_{G_\nu}(z_1, z_2)$ have been found in [306].

Example 3.1.5 (cf. [434]). Let $G := \{z \in \mathbb{C}^2 : |z_1 z_2| < 1 \text{ and } |z_2| < R\}$ with $R > 1$. Then, the following inequality is true:

$$\boldsymbol{\ell}_G((1,0),(0,1)) > \boldsymbol{\ell}_G((1,0),0) + \boldsymbol{\ell}_G(0,(0,1)).$$

The details are left to the reader (cf. Exercise 3.9.1).

Remark 3.1.6. Already here, we draw the attention of the reader to the fact that in the case of a convex domain $G \subset \mathbb{C}^n$, the triangle inequality for the Lempert function $\boldsymbol{\ell}_G$ is always valid. This is the consequence of a very deep result of L. Lempert [340, 341, 342], and H. L. Royden & P. M. Wong [460] (cf. Chapter 11), namely, that $\boldsymbol{\ell}_G = \boldsymbol{c}_G$ for such domains. There also exists a direct proof due to L. Lempert [340]; see Exercise 3.9.10.

To overcome the difficulty connected with the triangle inequality we modify the function $\boldsymbol{\ell}_G$ in such a way that the new function becomes a pseudodistance.

For $z', z'' \in G$ we put

$$\boldsymbol{k}_G(z', z'') := \inf\left\{\sum_{j=1}^{N} \boldsymbol{\ell}_G(z_{j-1}, z_j) : N \in \mathbb{N}, z_0 = z', \ z_1, \dots, z_{N-1} \in G, \ z_N = z''\right\},$$

$\boldsymbol{k}_G^* := \tanh \boldsymbol{k}_G$. The function $\boldsymbol{k}_G$ is called the *Kobayashi pseudodistance for G*.

Remark 3.1.7. Notice that the following properties hold for the system $(\boldsymbol{k}_G)_G$:

(a) $\boldsymbol{k}_G$ is the largest minorant of $\boldsymbol{\ell}_G$ that satisfies the triangle inequality;

(b) $\boldsymbol{k}_G$ is a pseudodistance on G;

(c) if $F \in \mathcal{O}(G, D)$, then $\boldsymbol{k}_D(F(z'), F(z'')) \le \boldsymbol{k}_G(z', z'')$, i.e., the system $(\boldsymbol{k}_G)_G$ is contractible with respect to holomorphic mappings;

(d) $\boldsymbol{k}_{\mathbb{D}} = \boldsymbol{\ell}_{\mathbb{D}} = \boldsymbol{p}$. Even more, we have:

(e) if $(d_G)_G$ is a system of pseudodistances $d_G : G \times G \longrightarrow [0, +\infty)$ with the properties stated in (c) and (d), then $d_G \le \boldsymbol{k}_G$;

(f) in particular, $\boldsymbol{c}_G \le \boldsymbol{k}_G$.

To be able to continue the discussion on the Lempert function and the Kobayashi pseudodistance, we need at least a few examples for which these objects can be calculated.

Example 3.1.8. Let q be a seminorm on $\mathbb{C}^n$. Denote by $G := \{z \in \mathbb{C}^n : q(z) < 1\}$ the associated open unit q-ball. For $z \in G$, we claim that the following formulas are true:

$$\boldsymbol{k}_G(0, z) = \boldsymbol{\ell}_G(0, z) = \boldsymbol{p}(0, q(z)).$$

To prove these equalities, we remind the reader of the similar formula for the Carathéodory pseudodistance. Thus, it suffices to verify that $\boldsymbol{\ell}_G(0, z) \le \boldsymbol{p}(0, q(z))$. In the case $q(z) \ne 0$, defining $\varphi(\lambda) := \lambda z / q(z)$, we obtain an analytic disc $\varphi \in \mathcal{O}(\mathbb{D}, G)$ with $\varphi(0) = 0$ and $\varphi(q(z)) = z$, which implies $\boldsymbol{\ell}_G(0, z) \le \boldsymbol{p}(0, q(z))$. In the case $q(z) = 0$, we consider the family of analytic discs $\varphi_t \in \mathcal{O}(\mathbb{D}, G)$, $\varphi_t(\lambda) := \lambda t z$ for $t > 1$ with $\varphi_t(0) = 0$ and $\varphi_t(1/t) = z$. Hence, we find that $\boldsymbol{\ell}_G(0, z) \le \boldsymbol{p}(0, 1/t) \underset{t \to \infty}{\longrightarrow} 0$.

In particular, we mention the following special cases.

Example 3.1.9. (a) $\boldsymbol{k}_{\mathbb{D}^n}(0, z) = \boldsymbol{\ell}_{\mathbb{D}^n}(0, z) = \max\{\boldsymbol{p}(0, |z_j|) : 1 \le j \le n\}$,
(b) $\boldsymbol{k}_{\mathbb{B}_n}(0, z) = \boldsymbol{\ell}_{\mathbb{B}_n}(0, z) = \boldsymbol{p}(0, \|z\|)$.

As a consequence of this example we obtain

Proposition 3.1.10. *The function* $\boldsymbol{k}_G : G \times G \longrightarrow [0, +\infty)$ *is continuous.*

Proof. In view of the triangle inequality, it suffices to prove that $\boldsymbol{k}_G(z_0, \cdot) : G \longrightarrow \mathbb{R}_+$ is continuous for fixed $z_0 \in G$. To see this, we take $w_0 \in G$ and $w \in \mathbb{B}(w_0, R) \subset G$

and we estimate

$$|\boldsymbol{k}_G(z_0, w) - \boldsymbol{k}_G(z_0, w_0)| \le \boldsymbol{k}_G(w_0, w) \le \boldsymbol{k}_{\mathbb{B}(w_0,R)}(w_0, w)$$
$$= \boldsymbol{p}\left(0, \frac{\|w - w_0\|}{R}\right) \le C\|w - w_0\| \quad \text{if} \quad \|w - w_0\| < \frac{R}{2}.$$

Here, we have used the contractibility of the Kobayashi pseudodistance under the embedding $\mathbb{B}(w_0, R) \longrightarrow G$ and Example 3.1.9. □

Our next aim is to present an example that shows that the Lempert function is in general not continuous.

Proposition 3.1.11. *Let $G := \{z \in \mathbb{C}^n : \mathfrak{h}(z) < 1\}$ denote a balanced pseudoconvex domain with Minkowski function $\mathfrak{h}$, i.e., $\mathfrak{h} : \mathbb{C}^n \longrightarrow [0, +\infty)$ is a psh function with $\mathfrak{h}(\lambda z) = |\lambda|\mathfrak{h}(z)$ $(\lambda \in \mathbb{C},\ z \in \mathbb{C}^n)$; cf. Appendix* B.7.6. *Then, the Lempert function of G is given by $\boldsymbol{\ell}_G(0, z) = \boldsymbol{p}(0, \mathfrak{h}(z))$, $z \in G$.*

Observe that in the case where $\mathfrak{h}$ is a seminorm, this formula was already derived in Example 3.1.8. Similarly, we also obtain here that $\boldsymbol{\ell}_G(0, \cdot) \le \boldsymbol{p}(0, \mathfrak{h}(\cdot))$ without the assumption that G is pseudoconvex.

Proof. Assume $\varphi : \mathbb{D} \longrightarrow G$ to be a holomorphic map with $\varphi(0) = 0$ and $\varphi(\sigma) = z$ $(0 \le \sigma < 1)$. Then, φ can be written in the form $\varphi(\lambda) = \lambda \cdot \widetilde{\varphi}(\lambda)$ with $\widetilde{\varphi} \in \mathcal{O}(\mathbb{D}, \mathbb{C}^n)$ and hence we have $\mathfrak{h} \circ \varphi(\lambda) = |\lambda|\mathfrak{h} \circ \widetilde{\varphi}(\lambda) < 1$. By the maximum principle for the subharmonic function $\mathfrak{h} \circ \widetilde{\varphi}$, we obtain $\mathfrak{h} \circ \widetilde{\varphi} \le 1$, hence $\mathfrak{h}(z) = \mathfrak{h} \circ \varphi(\sigma) = \sigma\mathfrak{h} \circ \widetilde{\varphi}(\sigma) \le \sigma$, and therefore $\boldsymbol{\ell}_G(0, z) \ge \boldsymbol{p}(0, \mathfrak{h}(z))$. □

Remark 3.1.12. Before we continue the discussion of the Lempert function, we want to point out that it does not seem to be known how to calculate the Kobayashi pseudodistance $\boldsymbol{k}_G(0, \cdot)$ in the situation of Proposition 3.1.11. The following example shows that for $\boldsymbol{k}_G(0, \cdot)$ there is no formula like the one in Proposition 3.1.11.

For $0 < \varepsilon < 1/4$, we put $G := \{z \in \mathbb{C}^2 : |z_1| < 1,\ |z_2| < 1,\ |z_1 z_2| < \varepsilon\}$, so that $\mathfrak{h}(z) = \max\{|z_1|, |z_2|, \sqrt{|z_1 z_2|/\varepsilon}\}$. By Proposition 3.1.11 we see that if $\varepsilon < t < \sqrt{\varepsilon}$, then

$$\boldsymbol{\ell}_G(0, (t, t)) = \boldsymbol{p}\left(0, \frac{t}{\sqrt{\varepsilon}}\right).$$

On the other hand, the definition easily implies the following inequality:

$$\boldsymbol{k}_G(0, (t, t)) \le \boldsymbol{\ell}_G(0, (t, 0)) + \boldsymbol{\ell}_G((t, 0), (t, t)) \le \boldsymbol{p}(0, t) + \boldsymbol{p}\left(0, \frac{t^2}{\varepsilon}\right)$$
$$= \frac{1}{2}\log\left(\frac{1+t}{1-t} \cdot \frac{\varepsilon + t^2}{\varepsilon - t^2}\right) < \frac{1}{2}\log\frac{\sqrt{\varepsilon} + t}{\sqrt{\varepsilon} - t} = \boldsymbol{\ell}_G(0, (t, t)) \qquad \text{if } t \text{ is near } \varepsilon.$$

Observe that this is yet another example showing that the Lempert function does not satisfy the triangle inequality.

Because of the existence of many bounded balanced pseudoconvex domains whose Minkowski functions are not continuous, Proposition 3.1.11 shows that the Lempert function is not continuous (in general), not even as a function of one variable. Nevertheless, $\boldsymbol{\ell}_G$ is always upper semicontinuous (cf. Proposition 3.1.14).

To give the reader some feeling of how bad the behavior of $\boldsymbol{\ell}_G$ may be, we present the following example of a balanced domain, due to J. Siciak (cf. [477]).

Example 3.1.13. We claim that if $n \geq 2$, then there exists a psh function $\psi : \mathbb{C}^n \longrightarrow [0, +\infty)$ with $\psi(\lambda z) = |\lambda|\psi(z)$ ($\lambda \in \mathbb{C},\ z \in \mathbb{C}^n$) such that $\psi \not\equiv 0$ but $\psi = 0$ on a dense subset of $\mathbb{C}^n$.

If for a while we assume the existence of such a ψ, then $\mathfrak{h}(z) := \psi(z) + \|z\|$ defines a bounded balanced pseudoconvex domain $G = G_{\mathfrak{h}} := \{z \in \mathbb{C}^n : \mathfrak{h}(z) < 1\}$ whose Lempert function $\boldsymbol{\ell}_G(0, \cdot) = \boldsymbol{p}(0, \mathfrak{h}(\cdot))$ behaves very irregularly.

What remains is the construction of the function ψ.

We write $\mathbb{Q}^{2n-2} \cap \mathbb{C}^{n-1} = \{r_j : j \in \mathbb{N}\}$ and we define the linear functionals $l_j : \mathbb{C}^n \longrightarrow \mathbb{C}$ by $l_j(z) := \langle z, (1, r_j)\rangle$. Denote by $L_j := \ker l_j$ and set $L := \bigcup_{j=1}^{\infty} L_j$. We then construct a sequence of psh functions by

$$\psi_j(z) := \left(\frac{|l_1(z) \ldots l_j(z)|}{\sup\{|l_1(w) \ldots l_j(w)| : \|w\| \leq 1\}}\right)^{1/j}.$$

Observe that

$$\psi_j \geq 0, \quad \psi_j(\lambda z) = |\lambda|\psi_j(z), \quad \psi_j|_{\bigcup_{\nu=1}^{j} L_\nu} \equiv 0.$$

Moreover, by the maximum principle, there are points z_j, $\|z_j\| = 1$, such that $\psi_j(z_j) = 1$.

By the Hartogs lemma for psh functions (cf. Appendix B.4.21), it turns out that there is a point z^*, $\|z^*\| < 2$, with $\limsup_{j\to\infty} \psi_j(z^*) \geq 2/3$. So ,taking an appropriate subsequence $(\psi_{j_\nu})_\nu \subset (\psi_j)_j$ with $\psi_{j_\nu}(z^*) \geq 1/2$, and defining

$$\psi(z) := \prod_{\nu=1}^{\infty} (\psi_{j_\nu}(z))^{2^{-\nu}},$$

we obtain a psh function on $\mathbb{C}^n$ with

$$\psi|_L \equiv 0, \quad \psi(z^*) \geq \frac{1}{2}, \quad \text{and} \quad \psi(\lambda z) = |\lambda|\psi(z)\ (\lambda \in \mathbb{C},\ z \in \mathbb{C}^n).$$

Proposition 3.1.14. *The Lempert function $\boldsymbol{\ell}_G$ is upper semicontinuous.*

Proof. Fix two points $z_0, w_0 \in G$. Then, for a positive number ε, we can choose an analytic disc $\varphi \in \mathcal{O}(\overline{\mathbb{D}}, G)$ with

$$\varphi(0) = z_0, \ \varphi(\sigma) = w_0 \ (0 \le \sigma < 1), \quad \text{and} \quad \boldsymbol{p}(0,\sigma) < \boldsymbol{\ell}_G(z_0, w_0) + \varepsilon.$$

The compact subset $\varphi(\overline{\mathbb{D}})$ has a positive boundary distance, say η.

If we take two points $z, w \in G$ with $\|z - z_0\| < \eta\sigma/6$ and $\|w - w_0\| < \eta\sigma/2$, we are able to define a holomorphic map $h : \mathbb{D} \to \mathbb{C}^n$ by

$$h(\lambda) := \varphi(\lambda) + [(z - z_0)(\sigma - \lambda) + \lambda(w - w_0)] \cdot \frac{1}{\sigma}.$$

Simple estimates yield $h \in \mathcal{O}(\mathbb{D}, G)$ with $h(0) = z$ and $h(\sigma) = w$, which implies that

$$\boldsymbol{\ell}_G(z, w) \le \boldsymbol{p}(0,\sigma) < \boldsymbol{\ell}_G(z_0, w_0) + \varepsilon.$$

In the remaining case where $\sigma = 0$, i.e., $z_0 = w_0$, we choose $R > 0$, $\alpha > 1$ such that $\mathbb{B}(z_0, R) \subset\subset G$ and $\boldsymbol{p}(0, 1/\alpha) < \varepsilon$. Now, if $\|z - z_0\| < R/(4\alpha)$ and $\|w - w_0\| < R/(4\alpha)$, we get a holomorphic map $\psi \in \mathcal{O}(\mathbb{D}, G)$ defined by $\psi(\lambda) := z + \lambda\alpha(w - z)$ with $\psi(0) = z$, $\psi(1/\alpha) = w$; thus, we obtain $\boldsymbol{\ell}_G(z, w) \le \boldsymbol{p}(0, 1/\alpha) < \varepsilon$.

Hence, the upper semicontinuity of $\boldsymbol{\ell}_G$ has been established. □

3.2 Tautness

Although, in general, the Lempert function is not continuous, there is a sufficiently rich family of pseudoconvex domains whose Lempert functions are continuous.

Let Ω be an open set in $\mathbb{C}^n$. Then, Ω is called *taut* if the space $\mathcal{O}(\mathbb{D}, \Omega)$ is normal, i.e., whenever we start with a sequence $(f_j)_{j\in\mathbb{N}} \subset \mathcal{O}(\mathbb{D}, \Omega)$, there exists a subsequence (f_{j_ν}) with $f_{j_\nu} \overset{K}{\underset{\nu\to\infty}{\Longrightarrow}} f \in \mathcal{O}(\mathbb{D}, \Omega)$ or there exists a subsequence (f_{j_ν}) that diverges uniformly on compact sets, i.e., for any two compact sets $K \subset \mathbb{D}$, $L \subset \Omega$ there is an index ν_0 such that $f_{j_\nu}(K) \cap L = \varnothing$ if $\nu \ge \nu_0$. Note that Ω is taut if and only if every connected component of Ω is taut.

Ω is called *locally taut* if any boundary point of Ω has an open neighborhood U such that $\Omega \cap U$ is taut.

The notion of taut domains was introduced by H. Wu (cf. [534]). It can be reformulated in terms similar to the Kobayashi pseudodistance (cf. [459]).

For a natural number m we introduce

$$\begin{aligned}
\boldsymbol{k}_G^{(m)}(z', z'') &:= \inf\Big\{\sum_{j=1}^{m} \boldsymbol{\ell}_G(z_{j-1}, z_j) : z' = z_0,\ z_1 \in G, \dots, z_{m-1} \in G,\ z_m = z''\Big\} \\
&= \inf\Big\{\sum_{j=1}^{m} p(0, \sigma_j) : \sigma_j \in \mathbb{D} : \exists_{\varphi_j \in \mathcal{O}(\mathbb{D}, G)} \text{ with } \varphi_1(0) = z', \\
&\qquad \varphi_j(\sigma_j) = \varphi_{j+1}(0),\ 1 \le j < m,\ \varphi_m(\sigma_m) = z''\Big\}, \quad z', z'' \in G.
\end{aligned}$$

Observe that

$$\ell_G = k_G^{(1)} \geq k_G^{(m)} \geq k_G^{(m+1)} \geq \lim_{l\to\infty} k_G^{(l)} = k_G.$$

Define

$$K(G) := \inf\{m \in \mathbb{N} : k_G^{(m)} \equiv k_G\} \in \mathbb{N} \cup \{+\infty\}.$$

We know that, in general, $K(G) = +\infty$ (cf. Exercise 3.9.1). It seems to be an open problem to find a large class of domains G (e.g., strongly pseudoconvex) for which $K(G) < +\infty$. ?

Moreover, it is not known whether there exist G and $a, b \in G$ such that $k_G^{(m)}(a, b)$ is strictly decreasing. ?

With the above notation in mind, we have the following characterization of taut domains:

Proposition 3.2.1. *The following statements are equivalent:*

(i) *G is a taut domain;*

(ii) *for any $m \in \mathbb{N}$, $R > 0$, and $z_0 \in G$ the set $\{z \in G : k_G^{(m)}(z_0, z) < R\}$ is a relatively compact subset of G;*

(iii) *for any $R > 0$ and $z_0 \in G$ the set $\{z \in G : k_G^{(2)}(z_0, z) < R\}$ is a relatively compact subset of G.*

Remark 3.2.2.

(a) In particular, the proposition says that if all k_G-balls (with finite radii) are relatively compact subsets of G, then G is a taut domain.

(b) The converse statement to (a) is false, as we will see in Chapter 13.

Proof of Proposition 3.2.1. (i) $\Longrightarrow$ (ii). Suppose (ii) is not true, i.e., we can find $m \in \mathbb{N}$, $R > 0$, $z_0 \in G$, and a sequence $(z_\nu)_{\nu\in\mathbb{N}}$ of points $z_\nu \in G$ with the following properties:

$$k_G^{(m)}(z_0, z_\nu) < R \quad \text{and} \quad \Big(z_\nu \underset{\nu\to\infty}{\longrightarrow} \hat{z} \in \partial G \quad \text{or} \quad z_\nu \underset{\nu\to\infty}{\longrightarrow} \infty\Big). \tag{*}$$

By definition, we can choose functions $\varphi_{\nu,j} \in \mathcal{O}(\mathbb{D}, G)$, $1 \leq j \leq m$, and points $\sigma_{\nu,j} \in \mathbb{D}$, $1 \leq j \leq m$, that share the following properties

$$\varphi_{\nu,1}(0) = z_0,\ \varphi_{\nu,j}(\sigma_{\nu,j}) = \varphi_{\nu,j+1}(0),\ 1 \leq j < m,$$

$$\varphi_{\nu,m}(\sigma_{\nu,m}) = z_\nu \ \text{ and } \sum_{j=1}^{m} p(0, \sigma_{\nu,j}) < R.$$

Since $\varphi_{\nu,1}(0) = z_0$, there is a subsequence $(\varphi_{1\nu,1}) \subset (\varphi_{\nu,1})$ with $\varphi_{1\nu,1} \underset{\nu\to\infty}{\overset{K}{\Longrightarrow}} \varphi_1 \in \mathcal{O}(\mathbb{D}, G)$ and $\varphi_1(0) = z_0$. We may assume that $\sigma_{1\nu,1} \longrightarrow \sigma_1 \in \mathbb{D}$, and we arrive at

$$\varphi_1(0) = z_0 \quad \text{and} \quad \varphi_1(\sigma_1) = \lim_{\nu\to\infty} \varphi_{1\nu,1}(\sigma_{1\nu,1}) = \lim_{\nu\to\infty} \varphi_{1\nu,2}(0).$$

Repeating this argument several times, we end up with subsequences $(\varphi_{m\nu,1}) \subset (\varphi_{\nu,1}), \dots, (\varphi_{m\nu,m}) \subset (\varphi_{\nu,m})$ and $(\sigma_{m\nu,1}) \subset (\sigma_{\nu,1}), \dots, (\sigma_{m\nu,m}) \subset (\sigma_{\nu,m})$ with

$$\varphi_{m\nu,j} \underset{\nu\to\infty}{\overset{K}{\Longrightarrow}} \varphi_j \in \mathcal{O}(\mathbb{D}, G), \quad \sigma_{m\nu,j} \longrightarrow \sigma_j \in \mathbb{D}$$

and

$$\varphi_1(0) = z_0, \ \varphi_j(\sigma_j) = \varphi_{j+1}(0), \ 1 \le j < m, \ \varphi_m(\sigma_m) = \lim_{\nu\to\infty} z_{m\nu},$$

which contradicts $(*)$.

What remains is to prove that (iii) is sufficient for G to be taut.

(iii) $\Longrightarrow$ (i). Now, we have to begin with a sequence $(f_j)_j \subset \mathcal{O}(\mathbb{D}, G)$. We assume that this sequence is not compactly uniformly divergent, i.e., there exist compact sets $K \subset \mathbb{D}$ and $L \subset G$ and a subsequence $(f_{1j}) \subset (f_j)$ with $f_{1j}(K) \cap L \neq \varnothing$ for all j.

Take points $\zeta_j \in K$ with $f_{1j}(\zeta_j) =: z_j \in L$. Because of the compactness, we may assume that

$$\zeta_j \longrightarrow \zeta^* \in K \quad \text{and} \quad z_j \to z^* \in L.$$

Now, for suitable $r > 0$ and $j_0 \in \mathbb{N}$, we have

$$\mathbb{B}(z^*, r) \subset G, \quad \boldsymbol{p}(\zeta_j, \zeta^*) < 1, \ z_j \in \mathbb{B}(z^*, r/4) \text{ if } j \ge j_0.$$

Defining $g_j(\lambda) := z^* + 3\lambda(z_j - z^*)$ gives a sequence of holomorphic mappings $g_j \in \mathcal{O}(\mathbb{D}, G)$ with $g_j(0) = z^*$ and $g_j(1/3) = z_j = f_{1j}(\zeta_j)$.

Therefore, for $\theta \in \mathbb{D}$ with $\boldsymbol{p}(\zeta^*, \theta) < R$ (R arbitrary) this leads to the following inequalities:

$$\begin{aligned} \boldsymbol{k}_G^{(2)}(z^*, f_{1j}(\theta)) &\le \boldsymbol{k}_G^{(1)}(g_j(0), g_j(1/3)) + \boldsymbol{k}_G^{(1)}(f_{1j}(\zeta_j), f_{1j}(\theta)) \\ &\le \boldsymbol{p}(0, 1/3) + \boldsymbol{p}(\zeta_j, \theta) \le \boldsymbol{p}(0, 1/3) + \boldsymbol{p}(\zeta^*, \zeta_j) + \boldsymbol{p}(\zeta^*, \theta) \\ &< \boldsymbol{p}(0, 1/3) + 1 + R =: \widehat{R}. \end{aligned} \tag{**}$$

By our assumption, $(**)$ implies that the sequence (f_{1j}) is locally uniformly bounded. Thus, Montel's theorem can be applied and we obtain a further subsequence $(f_{2j}) \subset (f_{1j})$ with $f_{2j} \underset{j\to\infty}{\overset{K}{\Longrightarrow}} f \in \mathcal{O}(\mathbb{D}, \mathbb{C}^n)$. Moreover, $(**)$ shows that if $\boldsymbol{p}(\zeta^*, \theta) < R$, then $f(\theta)$ belongs to the $\mathbb{C}^n$-closure of $\{w \in G : \boldsymbol{k}^{(2)}(z^*, w) < \widehat{R}\}$ which, by the assumption, is contained in G. Hence, $f \in \mathcal{O}(\mathbb{D}, G)$, which completes the proof. □

Before we continue our investigations of the Kobayashi pseudodistance, we like to make a digression about tautness. We will collect only the most important results on this subject.

We first recall the Kontinuitätssatz of Complex Analysis (cf. [327]). Applying this theorem, it is easy to see that a taut domain in $\mathbb{C}^n$ is necessarily a pseudoconvex domain. We will see that inside the class of pseudoconvex domains there is a rather large subfamily of taut domains.

Remark 3.2.3.

(a) Any bounded *hyperconvex* domain $G \subset \mathbb{C}^n$ (cf. Appendix B.7), i.e., any bounded domain G for which there exists a continuous negative plurisubharmonic exhaustion function $\varphi : G \longrightarrow (-\infty, 0)$, is a taut domain.

(b) Even more is true (cf. [295]). If G is a bounded pseudoconvex domain fulfilling the following boundary regularity:

$$\text{for any } z_0 \in \partial G \text{ there exist a unit vector } v \in \mathbb{C}^n, \text{ a neighborhood } U \text{ of } z_0, \\ \text{and a positive number } \varepsilon \text{ such that} : (U \cap \overline{G}) + (0,\varepsilon)\cdot v \subset G,$$

then G is a taut domain.

Observe that any bounded domain with a $\mathcal{C}^1$-boundary satisfies this boundary regularity. Hence, any bounded pseudoconvex domain with a $\mathcal{C}^1$-boundary is taut. We point out that these domains are also hyperconvex (cf. [295]), so that (a) yields tautness too. For a domain with a Lipschitz boundary, a similar result may be found in [122].

(c) In the case of a balanced pseudoconvex domain $G := \{z \in \mathbb{C}^n : \mathfrak{h}(z) < 1\}$ with Minkowski function $\mathfrak{h}$, there is even a complete characterization of tautness (cf. [36]), namely:

G is taut iff $\mathfrak{h}$ is continuous with $\mathfrak{h} \geq C\|\ \|$ for a suitable $C > 0$.

In particular, any bounded complete Reinhardt domain of holomorphy is taut.

(d) (cf. [32]) Let $\Pi : D \longrightarrow G$ be a holomorphic covering. Then, D is taut iff G is taut.

Indeed, first assume that D is taut and let $(f_k)_{k=1}^{\infty} \subset \mathcal{O}(\mathbb{D}, G)$. Suppose that $(f_k)_{k=1}^{\infty}$ is not divergent locally uniformly. Then, there exist compact sets $K \subset\subset \mathbb{D}$, $L \subset\subset G$, and a subsequence $(f_{k_s})_{s=1}^{\infty}$ such that $f_{k_s}(K) \cap L \neq \varnothing$, $s \in \mathbb{N}$. To simplify notation, suppose that $f_k(K) \cap L \neq \varnothing$, $k \in \mathbb{N}$. Let $\lambda_k \in K$ be such that $f_k(\lambda_k) \in L$, $k \in \mathbb{N}$. We may assume that $\lambda_k \longrightarrow \lambda_0 \in K$ and $f_k(\lambda_k) \longrightarrow a \in L$. Fix a point $\widetilde{a} \in \Pi^{-1}(a)$ and a sequence $\Pi^{-1}(f_k(\lambda_k)) \ni \widetilde{a}_k \longrightarrow \widetilde{a}$. Let $\widetilde{f}_k \in \mathcal{O}(\mathbb{D}, D)$ be the holomorphic lifting of f_k with $\widetilde{f}_k(\lambda_k) = \widetilde{a}_k$, $k \in \mathbb{N}$. Since D is taut and the sequence $(\widetilde{a}_k)_{k=1}^{\infty}$ is convergent, we may assume that

$\widetilde{f}_k \longrightarrow \widetilde{f}$ locally uniformly in $\mathbb{D}$, where $\widetilde{f} \in \mathcal{O}(\mathbb{D}, D)$ and $\widetilde{f}(\lambda_0) = \widetilde{a}$. Hence, $f_k \longrightarrow f := \Pi \circ \widetilde{f}$ locally uniformly in $\mathbb{D}$.

Now, let G be taut and let $(g_k)_{k=1}^{\infty} \subset \mathcal{O}(\mathbb{D}, D)$. Suppose that $(g_k)_{k=1}^{\infty}$ is not divergent locally uniformly. Then, as above, we may assume that there exist compact sets $K \subset\subset \mathbb{D}$, $L \subset\subset D$, and a sequence $(\lambda_k)_{k=1}^{\infty} \subset K$ such that $g_k(\lambda_k) \in L$, $k \in \mathbb{N}$, $\lambda_k \longrightarrow \lambda_0 \in K$, and $g_k(\lambda_k) \longrightarrow b \in L$. Consequently, since G is taut, we may assume that $\Pi \circ g_k \longrightarrow f \in \mathcal{O}(\mathbb{D}, G)$ locally uniformly in $\mathbb{D}$. Observe that $f(\lambda_0) = \Pi(b)$. Let $g : \mathbb{D} \longrightarrow D$ be the lifting of f with $g(\lambda_0) = b$. Let

$$\Delta := \{\lambda \in \mathbb{D} : \exists_{U \subset \mathbb{D}} : U \text{ is a neighborhood of } \lambda,\ g_k \longrightarrow g \text{ uniformly on } U\}.$$

Our problem is to show that $\Delta = \mathbb{D}$. Observe that if $g_k(\lambda^*) \longrightarrow g(\lambda^*)$ for some $\lambda^* \in \mathbb{D}$, then $\lambda^* \in \Delta$. In particular, $\lambda_0 \in \Delta$. Hence, Δ is non-empty, open, and closed in $\mathbb{D}$, which finishes the proof.

(e) The above property and Corollary 1.2.9 immediately imply the following result: *a domain $G \subset \mathbb{C}$ is taut iff $\#(\mathbb{C} \setminus G) \geq 2$ iff there exists a holomorphic covering $\Pi : \mathbb{D} \longrightarrow G$.*

The property of being taut is even a local one. Namely, there is the following result due to H. Gaussier ([189], see also [380]): let $G \subset \mathbb{C}^n$ be an unbounded domain: Let $U := \mathbb{C}^n \setminus \overline{\mathbb{B}}(s)$. A function $u \in \mathcal{PSH}(U)$ is called a *local psh peak function at* ∞ if $u \in \mathcal{C}(U \cap \overline{G})$, $u < 0$ on $U \cap \overline{G}$, and $\lim_{G \ni z \to \infty} u(z) = 0$. Moreover, a u is called a *local psh antipeak function at* ∞ if $u \in \mathcal{C}(U \cap \overline{G})$, $0 > u > -\infty$ on $U \cap \overline{G}$, and $\lim_{G \ni z \to \infty} u(z) = -\infty$.

Lemma 3.2.4. *Let $G \subset \mathbb{C}^n$ be an unbounded domain that allows a local psh peak function and a local psh antipeak function at ∞. Then, for any $R > 0$ there exists an $R' > R$ such that every analytic disc $f \in \mathcal{O}(\mathbb{D}, G)$ with $|f(0)| > R'$ fulfills $|f(\lambda)| > R$, $|\lambda| < 1/2$.*

Proof. Take a local psh peak function $\varphi \in \mathcal{C}(\overline{G} \setminus \mathbb{B}(s)) \cap \mathcal{PSH}(G \setminus \overline{\mathbb{B}}(s))$ and a local psh antipeak function $\psi \in \mathcal{C}(\overline{G} \setminus \mathbb{B}(s)) \cap \mathcal{PSH}(G \setminus \overline{\mathbb{B}}(s))$. Note that there exists a positive c such that $\varphi(z) < -c$, if $z \in \overline{G}$ and $s \leq \|z\| \leq s+1$. By the property of a local peak function, one can choose an $s' > s+1$ such that $\varphi(z) > -c/2$ when $z \in \overline{G} \setminus \mathbb{B}(s')$. Using Appendix B.4.18, it is easy to see that there is a negative $\widetilde{\varphi} \in \mathcal{C}(\overline{G}) \cap \mathcal{PSH}(G)$ with $\varphi = \widetilde{\varphi}$ on $\overline{G} \setminus \mathbb{B}(s')$. Moreover, one may find a positive ε such that $\varphi(z) + \varepsilon\psi(z) > -3c/4$ whenever $z \in \overline{G}$ and $s' \leq \|z\| \leq s'+1$. Note that $\varphi(z) + \varepsilon\psi(z) < -c$ for $z \in \overline{G}$, $s \leq \|z\| \leq s+1$. As above, one constructs a negative function $\varrho \in \mathcal{C}(\overline{G}) \cap \mathcal{PSH}(G)$ with $\varrho = \varphi + \varepsilon\psi$ on $\overline{G} \setminus \mathbb{B}(s')$. Recall that $\varrho(z) \underset{G \ni z \to \infty}{\longrightarrow} -\infty$.

Now fix an $R > s'+1$. Then, $\varrho \geq -L$ for $z \in \overline{G}$, $\|z\| \leq R$. Then, one finds a negative α such that, if $z \in G$, $\widetilde{\varphi}(z) \geq 2\alpha$, then $\varrho(z) < -L$; in particular, $\|z\| > R$.

In virtue of the properties of φ and ϱ, there exists a negative α such that for every $z \in \overline{G}$, $\widetilde{\varphi}(z) \geq 2\alpha$ one has $\varrho(z) < -L$. Using the fact that $\widetilde{\varphi}(z) = \varphi(z) \underset{\|z\|\to\infty}{\longrightarrow} 0$ one can choose an $R' > R$ such that if $z \in G \setminus \overline{\mathbb{B}}(R')$, then $\widetilde{\varphi}(z) \geq \alpha$.

Now, take an analytic disc $f \in \mathcal{O}(\mathbb{D}, G)$ with $|f(0)| > R'$. Then, $u := \widetilde{\varphi} \circ f \in \mathcal{SH}(\mathbb{D})$. For an $r \in (1/2, 1)$ put

$$E_{\alpha,r} := \{t \in [0, 2\pi] : u(re^{it}) \geq 2\alpha\}.$$

Then,

$$\alpha \leq u(0) \leq 1/(2\pi) \int_0^{2\pi} u(re^{it})dt \leq \frac{2\alpha}{2\pi}(2\pi - \mathcal{L}^1(E_{\alpha,r})),$$

which gives $\mathcal{L}^1(E_{\alpha,r}) \geq \pi$. Finally, fix a $\lambda \in \mathbb{D}(1/2)$. Then, using the Poisson integral formula, it follows that

$$u(\lambda) \leq \frac{r + |\lambda|}{2\pi(r - |\lambda|)} \int_0^{2\pi} u(re^{it})dt \leq -L\frac{3}{2\pi}\mathcal{L}^1(E_{\alpha,r}) \leq -3L/2 < -L.$$

Therefore, $|f(\lambda)| > R$. □

Remark 3.2.5. Note that for any compact set $K \subset \mathbb{D}$ there exists a positive r such that if $\boldsymbol{h}_a(0) \in K$, $a \in \mathbb{D}$, then $\mathbb{D}(\boldsymbol{h}_a(0), r) \subset \boldsymbol{h}_a(\mathbb{D}(1/2))$.

With these results at hand one may derive the following proposition.

Proposition 3.2.6. *Let $G \subset \mathbb{C}^n$ be a domain and in the case where G is unbounded we assume that there exist a local psh peak function and a local psh antipeak function. Then, G is taut if and only if G is locally taut.*

Proof. Let $(f_j)_{j\in\mathbb{N}} \subset \mathcal{O}(\mathbb{D}, G)$ be given.

Case 1. Assume there are a point $a \in \mathbb{D}$ and a subsequence $(f_{j_k})_{k\in\mathbb{N}}$ of $(f_j)_{j\in\mathbb{N}}$ with $f_{j_k}(a) \longrightarrow \infty$. Put

$$M := \{b \in \mathbb{D} : f_{j_k}(b) \longrightarrow \infty\}.$$

We will prove that M is open and closed. First, fix a $b \in M$. Then, there is a positive r such that $\mathbb{D}(b, r) \subset \boldsymbol{h}_{-b}(\mathbb{D}(1/2))$ (use Remark 3.2.5). Now let $R > 0$. Then, in virtue of Lemma 3.2.4, there are $R' > R$ and a k_R such that $|f_{j_k} \circ \boldsymbol{h}_{-b}(0)| = |f_{j_k}(b)| > R'$ for all $k \geq k_R$, and therefore $f_{j_k} \circ \boldsymbol{h}_{-b}(\mathbb{D}(1/2)) \cap \mathbb{B}(R) = \varnothing$. Applying the above observation, we have $f_{j_k}(\mathbb{D}((b, r)) \cap \mathbb{B}(R) = \varnothing$. Hence, $f_{j_k}|_{\mathbb{D}(b,r)}$ converges uniformly to ∞. In particular, $\mathbb{D}(b, r) \subset M$, i.e., M is open.

Now, let $b \in \mathbb{D}$ be such that there are points $b_j \in M$ with $b_j \longrightarrow b$. Note that the r from above can be chosen independent of b_j. So if j_0 is sufficiently large, we have $b \subset \mathbb{D}(b_{j_0}, r)$. Since $(f_{j_k})_{k\in\mathbb{N}}$ converges uniformly on $\mathbb{D}(b_{j_0}, r)$ to ∞, then it also does so at the point b. Hence, $b \in M$. Therefore, $M = \mathbb{D}$ and, as above, it even follows that $(f_{j_k})_{k\in\mathbb{N}}$ converges locally uniformly to ∞.

Case 2. Assume that $(f_j(a))_{j\in\mathbb{N}}$ is bounded for all $a \in \mathbb{D}$. Then, $(f_j)_j$ is locally bounded (use the reasoning from the first case). Then, the Montel theorem gives a subsequence $(f_{j_k})_k$ that converges locally uniformly to an $f \in \mathcal{O}(\mathbb{D}, \overline{G})$. Put

$$M := \{b \in \mathbb{D} : f(b) \in \partial G\}.$$

Assume that $M \neq \varnothing$. Obviously, M is closed in $\mathbb{D}$. Take a $b \in M$. Since G is locally taut, there is a open neighborhood U of $f(b)$ such that $G \cap U$ is taut. Then, $f(\mathbb{D}(b,r)) \subset\subset U$ for a small $r > 0$. And so, if j is large, then $f_j|_{\mathbb{D}(b,r)} \subset G \cap U$. Therefore, $f(\mathbb{D}(b,r)) \subset \partial G$. Hence, $M = \mathbb{D}$, which means that $(f_j)_j$ is locally uniformly divergent. □

We hope that this series of remarks conveyed the idea how rich the family of taut domains is.

Now, we come back to our discussion on the Lempert function. Namely, we want to study this function on taut domains. Our first result is the following one:

Proposition 3.2.7. *Let z', z'' be two points of a taut domain $G \subset \mathbb{C}^n$. Then, there exist a holomorphic map $\varphi \in \mathcal{O}(\mathbb{D}, G)$ and a number $\sigma \in [0,1)$ with $\varphi(0) = z'$, $\varphi(\sigma) = z''$, and $\boldsymbol{\ell}_G(z', z'') = \boldsymbol{p}(0, \sigma)$.*

Proof. By definition we find a sequence $(\varphi_j)_j \subset \mathcal{O}(\mathbb{D}, G)$ satisfying

$$\varphi_j(0) = z',\ \varphi_j(\sigma_j) = z''\ (0 \le \sigma_j < 1), \quad \text{and} \quad \boldsymbol{p}(0,\sigma_j) \searrow \boldsymbol{\ell}_G(z', z'').$$

Since G is a taut domain and $\varphi_j(0) = z'$, we can choose subsequences $(\varphi_{1j}) \subset (\varphi_j)$ and $(\sigma_{1j}) \subset (\sigma_j)$ with $\varphi_{1j} \underset{j\to\infty}{\overset{K}{\Longrightarrow}} \varphi \in \mathcal{O}(\mathbb{D}, G)$ and $\sigma_{1j} \longrightarrow \sigma \in [0,1)$. From this we conclude that $\varphi(\sigma) = z''$, $\varphi(0) = z'$, and $\boldsymbol{p}(0,\sigma) = \boldsymbol{\ell}_G(z', z'')$. Hence, we have proved that there always exist *extremal discs* through two given points. □

Remark 3.2.8.

(a) We note that the claim of Proposition 3.2.7 is no longer true if G is not taut. For example, take $G_0 := \mathbb{B}_2 \setminus \{(1/2, 0)\}$ and $z' := (0,0)$, $z'' := (1/4, 0)$. Using the analytic maps $\varphi_R(\lambda) := (R\lambda, s(R)\cdot\lambda(\lambda - 1/(4R)))$, $R < 1$, $s(R) \ll 1$, one can easily deduce that $\boldsymbol{\ell}_{G_0}(z', z'') \le p(0, 1/4)$.

Now suppose that there exists $\varphi = (\varphi_1, \varphi_2) \in \mathcal{O}(\mathbb{D}, G_0)$, $\varphi(0) = z'$, $\varphi(\sigma) = z''$ such that $\boldsymbol{\ell}_G(z', z'') = p(0,\sigma)$. Thus, $\sigma \le 1/4$. On the other hand, the Schwarz lemma implies that $1/4 \le \sigma$. Hence, $\varphi_1(1/4) = 1/4$, i.e., $\varphi_1(\lambda) \equiv \lambda$. Since $\varphi \in \mathcal{O}(\mathbb{D}, \mathbb{B}_2)$, it turns out that $\varphi_2 \equiv 0$. In particular, $\varphi(1/2) = (1/2, 0)$, which contradicts the definition of G.

(b) Let $\varphi \in \mathcal{O}(\mathbb{D}, G)$ be an analytic disc through the points $z', z'' \in G$ such that $\varphi(0) = z'$, $\varphi(\sigma) = z''$ $(0 \le \sigma < 1)$, and $\boldsymbol{k}_G(z', z'') = \boldsymbol{p}(0, \sigma)$. Then, for all $0 \le \sigma' < \sigma'' \le \sigma$, we also have

$$\boldsymbol{k}_G(\varphi(\sigma'), \varphi(\sigma'')) = \boldsymbol{p}(\sigma', \sigma'').$$

For the proof, observe that

$$\begin{aligned}\boldsymbol{k}_G(z', z'') &\le \boldsymbol{k}_G(\varphi(0), \varphi(\sigma')) + \boldsymbol{k}_G(\varphi(\sigma'), \varphi(\sigma'')) + \boldsymbol{k}_G(\varphi(\sigma''), \varphi(\sigma)) \\ &\le \boldsymbol{p}(0, \sigma') + \boldsymbol{p}(\sigma', \sigma'') + \boldsymbol{p}(\sigma'', \sigma) = \boldsymbol{p}(0, \sigma).\end{aligned}$$

Proposition 3.2.9. *If G is a taut domain in $\mathbb{C}^n$, then the Lempert function $\boldsymbol{\ell}_G$ is continuous on $G \times G$.*

Proof. Since we already know that $\boldsymbol{\ell}_G$ is upper semicontinuous, suppose that $\boldsymbol{\ell}_G$ is not lower semicontinuous at $(z_0, w_0) \in G \times G$, say. This means that $\boldsymbol{\ell}_G(z_0, w_0) > 0$ and, moreover, there are sequences $(z_\nu)_\nu$, $(w_\nu)_\nu \subset G$ with $\lim_{\nu\to\infty} z_\nu = z_0$, $\lim_{\nu\to\infty} w_\nu = w_0$, and $\boldsymbol{\ell}_G(z_\nu, w_\nu) \le \boldsymbol{\ell}_G(z_0, w_0) - \varepsilon \in (0, +\infty)$ for a suitable $\varepsilon > 0$. Proceeding as usual, we find holomorphic mappings $\varphi_\nu \in \mathcal{O}(\mathbb{D}, G)$ with

$$\varphi_\nu(0) = z_\nu,\ \varphi_\nu(\sigma_\nu) = w_\nu\ (0 \le \sigma_\nu < 1), \quad \text{and} \quad \boldsymbol{p}(0, \sigma_\nu) < \boldsymbol{\ell}_G(z_0, w_0) - \frac{\varepsilon}{2}.$$

On the other hand, since G is taut, there are subsequences $(\varphi_{1\nu}) \subset (\varphi_\nu)$ and $(\sigma_{1\nu}) \subset (\sigma_\nu)$ with $\varphi_{1\nu} \underset{\nu\to\infty}{\overset{K}{\Longrightarrow}} \varphi \in \mathcal{O}(\mathbb{D}, G)$, $\sigma_{1\nu} \underset{\nu\to\infty}{\longrightarrow} \sigma_0 \in [0, 1)$, and therefore $\varphi(0) = z_0$, $\varphi(\sigma_0) = w_0$. Now, we can conclude that

$$\boldsymbol{\ell}_G(z_0, w_0) \le \boldsymbol{p}(0, \sigma_0) = \lim_{\nu\to\infty} \boldsymbol{p}(0, \sigma_{1\nu}) \le \boldsymbol{\ell}_G(z_0, w_0) - \varepsilon/2,$$

and this contradiction finishes the proof. □

3.3 General properties of $\boldsymbol{k}$

We already know that the $\| \ \|$-topology of a domain $G \subset \mathbb{C}^n$ is stronger than the $\boldsymbol{k}_G$-topology on G. We remember that, in the case of the Carathéodory distance, the $\| \ \|$-topology can be different from the $\boldsymbol{c}_G$-topology. To discuss the analogous question for the Kobayashi distance, we need the following observation; see [313].

Proposition 3.3.1. *The Kobayashi pseudodistance is inner, i.e., if $z', z'' \in G$, then*

$$\begin{aligned}\boldsymbol{k}_G(z', z'') = \inf\{L_{\boldsymbol{k}_G}(\alpha) : \alpha : [0,1] \longrightarrow G \\ \text{is continuous and } \| \ \|\text{-rectifiable with } \alpha(0) = z',\ \alpha(1) = z''\},\ \text{where}\end{aligned}$$

$$L_{\boldsymbol{k}_G}(\alpha) := \sup\left\{\sum_{j=1}^{N} \boldsymbol{k}_G(\alpha(t_{j-1}), \alpha(t_j)) : N \in \mathbb{N},\ 0 = t_0 < t_1 < \cdots < t_N = 1\right\}$$

*denotes the $\boldsymbol{k}_G$-*length of α.

Proof. First, observe that $L_{\boldsymbol{k}_G}(\alpha) \geq \boldsymbol{k}_G(z', z'')$ for any such competing curve α. To verify the opposite inequality, fix $\varepsilon > 0$ and choose points $\sigma_1, \dots, \sigma_k \in [0, 1)$ and maps $\varphi_1, \dots, \varphi_k \in \mathcal{O}(\mathbb{D}, G)$ with

$$\varphi_1(0) = z', \ \varphi_j(\sigma_j) = \varphi_{j+1}(0), \ 1 \leq j < k, \ \varphi_k(\sigma_k) = z'',$$
$$\text{and} \sum_{j=1}^{k} \boldsymbol{p}(0, \sigma_j) < \boldsymbol{k}_G(z', z'') + \varepsilon.$$

Of course, we may assume $\sigma_j > 0$. Then, we are able to define a piecewise $\mathcal{C}^1$-curve in G connecting z' and z'' by

$$\alpha(t) := \varphi_j\left(\left(t - \frac{j-1}{k}\right) \cdot k\sigma_j\right) \quad \text{if} \quad t \in \left[\frac{j-1}{k}, \frac{j}{k}\right].$$

Therefore, we obtain

$$L_{\boldsymbol{k}_G}(\alpha) \leq \sum_{j=1}^{k} L_{\boldsymbol{k}_{\mathbb{D}}}([0, \sigma_j]) = \sum_{j=1}^{k} \boldsymbol{p}(0, \sigma_j) < \boldsymbol{k}_G(z', z'') + \varepsilon,$$

which concludes the proof. □

Remark 3.3.2.

(a) Recall that, in general, the formula analogous to that of Proposition 3.3.1 fails to hold for the Carathéodory pseudodistance.

(b) Later (see § 3.6) we will have another method to prove Proposition 3.3.1 (cf. Exercise 3.9.7).

It is obvious that a necessary condition for the $\| \ \|$-topology and the $\boldsymbol{k}_G$-topology to be equal is that $\boldsymbol{k}_G$ is a distance. We say that a domain $G \subset \mathbb{C}^n$ is *$\boldsymbol{k}$-hyperbolic*, if its Kobayashi pseudodistance is a distance.

Remark 3.3.3. By the well-known inequality $\boldsymbol{c}_G \leq \boldsymbol{k}_G$, it is clear that any $\boldsymbol{c}$-hyperbolic domain is also $\boldsymbol{k}$-hyperbolic. Therefore, any bounded domain G is $\boldsymbol{k}$-hyperbolic.

So far, we have shown that the Kobayashi pseudodistance is inner and continuous. These conditions suffice to prove the following comparison property of the topologies (cf. [33]):

Proposition 3.3.4. *If G is a $\boldsymbol{k}$-hyperbolic domain in $\mathbb{C}^n$, then its $\| \ \|$-topology is equal to the $\boldsymbol{k}_G$-topology.*

Proof. Cf. Remark 2.7.7. □

For the rest of this section, we turn to the question of how the Kobayashi pseudodistance behaves under certain set-theoretic operations.

Proposition 3.3.5.

(a) *Let G be a domain in $\mathbb{C}^n$, $G = \bigcup_{\nu=1}^{\infty} G_\nu$, where $(G_\nu)_\nu$ is an increasing sequence of subdomains. Then, for z', $z'' \in G$ we have*

$$\boldsymbol{\ell}_G(z',z'') = \lim_{\nu\to\infty} \boldsymbol{\ell}_{G\nu}(z',z'') \quad \textit{and} \quad \boldsymbol{k}_G(z',z'') = \lim_{\nu\to\infty} \boldsymbol{k}_{G_\nu}(z',z'').$$

(b) *If a domain $G \subset \mathbb{C}^n$ is taut and has a $\mathcal{C}^1$-boundary, and if $(G_j)_j$ is a sequence of domains $G_j \subset \mathbb{C}^n$ with $G_j \supset\supset G_{j+1}$ for all $j \in \mathbb{N}$, $\overline{G} = \bigcap_{j=1}^{\infty} G_j$, then, for every pair of points z', $z'' \in G$, the following is true:*

$$\lim_{j\to\infty} \boldsymbol{\ell}_{G_j}(z',z'') = \boldsymbol{\ell}_G(z',z'').$$

Remark 3.3.6.

(a) The assumptions in (b) can be reformulated by saying that G is a bounded pseudoconvex domain with $\mathcal{C}^1$-boundary.

(b) Exercise 3.9.3 will give examples showing that the assumptions in Proposition 3.3.5(b) cannot be considerably weakened. If we replace the Lempert function by the Kobayashi pseudodistance, we do not know whether Proposition 3.3.5(b) remains true.

Proof of Proposition 3.3.5. (a) For given points z', $z'' \in G$, there is a sufficiently large index ν_0 such that z', $z'' \in G_\nu$ if $\nu \geq \nu_0$. For these ν, the following inequalities are obvious:

$$\boldsymbol{\ell}_{G_\nu}(z',z'') \geq \boldsymbol{\ell}_{G_{\nu+1}}(z',z'') \geq \boldsymbol{\ell}_G(z',z''),$$
$$\boldsymbol{k}_{G_\nu}(z',z'') \geq \boldsymbol{k}_{G_{\nu+1}}(z',z'') \geq \boldsymbol{k}_G(z',z'').$$

If we now assume that $\lim_{\nu\to\infty} \boldsymbol{\ell}_{G_\nu}(z',z'') > A > \boldsymbol{\ell}_G(z',z'')$, we are able to select a function $\varphi \in \mathcal{O}(\overline{\mathbb{D}}, G)$ with

$$\varphi(0) = z',\ \varphi(\sigma) = z''\ (0 \leq \sigma < 1), \quad \text{and} \quad \boldsymbol{p}(0,\sigma) < A.$$

Then, $\varphi(\overline{\mathbb{D}}) \subset G_\nu$ if $\nu \gg \nu_0$, and so $\boldsymbol{\ell}_{G_\nu}(z',z'') \leq \boldsymbol{p}(0,\sigma) < \lim_{\nu\to\infty} \boldsymbol{\ell}_{G_\nu}(z',z'')$, which leads to the expected contradiction.

Recall that $\boldsymbol{k}_G$ is the largest pseudodistance below $\boldsymbol{\ell}_G$; hence,

$$\boldsymbol{k}_G(z',z'') = \lim_{\nu\to\infty} \boldsymbol{k}_{G_\nu}(z',z'').$$

(b) Contrary to the situation just discussed, the sequence $(\boldsymbol{\ell}_{G_j}(z',z''))_j$ is increasing with $\lim_{j\to\infty} \boldsymbol{\ell}_{G_j}(z',z'') \leq \boldsymbol{\ell}_G(z',z'')$.

Now, let us suppose that $\lim_{j\to\infty} \boldsymbol{\ell}_{G_j}(z',z'') < A < \boldsymbol{\ell}_G(z',z'')$. By definition, we are able to select holomorphic functions $\varphi_j \in \mathcal{O}(\mathbb{D}, G_j)$ with

$$\varphi_j(0) = z',\ \varphi_j(\sigma_j) = z''\ (0 \leq \sigma_j < 1), \quad \text{and} \quad \boldsymbol{p}(0,\sigma_j) < A.$$

In view of Montel's theorem we may assume that

$$\varphi_j \underset{j\to\infty}{\overset{K}{\Longrightarrow}} \varphi \in \mathcal{O}(\mathbb{D}, \mathbb{C}^n), \quad \sigma_j \longrightarrow \sigma \in [0,1), \quad \varphi(0) = z', \quad \text{and } \varphi(\sigma) = z''.$$

Since the sequence $(G_j)_j$ is strictly decreasing, it follows that $\varphi(\mathbb{D}) \subset \bigcap G_j = \overline{G}$. But, on the other hand, the assumptions of tautness and $\mathcal{C}^1$-boundary force the analytic disc $\varphi(\mathbb{D})$ to be inside G. Thus, φ can be considered a competitor in the definition of $\boldsymbol{\ell}_G$, which leads to $\boldsymbol{\ell}_G(z', z'') \le \boldsymbol{p}(0, \sigma) \le A$; a contradiction. □

In calculations of the Kobayashi pseudodistance, holomorphic coverings often play an important roledue to the following result of S. Kobayashi (cf. [317]):

Theorem 3.3.7. *Let $\Pi : \widetilde{G} \longrightarrow G$ be a holomorphic covering. Then, for $x, y \in G$ and $\widetilde{x} \in \widetilde{G}$, $\Pi(\widetilde{x}) = x$, the Lempert function and the Kobayashi pseudodistance for G satisfy the following formulas:*

$$\boldsymbol{\ell}_G(x, y) = \inf\{\boldsymbol{\ell}_{\widetilde{G}}(\widetilde{x}, \widetilde{y}) : \widetilde{y} \in \widetilde{G},\ \Pi(\widetilde{y}) = y\}, \tag{3.3.1}$$

$$\boldsymbol{k}_G(x, y) = \inf\{\boldsymbol{k}_{\widetilde{G}}(\widetilde{x}, \widetilde{y}) : \widetilde{y} \in \widetilde{G},\ \Pi(\widetilde{y}) = y\}. \tag{3.3.2}$$

Proof. By the contractibility of $(\boldsymbol{\ell}_G)_G$, we only have to verify the inequality "$\ge$".
Thus, for a given $\varepsilon > 0$, we take a holomorphic map $\varphi \in \mathcal{O}(\mathbb{D}, G)$ with the following properties:

$$\varphi(0) = x, \ \varphi(\sigma) = y \ (0 \le \sigma < 1), \quad \text{and} \quad \boldsymbol{p}(0, \sigma) < \boldsymbol{\ell}_G(x, y) + \varepsilon.$$

Remembering the properties of holomorphic coverings, we lift φ to $\widetilde{\varphi} \in \mathcal{O}(\mathbb{D}, \widetilde{G})$ with $\widetilde{\varphi}(0) = \widetilde{x}$ and $\Pi \circ \widetilde{\varphi} = \varphi$, in particular, $\Pi(\widetilde{\varphi}(\sigma)) = y$. Therefore, we conclude that

$$\boldsymbol{p}(0, \sigma) \ge \boldsymbol{\ell}_{\widetilde{G}}(\widetilde{x}, \widetilde{\varphi}(\sigma)) \ge \inf\{\boldsymbol{\ell}_{\widetilde{G}}(\widetilde{x}, \widetilde{y}) : \widetilde{y} \in \widetilde{G} \text{ with } \Pi(\widetilde{y}) = y\},$$

which proves (3.3.1). Formula (3.3.2) may be proved similarly. □

Remark 3.3.8.

(a) From the previous theorem, we conclude that, if all $\boldsymbol{k}_{\widetilde{G}}$-balls with finite radii are relatively compact subsets of $\widetilde{G}$, then for $x, y \in G$, and $\widetilde{x} \in \widetilde{G}$, $\Pi(\widetilde{x}) = x$, there exists a point $\widetilde{y} \in \widetilde{G}$ with $\Pi(\widetilde{y}) = y$ and $\boldsymbol{k}_G(x, y) = \boldsymbol{k}_{\widetilde{G}}(\widetilde{x}, \widetilde{y})$ (the same statement is also true for the Lempert functions, provided that $\widetilde{G}$ is taut).

(b) Notice that the above sharp form of Theorem 3.3.7 is not true in general – cf. Remark 10.1.2.

(c) In Chapter 9, which is devoted to the annulus, we will see how Theorem 3.3.7 is applied in concrete situations.

(d) Observe that if $P = \{\lambda \in \mathbb{C} : 1/2 < |\lambda| < 2\}$ denotes an annulus, then there is the universal covering $\Pi : \mathbb{D} \longrightarrow P$, and we can therefore find two points $\lambda', \lambda'' \in \mathbb{D}$ with $\pi(\lambda') = 1$, $\Pi(\lambda'') = -1$, and $\boldsymbol{k}_P(1, -1) = \boldsymbol{p}(\lambda', \lambda'')$. But it is clear that Π is not a $\boldsymbol{k}$-isometry.

(e) Since any domain $G \subset \mathbb{C}$ has $\mathbb{D}$ or $\mathbb{C}$ as its universal covering, Theorem 3.3.7 also implies that $\boldsymbol{\ell}_G$ satisfies the triangle inequality; hence, $\boldsymbol{k}_G = \boldsymbol{\ell}_G$.

(f) We also mention that for $G := \mathbb{C} \setminus \{0, 1\}$ we have $\boldsymbol{c}_G \equiv 0$, whereas $\boldsymbol{k}_G(z', z'') > 0$ if $z' \neq z''$. The latter fact follows due to Theorem 3.3.7 and the well-known result that $\mathbb{D}$ is the universal covering of G.

So far, we know only few examples of domains G for which $\boldsymbol{c}_G \not\equiv \boldsymbol{k}_G$ (recall, for example, the domain constructed in Theorem 2.6.3). For plane domains we have the following complete characterization of such domains.

Proposition 3.3.9.

(a) *Let G be a $\boldsymbol{c}$-hyperbolic domain in $\mathbb{C}$ and let us suppose that there is at least one pair of different points z', $z'' \in G$ with $\boldsymbol{k}_G(z', z'') = \boldsymbol{c}_G(z', z'')$. Then, G is biholomorphically equivalent to $\mathbb{D}$ and so $\boldsymbol{k}_G \equiv \boldsymbol{c}_G$.*

(b) *If a plane domain G is not $\boldsymbol{c}$-hyperbolic, then $\boldsymbol{c}_G \equiv 0$ and either $\boldsymbol{k}_G \equiv 0$ or G is $\boldsymbol{k}$-hyperbolic.*

Proof. In the case when G is $\boldsymbol{c}$-hyperbolic, there is the holomorphic covering $\Pi : \mathbb{D} \longrightarrow G$. According to the above remark, we choose points λ', λ'' in the unit disc such that $\Pi(\lambda') = z'$, $\Pi(\lambda'') = z''$, and

$$\boldsymbol{k}_G(z', z'') = \boldsymbol{k}_{\mathbb{D}}(\lambda', \lambda'') = p(\lambda', \lambda'').$$

On the other hand, $\boldsymbol{c}_G(z', z'')$ can be written as $\boldsymbol{c}_G(z', z'') = p(f(z'), f(z''))$ for a suitable $f \in \mathcal{O}(G, \mathbb{D})$. For the function $f \circ \Pi \in \mathcal{O}(\mathbb{D}, \mathbb{D})$ this implies

$$\boldsymbol{p}(f \circ \Pi(\lambda'), f \circ \Pi(\lambda'')) = \boldsymbol{c}_G(z', z'') = \boldsymbol{k}_G(z', z'') = p(\lambda', \lambda'').$$

Now, the Schwarz–Pick lemma tells us that $f \circ \Pi$ is a biholomorphic map, and therefore π is biholomorphic.

We turn to the proof of claim (b). Since G is not $\boldsymbol{c}$-hyperbolic, we have $\boldsymbol{c}_G \equiv 0$ (cf. Proposition 2.5.1). In the case where the universal covering of G is given by $\mathbb{C}$, it is clear that $\boldsymbol{k}_G \equiv 0$, so we may assume that $\Pi : \mathbb{D} \longrightarrow G$ is the universal covering. Hence, by Theorem 3.3.7 and Remark 3.3.8, we conclude that whenever z', $z'' \in G$, $z' \neq z''$, there are points λ', $\lambda'' \in \mathbb{D}$, $\Pi(\lambda') = z'$, $\Pi(\lambda'') = z''$ with

$$\boldsymbol{k}_G(z', z'') = p(\lambda', \lambda'') > 0. \qquad \square$$

Corollary 3.3.10. *Let $P := \{\lambda : 1/R < |\lambda| < R\}$ $(R > 1)$. Then, for z', $z'' \in P$, $z' \neq z''$, we have*

$$\boldsymbol{c}_P(z', z'') < \boldsymbol{k}_P(z', z'').$$

3.4 An extension theorem

The discussion in the previous chapter has shown that the Carathéodory pseudodistance is preserved, if a "small" set is removed from the considered domain. We will see that in the case of the Kobayashi pseudodistance a similar result is true. Nevertheless, the reader should note the following fact:

Example 3.4.1. $\boldsymbol{k}_{\mathbb{D}_*} \not\equiv \boldsymbol{k}_{\mathbb{D}}|_{\mathbb{D}_*\times\mathbb{D}_*}$ ($\mathbb{D}_* := \mathbb{D} \setminus \{0\}$).

To prove this non-identity, we begin with two different points λ', λ'' in $\mathbb{D}_*$. It is clear that the equality would lead to the following equation:

$$\boldsymbol{c}_{\mathbb{D}_*}(\lambda', \lambda'') = \boldsymbol{c}_{\mathbb{D}}(\lambda', \lambda'') = \boldsymbol{k}_{\mathbb{D}}(\lambda', \lambda'') = \boldsymbol{k}_{\mathbb{D}_*}(\lambda', \lambda''),$$

which would imply that $\mathbb{D}_*$ is simply connected (use Proposition 3.3.9(a)).

Theorem 3.4.2. *Let G be a domain in $\mathbb{C}^n$ ($n \geq 2$) containing a relatively closed subset A with $H^{2n-2}(A) = 0$, where H^{2n-2} denotes the $(2n-2)$-dimensional Hausdorff measure. Then, the following statements hold:*

(a) *any $g \in \mathcal{O}(\overline{\mathbb{D}}, G)$ with $g(0) \notin A$ can be uniformly approximated on $\overline{\mathbb{D}}$ by a sequence of analytic discs $(g_\nu)_\nu$, $g_\nu \in \mathcal{O}(\overline{\mathbb{D}}, G \setminus A)$;*

(b) $\boldsymbol{k}_G|_{(G\setminus A)\times(G\setminus A)} = \boldsymbol{k}_{G\setminus A}$.

A weaker version of this theorem can be found in [81]; the formulation above is taken from [434].

Proof. Taking (a) for granted, we are going to deduce (b).

First, observe that for arbitrary $z', z'' \in G \setminus A$ we have

$$\begin{aligned}&\boldsymbol{k}_G(z', z'')\\ &= \inf\left\{\sum_{j=1}^{N} \boldsymbol{\ell}_G(z_{j-1}, z_j) : N \in \mathbb{N}, z_0 = z',\ z_1, \dots, z_{N-1} \in G \setminus A,\ z_N = z''\right\}.\end{aligned}$$

This is a simple consequence of the upper semicontinuity of $\boldsymbol{\ell}_G$ (cf. Proposition 3.1.14). (Note that the formula remains true for arbitrary, nowhere dense subsets $A \subset G$.) In particular, if $\boldsymbol{k}_{G\setminus A} \leq \boldsymbol{\ell}_G$ on $(G \setminus A) \times (G \setminus A)$, then (b) is true.

Fix $z', z'' \in G \setminus A$ and let $\varphi \in \mathcal{O}(\overline{\mathbb{D}}, G)$ be such that $z' = \varphi(0)$, $z'' = \varphi(\sigma)$ for some $0 \leq \sigma < 1$. By (a) there exists $(\varphi_\nu)_{\nu=1}^{\infty} \subset \mathcal{O}(\overline{\mathbb{D}}, G \setminus A)$ with $\varphi_\nu \longrightarrow \varphi$ uniformly on $\overline{\mathbb{D}}$. Hence, $\boldsymbol{k}_{G\setminus A}(\varphi_\nu(0), \varphi_\nu(\sigma)) \leq \boldsymbol{p}(0, \sigma)$, $\nu \in \mathbb{N}$. Consequently, by the continuity of $\boldsymbol{k}_{G\setminus A}$ (cf. Proposition 3.1.10) it follows that $\boldsymbol{k}_{G\setminus A}(z', z'') \leq \boldsymbol{p}(0, \sigma)$. Since φ is arbitrary, we get $\boldsymbol{k}_{G\setminus A}(z', z'') \leq \boldsymbol{\ell}_G(z', z'')$. Hence, (b) is verified under the assumption that (a) holds.

Before we proceed, we recall some results from geometric measure theory (cf. [164]).

1) Let D denote a disc in $\mathbb{C}$ and let F be a subset of $D\times\mathbb{C}^{n-1}$ with $(2n-2)$-Hausdorff measure zero. Then, for any $\varepsilon > 0$ there exists a vector $v \in \mathbb{C}^{n-1}$, $0 < \|v\| < \varepsilon$, such that the section $D \times \{v\}$ does not intersect F.

2) If $F : G_1 \longrightarrow G_2$ is a diffeomorphism, where G_j are open sets in $\mathbb{R}^n$, then any subset $M \subset G_1$ of k-Hausdorff measure zero is mapped onto the set $F(M)$ of k-Hausdorff measure zero.

Now, we come to the proof of claim (a).

We start with a holomorphic map $g : U \longrightarrow G$, where U denotes an open neighborhood of $\overline{\mathbb{D}}$ with $g(0) \notin A$.

First, we approximate g uniformly by regular holomorphic maps $g_t : U \longrightarrow \mathbb{C}^n$ ($t > 0$), where all these functions share the following properties:

(i) if $0 < t \leq t_0$, then $g_t(U') \subset G$, for a suitable open neighborhood $U' \subset U$ of $\overline{\mathbb{D}}$;

(ii) $\|g_t - g\|_{U'} \leq t$ and $g_t(0) \notin A$ if $0 < t \leq t_0$;

(iii) $g_t'(\lambda) \neq 0$ if $0 < t \leq t_0$ and $\lambda \in U$.

Observe that the $\mathcal{C}^1$-map $\widetilde{g} : U \times \mathbb{R} \longrightarrow \mathbb{C}^n$, defined by $\widetilde{g}(\lambda, t) := tg'(\lambda)$, has the image in $\mathbb{C}^n$ of zero measure. Thus, there exists a vector $v \neq 0$, $v \notin \widetilde{g}(U \times \mathbb{R})$, $\|v\| \leq 1/2$. Bearing this in mind, we obtain our desired functions by putting $g_t(\lambda) := g(\lambda) + t\lambda v$. Therefore, without loss of generality, we may assume that the map g we want to approximate is a regular one. Since $g(0) \notin A$, we see that $g(r\overline{\mathbb{D}}) \cap A = \varnothing$ if $0 < r < 1$ is suitably chosen.

Now, we fix a point λ_0 with $r \leq |\lambda_0| \leq 1$. Since $g'(\lambda_0) \neq 0$, we can select $(n-1)$ orthonormal vectors $v_2, \dots, v_n$ in $\mathbb{C}^n$, all of them orthogonal to $g'(\lambda_0)$. Then, we consider the following map:

$$U \times \mathbb{C}^{n-1} \ni (\lambda, \lambda_2, \dots, \lambda_n) \overset{\widetilde{g}}{\longmapsto} g(\lambda) - \sum_{j=2}^{n} \lambda_j v_j.$$

Observe that the Jacobian matrix of $\widetilde{g}$ at $(\lambda_0, 0, \dots, 0)$ is non-singular. Hence, we find a small disc D around λ_0 and a positive number $\varepsilon(\lambda_0)$ such that $\widetilde{g}$ maps biholomorphically $D \times \{\widetilde{\lambda} \in \mathbb{C}^{n-1} : \|\widetilde{\lambda}\| < \varepsilon(\lambda_0)\}$ into G.

By a compactness argument, we cover the annulus $\{\lambda \in \mathbb{C} : r \leq |\lambda| \leq 1\}$ by a finite number of discs $D_1, \dots, D_k$, such that for some vectors $\{v_2^{(j)}, \dots, v_n^{(j)}\}$, and a suitable $\varepsilon > 0$, the mappings

$$\widetilde{g}_j : \overline{D}_j \times \{\widetilde{\lambda} \in \mathbb{C}^{n-1} : \|\widetilde{\lambda}\| < \varepsilon\} \longrightarrow G,$$

defined by

$$\widetilde{g}_j(\lambda, \lambda_2, \ldots, \lambda_n) = g(\lambda) - \sum_{\nu=2}^{n} \lambda_\nu v_\nu^{(j)},$$

are biholomorphic into G.

Since $\widetilde{g}_1$ is biholomorphic, we conclude that

$$H^{2n-2}(\{(\lambda, \widetilde{\lambda}) \in \overline{D}_1 \times \mathbb{C}^{n-1} : \|\widetilde{\lambda}\| < \varepsilon,\ g_1(\lambda, \widetilde{\lambda}) \in A\}) = 0.$$

Hence, according to our initial remarks, we find a small vector $\widetilde{\lambda}^{(1)} \in \mathbb{C}^{n-1}$ fulfilling the following conditions:

$$\widetilde{g}_1(\lambda, \widetilde{\lambda}^{(1)}) \notin A \text{ if } \lambda \in r\overline{\mathbb{D}} \cup \overline{D}_1, \text{ and } \|g - \widetilde{g}_1(\cdot, \widetilde{\lambda}^{(1)})\|_{\overline{\mathbb{D}}} \text{ is arbitrarily small.}$$

Now, we proceed with the second disc. For D_2 we only mention, that

$$\widetilde{g}_2(\lambda, \lambda_2, \ldots, \lambda_n) - \sum_{\nu=2}^{n} \lambda_\nu^{(1)} v_\nu^{(1)}$$

plays the role of $\widetilde{g}_1$ above.

So, we obtain small $(\lambda_\nu^{(2)})_{2 \le \nu \le n}$ such that the image of the set $r\overline{\mathbb{D}} \cup \overline{D}_1 \cup \overline{D}_2$ by the function

$$\widetilde{g}_2(\cdot, \lambda_2^{(2)}, \ldots, \lambda_n^{(2)}) - \sum_{\nu=2}^{n} \lambda_\nu^{(1)} v_\nu^{(1)}$$

does not intersect A. After a finite number of steps, we reach the desired function $\widetilde{g} \in \mathcal{O}(\overline{\mathbb{D}}, G \setminus A)$ that is near to g.

Thus, Theorem 3.4.2 is completely proven. □

Corollary 3.4.3. *If A denotes an analytic subset of the domain G of codimension at least two, then the following formula is true:*

$$\boldsymbol{k}_{G \setminus A} = \boldsymbol{k}_G|_{(G \setminus A) \times (G \setminus A)}.$$

Proof. A standard argument from measure theory leads to the conclusion that A has $(2n-3)$-Hausdorff measure zero. Hence, the corollary is a particular case of Theorem 3.4.2. □

3.5 The Kobayashi–Royden pseudometric

In Chapter 2, we have already learned that for the Carathéodory pseudodistance there is an infinitesimal version, the Carathéodory–Reiffen pseudometric, which measures the lengths of tangent vectors. A similar notion with respect to the Kobayashi pseudodistance will be introduced and investigated in this section.

Let G be a domain in $\mathbb{C}^n$. The function $\varkappa_G : G \times \mathbb{C}^n \longrightarrow [0, +\infty)$, defined by

$$\varkappa_G(z; X) := \inf\{\gamma(\lambda)|\alpha| : \exists_{\varphi\in\mathcal{O}(\mathbb{D},G)}\ \exists_{\lambda\in\mathbb{D}} : \varphi(\lambda) = z,\ \alpha\varphi'(\lambda) = X\},$$

is called the *Kobayashi–Royden pseudometric*.

Observe that

(a) $\varkappa_G(z; X) = \inf\{\alpha > 0 : \exists_{\varphi\in\mathcal{O}(\mathbb{D},G)} : \varphi(0) = z,\ \alpha\varphi'(0) = X\}$
$\qquad = \inf\{\alpha > 0 : \exists_{\varphi\in\mathcal{O}(\overline{\mathbb{D}},G)} : \varphi(0) = z,\ \alpha\varphi'(0) = X\};$

(b) $\varkappa_G(z; \lambda X) = |\lambda|\varkappa_G(z; X),\ \lambda \in \mathbb{C},\ X \in \mathbb{C}^n,\ z \in G \subset \mathbb{C}^n;$

(c) $\varkappa_D(F(z); F'(z)X) \le \varkappa_G(z; X),\ \ F \in \mathcal{O}(G, D),\ z \in G \subset \mathbb{C}^n,\ X \in \mathbb{C}^n.$

Hence, $\varkappa_G(z; \cdot)$ assigns a length to any tangent vector in z and, moreover, (c) shows that the system $(\varkappa_G)_G$ is contractible with respect to holomorphic mappings. In particular, if $F : G \longrightarrow D$ is a biholomorphic map, then

(d) $\varkappa_D(F(z); F'(z)X) = \varkappa_G(z; X),\ z \in G,\ X \in \mathbb{C}^n;$

(e) $\varkappa_{\mathbb{D}}(\lambda; X) \le \gamma(\lambda)|X|,\ \lambda \in \mathbb{D},\ X \in \mathbb{C}.$

Applying the Schwarz–Pick lemma, we obtain the following comparison result:

Lemma 3.5.1. *For any domain $G \subset \mathbb{C}^n$ we have $\gamma_G \le \varkappa_G$.*

Proof. For a point $z \in G \subset \mathbb{C}^n$ and a vector $X \in \mathbb{C}^n$, we choose a function $f \in \mathcal{O}(G, \mathbb{D})$ with $f(z) = 0$ and $|f'(z)X| = \gamma_G(z; X)$. Then, for every $\varphi \in \mathcal{O}(\mathbb{D}, G)$ with $\varphi(0) = z$ the composition $f \circ \varphi \in \mathcal{O}(\mathbb{D}, \mathbb{D})$ has the origin as a fixed point. Thus, $|(f \circ \varphi)'(0)| \le 1$. In the case $\alpha\varphi'(0) = X$ $(\alpha > 0)$ we obtain

$$\alpha \ge |(f \circ \varphi)'(0)| \cdot \alpha = |f'(z)X| = \gamma_G(z; X).$$

Since φ is arbitrary, the claim follows. □

Corollary 3.5.2. $\varkappa_{\mathbb{D}}(\lambda; X) = \gamma(\lambda)|X|,\ \lambda \in \mathbb{D},\ X \in \mathbb{C}.$

Moreover, it turns out that whenever there is a system $(\delta_G)_G$ of functions $\delta_G : G \times \mathbb{C}^n \longrightarrow [0, +\infty)$ $(G \subset \mathbb{C}^n)$ with the properties (b), (c), and $\delta_{\mathbb{D}} = \gamma_{\mathbb{D}}$, then $\delta_G \le \varkappa_G$ for any G. We recall that we already know that $\gamma_G \le \delta_G$ is also true.

To see at least a few concrete examples, we calculate the Kobayashi–Royden pseudometric for balanced pseudoconvex domains.

Proposition 3.5.3. *Let G be a balanced pseudoconvex domain in $\mathbb{C}^n$ given by $G := \{z \in \mathbb{C}^n : \mathfrak{h}(z) < 1\}$, where h is its Minkowski function. Then,*

$$\varkappa_G(0; X) = \mathfrak{h}(X), \quad X \in \mathbb{C}^n.$$

Proof. First of all, observe that if $\mathfrak{h}(X) \neq 0$, then $\mathbb{D} \ni \lambda \overset{\varphi}{\longmapsto} \lambda X/\mathfrak{h}(X)$ is an analytic disc in G with $\varphi(0) = 0$, $\varphi'(0) = X/\mathfrak{h}(X)$; hence, we obtain $\varkappa_G(0; X) \leq \mathfrak{h}(X)$.

Proof of the fact that the same inequality is also true if $\mathfrak{h}(X) = 0$ is left to the reader (cf. Example 3.1.8). On the other hand, let $\varphi \in \mathcal{O}(\mathbb{D}, G)$ with $\varphi(0) = 0$, $\alpha\varphi'(0) = X$ ($\alpha > 0$). As in Proposition 3.1.11, we observe that $\mathfrak{h}(\varphi(\lambda)) = \mathfrak{h}(\lambda\widetilde{\varphi}(\lambda)) < 1$, and therefore $\mathfrak{h} \circ \widetilde{\varphi} \leq 1$.

Thus, we end up with $\mathfrak{h}(X) = \mathfrak{h}(\alpha\varphi'(0)) = \alpha\mathfrak{h} \circ \widetilde{\varphi}(0) \leq \alpha$, which guarantees the missing inequality. □

We emphasize that the proof above is based on the information that G is pseudoconvex, i.e., that the Minkowski function is plurisubharmonic. Indeed, the following example, partially due to N. Sibony (cf. [174]), shows that the above formula is false if G is not pseudoconvex.

Example 3.5.4. Let G_ε, $0 < \varepsilon < 1$, be the following complete Reinhardt domain:

$$G_\varepsilon := \{z = (z_1, z_2) \in \mathbb{C}^2 : |z_1| < 1,\ |z_2| < \varepsilon \quad \text{or} \quad |z_1| < \varepsilon,\ |z_2| < 1\}.$$

Note that G_ε is not pseudoconvex, and its envelope of holomorphy $\widetilde{G}_\varepsilon$ is given by

$$\widetilde{G}_\varepsilon := \{z \in \mathbb{C}^2 : |z_1| < 1,\ |z_2| < 1,\ |z_1 z_2| < \varepsilon\}.$$

By $\mathfrak{h}_\varepsilon$ (resp. $\widetilde{\mathfrak{h}}_\varepsilon$), we denote the Minkowski function of G_ε (resp. $\widetilde{G}_\varepsilon$).

Then, the *$\varkappa$-indicatrix of G_ε at* 0,

$$I(\varkappa_{G_\varepsilon}; 0) := \{z \in \mathbb{C}^2 : \varkappa_{G_\varepsilon}(0; z) < 1\},$$

is a balanced domain in $\mathbb{C}^2$ with

$$G_\varepsilon \subset I(\varkappa_{G_\varepsilon}; 0) \subset I(\varkappa_{\widetilde{G}_\varepsilon}; 0) = \widetilde{G}_\varepsilon \subsetneq \operatorname{conv} G_\varepsilon = \operatorname{conv} \widetilde{G}_\varepsilon.$$

First, we show that $I(\varkappa_{G_\varepsilon}; 0) \subsetneq \widetilde{G}_\varepsilon$, which implies that $I(\varkappa_{G_\varepsilon}; 0)$ is not pseudoconvex, and, therefore, that $\varkappa_{G_\varepsilon}(0; \cdot)$ is not plurisubharmonic.

Fix $t_0 \in (\varepsilon, 1)$ and put $X_0 := (t_0, \varepsilon/t_0)$. Now, suppose that $\varkappa_{G_\varepsilon}(0; X_0) = 1$. This implies that there exists a holomorphic map $\varphi \in \mathcal{O}(\mathbb{D}, \overline{G}_\varepsilon)$ satisfying $\varphi(0) = 0$ and $\varphi'(0) = X_0 = (t_0, \varepsilon/t_0)$.

Observe that $|\varphi_1\varphi_2| \leq \varepsilon$, where $\varphi = (\varphi_1, \varphi_2)$. Since $\varphi_1'(0)\varphi_2'(0) = \varepsilon$, the maximum principle applied to $\varphi_1\varphi_2/\lambda^2$ implies that $(\varphi_1\varphi_2)(\lambda) = \varepsilon\lambda^2$, $\lambda \in \mathbb{D}$. Therefore, the non-tangential boundary values φ_j^* of φ_j, $j = 1, 2$, satisfy $|\varphi_1^*\varphi_2^*| = \varepsilon$ almost everywhere on $\mathbb{T}$. This can be read as

$$\varphi^*(e^{i\theta}) \in \{z \in \mathbb{C}^2 : |z_1| = 1,\ |z_2| = \varepsilon \quad \text{or} \quad |z_1| = \varepsilon,\ |z_2| = 1\}$$

whenever $\varphi^*(e^{i\theta}) = (\varphi_1^*(e^{i\theta}), \varphi_2^*(e^{i\theta}))$ exists.

Now, suppose that there are two different points $e^{i\theta_1}, e^{i\theta_2} \in \mathbb{T}$ at which φ^* exists, and such that

$$|\varphi_1^*(e^{i\theta_1})| = 1,\ |\varphi_2^*(e^{i\theta_1})| = \varepsilon \quad \text{and} \quad |\varphi_1^*(e^{i\theta_2})| = \varepsilon,\ |\varphi_2^*(e^{i\theta_2})| = 1.$$

Choosing curves α_ν in $\mathbb{D} \cup \{e^{i\theta_1}, e^{i\theta_2}\}$ with $\alpha_\nu(0) = e^{i\theta_1}$, $\alpha_\nu(1) = e^{i\theta_2}$, and

$$\sup_{0 \le t \le 1} \operatorname{dist}(\alpha_\nu(t), \mathbb{T}) \underset{\nu\to\infty}{\longrightarrow} 0,$$

leads to curves $\varphi \circ \alpha_\nu$ in $\overline{G}_\varepsilon$ connecting $\varphi^*(e^{i\theta_1})$ with $\varphi^*(e^{i\theta_2})$. Thus, there exist points $\lambda_\nu \in \alpha_\nu((0,1))$, $|\lambda_\nu| \longrightarrow 1$ with $|\varphi_1(\lambda_\nu)| = |\varphi_2(\lambda_\nu)| \le \varepsilon$. So, we have

$$\varepsilon = \lim_{\nu\to\infty} \varepsilon|\lambda_\nu|^2 = \lim_{\nu\to\infty} |(\varphi_1\varphi_2)(\lambda_\nu)| \le \varepsilon^2;$$

a contradiction.

So, we have $|\varphi_1^*| = 1$, $|\varphi_2^*| = \varepsilon$ a.e. (or conversely). Then, the maximum principle implies that $|\varphi_2| \le \varepsilon$ (resp. $|\varphi_1| \le \varepsilon$), and so the Schwarz lemma gives $|\varphi_2'(0)| \le \varepsilon$ (resp. $|\varphi_1'(0)| \le \varepsilon$) contradicting $\varphi_2'(0) = \varepsilon/t_0 > \varepsilon$ (resp. $\varphi_1'(0) = t_0 > \varepsilon$). Hence, $\varkappa_{G_\varepsilon}((0;(t,\varepsilon/t)) > \widetilde{\mathfrak{h}}_\varepsilon(t,\varepsilon/t)$ whenever $\varepsilon < t < 1$.

For the final discussion of this example, we will restrict ourselves to the case where $0 < \varepsilon < 2/(1+e^2)$:

Let $\beta \in [\varepsilon, 1)$ with $\varepsilon < 2/(e^{2\beta/\varepsilon} + 1)$. Under these assumptions, we find positive numbers α and A such that

$$1 < A\alpha < \min\{\alpha\varepsilon/2, \alpha/(e^{\alpha\beta} + 1)\}.$$

Then, it is easy to see that the holomorphic map $\varphi : \mathbb{D} \longrightarrow \mathbb{C}^2$, defined by

$$\varphi(\lambda) := A(e^{\varepsilon\alpha\lambda} - 1, e^{-\beta\alpha\lambda} - 1),$$

has its image inside G_ε. Since $\varphi(0) = 0$ and $\varphi'(0) = A\alpha(\varepsilon, -\beta)$, it follows that

$$\varkappa_{G_\varepsilon}(0;(\varepsilon,\beta)) \le \frac{1}{A\alpha} < 1 = \mathfrak{h}_\varepsilon(\varepsilon,\beta),$$

i.e., in general, the conclusion of Proposition 3.5.3 becomes false if the balanced domain is not pseudoconvex.

We want to emphasize that we do not know any formula for $\varkappa_{G_\varepsilon}(0;\cdot)$. ?

Example 3.5.4 shows that, in general, the $\varkappa$-indicatrix is not a pseudoconvex domain. This phenomenon also occurs for very regular domains as the following example, due to N. Sibony, illustrates.

Example 3.5.5 (cf. [174]). For $\varepsilon > 0$, we set

$$G_\varepsilon := \{z = (z_1, z_2) \in \mathbb{C}^2 : \varepsilon(|z_1|^2 + |z_2|^2) + |z_1^2 - z_2^3|^2 < 1\}.$$

Observe that G_ε is a strongly pseudoconvex domain with smooth $\mathcal{C}^\infty$-boundary; cf. Appendix B.7. Moreover, G_ε is contractible. For example, take $z \longmapsto (t^3 z_1, t^2 z_2)$, $0 \le t \le 1$, $z \in G_\varepsilon$.

Since the map $z \longmapsto (e^{3it} z_1, e^{2it} z_2)$ is a biholomorphic automorphism of G_ε, the $\varkappa$-indicatrix of G_ε at 0 is invariant under the rotations $z \longmapsto (e^{3it} z_1, e^{2it} z_2)$ and $z \longmapsto (e^{it} z_1, e^{it} z_2)$, $t \in \mathbb{R}$, and, therefore, $I(\varkappa_{G_\varepsilon}; 0)$ is a Reinhardt domain containing the origin.

Now, we claim that the point $(2/3, 4/3) \in \overline{I(\varkappa_{G_\varepsilon}; 0)}$ for small ε. To verify this, we study the following holomorphic map $\varphi : \mathbb{D} \longrightarrow \mathbb{C}^2$,

$$\varphi(\lambda) := (2\lambda/3 + 2\lambda^2 + \lambda^3, 4\lambda/3 + \lambda^2),$$

with $\varphi(0) = 0$ and $\varphi'(0) = (2/3, 4/3)$. Because of

$$\varphi_1(\lambda)^2 - \varphi_2(\lambda)^3 = (4/9)\lambda^2 + (8/27)\lambda^3,$$

we obtain

$$\begin{aligned}\varepsilon(|\varphi_1(\lambda)|^2 + |\varphi_2(\lambda)|^2) + |\varphi_1(\lambda)^2 - \varphi_2(\lambda)^3|^2 \\ \le \varepsilon(|\varphi_1(\lambda)|^2 + |\varphi_2(\lambda)|^2) + (20/27)^2 < 1\end{aligned}$$

if ε is sufficiently small, i.e., $\varphi(\mathbb{D}) \subset G_\varepsilon$. Therefore, $\varkappa_{G_\varepsilon}(0; (2/3, 4/3)) \le 1$.

On the other hand, we will prove that the point $(0, 4/3)$ does not belong to $\overline{I(\varkappa_{G_\varepsilon}; 0)}$. Then, this implies that $I(\varkappa_{G_\varepsilon}; 0)$ is not a pseudoconvex (Reinhardt) domain, and therefore $\varkappa_{G_\varepsilon}(0; \cdot)$ is not a psh function if $\varepsilon \ll 1$.

Now let us assume that $(0, 4/3) \in \overline{I(\varkappa_{G_\varepsilon}; 0)}$. Then, there exist points $(\lambda'_\nu, \lambda''_\nu) \in I(\varkappa_{G_\varepsilon}; 0)$ with $(\lambda'_\nu, \lambda''_\nu) \underset{\nu\to\infty}{\longrightarrow} (0, 4/3)$ and analytic discs $\varphi_\nu \in \mathcal{O}(\mathbb{D}, G_\varepsilon)$ with $\varphi_\nu(0) = 0$, $\sigma_\nu \varphi'_\nu(0) = (\lambda'_\nu, \lambda''_\nu)$, where $0 < \sigma_\nu < 1$. Taking a suitable subsequence, we obtain a map $\varphi \in \mathcal{O}(\mathbb{D}, \overline{G}_\varepsilon)$ with $\varphi(0) = 0$, $\sigma\varphi'(0) = (0, 4/3)$, where $0 < \sigma \le 1$. Therefore, φ can be represented as

$$\varphi(\lambda) = (\lambda^2 h(\lambda), 4\lambda g(\lambda)/(3\sigma)), \quad g, h \in \mathcal{O}(\mathbb{D}) \text{ with } g(0) = 1.$$

The maximum principle leads to $|\lambda h^2(\lambda) - (4/3\sigma)^3 g^3(\lambda)|^2 \le 1$, $\lambda \in \mathbb{D}$. In particular, for $\lambda = 0$ it follows that $(4/3\sigma)^6 \le 1$; a contradiction.

To summarize: there exist simply connected, strongly pseudoconvex domains $G \subset \mathbb{C}^2$, $0 \in G$, such that $\varkappa_G(0; \cdot)$ is not a psh function.

Now, we come back to Proposition 3.5.3 and apply it to the cases of the unit ball and the unit polydisc.

Example 3.5.6.

(a) For $z \in \mathbb{B}_n$ and $X \in \mathbb{C}^n$, we claim that the following formula is true.

$$\varkappa_{\mathbb{B}_n}(z;X) = \left[\frac{\|X\|^2}{1-\|z\|^2} + \frac{|\langle z,X\rangle|^2}{(1-\|z\|^2)^2}\right]^{1/2}. \tag{3.5.1}$$

For the proof, we recall that $\varkappa_{\mathbb{B}_n}$ is invariant under $\operatorname{Aut}(\mathbb{B}_n)$ (cf. Corollary 2.3.5).

(b) A similar argument leads to the corresponding formula for the unit polycylinder.

$$\varkappa_{\mathbb{D}^n}(z;X) = \max\left\{\frac{|X_1|}{1-|z_1|^2},\dots,\frac{|X_n|}{1-|z_n|^2}\right\}. \tag{3.5.2}$$

Remark 3.5.7. In the context of Remark 2.3.8, one could expect to also get a formula for $\varkappa_{\mathbb{L}_n}(z;X)$, where $\mathbb{L}_n$ is the unit Lie ball. Unfortunately, in the case of the Lie ball the formula for the automorphism $h_a \in \operatorname{Aut}(\mathbb{L}_n)$ with $h_a(a) = 0$ is rather complicated. The reader may try to get a formula for $\varkappa_{\mathbb{L}_n}(z;X)$ using [237].

From these examples, we deduce the following consequences; cf. Corollary 2.3.3.

Corollary 3.5.8.

(a) *Let $G_j = G_{\mathfrak{h}_j} \subsetneq \mathbb{C}^n$ be pseudoconvex balanced domains with Minkowski functions $\mathfrak{h}_j$, $j = 1,2$. Then, the following conditions are equivalent:*

(i) *there exists a biholomorphic mapping $F : G_1 \longrightarrow G_2$ with $F(0) = 0$;*

(ii) *there exists a $\mathbb{C}$-linear isomorphism $L : \mathbb{C}^n \longrightarrow \mathbb{C}^n$ such that $\mathfrak{h}_2 \circ L = \mathfrak{h}_1$, i.e., e_1 and G_2 are linearly equivalent.*

(b) *(Cf. [411]) Let $G = G_{\mathfrak{h}}$ be a bounded balanced domain in $\mathbb{C}^n$. Then, the following conditions are equivalent:*

(i) *there exists a biholomorphic mapping $F : G_{\mathfrak{h}} \longrightarrow \mathbb{B}_n$ with $F(0) = 0$;*

(ii) *$G_{\mathfrak{h}}$ and $\mathbb{B}_n$ are linearly equivalent;*

(iii) *$\mathfrak{h}^2 \in \mathcal{C}^2(\mathbb{C}^n)$.*

Corollary 3.5.9.

(a) *For any compact subset K of any domain $G \subset \mathbb{C}^n$ there is a suitable constant $C > 0$ such that the following inequality holds on $K \times \mathbb{C}^n$:*

$$\varkappa_G(z;X) \le C\|X\|, \quad z \in K,\ X \in \mathbb{C}^n.$$

(b) *If, in addition, G is bounded, then there exists a constant $C > 0$ such that for any $z \in G$, $X \in \mathbb{C}^n$ we have $\varkappa_G(z;X) \ge C\|X\|$.*

Proof. Use (3.5.1) and the contractibility of the system $(\varkappa_G)_G$. □

Observe that the second part of Corollary 3.5.9 can be read as follows: $\varkappa_G$ is positive definite if G is bounded.

In general, it turns out that $\varkappa_G(z;\cdot)$ need not be a norm on $\mathbb{C}^n$.

Example 3.5.10. Put $G := \{z \in \mathbb{C}^2 : |z_1| < 1, |z_2| < 1, |z_1 z_2| < 1/2\}$. Of course, G is a bounded balanced pseudoconvex domain with Minkowski function

$$\mathfrak{h}(z) = \max\{|z_1|, |z_2|, \sqrt{2|z_1||z_2|}\}.$$

Therefore, we know that $\varkappa_G(0; X) = \mathfrak{h}(X)$. In particular, we have

$$\varkappa_G\left(0; \left(\frac{3}{2}, \frac{3}{2}\right)\right) = \frac{3}{\sqrt{2}} > 2 = \varkappa_G\left(0; \left(1, \frac{1}{2}\right)\right) + \varkappa_G\left(0; \left(\frac{1}{2}, 1\right)\right).$$

Moreover, Proposition 3.5.3 shows that, in general, $\varkappa_G(z;\cdot)$ is not continuous as a function of the second variable. For more details, the reader should compare the analogous situation for the Lempert function in § 3.1.

The next example shows that the Kobayashi–Royden pseudometric is, in general, not continuous even as a function of the first variable.

Example 3.5.11 (cf. [133]). We will construct a pseudoconvex domain $G \subset \mathbb{C}^2$ satisfying the following two properties:

(a) there is a dense subset $M \subset \mathbb{C}$ such that $(M \times \mathbb{C}) \cup (\mathbb{C} \times \{0\}) \subset G$; in particular, $\varkappa_G(z; (0, 1)) = 0$ for all $z \in A := M \times \mathbb{C}$ and $\boldsymbol{k}_G \equiv 0$;

(b) there exists a point $z^0 \in G \setminus A$ such that $\varkappa_G(z^0; \cdot)$ is positive definite, i.e., $\varkappa_G(z^0; X) \geq C\|X\|$ (for some $C > 0$).

Construction of G.

(i) As a first step, we claim the existence of a subharmonic function $u : \mathbb{C} \longrightarrow \mathbb{R}_{-\infty}$ with $u \not\equiv -\infty$ but such that the set $M := \{\lambda \in \mathbb{C} : u(\lambda) = -\infty\}$ is dense in $\mathbb{C}$.

We construct u as an infinite sum. So, we start by choosing a dense sequence $(a_j^{(k)})_j \subset \mathbb{B}(k) \setminus \{0\}$. Then we put

$$u_k(\lambda) := \sum_{j=1}^{\infty} \lambda_j^{(k)} \log \frac{|\lambda - a_j^{(k)}|}{2(j+k)},$$

where the numbers $\lambda_j^{(k)} > 0$ are chosen in such a way that

$$\sum_{j=1}^{\infty} \lambda_j^{(k)} \log \frac{|a_j^{(k)}|}{2(j+k)} \in \mathbb{R}.$$

Observe that, locally uniformly, almost all of the summands become negative, which shows that u_k is a subharmonic function on $\mathbb{C}$; cf. Appendix B.4.14. Moreover, by definition we have

$$u_k \le 0 \text{ on } \mathbb{B}(k), \quad u_k(a_j^{(k)}) = -\infty, \quad \text{and} \quad u_k(0) > -\infty.$$

Now, we repeat the same procedure.

We choose new positive numbers λ_k such that $\sum_{k=1}^{\infty} \lambda_k u_k(0)$ is a real number. Then, we obtain our function u by the formula

$$u(\lambda) := \sum_{k=1}^{\infty} \lambda_k u_k(\lambda).$$

The same argument as before leads to the conclusion that u is subharmonic on $\mathbb{C}$, $u(0) \neq -\infty$ but $u = -\infty$ on the dense set $M := \bigcup_k \{a_j^{(k)} : j \in \mathbb{N}\}$.

(ii) We use the function u just constructed to define a new plurisubharmonic function $\psi : \mathbb{C}^2 \longrightarrow \mathbb{R}_{-\infty}$ by

$$\psi(z) := |z_2| \exp(|z_1|^2 + |z_2|^2) \cdot \exp(u(z_1)).$$

Then, our domain G is obtained as

$$G := \{z \in \mathbb{C}^2 : \psi(z) < 1\}.$$

We observe that the set $A := \{z \in \mathbb{C}^2 : z_1 \in M\}$ is a dense subset of G.

Now, we are in a position to verify our claims (a) and (b).

By construction, it is clear that if $z \in A$, then $\{z_1\} \times \mathbb{C} \subset G$, which implies that

$$\varkappa_G(z; (0,1)) = \varkappa_G(\Phi(0); \Phi'(0)1) \le \varkappa_{\mathbb{C}}(0; 1) = 0,$$

where $\Phi(\lambda) := (z_1, z_2 + \lambda)$. Moreover, using the triangle inequality, it is clear that $\boldsymbol{k}_G \equiv 0$.

To prove (b) we select a point $z_1^0 \in \mathbb{C}$ with $u(z_1^0) > -\infty$ and $t > 0$ such that $z^0 := (z_1^0, t) \in G$.

Now, we observe that whenever $\varphi \in \mathcal{O}(\mathbb{D}, G)$ is an analytic disc in G with $\varphi(0) = z^0$, $\|\varphi'(0)\| \le C$ for a suitable positive C which can be chosen independently of φ.

Let φ be as above. Then, the Cauchy integral leads to the following representation of $\varphi_j'(0)$ ($\varphi = (\varphi_1, \varphi_2)$):

$$\varphi_j'(0) = \frac{1}{2\pi i} \int_{|\zeta|=1/2} \frac{\varphi_j(\zeta)}{\zeta^2} d\zeta.$$

Hence,

$$\|\varphi'(0)\|^2 \le \frac{2}{\pi} \int_0^{2\pi} \left\| \varphi\left(\frac{1}{2} e^{i\theta}\right) \right\|^2 d\theta.$$

Here, we have used the Hölder inequality.

To continue the estimate, we introduce the plurisubharmonic function

$$\widehat{\psi}(z) := \log \psi(z) - \frac{1}{2}\|z\|^2,$$

for which the following inequality holds on G: $\widehat{\psi}(z) < -\|z\|^2/2$. Then, the mean value inequality for subharmonic functions shows that

$$\begin{aligned} -\infty < \widehat{\psi}(z^0) = \widehat{\psi} \circ \varphi(0) &\leq \frac{1}{2\pi}\int_0^{2\pi} \widehat{\psi} \circ \varphi\left(\frac{1}{2}e^{i\theta}\right) d\theta \\ &\leq -\frac{1}{4\pi}\int_0^{2\pi} \left\|\varphi\left(\frac{1}{2}e^{i\theta}\right)\right\|^2 d\theta. \end{aligned}$$

Finally, combining the information obtained, we find that $\|\varphi'(0)\|^2 \leq -8\widehat{\psi}(z^0)$. Hence, we have also proved part (b) of our claim.

Remark 3.5.12. We emphasize that the above example can be modified to obtain a pseudoconvex domain $G \subset \mathbb{C}^2$ that is not $\boldsymbol{k}$-hyperbolic, but such that $\boldsymbol{\varkappa}_G(z;\cdot)$ is positive definite for every $z \in G$.

In fact, define

$$u(\lambda) := \sum_{j=2}^{\infty} \frac{1}{j^2} \max\left\{\log\frac{|\lambda - 1/j|}{2}, -j^3\right\}, \quad \lambda \in \mathbb{D},$$
$$G := \{z \in \mathbb{D} \times \mathbb{C} : \psi(z) := |z_2| \cdot (\exp\|z\|^2) \cdot \exp u(z_1) < 1\}.$$

Note that u is subharmonic on $\mathbb{D}$, $u(\lambda) \neq -\infty$, $\lambda \in \mathbb{D}$. Hence, G is a pseudoconvex domain containing larger and larger discs

$$\{(1/k, z_2) : \psi(1/k, z_2) < 1\}.$$

Thus, the continuity of the Kobayashi pseudodistance implies $\boldsymbol{k}_G((0,0),(0,z_2)) = 0$, whenever $(0, z_2) \in G$, i.e., G is not $\boldsymbol{k}$-hyperbolic.

On the other hand, following the argument of (b) we easily derive

$$\boldsymbol{\varkappa}_G(z;X) \geq \frac{\|X\|}{-8\left(\log\psi(z) - \frac{1}{2}\|z\|^2\right)}, \quad z \in G.$$

To summarize, the last two examples show that, in general, the Kobayashi–Royden pseudometric is continuous neither in the first nor in the second variable. Nevertheless, we will see that this function is always upper semicontinuous even in both variables simultaneously.

Proposition 3.5.13. *For any domain $G \subset \mathbb{C}^n$ the Kobayashi–Royden pseudometric $\boldsymbol{\varkappa}_G : G \times \mathbb{C}^n \longrightarrow [0, +\infty)$ is upper semicontinuous.*

Proof. We start with a fixed point $z^0 \in G$ and a tangent vector $X^0 \in \mathbb{C}^n$, and we suppose that $\varkappa_G(z^0; X^0) < A$. By definition, we find an analytic disc $\varphi \in \mathcal{O}(\mathbb{D}, G)$ with $\varphi(0) = z^0$ and $\alpha\varphi'(0) = X^0$, where α is a positive number, $\alpha < A$. Without loss of generality, we may assume that φ is holomorphic on $\overline{\mathbb{D}}$. Thus, an ε-neighborhood of $\varphi(\overline{\mathbb{D}})$ remains inside G.

Now, we take $z \in G$ with $\|z-z^0\| < \varepsilon/4$ and $X \in \mathbb{C}^n$ with $(1/\alpha)\|X-X^0\| < \varepsilon/4$. The following mapping $\psi : \mathbb{D} \longrightarrow \mathbb{C}^n$,

$$\psi(\lambda) := \varphi(\lambda) + \left(z - z^0\right) + (\lambda/\alpha)\left(X - X^0\right),$$

has its image in G, and it satisfies

$$\psi(0) = z, \ \alpha\psi'(0) = \alpha\varphi'(0) + X - X^0 = X,$$

which means that $\varkappa_G(z; X) \le \alpha < A$, i.e., the upper semicontinuity of $\varkappa_G$ at $(z^0; X^0)$ is verified. □

Similarly to the case of the Lempert function (cf. § 3.2), we also obtain better results for the Kobayashi–Royden pseudometric on taut domains. Since the argument here is more or less the same, we only formulate the result. The proof is left to the reader as an exercise.

Proposition 3.5.14. *Let G be a taut domain in $\mathbb{C}^n$. Then,*

(a) *for any $z \in G$ and for any $X \in \mathbb{C}^n$ there exists an extremal analytic disc $\varphi \in \mathcal{O}(\mathbb{D}, G)$, i.e., $\varphi(0) = z$ and $\varkappa_G(z; X)\varphi'(0) = X$;*

(b) *the Kobayashi–Royden pseudometric is continuous on $G \times \mathbb{C}^n$.*

Remark 3.5.15. Taking only injective holomorphic discs $\varphi : \mathbb{D} \longrightarrow G$, K. T. Hahn ([207]) has introduced a family $(\boldsymbol{\eta}_G)_G$ of pseudometrics and a family $(\boldsymbol{h}_G)_G$ of pseudodistances that are contractible with respect to injective holomorphic mappings – see § 8.1.

3.6 The Kobayashi–Buseman pseudometric

We recall that, in general, the Kobayashi–Royden pseudometric $\varkappa_G(z_0; \cdot)$ does not give a seminorm on $\mathbb{C}^n$ if $G \subset \mathbb{C}^n$. Thus, for example, it may happen that the zero set of $\varkappa_G(z_0; \cdot)$ is not a linear subspace of the tangent space $\mathbb{C}^n$. To overcome this unpleasant fact S. Kobayashi [315] has introduced a new infinitesimal pseudometric, following old ideas of Buseman (cf. [80]).

Let G be a domain in $\mathbb{C}^n$. Then, the function $\hat{\varkappa}_G : G \times \mathbb{C}^n \longrightarrow [0, +\infty)$ given by

$$\hat{\varkappa}_G(z; X) := \widehat{\varkappa_G(z; \cdot)}(X), \quad z \in G, \ X \in \mathbb{C}^n,$$

where the right hand side is taken in the sense of § 2.2.2, is called the *Kobayashi–Buseman pseudometric for* G. Notice that (cf. Remark 2.2.4(f))

$$\hat{\varkappa}_G(z;X) = \sup\{|Y \bullet X| : Y \in \Gamma_G(z)\},$$

where

$$\Gamma_G(z) := \{Y \in \mathbb{C}^n : |Y \bullet Z| \le 1 \text{ for all } Z \in \mathbb{C}^n \text{ with } \varkappa_G(z;Z) < 1\}$$

denotes the polar of the unit Kobayashi–Royden ball

$$\Delta(z) := \{X \in \mathbb{C}^n : \varkappa_G(z;X) < 1\}.$$

Remark 3.6.1. At first, we collect simple properties of this new function $\hat{\varkappa}_G$ (cf. § 2.2.1):

(a) $\hat{\varkappa}_G \le \varkappa_G$;

(b) $\hat{\varkappa}_G(z;\cdot)$ is a complex seminorm on $\mathbb{C}^n$;

(c) $\hat{\varkappa}_G(z;\cdot)$ is continuous on $\mathbb{C}^n$;

(d) $\{X \in \mathbb{C}^n : \hat{\varkappa}_G(z;X) \le 1\}$ is nothing other than the bipolar set of $\Delta(z)$, and by a standard result from functional analysis it is the closed absolutely convex hull of $\Delta(z)$;

(e) $\{X \in \mathbb{C}^n : \hat{\varkappa}_G(z;X) < 1\}$ coincides with the convex hull of $\Delta(z)$;

(f) the system $(\hat{\varkappa}_G)_G$ of the Kobayashi–Buseman pseudometrics is contractible with respect to holomorphic mappings and $\hat{\varkappa}_{\mathbb{D}} = \varkappa_{\mathbb{D}}$.

We also have to mention that, in general, $\hat{\varkappa}_G$ is not a continuous function of both variables as Example 3.5.11 can easily show. Nevertheless, the result analogous to Theorem 3.5.13 remains true (cf. [315]).

Proposition 3.6.2. *For any domain* $G \subset \mathbb{C}^n$*, the Kobayashi–Buseman pseudometric is an upper semicontinuous function on* $G \times \mathbb{C}^n$.

Proof. We begin with the following observation: let $z^0 \in G$ and $X^0 \in \mathbb{C}^n$ be fixed. Then, for $z \in G$, $X \in \mathbb{C}^n$, it turns out that

$$\begin{aligned}\hat{\varkappa}_G\left(z;X\right) &\le \hat{\varkappa}_G\left(z;X^0\right) + \hat{\varkappa}_G\left(z;X-X^0\right)\\ &\le \hat{\varkappa}_G\left(z;X^0\right) + \sum_{j=1}^{n} |X_j - X_j^0|\hat{\varkappa}_G\left(z;e_j\right),\end{aligned}$$

where e_j denotes the j-th standard unit vector. Hence, it suffices to prove that for a fixed vector, for example for X^0, the function $\hat{\varkappa}_G(\cdot;X^0)$ is upper semicontinuous at z^0.

Assuming the contrary, there exists a sequence $(z^\nu)_\nu \subset G$ with $z^\nu \underset{\nu\to\infty}{\longrightarrow} z^0$ and $\widehat{\varkappa}_G(z^\nu; X^0) > A > \widehat{\varkappa}_G(z^0; X^0)$. By definition of $\widehat{\varkappa}_G$, we find vectors $Y^\nu \in \mathbb{C}^n$ such that $|Y^\nu \bullet Z| \le 1$ for all Z with $\varkappa_G(z^\nu; Z) < 1$ and $|Y^\nu \bullet X^0| \ge A$. Hence, $|Y^\nu \bullet Z| \le \varkappa_G(z^\nu; Z) \le M\|Z\|$ for $Z \in \mathbb{C}^n$ and $\nu \ge 1$, where M is a suitable constant. In particular, the sequence $(Y^\nu)_\nu$ is bounded, so we may assume that $Y^\nu \underset{\nu\to\infty}{\longrightarrow} Y^0$. Hence, we have $|Y^0 \bullet X^0| \ge A$.

Now, fix an arbitrary vector Z with $\varkappa_G(z^0; Z) < 1$. The upper semicontinuity of $\varkappa_G$ ensures that $\varkappa_G(z^\nu; Z) < 1$ if ν is sufficiently large. Thus, we arrive at $|Y^\nu \bullet Z| \le 1$, and therefore $|Y^0 \bullet Z| \le 1$. Hence, we have: $\widehat{\varkappa}_G(z^0; X^0) \ge A > \widehat{\varkappa}_G(z^0; X^0)$; a contradiction. □

So far, we know that the Kobayashi–Royden pseudometric and Kobayashi–Buseman pseudometrics are upper semicontinuous functions. Therefore, they can be used to define the length of a piecewise $\mathcal{C}^1$-curve, and then the minimal length of all such curves connecting two fixed points will yield a new pseudodistance.

For a piecewise $\mathcal{C}^1$-curve $\alpha : [0,1] \longrightarrow G$ we set:

$$L_{\varkappa_G}(\alpha) := \int_0^1 \varkappa_G(\alpha(t); \alpha'(t))dt, \quad L_{\widehat{\varkappa}_G}(\alpha) := \int_0^1 \widehat{\varkappa}_G(\alpha(t); \alpha'(t))dt.$$

The numbers $L_{\varkappa_G}(\alpha)$, $L_{\widehat{\varkappa}_G}(\alpha)$ are called the *$\varkappa_G$-length* and the *$\widehat{\varkappa}_G$-length* of the curve α, respectively.

For points $z', z' \in G$ we define:

$$(\textstyle\int \varkappa_G)(z, z'') := \inf\{L_{\varkappa_G}(\alpha) : \alpha \text{ is a piecewise } \mathcal{C}^1\text{-curve in } G \text{ from } z' \text{ to } z''\},$$
$$(\textstyle\int \widehat{\varkappa}_G)(z', z'') := \inf\{L_{\widehat{\varkappa}_G}(\alpha) : \alpha \text{ is a piecewise } \mathcal{C}^1\text{-curve in } G \text{ from } z' \text{ to } z''\}.$$

$\int \varkappa_G$ (resp. $\int \widehat{\varkappa}_G$) is called the *integrated form* of $\varkappa_G$ (resp. of $\widehat{\varkappa}_G$).

Remark 3.6.3. Observe that the systems $(\int \varkappa_G)_G$ and $(\int \widehat{\varkappa}_G)_G$ of pseudodistances are contractible with respect to holomorphic mappings, and that $\int \varkappa_{\mathbb{D}} = \int \widehat{\varkappa}_{\mathbb{D}} = \boldsymbol{p}$. Therefore, we obtain the following chain of inequalities: $\int \widehat{\varkappa}_G \le \int \varkappa_G \le \boldsymbol{k}_G$.

The main result, according to the integrated forms introduced above, is that they coincide with the Kobayashi pseudodistance.

Theorem 3.6.4. *If G is a domain in $\mathbb{C}^n$, then $\boldsymbol{k}_G = \int \varkappa_G = \int \widehat{\varkappa}_G$.*

Before we begin the proof, we recall the following result due to Harris (cf. [215]).

Lemma 3.6.5. *Let $d : G \times G \longrightarrow [0, +\infty)$ be a pseudodistance that is locally uniformly majorized by the Euclidean distance. Moreover, suppose that $\delta : G \times \mathbb{C}^n \longrightarrow [0, +\infty)$ is an upper semicontinuous function with $\delta(a; \lambda X) = |\lambda|\delta(a; X)$*

for all $a \in G$, $\lambda \in \mathbb{C}$, and $X \in \mathbb{C}^n$. Set

$$(\textstyle\int \delta)(z', z'') := \inf\Big\{ \int_0^1 \delta(\alpha(t); \alpha'(t))dt : \alpha : [0,1] \longrightarrow G \text{ is a piecewise } \mathcal{C}^1\text{-curve joining } z', z'' \Big\}.$$

Then, $d \le \int \delta$ if the following relation is true:

$$\limsup_{t \to 0+} \frac{d(a, a+tX)}{t} \le \delta(a; X), \quad a \in G, \ X \in \mathbb{C}^n.$$

Proof. We fix two points z', z'' in G and we take a $\mathcal{C}^1$-curve $\alpha : [0,1] \longrightarrow G$ connecting these two points. With $f(t) := d(z', \alpha(t))$, we observe that if s, t are near $t_0 \in [0,1]$, then

$$\begin{aligned} |f(t) - f(s)| &= |d(z', \alpha(t)) - d(z', \alpha(s))| \\ &\le d(\alpha(s), \alpha(t)) \le C\,\|\alpha(s) - \alpha(t)\| \le \widetilde{C}\,|s - t|. \end{aligned}$$

Here, we have used the fact that d is majorized by the Euclidean distance. Thus, we know that the function f is locally Lipschitz. Hence, f' exists almost everywhere and $d(z', z'') = \int_0^1 f'(t)dt$. It remains to estimate f':

$$\begin{aligned} |f'(\tau)| &= \lim_{h \to 0+} \frac{|f(\tau + h) - f(\tau)|}{h} \le \limsup_{h \to 0+} \frac{d(\alpha(\tau + h), \alpha(\tau))}{h} \\ &\le \limsup_{h \to 0+} \frac{d(\alpha(\tau) + h\alpha'(\tau), \alpha(\tau))}{h} + \limsup_{h \to 0+} \frac{d(\alpha(\tau) + h\alpha'(\tau), \alpha(\tau + h))}{h} \\ &\le \delta(\alpha(\tau); \alpha'(\tau)) + \limsup_{h \to 0+} \frac{C\,\|\alpha(\tau) + h\alpha'(\tau) - \alpha(\tau + h)\|}{h} = \delta(\alpha(\tau); \alpha'(\tau)) \end{aligned}$$

for almost all $\tau \in [0,1]$. Finally, we obtain

$$d(z', z'') \le \int_0^1 \delta(\alpha(\tau); \alpha'(\tau))d\tau.$$

By means of the triangle inequality for d, the case of piecewise $\mathcal{C}^1$-curves is clear. □

Proof of Theorem 3.6.4. First, recall that $\boldsymbol{\varkappa}_G$ is upper semicontinuous and that $\boldsymbol{k}_G$ is locally uniformly majorized by the Euclidean distance, i.e., all assumptions of Lemma 3.6.5 are fulfilled.

So, by Remark 3.6.3, to obtain $\boldsymbol{k}_G = \int \boldsymbol{\varkappa}_G$, we have to verify the inequality of Lemma 3.6.5 for $\boldsymbol{k}_G$ and $\boldsymbol{\varkappa}_G$.

Let us fix $z^0 \in G$ and $X^0 \in \mathbb{C}^n$. If $\varepsilon > 0$, then we are able to choose an analytic disc $\varphi \in \mathcal{O}(\mathbb{D}, G)$ with $\varphi(0) = z^0$, $\alpha\varphi'(0) = X^0$, and $0 < \alpha < \boldsymbol{\varkappa}_G(z^0; X^0) + \varepsilon$. Observe that φ can be written as $\varphi(\lambda) = z^0 + (\lambda/\alpha)X^0 + \lambda^2 H(\lambda)$, where $H \in \mathcal{O}(\mathbb{D}, \mathbb{C}^n)$.

Then, for positive t, $t\alpha < 1$, $z^0 + tX^0 \in G$, we obtain

$$\begin{aligned}\frac{\boldsymbol{k}_G\left(z^0, z^0 + tX^0\right)}{t} &\le \frac{\boldsymbol{k}_G\left(z^0, \varphi(t\alpha)\right)}{t} + \frac{\boldsymbol{k}_G\left(\varphi(t\alpha), z^0 + tX^0\right)}{t} \\ &\le \frac{\boldsymbol{k}_G\left(\varphi(0), \varphi(t\alpha)\right)}{t} + \frac{C}{t}\|t^2\alpha^2 H(t\alpha)\| \\ &\le \frac{1}{2t}\log\frac{1+t\alpha}{1-t\alpha} + C\alpha^2 t\|H(t\alpha)\|.\end{aligned}$$

Thus, in the limit case, we get

$$\limsup_{t\to 0+}\frac{\boldsymbol{k}_G\left(z^0, z^0 + tX^0\right)}{t} \le \alpha < \boldsymbol{\varkappa}_G\left(z^0; X^0\right) + \varepsilon.$$

Since ε was arbitrary, we get $\boldsymbol{k}_G = \int \boldsymbol{\varkappa}_G$ by Lemma 3.6.5.

To prove the remaining inequality $\boldsymbol{k}_G \le \int \widehat{\boldsymbol{\varkappa}}_G$, we will use Lemma 3.6.5 again. As above, we begin with a fixed point $z^0 \in G$ and a vector $X^0 \in \mathbb{C}^n$. From what we have mentioned about $\{Y \in \mathbb{C}^n : \widehat{\boldsymbol{\varkappa}}_G(z^0; Y) < 1\}$, it is easy to conclude that for any $\varepsilon > 0$ the vector X^0 can be represented as $X^0 = \sum_{j=1}^m X^j$ with vectors X^j satisfying

$$\sum_{j=1}^m \boldsymbol{\varkappa}_G\left(z^0; X^j\right) \le \widehat{\boldsymbol{\varkappa}}_G\left(z^0; X^0\right) + \varepsilon/2.$$

Moreover, since $\boldsymbol{\varkappa}_G$ is upper semicontinuous, we have $\boldsymbol{\varkappa}_G(z; X^j) \le \boldsymbol{\varkappa}_G(z^0; X^j) + \varepsilon/2m$ if z belongs to an appropriate neighborhood $U = U(z^0)$ of z^0. Putting the above together, we get

$$\begin{aligned}&\limsup_{t\to 0+}\frac{1}{t}\boldsymbol{k}_G\left(z^0, z^0 + tX^0\right) \le \limsup_{t\to 0+}\frac{1}{t}\boldsymbol{k}_G\left(z^0, z^0 + tX^1\right) \\ &\qquad + \sum_{j=1}^{m-1}\limsup_{t\to 0+}\frac{1}{t}\boldsymbol{k}_G\left(z^0 + t\left(X^1 + \dots + X^j\right), z^0 + t\left(X^1 + \dots + X^{j+1}\right)\right) \\ &\quad \le \boldsymbol{\varkappa}_G\left(z^0; X^1\right) + \sum_{j=1}^{m-1}\limsup_{t\to 0+}\frac{1}{t}\int_0^t \boldsymbol{\varkappa}_G\Big(z^0 + t\left(X^1 + \dots + X^j\right) \\ &\quad + \tau X^{j+1}; X^{j+1}\Big)\,d\tau \\ &\quad \le \boldsymbol{\varkappa}_G\left(z^0; X^1\right) + \dots + \boldsymbol{\varkappa}_G\left(z^0; X^m\right) + \frac{\varepsilon}{2} \le \widehat{\boldsymbol{\varkappa}}_G\left(z^0; X^0\right) + \varepsilon,\end{aligned}$$

which concludes the proof of Theorem 3.6.4. □

Remark 3.6.6. We emphasize that the argument used to prove Theorem 3.6.4 is also valid in a more general context, which will be discussed later in Chapter 4.

Before we proceed to apply the previous result, we want to recall that the Carathéodory–Reiffen pseudometric has appeared as a kind of a strong derivative of the Carathéodory pseudodistance (cf. Proposition 2.7.1(d)). The following example (see [508]) shows that there is no analogous relation between the Kobayashi pseudodistance and the Kobayashi–Royden pseudometric.

Example 3.6.7. Let us consider the following bounded balanced pseudoconvex domain in $\mathbb{C}^2$:

$$G := \{z \in \mathbb{C}^2 : |z_1| < 1, |z_2| < 1, |z_1 z_2| < a^2\} \quad (0 < a < 1/2).$$

For $z' = (z'_1, z'_2)$ and $z'' = (z''_1, z''_2)$ with $|z'_j| < a^2$ and $|z''_j| < a^2$ we easily derive

$$\boldsymbol{k}_G(z', z'') \le \boldsymbol{k}_G(z', (z''_1, z'_2)) + \boldsymbol{k}_G((z''_1, z'_2), z'') \le \boldsymbol{p}(z'_1, z''_1) + \boldsymbol{p}(z'_2, z''_2).$$

Estimating the following general differential quotient, we obtain

$$\limsup_{\substack{a' \to 0 \\ X' \to (a,a) \\ \lambda \twoheadrightarrow 0}} \frac{\boldsymbol{k}_G(a', a' + \lambda X')}{|\lambda|} \le \limsup_{\substack{a' \to 0 \\ X' \to (a,a) \\ \lambda \twoheadrightarrow 0}} \frac{\boldsymbol{p}(a'_1, a'_1 + \lambda X'_1)}{|\lambda|} + \limsup_{\substack{a' \to 0 \\ X' \to (a,a) \\ \lambda \twoheadrightarrow 0}} \frac{\boldsymbol{p}(a'_2, a'_2 + \lambda X'_2)}{|\lambda|} = 2a.$$ [1]

On the other hand, we know that $\boldsymbol{\varkappa}_G(0; (a, a)) = 1$. Hence, the general differential quotient differs from the Kobayashi–Royden metric. The full discussion of differential quotients of pseudodistances will be given in Chapters 4 and 5. But already here, we would like to see what happens for plane domains.

Remark 3.6.8. Let G be a domain in $\mathbb{C}$. Then, we claim that

$$\boldsymbol{\varkappa}_G(a; 1) = \lim_{\substack{z', z'' \to a \\ z' \ne z''}} \frac{\boldsymbol{k}_G(z', z'')}{|z' - z''|} = \lim_{\substack{z', z'' \to a \\ z' \ne z''}} \frac{\boldsymbol{\ell}_G(z', z'')}{|z' - z''|}, \qquad a \in G.$$

Without loss of generality, we may assume that G is a taut domain. Suppose that $(z'_\nu)_{\nu\in\mathbb{N}}, (z''_\nu)_{\nu\in\mathbb{N}} \subset G$ with $z'_\nu \ne z''_\nu$, $\nu \in \mathbb{N}$, $\lim_{\nu\to\infty} z'_\nu = a = \lim_{\nu\to\infty} z''_\nu$ are such that $\lim \frac{\boldsymbol{k}_G(z'_\nu, z''_\nu)}{|z'_\nu - z''_\nu|}$ exists. Then, there are holomorphic mappings $\varphi_\nu : \mathbb{D} \longrightarrow G$, $\varphi_\nu(0) = z'_\nu$, $\varphi_\nu(\sigma_\nu) = z''_\nu$ with $0 < \sigma_\nu < 1$, and $\boldsymbol{k}_G(z'_\nu, z''_\nu) = \boldsymbol{\ell}_G(z'_\nu, z''_\nu) =$

[1] $\limsup_{\lambda \twoheadrightarrow 0} := \limsup_{\substack{\lambda \to 0 \\ \lambda \ne 0}}$, cf. § C.

$p(0,\sigma_\nu) \underset{\nu\to\infty}{\longrightarrow} 0$. Therefore, taking an appropriate subsequence, we may assume that $\varphi_\nu \overset{K}{\underset{\nu\to\infty}{\Longrightarrow}} \varphi$ with $\varphi \in \mathcal{O}(\mathbb{D}, G)$, $\varphi(0) = a$. Hence, we obtain

$$\lim_{\nu\to\infty} \frac{\boldsymbol{k}_G(z'_\nu, z''_\nu)}{|z'_\nu - z''_\nu|} = \lim_{\nu\to\infty} \left(\frac{\boldsymbol{p}(0,\sigma_\nu)}{\sigma_\nu} \cdot \frac{\sigma_\nu}{|\varphi_\nu(0) - \varphi_\nu(\sigma_\nu)|}\right) = \frac{1}{|\varphi'(0)|} \geq \boldsymbol{\varkappa}_G(a;1).$$

On the other hand, take $\varphi \in \mathcal{O}(\mathbb{D}, G)$ with $\varphi(0) = a$, $\varphi'(0) \neq 0$. Since $\varphi'(0) \neq 0$, we may assume that there are $\sigma'_\nu \in \mathbb{D}$, $\sigma''_\nu \in \mathbb{D}$, $\sigma'_\nu \neq \sigma''_\nu$, with $\lim_{\nu\to\infty} \sigma'_\nu = \lim_{\nu\to\infty} \sigma''_\nu = 0$, such that $\varphi(\sigma'_\nu) = z'_\nu$, $\varphi(\sigma''_\nu) = z''_\nu$. Hence, we get

$$\lim_{\nu\to\infty} \frac{\boldsymbol{k}_G(z'_\nu, z''_\nu)}{|z'_\nu - z''_\nu|} \leq \limsup_{\nu\to\infty} \frac{\boldsymbol{p}(\sigma'_\nu, \sigma''_\nu)}{|\sigma'_\nu - \sigma''_\nu|} \cdot \frac{|\sigma'_\nu - \sigma''_\nu|}{|\varphi(\sigma'_\nu) - \varphi(\sigma''_\nu)|} = \frac{1}{|\varphi'(0)|},$$

which implies the remaining inequality.

3.7 Product formula

We conclude this chapter with a discussion of the way to calculate the objects studied before on product domains. We begin with information that can be derived directly from the original definitions.

Proposition 3.7.1. *Suppose that two domains $G \subset \mathbb{C}^n$ and $D \subset \mathbb{C}^m$ are given. Then, the following formulas hold:*

(a) $\boldsymbol{\ell}_{G\times D}((z',w'),(z'',w'')) = \max\{\boldsymbol{\ell}_G(z',z''), \boldsymbol{\ell}_D(w',w'')\}$,
$$z',z'' \in G,\ w',w'' \in D;$$

(b) $\boldsymbol{\varkappa}_{G\times D}((z,w);(X,Y)) = \max\{\boldsymbol{\varkappa}_G(z;X), \boldsymbol{\varkappa}_D(w;Y)\}$,
$$z \in G,\ w \in D,\ X \in \mathbb{C}^n,\ Y \in \mathbb{C}^m;$$

(c) $\widehat{\boldsymbol{\varkappa}}_{G\times D}((z,w);(X,Y)) = \max\{\widehat{\boldsymbol{\varkappa}}_G(z;X), \widehat{\boldsymbol{\varkappa}}_D(w;Y)\}$, *$z, w, X, Y$ as above.*

Proof. By the contractibility property with respect to holomorphic mappings (here for projections), the inequalities "$\geq$" are obvious.

Now, we proceed to prove the inverse inequalities.

(a) Suppose that $\boldsymbol{\ell}_{G\times D}((z',w'),(z'',w'')) > A > \max\{\boldsymbol{\ell}_G(z',z''), \boldsymbol{\ell}_D(w',w'')\}$.

Then, by definition, we find holomorphic maps $\varphi \in \mathcal{O}(\mathbb{D}, G)$ and $\psi \in \mathcal{O}(\mathbb{D}, D)$ sharing the following properties:

$$\begin{aligned} &\varphi(0) = z', &&\varphi(\sigma) = z'' \quad (0 \leq \sigma < 1), &&\boldsymbol{p}(0,\sigma) < A,\\ &\psi(0) = w', &&\psi(\tau) = w'' \quad (0 \leq \tau < 1), &&\boldsymbol{p}(0,\tau) < A. \end{aligned}$$

Without loss of generality, we may assume $0 \le \sigma \le \tau$, which opens the possibility of defining a new map $F \in \mathcal{O}(\mathbb{D}, G \times D)$ by

$$F(\lambda) := \left(\varphi\left(\frac{\sigma}{\tau}\lambda\right), \psi(\lambda)\right)$$

with $F(0) = (z', w')$ and $F(\tau) = (z'', w'')$. Hence, $\boldsymbol{\ell}_{G\times D}((z', w'), (z'', w'')) \le \boldsymbol{p}(0, \tau) < A$, contrary to our assumption.

(b) The argument here exactly follows the one used just before, and is hence left to the reader.

(c) Assertion (c) is an immediate consequence of the formula (b) and the properties of $\widehat{\boldsymbol{\varkappa}}_G$ that were formulated in Remark 3.6.1. □

Now, we combine the results given in Theorem 3.6.4 and Proposition 3.7.1 to obtain the following product formula.

Theorem 3.7.2. *Suppose $G_j \subset \mathbb{C}^{n_j}$ is a domain, $j = 1, 2$. Then, the following formula is true on $G_1 \times G_2$*

$$\boldsymbol{k}_{G_1\times G_2}((z_1', z_2'), (z_1'', z_2'')) = \max\{\boldsymbol{k}_{G_1}(z_1', z_1''), \boldsymbol{k}_{G_2}(z_2', z_2'')\}.$$

Proof. Obviously, it suffices to prove the inequality "$\le$". Let us suppose (without loss of generality) that $\boldsymbol{k}_{G_1}(z_1', z_1'') \ge \boldsymbol{k}_{G_2}(z_2', z_2'')$. For any $\varepsilon > 0$, Theorem 3.6.4 guarantees the existence of $\mathcal{C}^1$-curves

$$\gamma_j : [0,1] \longrightarrow G_j, \quad \gamma_j(0) = z_j', \quad \gamma_j(1) = z_j'',$$
$$\int_0^1 \boldsymbol{\varkappa}_{G_j}(\gamma_j(t); \gamma_j'(t))dt < \boldsymbol{k}_{G_j}(z_j', z_j'') + \varepsilon, \quad j = 1, 2.$$

Then, we choose continuous functions $h_j > 0$ on $[0, 1]$ with

$$\boldsymbol{\varkappa}_{G_j}(\gamma_j(t); \gamma_j'(t)) \le h_j(t), \quad \int_0^1 h_j(t)dt < \boldsymbol{k}_{G_j}(z_j', z_j'') + \varepsilon, \quad j = 1, 2.$$

It is easy to modify h_j in such a way that we have $\int_0^1 h_1(t)dt = \int_0^1 h_2(t)dt$. Thus, we obtain two strictly increasing functions

$$F_j(t) := \int_0^t h_j(\tau)d\tau, \quad j = 1, 2,$$

with $F_1(0) = F_2(0) = 0$ and $F_1(1) = F_2(1)$. Therefore, $\psi := F_2^{-1} \circ F_1$ gives a parameter transform of the unit interval with $\psi'(t) = h_1(t)/h_2(\psi(t))$.

Then, we observe that the curve $\boldsymbol{\gamma} : [0,1] \longrightarrow G_1 \times G_2$, $\boldsymbol{\gamma}(t) := (\boldsymbol{\gamma}_1(t), \boldsymbol{\gamma}_2 \circ \psi(t))$ connects (z_1', z_2') with (z_1'', z_2''). Moreover, we obtain

$$\boldsymbol{k}_{G_1 \times G_2}((z_1', z_2'), (z_1'', z_2'')) \le \int_0^1 \boldsymbol{\varkappa}_{G_1 \times G_2}(\boldsymbol{\gamma}(t); \boldsymbol{\gamma}'(t))dt$$

$$= \int_0^1 \max\{\boldsymbol{\varkappa}_G(\boldsymbol{\gamma}_1(t); \boldsymbol{\gamma}_1'(t)), \boldsymbol{\varkappa}_D(\boldsymbol{\gamma}_2 \circ \psi(t); \boldsymbol{\gamma}_2'(\psi(t))\psi'(t))\}dt$$

$$\le \int_0^1 \max\{h_1(t), h_2 \circ \psi(t) \cdot \psi'(t)\}dt = \int_0^1 h_1(t)dt < \boldsymbol{k}_{G_1}(z_1', z_1'') + \varepsilon,$$

which implies the desired formula. □

Remark 3.7.3. In Chapter 18, we will discover that the analogous product formula is also true for the Carathéodory distance.

3.8 Higher-order Lempert functions and Kobayashi–Royden pseudometrics

Definition 3.8.1. For a domain $G \subset \mathbb{C}^n$ and $m \in \mathbb{N}$ we define the m-th *Lempert function*

$$\boldsymbol{\ell}_G^{(m)*}(a,z) := \inf\{\sigma^m : \sigma \in [0,1),\ \exists_{\varphi = a + \lambda^m \psi \in \mathcal{O}(\mathbb{D},G)}{}^2 : \varphi(\sigma) = z\}, \quad a, z \in G.$$

Similarly, we define the m-th *Kobayashi–Royden pseudometric*

$$\boldsymbol{\varkappa}_G^{(m)}(a;X) := \inf\{\sigma : \sigma \in \mathbb{R}_+,\ \exists_{\varphi = a + \lambda^m \psi \in \mathcal{O}(\mathbb{D},G)} : \sigma\psi(0) = X\,{}^3\},$$
$$a \in G,\ X \in \mathbb{C}^n;$$

cf. [542, 543, 296, 297].

Notice that the function $\boldsymbol{\ell}_G^{(m)*}$ is well defined. In fact, we know that for arbitrary $a, z \in G$ there exists a disc $\varphi = a + \lambda\psi \in \mathcal{O}(\overline{\mathbb{D}}, G)$ with $\varphi(\sigma) = z$ for a $\sigma \in [0,1)$ (cf. Remark 3.1.1). Thus, it suffices to take $\mathbb{D} \ni \lambda \longmapsto \varphi(\lambda^m) \in G$.

To have a complete system of notation, we put $\boldsymbol{\ell}_G^{(m)} := \tanh^{-1}\left(\boldsymbol{\ell}_G^{(m)*}\right) = \boldsymbol{p}\left(0, \boldsymbol{\ell}_G^{(m)*}\right)$.

[2] Equivalently: $\varphi \in \mathcal{O}(\mathbb{D}, G)$, $\varphi(0) = a$, and $\operatorname{ord}_0(\varphi - a) \ge m$.

[3] Equivalently: $(\sigma/m!)\varphi^{(m)}(0) = X$.

Remark 3.8.2 (Basic properties of $\boldsymbol{\ell}_G^{(m)*}$ and $\boldsymbol{\varkappa}_G^{(m)}$).

(a) $$\boldsymbol{\ell}_G^{(m)*}(a,z) = \inf\{(\boldsymbol{m}_{\mathbb{D}}(\lambda_0', \lambda_0''))^m : \lambda_0', \lambda_0'' \in \mathbb{D} : \exists_{\varphi\in\mathcal{O}(\mathbb{D},G)} : \varphi(\lambda_0') = a,\ \operatorname{ord}_{\lambda_0'}(\varphi - a) \ge m,\ \varphi(\lambda_0'') = z\}, \quad a, z \in G,$$

$$\boldsymbol{\varkappa}_G^{(m)}(a;X) = \inf\left\{\frac{|\sigma|}{(1-|\lambda_0|^2)^m} : \sigma \in \mathbb{C},\ \lambda_0 \in \mathbb{D} : \exists_{\varphi\in\mathcal{O}(\mathbb{D},G)} : \varphi(\lambda_0) = a,\ \operatorname{ord}_{\lambda_0}(\varphi - a) \ge m,\ \sigma\varphi_{(m)}(\lambda_0) = X\right\}, \quad a \in G,\ X \in \mathbb{C}^n,$$

where $\varphi_{(m)}(\lambda_0) := \frac{1}{m!}\varphi^{(m)}(\lambda_0)$.

(b) In the above definitions one may additionally require $\varphi \in \mathcal{O}(\overline{\mathbb{D}}, G)$.

(c) $\boldsymbol{\ell}_{\mathbb{D}}^{(m)*}(0,\lambda) = |\lambda|$, $\lambda \in \mathbb{D}$, and $\boldsymbol{\varkappa}_{\mathbb{D}}^{(m)}(0;1) = 1$.
Indeed, this follows directly from the Schwarz lemma.

(d) The systems $(\boldsymbol{\ell}_G^{(m)*})_G$, $(\boldsymbol{\ell}_G^{(m)})_G$, $(\boldsymbol{\varkappa}_G^{(m)})_G$ are holomorphically contractible.
Indeed, if $F : G \longrightarrow D$ is holomorphic, then, for every φ from the definition, we have $\operatorname{ord}_0(F\circ\varphi - F(a)) \ge m$ and $(F\circ\varphi)^{(m)}(0) = F'(a)(\varphi^{(m)}(0))$.

(e) $\boldsymbol{\ell}_G^{(m\nu)*} \le \boldsymbol{\ell}_G^{(m)*}$ and $\boldsymbol{\varkappa}_G^{(m\nu)} \le \boldsymbol{\varkappa}_G^{(m)}$, $\nu \in \mathbb{N}$.
Indeed, let $\varphi = a + \lambda^m\psi \in \mathcal{O}(\mathbb{D},G)$, $\varphi(\sigma) = z$ (resp. $\sigma\psi(0) = X$). Put $\varphi_0(\lambda) := \varphi(\lambda^\nu) = a + \lambda^{m\nu}\psi(\lambda^\nu)$. Then, $\varphi_0(\sqrt[\nu]{\sigma}) = z$. Thus, $\boldsymbol{\ell}_G^{(m\nu)*}(a,z) \le \sqrt[\nu]{\sigma}^{m\nu} = \sigma^m$ (resp. $\boldsymbol{\varkappa}_G^{(m\nu)}(a;X) \le \sigma$).

(f) If $G_\nu \nearrow G$, then $\boldsymbol{\ell}_{G_\nu}^{(m)*} \searrow \boldsymbol{\ell}_G^{(m)*}$ and $\boldsymbol{\varkappa}_{G_\nu}^{(m)} \searrow \boldsymbol{\varkappa}_G^{(m)}$, $\nu \in \mathbb{N}$.

? What about the symmetry of the function $\boldsymbol{\ell}_G^{(m)*}$ for $m \ge 2$?

Before presenting more properties for the higher order Kobayashi–Royden pseudometrics, we discuss examples showing that, in general, $\boldsymbol{\varkappa}_G \ne \boldsymbol{\varkappa}_G^{(m)}$.

Example 3.8.3.

(a) Put

$$D := \{z \in \mathbb{C}^3 : \varrho(z) := \operatorname{Re} z_3 + |z_1^2 - z_2^3|^2 < 0\}.$$

Obviously, D is given by the psh function ϱ and is, therefore, a pseudoconvex domain. Let $z_t := (0,0,-t)$, $t \in (0,1)$, and $X_0 := (a,b,0)$ with $|a|^2+|b|^2 = 1$. Assume that $a \ne 0$ and set $\varphi(\lambda) := (at^{1/4}\lambda, bt^{1/4}\lambda, -t)$, $\lambda \in \mathbb{D}$. Then, $\varphi \in \mathcal{O}(\mathbb{D}, D)$ with $\varphi(0) = z_t$ and $\varphi'(0) = t^{1/4}X_0$. Hence, we have $\boldsymbol{\varkappa}_D(z_t;X_0) \le t^{-1/4}$.

To get a lower estimate, now let $\varphi \in \mathcal{O}(\mathbb{D}, D)$ be given with $\varphi(0) = z_t$, $\alpha\varphi'(0) = X_0$ with a positive α. Then, $\varphi(\lambda) = (\lambda\widetilde{\varphi}_1(\lambda), \lambda\widetilde{\varphi}_2(\lambda), \varphi_3(\lambda))$ with $\alpha\widetilde{\varphi}_1(0) = a$ and $\alpha\widetilde{\varphi}_2(0) = b$. Since φ maps $\mathbb{D}$ into D, it follows for $r \in (0,1)$ (using the mean value (in)equality for (sub)harmonic functions):

$$\begin{aligned}0 &> -t + \frac{1}{2\pi}\int_0^{2\pi} \left|\varphi_1^2(re^{i\theta}) - \varphi_2^3(re^{i\theta})\right|^2 d\theta \\ &= -t + \frac{1}{2\pi}\int_0^{2\pi} r^4\left|\widetilde{\varphi}_1^2(re^{i\theta}) - re^{i\theta}\widetilde{\varphi}_2^3(re^{i\theta})\right|^2 d\theta \\ &\geq -t + r^4|\widetilde{\varphi}_1^2(0)|^2 = -t + \frac{|a|^4 r^4}{\alpha^4} \underset{r\nearrow 1}{\longrightarrow} -t + \left(\frac{|a|}{\alpha}\right)^4;\end{aligned}$$

hence, $\alpha \geq \frac{|a|}{t^{1/4}}$.

Summarizing, if $a \neq 0$, then we get

$$|a|(1/t)^{1/4} \leq \varkappa_D(z_t; X_0) \leq (1/t)^{1/4}.$$

What is the exact value of $\varkappa_D(z_t; X_0)$? **?**

Now, assume that $X_0 = (0, b, 0)$, i.e., $|b| = 1$. Using the analytic disc $\varphi(\lambda) := (0, bt^{1/6}\lambda, -t)$, $\lambda \in \mathbb{D}$, gives the following upper estimate $\varkappa_D(z_t; X_0) \leq (1/t)^{1/6}$. So it remains to discuss the lower estimate. Let $\varphi \in \mathcal{O}(\mathbb{D}, D)$ with $\varphi(0) = z_t$, and $\alpha\varphi'(0) = X_0$ with positive α. Observe that now $\varphi_1(\lambda) = \lambda^2\widetilde{\varphi}_1(\lambda)$ and $\varphi_2(\lambda) = \lambda\widetilde{\varphi}_2(\lambda)$, $\lambda \in \mathbb{D}$, with $\alpha\widetilde{\varphi}_2(0) = b$. Then, following the calculation from above, we have for $r \in (0,1)$:

$$\begin{aligned}0 &> -t + \frac{1}{2\pi}\int_0^{2\pi} \left|\varphi_1^2(re^{i\theta}) - \varphi_2^3(re^{i\theta})\right|^2 d\theta \\ &= -t + \frac{1}{2\pi}\int_0^{2\pi} r^6\left|re^{i\theta}\widetilde{\varphi}_1^2(re^{i\theta}) - \widetilde{\varphi}_2^3(re^{i\theta})\right|^2 d\theta \\ &\geq -t + r^6|\widetilde{\varphi}_2^3(0)|^2 = -t + \frac{|b|^6 r^6}{\alpha^6} \underset{r\nearrow 1}{\longrightarrow} -t + \left(\frac{1}{\alpha}\right)^6;\end{aligned}$$

hence, $\alpha \geq \frac{1}{t^{1/6}}$. So, we end up with precise values for the case where $a = 0$, namely: $\varkappa_D(z_t; X_0) = (1/t)^{1/6}$.

On the other hand, taking the following analytic discs $\varphi_k(\lambda) := (k^3\lambda^3, bk^2\lambda^2, -t)$, $\lambda \in \mathbb{D}$, $k \in \mathbb{N}$, with values in D gives $\varphi_k(0) = z_t$, $\varphi_k'(0) = 0$, and $\varphi_k''(0)/2 = k^2(0,b,0)$. Therefore, $\varkappa_D^{(2)}(z_t, (0,b,0)) \leq 1/k^2 \underset{k\to\infty}{\longrightarrow} 0$, i.e.,

$$\varkappa_D^{(2m)}(z_t; (0,b,0)) \leq \varkappa_D^{(2)}(z_t; (0,b,0)) = 0 \neq \varkappa_D(z_t; (0,b,0)).$$

Notice that estimates of $\varkappa_D^{(k)}(z_t; X_0)$ with (k even, $a \neq 0$) or (k odd, $k \geq 3$) are not known. **?**

(b) Note that Theorem 20.1.1 shows that there are also examples of domains where all the higher Kobayashi–Royden metrics are pairwise different.

The following basic properties of $\boldsymbol{\ell}_G^{(m)}$ and $\boldsymbol{\varkappa}_G^{(m)}$ may be proved similar to $\boldsymbol{\ell}_G$ and $\boldsymbol{\varkappa}_G$.

Proposition 3.8.4. *The functions $\boldsymbol{\ell}_G^{(m)}$, $\boldsymbol{\varkappa}_G^{(m)}$ are upper semicontinuous.*

Proof. Cf. the proofs of Propositions 3.1.14 and 3.5.13. □

Proposition 3.8.5. *Assume that G is taut and fix an $m \in \mathbb{N}$. Then,*

(a) *for every $a, z \in G$ (resp. $(a, X) \in G \times \mathbb{C}^n$) there exists an* extremal disc $\varphi = a + \lambda^m \psi \in \mathcal{O}(\mathbb{D}, G)$ *such that* $\varphi\left(\sqrt[m]{\boldsymbol{\ell}_G^{(m)*}(a,z)}\right) = z$ *(resp.* $\boldsymbol{\varkappa}_G^{(m)}(a; X)\psi(0) = X$*).*

(b) *The functions $\boldsymbol{\ell}_G^{(m)}$ and $\boldsymbol{\varkappa}_G^{(m)}$ are continuous.*

Proof. Cf. the proofs of Propositions 3.2.7, 3.2.9, and 3.5.14. □

Remark 3.8.6. Suppose that $\varphi \in \mathcal{O}(\mathbb{D}, G)$ is an extremal disc for $\boldsymbol{\ell}_G^{(m)}(a,b)$ (resp. $\boldsymbol{\varkappa}_G^{(m)}(a; X)$) with $a \neq b$ (resp. $X \neq 0$), i.e., $\varphi(0) = a$, $\operatorname{ord}_0(\varphi - a) \geq m$, $\varphi(\sigma) = b$ with $\sigma := \sqrt[m]{\boldsymbol{\ell}_G^{(m)*}(a,b)}$ (resp. $(\sigma/m!)\varphi^{(m)}(0) = X$ with $\sigma := \boldsymbol{\varkappa}_G^{(m)}(a; X)$). Then, $\varphi(\mathbb{D})$ cannot be relatively compact in G.

Indeed, suppose that $\varphi(\mathbb{D}) \subset\subset G$. For $0 < \theta < 1$ define

$$\widetilde{\varphi}(\lambda) := \varphi(\lambda) + \left(\frac{\lambda}{\theta\sigma}\right)^m (\varphi(\sigma) - \varphi(\theta\sigma))$$
$$\left(\text{resp. } \widetilde{\varphi}(\lambda) := \varphi(\lambda) + \frac{\lambda^m(1-\theta)}{\theta\sigma} X\right), \quad \lambda \in \mathbb{D}.$$

Then, $\widetilde{\varphi}(0) = a$, $\operatorname{ord}_0(\widetilde{\varphi} - a) \geq m$, $\widetilde{\varphi}(\theta\sigma) = b$ (resp. $(\theta\sigma/m!)\widetilde{\varphi}^{(m)} = X$). Taking θ sufficiently near 1, we may easily obtain $\widetilde{\varphi}(\mathbb{D}) \subset G$; a contradiction.

Proposition 3.8.7 (Product property). *Let $G_j \subset \mathbb{C}^{n_j}$ be a domain, $j = 1, 2$. Then,*

$$\boldsymbol{\ell}_{G_1 \times G_2}^{(m)}((a_1, a_2), (z_1, z_2)) = \max\{\boldsymbol{\ell}_{G_1}^{(m)}(a_1, z_1), \boldsymbol{\ell}_{G_2}^{(m)}(a_2, z_2)\}, \; a_j, z_j \in G_j,$$
$$\boldsymbol{\varkappa}_{G_1 \times G_2}^{(m)}((a_1, a_2); (X_1, X_2)) = \max\{\boldsymbol{\varkappa}_{G_1}^{(m)}(a_1; X_1), \boldsymbol{\varkappa}_{G_2}^{(m)}(a_2; X_2)\},$$
$$(a_j, X_j) \in G_j \times \mathbb{C}^{n_j}, \; j = 1, 2.$$

Proof. Cf. the proof of Proposition 3.7.1. □

Proposition 3.8.8. *Let $\Pi : \widetilde{G} \longrightarrow G \subset \mathbb{C}^n$ be a holomorphic covering. Fix points $a \in G$, $\widetilde{a} \in \Pi^{-1}(a)$. Then,*

$$\boldsymbol{\ell}_G^{(m)}(a,z) = \inf\{\boldsymbol{\ell}_{\widetilde{G}}^{(m)}(\widetilde{a},\widetilde{z}) : \widetilde{z} \in \Pi^{-1}(z)\}, \quad z \in G,$$
$$\boldsymbol{\varkappa}_G^{(m)}(a;X) = \boldsymbol{\varkappa}_{\widetilde{G}}^{(m)}(\widetilde{a};(\Pi'(\widetilde{a}))^{-1}(X)), \quad X \in \mathbb{C}^n,\ m \in \mathbb{N}.$$

Proof. Cf. the proof of Theorem 3.3.7 (see also Exercise 3.9.8). □

Corollary 3.8.9. *If $G \subset \mathbb{C}$, then $\boldsymbol{\ell}_G^{(m)} = \boldsymbol{\ell}_G = \boldsymbol{k}_G$, and $\boldsymbol{\varkappa}_G^{(m)} = \boldsymbol{\varkappa}_G$, $m \in \mathbb{N}$.*

3.9 Exercises

Exercise 3.9.1.

(a) Prove the inequality in Example 3.1.5.

(b) For $N \in \mathbb{N}$, construct N holomorphic functions $f_j \in \mathcal{O}(\overline{\mathbb{D}},\mathbb{D})$ with the following properties:

(i) $\{\lambda \in \overline{\mathbb{D}} : f_j(\lambda) = f_{j+1}(\lambda)\} = \{\lambda_j\} \subset \mathbb{D}$ for $1 \le j \le N-1$;

(ii) $\{\lambda \in \overline{\mathbb{D}} : f_j(\lambda) = f_k(\lambda)\} = \varnothing$ for $1 \le j,k \le N$ with $|j-k| \ge 2$.

Define $G_\varepsilon := \{z \in \mathbb{D} \times \mathbb{D} : \prod_{j=1}^N |z_2 - f_j(z_1)| < \varepsilon\}$ and set $z^{(j)} := (\lambda_0, f_j(\lambda_0))$ with $\lambda_0 \in \mathbb{D} \setminus \{\lambda_1,\dots,\lambda_N\}$.

Prove that there is a sufficiently small ε such that

$$\boldsymbol{k}_{G_\varepsilon}^{(j-1)}(z^{(1)},z^{(j)}) > \boldsymbol{k}_{G_\varepsilon}^{(j)}(z^{(1)},z^{(j)}) \quad \text{for} \quad 2 \le j \le N. \qquad (*)$$

Use exhaustion to provide a strongly pseudoconvex domain $G \subset G_\varepsilon$ with real analytic boundary such that $(*)$ remains true for G.

Compare Theorems 14.5.7 and 14.5.9 for more examples showing that, in general, $\boldsymbol{k}_G^{(N)} \not\equiv \boldsymbol{k}_G$.

Exercise 3.9.2. Let G be a plane domain and let $f \in \mathcal{O}(\mathbb{D},G)$ be such that for two different points $\lambda',\lambda'' \in \mathbb{D}$ we have $\boldsymbol{k}_G(f(\lambda'),f(\lambda'')) = \boldsymbol{k}_{\mathbb{D}}(\lambda',\lambda'')$. Prove that f is a holomorphic covering.

Exercise 3.9.3.

(a) Note that $G := \{z \in \mathbb{C}^2 : |z_1| < |z_2| < 1\}$ is a taut domain without $\mathcal{C}^1$-boundary. Let $G_\nu := \{z \in \mathbb{C}^2 : |z_1| < |z_2| + 1/\nu < 1 + 2/\nu\}$. Prove that

$$\lim_{\nu\to\infty} \boldsymbol{\ell}_{G_\nu}((0,1/2),(0,-1/2)) < \boldsymbol{\ell}_G((0,1/2),(0,-1/2)).$$

(b) Construct a domain G with $\mathcal{C}^\infty$-boundary, which is not taut, and a sequence $(G_\nu)_\nu$ as in Proposition 3.3.5(b) such that

$$\lim_{\nu\to\infty} \boldsymbol{\ell}_{G_\nu}((1/2,1/2),(-1/2,1/2)) < \boldsymbol{\ell}_G((1/2,1/2),(-1/2,1/2)).$$

Hint. Start with

$$G' := \{z \in \mathbb{C}^2 : |z_1| \le 1/2,\ |z_2| < 1 \text{ or } 1/2 < |z_1| < 1,\ |z_2| < 1/2\}$$

and smooth it out.

(c) To disprove the claim in Proposition 3.3.5(b), construct a sequence of domains $(G_\nu)_\nu \subset \mathbb{C}^3$ with $G_\nu \supset G_{\nu+1} \supset \mathbb{B}(\frac{\nu+1}{\nu})$, but not with $G_\nu \supset\supset G_{\nu+1}$, and such that $\bigcap G_\nu = \overline{\mathbb{B}}_3$.

Hint. Build G_ν via thickening

$$G'_\nu := \mathbb{B}\left(\frac{\nu+1}{\nu}\right) \cup \bigcup_{j\ge\nu}\{(\lambda,\lambda,j\lambda(\lambda-1/2)) : \lambda \in \overline{\mathbb{D}}\}$$

and consider $\boldsymbol{\ell}_{\mathbb{B}_3}((0,0,0),(1/2,1/2,0))$.

(The ideas of the examples above are due to W. Zwonek.)

Exercise 3.9.4.

(a) Prove that $\boldsymbol{\varkappa}_G(0;\cdot) \equiv 0$ if

$$G := \{z = (z_1,z_2) \in \mathbb{C}^2 : |z_1| < 1 \text{ or } |z_2| < 1\}.$$

(b) Let $G := \{z \in \mathbb{C}^2 : |z_1| < 1,\ |z_2| < 1 \text{ and if } |z_1| = |z_2|, \text{ then } |z_1| < 1/2\}$. Prove that $1 < \boldsymbol{\varkappa}_G(0;(1,1))$. We don't know the exact value of $\boldsymbol{\varkappa}_G(0;(1,1))$.

Exercise 3.9.5.

(a) How does the Kobayashi–Royden pseudometric behave under an increasing sequence of domains?

(b) Prove that if $(G_j)_{j=1}^\infty \subset \mathbb{C}^n$ is a sequence of bounded taut domains such that $G_{j+1} \subset G_j$, $j \in \mathbb{N}$, and if $G := \bigcap_{j=1}^\infty G_j$ is a domain, then $\boldsymbol{\varkappa}_{G_j} \nearrow \boldsymbol{\varkappa}_G$. Show for $n = 1$ that the result remains true if $G := \operatorname{int}\bigcap_{j=1}^\infty G_j$ is a domain.

(c) Let $(G_j)_{j=1}^\infty \subset \mathbb{C}$ be a sequence of simply connected domains such that $G_{j+1} \subset G_j$, $j \in \mathbb{N}$. Assume that $G := \operatorname{int}\bigcap_{j=1}^\infty G_j$ is connected and $\mathbb{C} \setminus \overline{G_1} \neq \varnothing$. Prove that $\boldsymbol{\gamma}_{G_j} \nearrow \boldsymbol{\gamma}_G$.

Exercise 3.9.6. Complete the argument to show that, in general, the Kobayashi–Buseman pseudometric is not continuous (cf. Example 3.5.11).

Exercise 3.9.7. Use Theorem 3.6.4 to find an alternative proof of the fact that the Kobayashi pseudodistance is inner.

Exercise 3.9.8. Prove the following analogue of Theorem 3.3.7 for the Kobayashi–Royden pseudometric: let $\Pi : \widetilde{G} \longrightarrow G$ be a holomorphic covering; then $\varkappa_G(z; X) = \varkappa_{\widetilde{G}}(\widetilde{z}; Y)$ if $\widetilde{z} \in \widetilde{G}$ with $\Pi(\widetilde{z}) = z$ and $\Pi'(\widetilde{z})Y = X$.

Exercise 3.9.9. Prove the extension theorem for the Kobayashi–Royden pseudometric across closed sets with $(2n-2)$-Hausdorff measure zero (cf. Theorem 3.4.2) and next, prove Theorem 3.4.2, using Theorem 3.6.4.

Exercise 3.9.10 (cf. [340]). Prove that if G is a convex domain in $\mathbb{C}^n$, then the Lempert function $\boldsymbol{\ell}_G$ satisfies the triangle inequality.

Exercise 3.9.11. Let $D \subset \mathbb{B}_n$ be a domain, $a \in D$, and $X \in \mathbb{C}^n$, $X \neq 0$. Assume that $(a + \mathbb{C}X) \cap \mathbb{B}_n = (a + \mathbb{C}X) \cap D$. Prove that $\varkappa_D(a; X) = \varkappa_{\mathbb{B}_n}(a; X)$.

Exercise 3.9.12. Prove that:

(a) $\boldsymbol{k}_G^{(m)}$ is upper semicontinuous for an arbitrary domain G,

(b) the system $(\boldsymbol{k}_G^{(m)})_G$ is holomorphically contractible.

Exercise 3.9.13. Assume that $G \subset \mathbb{C}^n$ is a domain such that for any $a \in G$, $R > 0$, the set $\{z \in G : \boldsymbol{k}_G(a, z) < R\}$ is relatively compact in G (in particular, G is taut – cf. Remark 3.2.2(a)). Prove that for every $m \in \mathbb{N}_2$ we have:

(a) For any $z', z'' \in G$ there exist $z' = z_0, z_1, \dots, z_{m-1}, z_m = z'' \in G$ such that $\boldsymbol{k}_G^{(m)}(z', z'') = \sum_{j=1}^{m} \boldsymbol{\ell}_G(z_{j-1}, z_j)$.

(b) The function $\boldsymbol{k}_G^{(m)} : G \times G \longrightarrow \mathbb{R}_+$ is continuous.

3.10 List of problems

Chapter 4

Contractible systems

Summary. Most of the material presented in this chapter may be found in [137, 178], and [215]. The complex Green function has been introduced by M. Klimek in [303], the Azukawa pseudometric by K. Azukawa in [26], and the Sibony pseudometric by N. Sibony in [474] (cf. § 4.2). Another definition of a (symmetric) complex Green function has been proposed by U. Cegrell in [89]. Higher order Möbius functions and Reiffen pseudometrics were studied in [257, 258].

Introduction. In Chapters 2 and 3 we proved that $(\boldsymbol{c}_G)_G$, $(\boldsymbol{k}_G)_G$ (resp. $(\boldsymbol{\gamma}_G)_G$, $(\boldsymbol{\varkappa}_G)_G$) are "extremal" contractible systems of pseudodistances (resp. of pseudometrics). It is known that $(\boldsymbol{c}_G)_G$ and $(\boldsymbol{\ell}_G)_G$ are extremal contractible systems of functions. We also constructed some intermediate systems, e.g., $(\boldsymbol{c}^i_G)_G$, $(\boldsymbol{\varkappa}^{(m)}_G)_G$, or $(\widehat{\boldsymbol{\varkappa}}_G)_G$. Observe that there are many such intermediate contractible systems, e.g., $(d^{(t)}_G)_G$, $(\delta^{(t)}_G)_G$, where $d^{(t)}_G := (1-t)\boldsymbol{c}_G + t\boldsymbol{k}_G$, $\delta^{(t)}_G := (1-t)\boldsymbol{\gamma}_G + t\boldsymbol{\varkappa}_G$ $(0 < t < 1)$. We are only interested in the systems that appear in complex analysis in a natural way; most of them arise from certain extremal problems.

4.1 Abstract point of view

Let $\mathfrak{G}$ denote the family of all domains in all $\mathbb{C}^n$'s and let $\mathfrak{G}_0$ be a subfamily of $\mathfrak{G}$ with the following two properties:

(a) $\mathbb{D} \in \mathfrak{G}_0$,

(b) if $G \in \mathfrak{G}_0$ and D is a domain biholomorphic to G, then $D \in \mathfrak{G}_0$, i.e., $\mathfrak{G}_0$ *is invariant under biholomorphic mappings*.

Standard examples:

$$\mathfrak{G}_0 = \mathfrak{G},$$

$$\mathfrak{G}_0 = \mathfrak{G}_h := \{G \in \mathfrak{G} : G \text{ is a domain of holomorphy}\},$$

$$\mathfrak{G}_0 = \mathfrak{G}_c := \{G \in \mathfrak{G} : G \text{ is biholomorphic to a convex domain}\}, \tag{4.1.1}$$

$$\mathfrak{G}_0 = \mathfrak{G}_b := \{G \in \mathfrak{G} : G \text{ is biholomorphic to a bounded domain}\}. \tag{4.1.2}$$

Let $\underline{d} = (d_G)_{G\in\mathfrak{G}_0}$ be a system of functions $d_G : G \times G \longrightarrow \mathbb{R}_+$ ($G \in \mathfrak{G}_0$). We say that $\underline{d}$ is a (holomorphically) *contractible family of functions* if

$$d_{\mathbb{D}} = \boldsymbol{p}, \tag{4.1.3}$$

$$d_D(F(z'), F(z'')) \le d_G(z', z''), \quad z', z'' \in G, \tag{4.1.4}$$

for every $F \in \mathcal{O}(G, D)$ ($G, D \in \mathfrak{G}_0$). If, moreover, for any $G \in \mathfrak{G}_0$ the function d_G is a pseudodistance, then we say that $\underline{d}$ is a *contractible family of pseudodistances.*

Now let $\underline{\delta} = (\delta_G)_{G\in\mathfrak{G}_0}$ be a system of *pseudometrics* $\delta_G : G \times \mathbb{C}^n \longrightarrow \mathbb{R}_+$ ($G \in \mathfrak{G}_0$, G a domain in $\mathbb{C}^n$), i.e., $\delta_G(a; \lambda X) = |\lambda|\delta_G(a; X)$, $a \in G$, $\lambda \in \mathbb{C}$, $X \in \mathbb{C}^n$. We say that $\underline{\delta}$ is a *contractible family of pseudometrics* if

$$\delta_{\mathbb{D}} = \boldsymbol{\gamma}_{\mathbb{D}},$$

and if for any $F \in \mathcal{O}(G, D)$ ($G, D \in \mathfrak{G}_0$) we have

$$\delta_D(F(z); F'(z)X) \le \delta_G(z; X), \quad z \in G, \ X \in \mathbb{C}^n. \tag{4.1.5}$$

(We use the following convention: when we write "$\delta_G(a; X)$, $a \in G$, $X \in \mathbb{C}^n$", then it is automatically assumed that G is a domain in $\mathbb{C}^n$.)

Notice that we do not require that $\delta_G(a; \cdot)$ be a seminorm nor that δ_G be upper semicontinuous.

It is clear that if F is biholomorphic, then equalities hold in (4.1.4) and (4.1.5).

If we substitute (4.1.3) with

$$d_{\mathbb{D}} = \boldsymbol{m}, \tag{4.1.6}$$

then we say that $\underline{d}$ is an $\boldsymbol{m}$*-contractible family of functions* (resp. $\boldsymbol{m}$*-contractible family of pseudodistances*).

Remark 4.1.1.

(a) $\boldsymbol{c}, \boldsymbol{c}^i, \boldsymbol{\ell}^{(m)}, \boldsymbol{k}^{(m)}, \boldsymbol{k}$ are contractible families of functions; $\boldsymbol{c}, \boldsymbol{c}^i, \boldsymbol{k}$ are contractible families of pseudodistances.

(b) $\boldsymbol{c}^*, \boldsymbol{\ell}^{(m)*}, \boldsymbol{k}^*$ are $\boldsymbol{m}$-contractible families of functions; $\boldsymbol{c}^*, \boldsymbol{k}^*$ are $\boldsymbol{m}$-contractible families of pseudodistances.

(c) $\boldsymbol{\gamma}, \hat{\boldsymbol{\varkappa}}, \boldsymbol{\varkappa}^{(m)}$ are contractible families of pseudometrics.

Remark 4.1.2.

(a) If $\underline{d} = (d_G)_{G\in\mathfrak{G}_0}$ is a contractible family, then $\boldsymbol{c}_G \le d_G \le \boldsymbol{\ell}_G$, $G \in \mathfrak{G}_0$. If, moreover, d_G is a pseudodistance, then $\boldsymbol{c}_G \le d_G \le \boldsymbol{k}_G \le \boldsymbol{\ell}_G$.

(b) If $\underline{d}$ is an $\boldsymbol{m}$-contractible family, then $\boldsymbol{c}^*_G \le d_G \le \boldsymbol{\ell}^*_G$, $G \in \mathfrak{G}_0$.

(c) If $\underline{\delta} = (\delta_G)_{G\in\mathfrak{G}_0}$ is a contractible family of pseudometrics, then $\boldsymbol{\gamma}_G \le \delta_G \le \boldsymbol{\varkappa}_G$, $G \in \mathfrak{G}_0$. If, moreover, $\delta_G(a; \cdot)$ is a seminorm, then $\boldsymbol{\gamma}_G(a; \cdot) \le \delta_G(a; \cdot) \le \hat{\boldsymbol{\varkappa}}_G(a; \cdot) \le \boldsymbol{\varkappa}_G(a; \cdot)$.

Remark 4.1.3. In view of Remark 4.1.2(b), if $\underline{d}$ is an $\boldsymbol{m}$-contractible family, then $d_G < 1$. If $\underline{d} = (d_G)_{G\in\mathfrak{G}_0}$ is a contractible family, then we put $d_G^* := \tanh d_G$, $\underline{d}^* := (d_G^*)_{G\in\mathfrak{G}_0}$. Observe that the operator

$$\underline{d} \longmapsto \underline{d}^*$$

is a bijection between the class of contractible families of functions and the class of $\boldsymbol{m}$-contractible families of functions. From now on, $\boldsymbol{m}$-contractible objects will be marked by "$*$", e.g., d_G^* (this agrees with $\boldsymbol{c}^*$, $\boldsymbol{\ell}^*$, and $\boldsymbol{k}^*$). Observe that if $\underline{d}$ is a contractible family of pseudodistances, then so is $\underline{d}^*$. We have introduced the notion of $\boldsymbol{m}$-contractible families for technical reasons only (they are much simpler in calculations; cf. Chapter 2).

Remark 4.1.4. In Chapter 11 we will prove that for any $G \in \mathfrak{G}_c$ (cf. (4.1.1)) we have

$$\boldsymbol{c}_G = \boldsymbol{\ell}_G = \boldsymbol{k}_G,\ \boldsymbol{c}_G^* = \boldsymbol{\ell}_G^* = \boldsymbol{k}_G^*,\ \text{and}\ \boldsymbol{\gamma}_G = \boldsymbol{\varkappa}_G.$$

Thus, in the class $\mathfrak{G}_c$ the theory of contractible objects reduces to the Carathéodory case; the reader should always remember this fact.

To complete these (tedious) general preliminaries, observe that, unfortunately, our notions of contractible objects are too restrictive. They do not cover, for instance, the Bergman distance and metric (cf. Chapter 12), nor the Hahn pseudodistance and pseudometric (cf. § 8.1). Problems appear because of the requirement that $\underline{d}$ (or $\underline{\delta}$) be contractible with respect to *all* holomorphic mappings. To avoid these difficulties, one could proceed as follows.

First, for any $G, D \in \mathfrak{G}_0$ we fix a family $\mathcal{F}(G, D) \subset \mathcal{O}(G, D)$ such that:

(a) if $F : G \longrightarrow D$ is biholomorphic, then $F \in \mathcal{F}(G, D)$,

(b) if $F_1 \in \mathcal{F}(G_1, G_2)$, $F_2 \in \mathcal{F}(G_2, G_3)$, then $F_2 \circ F_1 \in \mathcal{F}(G_1, G_3)$.

Next, in (4.1.4) (or (4.1.5)) we only take $F \in \mathcal{F}(G, D)$. For example:

$\mathcal{F}(G, D) = \mathcal{F}_i(G, D) := \{F \in \mathcal{O}(G, D)\colon F \text{ is injective}\}$; this family is good for the Hahn pseudodistance and pseudometric;

$\mathcal{F}(G, D) = \mathcal{F}_b(G, D) := \{F \in \mathcal{O}(G, D) : F \text{ is biholomorphic})\}$; this family is good for the Bergman pseudodistance and pseudometric.

With the exception of Chapter 8, we will not go into this generalization, but we advise the reader to verify which of the results may be extended to the above general case.

Let us come back to the standard situation.

Remark 4.1.5. The functions d_G and δ_G in the definitions of contractible families of functions and pseudometrics look very arbitrary, but this is not so. They must satisfy a lot of regularity properties. For instance, since $\boldsymbol{k}_G$ is continuous, we get the following properties:

(a) If $\underline{d} = (d_G)_{G \in \mathfrak{G}_0}$ is a contractible family of pseudodistances, then for each $G \in \mathfrak{G}_0$ the pseudodistance d_G is continuous (use Remark 4.1.2(a)).

(b) Similarly, if $\underline{d}^*$ is an $\boldsymbol{m}$-contractible family and $\mathbb{B}(a, 3r) \subset G \in \mathfrak{G}_0$, then

$$d_G^*(z', z'') \le \boldsymbol{\ell}_G^*(z', z'') \le \boldsymbol{\ell}_{\mathbb{B}(z',2r)}^*(z', z'') = \frac{\|z' - z''\|}{2r}, \quad z', z'' \in \mathbb{B}(a, r).$$

If $\underline{d}$ is a contractible family, then, of course, we have analogous inequalities with $\tanh^{-1}\left(\frac{\|z'-z''\|}{2r}\right)$ at the end. This last term is not handy in calculations but, since $\tanh^{-1}(t) \sim t$ as $t \longrightarrow 0_+$, we can reformulate the estimate as follows:

(c) If $\underline{d}$ is a contractible family, then for any $a \in G \in \mathfrak{G}_0$ there exist $M, r > 0$ such that

$$d_G(z', z'') \le M\|z' - z''\|, \quad z', z'' \in \mathbb{B}(a, r) \subset G;$$

we briefly say that d_G is *locally bounded by the Euclidean distance.*

(d) If $\underline{\delta} = (\delta_G)_{G \in \mathfrak{G}_0}$ is a contractible family of pseudometrics and if, moreover, $\mathbb{B}(a, 2r) \subset G \in \mathfrak{G}_0$, then

$$\delta_G(z; X) \le \boldsymbol{\varkappa}_G(z; X) \le \boldsymbol{\varkappa}_{\mathbb{B}(z,r)}(z; X) = \frac{\|X\|}{r}, \quad z \in \mathbb{B}(a, r), \ X \in \mathbb{C}^n.$$

Let $\underline{d}$ be a contractible family of pseudodistances. We say that G is *$\underline{d}$-hyperbolic* if d_G is a distance.

Let $\underline{\delta}$ be a contractible family of pseudometrics. We say that

(a) G is *$\underline{\delta}$-hyperbolic* if

$$\forall_{a \in G} \ \exists_{M,r>0} : \ \delta_G(z; X) \ge M\|X\|, \quad z \in \mathbb{B}(a, r) \subset G;$$

(b) G is *pointwise $\underline{\delta}$-hyperbolic* (briefly, *p. $\underline{\delta}$-hyperbolic*) if

$$\forall_{a \in G} : \ \delta_G(a; X) > 0, \quad X \in (\mathbb{C}^n)_*.$$

Notice that

$$\begin{array}{ccccc}
[G \text{ is } \boldsymbol{c}\text{-hyperbolic}] & \Longrightarrow & [G \text{ is } \underline{d}\text{-hyperbolic}] & \Longrightarrow & [G \text{ is } \boldsymbol{k}\text{-hyperbolic}] \\
 & & & & \Updownarrow \text{ (cf. Theorem 7.2.2)} \\
[G \text{ is } \boldsymbol{\gamma}\text{-hyperbolic}] & \Longrightarrow & [G \text{ is } \underline{\delta}\text{-hyperbolic}] & \Longrightarrow & [G \text{ is } \boldsymbol{\varkappa}\text{-hyperbolic}] \\
\Updownarrow & & \Downarrow & & \Downarrow \text{ (cf. Remark 3.5.12)} \\
[G \text{ is p. } \boldsymbol{\gamma}\text{-hyperbolic}] & \Longrightarrow & [G \text{ is p. } \underline{\delta}\text{-hyperbolic}] & \Longrightarrow & [G \text{ is p. } \boldsymbol{\varkappa}\text{-hyperbolic}]
\end{array}$$

In particular, if $G \in \mathfrak{G}_b \cap \mathfrak{G}_0$ (cf. (4.1.2)), then G is $\underline{d}$- and $\underline{\delta}$-hyperbolic for any $\underline{d}$ and $\underline{\delta}$; moreover, $\operatorname{top} d_G = \operatorname{top} G$. In general,
$[\operatorname{top} \boldsymbol{c}_G = \operatorname{top} G] \Longrightarrow [\operatorname{top} d_G = \operatorname{top} G]$ (cf. Proposition 2.6.1(b)).

To justify the above general definitions of contractible families of functions and pseudometrics we need examples of such objects (different from the Carathéodory and Kobayashi cases). This will be done in the following sections.

4.2 Extremal problems for plurisubharmonic functions

Let us begin with the following trivial remarks:

$$c_G^*(a,z) = \sup\{u(z) : u \in \mathcal{M}_G^{(1)}(a)\}, \tag{4.2.1}$$

$$\gamma_G(a;X) = \sup\left\{\lim_{\lambda \nrightarrow 0} \frac{1}{|\lambda|} u(a+\lambda X) : u \in \mathcal{M}_G^{(1)}(a)\right\}, \tag{4.2.2}$$

where $\mathcal{M}_G^{(1)}(a) := \{|f| : f \in \mathcal{O}(G,\mathbb{D}),\ f(a)=0\}$. Observe that

$$\begin{aligned}\mathcal{M}_G^{(1)}(a) \subset \mathcal{K}_G(a) := \{u : G \longrightarrow [0,1) : u \text{ is log-psh},&\\ \exists_{M,r>0} : u(z) \le M\|z-a\|,\ z \in \mathbb{B}(a,r) \subset G\},&\quad a \in G \in \mathfrak{G}.\end{aligned}$$

This suggests considering systems $(\mathcal{L}_G(a))_{a\in G\in\mathfrak{G}_0}$ of subclasses $\mathcal{L}_G(a) \subset \mathcal{K}_G(a)$ and defining corresponding contractible objects via certain extremal problems for $\mathcal{L}_G(a)$. More precisely, we assume that such a system $(\mathcal{L}_G(a))_{a\in G\in\mathfrak{G}_0}$ satisfies the following conditions:

$$\text{the function } \mathbb{D} \ni \lambda \longmapsto |\lambda| \text{ belongs to } \mathcal{L}_{\mathbb{D}}(0), \tag{4.2.3}$$

$$\text{if } F \in \mathcal{O}(G,D) \text{ and } u \in \mathcal{L}_D(F(a)), \text{ then } u \circ F \in \mathcal{L}_G(a). \tag{4.2.4}$$

Observe that (4.2.3) and (4.2.4) imply that $\mathcal{M}_G^{(1)}(a) \subset \mathcal{L}_G(a)$; in other words, $(\mathcal{M}_G^{(1)}(a))_{a\in G\in\mathfrak{G}}$ is the minimal "admissible" family.

Remark 4.2.1. One can easily prove that the following families $\mathcal{L} = (\mathcal{L}_G(a))_{a\in G\in\mathfrak{G}}$ are "admissible":

(a) $\mathcal{L}_G(a) = \mathcal{K}_G(a)$ (and so $(\mathcal{K}_G(a))_{a\in G\in\mathfrak{G}}$ is the maximal "admissible" family);

(b) $\mathcal{L}_G(a) = \mathcal{S}_G(a) := \{\sqrt{u} : u : G \to [0,1),\ u$ is log-psh, $u(a)=0$, and u is $\mathcal{C}^2$ near $a\}$;

(c) $\mathcal{L}_G(a) = \mathcal{M}_G^{(k)}(a) := \{|f|^{\frac{1}{k}} : f \in \mathcal{O}(G,\mathbb{D}),\ \operatorname{ord}_a f \ge k\}$ ($k \in \mathbb{N}$), where $\operatorname{ord}_a f$ denotes the order of the zero of f at a.

Clearly,

$$\mathcal{M}_G^{(1)}(a) \subset \mathcal{M}_G^{(k)}(a) \subset \mathcal{K}_G(a), \quad \mathcal{M}_G^{(1)}(a) \subset \mathcal{S}_G(a) \subset \mathcal{K}_G(a), \quad a \in G \in \mathfrak{G}.$$

Moreover, if $G \subset \mathbb{C}^1$, then

$$\{|f|^{\frac{1}{2}} : f \in \mathcal{O}(G,\mathbb{D}),\ \operatorname{ord}_a f = 2\} \subset \mathcal{S}_G(a), \quad a \in G.$$

If $\mathcal{L} = (\mathcal{L}_G(a))_{a\in G\in\mathfrak{G}_0}$ satisfies (4.2.3) and (4.2.4), then we define

$$d_G^{(\mathcal{L})}(a,z) := \sup\{u(z) : u \in \mathcal{L}_G(a)\}, \quad a,z \in G \in \mathfrak{G}_0,$$
$$\delta_G^{(\mathcal{L})}(a;X) := \sup\left\{\limsup_{\lambda\to 0}\frac{1}{|\lambda|}u(a+\lambda X) : u \in \mathcal{L}_G(a)\right\}, \quad a \in G \in \mathfrak{G}_0, X \in \mathbb{C}^n.$$

In the special cases we set

$$\begin{aligned} g_G &:= d_G^{(\mathcal{K})}, & A_G &:= \delta_G^{(\mathcal{K})}, \\ s_G &:= d_G^{(\mathcal{S})}, & S_G &:= \delta_G^{(\mathcal{S})}, \\ m_G^{(k)} &:= d_G^{(\mathcal{M}^{(k)})}, & \gamma_G^{(k)} &:= \delta_G^{(\mathcal{M}^{(k)})}, \quad G \in \mathfrak{G}. \end{aligned}$$

By (4.2.1) and (4.2.2) we have

$$m_G^{(1)} = c_G^*, \quad \gamma_G^{(1)} = \gamma_G, \quad G \in \mathfrak{G}.$$

Moreover,

$$\begin{aligned} c_G^* &\le m_G^{(k)} \le g_G, & \gamma_G &\le \gamma_G^{(k)} \le A_G, \\ c_G^* &\le s_G \le g_G, & \gamma_G &\le S_G \le A_G, \end{aligned} \tag{4.2.5}$$

and in the case where $G \subset \mathbb{C}^1$, we get

$$\gamma_G \le \gamma_G^{(2)} \le S_G \le A_G.$$

Proposition 4.2.2. *For any $\mathcal{L}$ the system $\underline{d}^{(\mathcal{L})} := (d_G^{(\mathcal{L})})_{G\in\mathfrak{G}_0}$ is an $\boldsymbol{m}$-contractible family.*

Proof. In view of (4.2.4), the only difficulty is to show that $d_{\mathbb{D}}^{(\mathcal{L})} = \boldsymbol{m}$. It suffices to prove that $d_{\mathbb{D}}^{(\mathcal{L})}(0,\lambda) = |\lambda|$, $\lambda \in \mathbb{D}$. Since $\mathcal{M}_{\mathbb{D}}^{(1)}(0) \subset \mathcal{L}_{\mathbb{D}}(0)$, we get "$\ge$". The inequality "$\le$" follows from the Schwarz lemma for subharmonic functions (cf. Appendix B.4.24). □

Lemma 4.2.3 (cf. [304]). *We have $g_G(a,\cdot) \in \mathcal{K}_G(a)$, $a \in G \in \mathfrak{G}$. Consequently,*

$$A_G(a;X) = \limsup_{\lambda\to 0}\frac{1}{|\lambda|}g_G(a,a+\lambda X), \quad a \in G,\ X \in \mathbb{C}^n.$$

(In particular, $\delta_G^{(\mathcal{L})} \le A_G < +\infty$.)

Proof. Fix an $a \in G \in \mathfrak{G}$ and let u^* denote the upper semicontinuous regularization of the function $u := \boldsymbol{g}_G(a, \cdot)$. Obviously u^* is log-psh on G. According to Proposition 4.2.2 (and Remark 4.1.5(b)) we have

$$u(z) = \boldsymbol{g}_G(a, z) \le \frac{\|z - a\|}{r}, \quad z \in \mathbb{B}(a, r) \subset G.$$

Hence

$$u^*(z) \le \frac{\|z - a\|}{r}, \quad z \in \mathbb{B}(a, r),$$

and therefore $u^* \in \mathcal{K}_G(a)$. Since u is extremal, we get $u^* = u$. □

Lemma 4.2.4. $\log \boldsymbol{A}_G(a; \cdot) \in \mathcal{PSH}(\mathbb{C}^n)$, $a \in G$.

Proof. Fix an $a \in G$ and define

$$u(\lambda, X) := \log \boldsymbol{g}_G(a, a + \lambda X) - \log|\lambda|, \quad (\lambda, X) \in \Omega_0,$$

where $\Omega_0 := \Omega \setminus (\{0\} \times \mathbb{C}^n)$, $\Omega := \{(\lambda, X) \in \mathbb{C} \times \mathbb{C}^n : a + \lambda X \in G\}$. Observe that $u \in \mathcal{PSH}(\Omega_0)$. Moreover, since $\boldsymbol{g}_G(a, \cdot) \in \mathcal{K}_G(a)$, we easily conclude that u is locally bounded from above in Ω. Hence, by Appendix B.4.23(a), u extends to a $\widetilde{u} \in \mathcal{PSH}(\Omega)$ with

$$\widetilde{u}(0, X) := \limsup_{\Omega_0 \ni (\lambda, Y) \to (0, X)} u(\lambda, Y).$$

In particular, $\widetilde{u}(0, \cdot) \in \mathcal{PSH}(\mathbb{C}^n)$ and, by Appendix B.4.23(c) and Lemma 4.2.3,

$$\widetilde{u}(0, X) = \limsup_{\lambda \not\to 0} u(\lambda, X) = \log \boldsymbol{A}_G(a; X), \quad X \in \mathbb{C}^n. \qquad \square$$

Proposition 4.2.5. *For any $\mathcal{L}$ the system $\underline{\delta}^{(\mathcal{L})} := (\delta_G^{(\mathcal{L})})_{G \in \mathfrak{G}_0}$ is a contractible family of pseudometrics.*

Proof. It is clear that $\delta_G^{(\mathcal{L})}(a; \cdot)$ is a pseudometric. Using the same methods as in the proof of Proposition 4.2.2, we get $\delta_{\mathbb{D}}^{(\mathcal{L})}(0; \cdot) = \boldsymbol{\gamma}_{\mathbb{D}}(0; \cdot)$. It remains to verify the holomorphic contractibility. More precisely, it suffices to prove that for any $F \in \mathcal{O}(G, D)$ and for any $u \in \mathcal{L}_D(F(a))$ we have

$$\limsup_{\lambda \not\to 0} \frac{1}{|\lambda|} u(F(a) + \lambda F'(a)X) \le \limsup_{\lambda \not\to 0} \frac{1}{|\lambda|} (u \circ F)(a + \lambda X). \qquad \square$$

This is a direct consequence of the following

Lemma 4.2.6 (cf. [26, 27]). *Let $\varphi_1, \varphi_2 : \overline{\mathbb{D}}(\eta) \longrightarrow G$ ($\eta > 0$) be holomorphic mappings with $\varphi_1(0) = \varphi_2(0) =: a$ and $\varphi_1'(0) = \varphi_2'(0) =: X$. Then, for any $u \in \mathcal{K}_G(a)$ we have*

$$\limsup_{\lambda \nrightarrow 0} \frac{1}{|\lambda|} u(\varphi_1(\lambda)) = \limsup_{\lambda \nrightarrow 0} \frac{1}{|\lambda|} u(\varphi_2(\lambda)).$$

Proof. Define a mapping Φ by

$$\mathbb{D}(\eta) \times \mathbb{C} \ni (\lambda, \xi) \overset{\Phi}{\longmapsto} (1 - \xi)\varphi_1(\lambda) + \xi\varphi_2(\lambda) \in \mathbb{C}^n.$$

Note that $\Phi(\cdot, 0) = \varphi_1$, $\Phi(\cdot, 1) = \varphi_2$. Put $\Omega := \Phi^{-1}(G)$, $\Omega_0 := \Omega \setminus (\{0\} \times \mathbb{C})$, and let

$$v(\lambda, \xi) := \frac{1}{|\lambda|} u(\Phi(\lambda, \xi)), \quad (\lambda, \xi) \in \Omega_0.$$

Then, v is psh in Ω_0. Let $M, r > 0$ be such that

$$u(z) \le M\|z - a\|, \quad z \in \mathbb{B}(a, r) \subset G.$$

Write $\varphi_j(\lambda) = a + \lambda X + \lambda^2 \psi_j(\lambda)$, $\lambda \in \overline{\mathbb{D}}(\eta)$ ($j = 1, 2$) and set

$$M_0 := \max\{\|\psi_1\|_{\overline{\mathbb{D}}(\eta)}, \|\psi_2\|_{\overline{\mathbb{D}}(\eta)}\}.$$

Then, for $(\lambda, \xi) \in \Phi^{-1}(\mathbb{B}(a, r))$, $\lambda \neq 0$, we have

$$v(\lambda, \xi) \le M[\|X\| + |\lambda| M_0(|1 - \xi| + |\xi|)].$$

This shows that v is locally bounded in Ω and, therefore, by putting

$$v(0, \xi) := \limsup_{\substack{\lambda \nrightarrow 0 \\ \xi' \to \xi}} v(\lambda, \xi'), \quad \xi \in \mathbb{C},$$

we extend v to a psh function on the whole Ω (cf. Appendix B.4.23). By the above inequality, we get

$$v(0, \xi) \le M\|X\|, \quad \xi \in \mathbb{C}.$$

Consequently, $v(0, \cdot) = \text{const}$; cf. Appendix B.4.27. Finally,

$$\begin{aligned} \limsup_{\lambda \nrightarrow 0} \frac{1}{|\lambda|} u(\varphi_1(\lambda)) &= \limsup_{\lambda \nrightarrow 0} v(\lambda, 0) = v(0, 0) = v(0, 1) \\ &= \limsup_{\lambda \nrightarrow 0} v(\lambda, 1) = \limsup_{\lambda \nrightarrow 0} \frac{1}{|\lambda|} u(\varphi_2(\lambda)). \qquad \square \end{aligned}$$

Using the same method as above, one can easily prove the following slight generalization of Lemma 4.2.6.

Lemma 4.2.7. *Let $a \in G \subset \mathbb{C}^n \ni X$ and $m \in \mathbb{N}$. Take $\varphi_j \in \mathcal{O}(\mathbb{D}(\eta), G)$, $\varphi_j = a + \lambda^m \psi_j$, $j = 1, 2$, such that $\psi_1(0) = \psi_2(0) = X$. Then, for any $u \in \mathcal{K}_G(a)$, we have*

$$\limsup_{\lambda \to 0} \frac{1}{|\lambda|^m} u(\varphi_1(\lambda)) = \limsup_{\lambda \to 0} \frac{1}{|\lambda|^m} u(\varphi_2(\lambda)).$$

In particular,

$$\limsup_{\lambda \to 0} \frac{1}{|\lambda|^m} \boldsymbol{g}_G(a, \varphi_j(\lambda)) = \boldsymbol{A}_G(a; X), \quad j = 1, 2.$$

The function $\boldsymbol{g}_G$ is called the *complex Green function for* G, $\boldsymbol{A}_G$ is the *Azukawa pseudometric*, $\boldsymbol{S}_G$ is the *Sibony pseudometric*, $\boldsymbol{m}_G^{(k)}$ is the k-th *Möbius function*, and $\boldsymbol{\gamma}_G^{(k)}$ is the k-th *Reiffen pseudometric*.

Remark 4.2.8.

(a) If $G \subset \mathbb{C}^1$, then $-\log \boldsymbol{g}_G(a, \cdot)$ coincides with the classical Green function for G with pole at a; in particular, $\boldsymbol{g}_G^2(a, \cdot)$ is of class $\mathcal{C}^2$ on G; cf. Appendix B.5.

(b) If $\boldsymbol{g}_G^2(a, \cdot)$ is of class $\mathcal{C}^2$ near a (e.g., $G \subset \mathbb{C}^1$), then

$$\boldsymbol{s}_G(a, \cdot) = \boldsymbol{g}_G(a, \cdot) \quad \text{and} \quad \boldsymbol{S}_G(a; \cdot) = \boldsymbol{A}_G(a; \cdot).$$

Remark 4.2.9.

(a) Let $D \subset \mathbb{C}^m$ be a Liouville domain (cf. Remark 2.1.5). Then, for every domain $G \subset \mathbb{C}^n$ and $k \in \mathbb{N}$, we have:

- $\boldsymbol{m}_{G \times D}^{(k)}((a, b), (z, w)) = \boldsymbol{m}_G^{(k)}(a, z), \quad (a, b), (z, w) \in G \times D,$
- $\boldsymbol{\gamma}_{G \times D}^{(k)}((a, b); (X, Y)) = \boldsymbol{\gamma}_G^{(k)}(a; X), \quad (a, b) \in G \times D, (X, Y) \in \mathbb{C}^n \times \mathbb{C}^m.$

(b) Let $D \subset \mathbb{C}^m$ be a *psh Liouville domain*, i.e., each bounded from above function $u \in \mathcal{PSH}(D)$ is constant. Then, for every domain $G \subset \mathbb{C}^n$, we have:

- $\boldsymbol{s}_{G \times D}((a, b), (z, w)) = \boldsymbol{s}_G(a, z), \quad (a, b), (z, w) \in G \times D,$
- $\boldsymbol{S}_{G \times D}((a, b); (X, Y)) = \boldsymbol{S}_G(a; X), \quad (a, b) \in G \times D, (X, Y) \in \mathbb{C}^n \times \mathbb{C}^m,$
- $\boldsymbol{g}_{G \times D}((a, b), (z, w)) = \boldsymbol{g}_G(a, z), \quad (a, b), (z, w) \in G \times D,$
- $\boldsymbol{A}_{G \times D}((a, b); (X, Y)) = \boldsymbol{A}_G(a; X), \quad (a, b) \in G \times D, (X, Y) \in \mathbb{C}^n \times \mathbb{C}^m.$

General product properties (with arbitrary D) of the above invariant functions and pseudometrics will be discussed in Chapter 18.

Now, we are going to present the basic properties of the above contractible families of functions and pseudometrics.

4.2.1 Properties of $\boldsymbol{g}_G$ and $\boldsymbol{A}_G$

Proposition 4.2.10.

(a) *If* $G_\nu \nearrow G$, *then* $\boldsymbol{g}_{G_\nu} \searrow \boldsymbol{g}_G$ *and* $\boldsymbol{A}_{G_\nu} \searrow \boldsymbol{A}_G$.

(b) *If* $G = G_{\mathfrak{h}} \subset \mathbb{C}^n$ *is a balanced domain, then the following conditions are equivalent:*

(i) $\boldsymbol{g}_G(0,\cdot) = \mathfrak{h}$ *on* G;

(ii) $\boldsymbol{A}_G(0;\cdot) = \mathfrak{h}$ *on* $\mathbb{C}^n$;

(iii) G *is pseudoconvex.*

In particular, if G *is pseudoconvex, then* $\boldsymbol{g}_G(0,\cdot) = \boldsymbol{\ell}^*_G(0,\cdot)$ *and* $\boldsymbol{A}_G(0;\cdot) = \boldsymbol{\varkappa}_G(0;\cdot)$; *cf. Propositions* 3.1.11 *and* 3.5.3.

(c) *If* $G = G_{\mathfrak{h}}$ *is a balanced domain, then* $\boldsymbol{A}_G(0;\cdot) = \widetilde{\mathfrak{h}}$ (*cf.* § 2.2.2).

(d) (*Cf.* [303]) *If* $P \subset G$ *is a closed pluripolar set, then*

$$\boldsymbol{g}_{G\setminus P} = \boldsymbol{g}_G|_{(G\setminus P)\times(G\setminus P)}, \qquad \boldsymbol{A}_{G\setminus P} = \boldsymbol{A}_G|_{(G\setminus P)\times\mathbb{C}^n}.$$

(e) (*Cf.* [122, 305]) *If* G *is bounded, then for any ball* $\mathbb{B}(a,r) \subset G$ *and for any* $\varepsilon > 0$ *there exists a* $\delta \in (0,r)$ *such that*

$$[\boldsymbol{g}_G(x,z)]^{1+\varepsilon} \le \boldsymbol{g}_G(y,z), \quad x,y \in \mathbb{B}(a,\delta),\ z \in G \setminus \mathbb{B}(a,r). \tag{4.2.6}$$

(f) *If* G *is bounded, then for any* $z_0 \in G$ *the function* $G \ni z \longmapsto \boldsymbol{g}_G(z,z_0)$ *is continuous. Note that for unbounded domains* G *the function* $G \ni z \longmapsto \boldsymbol{g}_G(z,z_0)$ *need not be continuous; cf. Remark* 6.1.4.

(g) (*Cf.* [261]) *For any* G *the function* $\boldsymbol{g}_G$ *is upper semicontinuous on* $G \times G$. *Note that the function* $G \ni z \longmapsto \boldsymbol{g}_G(a,z)$ *need not be continuous (even for bounded domains of holomorphy); cf.* (b). *If the function* $\boldsymbol{g}_G(a,\cdot)$ *is continuous at* b, *then the function* $\boldsymbol{g}_G$ *is continuous at* (a,b).

(h) (*Cf.* [303]) *If* G *is hyperconvex, then* $\lim_{z\to\partial G} \boldsymbol{g}_G(a,z) = 1$, $a \in G$.

(i) (*Cf.* [303]) *For each* $a \in G$, *the function* $\log \boldsymbol{g}_G(a,\cdot)$ *is a maximal plurisubharmonic function on* $G \setminus \{a\}$. *In particular, if* $\log \boldsymbol{g}_G(a,\cdot) \in L^\infty_{\mathrm{loc}}(G \setminus \{a\})$ (*e.g. if the set* G *is bounded*), *then* $(dd^c \log \boldsymbol{g}_G(a,\cdot))^n = 0$ *in* $G\setminus\{a\}$; *cf. Appendix* B.6.

(j) (*Cf.* [122, 305]) *If* G *is hyperconvex, then the function* $\boldsymbol{g}_G$ *is continuous on* $G \times \overline{G}$, *where* $\boldsymbol{g}_G|_{G\times\partial G} := 1$; *cf.* (h).

(k) (*Cf.* [261]) *For any* G *the function* $\boldsymbol{A}_G$ *is upper semicontinuous on* $G \times \mathbb{C}^n$. *Note that the function* $\mathbb{C}^n \ni X \longmapsto \boldsymbol{A}_G(a;X)$ *need not be continuous (even for bounded domains of holomorphy); cf.* (b).

(l) *The function* $\boldsymbol{g}_G : G \times G \longrightarrow [0,1)$ *is symmetric iff for any* $z_0 \in G$ *the function* $G \ni z \longmapsto \boldsymbol{g}_G(z, z_0)$ *is log-psh.*

Note that, in general, for $n \geq 2$ *the function* $\boldsymbol{g}_G$ *need not be symmetric; cf.* [38]*; see also Proposition* 6.1.3 *and Exercise* 4.4.5.

? What are the relations between $\boldsymbol{g}_G(0,\cdot)$ and $\widetilde{\mathfrak{h}}$ for balanced domains $G = G_{\mathfrak{h}}$?

Proof. (a) The holomorphic contractibility implies that $\boldsymbol{g}_G \leq \boldsymbol{g}_{G_{\nu+1}} \leq \boldsymbol{g}_{G_\nu}$ and $\boldsymbol{A}_G \leq \boldsymbol{A}_{G_{\nu+1}} \leq \boldsymbol{A}_{G_\nu}$, $\nu \in \mathbb{N}$.

Fix an $a \in G$ and let $u := \lim_{\nu\to+\infty} \boldsymbol{g}_{G_\nu}(a,\cdot)$. Then, $u : G \to [0,1)$, $u \geq \boldsymbol{g}_G(a,\cdot)$, and $\log u \in \mathcal{PSH}(G)$ (cf. Appendix B.4.14). Let $\mathbb{B}(a,r) \subset\subset G$. Then $\mathbb{B}(a,r) \subset\subset G_\nu$, $\nu \geq \nu_0$. Using the holomorphic contractibility gives

$$\boldsymbol{g}_{G_\nu}(a,z) \leq \boldsymbol{g}_{\mathbb{B}(a,r)}(a,z) = \frac{\|z-a\|}{r}, \quad z \in \mathbb{B}(a,r),\ \nu \geq \nu_0. \tag{4.2.7}$$

Consequently, $u(z) \leq \frac{\|z-a\|}{r}$, $z \in \mathbb{B}(a,r)$. Hence, $u \in \mathcal{K}_G(a)$ and, therefore, $u \leq \boldsymbol{g}_G(a,\cdot)$.

Fix an $X \in \mathbb{C}^n$ and let $R > 0$ be such that $a + \mathbb{D}(R)X \subset\subset G$. Then, $a + \mathbb{D}(R)X \subset\subset G_\nu$, $\nu \geq \nu_0$. Define

$$u_\nu(\lambda) := \begin{cases} \frac{1}{|\lambda|}\boldsymbol{g}_{G_\nu}(a, a+\lambda X), & \text{if } 0 < |\lambda| < R \\ \boldsymbol{A}_{G_\nu}(a;X), & \text{if } \lambda = 0 \end{cases}, \quad \nu \geq \nu_0,$$

$$u(\lambda) := \begin{cases} \frac{1}{|\lambda|}\boldsymbol{g}_{G}(a, a+\lambda X), & \text{if } 0 < |\lambda| < R \\ \boldsymbol{A}_{G}(a;X), & \text{if } \lambda = 0 \end{cases}.$$

Observe that $u_\nu, u \in \mathcal{SH}(\mathbb{D}(R))$, $u \leq u_{\nu+1} \leq u_\nu$, and, by the first part of the proof, $u_\nu \searrow u$ in $\mathbb{D}_*(R)$. Let $\widetilde{u} := \lim_{\nu\to+\infty} u_\nu$. Then, $\widetilde{u} \in \mathcal{SH}(\mathbb{D}(R))$ and $\widetilde{u} = u$ on $\mathbb{D}_*(R)$. Consequently, $\widetilde{u} \equiv u$. In particular, $\boldsymbol{A}_{G_\nu}(a;X) = u_\nu(0) \searrow u(0) = \boldsymbol{A}_G(a;X)$.

(b) The inequalities "$\leq$" in (i) and (ii) are obvious. If $\mathfrak{h} \in \mathcal{PSH}(\mathbb{C}^n)$, then $\mathfrak{h}|_G \in \mathcal{K}_G(0)$ (cf. Appendix B.7.6), which shows that (iii) $\Longrightarrow$ (ii) and (iii) $\Longrightarrow$ (i). To get the remaining implications use Lemmas 4.2.3, 4.2.4, and Appendix B.7.6.

(c) Since $\boldsymbol{A}_G(0;\cdot) \leq \boldsymbol{\varkappa}_G(0;\cdot) \leq \mathfrak{h}_G$ and $\log \boldsymbol{A}_G(0;\cdot) \in \mathcal{PSH}(\mathbb{C}^n)$ (Lemma 4.2.4), we get $\boldsymbol{A}_G(0;\cdot) \leq \widetilde{\mathfrak{h}}_G$.

Let $\widetilde{G} := G_{\widetilde{\mathfrak{h}}}$. Recall (Remark 2.2.7(e)) that $\widetilde{G}$ is the envelope of holomorphy of G. Thus, by (b), $\widetilde{\mathfrak{h}}_G = \mathfrak{h}_{\widetilde{G}} = \boldsymbol{A}_{\widetilde{G}}(0;\cdot) \leq \boldsymbol{A}_G(0;\cdot)$.

(d) Since pluripolar sets are removable for upper bounded psh functions (cf. Appendix B.4.23), we obtain

$$\mathcal{K}_{G\setminus P}(a) = \mathcal{K}_G(a)|_{G\setminus P}, \quad a \in G \setminus P.$$

(e) Fix $\mathbb{B}(a,r) \subset G$ and $\varepsilon > 0$. We may assume that $\mathbb{B}(a,3r) \subset G$. Put $R :=$ diam G. Recall (cf. Remark 4.1.5(b)) that

$$\boldsymbol{g}_G(z,w) \le \frac{\|z-w\|}{2r}, \quad z,w \in \mathbb{B}(a,r). \tag{4.2.8}$$

Take $\delta \in (0, r/3)$ such that

$$\left(\frac{3\delta}{2r}\right)^{1+\varepsilon} < \frac{\delta}{R}. \tag{4.2.9}$$

Fix $x_0, y_0 \in \mathbb{B}(a,\delta)$ and define

$$u(z) := \begin{cases} \frac{\|z-y_0\|}{R}, & \text{if } z \in \overline{\mathbb{B}}(a,2\delta) \\ \max\{[\boldsymbol{g}_G(x_0,z)]^{1+\varepsilon}, \frac{\|z-y_0\|}{R}\}, & \text{if } z \in G \setminus \overline{\mathbb{B}}(a,2\delta) \end{cases}.$$

Observe that if $\|z-a\| = 2\delta$, then by (4.2.8) and (4.2.9) we get

$$[\boldsymbol{g}_G(x_0,z)]^{1+\varepsilon} < \frac{\|z-y_0\|}{R}.$$

Thus u is log-psh on G, and therefore $u \in \mathcal{K}_G(y_0)$. Hence

$$\boldsymbol{g}_G(y_0,z) \ge u(z) \ge [\boldsymbol{g}_G(x_0,z)]^{1+\varepsilon}, \quad z \in G \setminus \mathbb{B}(a,2\delta) \supset G \setminus \mathbb{B}(a,r).$$

(f) First, observe that, by (4.2.8), the function $\boldsymbol{g}_G$ is continuous on the diagonal of $G \times G$ (for arbitrary G). Next, for $z_0 \notin \mathbb{B}(a,r)$, take an arbitrary $\varepsilon > 0$ and let δ be as in (4.2.6). Then, we get

$$\limsup_{x \to a} [\boldsymbol{g}_G(x,z_0)]^{1+\varepsilon} \le \boldsymbol{g}_G(a,z_0), \quad [\boldsymbol{g}_G(a,z_0)]^{1+\varepsilon} \le \liminf_{y \to a} \boldsymbol{g}_G(y,z_0).$$

(g) By (a) we may assume that G is bounded. Recall that $\boldsymbol{g}_G$ is continuous at the diagonal. Take $(a,b) \in G \times G$, $b \notin \mathbb{B}(a,r)$. Using (4.2.6) and the fact that $\boldsymbol{g}_G(a,\cdot)$ is upper semicontinuous, we get

$$\limsup_{(x,z) \to (a,b)} [\boldsymbol{g}_G(x,z)]^{1+\varepsilon} \le \limsup_{z \to b} \boldsymbol{g}_G(a,z) = \boldsymbol{g}_G(a,b).$$

If $\boldsymbol{g}_G(a,\cdot)$ is continuous at $b \notin \mathbb{B}(a,r)$, then we get

$$[\boldsymbol{g}_G(a,b)]^{1+\varepsilon} = \lim_{z \to b} [\boldsymbol{g}_G(a,z)]^{1+\varepsilon} \le \liminf_{(y,z) \to (a,b)} \boldsymbol{g}_G(y,z).$$

(h) Let $\psi : G \longrightarrow (-\infty, 0)$ be a continuous psh function such that $\{\psi < -t\} \subset\subset G$ whenever $t > 0$. If $G \subset \mathbb{B}(R)$, then for $z \in G$ we define

$$\varphi(z) := \sup\Big\{\nu\psi(z) + \frac{1}{\nu}(\|z\|^2 - R^2) : \nu \in \mathbb{N}\Big\}.$$

Observe that φ is locally the maximum of a finite number of negative continuous strictly psh functions (cf. Appendix B.4), therefore φ itself is a negative continuous strictly psh function. Moreover, $\{\varphi < -t\} \subset\subset G$ for all $t > 0$.

Fix $a \in G$ and $0 < r < 1/2$ such that $\mathbb{B}(a, 2r) \subset\subset G$. Let $\chi \in \mathcal{C}_0^\infty(G, [0, 1])$ be such that $\chi = 1$ on $\mathbb{B}(a, r)$ and $\operatorname{supp} \chi \subset \mathbb{B}(a, 2r)$. For $C > 0$, let

$$u(z) := \exp(C\varphi(z) + \chi(z) \log \|z - a\|), \quad z \in G.$$

Obviously, $0 \le u < 1$ and $u(z) \le \|z - a\|$, $z \in \mathbb{B}(a, r)$. Moreover, since φ is strictly psh, we easily conclude that for $C \gg 0$ the function $\log u$ is psh on G. Let us fix such a C. Then, $u \in \mathcal{K}_G(a)$, and consequently

$$\boldsymbol{g}_G(a, z) \ge \exp(C\varphi(z)), \quad z \in G \setminus \mathbb{B}(a, 2r),$$

which directly implies the required result.

(i) Let G_0 be a relatively compact open subset of $G \setminus \{a\}$ and let v be a function upper semicontinuous on $\overline{G_0}$ and psh in G_0 such that $v \le \log \boldsymbol{g}_G(a, \cdot)$ on ∂G_0. Define

$$u(z) := \begin{cases} \max\{\boldsymbol{g}_G(a, z), \exp(v(z))\}, & z \in G_0 \\ \boldsymbol{g}_G(a, z), & z \in G \setminus G_0 \end{cases}.$$

Then, $u \in \mathcal{K}_G(a)$ (cf. Appendix B.4.18) and, therefore, by the definition of $\boldsymbol{g}_G$, we have $u \le \boldsymbol{g}_G(a, \cdot)$. In particular, $v \le \log \boldsymbol{g}_G(a, \cdot)$ in G_0, which finishes the proof of maximality. It remains to observe that if G is bounded, then

$$\frac{\|z - w\|}{\operatorname{diam} G} \le \boldsymbol{g}_G(z, w), \quad z, w \in G. \tag{4.2.10}$$

In particular, if G is bounded, then the function $\log \boldsymbol{g}_G(a, \cdot)$ is locally bounded in $G \setminus \{a\}$ and, therefore, using Appendix B.6.2, $(dd^c \log \boldsymbol{g}_G(a, \cdot))^n = 0$ in $G \setminus \{a\}$.

(j) Fix $(a, b) \in G \times \overline{G}$. Fix $r > 0$ with $\mathbb{B}(a, 3r) \subset G$ and put $R := \operatorname{diam} G$. We may assume that $a \ne b$ (cf. the proof of (f)) and that $b \notin \mathbb{B}(a, 2r)$. Let us assume for a while that

$$\text{the function } \boldsymbol{g}_G(a, \cdot) \text{ is continuous on } \overline{G}. \tag{4.2.11}$$

Take an arbitrary $\varepsilon > 0$ and let $\delta = \delta(\varepsilon)$ be as in (e). Note that, by (h), condition (4.2.6) holds for $x, y \in \mathbb{B}(a, \delta)$, $z \in \overline{G} \setminus \mathbb{B}(a, r)$. Now, we may argue as in the proof (g):

$$\begin{aligned}
[\boldsymbol{g}_G(a, b)]^{1+\varepsilon} &= \lim_{z \to b} [\boldsymbol{g}_G(a, z)]^{1+\varepsilon} \le \liminf_{(y,z) \to (a,b)} \boldsymbol{g}_G(y, z) \\
&\le \limsup_{(x,z) \to (a,b)} \boldsymbol{g}_G(x, z) \le \lim_{z \to b} [\boldsymbol{g}_G(a, z)]^{\frac{1}{1+\varepsilon}} = [\boldsymbol{g}_G(a, b)]^{\frac{1}{1+\varepsilon}}.
\end{aligned}$$

It remains to prove (4.2.11). By virtue of (h) and Lemma 4.2.3 it suffices, to prove that $\boldsymbol{g}_G(a,\cdot)$ is lower semicontinuous on G.

Since $\boldsymbol{g}_G(a,\cdot)$ is log-psh on G, we can find a sequence $(G_\nu)_{\nu=1}^{\infty}$ of subdomains of G and a sequence $(g_\nu)_{\nu=1}^{\infty}$ of $\mathcal{C}^\infty$ log-psh functions g_ν on G_ν with $G_\nu \nearrow G$ and $g_\nu \searrow \boldsymbol{g}_G(a,\cdot)$; cf. Appendix B.4.19.

Let $\varphi : G \longrightarrow (-\infty, 0)$ be a continuous psh function such that $\{\varphi < -t\} \subset\subset G$ whenever $t > 0$. We may assume that

$$\varphi(z) < -r \;\text{ for } z \in \overline{\mathbb{B}}(a,r). \tag{4.2.12}$$

Fix $\varepsilon_0 \in (0,1),\; \varepsilon_0 < r$, such that

$$\frac{\varepsilon}{R} > \varepsilon^{2(1-\varepsilon)} \geq e^{\varepsilon - 1/\varepsilon}, \quad 0 < \varepsilon < \varepsilon_0. \tag{4.2.13}$$

Since $\varepsilon_0 < r$, condition (4.2.12) implies that

$$\varphi(z) < -\varepsilon \;\text{ for } z \in \overline{\mathbb{B}}(a,\varepsilon), \quad 0 < \varepsilon < \varepsilon_0. \tag{4.2.14}$$

For an arbitrary $\varepsilon \in (0, \varepsilon_0)$ let $\eta = \eta(\varepsilon) \in (0, \varepsilon)$ be such that

$$\frac{\eta}{2r} < (\varepsilon\eta)^{1-\varepsilon}. \tag{4.2.15}$$

Then, using Dini's theorem, we choose $\nu(\varepsilon)$ with $\nu(\varepsilon) \to \infty$ as $\varepsilon \longrightarrow 0$ such that for $v_\varepsilon := g_{\nu(\varepsilon)}$ the following properties hold:

$$\{\varphi \leq -\varepsilon^3\} \subset G_{\nu(\varepsilon)},$$

$$v_\varepsilon < (\varepsilon\eta)^{1-\varepsilon} \;\text{ on } \overline{\mathbb{B}}(a,\eta) \;\text{ (use (4.2.8) and (4.2.15))}, \tag{4.2.16}$$

$$v_\varepsilon(z) < 1 \;\text{ if } \varphi(z) \leq -\varepsilon^3. \tag{4.2.17}$$

Now, we define $u_\varepsilon : G \longrightarrow \mathbb{R}$ by

$$u_\varepsilon(z) := \begin{cases} e^{-\varepsilon}(\varepsilon\|z-a\|)^{1-\varepsilon}, & \text{if } z \in \overline{\mathbb{B}}(a,\eta) \\ \max\{e^{-\varepsilon}(\varepsilon\|z-a\|)^{1-\varepsilon}, e^{-\varepsilon}v_\varepsilon(z)\}, & \text{if } z \in \overline{\mathbb{B}}(a,\varepsilon) \setminus \overline{\mathbb{B}}(a,\eta) \\ e^{-\varepsilon}v_\varepsilon(z), & \text{if } \varphi(z) \leq -\varepsilon \text{ and } z \notin \overline{\mathbb{B}}(a,\varepsilon)\;. \\ \max\{e^{-\varepsilon}v_\varepsilon(z), e^{\varphi(z)/\varepsilon^2}\}, & \text{if } -\varepsilon < \varphi(z) \leq -\varepsilon^3 \\ e^{\varphi(z)/\varepsilon^2}, & \text{if } \varphi(z) > -\varepsilon^3 \end{cases}$$

Note that, by (4.2.14), the function u_ε is well-defined. Then, (4.2.8), (4.2.10), (4.2.13), (4.2.16), and (4.2.17) show that u_ε is a continuous log-psh function on G with values in $[0,1)$. In particular, $u_\varepsilon^{1/(1-\varepsilon)} \in \mathcal{K}_G(a)$. Moreover, $u_\varepsilon \to \boldsymbol{g}_G(a,\cdot)$ as $\varepsilon \longrightarrow 0$, i.e.,

$$\boldsymbol{g}_G(a,\cdot) = \sup\{u_\varepsilon^{1/(1-\varepsilon)} : 0 < \varepsilon < \varepsilon_0\}.$$

Hence, $\boldsymbol{g}_G(a,\cdot)$ is a lower semicontinuous function.

(k) Suppose that $\boldsymbol{A}_G(a;X_0) < M$. By Lemma 4.2.3 there exists an $R > 0$ such that

$$\boldsymbol{g}_G(a, a+\lambda X_0) < M|\lambda|, \quad 0 < |\lambda| \le R.$$

By (g), there exists a neighborhood V of (a, X_0) in $G \times \mathbb{C}^n$ such that

$$\{z + \lambda X : (z,X) \in V,\ |\lambda| \le R\} \subset G,$$
$$\boldsymbol{g}_G(z, z+\lambda X) < MR, \quad |\lambda| = R,\ (z,X) \in V.$$

Fix $(z,X) \in V$. Observe that the function u, defined by

$$u(\lambda) := \frac{1}{|\lambda|}\boldsymbol{g}_G(z, z+\lambda X), \quad 0 < |\lambda| \le R,$$
$$u(0) := \boldsymbol{A}_G(z;X),$$

is subharmonic (Lemma 4.2.3). Since $u(\lambda) < M$ for $|\lambda| = R$, the maximum principle gives $u(0) = \boldsymbol{A}_G(z;X) < M$.

(l) If $\boldsymbol{g}_G$ is symmetric, then by Lemma 4.2.3, for any $z_0 \in G$, the function $u_{z_0}(z) := \boldsymbol{g}_G(z, z_0)$, $z \in G$, belongs to $\mathcal{K}_G(z_0)$. In particular, u_{z_0} is log-psh. Now suppose that for each $z_0 \in G$ the function u_{z_0} is log-psh. Fix z_0 and let $\mathbb{B}(z_0, 2r) \subset G$. Then (cf. Remark 4.1.5(b)), we have $u_{z_0}(z) \le \|z - z_0\|/r$, $z \in \mathbb{B}(z_0, r)$. Consequently, $u_{z_0} \in \mathcal{K}_G(z_0)$, and finally $\boldsymbol{g}_G(z, z_0) = u_{z_0}(z) \le \boldsymbol{g}_G(z_0, z)$, $z, z_0 \in G$. □

Basic relations between the Green function (resp. Azukawa pseudometric) and the higher order Lempert functions (resp. higher-order Kobayashi–Royden pseudometrics) (cf. § 3.8) are described by the following inequalities:

Proposition 4.2.11. $\boldsymbol{g}_G \le \boldsymbol{\ell}_G^{(k)*}$, $\boldsymbol{A}_G \le \boldsymbol{\varkappa}_G^{(k)}$.

Proof. Fix an $a \in G$ and let $u := \boldsymbol{g}_G(a,\cdot)$. Recall that $u(z) \le M\|z - a\|$, $z \in G$. Take a $\varphi = a + \lambda^k \psi \in \mathcal{O}(\mathbb{D}, G)$. Then, the function

$$\mathbb{D}_* \ni \lambda \overset{v}{\longmapsto} \frac{1}{|\lambda|^k} u(\varphi(\lambda))$$

is log-sh and locally bounded near $\lambda = 0$. Thus, v extends subharmonically to $\mathbb{D}$ and $v(0) = \limsup_{\lambda \to\!\!\!\to 0} v(\lambda)$. The maximum principle implies that $v \le 1$. In particular, if $\varphi(\sigma) = z$, then $u(z) \le \sigma^k$, which gives $\boldsymbol{g}_G(a,z) \le \boldsymbol{\ell}_G^{(k)*}(a,z)$. If $\sigma\psi(0) = X$, then, using Lemma 4.2.7, we get $1 \ge v(0) = \boldsymbol{A}_G(a;\psi(0))$, which implies that $\boldsymbol{A}_G(a;X) \le \boldsymbol{\varkappa}_G^{(k)}(a;X)$. □

In particular, using Propositions 3.1.11, 3.5.3, and 4.2.10(b), we get the following result:

Corollary 4.2.12. *Let* $G = \{z \in \mathbb{C}^n : \mathfrak{h}(z) < 1\}$ *be a balanced pseudoconvex domain. Then, we have* $\boldsymbol{g}_G(0,\cdot) \equiv \boldsymbol{\ell}_G^{(k)*}(0,\cdot) \equiv \mathfrak{h}$ *and* $\boldsymbol{A}_G(0;\cdot) \equiv \boldsymbol{\varkappa}_G^{(k)}(0;\cdot) \equiv \mathfrak{h}$, $k \in \mathbb{N}$.

Before discussing examples of pluricomplex Green functions $\boldsymbol{g}_G$, we introduce functions without singularities that can be used to approximate $\boldsymbol{g}_G$ from above. These functions will be useful in further investigations of $\boldsymbol{g}_G$.

Let $G \subset \mathbb{C}^n$ be a bounded domain and let $K \subset G$ be compact. Fix positive numbers r, R such that

$$\bigcup_{a \in K} \mathbb{B}(a,r) \subset\subset G \text{ and } K \subset \bigcap_{a \in K} \mathbb{B}(a,R).$$

Then, for $\varepsilon \in (0,r)$, we define

$$\boldsymbol{g}_G^{\varepsilon}(z,w) := \sup\{v(w) : 0 \le v < 1,\ \log v \in \mathcal{PSH}(G),\ v|_{\mathbb{B}(z,\varepsilon)} \le \varepsilon/r\}, \qquad (z,w) \in K \times G.$$

The following properties of $\boldsymbol{g}_G^{\varepsilon}$ may be found in [57, 59]:

Proposition 4.2.13. *Let* $G, K, r, R, \varepsilon \in (0,r)$ *be as above. Then,*

(a) $\log \boldsymbol{g}_G^{\varepsilon}(z,\cdot) \in \mathcal{PSH}(G)$ *and* $\varepsilon/r \le \boldsymbol{g}_G^{\varepsilon} < 1$ *on* $K \times G$.

(b) $\max\{\varepsilon/r, \boldsymbol{g}_G(z,w)\} \le \boldsymbol{g}_G^{\varepsilon}(z,w) \le \max\{\|z-w\|/r, \varepsilon/r\}$ *on* $K \times G$.

(c) $\boldsymbol{g}_G^{\varepsilon'} \le \boldsymbol{g}_G^{\varepsilon}$ *on* $K \times G$ *for* $0 < \varepsilon' < \varepsilon < r$.

(d) *If, in addition, G is hyperconvex, then,*

$$\boldsymbol{g}_G^{\varepsilon}(z,\cdot) \le (\boldsymbol{g}_G(z,\cdot))^{\frac{\log(r/\varepsilon)}{\log(R/\varepsilon)}} \text{ on } G \setminus \overline{\mathbb{B}}(z,\varepsilon) \text{ for all } z \in K.$$

(e) *If G is hyperconvex, then* $\boldsymbol{g}_G^{\varepsilon}(z,\cdot) \underset{\varepsilon \searrow 0}{\searrow} \boldsymbol{g}_G(z,\cdot)$ *on* $G \setminus \{z\}$ *for all* $z \in K$.

(f) *If G is hyperconvex and* $a \in G$, *then* $\boldsymbol{g}_G^{\varepsilon}(a,\cdot)$ *is continuous on* G.

(g) *For* $z \in K$ *the function* $\log \boldsymbol{g}_G^{\varepsilon}(z,\cdot)$ *is maximal on* $G \setminus \overline{\mathbb{B}}(z,\varepsilon)$.

Proof. (a) Fix $z \in K$ and $\varepsilon \in (0,r)$ and put $u := (\boldsymbol{g}_G^{\varepsilon}(z,\cdot))^* \ge \boldsymbol{g}_G^{\varepsilon}(z,\cdot)$. Then, $\log u = (\log \boldsymbol{g}_G^{\varepsilon}(z,\cdot))^* \in \mathcal{PSH}(G)$, $u \ge \varepsilon/r$ on G, and $u = \varepsilon/r$ on $\mathbb{B}(z,\varepsilon)$. Moreover, by the maximum principle we get $u < 1$ on G. Now, applying the Oka Theorem (see Appendix B.4.26), we see that $u \le \varepsilon/r$ also on $\overline{\mathbb{B}}(z,\varepsilon)$. Hence, u is a competitor in the definition of $\boldsymbol{g}_G^{\varepsilon}(z,\cdot)$, which gives $u = \boldsymbol{g}_G^{\varepsilon}(z,\cdot)$. In particular, $\log \boldsymbol{g}_G^{\varepsilon}(z,\cdot) \in \mathcal{PSH}(G)$.

(b) Note that the logarithm of the function on the left side is psh, and negative on G. Call it, for the moment, u. If $w \in \overline{\mathbb{B}}(z,\varepsilon) \subset \mathbb{B}(z,r)$, then

$$\boldsymbol{g}_G(z,w) \le \boldsymbol{g}_{\mathbb{B}(z,r)}(z,w) \le \|w-z\|/r \le \varepsilon/r.$$

Hence, u is a competitor for the definition of $\boldsymbol{g}_G^{\varepsilon}(z,\cdot)$ and the left inequality is verified.

For the right inequality, as long as $\|w-z\| \ge r$ or $\|w-z\| \le \varepsilon$, there is nothing to do. So, take a w with $\varepsilon < \|w-z\| < r$. Put

$$\mathbb{D}_*(r) \ni \lambda \overset{u}{\longmapsto} \log \boldsymbol{g}_G^{\varepsilon}\left(z, z+\lambda\frac{w-z}{\|w-z\|}\right) - \log\frac{|\lambda|}{r}.$$

Then, $u \le 0$ if $|\lambda| = \varepsilon$ or $|\lambda| = r$. Applying the maximum principle for subharmonic functions, it follows that $u \le 0$ on $\mathbb{A}(\varepsilon, r)$, which gives the remaining part of the right inequality for $\lambda = \|w-z\|$.

(c) Fix $0 < \varepsilon' < \varepsilon < r$ and $z \in K$. Note that $\log \boldsymbol{g}_G^{\varepsilon'}(z,\cdot)$ is negative and psh on G. Moreover, if $w \in \overline{\mathbb{B}}(z,\varepsilon)$, then

$$\boldsymbol{g}_G^{\varepsilon'}(z,w) \le \max\{\varepsilon'/r, \|w-z\|/r\} \le \varepsilon/r,$$

i.e., $\boldsymbol{g}_G^{\varepsilon'}(z,\cdot)$ is a competitor in the definition of $\boldsymbol{g}_G^{\varepsilon}(z,\cdot)$. Therefore, $\boldsymbol{g}_G^{\varepsilon'}(z,\cdot) \le \boldsymbol{g}_G^{\varepsilon}(z,\cdot)$.

(d) Fix an ε and a $z \in K$ as in the proposition. If $w \in G$ with $\|w-z\| = \varepsilon$, then

$$\frac{\log(R/\varepsilon)}{\log(r/\varepsilon)} \log \boldsymbol{g}_G^{\varepsilon}(z,w) = \log(\varepsilon/R) = \log \boldsymbol{g}_{\mathbb{B}(z,R)}(z,w) \le \log \boldsymbol{g}_G(z,w).$$

If the point w now tends to the boundary of G, then $\lim_{w\to\partial G} \log \boldsymbol{g}_G(z,w) = 0$, see Proposition 4.2.10(h). Applying the maximality of $\log \boldsymbol{g}_G(z,\cdot)$ (see Proposition 4.2.10(i)) delivers the claimed inequality on the domain $G \setminus \overline{\mathbb{B}}(z,\varepsilon)$.

(e) This property is a direct consequence of (c) and (d).

(f) It suffices to prove that $u := \boldsymbol{g}_G^{\varepsilon}(a,\cdot)$ is continuous on $G \setminus \overline{\mathbb{B}}(a,\varepsilon)$. So, let us fix a point $w^0 \in G \setminus \overline{\mathbb{B}}(a,\varepsilon)$ and choose a positive η. By the hyperconvexity, one finds a positive δ such that if $\operatorname{dist}(w,\partial D) < \delta$, then $u(w) > 1-\eta$. Moreover, fix $\varepsilon' > \varepsilon$ such that $u \le \varepsilon/r + \eta$ on $\overline{\mathbb{B}}(a,\varepsilon') \subset G$. Put $s := \min\{\delta, \varepsilon' - \varepsilon\}$. Now fix a point $w' \in G$ with $\|w' - w^0\| < s$. Let

$$G' := \{w \in G : w + w^0 - w' \in G\}.$$

Obviously, $G' \subset G$ is open and if $w \in \overline{\mathbb{B}}(a,\varepsilon)$, then $w + w^0 - w' \in \overline{\mathbb{B}}(a,\varepsilon')$ and $u(w + w^0 - w') - \eta \le \varepsilon/r$. Moreover, if $w \in G'$ tends to a $w^* \in \partial G' \cap G$, then $w + w^0 - w'$ tends to a point in ∂G, implying that

$$\limsup_{w \to w^*}(u(w + w^0 - w') - \eta) = 1 - \eta \le u(w^*).$$

Therefore, setting

$$v(w) := \begin{cases} \max\{u(w), u(w + w^0 - w') - \eta\}, & \text{if } w \in G' \\ u(w), & \text{if } w \in G \setminus G' \end{cases}$$

gives $\log v \in \mathcal{PSH}(G)$ and the fact that v is a competitor in the definition of $\boldsymbol{g}_G^{\varepsilon}(a,\cdot)$. Hence, $u(w') \ge v(w') \ge u(w^0) - \eta$, meaning that u is semicontinuous from below at the point w^0.

(g) Fix ε and $z \in K$, a relatively compact open subset $G_0 \subset G \setminus \overline{\mathbb{B}}(z,\varepsilon)$, and a function $v \in \mathcal{C}(\overline{G_0}) \cap \mathcal{PSH}(G_0)$ with $v \le \log \boldsymbol{g}_G^{\varepsilon}(z,\cdot)$ on ∂G_0. Put

$$u(w) := \begin{cases} \max\{\log \boldsymbol{g}_G^{\varepsilon}(z,w), v(w)\}, & \text{if } w \in G_0 \\ \log \boldsymbol{g}_G^{\varepsilon}(z,w), & \text{if } w \in G \setminus G_0 \end{cases}, \quad w \in G.$$

Then, by Appendix B.4.18, $u \in \mathcal{PSH}(G)$. Moreover, $u < 0$ and $u|_{\overline{\mathbb{B}}(z,\varepsilon)} = \log \boldsymbol{g}_G^{\varepsilon}(z,\cdot)|_{\overline{\mathbb{B}}(z,\varepsilon)} = \log(\varepsilon/r)$. Thus, $u \le \log \boldsymbol{g}_G^{\varepsilon}(z,\cdot)$. In particular, $v \le \log \boldsymbol{g}_G^{\varepsilon}(z,\cdot)$ on G_0. $\square$

Remark 4.2.14. For any domain $G \subset \mathbb{C}^n$, define

$$\tilde{\boldsymbol{g}}_G(a,z) := \inf_{\substack{\varphi \in \mathcal{O}(\mathbb{D},G) \\ \varphi(0)=z \\ a \in \varphi(\mathbb{D})}} \Big\{ \prod_{\lambda \in \varphi^{-1}(a)} |\lambda|^{\operatorname{ord}_\lambda(\varphi - a)} \Big\}, \quad a, z \in G.$$

Then, the following properties are true:

(a)

$$\tilde{\boldsymbol{g}}_G(a,z) = \inf_{\substack{\varphi \in \mathcal{O}(\overline{\mathbb{D}},G) \\ \varphi(0)=z \\ a \in \varphi(\mathbb{D})}} \Big\{ \prod_{\lambda \in \varphi^{-1}(a)} |\lambda|^{\operatorname{ord}_\lambda(\varphi - a)} \Big\}, \quad a, z \in G;$$

(b) $\boldsymbol{g}_G \le \tilde{\boldsymbol{g}}_G$;

(c) the system $(\tilde{\boldsymbol{g}}_G)_{G \in \mathfrak{G}}$ is an $\boldsymbol{m}$-contractible family of functions;

(d) the function $\tilde{\boldsymbol{g}}_G$ is upper semicontinuous;

(e) if $\log \tilde{\boldsymbol{g}}(a,\cdot) \in \mathcal{PSH}(G)$, then $\tilde{\boldsymbol{g}}_G(a,\cdot) \equiv \boldsymbol{g}_G(a,\cdot)$.

Notice that, in fact, $\tilde{\boldsymbol{g}}_G \equiv \boldsymbol{g}_G$ *for any domain G (cf. Theorem* 17.4.3). *The proof of this identity requires more advanced tools (the analytic disc method), which will be developed in Chapter* 17.

Add. (see Section 11.6) (a) Fix G, $a \in G$, and $z_0 \in G$, $a \neq z_0$. Suppose that $\tilde{\boldsymbol{g}}_G(a, z_0) < C' < C$. Let $\varphi \in \mathcal{O}(\mathbb{D}, G)$ be such that $\varphi(0) = z_0, a \in \varphi(\mathbb{D})$, and $\prod |\lambda_j|^{k_j} < C'$, where $(\lambda_j)_j$ denote all zeros of $\varphi - a$ and $k_j := \operatorname{ord}_{\lambda_j}(\varphi - a)$. We may assume that $|\lambda_j| \leq |\lambda_{j+1}|$. Take N such that $\prod_{j=1}^N |\lambda_j|^{k_j} < C'$. Choose $R \in (|\lambda_N|, 1)$ sufficiently near to 1. Define $\varphi_R(\lambda) := \varphi(R\lambda)$. Observe that $\varphi_R \in \mathcal{O}(\overline{\mathbb{D}}, G)$, $\varphi_R(0) = z_0$, and $a \in \varphi_R(\mathbb{D})$. Because $\varphi_R(\lambda_j/R) = a$ for $|\lambda_j| < R$, we get

$$\prod_{|\lambda_j|<R} \left(\frac{|\lambda_j|}{R}\right)^{k_j} < C' R^{-\sum_{j=1}^N k_j} < C.$$

Since C is arbitrary, we are done.

Add. (b) Fix G and $a, z_0 \in G$, $z_0 \neq a$. Let $u := \boldsymbol{g}_G(a, \cdot)$. Recall (cf. Lemma 4.2.3) that $u \in \mathcal{K}_G(a)$. In particular, u is log-psh on G and

$$u(z) \leq M\|z - a\| \quad \text{for } z \text{ near } a, \tag{4.2.18}$$

where M is a positive constant. Take an arbitrary $\varphi \in \mathcal{O}(\overline{\mathbb{D}}, G)$ with $\varphi(0) = z_0$ and $a \in \varphi(\mathbb{D})$. Let λ_j, $j = 1, \dots, N$, denote the solutions in $\mathbb{D}$ of the equation $\varphi(\lambda) = a$, counted with multiplicities. Define $f := \boldsymbol{h}_{\lambda_1} \cdot \dots \cdot \boldsymbol{h}_{\lambda_N}$, where $\boldsymbol{h}_{\lambda_0}(\lambda) := (\lambda - \lambda_0)/(1 - \overline{\lambda}_0\lambda)$. Put $v := u \circ \varphi / |f|$. It is clear that v is sh on $\mathbb{D} \setminus \{\lambda_1, \dots, \lambda_N\}$ and v is locally bounded on $\mathbb{D}$ (use (4.2.18)). Consequently, v extends subharmonically to $\mathbb{D}$. Since $\limsup_{|\lambda| \to 1} v(\lambda) \leq 1$, the maximum principle shows that $v \leq 1$. In particular, $u(z_0) = u(\varphi(0)) \leq |f(0)| = \prod_{j=1}^N |\lambda_j|$. Now, by (a), $\boldsymbol{g}_G(a, z_0) \leq \tilde{\boldsymbol{g}}_G(a, z_0)$.

Add. (c) Directly from the definition, it follows that the system $(\tilde{\boldsymbol{g}}_G)_{G \in \mathfrak{G}}$ is holomorphically contractible. Observe that if $G = \mathbb{D}$, $a, z_0 \in \mathbb{D}$, then the function $\varphi = \boldsymbol{h}_{-z_0}$ is a competitor in the infimum, which defines $\tilde{\boldsymbol{g}}_{\mathbb{D}}(a, z_0)$. Hence, by (b), $\boldsymbol{m}(a, z_0) = \boldsymbol{g}_{\mathbb{D}}(a, z_0) \leq \tilde{\boldsymbol{g}}_{\mathbb{D}}(a, z_0) \leq |\boldsymbol{h}_{-z_0}^{-1}(a)| = \boldsymbol{m}(a, z_0)$.

Add. (d) Let $a_0, z_0 \in G$ be fixed. Because of (c), the function $\tilde{\boldsymbol{g}}_G$ is continuous on the diagonal, and therefore we may assume that $a_0 \neq z_0$. Take $C > \tilde{\boldsymbol{g}}_G(a_0, z_0)$ and let $\varphi \in \mathcal{O}(\overline{\mathbb{D}}, G)$ be such that $\varphi(0) = z_0$, $a_0 \in \varphi(\mathbb{D})$, and $\prod_{j=1}^N |\lambda_j| \leq C$, where $\lambda_1, \dots, \lambda_N$ denote the zeros of $\varphi - a_0$ in $\mathbb{D}$ with multiplicities. Define

$$\varphi_{a,z}(\lambda) := \varphi(\lambda) + (z - z_0) \prod_{j=1}^N (1 - \lambda/\lambda_j) + (a - a_0)\left(1 - \prod_{j=1}^N (1 - \lambda/\lambda_j)\right), \quad \lambda \in \overline{\mathbb{D}}.$$

Then, $\varphi_{a,z}(\overline{\mathbb{D}}) \subset G$ for (a, z) near (a_0, z_0) and, therefore, $\varphi_{a,z}$ is a competitor in (a). Thus, $\tilde{\boldsymbol{g}}_G(a, z) \leq C$ for (a, z) near (a_0, z_0).

Add. (e) This follows directly from (b), (c), and the definition of $\boldsymbol{g}_G(a, \cdot)$.

4.2.2 Examples

The following example shows that, in general, there are no relations between $\boldsymbol{g}_G$ and $\boldsymbol{g}_{\widetilde{G}}|_{G\times G}$, where $\widetilde{G}$ is the envelope of holomorphy of G; see also Exercise 9.2.7.

Example 4.2.15 (cf. [257]). Given $0 < \alpha, \beta < 1$, we define

$$G_+ = G_+(\alpha,\beta) := \mathbb{D}^2 \setminus \{(z_1,z_2) \in \mathbb{D}^2 : |z_1| \geq \alpha,\ |z_2| \geq \beta\},$$
$$G_- = G_-(\alpha,\beta) := \mathbb{D}^2 \setminus \{(z_1,z_2) \in \mathbb{D}^2 : |z_1| \leq \alpha,\ |z_2| \geq \beta\}.$$

Observe that G_+ and G_- are Reinhardt domains, G_+ is balanced, but G_- is not balanced. They are not domains of holomorphy. The envelope of holomorphy of G_+ is given by the formula

$$\widetilde{G}_+ = \widetilde{G}_+(\alpha,\beta) := \{(z_1,z_2) \in \mathbb{D}^2 : |z_1|^{-\log\beta}|z_2|^{-\log\alpha} < e^{-\log\alpha\log\beta}\};$$

the envelope of holomorphy $\widetilde{G}_-$ of G_- is just the unit bidisc $\mathbb{D}^2$. Our aim is to prove the following:

(a) $\boldsymbol{g}_{G_+}(0,z) = \boldsymbol{g}_{\widetilde{G}_+}(0,z)$

$$= \max\left\{|z_1|, |z_2|, \left(e^{\log\alpha\log\beta}|z_1|^{-\log\beta}|z_2|^{-\log\alpha}\right)^{-\frac{1}{\log\alpha\beta}}\right\}$$

$$=: \mathfrak{h}_{\widetilde{G}_+}(z), \quad z = (z_1,z_2) \in G_+;$$

(b) if $\beta \geq \alpha$, then $\boldsymbol{g}_{G_-}(0,z) = \boldsymbol{g}_{\widetilde{G}_-}(0,z) = \max\{|z_1|,|z_2|\}, \quad z = (z_1,z_2) \in G_-$;

(c) if $\beta < \alpha$, then $\boldsymbol{g}_{G_-}(0,z) = \left.\begin{cases} \max\{|z_1|, \frac{\alpha}{\beta}|z_2|\}, & \text{if } |z_2| < \beta \\ \max\{|z_1|, |z_2|^{\frac{\log\alpha}{\log\beta}}\}, & \text{if } |z_1| > \alpha \end{cases}\right\} \not\equiv \boldsymbol{g}_{\widetilde{G}_-}(0,z),$

$$z = (z_1,z_2) \in G_-.$$

It is unclear whether $\boldsymbol{g}_G(0,\cdot) = \boldsymbol{g}_{\widetilde{G}}(0,\cdot)$ on G for balanced Reinhardt domains. ?

Proof of (a). (Cf. Figure 4.1)

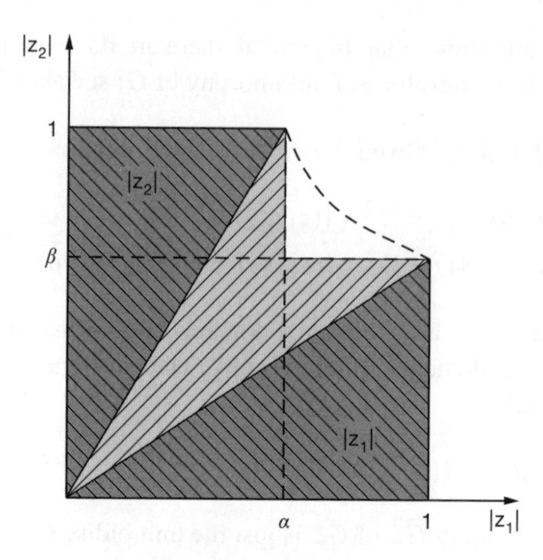

Figure 4.1. $\boldsymbol{g}_{G_+}(0,\cdot) = \boldsymbol{g}_{\widetilde{G}_+}(0,\cdot) = \mathfrak{h}_{\widetilde{G}_+}$.

Let

$$h(z) := \max\left\{|z_1|, |z_2|, \min\left\{\frac{|z_1|}{\alpha}, \frac{|z_2|}{\beta}\right\}\right\}.$$

By Proposition 4.2.10(b) and Remark 3.1.12, we have

$$\mathfrak{h}_{\widetilde{G}_+} = \boldsymbol{g}_{\widetilde{G}_+}(0,\cdot) \le \boldsymbol{g}_{G_+}(0,\cdot) \le \boldsymbol{\ell}^*_{G_+}(0,\cdot) \le h.$$

In particular,

$$\mathfrak{h}_{\widetilde{G}_+}(z) = \boldsymbol{g}_{\widetilde{G}_+}(0,z) = \boldsymbol{g}_{G_+}(0,z) = h(z) = \max\{|z_1|, |z_2|\} \tag{4.2.19}$$

for all $z = (z_1, z_2) \in G_+$ such that $|z_2| \le \beta|z_1|$ or $|z_1| \le \alpha|z_2|$. It remains to prove that $\boldsymbol{g}_{G_+}(0,z) \le \mathfrak{h}_{\widetilde{G}_+}(z)$ for all $z = (z_1, z_2) \in G_+$ with

$$\left(0 < |z_1| < \alpha \text{ and } \beta|z_1| < |z_2| < \frac{1}{\alpha}|z_1|\right)$$
$$\text{or } \left(0 < |z_2| < \beta \text{ and } \alpha|z_2| < |z_1| < \frac{1}{\beta}|z_2|\right).$$

Fix $0 < |z_1^0| < \alpha$ and define the function u by

$$U := \left\{\lambda \in \mathbb{C} : \beta|z_1^0| \le |\lambda| \le \frac{1}{\alpha}|z_1^0|\right\} \ni \lambda \overset{u}{\longmapsto} \boldsymbol{g}_{G_+}(0,(z_1^0,\lambda)).$$

Then, u is subharmonic and by (4.2.19)

$$u(\lambda) \le \begin{cases} |z_1^0|, & \text{if } |\lambda| = \beta|z_1^0| \\ \frac{1}{\alpha}|z_1^0|, & \text{if } |\lambda| = \frac{1}{\alpha}|z_1^0| \end{cases}.$$

Hence, by the Hadamard three-circles-theorem for subharmonic functions (cf. Appendix B.4.25), we get

$$u(\lambda) = \boldsymbol{g}_{G_+}(0,(z_1^0,\lambda)) \le \mathfrak{h}_{\widetilde{G}_+}(z_1^0,\lambda), \quad \lambda \in U.$$

The case where $0 < |z_2| < \beta$ is analogous. □

Proof of (*b*). (Cf. Figure 4.2)

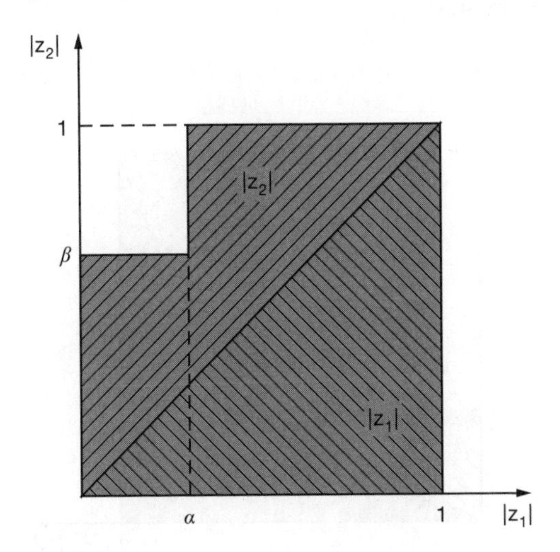

Figure 4.2. If $\beta \ge \alpha$, then $\boldsymbol{g}_{G_-}(0,z) = \boldsymbol{g}_{\widetilde{G}_-}(0,z) = \max\{|z_1|,|z_2|\}$.

Obviously,

$$\widetilde{h}_-(z) := \max\{|z_1|,|z_2|\} = \boldsymbol{g}_{\widetilde{G}_-}(0,z) \le \boldsymbol{g}_{G_-}(0,z), \quad z = (z_1,z_2) \in G_-.$$

Take $\zeta = (\zeta_1,\zeta_2) \in \partial G_-$ with $|\zeta_1| = 1$. Then, by Appendix B.4.24, we have

$$\boldsymbol{g}_{G_-}(0,\lambda\zeta) \le |\lambda| = \widetilde{h}_-(\lambda\zeta), \quad \lambda \in \mathbb{D}.$$

It remains to prove that

$$\boldsymbol{g}_{G_-}(0,z) \le \widetilde{h}_-(z) \tag{4.2.20}$$

for all $z = (z_1,z_2) \in G_-$ with $|z_2| \ge |z_1|$. Fix $0 < |z_2^0| < \beta$ and consider the function u given by

$$U := \{\lambda \in \mathbb{C} : |\lambda| \le |z_2^0| + \varepsilon\} \ni \lambda \overset{u}{\longmapsto} \boldsymbol{g}_{G_-}(0,(\lambda,z_2^0))$$

($\varepsilon > 0$ small). Then, by the maximum principle, we have

$$u(\lambda) \le |z_2^0| + \varepsilon, \quad \lambda \in U.$$

Letting $\varepsilon \longrightarrow 0$, we get (4.2.20).

Now, let $\alpha < |z_1^0| < 1$. We prove (4.2.20) by using the three-circles-theorem for the function

$$\{\lambda \in \mathbb{C} : |z_1^0| - \varepsilon < |\lambda| < 1\} \ni \lambda \longmapsto \boldsymbol{g}_{G_-}(0, (z_1^0, \lambda))$$

and then letting $\varepsilon \longrightarrow 0$. □

Proof of (c). (Cf. Figure 4.3)

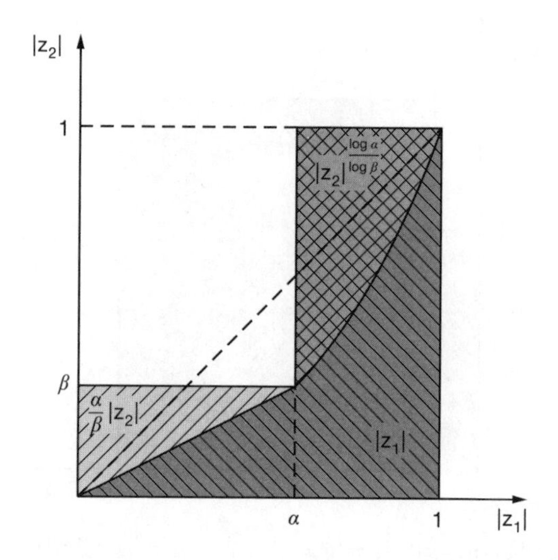

Figure 4.3. If $\beta < \alpha$, then $\boldsymbol{g}_{G_-}(0,\cdot) \not\equiv \boldsymbol{g}_{\tilde{G}_-}(0,\cdot)$.

Put

$$u(z) := \begin{cases} \max\{|z_1|, \frac{\alpha}{\beta}|z_2|\}, & \text{if } |z_2| < \beta \\ \max\{|z_1|, |z_2|^{\frac{\log\alpha}{\log\beta}}\}, & \text{if } |z_1| > \alpha \end{cases}.$$

One can see that u is well-defined and that $u \in \mathcal{K}_{G_-}(0)$. Hence, $\boldsymbol{g}_{G_-}(0,\cdot) \ge u$. By the Schwarz lemma (cf. the proof of (b)), we have $\boldsymbol{g}_{G_-}(0,z) \le u(z) = |z_1|$ for all $z = (z_1, z_2) \in G_-$ with $|z_2| < \frac{\beta}{\alpha}|z_1|$. So, it remains to prove that

$$\boldsymbol{g}_{G_-}(0,z) \le u(z) \text{ for all } z = (z_1, z_2) \in G_- \text{ with} \tag{4.2.21}$$

$$\left(|z_1| \le \alpha \text{ and } |z_2| \ge \frac{\beta}{\alpha}|z_1|\right) \text{ or } \alpha < |z_1| < 1.$$

In the former case one can apply the maximum principle (cf. the proof of (b)) to the function

$$\left\{\lambda \in \mathbb{C} : |\lambda| \le \frac{\alpha}{\beta}|z_2^0| + \varepsilon\right\} \ni \lambda \longmapsto g_{G_-}(0, (\lambda, z_2^0)).$$

The latter case is more complicated. It is sufficient to prove that (4.2.21) holds for $\alpha < |z_1| < 1$ and $|z_2| < |z_1|^{\frac{\log\beta}{\log\alpha}}$. Then, one can apply the three circles theorem to the function

$$\{\lambda \in \mathbb{C} : |z_1^0|^{\frac{\log\beta}{\log\alpha}} - \varepsilon \le |\lambda| < 1\} \ni \lambda \longmapsto g_{G_-}(0, (z_1^0, \lambda))$$

($\alpha < |z_1^0| < 1$, $\varepsilon > 0$, ε small). Take $r, s \in \mathbb{N}$ with $r/s \ge \log\beta/\log\alpha$, and consider the function

$$\mathbb{D} \ni \lambda \longmapsto g_{G_-}(0, (\lambda^s, \lambda^r)).$$

Then, by the Schwarz lemma for subharmonic functions,

$$g_{G_-}(0, (\lambda^s, \lambda^r)) \le |\lambda|^s = u(\lambda^s, \lambda^r), \quad \lambda \in \mathbb{D}.$$

Hence, by the maximum principle, we get $g_{G_-}(0, z) \le |z_1| = u(z)$ for all $\alpha < |z_1| < 1$ and $|z_2| \le |z_1|^{r/s}$. We conclude the proof by taking $\frac{r_\nu}{s_\nu} \searrow \frac{\log\beta}{\log\alpha}$. □

4.2.3 Properties of $\boldsymbol{S}_G$

Proposition 4.2.16 (cf. [304]).

$$\begin{aligned} \boldsymbol{S}_G(a; X) = \sup\{&[(\mathcal{L}u)(a; X)]^{\frac{1}{2}} : \\ &u : G \to [0,1), u \text{ log-psh}, u(a) = 0, \text{ and } u\ \mathcal{C}^2 \text{ near } a\}, \quad a \in G,\ X \in \mathbb{C}^n. \end{aligned}$$

In particular, $\boldsymbol{S}_G(a; \cdot)$ is a seminorm for any $a \in G \in \mathfrak{G}$.

Proof. It suffices to prove that if u is as above, then

$$\limsup_{\lambda \to\!\!\!\!\to 0} \frac{1}{|\lambda|^2} u(a + \lambda X) = (\mathcal{L}u)(a; X).$$

Fix a u and put $v(\lambda) := u(a + \lambda X)$. Since $v'(0) = 0$, we obtain

$$\begin{aligned} v(\lambda) = v(x, y) &= \frac{1}{2}\frac{\partial^2 v}{\partial x^2}(0)x^2 + \frac{\partial^2 v}{\partial x \partial y}(0)xy + \frac{1}{2}\frac{\partial^2 v}{\partial y^2}(0)y^2 + o(|\lambda|^2) \\ &=: P(x, y) + o(|\lambda|^2) \text{ in a neighborhood of } 0 \in \mathbb{C}. \end{aligned}$$

We already know that $u(z) \le M\|z - a\|^2$ for z near a. Hence the function $0 \ne \lambda \overset{\widetilde{v}}{\longmapsto} \frac{1}{|\lambda|^2} v(\lambda)$ (defined near 0) extends to a subharmonic function at 0 with $\widetilde{v}(0) := \limsup_{\lambda \to\!\!\!\!\to 0} \widetilde{v}(\lambda)$. In particular, by the Oka theorem (cf. Appendix B.4.26), we conclude that $\limsup_{t\to 0+} \widetilde{v}(tx, ty)$ is independent of (x, y). Consequently, the function $\frac{P(x,y)}{x^2+y^2}$ is constant, and therefore $\frac{\partial^2 v}{\partial x^2}(0) = \frac{\partial^2 v}{\partial y^2}(0)$ and $\frac{\partial^2 v}{\partial x \partial y}(0) = 0$. Finally,

$$\limsup_{\lambda\to 0} \frac{1}{|\lambda|^2} v(\lambda) = \frac{1}{4}(\Delta v)(0) = (\mathcal{L}u)(a; X). \qquad \square$$

Remark 4.2.17. Let $G \subset \mathbb{C}^n$ be a balanced pseudoconvex domain that is not convex. Then, Proposition 4.2.10(b) and the fact that $\boldsymbol{S}_G(0;\cdot)$ is a seminorm, imply that $\boldsymbol{A}_G(0;\cdot) \not\equiv \boldsymbol{S}_G(0;\cdot)$ (cf. (4.2.5)). Note that there are a lot of such domains, which may even be bounded. What are the relations between $\boldsymbol{S}_G(0;\cdot)$ and $\mathfrak{h}_{\widehat{G}}$?

Is the function $G \times \mathbb{C}^n \ni (z, X) \longmapsto \boldsymbol{S}_G(z; X)$ Borel measurable?

Example 4.2.18 (cf. [257]). Let $\boldsymbol{S}_G^*$ denote the upper semicontinuous regularization of $\boldsymbol{S}_G$. We show that the system $(\boldsymbol{S}_G^*)_{G\in\mathfrak{G}_h}$ is not holomorphically contractible. In particular, there are domains of holomorphy G for which $\boldsymbol{S}_G$ is not upper semicontinuous; see also Exercise 4.4.2.

Let

$$\varphi(\xi,\eta) := \sum_{k=1}^{\infty} \lambda_k \log\left(\frac{|\xi - a_k|^2 + |\eta|}{k}\right), \quad (\xi,\eta) \in \mathbb{C}^2,$$

where $(a_k)_{k=1}^{\infty} \subset \mathbb{D}_*$ is a dense subset of $\mathbb{D}$ and $\lambda_k > 0$, $k \geq 1$, are such that $\varphi(0,0) > -\infty$, $\varphi \in \mathcal{C}^2(\mathbb{C} \times \mathbb{C}_*) \cap \mathcal{PSH}(\mathbb{C} \times \mathbb{C})$. Define

$$G := \{(z_1, z_2) \in \mathbb{C}^2 : |z_1| e^{\varphi(z_2,0)} < 1\},$$
$$D := \{(z_1, z_2, z_3) \in \mathbb{C}^3 : |z_1| e^{\varphi(z_2,z_3)} < 1\},$$
$$G \ni (z_1, z_2) \overset{F}{\longmapsto} (z_1, z_2, 0) \in D.$$

Note that G and D are Hartogs domains of holomorphy. Take $(z_1^0, z_2^0) \in G \cap (\mathbb{C} \times \mathbb{D})$ and let $u : G \to [0, 1)$ be a log-psh function such that $u(z_1^0, z_2^0) = 0$ and u is $\mathcal{C}^2$ near (z_1^0, z_2^0). Since $\mathbb{C} \times \{a_k\} \subset G$, $k \geq 1$, we get $u(z_1, a_k) = \text{const}(k)$, $z_1 \in \mathbb{C}$, $k \geq 1$. Hence, since $\{0\} \times \mathbb{C} \subset G$, we have $u(z_1, a_k) = \text{const}$, $z_1 \in \mathbb{C}$, $k \geq 1$. Recall that $(a_k)_{k=1}^{\infty}$ is dense in $\mathbb{D}$ and u is $\mathcal{C}^2$ near (z_1^0, z_2^0). Therefore, $u \equiv \text{const} = 0$ near (z_1^0, z_2^0). This shows that $\boldsymbol{S}_G = 0$ in $(G \cap (\mathbb{C} \times \mathbb{D})) \times \mathbb{C}^2$, and consequently $\boldsymbol{S}_G^* = 0$ on $(G \cap (\mathbb{C} \times \mathbb{D})) \times \mathbb{C}^2$.

On the other hand, the function $D \ni (z_1, z_2, z_3) \longmapsto |z_1|^2 e^{2\varphi(z_2,z_3)}$ is a log-psh function on D that is $\mathcal{C}^2$ near any point $(0, 0, t) \in D$, $t > 0$. Hence,

$$\boldsymbol{S}_D((0,0,t);(1,0,0)) \geq e^{\varphi(0,t)} \geq e^{\varphi(0,0)} > 0.$$

Finally, we obtain that

$$\begin{aligned} \boldsymbol{S}_D^*(F(0,0); F'(0,0)(1,0)) &= \boldsymbol{S}_D^*((0,0,0);(1,0,0)) \\ &\geq \limsup_{t\to 0+} \boldsymbol{S}_D((0,0,t);(1,0,0)) \geq e^{\varphi(0,0)} > 0 = \boldsymbol{S}_G^*((0,0);(1,0)). \end{aligned}$$

The above example shows that there are contractible families of pseudometrics that are not upper semicontinuous. Moreover, if one forces these pseudometrics to be upper semicontinuous, then the new system is no longer holomorphically contractible. What are good sufficient conditions on domains of holomorphy G for $\boldsymbol{S}_G$ to be upper semicontinuous is an open question. ?

4.2.4 Properties of $\boldsymbol{m}_G^{(k)}$ and $\boldsymbol{\gamma}_G^{(k)}$

Proposition 4.2.19.

(a) *For any $a, z \in G$, and $k \in \mathbb{N}$ there exists an $f \in \mathcal{O}(G, \mathbb{D})$ with* $\operatorname{ord}_a f \geq k$ *such that $|f(z)|^{\frac{1}{k}} = \boldsymbol{m}_G^{(k)}(a, z)$ (we say that f is* extremal *for $\boldsymbol{m}_G^{(k)}(a, z)$).*

(b) *The function $\boldsymbol{m}_G^{(k)}(a, \cdot)$ is continuous and belongs to $\mathcal{K}_G(a)$.*

(c) *The function $\boldsymbol{m}_G^{(k)}$ is upper semicontinuous.*

(d) *If $G \in \mathfrak{G}_b$, then $\boldsymbol{m}_G^{(k)}$ is continuous.*

(e) *If $G_\nu \nearrow G$, then $\boldsymbol{m}_{G_\nu}^{(k)} \searrow \boldsymbol{m}_G^{(k)}$.*

Note that for $k \geq 2$ the function $\boldsymbol{m}_G^{(k)}$ is, in general, neither symmetric nor continuous; cf. Proposition 6.1.1.

Proof. (a) Use Montel's argument.

(b) The family $\{|f| : f \in \mathcal{O}(G, \mathbb{D}),\ \operatorname{ord}_a f \geq k\}$ is equicontinuous; cf. the proof of Proposition 2.6.1(a).

(c) Let $G \ni z_\nu \longrightarrow z_0 \in G$, $G \ni w_\nu \longrightarrow w_0 \in G$, and let f_ν be an extremal function for $\boldsymbol{m}_G^{(k)}(z_\nu, w_\nu)$, $\nu \geq 1$. We can find a subsequence $f_{\nu_\mu} \overset{K}{\underset{\mu\to\infty}{\Longrightarrow}} f_0 \in \mathcal{O}(G, \overline{\mathbb{D}})$ (use the Montel argument). Clearly, $\operatorname{ord}_{z_0} f_0 \geq k$ (hence, $f_0 : G \longrightarrow \mathbb{D}$), and consequently

$$\boldsymbol{m}_G^{(k)}(z_0, w_0) \geq |f_0(w_0)|^{\frac{1}{k}} = \lim_{\mu\to\infty} |f_{\nu_\mu}(w_{\nu_\mu})|^{\frac{1}{k}} = \lim_{\mu\to\infty} \boldsymbol{m}_G^{(k)}(z_{\nu_\mu}, w_{\nu_\mu}).$$

This proves that $\limsup_{(z,w)\to(z_0,w_0)} \boldsymbol{m}_G^{(k)}(z, w) = \boldsymbol{m}_G^{(k)}(z_0, w_0)$.

(d) We may assume that G is bounded. Fix $z_0, w_0 \in G$ and let f_0 be extremal for $\boldsymbol{m}_G^{(k)}(z_0, w_0)$. Define

$$\widetilde{f}_z(w) = f_0(w) - \sum_{|\alpha|<k} \frac{1}{\alpha!} D^\alpha f_0(z)(w - z)^\alpha, \quad z, w \in G.$$

Then, $\widetilde{f}_z \in \mathcal{H}^\infty(G)$ and $\operatorname{ord}_z \widetilde{f}_z \geq k$. Let

$$M(z) := \max\{1, \|\widetilde{f}_z\|_G\}.$$

Since $\operatorname{ord}_{z_0} f_0 \geq k$, $M(z) \longrightarrow 1$ as $z \longrightarrow z_0$. Put $f_z := \widetilde{f}_z/M(z)$. Then,

$$\boldsymbol{m}_G^{(k)}(z,w) \geq |f_z(w)|^{\frac{1}{k}},$$

and therefore

$$\liminf_{(z,w)\to(z_0,w_0)} \boldsymbol{m}_G^{(k)}(z,w) \geq |f_0(w_0)|^{\frac{1}{k}} = \boldsymbol{m}_G^{(k)}(z_0,w_0).$$

(e) Use the Montel argument; cf. Proposition 2.7.1(a). □

Proposition 4.2.20.

(a) $\boldsymbol{\gamma}_G^{(k)}(a;X) = \sup\{|f_{(k)}(a)X|^{\frac{1}{k}} : f \in \mathcal{O}(G,\mathbb{D}),\ \operatorname{ord}_a f \geq k\}$, *where* $f_{(k)}(a)X := \dfrac{1}{k!}f^{(k)}(a)X = \sum_{|\alpha|=k} \dfrac{1}{\alpha!}D^\alpha f(a)X^\alpha$

(note that $f_{(k)}(a) : \mathbb{C}^n \longrightarrow \mathbb{C}$ *is a homogeneous polynomial of degree* k*).*

(b) $\boldsymbol{\gamma}_G^{(k)}(a;X) = \lim_{\lambda\not\to 0} \dfrac{1}{|\lambda|}\boldsymbol{m}_G^{(k)}(a, a+\lambda X)$.

(c) *If* $G \in \mathfrak{G}_b$, *then*

$$\boldsymbol{\gamma}_G^{(k)}(a;X) = \lim_{\substack{z',z''\to a,\ z'\neq z''\\ \frac{z'-z''}{\|z'-z''\|}\to X}} \frac{\boldsymbol{m}_G^{(k)}(z',z'')}{\|z'-z''\|}, \quad a \in G,\ \|X\| = 1.$$

(d) *For any* $a \in G$, $X \in \mathbb{C}^n$ *there exists an* $f \in \mathcal{O}(G,\mathbb{D})$ *with* $\operatorname{ord}_a f \geq k$ *such that* $|f_{(k)}(a)X|^{\frac{1}{k}} = \boldsymbol{\gamma}_G^{(k)}(a;X)$.

(e) *The function* $\boldsymbol{\gamma}_G^{(k)}(a;\cdot)$ *is continuous and* log-*psh on* $\mathbb{C}^n$.

(f) *The function* $\boldsymbol{\gamma}_G^{(k)}$ *is upper semicontinuous.*

(g) *If* $G \in \mathfrak{G}_b$, *then* $\boldsymbol{\gamma}_G^{(k)}$ *is continuous.*

(h) *If* $G_\nu \nearrow G$, *then* $\boldsymbol{\gamma}_{G_\nu}^{(k)} \searrow \boldsymbol{\gamma}_G^{(k)}$.

Note that, for $k \geq 2$, the function $\boldsymbol{\gamma}_G^{(k)}$ need not be continuous in general; cf. Proposition 6.1.1.

Proof. (a) If $f \in \mathcal{O}(G,\mathbb{D})$ and $\operatorname{ord}_a f \geq k$, then $f(a+\lambda X) = \lambda^k f_{(k)}(a)X + o(|\lambda|^k)$ as $\lambda \longrightarrow 0$.

(b) By (a) we have

$$\gamma_G^{(k)}(a;X) \leq \liminf_{\lambda \to 0} \frac{1}{|\lambda|} m_G^{(k)}(a, a+\lambda X).$$

Take $0 \neq \lambda_\nu \longrightarrow 0$ and let f_ν be extremal for $m_G^{(k)}(a, a+\lambda_\nu X)$, $\nu \geq 1$. By the Montel argument, we get $f_{\nu_\mu} \overset{K}{\underset{\mu\to\infty}{\Longrightarrow}} f_0$. Hence,

$$\begin{aligned}\gamma_G^{(k)}(a;X) \geq |f_{0(k)}(a)X|^{\frac{1}{k}} &= \lim_{\mu\to+\infty} \frac{1}{|\lambda_{\nu_\mu}|} |f_{\nu_\mu}(a+\lambda_{\nu_\mu}X)|^{\frac{1}{k}} \\ &= \lim_{\mu\to\infty} \frac{1}{|\lambda_{\nu_\mu}|} m_G^{(k)}(a, a+\lambda_{\nu_\mu}X).\end{aligned}$$

(c) As in (b), one can prove that

$$\limsup_{\substack{z',z''\to a,\, z'\neq z'' \\ \frac{z'-z''}{\|z'-z''\|}\to X}} \frac{m_G^{(k)}(z',z'')}{\|z'-z''\|} \leq \gamma_G^{(k)}(a;X).$$

By methods analogous to those of the proof of Proposition 4.2.19(d), we can easily show that $\liminf \geq \gamma_G^{(k)}(a;X)$.

(d), (h) Use the Montel argument.

(e) The family of functions $\mathbb{C}^n \ni X \longmapsto f_{(k)}(a)X$, $f \in \mathcal{O}(G,\mathbb{D})$, $\operatorname{ord}_a f \geq k$, is equicontinuous on $\mathbb{C}^n$.

(f), (g) Cf. the proofs of Proposition 4.2.19(c, d). □

Observe that, directly from the definitions, we get

$$(m_G^{(k)})^k \cdot (m_G^{(\ell)})^\ell \leq (m_G^{(k+\ell)})^{k+\ell}, \quad (\gamma_G^{(k)})^k \cdot (\gamma_G^{(\ell)})^\ell \leq (\gamma_G^{(k+\ell)})^{k+\ell}.$$

Consequently, the limits

$$m_G^\infty := \lim_{k\to+\infty} m_G^{(k)}, \quad \gamma_G^\infty := \lim_{k\to+\infty} \gamma_G^{(k)}$$

exist and $m_G^\infty = \sup_{k\in\mathbb{N}} m_G^{(k)}$, $\gamma_G^\infty = \sup_{k\in\mathbb{N}} \gamma_G^{(k)}$.

We say that m_G^∞ (resp. γ_G^∞) is the *singular Carathéodory function* (resp. *pseudometric*); cf. [395]. Notice that

$$m_G^{(k)} \leq m_G^\infty \leq g_G, \quad \gamma_G^{(k)} \leq \gamma_G^\infty \leq A_G.$$

Remark 4.2.21.

(a) One can prove (cf. [374]) that if $\boldsymbol{\gamma}_G(a;X) > 0$ for all $X \in (\mathbb{C}^n)_*$, then for every $k \in \mathbb{N}$ there exist a constant $M > 0$ and an open neighborhood $U \subset G$ of a such that

$$|\boldsymbol{\gamma}_G^{(k)}(z',X') - \boldsymbol{\gamma}_G^{(k)}(z'',X'')| \le M\big(\|X'-X''\| + (\|X'\|+\|X''\|)\|z'-z''\|\big),$$
$$z',z'' \in U,\ X',X'' \in \mathbb{C}^n.$$

In particular, if $\boldsymbol{\gamma}_G(a;X) > 0$ for all $a \in G$ and $X \in (\mathbb{C}^n)_*$, then for every $k \in \mathbb{N}$ the function $\boldsymbol{\gamma}_G^{(k)}$ is continuous.

(b) If $G \subset \mathbb{C}^n$ is a *strictly hyperconvex domain* (i.e., there exist a larger domain $G' \supset\supset G$ and a function $u \in \mathcal{PSH}(G') \cap \mathcal{C}(G')$ such that $G = \{z \in G' : u(z) < 0\}$), then $\boldsymbol{\gamma}_G^{\infty}$ is continuous and $\boldsymbol{\gamma}_G^{\infty} = \boldsymbol{A}_G$ (see [374]).

We end this section with a list of relations between $\boldsymbol{m}_G^{(k)}$, $\boldsymbol{g}_G$, and $\boldsymbol{\ell}_G^*$, and $\boldsymbol{\gamma}_G^{(k)}$, $\boldsymbol{A}_G, \boldsymbol{\varkappa}_G$ for $G \subset \mathbb{C}^1$. Recall that if $G \subset \mathbb{C}^1$, then

$$\boldsymbol{m}_G^{(1)} = \boldsymbol{c}_G^* \le \boldsymbol{m}_G^{(k)} \le \boldsymbol{s}_G = \boldsymbol{g}_G \le \boldsymbol{\ell}_G^{(m)*} = \boldsymbol{\ell}_G^* = \tanh(\boldsymbol{k}_G),$$
$$\boldsymbol{\gamma}_G^{(1)} = \boldsymbol{\gamma}_G \le \boldsymbol{\gamma}_G^{(k)} \le \boldsymbol{S}_G = \boldsymbol{A}_G \le \boldsymbol{\varkappa}_G^{(m)} = \boldsymbol{\varkappa}_G.$$

Moreover, if G is biholomorphic to $\mathbb{D}$, then

$$\boldsymbol{c}_G^* = \boldsymbol{m}_G^{(k)} = \boldsymbol{g}_G = \boldsymbol{\ell}_G^*, \qquad \boldsymbol{\gamma}_G = \boldsymbol{\gamma}_G^{(k)} = \boldsymbol{A}_G = \boldsymbol{\varkappa}_G.$$

Proposition 4.2.22.

(a) *Let $G \subset \mathbb{C}^1$ be a taut domain (i.e., a hyperbolic domain). Then the following conditions are equivalent.*

(i) *There exist $k \in \mathbb{N}$ and $z_0', z_0'' \in G$, $z_0' \ne z_0''$, such that $\boldsymbol{m}_G^{(k)}(z_0',z_0'') = \boldsymbol{\ell}_G^*(z_0',z_0'')$.*

(ii) *There exist $k \in \mathbb{N}$ and $z_0' \in G$ such that $\boldsymbol{\gamma}_G^{(k)}(z_0';1) = \boldsymbol{\varkappa}_G(z_0';1)$.*

(iii) *There exist $z_0', z_0'' \in G$, $z_0' \ne z_0''$, such that $\boldsymbol{g}_G(z_0',z_0'') = \boldsymbol{\ell}_G^*(z_0',z_0'')$.*

(iv) *There exists a $z_0' \in G$ such that $\boldsymbol{A}_G(z_0';1) = \boldsymbol{\varkappa}_G(z_0';1)$.*

(v) *G is biholomorphic with $\mathbb{D}$.*

(b) *Let $G \subset \mathbb{C}^1$ be a domain biholomorphic to a bounded domain regular with respect to the Dirichlet problem; cf. Appendix B.5. Then, the following conditions are equivalent.*

(i) *There exist $z_0', z_0'' \in G$, $z_0' \ne z_0''$, such that $\boldsymbol{c}_G^*(z_0',z_0'') = \boldsymbol{g}_G(z_0',z_0'')$.*

(ii) *There exists a* $z_0' \in G$ *such that* $\boldsymbol{\gamma}_G(z_0'; 1) = \boldsymbol{A}_G(z_0'; 1)$.

(iii) *G is biholomorphic with* $\mathbb{D}$.

Proof. (a) The only difficulty is to prove that (iii) $\Longrightarrow$ (v), and (iv) $\Longrightarrow$ (v). Suppose that (iii) (resp. (iv)) is fulfilled and let $\varphi \in \mathcal{O}(\mathbb{D}, G)$ be such that $\varphi(0) = z_0'$ and $\varphi(\sigma) = z_0''$, where $\sigma := \boldsymbol{\ell}_G^*(z_0', z_0'')$ (resp. $\varphi(0) = z_0'$ and $\boldsymbol{\varkappa}_G(z_0'; 1)\varphi'(0) = 1$); cf. Propositions 3.2.7 and 3.5.14. We will prove that $\varphi : \mathbb{D} \longrightarrow G$ is biholomorphic. Let

$$u(\lambda) := \frac{\boldsymbol{g}_G(z_0', \varphi(\lambda))}{|\lambda|}, \quad \lambda \in \mathbb{D}_*.$$

Then u is subharmonic and $u \le 1$ (by the holomorphic contractibility of $\boldsymbol{g}_G$). Put

$$u(0) := \limsup_{\lambda \nrightarrow 0} u(\lambda) \ (= \boldsymbol{A}_G(z_0'; \varphi'(0))).$$

Clearly, u is subharmonic on the whole $\mathbb{D}$. Since $u(\sigma) = 1$ (resp. $u(0) = 1$), we get $u \equiv 1$. Let $\lambda_0 \in \mathbb{D}_*$. Define

$$v(\lambda) := \frac{\boldsymbol{g}_G(\varphi(\lambda_0), \varphi(\lambda))}{\boldsymbol{m}(\lambda_0, \lambda)}, \quad \lambda \in \mathbb{D} \setminus \{\lambda_0\}.$$

Then, as above, v is subharmonic and $v \le 1$. Since $\boldsymbol{g}_G$ is symmetric ($n = 1$!), we get $v(0) = u(\lambda_0) = 1$. Thus, $v \equiv 1$ and, therefore,

$$\boldsymbol{g}_G(\varphi(\lambda'), \varphi(\lambda'')) = \boldsymbol{m}(\lambda', \lambda''), \quad \lambda', \lambda'' \in \mathbb{D}.$$

This implies that φ is injective and proper. Consequently, φ is biholomorphic.

(b) Without loss of generality, we may assume that G itself is bounded and regular with respect to the Dirichlet problem. In particular,

$$\begin{aligned} &\boldsymbol{g}_G(z', z'') \ge \boldsymbol{c}_G^*(z', z'') > 0, \quad z', z'' \in G,\ z' \ne z'', \\ &\boldsymbol{A}_G(z'; 1) \ge \boldsymbol{\gamma}_G(z'; 1) > 0, \quad \lim_{z \to \partial G} \boldsymbol{g}_G(z', z) = 1, \quad z' \in G. \end{aligned}$$

Nontrivial implications are (i) $\Longrightarrow$ (iii), and (ii) $\Longrightarrow$ (iii). Assume that (i) (resp. (ii)) is satisfied and let $f \in \mathcal{O}(G, \mathbb{D})$ denote an extremal function for $\boldsymbol{c}_G^*(z_0', z_0'')$ (resp. $\boldsymbol{\gamma}_G(z_0'; 1)$), i.e., $f(z_0') = 0, |f(z_0'')| = \boldsymbol{c}_G^*(z_0', z_0'')$ (resp. $f(z_0') = 0, |f'(z_0')| = \boldsymbol{\gamma}_G(z_0'; 1)$). We will prove that $f : G \longrightarrow \mathbb{D}$ is biholomorphic. Put

$$u(z) := \frac{|f(z)|}{\boldsymbol{g}_G(z_0', z)}, \quad z \in G \setminus \{z_0'\}, \qquad u(z_0') := \limsup_{z \nrightarrow z_0'} u(z) = \frac{|f'(z_0')|}{\boldsymbol{A}_G(z_0'; 1)}.$$

Then, u is subharmonic (since $\log \boldsymbol{g}_G(z_0', \cdot)$ is harmonic in $G \setminus \{z_0'\}$) and $u \le 1$. Since $u(z_0'') = 1$ (resp. $u(z_0') = 1$), we have $u \equiv 1$. Take $z_0 \in G \setminus \{z_0'\}$ and put

$$v(z) := \frac{\boldsymbol{m}(f(z_0), f(z))}{\boldsymbol{g}_G(z_0, z)}, \quad z \in G \setminus \{z_0\}.$$

Then, v is subharmonic, $v \le 1$, and $v(z_0') = u(z_0) = 1$. Hence, $v \equiv 1$, and finally

$$\boldsymbol{m}(f(z'), f(z'')) = \boldsymbol{g}_G(z', z''), \quad z', z'' \in G.$$

In particular, f is injective and proper, and therefore f is biholomorphic. □

Remark 4.2.23.

(a) If $G \subset \mathbb{C}^1$ is not taut, then assertion (a) in Proposition 4.2.22 is not true, e.g., $G = \mathbb{C}_*$ ($\boldsymbol{c}_G^* = \boldsymbol{\ell}_G^* \equiv 0$).

(b) If $G \subset \mathbb{C}^1$ is not regular with respect to the Dirichlet problem, then Proposition 4.2.22(b) is false, e.g., $G = \mathbb{D}_*$ ($\boldsymbol{c}_{\mathbb{D}_*}^* = \boldsymbol{g}_{\mathbb{D}_*} = \boldsymbol{c}_{\mathbb{D}}^*|_{\mathbb{D}_* \times \mathbb{D}_*}$).

(c) If $G = P = \{z \in \mathbb{C} : 1/R < |z| < R\}$ (note that P is taut and regular with respect to the Dirichlet problem), then, for $k \ge 2$, there are points $z_0', z_0'' \in P$, $z_0' \ne z_0''$, such that $\boldsymbol{m}_P^{(k)}(z_0', z_0'') = \boldsymbol{g}_P(z_0', z_0'')$ and $\boldsymbol{\gamma}_P^{(k)}(z_0'; 1) = \boldsymbol{A}_G(z_0'; 1)$; cf. Proposition 9.1.12(c).

(d) If $G \subset \mathbb{C}$ such that the set of one-point connected components of $\mathbb{C} \setminus G$ is a polar set, then $\boldsymbol{g}_G \equiv \boldsymbol{m}_G^\infty$ and $\boldsymbol{A}_G \equiv \boldsymbol{\gamma}_G^\infty$ (see [395]).

(e) If S denotes the standard ternary Cantor set in $[0, 1]$, then $G := \mathbb{C} \setminus S$ is hyperconvex, $\boldsymbol{g}_G \not\equiv \boldsymbol{m}_G^\infty$, and $\boldsymbol{A}_G \not\equiv \boldsymbol{\gamma}_G$ (see [395]).

(f) Observe that $\log \boldsymbol{\ell}_G^*(a, \cdot)$ is psh iff $\boldsymbol{g}_G(a, \cdot) = \boldsymbol{\ell}_G^*(a, \cdot)$. Hence, by Proposition 4.2.22, if $G \subset \mathbb{C}^1$ is a multi-connected taut domain, then, for any $a \in G$, the function $\log \boldsymbol{\ell}_G^*(a, \cdot)$ is not subharmonic; cf. Proposition 2.7.1(b).

(g) We will see in Lemma 20.3.1 that if $G \subset \mathbb{C}$ is a bounded domain with a smooth boundary at $p \in \partial G$, then $\lim_{G \ni \lambda \to p} \frac{\boldsymbol{\gamma}_G(\lambda;1)}{\boldsymbol{\varkappa}_G(\lambda;1)} = 1$, which may be considered an extension of Proposition 4.2.22(a); see also [123].

4.3 Inner pseudodistances. Integrated forms. Derivatives. Buseman pseudometrics. $\mathcal{C}^1$-pseudodistances

Roughly speaking, the aim of this section is to define abstract versions of the following operations:

$$\begin{aligned}
\boldsymbol{c}_G &\longmapsto \boldsymbol{c}_G^i && \text{(cf. § 2.7)},\\
\boldsymbol{\gamma}_G &\longmapsto \textstyle\int \boldsymbol{\gamma}_G && \text{(cf. § 2.7)},\\
\boldsymbol{c}_G &\longmapsto \boldsymbol{\gamma}_G && \text{(cf. § 2.7)},\\
\boldsymbol{\varkappa}_G &\longmapsto \widehat{\boldsymbol{\varkappa}}_G && \text{(cf. § 3.6)}.
\end{aligned}$$

Fix a domain $G \subset \mathbb{C}^n$. Denote by $\mathfrak{D}(G)$ the family of all pseudodistances $d : G \times G \longrightarrow \mathbb{R}_+$ such that

$$\forall_{a \in G}\ \exists_{M,r>0} : d(z', z'') \le M\|z' - z''\|, \quad z', z'' \in \mathbb{B}(a, r) \subset G. \tag{4.3.1}$$

Note that:

- if $\underline{d} = (d_G)_G$ is a contractible family of pseudodistances, then $d_G \in \mathfrak{D}(G)$ for every G (cf. Remark 4.1.5(c));
- if $d \in \mathfrak{D}(G)$, then d is continuous on $G \times G$;
- if $d \in \mathfrak{D}(G)$, then $\tanh d \in \mathfrak{D}(G)$.

Let $\mathfrak{M}(G)$ denote the family of all *pseudometrics* $\delta : G \times \mathbb{C}^n \longrightarrow \mathbb{R}_+$, i.e.,

$$\delta(z; \lambda X) = |\lambda|\delta(z; X), \quad z \in G,\ \lambda \in \mathbb{C},\ X \in \mathbb{C}^n,$$

such that

$$\forall_{a \in G}\ \exists_{M,r>0} : \delta(z; X) \le M\|X\|, \quad z \in \mathbb{B}(a, r) \subset G,\ X \in \mathbb{C}^n. \tag{4.3.2}$$

Observe that:

- if $\delta \in \mathfrak{M}(G)$ is upper semicontinuous, then (4.3.2) is automatically satisfied;
- if $(\delta_G)_G$ is a contractible family of pseudometrics, then $\delta_G \in \mathfrak{M}(G)$ for every G (cf. Remark 4.1.5(d)).

We will define abstract operators

$$\begin{array}{ccc} \mathfrak{D}(G) & \overset{\int}{\longrightarrow} & \mathfrak{M}(G) \\ i \downarrow & & \sim\downarrow\wedge \\ \mathfrak{D}(G) & \underset{\mathcal{D}}{\longrightarrow} & \mathfrak{M}(G) \end{array}$$

and to study interrelations between them.

4.3.1 Operator $d \longmapsto d^i$

Let $d \in \mathfrak{D}(G)$. For a curve $\alpha : [0,1] \longrightarrow G$ put

$$L_d(\alpha) := \sup\left\{\sum_{j=1}^{N} d(\alpha(t_{j-1}), \alpha(t_j)) : N \in \mathbb{N},\ 0 = t_0 < \cdots < t_N = 1\right\}.$$

The number $L_d(\alpha) \in [0, +\infty]$ is called the *d-length of α*. If $L_d(\alpha) < +\infty$, then we say that α is *d-rectifiable*. Note that:

- $L_d(\alpha) = L_{\tanh d}(\alpha)$,
- if α is rectifiable in the Euclidean sense, then α is d-rectifiable (cf. the proof of Lemma 2.7.3(a)).

Let

- $\mathfrak{C}_{in}(G) :=$ the family of all curves in G,
- $\mathfrak{C}_i(G) :=$ the family of all curves in G that are rectifiable in the Euclidean sense,
- $\mathfrak{C}_{ic}(G) :=$ the family of all piecewise $\mathcal{C}^1$-curves in G,

and let $\mathfrak{C} \in \{\mathfrak{C}_{in}, \mathfrak{C}_i, \mathfrak{C}_{ic}\}$.

Obviously:

- $\mathfrak{C}_{in}(G) \subsetneq \mathfrak{C}_i(G) \subsetneq \mathfrak{C}_{in}(G)$;
- if $\alpha, \beta \in \mathfrak{C}(G)$ and $\alpha(1) = \beta(0)$, then $\alpha \cup \beta \in \mathfrak{C}(G)$;
- if $\alpha \in \mathfrak{C}(G)$, $F \in \mathcal{O}(G, D)$, then $F \circ \alpha \in \mathfrak{C}(D)$.

We define the *inner pseudodistance for d with respect to* $\mathfrak{C}$:

$$d^{\mathfrak{C}}(a,z) := \inf\{L_d(\alpha) : \alpha \in \mathfrak{C}(G),\ \alpha(0) = a,\ \alpha(1) = z\}, \quad (a,z) \in G \times G.$$

Note that $d^{\mathfrak{C}} = (\tanh d)^{\mathfrak{C}}$. Put:

- $d^{in} := d^{\mathfrak{C}_{in}}$ (cf. [452]),
- $d^i := d^{\mathfrak{C}_i}$ (cf. [260]),
- $d^{ic} := d^{\mathfrak{C}_{ic}}$ (cf. [508]).

We say that d is *inner* if $d = d^{ic}$; see also [37]. We have:

- $d \le d^{in} \le d^i \le d^{ic}$,
- if d is inner, then $d = d^{in} = d^i = d^{ic}$;
- $\boldsymbol{c}^i_G = \boldsymbol{c}^{ic}_G$ for any G (cf. Lemma 2.7.3(d));
- if G is γ-hyperbolic (e.g., G is biholomorphic to a bounded domain) or $G \subset \mathbb{C}^1$, then $\boldsymbol{c}^{in}_G = \boldsymbol{c}^i_G = \boldsymbol{c}^{ic}_G$ (cf. Lemma 2.7.3(b)); it is not known whether $\boldsymbol{c}^{in}_G = \boldsymbol{c}^i_G$ for arbitrary domain $G \subset \mathbb{C}^n$; ?
- $\boldsymbol{m}^i_{\mathbb{D}} = \boldsymbol{p}_{\mathbb{D}} = \boldsymbol{p}^i_{\mathbb{D}}$;
- in general $\boldsymbol{c}^i_G \not\equiv \boldsymbol{c}_G$ (cf. Example 2.7.9, see also [259]);
- $\boldsymbol{k}_G = \boldsymbol{k}^{ic}_G$ for any G (cf. Proposition 3.3.1), i.e., $\boldsymbol{k}_G$ is inner.

Proposition 4.3.1. *Let $d \in \mathfrak{D}(G)$ and $\mathfrak{C} \in \{\mathfrak{C}_{in}, \mathfrak{C}_i, \mathfrak{C}_{ic}\}$. Then,*

(a) $d^{\mathfrak{C}} \in \mathfrak{D}(G)$.

(b) $L_{d^{\mathfrak{C}}} = L_d$*; consequently,* $(d^{\mathfrak{C}})^{\mathfrak{C}} = d^{\mathfrak{C}}$*. In particular,* d^{ic} *is inner.*

(c) *Any $d^{\mathfrak{C}}$-ball $B_{d^{\mathfrak{C}}}(a,r)$ is arcwise connected in* $\operatorname{top} G$*; in particular, if $d^{\mathfrak{C}}$ is a distance, then* $\operatorname{top} d^{\mathfrak{C}} = \operatorname{top} G$.

(d) *If $(d_G)_G$ is a holomorphically contractible family of pseudodistances with $d_{\mathbb{D}} \in \{\boldsymbol{m}, \boldsymbol{p}\}$, then the family $(d_G^{\mathfrak{C}})_G$ is holomorphically contractible with $d_{\mathbb{D}}^{\mathfrak{C}} = \boldsymbol{p}$.*

Proof. (a) It is clear that $d^{\mathfrak{C}}$ is a pseudodistance. It remains to verify (4.3.1) for $d^{\mathfrak{C}}$. Fix an $a \in G$ and let M, r be as in (4.3.1). Then, for any $z', z'' \in \mathbb{B}(a,r)$ we have

$$d^{\mathfrak{C}}(z',z'') \le L_d([z',z'']) \le M L_{\|\,\|}([z',z'']) = M\|z'-z''\|.$$

(b) Cf. Remark 2.7.5(c).

(c) Cf. Remark 2.7.7.

(d) The proof is left to the reader. □

4.3.2 Operator $\delta \longmapsto \int \delta$

Let $\delta \in \mathcal{M}(G)$ be Borel measurable. For $\alpha \in \mathfrak{C}_{ic}(G)$ put

$$L_\delta(\alpha) := \int_0^1 \delta(\alpha(t);\alpha'(t))dt.$$

The number $L_\delta(\alpha) \in \mathbb{R}_+$ is called the *δ-length of* α. Define

$$(\textstyle\int \delta)(z',z'') := \inf\{L_\delta(\alpha) : \alpha \in \mathfrak{C}_{ic}(G),\ \alpha(0)=z',\ \alpha(1)=z''\}, \quad z',z'' \in G.$$

We say that $\int \delta$ is the *integrated form of* δ. Recall that, for any domain $G \subset \mathbb{C}^n$, we have:

- $\int \gamma_G = \boldsymbol{c}_G^i$ (cf. Lemma 2.7.3(c)),
- $\int \varkappa_G = \int \hat{\varkappa}_G = \boldsymbol{k}_G$ (cf. Theorem 3.6.4).

Proposition 4.3.2. *Let $\delta \in \mathcal{M}(G)$ be Borel measurable. Then,*

(a) $\int \delta \in \mathcal{D}(G)$.

(b) *$L_{\int \delta}(\alpha) \le L_\delta(\alpha)$ for any $\alpha \in \mathfrak{C}_{ic}(G)$; in particular, $(\int \delta)^{ic} = \int \delta$, i.e., $\int \delta$ is inner.*

(c) *If $(\delta_G)_G$ is a holomorphically contractible family of pseudometrics such that δ_G is Borel measurable for every G, then the family $(\int \delta_G)_G$ is a holomorphically contractible family of pseudodistances.*

Proof. (a) It is clear that $\int\delta$ is a pseudodistance. We only need to verify (4.3.1). Fix an $a \in G$ and let M, r be as in (4.3.2). Then, for any $z', z'' \in \mathbb{B}(a,r)$ we have

$$(\textstyle\int\delta)(z',z'') \le L_\delta([z',z'']) = \int_0^1 \delta(z' + t(z''-z'); z''-z')dt \le M\|z'-z''\|.$$

(b) Let $\alpha \in \mathfrak{C}_{ic}(G)$ and let $0 = t_0 < \cdots < t_N = 1$. Then,

$$\sum_{j=1}^{N} (\textstyle\int\delta)(\alpha(t_{j-1}),\alpha(t_j)) \le \sum_{j=1}^{N} L_\delta(\alpha|_{[t_{j-1},t_j]}) = L_\delta(\alpha).$$

(c) The proof is left to the reader. □

4.3.3 Operator $d \longmapsto \mathfrak{D}d$

For a function $d : G \times G \longrightarrow \mathbb{R}_+$ with (4.3.1), we put

$$(\mathfrak{D}d)(a;X) = \limsup_{\substack{\mathbb{C}_* \ni \lambda \to 0 \\ z \to a}} \frac{1}{|\lambda|} d(z, z+\lambda X), \quad a \in G,\ X \in \mathbb{C}^n.$$

Note that:

- according to (4.3.1), the above limit is finite,
- $\mathfrak{D}d = \mathfrak{D}(\tanh d)$,
- $\mathfrak{D}c_G = \mathfrak{D}c_G^i = \gamma_G$ – cf. Propositions 2.7.1(d), 2.7.6.

Proposition 4.3.3. *Let $d \in \mathfrak{D}(G)$. Then:*

(a) $\mathfrak{D}d \in \mathcal{M}(G)$.

(b) *For each $a \in G$, the function $(\mathfrak{D}d)(a;\cdot)$ is a $\mathbb{C}$-seminorm.*

(c) *For every $a \in G$, we have*

$$(\mathfrak{D}d)(a;X) = \limsup_{\substack{\mathbb{C}_* \ni \lambda \to 0 \\ z \to a \\ X' \to X}} \frac{1}{|\lambda|} d(z, z+\lambda X'), \quad X \in \mathbb{C}^n, \tag{4.3.3}$$

$$(\mathfrak{D}d)(a;X) = \limsup_{\substack{z',z'' \to a,\ z' \neq z'' \\ \frac{z'-z''}{\|z'-z''\|} \to X}} \frac{d(z',z'')}{\|z'-z''\|}, \quad X \in \partial\mathbb{B}_n. \tag{4.3.4}$$

(d) *$\mathfrak{D}d$ is upper semicontinuous.*

(e) $L_d(\alpha) \le L_{\mathfrak{D}d}(\alpha)$ *for any* $\alpha \in \mathfrak{C}_{ic}(G)$. *In particular,* $d \le d^{ic} \le \int \mathfrak{D}d$.

(f) *Let* $\delta \in \mathfrak{M}(G)$ *be such that, for each* $X \in \mathbb{C}^n$, *the function* $\delta(\cdot; X)$ *is upper semicontinuous. Then,* $\mathfrak{D}(\int \delta) \le \delta$. *In particular,* $\mathfrak{D}\boldsymbol{k}_G \le \hat{\boldsymbol{\varkappa}}_G$ (*cf. Proposition* 3.6.2, *Theorem* 3.6.4).

We point out that it is not know whether the system $(\mathfrak{D}\boldsymbol{k}_G)_G$ is holomorphically contractible. Notice that the system $(\mathfrak{D}\boldsymbol{k}_G)_{G \text{ is taut}}$ *is* holomorphically contractible – cf. Proposition 5.3.6.

?

Proof. (a) Directly from the definition, we get

$$(\mathfrak{D}d)(a; \lambda X) = |\lambda|(\mathfrak{D}d)(a; X), \quad a \in G,\ \lambda \in \mathbb{C},\ X \in \mathbb{C}^n.$$

Take an $a \in G$ and let M, r be as in (4.3.1). Then, for any $z \in \mathbb{B}(a, r)$, we get

$$\begin{aligned}(\mathfrak{D}d)(z; X) &\le \limsup_{\substack{\mathbb{C}_* \ni \lambda \to 0\\ z' \to z}} \frac{1}{|\lambda|} d(z', z' + \lambda X)\\ &\le \limsup_{\substack{\mathbb{C}_* \ni \lambda \to 0\\ z' \to z}} \frac{1}{|\lambda|} M\|\lambda X\| = M\|X\|, \quad X \in \mathbb{C}^n.\end{aligned}$$

(b)
$$\begin{aligned}(\mathfrak{D}d)(a; X' + X'') &= \limsup_{\substack{\mathbb{C}_* \ni \lambda \to 0\\ z \to a}} \frac{1}{|\lambda|} d(z, z + \lambda(X' + X''))\\ &\le \limsup_{\substack{\mathbb{C}_* \ni \lambda \to 0\\ z \to a}} \frac{1}{|\lambda|} d(z, z + \lambda X')\\ &\quad + \limsup_{\substack{\mathbb{C}_* \ni \lambda \to 0\\ z \to a}} \frac{1}{|\lambda|} d(z + \lambda X', z + \lambda X' + \lambda X'')\\ &\le (\mathfrak{D}d)(a; X') + (\mathfrak{D}d)(a; X''), \quad a \in G,\ X', X'' \in \mathbb{C}^n.\end{aligned}$$

(c) Fix $a \in G$ and $X \in \mathbb{C}^n$. Let M, r be as in (4.3.1). Then, for $z \in \mathbb{B}(a, r/2)$, $|\lambda| < (r/2)/(\|X\| + (r/2))$, and $X' \in \mathbb{B}(X, r/2)$ we have

$$\begin{aligned}d(z, z + \lambda X') &\le d(z, z + \lambda X) + d(z + \lambda X, z + \lambda X')\\ &\le d(z, z + \lambda X) + M|\lambda|\|X' - X\|,\end{aligned}$$

which implies (4.3.3). Now, (4.3.4) follows from the relation

$$\frac{d(z', z'')}{\|z' - z''\|} = \frac{1}{|\lambda|} d(z, z + \lambda X') \text{ with } z = z'',\ \lambda := \|z' - z''\|,\ X' := \frac{z' - z''}{\|z' - z''\|}.$$

(d) Fix $a \in G$, $X \in \mathbb{C}^n$ and suppose that $(\mathcal{D}d_G)(a;X) < A$. By (c), there exists an $\eta > 0$ such that

$$\frac{1}{|\lambda|} d_G(z, z + \lambda X') < A, \quad 0 < |\lambda| < \eta,\ z \in \mathbb{B}(a,\eta),\ X' \in \mathbb{B}(X,\eta).$$

Hence, again using (c), we get

$$(\mathcal{D}d_G)(z;X') \le A, \quad z \in \mathbb{B}(a,\eta),\ X' \in \mathbb{B}(X,\eta).$$

(e) Cf. the proof of Lemma 3.6.5.

(f) Fix $a \in G$ and $X \in \mathbb{C}^n$. Then,

$$\begin{aligned}
(\mathcal{D}(\textstyle\int \delta))(a;X) &= \limsup_{\substack{\mathbb{C}_* \ni \lambda \to 0 \\ z \to a}} \frac{1}{|\lambda|} (\textstyle\int \delta)(z, z+\lambda X) \le \limsup_{\substack{\mathbb{C}_* \ni \lambda \to 0 \\ z \to a}} \frac{1}{|\lambda|} L_\delta([z, z+\lambda X]) \\
&\le \limsup_{\substack{\mathbb{C}_* \ni \lambda \to 0 \\ z \to a}} \frac{1}{|\lambda|} \int_0^1 \delta(z + t\lambda X; \lambda X) dt \\
&= \limsup_{\substack{\mathbb{C}_* \ni \lambda \to 0 \\ z \to a}} \int_0^1 \delta(z + t\lambda X; X) dt \\
&\overset{\text{Fatou}}{\le} \int_0^1 \Big(\limsup_{\substack{\mathbb{C}_* \ni \lambda \to 0 \\ z \to a}} \delta(z + t\lambda X; X) \Big) dt \overset{\text{semicontinuity}}{\le} \delta(a;X).
\end{aligned}$$

□

4.3.4 Operator $\delta \longmapsto \widehat{\delta}$

Let $G \subset \mathbb{C}^n$ be a domain and let $\delta \in \mathcal{M}(G)$ be such that $\delta(a;\cdot)$ is upper semicontinuous for each $a \in G$. Notice that $\delta(a;\cdot) = \mathfrak{h}_{D_a}$, where $D_a := \{X \in \mathbb{C}^n : \delta(a;X) < 1\}$; D_a is a balanced domain. We define

$$\widehat{\delta}^{(m)}(a;X) := \mathfrak{h}^{(m)}_{D_a}(X), \quad m \in \mathbb{N},\ a \in G,\ X \in \mathbb{C}^n,$$

and the *Buseman pseudometric for* δ

$$\widehat{\delta}(a;X) := \mathfrak{h}_{\widehat{D}_a}(X), \quad a \in G,\ X \in \mathbb{C}^n;$$

see also [80].

Remark 4.3.4. Let $G \subset \mathbb{C}^n$ be a domain and let $\delta \in \mathcal{M}(G)$ be such that $\delta(a;\cdot)$ is upper semicontinuous for each $a \in G$.

(a) $\widehat{\delta} \le \widehat{\delta}^{(m)} \le \widehat{\delta}^{(1)} = \delta$. If $n = 1$, then $\widehat{\delta} = \widehat{\delta}^{(m)} = \delta$.

(b) $\widehat{\delta}^{(m)} \searrow \widehat{\delta}$.

(c) $\widehat{\delta}^{(m)}, \widehat{\delta} \in \mathcal{M}(G)$.

(d) $\widehat{\delta} = \widehat{\delta}^{(2n-1)}$ (cf. Lemma 2.2.5(b)).

(e) The above definitions may be applied to $\delta \in \{\boldsymbol{\gamma}_G^{(k)}, \boldsymbol{S}_G, \boldsymbol{A}_G, \boldsymbol{\varkappa}_G^{(k)}\}$. Observe that $\widehat{\boldsymbol{\varkappa}}_G$ agrees with the definition in § 3.6 (cf. Remark 3.6.1).

(f) If δ is upper semicontinuous, then so is $\widehat{\delta}^{(m)}$ and, consequently, $\widehat{\delta}$ is also upper semicontinuous.

In fact, fix $m \in \mathbb{N}$, $a \in G$, $X_0 \in \mathbb{C}^n$, and let $\widehat{\delta}^{(m)}(a; X_0) < A$. Then, there exist $X_1, \dots, X_m \in \mathbb{C}^n$ such that $X_0 = \sum_{j=1}^m X_j$ and $\sum_{j=1}^m \delta(a; X_j) < A$. Using the upper semicontinuity of δ, we find an $r > 0$ such that $\sum_{j=1}^m \delta(z; X_j') < A$ for all $z \in \mathbb{B}(a,r) \subset G$ and $X_j' \in \mathbb{B}(X_j, r)$, $j = 1, \dots, m$. Thus, $\widehat{\delta}^{(m)}(z; X') < A$ for all $z \in \mathbb{B}(a,r)$ and $X' \in \mathbb{B}(X_0, r)$.

(g) If δ is continuous and $\delta(a; X) > 0$ for all $a \in G$, $X \neq 0$, then $\widehat{\delta}^{(m)}$ is continuous. In particular, if G is taut, then $\widehat{\boldsymbol{\varkappa}}_G^{(m)}$ is continuous.

In fact, fix $(a, X_0) \in G \times \mathbb{C}^n$ and let $z_k \longrightarrow a$, $X_k' \longrightarrow X_0$, $\widehat{\delta}^{(m)}(z_k; X_k') \longrightarrow \alpha$. By (f), we only need to show that $\alpha \ge \delta^{(m)}(a; X_0)$. By Remark 2.2.4(h), $\widehat{\delta}^{(m)}(z_k; X_k') = \sum_{j=1}^m \delta(z_k; X_{k,j})$ with $X_k' = \sum_{j=1}^m X_{k,j}$. Observe that there exists a constant $c > 0$ such that $\delta(z_k; X) \ge c\|X\|$, $k \in \mathbb{N}$, $X \in \mathbb{C}^n$. Consequently, the sequences $(X_{k,j})_{k=1}^\infty$, $j = 1, \dots, m$, are bounded and, therefore, we may assume that $X_{k,j} \longrightarrow X_j$ when $k \longrightarrow +\infty$, $j = 1, \dots, m$. Observe that $X_0 = \sum_{j=1}^m X_j$. Using the continuity of δ, we get $\alpha = \sum_{j=1}^m \delta(a; X_j) \ge \widehat{\delta}^{(m)}(a; X_0)$.

(h) The proof in (g) shows that if $\delta(a; \cdot)$ is continuous and $\delta(a; X) > 0$ for all $X \neq 0$, then $\widehat{\delta}^{(m)}(a; \cdot)$ is continuous.

(i) If δ is upper semicontinuous, then $\int \widehat{\delta} = \int \widehat{\delta}^{(m)} = \int \delta$.

In fact, the inequalities "$\le$" are obvious. By Proposition 4.3.3(b, f), we have $\mathcal{D}(\int \delta) \le \widehat{\delta}$ and hence, by Proposition 4.3.3(e), we get

$$\int \widehat{\delta} \ge \int \mathcal{D}(\int \delta) \ge (\int \delta)^{ic} = \int \delta.$$

(j) If $(\delta_G)_G$ is a holomorphically contractible family of pseudometrics such that for any domain G and for any point $a \in G$, the function $\delta_G(a; \cdot)$ is upper semicontinuous, then the systems $(\widehat{\delta}_G^{(m)})_G$, $(\widehat{\delta}_G)_G$ are holomorphically contractible.

In fact, if $F \in \mathcal{O}(G, D)$, then for $a \in G$ and $X = \sum_{j=1}^{m} X_j$, we have

$$\widehat{\delta}_D^{(m)}(F(a); F'(a)X) \le \sum_{j=1}^{m} \delta_D(F(a); F'(a)X_j) \le \sum_{j=1}^{m} \delta_G(a; X_j),$$

which implies $\widehat{\delta}_D^{(m)}(F(a); F'(a)X) \le \widehat{\delta}_G^{(m)}(a; X)$.

4.3.5 Operator $\delta \longmapsto \widetilde{\delta}$

Let $\delta \in \mathcal{M}(G)$ be such that $\delta(a; \cdot)$ is upper semicontinuous for each $a \in G$. Recall that, for any fixed $a \in G$, the function $\delta(a; \cdot)$ is the Minkowski function of the balanced indicatrix of $\delta(a; \cdot)$. In light of subsection 2.2.2 one defines the *DNT-pseudometric*

$$\widetilde{\delta}(a; X) := \widetilde{\delta(a; \cdot)}(X), \quad a \in G,\ X \in \mathbb{C}^n.$$

Obviously, $\widetilde{\delta} : G \times \mathbb{C}^n \longrightarrow [0, +\infty)$ is again a pseudometric on G, which was introduced in [134]. Note that $\widehat{\delta} \le \widetilde{\delta} \le \delta$ and $\widetilde{\delta}(a; \cdot) \in \mathcal{PSH}(\mathbb{C}^n)$.

The above operator may be applied to $\delta \in \{\boldsymbol{\gamma}_G^{(k)}, \boldsymbol{S}_G, \boldsymbol{A}_G, \boldsymbol{\varkappa}_G^{(k)}\}$.

Remark 4.3.5. Let $\delta \in \mathcal{M}(G)$ be such that $\delta(a; \cdot)$ is upper semicontinuous for each $a \in G$.

(a) If δ is upper semicontinuous, so is $\widetilde{\delta}$.

In fact, fix a point $a \in G$ and a vector $X \in \mathbb{C}^n$. Suppose that $\widetilde{\delta}$ is not upper semicontinuous at $(a; X)$. Then, there are a positive ε_0 and sequences $G \ni a_j \longrightarrow a$, $\mathbb{C}^n \ni X_j \longrightarrow X$ such that $\widetilde{\delta}(a_j; X_j) \ge \widetilde{\delta}(a; X) + \varepsilon_0 =: A$ for all j.

Let $Z \in \mathbb{C}^n$ with $\delta(a; Z) < 1$, i.e., Z belongs to the indicatrix

$$I := \{Y \in \mathbb{C}^n : \delta(a; Y) < 1\}$$

of $\delta(a; \cdot)$. Then, upper semicontinuity implies that there exist a $j_Z \in \mathbb{N}$ and a neighborhood $V_Z = V_Z(0)$ such that $\delta(a_j; Z + Y) < 1$, $j \ge j_Z$ and $Y \in V_Z$. In other words, $Z \in \operatorname{int} \bigcap_{j \ge j_Z} I_j$, where $I_j := \{Y \in \mathbb{C}^n : \delta(a_j; Y) < 1\}$ denotes the indicatrix of $\delta(a_j; \cdot)$. Hence,

$$I \subset \bigcup_{k \in \mathbb{N}} \operatorname{int} \bigcap_{j \ge k} I_j \subset \bigcup_{k \in \mathbb{N}} \operatorname{int} \bigcap_{j \ge k} \widetilde{I}_j =: D,$$

where $\widetilde{I}_j$ denotes the envelope of holomorphy of I_j. Since D is a balanced pseudoconvex domain, the envelope of holomorphy $\widetilde{I}$ of I is contained in D.

Observe that $X/A \in \widetilde{I}$. Therefore, one may choose a $k \in \mathbb{N}$ and a positive r_k such that $X/A + \mathbb{B}(r_k) \subset \bigcap_{j \geq k} \widetilde{I}_j$. Put $Y_j := (X_j - X)/A$. Then, $\|Y_j\| < r_k$ for all $j \geq j_k > k$. So we have

$$\widetilde{\delta}(a_j; X_j) = \widetilde{\delta}(a_j; X + (X_j - X)) < A, \quad j \geq j_k;$$

a contradiction.

(b) Let δ be as above and let $a \in G$. Assume that $\delta(a; \cdot) > 0$ on $\mathbb{C}^n \setminus \{0\}$ and that δ is continuous at each point (a, X), $X \in \mathbb{C}^n$. Then, $\widetilde{\delta}$ is continuous at (a, X), $X \in \mathbb{C}^n$. In particular, if δ is continuous and $\delta(a; X) > 0$, $a \in G$, $X \in (\mathbb{C}^n)_*$, then $\widetilde{\delta}$ is continuous. For example, if G is taut, then $\widetilde{\varkappa}_G^{(k)}$ is continuous.

In fact, using a compactness argument, it follows that there exist an $M > 1$ and a neighborhood $V = V(a) \subset G$ such that $\|X\|/M \leq \delta(a; X) \leq M\|X\|$, $X \in \mathbb{C}^n$. Moreover, for a positive $\varepsilon < 1/M$, we get a smaller neighborhood $U_\varepsilon = U_\varepsilon(a) \subset V$ such that $|\delta(b; X) - \delta(a; X)| \leq \varepsilon\|X\|$, $X \in \mathbb{C}^n$ and $b \in U_\varepsilon$. Hence, if $b \in U_\varepsilon$, then

$$(1 - \varepsilon M)\delta(a; X) \leq \delta(b; X) \leq \varepsilon\|X\| + \delta(a; X) \leq (1 + \varepsilon M)\delta(a; X).$$

Therefore, $(1 - \varepsilon M)I(b) \subset I(a) \subset (1 + \varepsilon M)I(b)$, where $I(z)$ denotes the indicatrix of $\delta(z; \cdot)$, $z \in G$. Now turning to the envelopes of holomorphy gives

$$(1 - \varepsilon M)\widetilde{I}(b) \subset \widetilde{I}(a) \subset (1 + \varepsilon M)\widetilde{I}(b), \quad b \in U_\varepsilon.$$

Rephrasing this inclusion leads to the following inequality:

$$(1 - \varepsilon M)\widetilde{\delta}(b; X) \leq \widetilde{\delta}(a; X) \leq (1 + \varepsilon M)\widetilde{\delta}(b; X), \quad b \in U_\varepsilon.$$

Now let $G \times \mathbb{C}^n \ni (a_j, X_j) \longrightarrow (a, X)$, $X \neq 0$. Then the above estimate has shown that the sequence $\widetilde{\delta}(a_j; \cdot)$ converges uniformly on $\partial\mathbb{B}_n$ to $\widetilde{\delta}(a; \cdot)$. Taking this into account, it easily follows that $\widetilde{\delta}(a_j; X_j) \longrightarrow \widetilde{\delta}(a; X)$.

(c) If $\delta_j \searrow \delta$, then $\widetilde{\delta}_j \searrow \widetilde{\delta}$.

Use Remark 2.2.7.

(d) If δ is upper semicontinuous, then $\smallint \widehat{\delta} = \smallint \widetilde{\delta} = \smallint \delta$.

Use Remark 4.3.4(i).

(e) If $(\delta_G)_G$ is a holomorphically contractible family of pseudometrics such that for any domain G and for any point $a \in G$, the function $\delta_G(a; \cdot)$ is upper semicontinuous, then the family $(\widetilde{\delta}_G)_G$ is also a contractible family. In particular, $(\widetilde{\varkappa}_G^{(k)})_{G \in \mathfrak{G}}$ is a contractible family of upper semicontinuous pseudometrics.

(f) Let $(\delta_G)_G$ be as before. Let $(G_j)_j$ be an increasing sequence of domains in $\mathbb{C}^n$ with $G_j \nearrow G$. If $\delta_{G_j} \searrow \delta_G$, then $\widetilde{\delta}_{G_j} \searrow \widetilde{\delta}_G$. In particular, $\widetilde{\varkappa}^{(k)}_{G_j} \searrow \widetilde{\varkappa}^{(k)}_G$.

(g) $A_G \leq \widetilde{\varkappa}_G$.

Remark 4.3.6. Let $G \subset \mathbb{C}^n$ be a domain. Let us summarize the relations between the operators ^{i}, $\int$, $\mathfrak{D}$, $\hat{}$, and $\tilde{}$ acting on the families $\mathfrak{D}(G)$ and $\mathfrak{M}(G) \cap \mathcal{C}^{\uparrow}(G \times \mathbb{C}^n) = \{\delta \in \mathfrak{M}(G) : \delta \text{ is upper semicontinuous}\}$:

Table 4.1

	i	$\int$	$\mathfrak{D}$	$\hat{}$	$\tilde{}$
i	$i \circ i = i$	$i \circ \int = \int$	$\times$	$\times$	$\times$
$\int$	$\times$	$\times$	$\int \circ \mathfrak{D} \geq i$	$\int \circ \hat{} = \int$	$\int \circ \tilde{} = \int$
$\mathfrak{D}$	$\mathfrak{D} \circ i = \mathfrak{D}$	$\mathfrak{D} \circ \int \leq \hat{}$	$\times$	$\times$	$\times$
$\hat{}$	$\times$	$\times$	$\hat{} \circ \mathfrak{D} = \mathfrak{D}$	$\hat{} \circ \hat{} = \hat{}$	$\hat{} \circ \tilde{} = \hat{}$
$\tilde{}$	$\times$	$\times$	$\tilde{} \circ \mathfrak{D} = \mathfrak{D}$	$\tilde{} \circ \hat{} = \hat{}$	$\tilde{} \circ \tilde{} = \tilde{}$

Remark 4.3.7.

(a) In general, the inequality $\mathfrak{D} \circ \int \leq \hat{}$ may be strict.

For, by Example 3.5.11 there exists a pseudoconvex Hartogs domain $G_0 \subset \mathbb{C}^2$ such that $\boldsymbol{k}_{G_0} \equiv 0$ and $\hat{\boldsymbol{\varkappa}}_{G_0} \not\equiv 0$; recall that $\boldsymbol{k}_G = \int \boldsymbol{\varkappa}_G$ (cf. Theorem 3.6.4). In particular, $\mathfrak{D}(\int \boldsymbol{\varkappa}_{G_0}) = \mathfrak{D}\boldsymbol{k}_{G_0} \equiv 0$. Thus, $\mathfrak{D}(\int \boldsymbol{\varkappa}_{G_0}) \not\equiv \hat{\boldsymbol{\varkappa}}_{G_0}$.

(b) The same is true for the inequality $\int \circ \mathfrak{D} \geq i$.

For example, take $G = \mathbb{C}$, $d(z', z'') := |\operatorname{Re} z' - \operatorname{Re} z''|$, $z', z'' \in \mathbb{C}$ (it is clear that (4.3.1) is satisfied). Then, $(\mathfrak{D}d)(a; X) = |X|$, $a, X \in \mathbb{C}$. Hence, $(\int \mathfrak{D}d)(z', z'') = |z' - z''|$, $z', z'' \in \mathbb{C}$. In particular, $(\int \mathfrak{D}d)(0, i) = 1$. On the other hand, $d^i(0, i) = L_d([0, i]) = 0$.

4.3.6 $\mathcal{C}^1$-pseudodistances

We say that d is a $\mathcal{C}^1$*-pseudodistance* if

$$\forall_{K \subset\subset G}\ \forall_{\varepsilon > 0}\ \exists_{\eta > 0} :$$
$$|d(z', z'') - (\mathfrak{D}d)(z; z' - z'')| \leq \varepsilon \|z' - z''\|, \quad z \in K,\ z', z'' \in \mathbb{B}(z, \eta);$$

cf. [447, 448].

Remark 4.3.8. One can easily prove that for $d \in \mathfrak{D}(G)$ the following conditions are equivalent:

(i) d is a $\mathcal{C}^1$-pseudodistance;

(ii) $\mathfrak{D}d$ is continuous on $G \times \mathbb{C}^n$, and

$$(\mathfrak{D}d)(a; X) = \lim_{\substack{\mathbb{C}_* \ni \lambda \to 0 \\ z \to a \\ X' \to X}} \frac{1}{|\lambda|} d(z, z + \lambda X'), \quad a \in G, \ X \in \mathbb{C}^n.$$

(iii) $\mathfrak{D}d$ is continuous, and

$$(\mathfrak{D}d)(a; X) = \lim_{\substack{z', z'' \to a, \, z' \neq z'' \\ \frac{z' - z''}{\|z' - z''\|} \to X}} \frac{d(z', z'')}{\|z' - z''\|}, \quad a \in G, \ X \in \partial \mathbb{B}_n.$$

Proposition 4.3.9. *Let d be a $\mathcal{C}^1$-pseudodistance. Then,*

(a) $L_d(\alpha) = L_{\mathfrak{D}d}(\alpha)$ *for any* $\alpha \in \mathfrak{C}_{ic}(G)$*;*

(b) $d^i = \int(\mathfrak{D}d)$ *(cf. Remark* 4.3.7(b)*);*

(c) d^i *is a* $\mathcal{C}^1$*-pseudodistance (recall that* $\mathfrak{D}d^i = \mathfrak{D}d$*, cf. Remark* 4.3.6).

Recall that $\boldsymbol{c}_G$ and $\boldsymbol{c}_G^i$ are $\mathcal{C}^1$-pseudodistances (cf. Propositions 2.7.1(d'), 2.7.6).

Proof. Use the same methods as in the proofs of Lemma 2.7.3(c, d) and Proposition 2.7.6. □

As an application of the above results, we present the following proposition (clarifying Remark 3.1.12 and Example 3.6.7).

Proposition 4.3.10 (cf. [509]). *Let $G = G_{\mathfrak{h}} \subset \mathbb{C}^n$ be a balanced domain with Minkowski function $\mathfrak{h}$. Then,*

(a) $\widehat{\mathfrak{h}} = \boldsymbol{\gamma}_G(0; \cdot) = (\mathfrak{D}\boldsymbol{k}_G)(0; \cdot) = \widehat{\boldsymbol{\varkappa}}_G(0; \cdot)$*, cf. Proposition* 2.3.1(d)*;*

(b) *if G is pseudoconvex, then for any $a \in G$ the following conditions are equivalent:*

  (i) $\widehat{\mathfrak{h}}(a) = \mathfrak{h}(a)$,

  (ii) $\boldsymbol{\gamma}_G(0; a) = \mathfrak{h}(a)$,

  (iii) $\boldsymbol{c}_G^*(0, a) = \mathfrak{h}(a)$,

  (iv) $\widehat{\boldsymbol{\varkappa}}_G(0; a) = \boldsymbol{\varkappa}_G(0; a)$,

  (v) $\boldsymbol{k}_G(0, a) = \boldsymbol{\ell}_G(0, a)$.

Proof. (a) By Proposition 2.3.1(d), Theorem 4.3.3(g), and the proof of Proposition 3.5.3 we get

$$\widehat{\mathfrak{h}} = \boldsymbol{\gamma}_G(0;\cdot) = (\mathcal{D}\boldsymbol{c}_G)(0;\cdot) \le (\mathcal{D}\boldsymbol{k}_G)(0;\cdot) \le \hat{\boldsymbol{\varkappa}}_G(0;\cdot) \le \boldsymbol{\varkappa}_G(0;\cdot) = \mathfrak{h}.$$

Since $\hat{\boldsymbol{\varkappa}}_G(0;\cdot)$ is a seminorm, we obtain $\hat{\boldsymbol{\varkappa}}_G(0;\cdot) \le \widehat{\mathfrak{h}}$, which completes the proof of (a).

(b) By Propositions 2.3.1(b, d), 3.1.11, 3.5.3, and (a), we only need to prove that $\widehat{\mathfrak{h}}(a) = \mathfrak{h}(a)$ if $\boldsymbol{k}_G^*(0,a) = \mathfrak{h}(a)$. By Remark 3.2.8(b), if $\boldsymbol{k}_G^*(0,a) = \mathfrak{h}(a)$, then $\boldsymbol{k}_G^*(0,ta) = t\mathfrak{h}(a)$, $0 \le t \le 1$. Hence, $(\mathcal{D}\boldsymbol{k}_G)(0;a) \ge \mathfrak{h}(a)$, which by (a) gives $\widehat{\mathfrak{h}}(a) = \mathfrak{h}(a)$. □

Remark 4.3.11. For the real case, a similar discussion as in § 4.3 may be found in [73].

4.4 Exercises

Exercise 4.4.1. Let $(d_G)_G$ (resp. $(\delta_G)_G$) be a contractible family of functions (resp. pseudometrics). Fix a domain G and assume that there exists a sequence $(G_j^0)_{j=1}^\infty$ of subdomains of G such that $G_j^0 \subset G_{j+1}^0 \subset\subset G$, $G = \bigcup_{j=1}^\infty G_j^0$, and $d_{G_j^0} \searrow d_G$ (resp. $\delta_{G_j^0} \searrow \delta_G$) – cf. Propositions 2.7.1(a), 3.3.5(a). Let $G = \bigcup_{i \in I} G_i$, where each G_i is a subdomain of G, and for any compact $K \subset\subset G$ there exists an $i_0 \in I$ with $K \subset G_{i_0}$.

Prove that $\inf_{i\in I} d_{G_i} = d_G$ (resp. $\inf_{i \in I} \delta_{G_i} = \delta_G$). In particular, for any sequence $(G_j)_{j=1}^\infty$ of subdomains of G with $G_j \nearrow G$ we have $d_{G_j} \searrow d_G$ (resp. $\delta_{G_j} \searrow \delta_G$).

Exercise 4.4.2 (cf. Example 4.2.18). Let

$$\varphi(\xi) := \sum_{k=2}^\infty \lambda_k \log \frac{|\xi - 1/k|}{2}, \quad \xi \in \mathbb{C},$$

where $\lambda_k > 0$ are such that $\varphi(0) > -\infty$. Define

$$G = \{(z_1, z_2) \in \mathbb{C}^2 : |z_1| e^{\varphi(z_2)} < 1\}.$$

Prove that $\limsup_{t\to 0+} \boldsymbol{S}_G((-t,0);(0,1)) > \boldsymbol{S}_G((0,0);(0,1))$.

Exercise 4.4.3. Let $F : G \longrightarrow D$ be holomorphic ($G \subset \mathbb{C}^n$, $D \subset \mathbb{C}^m$). Let $a \in G$ and let $r := \operatorname{ord}_a(F - F(a)) = \min\{\operatorname{ord}_a(F_j - F_j(a)) : j = 1,\dots,m\}$. Prove that

$$\boldsymbol{m}_D^{(k)}(F(a), F(z)) \le [\boldsymbol{m}_G^{(kr)}(a,z)]^r, \quad z \in G.$$

Exercise 4.4.4. Let $G = G_{\mathfrak{h}}$ be a balanced domain in $\mathbb{C}^n$ ($\mathfrak{h}$ is the Minkowski function of G). Fix $a \in G$ and $k \in \mathbb{N}$. Prove that the following conditions are equivalent (cf. Proposition 2.3.1(b)):

(i) $\boldsymbol{m}_G^{(k)}(0, a) = \mathfrak{h}(a)$;

(ii) $\boldsymbol{m}_G^{(k)}(0, \cdot) = \mathfrak{h}$ on $G \cap (\mathbb{C}a)$;

(iii) $\boldsymbol{\gamma}_G^{(k)}(0; a) = \mathfrak{h}(a)$;

(iv) $\boldsymbol{\gamma}_G^{(k)}(0; \cdot) = \mathfrak{h}$ on $\mathbb{C}a$;

(v) there exists a homogeneous polynomial Q of degree k, such that $|Q(a)|^{1/k} = |\mathfrak{h}(a)$ and $|Q|^{1/k} \leq \mathfrak{h}$.

Exercise 4.4.5 (cf. [433]). Let $\varphi \in \mathcal{O}(\mathbb{D}, \mathbb{C}^2)$ be defined by

$$\varphi(\lambda) := (\lambda^2 - 1/4, \lambda(\lambda^2 - 1/4)).$$

Fix $a := (0, 0) = \varphi(\pm 1/2)$ and $b := (-1/4, 0) = \varphi(0)$. Let $D \subset \mathbb{C}^2$ be the connected component of the set $\{z \in \mathbb{C}^2 : |-16(z_1 + 1/4)^2(z_1 - 1/4) + 16z_2^2| < 1\}$ that contains $\varphi(\mathbb{D})$. Prove that $\boldsymbol{g}_D(a, b) \leq 1/4 < 1/2 \leq \boldsymbol{g}_D(b, a)$.

Hint. Use the log-plurisubharmonic functions $u_1(z) := |16z_1/(15 - 4z_1)|$ and $u_2(z) := |-16(z_1 + 1/4)^2(z_1 - 1/4) + 16z_2^2|^{1/2}$.

Exercise 4.4.6. Prove that $\int(\mathfrak{D}\boldsymbol{k}_G) = \boldsymbol{k}_G$.

Exercise 4.4.7. Prove the following theorems (cf. [134]).
(a) *If* $(D_j)_{j=1}^{\infty}$ *is an increasing sequence of domains in* $\mathbb{C}^n$ *with* $D = \bigcup_{j=1}^{\infty} D_j$ *and* $D_j \times \mathbb{C}^n \ni (a_j, X_j) \longrightarrow (a, X) \in D \times \mathbb{C}^n$, *then* $\limsup_{j \to +\infty} \widetilde{\boldsymbol{\varkappa}}_{D_j}(a_j; X_j) \leq \widetilde{\boldsymbol{\varkappa}}_D(a; X)$.
(b) *Let* $D \subset\subset \mathbb{C}^n$ *be a pseudoconvex domain with a* $\mathcal{C}^1$*-boundary. Assume that* $D \subset \bigcap_{j=1}^{\infty} D_j \subset \overline{D}$, *where* $(D_j)_{j=1}^{\infty}$ *is a decreasing sequence of bounded domains. If* $D_j \times \mathbb{C}^n \ni (a_j, X_j) \longrightarrow (a, X) \in D \times \mathbb{C}^n$, *then* $\widetilde{\boldsymbol{\varkappa}}_{D_j}(a_j; X_j) \longrightarrow \widetilde{\boldsymbol{\varkappa}}_D(a; X)$.

4.5 List of problems

4.1. What are relations between $\boldsymbol{g}_G(0, \cdot)$ and $\widetilde{\mathfrak{h}}$ for balanced domains $G = G_{\mathfrak{h}}$? 160
4.2. Decide whether $\boldsymbol{g}_G(0, \cdot) = \boldsymbol{g}_{\widehat{G}}(0, \cdot)$ on G for balanced Reinhardt domains 169
4.3. What are relations between $\boldsymbol{S}_G(0; \cdot)$ and $\mathfrak{h}_{\widehat{G}}$? 174
4.4. Is the function $G \times \mathbb{C}^n \ni (z, X) \longmapsto \boldsymbol{S}_G(z; X)$ Borel measurable? . . 174
4.5. What are good sufficient conditions on domains of holomorphy G, for $\boldsymbol{S}_G$ to be upper semicontinuous? 174
4.6. Does the equality $\boldsymbol{c}_G^{in} = \boldsymbol{c}_G^{i}$ hold for arbitrary domain G? 182
4.7. Decide whether the system $(\mathfrak{D}\boldsymbol{k}_G)_G$ is holomorphically contractible . . 185

Chapter 5

Properties of standard contractible systems

Summary. Section 5.1 (based on [567] and [565]) presents more advanced regularity properties of the Green function and the Azukawa pseudometric for hyperconvex domains. The Lipschitz continuity of some invariant pseudodistances and pseudometrics is discussed in § 5.2 (based on [386]). Finally, § 5.3 (based on [384]) completes the presentation of the properties of the derivative $\mathcal{D}$ from § 4.3.3.

5.1 Regularity properties of g_G and A_G

This section is based on [567] and [565]. We will discuss various regularity properties of the Green function and the Azukawa pseudometric.

For a domain $G \subset \mathbb{C}^n$, define the function $\mathfrak{b}_G : G \longrightarrow [0, 1]$,

$$\mathfrak{b}_G(a) := \liminf_{z \to \partial G} g_G(a, z);$$

in the case where G is unbounded, we assume that $\infty \in \partial G$. Put

$$G_\theta(a) := \{z \in G : g_G(a, z) < \theta\}, \quad 0 < \theta < 1.$$

Remark 5.1.1.

(a) If G is bounded, then $\mathfrak{b}_G(a) > 0$, $a \in G$ (cf. (4.2.10)).

(b) Let $G = G_{\mathfrak{h}} \subset$ be a balanced pseudoconvex domain with $\mathfrak{h}^{-1}(0) \neq \{0\}$. Then, $\mathfrak{b}_G(0) = 0$ (cf. Proposition 4.2.10(b)).

(c) If G is hyperconvex, then $\mathfrak{b}_G \equiv 1$ (cf. Proposition 4.2.10(h)).

(d) If $0 < \theta < \mathfrak{b}_G(a)$, then $G_\theta(a) \subset\subset G$.

Indeed, suppose that there exists a sequence $G_\theta(a) \ni z_\nu \longrightarrow \zeta \in \partial G$ (recall that if G is unbounded, then $\infty \in \partial G$). Then, $\mathfrak{b}_G(a) \le \theta$; a contradiction.

(e) $G_\theta(a)$ is a domain.

Indeed, suppose that U is a connected component of $G_\theta(a)$ such that $a \notin U$. Define

$$u(z) := \begin{cases} g_G(a, z), & \text{if } z \in G \setminus U \\ \theta, & \text{if } z \in U \end{cases}.$$

Then $u \in \mathcal{K}_G(a)$ (cf. Appendix B.4.18). In particular, $g_G(a, z) \ge u(z) = \theta$ in U; a contradiction.

(f) $\boldsymbol{g}_{G_\theta(a)}(a,z) = (1/\theta)\boldsymbol{g}_G(a,z)$, $z \in G_\theta(a)$.

Indeed, first observe that $(1/\theta)\boldsymbol{g}_G(a,\cdot) \in \mathcal{K}_{G_\theta(a)}(a)$, which gives $\boldsymbol{g}_{G_\theta(a)}(a,\cdot) \geq (1/\theta)\boldsymbol{g}_G(a,\cdot)$. Define

$$u(z) := \begin{cases} \boldsymbol{g}_G(a,z), & \text{if } z \in G \setminus G_\theta(a) \\ \theta \boldsymbol{g}_{G_\theta(a)}(a,z), & \text{if } z \in G_\theta(a) \end{cases}.$$

Then, $u \in \mathcal{K}_G(a)$ (cf. Appendix B.4.18). Consequently,

$$\boldsymbol{g}_G(a,\cdot) \geq u = \theta \boldsymbol{g}_{G_\theta(a)}(a,\cdot) \text{ on } G_\theta(a).$$

(g) $\boldsymbol{A}_{G_\theta(a)}(a;X) = (1/\theta)\boldsymbol{A}_G(a;X)$, $X \in \mathbb{C}^n$.

Indeed, it suffices to use (f) and Lemma 4.2.3.

(h) If $\mathfrak{b}_G(a) > 0$, then $\boldsymbol{g}_G(a,z) > 0$, $z \in G \setminus \{a\}$, and $\boldsymbol{A}_G(a;X) > 0$, $X \in (\mathbb{C}^n)_*$.

Indeed, suppose that $\boldsymbol{g}_G(a,z_0) = 0$ for some $z_0 \neq a$. Take $0 < \theta < \mathfrak{b}_G(a)$. Observe that $z_0 \in G_\theta(a)$. Then, using (d) and (f), we get $0 = \boldsymbol{g}_G(a,z_0) = \theta \boldsymbol{g}_{G_\theta(a)}(a,z_0) > 0$; a contradiction. Moreover, by (d) and (g) we get $\boldsymbol{A}_G(a;X) = \theta \boldsymbol{A}_{G_\theta(a)}(a;X) > 0$, $X \neq 0$.

(i) If $0 < \theta < \mathfrak{b}_G(a)$ and $\boldsymbol{g}_G(a,\cdot)$ is continuous, then $G_\theta(a)$ is hyperconvex.

In view (d), it suffices to check that $\log \boldsymbol{g}_{G_\theta(a)}(a,\cdot)$ is a psh exhaustion function. For, using (f), for $\zeta \in \partial G_\theta(a) \subset G$, we get

$$\liminf_{z\to\zeta} \boldsymbol{g}_{G_\theta(a)}(a,z) = (1/\theta) \liminf_{z\to\zeta} \boldsymbol{g}_G(a,z) = (1/\theta)\boldsymbol{g}_G(a,\zeta) = 1.$$

(j) If $\mathfrak{b}_G(a) > 0$ and $\boldsymbol{g}_G$ is continuous, then for any $0 < \theta_2 < \eta'' < \eta' < \theta_1 < \mathfrak{b}_G(a)$ there exists a $\delta > 0$ such that

$$G_{\theta_2}(x) \subset G_{\eta''}(a) \subset\subset G_{\eta'}(a) \subset G_{\theta_1}(y), \quad x, y \in \mathbb{B}(a,\delta).$$

Indeed, by (d) we have to check that

$$G_{\theta_2}(x) \subset G_{\eta''}(a), \quad G_{\eta'}(a) \subset G_{\theta_1}(y), \quad x, y \in \mathbb{B}(a,\delta), \quad x, y \in \mathbb{B}(a,\delta).$$

First, observe that the function $\boldsymbol{g}_G$ is uniformly continuous on $G_{\theta_1}(a) \times G_{\theta_1}(a)$. In particular, there exists a $\delta > 0$ such that:

- $\boldsymbol{g}_G(x,a) < \eta''$ for all $x \in B(a,\delta)$,
- $\boldsymbol{g}_G(x,z) > \eta''$ for all $x \in B(a,\delta)$ and $z \in G$ with $\boldsymbol{g}_G(a,z) = \eta'$.

Fix an $x \in \mathbb{B}(a,\delta)$ and take an arbitrary $z \in G_{\eta''}(x)$. Let $\gamma : [0,1] \longrightarrow G_{\eta''}(x)$ be a curve with $\gamma(0) = a$, $\gamma(1) = z$ (cf. (e)). Suppose that $z \notin G_{\eta''}(a)$, i.e., $\boldsymbol{g}_G(a,z) \geq \eta''$. By continuity, there exists a $t_0 \in (0,1]$ such that $\boldsymbol{g}_G(a,\gamma(t_0)) = \eta'$; a contradiction.

To prove that $G_{\eta'}(a) \subset G_{\theta_1}(y)$, $y \in \mathbb{B}(a,\delta)$, suppose that $\boldsymbol{g}_G(a,z_\nu) < \eta'$ and $\boldsymbol{g}_G(y_\nu,z_\nu) \geq \theta_1$ with $y_\nu \longrightarrow a$. Since $G_{\eta'}(a) \subset\subset G$ (cf. (d)) and $\boldsymbol{g}_G$ is continuous, we get a contradiction.

Theorem 5.1.2 (cf. [567]). *Assume that $\mathfrak{b}_G(a) > 0$, $a \in G$, and $\boldsymbol{g}_G$ is continuous (e.g. G is hyperconvex). Then,*

(a) *$\boldsymbol{A}_G$ is continuous.*

(b) $$\boldsymbol{A}_G(a;X) = \lim_{\lambda\to 0}\frac{1}{|\lambda|}\boldsymbol{g}_G(a,a+\lambda X), \quad a \in G,\ X \in \mathbb{C}^n.$$

(c) $$\lim_{\substack{z',z'',w',w''\to a,\ z'\neq z'',\ w'\neq w''\\ \frac{w'-w''}{\|w'-w''\|}-\frac{z'-z''}{\|z'-z''\|}\to 0,\ \frac{\|w'-w''\|}{\|z'-z''\|}\to 1}}\frac{\boldsymbol{g}_G(z',z'')}{\boldsymbol{g}_G(w',w'')} = 1, \quad a \in G.$$

(d) $$\boldsymbol{A}_G(a;X) = \lim_{\substack{z',z''\to a,\ z'\neq z''\\ \frac{z'-z''}{\|z'-z''\|}\to X}}\frac{\boldsymbol{g}_G(z',z'')}{\|z'-z''\|}, \quad a \in G,\ \|X\| = 1.$$

(e) $$\lim_{z',z''\to a,\ z'\neq z''}\frac{\boldsymbol{g}_G(z',z'')}{\boldsymbol{g}_G(z'',z')} = 1, \quad a \in G.$$

Proof. (a) Obviously, $\boldsymbol{A}_G$ is continuous at $(a,0)$. Let $(z_\nu,X_\nu) \longrightarrow (a,X)$, $X \neq 0$. Take arbitrary $0 < \theta_2 < \theta_1 < \mathfrak{b}_G(a)$. Let $\Phi_\nu : \mathbb{C}^n \longrightarrow \mathbb{C}^n$ be an affine isomorphism with $\Phi_\nu(z_\nu) = a$, $\Phi_\nu'(z_\nu)X_\nu = X$, $\nu \in \mathbb{N}$, and $\Phi_\nu \longrightarrow \mathrm{id}$. By Remark 5.1.1(j) we get

$$\Phi_\nu(G_{\theta_2}(z_\nu)) \subset\subset G_{\theta_1}(a), \quad G_{\theta_2}(a) \subset\subset \Phi_\nu(G_{\theta_1}(z_\nu)), \quad \nu \gg 1.$$

Hence, by Remark 5.1.1(g), we obtain

$$\begin{aligned}(1/\theta_2)\boldsymbol{A}_G(z_\nu;X_\nu) &= \boldsymbol{A}_{G_{\theta_2}(z_\nu)}(z_\nu;X_\nu) = \boldsymbol{A}_{\Phi_\nu(G_{\theta_2}(z_\nu'))}(\Phi_\nu(z_\nu);\Phi_\nu'(z_\nu')X_\nu)\\ &= \boldsymbol{A}_{\Phi_\nu(G_{\theta_2}(z_\nu'))}(a;X) \geq \boldsymbol{A}_{G_{\theta_1}(a)}(a;X) = (1/\theta_1)\boldsymbol{A}_G(a;X), \quad \nu \gg 1.\end{aligned}$$

Similarly,

$$\begin{aligned}(1/\theta_1)\boldsymbol{A}_G(z_\nu;X_\nu) &= \boldsymbol{A}_{G_{\theta_1}(z_\nu)}(z_\nu;X_\nu) = \boldsymbol{A}_{\Phi_\nu(G_{\theta_1}(z_\nu))}(\Phi_\nu(z_\nu);\Phi_\nu'(z_\nu)X_\nu)\\ &= \boldsymbol{A}_{\Phi_\nu(G_{\theta_1}(z_\nu))}(a;X) \leq \boldsymbol{A}_{G_{\theta_2}(a)}(a;X) = (1/\theta_2)\boldsymbol{A}_G(a;X), \quad \nu \gg 1,\end{aligned}$$

which completes the proof.

(b) Fix an $a \in G$ and $0 < \theta < \mathfrak{b}_G(a)$. Then, by Remark 5.1.1(f),

$$\liminf_{\lambda \to 0} \frac{1}{|\lambda|} g_G(a, a + \lambda X) = \theta \liminf_{\lambda \to 0} \frac{1}{|\lambda|} g_{G_\theta(a)}(a, a + \lambda X).$$

Thus, if the result is true for $G_\theta(a)$, then it holds for G (cf. Remark 5.1.1(g)). Consequently, we may assume that G is hyperconvex.

Suppose that for some $a \in G$ and $X \neq 0$ there exist a sequence $(\lambda_\nu)_{\nu=1}^\infty \subset \mathbb{D}_*$ and $0 < \theta < 1$ such that $\lambda_\nu \longrightarrow 0$ and $\frac{1}{|\lambda_\nu|} g_G(a, a + \lambda_\nu X) < \theta^2 A_G(a; X)$, $\nu \in \mathbb{N}$. Let $a + \mathbb{D}(R)X \subset G$. We may assume that $|\lambda_\nu| < R$, $\nu \in \mathbb{N}$. Since G is hyperconvex, we have $G_\theta(a) \subset\subset G$. To simplify notation, assume that $a = 0$. Fix a $\delta \in (0, \pi)$ such that $e^{it} G_\theta(0) \subset\subset G$ for $|t| \le \delta$. Hence,

$$\begin{aligned} g_G(0, e^{it}\lambda_\nu X) &\le g_{e^{it}G_\theta(0)}(0, e^{it}\lambda_\nu X) = g_{G_\theta(0)}(0, \lambda_\nu X) \\ &= (1/\theta) g_G(0, \lambda_\nu X) < |\lambda_\nu| \theta A_G(a; X), \quad |t| \le \delta,\ \nu \in \mathbb{N}. \end{aligned}$$

Define the sh function

$$u(\lambda) := \begin{cases} \frac{1}{|\lambda|} g_G(0, \lambda X), & \text{if } 0 < |\lambda| < R \\ A_G(0; X), & \text{if } \lambda = 0 \end{cases}.$$

Then,

$$\begin{aligned} A_G(0; X) = u(0) &\le \frac{1}{2\pi} \int_0^{2\pi} u(e^{it}\lambda_\nu X) dt \\ &= \frac{1}{2\pi} \left(\int_{|t| \le \delta} u(e^{it}\lambda_\nu X) dt + \int_{|t| > \delta} u(e^{it}\lambda_\nu X) dt \right) \\ &\le \frac{1}{2\pi} \left(2\delta\theta A_G(0; X) + \int_{|t| > \delta} u(e^{it}\lambda_\nu X) dt \right). \end{aligned}$$

Hence, using Fatou's lemma and the upper semicontinuity of u, we have

$$A_G(0; X) \le \limsup_{\nu \to +\infty} \frac{1}{2\pi} (2\delta\theta A_G(0; X) + (2\pi - 2\delta) u(0)) < A_G(0; X);$$

a contradiction.

(c) Fix sequences $z'_\nu, z''_\nu, w'_\nu, w''_\nu$ with the properties from (c). Take arbitrary $0 < \theta_2 < \theta_1 < \mathfrak{b}_G(a)$. Similar to (a) we find affine isomorphisms $\Phi_\nu : \mathbb{C}^n \longrightarrow \mathbb{C}^n$, $\nu \in \mathbb{N}$, such that $\Phi_\nu(w'_\nu) = z'_\nu$, $\Phi_\nu(w''_\nu) = z''_\nu$, $\varphi_\nu \longrightarrow \text{id}$, and $\Phi_\nu(G_{\theta_2}(w'_\nu)) \subset\subset G_{\theta_1}(z'_\nu)$, $\nu \gg 1$. By Remark 5.1.1(f), we get

$$\begin{aligned} (1/\theta_1) g_G(z'_\nu, z''_\nu) &= g_{G_{\theta_1}(z'_\nu)}(z'_\nu, z''_\nu) \le g_{\Phi_\nu(G_{\theta_2}(w'_\nu))}(\Phi_\nu(w'_\nu), \Phi_\nu(w''_\nu)) \\ &= g_{G_{\theta_2}(w'_\nu)}(w'_\nu, w''_\nu) = (1/\theta_2) g_G(w'_\nu, w''_\nu), \quad \nu \gg 1. \end{aligned}$$

Similarly, we get $(1/\theta_1)\boldsymbol{g}_G(w'_\nu, w''_\nu) \le (1/\theta_2)\boldsymbol{g}_G(z'_\nu, z''_\nu)$, $\nu \gg 1$, which finishes the proof.

(d) Take $z'_\nu, z''_\nu \longrightarrow a$, $z'_\nu \neq z''_\nu$, $\frac{z'_\nu - z''_\nu}{\|z'_\nu - z''_\nu\|} \to X$. Put $w'_\nu := a$, $w''_\nu := a - \|z'_\nu - z''_\nu\|X$. Then, by (b, c), we get

$$\begin{aligned}\lim_{\nu\to+\infty} \frac{\boldsymbol{g}_G(z'_\nu, z''_\nu)}{\|z'_\nu - z''_\nu\|} &= \lim_{\nu\to+\infty} \frac{\boldsymbol{g}_G(z'_\nu, z''_\nu)}{\boldsymbol{g}_G(w'_\nu, w''_\nu)} \frac{\boldsymbol{g}_G(w'_\nu, w''_\nu)}{\|z'_\nu - z''_\nu\|} \\ &= \lim_{\nu\to+\infty} \frac{\boldsymbol{g}_G(a, a - \|z'_\nu - z''_\nu\|X)}{\|z'_\nu - z''_\nu\|} \\ &= \boldsymbol{A}_G(a; -X) = \boldsymbol{A}_G(a; X).\end{aligned}$$

(e) Take $z'_\nu, z''_\nu \longrightarrow a$, $z'_\nu \neq z''_\nu$,

$$\lim_{\nu\to+\infty} \frac{\boldsymbol{g}_G(z'_\nu, z''_\nu)}{\boldsymbol{g}_G(z''_\nu, z'_\nu)} = \alpha \in [0, +\infty].$$

We may assume that $\frac{z'_\nu - z''_\nu}{\|z'_\nu - z''_\nu\|} \to X$. Then, by (d), we have

$$\alpha = \lim_{\nu\to+\infty} \frac{\boldsymbol{g}_G(z'_\nu, z''_\nu)}{\|z'_\nu - z''_\nu\|} \frac{\|z''_\nu - z'_\nu\|}{\boldsymbol{g}_G(z''_\nu, z'_\nu)} = \boldsymbol{A}_G(a; X)/\boldsymbol{A}_G(a; -X) = 1. \qquad \square$$

Remark 5.1.3. Notice that the proof of Theorem 5.1.2(b) shows that property (b) holds for a fixed $a = a_0 \in G$ if $\mathfrak{b}_G(a_0) > 0$ and $\boldsymbol{g}_G(a_0, \cdot)$ is continuous.

Example 5.1.4. There exists a bounded pseudoconvex domain $G \subset \mathbb{C}^2$, $0 \in G$, such that $\lim_{\lambda\to 0} \frac{1}{|\lambda|}\boldsymbol{g}_G(0, \lambda\boldsymbol{1})$ does not exist.

Let $G_\mathfrak{h}$ be a bounded balanced pseudoconvex domain such that $\mathfrak{h}(\boldsymbol{1}) = 1$ and there exist sequences $\mathbb{D} \ni a_k \longrightarrow 0$, $\mathbb{D} \ni b_k \longrightarrow 0$ such that $\mathfrak{h}(1, e^{a_k}) \longrightarrow \delta < 1$ and $\mathfrak{h}(1, e^{b_k}) \longrightarrow 1$ (such a function $\mathfrak{h}$ may be easily constructed using Example 3.1.13). Consider the biholomorphism

$$\mathbb{C}^2 \ni (z_1, z_2) \overset{\Phi}{\longmapsto} (z_1, z_2 e^{z_1}) \in \mathbb{C}^2$$

and let $G := \Phi^{-1}(G_\mathfrak{h})$. Then,

$$\frac{\boldsymbol{g}_G(0, (a_k, a_k))}{|a_k|} = \frac{\boldsymbol{g}_{G_\mathfrak{h}}(0, (a_k, a_k e^{a_k}))}{|a_k|} = \frac{\mathfrak{h}(a_k, a_k e^{a_k})}{|a_k|} = \mathfrak{h}(1, e^{a_k}) \longrightarrow \delta,$$

$$\frac{\boldsymbol{g}_G(0, (b_k, b_k))}{|b_k|} = \frac{\boldsymbol{g}_{G_\mathfrak{h}}(0, (b_k, b_k e^{b_k}))}{|b_k|} = \frac{\mathfrak{h}(b_k, b_k e^{b_k})}{|b_k|} = \mathfrak{h}(1, e^{b_k}) \longrightarrow 1.$$

In the case where $n = 1$, the situation is simpler. We have the following result:

Proposition 5.1.5 (cf. [567]). *Assume that $G \subset \mathbb{C}$. Then,*

(a) $\boldsymbol{A}_G$ *is continuous,*

(b) $\displaystyle \boldsymbol{A}_G(a;1) = \lim_{z',z''\to a,\, z'\neq z''} \frac{\boldsymbol{g}_G(z',z'')}{|z'-z''|},\ a \in G;$

in particular, $\displaystyle \boldsymbol{A}_G(a;1) = \lim_{\lambda\to 0} \frac{1}{|\lambda|}\boldsymbol{g}_G(a, a+\lambda),\ a \in G.$

Proof. If ∂G is polar, then $\boldsymbol{g}_G \equiv 0$ and $\boldsymbol{A}_G \equiv 0$ (cf. Proposition 4.2.10(d)), so the result in this case is trivial. Consequently, we may assume that ∂G is not polar. Observe that now $\boldsymbol{g}_G = e^{-\mathfrak{g}_G}$, where $\mathfrak{g}_G$ is the classical Green function of G (cf. Appendix B.5.7). Consequently, $\boldsymbol{g}_G$ is continuous and $\boldsymbol{g}_G(a,z) > 0$, $z \neq a$ (cf. Appendix B.5.5). Thus, by Theorem 5.1.2, we only need to check that $\mathfrak{b}_G(a) > 0$, $a \in G$. Take a point $\zeta \in \mathbb{C} \cap \partial G$. Let $r > 0$ be such that $(\partial G) \setminus \overline{\mathbb{D}}(\zeta, r)$ is not polar. Define $D := G \cup \mathbb{D}(\zeta, r)$. Then, D is a domain and ∂D is not polar. In particular, we have

$$\mathfrak{b}_G(a) = \liminf_{z\to\zeta} \boldsymbol{g}_G(a,z) \geq \liminf_{z\to\zeta} \boldsymbol{g}_D(a,z) = \boldsymbol{g}_D(a,\zeta) > 0, \quad a \in G.$$

In the case where $\zeta = \infty$, we fix a $b \in \mathbb{C} \cap \partial G$ and we use the transformation $\overline{\mathbb{C}} \ni z \longmapsto \frac{1}{z-b} \in \overline{\mathbb{C}}$. □

5.2 Lipschitz continuity of $\boldsymbol{\ell}^*$, $\boldsymbol{\varkappa}$, $\boldsymbol{g}$, and A

This section is mainly based on [386]. It discusses better regularity like the Lipschitz property of invariant functions.

Recall that $\boldsymbol{\ell}_G^*(0,z) = \boldsymbol{g}_G(0,z) = \mathfrak{h}(z)$, $z \in G$, and $A_G(0;X) = \mathfrak{h}(X)$, $X \in \mathbb{C}^n$, if $G = G_{\mathfrak{h}} \subset \mathbb{C}^n$ is a bounded pseudoconvex balanced domain with Minkowski function $\mathfrak{h}$ (see Proposition 4.2.10(b)). In particular, if $\mathfrak{h}_0(z) := |z_1|+|z_2|+\sqrt{|z_1z_2|}$, $z \in \mathbb{C}^2$, then $\boldsymbol{\ell}^*_{G_{\mathfrak{h}_0}}(0,\cdot) = \boldsymbol{g}_{G_{\mathfrak{h}_0}}(0,\cdot)$ is not Lipschitz continuous on $G_{\mathfrak{h}_0}$. Let $G = G_{\mathfrak{h}}$ be as above. Analyzing the Lipschitz continuity of the Minkowski function $\mathfrak{h}$ leads to the following equivalent statements:

(i) there exists a $c > 0$ such that $|\mathfrak{h}(z') - \mathfrak{h}(z'')| \leq c\|z'-z''\|$, $z', z'' \in G$;

(ii) there exists a $c' > 0$ such that $1 - \mathfrak{h}(z) \leq c' \operatorname{dist}(z, \partial G)$, $z \in G$.

The proof is left to the reader as an exercise. So, we see that the condition in (ii) is necessary. In fact the following result is true.

Theorem 5.2.1 (cf. [386]). *Let $D \subset \mathbb{C}^n$ be a $\boldsymbol{\varkappa}$-hyperbolic domain and $K \subset D$ a compact subset such that*

$$\sup_{z\in K,\, w\in D} \frac{1-\boldsymbol{\ell}_D^*(z,w)}{\operatorname{dist}(w,\partial D)} < +\infty. \tag{5.2.1}$$

Then the following estimates hold:

(a) $\boldsymbol{\ell}_D^*$ *is a Lipschitz function on* $K \times D$*;*

(b) *there is a* $c > 0$ *such that*

$$|\boldsymbol{\varkappa}_D(z'; X') - \boldsymbol{\varkappa}_D(z''; X'')| \leq c\left((\|X'\| + \|X''\|) \cdot \|z' - z''\| + \|X' - X''\|\right), \quad z', z'' \in K,\ X', X'' \in \mathbb{C}^n.$$

We point out that $\boldsymbol{\varkappa}$-hyperbolicity is equivalent to $\boldsymbol{k}$-hyperbolicity (see Theorem 13.2.2).

Remark 5.2.2.

(a) By symmetry it is clear that $\boldsymbol{\ell}_D^*$ is also a Lipschitz function on $D \times K$ (under the assumption (5.2.1)). On the other hand, it is an easy exercise to see that $\boldsymbol{\ell}_{\mathbb{D}}^*$ is not Lipschitz on $\mathbb{D} \times \mathbb{D}$.

(b) For the Carathéodory–Reiffen pseudometric the estimate in (b) is true for any domain D in $\mathbb{C}^n$ (see Theorem 2.7.1(c)).

(c) Later, we will see that any compact subset K of a strongly pseudoconvex domain satisfies the assumption of Theorem 5.2.1.

(d) A result similar to (b) can be found in [328].

We emphasize that for a certain class of domains, the assumption in Theorem 5.2.1 is also necessary for $\boldsymbol{\ell}_D^*$ to be a Lipschitz function (see the introduction above).

Corollary 5.2.3. *Let* $D \subset \mathbb{C}^n$ *be a taut domain and* $K \subset\subset D$*. Then* $\boldsymbol{\ell}_D^*$ *is Lipschitz on* $K \times D$ *if and only if equation* (5.2.1) *holds.*

Proof. By Theorem 5.2.1(a), it remains to show that (5.2.1) is a consequence of the Lipschitz property of $\boldsymbol{\ell}_D^*$. Suppose this equation does not hold. Then, there are sequences $(z_j)_j \subset K$ and $(w_j)_j \subset D$ such that $1 - \boldsymbol{\ell}_D^*(z_j, w_j) \geq j\operatorname{dist}(w_j, \partial D)$, $j \in \mathbb{N}$. Choose points $b_j \in \partial D$ with $\|b_j - w_j\| = \operatorname{dist}(w_j, \partial D)$, $j \in \mathbb{N}$, and sequences $(b_{j,k})_k \subset D$ with $b_{j,k} \underset{k\to\infty}{\longrightarrow} b_j$ for all j. Then,

$$\begin{aligned} j &\leq \frac{1 - \boldsymbol{\ell}_D^*(z_j, b_{j,k}) + \boldsymbol{\ell}_D^*(z_j, b_{j,k}) - \boldsymbol{\ell}_D^*(z_j, w_j)}{\|w_j - b_j\|} \\ &\leq \frac{1 - \boldsymbol{\ell}_D^*(z_j, b_{j,k})}{\|w_j - b_j\|} + c\frac{\|w_j - b_{j,k}\|}{\|w_j - b_j\|}, \quad j, k \in \mathbb{N}. \end{aligned}$$

Recall Proposition 3.2.1. Therefore, for each $j \in \mathbb{N}$ we can choose an index k_j sufficiently large, such that for these $k = k_j$ and all j the right hand side of the above inequality is bounded by $1 + 2c$; a contradiction. □

Now, we turn to the proof of Theorem 5.2.1.

Proof of Theorem 5.2.1. First, note that assumption (5.2.1) means that there is a $c > 0$ such that for any $r \in (0,1)$ and $\varphi \in \mathcal{O}(\mathbb{D}, D)$ with $\varphi(0) \in K$ one has

$$c \operatorname{dist}(\varphi(\mathbb{D}(r)), \partial D) \geq 1 - r. \tag{5.2.2}$$

(a) Since D is bounded, there exists a $c_1 > 1 > 1/2 > c_2$ such that

$$c_2\|z - w\| \leq \boldsymbol{\ell}_D^*(z, w) \leq c_1\|z - w\|, \quad z, w \in D.$$

(Use the holomorphic contractibility of $\boldsymbol{\ell}^*$.) Put $c_3 := c_1(1 + 2c/c_2)$.

Fix points $z', z'' \in K$ and $w', w'' \in D$. Then, it suffices to prove the following two inequalities:

$$|\boldsymbol{\ell}_D^*(z', w') - \boldsymbol{\ell}_D^*(z', w'')| \leq c_3\|w' - w''\|, \tag{5.2.3}$$

$$|\boldsymbol{\ell}_D^*(z', w'') - \boldsymbol{\ell}_D^*(z'', w'')| \leq 2c_3\|z' - z''\|. \tag{5.2.4}$$

To prove (5.2.3) we may assume that $0 \neq \alpha := \boldsymbol{\ell}_D^*(z', w') \leq \boldsymbol{\ell}_D^*(z', w'')$. Put $r := 1 - c\|w' - w''\|/\alpha$. Now, we discuss the following three cases.

Case 1°. $r > \max\{\alpha, c_2\}$: Then, for any $\alpha' \in (\alpha, r)$ there exists a $\varphi \in \mathcal{O}(\mathbb{D}, D)$ with $\varphi(0) = z'$ and $\varphi(\alpha') = w'$. Put $\varphi(\lambda) := \varphi(r\lambda) + \frac{r\lambda}{\alpha'}(w'' - w')$, $\lambda \in \mathbb{D}$. Applying (5.2.2) it follows that $\psi \in \mathcal{O}(\mathbb{D}, D)$ with $\psi(0) = z'$ and $\psi(\alpha'/r) = w''$. Therefore, $\boldsymbol{\ell}_D^*(z', w'') \leq \alpha/r$. Hence,

$$\boldsymbol{\ell}_D^*(z', w'') - \boldsymbol{\ell}_D^*(z', w') \leq \boldsymbol{\ell}_D^*(z', w')\frac{1-r}{r} = \frac{c}{r}\|w' - w''\| \leq \frac{c}{c_2}\|w' - w''\|.$$

Case 2°. $\alpha \geq \max\{r, c_2\}$: Then,

$$\boldsymbol{\ell}_D^*(z', w'') - \boldsymbol{\ell}_D^*(z', w') \leq 1 - \alpha \leq 1 - r \leq \frac{c}{c_2}\|w' - w''\|.$$

Case 3°. $c_2 \geq \max\{r, \alpha\}$: Then, we have

$$\|w' - w''\| = \frac{\alpha}{c}(1 - r) \geq \frac{\alpha}{c}(1 - c_2) \geq \frac{c_2}{c}(1 - c_2)\|z' - w'\|.$$

Using the triangle inequality, we get

$$\|z' - w''\| \leq \left(\frac{c}{c_2(1 - c_2)} + 1\right)\|w' - w''\| \leq \left(1 - \frac{2c}{c_2}\right)\|w' - w''\|.$$

Hence,

$$\ell^*_D(z', w'') - \ell^*_D(z', w') \le \ell^*_D(z', w'') \le c_1\|z' - w''\| \le c_1\left(1 + \frac{2c}{c_2}\right)\|w' - w''\|,$$

which gives (5.2.3).

It remains to prove (5.2.4). We may assume that

$$0 < \beta := \ell^*_D(z'', w'') \le \ell^*_D(z', w'').$$

Put $s := 1 - 2c\|w' - w''\|/\beta$. Then, the proof follows the steps from the one before. Therefore, it is left to the reader as an exercise. This completes the proof of part (a).

(b) Put

$$c_4 := \sup_{u\in K,\, U\in\partial\mathbb{B}_n} \varkappa_D(u; U), \quad c_5 := \inf_{u\in K,\, U\in\partial\mathbb{B}_n} \varkappa_D(u; U), \quad c_6 := c_4(1 + 2c/c_5).$$

Since D is assumed to be $\varkappa$-hyperbolic, we have $c_5 > 0$.

Now fix $z', z'' \in K$ and $X', X'' \in \mathbb{C}^n$. Using the triangle inequality, it is enough to verify the following two estimates:

$$|\varkappa_D(z'; X'') - \varkappa_D(z''; X'')| \le 4cc_4\|X''\| \cdot \|z' - z''\|, \tag{5.2.5}$$

$$|\varkappa_D(z'; X') - \varkappa_D(z'; X'')| \le c_6\|X' - X''\|. \tag{5.2.6}$$

To prove (5.2.5), observe that $|\varkappa_D(z'; X'') - \varkappa_D(z''; X'')| \le 2c_4\|X''\|$ which immediately implies (b), if $p := 1 - c\|z' - z''\| \le 1/2$. Now assume that $p > 1/2$, and also that $\varkappa_D(z'; X'') \le \varkappa_D(z''; X'')$. Let $\varphi \in \mathcal{O}(\mathbb{D}, D)$ with $\varphi(0) = z'$ and $s\varphi'(0) = X''$. Put $\psi(\lambda) := \varphi(p\lambda) + z'' - z'$. Then, $\psi(0) = z''$ and $(s/p)\psi'(0) = X''$. Moreover, using (5.2.2), we get $\psi \in \mathcal{O}(\mathbb{D}, D)$, which implies that $p\varkappa_D(z''; X'') \le \varkappa_D(z'; X'')$. Hence,

$$\varkappa_D(z''; X') - \varkappa_D(z'; X') \le (1/p - 1)c_4\|X''\| \le 2cc_4\|X''\| \cdot \|z' - z''\|.$$

To verify property (5.2.6), we may assume that $X' \ne 0$ and

$$0 < s := \varkappa_D(z'; X') \le \varkappa_D(z'; X'')$$

(here, we have used the $\varkappa$-hyperbolicity of D). Put $q := 1 - c\|X' - X''\|/s$.

In a first step, we assume that $q > 1/2$. Let $\varphi \in \mathcal{O}(\mathbb{D}, D)$ with $\varphi(0) = z'$ and $s'\varphi'(0) = X$ for some positive s'. Put $\psi(\lambda) := \varphi(q\lambda) + q\lambda(X'' - X')/s'$. Then (as above), $\psi \in \mathcal{O}(\mathbb{D}, D)$ with $\psi(0) = z'$ and $s'\psi'(0) = qX''$. Therefore, $\varkappa_D(z'; X'') \le s/q$. Hence,

$$\varkappa_D(z'; X'') - \varkappa_D(z'; X') \le s(1 - q)/q = c\|X' - X''\|/q \le c_6\|X' - X''\|.$$

Now let $q \le 1/2$. Then, $\|X' - X''\| = s(1 - q)/c \ge c_5\|X'\|/(2c)$, and, by the triangle inequality, $\|X''\| \le (1 + \frac{2c}{c_5})\|X' - X''\|$. Therefore, it follows that

$$\varkappa_D(z'; X'') - \varkappa_D(z'; X') \le \varkappa_D(z'; X'') \le c_4\|X''\| \le c_6\|X' - X''\|.$$

This completes the proof of part (b). □

Now, we turn to discuss the Lipschitz property for the Buseman pseudometric.

Theorem 5.2.4. *Let $D \subset \mathbb{C}^n$ and K be as in Proposition* 5.2.1. *Then, there exists a $C > 0$ such that for every $z', z'' \in K$ and $X', X'' \in \mathbb{C}^n$ the following inequality holds:*

$$|\hat{\varkappa}_D(z'; X') - \hat{\varkappa}_D(z''; X'')| \le C\left((\|X'\| + \|X''\|)\|z' - z''\| + \|X' - X''\|\right).$$

Proof. In virtue of Proposition 5.2.1, there exists a constant $d_1 > 0$ such that

$$|\varkappa_D(w'; X) - \varkappa_D(w''; X)| \le d_1\|X\| \cdot \|w' - w''\|, \quad w', w'' \in K,\ X \in \mathbb{C}^n.$$

Now fix points $z', z'' \in K$ and vectors $X', X'' \in \mathbb{C}^n$. Since $\hat{\varkappa}_D(z; \cdot)$ is a norm, $z \in D$, and D is bounded, it suffices to find a $d_2 > 0$ such that

$$|\hat{\varkappa}_D(z'; X) - \hat{\varkappa}_D(z''; X)| \le d_2\|X\| \cdot \|z' - z''\|, \quad X \in \mathbb{C}^n.$$

To verify this inequality, we may assume that $X \neq 0$, $z' \neq z''$, and $\hat{\varkappa}_D(z'; X) \le \hat{\varkappa}_D(z''; X)$. Then, by Lemma 2.2.5(b), there are $(2n-1)$ vectors $X_1, \dots, X_{2n-1}$ with $X = \sum_{j=1}^{2n-1} X_j$ such that

$$\sum_{j=1}^{2n-1} \varkappa_D(z'; X_j) \le \hat{\varkappa}_D(z'; X) + \|X\| \cdot \|z' - z''\|.$$

Therefore,

$$\begin{aligned}
0 &\le \hat{\varkappa}_D(z''; X) - \hat{\varkappa}_D(z'; X) \\
&\le \left(\sum_{j=1}^{2n-1} \varkappa_D(z''; X_j) - \varkappa_D(z'; X_j)\right) + \|X\| \cdot \|z' - z''\| \\
&\le \|z' - z''\|\left(d_1 \sum_{j=1}^{2n-1} \|X_j\| + \|X\|\right).
\end{aligned}$$

Note that $D \subset \mathbb{B}(z', R)$, where the R can be chosen independently of z'. Then,

$$\begin{aligned}
\sum_{j=1}^{2n-1} \|X_j\| &\le R \sum_{j=1}^{2n-1} \varkappa_{\mathbb{B}(z',R)}(z'; X_j) \le R \sum_{j=1}^{2n-1} \varkappa_D(z'; X_j) \\
&\le R\left(\varkappa_D(z'; X) + \|X\| \cdot \|z' - z''\|\right) \le d_3\|X\|,
\end{aligned}$$

where $d_3 = R\left(\operatorname{diam}(D) + \sup_{u \in K,\, U \in \partial\mathbb{B}_n} \varkappa_D(u; U)\right)$ is independent of the point in K and the vector. □

Finally, we start to discuss the Lipschitz property for the Green function and the Azukawa pseudometric. Coming back to the introduction of this section, the second condition (ii) there may be read as

$$\log g_G(0,\cdot) \in \mathcal{PSH}(G),$$
$$\log g_G(0,z) \geq -d \operatorname{dist}(z,\partial G), \quad z \in G, \qquad c' \operatorname{dist}(z,\partial G) < 1;$$

here, d is a suitable chosen positive number, i.e., there exists a negative psh function on G with a "linear" lower bound near ∂G. In general, the following is true (see [386]):

Proposition 5.2.5 (cf. [386]). *Let $D \subset \mathbb{C}^n$ be a bounded domain. Then, the following conditions are equivalent:*

(i) *there exists a $u \in \mathcal{PSH}(D)$ with $u < 0$ on D and $\inf_{z\in D} u(z)/\operatorname{dist}(z,\partial D) > -\infty$;*

(ii) *D is bounded hyperconvex and there are a point $z_0 \in D$ and a number $C > 0$ such that*

$$|\log g_D(z_0,w_1) - \log g_D(z_0,w_2)| \leq C \frac{\|w_1 - w_2\|}{\min\{\|z_0 - w_1\|, \|z_0 - w_2\|\}},$$
$$w_1, w_2 \in D \setminus \{z_0\};$$

(iii) *D is hyperconvex and for any compact set $K \subset D$ there exists a $C > 0$ such that*

$$|\log g_D(z,w_1) - \log g_D(z,w_2)| \leq C \frac{\|w_1 - w_2\|}{\min\{\|z - w_1\|, \|z - w_2\|\}},$$
$$z \in K, w_1, w_2 \in D \setminus \{z\}.$$

Assuming Proposition 5.2.5, the following result follows:

Corollary 5.2.6. *Let D and u be as in Proposition 5.2.5(i) and $K \subset\subset D$. Then, there exists a $C > 0$ such that*

$$|g_D(z,w_1) - g_D(z,w_2)| \leq C\|w_1 - w_2\|, \quad z \in K,\ w_1, w_2 \in D.$$

Proof. Recall that there is a $c > 0$ such that $g_D(z,w) \leq c\|z - w\|$, $z \in K$, $w \in D$. Therefore, we may assume that $w_1 \neq z \neq w_2$. Moreover, without loss of generality, let $\|z - w_2\| \leq \|z - w_1\|$. Two cases are possible:

Case 1°. $|\log g_D(z, w_1) - \log g_D(z, w_2)| < 1$. Then,

$$\begin{aligned}|g_D(z, w_1) - g_D(z, w_2)| &= g_D(z, w_2)|\exp(\log g_D(z, w_1) - \log g_D(z, w_2)) - 1| \\ &< ec\|z - w_2\| \cdot |\log g_D(z, w_1) - \log g_D(z, w_2)| \\ &\le ecC\|w_1 - w_2\|,\end{aligned}$$

where C is the constant from Proposition 5.2.5(ii).

Case 2°. $|\log g_D(z, w_1) - \log g_D(z, w_2)| \ge 1$. Then, by Proposition 5.2.5,

$$C\|w_1 - w_2\| \ge \|z - w_2\| \ge \|z - w_1\| - \|w_1 - w_2\|,$$

and therefore, $(C + 1)\|w_1 - w_2\| \ge \|z - w_1\|$. Hence, it follows that

$$\begin{aligned}|g_D(z, w_1) - g_D(z, w_2)| &< \max\{g_D(z, w_1), g_D(z, w_2)\} \\ &\le c\|z - w_1\| \le c(C + 1)\|w_1 - w_2\|. \qquad \square\end{aligned}$$

Remark 5.2.7. Assume that D is a bounded domain with a smooth boundary. Then, the Hopf Lemma shows that the u in Proposition 5.2.5 satisfies also a "linear" upper estimate, namely $u(z) \le -c \operatorname{dist}(z, \partial D)$.

Proof of Proposition 5.2.5. (ii) $\Longrightarrow$ (i): Put $u := g_D(z_0, \cdot) - 1$. Since D is assumed to be hyperconvex, one concludes $\lim_{w \to \partial D} u(w) = 0$. Assume that the estimate for u is not correct. Then, there is a sequence $(z_j)_j \subset D$ such that $-1 \le u(z_j) \le -j \operatorname{dist}(z_j, \partial D)$, $j \in \mathbb{N}$; in particular, we may assume that $z_j \underset{j\to\infty}{\longrightarrow} z' \in \partial D$. Choose points $w_j \in \partial D$ with $\operatorname{dist}(z_j, \partial D) = \|z_j - w_j\|$ and sequences $(w_{j,k})_k \subset D$ with $w_{j,k} \underset{j\to\infty}{\longrightarrow} w_j$, $j \in \mathbb{N}$. Then, there is a constant $c > 0$ such that

$$\begin{aligned}j &\le \frac{1 - g_D(z_0, w_{j,k}) + g_D(z_0, w_{j,k}) - g_D(z_0, z_j)}{\|z_j - w_j\|} \\ &\le \frac{1 - g_D(z_0, w_{j,k})}{\|z_j - w_j\|} + C\frac{\|w_{j,k} - z_j\|}{\min\{\|z_0 - w_{j,k}\|, \|z_0 - z_j\|\}} \le 1 + Cc, \quad j \gg 1,\end{aligned}$$

if $k = k_j$ is sufficiently large for every $j \in \mathbb{N}$; note that such k_j always exists since D is assumed to be hyperconvex (see Proposition 4.2.10(i)).

Since (iii) $\Longrightarrow$ (ii) is trivial, it remains to prove (i) $\Longrightarrow$ (iii): Observe that, since D is bounded, u is a psh exhaustion function; hence, D is bounded hyperconvex.

Fix a compact set $K \subset D$. Let $r \in (0, 1)$ be such that $\bigcup_{a\in K} \mathbb{B}(a, r) \subset\subset D$ and let $R > r$ be such that $D \subset \bigcap_{a\in K} \mathbb{B}(a, R)$. Then choose a domain $D_1 \subset\subset D$ containing K. Put $\delta := 2\sup_{z\in D_1} u(z) < 0$. Then, by the assumption on u, there exists an open set $D_2 \subset\subset D$ with $D_1 \cup \bigcup_{a\in K} \overline{\mathbb{B}}(a, r) \subset\subset D_2$ such that $\inf_{z\in D\setminus D_2} u(z) \ge \delta$. Then, $\inf_{w\in D\setminus D_2} \operatorname{dist}(w, \partial D) \ge s > 0$ for a certain s.

For an $a \in K$ set $\varphi(a,\cdot) := \log(\|\cdot - a\|/\operatorname{diam} D_2)$, $m := \inf_{K\times D_1}\varphi$, and

$$v_a := \begin{cases} \varphi(a,\cdot)+m & \text{on } D_1 \\ \max\{\varphi(a,\cdot)+m, mu/\delta\} & \text{on } D_2\setminus D_1 \\ mu/\delta & \text{on } D\setminus D_2 \end{cases}.$$

Then, $v_a \in \mathcal{PSH}(D)$. Therefore, $\log \boldsymbol{g}_D(a,\cdot) \ge v_a$ and so

$$\log \boldsymbol{g}_D(a,w) \ge mu/\delta \ge -mM/\delta \operatorname{dist}(w,\partial D), \quad w \in D\setminus D_2,$$

where $M := \inf_D(u/\operatorname{dist}(\cdot,\partial D)) > -\infty$. Moreover, if $w \in D\setminus\overline{\mathbb{B}}(a,r)$, then

$$\log \boldsymbol{g}_D(a,w) \ge \log \boldsymbol{g}_{\mathbb{B}(a,R)}(a,w) = \log(\|w-a\|/R) \ge \log(r/(2R)), \quad w \in D\setminus\overline{\mathbb{B}}(a,r/2).$$

Hence, we get for all $a \in K$

$$\frac{\log \boldsymbol{g}_D(a,z)}{\operatorname{dist}(z,\partial D)} > \min\left\{\frac{-mM}{\delta}, \frac{\log(r/(2R))}{s}\right\} =: -c_4 > -\infty, \quad w \in D\setminus\overline{\mathbb{B}}(a,r/2). \tag{5.2.7}$$

To continue, let us first recall the following functions and their properties (see Proposition 4.2.13):

$$\boldsymbol{g}_D^{\varepsilon}(z,w) := \sup\{v(w) : 0 \le v < 1,\ \log v \in \mathcal{PSH}(D),\ v|_{\mathbb{B}(z,\varepsilon)} \le \varepsilon/r\},$$

$z \in K$, $\varepsilon \in (0,r)$, and $w \in D$.

Our main goal (*) is to find positive numbers c_1, c_2 such that if $z \in K$, $w_1, w_2 \in D\setminus\{z\}$, and ε are satisfying

$$\max\{\varepsilon, c_1\|w_1-w_2\|\} < \min\{r/2, \|z-w_1\|, \|z-w_2\|\}, \tag{5.2.8}$$

then

$$|\log \boldsymbol{g}_D^{\varepsilon}(z,w_1) - \log \boldsymbol{g}_D^{\varepsilon}(z,w_2| \le c_2\frac{\|w_1-w_2\|}{\min\{\|z-w_1\|, \|z-w_2\|\}}. \tag{5.2.9}$$

Let us assume for a moment that equation (5.2.9) has been proved under the assumptions of (5.2.8). Fix an $\varepsilon < r$. Take points $z \in K$ and $w_1, w_2 \in D\setminus\{z\}$. One may assume that $\boldsymbol{g}_D^{\varepsilon}(z,w_1) \le \boldsymbol{g}_D^{\varepsilon}(z,w_2)$. There is a semicircle γ with diameter $[w_1,w_2]$, such that $\gamma(0) = w_1$, $\gamma(\pi) = w_2$, and $\operatorname{dist}(z,\gamma([0,\pi])) = \min\{\|z-w_1\|, \|z-w_2\|\}$. Put $t' := \sup\{t \in (0,\pi] : \gamma(\tau) \in D \text{ for all } \tau \in [0,t)\}$. If $t' = \pi$, then “integration” along γ leads to

$$\log \boldsymbol{g}_D^{\varepsilon}(z,w_2) - \log \boldsymbol{g}_D^{\varepsilon}(z,w_2) \le \pi c_2\frac{\|w_1-w_2\|}{\min\{\|z-w_1\|, \|z-w_2\|\}}. \tag{5.2.10}$$

Now, we deal with the case when $t' < \pi$. Then, obviously $\gamma(t') \in \partial D$. Since

$$\lim_{w \to \partial D} \log \boldsymbol{g}_D^\varepsilon(z, w) = 0 > \log \boldsymbol{g}_D^\varepsilon(z, w_2) \geq \log \boldsymbol{g}_D^\varepsilon(z, w_1)$$

and $\boldsymbol{g}_D^\varepsilon$ is continuous, we find a $t^* \in [0, t')$ with $\boldsymbol{g}_D^\varepsilon(z, \gamma(t^*)) = \boldsymbol{g}_D^\varepsilon(z, w_1)$. Repeating the "integration" argument from above together with equation (5.2.9), we again get the estimate (5.2.10). So, it remains to let ε run to 0 to receive (iii) with $C = \pi c_2$.

Finally, we will prove equation (5.2.9) under the assumption of (5.2.8). Let z, w_1, w_2 and ε be as in (*). We may assume that $\boldsymbol{g}_D^\varepsilon(z, w_1) < \boldsymbol{g}_D^\varepsilon(z, w_2)$. Take an $f \in \mathcal{O}(D, \mathbb{D})$ that is an extremal for the Carathéodory distance $\boldsymbol{c}_D^*(z, w_1)$ with $f(z) = 0$. Put $h := f/f(w_1)$ on D. Then,

$$|h(\zeta)| \leq \frac{\boldsymbol{c}_D^*(z, \zeta)}{\boldsymbol{c}_D^*(z, w_1)} \leq \frac{R\|\zeta - z\|}{r\|w_1 - z\|} \leq \frac{R}{\|w_1 - z\|}, \quad \zeta \in \mathbb{B}(z, r); \tag{5.2.11}$$

use the fact that $\mathbb{B}(z, r) \subset D \subset \mathbb{B}(z, R)$. Set $c_1 := R/r$. Put

$$D' := \{\zeta \in D : \zeta + h(\zeta)(w_2 - w_1) \in D\}, \text{ and } D'' := D' \setminus \overline{\mathbb{B}}(z, \varepsilon).$$

Note that if $\zeta \in \overline{\mathbb{B}}(z, r/2)$, then $|h(\zeta)|\|w_2 - w_1\| \leq c_1 \frac{\|w_1 - w_2\|}{\|w_1 - z\|} \frac{r}{2} < \frac{r}{2}$. Hence, $\overline{\mathbb{B}}(z, \varepsilon) \subset \overline{\mathbb{B}}(z, r/2) \subset D'$. Note that $w_1 \in D''$. Finally, define

$$\hat{g}(\zeta) := \log \boldsymbol{g}_D^\varepsilon(z, \zeta + (w_2 - w_1)h(\zeta)), \quad \zeta \in D'.$$

By equation (5.2.7), we know that

$$\log \boldsymbol{g}_D(z, \zeta) \geq -c_4 \operatorname{dist}(\zeta, \partial D), \quad \zeta \in D \setminus \mathbb{B}(z, r/2).$$

Thus, (5.2.7) and (5.2.11) imply that for $\zeta \in D'$ we have

$$\log \boldsymbol{g}_D^\varepsilon(z, \zeta) \geq \log \boldsymbol{g}_D(z, \zeta) \geq -c_4 \operatorname{dist}(\zeta, \partial D) \geq -Rc_4 \frac{\|w_2 - w_1\|}{\|w_1 - z\|}.$$

Put $v(\zeta) := \hat{g}(\zeta) - G_D^\varepsilon(z, \zeta)$, $\zeta \in D'$. Then,

$$\limsup_{D' \ni \zeta \to \partial D'} v(\zeta) \leq Rc_4 \frac{\|w_2 - w_1\|}{\|w_1 - z\|}.$$

On the other hand, for $\zeta \in \partial\mathbb{B}(z, \varepsilon)$, it follows from Proposition 4.2.13(b) and (5.2.11) that

$$\begin{aligned}
v(\zeta) &\leq \log \frac{\max\{\varepsilon, \|\zeta + (w_2 - w_1)h(\zeta) - z\|}{r} - \log \frac{\varepsilon}{r} \\
&= \log^+ \frac{\|\zeta + (w_2 - w_1)h(\zeta) - z\|}{\varepsilon} \leq \log\left(1 + c_1 \frac{\|w_2 - w_1\|}{\|w_1 - z\|}\right).
\end{aligned}$$

Since $\log \boldsymbol{g}_D^{\varepsilon}(z,\cdot)$ is a maximal psh function on D'', the domination principle implies that

$$v(\zeta) \le c_2 \frac{\|w_2 - w_1\|}{\|w_1 - z\|}, \qquad \zeta \in D'',$$

where $c_2 := \max\{c_1, Rc_4\}$. Applying this information for $\zeta = w_1$ gives the desired equation (5.2.9). □

Starting with Corollary 5.2.6, we get under the same assumptions even more, namely that $\boldsymbol{g}_D$ is Lipschitz in both variables.

Theorem 5.2.8. *Let D and u be as in Proposition* 5.2.5 *and let K be a compact subset of D. Then, there exists a positive constant c such that*

$$|\boldsymbol{g}_D(z_1, w_1) - \boldsymbol{g}_D(z_2, w_2)| \le C\,(\|z_1 - z_2\| + \|w_1 - w_2\|),$$
$$z_1, z_2 \in K,\ w_1, w_2 \in D.$$

Proof. First, we enlarge K, or more precisely, we may assume that $K \subset D_1 \subset\subset D$ with a suitable domain D_1 that has a $\mathcal{C}^\infty$-smooth boundary. Then, one may find a positive number c_2 such that all pairs of points $a_1, a_2 \in D_1$ can be connected by a curve γ inside of D_1 with $L(\gamma) \le c_2\|a_1 - a_2\|$. Put $K_1 = \overline{D}_1$.

Note that D is bounded. Therefore, $\mathfrak{b}_D(a) > 0$ for all $a \in D$. Put

$$D_\theta(a) = \{w \in D : \boldsymbol{g}_D(a, w) < \theta\}, \quad 0 < \theta < 1.$$

Recall from the proof of Proposition 5.2.5, (i) $\Longrightarrow$ (iii), equation (5.2.7), that there are positive numbers r, M such that, if $w \in D$ with $\operatorname{dist}(w, \partial D) < r$ and $z \in K_1$, then $\log \boldsymbol{g}_D(z, w) \ge -M \operatorname{dist}(w, \partial D)$. Using this fact, one finds a constant $c > 0$ such that

$$\operatorname{dist}(D_\theta(a), \partial D) \ge c(1 - \theta), \quad a \in K_1,\ \theta \in (0,1).$$

In virtue of Corollary 5.2.6 it is enough to find a $c_1 > 0$ such that

$$|\boldsymbol{g}_D(z_1, w) - \boldsymbol{g}_D(z_2, w)| \le c_1\|z_1 - z_2\|, \quad z_1, z_2 \in K,\ w \in D.$$

Fix $z_1, z_2 \in K$ and $w \in D$, and connect the first two points via a curve γ inside of D_1, such that $L(\gamma) \le c_1\|z_1 - z_2\|$. Put $\varepsilon(w) := \sup_{z \in \overline{D}_1} \boldsymbol{g}_D(z, w)$. Then, $\varepsilon(w) > 0$. Now, we take a partition $t_0 = 0, t_1, \dots, t_N = 1$ of $[0, 1)$ such that

$$\theta_j := 1 - \|a_j - a_{j+1}\|/c > \varepsilon(w), \quad j = 0, \dots, N-1,$$

where $a_j := \gamma(t_j)$, $j = 0, \dots, N$. Note that $w \in D_{\theta_j}(a_j)$ and $D \subset\subset \widetilde{D}_j := D_{\theta_j}(a_j) + a_{j+1} - a_j$. So, one gets

$$\begin{aligned}\boldsymbol{g}_D(a_j, w) &= \theta_j \boldsymbol{g}_{D_{\theta_j}(a_j)}(a_j, w) = \theta_j \boldsymbol{g}_{\widetilde{D}}(a_{j+1}, w + a_{j+1} - a_j)\\ &\ge \theta_j \boldsymbol{g}_D(a_{j+1}, w + a_{j+1} - a_j), \quad j = 0, \dots, N-1,\end{aligned}$$

where the first equality follows from Remark 5.1.1(f). Therefore,

$$\begin{aligned} \boldsymbol{g}_D(a_{j+1}, w) - \boldsymbol{g}_D(a_j, w) &\le \boldsymbol{g}_D(a_{j+1}, w) - \theta_j \boldsymbol{g}_D(a_{j+1}, w + a_{j+1} - a_j) \\ &\le C\|a_j - a_{j+1}\| + (1 - \theta_j) \\ &= (C + 1/c)\|a_{j+1} - a_j\|, \quad j = 0, \dots, N-1, \end{aligned}$$

where C may denote the constant from Corollary 5.2.6. By symmetry, one also has $\boldsymbol{g}_D(a_j, w) - \boldsymbol{g}_D(a_{j+1}, w) \le (C + 1/c)\|a_{j+1} - a_j\|$, $j = 0, \dots, N-1$.

Finally, using the triangle inequality, we get

$$\begin{aligned} |\boldsymbol{g}_D(z_1, w) - \boldsymbol{g}_D(z_2, w)| &\le \sum_{j=0}^{N-1} |\boldsymbol{g}_D(a_{j+1}, w) - \boldsymbol{g}_D(a_j, w)| \\ &\le (C + 1/c) \sum_{j=0}^{N-1} \|a_{j+1} - a_j\| \\ &\le (C + 1/c) L(\gamma) \le c_2 (C + 1/c)\|z_1 - z_2\|, \end{aligned}$$

which finishes the proof. □

It is not clear whether $\boldsymbol{g}_D$, under the assumptions of Theorem 5.2.8, is also a Lipschitz function on $D \times K$. ?

Moreover, there is the following result on the Lipschitz property of the Azukawa pseudometric.

Theorem 5.2.9. *Let D and u be as in Proposition 5.2.5. Moreover, let $K \subset D$ be compact. Then, there is a $C > 0$ such that*

$$|\boldsymbol{A}_D(z_1; X) - \boldsymbol{A}_D(z_2; Y)| \le C\left((\|X\| + \|Y\|) \cdot \|z_1 - z_2\| + \|X - Y\|\right), \quad z_1, z_2 \in K, \; X, Y \in \mathbb{C}^n.$$

Proof. To prove this theorem, it suffices to verify the following statements:

(a) there is a $c_3 > 0$ such that

$$|\boldsymbol{A}_D(z; X) - \boldsymbol{A}_D(z; Y)| \le c_1\|X - Y\|, \quad z \in K, \; X, Y \in \mathbb{C}^n;$$

(b) if $c_4 := \sup_{z \in \overline{D}_1, \|X\|=1} \boldsymbol{A}_D(z; X)$, then

$$|\boldsymbol{A}_D(z_1; X) - \boldsymbol{A}_D(z_2; X)| \le c_4\|X\| \cdot \|z_1 - z_2\|,$$

where D_1 is chosen as in the proof of the former theorem.

To verify (a) recall (see Corollary 5.2.6) that there is $c_3 > 0$ such that

$$\left| \boldsymbol{g}_D(z, t + \lambda X)/|\lambda| - \boldsymbol{g}_D(z, z + \lambda Y)/|\lambda| \right| \le c_3 \|X - Y\|,$$

whenever $z \in K$, $X, Y \in \mathbb{C}^n$, and $\lambda \in \mathbb{C}_*$ sufficiently small. If $\lambda \longrightarrow 0$, then Theorem 5.1.2(b) gives (a).

To show (b), one may assume that $\boldsymbol{A}_D(z_1; X) \le \boldsymbol{A}_D(z_1; X)$. Now, we copy the proof of Theorem 5.2.8, i.e., take γ, c, a_j, θ_j, and $\widetilde{D}_j$ as there. Then,

$$\boldsymbol{A}_D(a_j; X) = \theta_j \boldsymbol{A}_{D_{\theta_j}(a_j)}(a_j; X) = \theta_j \boldsymbol{A}_{\widetilde{D}_j}(a_{j+1}; X) \ge \theta_j \boldsymbol{A}_D(a_{j+1}; X),$$

where the first equality follows from Remark 5.1.1(g). Then,

$$\begin{aligned} \boldsymbol{A}_D(z_2; X) - \boldsymbol{A}_D(z_1; X) &\le \sum_{j=0}^{N-1} |\boldsymbol{A}_D(a_{j+1}; X) - \boldsymbol{A}_D(a_j; X)| \\ &\le \sum_{j=0}^{N-1} (1 - \theta_j) \boldsymbol{A}_D(a_{j+1}; X) \\ &\le c_4 \|X\| L(\gamma)/c \le c_4 \|z_1 - z_2\| \cdot \|X\|, \end{aligned}$$

which finishes the proof of (b). □

5.3 Derivatives

We need the following auxiliary notion: we say that a domain $G \subset \mathbb{C}^n$ is ***k**-hyperbolic* at a point $a \in G$ if $\boldsymbol{k}_G(a, z) > 0$ for all $z \in G \setminus \{a\}$ (cf. § 13.2).

Proposition 5.3.1 (cf. Theorem 13.2.2). *Let $G \subset \mathbb{C}^n$ be a domain and let $a \in G$. Then, the following conditions are equivalent:*

(i) *G is $\boldsymbol{k}$-hyperbolic at a;*

(ii) *for each ball $\mathbb{B}(a, r)$ there exists a $\delta > 0$ such that $B_{\boldsymbol{k}_G}(a, \delta) \subset \mathbb{B}(a, r)$; equivalently, for every sequence $(z_s)_{s=1}^{\infty} \subset G$, if $\boldsymbol{k}_G(a, z_s) \longrightarrow 0$, then $z_s \longrightarrow a$ in top G;*

(iii) *for each ball $\mathbb{B}(a, r) \subset G$ there exists a $\delta \in (0, 1)$ such that, for any $m \in \mathbb{N}$ and for any $F \in \mathcal{O}(\mathbb{B}_m, G)$, if $F(0) \in \mathbb{B}(a, \delta)$, then $F(\mathbb{B}_m(\delta)) \subset \mathbb{B}(a, r)$;*

(iv) *for each ball $\mathbb{B}(a, r) \subset G$ there exists a $\delta \in (0, 1)$ such that, for any $\varphi \in \mathcal{O}(\mathbb{D}, G)$, if $\varphi(0) \in \mathbb{B}(a, \delta)$, then $\varphi(\mathbb{D}(\delta)) \subset \mathbb{B}(a, r)$;*

(v) *there exist $r > 0$ and $M > 0$ such that $\boldsymbol{\varkappa}_G(z; X) \ge M \|X\|$, $z \in \mathbb{B}(a, r) \subset G$, $X \in \mathbb{C}^n$;*

(vi) *there exist* $r > 0$, $R > 0$ *such that* $B_{\boldsymbol{k}_G}(a, r) \subset \mathbb{B}(R)$;

(vii) *there exist* $r > 0$ *and* $M > 0$ *such that* $\boldsymbol{k}_G(z', z'') \geq M\|z' - z''\|$, $z', z'' \in \mathbb{B}(a, r) \subset G$;

(viii) *for each ball* $B(a, \tau) \subset G$ *there exist* $\varrho \in (0, \tau)$ *and* $\varepsilon > 0$ *such that* $\boldsymbol{\ell}_G(z', z'') \geq \varepsilon$ *for* $z' \in \mathbb{B}(a, \varrho)$ *and* $z'' \in G \setminus \mathbb{B}(a, \tau)$.

Proof. (i) $\Longrightarrow$ (ii) (cf. Remark 2.6.7): We may assume that $\mathbb{B}(a, r) \subset\subset G$. Put $\delta := \min\{\boldsymbol{k}_G(a, z) : z \in \partial\mathbb{B}(a, r)\}$. Condition (i) implies that $\delta > 0$. Obviously, $B_{\boldsymbol{k}_G}(a, \delta) \cap \partial\mathbb{B}(a, r) = \varnothing$. Since $\boldsymbol{k}_G$ is an inner pseudodistance (Proposition 3.3.1), the ball $B_{\boldsymbol{k}_G}(a, \delta)$ is connected. Thus $B_{\boldsymbol{k}_G}(a, \delta) \subset \mathbb{B}(a, r)$.

(ii) $\Longrightarrow$ (iii): Suppose that for every $s \gg 1$ there exist $m_s \in \mathbb{N}$, $F_s \in \mathcal{O}(\mathbb{B}_{m_s}, G)$, and $w_s \in \mathbb{B}_{m_s}(1/s)$ such that $F_s(0) \in \mathbb{B}(1/s)$ and $\|F(w_s) - a\| \geq r$. Then,

$$\begin{aligned}\boldsymbol{k}_G(a, F_s(w_s)) &\leq \boldsymbol{k}_G(a, F_s(0)) + \boldsymbol{k}_G(F_s(0), F_s(w_s)) \\ &\leq \boldsymbol{k}_G(a, F_s(0)) + \boldsymbol{k}_{\mathbb{B}_{m_s}}(0, w_s) \longrightarrow 0.\end{aligned}$$

Consequently, by (ii), we get $F_s(w_s) \longrightarrow a$; a contradiction.

(iv) is a special case of (iii).

(iv) $\Longrightarrow$ (v): Take a ball $\mathbb{B}(a, 3r) \subset G$ and let δ be as in (iv). Fix $z \in \mathbb{B}(a, \delta)$ and $X \in (\mathbb{C}^n)_*$. Let $\varphi \in \mathcal{O}(\mathbb{D}, G)$ be such that $\varphi(0) = z$ and $\alpha\varphi'(0) = X$. Then, using (iv), we get $\varphi(\mathbb{D}(\delta)) \subset \mathbb{B}(a, r) \subset \mathbb{B}(z, r + \delta) \subset G$. Put $\psi(\lambda) := \varphi(\delta\lambda)$, $\lambda \in \mathbb{D}$. Then, $\psi \in \mathcal{O}(\mathbb{D}, \mathbb{B}(z, r + \delta))$, $\psi(0) = z$, and $\alpha\psi'(0) = \delta X$. Consequently, $\boldsymbol{\varkappa}_G(z; X) \geq \delta\boldsymbol{\varkappa}_{\mathbb{B}(z, r+\delta)}(z; X) = \delta\|X\|/(r + \delta)$.

(v) $\Longrightarrow$ (vi): We will show that $B_{\boldsymbol{k}_G}(a, rM) \subset \mathbb{B}(a, r)$. Suppose that there exists a $z \in B_{\boldsymbol{k}_G}(a, rM) \setminus \mathbb{B}(a, r)$. By Theorem 3.6.4, there exists a piecewise $\mathcal{C}^1$-curve $\alpha : [0, 1] \longrightarrow G$, $\alpha(0) = a$, $\alpha(1) = z$, and $\int_0^1 \boldsymbol{\varkappa}_G(\alpha(t); \alpha'(t))dt < rM$. Let $t_0 \in (0, 1]$ be such that $z_0 = \alpha(t_0) \in \partial\mathbb{B}(a, r)$ and $\alpha([0, t_0)) \subset \mathbb{B}(a, r)$. Then, $rM > \int_0^{t_0} \boldsymbol{\varkappa}_G(\alpha(t); \alpha'(t))dt \geq M\|a - z_0\| = Mr$; a contradiction.

(vi) $\Longrightarrow$ (iv): Let $\delta \in (0, 1)$ be such that $\boldsymbol{p}(0, \delta) < r/2$ and $\mathbb{B}(a, \delta) \subset B_{\boldsymbol{k}_G}(a, r/2)$. Fix $z \in \mathbb{B}(a, \delta)$ and $X \in (\mathbb{C}^n)_*$. Let $\varphi \in \mathcal{O}(\mathbb{D}, G)$ be such that $\varphi(0) = z$ and $\alpha\varphi'(0) = X$. Then for $\lambda \in \mathbb{D}(\delta)$ we get

$$\boldsymbol{k}_G(a, \varphi(\lambda)) \leq \boldsymbol{k}_G(a, \varphi(0)) + \boldsymbol{k}_G(\varphi(0), \varphi(\lambda)) \leq \boldsymbol{k}_G(a, \varphi(0)) + \boldsymbol{p}(0, \lambda) < r.$$

Hence, by (vi), $\varphi(\lambda) \in \mathbb{B}(\varphi(0), 2R)$ for $\lambda \in \mathbb{D}(\delta)$. Consequently, by the Schwarz lemma, we get $\|\varphi'(0)\| \leq 2R/\delta$. Thus, $\boldsymbol{\varkappa}_G(z; X) \geq (\delta/(2R))\|X\|$.

(v) $\Longrightarrow$ (vii): We may assume that $\mathbb{B}(a, r) \subset\subset G$. Fix $z', z'' \in \mathbb{B}(a, r/2)$. Let $\alpha : [0, 1] \longrightarrow G$ be a piecewise $\mathcal{C}^1$-curve with $\alpha(0) = z'$, $\alpha(1) = z''$. If $\alpha([0, 1]) \subset \mathbb{B}(a, r)$, then condition (v) implies that $\int_0^1 \boldsymbol{\varkappa}_G(\alpha(t); \alpha'(t))dt \geq M\|z' - z''\|$.

If $b_1 := \alpha(t_1) \in \partial\mathbb{B}(a,r)$, $b_2 := \alpha(t_2) \in \partial\mathbb{B}(a,r)$, $\alpha([0,t_1) \cup (t_2,1]) \subset \mathbb{B}(a,r)$ for some $0 < t_1 < t_2 < 1$, then

$$\int_0^1 \varkappa_G(\alpha(t);\alpha'(t))dt \geq M(\|z' - b_1\| + \|z'' - b_2\|) \geq Mr \geq M\|z' - z''\|.$$

(vii) $\Longrightarrow$ (i): Fix a $z \in G \setminus \{a\}$. The case where $z \in \mathbb{B}(a,r)$ is obvious. Thus, assume that $z \notin \mathbb{B}(a,r)$. Let $\alpha : [0,1] \longrightarrow G$ be a piecewise $\mathcal{C}^1$-curve with $\alpha(0) = a$ and $\alpha(1) = z$. Let $t_0 \in (0,1]$ be such that $z_0 := \alpha(t_0) \in \partial\mathbb{B}(a,r)$ and $\alpha([0,t_0)) \subset \mathbb{B}(a,r)$. Then $\int_0^1 \varkappa_G(\alpha(t);\alpha'(t))dt \geq \int_0^{t_0} \varkappa_G(\alpha(t);\alpha'(t))dt \geq \boldsymbol{k}_G(a,z_0) \geq Mr$. Hence, by Theorem 3.6.4, we conclude that $\boldsymbol{k}_G(a,z) \geq Mr$.

(vii) $\Longrightarrow$ (viii): We may assume that $\tau \leq r$, where r is as in (vii). Fix $z' \in \mathbb{B}(a,\tau/2)$ and $z'' \in G \setminus \mathbb{B}(a,\tau)$. Then,

$$\begin{aligned}\boldsymbol{\ell}_G(z',z'') &\geq \boldsymbol{k}_G(z',z'') \geq \boldsymbol{k}_G(a,z'') - \boldsymbol{k}_G(a,z') \\ &\geq M\tau - M\|z' - a\| \geq M\tau/2 > 0.\end{aligned}$$

(viii) $\Longrightarrow$ (iv): Fix an $r > 0$ and let $\varepsilon > 0$ be as in (viii) with $\tau := r$. Suppose that (iv) does not hold. Then, for $s \gg 1$ there exist $\varphi_s \in \mathcal{O}(\mathbb{D},G)$ and $\lambda_s \in \mathbb{D}(1/s)$ such that $\varphi_s(0) \in \mathbb{B}(a,1/s)$ and $\|\varphi_s(\lambda_s) - a\| \geq r$. Thus, we have

$$0 \longleftarrow \boldsymbol{p}(0,\lambda_s) \geq \boldsymbol{\ell}_G(\varphi_s(0),\varphi_s(\lambda_s)) \geq \varepsilon, \quad s \gg 1;$$

a contradiction. □

Using Propositions 3.5.14 and 5.3.1(v), we get the following important corollary.

Corollary 5.3.2. *If G is taut, then G is $\boldsymbol{k}$-hyperbolic at each point.*

We are going to discuss more subtle interrelations between $\mathcal{D}\boldsymbol{k}_G$, $\mathcal{D}\boldsymbol{k}_G^{(m)}$, $\mathcal{D}\boldsymbol{\ell}_G$, $\hat{\varkappa}_G^{(m)}$, and $\hat{\varkappa}_G$. Recall that $\mathcal{D}\boldsymbol{k}_G \leq \hat{\varkappa}_G$. The main result is the following proposition.

Proposition 5.3.3 (cf. [384]). *Let $G \subset \mathbb{C}^n$ be a domain and let $(a,X) \in G \times \mathbb{C}^n$. Assume that*

- *$G \subset \mathbb{C}^n$ is $\boldsymbol{k}$-hyperbolic at a,*
- *$\varkappa_G$ is continuous at (a,X) (e.g., G is taut).*

Then,

$$\varkappa_G(a;X) = \lim_{\substack{\mathbb{C}_* \ni \lambda \to 0 \\ z \to a \\ X' \to X}} \frac{1}{|\lambda|}\boldsymbol{\ell}_G(z, z + \lambda X') = (\mathcal{D}\boldsymbol{\ell}_G)(a;X).$$

(Notice that the above result is not true in general – Example 5.3.7(c).) Consequently,

- *if* $\|X\| = 1$, *then*

$$\boldsymbol{\varkappa}_G(a;X) = \lim_{\substack{z',z''\to a,\ z'\neq z'' \\ \frac{z'-z''}{\|z'-z''\|}\to X}} \frac{\boldsymbol{\ell}_G(z',z'')}{\|z'-z''\|};$$

- *if* $\boldsymbol{\varkappa}_G$ *is continuous at each point from* $\{a\} \times \partial\mathbb{B}_n$, *then*

$$\lim_{z',z''\to a,\ z'\neq z''} \frac{\boldsymbol{\ell}_G(z',z'') - \boldsymbol{\varkappa}_G(a;z'-z'')}{\|z'-z''\|} = 0;$$

- *if* $\boldsymbol{\varkappa}_G$ *is continuous at each point from* $\{a\} \times \partial\mathbb{B}_n$ *and* $\min_{\{a\}\times\partial\mathbb{B}_n} \boldsymbol{\varkappa}_G > 0$, *then*

$$\lim_{z',z''\to a,\ z'\neq z''} \frac{\boldsymbol{\ell}_G(z',z'')}{\boldsymbol{\varkappa}_G(a;z'-z'')} = 1.$$

Notice that in the case where G is taut, all the above assumptions are satisfied globally. In the taut case, Proposition 5.3.3 was first proved in [407] (see Remark 5.3.5).

Lemma 5.3.4. *Assume that* $G \subset \mathbb{C}^n$ *is* $\boldsymbol{k}$*-hyperbolic at* a *and let* $U \subset G$ *be an open neighborhood of* a. *Then, there exists an open neighborhood* $U_0 \subset U$ *of* a, *such that:*

$$\boldsymbol{k}_G^{(m)}(z',z'') = \inf\left\{\sum_{j=1}^{m} \boldsymbol{\ell}_G(z_{j-1},z_j) : z_0,\dots,z_m \in U,\ z_0 = z',\ z_m = z''\right\},$$
$$z',z'' \in U_0,\ m \in \mathbb{N}.$$

Proof. Since G is $\boldsymbol{k}$-hyperbolic at a, we find $0 < r_4 < r_3 < r_2 < r_1$ such that $U_0 := \mathbb{B}(a,r_4) \subset B_{\boldsymbol{k}_G}(a,r_3) \subset B_{\boldsymbol{k}_G}(a,r_2) \subset \mathbb{B}(a,r_1) \subset U$ with $2\boldsymbol{p}(0,r_4/r_1) < r_2 - r_3$. Take $z',z'' \in U_0$. Then,

$$\begin{aligned}\boldsymbol{k}_G^{(m)}(z',z'') &\le \boldsymbol{k}_{\mathbb{B}(a,r_1)}^{(m)}(z',z'') = \boldsymbol{k}_{\mathbb{B}(a,r_1)}(z',z'') \\ &\le \boldsymbol{k}_{\mathbb{B}(a,r_1)}(z',a) + \boldsymbol{k}_{\mathbb{B}(a,r_1)}(z'',a) \le 2\boldsymbol{p}(0,r_4/r_1).\end{aligned}$$

On the other hand, if $z' = z_0,\dots,z_k \in G$ and $z_k \notin U$, then

$$\begin{aligned}\sum_{j=1}^{k} \boldsymbol{\ell}_G(z_{j-1},z_j) &\ge \sum_{j=1}^{k} \boldsymbol{k}_G(z_{j-1},z_j) \ge \boldsymbol{k}_G(z',z_k) \\ &\ge \boldsymbol{k}_G(a,z_k) - \boldsymbol{k}_G(z',a) \ge r_2 - r_3.\end{aligned}$$ □

Proof of Proposition 5.3.3. Let $\mathbb{C}_* \ni \lambda_k \longrightarrow 0$, $z_k \longrightarrow a$, $X_k' \longrightarrow X$ be such that $\frac{1}{|\lambda_k|}\boldsymbol{\ell}_G^*(z_k, z_k + \lambda_k X_k') \longrightarrow \alpha$.

Step 1°. Additionally assume that $X \neq 0$. Let $\varphi_k \in \mathcal{O}(\mathbb{D}, G)$, $\sigma_k \in (0,1)$ be such that $\varphi_k(0) = z_k$, $\varphi_k(\sigma_k) = z_k + \lambda_k X'_k$, $\sigma_k \leq \boldsymbol{\ell}^*_G(z_k, z_k + \lambda_k X'_k) + |\lambda_k|/k$, $k \in \mathbb{N}$. By Remark 4.1.5(c), we have

$$\sigma_k \leq M \|\lambda_k X'_k\| + |\lambda_k|/k \leq \text{const}\,|\lambda_k|, \quad k \gg 1.$$

Let $\mathbb{B}(a, R) \subset\subset G$. By Proposition 5.3.1(iv), there exists a $\delta \in (0,1)$ such that, if $z_k \in \mathbb{B}(a, \delta)$, then $\varphi_k(\overline{\mathbb{D}}(\delta)) \subset \mathbb{B}(a, R)$. Write $\varphi_k(\lambda) = \sum_{\nu=0}^{\infty} c_{k,\nu}\lambda^\nu$. Then, by the Cauchy inequalities, $\|c_{k,\nu}\| \leq \text{const}/\delta^\nu$, $\nu \in \mathbb{Z}_+$, $k \gg 1$. Put

$$R_k := \varphi_k(\sigma_k) - \varphi_k(0) - \varphi'_k(0)\sigma_k = \lambda_k X'_k - \varphi'_k(0)\sigma_k.$$

We have $\|R_k\| \leq \sum_{\nu=2}^{\infty} \|c_{k,\nu}\|\sigma_k^\nu \leq \text{const}\,\sigma_k^2$, $k \gg 1$. Hence, $R_k/\lambda_k \longrightarrow 0$. We get

$$\boldsymbol{\varkappa}_G(z_k; X'_k - R_k/\lambda_k) \leq \sigma_k/|\lambda_k| \leq \frac{1}{|\lambda_k|}\boldsymbol{\ell}^*_G(z_k, z_k + \lambda_k X'_k) + 1/k, \quad k \gg 1.$$

Thus, using the continuity of $\boldsymbol{\varkappa}_G$ at (a, X), we get

$$\boldsymbol{\varkappa}_G(a; X) \leq \liminf_{k\to+\infty} \frac{1}{|\lambda_k|}\boldsymbol{\ell}^*_G(z_k, z_k + \lambda_k X'_k).$$

The above inequality is obviously true also for $X = 0$.

Step 2°. It remains to prove that

$$\limsup_{k\to+\infty} \frac{1}{|\lambda_k|}\boldsymbol{\ell}^*_G(z_k, z_k + \lambda_k X'_k) \leq \boldsymbol{\varkappa}_G(a; X).$$

First observe that the case $X = 0$ is elementary (use by Remark 4.1.5(b)). Suppose that $X \neq 0$. Let $\varphi \in \mathcal{O}(\overline{\mathbb{D}}, G)$ be such that $\varphi(0) = a$ and $\sigma\varphi'(0) = X$. Define

$$\sigma_k := \sigma\lambda_k, \quad R_k := \frac{\varphi(\sigma_k) - a - \lambda_k X'_k}{\sigma_k}, \quad k \gg 1.$$

Observe that $R_k \longrightarrow 0$. Put

$$\varphi_k(\zeta) := \varphi(\zeta) - \zeta R_k + z_k - a.$$

Then, for $k \gg 1$ we have $\varphi_k(\mathbb{D}) \subset G$, $\varphi_k(0) = z_k$, and $\varphi_k(\sigma_k) = z_k + \lambda_k X'_k$. Hence,

$$\frac{1}{|\lambda_k|}\boldsymbol{\ell}^*_G(z_k, z_k + \lambda_k X'_k) \leq \frac{1}{|\lambda_k|}\sigma_k = \sigma. \qquad \square$$

Remark 5.3.5.

(a) Step 2° of the proof of Proposition 5.3.3 shows that the inequality

$$(\mathcal{D}\boldsymbol{\ell}_G)(a;X) = \limsup_{\substack{\mathbb{C}_* \ni \lambda \to 0 \\ z \to a \\ X' \to X}} \frac{1}{|\lambda|} \boldsymbol{\ell}_G(z, z + \lambda X') \le \boldsymbol{\varkappa}_G(a;X), \quad a \in G,\ X \in \mathbb{C}^n.$$

is true for an *arbitrary* domain $G \subset \mathbb{C}^n$.

(b) Observe that in the case where G is taut, Step 1° of the proof of Proposition 5.3.3 may be simplified.

Assume that $X \neq 0$. Since G is taut, there exist extremal discs $\varphi_k \in \mathcal{O}(\mathbb{D}, G)$ such that $\varphi_k(0) = z_k$, $\varphi_k(\sigma_k) = z_k + \lambda_k X'_k$ with $\sigma_k := \boldsymbol{\ell}^*_G(z_k, z_k + \lambda_k X'_k)$, $k \in \mathbb{N}$. We may assume that $\varphi_k \longrightarrow \varphi$ locally uniformly in $\mathbb{D}$ with $\varphi \in \mathcal{O}(\mathbb{D}, G)$, $\varphi(0) = a$. Observe that

$$\frac{\lambda_k X'_k}{\sigma_k} = \frac{\varphi_k(\sigma_k) - \varphi_k(0)}{\sigma_k} \longrightarrow \varphi'(0).$$

Since $\sigma_k/|\lambda_k| \longrightarrow \alpha$, we conclude that $\alpha > 0$. We may assume that $\sigma_k/\lambda_k \longrightarrow e^{i\theta}\alpha$. Thus, $e^{i\theta}\alpha\varphi'(0) = X$, which implies that $\boldsymbol{\varkappa}_G(a;X) \le \alpha$ and hence,

$$\boldsymbol{\varkappa}_G(a;X) \le \liminf_{k \to +\infty} \frac{1}{|\lambda_k|} \boldsymbol{\ell}^*_G(z_k, z_k + \lambda_k X'_k).$$

Proposition 5.3.6 (cf. [384]). *Let $G \subset \mathbb{C}^n$ be a domain and let $a \in G$. Assume that,*

- *G is $\boldsymbol{k}$-hyperbolic at a,*
- *$\boldsymbol{\varkappa}_G$ is continuous at each point from $\{a\} \times \partial\mathbb{B}_n$ and $\min_{\{a\} \times \partial\mathbb{B}_n} \boldsymbol{\varkappa}_G > 0$ (e.g., G is taut).*

Then,

$$\widehat{\boldsymbol{\varkappa}}^{(m)}_G(a;X) = \lim_{\substack{\mathbb{C}_* \ni \lambda \to 0 \\ z \to a \\ X' \to X}} \frac{1}{|\lambda|} \boldsymbol{k}^{(m)}_G(z, z + \lambda X') = (\mathcal{D}\boldsymbol{k}^{(m)}_G)(a;X),$$
$$X \in \mathbb{C}^n,\ m \in \mathbb{N} \cup \{\infty\}, \quad (5.3.1)$$

where $\widehat{\varkappa}_G^{(\infty)} := \widehat{\varkappa}_G$ *and* $\boldsymbol{k}_G^{(\infty)} := \boldsymbol{k}_G$. *Consequently, using Remark* 4.3.4(h) *and Lemma* 2.2.5(b), *we get*

$$\widehat{\varkappa}_G^{(m)}(a;X) = \lim_{\substack{z',z''\to a,\, z'\neq z''\\ \frac{z'-z''}{\|z'-z''\|}\to X}} \frac{\boldsymbol{k}_G^{(m)}(z',z'')}{\|z'-z''\|}, \quad X \in \partial\mathbb{B}_n,\ m \in \mathbb{N}\cup\{\infty\},$$

$$\lim_{z',z''\to a,\, z'\neq z''} \frac{\boldsymbol{k}_G^{(m)}(z',z'') - \widehat{\varkappa}_G^{(m)}(a;z'-z'')}{\|z'-z''\|} = 0, \quad m \in \mathbb{N},$$

$$\lim_{z',z''\to a,\, z'\neq z''} \frac{\boldsymbol{k}_G^{(m)}(z',z'')}{\widehat{\varkappa}_G^{(m)}(a;z'-z'')} = 1, \quad m \in \mathbb{N},$$

$$\lim_{z',z''\to a,\, z'\neq z''} \frac{\boldsymbol{k}_G(z',z'')}{\widehat{\varkappa}_G^{(2n-1)}(a;z'-z'')} = 1. \tag{5.3.2}$$

Notice that the assumptions are satisfied if G is taut. In the taut case with $m = \infty$, property (5.3.1) was first proved in [308] and property (5.3.2) – in [382].

Proof of Proposition 5.3.6. Step 1°. The case where $m = 1$ reduces to Proposition 5.3.3.

Step 2°. The case where $m \in \mathbb{N}_2$.

Let $\mathbb{C}_* \ni \lambda_k \longrightarrow 0$, $z_k \longrightarrow a$, $X_k' \longrightarrow X$. Suppose that $X = X_1 + \cdots + X_m$. Put $z_{k,0} := z_k$, $z_{k,j} := z_k + \lambda_k(X_1 + \cdots + X_j)$, $j = 1,\dots,m-1$, $z_{k,m} := z_k + \lambda_k X_k'$. Observe that $X_k' - (X_1 + \cdots + X_{m-1}) \longrightarrow X_m$ and $z_{k,j} \longrightarrow a$, when $k \longrightarrow +\infty$. Then, by Proposition 5.3.3,

$$\begin{aligned}\sum_{j=1}^m \varkappa_G(a;X_j) &= \sum_{j=1}^m \lim_{k\to+\infty} \frac{1}{|\lambda_k|}\boldsymbol{\ell}_G(z_{k,j-1},z_{k,j})\\ &\geq \limsup_{k\to+\infty} \frac{1}{|\lambda_k|}\boldsymbol{k}_G^{(m)}(z_k, z_k+\lambda_k X_k'),\end{aligned}$$

which implies that

$$\widehat{\varkappa}_G^{(m)}(a;X) \geq \limsup_{k\to+\infty} \frac{1}{|\lambda_k|}\boldsymbol{k}_G^{(m)}(z_k, z_k+\lambda_k X_k').$$

Conversely, by Lemma 5.3.4, there exist $z_k = z_{k,0}, z_{k,1}, \dots, z_{k,m-1}, z_{k,m} = z_k + \lambda_k X_k' \in G$, such that

$$\boldsymbol{k}_G^{(m)}(z_k, z_k+\lambda_k X_k') \geq -|\lambda_k|/k + \sum_{j=1}^m \boldsymbol{\ell}_G(z_{k,j-1},z_{k,j})$$

and $z_{k,j} \longrightarrow a$ when $k \longrightarrow +\infty$. Take a $\theta \in (0,1)$. By Proposition 5.3.3

$$\boldsymbol{\ell}_G(z_{k,j-1}, z_{k,j}) \geq \theta \boldsymbol{\varkappa}_G(a; z_{k,j} - z_{k,j-1}), \quad j = 1, \dots, m,\ k \gg 1.$$

Hence,

$$\begin{aligned} \boldsymbol{k}_G^{(m)}(z_k, z_k + \lambda_k X_k') &\geq -|\lambda_k|/k + \theta \sum_{j=1}^{m} \boldsymbol{\varkappa}_G(a; z_{k,j} - z_{k,j-1}) \\ &\geq -|\lambda_k|/k + \theta \widehat{\boldsymbol{\varkappa}}_G^{(m)}(a; \lambda_k X_k') \\ &= |\lambda_k|(-1/k + \theta \widehat{\boldsymbol{\varkappa}}_G^{(m)}(a; X_k')), \quad k \gg 1. \end{aligned}$$

Thus, by Remark 4.3.4(h),

$$\liminf_{k\to+\infty} \frac{1}{|\lambda_k|} \boldsymbol{k}_G^{(m)}(z_k, z_k + \lambda_k X_k') \geq \widehat{\boldsymbol{\varkappa}}_G^{(m)}(a; X).$$

Step 3°. The case $m = \infty$.

We already know that

$$\widehat{\boldsymbol{\varkappa}}_G(a; X) \geq \limsup_{k\to+\infty} \frac{1}{|\lambda_k|} \boldsymbol{k}_G(z_k, z_k + \lambda_k X_k');$$

cf. Proposition 4.3.3(g).

Conversely, by Lemma 5.3.4, there exist $m_k \in \mathbb{N}$, $z_k = z_{k,0}, z_{k,1}, \dots, z_{k,m_k-1}$, $z_{k,m_k} = z_k + \lambda_k X_k' \in G$ such that

$$\boldsymbol{k}_G(z_k, z_k + \lambda_k X_k') \geq -|\lambda_k|/k + \sum_{j=1}^{m_k} \boldsymbol{\ell}_G(z_{k,j-1}, z_{k,j})$$

and $z_{k,j} \longrightarrow a$ when $k \longrightarrow +\infty$. Take a $\theta \in (0,1)$. By Proposition 5.3.3

$$\boldsymbol{\ell}_G(z_{k,j-1}, z_{k,j}) \geq \theta \boldsymbol{\varkappa}_G(a; z_{k,j} - z_{k,j-1}), \quad j = 1, \dots, m_k,\ k \gg 1.$$

Hence,

$$\begin{aligned} \boldsymbol{k}_G(z_k, z_k + \lambda_k X_k') &\geq -|\lambda_k|/k + \theta \sum_{j=1}^{m_k} \boldsymbol{\varkappa}_G(a; z_{k,j} - z_{k,j-1}) \\ &\geq -|\lambda_k|/k + \theta \widehat{\boldsymbol{\varkappa}}_G^{(m_k)}(a; \lambda_k X_k') \\ &\geq |\lambda_k| \left(-1/k + \theta \widehat{\boldsymbol{\varkappa}}_G(a; X_k')\right), \quad k \gg 1. \end{aligned}$$

Thus, by Remark 4.3.4(h),

$$\liminf_{k\to+\infty} \frac{1}{|\lambda_k|} \boldsymbol{k}_G(z_k, z_k + \lambda_k X_k') \geq \widehat{\boldsymbol{\varkappa}}_G(a; X). \qquad \square$$

Example 5.3.7 (cf. [384]).

(a) Let $G \subset \mathbb{C}^2$ be the domain from Example 3.5.11. Then, $\boldsymbol{k}_G^{(3)} \equiv 0$ and $\boldsymbol{\varkappa}_G(z^0; X) \geq C\|X\|$, $X \in \mathbb{C}^2$, for some $C > 0$. In particular, $\hat{\boldsymbol{\varkappa}}_G^{(m)}(z^0; X) = \hat{\boldsymbol{\varkappa}}_G^{(3)}(z^0; X) \geq C\|X\|$, $X \in \mathbb{C}^2$. Thus,

$$0 = \mathfrak{D}\boldsymbol{k}_G(z^0; X) \leq \mathfrak{D}\boldsymbol{k}_G^{(3)}(z^0; X) < \hat{\boldsymbol{\varkappa}}_G^{(3)}(z^0; X) = \hat{\boldsymbol{\varkappa}}_G(z^0; X), \quad X \neq 0.$$

(b) Example 5.1.4 gives a bounded pseudoconvex domain $G \subset \mathbb{C}^2$, $0 \in G$, such that $\lim_{\lambda \to 0} \frac{1}{|\lambda|} \boldsymbol{\ell}_G(0, \lambda \mathbf{1})$ does not exist.

(c) Let $G \subset \mathbb{C}^n$ be a pseudoconvex balanced domain. We know that $\boldsymbol{g}_G(0, \cdot) = \boldsymbol{\ell}_G^*(0, \cdot) = \boldsymbol{\varkappa}_G(0; \cdot) = \mathfrak{h}_G$. Thus, by Remark 5.3.5(a),

$$\mathfrak{h}_G \leq \mathfrak{D}\boldsymbol{\ell}_G(0; \cdot) \leq \boldsymbol{\varkappa}_G(0; \cdot) = \mathfrak{h}_G.$$

Observe that, if $\mathfrak{h}_G$ is not continuous at X and $\alpha := \liminf_{X' \to X} \mathfrak{h}_G(X') < \mathfrak{h}_G(X)$, then

$$\liminf_{\substack{\mathbb{C}_* \ni \lambda \to 0 \\ z \to 0 \\ X' \to X}} \frac{1}{|\lambda|} \boldsymbol{\ell}_G(z, z + \lambda X') \leq \alpha < \boldsymbol{\varkappa}_G(0; X)$$

and therefore, the limit

$$\lim_{\substack{\mathbb{C}_* \ni \lambda \to 0 \\ z \to 0 \\ X' \to X}} \frac{1}{|\lambda|} \boldsymbol{\ell}_G(z, z + \lambda X')$$

does not exist (cf. Proposition 5.3.3).

(d) There exists a pseudoconvex domain $G \subset \mathbb{C}^2$ such that

- $I := \{(it/2, 1/2) : t \in [0, 1]\} \subset G$,
- $\boldsymbol{k}_G(z', z'') = 0$, $z', z'' \in I$,
- there exists a $C > 0$ such that $\boldsymbol{k}_G(z, z - te_1) \geq Ct$, $z \in I$, $0 \leq t \leq 1/2 - 1/e$.

Put $[0, 1] \ni t \overset{\alpha}{\longmapsto} (it/2, 1/2) \in G$. Observe that, if the above conditions are satisfied, then:

- $L_{\boldsymbol{k}_G}(\alpha) = 0$,
- $(\mathfrak{D}\boldsymbol{k}_G)(a; e_1) \geq C$, $a \in I$,

- $$L_{\mathcal{D}k_G}(\alpha) = \int_0^1 (\mathcal{D}k_G)((it/2, 1/2); (i/2, 0))dt \geq C/2,$$

- $$\liminf_{\substack{\mathbb{C}_* \ni \lambda \to 0 \\ z \to a \\ X' \to e_1}} \frac{1}{|\lambda|} \ell_G(z, z + \lambda X') \leq \liminf_{\mathbb{R}_* \ni t \to 0} \frac{1}{|t|} k_G(a, a + ite_1) = 0, \quad a \in I.$$

To construct such a domain G, we use a modification of Example 3.5.11. Let $A > 1$ and put

$$u(\lambda) := \sum_{k=1}^{\infty} \frac{1}{k^2} \log \frac{|\lambda - 1/k|}{A}, \quad \lambda \in \mathbb{C}.$$

Then:

- $u(\lambda) \geq u(0) = \sum_{k=1}^{\infty} \frac{1}{k^2} \log \frac{1}{Ak} > -\infty$ for $\operatorname{Re} \lambda \leq 0$,
- $u(\lambda) \leq \sum_{k=1}^{\infty} \frac{1}{k^2} \log \frac{R+1}{A} = \frac{\pi^2}{6} \log \frac{R+1}{A}$, $\lambda \in \mathbb{D}(R)$,
- $u \in \mathcal{SH}(\mathbb{C})$.

Let $(r_j)_{j=1}^{\infty} \subset [0, 1/2]$ be dense in $[0, 1/2]$. Define

$$v(\lambda) := \sum_{j=1}^{\infty} \frac{u(\lambda - ir_j)}{j^2}, \quad \lambda \in \mathbb{C}.$$

Then:

- $v(\lambda) \geq v(0) \geq \frac{\pi^2}{6} u(0) > -\infty$ for $\operatorname{Re} \lambda \leq 0$,
- $v(\lambda) < \frac{\pi^4}{36} \log \frac{3}{A} < -1$, $\lambda \in \mathbb{D}$, provided that $A \gg 1$,
- $v(1/k + ir_j) = -\infty$ for any $(j, k) \in \mathbb{N}^2$,
- $v \in \mathcal{SH}(\mathbb{C})$.

Put

$$\psi(z) := |z_2| \exp(\|z\|^2) \exp(v(z_1)), \quad z = (z_1, z_2) \in \mathbb{C}^2.$$

Observe that:

- $\psi \in \mathcal{PSH}(\mathbb{C}^2)$,
- $\mathbb{B}_2 \subset G := \{z \in \mathbb{C}^2 : \psi(z) < 1\}$, G is a pseudoconvex domain,
- $\mathbb{C} \times \{0\} \subset G$,

- $\{1/k + ir_j\} \times \mathbb{C} \subset G$ for any $(j,k) \in \mathbb{N}^2$,
- $\boldsymbol{k}_G((1/k + ir_j, 1/2), (1/k' + ir_{j'}, 1/2)) = 0$ for any $(j,k), (j',k') \in \mathbb{N}^2$; hence, in view of the continuity of $\boldsymbol{k}_G$, we have $\boldsymbol{k}_G(z', z'') = 0$, $z', z'' \in I$,
- the function $\mathbb{C}^2 \ni z \longmapsto \log \psi(z) - \frac{1}{2}\|z\|^2$ is psh,
- for $z = (z_1, z_2) \in \mathbb{B}_2$ with $\operatorname{Re} z_1 \le 0$ and $|z_2| \ge 1/e$, we have

$$\frac{1}{2}\|z\|^2 - \log \psi(z) = -\frac{1}{2}\|z\|^2 - \log|z_2| - v(z_1) \le 1 - v(0),$$

- for $z = (z_1, z_2) \in \mathbb{B}_2$ with $\operatorname{Re} z_1 \le 0$ and $|z_2| \ge 1/e$, we have

$$\boldsymbol{\varkappa}_G(z; X) \ge \frac{\|X\|}{8(\frac{1}{2}\|z\|^2 - \log \psi(z))} \ge \frac{\|X\|}{8(1 - v(0))} = C\|X\|, \quad X \in \mathbb{C}^2,$$

- If $z = (z_1, z_2) \in I$ and $0 < t < 1/2$, then $\|z - te_1\| < 1$, $\operatorname{Re}(z_1 - t) < 0$, and $\|z_2\| = 1/2 > 1/e$,
- $$\boldsymbol{k}_G(z, z - te_1) \ge \int_0^t \boldsymbol{\varkappa}_G(z - se_1; -e_1)ds \ge Ct.$$

Example 5.3.8 (cf. [385]). Denote by $\mathbb{M}(3\times 3; \mathbb{C})$ the set of all 3×3 complex matrices and by Ω_3 the *spectral unit ball*, i.e., the set of all matrices from $\mathbb{M}(3 \times 3; \mathbb{C})$ with all their eigenvalues in $\mathbb{D}$ (cf. Miscellanea § A.9). Ω_3 can be seen as a domain in $\mathbb{C}^9$. Put

$$A := \begin{pmatrix} 0 & 0 & 0 \\ 0 & 0 & 1 \\ 0 & 0 & 0 \end{pmatrix} \text{ and } B_t := \begin{pmatrix} 1 & 0 & 0 \\ 0 & \omega & 0 \\ 0 & 3t & \omega^2 \end{pmatrix}, \ t \in \mathbb{C},$$

where $\omega := e^{2\pi i/3}$, and $B := B_0$. Then, we have the following results.

(a) $1 = \boldsymbol{\varkappa}_{\Omega_3}(A; B) > 0 = \lim_{\mathbb{C}_* \ni \lambda \to 0} \frac{1}{|\lambda|} \boldsymbol{\ell}^*_{\Omega_3}(A, A + \lambda B)$.

(b) Let $(t_j)_j \subset \mathbb{C}_*$ and $(C_j)_j \subset \mathbb{M}(3 \times 3; \mathbb{C})$ $(C_j = (c^j_{k,l}))$ be sequences such that $t_j \to 0$, $C_j \to B$, and $\liminf_{j\to\infty} |c^j_{3,2}/t_j - 3| > 0$. Then

$$\lim_{j\to\infty} \frac{\boldsymbol{\ell}^*_{\Omega_3}(A, A + t_j C_j)}{|t_j|} = 0.$$

(c) $\varkappa_{\Omega_3}(A;B) = \lim_{\mathbb{D}_* \ni \lambda \to 0} \dfrac{\ell^*_{\Omega_3}(A, A+\lambda B_\lambda)}{|\lambda|}$ (compare the condition in (b) for $c^j_{3,2}$).

(d) Put $\widetilde{\Omega}_3 := \{C \in \Omega_3 : \operatorname{trace} C = 0\}$. The set $\widetilde{\Omega}_3$ may be considered a pseudoconvex domain in $\mathbb{C}^8$. Observe that $A, B \in \widetilde{\Omega}_3$. Then,

$$\varkappa_{\widetilde{\Omega}_3}(A;B) > 0 = \lim_{\mathbb{C}_* \ni \lambda \to 0} \frac{1}{|\lambda|} \ell^*_{\widetilde{\Omega}_3}(A, A+\lambda B).$$

For proof, the reader is asked to consult the original paper. We only point out that Ω_3 is not taut. We emphasize that we do not know whether

- $\varkappa_{\Omega_3}(A;B) > \mathcal{D}\ell^*_{\Omega_3}(A;B)$;
- there exist a domain $D \subset \mathbb{C}^n$, a point $a \in D$, and a vector $X \in \mathbb{C}^n$, such that $\varkappa_D(a;X) > \mathcal{D}\ell^*_D(a;X)$.

5.4 List of problems

Chapter 6

Elementary Reinhardt domains

Summary. We will discuss basic holomorphically invariant pseudodistances and pseudometrics for elementary Reinhardt domains $\boldsymbol{D}_\alpha$. We point out that, so far, the class of elementary Reinhardt domains is the only category of domains in $\mathbb{C}^n$ for which holomorphically invariant objects may be given by effective formulas. Notice that, for finite intersections of elementary Reinhardt domains, the problem of finding effective formulas is, in general, not solved. § 6.1 is based on [257] and [258], § 6.2 is based on [263], and § 6.3 – on [422] and [565].

6.1 Elementary n-circled domains

Let us begin with the following elementary Reinhardt domain:

$$G = \boldsymbol{D}_\alpha = \{z \in \mathbb{C}^n : |z^\alpha| < 1\},$$

where $\alpha = (\alpha_1, \dots, \alpha_n) \in \mathbb{N}^n$, $\alpha_1, \dots, \alpha_n$ are relatively prime ($n \geq 2$); cf. Proposition 2.10.2.

More general Reinhardt domains will be studied in §§ 6.2 *and* 6.3, *and in Chapter* 10.

Put $\Phi(z) := z^\alpha$ and $r(a) := \operatorname{ord}_a[\Phi - \Phi(a)]$. Moreover, we define

$$G_0 := \{(z_1, \dots, z_n) \in G : z_1 \dots z_n \neq 0\}.$$

Our first aim is to find effective formulas for $\boldsymbol{m}_G^{(k)}$, $\boldsymbol{s}_G$, $\boldsymbol{g}_G$, $\boldsymbol{\gamma}_G^{(k)}$, $\boldsymbol{S}_G$, and $\boldsymbol{A}_G$.

Proposition 6.1.1. *Let $a \in G$ and let $r := r(a)$. Then, for $k \in \mathbb{N}$, we have*

$$\boldsymbol{m}_G^{(k)}(a, z) = [\boldsymbol{m}(\Phi(a), \Phi(z))]^{\frac{1}{k}\lceil\frac{k}{r}\rceil}, \quad z \in G, \tag{6.1.1}$$

$$\boldsymbol{\gamma}_G^{(k)}(a; X) = \begin{cases} [\boldsymbol{\gamma}_{\mathbb{D}}(\Phi(a); \Phi_{(r)}(a)X)]^{\frac{1}{r}} & \text{if } r \mid k \\ 0 & \text{otherwise} \end{cases}, \quad X \in \mathbb{C}^n, \tag{6.1.2}$$

where $\Phi_{(r)}(a)X := \sum_{|\beta|=r} \frac{1}{\beta!} D^\beta \Phi(a) X^\beta$.

Remark 6.1.2. If $k = 1$, then the assertion reduces to Proposition 2.10.2(b).
If $r(a) = 1$ (e.g., $a \in G_0$), then,

$$\boldsymbol{m}_G^{(k)}(a,z) = \boldsymbol{m}(\Phi(a),\Phi(z)), \quad z \in G,\ k \geq 1,$$
$$\boldsymbol{\gamma}_G^{(k)}(a;X) = \boldsymbol{\gamma}_{\mathbb{D}}(\Phi(a);\Phi'(a)X), \quad X \in \mathbb{C}^n,\ k \geq 1.$$

Fix $k \geq 2$ and $b \in G_0$. By (6.1.1), we have

$$\boldsymbol{m}_G^{(k)}(0,b) = |\Phi(b)|^{\frac{1}{k}\lceil\frac{k}{|\alpha|}\rceil} > |\Phi(b)| = \boldsymbol{m}_G^{(k)}(b,0).$$

This shows that the function $\boldsymbol{m}_G^{(k)}$ ($k \geq 2$) is not symmetric. If D is a subdomain of G (with $0, b \in D$), then, by Proposition 4.2.19(e), for any $k \geq 2$ we have $\boldsymbol{m}_D^{(k)}(0,b) > \boldsymbol{m}_D^{(k)}(b,0)$ provided that D is sufficiently close to G. Thus, there exist "very regular" domains in $\mathbb{C}^n$ (e.g., bounded Reinhardt domains with real analytic boundary) such that $\boldsymbol{m}_D^{(k)}$ is not symmetric ($k \geq 2$).

Let $G_0 \ni a_\nu \longrightarrow 0$. Then (for $k \geq 2$),

$$\boldsymbol{m}_G^{(k)}(a_\nu,b) = \boldsymbol{m}(\Phi(a_\nu),\Phi(b)) \longrightarrow |\Phi(b)| < \boldsymbol{m}_G^{(k)}(0,b).$$

Consequently, $\boldsymbol{m}_G^{(k)}$ is not continuous in the first variable; cf. Proposition 4.2.19.

Moreover, for $X \in \mathbb{C}^n$ we get

$$\lim_{G_0 \ni a \to 0} \boldsymbol{\gamma}_G^{(|\alpha|)}(a;X) = \lim_{G_0 \ni a \to 0} \boldsymbol{\gamma}_{\mathbb{D}}(\Phi(a);\Phi'(a)X) = |\Phi'(0)X|.$$

On the other hand, $\boldsymbol{\gamma}_G^{(|\alpha|)}(0;X) = |X^\alpha|^{1/|\alpha|}$. Comparing the zero sets of the functions $X \longmapsto \Phi'(0)X$ and $X \longmapsto X^\alpha$, we see that there exists an X_0 such that $|\Phi'(0)X_0| \neq |X_0^\alpha|^{1/|\alpha|}$. Therefore, the function $\boldsymbol{\gamma}_G^{|\alpha|}(\cdot;X_0)$ is not continuous at 0; cf. Proposition 4.2.20.

Proof of Proposition 6.1.1. Let

$$f(z) := \left(\frac{\Phi(z)-\Phi(a)}{1-\Phi(z)\overline{\Phi(a)}}\right)^{\lceil\frac{k}{r}\rceil}, \quad z \in G.$$

Then, $f \in \mathcal{O}(G,\mathbb{D})$ and $\operatorname{ord}_a f = r\lceil k/r\rceil \geq k$. Hence, $\boldsymbol{m}_G^{(k)}(a,z) \geq |f(z)|^{1/k}$, $z \in G$, which gives the inequality "$\geq$" (in (6.1.1)). Similarly, $\boldsymbol{\gamma}_G^{(k)}(a;X) \geq |f_{(k)}(a)X|^{1/k}$, which is the right hand side of (6.1.2).

Now let $f \in \mathcal{O}(G,\mathbb{D})$ with $\operatorname{ord}_a f \geq k$ be arbitrary. Then, $f = \varphi \circ \Phi$ (cf. § 2.10), where $\varphi \in \mathcal{O}(\mathbb{D},\mathbb{D})$ and $\operatorname{ord}_a f = r \operatorname{ord}_{\Phi(a)} \varphi \geq k$. Hence, $\operatorname{ord}_{\Phi(a)} \varphi \geq \lceil\frac{k}{r}\rceil$, and, therefore, $|f(z)|^{1/k} \leq [\boldsymbol{m}(\Phi(a),\Phi(z))]^{\frac{1}{k}\lceil\frac{k}{r}\rceil}$ ($\boldsymbol{m} = \boldsymbol{m}_{\mathbb{D}}^{(l)}$, $l \geq 1$). The proof of (6.1.1) is finished.

To prove (6.1.2), observe that, if $r \not| k$, then $\operatorname{ord}_a f > k$ and so $f_{(k)}(a)X = 0$. If $k = lr$, then $f_{(k)}(a)X = \varphi_{(l)}(\Phi(a))\Phi_{(r)}(a)X$, which directly implies the required result. □

Proposition 6.1.3. *Let $a \in G$, $r := r(a)$. Then,*

$$\boldsymbol{g}_G(a,z) = [\boldsymbol{m}(\Phi(a), \Phi(z))]^{\frac{1}{r}}, \quad z \in G,$$
$$\boldsymbol{A}_G(a;X) = [\boldsymbol{\gamma}_{\mathbb{D}}(\Phi(a); \Phi_{(r)}(a)X)]^{\frac{1}{r}}, \quad X \in \mathbb{C}^n.$$

Remark 6.1.4. $\boldsymbol{g}_G(a,\cdot) = \boldsymbol{m}_G^{(k)}(a,\cdot)$ iff $r|k$ iff $\boldsymbol{A}_G(a;\cdot) = \boldsymbol{\gamma}_G^{(k)}(a;\cdot)$. In particular, the equalities hold for $a \in G_0$.

If $b \in G_0$, then $\boldsymbol{g}_G(0,b) = |\Phi(b)|^{1/|\alpha|} > |\Phi(b)| = \boldsymbol{g}_G(b,0)$. Thus, $\boldsymbol{g}_G$ is not symmetric and, by the same methods as in the case of the Möbius functions, we can find "very regular" subdomains D of G for which $\boldsymbol{g}_D$ is not symmetric (cf. Proposition 4.2.10). An argument as in Remark 6.1.2 shows that the functions $\boldsymbol{g}_G$ and $\boldsymbol{A}_G$ are not continuous in the first variable.

Proof of Proposition 6.1.3. Proposition 6.1.1 gives the inequalities "$\geq$". The converse inequalities are immediate consequences of the following lemma. □

Lemma 6.1.5. *For any $u \in \mathcal{K}_G(a)$, there exists a function $v : \mathbb{D} \longrightarrow [0,1)$ such that $u = v \circ \Phi$ and $v^r \in \mathcal{K}_{\mathbb{D}}(\Phi(a))$.*

Proof. For $\lambda \in \mathbb{D}$, let $V_\lambda := \{z \in \mathbb{C}^n : z^\alpha = \lambda\}$. Fix $\lambda \in \mathbb{D}_*$ and let μ be a fixed α_n-th root of λ. Then, the mapping

$$(\mathbb{C}_*)^{n-1} \ni (w_1,\dots,w_{n-1}) \overset{\Psi}{\longmapsto} (w_1^{\alpha_n},\dots,w_{n-1}^{\alpha_n}, \mu w_1^{-\alpha_1} \cdot \dots \cdot w_{n-1}^{-\alpha_{n-1}})$$

is a holomorphic surjection of $(\mathbb{C}_*)^{n-1}$ onto V_λ. The psh function $u \circ \Psi$ extends to a bounded psh function on $\mathbb{C}^{n-1}$, which implies (cf. Appendix B.4.27) that $u \circ \Psi =$ const and, hence, $u|_{V_\lambda} =$ const $=: v(\lambda)$. It is clear that $u =$ const $=: v(0)$ on V_0. Thus, we have constructed a function $v : \mathbb{D} \longrightarrow [0,1)$ such that $u = v \circ \Phi$. Now, we prove that $v^r \in \mathcal{K}_{\mathbb{D}}(\Phi(a))$. For we take $\lambda_0 = \Phi(b) \in \mathbb{D}_*$ and we observe that

$$v(\lambda) = u\left(b_1,\dots,b_{n-1}, \left(\frac{\lambda}{b_1^{\alpha_1} \cdot \dots \cdot b_{n-1}^{\alpha_{n-1}}}\right)^{\frac{1}{\alpha_n}}\right) \text{ for } \lambda \text{ near } \lambda_0,$$

where the branch of the α_n-th root is chosen in such a way that $(b_n^{\alpha_n})^{1/\alpha_n} = b_n$. This shows that v is log-subharmonic near λ_0, and, consequently, in $\mathbb{D}_*$. Moreover, if $\Phi(a) \neq 0$, then, taking $b = a$, we get

$$v(\lambda) \leq M\left|\left(\frac{\lambda}{a_1^{\alpha_1} \cdot \dots \cdot a_{n-1}^{\alpha_{n-1}}}\right)^{\frac{1}{\alpha_n}} - a_n\right| \leq M_1|\lambda - \Phi(a)| \text{ for } \lambda \text{ near } \Phi(a).$$

It remains to prove that v is log-subharmonic at $0 \in \mathbb{D}$ (since v is bounded, it suffices to show that $v(0) = \limsup_{\lambda \not\to 0} v(\lambda)$, cf. Appendix B.4.23) and that $v^r \in \mathcal{K}_{\mathbb{D}}(0)$ in the case where $\Phi(a) = 0$.

We have

$$\limsup_{\lambda \not\to 0} v(\lambda) \geq \limsup_{G_0 \ni z \to 0} v(\Phi(z)) = \limsup_{z\to 0} u(z) = u(0) = v(0).$$

On the other hand,

$$\limsup_{\lambda \not\to 0} v(\lambda) = \limsup_{\lambda \not\to 0} u(\lambda^{\frac{1}{|\alpha|}}, \dots, \lambda^{\frac{1}{|\alpha|}}) \leq \limsup_{z\to 0} u(z) = u(0).$$

Now suppose that $\Phi(a) = 0$. We may assume that $a_1, \dots, a_s \neq 0$, $a_{s+1} = \dots = a_n = 0$, $0 \leq s \leq n-1$ ($r = \alpha_{s+1} + \dots + \alpha_n$). Then, as above, for λ near 0

$$v(\lambda) = u\left(a_1, \dots, a_s, \left(\frac{\lambda}{a_1^{\alpha_1} \cdots a_s^{\alpha_s}}\right)^{\frac{1}{r}}, \dots, \left(\frac{\lambda}{a_1^{\alpha_1} \cdots a_s^{\alpha_s}}\right)^{\frac{1}{r}}\right) \leq M_1 |\lambda|^{\frac{1}{r}}. \qquad \square$$

Proposition 6.1.6.

$$s_G(a,z) = \begin{cases} g_G(a,z), & \textit{if } g_G^2(a,\cdot) \textit{ is } \mathcal{C}^2 \textit{ near } a \\ ?, & \textit{otherwise} \end{cases},$$

$$S_G(a;X) = \begin{cases} A_G(a;X), & \textit{if } g_G^2(a,\cdot) \textit{ is } \mathcal{C}^2 \textit{ near } a \\ 0, & \textit{otherwise} \end{cases}.$$

?

Remark 6.1.7. The function $g_G^2(a,\cdot)$ is $\mathcal{C}^2$ near a iff $\#\{j : a_j = 0\} \leq 1$ (in particular, $g_G^2(a,\cdot)$ is of class $\mathcal{C}^2$ near a if $a \in G_0$).

Proof of Proposition 6.1.6. In view of Proposition 6.1.3, we only need to show that if $a_1, \dots, a_s \neq 0$, $a_{s+1} = \dots = a_n = 0$, $0 \leq s \leq n-2$, then $q := S_G(a;\cdot) \equiv 0$. Recall that q is a $\mathbb{C}$-seminorm and (by Proposition 4.2.16)

$$q(X) \leq A_G(a;X) = |a_1^{\alpha_1} \cdot \dots \cdot a_s^{\alpha_s} X_{s+1}^{\alpha_{s+1}} \cdot \dots \cdot X_n^{\alpha_n}|^{\frac{1}{r}} =: M|X_{s+1}^{\alpha_{s+1}} \cdot \dots \cdot X_n^{\alpha_n}|^{\frac{1}{r}},$$

where $r = r(a) = \alpha_{s+1} + \dots + \alpha_n$. In particular,

$$\{X \in \mathbb{C}^n : M|X_{s+1}^{\alpha_{s+1}} \cdot \dots \cdot X_n^{\alpha_n}|^{\frac{1}{r}} < 1\} \subset B_q(0,1) = \{X \in \mathbb{C}^n : q(X) < 1\}.$$

Since $B_q(0,1)$ is convex and $s \leq n-2$, we get $B_q(0,1) = \mathbb{C}^n$, i.e., $q \equiv 0$. $\square$

6.2 General point of view

In the context of § 6.1, one may consider the following general situation: let $G \subset \mathbb{C}^n$, $B \subset \mathbb{C}^m$ be domains, $m < n$, and let $\Phi = (\Phi_1, \dots, \Phi_m) : G \longrightarrow B$ be holomorphic. We are interested in cases where formulas similar to those in Propositions 6.1.1 and 6.1.3 are true. Define $r = r(a) := \operatorname{ord}_a(\Phi - \Phi(a))$,

$$\Phi_{(r)}(a)X := \frac{1}{r!}\Phi^{(r)}(a)X = \sum_{\beta \in \mathbb{Z}_+^n,\, |\beta| = r} \frac{1}{\beta!} D^\beta \Phi(a) X^\beta, \quad a \in G,\ X \in \mathbb{C}^n;$$

the mapping $\mathbb{C}^n \ni X \longmapsto \Phi_{(r)}(a)X \in \mathbb{C}^m$ is a homogeneous polynomial of degree r.

Proposition 6.2.1 (cf. [263]). *Assume that there exists a relatively closed pluripolar set $\Sigma \subset B$, such that for each $\xi \in B \setminus \Sigma$:*

(A) *the analytic set $V_\xi := \Phi^{-1}(\xi)$ is non-empty and has the* psh Liouville property, *i.e., if $u \in \mathcal{PSH}(V_\xi)$ is bounded from above, then $u \equiv$ const (cf. § B.4),*

(B) *there exists a $b \in V_\xi$ such that* $\operatorname{rank} \Phi'(b) = m$.

Then $\boldsymbol{m}_G(a,z) = \boldsymbol{m}_B(\Phi(a), \Phi(z)), \quad a, z \in G.$ (6.2.1)

Moreover, if $a \in G$ is such that

(C) $\dim\{X \in \mathbb{C}^n : \Phi^{(r)}(a)X = 0\} = n - m$ *(cf. § B.1.1), where $r := r(a)$, then for $z \in G$, $X \in \mathbb{C}^n$, and $k \in \mathbb{N}$, we have:*

$$\boldsymbol{m}_G^{(k)}(a,z) = \left(\boldsymbol{m}_B^{(\lceil k/r \rceil)}(\Phi(a), \Phi(z))\right)^{\lceil k/r \rceil / k}, \tag{6.2.2}$$

$$\boldsymbol{\gamma}_G^{(k)}(a;X) = \begin{cases} \left(\boldsymbol{\gamma}_B^{(k/r)}(\Phi(a); \Phi_{(r)}(a)X)\right)^{1/r}, & \text{if } r \mid k \\ 0, & \text{if } r \nmid k \end{cases}, \tag{6.2.3}$$

$$\boldsymbol{g}_G(a,z) = \left(\boldsymbol{g}_B(\Phi(a), \Phi(z))\right)^{1/r}, \tag{6.2.4}$$

$$\boldsymbol{A}_G(a;X) = \left(\boldsymbol{A}_B(\Phi(a); \Phi_{(r)}(a)X)\right)^{1/r}. \tag{6.2.5}$$

Remark 6.2.2.

(a) Condition (A) implies that the fiber V_ξ is connected for any $\xi \in B \setminus \Sigma$.

(b) If $r(a) = 1$, then condition (C) says that $\operatorname{rank} \Phi'(a) = m$.

(c) If $m = 1$, then condition (C) is automatically satisfied.

(d) If (C) is not satisfied, then formulas (6.2.2–6.2.5) need not be true; cf. Example 6.2.9.

Proof of Proposition 6.2.1. First, observe that if $a \in G$, $g \in \mathcal{O}(B, \mathbb{D})$, and $\operatorname{ord}_{\Phi(a)} g \geq \ell := \lceil k/r \rceil$ with $r := r(a)$, then $\operatorname{ord}_a(g \circ \Phi) \geq k$. Consequently, $\boldsymbol{m}_G^{(k)}(a,z) \geq (\boldsymbol{m}_B^{(\ell)}(\Phi(a), \Phi(z)))^{\ell/k}$.

Let $f \in \mathcal{O}(G, \mathbb{D})$, $\operatorname{ord}_a f \geq k$. By (A) there exists a function $g : B \setminus \Sigma \longrightarrow \mathbb{D}$ such that $f = g \circ \Phi$ on $G \setminus \Phi^{-1}(\Sigma)$. Take a $\xi_0 \in B \setminus \Sigma$ and let $b_0 \in V_{\xi_0}$ be such that $\operatorname{rank} \Phi'(b_0) = m$ (cf. (B)). After a permutation of variables we may assume that the matrix $\left[\frac{\partial \Phi_j}{\partial z_k}(b_0)\right]_{j,k=1,\dots,m}$ is non-singular. Write $z = (z', z'') \in \mathbb{C}^m \times \mathbb{C}^{n-m}$. By the implicit mapping theorem, there exist open neighborhoods U, W of b_0' and (b_0'', ξ_0), respectively, and a holomorphic mapping $\varphi : W \longrightarrow U$ such that $\varphi(b_0'', \xi_0) = b_0'$ and

$$\{(z', z'', \xi) \in U \times W : \Phi(z', z'') = \xi\} = \{(\varphi(z'', \xi), z'', \xi) : (z'', \xi) \in W\}.$$

In particular, $g(\xi) = f(\varphi(z'', \xi), z'')$, $(z'', \xi) \in W$, which shows that $g \in \mathcal{O}(B \setminus \Sigma)$. Since Σ is pluripolar, Appendix B.4.23(d) implies that g extends holomorphically to a $\widetilde{g} \in \mathcal{O}(B, \mathbb{D})$. By the identity principle for holomorphic functions, we get $\widetilde{g} \circ \Phi = f$ in G. In particular, $\widetilde{g}(\Phi(a)) = f(a) = 0$. Thus, $\boldsymbol{m}_B(\Phi(a), \Phi(z)) \geq \boldsymbol{m}_G(a, z)$, $z \in G$, which gives (6.2.1).

Assume that (C) is satisfied. Let L be an m-dimensional vector subspace of $\mathbb{C}^n$, such that

$$L \cap \{X \in \mathbb{C}^n : \Phi^{(r)}(a)(X) = 0\} = \{0\}$$

(cf. [349], Ch. VII, § 7). In particular,

$$\|\Phi_{(r)}(a)X\| \geq C_0 \|X\|^r, \quad X \in L,$$

with a constant $C_0 > 0$. The Taylor formula for Φ shows that there exist $0 < \varrho < \operatorname{dist}(a, \partial G)$ and $C > 0$, such that

$$\|\Phi(a + X) - \Phi(a)\| \geq C\|X\|^r, \quad X \in L \cap \mathbb{B}(\varrho). \tag{6.2.6}$$

We chose neighborhoods $U \subset \mathbb{B}(\varrho)$ of 0 and $W \subset \mathbb{C}^m$ of $\Phi(a)$ such that the mapping $L \cap U \ni X \longmapsto \Phi(a + X) \in W$ is proper (in particular, surjective). For $\xi \in W$ let $X(\xi) \in L \cap U$ be such that $\Phi(a + X(\xi)) = \xi$. Then, by (6.2.6), we get

$$\begin{aligned}|\widetilde{g}(\xi)| &= |f(a + X(\xi))| \leq \text{const}\, \|X(\xi)\|^k \\ &\leq \text{const}\, \|\Phi(a + X(\xi)) - \Phi(a)\|^{k/r} = \text{const}\, \|\xi - \Phi(a)\|^{k/r}, \quad \xi \in W.\end{aligned}$$

Hence, $\operatorname{ord}_{\Phi(a)} \widetilde{g} \geq \ell$ and, therefore,

$$\boldsymbol{m}_B^{(\ell)}(\Phi(a), \Phi(z)) \geq |\widetilde{g}(\Phi(z))|^{1/\ell} = |f(z)|^{1/\ell},$$

which implies that

$$\boldsymbol{m}_B^{(\ell)}(\Phi(a), \Phi(z)) \geq (\boldsymbol{m}_G^{(k)}(a, z))^{k/\ell}.$$

The proof of (6.2.2) is completed.

Using Lemma 4.2.7 and Propositions 4.2.19(b), 4.2.20(b), we get

$$\begin{aligned}\boldsymbol{\gamma}_G^{(k)}(a;X) &= \lim_{\lambda\to 0}\frac{\boldsymbol{m}_G^{(k)}(a,a+\lambda X)}{|\lambda|} = \lim_{\lambda\to 0}\frac{(\boldsymbol{m}_B^{(\ell)}(\Phi(a),\Phi(a+\lambda X)))^{\ell/k}}{|\lambda|}\\ &= \limsup_{\lambda\nrightarrow 0}\left(\frac{\boldsymbol{m}_B^{(\ell)}(\Phi(a),\Phi(a)+\lambda^r\Phi_{(r)}(a)X))}{|\lambda|^r}\right)^{\ell/k}|\lambda|^{r\ell/k-1}\\ &= \left(\boldsymbol{\gamma}_B^{(\ell)}(\Phi(a);\Phi_{(r)}(a)X)\right)^{\ell/k}\cdot 0^{r\ell/k-1},\end{aligned}$$

which gives (6.2.3).

We move to (6.2.4). If $v \in \mathcal{K}_B(\Phi(a))$, then,

$$v\circ\Phi(z) \le \text{const}\,\|\Phi(z)-\Phi(a)\| \le \text{const}\,\|z-a\|^r$$

for z from a neighborhood of a. This proves that $(v\circ\Phi)^{1/r} \in \mathcal{K}_G(a)$. Thus $\boldsymbol{g}_G(a,z) \ge (v\circ\Phi(z))^{1/r}$ and consequently, $\boldsymbol{g}_G(a,z) \ge (\boldsymbol{g}_B(\Phi(a),\Phi(z)))^{1/r}$.

Now, let $u \in \mathcal{K}_G(a)$. By (A), there exists a function $v : B\setminus\Sigma \longrightarrow [0,1)$ such that $u = v\circ\Phi$ on $G\setminus\Phi^{-1}(\Sigma)$. Similar to the holomorphic case, condition (B) and the implicit mapping theorem imply that $\log v \in \mathcal{PSH}(B\setminus\Sigma)$. Now, by the Riemann type theorem for psh functions, v extends to a log-psh function $\widetilde{v}$ on B (cf. Appendix B.4.23(a)). By the identity principle for psh functions, we get $u = \widetilde{v}\circ\Phi$ in G.

Let U, W, $X(\xi)$ be as above. Then,

$$\begin{aligned}\widetilde{v}(\xi) &= u(a+X(\xi)) \le \text{const}\,\|X(\xi)\|\\ &\le \text{const}\,\|\Phi(a+X(\xi))-\Phi(a)\|^{1/r} = \text{const}\,\|\xi-\Phi(a)\|^{1/r},\quad \xi\in W.\end{aligned}$$

Hence, $\widetilde{v}^r \in \mathcal{K}_B(\Phi(a))$. Consequently, $\boldsymbol{g}_B(\Phi(a),\Phi(z)) \ge \widetilde{v}^r(\Phi(z))$, which gives $(\boldsymbol{g}_B(\Phi(a),\Phi(z)))^{1/r} \ge \boldsymbol{g}_G(a,z)$. The proof of (6.2.4) is completed.

Finally, using Lemma 4.2.7, we get

$$\begin{aligned}\boldsymbol{A}_G(a;X) &= \limsup_{\lambda\nrightarrow 0}\frac{\boldsymbol{g}_G(a,a+\lambda X)}{|\lambda|} = \limsup_{\lambda\nrightarrow 0}\frac{(\boldsymbol{g}_B(\Phi(a),\Phi(a+\lambda X)))^{1/r}}{|\lambda|}\\ &= \left(\limsup_{\lambda\nrightarrow 0}\frac{\boldsymbol{g}_B(\Phi(a),\Phi(a)+\lambda^r\Phi_{(r)}(a)X))}{|\lambda|^r}\right)^{1/r}\\ &= \left(\boldsymbol{A}_B(\Phi(a);\Phi_{(r)}(a)X)\right)^{1/r},\end{aligned}$$

which gives (6.2.5). □

Proposition 6.2.3. *Let $V \subset \mathbb{C}^n$ be a connected algebraic set. Then, V has the psh Liouville property.*

Proof. We may assume that V is irreducible and $k := \dim V \in \{1, \dots, n-1\}$. It is known that, after a linear change of coordinates, the projection

$$\mathbb{C}^k \times \mathbb{C}^{n-k} \supset V \ni (z, w) \stackrel{\pi}{\longmapsto} z \in \mathbb{C}^k$$

is proper (cf. [349], Ch. VII,§ 7). Let $S \subset \mathbb{C}^k$ be an analytic set such that

$$\pi|_{V \setminus \pi^{-1}(S)} : V \setminus \pi^{-1}(S) \longrightarrow \mathbb{C}^k \setminus S$$

is a holomorphic covering. Obviously, $V_0 := V \setminus \pi^{-1}(S) \subset \operatorname{Reg}(V) =$ the set of all regular points of V. Observe that V_0 is connected.

Let $u \in \mathcal{PSH}(V)$ be bounded from above. Define

$$v(z) := \max\{u(z, w) : (z, w) \in V\}, \quad z \in \mathbb{C}^k \setminus S.$$

Then $v \in \mathcal{PSH}(\mathbb{C}^k \setminus S)$ and v is bounded from above. Consequently, v extends to a psh function on $\mathbb{C}^k$ (cf. Appendix B.4.23), which implies that $v \equiv C = \text{const}$ (cf. Appendix B.4.27). Obviously, $u \le C$ on V_0. Let $\widetilde{V}_0 := \{(z, w) \in V_0 : u(z, w) = C\}$. Then, $\widetilde{V}_0 \ne \varnothing$ and $\widetilde{V}_0$ is closed in V_0. Moreover, by the maximum principle, $\widetilde{V}_0$ is open. Thus, $\widetilde{V}_0 = V_0$, i.e., $u = C$ on V_0. Hence, $u = C$ on $\operatorname{Reg}(V)$. Take an $a \in \operatorname{Sing}(V)$ and let $\varphi : \mathbb{D} \longrightarrow V$ be a holomorphic disc such that $\varphi(0) = 0$ and $\varphi(\mathbb{D}) \not\subset \operatorname{Sing}(V)$. Then, 0 is an isolated point of $\varphi^{-1}(\operatorname{Sing}(V))$. Thus, $u(a) = u(\varphi(0)) = \limsup_{\mathbb{D}_* \ni \lambda \to 0} u(\varphi(\lambda)) = C$ (cf. e.g., [121], the first part of the proof of Theorem 1.7). □

Example 6.2.4 (Primitive polynomials). A polynomial P of n-complex variables is *primitive* if P cannot be written in the form $P = f(\widetilde{P})$, where f is a polynomial of one complex variable of degree ≥ 2 and $\widetilde{P}$ is a polynomial of n-complex variables (cf. [116]).

Observe that a homogeneous polynomial Q is primitive iff Q cannot be written as $Q = \widetilde{Q}^p$, where $p \ge 2$ and $\widetilde{Q}$ is a homogeneous polynomial. In particular, a monomial z^α, where $\alpha = (\alpha_1, \dots, \alpha_n) \in \mathbb{N}^n$, $n \ge 2$, is primitive iff the numbers $\alpha_1, \dots, \alpha_n$ are relatively prime.

It is known (cf. [116]) that if P is a primitive polynomial, then the fibers $P^{-1}(\xi)$ are connected except for a finite number of $\xi \in \mathbb{C}$.

Observe that if P is a polynomial, then the set $P(\{z \in \mathbb{C}^n : P'(z) = 0\})$ is finite.

Indeed, let $V := \{z \in \mathbb{C}^n : P'(z) = 0\}$. We will show that the number of elements of $P(V)$ is at most equal to the number of irreducible components of V. Let V_0 be an irreducible component of V. Then, the set $\operatorname{Reg}(V_0)$ is a connected complex manifold and $d(P|_{\operatorname{Reg}(V_0)}) \equiv 0$. Hence $P|_{\operatorname{Reg}(V_0)}$ is constant and, consequently, $P|_{V_0}$ is constant.

Thus, using Proposition 6.2.3, we conclude that *if Φ is a primitive polynomial, then* (A), (B) *are satisfied with a finite set* $\Sigma \subset B$.

In particular, *Proposition* 6.2.1 *generalizes Propositions* 6.1.1 *and* 6.1.3.

Before the next example, we need to state the following lemma.

Lemma 6.2.5. *Let*

$$\alpha^j = (\alpha_1^j, \dots, \alpha_n^j) \in \mathbb{Z}^n, \quad j = 1, \dots, m, \ 1 \le m \le n-1,$$

be such that $\operatorname{rank} A = m$, *where* $A := [\alpha_k^j]_{j=1,\dots,m,\ k=1,\dots,n}$. *Put*

$$\mathbb{C}^n(\alpha^1) \cap \dots \cap \mathbb{C}^n(\alpha^m) \ni z \overset{\Phi}{\longmapsto} (z^{\alpha^1}, \dots, z^{\alpha^m}) \in \mathbb{C}^m.$$

Then,

(a) $\operatorname{rank} \Phi'(a) = m$, $a \in \mathbb{C}_*^n$. *In particular,* $V_\xi := \Phi^{-1}(\xi)$ *is an* $(n-m)$-*dimensional complex manifold for* $\xi \in \mathbb{C}^m \setminus V_0$, *where*

$$V_0 = V_0^m := \{(\xi_1, \dots, \xi_m) \in \mathbb{C}^m : \xi_1 \cdots \xi_m = 0\}.$$

(b) *The following conditions are equivalent:*

- (i) *for every* $\xi \in \mathbb{C}^m \setminus V_0$, *the manifold* V_ξ *is connected;*
- (ii) *there exists a* $\xi \in \mathbb{C}^m \setminus V_0$ *such that the manifold* V_ξ *is connected;*
- (iii) $A\mathbb{Z}^n = \mathbb{Z}^m$.

Remark 6.2.6. One may prove that the above conditions are equivalent to the following one:

(iv) *the greatest common divisor of all determinants of* $m \times m$ *submatrices of* A *equals* 1;

cf. Exercise 6.4.1. Note that in the case $m = 1$, condition (iv) states that $\alpha_1^1, \dots, \alpha_n^1$ are relatively prime.

Proof of Lemma 6.2.5. (a) We have

$$\frac{\partial \Phi_j}{\partial z_k}(a) = a^{\alpha^j} \frac{\alpha_k^j}{a_k}, \quad j = 1, \dots, m, \ k = 1, \dots, n.$$

Thus, $\operatorname{rank} \Phi'(a) = \operatorname{rank} A = m$.

(b) (due to W. Zwonek) The implication (i) $\Longrightarrow$ (ii) is trivial.

(ii) $\Longrightarrow$ (iii) $\Longrightarrow$ (i): Let $\xi = (u_1 e^{2\pi i \theta_1}, \dots, u_m e^{2\pi i \theta_m}) \in \mathbb{C}^m \setminus V_0$. Take arbitrary two points $a, b \in V_\xi$,

$$a = (r_1 e^{2\pi i \varphi_1}, \dots, r_n e^{2\pi i \varphi_n}), \quad b = (s_1 e^{2\pi i \psi_1}, \dots, s_n e^{2\pi i \psi_n}).$$

We have $\Phi(r) = \Phi(s) = u$, $A\varphi = \theta \bmod \mathbb{Z}^m$, $A\psi = \theta \bmod \mathbb{Z}^m$.

Observe that V_ξ is connected iff for arbitrary $a, b \in V_\xi$ there exists a curve $\gamma : [0,1] \longrightarrow V_\xi$ such that $\gamma(0) = a$, $\gamma(1) = b$. Write

$$\gamma(t) = (R_1(t)e^{2\pi i(\varphi_1+\sigma_1(t))}, \dots, R_n(t)e^{2\pi i(\varphi_n+\sigma_n(t))}),$$

where $R : [0,1] \longrightarrow \mathbb{R}^n_{>0}$ is continuous, $\sigma : [0,1] \longrightarrow \mathbb{R}^n$ is such that the mapping $[0,1] \ni t \longmapsto (e^{2\pi i\sigma_1(t)}, \dots, e^{2\pi i\sigma_n(t)})$ is continuous, $\Phi(R(t)) = u$, $A\sigma(t) = 0 \bmod \mathbb{Z}^m$, $t \in [0,1]$, $R(0) = r$, $R(1) = s$, $\sigma(0) = 0 \bmod \mathbb{Z}^n$, $\sigma(1) = \psi - \varphi \bmod \mathbb{Z}^n$. Note that the set $\{x \in \mathbb{R}^n_{>0} : \Phi(x) = u\}$ is connected (its logarithmic image is an affine subspace of $\mathbb{R}^n$). Hence, we can always find an R with the required properties.

Let $\mathbb{R}^n \ni x \overset{\boldsymbol{q}}{\longmapsto} [x] \in \mathbb{R}^n/\mathbb{Z}^n \simeq \mathbb{T}^n$, $X := \boldsymbol{q}(A^{-1}(\mathbb{Z}^m)) = \bigcup_{p\in\mathbb{Z}^m} \boldsymbol{q}(A^{-1}(p))$ (X is a compact space).

The mapping σ may be identified with a curve $\widetilde{\sigma} : [0,1] \longrightarrow X$ such that $\widetilde{\sigma}(0) = [0]$, $\widetilde{\sigma}(1) = [\psi - \varphi]$. Thus, V_ξ is connected iff X is arcwise connected. Since $A^{-1}(p)$ is an affine subspace of $\mathbb{R}^n$, each set $\boldsymbol{q}(A^{-1}(p))$ is arcwise connected (and compact). Note that for $p \in \mathbb{Z}^m$ we have: $\boldsymbol{q}(A^{-1}(p)) \cap \boldsymbol{q}(A^{-1}(0)) \neq \varnothing$ iff $p \in A\mathbb{Z}^n$. Consequently, X is arcwise connected iff $A\mathbb{Z}^n = \mathbb{Z}^m$. □

Now, we are able to present the following generalization of Propositions 6.1.1, 6.1.3, and 6.1.6:

Proposition 6.2.7. *Let* $\alpha = (\alpha_1, \dots, \alpha_n) \in (\mathbb{Z}^n)_*$ *be such that* $\alpha_1, \dots, \alpha_n$ *are relatively prime. Put* $\Phi(z) = z^\alpha$, $z \in \mathbb{C}^n(\alpha)$. *Let* $a = (a_1, \dots, a_n) \in \boldsymbol{D}_\alpha$, $r := r(a)$, $s = s(a) := \#\{j \in \{1, \dots, n\} : \alpha_j a_j \neq 0\}$. *Then, for* $z \in \boldsymbol{D}_\alpha$, $X \in \mathbb{C}^n$, *and* $k \in \mathbb{N}$, *we have:*

$$\boldsymbol{m}^{(k)}_{\boldsymbol{D}_\alpha}(a,z) = \left(\boldsymbol{m}^{(\lceil k/r\rceil)}_{\mathbb{D}}(\Phi(a), \Phi(z))\right)^{\lceil k/r\rceil/k},$$

$$\boldsymbol{\gamma}^{(k)}_{\boldsymbol{D}_\alpha}(a;X) = \begin{cases} \left(\boldsymbol{\gamma}^{(k/r)}_{\mathbb{D}}(\Phi(a); \Phi_{(r)}(a)X)\right)^{1/r}, & r \mid k \\ 0, & r \nmid k \end{cases},$$

$$\boldsymbol{g}_{\boldsymbol{D}_\alpha}(a,z) = (\boldsymbol{m}_{\mathbb{D}}(\Phi(a), \Phi(z)))^{1/r},$$

$$\boldsymbol{A}_{\boldsymbol{D}_\alpha}(a;X) = (\boldsymbol{\gamma}_{\mathbb{D}}(\Phi(a); \Phi_{(r)}(a)X))^{1/r},$$

$$\boldsymbol{s}_{\boldsymbol{D}_\alpha}(a,z) = \begin{cases} ?, & s \le n-2 \\ (\boldsymbol{m}_{\mathbb{D}}(\Phi(a), \Phi(z)))^{1/r}, & s \ge n-1 \end{cases},$$

$$\boldsymbol{S}_{\boldsymbol{D}_\alpha}(a;X) = \begin{cases} 0, & s \le n-2 \\ (\boldsymbol{\gamma}_{\mathbb{D}}(\Phi(a); \Phi_{(r)}(a)X))^{1/r}, & s \ge n-1 \end{cases}.$$

The case where $\boldsymbol{D}_\alpha$ is of irrational type will be discussed in § 6.3.

?

Proof. The formulas for $\boldsymbol{m}^{(k)}_{\boldsymbol{D}_\alpha}$, $\boldsymbol{\gamma}^{(k)}_{\boldsymbol{D}_\alpha}$, $\boldsymbol{g}_{\boldsymbol{D}_\alpha}$, and $\boldsymbol{A}_{\boldsymbol{D}_\alpha}$ follow immediately from Propositions 6.2.1, 6.2.3, and Lemma 6.2.5.

Observe that $\boldsymbol{g}^2_{\boldsymbol{D}_\alpha}(a,\cdot)$ is of class $\mathcal{C}^2$ in a neighborhood of a iff $s \geq n-1$. This implies the formulas for $s \geq n-1$ (cf. Remark 4.2.8). To show that $\boldsymbol{S}_{\boldsymbol{D}_\alpha}(a;\cdot) \equiv 0$ in the case where $s \leq n-2$, we may argue as in the proof of Proposition 6.1.6. □

The case of a finite intersection $\boldsymbol{D}_{\alpha^1} \cap \cdots \cap \boldsymbol{D}_{\alpha^m}$ is essentially more complicated – only certain special cases are completely understood (cf. Examples 6.2.8, 6.2.9, and Proposition 6.2.10).

Example 6.2.8. Let $\alpha^1, \dots, \alpha^m, A, \Phi$ be as in Lemma 6.2.5 (rank $A = m$). Take $B := \mathbb{D}^m$ and define

$$G := \Phi^{-1}(B) = \boldsymbol{D}_{\alpha^1} \cap \cdots \cap \boldsymbol{D}_{\alpha^m}.$$

Observe that G is connected. *Assume* that $A\mathbb{Z}^n = \mathbb{Z}^m$. For example:

$$A = \begin{bmatrix} \alpha_1^1 & \dots & \alpha_{n-m}^1 & 1 & 0 & \dots & 0 \\ \alpha_1^2 & \dots & \alpha_{n-m}^2 & 0 & 1 & \dots & 0 \\ \vdots & & & & & & \\ \alpha_1^m & \dots & \alpha_{n-m}^m & 0 & 0 & \dots & 1 \end{bmatrix}.$$

Then, Proposition 6.2.3 and Lemma 6.2.5 imply that conditions (A, B) from Proposition 6.2.1 are satisfied with $\Sigma := \boldsymbol{V}_0 \cap \mathbb{D}^m$.

Consequently, *if $a \in G$ satisfies* (C), *then Proposition* 6.2.1 *gives effective formulas for $\boldsymbol{m}^{(k)}_G(a,\cdot)$, $\boldsymbol{\gamma}^{(k)}_G(a;\cdot)$, $\boldsymbol{g}_G(a,\cdot)$, and $\boldsymbol{A}_G(a;\cdot)$.*

Example 6.2.9. Consider a more concrete situation: $n = 3$, $m = 2$,

$$A = \begin{bmatrix} 1 & 1 & 0 \\ 1 & 0 & 1 \end{bmatrix}, \quad G = \{(z_1, z_2, z_3) \in \mathbb{C}^3 : |z_1 z_2| < 1,\ |z_1 z_3| < 1\}.$$

Observe that for $a = (a_1, a_2, a_3) \in G$ we have:

- $r(a) = 1$ iff $a \neq 0$,
- $\operatorname{rank} \Phi'(a) = 2$ iff $a_1 \neq 0$,
- $r(0) = 2$,
- $\dim\{X \in \mathbb{C}^3 : \Phi^{(2)}(0)X = 0\} = \dim\{X \in \mathbb{C}^3 : X_1X_2 = 0,\ X_1X_3 = 0\} = 2$.

Thus, by Proposition 6.2.1, if $a_1 \neq 0$, then

$$\begin{aligned} \boldsymbol{m}^{(k)}_G(a,z) = \boldsymbol{g}_G(a,z) &= \boldsymbol{m}_{\mathbb{D}^2}(\Phi(a), \Phi(z)) \\ &= \max\{\boldsymbol{m}_{\mathbb{D}}(a_1a_2, z_1, z_2),\ \boldsymbol{m}_{\mathbb{D}}(a_1a_3, z_1, z_3)\}, \quad z \in G. \end{aligned}$$

The points $(0, a_2, a_3)$ are not covered by Proposition 6.2.1. Let us consider the following three cases:

- $a = 0$: Then,

$$\boldsymbol{m}_G^{(2p)}(0, z) = \boldsymbol{g}_G(0, z) = (\max\{|z_1 z_2|, |z_1 z_3|\})^{1/2}, \quad p \in \mathbb{N},$$
$$\boldsymbol{m}_G^{(2p+1)}(0, z) = (\max\{|z_1 z_2|, |z_1 z_3|\})^{\frac{p+1}{2p+1}}, \quad p \in \mathbb{Z}_+,\ z \in G;$$

thus formulas (6.2.2) and (6.2.4) (and, consequently, (6.2.3) and (6.2.5)) are still valid.

Indeed, the formula for $\boldsymbol{g}_G(0, \cdot)$ follows from Proposition 4.2.10(b). If $f(z) := (z_1 z_2)^p$ or $f(z) := (z_1 z_3)^p$, then $\operatorname{ord}_0 f = 2p$. Consequently,

$$(\max\{|z_1 z_2|, |z_1 z_3|\})^{1/2} \le \boldsymbol{m}_G^{(2p)}(0, z) \le \boldsymbol{g}_G(0, z), \quad z \in G,$$

which gives the formula for $\boldsymbol{m}_G^{(2p)}(0, \cdot)$. If $f(z) := (z_1 z_2)^{p+1}$ or $f(z) := (z_1 z_3)^{p+1}$, then $\operatorname{ord}_0 f = 2p + 2$, which gives the inequality

$$\boldsymbol{m}_G^{(2p+1)}(0, z) \ge (\max\{|z_1 z_2|, |z_1 z_3|\})^{\frac{p+1}{2p+1}}, \quad z \in G.$$

The proof of Proposition 6.2.1 shows that every function $f \in \mathcal{O}(G, \mathbb{D})$ is of the form $f = \widetilde{g} \circ \Phi$, where $\widetilde{g} \in \mathcal{O}(\mathbb{D}^2, \mathbb{D})$. If, moreover, $\operatorname{ord}_0 f \ge 2p + 1$, then

$$|f(z)|^{\frac{1}{2p+1}} \le \boldsymbol{m}_G^{(2p+1)}(0, z) \le \boldsymbol{g}_G(0, z) = (\max\{|z_1 z_2|, |z_1 z_3|\})^{1/2}.$$

Hence,

$$|\widetilde{g}(\xi)| \le (\max\{|\xi_1|, |\xi_2|\})^{p+\frac{1}{2}}, \quad \xi = (\xi_1, \xi_2) \in \mathbb{D}^2,$$

or,

$$|\widetilde{g}(\xi)| \le (\max\{|\xi_1|, |\xi_2|\})^{p+1}, \quad \xi = (\xi_1, \xi_2) \in \mathbb{D}^2.$$

Thus,

$$|f(z)| \le (\max\{|z_1 z_2|, |z_1 z_3|\})^{p+1},$$

which finally gives the formula for $\boldsymbol{m}_G^{(2p+1)}(0, \cdot)$.

- $a = (0, 0, a_3)$, $a_3 \ne 0$: Then,

$$\boldsymbol{g}_G((0,0,a_3), z) = \boldsymbol{m}_G^{(2p)}((0,0,a_3), z) = \max\{|z_1 z_2|^{1/2}, |z_1 z_3|\}, \quad p \in \mathbb{N},$$
$$\boldsymbol{m}_G^{(2p+1)}((0,0,a_3), z) = \max\{|z_1 z_2|^{\frac{p+1}{2p+1}}, |(z_1 z_2)^p z_1 z_3|^{\frac{1}{2p+1}}, |z_1 z_3|\}, \quad p \in \mathbb{Z}_+,\ z \in G;$$

thus, formulas (6.2.2) and (6.2.4) are not valid.

Indeed, the inequalities "$\geq$" are elementary (if $f(z) := (z_1z_2)^p$ or $f(z) := (z_1z_3)^{2p}$, then $\operatorname{ord}_a f \geq 2p$; if $f(z) := (z_1z_2)^{p+1}$ or $f(z) := (z_1z_2)^p z_1z_3$ or $f(z) := (z_1z_3)^{2p+1}$, then $\operatorname{ord}_a f \geq 2p+1$). It suffices to check the formulas only for $z \in G_0 := G \cap (\mathbb{C}^2 \times \mathbb{C}_*)$. Define the biholomorphic mapping

$$F : \mathbb{C}^2 \times \mathbb{C}_* \longrightarrow \mathbb{C}^2 \times \mathbb{C}_*, \quad F(w) := (w_1/w_3, w_2w_3, w_3),$$

and observe $G_0 = F(D \times \mathbb{C}_*)$, where

$$D := \{(w_1, w_2) \in \mathbb{C}^2 : |w_1| < 1,\ |w_1w_2| < 1\}.$$

Thus, if $(z_1, z_2, z_3) \in G_0$, then (using Remark 4.2.9) we get

$$\begin{aligned} \boldsymbol{m}_G^{(k)}((0,0,a_3),(z_1,z_2,z_3)) &= \boldsymbol{m}_G^{(k)}(F(0,0,a_3), F(z_1z_3, z_2/z_3, z_3)) \\ &\leq \boldsymbol{m}_{D\times\mathbb{C}_*}^{(k)}((0,0,a_3),(z_1z_3, z_2/z_3, z_3)) \\ &= \boldsymbol{m}_D^{(k)}((0,0),(z_1z_3, z_2/z_3)). \end{aligned}$$

Observe that

$$\boldsymbol{m}_D^{(k)}(0,w) \leq \boldsymbol{g}_D(0,w) = \mathfrak{h}_D(w) = \max\{|w_1|, |w_1w_2|^{1/2}\}, \quad k \in \mathbb{N},\ w \in D;$$

cf. Proposition 4.2.10(b). Thus, we only need to check that

$$\boldsymbol{m}_D^{(2p+1)}(0,w) \leq \max\{|w_1|, |w_1(w_1w_2)^p|^{\frac{1}{2p+1}}, |w_1w_2|^{\frac{p+1}{2p+1}}\}, \quad p \in \mathbb{Z}_+,\ w \in D.$$

Note that $\{0\}\times\mathbb{C} \subset D$. In particular, by the Liouville theorem, $\boldsymbol{m}_D^{(k)}((0,0),(0,w_2)) = 0$, $w_2 \in \mathbb{C}$. So, we may assume that $w_1 \neq 0$. Fix a point $w_1^0 \in \mathbb{D}_*$. We like to show that

$$\begin{aligned} L(w_2) &:= \boldsymbol{m}_D^{(2p+1)}(0,(w_1^0, w_2)) \\ &\leq \begin{cases} |w_1^0|, & \text{if } |w_2| \leq |w_1^0| \\ |w_1^0(w_1^0w_2)^p|^{\frac{1}{2p+1}}, & \text{if } |w_1^0| \leq |w_2| \leq 1 \\ |w_1^0w_2|^{\frac{p+1}{2p+1}}, & \text{if } 1 \leq |w_2| < 1/|w_1^0| \end{cases} =: R(w_2). \end{aligned}$$

Since the function L is log-sh, the Hadamard three circles theorem (Appendix B.4.25) implies that it is enough to prove that $L(w_2) \leq R(w_2)$ on the "jumping lines" $|w_2| = |w_1^0|$ and $|w_2| = 1$.

If $|w_2| = |w_1^0|$, then $L(w_2) \leq \boldsymbol{g}_D((0,0),(w_1^0, w_2)) = |w_1^0| = R(w_2)$.

Take an $f \in \mathcal{O}(D, \mathbb{D})$ with $\operatorname{ord}_0 f \ge 2p+1$. Then,

$$|f(\lambda, e^{i\theta})| \le (\boldsymbol{g}_D((0,0),(\lambda, e^{i\theta}))^{2p+1} = |\lambda|^{p+\frac{1}{2}}, \quad \lambda \in \mathbb{D}.$$

Hence, $|f(\lambda, e^{i\theta})| \le |\lambda|^{p+1}$, $\lambda \in \mathbb{D}$, which implies that for $|w_2| = 1$ we get $L(w_2) \le |w_1^0|^{\frac{p+1}{2p+1}} = R(w_2)$.

- $a = (0, a_2, a_3)$, $a_2 a_3 \neq 0$: Define

$$f(z) := \frac{a_3 z_1 z_2 - a_2 z_1 z_3}{|a_2| + |a_3|}, \quad z = (z_1, z_2, z_3) \in G.$$

Then, $f \in \mathcal{O}(G, \mathbb{D})$ and $\operatorname{ord}_a f = 2$. Hence,

$$\begin{aligned} \boldsymbol{g}_G(a,z) &\ge \boldsymbol{m}_G^{(k)}(a,z) \\ &\ge \max\left\{|z_1 z_2|, |z_1 z_3|, \left(\frac{|a_3 z_1 z_2 - a_2 z_1 z_3|}{|a_2| + |a_3|}\right)^{\lceil k/2 \rceil / k}\right\}, \quad z \in G. \end{aligned}$$

In particular, formulas (6.2.2) and (6.2.4) are not valid.

Formulas for $\boldsymbol{m}_G^{(k)}((0, a_2, a_3), \cdot)$ and $\boldsymbol{g}_G((0, a_2, a_3), \cdot)$ are not known; see also Exercise 6.4.2 ?

The main idea of Example 6.2.9 may be extended to the following general result.

Proposition 6.2.10 (cf. [263]). *Let* $\alpha^j = (\alpha_1^j, \dots, \alpha_j^j) \in (\mathbb{Z}_+^s)_* \times \mathbb{Z}^{n-s}$, $j = 1, \dots, m$, $m \ge 2$, $1 \le s \le n-1$,

$$G := \boldsymbol{D}_{\alpha^1} \cap \dots \cap \boldsymbol{D}_{\alpha^m}$$

(we do not assume that $m \le n-1$, $\operatorname{rank} A = m$, *or* $A\mathbb{Z}^n = \mathbb{Z}^m$*). Fix an* $a = (0, \dots, 0, a_{s+1}, \dots, a_n)$ *with* $a_{s+1} \cdots a_n \neq 0$. *Put*

$$A := [\alpha_k^j]_{\substack{j=1,\dots,m \\ k=1,\dots,n}} \in \mathbb{M}(m \times n; \mathbb{Z}),$$

$$\widetilde{A} := [\alpha_k^j]_{\substack{j=1,\dots,m \\ k=1,\dots,s}} = \begin{bmatrix} \beta^1 \\ \vdots \\ \beta^m \end{bmatrix} \in \mathbb{M}(m \times s; \mathbb{Z}_+).$$

Let $r_j := \operatorname{ord}_a z^{\alpha^j} = |\beta^j| > 0$, $j = 1, \dots, m$. *Then, the following conditions are equivalent:*

(i) $\operatorname{rank} A = \operatorname{rank} \widetilde{A}$;

(ii) $\boldsymbol{g}_G(a,z) = \max\{|z^{\alpha^j}|^{1/r_j} : j = 1, \dots, m\}$, $z \in G$;

(iii) $\boldsymbol{g}_G(a, (\lambda z', z'')) = |\lambda| \boldsymbol{g}_G(a,z)$, $z = (z', z'') \in G \subset \mathbb{C}^s \times \mathbb{C}^{n-s}$;

(iv) *for every* $k \in \mathbb{N}$ *the set*

$$\Big\{(z', z'') \in G \subset \mathbb{C}^s \times \mathbb{C}^{n-s} : \limsup_{\theta \to 0+} \frac{1}{\theta} \boldsymbol{m}_G^{(k)}(a, (\lambda z', z'')) < +\infty\Big\}$$

is not pluripolar.

Note that (ii) gives an effective formula for $\boldsymbol{g}_G(a, \cdot)$, which does not look like the formulas in Proposition 6.2.1 unless $r_1 = \dots = r_m$.

Proof. (i) $\Longrightarrow$ (ii): Let

$$L(z) := \boldsymbol{g}_G(a, z), \quad R(z) := \max\{|z^{\alpha^j}|^{1/r_j} : j = 1, \dots, m\}, \qquad z \in G.$$

The inequality $L \geq R$ follows from the definition of $\boldsymbol{g}_G$. To prove that $L \leq R$ it suffices to show that $L(z) \leq R(z)$ for any $z \in G_0 := G \cap (\mathbb{C}^s \times \mathbb{C}_*^{n-s})$.

By virtue of (i), for any $k \in \{s+1, \dots, n\}$, the system of equations

$$\alpha_1^j x_1 + \dots + \alpha_s^j x_s = -\alpha_k^j, \quad j = 1, \dots, m,$$

has a rational solution $(Q_1^k/\mu_k, \dots, Q_s^k/\mu_k)$ with $Q_1^k, \dots, Q_s^k \in \mathbb{Z}$, $\mu_k \in \mathbb{N}$. Put $Q_k^k := \mu_k$ and $Q_j^k := 0$, $j = s+1, \dots, n$, $j \neq k$. Then,

$$\alpha_1^j Q_1^k + \dots + \alpha_n^j Q_n^k = 0, \quad j = 1, \dots, m, \ k = s+1, \dots, n. \tag{6.2.7}$$

Let $Q_j := (Q_j^{s+1}, \dots, Q_j^n) \in \mathbb{Z}^{n-s}$, $j = 1, \dots, n$. Define

$$F : \mathbb{C}^s \times \mathbb{C}_*^{n-s} \longrightarrow \mathbb{C}^s \times \mathbb{C}_*^{n-s},$$

$$\begin{aligned} F(\eta, \xi) &:= (\xi^{Q_1}\eta_1, \dots, \xi^{Q_s}\eta_s, \xi^{Q_{s+1}}, \dots, \xi^{Q_n}) \\ &= (\xi^{Q_1}\eta_1, \dots, \xi^{Q_s}\eta_s, \xi_{s+1}^{\mu_{s+1}}, \dots, \xi_n^{\mu_n}), \\ \eta &= (\eta_1, \dots, \eta_s) \in \mathbb{C}^s, \ \xi = (\xi_{s+1}, \dots, \xi_n) \in \mathbb{C}_*^{n-s}. \end{aligned}$$

Observe that F is surjective. Indeed, for $z = (z_1, \dots, z_n) \in \mathbb{C}^s \times \mathbb{C}_*^{n-s}$, take an arbitrary $\xi_j \in \sqrt[\mu_j]{z_j}$, $j = s+1, \dots, n$, and define $\eta_j := z_j/\xi^{Q_j}$, $j = 1, \dots, s$.

Moreover, if $z = F(\eta, \xi)$, then, by (6.2.7), we get

$$z^{\alpha^j} = \xi^{\alpha_1^j Q_1 + \dots + \alpha_n^j Q_n}\eta^{\beta^j} = \eta^{\beta^j}, \quad j = 1, \dots, m. \tag{6.2.8}$$

Let

$$D := \{\eta \in \mathbb{C}^s : |\eta^{\beta^j}| < 1, \ j = 1, \dots, m\}.$$

Using (6.2.8) we get $F(D \times \mathbb{C}_*^{n-s}) = G_0$. Fix a $\xi_0 \in \mathbb{C}_*^{n-s}$ such that $a = F(0, \xi_0)$. Then, for any $z = F(\eta, \xi) \in G_0$, we have

$$\begin{aligned} \boldsymbol{g}_G(a, z) &= \boldsymbol{g}_G(F(0, \xi_0), F(\eta, \xi)) \le \boldsymbol{g}_{D \times \mathbb{C}_*^{n-s}}((0, \xi_0), (\eta, \xi)) \overset{(*)}{=} \boldsymbol{g}_D(0, \eta) \\ &= \mathfrak{h}_D(\eta) = \max\left\{|\eta^{\beta^j}|^{1/r_j} : j = 1, \dots, m\right\} \\ &= \max\left\{|z^{\alpha^j}|^{1/r_j} : j = 1, \dots, m\right\}, \end{aligned}$$

where (*) follows from Remark 4.2.9.

The implications (ii) $\Longrightarrow$ (iii) $\Longrightarrow$ (iv) are obvious.

(iv) $\Longrightarrow$ (i): Suppose that $\operatorname{rank} \widetilde{A} < \operatorname{rank} A$. We may assume that

$$2 \le t := \operatorname{rank} A = \operatorname{rank} \begin{bmatrix} \alpha^1 \\ \vdots \\ \alpha^t \end{bmatrix}, \quad \operatorname{rank} \begin{bmatrix} \beta^1 \\ \vdots \\ \beta^t \end{bmatrix} < t.$$

Then, there exist $c_1, \dots, c_t \in \mathbb{Z}$ such that $c_1\beta^1 + \dots + c_t\beta^t = 0$ and $|c_1| + \dots + |c_t| > 0$. To simplify notation, assume that $c_1, \dots, c_u \ge 0$, $c_{u+1}, \dots, c_t < 0$ for some $1 \le u \le t - 1$. Let

$$d := a^{c_1\alpha^1 + \dots + c_t\alpha^t}, \quad r := c_1 r_1 + \dots + c_u r_u = -(c_{u+1} r_{u+1} + \dots + c_t r_t),$$

$$f(z) := \frac{z^{c_1\alpha^1 + \dots + c_u\alpha^u} - d z^{-(c_{u+1}\alpha^{u+1} + \dots + c_t\alpha^t)}}{1 + |d|}, \quad z \in G.$$

Observe that f is well-defined, $f \in \mathcal{O}(G, \mathbb{D})$, $\operatorname{ord}_a f \ge r + 1$, and $f \not\equiv 0$ (because $\alpha^1, \dots, \alpha^t$ are linearly independent). Take a $b = (b', b'') \in G \subset \mathbb{C}^s \times \mathbb{C}^{n-s}$ with $f(b) \ne 0$. Observe that $f(\theta b', b'') = \theta^r f(b)$, $0 \le \theta \le 1$. Hence,

$$\begin{aligned} \frac{1}{\theta} \boldsymbol{m}_G^{(r+1)}(a, (\theta b', b'')) &\ge \frac{1}{\theta} |f(\theta b', b'')|^{1/(r+1)} \\ &= \theta^{-1/(r+1)} |f(b)|^{1/(r+1)} \underset{\theta \to 0+}{\longrightarrow} +\infty; \end{aligned}$$

a contradiction. □

6.3 Elementary n-circled domains II

Our aim is to complete the discussion from §§ 6.1 and 6.2 and to establish effective formulas for

$$d_{D_\alpha} \in \{\boldsymbol{m}_{D_\alpha}^{(k)}, \boldsymbol{s}_{D_\alpha}, \boldsymbol{g}_{D_\alpha}, \boldsymbol{\ell}_{D_\alpha}^{(k)*}, \boldsymbol{k}_{D_\alpha}^*\}, \quad \delta_{D_\alpha} \in \{\boldsymbol{\gamma}_{D_\alpha}^{(k)}, \boldsymbol{S}_{D_\alpha}, \boldsymbol{A}_{D_\alpha}, \boldsymbol{\varkappa}_{D_\alpha}^{(k)}\},$$

where

$$\boldsymbol{D}_\alpha = \{(z_1, \dots, z_n) \in \mathbb{C}^n(\alpha) : |z^\alpha| < 1\},$$

$\alpha \in (\mathbb{R}^n)_*$, $|z^\alpha| := |z_1|^{\alpha_1} \cdots |z_n|^{\alpha_n}$ (cf. § 2.10). This section is based on [422] and [565].

Remark 6.3.1.

(a) Recall that the general elementary Reinhardt domain $\boldsymbol{D}_{\alpha,C}$ is biholomorphic to $\boldsymbol{D}_\alpha$. Thus, we are in fact, studying holomorphically contractible functions and/or pseudometrics for $\boldsymbol{D}_{\alpha,C}$.

(b) If $\alpha_{s+1} = \dots = \alpha_n = 0$ for some $s \in \{1, \dots, n-1\}$, then $\boldsymbol{D}_\alpha^n = \boldsymbol{D}_\beta^s \times \mathbb{C}^{n-s}$ with $\beta := (\alpha_1, \dots, \alpha_s)$. Consequently,

$$d_{\boldsymbol{D}_\alpha^n}((a,b),(z,w)) = d_{\boldsymbol{D}_\beta^s}(a,z), \ (a,b),(z,w) \in \boldsymbol{D}_\beta^s \times \mathbb{C}^{n-s},$$
$$\delta_{\boldsymbol{D}_\alpha^n}((a,b);(X,Y)) = \delta_{\boldsymbol{D}_\beta^s}(a;X), \ (a,b) \in \boldsymbol{D}_\beta^s \times \mathbb{C}^{n-s}, \ (X,Y) \in \mathbb{C}^s \times \mathbb{C}^{n-s};$$

cf. Remark 4.2.9 and Propositions 3.7.1, 3.8.7. Thus, *we may assume that* $\alpha \in \mathbb{R}_*^n$.

(c) If $n = 1$, then either $\boldsymbol{D}_\alpha = \mathbb{D}$ (for $\alpha > 0$) or $\boldsymbol{D}_\alpha = \mathbb{A}(1, +\infty)$ (for $\alpha < 0$). The first case is obvious ($d_\mathbb{D} = \boldsymbol{m}$, $\delta_\mathbb{D} = \boldsymbol{\gamma}$). In the second case, observe that the mapping $\mathbb{A}(1, +\infty) \ni z \longmapsto 1/z \in \mathbb{D}_*$ is biholomorphic. Thus, the problem reduces to $\mathbb{D}_*$. Recall that

$$\boldsymbol{m}_{\mathbb{D}_*}^{(k)} = \boldsymbol{s}_{\mathbb{D}_*} = \boldsymbol{g}_{\mathbb{D}_*} = \boldsymbol{m}|_{\mathbb{D}_* \times \mathbb{D}_*}, \quad \boldsymbol{\ell}_{\mathbb{D}_*}^{(k)*} = \boldsymbol{\ell}_{\mathbb{D}_*}^* = \boldsymbol{k}_{\mathbb{D}_*}^*,$$
$$\boldsymbol{\gamma}_{\mathbb{D}_*}^{(k)} = \boldsymbol{S}_{\mathbb{D}_*} = \boldsymbol{A}_{\mathbb{D}_*} = \boldsymbol{\gamma}_\mathbb{D}|_{\mathbb{D}_* \times \mathbb{C}}, \quad \boldsymbol{\varkappa}_{\mathbb{D}_*}^{(k)} = \boldsymbol{\varkappa}_{\mathbb{D}_*}.$$

(d) Using rotations R_θ and permutations of variables, we see that it is enough to determine $d_{\boldsymbol{D}_\alpha}(a, \cdot)$ and $\delta_{\boldsymbol{D}_\alpha}(a, \cdot)$, with a fixed point $a \in \boldsymbol{D}_\alpha$, under the following additional assumptions (AS):

- $n \geq 2$;
- $\alpha_1, \dots, \alpha_q < 0$ *and* $\alpha_{q+1}, \dots, \alpha_n > 0$ *for a* $q = q(\alpha) \in \{0, 1, \dots, n\}$;
- *if* $q < n$, *then we set* $t = t(\alpha) := \min\{\alpha_{q+1}, \dots, \alpha_n\}$;
- $a = (a_1, \dots, a_n) \in \boldsymbol{D}_\alpha$, $a_1, \dots, a_s > 0$, $a_{s+1} = \dots = a_n = 0$ *for an* $s = s(a) \in \{q, \dots, n\}$;
- *if* $s < n$, *then we set* $r = r(a) = r_\alpha(a) := \alpha_{s+1} + \dots + \alpha_n$;
- *if* $s = n$ *(in particular, if* $q = n$*), then* $r = r(a) = r_\alpha(a) := 1$;

- *if $\boldsymbol{D}_\alpha$ is of rational type, then we assume that $\alpha \in \mathbb{Z}^n$ and $\alpha_1, \dots, \alpha_n$ are relatively prime;*
- *if $\boldsymbol{D}_\alpha$ is of irrational type and $q < n$, then we assume that $t(\alpha) = 1$.*

Observe that if $\alpha \in \mathbb{Z}^n$, then $r(a) = \operatorname{ord}_a(\Phi - \Phi(a))$ with $\Phi(z) := z^\alpha$.

We begin with

$$d_{\boldsymbol{D}_\alpha} \in \left\{\boldsymbol{m}^{(k)}_{\boldsymbol{D}_\alpha}, \boldsymbol{s}_{\boldsymbol{D}_\alpha}, \boldsymbol{g}_{\boldsymbol{D}_\alpha}\right\}, \quad \delta_{\boldsymbol{D}_\alpha} \in \left\{\boldsymbol{\gamma}^{(k)}_{\boldsymbol{D}_\alpha}, \boldsymbol{S}_{\boldsymbol{D}_\alpha}, \boldsymbol{A}_{\boldsymbol{D}_\alpha}\right\}.$$

The remaining objects will be discussed in Chapter 10.

Recall that the case where $\boldsymbol{D}_\alpha$ is of rational type was presented in Proposition 6.2.7.

Using Lemma 2.10.1 (see also Proposition 2.10.2), we immediately get the following result:

Proposition 6.3.2. *Assume that $\boldsymbol{D}_\alpha$ is of irrational type. Then $\boldsymbol{m}^{(k)}_{\boldsymbol{D}_\alpha} \equiv 0$ and $\boldsymbol{\gamma}^{(k)}_{\boldsymbol{D}_\alpha} \equiv 0$, $k \in \mathbb{N}$.*

The case of the Green function is more complicated.

Proposition 6.3.3. *Assume that $\boldsymbol{D}_\alpha$ is of irrational type and conditions* (AS) *are satisfied. Then, for $z \in \boldsymbol{D}_\alpha$ and $X \in \mathbb{C}^n$, we have:*

$$\boldsymbol{g}_{\boldsymbol{D}_\alpha}(a, z) = \begin{cases} \left(\prod_{j=1}^{n} |z_j|^{\alpha_j}\right)^{1/r}, & s < n \\ 0, & s = n \end{cases},$$

$$\boldsymbol{A}_{\boldsymbol{D}_\alpha}(a; X) = \begin{cases} \left(\prod_{j=1}^{s} a_j^{\alpha_j} \prod_{j=s+1}^{n} |X_j|^{\alpha_j}\right)^{1/r}, & s < n \\ 0, & s = n \end{cases}.$$

We need some auxiliary results. Proposition 4.2.10(d) implies the following lemma:

Lemma 6.3.4. *Suppose that $q > 0$, put $\beta := (-\alpha_1, \dots, -\alpha_q, \alpha_{q+1}, \dots, \alpha_n)$, $\Sigma := \{(z_1, \dots, z_n) \in \mathbb{C}^n : z_1 \cdots z_q = 0\}$, and let*

$$\boldsymbol{D}_\alpha \ni (\zeta_1, \dots, \zeta_n) \overset{F}{\longmapsto} (1/\zeta_1, \dots, 1/\zeta_q, \zeta_{q+1}, \dots, \zeta_n) \in \boldsymbol{D}_\beta \setminus \Sigma.$$

Then, F is biholomorphic and

$$\boldsymbol{g}_{\boldsymbol{D}_\alpha}(a, z) = \boldsymbol{g}_{\boldsymbol{D}_\beta}(F(a), F(z)).$$

Note that $r_\beta(F(a)) = r_\alpha(a)$.

Definition 6.3.5. For $\zeta = (\zeta_1, \dots, \zeta_n) \in \mathbb{C}^n$, define

$$\mathbb{T}_\zeta := \{R_\theta(\zeta) : \theta \in \mathbb{R}^n\} = \{(e^{i\theta_1}\zeta_1, \dots, e^{i\theta_n}\zeta_n) : \theta_1, \dots, \theta_n \in \mathbb{R}\};$$

we consider $\mathbb{T}_\zeta$ as a commutative group. Let $\mathbb{T}_\zeta(\alpha)$ denote the subgroup of $\mathbb{T}_\zeta$, generated by

$$\left\{\left(e^{\frac{\alpha_{j_1}}{\alpha_1}2s_1\pi i}\zeta_1, \dots, e^{\frac{\alpha_{j_n}}{\alpha_n}2s_n\pi i}\zeta_n\right) : j_1, \dots, j_n \in \{1, \dots, n\},\ s_1, \dots, s_n \in \mathbb{Z}\right\}. \tag{6.3.1}$$

Remark 6.3.6.

(a) If $\alpha \in \mathbb{Z}_*^n$ and $\alpha_1, \dots, \alpha_n$ are relatively prime, then

$$\mathbb{T}_\zeta(\alpha) := \{(\varepsilon_1\zeta_1, \dots, \varepsilon_n\zeta_n) : \varepsilon_j^{\alpha_j} = 1,\ j = 1, \dots, n\}.$$

Indeed, the inclusion "$\subset$" is obvious. We know that $k_1\alpha_1 + \dots + k_n\alpha_n = 1$ for some $k_1, \dots, k_n \in \mathbb{Z}$. Let $\varepsilon_j = e^{\frac{\ell_j}{\alpha_j}2\pi i}$, where $\ell_j \in \mathbb{Z}$, $j = 1, \dots, n$. Then,

$$\frac{\ell_j}{\alpha_j} = \frac{\alpha_1}{\alpha_j}k_1\ell_j + \dots + \frac{\alpha_n}{\alpha_j}k_n\ell_j, \quad j = 1, \dots, n.$$

Thus, $(\varepsilon_1\zeta_1, \dots, \varepsilon_n\zeta_n)$ may be obtained by superposition of n elementary operations of type (6.3.1), generated by $(j_1, \dots, j_n, s_1, \dots, s_n) = (\nu, \dots, \nu, k_\nu\ell_1, \dots, k_\nu\ell_n)$, $\nu = 1, \dots, n$.

(b) If $\alpha \in \mathbb{R}_*^n$ is of irrational type, then, by the Kronecker theorem (cf. B.10.1), $\overline{\mathbb{T}_\zeta(\alpha)} = \mathbb{T}_\zeta$.

(c) $\widetilde{\zeta} \in \mathbb{T}_\zeta(\alpha) \iff \zeta \in \mathbb{T}_{\widetilde{\zeta}}(\alpha)$.

Lemma 6.3.7.

(a) *Let $\widetilde{z} \in \mathbb{T}_z(\alpha)$, $\varphi \in \mathcal{O}(\mathbb{D}, \boldsymbol{D}_\alpha)$, $\varphi(0) = a$, $\varphi(\sigma) = z$, $\sigma \neq 0$. Then, there exists a $\widetilde{\varphi} \in \mathcal{O}(\mathbb{D}, \boldsymbol{D}_\alpha)$ such that $\widetilde{\varphi}(0) = a$ and $\widetilde{\varphi}(\sigma) = \widetilde{z}$. Consequently, $\boldsymbol{\ell}_{\boldsymbol{D}_\alpha}(a, \widetilde{z}) = \boldsymbol{\ell}_{\boldsymbol{D}_\alpha}(a, z)$.*

(b) *Assume that $\alpha \in \mathbb{Z}_*^n$ and $\alpha_1, \dots, \alpha_n$ are relatively prime. Let $\varphi \in \mathcal{O}(\mathbb{D}, \boldsymbol{D}_\alpha \setminus V_0)$, $\varphi(0) = a$, and $z_j^{\alpha_j} = \varphi_j^{\alpha_j}(\sigma)$, $j = 1, \dots, n$, for some $\sigma \geq \delta > 0$. Then, there exists a $\widetilde{\varphi} \in \mathcal{O}(\mathbb{D}, \boldsymbol{D}_\alpha \setminus V_0)$ such that*

- $\widetilde{\varphi}(0) = a$, $\widetilde{\varphi}(\sigma) = z$,
- $\widetilde{\varphi}_1^{\alpha_1} \cdots \widetilde{\varphi}_n^{\alpha_n} = \varphi_1^{\alpha_1} \cdots \varphi_n^{\alpha_n}$,
- $(1/c)|\varphi_j| \leq |\widetilde{\varphi}_j| \leq c|\varphi_j|$, $j = 1, \dots, n$, *where $c > 0$ depends only on α and δ.*

Proof. (a) Let $\mathbb{H} := (-1, 1) \times \mathbb{R}$. Observe that, instead of holomorphic discs $\mathbb{D} \longrightarrow \boldsymbol{D}_\alpha$, we may consider holomorphic strips $\mathbb{H} \longrightarrow \boldsymbol{D}_\alpha$. We may also assume that $\sigma = i\tau$, $\tau > 0$. Thus, $\varphi \in \mathcal{O}(\mathbb{H}, \boldsymbol{D}_\alpha)$, $\varphi(0) = a$, $\varphi(i\tau) = z$. Define

$$\begin{aligned}\widetilde{\varphi}(\lambda) &= (\widetilde{\varphi}_1(\lambda), \dots, \widetilde{\varphi}_n(\lambda)) \\ &:= \left(\varphi_1(\lambda), \dots, \varphi_{n-2}(\lambda), e^{-2s_n\pi\lambda/\tau}\varphi_{n-1}(\lambda), e^{\frac{\alpha_{n-1}}{\alpha_n}2s_n\pi\lambda/\tau}\varphi_n(\lambda)\right), \ \lambda \in \mathbb{H}.\end{aligned}$$

Then, $\widetilde{\varphi} : \mathbb{H} \longrightarrow \boldsymbol{D}_\alpha$, $\widetilde{\varphi}(0) = \varphi(0) = a$,

$$\widetilde{\varphi}(i\tau) = \left(z_1, \dots, z_{n-1}, e^{\frac{\alpha_{n-1}}{\alpha_n}2s_n\pi i}z_n\right),$$

and $\widetilde{\varphi}_1^{\alpha_1} \cdots \widetilde{\varphi}_n^{\alpha_n} = \varphi_1^{\alpha_1} \cdots \varphi_n^{\alpha_n}$. Observe that

$$\max\{|\varphi_j/\widetilde{\varphi}_j|, |\widetilde{\varphi}_j/\varphi_j|\} \le e^{c/\tau}, \quad j = 1, \dots, n,$$

where c depends only on α and $|s_n|$.

Repeating the same procedure with respect to other coordinates of α and/or z, we get the required result.

(b) follows directly from the proof of (a) and Remark 6.3.6(a). □

Corollary 6.3.8. *Assume that α is of irrational type and $z'' \in \mathbb{T}_{z'}$. Then, $\boldsymbol{\ell}_{\boldsymbol{D}_\alpha}(a, z') = \boldsymbol{\ell}_{\boldsymbol{D}_\alpha}(a, z'')$. In particular, if $z \in \mathbb{T}_a$, then $\boldsymbol{\ell}_{\boldsymbol{D}_\alpha}(a, z) = 0$.*

Proof. By Lemma 6.3.7, we have

$$\boldsymbol{\ell}_{\boldsymbol{D}_\alpha}(a, z'') = \boldsymbol{\ell}_{\boldsymbol{D}_\alpha}(a, \widetilde{z}), \quad \widetilde{z} \in \mathbb{T}_{z''}(\alpha) \subset \mathbb{T}_{z''} = \mathbb{T}_{z'}.$$

Since $\overline{\mathbb{T}_{z''}(\alpha)} = \mathbb{T}_{z'}$ (Remark 6.3.6(b)) and $\boldsymbol{\ell}_{\boldsymbol{D}_\alpha}$ is upper semicontinuous (Proposition 3.1.14), we get $\boldsymbol{\ell}_{\boldsymbol{D}_\alpha}(a, z') \ge \limsup_{\mathbb{T}_{z''}(\alpha) \ni \widetilde{z} \to z'} \boldsymbol{\ell}_{\boldsymbol{D}_\alpha}(a, \widetilde{z}) = \boldsymbol{\ell}_{\boldsymbol{D}_\alpha}(a, z'')$. □

Lemma 6.3.9.

(a) *For arbitrary $\lambda_0', \lambda_0'' \in \mathbb{D}$, $b_1, \dots, b_n, c_1, \dots, c_\mu \in \mathbb{C}_*$ with $\lambda_0' \neq \lambda_0''$, $|b_1|^{\alpha_1} \cdots |b_n|^{\alpha_n} = 1$, and $\mu < n$, there exist $\psi_1, \dots, \psi_n \in \mathcal{O}(\mathbb{D}, \mathbb{C}_*)$ such that*

- $\psi_j(\lambda_0') = b_j$, $j = 1, \dots, n$,
- $\psi_j(\lambda_0'') = c_j$, $j = 1, \dots, \mu$,
- $|\psi_1|^{\alpha_1} \cdots |\psi_n|^{\alpha_n} \equiv 1$.

(b) *For arbitrary $\lambda_0 \in \mathbb{D}$, $\sigma \in \mathbb{C}_*$, $b_1, \dots, b_n \in \mathbb{C}_*$, $X_1, \dots, X_\mu \in \mathbb{C}$ with $|b_1|^{\alpha_1} \cdots |b_n|^{\alpha_n} = 1$, and $\mu < n$, there exist $\psi_1, \dots, \psi_n \in \mathcal{O}(\mathbb{D}, \mathbb{C}_*)$ such that*

- $\psi_j(\lambda_0) = b_j$, $j = 1, \dots, n$,
- $\sigma\psi_j'(\lambda_0) = X_j$, $j = 1, \dots, \mu$,
- $|\psi_1|^{\alpha_1} \cdots |\psi_n|^{\alpha_n} \equiv 1$.

(c) *Assume that $\alpha \in \mathbb{Z}_*^n$ and $\alpha_1, \dots, \alpha_n$ are relatively prime. Let $L_0', L_0'' \subset\subset \mathbb{D}$, $L \subset\subset \mathbb{C}_*$ be such that*

$$\min\{\boldsymbol{m}(\lambda_1, \lambda_2) : \lambda_1 \in L_0',\ \lambda_2 \in L_0''\} \geq \delta > 0.$$

Then, there exists a compact $K \subset\subset \mathbb{C}_*$, depending only on α, δ and L, such that for any $\lambda_0' \in L_0'$, $\lambda_0'' \in L_0''$, $b, c \in L^n$ with $b^\alpha = c^\alpha = 1$, there exists a $\widetilde{\psi} \in \mathcal{O}(\mathbb{D}, K^n)$ such that $\widetilde{\psi}(\lambda_0') = b$, $\widetilde{\psi}(\lambda_0'') = c$, and $\widetilde{\psi}_1^{\alpha_1} \cdots \widetilde{\psi}_n^{\alpha_n} \equiv 1$.

Proof. (a), (b) We may assume $\lambda_0' = 0$, $\lambda_0'' = \sigma \in (0, 1)$ (resp. $\lambda_0 = 0$, $\sigma > 0$) and $\mu = n - 1$. Let $q_j := \operatorname{Log} b_j$, $j = 1, \dots, n - 1$. Put

$$\psi_j(\lambda) := \exp(p_j\lambda + q_j), \quad j = 1, \dots, n - 1,$$
$$\psi_n(\lambda) := e^{i\theta} \exp\Big(-\frac{1}{\alpha_n} \sum_{j=1}^{n-1} \alpha_j(p_j\lambda + q_j)\Big).$$

Then,

- $\psi_j(0) = b_j$, $j = 1, \dots, n - 1$,
- $|\psi_1|^{\alpha_1} \cdots |\psi_n|^{\alpha_n} \equiv 1$,
- $|\psi_n(0)| = (|b_1|^{\alpha_1} \cdots |b_{n-1}|^{\alpha_{n-1}})^{-1/\alpha_n} = |b_n|$, so for arbitrary $p_1, \dots, p_{n-1}$ we may find a $\theta \in \mathbb{R}$ such that $\psi_n(0) = b_n$. Finally, we put $p_j := \frac{1}{\sigma} \operatorname{Log} \frac{c_j}{b_j}$ (resp. $p_j := \frac{X_j}{\sigma b_j}$), $j = 1, \dots, n - 1$.

(c) Using (a), we get a compact $K_0 = K_0(\alpha, \delta, L) \subset\subset \mathbb{C}_*$ and $\psi_j \in \mathcal{O}(\mathbb{D}, K_0)$ such that $\psi_j(\lambda_0') = b_j$, $\psi_j(\lambda_0'') = c_j$, $j = 1, \dots, n-1$. Put $\psi_n := (\psi^{\alpha_1} \cdots \psi^{\alpha_{n-1}})^{-1/\alpha_n}$, where the power $(-1/\alpha_n)$ is chosen so that $\psi_n(\lambda_0') = b_n$. We have $\psi_n^{\alpha_n}(\lambda_0'') = c_n^{\alpha_n}$. Now, using Lemma 6.3.7(b), we may modify ψ to get the required K and $\widetilde{\psi}$. □

Proof of Proposition 6.3.3. The formulas for $\boldsymbol{A}_{\boldsymbol{D}_\alpha}(a; X)$ follow from the formulas for $\boldsymbol{g}_{\boldsymbol{D}_\alpha}(a, \cdot)$ and Lemma 4.2.7.

In order to determine $\boldsymbol{g}_{\boldsymbol{D}_\alpha}(a, \cdot)$ in the case where $q > 0$, we first apply Lemma 6.3.4 and reduce the problem to $q = 0$.

In the case $s < n$, the function $\boldsymbol{D}_\alpha \ni \zeta \overset{u}{\longmapsto} |\zeta^\alpha|^{1/r}$ belongs to $\mathcal{K}_{\boldsymbol{D}_\alpha}(a)$. Indeed, if ζ is in a small neighborhood of a, then

$$u(\zeta) \leq \text{const}\Big(\prod_{j=s+1}^{n} |\zeta_j|^{\alpha_j}\Big)^{1/r} \leq \text{const}\,\|(\zeta_{s+1}, \dots, \zeta_n)\| \leq \text{const}\,\|\zeta - a\|.$$

Thus, $\boldsymbol{g}_{\boldsymbol{D}_\alpha}(a,z) \geq u(z)$. To prove the opposite inequality, first observe that

$$\boldsymbol{V}_0 = \{(z_1,\dots,z_n) \in \mathbb{C}^n : z_1 \cdots z_n = 0\} \subset \boldsymbol{D}_\alpha$$

(because $q = 0$) and that $\boldsymbol{g}_{\boldsymbol{D}_\alpha}(a,\cdot) \equiv 0$ on $\boldsymbol{V}_0$ (because $a \in \boldsymbol{V}_0$). This finishes the proof in the case where $z \in \boldsymbol{V}_0$. In the case where $z \in \boldsymbol{D}_\alpha \setminus \boldsymbol{V}_0$ put $\sigma := |z^\alpha|^{1/r} > 0$. By Lemma 6.3.9(a), there exist $\psi_1,\dots,\psi_n \in \mathcal{O}(\mathbb{D},\mathbb{C}_*)$ such that

- $\psi_j(\sigma) = z_j$, $j = 1,\dots,s$, $\varphi_j(\sigma) = z_j/\sigma$, $j = s+1,\dots,n$,
- $\psi_j(0) = a_j$, $j = 1,\dots,s$,
- $|\psi_1|^{\alpha_1} \cdots |\psi_n|^{\alpha_n} \equiv 1$.

Define $\varphi := (\psi_1,\dots,\psi_s,\lambda\psi_{s+1},\dots,\lambda\psi_n)$. Then, $|\varphi_1|^{\alpha_1} \cdots |\varphi_n|^{\alpha_n} \equiv |\lambda|^{n-s}$. In particular, $\varphi : \mathbb{D} \longrightarrow \boldsymbol{D}_\alpha$. Moreover, $\varphi(0) = a$ and $\varphi(\sigma) = z$. Consequently,

$$\boldsymbol{g}_{\boldsymbol{D}_\alpha}(a,z) \leq \boldsymbol{\ell}^*_{\boldsymbol{D}_\alpha}(a,z) \leq \boldsymbol{m}_{\mathbb{D}}(0,\sigma) = |z^\alpha|^{1/r}.$$

It remains to consider the case where $q = 0$, $s = n$. We would like to show that $\boldsymbol{g}_{\boldsymbol{D}_\alpha}(a,\cdot) \equiv 0$. By Corollary 6.3.8, we know that $\boldsymbol{g}_{\boldsymbol{D}_\alpha}(a,\zeta) = 0$, $\zeta \in \mathbb{T}_a$. Hence, by the maximum principle for psh functions, we get $\boldsymbol{g}_{\boldsymbol{D}_\alpha}(a,\cdot) = 0$ on $\mathbb{P}(a)$. Recall that $\boldsymbol{g}_{\boldsymbol{D}_\alpha}(a,\cdot) = 0$ is log-psh function. Thus, $\boldsymbol{g}_{\boldsymbol{D}_\alpha}(a,\cdot) \equiv 0$. □

Proposition 6.3.10. *Assume that $\boldsymbol{D}_\alpha$ is of irrational type, and the conditions* (AS) *are satisfied. Then, for $z \in \boldsymbol{D}_\alpha$ and $X \in \mathbb{C}^n$ we have:*

$$\boldsymbol{s}_{\boldsymbol{D}_\alpha}(a,z) = \begin{cases} ?, & s \leq n-2 \\ \left(\prod_{j=1}^{n} |z_j|^{\alpha_j}\right)^{1/\alpha_n}, & s = n-1\,, \\ 0, & s = n \end{cases}$$

$$\boldsymbol{S}_{\boldsymbol{D}_\alpha}(a;X) = \begin{cases} 0, & s \leq n-2 \\ \left(\prod_{j=1}^{n-1} a_j^{\alpha_j}\right)^{1/\alpha_n} |X_n|, & s = n-1\,. \\ 0, & s = n \end{cases}$$

Proof. By Proposition 6.3.3, we know that $\boldsymbol{g}^2_{\boldsymbol{D}_\alpha}(a,\cdot)$ is of class $\mathcal{C}^2$ in a neighborhood of a iff $s \geq n-1$. This implies the formulas for $s \geq n-1$ (cf. Remark 4.2.8). To show that $\boldsymbol{S}_{\boldsymbol{D}_\alpha}(a;\cdot) \equiv 0$ in the case where $s \leq n-2$, we may argue as in the proof of Proposition 6.1.6. □

6.4 Exercises

Exercise 6.4.1. Complete the details of the proof of the following result:
Let $A \in \mathbb{M}(m \times n; \mathbb{Z})$ *be an* $m \times n$ *matrix with integer entries,* $m \leq n$*. Assume that* $\operatorname{rank} A = m$*. Let* $\mathfrak{g}(A)$ *stand for the greatest common divisor of all determinants of* $m \times m$ *submatrices of* A*. Then,* $A\mathbb{Z}^n = \mathbb{Z}^m$ *iff* $\mathfrak{g}(A) = 1$*.*

Proof. (Due to W. Jarnicki) Let $P, Q \in \mathbb{M}(k \times \ell; \mathbb{Z})$. We say that $P \simeq Q$ if there exist $N \in \mathbb{Z}_+$, $P_0, \dots, P_N \in \mathbb{M}(k \times \ell; \mathbb{Z})$, and $\lambda_1, \dots, \lambda_N \in \mathbb{Z}$ such that $P = P_0$, $Q = P_N$, $j = 1, \dots, N$, P_j is obtained from P_{j-1} by one of the following procedures:

- adding a column, multiplied by λ_j, to another column,
- exchanging two columns or two rows,
- negating a column or a row.

If $A \simeq B$, then $\mathfrak{g}(A) = \mathfrak{g}(B)$, and ($A\mathbb{Z}^n = \mathbb{Z}^m \iff B\mathbb{Z}^n = \mathbb{Z}^m$).

Step 1°. *Let* $P \in \mathbb{M}(k \times \ell; \mathbb{Z})$ *and let* $v = (v_1, \dots, v_k) \in \mathbb{Z}^k$ *be such that* $\operatorname{GCD}(v_1, \dots, v_k) = 1$*. Let* $g \in \mathbb{Z}_*$ *be the generator of the group*

$$\left\{x \in \mathbb{Z} : xv \in P\mathbb{Z}^\ell\right\}.$$

Then, there exists a $Q \in \mathbb{M}(k \times \ell; \mathbb{Z})$ *such that* $P \simeq Q$ *and* $gv = Qe_1$*, where* $e_1, \dots, e_k$ *is the canonical basis of* $\mathbb{Z}^k$*.*

Indeed, for $\ell = 1$ the result is trivial. Assume that $\ell \geq 2$. Let $w \in \mathbb{Z}^\ell$ be such that $Pw = gv$. We proceed by induction on $|w| := |w_1| + \dots + |w_\ell|$. The case $|w| = 1$ is elementary. Assume now that $|w| > 1$. Obviously $\operatorname{GCD}(w_1, \dots, w_\ell) = 1$. Thus, we may assume that $|w_2| \geq |w_1| > 0$. By adding/subtracting column 2 to/from column 1, we get a matrix Q such that $P \simeq Q$ and $Q\widetilde{w} = gv$, where $\widetilde{w} = (w_1, w_2 \pm w_1, w_3, \dots, w_\ell)$. Observe that $|\widetilde{w}| < |w|$, and we may continue inductively.

Step 2°. *If* $A\mathbb{Z}^n = \mathbb{Z}^m$*, then* $\mathfrak{g}(A) = 1$*.*

Indeed, it is enough to find a $B \in \mathbb{M}(m \times n; \mathbb{Z})$ that $A \simeq B$ and B has an $m \times m$ submatrix with determinant 1.

For $s = 1, \dots, m$, let $A_{(s)}$ be the submatrix of A obtained by selecting columns $s, s+1, \dots, n$. We proceed in m steps.

First, we apply Step 1° to $P = A_{(1)}$, $v = e_1$, and $g = 1$. Thus, we may assume that $a_{1,1} = 1$ and $a_{i,1} = 0$ for $i > 1$.

In step s take $w = (w_1, \dots, w_n) \in \mathbb{Z}^n$ such that $e_s = Aw$. By step $(s-1)$, we have

$$A_{(s)}(w_s, w_{s+1}, \dots, w_n) = (u_1, \dots, u_{s-1}, 1, 0, \dots, 0) =: v$$

for some $u_1, \dots, u_{s-1} \in \mathbb{Z}$. Now, we apply Step 1° to $P = A_{(s)}$ and $g = 1$, and we may assume that $a_{j,j} = 1$ for $j = 1, \dots, s$ and that $a_{i,j} = 0$ for $j = 1, \dots, s$ and $i > j$.

Step 3°. *If* $\mathfrak{g}(A) = 1$, *then* $A\mathbb{Z}^n = \mathbb{Z}^m$.

Indeed, suppose $A\mathbb{Z}^n \subsetneq \mathbb{Z}^m$. Then, there exists a $j \in \{1, \dots, m\}$ such that $e_j \notin A\mathbb{Z}^n$. We may assume that $j = 1$. Let $g > 1$ be the generator of $\{x \in \mathbb{Z} : xe_1 \in A\mathbb{Z}^n\}$. It is enough to show that the determinant of any $m \times m$ submatrix of A is divisible by g. Let S be an $m \times m$ submatrix of A. We may assume that $\det S \neq 0$. Let g' be the generator of $\{x \in \mathbb{Z} : xe_1 \in S\mathbb{Z}^m\}$. Observe that $g|g'$. It is therefore enough to prove that $g'|\det S$. By Step 1° there exists a $T \in \mathbb{M}(m \times m; \mathbb{Z})$ such that $S \simeq T$ and $Te_1 = g'e_1$. Applying the Laplace expansion, we get $g'|\det T$. Now it suffices to observe that $|\det T| = \mathfrak{g}(T) = \mathfrak{g}(S) = |\det S|$. □

Exercise 6.4.2. Let $G := \{(z_1, z_2, z_3) \in \mathbb{C}^3 : |z_1 z_2| < 1,\ |z_1 z_3| < 1\}$ be as in Example 6.2.9 and let $a = (0, a_2, a_3)$ with $a_2 a_3 \neq 0$. Prove that

$$\boldsymbol{m}_G^{(k)}(a,z) = \boldsymbol{g}_G(a,z) = \max\{|z_1 z_2|, |z_1 z_3|\}, \quad z \in G \cap \{a_3 z_2 - a_2 z_3 = 0\},$$

$$\boldsymbol{m}_G^{(k)}(a,z) \not\equiv \max\left\{|z_1 z_2|, |z_1 z_3|, \left(\frac{|a_3 z_1 z_2 - a_2 z_1 z_3|}{|a_2| + |a_3|}\right)^{\lceil k/2 \rceil / k}\right\},$$

$$\boldsymbol{m}_G^{(k)}(a, (\xi z_1, \eta z_2, \eta z_3)) = \boldsymbol{m}_G^{(k)}(a,z),$$

$$\boldsymbol{g}_G(a, (\xi z_1, \eta z_2, \eta z_3)) = \boldsymbol{g}_G(a,z), \quad z \in G,\ \xi, \eta \in \mathbb{T}.$$

6.5 List of problems

Chapter 7

Symmetrized polydisc

Summary. A famous result by L. Lempert (see Theorem 11.2.1) states that $\boldsymbol{c}_G = \boldsymbol{\ell}_G$ for convex domains G and, therefore, also for all domains that can be exhausted by domains which are biholomorphically equivalent to convex ones. In the first section below, we discuss a domain, the symmetrized bidisc $\mathbb{G}_2$, originating from investigations on interpolation problems on the so called spectral ball. It cannot be exhausted by domains biholomorphic to convex ones, but, surprisingly, it turned out that nevertheless $\boldsymbol{c}_{\mathbb{G}_2} = \boldsymbol{\ell}_{\mathbb{G}_2}$. The second section presents results for the higher-dimensional analogue of $\mathbb{G}_2$ showing that here the situation with respect to invariant distances is totally different.

7.1 Symmetrized bidisc

The aim of this section is to construct a bounded hyperconvex $(1,2)$-balanced domain $\mathbb{G}_2 \subset \mathbb{C}^n$ such that $\boldsymbol{c}_{\mathbb{G}_2} \equiv \boldsymbol{\ell}_{\mathbb{G}_2}$, but $\mathbb{G}_2$ cannot be exhausted by domains biholomorphic to convex domains. The role played by such an example will be discussed in detail in Chapter 11. Let us only mention that the problem of existence of such domain was open for more than 20 years. The example was constructed in a series of papers by J. Agler, C. Costara, and N. J. Young [11, 12, 112, 111, 113] and [212] when they investigated the 2×2-spectral Nevanlinna–Pick problem.

Let

$$\begin{aligned}
\pi &: \mathbb{C}^2 \longrightarrow \mathbb{C}^2,\\
\pi(\lambda_1,\lambda_2) &:= (\lambda_1+\lambda_2, \lambda_1\lambda_2),\\
\mathbb{G}_2 &:= \pi(\mathbb{D}^2) = \{(\lambda_1+\lambda_2, \lambda_1\lambda_2) : \lambda_1,\lambda_2 \in \mathbb{D}\}, \quad \sigma_2 := \pi(\mathbb{T}^2) \subset \partial\mathbb{G}_2,\\
\Delta_2 &:= \{(\lambda,\lambda) : \lambda \in \mathbb{D}\}, \quad \Sigma_2 := \pi(\Delta_2) = \{(2\lambda,\lambda^2) : \lambda \in \mathbb{D}\},\\
F_a(s,p) &:= \frac{2ap-s}{2-as}, \quad a \in \overline{\mathbb{D}},\ (s,p) \in (\mathbb{C} \setminus \{2/a\}) \times \mathbb{C}.
\end{aligned}$$

In this section, we will use the variables $(s,p) \in \mathbb{C}^2$, recalling that s corresponds to the sum in the mapping π and p to the product in π, respectively.

Note that:

- π is symmetric;
- $\pi : \mathbb{C}^2 \longrightarrow \mathbb{C}^2$ is proper;
- $\mathbb{G}_2$ is open (cf. Appendix B.2.1(a));

- $\mathbb{G}_2$ is an $(1,2)$-balanced domain;
- $\pi|_{\mathbb{D}^2} : \mathbb{D}^2 \longrightarrow \mathbb{G}_2$ is proper with multiplicity 2 (cf. Appendix B.2.1(b)) – indeed, it suffices to check that, if $\pi(\mu) = \pi(\lambda)$ and $\lambda \in \mathbb{D}^2$, then $\mu \in \mathbb{D}^2$; since π is symmetric, the implication is clear for $\lambda_1 \neq \lambda_2$; the remaining case where $\lambda_1 = \lambda_2$ is obvious;
- $\Delta_2 = \{\lambda \in \mathbb{D}^2 : \det \pi'(\lambda) = 0\}$;
- $\pi|_{\mathbb{D}^2 \setminus \Delta_2} : \mathbb{D}^2 \setminus \Delta_2 \longrightarrow \mathbb{G}_2 \setminus \Sigma_2$ is a holomorphic covering (cf. Appendix B.2.1(b)).

The domain $\mathbb{G}_2$ is called the *symmetrized bidisc*. One can prove (cf. Lemma 7.1.3) that $|s| < 2$ and $|\boldsymbol{F}_a| < 1$ on $\mathbb{G}_2$ and that (Remark 7.1.6) $\mathbb{G}_2$ is hyperconvex.

Theorem 7.1.1. *We have*

$$\begin{aligned} \boldsymbol{c}^*_{\mathbb{G}_2}((s_1,p_1),(s_2,p_2)) &= \boldsymbol{\ell}^*_{\mathbb{G}_2}((s_1,p_1),(s_2,p_2)) \\ &= \max\{\boldsymbol{m}_{\mathbb{D}}(\boldsymbol{F}_z(s_1,p_1),\boldsymbol{F}_z(s_2,p_2)) : z \in \overline{\mathbb{D}}\} \\ &= \max\{\boldsymbol{m}_{\mathbb{D}}(\boldsymbol{F}_z(s_1,p_1),\boldsymbol{F}_z(s_2,p_2)) : z \in \mathbb{T}\}, \\ &\qquad\qquad (s_1,p_1),(s_2,p_2) \in \mathbb{G}_2. \end{aligned}$$

Moreover, $\mathbb{G}_2$ cannot be written in the form $\mathbb{G}_2 = \bigcup_{i \in I} G_i$, where each domain G_i is biholomorphic to a convex domain and, for any compact $K \subset\subset \mathbb{G}_2$, there exists an $i_0 \in I$ with $K \subset G_{i_0}$.

Remark 7.1.2. In Chapter 11, we will see that $\boldsymbol{c}_G = \boldsymbol{\ell}_G$ for any convex domain G. Therefore, if a domain is the increasing union of a sequence of convex domains, then, immediately, $\boldsymbol{c}_G = \boldsymbol{\ell}_G$. Therefore, the above theorem is definitely not a consequence of the results in Chapter 11.

The proof will be given after presenting some auxiliary lemmas.

Lemma 7.1.3 (cf. [12]). *For $(s,p) \in \mathbb{C}^2$, the following conditions are equivalent:*

(i) $(s,p) \in \mathbb{G}_2$;

(ii) $|s - \bar{s}p| + |p|^2 < 1$;

(iii) $|s| < 2$, $|s - \bar{s}p| + |p|^2 < 1$;

(iv) $|\frac{2zp-s}{2-zs}| < 1$, $z \in \overline{\mathbb{D}}$ (*i.e.*, $|\boldsymbol{F}_z(s,p)| < 1$, $z \in \overline{\mathbb{D}}$);

(v) $|\frac{2p-\bar{z}s}{2-zs}| < 1$, $z \in \overline{\mathbb{D}}$;

(vi) $2|s - \bar{s}p| + |s^2 - 4p| + |s|^2 < 4$.

In particular, $\boldsymbol{F}_a \in \mathcal{O}(\mathbb{G}_2, \mathbb{D})$ $(a \in \overline{\mathbb{D}})$.

Proof. It is clear that $(s,p) \in \mathbb{G}_2$ iff both roots of the polynomial $f(z) = z^2 - sz + p$ belong to $\mathbb{D}$. By the Cohn criterion (cf. [441]) $f^{-1}(0) \subset \mathbb{D}$ iff $|p| < 1$, and the root of the polynomial

$$g(z) := \frac{1}{z}\left(f(z) - pz^2\overline{f(1/\bar{z})}\right) = (1-|p|^2)z - (s - p\bar{s})$$

belongs to $\mathbb{D}$. Thus, (i) $\iff$ (ii).

The equivalence (iv) $\iff$ (v) follows from the maximum principle:

$$\begin{aligned}\max\left\{\left|\frac{2zp-s}{2-zs}\right| : z \in \overline{\mathbb{D}}\right\} &= \max\left\{\left|\frac{2zp-s}{2-zs}\right| : z \in \mathbb{T}\right\} \\ &= \max\left\{\left|\frac{2p-\bar{z}s}{2-zs}\right| : z \in \mathbb{T}\right\} \\ &= \max\left\{\left|\frac{2p-\bar{z}s}{2-zs}\right| : z \in \overline{\mathbb{D}}\right\}.\end{aligned}$$

Observe that (iv) with $z = 0$ gives $|s| < 2$ and, moreover,

$$\max\left\{\left|\frac{2p-\bar{z}s}{2-zs}\right|^2 : z \in \mathbb{T}\right\} = \max\left\{\frac{4|p|^2 + |s|^2 - 4\operatorname{Re}(zp\bar{s})}{4 + |s|^2 - 4\operatorname{Re}(zs)} : z \in \mathbb{T}\right\}.$$

Thus, (iv) $\Longrightarrow$ (iii) $\Longrightarrow$ (ii) $\Longrightarrow$ (iv).

Note that for $|s| < 2$ the mapping $\overline{\mathbb{D}} \ni z \longmapsto \boldsymbol{F}_z(s,p)$ maps $\overline{\mathbb{D}}$ onto $\overline{\mathbb{D}}(a,r) \subset \mathbb{C}$ with $a := 2\frac{\bar{s}p - s}{4-|s|^2}$, $r := \frac{|s^2-4p|}{4-|s|^2}$. We have $\mathbb{D}(a,r) \subset \mathbb{D}$ iff $|a| + r < 1$. Thus, (vi) $\iff$ (iv). □

Lemma 7.1.4 (cf. [12]). *For $(s,p) \in \mathbb{C}^2$, the following conditions are equivalent:*

(i) $(s,p) \in \overline{\mathbb{G}}_2$;

(ii) $|s| \le 2$, $|s - \bar{s}p| + |p|^2 \le 1$;

(iii) $\left|\frac{2zp-s}{2-zs}\right| \le 1$, $z \in \overline{\mathbb{D}}$;[1]

(iv) $\left|\frac{2p-\bar{z}s}{2-zs}\right| \le 1$, $z \in \overline{\mathbb{D}}$.[2]

Notice that the condition $|s - \bar{s}p| + |p|^2 \le 1$ does not imply that $(s,p) \in \overline{\mathbb{G}}_2$ (e.g. $(s,p) = (5/2, 1)$).

Proof. Using Lemma 7.1.3, we see that (i) $\Longrightarrow$ (ii). Moreover, (ii) $\iff$ (iii) $\iff$ (iv). It remains to observe that, if (ii) is satisfied and $s \ne p\bar{s}$, then $(ts, p) \in \mathbb{G}_2$, $0 < t < 1$. If (ii) is satisfied and $s = p\bar{s}$, then $(ts, t^2p) \in \mathbb{G}_2$, $0 < t < 1$. □

[1] Notice that, if $(s,p) \in \overline{\mathbb{G}}_2$ and $|s| = 2$, then $(s,p) = (2\eta, \eta^2)$ for some $\eta \in \mathbb{T}$ and, consequently, $\frac{2zp-s}{2-zs} = -\eta$, which implies that the function $z \longmapsto \frac{2zp-s}{2-zs}$ has no essential singularities.

[2] As above, notice that, if $(s,p) = (2\eta, \eta^2)$ for some $\eta \in \mathbb{T}$, then $\frac{2p-\bar{z}s}{2-\bar{z}s} = \eta^2$ and, consequently, the function $z \longmapsto \frac{2p-\bar{z}s}{2-\bar{z}s}$ has no essential singularities.

Corollary 7.1.5. *For $(s, p) \in \mathbb{C}^2$, the following conditions are equivalent:*

(i) $(s, p) \in \sigma_2$;

(ii) $s = \bar{s}p$, $|p| = 1$, *and* $|s| \le 2$.

In particular, $|\boldsymbol{F}_a| = 1$ *on* σ_2 *for* $a \in \mathbb{T}$.[3]

Remark 7.1.6. Observe that

$$\mathfrak{h}_{\mathbb{G}_2}(s, p) = \max\left\{\max\{|\lambda_1|, |\lambda_2|\} : (\lambda_1, \lambda_2) \in \pi^{-1}(s, p)\right\}, \quad (s, p) \in \mathbb{C}^2.$$

is a continuous plurisubharmonic function such that

$$\overline{\mathbb{G}_2} = \{(s, p) \in \mathbb{C}^2 : \mathfrak{h}_{\mathbb{G}_2}(s, p) \le 1\};$$

cf. § 2.2.4. In particular, $\mathbb{G}_2$ is hyperconvex.

The maximum principle for plurisubharmonic functions gives the following result:

Lemma 7.1.7. *Let $\varphi : \overline{\mathbb{D}} \longrightarrow \mathbb{C}^2$ be a continuous mapping, holomorphic in $\mathbb{D}$ such that $\varphi(\mathbb{T}) \subset \overline{\mathbb{G}_2}$. Then, $\varphi(\overline{\mathbb{D}}) \subset \overline{\mathbb{G}_2}$. If $\varphi(\mathbb{D}) \cap \mathbb{G}_2 \neq \varnothing$, then $\varphi(\mathbb{D}) \subset \mathbb{G}_2$. If $\varphi(\mathbb{D}) \cap \mathbb{G}_2 = \varnothing$, then $\varphi(\mathbb{D}) \subset \partial\mathbb{G}_2$.*

Remark 7.1.8. If $f \in \mathcal{O}(\mathbb{D}^2)$ is symmetric, then the relation $F(\pi(\lambda_1, \lambda_2)) = f(\lambda_1, \lambda_2)$ defines a function $F \in \mathcal{O}(\mathbb{G}_2)$.

In particular, if $h \in \mathcal{O}(\mathbb{D}, \mathbb{D})$, then the relation $H_h(\pi(\lambda_1, \lambda_2)) = \pi(h(\lambda_1), h(\lambda_2))$ defines a holomorphic mapping $H_h : \mathbb{G}_2 \longrightarrow \mathbb{G}_2$ with $H_h(\Sigma_2) \subset \Sigma_2$.

Observe that if $h \in \operatorname{Aut}(\mathbb{D})$, then $H_h \in \operatorname{Aut}(\mathbb{G}_2)$, $H_h^{-1} = H_{h^{-1}}$, $H_h(\Sigma_2) = \Sigma_2$, and $H_h(\sigma_2) = \sigma_2$.

In particular, if $h(\lambda) := \tau\lambda$ for some $\tau \in \mathbb{T}$, then we get the "rotation" $R_\tau(s, p) := H_h(s, p) = (\tau s, \tau^2 p)$.

Remark 7.1.9. For any point $(s_0, p_0) = (2a, a^2) \in \Sigma_2$, we get $H_{\boldsymbol{h}_a}(s_0, p_0) = (0, 0)$.

Lemma 7.1.10. *σ_2 is the Shilov (and Bergman) boundary of $\mathbb{G}_2$.*

Proof. It is clear that the modulus of any function $f \in \mathcal{C}(\overline{\mathbb{G}_2}) \cap \mathcal{O}(\mathbb{G}_2)$ attains its maximum on σ_2. We have to prove that σ_2 is minimal. First, observe that the function $f_0(s, p) := s + 2$ is a peak function at $(2, 1) \in \sigma_2$. Take any other point $(s_0, p_0) = \pi(\lambda_1^0, \lambda_2^0) \in \sigma_2$. The case where $\lambda_1^0 = \lambda_2^0$ reduces (via a rotation R_τ) to the previous one. Thus, assume that $\lambda_1^0 \neq \lambda_2^0$. We will find a Blaschke product B of order 2, such that

$$\{\lambda \in \mathbb{T} : B(\lambda) = 1\} = \{\lambda_1^0, \lambda_2^0\}.$$

[3] Notice that for $(s, p) \in \sigma_2$, $|s| < 2$, we have $\boldsymbol{F}_a(s, p) = ap\frac{2-\bar{a}\bar{s}}{2-as}$.

Suppose for a moment that such a B is already constructed. Then, $f_0 \circ H_B$ is a peak function for (s_0, p_0).

We move to the construction of B. Using a rotation we may reduce the proof to the case $\lambda_2^0 = \overline{\lambda_1^0}$. Then, using the fact that the mapping $(-1,1) \ni a \longmapsto \frac{2a}{1+a^2} \in (-1,1)$ is bijective, we see that there exists an $a \in (-1,1)$ such that $\boldsymbol{h}_a(\lambda_1^0) = -\boldsymbol{h}_a(\overline{\lambda_1^0})$. Finally, we take $B := \tau \boldsymbol{h}_a^2$ with $\tau \in \mathbb{T}$ such that $\tau \boldsymbol{h}_a^2(\lambda_1^0) = 1$. $\square$

Proposition 7.1.11 (cf. [112, 153]). *The domain $\mathbb{G}_2$ cannot be written in the form $\mathbb{G}_2 = \bigcup_{i \in I} G_i$, where each domain G_i is biholomorphic to a convex domain and for any compact $K \subset\subset \mathbb{G}_2$ there exists an $i_0 \in I$ with $K \subset G_{i_0}$.*

Proof. First, observe that $\overline{\mathbb{G}}_2$ is not convex: for example, $(2,1), (2i,-1) \in \overline{\mathbb{G}}_2$, but $(1+i, 0) \notin \overline{\mathbb{G}}_2$. Consequently, $\mathbb{G}_2$ is also not convex.

Suppose that $\mathbb{G}_2 = \bigcup_{i \in I} G_i$, where each domain G_i is biholomorphic to a convex domain and for any compact $K \subset\subset \mathbb{G}_2$ there exists an $i_0 \in I$ with $K \subset G_{i_0}$. For any $0 < \varepsilon < 1$, take an $i = i(\varepsilon) \in I$ such that $\{(s,p) \in \mathbb{C}^2 : \mathfrak{h}_{\mathbb{G}_2}(s,p) \le 1-\varepsilon\} \subset G_{i(\varepsilon)}$, and let $f_\varepsilon = (g_\varepsilon, h_\varepsilon) : G_{i(\varepsilon)} \longrightarrow D_\varepsilon$ be a biholomorphic mapping onto a convex domain $D_\varepsilon \subset \mathbb{C}^n$ with $f_\varepsilon(0,0) = (0,0)$ and $f_\varepsilon'(0,0) = \mathbb{I}_2$.

Take two arbitrary points $(s_1, p_1), (s_2, p_2) \in \mathbb{C}^2$ and put

$$C := \max\{\mathfrak{h}_{\mathbb{G}_2}(s_1, p_1), \mathfrak{h}_{\mathbb{G}_2}(s_2, p_2)\}.$$

Our aim is to prove that $\mathfrak{h}_{\mathbb{G}_2}(t(s_1,p_1) + (1-t)(s_2,p_2)) \le C$, $t \in [0,1]$, which, in particular, shows that $\mathbb{G}_2$ is convex; a contradiction.

Observe that for $|\lambda| < (1-\varepsilon)/C$ we have $\mathfrak{h}_{\mathbb{G}_2}(\lambda s_j, \lambda^2 p_j) = |\lambda| \mathfrak{h}_{\mathbb{G}_2}(s_j, p_j) < 1-\varepsilon$, $j = 1,2$. Consequently, for any $t \in [0,1]$, the mapping $\varphi_{\varepsilon,t} : \mathbb{D}(\frac{1-\varepsilon}{C}) \longrightarrow \mathbb{G}_2$,

$$\varphi_{\varepsilon,t}(\lambda) = (\psi_{\varepsilon,t}(\lambda), \chi_{\varepsilon,t}(\lambda)) := f_\varepsilon^{-1}(t f_\varepsilon(\lambda s_1, \lambda^2 p_1) + (1-t) f_\varepsilon(\lambda s_2, \lambda^2 p_2))$$

is well defined. We have $\varphi_{\varepsilon,t}(0) = (0,0)$ and

$$\psi_{\varepsilon,t}'(0) = t s_1 + (1-t) s_2, \quad \chi_{\varepsilon,t}'(0) = 0,$$

$$\frac{1}{2}\chi_{\varepsilon,t}''(0) = t p_1 + (1-t) p_2 + \mu_\varepsilon t(1-t)(s_1 - s_2)^2, \text{ where } \mu_\varepsilon := \frac{1}{2}\frac{\partial^2 h_\varepsilon}{\partial s^2}(0,0).$$

Define $\Phi_{\varepsilon,t} : \mathbb{D}(\frac{1-\varepsilon}{C}) \longrightarrow \mathbb{C}^2$,

$$\Phi_{\varepsilon,t}(\lambda) := \begin{cases} (\frac{1}{\lambda}\psi_{\varepsilon,t}(\lambda), \frac{1}{\lambda^2}\chi_{\varepsilon,t}(\lambda)), & \lambda \neq 0 \\ (t s_1 + (1-t) s_2, t p_1 + (1-t) p_2 + \mu_\varepsilon t(1-t)(s_1 - s_2)^2), & \lambda = 0 \end{cases}.$$

Then, $\Phi_{\varepsilon,t}$ is holomorphic and, by the maximum principle, we get

$$\mathfrak{h}_{\mathbb{G}_2}(\Phi_{\varepsilon,t}(0)) \le \limsup_{s \longrightarrow \frac{1-\varepsilon}{C}} \max_{|\lambda|=s} \mathfrak{h}_{\mathbb{G}_2}(\Phi_{\varepsilon,t}(\lambda)) = \limsup_{s \longrightarrow \frac{1-\varepsilon}{C}} \frac{1}{s} \max_{|\lambda|=s} \mathfrak{h}_{\mathbb{G}_2}(\varphi_{\varepsilon,t}(\lambda)) \le \frac{C}{1-\varepsilon},$$

$$\text{i.e.,}\quad \mathfrak{h}_{\mathbb{G}_2}(ts_1 + (1-t)s_2, tp_1 + (1-t)p_2 + \mu_\varepsilon t(1-t)(s_1 - s_2)^2) \le \frac{C}{1-\varepsilon}.$$

We only need to prove that $\mu_\varepsilon \longrightarrow 0$.

Taking $t = 1/2$, we get

$$\mathfrak{h}_{\mathbb{G}_2}\left(\frac{1}{2}(s_1 + s_2), \frac{1}{2}(p_1 + p_2) + \frac{1}{4}\mu_\varepsilon(s_1 - s_2)^2\right) \le \frac{C}{1-\varepsilon}.$$

For $\alpha \in \mathbb{T}$, take $(s_1, p_1) := \pi(\alpha, -1) = (\alpha - 1, -\alpha)$, $(s_2, p_2) := \pi(\alpha, 1) = (\alpha + 1, \alpha)$. Then, $C = 1$ and

$$\mathfrak{h}_{\mathbb{G}_2}(\alpha, \mu_\varepsilon) \le \frac{1}{1-\varepsilon}.$$

Hence, $((1-\varepsilon)\alpha, (1-\varepsilon)^2\mu_\varepsilon) \in \overline{\mathbb{G}}_2$ and so, by Lemma 7.1.4,

$$|(1-\varepsilon)\alpha - (1-\varepsilon)^2\mu_\varepsilon(1-\varepsilon)\overline{\alpha}| + (1-\varepsilon)^4|\mu_\varepsilon|^2 \le 1, \quad \alpha \in \mathbb{T}.$$

It follows that

$$(1-\varepsilon) + (1-\varepsilon)^3|\mu_\varepsilon| + (1-\varepsilon)^4|\mu_\varepsilon|^2 \le 1$$

and, finally, $|\mu_\varepsilon| \le \frac{\varepsilon}{(1-\varepsilon)^3} \longrightarrow 0$. □

Remark 7.1.12. Let $G \subset \mathbb{C}^m$ be a balanced domain. Then, $D := \mathbb{G}_2 \times G$ cannot be exhausted by domains biholomorphic to convex domains. For a proof of this fact see [389]. If, on the other hand, in addition, G is convex, then the Lempert Theorem A.5.5 and the product property (see Chapter 18.1) will lead to $\boldsymbol{c}_D = \boldsymbol{\ell}_D$.

Lemma 7.1.13 (cf. [111, 113]). *Let $\varphi : \mathbb{D} \longrightarrow \mathbb{G}_2$ be a mapping of the form*

$$\varphi = (S, P) = \left(\frac{\widetilde{S}}{P_0}, \frac{\widetilde{P}}{P_0}\right), \tag{7.1.1}$$

where P_0, $\widetilde{P}$, $\widetilde{S}$ are polynomials of degree ≤ 2 with $P_0^{-1}(0) \cap \overline{\mathbb{D}} = \varnothing$. Assume that $\varphi(\mathbb{T}) \subset \sigma_2$ and $\varphi(\xi) = (2\eta, \eta^2)$ for some $\xi, \eta \in \mathbb{T}$. Then, $h := \boldsymbol{F}_{\overline{\eta}} \circ \varphi \in \operatorname{Aut}(\mathbb{D})$. In particular, if $a' := \varphi(\lambda')$, $a'' := \varphi(\lambda'')$, then,

$$\begin{aligned}\boldsymbol{m}_{\mathbb{D}}(\lambda', \lambda'') &= \boldsymbol{m}_{\mathbb{D}}(h(\lambda'), h(\lambda'')) = \boldsymbol{m}_{\mathbb{D}}(\boldsymbol{F}_{\overline{\eta}}(a'), \boldsymbol{F}_{\overline{\eta}}(a''))\\ &\le \max\{\boldsymbol{m}_{\mathbb{D}}(\boldsymbol{F}_z(a'), \boldsymbol{F}_z(a'')) : z \in \mathbb{T}\}\\ &\le \max\{\boldsymbol{m}_{\mathbb{D}}(\boldsymbol{F}_z(a'), \boldsymbol{F}_z(a'')) : z \in \overline{\mathbb{D}}\} \le \boldsymbol{c}^*_{\mathbb{G}_2}(a', a'')\\ &\le \boldsymbol{\ell}^*_{\mathbb{G}_2}(a', a'') = \boldsymbol{\ell}^*_{\mathbb{G}_2}(\varphi(\lambda'), \varphi(\lambda'')) \le \boldsymbol{m}_{\mathbb{D}}(\lambda', \lambda'').\end{aligned}$$

Consequently, the formulas from Theorem 7.1.1 *hold for all* $(s_1, p_1), (s_2, p_2) \in \varphi(\mathbb{D})$, *and* φ *is a* complex geodesic, *i.e.,* $\boldsymbol{m}_{\mathbb{D}}(\lambda', \lambda'') = \boldsymbol{\ell}_{\mathbb{G}_2}(\varphi(\lambda'), \varphi(\lambda''))$, $\lambda', \lambda'' \in \mathbb{D}$.

Proof. Put

$$h := \boldsymbol{F}_{\overline{\eta}} \circ \varphi = \frac{2\overline{\eta}P - S}{2 - \overline{\eta}S} = \frac{2\overline{\eta}\widetilde{P} - \widetilde{S}}{2P_0 - \overline{\eta}\widetilde{S}}.$$

First, observe that $h(\mathbb{D}) \subset \mathbb{D}$ and $h(\mathbb{T}) \subset \mathbb{T}$. It is clear that h is a rational function of degree ≤ 2. Notice that $2\overline{\eta}P(\xi) - S(\xi) = 0 = 2 - \overline{\eta}S(\xi)$. Consequently, h is a rational function of degree ≤ 1 and, therefore, h must be an automorphism of the unit disc. □

Lemma 7.1.14. *If* φ *satisfies the assumptions of Lemma* 7.1.13, *then for any* $g \in \operatorname{Aut}(\mathbb{D})$, *the mapping* $\psi := H_g \circ \varphi$ *satisfies the same assumptions.*

Proof. The only problem is to check that ψ has the form (7.1.1). Let $g = \tau \boldsymbol{h}_a$ for some $\tau \in \mathbb{T}$, $a \in \mathbb{D}$. Fix a λ and let $\varphi(\lambda) = (S(\lambda), P(\lambda)) = \pi(z_1, z_2)$. Then,

$$\begin{aligned}\psi(\lambda) &= \pi(g(z_1), g(z_2)) = (\tau(\boldsymbol{h}_a(z_1) + \boldsymbol{h}_a(z_2)), \tau^2\boldsymbol{h}_a(z_1)\boldsymbol{h}_a(z_2)) \\ &= \left(\tau\frac{(1+|a|^2)(z_1+z_2) - 2\overline{a}z_1z_2 - 2a}{1 - \overline{a}(z_1+z_2) + \overline{a}^2z_1z_2}, \tau^2\frac{z_1z_2 - a(z_1+z_2) + a^2}{1 - \overline{a}(z_1+z_2) + \overline{a}^2z_1z_2}\right).\end{aligned}$$

Consequently,

$$\begin{aligned}\psi &= \left(\tau\frac{(1+|a|^2)S - 2\overline{a}P - 2a}{1 - \overline{a}S + \overline{a}^2P}, \tau^2\frac{P - aS + a^2}{1 - \overline{a}S + \overline{a}^2P}\right) \\ &= \left(\tau\frac{(1+|a|^2)\widetilde{S} - 2\overline{a}\widetilde{P} - 2aP_0}{P_0 - \overline{a}\widetilde{S} + \overline{a}^2\widetilde{P}}, \tau^2\frac{\widetilde{P} - a\widetilde{S} + a^2P_0}{P_0 - \overline{a}\widetilde{S} + \overline{a}^2\widetilde{P}}\right).\end{aligned}$$

□

Proof of Theorem 7.1.1. We already know (Proposition 7.1.11) that $\mathbb{G}_2$ cannot be exhausted by domains biholomorphic to convex domains.

Step 1°. First, consider the case where $s_1 = 0$. The case where $s_2 = 0$ is simple. Consider the embedding $\mathbb{D} \ni \lambda \overset{\varphi}{\longmapsto} (0, \lambda) \in \mathbb{G}_2$ and the projection $\mathbb{G}_2 \ni (s, p) \overset{F}{\longmapsto} p \in \mathbb{D}$. Then,

$$\begin{aligned}\boldsymbol{m}_{\mathbb{D}}(p_1, p_2) &= \max\{\boldsymbol{m}_{\mathbb{D}}(zp_1, zp_2) : z \in \mathbb{T}\} = \max\{\boldsymbol{m}_{\mathbb{D}}(zp_1, zp_2) : z \in \overline{\mathbb{D}}\} \\ &= \boldsymbol{m}_{\mathbb{D}}(\boldsymbol{F}_z(0, p_1), \boldsymbol{F}_z(0, p_2)) \leq \boldsymbol{c}^*_{\mathbb{G}_2}((0, p_1), (0, p_2)) \\ &\leq \boldsymbol{\ell}^*_{\mathbb{G}_2}((0, p_1), (0, p_2)) = \boldsymbol{\ell}^*_{\mathbb{G}_2}(\varphi(p_1), \varphi(p_2)) \leq \boldsymbol{m}_{\mathbb{D}}(p_1, p_2),\end{aligned}$$

which completes the proof.

Step 2°. Assume that $s_1 = 0$, $s_2 \neq 0$. Let $t_0 \in (0,1)$ be defined by the formula

$$t_0 := \max\left\{\boldsymbol{m}_{\mathbb{D}}\left(p_1, \frac{2p_2 - \overline{z}s_2}{2 - zs_2}\right) : z \in \overline{\mathbb{D}}\right\} = \boldsymbol{m}_{\mathbb{D}}\left(p_1, \frac{2p_2 - \overline{\xi}s_2}{2 - \xi s_2}\right),$$

where $\xi \in \overline{\mathbb{D}}$.

Our aim is to construct a mapping $\varphi : \mathbb{D} \longrightarrow \mathbb{G}_2$ satisfying all the assumptions of Lemma 7.1.13, such that $\varphi(t_0) = (0, p_1)$, $\varphi(0) = (s_2, p_2)$.

First, we prove that

$$\boldsymbol{m}_{\mathbb{D}}\left(p_1, \frac{2p_2 - \overline{z}s_2}{2 - zs_2}\right) < t_0, \quad z \in \mathbb{D},$$

and so $\xi \in \mathbb{T}$.

Indeed, let $L : \mathbb{C} \longrightarrow \mathbb{C}$, $L(z) := p_1 z - \overline{z}$. Then L is an $\mathbb{R}$-linear isomorphism. In particular, if $D := L(\mathbb{D})$, then $\partial D = L(\mathbb{T})$. Observe that

$$t_0 = \max\{|\Phi(L(z))| : z \in \overline{\mathbb{D}}\} = \max\{|\Phi(w)| : w \in \overline{D}\},$$

where

$$\Phi(w) := \frac{2(p_2 - p_1) + s_2 w}{2(1 - p_1\overline{p}_2) + \overline{s}_2 w}.$$

Note that $\Phi \not\equiv \text{const}$. Now, the required result easily follows from the maximum principle.

In particular, $\boldsymbol{m}_{\mathbb{D}}(p_1, p_2) < t_0$.

Using the automorphism R_ξ, we may reduce the problem to the case of $\xi = 1$.

Step 3°. Put

$$a_0 := \boldsymbol{F}_1(s_2, p_2) = \frac{2p_2 - s_2}{2 - s_2} \in \mathbb{D}$$

and let $h \in \operatorname{Aut}(\mathbb{D})$ be such that $h(t_0) = p_1$, $h(0) = a_0$. Let $\tau \in \mathbb{T}$ be such that $h(\tau) = 1$.

There exists a Blaschke product P of order 2, such that $P(t_0) = p_1$, $P(0) = p_2$, and $P(\tau) = 1$.

Indeed, first observe that it suffices to find a Blaschke product Q of order 2 with $Q(t_0) = \boldsymbol{h}_{p_2}(p_1) =: p_1'$, $Q(0) = 0$, and $Q(\tau) = \boldsymbol{h}_{p_2}(1) =: \tau' \in \mathbb{T}$ (having such a Q we put $P := \boldsymbol{h}_{-p_2} \circ Q$).

We have $Q(\lambda) = \lambda g(\lambda)$, where $g \in \operatorname{Aut}(\mathbb{D})$ is such that $g(t_0) = p_1'/t_0 =: a \in \mathbb{D}$ (recall that $t_0 > \boldsymbol{m}_{\mathbb{D}}(p_1, p_2) = |p_1'|$) and $g(\tau) = \tau'/\tau =: \tau'' \in \mathbb{T}$. Define

$$g := \boldsymbol{h}_{-a} \circ (\zeta \cdot \boldsymbol{h}_{t_0}),$$

where $\zeta := \boldsymbol{h}_a(\tau'')\overline{\boldsymbol{h}_{t_0}(\tau)}$. Then, $g(t_0) = \boldsymbol{h}_{-a}(0) = a$ and $g(\tau) = \boldsymbol{h}_{-a}(\zeta \cdot \boldsymbol{h}_{t_0}(\tau)) = \tau''$.

Step 4°. Define

$$S := 2\frac{P-h}{1-h}, \quad \varphi := (S, P).$$

First observe that φ has the form (7.1.1). Indeed, let $P = \widetilde{P}/P_0$. The only problem is to show that $S = \widetilde{S}/P_0$ with $\widetilde{S}$ being a polynomial of degree ≤ 2. Let $h = \widetilde{h}/h_0$. Then,

$$S = 2\frac{\widetilde{P}h_0 - \widetilde{h}P_0}{P_0(h_0 - \widetilde{h})}.$$

Since $h(\tau) = P(\tau) = 1$, the polynomial $\widetilde{P}h_0 - \widetilde{h}P_0$ is divisible by $\widetilde{h} - h_0$.

Observe that

- $\varphi(t_0) = (S(t_0), P(t_0)) = \left(2\frac{P(t_0) - h(t_0)}{1 - h(t_0)}, p_1\right) = (0, p_1) = (s_1, p_1);$
- $\varphi(0) = (S(0), P(0)) = \left(2\frac{p_2 - a_0}{1 - a_0}, p_2\right) = \left(2\frac{p_2 - \frac{2p_2 - s_2}{2 - s_2}}{1 - \frac{2p_2 - s_2}{2 - s_2}}, p_2\right) = (s_2, p_2);$
- $\boldsymbol{F}_1 \circ \varphi = \dfrac{2P - S}{2 - S} = \dfrac{2P - 2\frac{P-h}{1-h}}{2 - 2\frac{P-h}{1-h}} = h;$
- on $\mathbb{T}$ we get $\overline{S}P = 2\dfrac{\overline{P} - \overline{h}}{1 - \overline{h}}P = 2\dfrac{1 - P/h}{1 - 1/h} = S.$

We have

$$h = \frac{2P - S}{2 - S} = \frac{2\widetilde{P} - \widetilde{S}}{2P_0 - \widetilde{S}}.$$

Note that $\deg(2\widetilde{P} - \widetilde{S}) = 2$ or $\deg(2P_0 - \widetilde{S}) = 2$. Thus, the polynomials $2\widetilde{P} - \widetilde{S}$, $2P_0 - \widetilde{S}$ must have a common zero, say z_0. We have $2\widetilde{P}(z_0) = \widetilde{S}(z_0) = 2P_0(z_0)$. Thus, $P(z_0) = 1$, which implies that $z_0 \in \mathbb{T}$ and $S(z_0) = 2$.

Put $C := \max\{|S(\lambda)| : \lambda \in \mathbb{T}\}$ (we already know that $C \geq 2$). Define $\psi := (2S/C, P)$. Then, ψ satisfies all assumptions of Lemma 7.1.13 and, consequently,

Theorem 7.1.1 holds for points from $\psi(\mathbb{D})$. In particular, there exists an $\eta \in \overline{\mathbb{D}}$ such that

$$t_0 = \boldsymbol{m}_{\mathbb{D}}\left(\overline{\eta}p_1, \frac{2\overline{\eta}p_2 - 2s_2/C}{2 - \overline{\eta}2s_2/C}\right) = \boldsymbol{m}_{\mathbb{D}}\left(p_1, \frac{2p_2 - \eta 2s_2/C}{2 - \overline{\eta}2s_2/C}\right).$$

Hence, $C \le 2$, and finally, $C = 2$. Consequently, $\varphi = \psi$, which completes the proof of the theorem in the case $s_1 = 0$.

Step 5°. Now, let $(s_1, p_1), (s_2, p_2) \in \mathbb{G}_2$ be arbitrary. Suppose that $s_1 = \lambda_1^0 + \lambda_2^0$, $p_1 = \lambda_1^0\lambda_2^0$ with $\lambda_1^0, \lambda_2^0 \in \mathbb{D}$. One can easily prove that there exists an automorphism $g \in \operatorname{Aut}(\mathbb{D})$ such that $g(\lambda_1^0) + g(\lambda_2^0) = 0$. Then, $H_g(s_1, p_1) = (0, p_1')$. Put $(s_2', p_2') := H_2(s_2, p_2)$. We have the following two cases:

- $s_2' = 0$: We already know that the mapping $\varphi = (0, h)$ with suitable $h \in \operatorname{Aut}(\mathbb{D})$ $(h(t_0) = p_1'$, $t_0 := \boldsymbol{m}_{\mathbb{D}}(p_1', p_2')$, $h(0) = p_2')$ is a complex geodesic for $(0, p_1')$ and $(0, p_2')$. By an argument like the one in the proof of Lemma 7.1.14, we easily conclude that, if $g^{-1} = \tau \boldsymbol{h}_a$, then

$$\psi := H_{g^{-1}} \circ \varphi = \left(\tau\frac{-2\overline{a}h - 2a}{1 + \overline{a}^2 h}, \tau^2\frac{h + a^2}{1 + \overline{a}^2 h}\right) = (\overline{\beta}q + \beta, q),$$

where

$$\beta := -\tau\frac{2a}{1 + |a|^2} \in \mathbb{D}, \quad q := \tau^2 \boldsymbol{h}_{-a^2} \circ h \in \operatorname{Aut}(\mathbb{D}).$$

For any $\alpha \in \mathbb{T}$ we have:

$$\boldsymbol{F}_\alpha \circ \psi = \alpha\frac{2q - \overline{\alpha}(\overline{\beta}q + \beta)}{2 - \alpha(\overline{\beta}q + \beta)} = \frac{2 - \overline{\alpha\beta}}{2 + \alpha\beta} \cdot \frac{q - \frac{\overline{\alpha\beta}}{2 - \overline{\alpha\beta}}}{1 - \frac{\alpha\beta}{2 + \alpha\beta} \cdot q} =: q_\alpha \in \operatorname{Aut}(\mathbb{D}).$$

Hence,

$$\begin{aligned}
t_0 &= \boldsymbol{m}_{\mathbb{D}}(q_\alpha(t_0), q_\alpha(0)) = \boldsymbol{m}_{\mathbb{D}}(\boldsymbol{F}_\alpha(\psi(t_0)), \boldsymbol{F}_\alpha(\psi(0))) \\
&\le \max\left\{\boldsymbol{m}_{\mathbb{D}}\left(\frac{2p_1 - \overline{z}s_1}{2 - zs_1}, \frac{2p_2 - \overline{z}s_2}{2 - zs_2}\right) : z \in \mathbb{T}\right\} \\
&= \max\{\boldsymbol{m}_{\mathbb{D}}(\boldsymbol{F}_z(s_1, p_1), \boldsymbol{F}_z(s_2, p_2)) : z \in \mathbb{T}\} \\
&\le \max\{\boldsymbol{m}_{\mathbb{D}}(\boldsymbol{F}_z(s_1, p_1), \boldsymbol{F}_z(s_2, p_2)) : z \in \overline{\mathbb{D}}\} \\
&\le \boldsymbol{c}^*_{\mathbb{G}_2}((s_1, p_1), (s_2, p_2)) \le \boldsymbol{\ell}^*_{\mathbb{G}_2}((s_1, p_1), (s_2, p_2)) = \boldsymbol{\ell}^*_{\mathbb{G}_2}(\psi(t_0), \psi(0)) \le t_0.
\end{aligned}$$

- $s_2' \ne 0$: We know that there exists a mapping $\varphi : \mathbb{D} \longrightarrow \mathbb{G}_2$ as in Lemma 7.1.13 such that $H_g(s_j, p_j) \in \varphi(\mathbb{D})$, $j = 1, 2$. It remains to observe that, by Lemma 7.1.14, the mapping $H_{g^{-1}} \circ \varphi$ also satisfies all assumptions of Lemma 7.1.13. □

Corollary 7.1.15 (cf. [11]).

(a)
$$\begin{aligned}&\boldsymbol{c}^*_{\mathbb{G}_2}((s_1,p_1),(s_2,p_2)) = \boldsymbol{\ell}^*_{\mathbb{G}_2}((s_1,p_1),(s_2,p_2))\\ &= \max\left\{\left|\frac{(s_1p_2-p_1s_2)z^2+2(p_1-p_2)z+s_2-s_1}{(p_1\overline{s}_2-s_1)z^2+2(1-p_1\overline{p}_2)z+s_1\overline{p}_2-\overline{s}_2}\right| : z\in\mathbb{T}\right\},\\ &\qquad (s_1,p_1),\ (s_2,p_2)\in\mathbb{G}_2.\end{aligned}$$

In particular,

(b)
$$\begin{aligned}\boldsymbol{c}^*_{\mathbb{G}_2}((0,0),(s,p)) &= \boldsymbol{\ell}^*_{\mathbb{G}_2}((0,0),(s,p)) = \max\{|\boldsymbol{F}_z(s,p)| : z\in\mathbb{T}\}\\ &= \frac{2|s-\overline{s}p|+|s^2-4p|}{4-|s|^2},\quad (s,p)\in\mathbb{G}_2,\end{aligned}$$

(c)
$$\begin{aligned}&\boldsymbol{c}^*_{\mathbb{G}_2}((2\mu,\mu^2),(s,p)) = \boldsymbol{\ell}^*_{\mathbb{G}_2}((2\mu,\mu^2),(s,p))\\ &= \frac{2|s-\overline{s}p+\mu^2(\overline{s}-s\overline{p})-2\mu(1-|p|^2)|+(1-|\mu|^2)|s^2-4p|}{|2-\overline{\mu}|^2-|s-2\overline{\mu}p|^2},\\ &\qquad \mu\in\mathbb{D},\ (s,p)\in\mathbb{G}_2.\end{aligned}$$

Proof. (a) follows directly from Theorem 7.1.1.

(b) Fix $(s,p)\in\mathbb{G}_2$. The first two equalities follow from Theorem 7.1.1. The cases $s=0$ or $p=0$ are elementary. Let $f(z) := \boldsymbol{F}_z(s,p)$. Since f is a homography, the set $f(\mathbb{T})$ is a circle for which the points $f(0)=-s/2$ and $f(\infty)=-2p/s$ are symmetric. Hence,

$$f(\mathbb{T}) = \left\{w\in\mathbb{C} : \left|\frac{w+\frac{s}{2}}{w+\frac{2p}{s}}\right| = \tfrac{1}{2}|s|\right\} = \{w\in\mathbb{C} : |w-w_0|=r\},$$

where

$$w_0 := \frac{2(p\overline{s}-s)}{4-|s|^2},\quad r := \frac{|s^2-4p|}{4-|s|^2},$$

which directly implies the remaining equality.

(c) We have

$$\begin{aligned}\boldsymbol{c}^*_{\mathbb{G}_2}((2\mu,\mu^2),(s,p)) &= \boldsymbol{c}^*_{\mathbb{G}_2}(H_{\boldsymbol{h}_\mu}(2\mu,\mu^2), H_{\boldsymbol{h}_\mu}(s,p))\\ &= \boldsymbol{c}^*_{\mathbb{G}_2}\left((0,0),\left(\frac{(1+|\mu|^2)s-2\mu-2\overline{\mu}p}{1-\overline{\mu}s+\overline{\mu}^2p}, \frac{p-\mu s+\mu^2}{1-\overline{\mu}s+\overline{\mu}^2p}\right)\right).\end{aligned}$$

Now we only need to use (b). □

Theorem 7.1.16.

$$\begin{aligned}\boldsymbol{\gamma}_{\mathbb{G}_2}((s_0,p_0);(X,Y)) &= \boldsymbol{\varkappa}_{\mathbb{G}_2}((s_0,p_0);(X,Y))\\ &= \max\{\boldsymbol{\gamma}_{\mathbb{D}}(\boldsymbol{F}_z(s_0,p_0);\boldsymbol{F}_z'(s_0,p_0)(X,Y)) : z\in\overline{\mathbb{D}}\}\\ &= \max\{\boldsymbol{\gamma}_{\mathbb{D}}(\boldsymbol{F}_z(s_0,p_0);\boldsymbol{F}_z'(s_0,p_0)(X,Y)) : z\in\mathbb{T}\},\\ &\qquad (s_0,p_0)\in\mathbb{G}_2,\ (X,Y)\in\mathbb{C}^2.\end{aligned}$$

Proof. We already know that $\boldsymbol{c}_{\mathbb{G}_2}\equiv\boldsymbol{\ell}_{\mathbb{G}_2}$ (Theorem 7.1.1). For taut domains $G\subset\mathbb{C}^n$, we have a general implication $\boldsymbol{c}_G\equiv\boldsymbol{\ell}_G\Longrightarrow\boldsymbol{\gamma}_G\equiv\boldsymbol{\varkappa}_G$ (cf. Proposition 11.1.7[4]). Fix $(s_0,p_0)\in\mathbb{G}_2$ and $(X,Y)\neq(0,0)$. Obviously,

$$\begin{aligned}\boldsymbol{\gamma}_{\mathbb{G}_2}((s_0,p_0);(X,Y)) &\geq \max\{\boldsymbol{\gamma}_{\mathbb{D}}(\boldsymbol{F}_z(s_0,p_0);\boldsymbol{F}_z'(s_0,p_0)(X,Y)) : z\in\overline{\mathbb{D}}\}\\ &\geq \max\{\boldsymbol{\gamma}_{\mathbb{D}}(\boldsymbol{F}_z(s_0,p_0);\boldsymbol{F}_z'(s_0,p_0)(X,Y)) : z\in\mathbb{T}\}.\end{aligned}$$

By Theorem 7.1.1, there exists a $z_\nu\in\mathbb{T}$ such that

$$\begin{aligned}&\boldsymbol{m}_{\mathbb{D}}(\boldsymbol{F}_{z_\nu}(s_0,p_0),\boldsymbol{F}_{z_\nu}(s_0+\tfrac{1}{\nu}X,p_0+\tfrac{1}{\nu}Y))\\ &\qquad= \boldsymbol{c}^*_{\mathbb{G}_2}((s_0,p_0),(s_0+\tfrac{1}{\nu}X,p_0+\tfrac{1}{\nu}Y)),\quad \nu\gg1.\end{aligned}$$

We may assume that $z_\nu\longrightarrow z_0\in\mathbb{T}$. Consequently,

$$\begin{aligned}\boldsymbol{\gamma}_{\mathbb{G}_2}((s_0,p_0);(X,Y)) &= \lim_{\nu\to+\infty}\frac{\boldsymbol{c}^*_{\mathbb{G}_2}((s_0,p_0),(s_0+\frac{1}{\nu}X,p_0+\frac{1}{\nu}Y))}{\frac{1}{\nu}}\\ &= \lim_{\nu\to+\infty}\frac{1}{\nu}\left|\frac{\boldsymbol{F}_{z_\nu}(s_0+\frac{1}{\nu}X,p_0+\frac{1}{\nu}Y)-\boldsymbol{F}_{z_\nu}(s_0,p_0)}{1-\overline{\boldsymbol{F}_{z_\nu}(s_0,p_0)}\boldsymbol{F}_{z_\nu}(s_0+\frac{1}{\nu}X,p_0+\frac{1}{\nu}Y)}\right|\\ &= \frac{1}{1-|\boldsymbol{F}_{z_0}(s_0,p_0)|^2}\lim_{\nu\to+\infty}\left|\frac{\boldsymbol{F}_{z_\nu}(s_0+\frac{1}{\nu}X,p_0+\frac{1}{\nu}Y)-\boldsymbol{F}_{z_\nu}(s_0,p_0)}{\frac{1}{\nu}}\right|\\ &= \frac{1}{1-|\boldsymbol{F}_{z_0}(s_0,p_0)|^2}\left|\frac{-2+2z_0^2p_0}{(2-z_0s_0)^2}X+\frac{2z_0}{2-z_0s_0}Y\right|\\ &= \boldsymbol{\gamma}_{\mathbb{D}}(\boldsymbol{F}_{z_0}(s_0,p_0);\boldsymbol{F}'_{z_0}(s_0,p_0)(X,Y)).\end{aligned}$$

□

---

[4] For the reader's convenience, we here repeat the corresponding part of the proof of Proposition 11.1.7: take $z_0\in G$ and $X_0\in(\mathbb{C}^n)_*$. Let $\varphi_\nu:\mathbb{D}\longrightarrow G$ be an extremal disc for $\boldsymbol{\ell}^*_G(z_0,z_0+(1/\nu)X_0)$ with $\varphi_\nu(0)=z_0$, $\varphi_\nu(\sigma_\nu)=z_0+(1/\nu)X_0$ ($\sigma_\nu\in(0,1)$), $\nu\gg1$. We may assume that $\varphi_\nu\overset{K}{\underset{\nu\to\infty}{\Longrightarrow}}\varphi_0\in\mathcal{O}(\mathbb{D},G)$, $\varphi_0(0)=0$. Then, $\varphi_0'(0)=\lim_{\nu\to\infty}\frac{\varphi_\nu(\sigma_\nu)-\varphi_\nu(0)}{\sigma_\nu}=(\lim_{\nu\to\infty}\frac{1}{\nu\sigma_\nu})\cdot X_0$. On the other hand, $\lim_{\nu\to\infty}\nu\sigma_\nu=\lim_{\nu\to\infty}\frac{\boldsymbol{\ell}^*_G(z_0,z_0+\frac{1}{\nu}X_0)}{\frac{1}{\nu}}=\lim_{\nu\to\infty}\frac{\boldsymbol{c}^*_G(z_0,z_0+\frac{1}{\nu}X_0)}{\frac{1}{\nu}}=\boldsymbol{\gamma}_G(z_0;X_0)$. Hence, $X_0=\boldsymbol{\gamma}_G(z_0;X_0)\varphi_0'(0)$, which proves that $\boldsymbol{\gamma}_G(z_0;X_0)=\boldsymbol{\varkappa}_G(z_0;X_0)$.

Corollary 7.1.17.

$$\begin{aligned}\gamma_{\mathbb{G}_2}((s,p);(X,Y)) &= \varkappa_{\mathbb{G}_2}((s,p);(X,Y))\\ &= \max\left\{\frac{1}{2}\frac{|(sY-pX)z^2-2Yz+X|}{1-|p|^2-\operatorname{Re}(zs-z\overline{s}p)} : z\in\mathbb{T}\right\}\\ &= \max\left\{\left|\frac{(sY-pX)z^2-2Yz+X}{(p\overline{s}-s)z^2+2(1-|p|^2)z+s\overline{p}-\overline{s}}\right| : z\in\mathbb{T}\right\}\\ &\qquad (s,p)\in\mathbb{G}_2,\ (X,Y)\in\mathbb{C}^2.\end{aligned}$$

In particular,

$$\gamma_{\mathbb{G}_2}((0,0);(X,Y)) = \frac{1}{2}|X|+|Y|,\quad (X,Y)\in\mathbb{C}^2.$$

Theorem 7.1.18 (cf. [266, 113, 14]).

$$\operatorname{Aut}(\mathbb{G}_2) = \{H_h : h\in\operatorname{Aut}(\mathbb{D})\}.^5$$

Proof. Step 1°. First, observe that $\operatorname{Aut}(\mathbb{G}_2)$ does not act transitively on $\mathbb{G}_2$.

Otherwise, by the Cartan classification theorem (cf. [18, 184]), $\mathbb{G}_2$ would be biholomorphic to $\mathbb{B}_2$ or $\mathbb{D}^2$, which is, by Theorem 7.1.1, impossible.[6]

Step 2°. Next, observe that $F(\Sigma_2)=\Sigma_2$ for every $F\in\operatorname{Aut}(\mathbb{G}_2)$.

Indeed, let $V := \{F(0,0) : F\in\operatorname{Aut}(\mathbb{G}_2)\}$. By W. Kaup's theorem, V is a connected complex submanifold of $\mathbb{G}_2$ (cf. [290]). We already know that $\Sigma_2\subset V$ (Remark 7.1.9). Since $\operatorname{Aut}(\mathbb{G}_2)$ does not act transitively, we have $V\subsetneq\mathbb{G}_2$. Thus, $V=\Sigma_2$.

Take a point $(s_0,p_0)=H_h(0,0)\in\Sigma_2$ with $h\in\operatorname{Aut}(\mathbb{D})$ (Remark 7.1.9). Then, for every $F\in\operatorname{Aut}(\mathbb{G}_2)$, we get $F(s_0,p_0)=(F\circ H_h)(0,0)\in V=\Sigma_2$.

Step 3°. By Remark 7.1.9, we only need to show that every automorphism $F\in\operatorname{Aut}(\mathbb{G}_2)$ with $F(0,0)=(0,0)$ is equal to a "rotation" R_τ. Fix such an $F=(S,P)$.

First, observe that $F|_{\Sigma_2}\in\operatorname{Aut}(\Sigma_2)$. Hence, the mapping

$$\mathbb{D}\ni\lambda\longmapsto(2\lambda,\lambda^2)\longmapsto F(2\lambda,\lambda^2)\longmapsto\frac{1}{2}\operatorname{pr}_s(F(2\lambda,\lambda^2))\in\mathbb{D}$$

[5] See Theorem 7.2.10 for a more general result.

[6] Instead of Theorem 7.1.1, one can also argue as follows: in the case where $\mathbb{G}_2\simeq\mathbb{B}_2$, we use the Remmert–Stein theorem (cf. [370], p. 71) which says that there is no proper holomorphic mapping $\mathbb{D}^2\longrightarrow\mathbb{B}_2$. In the case where $\mathbb{G}_2\simeq\mathbb{D}^2$, we use the characterization of proper holomorphic mappings $F:\mathbb{D}^2\longrightarrow\mathbb{D}^2$ (cf. [370], p. 76), which says that any such a mapping has the form $F(z_1,z_2)=(F_1(z_1),F_2(z_2))$ up to a permutation of the variables.

must be a rotation, i.e., $F(2\lambda, \lambda^2) = (2\alpha\lambda, \alpha^2\lambda^2)$ for some $\alpha \in \mathbb{T}$. Taking $R_{1/\alpha} \circ F$ instead of F, we may assume that $\alpha = 1$. In particular, $F'(0,0)\left[\begin{smallmatrix}2\\0\end{smallmatrix}\right] = \left[\begin{smallmatrix}2\\0\end{smallmatrix}\right]$ and, therefore, $F'(0,0) = \left[\begin{smallmatrix}1 & b\\0 & d\end{smallmatrix}\right]$.

For $\tau \in \mathbb{T}$, put $G_\tau := F^{-1} \circ R_{1/\tau} \circ F \circ R_\tau \in \operatorname{Aut}(\mathbb{G}_2)$. Obviously, $G_\tau(0,0) = (0,0)$. Moreover, $G'_\tau(0,0) = \left[\begin{smallmatrix}1 & b(\tau-1)\\0 & 1\end{smallmatrix}\right]$. Let $G_\tau^n : \mathbb{G}_2 \longrightarrow \mathbb{G}_2$ be the n-th iterate of G_τ. We have $(G_\tau^n)'(0,0) = \left[\begin{smallmatrix}1 & nb(\tau-1)\\0 & 1\end{smallmatrix}\right]$. Using the Cauchy inequalities, we get

$$|nb(\tau - 1)| \le \text{const}, \quad n \in \mathbb{N},\ \tau \in \mathbb{T},$$

which implies that $b = 0$, i.e., $F'(0,0)$ is diagonal.

Step 4°. We have $G'_\tau(0,0) = \mathbb{I}_2$. Hence, by the Cartan theorem (cf. [370], p. 66), $G_\tau = \mathrm{id}$. Consequently, $R_\tau \circ F = F \circ R_\tau$, i.e.,

$$(\tau S(s,p), \tau^2 P(s,p)) = (S(\tau s, \tau^2 p), P(\tau s, \tau^2 p)), \quad (s,p) \in \mathbb{G}_2,\ \tau \in \mathbb{T}.$$

Hence, $F(s,p) = (s, p + Cs^2)$. Since $F(2\lambda, \lambda^2) = (2\lambda, \lambda^2)$, we have $(2\lambda, \lambda^2 + 4C\lambda^2) = (2\lambda, \lambda^2)$, which immediately implies that $C = 0$, i.e., $F = \mathrm{id}$. □

Remark 7.1.19. Recall (Proposition 2.3.1(a)) that if $G \subset \mathbb{C}^n$ is a balanced domain, then $\boldsymbol{\gamma}_G(0;\cdot) \le \mathfrak{h}_G$ and, hence, $G \subset \{X \in \mathbb{C}^n : \boldsymbol{\gamma}_G(0;X) < 1\}$. Notice that this is not true in the category of d-balanced domains. For example, Corollary 7.1.15 shows that for $(X,Y) = (2\lambda, \lambda^2)$, with $\lambda \in \mathbb{D}$, we have $(X,Y) \in \mathbb{G}_2$ and $\boldsymbol{\gamma}_{\mathbb{G}_2}((0,0);(X,Y)) = |\lambda| + |\lambda|^2$. Taking $\lambda \approx 1$, we get $\boldsymbol{\gamma}_{\mathbb{G}_2}((0,0);(X,Y)) \approx 2$.

It should be mentioned that a more general approach to calculating the Carathéodory distance for $\mathbb{G}_2$ may be found in [14].

To complete the geometric description of the symmetrized bidisc, we will show that $\mathbb{G}_2$ is $\mathbb{C}$-convex. Before proving this result, let us recall some notions that may be of interest in this context. A domain $G \subset \mathbb{C}^n$ is said to be

- *linearly convex*, if for any $a \notin G$ there exists a complex hyperplane H through a with $H \cap G = \varnothing$;
- *weakly linearly convex*, if the former property is satisfied for all $a \in \partial G$;
- *locally weakly linearly convex*, if for every $a \in \partial G$ there exist a neighborhood $U = U(a)$ and a complex hyperplane H through a such that $U \cap H \cap G = \varnothing$;
- $\mathbb{C}$*-convex*, if for any complex line $L = a + b\mathbb{C}$, $\mathbb{C}^n \ni a, b \neq 0$, with $L \cap G \neq \varnothing$ this intersection is connected and simply connected (when thought as a one-dimensional domain).

It is known that: $\mathbb{C}$-convex $\Longrightarrow$ linearly convex $\Longrightarrow$ weakly linearly convex $\Longrightarrow$ locally weakly linearly convex.

A detailed discussion of the above properties may be found in [235] and [20]. We only mention that, for bounded domains in $\mathbb{C}^n$, $n > 1$, with a $\mathcal{C}^1$-boundary, all these properties are equivalent. Moreover, in the case where G has even a $\mathcal{C}^2$-boundary, the following result is true:

G is $\mathbb{C}$-convex if and only if

$$\mathcal{L}r(a; X) \geq \Big| \sum_{j,k=1}^{n} \frac{\partial^2 r}{\partial z_j \partial z_k}(a) X_j X_k \Big| \tag{$*$}$$

whenever $a \in \partial G$ and $X \in \mathbb{C}^n$ with $\sum_{j=1}^{n} \frac{\partial r}{\partial z_j}(a) X_j = 0$; here r is a defining function for G.

Moreover, a bounded domain $G \subset \mathbb{C}^n$ is called *strongly linearly convex* if in $(*)$ the strong inequality holds as long as $X \neq 0$.

Put

$$\mathbb{G}_2(\varepsilon) := \{(s, p) \in \mathbb{C}^2 : \sqrt{|s - \bar{s}p|^2 + \varepsilon} + |p|^2 < 1\}. \tag{7.1.2}$$

Observe that $\mathbb{G}_2(0) = \mathbb{G}_2$ and $\mathbb{G}_2(\varepsilon) \nearrow \mathbb{G}_2$ as $\varepsilon \to 0+$. Moreover, $\overline{\mathbb{G}_2(\varepsilon)} \subset \mathbb{C} \times \mathbb{D}$, $\varepsilon \in (0, 1)$, and the mapping

$$\mathbb{C} \times \mathbb{D} \ni (s, p) \longmapsto (s - \bar{s}p, p) \in \mathbb{C}^2 \tag{7.1.3}$$

is an $\mathbb{R}$-diffeomorphism onto the image. It shows, in particular, that $\mathbb{G}_2(\varepsilon)$ is $\mathbb{R}$-diffeomorphic to the convex domain $G_\varepsilon = \{(w, z) \in \mathbb{C}^2 : \sqrt{|w|^2 + \varepsilon} + |z|^2 < 1\}$.

Now, we formulate the result which was announced above.

Theorem 7.1.20 (cf. [429]). *$\mathbb{G}_2(\varepsilon)$ is a strongly linearly convex domain with $\mathcal{C}^\omega$-boundary, $\varepsilon \in (0, 1)$. In particular, $\mathbb{G}_2$ is $\mathbb{C}$-convex.*

Remark 7.1.21.

(a) Another proof of the weaker fact that $\mathbb{G}_2$ is $\mathbb{C}$-convex may be found in [390].

(b) Any bounded $\mathbb{C}$-convex domain with a $\mathcal{C}^2$-boundary can be exhausted by a sequence $(G_j)_{j \in \mathbb{N}}$ of strongly linearly convex domains G_j with $\mathcal{C}^\infty$-boundaries (see [243]). Note that $\mathbb{G}_2$ does not have a $\mathcal{C}^2$-boundary.

? (c) Can any $\mathbb{C}$-convex bounded domain be exhausted by strongly linearly convex domains with $\mathcal{C}^k$-boundaries, $k \in \{2, \ldots, \infty, \omega\}$?

(d) Moreover, using Theorems A.5.5 and 7.1.20 gives another proof of Theorem 7.1.1.

Proof. Let $\varepsilon \in (0,1)$ be fixed. Choose the following defining C^ω functions for the domain $\mathbb{G}_2(\varepsilon)$:

$$r_\varepsilon(s,p) := |s - \bar{s}p|^2 + \varepsilon - (1-|p|^2)^2, \ (s,p) \in \mathbb{C} \times \mathbb{D}. \tag{7.1.4}$$

Note that r_ε is real analytic. Moreover, the gradient of r_ε does not vanish on $\partial\mathbb{G}_2(\varepsilon)$.

Fix a boundary point $(\xi,\eta) \in \partial\mathbb{G}_2(\varepsilon)$ and a holomorphic tangent vector $X = (X_1, X_2) \neq 0$ at $\partial\mathbb{G}_2(\varepsilon)$ in (ξ,η), i.e., $(r_\varepsilon)_s(\xi,\eta)X_1 + (r_\varepsilon)_p(\xi,\eta)X_2 = 0$.

First we mention that $r_\varepsilon(e^{it}s, e^{2it}p) = r_\varepsilon(s,p)$, $(s,p) \in \mathbb{C}\times\mathbb{D}$, $t \in \mathbb{R}$. Therefore, it suffices to deal only with a boundary point (ξ,η) with $\xi \geq 0$. Then, $\xi^2 = \frac{(1-|\eta|^2)^2-\varepsilon}{|1-\eta|^2}$; in particular, $\varepsilon \leq (1-|\eta|^2)^2$. Put

$$\varrho(\lambda) := r_\varepsilon((\xi,\eta) + \lambda X), \quad \lambda \in \mathbb{C}.$$

We have to show that $\varrho_{\lambda\bar{\lambda}}(0) > |\varrho_{\lambda\lambda}(0)|$.

Using Taylor expansion we get

$$\begin{aligned}\varrho(\lambda) = 2\operatorname{Re}\Big(\big((\bar{\xi} - \xi\bar{\eta})(X_1 - \bar{\xi}X_2) - (\xi - \bar{\xi}\eta)X_1\bar{\eta} + 2\bar{\eta}X_2 - 2|\eta|^2\bar{\eta}X_2\big)\lambda\Big)\\ + |\lambda|^2\Big(|X_1 - \bar{\xi}X_2|^2 + |X_1|^2|\eta|^2 - 2\operatorname{Re}((\bar{\xi} - \xi\bar{\eta})\overline{X}_1X_2) + 2|X_2|^2 - 2|\eta|^2|X_2|^2\Big)\\ - \operatorname{Re}\big(2(X_1 - \bar{\xi}X_2)X_1\bar{\eta}\lambda^2\big) - \big(\operatorname{Re}(2\bar{\eta}X_2\lambda)\big)^2 + o(\lambda^2).\end{aligned} \tag{7.1.5}$$

The above formula shows in particular that the tangent vector X is given by the formula

$$X_1(\bar{\xi} - \xi\bar{\eta} - \bar{\eta}(\xi - \bar{\xi}\eta)) = X_2(\bar{\xi}(\bar{\xi} - \xi\bar{\eta}) - 2\bar{\eta} + 2|\eta|^2\bar{\eta}). \tag{7.1.6}$$

It is elementary to see that for the $\mathcal{C}^2$-function $v(\lambda) = \operatorname{Re}(A\lambda) + a|\lambda|^2 + \operatorname{Re}(b\lambda^2) - (\operatorname{Re}(c\lambda))^2 + o(\lambda^2)$, where $a \in \mathbb{R}$, $A, b, c \in \mathbb{C}$, the condition for

$$v_{\lambda\bar{\lambda}}(0) > |v_{\lambda\lambda}(0)|$$

is equivalent to

$$a - \frac{|c|^2}{2} > \left|b - \frac{c^2}{2}\right|. \tag{7.1.7}$$

Applying this information to the function ϱ the following inequality

$$\begin{aligned}|X_1 - \bar{\xi}X_2|^2 + |X_1|^2|\eta|^2 - 2\operatorname{Re}((\bar{\xi} - \xi\bar{\eta})\overline{X}_1X_2)\\ + 2|X_2|^2 - 2|\eta|^2|X_2|^2 - \frac{|2\bar{\eta}X_2|^2}{2} > \left|2(X_1 - \bar{\xi}X_2)X_1\bar{\eta} + \frac{(2\bar{\eta}X_2)^2}{2}\right|\end{aligned} \tag{7.1.8}$$

has to be verified.

Note that if $\xi = 0$, then $X_2 = 0$ and the former inequality is obviously true. Thus, from now on, we may assume that $\xi > 0$.

Substitute condition (7.1.6) and divide both sides by $|X_2|^2$. After reductions we get the following inequality

$$\begin{aligned}
&\left|2|\eta|^2\overline{\eta} - 2\overline{\eta} + \overline{\xi}\overline{\eta}(\xi - \overline{\xi}\eta)\right|^2 + |\eta|^2|\overline{\xi}(\overline{\xi} - \xi\overline{\eta}) - 2\overline{\eta} + 2|\eta|^2\overline{\eta}|^2 \\
&\quad - 2\operatorname{Re}\left((\overline{\xi} - \xi\overline{\eta})(\xi(\xi - \overline{\xi}\eta) - 2\eta + 2|\eta|^2\eta)(\overline{\xi} - \xi\overline{\eta} - \overline{\eta}(\xi - \overline{\xi}\eta))\right) \\
&\quad + 2|\overline{\xi} - \xi\overline{\eta} - \overline{\eta}(\xi - \overline{\xi}\eta)|^2 - 4|\eta|^2|\overline{\xi} - \xi\overline{\eta} - \overline{\eta}(\xi - \overline{\xi}\eta)|^2 \\
&> \left|2(2|\eta|^2\overline{\eta} - 2\overline{\eta} + \overline{\xi}\overline{\eta}(\xi - \overline{\xi}\eta))(\overline{\xi}(\overline{\xi} - \xi\overline{\eta}) - 2\overline{\eta} + 2|\eta|^2\overline{\eta})\eta\right. \\
&\quad \left. + 2\overline{\eta}^2(\overline{\xi} - \xi\overline{\eta} - \overline{\eta}(\xi - \overline{\xi}\eta))^2\right|.
\end{aligned} \tag{7.1.9}$$

Then, elementary calculations lead to

$$\begin{aligned}
&|\eta|^2\left|2|\eta|^2 - 2 + \overline{\xi}(\xi - \overline{\xi}\eta)\right|^2 + |\eta|^2\left|\overline{\xi}(\xi - \xi\overline{\eta}) + 2|\eta|^2\overline{\eta} - 2\overline{\eta}\right|^2 \\
&\quad - 2\operatorname{Re}\left((\overline{\xi} - \xi\overline{\eta})(\xi(\xi - \overline{\xi}\eta) - 2\eta + 2|\eta|^2\eta)(\overline{\xi} - \xi\overline{\eta} - \overline{\eta}(\xi - \overline{\xi}\eta))\right) \\
&\quad + 2|\overline{\xi} - \xi\overline{\eta} - \overline{\eta}(\xi - \overline{\xi}\eta)|^2 - 4|\eta|^2|\overline{\xi} - \xi\overline{\eta} - \overline{\eta}(\xi - \overline{\xi}\eta)|^2 \\
&> 2|\eta|^2\left|(2|\eta|^2 - 2 + \overline{\xi}(\xi - \overline{\xi}\eta))(\overline{\xi}(\overline{\xi} - \xi\overline{\eta}) - 2\overline{\eta} + 2|\eta|^2\overline{\eta})\right. \\
&\quad \left. + (\overline{\xi} - \xi\overline{\eta} - \overline{\eta}(\xi - \overline{\xi}\eta))^2\right|.
\end{aligned} \tag{7.1.10}$$

Now using the special form of ξ, we are lead to verify

$$\begin{aligned}
&|\eta|^2\left|2(|\eta|^2 - 1)(1 - \overline{\eta}) + (1 - |\eta|^2)^2 - \varepsilon\right|^2 \\
&\quad + |\eta|^2\left|(1 - |\eta|^2)^2 - \varepsilon - 2\overline{\eta}(1 - |\eta|^2)(1 - \eta)\right|^2 - 2((1 - |\eta|^2)^2 - \varepsilon) \\
&\quad \cdot \operatorname{Re}\left((1 - \overline{\eta})\left(\frac{(1 - |\eta|^2)^2 - \varepsilon}{1 - \overline{\eta}} - 2\eta(1 - |\eta|^2)\right)(1 - 2\overline{\eta} + |\eta|^2)\right) \\
&\quad + 2((1 - |\eta|^2)^2 - \varepsilon)\left|1 - 2\overline{\eta} + |\eta|^2\right|^2 - 4|\eta|^2((1 - |\eta|^2)^2 - \varepsilon)\left|1 - 2\overline{\eta} + |\eta|^2\right|^2 \\
&> 2|\eta|^2\left|(2(|\eta|^2 - 1)(1 - \overline{\eta}) + (1 - |\eta|^2)^2 - \varepsilon)((1 - |\eta|^2)^2 - \varepsilon\right. \\
&\quad \left. - 2\overline{\eta}(1 - |\eta|^2)(1 - \eta)) + ((1 - |\eta|^2)^2 - \varepsilon)(1 - 2\overline{\eta} + |\eta|^2)^2\right|,
\end{aligned} \tag{7.1.11}$$

which is equivalent to

$$\begin{aligned}
&|1 - 2\eta + |\eta|^2|^2 2|\eta|^2\varepsilon + 2|\eta|^2\varepsilon^2 + 2\varepsilon((1 - |\eta|^2)^2 - \varepsilon)\operatorname{Re}(1 - 2\eta + |\eta|^2) \\
&\qquad > 2|\eta|^2|\varepsilon^2 - \varepsilon(1 - 2\eta + |\eta|^2)^2|.
\end{aligned} \tag{7.1.12}$$

Note that $\mathrm{Re}(1 - 2\eta + |\eta|^2) = |1 - \eta|^2 > 0$, which easily implies that the above inequality holds for all possible η (i.e., satisfying the inequality $(1 - |\eta|^2)^2 \geq \varepsilon$). $\square$

Corollary 7.1.22. *There exists a strongly linearly convex domain with $\mathcal{C}^\omega$-boundary, namely $\mathbb{G}_2(\varepsilon)$ for small ε, which cannot be exhausted by domains that are biholomorphic to convex domains.*

Proof. Use Proposition 7.1.11 and Theorem 7.1.20. $\square$

Remark 7.1.23. Recently, another domain entered the discussion (see [7]), namely the so called *tetrablock*

$$\mathbb{E} := \{z \in \mathbb{C}^3 : 1 - z_1\zeta_1 - z_2\zeta_2 + z_3\zeta_1\zeta_2 \neq 0,\ \zeta_1, \zeta_2 \in \overline{\mathbb{D}}\}.$$

For this domain, one can also prove that $\boldsymbol{\ell}_{\mathbb{E}} = \boldsymbol{c}_{\mathbb{E}}$, that it cannot be exhausted by domains biholomorphic to convex domains (see [158] and [154]), and that it is $\mathbb{C}$-convex (see [572]). For more details, see § 16.8.

7.2 Symmetrized polydisc

This section is in some sense a continuation of Section 7.1. Let $n \geq 1$ and let $\pi_n : \mathbb{C}^n \longrightarrow \mathbb{C}^n$,

$$\pi_n(\lambda_1, \dots, \lambda_n) := \Big(\sum_{1 \leq j_1 < \dots < j_k \leq n} \lambda_{j_1} \cdots \lambda_{j_k} \Big)_{k=1,\dots,n}$$

Put $\mathbb{G}_n := \pi_n(\mathbb{D}^n)$ (cf. Section 7.1); note that $\mathbb{G}_1 = \mathbb{D}$ and $\mathbb{G}_2$ is the symmetrized bidisc. Observe that:

- π_n is symmetric;
- π_n is proper with multiplicity $n!$ – cf. Appendix B.2.2;
- $\mathbb{G}_n$ is open (cf. Appendix B.2.1(b)) and obviously connected;
- $\Delta_n := \{\lambda \in \mathbb{C}^n : \det \pi_n'(\lambda) = 0\} = \{\lambda \in \mathbb{C}^n : \exists_{i \neq j} : \lambda_i = \lambda_j\}$ (the reader is also asked to calculate $\det \pi_n'(\lambda)$);
- if $\lambda \notin \Delta_n$, then

$$\pi_n^{-1}(\pi_n(\lambda)) = \{(\lambda_{\sigma(1)}, \dots, \lambda_{\sigma(n)}) : \sigma \text{ is a permutation of } \{1, \dots, n\}\};$$

- $\pi_n^{-1}(\mathbb{G}_n) = \mathbb{D}^n$ – indeed let $\lambda \in \mathbb{D}^n$ and $z = \pi_n(\lambda)$. In the case when $\lambda \notin \Delta_n$, then all preimages of z are given by the following $n!$ points $(\lambda_{\sigma(1)}, \dots, \lambda_{\sigma(n)}) \in \mathbb{D}^n$, with σ any of the above permutations. So, it remains to discuss the situation $z = \pi_n(\lambda)$ with $\lambda \in \mathbb{D}^n \cap \Delta_n$. Assume there also is a preimage $\lambda' \notin \mathbb{D}^n$. Since π_n is open, there exist open disjoint neighborhoods U and $V \subset \mathbb{C}^n$ of λ' and λ, respectively, $V \subset \mathbb{D}^n$, such that $W := \pi_n(U) = \pi_n(V) \ni z$. Take a point $\mu \in V \setminus \Delta_n$. Then $w := \pi_n(\mu)$ has the $n!$ preimages in $\mathbb{D}^n$, which are given by $(\mu_{\sigma(1)}, \dots, \mu_{\sigma(n)})$, with

σ a permutation of $\{1,\dots,n\}$. But there is one more preimage of w sitting in U; a contradiction.

- $\pi_n|_{\mathbb{D}^n} : \mathbb{D}^n \longrightarrow \mathbb{G}_n$ is proper.

Thus, $\mathbb{G}_n$ is holomorphically convex and hence pseudoconvex. The domain $\mathbb{G}_n$ is called the *symmetrized polydisc* or, more precisely, the *symmetrized n-disc*. It is $(1,2,\dots,n)$-balanced, and its $(1,\dots,n)$-Minkowski function is given by

$$\mathfrak{h}_{\mathbb{G}_n}(z) = \max\{|\lambda_1|,\dots,|\lambda_n| : \pi_n(\lambda_1,\dots,\lambda_n) = z\}.$$

$\mathfrak{h}_{\mathbb{G}_n}$ is obviously continuous and, moreover, using Proposition 2.2.15, it is even psh. Therefore, $\mathbb{G}_n$ is hyperconvex and hence also taut. In particular,

$$z \in \mathbb{G}_n \iff (-z_1, z_2, \dots, (-1)^{n-1} z_{n-1}, (-1)^n z_n) \in \mathbb{G}_n.$$

For a point $z \in \mathbb{C}^n$, put

$$P_z(\lambda) = \lambda^n + \sum_{j=1}^{n} z_j \lambda^{n-j}, \quad \lambda \in \mathbb{C}.$$

Then, it is clear that $z \in \mathbb{G}_n$ if and only if all the zeros of P_z belong to $\mathbb{D}$.

Moreover, set

$$f_z(\lambda) := \frac{\sum_{j=1}^{n} j z_j \lambda^{j-1}}{n + \sum_{j=1}^{n-1}(n-j) z_j \lambda^j} = \frac{Q_z(\lambda)}{R_z(\lambda)}, \quad \lambda \in A_n(z),$$

where $A_n(z) := \mathbb{C} \setminus \{\mu \in \mathbb{C} : R_z(\mu) \neq 0\}$. These rational functions can be used to describe $\mathbb{G}_n$ (see [114], Proposition 3.1; cf. also Lemma 7.1.3).

Proposition 7.2.1. *Let $z \in \mathbb{C}^n$. Then, the following properties are equivalent:*

(i) $z \in \mathbb{G}_n$*;*

(ii) $f_z \in \mathcal{O}(\overline{\mathbb{D}})$ *and* $\sup_{\lambda \in \overline{\mathbb{D}}} |f_z(\lambda)| < 1$.

Proof. (i) $\Longrightarrow$ (ii): Fix a point $z \in \mathbb{G}_n$. Then, $P := P_z$ has all its zeros in $\mathbb{D}$. Applying the Lucas theorem (cf. [441]) gives that all zeros of P' are also sitting in $\mathbb{D}$. Thus the function $\mathbb{C}_* \ni \lambda \longmapsto P'(1/\lambda) = R_z(\lambda)/\lambda^{n-1}$ is zero-free in $\overline{\mathbb{D}} \setminus \{0\}$ and, obviously, $0 \in A_n(z)$. Thus, $\overline{\mathbb{D}} \subset A_n(z)$. Hence, f_z is holomorphic in a neighborhood of $\overline{\mathbb{D}}$.

Because of the maximum principle, it suffices to discuss the estimate only on $\mathbb{T}$. So, fix a $\lambda \in \mathbb{T}$. Note that

$$f_z(\lambda) = \frac{n\lambda^{n-1} P(1/\lambda) - \lambda^{n-2} P'(1/\lambda)}{\lambda^{n-1} P'(1/\lambda)} = n\lambda \frac{nP(\zeta)}{\zeta P'(\zeta)} - 1, \tag{7.2.1}$$

where $\zeta := 1/\lambda$. So, we have to verify that $1 - \frac{nP(\zeta)}{\zeta P'(\zeta)} \in \mathbb{D}$. Recall the conformal mapping $\mathbb{D} \ni z \overset{g}{\longmapsto} \frac{1}{1-z} = w$ onto the halfplane $\{w \in \mathbb{C} : \operatorname{Re} w > 1/2\} =: H$. Using its inverse mapping $g^{-1} : H \to \mathbb{D}$, we have to show that $\frac{\zeta P'(\zeta)}{nP(\zeta)} \in H$.

Denote the zeros of P by λ_j, $j = 1, \dots, n$. By the assumption, we know that $\lambda_j \in \mathbb{D}$. Then,

$$\frac{\zeta P'(\zeta)}{nP(\zeta)} = \frac{1}{n} \sum_{j=1}^{n} \frac{1}{1 - \lambda_j/\zeta}.$$

Therefore, $\frac{\zeta P'(\zeta)}{nP(\zeta)} \in \operatorname{conv}\{\frac{1}{1-\lambda_j/\zeta} : j = 1, \dots, n\}$. Since $\lambda_j/\zeta \in \mathbb{D}$, we have $g(\lambda_j/\zeta) = \frac{1}{1-\lambda_j/\zeta} \in H$, $j = 1, \dots, n$. Hence, the above equation implies that also $\operatorname{Re} \frac{\zeta P'(\zeta)}{nP(\zeta)} > 1/2$, as we wanted to know.

(ii) $\Longrightarrow$ (i): Let $z \in \mathbb{C}^n$ be fixed and $P = P_z$. In a first step, we want to show that the denominator of f_z has no zeros in $\overline{\mathbb{D}}$. Let us assume the contrary. So, we may take a $\lambda_0 \in \overline{\mathbb{D}}$ with $R_z(\lambda_0) = 0$. Obviously, $\lambda_0 \neq 0$. Hence, the numerator of f_z vanishes at λ_0. Using (7.2.1), it follows that $P'(1/\lambda_0) = 0$ and $n\lambda_0^{n-1} P(1/\lambda_0) - \lambda_0^{n-2} P'(1/\lambda_0) = 0$. Therefore, P has a zero of order $m \geq 2$ at the point $1/\lambda_0$, i.e., $P(\lambda) = (\lambda - 1/\lambda_0)^m \widetilde{P}(\lambda)$ with $\widetilde{P}(1/\lambda_0) \neq 0$ near $1/\lambda_0$. Putting this representation into (7.2.1) we get, after a small calculation, $|f_z(\lambda_0)| = |1/\lambda_0| \geq 1$; a contradiction.

By assumption, we have $|f_z(\lambda)| < 1$, $\lambda \in \mathbb{T}$. Therefore, $|Q_z(\lambda)| < |R_z(\lambda)|$ or $|\lambda^{n-1} Q_z(1/\lambda)| < |\lambda^n R_z(1/\lambda)|$, $\lambda \in \mathbb{T}$. So,

$$\Big| \sum_{j=1}^{n} j z_j \lambda^{n-j} \Big| < |\lambda^n R_z(1/\lambda)| = |n\lambda^n + \sum_{j=1}^{n-1} (n-j) z_j \lambda^{n-j}|, \quad \lambda \in \mathbb{T}.$$

By virtue of the Rouché theorem, it follows that nP has as many zeros in $\mathbb{D}$ as the polynomial $\lambda \longmapsto \lambda^n R(1/\lambda)$. Hence, all zeros of P lie in $\mathbb{D}$ and, therefore, $z \in \mathbb{G}_n$. □

We point out that the former proof has also shown that $R_z(\lambda) \neq 0$ if $z \in \mathbb{G}_n$ and $\lambda \in \overline{\mathbb{D}}$. Therefore, the function $\mathbb{G}_n \ni z \longmapsto f_z(\lambda)$ is holomorphic for such λ's.

Corollary 7.2.2. *If $z, w \in \mathbb{G}_n$, then*

$$\boldsymbol{c}_{\mathbb{G}_n}(z, w) \geq p_n(z, w) := \sup_{\lambda \in \mathbb{T}} \{\boldsymbol{p}(f_z(\lambda), f_w(\lambda))\}.$$

Note that we also have $\boldsymbol{c}_{\mathbb{G}_n}(z, w) = \sup_{\lambda \in \overline{\mathbb{D}}} \{\boldsymbol{p}(f_z(\lambda), f_w(\lambda))\}$, since the function on the right hand side is sh in λ.

Observe that the function f_z may be read in the following form:

$$f_z(\lambda) = \frac{\sum_{j=1}^{n} j!(n-j)!z_j\binom{n-1}{j-1}\lambda^{j-1}}{n! + \sum_{j=1}^{n-1} j!(n-j)!z_j\binom{n-1}{j}\lambda^j}.$$

Taking into account that the point $(\binom{n-1}{1}\lambda, \dots, \binom{n-1}{n-1}\lambda^{n-1}) \in \Gamma_{n-1}$, where

$$\Gamma_n := \{w \in \mathbb{C}^n : \mathfrak{h}_{\mathbb{G}_n}(w) \le 1\},$$

we may introduce the following rational function on $\mathbb{C}^{n-1}$. For a fixed $z \in \mathbb{C}^n$, put

$$g_z(w) := \frac{\sum_{j=1}^{n} j!(n-j)!z_j w_{j-1}}{n! + \sum_{j=1}^{n-1} j!(n-j)!z_j w_j}, \qquad w \in \mathbb{C}^{n-1}.$$

Using this new function, there is another criterion to decide whether a point $z \in \mathbb{C}^n$ belongs to the symmetrized polydisc.

Proposition 7.2.3. *For a point $z \in \mathbb{C}^n$, $n \ge 2$, the following two conditions are equivalent:*

(i) $z \in \mathbb{G}_n$;

(ii) $g_z \in \mathcal{O}(\Gamma_{n-1})$ *and* $\sup_{w \in \Gamma_{n-1}} |g_z(w)| < 1$.

Proof. (ii) $\Longrightarrow$ (i): In view of the above remark, we have

$$f_z(\lambda) = g_z\left(\binom{n-1}{1}\lambda, \dots, \binom{n-1}{n-1}\lambda^{n-1}\right)$$

and, therefore, property (i) follows due to Proposition 7.2.1.

(i) $\Longrightarrow$ (ii): Now, we start with a $z \in \mathbb{G}_n$. Then, $\sup_{\lambda \in \overline{\mathbb{D}}} |f_z(\lambda)| < r < 1$ for a certain r. Then,

$$\frac{\sum_{j=1}^{n} jz_j\lambda^{j-1}}{n + \sum_{j=1}^{n-1}(n-j)z_j\lambda^j} \neq -r\mu, \quad |\lambda| \le 1 \le |\mu|;$$

or, equivalently,

$$\sum_{j=1}^{n-1}\left((j+1)z_{j+1} + (n-j)z_j\mu r\right)\lambda^j + nr\mu + z_1 \neq 0, \quad |\lambda| \le 1 \le |\mu|.$$

Then,

$$1 + \sum_{j=1}^{n-1} \frac{(j+1)z_{j+1} + (n-j)z_j\mu r}{nr\mu + z_1} \neq 0, \quad |\lambda| \le 1 \le |\mu|.$$

Note that the denominator is different from zero. Put

$$\widetilde{S}_j(\mu) := \frac{(j+1)z_{j+1} + (n-j)z_j\mu r}{nr\mu + z_1}, \quad j = 1, \dots, n-1.$$

Dividing by λ^{n-1} and setting $\xi = \lambda^{-1}$ leads to the polynomial

$$\xi^{n-1} + \sum_{j=1}^{n-1} \widetilde{S}_j(\mu)\xi^{n-1-j},$$

which has no zeros outside of $\mathbb{D}$. Thus, $(\widetilde{S}_1(\mu), \dots, \widetilde{S}_{n-1}(\mu)) \in \mathbb{G}_{n-1}$.

Now, we need the following well known result on zeros of polynomials (see [356], Corollary 16.1(a)):

Lemma* 7.2.4. *Assume that the polynomials $A(\lambda) = \sum_{j=1} n a_j \lambda^j / \binom{n}{j}$ and $B(\lambda) = \sum_{j=1}^n b_j \lambda^j / \binom{n}{j}$ have all their zeros in $\mathbb{D}(r)$ and $\overline{\mathbb{D}}(R)$, respectively. Then, the polynomial $C(\lambda) = \sum_{j=1}^n a_j b_j \lambda^j / \binom{n}{j}$ has all its zeros inside of $\mathbb{D}(rR)$.*

From this lemma, we may conclude that, if $\zeta \in \mathbb{G}_{n-1}$ and if $\mathfrak{h}_{\mathbb{G}_{n-1}}(\eta) \le 1$, $\zeta, \eta \in \mathbb{C}^{n-1}$, then

$$\left(\frac{\zeta_1\eta_1}{\binom{n-1}{1}}, \dots, \frac{\zeta_{n-1}\eta_{n-1}}{\binom{n-1}{n-1}} \right) \in \mathbb{G}_{n-1}.$$

Applying this result shows that

$$\left(\frac{w_1\widetilde{S}_1(\mu)}{\binom{n-1}{1}}, \dots, \frac{w_{n-1}\widetilde{S}_{n-1}(\mu)}{\binom{n-1}{n-1}} \right) \in \mathbb{G}_{n-1}.$$

Thus, $1 + \sum_{j=1}^{n-1} w_j \widetilde{S}_j(\mu)/\binom{n-1}{j} \neq 0$. Now reversing the former calculations leads to

$$\frac{\sum_{j=1}^n z_j w_{j-1}/\binom{n}{j}}{1 + \sum_{j=1}^{n-2} z_j w_{j-1}/\binom{n}{j}} \neq r\mu, \quad |\mu| \ge 1, w \in \Gamma_{n-1}.$$

Hence, we end up with

$$\sup_{w \in \Gamma_{n-1}} |g_z(w)| = \sup_{w \in \Gamma_{n-1}} \left| \frac{\sum_{j=1}^n z_j w_{j-1}/\binom{n}{j}}{1 + \sum_{j=1}^{n-2} z_j w_{j-1}/\binom{n}{j}} \right| \le r < 1. \qquad \square$$

In the future we will also need the following rational mappings. For $\lambda \in \overline{\mathbb{D}}$, we put

$$\Phi_{n,\lambda}(z) := \left(\frac{(n-j)z_j + \lambda(j+1)z_{j+1}}{n + \lambda z_1} \right)_{j=1,\dots,n-1},$$

which is holomorphic on $\mathbb{D}(n) \times \mathbb{C}^{n-1}$. Its main property is the following one:

Proposition 7.2.5. *For a point $z \in \mathbb{C}^n$, $n \geq 2$, the following properties are equivalent:*

(i) $z \in \mathbb{G}_n$;

(ii) $\Phi_{n,\lambda}(z) \in \mathbb{G}_{n-1}$ *for all* $\lambda \in \overline{\mathbb{D}}$.

Proof. Let us begin by observing that

$$\Gamma_{n-1} = \{(t_1 + \mu, t_2 + \mu t_1, \dots, t_{n-2} + \mu t_{n-3}, \mu t_{n-2}) : t \in \Gamma_{n-2},\ \mu \in \overline{\mathbb{D}}\}.$$

Therefore, one has the following equivalence:

(i) $z \in \mathbb{G}_n$;

(ii) $g_z \in \mathcal{O}(\Gamma_{n-1})$ and $\sup_{w \in \Gamma_{n-1}} |g_z(w)| < 1$;

(iii) $g_z \in \mathcal{O}(\Gamma_{n-1})$
and $\sup_{\lambda \in \overline{\mathbb{D}}} \sup_{t \in \Gamma_{n-2}} |g_z(t_1 + \lambda, t_2 + \lambda t_1, \dots, t_{n-2} + \lambda t_{n-3}, \lambda t_{n-2})| < 1$;

(iv) $t \longmapsto g_z(t_1 + \lambda, t_2 + \lambda t_1, \dots, t_{n-2} + \lambda t_{n-3}, \lambda t_{n-2})$ is holomorphic on Γ_{n-2} and $\sup_{t \in \Gamma_{n-2}} |g_z(t_1 + \lambda, t_2 + \lambda t_1, \dots, t_{n-2} + \lambda t_{n-3}, \lambda t_{n-2})| < 1$ for any $\lambda \in \overline{\mathbb{D}}$;

(v) $g_{\Phi_{n,\lambda}(z)} \in \mathcal{O}(\Gamma_{n-2})$ and $\sup_{t \in \Gamma_{n-2}} |g_{\Phi_{n,\lambda}(z)}(t)| < 1$ for all $\lambda \in \overline{\mathbb{D}}$;

(vi) $\Phi_{n,\lambda}(z) \in \mathbb{G}_{n-1}$ for all $\lambda \in \overline{\mathbb{D}}$. □

Remark 7.2.6. If $z \in \mathbb{G}_n$, $n \geq 2$, and $\lambda \in \overline{\mathbb{D}}$, then $f_z(\lambda) = \Phi_{2,\lambda} \circ \dots \circ \Phi_{n,\lambda}(z)$.

For later purposes, we here add the following consequence of the former result on a sort of peak-points, due to Kosiński and Zwonek (see [323]):

Corollary 7.2.7. *Let $a \in \partial\mathbb{G}_n$, $n \geq 1$. Then, there exist an open neighborhood U of a and an $f \in \mathcal{C}(\mathbb{G}_n \cup U) \cap \mathcal{O}(\mathbb{G}_n)$ such that $f(a) = 1$ and $|f| < 1$ on $\mathbb{G}_n$.*

Proof. We proceed via induction on n, and observe that the case where $n = 1$ is trivial. Now assume that $n \geq 2$, and the result is known for $n - 1$. In case that $|a_1| = n$, one may take simply the function $f(z) := z_1 \frac{|a_1|}{n a_1}$. So it remains to discuss the case $|a_1| < n$. By virtue of Proposition 7.2.5, there is a $\lambda_0 \in \overline{\mathbb{D}}$ such that $b := \Phi_{n,\lambda_0}(a) \in \partial\mathbb{G}_{n-1}$. Therefore, by the induction assumption, we find an open neighborhood V of b and a $g \in \mathcal{C}(\mathbb{G}_{n-1} \cup V) \cap \mathcal{O}(\mathbb{G}_{n-1})$ satisfying $g(b) = 1$ and $|g| < 1$ on $\mathbb{G}_{n-1}$. Then, $f := g \circ \Phi_{n,\lambda_0}$ fulfills all the desired properties for a correctly chosen $U = U(a)$. □

Recall Theorem 7.1.1, which even gives $\ell_{\mathbb{G}_2} = c_{\mathbb{G}_2} = p_2$. Before we deal with some of the invariant metrics on $\mathbb{G}_n$ and the question whether $\ell_{\mathbb{G}_n} = c_{\mathbb{G}_n}$, we first discuss some of the geometric properties of $\mathbb{G}_n$, which may be found in [377] and [390].

Proposition 7.2.8. (a) *$\mathbb{G}_n$, $n \geq 3$, is linearly convex.*

(b) *$\mathbb{G}_n$ is not $\mathbb{C}$-convex.*

(c) *$\mathbb{G}_n$ cannot be exhausted by domains biholomorphic to convex ones.*

Proof. (a) Fix a point $z = \pi_n(\lambda) \notin \mathbb{G}_n$. We we may assume that $|\lambda_1| \geq 1$. Then, the set

$$B := \{\pi_n(\lambda_1, \mu_2, \dots, \mu_n) : \mu_2, \dots, \mu_n \in \mathbb{C}\}$$

is disjoint to $\mathbb{G}_n$. Moreover,

$$B = \{(\lambda_1 + z_1, \lambda_1 z_1 + z_2, \dots, \lambda_1 z_{n-2} + z_{n-1}, \lambda_1 z_{n-1}) : z_1, \dots, z_{n-1} \in \mathbb{C}\};$$

thus, B is a complex affine hyperplane. Hence, $\mathbb{G}_n$ is linearly convex.

(b) Consider the points

$$a_t := \pi_n(t, t, t, 0, \dots, 0) = (3t, 3t^2, t^3, 0, \dots, 0) \in \mathbb{G}_n,$$
$$b_t := \pi_n(-t, -t, -t, 0, \dots, 0) = (-3t, 3t^2, -t^3, 0, \dots, 0) \in \mathbb{G}_n,$$

where $t \in (0, 1)$. Denote by L_t the complex line passing through a_t and b_t, that is,

$$L_t = \{c_{t,\lambda} = (3t(1 - 2\lambda), 3t^2, t^3(1 - 2\lambda), 0, \dots, 0) : \lambda \in \mathbb{C}\}.$$

Note that $c_{t,0} = a_t$ and $c_{t,1} = b_t$. Assume now that $\mathbb{G}_n \cap L_t$ is connected. Then, there exists a $\tau \in \mathbb{R}$ such that $c_{t,\lambda} \in \mathbb{G}_n$ with $\lambda := 1/2 + i\tau$, i.e.,

$$c_{t,\lambda} = (-6\tau t i, 3t^2, -2\tau t^3 i, 0, \dots, 0) \in \mathbb{G}_n,$$

or $c_{t,\lambda} = \pi_n(\mu)$, where $\mu = (\mu_1, \mu_2, \mu_3, 0, \dots, 0) \in \mathbb{D}^n$, which implies $-36\tau^2 t^2 = (\mu_1 + \mu_2 + \mu_3)^3 = \mu_1^2 + \mu_2^2 + \mu_3^2 + 6t^2$. Therefore,

$$t^2 = \frac{|\mu_1^2 + \mu_2^2 + \mu_3^2|}{36\tau^2 + 6} < \frac{3}{36\tau^2 + 6} \leq \frac{1}{2},$$

which gives a contradiction for $t \in [1/\sqrt{2}, 1)$. Hence, $\mathbb{G}_n$ is not $\mathbb{C}$-convex.

(c) A proof of this remaining claim, similar to the one of Lemma 7.1.11, can be found in [377]. Since this result will also be a direct consequence of further discussions and Lempert's Theorem 11.2.1, we will omit its direct proof here. □

Remark 7.2.9.

(a) It should be mentioned that $\mathbb{G}_2$ is starlike with respect to the origin, but $\mathbb{G}_n$, $n \geq 3$, is not.

(b) The above result shows that there are d-balanced, linearly convex domains ($d \neq (1, \dots, 1)$) that are not convex. It is interesting to mention that any linearly convex balanced domain is automatically convex. For a proof see [390].

(c) One may, in general, study images $\pi(G)$ of convex domains $G \subset \mathbb{C}^n$ and ask whether they are convex, $\mathbb{C}$-convex, or can be exhausted by domains biholomorphic to convex ones. Particular cases are discussed in [548] – cf. § A.3.

Much more is known about the structure of the symmetrized polydisc. For completeness, we present (without giving the proof) the following result due to A. Edigarian and W. Zwonek (see [157]):

Theorem* 7.2.10. *Any proper holomorphic mapping $F : \mathbb{G}_n \longrightarrow \mathbb{G}_n$ is of the form*

$$F(\pi_n(\lambda_1, \dots, \lambda_n)) = \pi_n(B(\lambda_1), \dots, B(\lambda_n)),$$

where B is a finite Blaschke product. In particular,

$$\operatorname{Aut}(\mathbb{G}_n) = \{H_h : h \in \operatorname{Aut}(\mathbb{D})\},$$

where $H_h(\pi_n(\lambda_1, \dots, \lambda_n)) = \pi_n(h(\lambda_1), \dots, h(\lambda_n))$.

The reader who is interested in a proof may consult the original paper [157].

In the following, we will discuss the Carathéodory distance and the Lempert function on $\mathbb{G}_n$ and show that they do not coincide which is quite different from the 2-dimensional case. The arguments given here are taken from [389] and [388].

For this discussion, we introduce the following objects:

- $\widetilde{f_\lambda}(X) := \frac{\sum_{j=1}^n jX_j\lambda^{j-1}}{n}$, where $\lambda \in \mathbb{C}$ and $\mathbb{C}^n \ni X = \sum_{j=1}^n X_j e_j$;
- $\varrho_n(X) := \max\{|\widetilde{f_\lambda}(X)| : \lambda \in \mathbb{T}\}$, $X \in \mathbb{C}^n$.

Observe that

$$\gamma_{\mathbb{G}_n}(0; X) \geq \lim_{\mathbb{C}_* \ni t \to 0} \frac{p_n(0, tX)}{t} = \varrho_n(X), \quad X \in \mathbb{C}^n.$$

Denote the span of the standard unit vectors e_j, e_k by $L_{j,k}$. Thus, if $X \in L_{j,k}$, then $\varrho_n(X) = \frac{j|X_j| + k|X_k|}{n}$.

The following result gives some relations between certain pseudometrics on $\mathbb{G}_n$.

Proposition 7.2.11 (cf. [389]). *Let $n \geq 3$ and let $k, l \in \{1, \dots, n\}$. Then,*

(a) *If k does not divide n, then $\boldsymbol{\gamma}_{\mathbb{G}_n}(0; \boldsymbol{e}_k) > \varrho_n(\boldsymbol{e}_k)$;*

(b) *k divides n if and only if $\boldsymbol{\varkappa}_{\mathbb{G}_n}(0; \boldsymbol{e}_k) = \varrho_n(\boldsymbol{e}_k)$;*

(c) *if k and l divide n, then $\boldsymbol{\varkappa}^{(2)}_{\mathbb{G}_n}(0; X) = \widehat{\boldsymbol{\varkappa}}_{\mathbb{G}_n}(0; X) = \varrho_n(X)$, $X \in L_{k,l}$;*

(d) *if $X \in L_{1,n} \setminus (L_{1,1} \cup L_{n,n})$, then $\boldsymbol{\varkappa}_{\mathbb{G}_n}(0; X) > \varrho_n(X)$.*

As a consequence, we see that $\boldsymbol{\ell}_{\mathbb{G}_n}$ and $\boldsymbol{c}_{\mathbb{G}_n}$ are, in general, different on $\mathbb{G}_n \times \mathbb{G}_n$, $n \geq 3$.

Corollary 7.2.12. *If $n \geq 3$, then $\boldsymbol{\ell}_{\mathbb{G}_n} \not\equiv \boldsymbol{k}_{\mathbb{G}_n}$ and $\boldsymbol{c}_{\mathbb{G}_n} \not\equiv p_n$.*

Proof. Assume that $\boldsymbol{\ell}_{\mathbb{G}_n} \equiv \boldsymbol{k}_{\mathbb{G}_n}$. Then, by virtue of Propositions 5.3.3 and 5.3.6, one has $\boldsymbol{\varkappa}_{\mathbb{G}_n} \equiv \widehat{\boldsymbol{\varkappa}}_{\mathbb{G}_n}$ on $\mathbb{G}_n \times \mathbb{C}^n$. Therefore, if $X \in L_{1,1} \setminus (L_{1,1} \cup L_{n,n})$, then $\boldsymbol{\varkappa}_{\mathbb{G}_n}(0; X) > \varrho_n(X) = \widehat{\boldsymbol{\varkappa}}_{\mathbb{G}_n}(0; X)$; a contradiction. The second claim follows from (a) and also via differentiation. □

This result, together with Theorem 7.1.20 and Proposition 7.2.8, immediately leads to the following important problem: does $\boldsymbol{\ell}_G = \boldsymbol{c}_G$ hold, if the domain G is assumed to be $\mathbb{C}$-convex.

?

Proof of Proposition 7.2.11. (a) Let $\sqrt[k]{1} = \{\xi_1, \dots, \xi_k\} \subset \mathbb{T}$. Put

$$g_z(\lambda) = \lambda f_z(\lambda) \text{ and } g_{z,k}(\lambda) := \frac{\sum_{j=1}^k g_z(\xi_k \lambda)}{k\lambda^k}, \quad z \in \mathbb{G}_n,\ \lambda \in \overline{\mathbb{D}}.$$

Using that $\sum_{j=1}^k \xi_j^m = 0$ for $m = 1, \dots, k-1$, the Taylor expansion formula shows that $g_{z,k}$ can be holomorphically extended through 0 as $g_{z,k}(0) = P_k(z)$, where P_k is a polynomial in z with

!

- $\frac{\partial P_k}{\partial z_k}(0) = n/k = \varrho_n(\boldsymbol{e}_k)$;
- $t^k P_k(w) = P_k(tw_1, t^2 w_2, \dots, t^n w_n), \quad w \in \mathbb{C}^n, t \in \mathbb{C}$.

The maximum principle and Proposition 7.2.1 implies that $g_{z,k} \in \mathcal{O}(\mathbb{D}, \overline{\mathbb{D}})$; in particular, $|P_k(z)| \leq 1$ for all $z \in \mathbb{G}_n$. To prove the inequality claimed in (a) it suffices to show that even $|P_k| < 1$ on $\overline{\mathbb{G}_n}$. Assuming the contrary there exists a $z \in \partial\mathbb{G}_n$ with $P_k(z) = e^{i\theta}$. Applying the maximum principle together with the triangle inequality, it follows that $g_z(\xi_j \lambda) = e^{i\theta}\lambda^k$, $\lambda \in \mathbb{T}$ and $1 \leq j \leq n$. In particular, $g_z(\lambda) = e^{i\theta}\lambda^k$, i.e.,

$$\sum_{j=1} njz_j\lambda^j = e^{i\theta}\left(n\lambda^k + \sum_{j=1}^{n-1}(n-j)z_j\lambda^{k+j}\right).$$

Comparing the coefficients of these two polynomials in λ yields

- $z_k = e^{i\theta} n/k, z_{n+1-k} = \cdots z_{n-1} = 0$;
- $(k+j)z_{k+j} = e^{i\theta}(n-j)z_j, \quad 1 \le j \le n-k$. Hence $z_{kl} = e^{i\theta}\binom{n/k}{l}$, $1 \le l \le \lfloor n/k \rfloor$. On the other hand, since, by assumption, k does not divide n, one has $n-k < k\lfloor n/k \rfloor < n$, and so $z_{k\lfloor n/k \rfloor} = 0$; a contradiction.

(b) First assume that $\varkappa_{\mathbb{G}_n}(0; \boldsymbol{e}_k) = \varrho_n(\boldsymbol{e}_k)$. Then, $\boldsymbol{\gamma}_{\mathbb{G}_n}(0; \boldsymbol{e}_k) = \varrho_n(\boldsymbol{e}_k)$. So, by (a), k divides n.

Conversely, let k divide n, i.e., $ks = n$ for a certain $s \in \mathbb{N}$. Put

$$\varphi(\zeta) := \begin{cases} 0, & \text{if } j \notin k\mathbb{N}, 1 \le j \le n \\ \binom{s}{m}\zeta^m, & \text{if } 1 \le j = mk \le n \end{cases}.$$

Obviously, $\varphi \in \mathcal{O}(\mathbb{D}, \mathbb{C}^n)$ with $\varphi(0) = 0$. To prove that, in fact, $\varphi \in \mathcal{O}(\mathbb{D}, \mathbb{G}_n)$, consider the associated polynomial for $\zeta \in \mathbb{D}$

$$\begin{aligned} P_\zeta(\lambda) &= \lambda^n + \binom{s}{1}\zeta^1\lambda^{n-1k} + \cdots + \binom{s}{m}\zeta^m\lambda^{n-mk} + \cdots + \binom{s}{s}\zeta^s\lambda^{n-sk} \\ &= (\lambda^k + \zeta)^s. \end{aligned}$$

and observe that all of its zeros are contained in $\mathbb{D}$. It remains to mention that $(k/n)\varphi'(0) = \boldsymbol{e}_k$ to get $\varrho_n(\boldsymbol{e}_k) = k/n \le \varkappa_{\mathbb{G}_n}(0; \boldsymbol{e}_k) \le k/n$.

(c) Let $X \in L_{k,l}$. Then, using (b), one obtains

$$\begin{aligned} \varrho_n(X) &\le \widehat{\varkappa}_{\mathbb{G}_n}(0; X) \le \varkappa^{(2)}_{\mathbb{G}_n}(0; X) \le \varkappa_{\mathbb{G}_n}(0; X_k\boldsymbol{e}_k) + \varkappa_{\mathbb{G}_n}(0; X_l\boldsymbol{e}_l) \\ &= \varrho_n(X_k\boldsymbol{e}_k) + \varrho_n(X_l\boldsymbol{e}_l) = \varrho_n(X). \end{aligned}$$

(d) Fix an $X \in L_{1,n} \setminus (L_{1,1} \cup L_{n,n})$. Recall that

- $\varkappa_{\mathbb{G}_n}(0; Y) = \varkappa_{\mathbb{G}_n}(0; (\lambda Y_1, \lambda^2 Y_2, \dots, \lambda^n Y_n))$ (use the automorphism $\mathbb{G}_n \ni z \longmapsto (\lambda z_1, \dots, \lambda^n z_n)$);
- $\varkappa_{\mathbb{G}_n}(0; Y) = \varkappa_{\mathbb{G}_n}(0; \lambda Y)$, $Y \in \mathbb{C}^n$ and $\lambda \in \mathbb{T}$.

Therefore, we may assume that $X_1 > 0$ and $X_2 > 0$.

Using the mappings $\Phi_{n,1} : \mathbb{G}_n \longrightarrow \mathbb{G}_{n-1}$, it follows that

$$\varkappa_{\mathbb{G}_n}(0; X) \ge \varkappa_{\mathbb{G}_{n-1}}(\Phi_{n,1}(0); \Phi'_{n,1}(0)X) = \varkappa_{\mathbb{G}_{n-1}}\left(0; \frac{n-1}{n}X_1\boldsymbol{e}_1 + X_n\boldsymbol{e}_n\right).$$

Continuing via induction on n finally leads to

$$\varkappa_{\mathbb{G}_n}(0; X) \ge \varkappa_{\mathbb{G}_3}(0; Y) \text{ with } Y := \frac{3X_1}{n}\boldsymbol{e}_1 + X_n\boldsymbol{e}_3.$$

Suppose now that $\varkappa_{\mathbb{G}_n}(0; X) = \varrho_n(X)$. Then,

$$\varrho_n(X) \geq \varkappa_{\mathbb{G}_3}(0; Y) \geq \varrho_3(Y) = \varrho_n(X);$$

hence, $\varkappa_{\mathbb{G}_3}(0; Y) = \varrho_3(Y)$.

Since $\mathbb{G}_3$ is taut, there exists an extremal analytic disc for $\varkappa_{\mathbb{G}_3}(0; Y)$, i.e., there is a $\varphi \in \mathcal{O}(\mathbb{D}, \mathbb{G}_3)$ with $\varphi(0) = 0$ and $\varkappa_{\mathbb{G}_3}(0; Y)\varphi'(0) = Y$. So, we may write $\varphi(\zeta) = \zeta\widetilde{\varphi}(\zeta)$, $\zeta \in \mathbb{D}$. Thus, $\varrho_3(Y)\widetilde{\varphi}(0) = Y$ or, in more detail,

$$\widetilde{\varphi}_1(0) = \frac{Y_1}{3(Y_1 + 3Y_3)}, \quad \widetilde{\varphi}_2(0) = 0, \quad \widetilde{\varphi}_3(0) = \frac{Y_1}{3(Y_1 + 3Y_3)}.$$

Put $g_\lambda(\zeta) := f_{\varphi(\zeta)}(\lambda)/\zeta$, $\zeta \in \mathbb{D}$ and $\lambda \in \mathbb{T}$. Then, $g_\lambda \in \mathcal{O}(\mathbb{D}, \overline{\mathbb{D}})$. In particular, $g_{\pm 1}(0) = 1$, and so, by the maximum principle, one has $g_{\pm 1} \equiv 1$ on $\mathbb{D}$. Therefore,

$$\widetilde{\varphi}_1 \pm 2\widetilde{\varphi}_2 + 3\widetilde{\varphi}_3 = 3 \pm 2\,\mathrm{id}\,|_{\mathbb{D}}\widetilde{\varphi}_1 + \mathrm{id}\,|_{\mathbb{D}}\widetilde{\varphi}_2.$$

Exploiting these two equation gives $\widetilde{\varphi}_2 = \mathrm{id}\,|_{\mathbb{D}}\widetilde{\varphi}_1$ and $\widetilde{\varphi}_3 = 1 + \frac{\mathrm{id}\,|_{\mathbb{D}}^2 - 1}{3}\widetilde{\varphi}_1$ on $\mathbb{D}$.

Put $\psi = \widetilde{\varphi}_1/3$. Then $|g_\lambda(\zeta)| \leq 1$, $\zeta \in \mathbb{D}$, may be equivalently reformulated in the following way:

$$\left| \frac{\psi(\zeta) + 2\lambda\zeta\psi(\zeta) + \lambda^2(1 + (\zeta^2 - 1)\psi(\zeta))}{1 + 2\lambda\zeta\psi(\zeta) + \lambda^2\zeta^2\psi(\zeta)} \right| \leq 1,$$

or

$$\left| \frac{\psi(\zeta)(1 + \lambda\zeta)^2 + \lambda^2(1 - \psi(\zeta))}{\psi(\zeta)(1 + \lambda\zeta)^2 + 1 - \psi(\zeta)} \right| \leq 1,$$

or

$$\mathrm{Re}\left(\psi(\zeta)(1 - \overline{\psi(\zeta)})((\overline{\lambda} + \zeta)^2 - (1 + \lambda\zeta)^2)\right) \leq 0.$$

Let $\mathbb{T} \ni \lambda = x + iy$, $\zeta = ir \in \mathbb{D}$, $a := \mathrm{Re}\,\psi(\zeta) - |\psi(\zeta)|^2$, and $b := \mathrm{Im}\,\psi(\zeta)$. Then, the above condition may be rewritten as

$$y\left(a(2r - y(r^2 + 1)) + bx(1 - r^2)\right) \leq 0, \quad x^2 + y^2 = 1,\ r \in (-1, 1).$$

Setting $x = 0$ implies that $a \geq 0$ Then, letting $y \longrightarrow 0+$ gives $-2ar \geq (1 - r^2)|b|$. Hence, $a = b = 0$ for positive r. Then, using the identity principle leads to either $\psi \equiv 0$ or $\psi \equiv 1$ on $\mathbb{D}$. Therefore, either $X_1 = 0$ or $X_n = 0$; a contradiction. □

So far it is still not clear whether $\boldsymbol{k}_{\mathbb{G}_n} = \boldsymbol{c}_{\mathbb{G}_n}$, $n \geq 3$. In fact, this question is still open if $n \geq 4$.

For $n = 3$ we have the following result.

Proposition 7.2.13 (see [388]). $\hat{\boldsymbol{\varkappa}}_{\mathbb{G}_3}(0;\cdot) \neq \boldsymbol{\gamma}_{\mathbb{G}_3}(0;\cdot)$ *and then (by differentiation)* $\boldsymbol{k}_{\mathbb{G}_3}(0,\cdot) \neq \boldsymbol{c}_{\mathbb{G}_3}(0,\cdot)$.

The proof is an immediate consequence of the following two lemmas:

Lemma 7.2.14. $\boldsymbol{\gamma}_{\mathbb{G}_3}(0;\boldsymbol{e}_2) \leq C_0 := \sqrt{\frac{8}{13\sqrt{13}-35}} = 0.8208\ldots$.

Proof. By virtue of Proposition 2.3.2, it suffices for us to prove that, for any $c \in \mathbb{C}$, the following inequality holds:

$$\max_{z\in\partial\mathbb{G}_3} |z_2 - cz_1^2|^2 \geq 1/C_0^2.$$

First, observe that it is enough to prove this inequality for all real c. Indeed, take a $c \in \mathbb{C}$ and note that if $z \in \partial\mathbb{G}_n$, then also $\overline{z} \in \partial\mathbb{G}_n$. Then,

$$\begin{aligned}2\max_{z\in\partial\mathbb{G}_3} |z_2 - cz_1^2| &\geq \max_{z\in\partial\mathbb{G}_3} \left(|z_2 - cz_1^2| + |\overline{z}_2 - c\overline{z}_1^2|\right)\\ &\geq \max_{z\in\partial\mathbb{G}_3} |2z_2 - (c+\overline{c})z_1^2| = 2\max_{z\in\partial\mathbb{G}_3} |z_2 - \operatorname{Re}(c)z_1^2|.\end{aligned}$$

Now let $c \in \mathbb{R}$. Then, using that $(1,1,e^{i\varphi}) \in \partial\mathbb{G}_3$,

$$\begin{aligned}\max_{z\in\partial\mathbb{G}_3} |z_2 - cz_1^2|^2 &\geq \max_{\varphi\in[0,2\pi)} |1 + 2e^{i\varphi} - c(2+e^{i\varphi})^2|^2\\ &\geq \max_{\varphi\in[0,2\pi)} \left(4c(4c-1)\cos^2\varphi + 4(10c^2 - 7c + 1)\cos\varphi + 25c^2 - 22c + 5\right).\end{aligned}$$

Put

$$f_c(t) := 4c(4c-1)t^2 + 4(2c-1)(5c-1)t + 25c^2 - 22c + 5, \quad t \in [-1,1].$$

If $c \notin (\frac{1}{6}, \frac{5-\sqrt{17}}{4}) =: \Delta$, then

$$\max_{t\in[-1,1]} f_c(t) = \max\{f_c(-1), f_c(1)\} \geq \left(\frac{9-\sqrt{17}}{4}\right)^2 > C_0^{-2}.$$

Otherwise,

$$\max_{t\in[-1,1]} f_c(t) = f_c\left(\frac{10c^2 - 7c + 1}{2c(1-4c)}\right) = \frac{(3c-1)^3}{c(4c-1)} =: g(c).$$

It remains to mention that $\min_{c\in\Delta} g(c) = g(\frac{\sqrt{13}-1}{12}) = C_0^{-2}$. □

Recall that for an arbitrary domain $D \subset \mathbb{C}^n$ the following chain of inequalities is always true: $\boldsymbol{\gamma}_D \leq \widehat{\boldsymbol{\gamma}_D^{(k)}} \leq \hat{\boldsymbol{\varkappa}}_D$, where $\widehat{\boldsymbol{\gamma}_D^{(k)}}(a;X) := \widehat{\boldsymbol{\gamma}_D^{(k)}(a;\cdot)}(X)$.

Lemma 7.2.15. $\widehat{\boldsymbol{\gamma}^{(2)}_{\mathbb{G}_3}}(0; \boldsymbol{e}_2) \geq C_1 := \sqrt{0.675} = 0.8215\ldots$.

Proof. Put $g : \mathbb{C}^3 \to \mathbb{C}$, as

$$g(z) := 0.675z_2^2 - 0.291z_2z_1^2 + 0.033z_1^4, \quad z \in \mathbb{C}^3.$$

Assume we knew that $g(\mathbb{G}_3) \subset \mathbb{D}$. Then, we immediately have $\boldsymbol{\gamma}^{(2)}_{\mathbb{G}_3}(0; (X_1, 1, X_3)) \geq \sqrt{0.675}$, since f is a competitor for the definition of $\boldsymbol{\gamma}^{(2)}_{\mathbb{G}_3}(0; \cdot)$. Applying Lemma 2.2.5 it follows that also $\widehat{\boldsymbol{\gamma}^{(2)}_{\mathbb{G}_3}}(0; \boldsymbol{e}_2) \geq C_1$.

So it remains to estimate g on the symmetrized polydisc $\mathbb{G}_3$. It suffices to verify that $|g| < 1$ on $\pi_n(\mathbb{T}^n)$, since $g \circ \pi_n$ is holomorphic on $\mathbb{C}^n$. Put

- $\tilde{g}_1(\theta_1, \theta_2) := 1 + e^{i\theta_1} + e^{i\theta_2}$,
- $\tilde{g}_2(\theta_1, \theta_2) := e^{i(\theta_1+\theta_2)} + e^{i\theta_1} + e^{i\theta_2}$,

where $\theta_1, \theta_2 \in [0, 2\pi)$. So, the following function

$$\tilde{g}(\theta) := 0.675\tilde{g}_2^2(\theta) - 0.291\tilde{g}_2(\theta)\tilde{g}_1^2(\theta) + 0.033\tilde{g}_1^4(\theta), \quad \theta = (\theta_1, \theta_2) \in [0, 2\pi)^2,$$

should have absolute values less than 1 for all θ's, which can be verified by a computer program. □

Remark 7.2.16. It should be added that $\boldsymbol{\gamma}^{(2)}_{\mathbb{G}_3}(0; \cdot)$ is not a norm. For a proof see [388].

We close this section by asking for more concrete formulas of the invariant functions for the symmetrized polydisc. ?

7.3 List of problems

Chapter 8

Non-standard contractible systems

Summary. In § 8.1 we discuss modifications of the Lempert function and the Kobayashi–Royden pseudometric due to K. T. Hahn. They deal only with injective analytic discs. Nevertheless, it turns out that these new objects coincide with the Lempert function, respectively the Kobayashi–Royden pseudometric for $n \geq 3$ (Theorem 8.1.9). Using the pointwise operator $\mathbb{W}$ from § 2.2.3, the Wu pseudometric is studied in § 8.3. We put emphasis on regularity properties. The chapter concludes (see § 8.2) with the study of generalized Green functions with pole functions and generalized Möbius functions with weight functions.

8.1 Hahn function and pseudometric

K. T. Hahn has introduced ([207]) the following family $(\boldsymbol{\eta}_G)_G$ of pseudometrics:

$$\boldsymbol{\eta}_G : G \times \mathbb{C}^n \longrightarrow \mathbb{R}_+,$$

$$\begin{aligned}\boldsymbol{\eta}_G(a; X) :=& \inf\{\boldsymbol{\gamma}_{\mathbb{D}}(\lambda; \alpha) : \lambda \in \mathbb{D},\ \alpha \in \mathbb{C},\ \exists_{\varphi \in \mathcal{O}(\mathbb{D}, G)},\ \varphi \text{ is injective},\\ &\qquad\qquad \varphi(\lambda) = a,\ \alpha\varphi'(\lambda) = X\}\\ =& \inf\{\sigma \geq 0 : \exists_{\varphi \in \mathcal{O}(\overline{\mathbb{D}}, G)},\ \varphi \text{ is injective}, \varphi(0) = a,\ \sigma\varphi'(0) = X\}.\end{aligned}$$

Obviously, $\boldsymbol{\varkappa}_G \leq \boldsymbol{\eta}_G$. *The family $(\boldsymbol{\eta}_G)_G$ is contractible only with respect to injective holomorphic mappings* and $\boldsymbol{\eta}_{\mathbb{D}} = \boldsymbol{\gamma}_{\mathbb{D}}$. Moreover, if $(\delta_G)_G$ is a family of pseudometrics $\delta_G : G \times \mathbb{C}^n \longrightarrow \mathbb{R}_+$ ($G \subset \mathbb{C}^n$) that is contractible with respect to injective holomorphic mappings, and $\delta_{\mathbb{D}}(0; 1) = 1$, then $\delta_G \leq \boldsymbol{\eta}_G$ for an arbitrary domain G.

Along the same lines, K. T. Hahn also introduced a family $(\boldsymbol{h}^*_G)_G$ of functions $\boldsymbol{h}^*_G : G \times G \longrightarrow \mathbb{R}_+$,

$$\begin{aligned}\boldsymbol{h}^*_G(a, b) :=& \inf\{\boldsymbol{m}_{\mathbb{D}}(\lambda, \mu) : \lambda, \mu \in \mathbb{D},\ \exists_{\varphi \in \mathcal{O}(\mathbb{D}, G)} :\ \varphi \text{ is injective},\\ &\qquad\qquad \varphi(\lambda) = a,\ \varphi(\mu) = b\}\\ =& \inf\{\sigma \in [0, 1) : \exists_{\varphi \in \mathcal{O}(\overline{\mathbb{D}}, G)} :\ \varphi \text{ is injective},\ \varphi(0) = a,\ \varphi(\sigma) = b\}.\end{aligned}$$

Clearly, $\boldsymbol{\ell}^*_G \leq \boldsymbol{h}^*_G$. *The family $(\boldsymbol{h}^*_G)_G$ is holomorphically contractible with respect to injective holomorphic mappings* and $\boldsymbol{h}^*_{\mathbb{D}} = \boldsymbol{m}_{\mathbb{D}}$. Moreover, if $(d_G)_G$ is a family of functions $d_G : G \times G \longrightarrow \mathbb{R}_+$ that is contractible with respect to injective holomorphic mappings and $d_{\mathbb{D}}(0; t) = t$, $t \in [0, 1)$, then $d_G \leq \boldsymbol{h}^*_G$ for arbitrary domain G. See also [281, 512].

Remark 8.1.1. Observe that the infimum in the definition of $\boldsymbol{h}_G^*(a,b)$ is taken over a non-empty set.

Indeed, fix points $a, b \in G$, $a \neq b$. Then, there is an injective $\mathcal{C}^1$-curve $\alpha = (\alpha_1, \dots, \alpha_n) : [0,1] \longrightarrow G$ with $\alpha(0) = a$, $\alpha(1) = b$ such that $\alpha'(t) \neq 0$, $t \in [0,1]$. By the Weierstrass approximation theorem, we find a sequence $(p_j)_{j=1}^{\infty}$ of polynomial mappings $p_j = (p_{j,1}, \dots, p_{j,n}) : \mathbb{R} \longrightarrow \mathbb{C}^n$ such that:

- $p_j(0) = a$, $p_j(1) = b$,
- $p_j \longrightarrow \alpha$ and $p_j' \longrightarrow \alpha'$ uniformly on $[0,1]$,
- $p_j([0,1]) \subset G$, $j \in \mathbb{N}$.

Then, $p_j|_{[0,1]}$ must be injective for $j \gg 1$. Indeed, suppose the contrary. Then, passing to a subsequence, we may assume that there exist $t_j', t_j'' \in [0,1]$, $t_j' \neq t_j''$, with $p_j(t_j') = p_j(t_j'')$, $j \in \mathbb{N}$, and $t_j' \longrightarrow t'$ and $t_j'' \longrightarrow t''$. The uniform convergence $p_j \longrightarrow \alpha$ on $[0,1]$ implies that $\alpha(t') = \alpha(t'')$. Applying the fact that α is injective gives $t' = t'' =: t_0$. Suppose that $\operatorname{Re}\alpha_1'(t_0) \neq 0$. Then, $|\operatorname{Re}\alpha_1'(t)| \geq 2\varepsilon$ for $t \in [0,1] \cap [t_0 - \delta, t_0 + \delta]$ with sufficiently small $\varepsilon, \delta > 0$. We may assume that $t_j', t_j'' \in [t_0 - \delta, t_0 + \delta]$ for all $j \in \mathbb{N}$. Since $p_j' \longrightarrow \alpha'$ uniformly on $[0,1]$, we may also assume that $|\operatorname{Re} p_{j,1}'(t)| \geq \varepsilon$, $t \in [0,1] \cap [t_0 - \delta, t_0 + \delta]$ for all $j \in \mathbb{N}$. Then,

$$0 = \|p_j(t_j') - p_j(t_j'')\| \geq |\operatorname{Re} p_{j,1}(t_j') - \operatorname{Re} p_{j,1}(t_j'')| \geq \varepsilon|t_j' - t_j''|;$$

a contradiction.

Fix a j such that p_j is injective on $[0,1]$ and $p_j'(t) \neq 0$, $t \in [0,1]$. Consider p_j as a complex polynomial mapping $\mathbb{C} \longrightarrow \mathbb{C}^n$. Then, there exists a simply connected domain $U \subset \mathbb{C}$ such that

- $[0,1] \subset U$,
- $p_j(U) \subset G$,
- $p_j|_U$ injective.

Arguing as in the case of the Lempert function, we end up with an injective analytic disc in G, passing through a and b.

Remark 8.1.2.

(a) If G is biholomorphic to $\mathbb{C}^n$, then $\eta_G \equiv 0$, $\boldsymbol{h}_G^* \equiv 0$.

(b) $\boldsymbol{h}_{\mathbb{C}_*}^*(a,b) = \dfrac{|\frac{b}{a}-1| + 2 - 2\sqrt{1 + |\frac{b}{a}-1|}}{|\frac{b}{a}-1|}$, $a, b \in \mathbb{C}_*$, $|b| \leq |a|$ (recall that $\boldsymbol{\ell}_{\mathbb{C}_*}^* \equiv 0$).

Indeed, we may assume that $a = 1$, $b \in \overline{\mathbb{D}}$, $b \neq 1$. Let $\varphi \in \mathcal{O}(\mathbb{D}, \mathbb{C}_*)$ be injective with $\varphi(0) = 1$, $\varphi(\sigma) = b$ for a $\sigma \in (0,1)$. Applying the Koebe distortion theorem (cf. Appendix B.1.8), we have

$$\begin{aligned}|b-1| = |\varphi(\sigma)-\varphi(0)| &\le |\varphi'(0)|\frac{\sigma}{(1-\sigma)^2}\\ &\le 4\,\mathrm{dist}(1,\partial f(\mathbb{D}))\frac{\sigma}{(1-\sigma)^2} \le 4\frac{\sigma}{(1-\sigma)^2},\end{aligned}$$

which, after elementary calculations, gives the inequality

$$\boldsymbol{h}^*_{\mathbb{C}_*}(1,b) \ge \frac{|b-1|+2-2\sqrt{1+|b-1|}}{|b-1|} =: \sigma_0.$$

Now, fix a $\theta \in \mathbb{R}$ such that $e^{i\theta}(b-1) = |b-1|$. Consider the Koebe function

$$\varphi(\lambda) = 1 + \frac{4\lambda}{(1-e^{i\theta}\lambda)^2}, \quad \lambda \in \mathbb{D}.$$

Then, φ is injective, $\varphi(\mathbb{D}) \subset \mathbb{C}_*$, $\varphi(0) = 1$, and $\varphi(e^{i\theta}\sigma_0) = b$, which gives the opposite inequality.

(c) $\boldsymbol{\eta}_{\mathbb{C}_*}(a;1) = \frac{1}{4|a|}$, $a \in \mathbb{C}_*$ (recall that $\boldsymbol{\varkappa}_{\mathbb{C}_*} \equiv 0$); see Corollary 8.1.6 for a more general result.

Indeed, (cf. [363]), the inequality "$\ge$" follows from the Koebe distortion theorem. Using the biholomorphism $\mathbb{C}_* \ni z \longmapsto z/(4a) \in \mathbb{C}_*$, we see that to get "$\le$", we only need to show that $\boldsymbol{\eta}_{\mathbb{C}_*}(1/4;1) \le 1$. Let

$$\varphi(\lambda) := \frac{1}{4} + \frac{\lambda}{(1-\lambda)^2}, \quad \lambda \in \mathbb{D},$$

be the Koebe function. Then, $\varphi : \mathbb{D} \longrightarrow \mathbb{C}_*$ is injective, $\varphi(0) = 1/4$, and $\varphi'(0) = 1$.

(d) Let $G \subset \mathbb{C}$ be a taut domain. Let $a, b \in G$, $a \ne b$ (resp. $a \in G$) and put $\sigma := \boldsymbol{h}^*_G(a,b)$ (resp. $\sigma := \boldsymbol{\eta}_G(a;1)$). Then, there exists an extremal injective disc $\varphi \in \mathcal{O}(\mathbb{D}, G)$ such that $\varphi(0) = a$ and $\varphi(\sigma) = b$ (resp. $\varphi(0) = a$ and $\sigma\varphi'(0) = 1$).

Indeed, let $\varphi_s \in \mathcal{O}(\mathbb{D}, G)$ and $\sigma_s > 0$ be such that φ_s is injective, $\varphi_s(0) = a$, and $\varphi_s(\sigma_s) = b$, $\sigma_s \searrow \sigma$ (resp. $\sigma_s\varphi_s'(0) = 1$, $\sigma_s \searrow \sigma$). Since G is taut, we may assume that $\varphi_s \longrightarrow \varphi$ locally uniformly in $\mathbb{D}$, where $\varphi \in \mathcal{O}(\mathbb{D}, G)$, $\varphi(0) = a$, and $\varphi(\sigma) = b$ (resp. $\sigma\varphi'(0) = 1$). In particular, $\varphi \not\equiv$ const and therefore, by the Hurwitz theorem, φ must be injective.

(e) If $G \subset \mathbb{C}$ is a taut domain, then the following conditions are equivalent:

(i) G is simply connected;

(ii) $\boldsymbol{\ell}_G^* \equiv \boldsymbol{h}_G^*$ and $\boldsymbol{\varkappa}_G \equiv \boldsymbol{\eta}_G$;

(iii) there exist $a, b \in G$, $a \neq b$, such that $\boldsymbol{\ell}_G^*(a,b) = \boldsymbol{h}_G^*(a,b)$;

(iv) there exists an $a \in G$ such that $\boldsymbol{\varkappa}_G(a;1) = \boldsymbol{\eta}_G(a;1)$.

Indeed, the implications (i) $\Longrightarrow$ (ii) $\Longrightarrow$ (iii) and (ii) $\Longrightarrow$ (iv) are obvious. Suppose that $\boldsymbol{\ell}_G^*(a,b) = \boldsymbol{h}_G^*(a,b) =: \sigma$, $a \neq b$ (resp. $\boldsymbol{\varkappa}_G(a;1) = \boldsymbol{\eta}_G(a;1) =: \sigma$). By (d) we already know that there exists an extremal injective disc $\varphi \in \mathcal{O}(\mathbb{D}, G)$ such that $\varphi(0) = a$ and $\varphi(\sigma) = b$ (resp. $\varphi(0) = a$ and $\sigma\varphi'(0) = 1$). Let $\Pi : \mathbb{D} \longrightarrow G$ be the holomorphic covering with $\Pi(0) = a$ (cf. Remark 3.2.3). Let $\widetilde{\varphi} : \mathbb{D} \longrightarrow \mathbb{D}$ be the lifting of φ with $\widetilde{\varphi}(0) = 0$. Then,

$$\sigma = \boldsymbol{\ell}_G^*(a,b) = \boldsymbol{\ell}_G^*(\Pi \circ \widetilde{\varphi}(0), \Pi \circ \widetilde{\varphi}(\sigma)) \le \boldsymbol{m}_{\mathbb{D}}(0, \widetilde{\varphi}(\sigma)) \le \sigma.$$

Hence, by the Schwarz–Pick lemma, $\widetilde{\varphi} \in \operatorname{Aut}(\mathbb{D})$, which implies that Π is biholomorphic.

In the infinitesimal case we have

$$1 = \varphi'(0)\sigma = |\Pi'(0)\widetilde{\varphi}'(0)\boldsymbol{\varkappa}_G(a;1)| = |\widetilde{\varphi}'(0)|\boldsymbol{\varkappa}_G(\Pi(0);\Pi'(0)) \le |\widetilde{\varphi}'(0)| \le 1.$$

Hence, by the Schwarz lemma, $\widetilde{\varphi} \in \operatorname{Aut}(\mathbb{D})$, and consequently, Π is biholomorphic.

(f) If there exists an injective extremal mapping for $\boldsymbol{\ell}_G^*(a,b)$ (resp. $\boldsymbol{\varkappa}_G(a;X)$), then $\boldsymbol{h}_G^*(a,b) = \boldsymbol{\ell}_G^*(a,b)$ (resp. $\boldsymbol{\eta}_G(a;X) = \boldsymbol{\varkappa}_G(a;X)$). In particular,

- if G is a balanced pseudoconvex domain, then $\boldsymbol{h}_G^*(0,\cdot) = \boldsymbol{\ell}_G^*(0,\cdot)$ and $\boldsymbol{\eta}_G(0;\cdot) = \boldsymbol{\varkappa}_G(0;\cdot)$;
- if $G \in \{\mathbb{B}_n, \mathbb{D}^n\}$, then $\boldsymbol{h}_G^* \equiv \boldsymbol{\ell}_G^*$ and $\boldsymbol{\eta}_G \equiv \boldsymbol{\varkappa}_G$.

(g) The functions $\boldsymbol{h}_G^*$ and $\boldsymbol{\eta}_G$ are upper semicontinuous.

Indeed (the proof is due W. Jarnicki), take $a_0, b_0 \in G$, $a_0 \neq b_0$, $\boldsymbol{h}_G^*(a_0,b_0) < A$ and let $\varphi : \overline{\mathbb{D}} \longrightarrow G$ be an injective holomorphic mapping with $\varphi(0) = a_0$, $\varphi(\sigma) = b_0$, where $0 < \sigma < A$. Let $v_2, \dots, v_n \in \mathbb{C}^n$ be such that $b_0 - a_0, v_2, \dots, v_n$ are linearly independent. For $a, b \in \mathbb{C}^n$ such that $(b - a, v_2, \dots, v_n)$ are linearly independent, define $\Phi_{a,b} : \mathbb{C}^n \longrightarrow \mathbb{C}^n$,

$$\Phi_{a,b}(a_0 + \zeta_1(b_0 - a_0) + \zeta_2 v_2 + \dots + \zeta_n v_n) := a + \zeta_1(b - a) + \zeta_2 v_2 + \dots + \zeta_n v_n.$$

Obviously, $\Phi_{a,b}$ is biholomorphic and $\Phi_{a,b} \longrightarrow \mathrm{id}_{\mathbb{C}^n}$ locally uniformly in $\mathbb{C}^n$ when $(a,b) \longrightarrow (a_0,b_0)$. Observe that $\Phi_{a,b}(a_0) = a$, $\Phi_{a,b}(b_0) = b$. Define $\varphi_{a,b} := \Phi_{a,b} \circ \varphi$. Then, $\varphi_{a,b}$ is an injective holomorphic mapping, $\varphi_{a,b}(0) = a$, $\varphi_{a,b}(\sigma) = b$, and $\varphi_{a,b}(\overline{\mathbb{D}}) \subset G$ provided (a,b) is sufficiently near (a_0,b_0). This shows that $\boldsymbol{h}^*_G$ is upper semicontinuous at (a_0,b_0).

Let $\mathbb{B}(a_0,3r) \subset G$. Then, for $a,b \in \mathbb{B}(a_0,r)$ we have

$$\boldsymbol{h}^*_G(a,b) \le \boldsymbol{h}^*_{\mathbb{B}(a,2r)}(a,b) \overset{\text{(f)}}{=} \frac{\|a-b\|}{2r}.$$

Thus $\boldsymbol{h}^*_G$ is continuous at (a_0,a_0).

The infinitesimal case is left for the reader.

(h) If $G \subset \mathbb{C}$ is a taut domain, then the functions $\boldsymbol{h}^*_G$ and $\boldsymbol{\eta}_G$ are continuous.

Indeed, let $(a_k,b_k) \longrightarrow (a,b) \in G \times G$, $a \ne b$. By (d) we already know that for each $k \in \mathbb{N}$ there exists an extremal φ_k for $\boldsymbol{h}^*_G(a_k,b_k)$, i.e., $\varphi_k : \mathbb{D} \longrightarrow G$ is an injective holomorphic disc with $\varphi_k(0) = a_k$, $\varphi_k(\sigma_k) = b_k$, where $\sigma_k := \boldsymbol{h}^*_G(a_k,b_k)$. Since $\boldsymbol{h}^*_G$ is upper semicontinuous, we have $\limsup_{k\to+\infty} \sigma_k \le \sigma := \boldsymbol{h}^*_G(a,b)$. We have to show that $\lim_{k\to+\infty} \sigma_k = \sigma$. Passing to a subsequence, we may assume that $\sigma_k \longrightarrow \alpha \in [0,\sigma]$. Since G is taut, we may assume that $\varphi_k \longrightarrow \varphi$ locally uniformly in $\mathbb{D}$, where $\varphi \in \mathcal{O}(\mathbb{D},\mathbb{D})$, $\varphi(0) = a$, $\varphi(\alpha) = b$. In particular, $\varphi \not\equiv \mathrm{const}$. Thus, by the Hurwitz theorem, φ is injective and therefore, $\sigma \le \alpha$.

The infinitesimal case is left for the reader.

?

It seems to be unknown whether a similar result is true in dimension $n = 2$ (for $n \ge 3$ see Theorem 8.1.9).

(i) If $\mathbb{C}^n \supset G_k \nearrow G \subset \mathbb{C}^n$, then $\boldsymbol{h}^*_{G_k} \searrow \boldsymbol{h}^*_G$ and $\boldsymbol{\eta}_{G_k} \searrow \boldsymbol{\eta}_G$ (cf. the proof of Proposition 3.3.5(a)).

(j) If $G \subset \mathbb{C}^n$, $D \subset \mathbb{C}^m$ are domains, then

$$\boldsymbol{h}^*_{G\times D}((z',w'),(z'',w'')) \le \max\{\boldsymbol{h}^*_G(z',z''), \boldsymbol{h}^*_D(w',w'')\},$$
$$(z',w'),(z'',w'') \in G \times D,$$
$$\boldsymbol{\eta}_{G\times D}((z,w);(X,Y)) \le \max\{\boldsymbol{\eta}_G(z;X), \boldsymbol{\eta}_D(w;Y)\},$$
$$(z,w) \in G \times D,\ (X,Y) \in \mathbb{C}^n \times \mathbb{C}^m$$

(cf. the proof of Proposition 3.7.1).

(k) $\boldsymbol{h}^*_{\mathbb{C}_*\times\mathbb{C}^{n-1}} \equiv 0$, $\boldsymbol{\eta}_{\mathbb{C}_*\times\mathbb{C}^{n-1}} \equiv 0$ $(n \ge 2)$. In particular, $\boldsymbol{h}^*$ and $\boldsymbol{\eta}$ do not have product property (cf. Proposition 3.7.1(a, b)).

Indeed, (cf. [280]), it suffices to show that $\boldsymbol{h}^*_{\mathbb{C}_*\times\mathbb{C}^{n-1}}(\boldsymbol{e}_1, b) = 0$, $b \neq \boldsymbol{e}_1$, where $\boldsymbol{e}_1 := (1, 0, \dots, 0)$ (resp. $\boldsymbol{\eta}_{\mathbb{C}_*\times\mathbb{C}^{n-1}}(\boldsymbol{e}_1; X) = 0$, $X \neq 0$). Define

$$\varphi = (\varphi_1, \dots, \varphi_n) : \mathbb{C} \longrightarrow \mathbb{C}_* \times \mathbb{C}^{n-1}, \quad \varphi_1(\lambda) := \exp(\lambda \operatorname{Log} b_1),$$

$$\varphi_j(\lambda) := \begin{cases} \lambda b_j, & \text{if } b_j \neq 0 \\ \exp(\lambda \operatorname{Log} b_1) - 1 + (1 - b_1)\lambda, & \text{if } b_j = 0 \end{cases}, \quad j = 2, \dots, n.$$

Observe that φ is injective, $\varphi(0) = \boldsymbol{e}_1$, and $\varphi(1) = b$. Thus, $\boldsymbol{h}^*_{\mathbb{C}_*\times\mathbb{C}^{n-1}}(\boldsymbol{e}_1, b) \leq \boldsymbol{h}^*_{\mathbb{C}}(0, 1) = 0$.

In the infinitesimal case we take

$$\varphi_1(\lambda) : = \exp(\lambda X_1),$$

$$\varphi_j(\lambda) : = \begin{cases} \lambda X_j, & \text{if } X_j \neq 0 \\ \exp(\lambda X_1) - 1 + \lambda X_1, & \text{if } X_j = 0 \end{cases}, \quad j = 2, \dots, n.$$

Then φ is injective, $\varphi(0) = \boldsymbol{e}_1$, and $\varphi'(0) = X$. Thus $\boldsymbol{\eta}_{\mathbb{C}_*\times\mathbb{C}^{n-1}}(\boldsymbol{e}_1; X) \leq \boldsymbol{\eta}_{\mathbb{C}}(0; 1) = 0$.

In the context of Remark 8.1.2(k), one should mention that in fact the following result is true:

Proposition 8.1.3 (cf. [280]). *Let $G := \mathbb{C}^n \setminus F$, where $F \subsetneq \mathbb{C}^n$ is a convex closed set, $n \geq 2$. Then, the following conditions are equivalent:*

(i) *F contains at most one $(n-1)$-dimensional complex hyperplane;*

(ii) *for any $a \in G$, $X \in (\mathbb{C}^n)_*$, there exists an injective disc $\varphi \in \mathcal{O}(\mathbb{C}, G)$ such that $\varphi(0) = z$, $\varphi'(0) = X$;*

(iii) *for any $a, b \in G$, $a \neq b$, there exists an injective disc $\varphi \in \mathcal{O}(\mathbb{C}, G)$ such that $\varphi(0) = a$, $\varphi(1) = b$;*

(iv) $\boldsymbol{h}^*_G \equiv 0$;

(v) $\boldsymbol{\eta}_G \equiv 0$;

(vi) $\boldsymbol{\ell}^*_G \equiv 0$;

(vii) $\boldsymbol{\varkappa}_G \equiv 0$.

Proof. The implications (ii) $\Longrightarrow$ (iv) $\Longrightarrow$ (vi) and (iii) $\Longrightarrow$ (v) $\Longrightarrow$ (vii) are obvious. For the proof of the implications (vi) $\Longrightarrow$ (i) and (vii) $\Longrightarrow$ (i), suppose that F contains two different $(n-1)$-dimensional complex hyperplanes

$$H_j := \{\xi \in \mathbb{C}^n : \langle \xi, v^j \rangle = \alpha_j\}, \quad v^j \in \mathbb{C}^n, \ \|v^j\| = 1, \ \alpha_j \in \mathbb{C}, \ j = 1, 2.$$

If v^1, v^2 are $\mathbb{C}$-linearly dependent then, after an affine $\mathbb{C}$-isomorphism, we may assume that $v^1 = v^2 = \boldsymbol{e}_1 = (1, 0, \dots, 0)$, $\alpha_1 = 0$, $\alpha_2 = 1$. Then, $G \subset (\mathbb{C} \setminus \{0, 1\}) \times \mathbb{C}^{n-1})$. Hence, using the product properties for $\boldsymbol{\ell}^*$ and $\boldsymbol{\varkappa}$, we get

$$\boldsymbol{\ell}^*_G(a, b) \geq \boldsymbol{\ell}^*_{(\mathbb{C}\setminus\{0,1\})\times\mathbb{C}^{n-1}}(a, b) = \boldsymbol{\ell}^*_{\mathbb{C}\setminus\{0,1\}}(a_1, b_1) > 0 \text{ if } a_1 \neq b_1,$$
$$\boldsymbol{\varkappa}_G(a; X) \geq \boldsymbol{\varkappa}_{(\mathbb{C}\setminus\{0,1\})\times\mathbb{C}^{n-1}}(a; X) = \boldsymbol{\varkappa}_{\mathbb{C}\setminus\{0,1\}}(a_1; X_1) > 0 \text{ if } X_1 \neq 0;$$

a contradiction.

If v^1, v^2 are $\mathbb{C}$-linearly independent, then after an affine $\mathbb{C}$-isomorphism, we may assume that $v^1 = \boldsymbol{e}_1$, $v^2 = \boldsymbol{e}_2 = (0, 1, 0, \dots, 0)$, $\alpha_1 = \alpha_2 = 0$. Then, $(\{0\} \times \mathbb{C}^{n-1}) \cup (\mathbb{C} \times \{0\} \times \mathbb{C}^{n-2}) \subset F$, which implies that $F = \mathbb{C}^n$; a contradiction.

Now, we are going to prove that (i) $\Longrightarrow$ (ii) and (i) $\Longrightarrow$ (iii). Since the case where F contains an $(n-1)$-dimensional complex hyperplane was solved in Remark 8.1.2(k), we may assume that F contains no $(n-1)$-dimensional complex hyperplanes. Let $a \in G$. Since F coincides with its convex hull with respect to the real valued $\mathbb{R}$-linear functionals on $\mathbb{C}^n$, there are two affine $\mathbb{C}$-linear mappings $\ell_1, \ell_2 : \mathbb{C}^n \longrightarrow \mathbb{C}$ such that $\ell_1 - \ell_1(0)$ and $\ell_2 - \ell_2(0)$ are linearly independent and

$$\operatorname{Re}(\ell_1(a)) > 0 = \max_F \operatorname{Re}(\ell_1) = \max_F \operatorname{Re}(\ell_2).$$

Replacing ℓ_2 by $\ell_1 + \varepsilon\ell_2$ with $0 < \varepsilon \ll 1$, we may assume that

$$\operatorname{Re}(\ell_2(a)) > 0 \geq \max_F \operatorname{Re}(\ell_2).$$

Thus, if $G \ni a = (a_1, \dots, a_n)$ and $(\mathbb{C}^n)_* \ni X = (X_1, \dots, X_n)$, then after an affine $\mathbb{C}$-linear change of coordinates, we may assume that $\operatorname{Re} a_1 > 0$, $\operatorname{Re} a_2 > 0$, and

$$F \subset F_0 := \{\xi \in \mathbb{C}^n : \operatorname{Re}\xi_1 \leq 0,\ \operatorname{Re}\xi_2 \leq 0\}.$$

If $X_1 = X_2 = 0$, then the mapping $\mathbb{C} \ni \lambda \longmapsto a + \lambda X$ has the required properties. Otherwise, we may assume that $X_2 \neq 0$ and for $M > 0$, put

$$\varphi_j(\lambda) := a_j + \lambda X_j, \quad j = 2, 3, \dots, n,$$
$$\varphi_1(\lambda) := a_1 + \int_0^{2M} h(t) \exp(t\varphi_2(\lambda)) dt,$$

where

$$h(t) := \begin{cases} \frac{a_2 X_1}{X_2 M(1-\exp(a_2 M))}, & \text{if } t \in [0, M] \\ \frac{a_2 X_1}{X_2 M \exp(a_2 M)(\exp(a_2 M)-1)}, & \text{if } t \in (M, 2M] \end{cases}, \quad \lambda \in \mathbb{C}.$$

Then, $\varphi := (\varphi_1, \dots, \varphi_n) : \mathbb{C} \longrightarrow \mathbb{C}^n$ is an injective holomorphic mapping, $\varphi(0) = a$, and $\varphi'(0) = X$. We want to find an $M > 0$ such that $\varphi(\mathbb{C}) \subset \mathbb{C}^n \setminus F_0$. It suffices to find an M such that, for every $\lambda \in \mathbb{C}$, we have

$$(\operatorname{Re}\varphi_2(\lambda) \leq 0) \Longrightarrow (\operatorname{Re}\varphi_1(\lambda) > 0).$$

Since $\operatorname{Re} a_2 > 0$, we only need to observe that $\int_0^{2M} h(t) dt \longrightarrow 0$ when $M \longrightarrow +\infty$.

Let now $a, b \in G$ and $a \neq b$. As above, we may assume that $\operatorname{Re} a_2 > 0$, $\operatorname{Re} b_1 > 0$, and $F \subset F_0$. If $a_1 = b_1$ or $a_2 = b_2$, then the mapping $\mathbb{C} \ni \lambda \longmapsto a + \lambda(b - a)$ has the required properties. Otherwise, we may assume that $b_2 \neq a_2$ and for $m \in \mathbb{N}$, put

$$\varphi_j(\lambda) := a_j + \lambda(b_j - a_j), \quad j = 2, 3, \dots, n,$$
$$\varphi_1(\lambda) := b_1 + \int_0^{2M} h(t) \exp(t\varphi_2(\lambda)) dt,$$

where

$$M := \frac{(2m-1)\pi}{|a_2 - b_2|},$$
$$h(t) := \begin{cases} \frac{a_2(a_1 - b_1)\exp(b_2 M)}{(\exp(a_2 M) - 1)(\exp(b_2 M) - \exp(a_2 M))}, & \text{if } t \in [0, M] \\ \frac{a_2(a_1 - b_1)}{(\exp(a_2 M) - 1)(\exp(a_2 M) - \exp(b_2 M))}, & \text{if } t \in (M, 2M] \end{cases}, \quad \lambda \in \mathbb{C}.$$

As above, if $M \gg 1$, then $\varphi := (\varphi_1, \dots, \varphi_n) : \mathbb{C} \longrightarrow \mathbb{C}^n$ is injective with $\varphi(0) = a$ and $\varphi(1) = b$. □

To get more examples, we need some auxiliary results.

Definition 8.1.4. Let $D \subset \mathbb{C}_*$ be a domain. For $r > 0$ let $L_r := D \cap r\mathbb{T}$; L_r is either empty or is an at most countable union of pairwise disjoint open arcs, $L_r = \bigcup_{i \in I_r} L_{r,i}$. Define D^* to be the subset of $\mathbb{C}_*$ such that, for every $r > 0$, if we put $L_r^* := D^* \cap r\mathbb{T}$, then:

- if $L_r = r\mathbb{T}$, then $L_r^* = r\mathbb{T}$,
- if $L_r = \varnothing$, then $L_r^* = \varnothing$,
- if $\varnothing \neq L_r \subsetneq r\mathbb{T}$, $L_r = \bigcup_{i \in I_r} L_{r,i}$, then L_r^* is an open arc, L_r^* symmetric with respect to $\mathbb{R}_+$ (i.e., $L_r^* \cap \mathbb{R}_- = \varnothing$ and $z \in L_r^* \iff \bar{z} \in L_r^*$), and $\operatorname{length}(L_r^*) = \sum_{i \in I_r} \operatorname{length}(L_{r,i})$.

We say that D^* is the *circular symmetrization* of D.

One can prove that D^* is a domain. Moreover, if D is simply connected, then so is D^* (cf. [217], subsection 4.5.5) and $D^* \cap \mathbb{R}_- = \varnothing$.

Theorem* 8.1.5 (Symmetrization principle, cf. [217], Theorem 4.9). *Let $\varphi : \mathbb{D} \longrightarrow D \subset \mathbb{C}_*$ be a biholomorphic mapping such that $\varphi(0) \in (0, 1)$. Let $D_0 \subset \mathbb{C}$ be a simply connected domain with $D^* \subset D_0$. Let $\varphi_0 : \mathbb{D} \longrightarrow D_0$ be biholomorphic with $\varphi_0(0) = \varphi(0)$. Then, $|\varphi'(0)| \leq |\varphi_0'(0)|$.*

Corollary 8.1.6. *Let $0 \leq r < R < +\infty$ and let $\psi : \mathbb{A}(r, R) \setminus \mathbb{R}_- \longrightarrow \mathbb{D}$ be biholomorphic. Then,*

$$\eta_{\mathbb{A}(r,R)}(a; 1) = \eta_{\mathbb{A}(r,R) \setminus \mathbb{R}_-}(a; 1) = \frac{|\psi'(a)|}{1 - |\psi(a)|^2}, \quad a \in (r, R),$$

Remark 8.1.7. Notice that the biholomorphic mapping ψ from above may be explicitly given via theta-functions.

Corollary 8.1.8 (cf. [363]). $\boldsymbol{\eta}_{\mathbb{D}_*}(a;1) = \frac{1+|a|}{4|a|(1-|a|)}$, $a \in \mathbb{D}_*$.

Proof. It is clear that $\boldsymbol{\eta}_{\mathbb{D}_*}(a;1) = \boldsymbol{\eta}_{\mathbb{D}_*}(|a|;1)$. By Corollary 8.1.6, we only need to observe that the mapping

$$\mathbb{D} \setminus \mathbb{R}_- \ni z \overset{\psi}{\longmapsto} \frac{1-z-2\sqrt{z}}{1-z+2\sqrt{z}} \in \mathbb{D},$$

where $\sqrt{z} := \exp(\frac{1}{2}\operatorname{Log} z)$, is well defined and biholomorphic. □

? An effective formula for $\boldsymbol{h}^*_{\mathbb{A}(r,R)}$ seems to be unknown.

Recall (Remark 8.1.2(e)) that for $G := \mathbb{A}(r,R)$, $0 < r < R < +\infty$, we have $\boldsymbol{\ell}_G(a,b) \neq \boldsymbol{h}_G(a,b)$, $a,b \in G$, $a \neq b$, and $\boldsymbol{\varkappa}_G(a;1) \neq \boldsymbol{\eta}_G(a;1)$, $a \in G$. On the other hand, the following surprising result for $n \geq 3$ is due to M. Overholt:

Theorem 8.1.9 (cf. [406]). *If $G \subset \mathbb{C}^n$, $n \geq 3$, then $\boldsymbol{\ell}^*_G \equiv \boldsymbol{h}^*_G$ and $\boldsymbol{\varkappa}_G \equiv \boldsymbol{\eta}_G$.*

Proof. Fix $a, b \in G$, $a \neq b$. Without loss of generality, we may assume that $a = 0 \in G$, $b_1 \cdots b_n \neq 0$. Let $\varepsilon > 0$. Then, there is a disc $\varphi \in \mathcal{O}(\mathbb{D}, G)$ with $\varphi(0) = 0$, $\varphi(\sigma') = b$ for a $\sigma' \in (0,1)$ such that $\sigma' < \boldsymbol{\ell}^*_G(0,b)+\varepsilon$. We choose an $R \in (0,1)$ such that $\sigma := \sigma'/R \in (0,1)$ and $\sigma < \boldsymbol{\ell}^*_G(0,b)+\varepsilon$. Put $\varphi_R(\lambda) := \varphi(R\lambda)$, $|\lambda| < 1/R$. Obviously, $\varphi_R \in \mathcal{O}(\mathbb{D}(1/R), G)$ with $\varphi_R(0) = 0$ and $\varphi_R(\sigma) = b$. Since φ_R is continuous on $\overline{\mathbb{D}}$, we have $\operatorname{dist}(\varphi_R(\overline{\mathbb{D}}), \partial G) =: 2s > 0$.

Now, we take a polynomial mapping $\tilde{p} : \mathbb{C} \longrightarrow \mathbb{C}^n$ coming from the power series expansion of φ_R such that $\operatorname{dist}(\tilde{p}(\overline{\mathbb{D}}), \partial G) > s$ and

$$\|\varphi_R(\sigma) - \tilde{p}(\sigma)\|_{\mathbb{D}} < s\sigma.$$

Put

$$p(\lambda) := \tilde{p}(\lambda) + \frac{\lambda}{\sigma}(\varphi_R(\sigma) - \tilde{p}(\sigma)), \quad \lambda \in \mathbb{C}.$$

Hence, $p|_{\mathbb{D}} \in \mathcal{O}(\mathbb{D}, G)$ with $p(0) = 0$, $p(\sigma) = b$, and $\operatorname{dist}(p(\overline{\mathbb{D}}), \partial G) > 0$. Observe that p is a polynomial mapping with

$$p(\lambda) = \sum_{k=1}^{m} a_k \lambda^k, \quad \lambda \in \mathbb{C},$$

where $m > n$ is sufficiently large and $a_1, \dots, a_m \in \mathbb{C}^n$. Put $A := [a_2, \dots, a_m] \in \mathbb{M}(n \times (m-1); \mathbb{C}) =: \mathbb{M}$.

For every matrix $C = [c_2, \dots, c_m] \in \mathbb{M}$, define a polynomial mapping $p_C : \mathbb{C} \longrightarrow \mathbb{C}^n$,

$$p_C(\lambda) = \sum_{k=1}^{m} c_k \lambda^k, \quad \lambda \in \mathbb{C},$$

where

$$c_1 := \frac{1}{\sigma}\Big(b - \sum_{k=2}^{m} c_k \sigma^k\Big),$$

Observe that $p_C(0) = 0$ and $p_C(\sigma) = b$. Moreover, $p = p_A$. Let Ξ denote the set of all $C \in \mathbb{M}$ such that the mapping p_C is not injective. We are going to show that Ξ is nowhere dense in $\mathbb{M}$. Then, there exists a sequence $(C_s)_{s=1}^{\infty} \subset \mathbb{M} \setminus \Xi$ such that $C_s \longrightarrow A$. Since $\operatorname{dist}(p_A(\overline{\mathbb{D}}), \partial G) > 0$, we conclude that $p_{C_s}(\mathbb{D}) \subset G$ for $s \gg 1$, which will finish the proof.

First, observe that

$$\begin{aligned}
\Xi &= \{C \in \mathbb{M} : \exists_{\lambda_1, \lambda_2 \in \mathbb{C}} : \lambda_1 \neq \lambda_2,\ p_C(\lambda_1) = p_C(\lambda_2)\} \\
&= \Big\{C \in \mathbb{M} : \exists_{\lambda_1, \lambda_2 \in \mathbb{C}} : \lambda_1 \neq \lambda_2,\ \sum_{k=1}^{m} c_k \sum_{s=0}^{k-1} \lambda_1^s \lambda_2^{k-1-s} = 0\Big\} \\
&= \Big\{C \in \mathbb{M} : \exists_{\lambda_1, \lambda_2 \in \mathbb{C}} : \lambda_1 \neq \lambda_2,\ \sum_{k=2}^{m} c_k \Big(\sigma^{k-1} - \sum_{s=0}^{k-1} \lambda_1^s \lambda_2^{k-1-s}\Big) = \frac{1}{\sigma} b\Big\}.
\end{aligned}$$

Using the transformations $c_{k,j} \longrightarrow c_{k,j}\sigma^k / b_j$, $(\lambda_1, \lambda_2) \longrightarrow (\lambda_1/\sigma, \lambda_2/\sigma)$, we may assume that

$$\Xi = \Big\{C \in \mathbb{M} : \exists_{\lambda_1, \lambda_2 \in \mathbb{C}} : \lambda_1 \neq \lambda_2,\ \sum_{k=2}^{m} c_k \Big(1 - \sum_{s=0}^{k-1} \lambda_1^s \lambda_2^{k-1-s}\Big) = \mathbf{1}\Big\}.$$

Define $\Phi = (\Phi_1, \dots, \Phi_{m-1}) : \mathbb{C}^2 \longrightarrow \mathbb{C}^{m-1}$,

$$\Phi_j(\lambda) := 1 - \sum_{s=0}^{j} \lambda_1^s \lambda_2^{j-s}, \quad \lambda = (\lambda_1, \lambda_2) \in \mathbb{C}^2.$$

Observe that Φ is proper, and $\operatorname{rank} \Phi'(\lambda) = 2$, $\lambda \in \mathbb{C}^2$. In particular, $V := \Phi(\mathbb{C}^2)$ is an algebraic subset of $\mathbb{C}^{m-1}$ with $\dim V = 2$ (cf. [349], Ch. VIII, § 8).

For $C \in \mathbb{M}$, let

$$L(C) := \{(z_1, \dots, z_{m-1}) \in \mathbb{C}^{m-1} : c_2 z_1 + \dots + c_m z_{m-1} = \mathbf{1}\}.$$

Observe that $L(C)$ is an affine subspace of $\mathbb{C}^{m-1}$ and $\dim L(C) = m - 1 - \operatorname{rank} C$. In particular, $\dim L(C) = m - 1 - n < (m-1) - 2$ provided that $\operatorname{rank} C = n$. Consequently,

$$\{C \in \mathbb{M} : \operatorname{rank} C = n,\ L(C) \cap V = \varnothing\} \subset \mathbb{M} \setminus \Xi$$

is dense in $\mathbb{M}$ (cf. [349], Ch. VIII, § 11).

The infinitesimal case is left to the reader. □

So, the problem of comparison of the Lempert and Hahn functions remains only for the dimension 2. Here, we present an answer to what happens with the Hahn function for the product of two plane domains. It shows that both functions can be different also in the 2-dimensional case.

Before stating this result, we ask the reader to solve the following exercise, which will be important in the proof of the following proposition:

Exercise 8.1.10. Let $G \subset \mathbb{C}^n$ be a domain. Then, the following properties are equivalent:

(i) $\boldsymbol{\ell}_G^* \equiv \boldsymbol{h}_G^*$;

(ii) for any $\varphi \in \mathcal{O}(\mathbb{D}, G)$, $0 < \alpha < \delta < 1$ with $\varphi(0) \neq \varphi(\alpha)$, there exists an injective $\psi \in \mathcal{O}(\mathbb{D}, G)$ with $\psi(0) = \varphi(0)$ and $\psi(\delta) = \varphi(\alpha)$.

Theorem 8.1.11 (cf. [278]). *Let $G_j \subset \mathbb{C}$ be a domain, $j = 1, 2$.*

(a) *If at least one of the G_j's is simply connected, then $\boldsymbol{\ell}_{G_1 \times G_2}^* = \boldsymbol{h}_{G_1 \times G_2}^*$.*

(b) *If at least one of the G_j's is biholomorphically equivalent to $\mathbb{C}_*$, then $\boldsymbol{\ell}_{G_1 \times G_2}^* = \boldsymbol{h}_{G_1 \times G_2}^*$.*

(c) *Otherwise, $\boldsymbol{\ell}_{G_1 \times G_2}^* \not\equiv \boldsymbol{h}_{G_1 \times G_2}^*$.*

The proof of (c) will be based on the following lemma from classical complex analysis and the uniformization theorem (cf. Appendix B.1.7).

Lemma 8.1.12. *Let $G_j \subset \mathbb{C}$ be a non simply connected domain that is not biholomorphically equivalent to $\mathbb{C}_*$, $j = 1, 2$. Denote by $p_j : \mathbb{D} \longrightarrow G_j$ the universal covering mapping. Then, there are two different points $q_1, q_2 \in \mathbb{D}$ and automorphisms $f_j \in \operatorname{Aut}(\mathbb{D})$, $j = 1, 2$, such that $p_j(f_j(q_1)) = p_j(f_j(q_2))$, $j = 1, 2$, and*

$$\det \begin{bmatrix} (p_1 \circ f_1)'(q_1), & (p_1 \circ f_1)'(q_2) \\ (p_2 \circ f_2)'(q_1), & (p_2 \circ f_2)'(q_2) \end{bmatrix} \neq 0.$$

Proof. By assumption, the map p_j is not injective, $j = 1,2$. Therefore, there exists a $\psi_j \in \mathrm{Aut}(\mathbb{D})\setminus\{\mathrm{id}_{\mathbb{D}}\}$ such that $p_j \circ \psi_j = p_j$, $j = 1,2$; in particular, ψ_j is a lifting of p_j. Note that ψ_j has no fixed points in $\mathbb{D}$ (otherwise, applying the uniqueness of the lifting, it would be equal to $\mathrm{id}_{\mathbb{D}}$). Therefore, it has one or two fixed points on $\mathbb{T}$. Fix $\lambda' \in \mathbb{T}$ with $\psi_j(\lambda') \neq \lambda'$ for $j = 1,2$. Then, $\boldsymbol{m}(t\lambda', \psi_j(t\lambda')) \longrightarrow 1$ when $t \nearrow 1$, $j = 1,2$. Hence, we find $z_1, z_2 \in \mathbb{D}$ with

$$\boldsymbol{m}(z_1, \psi_1(z_1)) = \boldsymbol{m}(z_2, \psi_2(z_2)) \in (0,1).$$

Let $d \in (0,1)$ with $\boldsymbol{m}(-d,d) = \boldsymbol{m}(z_1, \psi_1(z_1))$. Then, there exists $h_j \in \mathrm{Aut}(\mathbb{D})$ with

$$h_j(-d) = z_j, \; h_j(d) = \psi_j(z_j), \quad j = 1,2.$$

Assume that $(p_j \circ h_j)'(-d) \neq \pm(p_j \circ h_j)'(d)$ for at least one of the j's, say for $j = 1$. Then, one of the following determinants does not vanish

$$\det\begin{bmatrix}(p_1\circ h_1)'(-d), & (p_1\circ h_1)'(d)\\ (p_2\circ h_2)'(-d), & (p_2\circ h_2)'(d)\end{bmatrix},$$

$$\det\begin{bmatrix}(p_1\circ h_1\circ(-\mathrm{id}_{\mathbb{D}}))'(-d), & (p_1\circ h_1\circ(-\mathrm{id}_{\mathbb{D}}))'(d)\\ (p_2\circ h_2)'(-d), & (p_2\circ h_2)'(d)\end{bmatrix}$$

(use that $(p_2 \circ h_2)'(d) \neq 0$).

So, we may put $f_1 = h_1$, $f_2 = h_2$ (resp. $f_1 = h_1 \circ (-\mathrm{id}_{\mathbb{D}})$, $f_2 = h_2$) and $q_1 = -d$, $q_2 = d$.

Now, for the remaining part of the proof, we may assume that

$$\big((p_j\circ h_j)'(d)\big)^2 = \big((p_j\circ h_j)'(-d)\big)^2, \quad j = 1,2. \tag{8.1.1}$$

Put $\widetilde{\psi}_j := h_j^{-1}\circ\psi_j\circ h_j$ and $\widetilde{p}_j := p_j \circ h_j$, $j = 1,2$. Then, $\widetilde{\psi}_j(-d) = d$ and $\widetilde{p}_j{}'(-d) = (\widetilde{p}_j \circ \widetilde{\psi}_j)'(-d) = \widetilde{p}_j{}'(\widetilde{\psi}_j(-d))\widetilde{\psi}_j'(-d)$. Taking the squares on both sides, we get $\big(\widetilde{\psi}_j'(-d)\big)^2 = 1$ (see (8.1.1)). Therefore, either $\widetilde{\psi}_j(-d) = d$, $\widetilde{\psi}_j'(-d) = -1$ or $\widetilde{\psi}_j(-d) = d$, $\widetilde{\psi}_j'(-d) = 1$.

It follows that

$$\psi := \widetilde{\psi}_1 = \widetilde{\psi}_2 = h_c \text{ with } c := \frac{-2d}{1+d^2}.$$

Now fix an $a \in \mathbb{D}$, and choose $\varphi \in \mathrm{Aut}(\mathbb{D})$ such that $\varphi(a) = \psi(a)$ and $\varphi(\psi(a)) = a$. Note that such a φ exists.

Suppose that $\varphi'(a) = \psi'(a)$. Then, $\varphi = \psi$ and, therefore, $\psi \circ \psi(a) = a$. So, $\psi \circ \psi$ has a fixed point in $\mathbb{D}$ and, therefore, it has none on $\mathbb{T}$. On the other hand, ψ is without fixed points on $\mathbb{D}$. So, it has at least one fixed point $b \in \mathbb{T}$. Then, $\psi \circ \psi(b) = b$; a contradiction.

Fix an $a_0 \in \mathbb{D} \cap \mathbb{R}$. Let $\varphi \in \operatorname{Aut}(\mathbb{D})$ with $\varphi(a_0) = \psi(a_0)$ and $\varphi(\psi(a_0)) = a_0$. Then, $\varphi = \boldsymbol{h}_{-a_0} \circ (-\operatorname{id}_{\mathbb{D}}) \circ \boldsymbol{h}_{h_{a_0}(\psi(a_0))} \circ \boldsymbol{h}_{a_0}$. By a direct calculation, it follows that $\varphi'(a_0) \neq -\psi'(a_0)$.

Summarizing, we know that, if $\varphi \in \operatorname{Aut}(\mathbb{D})$ is such that $\varphi(a_0) = \psi(a_0)$ and $\varphi(\psi(a_0)) = a_0$, then $\varphi'(a_0) \neq \pm\psi'(a_0)$. Then, $\varphi \circ \varphi = \operatorname{id}_{\mathbb{D}}$ (note that $\varphi \circ \varphi$ has two fixed points in $\mathbb{D}$) and so $\varphi'(\psi(a_0)) = \frac{1}{\varphi'(a_0)}$.

Finally, we put $q_1 := a_0$, $q_2 := \psi(a_0)$, $f_1 := h_1$, and $f_2 := h_2 \circ \varphi$. Then:

$$\begin{aligned} p_1(f_1(q_2)) &= (p_1 \circ h_1)(\psi(a_0)) \\ &= (p_1 \circ \psi_1)(h_1(a_0)) = (p_1 \circ h_1)(q_1) = (p_1 \circ f_1)(q_1), \\ p_2(f_2(q_2)) &= (p_2 \circ h_2)(\varphi(\psi(a_0))) = (p_2 \circ h_2)(a_0) = (p_2 \circ \psi_2)(h_2(a_0)) \\ &= (p_2 \circ h_2)(\psi(a_0)) = (p_2 \circ (h_2 \circ \varphi))(a_0) = (p_2 \circ f_2)(q_1). \end{aligned}$$

Moreover, we have

$$\begin{aligned} &\det \begin{bmatrix} (p_1 \circ f_1)'(q_1), & (p_1 \circ f_1)'(q_2) \\ (p_2 \circ f_2)'(q_1), & (p_2 \circ f_2)'(q_2) \end{bmatrix} \\ &= \det \begin{bmatrix} (p_1 \circ h_1)'(a_0), & (p_1 \circ h_1)'(\psi(a_0)) \\ (p_2 \circ h_2)'(\varphi(a_0))\varphi'(a_0), & (p_2 \circ h_2)'(\varphi(\psi(a_0)))\varphi'(\psi(a_0)) \end{bmatrix} \\ &= \det \begin{bmatrix} (p_1 \circ h_1)'(\psi(a_0))\psi'(a_0), & (p_1 \circ h_1)'(\psi(a_0)) \\ (p_2 \circ h_2)'(\psi(a_0))\varphi'(a_0), & (p_2 \circ h_2)'(\psi(a_0))/\varphi'(a_0) \end{bmatrix} \\ &= (p_1 \circ h_1)'(\psi(a_0))(p_2 \circ h_2)'(\psi(a_0)) \det \begin{bmatrix} \psi'(a_0), & 1 \\ \varphi'(a_0), & \psi'(a_0)/\varphi'(a_0) \end{bmatrix} \neq 0. \end{aligned}$$

Hence, this lemma is proved. □

Proof of Theorem 8.1.11. (a) Without loss of generality, we may assume that G_1 is simply connected. Our task is to apply Exercise 8.1.10. So, let $\varphi = (\varphi_1, \varphi_2) \in \mathcal{O}(\mathbb{D}, G_1 \times G_2)$ and $0 < \alpha < \delta < 1$ with $\varphi(0) \neq \varphi(\alpha)$.

Assume that $\varphi_1(0) \neq \varphi_1(\alpha)$. Recall that $\boldsymbol{\ell}^*_{G_1} = \boldsymbol{h}^*_{G_1}$. Hence, there exists an injective $\widetilde{\psi}_1 \in \mathcal{O}(\mathbb{D}, G_1)$ with $\widetilde{\psi}_1(0) = \varphi_1(0)$ and $\widetilde{\psi}_1(\delta) = \varphi_1(\alpha)$. Put

$$\psi(\lambda) := \left(\widetilde{\psi}_1(\lambda), \varphi_2\left(\frac{\alpha}{\delta}\lambda\right)\right), \quad \lambda \in \mathbb{D}.$$

Then, $\psi \in \mathcal{O}(\mathbb{D}, G_1 \times G_2)$, ψ is injective, and one has $\psi(0) = \varphi(0)$ and $\psi(\delta) = \varphi(\alpha)$.

Now, let $\varphi_1(0) = \varphi_1(\alpha)$ and $\varphi_2(0) \neq \varphi_2(\alpha)$. Take a $d \in (0, \operatorname{dist}(\varphi_1(0), \partial G_1))$ and put

$$h(\lambda) := \frac{\varphi_2(\frac{\alpha}{\delta}\lambda) - \varphi_2(0)}{\varphi_2(\alpha) - \varphi_2(0)}, \tag{8.1.2}$$

$$\psi_1(\lambda) := \varphi_1(0) + \frac{\delta d}{M\delta + 1}\left(h(\lambda) - \frac{\lambda}{\delta}\right), \quad \lambda \in \mathbb{D}, \tag{8.1.3}$$

where $M := \|h\|_{\overline{\mathbb{D}}}$. Observe that $\psi_1 \in \mathcal{O}(\mathbb{D}, G_1)$. Finally, define

$$\psi(\lambda) := \left(\psi_1(\lambda), \varphi_2\left(\frac{\alpha}{\delta}\lambda\right)\right), \quad \lambda \in \mathbb{D}.$$

Then, $\psi \in \mathcal{O}(\mathbb{D}, G_1 \times G_2)$ with $\psi(0) = \varphi(0)$ and $\psi(\delta) = \varphi(\alpha)$. Moreover, one easily sees that ψ is an injective analytic disc. Hence, (a) is proved.

(b) We may assume that $G_1 = \mathbb{C}_*$ and $G_2 \neq \mathbb{C}$. Let, as in (a), $\varphi = (\varphi_1, \varphi_2) \in \mathcal{O}(\mathbb{D}, G_1 \times G_2)$, $0 < \alpha < \delta < 1$, and $\varphi(0) \neq \varphi(\alpha)$. Moreover, applying a suitable automorphism of $\mathbb{C}_*$, we may even assume that $\varphi_1(0) = 1$.

In the case where $\varphi_2(0) = \varphi_2(\alpha)$ define $\widetilde{G}_2 := \varphi_2(0) + \operatorname{dist}(\varphi_2(0), \partial G_2)\mathbb{D}$. Obviously, $\widetilde{G}_2$ is a simply connected domain, $\widetilde{\varphi} = (\varphi_1, \widetilde{\varphi}_2) \in \mathcal{O}(\mathbb{D}, G_1 \times \widetilde{G}_2)$, where $\widetilde{\varphi}_2(\lambda) := \varphi_2(0)$, $\lambda \in \mathbb{D}$. By virtue of (a), there exists an injective analytic disc $\psi \in \mathcal{O}(\mathbb{D}, G_1 \times \widetilde{G}_2)$ with $\psi(0) = \varphi(0)$, $\psi(\delta) = \widetilde{\varphi}(\alpha) = \varphi(\alpha)$.

Next, we discuss the situation where $\varphi_2(0) \neq \varphi_2(\alpha)$. For the moment, we assume, in addition, that $\varphi_1(\alpha) = 1 + \delta$. Put

$$\psi(\lambda) := \left(1 + \lambda, \varphi_2\left(\frac{\alpha}{\delta}\lambda\right)\right), \quad \lambda \in \mathbb{D}.$$

Then, $\psi \in \mathcal{O}(\mathbb{D}, \mathbb{C}_* \times G_2)$ is injective and satisfies $\psi(0) = \varphi(0)$ and $\psi(\delta) = \varphi(\alpha)$.

Now, we turn to the remaining case $\varphi_1(\alpha) \neq 1 + \delta$. Then, for $k \in \mathbb{N}$, we choose numbers $d_k \in \mathbb{C} \setminus \{1\}$ such that $d_k^k = \frac{\varphi_1(\alpha)}{1+\delta}$ and $\operatorname{Arg}(d_k) \longrightarrow 0$ when $k \longrightarrow \infty$. Note that $d_k \longrightarrow 1$.

Put

$$c_k := \frac{\varphi_2(\alpha) - \varphi_2(0)}{1 - d_k}, \quad k \in \mathbb{N}.$$

Since $|c_k| \longrightarrow \infty$, we choose a k_0 such that $|c_{k_0}| > M := \sup\{|\varphi_1(\lambda)| : |\lambda| \leq \frac{\alpha}{\delta}\}$.

Define

$$\psi(\lambda) := \left((1 + \lambda)h^{k_0}(\lambda), \varphi_2\left(\frac{\alpha}{\delta}\lambda\right)\right), \quad \lambda \in \mathbb{D},$$

where

$$h(\lambda) := \frac{\varphi_2(\frac{\alpha}{\delta}\lambda) - c_{k_0}}{\varphi_2(0) - c_{k_0}}, \quad \lambda \in \mathbb{D}.$$

Then, $h \in \mathcal{O}(\mathbb{D}, \mathbb{C}_*)$ and so $\psi \in \mathcal{O}(\mathbb{D}, G_1 \times G_2)$ with $\psi(0) = (1, \varphi_2(0)) = \varphi(0)$. Moreover, a short calculation leads to $\psi(\delta) = \varphi(\alpha)$.

If $\psi(\lambda') = \psi(\lambda'')$, then $h(\lambda') = h(\lambda'')$, and, therefore, $\lambda' = \lambda''$, i.e., ψ is also injective. Hence, the proof of (b) is complete.

(c) Recall that the universal covering of G_j is $\mathbb{D}$ and that the covering mapping $p_j : \mathbb{D} \longrightarrow G_j$ is locally biholomorphic and surjective, but both are not injective,

$j = 1,2$. Applying Lemma 8.1.12, we find a point $q = (q_1, q_2) \in \mathbb{D}^2$, $q_1 \neq q_2$, and automorphisms $f_j \in \operatorname{Aut}(\mathbb{D})$, $j = 1,2$, such that with $\widetilde{p}_j := p_j \circ f_j$, $j = 1,2$, the following is true:

$$\widetilde{p}_j(q_1) = \widetilde{p}_j(q_2), \; j = 1,2, \text{ and } \det \begin{bmatrix} \widetilde{p}_1'(q_1), & \widetilde{p}_1'(q_2) \\ \widetilde{p}_2'(q_1), & \widetilde{p}_2'(q_2) \end{bmatrix} \neq 0.$$

Moreover, choose an $r \in (0,1)$ such that both mappings $\widetilde{p}_j$ are injective on $\overline{\mathbb{D}(r)}$ and put $a := (a_1, a_2) = (\widetilde{p}_1(0), \widetilde{p}_2(0))$, $b := (b_1, b_2) = (\widetilde{p}_1(r), \widetilde{p}_2(r)) \in G_1 \times G_2$. Note that $a_j \neq b_j$, $j = 1,2$.

Then, by virtue of Proposition 3.7.1, Theorem 3.3.7, and the choice of r, we have

$$\boldsymbol{\ell}^*_{G_1 \times G_2}(a,b) = \max\{\boldsymbol{\ell}^*_{G_1}(a_1,b_1), \boldsymbol{\ell}^*_{G_2}(a_2,b_2)\} = r.$$

Assume now that $\boldsymbol{\ell}^*_{G_1 \times G_2} = \boldsymbol{h}^*_{G_1 \times G_2}$; in particular,

$$r = \boldsymbol{\ell}^*_{G_1 \times G_2}(a,b) = \boldsymbol{h}^*_{G_1 \times G_2}(a,b).$$

Then, there exist a sequence of analytic discs $(\varphi_j)_{j=1}^\infty \subset \mathcal{O}(\mathbb{D}, G_1 \times G_2)$ and a sequence of numbers $(\alpha_j)_{j=1}^\infty \subset (1, 1/\sqrt{r})$ with $\alpha_j \searrow 1$ such that $\varphi_j(0) = a$ and $\varphi_j(\alpha_j r) = b$ for all j.

Then, we find $\psi_j = (\psi_{j,1}, \psi_{j,2}) \in \mathcal{O}(\mathbb{D}, G_1 \times G_2)$ injective such that $\psi_j(0) = a$ and $\psi_j(\alpha_j^2 r) = b$, $j \in \mathbb{N}$.

Recall that $\widetilde{p}_j$ are covering mappings. Therefore, we can lift the functions $\psi_{j,k}$, $k = 1,2$, i.e., there are holomorphic mappings $\widetilde{\psi}_{j,k} \in \mathcal{O}(\mathbb{D}, \mathbb{D})$ such that $\widetilde{p}_k \circ \widetilde{\psi}_{j,k} = \psi_{j,k}$ and $\widetilde{\psi}_{j,k}(0) = 0$. Note that $(\widetilde{p}_k \circ \widetilde{\psi}_{j,k})(\alpha_j^2 r) = \widetilde{p}_k(r)$. Recall that $\widetilde{p}_k$ is injective on $\overline{\mathbb{D}}(r)$ and therefore injective on $\mathbb{D}(r+\varepsilon)$, where $\varepsilon \in (0, 1-r)$ is sufficiently small. Then, for large j, we have that $\widetilde{\psi}_{j,k}(\alpha_j^2 r) = r$, $k = 1,2$.

By the Montel theorem, we may assume that $\widetilde{\psi}_{j,k} \longrightarrow \widetilde{\psi}_k \in \mathcal{O}(\mathbb{D}, \overline{\mathbb{D}})$ locally uniformly, $k = 1,2$. Since $\widetilde{\psi}(0) = 0$, it follows that, in fact, $\widetilde{\psi} \in \mathcal{O}(\mathbb{D}, \mathbb{D}^2)$. Moreover, because of the previous remark, $\widetilde{\psi}_k(r) = r$, $k = 1,2$. Then, by the Schwarz lemma, we have $\widetilde{\psi}_k = \mathrm{id}_{\mathbb{D}}$, $k = 1,2$.

Put

$$g = (g_1, g_2) : \mathbb{D}^2 \longrightarrow \mathbb{C}^2, \quad g_k(\lambda, \mu) := \widetilde{p}_k(\lambda) - \widetilde{p}_k(\mu).$$

Note that $g(q) = 0$ with $q := (q_1, q_2)$ and $\det g'(q) \neq 0$. Hence, we find neighborhoods $U = \mathbb{D}(q_1, s) \times \mathbb{D}(q_2, s) \subset \mathbb{D}^2$ of q and $V = V(0) \subset \mathbb{C}^2$ such that g maps U biholomorphically to V and $\mathbb{D}(q_1, s) \cap \mathbb{D}(q_2, s) = \varnothing$ (recall that $q_1 \neq q_2$).

Let now $g_j : \mathbb{D}^2 \longrightarrow \mathbb{C}^2$,

$$g_j(\lambda, \mu) := (\psi_{j,1}(\lambda) - \psi_{j,1}(\mu), \psi_{j,2}(\lambda) - \psi_{j,2}(\mu)), \quad (\lambda, \mu) \in \mathbb{D}^2, \; j \in \mathbb{N}.$$

By the result before we conclude that $g_j \longrightarrow g$ uniformly on U. Then, in virtue of the Hurwitz theorem, there exists a large index j_0 such that g_{j_0} vanishes in at least one point $(t_1, t_2) \in U$, i.e., $\psi_{j_0}(t_1) = \psi_{j_0}(t_2)$, which contradicts the injectivity of ψ_{j_0}. □

Using similar techniques, one can get the following result.

Theorem 8.1.13 (cf. [276]). *Let $G_j \subset \mathbb{C}$ be a domain, $j = 1, 2$.*

(a) *If at least one of the G_j's is simply connected, then $\varkappa_{G_1\times G_2} = \eta_{G_1\times G_2}$.*

(b) *If at least one of the G_j's is biholomorphically equivalent to $\mathbb{C}_*$, then $\varkappa_{G_1\times G_2} = \eta_{G_1\times G_2}$.*

(c) *Otherwise, $\varkappa_{G_1\times G_2} \not\equiv \eta_{G_1\times G_2}$.*

Proof. The proof is left to the reader. □

Any general characterization of those domains $G \subset \mathbb{C}^2$ for which $\boldsymbol{\ell}_G \equiv \boldsymbol{h}_G$ and/or $\varkappa_G \equiv \eta_G$ is not known.

8.2 Generalized Green, Möbius, and Lempert functions

Observe that in the definitions of $\boldsymbol{m}_G^{(k)}(a,z)$, $\boldsymbol{g}_G(a,z)$, and $\boldsymbol{\ell}_G^{(k)}(a,z)$ the roles played by the points $a, z \in G$ are not symmetric. The definitions distinguish the point a as a kind of a pole (center). From this point of view, one may propose the three definitions that follow.

Let $G \subset \mathbb{C}^n$ be a domain and let $\mathfrak{p} : G \longrightarrow \mathbb{R}_+$ be a function, $\mathfrak{p} \not\equiv 0$. Set $|\mathfrak{p}| := \{z \in G : \mathfrak{p}(z) > 0\}$.

Definition 8.2.1. Put

$$\boldsymbol{g}_G(\mathfrak{p}, z) := \sup\{u(z) :\ u : G \longrightarrow [0,1),\ \log u \in \mathcal{PSH}(G),\ \forall_{a\in|\mathfrak{p}|}\ \exists_{C=C(u,a)>0}\ \forall_{w\in G} : u(w) \le C\,\|w-a\|^{\mathfrak{p}(a)}\},\quad z \in G.$$ [1]

The function $\boldsymbol{g}_G(\mathfrak{p},\cdot)$ is called the *generalized pluricomplex Green function with pole function* $\mathfrak{p}$. Each point $a \in |\mathfrak{p}|$ is called *a pole with weight* $\mathfrak{p}(a)$; the set $|\mathfrak{p}|$ is called the *set of poles*. If $\mathfrak{p} = \chi_A$ = the characteristic function of a set $A \subset G$, then we put $\boldsymbol{g}_G(A,\cdot) := \boldsymbol{g}_G(\chi_A,\cdot)$.

For the case where the set $|\mathfrak{p}|$ is finite, the function $\boldsymbol{g}_G(\mathfrak{p},\cdot)$ was introduced by P. Lelong in [339]. The generalized pluricomplex Green function was studied, e.g., in [156, 337, 108, 84, 151, 247, 279].

Definition 8.2.2. Put

$$\boldsymbol{m}_G(\mathfrak{p}, z) := \sup\{|f(z)| : f \in \mathcal{O}(G,\mathbb{D}),\ \forall_{a\in|\mathfrak{p}|} : \operatorname{ord}_a f \ge \mathfrak{p}(a)\},\quad z \in G.$$ [2]

[1] Note that the growth condition may be equivalently formulated as follows: $\forall_{a\in|\mathfrak{p}|}\ \exists_{C,r>0}\ \forall_{w\in\mathbb{B}(a,r)\subset G}: u(w) \le C\,\|w-a\|^{\mathfrak{p}(a)}$.

The function $\boldsymbol{m}_G(\mathfrak{p}, \cdot)$ is called the *generalized Möbius function with weights* $\mathfrak{p}$. We put $\boldsymbol{m}_G(A, \cdot) := \boldsymbol{m}_G(\chi_A, \cdot)$, $A \subset G$.

Definition 8.2.3. Take a $z \in G$. Suppose that there exists a $\psi \in \mathcal{O}(\mathbb{D}, G)$ such that $\psi(0) = z$ and $|\mathfrak{p}| \subset \psi(\mathbb{D})$. (Notice that such a ψ exists if $|\mathfrak{p}|$ is at most countable (cf. Theorem 8.2.11).) Put

$$\boldsymbol{\ell}^*_G(\mathfrak{p}, z) := \inf\Big\{ \prod_{a \in |\mathfrak{p}|} |\mu_a|^{\mathfrak{p}(a)} : (\mu_a)_{a \in |\mathfrak{p}|} \subset \mathbb{D},$$

$$\exists_{\varphi \in \mathcal{O}(\mathbb{D}, G)} : \varphi(0) = z,\ \forall_{a \in |\mathfrak{p}|} : \varphi(\mu_a) = a \Big\}.^3 \quad (8.2.1)$$

For the case where the function ψ does not exist, we put

$$\boldsymbol{\ell}^*_G(\mathfrak{p}, z) := \inf\{\boldsymbol{\ell}^*_G(\mathfrak{p}_B, z) : \varnothing \neq B \subset |\mathfrak{p}|,\ \#B < +\infty\},$$

where $\mathfrak{p}_B := \mathfrak{p} \cdot \chi_B$. The function $\boldsymbol{\ell}^*_G(\mathfrak{p}, \cdot)$ is called the *generalized Lempert function with weights* $\mathfrak{p}$. As before, we put $\boldsymbol{\ell}^*_G(A, \cdot) := \boldsymbol{\ell}^*_G(\chi_A, \cdot)$, $A \subset G$.

The generalized Lempert function has been first introduced in [108] (see also [267]). The definition above is taken from [381] (see also [379]).

Directly from Definitions 8.2.1, 8.2.2, and 8.2.3 we get the following elementary properties of the generalized Möbius, Green, and Lempert functions (cf. [247]):

Remark 8.2.4.

(a) $\boldsymbol{m}_G(\mathfrak{p}, z) = 0$, $z \in |\mathfrak{p}|$; if $|\mathfrak{p}|$ is not (analytically) thin, then $\boldsymbol{m}_G(\mathfrak{p}, \cdot) \equiv 0$.

(b) $\boldsymbol{g}_G(\mathfrak{p}, z) = 0$, $z \in |\mathfrak{p}|$; if $|\mathfrak{p}|$ is not pluripolar, then $\boldsymbol{g}_G(\mathfrak{p}, \cdot) \equiv 0$.

(c) $\boldsymbol{m}_G(A, z) = \sup\{|f(z)| : f \in \mathcal{O}(G, \mathbb{D}),\ f|_A = 0\}$, $z \in G$.

(d) $\boldsymbol{c}^*_G(a, \cdot) = \boldsymbol{m}_G(a, \cdot) = \boldsymbol{m}_G(\{a\}, \cdot)$; $\boldsymbol{g}_G(a, \cdot) = \boldsymbol{g}_G(\{a\}, \cdot)$, $a \in G$.

(e) $\boldsymbol{m}_G(k\chi_{\{a\}}, \cdot) = (\boldsymbol{m}^{(k)}_G(a, \cdot))^k$, $k \in \mathbb{N}$.

(f) $\boldsymbol{m}_G(\mathfrak{p}, \cdot) \le \boldsymbol{g}_G(\mathfrak{p}, \cdot)$.

(g) $\boldsymbol{m}_G(k\mathfrak{p}, \cdot) \ge (\boldsymbol{m}_G(\mathfrak{p}, \cdot))^{\lceil k \rceil}$; $\boldsymbol{g}_G(k\mathfrak{p}, \cdot) = (\boldsymbol{g}_G(\mathfrak{p}, \cdot))^k$, $k > 0$.

---

[2] Note that $\operatorname{ord}_a f \ge \mathfrak{p}(a) \iff \operatorname{ord}_a f \ge \lceil \mathfrak{p}(a) \rceil$.

[3] Recall that for a function $f : |\mathfrak{p}| \longrightarrow [0, 1)$ we put

$$\prod_{a \in |\mathfrak{p}|} f(a) := \inf\Big\{ \prod_{a \in B} f(a) : \varnothing \neq B \subset |\mathfrak{p}|,\ \#B < +\infty \Big\}.$$

(h) In general, $\boldsymbol{m}_G(k\mathfrak{p},\cdot) \not\equiv (\boldsymbol{m}_G(\mathfrak{p},\cdot))^{\lceil k\rceil}$ – for instance, if $G := \mathbb{A}(1/R, R) \subset \mathbb{C}$ $(R > 1)$, then $\boldsymbol{m}_G(k\chi_{\{a\}},\cdot) = (\boldsymbol{m}_G^{(k)}(a,\cdot))^k \not\equiv (\boldsymbol{m}_G(a,\cdot))^k = (\boldsymbol{m}_G(\chi_{\{a\}},\cdot))^k$, $k \in \mathbb{N}_2$; cf. Proposition 9.1.5.

(i) If $\mathfrak{p}' \le \mathfrak{p}''$, then

$$\boldsymbol{m}_G(\mathfrak{p}',\cdot) \ge \boldsymbol{m}_G(\mathfrak{p}'',\cdot), \quad \boldsymbol{g}_G(\mathfrak{p}',\cdot) \ge \boldsymbol{g}_G(\mathfrak{p}'',\cdot);$$

in particular, if $A \subset B \subset G$, then

$$\boldsymbol{m}_G(A,\cdot) \ge \boldsymbol{m}_G(B,\cdot), \quad \boldsymbol{g}_G(A,\cdot) \ge \boldsymbol{g}_G(B,\cdot).$$

For arbitrary $\mathfrak{p}', \mathfrak{p}''$ we have

$$\boldsymbol{m}_G(\mathfrak{p}',\cdot)\boldsymbol{m}_G(\mathfrak{p}'',\cdot) \le \boldsymbol{m}_G(\mathfrak{p}'+\mathfrak{p}'',\cdot) \le \min\{\boldsymbol{m}_G(\mathfrak{p}',\cdot),\ \boldsymbol{m}_G(\mathfrak{p}'',\cdot)\},$$
$$\boldsymbol{g}_G(\mathfrak{p}',\cdot)\boldsymbol{g}_G(\mathfrak{p}'',\cdot) \le \boldsymbol{g}_G(\mathfrak{p}'+\mathfrak{p}'',\cdot) \le \min\{\boldsymbol{g}_G(\mathfrak{p}',\cdot),\ \boldsymbol{g}_G(\mathfrak{p}'',\cdot)\}.$$

In particular,

$$\boldsymbol{g}_G(\mathfrak{p},\cdot) \le \inf_{a\in G}(\boldsymbol{g}_G(a,\cdot))^{\mathfrak{p}(a)}.$$

(j) $\boldsymbol{g}_G(\mathfrak{p},z) = \sup\{u(z) : \log u \in \mathcal{PSH}(G),\ u \le \inf_{a\in G}(\boldsymbol{g}_G(a,\cdot))^{\mathfrak{p}(a)}\},\ z \in G.$

(k) $\log \boldsymbol{m}_G(\mathfrak{p},\cdot) \in \mathcal{C}(G) \cap \mathcal{PSH}(G)$, $\log \boldsymbol{g}_G(\mathfrak{p},\cdot) \in \mathcal{PSH}(G)$.

(l) If $|\mathfrak{p}|$ is finite, then

$$\boldsymbol{m}_G(\mathfrak{p},\cdot) \ge \prod_{a\in|\mathfrak{p}|}(\boldsymbol{m}_G(a,\cdot))^{\lceil \mathfrak{p}(a)\rceil}, \quad \boldsymbol{g}_G(\mathfrak{p},\cdot) \ge \prod_{a\in|\mathfrak{p}|}(\boldsymbol{g}_G(a,\cdot))^{\mathfrak{p}(a)}$$

(cf. Proposition 8.2.9(a)).

(m) If $F : G \longrightarrow D$ is holomorphic, then for any $\mathfrak{q} : D \longrightarrow \mathbb{R}_+$ with $|\mathfrak{q}| \cap F(G) \ne \varnothing$ we have

$$\boldsymbol{m}_D(\mathfrak{q}, F(z)) \le \boldsymbol{m}_G(\mathfrak{q}_F, z) \le \boldsymbol{m}_G(\mathfrak{q}\circ F, z),$$
$$\boldsymbol{g}_D(\mathfrak{q}, F(z)) \le \boldsymbol{g}_G(\mathfrak{q}_F, z) \le \boldsymbol{g}_G(\mathfrak{q}\circ F, z), \quad z \in G,$$

where

$$\mathfrak{q}_F(a) := \mathfrak{q}(F(a))\operatorname{ord}_a(F - F(a)), \quad a \in G.$$

In particular,

$$\boldsymbol{m}_D(B, F(z)) \le \boldsymbol{m}_G(F^{-1}(B), z),$$
$$\boldsymbol{g}_D(B, F(z)) \le \boldsymbol{g}_G(F^{-1}(B), z), \quad B \subset F(G),\ z \in G.$$

(n) For any $z_0 \in G$ there exists an *extremal function for* $\boldsymbol{m}_G(\mathfrak{p}, z_0)$, i.e., a function $f_{z_0} \in \mathcal{O}(G, \mathbb{D})$, $\operatorname{ord}_a f_{z_0} \geq \lceil \mathfrak{p}(a) \rceil$, $a \in G$, and $\boldsymbol{m}_G(\mathfrak{p}, z_0) = |f_{z_0}(z_0)|$.

(o) Let $P \subset G$ be a relatively closed thin set such that

$$\mathfrak{p}(a) \leq \limsup_{G \setminus P \ni z \to a} \mathfrak{p}(z), \quad a \in P.$$

Then, $\boldsymbol{m}_{G \setminus P}(\mathfrak{p}, \cdot) = \boldsymbol{m}_G(\mathfrak{p}, \cdot)$ on $G \setminus P$ (cf. Remark 2.1.3).

(p) Let $P \subset G$ be a relatively closed pluripolar set such that $\mathfrak{p} = 0$ on P. Then, $\boldsymbol{g}_{G \setminus P}(\mathfrak{p}, \cdot) = \boldsymbol{g}_G(\mathfrak{p}, \cdot)$ on $G \setminus P$ (cf. Proposition 4.2.10(d)).

(q) For arbitrary $\mathfrak{p}, \mathfrak{q} : G \longrightarrow \mathbb{R}_+$ with $\mathfrak{p} \leq \mathfrak{q}$ and $|\mathfrak{p}| = |\mathfrak{q}|$ we have $\boldsymbol{\ell}^*_G(\mathfrak{q}, \cdot) \leq \boldsymbol{\ell}^*_G(\mathfrak{p}, \cdot)$. Note that in fact the result holds for arbitrary $\mathfrak{p} \leq \mathfrak{q}$ (cf. Theorem 8.2.11).

(r) $\boldsymbol{\ell}^*_G(k\chi_{\{a\}}, \cdot) = (\boldsymbol{\ell}^*_G(a, \cdot))^k$, $k > 0$; in particular, $\boldsymbol{\ell}^*_G(\{a\}, \cdot) = \boldsymbol{\ell}^*_G(a, \cdot)$ $(a \in G)$.

(s) If $|\mathfrak{p}|$ is uncountable and a $\psi \in \mathcal{O}(\mathbb{D}, G)$ with $\psi(0) = z$, $|\mathfrak{p}| \subset \psi(\mathbb{D})$ exists, then $0 = \boldsymbol{\ell}^*_G(\mathfrak{p}, z) = \inf\{\boldsymbol{\ell}^*_G(\mathfrak{p}_B, z) : \varnothing \neq B \subset |\mathfrak{p}|,\ \#B < +\infty\}$.

Indeed, it suffices to apply a general property of summable families (cf. [233], Proposition 1.4) to the family $(\mathfrak{p}(a) \log |\mu_a|)_{a \in |\mathfrak{p}|}$.

Proposition 8.2.5. *If $G_k \nearrow G$ and $\mathfrak{p}_k \nearrow \mathfrak{p}$, then $\boldsymbol{m}_{G_k}(\mathfrak{p}_k, \cdot) \searrow \boldsymbol{m}_G(\mathfrak{p}, \cdot)$ and $\boldsymbol{g}_{G_k}(\mathfrak{p}_k, \cdot) \searrow \boldsymbol{g}_G(\mathfrak{p}, \cdot)$.*

Proof. The case of the generalized Möbius function follows from a Montel argument (based on Remark 8.2.4(n)). For the case of the generalized Green function, first recall that $\boldsymbol{g}_{G_k}(a, \cdot) \searrow \boldsymbol{g}_G(a, \cdot)$, $a \in G$; cf. Proposition 4.2.10(a). Let $u_k := \boldsymbol{g}_{G_k}(\mathfrak{p}_k, \cdot)$. Then $\log u_k \in \mathcal{PSH}(G_k)$ (by Remark 8.2.4(k)) and $\boldsymbol{g}_G(\mathfrak{p}, \cdot) \leq u_{k+1} \leq u_k$ on G_k (by Remark 8.2.4(i, m)). Let $u := \lim_{k \to +\infty} u_k$. Obviously, $u \geq \boldsymbol{g}_G(\mathfrak{p}, \cdot)$ and $\log u \in \mathcal{PSH}(G)$. Moreover, since $u_k \leq (\boldsymbol{g}_{G_k}(a, \cdot))^{\mathfrak{p}_k(a)}$, $a \in G_k$, we easily conclude that $u \leq (\boldsymbol{g}_G(a, \cdot))^{\mathfrak{p}(a)}$, $a \in G$. Hence, by Remark 8.2.4(j), $u = \boldsymbol{g}_G(\mathfrak{p}, \cdot)$. □

Proposition 8.2.6 (cf. [247]). *We have*

$$\boldsymbol{g}_G(\mathfrak{p}, \cdot) = \inf\{\boldsymbol{g}_G(\mathfrak{p}', \cdot) : \mathfrak{p}' \leq \mathfrak{p},\ \#|\mathfrak{p}'| < +\infty\} =: u.$$

Proof. Obviously $u \geq \boldsymbol{g}_G(\mathfrak{p}, \cdot)$. To prove the opposite inequality, we only need to show that $\log u$ is plurisubharmonic. Observe that $\boldsymbol{g}_G(\max\{\mathfrak{p}'_1, \dots, \mathfrak{p}'_N\}, \cdot) \leq \min\{\boldsymbol{g}_G(\mathfrak{p}'_1, \cdot), \dots, \boldsymbol{g}_G(\mathfrak{p}'_N, \cdot)\}$. Thus, we only need the following general result. □

Lemma 8.2.7. *Let $(v_i)_{i \in A} \subset \mathcal{PSH}(\Omega)$ $(\Omega \subset \mathbb{C}^n)$ be such that for any $i_1, \dots, i_N \in A$ there exists an $i_0 \in A$ such that $v_{i_0} \leq \min\{v_{i_1}, \dots, v_{i_N}\}$. Then, $v := \inf_{i \in A} v_i \in \mathcal{PSH}(\Omega)$.*

Proof. It suffices to consider only the case $n = 1$. Take a disc $\mathbb{D}(a,r) \subset\subset \Omega$, $\varepsilon > 0$, and a continuous function $w \in \mathcal{C}(\partial\mathbb{D}(a,r))$ such that $w \geq v$ on $\partial\mathbb{D}(a,r)$. We want to show that $v(a) \leq \frac{1}{2\pi}\int_0^{2\pi} w(a+re^{i\theta})d\theta + \varepsilon$. For any point $b \in \partial\mathbb{D}(a,r)$ there exists an $i = i(b) \in A$ such that $v_i(b) < w(b) + \varepsilon$. Hence, there exists an open arc $I = I(b) \subset \partial\mathbb{D}(a,r)$ with $b \in I$ such that $v_i(\lambda) < w(\lambda) + \varepsilon$, $\lambda \in I$. By a compactness argument, we find $b_1, \dots, b_N \in \partial\mathbb{D}(a,r)$ such that $\partial\mathbb{D}(a,r) = \bigcup_{j=1}^N I(b_j)$. By assumption, there exists an $i_0 \in A$ such that $v_{i_0} \leq \min\{v_{i(b_1)}, \dots, v_{i(b_N)}\}$. Then,

$$v(a) \leq v_{i_0}(a) \leq \frac{1}{2\pi}\int_0^{2\pi} v_{i_0}(a+re^{i\theta})d\theta \leq \frac{1}{2\pi}\int_0^{2\pi} w(a+re^{i\theta})d\theta + \varepsilon. \qquad \square$$

Proposition 8.2.8. *We have*

$$\boldsymbol{m}_G(\mathfrak{p},\cdot) = \inf\{\boldsymbol{m}_G(\mathfrak{p}',\cdot) : \mathfrak{p}' \leq \mathfrak{p},\ \#|\mathfrak{p}'| < +\infty\}.$$

Proof. The case where $|\mathfrak{p}|$ is finite is trivial. The case where the set $|\mathfrak{p}|$ is countable follows from Remark 8.2.4(i). In the general case, let $A_k := \{a \in G : \mathfrak{p}(a) = k\}$ and let B_k be a countable (or finite) dense subset of A_k, $k \in \mathbb{Z}_+$. Put $B := \bigcup_{k=0}^\infty B_k$, $\mathfrak{p}' := \mathfrak{p} \cdot \chi_B$. Then, $\mathfrak{p}' \leq \mathfrak{p}$, the set $|\mathfrak{p}'|$ is at most countable, and $\boldsymbol{m}_G(\mathfrak{p},\cdot) \equiv \boldsymbol{m}_G(\mathfrak{p}',\cdot)$. $\square$

Proposition 8.2.9.

(a)

$$\boldsymbol{m}_G(\mathfrak{p},\cdot) \geq \prod_{a\in G} (\boldsymbol{m}_G(a,\cdot))^{\lceil \mathfrak{p}(a)\rceil}, \quad \boldsymbol{g}_G(\mathfrak{p},\cdot) \geq \prod_{a\in G} (\boldsymbol{g}_G(a,\cdot))^{\mathfrak{p}(a)}.$$

(b) *If $G \subset \mathbb{C}$, then*

$$\boldsymbol{g}_G(\mathfrak{p},z) = \prod_{a\in G} (\boldsymbol{g}_G(a,z))^{\mathfrak{p}(a)}, \quad z \in G.$$

Notice that the formula in (b) is not true for $G \subset \mathbb{C}^n$, $n \geq 2$; cf. Example 8.2.27.

Proof. (a) Use Remark 8.2.4(l) and Propositions 8.2.6, 8.2.8.

(b) By Proposition 8.2.6 we may assume that the set $|\mathfrak{p}|$ is finite. Let

$$u := \prod_{a\in|\mathfrak{p}|} (\boldsymbol{g}_G(a,\cdot))^{\mathfrak{p}(a)}.$$

By (a), we only need to show that $\boldsymbol{g}_G(\mathfrak{p},\cdot) \leq u$. Now, by Proposition 8.2.5, we may assume that $G \subset\subset \mathbb{C}$ is regular with respect to the Dirichlet problem. Then, the function $\log u$ is subharmonic on G and harmonic on $G \setminus |\mathfrak{p}|$. The function $v := \log \boldsymbol{g}_G(\mathfrak{p},\cdot) - \log u$ is locally bounded from above in G and $\limsup_{z\to\zeta} v(z) \leq 0$, $\zeta \in \partial G$. Consequently, v extends to a subharmonic function on G and, by the maximum principle, $v \leq 0$ on G, i.e., $\boldsymbol{g}_G(\mathfrak{p},\cdot) \leq u$ on G. $\square$

Proposition 8.2.10 (cf. [156, 337]). *Let $G, D \subset \mathbb{C}^n$ be domains and let $F : G \longrightarrow D$ be a proper holomorphic mapping. Put $\Sigma := F(\{z \in G : \det F'(z) = 0\})$. Let $\mathfrak{q} : D \longrightarrow \mathbb{R}_+$, $\mathfrak{q} \not\equiv 0$, be such that $\Sigma \cap |\mathfrak{q}| = \varnothing$. Then,*

$$\boldsymbol{g}_D(\mathfrak{q}, F(z)) = \boldsymbol{g}_G(\mathfrak{q}_F, z) = \boldsymbol{g}_G(\mathfrak{q} \circ F, z), \quad z \in G.$$

In particular, if $B \subset D$ is such that $\Sigma \cap B = \varnothing$, then

$$\boldsymbol{g}_D(B, F(z)) = \boldsymbol{g}_G(F^{-1}(B), z), \quad z \in G.$$

Notice that the equality may be false if $\Sigma \cap |\mathfrak{q}| \neq \varnothing$ – cf. Example 8.2.24. Moreover, the analogue of the above result for Möbius functions is not true – cf. Example 9.1.8

For the behavior of the pluricomplex Green function under coverings see [29, 30].

Proof. We only need to show that $\boldsymbol{g}_D(\mathfrak{q}, F(z)) \geq \boldsymbol{g}_G(\mathfrak{q} \circ F, z)$, $z \in G$; cf. Remark 8.2.4(m). It is well-known that

$$F|_{G \setminus F^{-1}(\Sigma)} : G \setminus F^{-1}(\Sigma) \longrightarrow D \setminus \Sigma$$

is a holomorphic covering. Denote by N its multiplicity. Let $u : G \longrightarrow [0, 1)$ be a log-psh function such that

$$u(z) \leq C(a)\|z - a\|^{\mathfrak{q}(F(a))}, \quad a, z \in G.$$

Define $v : D \longrightarrow [0, 1)$,

$$v(w) := \max\{u(z) : z \in F^{-1}(w)\}, \quad w \in D.$$

Since F is proper, $\log v \in \mathcal{PSH}(D)$ (cf. [305], Proposition 2.9.26). Take a $b \in D$ with $\mathfrak{q}(b) > 0$ (recall that $b \notin \Sigma$) and let $F^{-1}(b) = \{a_1, \dots, a_N\}$ ($a_j \neq a_k$ for $j \neq k$). There exist open neighborhoods $U_1, \dots, U_N, V$ of $a_1, \dots, a_N, b$, respectively, such that $F|_{U_j} : U_j \longrightarrow V$ is biholomorphic, $j = 1, \dots, N$. Let $g_j := (F|_{U_j})^{-1}$, $j = 1, \dots, N$. Shrinking the neighborhoods, if necessary, we may assume that there is a constant $M > 0$ such that $\|g_j(w) - a_j\| \leq M\|w - b\|$, $w \in V$. Then, for $w \in V$, we get

$$\begin{aligned} v(w) &= \max\{u \circ g_j(w) : j = 1, \dots, N\} \\ &\leq \max\{C(a_j)\|g_j(w) - a_j\|^{\mathfrak{q}(b)} : j = 1, \dots, N\} \\ &\leq \max\{C(a_j) : j = 1, \dots, N\} M^{\mathfrak{q}(b)} \|w - b\|^{\mathfrak{q}(b)}. \end{aligned}$$

Consequently, $\boldsymbol{g}_D(\mathfrak{q}, \cdot) \geq v$ and, therefore, $\boldsymbol{g}_D(\mathfrak{q}, F(z)) \geq v(F(z)) \geq u(z)$, $z \in G$, which gives the required inequality. □

Now we will discuss properties of the generalized Lempert function.

Theorem 8.2.11 (cf. [381]). *Let $G \subset \mathbb{C}^n$ be a domain, let $\mathfrak{p} : G \longrightarrow \mathbb{R}_+$, $\mathfrak{p} \not\equiv 0$, be arbitrary, and let $z \in G$.*

(a) *If $|\mathfrak{p}|$ is at most countable, then there exists a $\psi \in \mathcal{O}(\mathbb{D}, G)$ such that $\psi(0) = z$ and $|\mathfrak{p}| \subset \psi(\mathbb{D})$.*

(b) *For arbitrary $\mathfrak{p}$ we have*

$$\boldsymbol{\ell}_G^*(\mathfrak{p}, z) = \inf\{\boldsymbol{\ell}_G^*(\mathfrak{p}_B, z) : \varnothing \neq B \subset |\mathfrak{p}|,\ \#B < +\infty\}.$$

Consequently, if $\mathfrak{p} \le \mathfrak{q}$, then

$$\boldsymbol{\ell}_G^*(\mathfrak{q}, \cdot) \le \boldsymbol{\ell}_G^*(\mathfrak{p}, \cdot). \tag{8.2.2}$$

Remark 8.2.12.

(a) For the case where G is convex and $|\mathfrak{p}|, |\mathfrak{q}|$ are finite, inequality (8.2.2) has been first proved by F. Wikström in [529]. On the other hand, in [530] he gave the following example of a complex analytic space for which (8.2.2) is not true:

The analytic space X we will discuss is given as an analytic subset of $\mathbb{B}_2$, namely $X = X_1 \cup X_2$, where $X_1 := \mathbb{D} \times \{0\}$,

$$X_2 := \Bigg\{(z_1, z_2) \in \mathbb{B}_2 : z_2\left(1 - \frac{9}{10} \cdot \frac{9}{5} z_1\right) - \sqrt{1 - \left(\frac{5}{9}\right)^2 \frac{9}{5} z_1 \left(\frac{9}{5} z_1 - \frac{9}{10}\right)} = 0\Bigg\},$$

Put

$$\psi(\lambda) := \left(\frac{5}{9}\lambda, \sqrt{1 - \left(\frac{5}{9}\right)^2} \lambda \frac{\lambda - \frac{9}{10}}{1 - \frac{9}{10}\lambda}\right), \quad \lambda \in \mathbb{D}.$$

Observe that $\psi : \mathbb{D} \longrightarrow X_2$ is biholomorphic, $\psi(0) = (0, 0)$, and $\psi(9/10) = (1/2, 0) =: a$. Let $b := \psi(3/4) \in X_2 \setminus X_1$ and define $A := \{a, b\} \subset X$. Then, $\boldsymbol{\ell}_X^*(A, (0, 0)) \le 27/40$. Moreover, if $\varphi \in \mathcal{O}(\mathbb{D}, X)$ with $\varphi(0) = (0, 0)$, $\varphi(\lambda_1) = (1/2, 0)$, and $\varphi(\lambda_2) = \psi(3/4)$, then $\varphi(\mathbb{D}) \subset X_2$. Put $\chi := \psi^{-1} \circ \varphi \in \mathcal{O}(\mathbb{D}, \mathbb{D})$. Then, $\chi(0) = 0$, $\chi(\lambda_1) = 9/10$, and $\chi(\lambda_2) = 3/4$. Using the Schwarz lemma one gets $|\lambda_1 \lambda_2| \ge 27/40$; in particular, $\boldsymbol{\ell}_X^*(A, (0, 0)) = 27/40 > 1/2 = \boldsymbol{\ell}_X^*(\{a\}, (0, 0))$, which gives the wanted counterexample.

(b) The proof of Theorem 8.2.11 will be based on Lemma 8.2.13 being a special version of the Arakelian theorem from [21]. Using an analogous method of proof, F. Forstnerič and J. Winkelmann [177] proved the following result:

Let M be a connected complex manifold. Then the set

$$\{\varphi \in \mathcal{O}(\mathbb{D}, M) : \varphi(\mathbb{D}) \text{ is dense in } M\}$$

is dense in $\mathcal{O}(\mathbb{D}, M)$ in the compact-open topology.

Lemma 8.2.13. *Let $0 < t < 1$, $E := \overline{\mathbb{D}}(t) \cup [t, 1)$, $f \in \mathcal{C}(E) \cap \mathcal{O}(\mathbb{D}(t))$, $\varepsilon > 0$. Then there exists a $g \in \mathcal{O}(\mathbb{D})$ such that $|g(z) - f(z)| < \varepsilon$, $z \in E$.*

Proof of Theorem 8.2.11. (a) will follow from the proof of (b) below.

(b) Fix a $z \in G$. We may assume that $|\mathfrak{p}|$ is at most countable. We will only consider the case where $|\mathfrak{p}|$ is countable (the case where $|\mathfrak{p}|$ is finite is left to the reader). We have to prove that for every non-empty finite set $B \subset |\mathfrak{p}|$ we have $\boldsymbol{\ell}_G^*(\mathfrak{p}, z) \le \boldsymbol{\ell}_G^*(\mathfrak{p}_B, z)$. Let $\varphi \in \mathcal{O}(\mathbb{D}, G)$ be such that $\varphi(0) = z$, $\varphi(\mu_j) = a_j$, $j = 1, \dots, m$. Notice that such a φ always exists if $m = 1$. Put $a_0 := z$, $\mu_0 := 0$. Let $t_0 := \max\{|\mu_j| : j = 1, \dots, m\}$ and fix an arbitrary $t \in (t_0, 1)$. Define $\mu_j = 1 - \frac{1-t}{j^2}$, $j \ge m + 1$. Since the sequence $(\mu_j)_{j=0}^{\infty}$ satisfies the Blaschke condition, we find a Blaschke product B_k with zero set $\{\lambda_0, \dots, \lambda_{k-1}, \lambda_{k+1}, \dots\}$, $k = 0, 1, 2, \dots$. Take a continuous curve $\gamma : [t, 1) \longrightarrow G$ with $\gamma(t) = \varphi(t)$ and $\gamma(\mu_j) = a_j$, $j \ge m + 1$. Let $E := \overline{\mathbb{D}}(t) \cup [t, 1)$, $f := \varphi|_{\overline{\mathbb{D}}(t)} \cup \gamma$, and $d(\lambda) := d_G(f(\lambda))$, $\lambda \in E$, where the distance to the boundary of G, $d_G : G \longrightarrow \mathbb{R}_+$, is taken in the maximum norm. Take two arbitrary functions $\eta_1, \eta_2 : E \longrightarrow \mathbb{R}$ such that

- $\eta_1(\lambda), \eta_2(\lambda) \le \log \frac{d(z)}{9}$, $\lambda \in E$,
- $\eta_1(\lambda) = \eta_2(\lambda) = \min_{\overline{\mathbb{D}}(t)} \log \frac{d}{9}$, $\lambda \in \overline{\mathbb{D}}(t)$,
- $\eta_1(\mu_j) - \eta_2(\mu_j) = \log \left(\frac{1}{2^{j+1}} |B_j(\mu_j)|\right)$, $j \ge m + 1$.

Put $\varepsilon := \min\{\frac{1}{2^{j+1}} |B_j(\mu_j)| : j = 0, \dots, m\} < 1$. Using Lemma 8.2.13, we find $\zeta_1, \zeta_2 \in \mathcal{O}(\mathbb{D})$ and $h \in \mathcal{O}(\mathbb{D}, \mathbb{C}^2)$ such that

- $|\zeta_j - \eta_j| \le 1$ on E, $j = 1, 2$,
- $\|h - f\|_\infty \le \varepsilon |e^{\zeta_1 - 1}| \le \varepsilon e^{\eta_1}$ on E (to get h, we apply Lemma 8.2.13 to the mapping $e^{1-\zeta_1} f$).

Put $\delta_j := h(\mu_j) - f(\mu_j)$, $j \in \mathbb{Z}_+$. Consequently,

- $\|h - f\|_\infty \le \frac{d}{9}$ on E,
- $\|\delta_j\|_\infty \le e^{\eta_1(\mu_j)} \frac{1}{2^{j+1}} |B_j(\mu_j)| = e^{\eta_2(\mu_j)} \frac{1}{2^{j+1}} |B_j(\mu_j)|$, $j = 0, \dots, m$,
- $\|\gamma_j\|_\infty \le e^{\eta_1(\mu_j)} = e^{\eta_2(\mu_j)} \frac{1}{2^{j+1}} |B_j(\mu_j)|$, $j \ge m + 1$.

Define

$$g(\lambda) := e^{\xi_2(\lambda)} \sum_{j=0}^{\infty} \frac{\delta_j}{e^{\xi_2(\mu_j)} B_j(\mu_j)} B_j(\lambda), \quad \lambda \in \mathbb{D}.$$

Observe that the above series is uniformly convergent in $\mathbb{D}$ (thus $g \in \mathcal{O}(\mathbb{D}, \mathbb{C}^2)$), $g(\mu_j) = \delta_j$, $j \in \mathbb{Z}_+$, and

$$\|g\|_\infty \leq e^{\operatorname{Re} \xi_2 + 1} \leq e^{\eta_2 + 2} \leq \frac{e^2}{9} d \quad \text{on } E.$$

Put $q := h - g$. Then, $q(\mu_j) = f(\mu_j) = a_j$, $j \in \mathbb{Z}_+$, and

$$\|q - f\|_\infty \leq \frac{e^2 + 1}{9} d < d \quad \text{on } E.$$

Consequently, $q \in \mathcal{O}(\mathbb{D}, G)$. *Notice that starting from $m = 1$ we here get a proof of* (*a*). Take a simply connected domain $\Delta \subset \mathbb{D}$ such that $E \subset \Delta$ and $q(\Delta) \subset G$. Let $\varrho_t : \mathbb{D} \longrightarrow \Delta$ be the Riemann conformal mapping with $\varrho_t(0) = 0$ and $\varrho_t'(0) > 0$. Define $\mu_{t,j} = \varrho_t^{-1}(\mu_j)$, $j \in \mathbb{N}$, and $\psi := q \circ \varrho_t : \mathbb{D} \longrightarrow G$. We have

$$\boldsymbol{\ell}_G^*(\mathfrak{p}, z) \leq \prod_{j=1}^{\infty} |\mu_{t,j}|^{\mathfrak{p}(a_j)} \leq \prod_{j=1}^{m} |\mu_{t,j}|^{\mathfrak{p}(a_j)}.$$

Now we let $t \longrightarrow 1$. Using the Carathéodory kernel theorem (cf. [195], Chapter II, § 5), we conclude that $\varrho_t \longrightarrow \mathrm{id}_{\mathbb{D}}$ locally uniformly in $\mathbb{D}$ when $t \longrightarrow 1$. In particular, $\prod_{j=1}^{m} |\mu_{t,j}|^{\mathfrak{p}(a_j)} \longrightarrow \prod_{j=1}^{m} |\mu_j|^{\mathfrak{p}(a_j)}$, which finishes the proof. □

Remark 8.2.14.

(a) $\boldsymbol{g}_G(\mathfrak{p}, \cdot) \leq \boldsymbol{\ell}_G^*(\mathfrak{p}, \cdot)$.

Indeed, by Proposition 8.2.6 and Theorem 8.2.11, we may assume that $|\mathfrak{p}|$ is finite. Fix a $z \in G \setminus |\mathfrak{p}|$, take an arbitrary function φ as in (8.2.1) and an arbitrary function u from the defining family of $\boldsymbol{g}_G(\mathfrak{p}, z)$ (cf. Definition 8.2.1). We will show that $u(z) = u(\varphi(0)) \leq \prod_{a \in |\mathfrak{p}|} |\mu_a|^{\mathfrak{p}(a)}$. Define

$$v(\lambda) := \log u(\varphi(\lambda)) - \sum_{a \in |\mathfrak{p}|} \mathfrak{p}(a) \log \boldsymbol{m}_{\mathbb{D}}(\lambda, \mu_a), \quad \lambda \in \mathbb{D} \setminus \{\mu_a : a \in |\mathfrak{p}|\}.$$

Observe that v is subharmonic. Moreover, v is bounded from above in a neighborhood of each point $a \in |\mathfrak{p}|$. Thus, v extends subharmonically to $\mathbb{D}$. It is clear that $\limsup_{\lambda \to \mathbb{T}} v(\lambda) \leq 0$. Consequently, by the maximum principle, $v(0) \leq 0$.

(b) $\prod_{a \in |\mathfrak{p}|} [\boldsymbol{m}_{\mathbb{D}}(a, \cdot)]^{\mathfrak{p}(a)} = \boldsymbol{g}_{\mathbb{D}}(\mathfrak{p}, \cdot) = \boldsymbol{\ell}_{\mathbb{D}}^*(\mathfrak{p}, \cdot)$.

Indeed, we only need to prove that $\boldsymbol{\ell}_{\mathbb{D}}^*(\mathfrak{p}, \cdot) \leq \prod_{a \in |\mathfrak{p}|} [\boldsymbol{m}_{\mathbb{D}}(a, \cdot)]^{\mathfrak{p}(a)}$. Fix a $z_0 \in \mathbb{D}$ and let $\varphi := \boldsymbol{h}_{-z_0}$, $\mu_a := \varphi^{-1}(a)$, $a \in |\mathfrak{p}|$. Then, $\varphi(0) = z_0$ and $\prod_{a \in |\mathfrak{p}|} |\mu_a|^{\mathfrak{p}(a)} = \prod_{a \in |\mathfrak{p}|} [\boldsymbol{m}_{\mathbb{D}}(\mu_a, 0)]^{\mathfrak{p}(a)} = \prod_{a \in |\mathfrak{p}|} [\boldsymbol{m}_{\mathbb{D}}(a, z_0)]^{\mathfrak{p}(a)}$.

(c) Let $F : G \longrightarrow D$ be a holomorphic mapping and let $\mathfrak{q} : D \longrightarrow \mathbb{R}_+$ be such that $\#F^{-1}(b) = 1$ for any $b \in |\mathfrak{q}|$ (e.g. F is bijective). Then,

$$\boldsymbol{\ell}_D^*(\mathfrak{q}, F(z)) \le \boldsymbol{\ell}_G^*(\mathfrak{q} \circ F, z), \quad z \in G.$$

Notice that the above inequality is not true for arbitrary holomorphic mappings – cf. Example 8.2.16.

Indeed, we may assume that $|\mathfrak{q}|$ is finite. We have

$$\begin{aligned}\boldsymbol{\ell}_D^*(\mathfrak{q}, F(z)) &= \inf\Big\{ \prod_{b\in|\mathfrak{q}|} |\mu_b|^{\mathfrak{q}(b)} : \exists_{\psi\in\mathcal{O}(\mathbb{D},D)} : \psi(0) = F(z),\ \psi(\mu_b) = b\Big\} \\ &\overset{\psi:=F\circ\varphi}{\le} \inf\Big\{ \prod_{a\in F^{-1}(|\mathfrak{q}|)} |\mu_a|^{\mathfrak{q}(F(a))} : \exists_{\varphi\in\mathcal{O}(\mathbb{D},G)} : \varphi(0) = z, \\ &\qquad\qquad \varphi(\mu_a) = a\Big\} \\ &= \boldsymbol{\ell}_G^*(\mathfrak{q} \circ F, z).\end{aligned}$$

(d) $\boldsymbol{\ell}_G^*(\mathfrak{p}, \cdot) \le \inf_{a\in|\mathfrak{p}|}(\boldsymbol{\ell}_G^*(a, \cdot))^{\mathfrak{p}(a)}$.

(e) The function $\boldsymbol{\ell}_G^*(\mathfrak{p}, \cdot)$ is upper semicontinuous.

Indeed, we may assume that $|\mathfrak{p}|$ is finite. Fix a $z_0 \in G$ and $A >$ such that $\boldsymbol{\ell}_G^*(\mathfrak{p}, z_0) < A$. We would like to find a ball $\mathbb{B}(z_0, \delta) \subset G$ such that $\boldsymbol{\ell}_G^*(\mathfrak{p}, z) < A$, $z \in \mathbb{B}(z_0, \delta)$. The case where $z_0 \in |\mathfrak{p}|$ follows easily from (d). Thus assume that $z_0 \notin |\mathfrak{p}|$. Let $\varphi \in \mathcal{O}(\mathbb{D}, G)$ be such that $\varphi(0) = z_0$, $\varphi(\mu_a) = a$, $a \in |\mathfrak{p}|$, and $\prod_{a\in|\mathfrak{p}|} |\mu_a|^{\mathfrak{p}(a)} < A$. We may assume that $\varphi \in \mathcal{O}(\overline{\mathbb{D}}, G)$. For $z \in \mathbb{C}^n$, define

$$\varphi_z(\lambda) := \varphi(\lambda) + (z - z_0) \prod_{a\in|\mathfrak{p}|} \Big(1 - \frac{\lambda}{\mu_a}\Big), \quad \lambda \in \mathbb{D}.$$

Then, $\varphi_z(0) = z$ and $\varphi_z(\mu_a) = a$, $a \in |\mathfrak{p}|$. Moreover, if $\delta \ll 1$, then $\varphi_z \in \mathcal{O}(\mathbb{D}, G)$ for $z \in \mathbb{B}(z_0, \delta)$. Thus, $\boldsymbol{\ell}_G^*(\mathfrak{p}, z) \le \prod_{a\in|\mathfrak{p}|} |\mu_a|^{\mathfrak{p}(a)} < A$, $z \in \mathbb{B}(z_0, \delta)$.

(f) If G is taut and $\mathfrak{p}$ is finite, then for every $z_0 \in G$ there exists an *extremal disc* for $\boldsymbol{\ell}_G^*(\mathfrak{p}, z_0)$, i.e., a function $\varphi \in \mathcal{O}(\mathbb{D}, G)$ such that $\varphi(0) = z_0$ and there exists a non-empty set $B \subset |\mathfrak{p}|$ with $\varphi(\mu_a) = a$ for $a \in B$ and $\prod_{a\in B} |\mu|^{\mathfrak{p}(a)} = \boldsymbol{\ell}_G^*(\mathfrak{p}, z_0)$.

Indeed, let $\varphi_s \in \mathcal{O}(\mathbb{D}, G)$ be such that $\varphi_s(0) = z_0$, $\varphi_s(\mu_{s,a}) = a$, $a \in |\mathfrak{p}|$, $\prod_{a\in|\mathfrak{p}|} |\mu_{s,a}|^{\mathfrak{p}(a)} \searrow \boldsymbol{\ell}_G^*(\mathfrak{p}, z_0)$ where $s \nearrow +\infty$. Since G is taut, we may assume that $\varphi_s \longrightarrow \varphi$ locally uniformly in $\mathbb{D}$ with $\varphi \in \mathcal{O}(\mathbb{D}, G)$, $\varphi(0) = z_0$. We may also assume that $\mu_{s,a} \longrightarrow \mu_a \in \overline{\mathbb{D}}$ when $s \longrightarrow +\infty$, $a \in |\mathfrak{p}|$. Let $B := \{a \in |\mathfrak{p}| : \mu_a \in \mathbb{D}\}$. Observe that $B \ne \varnothing$. We have

$$\boldsymbol{\ell}_G^*(\mathfrak{p}, z_0) \le \boldsymbol{\ell}_G^*(\mathfrak{p}_B, z_0) = \prod_{a\in B} |\mu_a|^{\mathfrak{p}(a)} = \lim_{s\to+\infty} \prod_{a\in|\mathfrak{p}|} |\mu_{s,a}|^{\mathfrak{p}(a)} = \boldsymbol{\ell}_G^*(\mathfrak{p}, z_0).$$

(g) If $|\mathfrak{p}|$ is finite, $z_0 \in G \setminus |\mathfrak{p}|$, and $\varphi \in \mathcal{O}(\mathbb{D}, G)$ is an extremal disc for $\boldsymbol{\ell}_G^*(\mathfrak{p}, z_0)$, then $\varphi(\mathbb{D})$ cannot be relatively compact in G (cf. Remark 3.8.6).

(h) If G is taut and $\mathfrak{p}$ is finite, then the function $\boldsymbol{\ell}_G^*(\mathfrak{p}, \cdot)$ is continuous (use the method from (f)).

(i) (Cf. [396], Proposition 2) Assume that $n = 1$, $z_0 \in G$, $|\mathfrak{p}|$ is at most countable, and $\Pi : \mathbb{D} \longrightarrow G$ is a holomorphic covering with $\Pi(0) = z_0$. Let $\mu_a \in \Pi^{-1}(a)$ be such that $|\mu_a| = \min\{|\mu| : \mu \in \Pi^{-1}(a)\}$. Then,

$$\boldsymbol{\ell}_G^*(\mathfrak{p}, z_0) = \prod_{a \in |\mathfrak{p}|} |\mu_a|^{\mathfrak{p}(a)}.$$

Moreover, if $|\mathfrak{p}|$ is finite and $\boldsymbol{\ell}_G^*(\mathfrak{p}, z_0) > 0$, then each extremal disc for $\boldsymbol{\ell}_G^*(\mathfrak{p}, z_0)$ is of the form $\lambda \longmapsto \Pi(e^{i\theta}\lambda)$ for a $\theta \in \mathbb{R}$.

Indeed, the inequality "$\leq$" follows directly from the definition. Let $\varphi \in \mathcal{O}(\mathbb{D}, G)$, $\varphi(0) = z_0, \varphi(\nu_a) = a, a \in |\mathfrak{p}|$. Take the lifting $\widetilde{\varphi} : \mathbb{D} \longrightarrow \mathbb{D}$ of φ with $\widetilde{\varphi}(0) = 0$ and $\Pi \circ \widetilde{\varphi} \equiv \varphi$. Note that by the Schwarz lemma we have $|\widetilde{\varphi}(\lambda)| \leq |\lambda|, \lambda \in \mathbb{D}$. Since $\Pi(\widetilde{\varphi}(\nu_0)) = \varphi(\nu_a) = a$, we get $|\mu_a| \leq |\widetilde{\varphi}(\nu_a)| \leq |\nu_a|$, which gives the inequality "$\geq$".

The form of the extremal discs follows easily from the Schwarz lemma.

Example 8.2.15. Notice that in [267] we used a different definition, namely, we set $\boldsymbol{\ell}_G^*(\mathfrak{p}, z) := 1$ in all the cases where the function ψ does not exist. Observe that both definitions are different. The following example (cf. [381]) shows that the definition from [381] is more flexible:

Let $G := \mathbb{D}^2$, $A \subset \mathbb{D}$, $b \in \mathbb{D}_*$. Assume that A is uncountable. It is easily seen that there is no $\psi \in \mathcal{O}(\mathbb{D}, G)$ such that $\psi(0) = (0, b)$ and $A \times \{0\} \subset \psi(\mathbb{D})$. On the other hand, using [135], Theorem 2.1 (see also [396], Corollary 6), for every non-empty finite set $B \subset A$ we get $\boldsymbol{\ell}_G^*(B \times \{0\}, (0, b)) = \max\{\boldsymbol{\ell}_{\mathbb{D}}^*(B, 0), \boldsymbol{\ell}_{\mathbb{D}}^*(0, b)\} = \max\{\prod_{\mu \in B} |\mu|, |b|\}$. Hence, $\boldsymbol{\ell}_G^*(A \times \{0\}, (0, b)) = |b|$.

Example 8.2.16. The following example (cf. [381]) shows that the system $(\boldsymbol{\ell}_G^*)_G$ is not holomorphically contractible in the sense of Remark 8.2.4(m) (even for $n = 1$ and proper holomorphic coverings):

Let $\mathbb{D}_* \ni \lambda \overset{F}{\longmapsto} \lambda^2 \in \mathbb{D}_*$. Observe that F is a proper holomorphic covering. Fix an $a \in \mathbb{D}_*$. By Exercise 3.9.8, we get

$$\boldsymbol{\ell}_{\mathbb{D}_*}^*(a^2, \lambda^2) = \min\{\boldsymbol{\ell}_{\mathbb{D}_*}^*(a, \lambda),\ \boldsymbol{\ell}_{\mathbb{D}_*}^*(-a, \lambda)\} > \boldsymbol{\ell}_{\mathbb{D}_*}^*(a, \lambda)\boldsymbol{\ell}_{\mathbb{D}_*}^*(-a, \lambda), \quad \lambda \in \mathbb{D}_* \setminus \{a, -a\}.$$

Fix a $\lambda \in \mathbb{D}_* \setminus \{a, -a\}$ and let $\Pi : \mathbb{D} \longrightarrow \mathbb{D}_*$ be a holomorphic covering with $\Pi(0) = \lambda$. Then, by Exercise 3.9.8 and Remark 8.2.14(i), we have

$$\begin{aligned}\boldsymbol{\ell}^*_{\mathbb{D}_*}(-a, \lambda) \cdot \boldsymbol{\ell}^*_{\mathbb{D}_*}(a, \lambda) &= \min\{|\eta| : \eta \in \Pi^{-1}(-a)\} \cdot \min\{|\xi| : \xi \in \Pi^{-1}(a)\} \\ &= \boldsymbol{\ell}^*_{\mathbb{D}_*}(\{-a, a\}, \lambda).\end{aligned}$$

Finally, taking $\mathfrak{q} := \chi_{\{a^2\}}$ gives

$$\boldsymbol{\ell}^*_{\mathbb{D}_*}(\mathfrak{q} \circ F, \lambda) = \boldsymbol{\ell}^*_{\mathbb{D}_*}(\{a, -a\}, \lambda) < \boldsymbol{\ell}^*_{\mathbb{D}_*}(a^2, \lambda^2) = \boldsymbol{\ell}^*_{\mathbb{D}_*}(\mathfrak{q}, F(\lambda)), \qquad \lambda \in \mathbb{D}_* \setminus \{a, -a\}.$$

Remark 8.2.17 (Coman conjecture). The *Coman conjecture* says that $\boldsymbol{g}_G(\mathfrak{p}, \cdot) \equiv \boldsymbol{\ell}^*_G(\mathfrak{p}, \cdot)$ for any convex bounded domain G and function $\mathfrak{p}$ with $\#|\mathfrak{p}| < +\infty$ (cf. [108]). D. Coman proved that his conjecture is true in the case where $G = \mathbb{B}_2$ is the unit ball in $\mathbb{C}^2$, $|\mathfrak{p}| = \{a_1, a_2\}$, and $\mathfrak{p}(a_1) = \mathfrak{p}(a_2)$ (cf. [108]).

The first counterexample was given by M. Carlehed and J. Wiegerinck in [84]. Let $G = \mathbb{D}^2$, $c_1, c_2, d \in \mathbb{D}_*$, $c_1 \neq c_2$, $|c_1 c_2| < |d| < |c_1|$, $\mathfrak{p}_{2,1} := 2\chi_{(c_1,0)} + \chi_{(c_2,0)}$. Then, $\boldsymbol{g}_{\mathbb{D}^2}(\mathfrak{p}_{2,1}, (0, d)) < \boldsymbol{\ell}^*_{\mathbb{D}^2}(\mathfrak{p}_{2,1}, (0, d))$ (cf. [267], Example 1.11.3).

P. J. Thomas and N. V. Trao [499] (see also [135]) found a counterexample with $G = \mathbb{D}^2$, $\mathfrak{p} = \chi_A$, where $A := \{-a, a\} \times \{-\varepsilon, \varepsilon\}$, $a \in (0, 1)$, $a^{3/2} < \gamma < a$, $0 < \varepsilon \ll 1$. Then, $\boldsymbol{g}_{\mathbb{D}^2}(A, (0, \gamma)) < \boldsymbol{\ell}^*_{\mathbb{D}^2}(A, (0, \gamma))$ (cf. [267], Example 1.11.4).

Finally, P. J. Thomas found in [498] a "minimal counterexample" for $\mathbb{B}_2$. He proved that there exists a set $A \subset \mathbb{B}_2$ with $\#A = 3$ such that $\boldsymbol{g}_{\mathbb{B}_2}(A, z_0) < \boldsymbol{\ell}^*_{\mathbb{B}_2}(A, z_0)$ for some $z_0 \in \mathbb{B}_2$ (see Theorem 8.2.20 for more details).

Similar to Remark 4.2.14, we may introduce the following function:

Definition 8.2.18. For $\mathfrak{p} : G \longrightarrow \mathbb{R}^+$, $\mathfrak{p} \not\equiv 0$, put

$$\tilde{\boldsymbol{g}}_G(\mathfrak{p}, z) := \inf\Big\{ \prod_{\lambda \in \mathbb{D}_*} |\lambda|^{\mathfrak{p}(\varphi(\lambda)) \operatorname{ord}_\lambda(\varphi - \varphi(\lambda))} : \varphi \in \mathcal{O}(\mathbb{D}, G),\ \varphi(0) = z \Big\}, \quad z \in G.$$

As always, we put $\tilde{\boldsymbol{g}}_G(A, \cdot) := \tilde{\boldsymbol{g}}_G(\chi_A, \cdot)$.

Lemma 8.2.19.

(a)
$$\tilde{\boldsymbol{g}}_G(\mathfrak{p}, z) = \inf\Big\{ \prod_{\lambda \in \mathbb{D}_*} |\lambda|^{\mathfrak{p}(\varphi(\lambda)) \operatorname{ord}_\lambda(\varphi - \varphi(\lambda))} : \varphi \in \mathcal{O}(\overline{\mathbb{D}}, G),\ \varphi(0) = z \Big\},\ z \in G.$$

(b) $\boldsymbol{g}_G(\mathfrak{p}, \cdot) \le \tilde{\boldsymbol{g}}_G(\mathfrak{p}, \cdot)$.

Notice that in fact $\tilde{\boldsymbol{g}}_G(\mathfrak{p}, \cdot) = \boldsymbol{g}_G(\mathfrak{p}, \cdot)$ (*cf. Theorem* 17.4.3).

Proof.
(a) The proof is left to the reader.

(b) Fix $z \in G$, $\varphi \in \mathcal{O}(\overline{\mathbb{D}}, G)$ with $\varphi(0) = z$, and a finite set $B \subset \mathbb{D}_* \cap \varphi^{-1}(|\mathfrak{p}|)$. Let u be as in the definition of $\boldsymbol{g}_G(\mathfrak{p}, z)$ (cf. Definition 8.2.1). We will prove that $u(\varphi(0)) \le \prod_{\lambda \in B} |\lambda|^{\mathfrak{p}(\varphi(\lambda)) \operatorname{ord}_\lambda(\varphi - \varphi(\lambda))}$, which implies that

$$u(\varphi(0)) \le \prod_{\lambda \in \mathbb{D}_*} |\lambda|^{\mathfrak{p}(\varphi(\lambda)) \operatorname{ord}_\lambda(\varphi - \varphi(\lambda))}$$

and, consequently, $\boldsymbol{g}_G(\mathfrak{p}, \cdot) \le \tilde{\boldsymbol{g}}_G(\mathfrak{p}, \cdot)$. Let

$$v(\xi) := \log u(\varphi(\xi)) - \sum_{\lambda \in B} \mathfrak{p}(\varphi(\lambda)) \operatorname{ord}_\lambda(\varphi - \varphi(\lambda)) \log \boldsymbol{m}_{\mathbb{D}}(\lambda, \xi).$$

Then, $v \in \mathcal{SH}(\mathbb{D}(r) \setminus B)$ for some $r > 1$ and $v = \log u \circ \varphi \le 0$ on $\mathbb{T}$. Moreover, one can easily check that v is locally bounded from above in $\mathbb{D}(r)$. Hence, v extends subharmonically to $\mathbb{D}(r)$ and, by the maximum principle, $v \le 0$ on $\mathbb{D}$. In particular, $v(0) \le 0$, which gives the required inequality. □

Next, we present the example of P. Thomas in detail. Its proof is partially based on Lemma 8.2.19.

Theorem 8.2.20 (cf. [498]). *Let $\varrho : (0,1) \longrightarrow \mathbb{R}_{>0}$ be such that $t \ge \varrho(t) \underset{t \to 0}{\longrightarrow} 0$. Put*

$$S(\varepsilon) := \{(0,0), (0,\varepsilon), (\varepsilon\varrho(\varepsilon), 0)\}.$$

Then, $\boldsymbol{g}_{\mathbb{B}_2}(S(\varepsilon), \cdot) \ne \boldsymbol{\ell}^*_{\mathbb{B}_2}(S(\varepsilon), \cdot)$ if ε is sufficiently small.

The proof of this result will be based on the following two lemmata:

Lemma 8.2.21. *Let ϱ and $S(\varepsilon)$ be as in Theorem 8.2.20. Then, for any $\delta \in (0, 1/4)$ there exists a positive $r_0 = r_0(\delta)$ such that for any $z \in \mathbb{D}^2$ with $\frac{1}{2}|z_2|^{3/2} \le |z_1| \le |z_2|^{3/2}$ and $0 < \|z\| \le r_0$, one can find an $\varepsilon_0 = \varepsilon_0(z) \in (0,1)$ such that*

$$\boldsymbol{\ell}^*_{\mathbb{D}^2}(S(\varepsilon), z) \ge |z_2|^{2-\delta}, \quad \varepsilon \in (0, \varepsilon_0].$$

Proof. Suppose that the lemma is not true. Then there exists a $\delta_0 \in (0, 1/4)$ and a sequence $(z^j)_{j \in \mathbb{N}}$ of points z^j satisfying

$$\frac{|z_2^j|^{3/2}}{2} \le |z_1^j| \le |z_2^j|^{3/2} \quad \text{and} \quad 0 < \|z^j\| \le 1/j,$$

and for each j there is a sequence $(\varepsilon_{j,m})_{m \in \mathbb{N}} \subset (0,1)$ such that $\varepsilon_{j,m} < 1/m$ and $\boldsymbol{\ell}^*_{\mathbb{D}^2}(S(\varepsilon_{j,m}), z^j) < |z_2^j|^{2-\delta_0}$.

Let j_1 be such that $8j_1^{2\delta_0-1/2} < 1$. Then, fix a $j \geq j_1$ and the corresponding point z^j. Then choose an $m_{j,0}$ satisfying

- $1/m_{j,0} < \frac{|z_2^j|^{3/2}}{4}$,
- $\varrho(\varepsilon_{j,m_{j,0}}) < |z_2^j|^{1-\delta_0}$.

For these data, we have $\boldsymbol{\ell}^*_{\mathbb{D}^2}(S(\varepsilon_{j,m_0}), z^j) < |z_2^j|^{2-\delta_0}$. Then, by the definition of the Lempert function, we find an analytic disc $\widetilde{\varphi} \in \mathcal{O}(\mathbb{D}, \mathbb{D}^2)$ and points $\zeta_0, \widetilde{\zeta}_l \in \mathbb{D}$, $l = 1, 2$, with

- $\widetilde{\varphi}(0) = z^j$,
- $\widetilde{\varphi}(\zeta_0) = (0, 0)$,
- $\widetilde{\varphi}(\widetilde{\zeta}_1) = (\varepsilon_{j,m_{j,0}}\varrho(\varepsilon_{j,m_{j,0}}), 0)$,
- $\widetilde{\varphi}(\widetilde{\zeta}_2) = (0, \varepsilon_{j,m_{j,0}})$,

such that $|\zeta_0\widetilde{\zeta}_1\widetilde{\zeta}_2| < |z_2^j|^{2-\delta_0}$.

Observe that $\zeta_0\widetilde{\zeta}_1\widetilde{\zeta}_2 \neq 0$ and $\widetilde{\zeta}_1 \neq \zeta_0 \neq \widetilde{\zeta}_2 \neq \widetilde{\zeta}_1$.

Now put $\varphi(\zeta) := \widetilde{\varphi}(-\boldsymbol{h}_{\zeta_0}(\zeta))$, $\zeta \in \mathbb{D}$. Then, $\varphi \in \mathcal{O}(\mathbb{D}, \mathbb{D}^2)$ and it shares the following properties:

- $\varphi(0) = (0, 0)$,
- $\varphi(\zeta_0) = z^j$,
- $\varphi(\zeta_1) = (\varepsilon_{j,m_{j,0}}\varrho(\varepsilon_{j,m_{j,0}}), 0)$, where $\zeta_1 := \boldsymbol{h}_{-\zeta_0}(-\widetilde{\zeta}_1) \neq 0$,
- $\varphi(\zeta_2) = (0, \varepsilon_{j,m_{j,0}})$, where $\zeta_2 := \boldsymbol{h}_{-\zeta_0}(-\widetilde{\zeta}_2) \neq 0$.

Moreover, we have

$$|\zeta_0\boldsymbol{h}_{\zeta_0}(\zeta_1)\boldsymbol{h}_{\zeta_0}(\zeta_2)| < |z_2^j|^{2-\delta_0}. \tag{8.2.3}$$

Note that φ_1, respectively φ_2, is vanishing at the points 0 and $\zeta_2 \neq 0$, respectively at 0 and $\zeta_1 \neq 0$. Therefore, there are functions $\psi_j \in \mathcal{O}(\mathbb{D}, \overline{\mathbb{D}})$, $j = 1, 2$, such that

$$\varphi_1(\zeta) = \zeta\boldsymbol{h}_{\zeta_2}(\zeta)\psi_1(\zeta), \quad \varphi_2(\zeta) = \zeta\boldsymbol{h}_{\zeta_1}(\zeta)\psi_2(\zeta), \quad \zeta \in \mathbb{D}.$$

Using this description, one obtains

- $\psi_1(\zeta_1) = \frac{\varepsilon_{j,m_{j,0}}\varrho(\varepsilon_{j,m_{j,0}})}{\zeta_1\boldsymbol{h}_{\zeta_2}(\zeta_1)} =: w_{1,1}$,
- $\psi_1(\zeta_0) = \frac{z_1^j}{\zeta_0\boldsymbol{h}_{\zeta_2}(\zeta_0)} =: w_{1,0} \neq 0$,
- $\psi_2(\zeta_2) = \frac{\varepsilon_{j,m_{j,0}}}{\zeta_2\boldsymbol{h}_{\zeta_1}(\zeta_2)} =: w_{2,2} \neq 0$,

- $\psi_2(\zeta_0) = \frac{z_2^j}{\zeta_0 \boldsymbol{h}_{\zeta_1}(\zeta_0)} =: w_{2,0} \neq 0$.

Then, we have the following two alternatives:

(a) either $|w_{1,1}| = |w_{1,0}| = 1$ or

$$\psi_1 \in \mathcal{O}(\mathbb{D}, \mathbb{D}) \text{ (in particular, } \max\{|w_{1,0}|, |w_{1,1}|\} < 1);$$

(b) either $|w_{2,2}| = |w_{2,0}| = 1$ or

$$\psi_2 \in \mathcal{O}(\mathbb{D}, \mathbb{D}) \text{ (in particular, } \max\{|w_{2,2}|, |w_{2,0}|\} < 1).$$

In other words, we have

(a) either $|w_{1,1}| = |w_{1,0}| = 1$ or $\boldsymbol{m}_{\mathbb{D}}(w_{1,0}, w_{1,1}) \le \boldsymbol{m}_{\mathbb{D}}(\zeta_0, \zeta_1)$;

(b) either $|w_{2,2}| = |w_{2,0}| = 1$ or $\boldsymbol{m}_{\mathbb{D}}(w_{2,0}, w_{2,2}) \le \boldsymbol{m}_{\mathbb{D}}(\zeta_0, \zeta_2)$.

Using the above equations, it follows that

$$\begin{aligned}\max\left\{\frac{1}{|w_{j,0}|} : j = 1, 2\right\} &\le \frac{1}{|w_{1,0} w_{2,0}|} = \frac{|\zeta_0 \boldsymbol{h}_{\zeta_2}(\zeta_0) \zeta_0 \boldsymbol{h}_{\zeta_1}(\zeta_0)|}{|z_1^j z_2^j|} \\ &\overset{(8.2.3)}{\le} \frac{|\zeta_0||z_2^j|^{2-\delta_0}}{|z_1^j z_2^j|} \le 2\frac{|\zeta_0||z_2^j|^{2-\delta_0}}{|z_2^j|^{3/2}|z_2^j|} = 2|\zeta_0||z_2^j|^{-\delta_0 - 1/2}\end{aligned}$$

or

$$\frac{|z_2^j|^{\delta_0 + 1/2}}{2|\zeta_0|} \le |w_{1,0} w_{2,0}| \le \min\{|w_{1,0}|, |w_{2,0}|\} \le 1. \tag{8.2.4}$$

Moreover, we have

$$|\boldsymbol{h}_{\zeta_2}(\zeta_0)| = \frac{|z_1^j|}{|w_{1,0}\zeta_0|} \le 2|z_2^j|^{3/2-\delta_0-1/2} = 2|z_2^j|^{1-\delta_0}, \tag{8.2.5}$$

$$\begin{aligned}2|z_2^j|^{1-\delta_0} &\overset{(8.2.4)}{\le} |w_{2,0}| 2|\zeta_0||z_2^j|^{1-\delta_0-\delta_0-1/2} = 2|w_{2,0}\zeta_0||z_2^j|^{1/2-2\delta_0} \\ &\le \min\{|\zeta_0|, |w_{2,0}|\}/2,\end{aligned} \tag{8.2.6}$$

where the condition on $j \ge j_1$ has been used. Hence, we have

$$|\boldsymbol{h}_{\zeta_2}(\zeta_0)| \le \tfrac{1}{2} \min\{|\zeta_0|, |w_{2,0}|\}.$$

In the case where $\psi_2 \in \mathcal{O}(\mathbb{D}, \mathbb{D})$, it follows that

$$|w_{2,0}| \le \boldsymbol{m}_{\mathbb{D}}(0, w_{2,2}) + \boldsymbol{m}_{\mathbb{D}}(w_{2,0}, w_{2,2}) \le |w_{2,2}| + |\boldsymbol{h}_{\zeta_2}(\zeta_0)| \le |w_{2,2}| + \frac{1}{2}|w_{2,0}|,$$

which gives $\frac{|w_{2,0}|}{2} \le |w_{2,2}|$. Note that the last inequality remains true if $\psi_2 \in \mathcal{O}(\mathbb{D}, \mathbb{T})$.

Moreover, we have

$$\frac{|\zeta_0|}{2} \geq |\boldsymbol{h}_{\zeta_2}(\zeta_0)| \geq \boldsymbol{m}_{\mathbb{D}}(\zeta_0, 0) - \boldsymbol{m}_{\mathbb{D}}(0, \zeta_2) = |\zeta_0| - |\zeta_2|,$$
$$|\zeta_2| \leq |\zeta_0| + |\boldsymbol{h}_{\zeta_2}(\zeta_0)| \leq \tfrac{3}{2}|\zeta_0|,$$

which implies that $\frac{|\zeta_0|}{2} \leq |\zeta_2| \leq \frac{3}{2}|\zeta_0|$.

Now we start to discuss an estimate for $|\boldsymbol{h}_{\zeta_1}(\zeta_2)|$:

$$|\boldsymbol{h}_{\zeta_1}(\zeta_2)| = \frac{\varepsilon_{j,m_{j,0}}}{|\zeta_2 w_{2,2}|} \leq 4\frac{\varepsilon_{j,m_{j,0}}}{|\zeta_0 w_{2,0}|} \overset{(8.2.6)}{\leq} \frac{4\varepsilon_{j,m_{j,0}}}{|z_2^j|^{\delta_0+1/2}}.$$

Now, using the choice of $m_{j,0}$ one gets $|\boldsymbol{h}_{\zeta_1}(\zeta_2)| \leq |z_2^j|^{3/2-\delta_0-1/2} = |z_2^j|^{1-\delta_0}$.

Finally, applying the triangle inequality, we put the information that we gained so far together and get

$$|\boldsymbol{h}_{\zeta_1}(\zeta_0)| = \boldsymbol{m}_{\mathbb{D}}(\zeta_0, \zeta_1) \leq |\boldsymbol{h}_{\zeta_0}(\zeta_2)| + |\boldsymbol{h}_{\zeta_1}(\zeta_2)| \overset{(8.2.5)}{\leq} 3|z_2^j|^{1-\delta_0}.$$

Now we are able to get two estimates for $w_{1,0}$, which shall finally lead to a contradiction.

Recall that $|w_{1,0}| \geq \frac{1}{2}|z_2^j|^{1/2+\delta_0}$ (see (8.2.4)). On the other hand, the triangle inequality for $\boldsymbol{m}_{\mathbb{D}}$ gives

$$|\zeta_1| \geq |\zeta_2| - |\boldsymbol{h}_{\zeta_1}(\zeta_2)| \geq |\zeta_0|/2 - |z_2^j|^{1-\delta_0}$$
$$\overset{(8.2.4)}{\geq} |\zeta_0|/2 - 2|\zeta_0||z_2^j|^{1-\delta_0-(\delta_0+1/2)} = |\zeta_0|/2 - 2|\zeta_0||z_2^j|^{1/2-2\delta_0} \geq |\zeta_0|/4,$$

using the choice of j_1. Hence, we have

$$|w_{1,1}| \leq \frac{|w_{1,1}|}{|w_{2,2}|} \leq \varrho(\varepsilon_{j,m_{j,0}})\frac{|\zeta_2|}{|\zeta_1|} \leq 6\varrho(\varepsilon_{j,m_{j,0}}) \leq |z_2^j|^{1-\delta_0} < 1.$$

Therefore, $\psi_1 \in \mathcal{O}(\mathbb{D}, \mathbb{D})$ and so the triangle inequality for $\boldsymbol{m}_{\mathbb{D}}$ leads to

$$\tfrac{1}{2}|z_2^j|^{1/2+\delta_0} \leq |w_{1,0}| \leq |w_{1,1}| + |\boldsymbol{h}_{\zeta_1}(\zeta_0)| \leq |z_2^j|^{1-\delta_0} + 3|z_2^j|^{1-\delta_0} = 4|z_2^j|^{1-\delta_0}.$$

Combining both estimates for $w_{1,0}$ we get $16|z_2^j|^{1-\delta_0} \geq |z_2^j|^{1/2+\delta_0}/2$, which is definitely not true for large j; a contradiction. □

Lemma 8.2.22. *Let ϱ and $S(\varepsilon)$ be as in Theorem* 8.2.20 *and let $\delta, \eta < 1/2$ be positive numbers with $(2\delta)^{2/3} < (1-\eta)^2 - \eta$. Then there exists an $\varepsilon_1 = \varepsilon_1(\delta, \eta) > 0$ such that*

$$g_{\mathbb{D}^2}(S(\varepsilon), z) \leq 8|z_2|^2, \quad \varepsilon \leq \varepsilon_1, \quad \delta \leq \tfrac{1}{2}|z_2|^{3/2} \leq |z_1| \leq |z_2|^{3/2} < 1,$$
$$|z_2| < (1-\eta)^2 - \eta.$$

Proof. The idea is to find an analytic disc passing through all the poles given by $S(\varepsilon)$, and twice through one of them, to get an upper estimate for $\widetilde{\boldsymbol{g}}_{\mathbb{D}^2}(S(\varepsilon), z)$. Fix δ and η and take $\eta' > 0$ and $k \in \mathbb{N}$, $k \geq 13$, such that $(1+\eta')^3 = 1+\eta$ and $\frac{k}{k-1} \leq 1+\eta'$. Finally, put $\varepsilon_1 := \frac{1}{18}\min\{\frac{\eta}{k}(2\delta)^{2/3}, \eta^2(1-2\eta)\}$. Now let $\varepsilon \in (0, \varepsilon_1)$ and $z \in \mathbb{D}^2$ with $\delta \leq \frac{1}{2}|z_2|^{3/2} \leq |z_1| \leq |z_2|^{3/2}$ and $|z_2| < (1-\eta)^2 - \eta$. Then,

$$\begin{aligned} 0 < |z_2|/2 \leq |z_2|(1-1/k) \leq |z_2| - \varepsilon_1 \leq |z_2 - \varepsilon| \\ \leq |z_2| + \varepsilon_1 \leq |z_2|(1+1/k) \leq 2|z_2|, \end{aligned}$$

and therefore $\frac{|z_2|}{|z_2-\varepsilon|} \leq \frac{k}{k-1} \leq 1+\eta'$.

Choose a $\lambda \in \mathbb{C}$ such that

$$\lambda^2 = \frac{z_1}{z_2(z_2-\varepsilon)}\left(\frac{z_1}{z_2-\varepsilon} + \varrho(\varepsilon)\right).$$

Then,

$$|\lambda|^2 \geq \frac{|z_1|}{|z_2(z_2-\varepsilon)|}\left(\frac{|z_1|}{|z_2-\varepsilon|} - \varrho(\varepsilon)\right) \geq \frac{|z_1|}{2|z_2|^2}\left(\frac{|z_1|}{2|z_2|} - \varrho(\varepsilon)\right) \geq \frac{|z_1|^2}{8|z_2|^3} \geq \frac{1}{32};$$

in particular, $\lambda \neq 0$. Moreover, we have

$$|\lambda|^2 \leq \frac{|z_1|}{|z_2(z_2-\varepsilon)|}\left(\frac{|z_1|}{|z_2-\varepsilon|} + \varrho(\varepsilon)\right) \leq (1+\eta')^2\frac{|z_1|^2}{|z_2|^3}\left(1 + \frac{|z_2|\varrho(\varepsilon)}{|z_1|(1+\eta')}\right).$$

Applying the choice of ε_1 leads finally to $|\lambda|^2 \leq (1+\eta')^3 = 1+\eta$.

Take a $\mu \in \mathbb{C}$ such that $\mu^2 = \varepsilon + \left(\frac{\varrho(\varepsilon)}{2\lambda}\right)^2$. Then,

$$|\mu^2| \leq \varepsilon + \frac{\varrho(\varepsilon)^2}{4|\lambda|^2} \leq \varepsilon + 8\varrho(\varepsilon)^2 \leq 9\varepsilon < (1-\eta)^2/2.$$

Put

$$\psi(\zeta) = \psi_{\lambda,\mu}(\zeta) := \left((\lambda\zeta - \varrho(\varepsilon)/2)(\zeta^2 - \mu^2), \zeta^2 - \left(\frac{\varrho(\varepsilon)}{2\lambda}\right)^2\right).$$

Note that this analytic disc is a modification of the Neil parabola. We have

$$\psi(\mu) = \psi(-\mu) = (0,\varepsilon), \quad \psi\left(\frac{\varrho(\varepsilon)}{2\lambda}\right) = (0,0), \quad \psi\left(-\frac{\varrho(\varepsilon)}{2\lambda}\right) = (\varrho(\varepsilon)\varepsilon, 0),$$

in particular, ψ passes through all the poles and even twice through $(0,\varepsilon)$. Observe that $\left|\frac{\varepsilon}{2\lambda}\right| < 1-\eta$.

Put $\zeta_z := \frac{1}{\lambda}\left(\frac{z_1}{z_2-\varepsilon} + \frac{\varrho(\varepsilon)}{2}\right)$. Then $\psi(\zeta_z) = z$. Moreover, there is the following estimate:

$$|\zeta_z^2 - z_2| = |z_2|\frac{|z_2-\varepsilon|\varrho(\varepsilon)^2}{4|z_1||\frac{z_1}{z_2-\varepsilon}+\varrho(\varepsilon)|} \le \frac{|z_2|^{1/2}\varrho(\varepsilon)^2}{|\frac{|z_2|^{1/2}}{4} - \varrho(\varepsilon)|}$$
$$\le \frac{4|z_2|^{1/2}\varrho(\varepsilon)^2}{|(2\delta)^{1/3} - 4\varrho(\varepsilon)|} \le \frac{8|z_2|^{1/2}\varrho(\varepsilon)^2}{(2\delta)^{1/3}} < \eta,$$

or $|\zeta_z|^2 \le |z_2| + 8\frac{\varrho(\varepsilon)^2}{(2\delta)^{1/3}} \le |z_2| + \eta < (1-\eta)^2 < 1-\eta$.

Now we discuss the image of our disc ψ for $|\zeta| < 1-\eta$. We get

$$|\psi_2(\zeta)| \le (1-\eta)^2 + \frac{\varrho(\varepsilon)^2}{4|\lambda|^2} \le (1-\eta)^2 + 8\varrho(\varepsilon)^2 < 1 \quad \text{and}$$
$$|\psi_1(\zeta)| \le (|\lambda|(1-\eta) + \varrho(\varepsilon)/2)\left((1-\eta)^2 + 10\varepsilon\right)$$
$$\le ((1-\eta)(1+\eta)+\varepsilon)\left((1-\eta)^2 + 10\varepsilon\right) < 1.$$

Put $\widetilde{\psi}(\zeta) := \psi((1-\eta)\zeta)$. Then, $\widetilde{\psi} \in \mathcal{O}(\mathbb{D}^2)$. Let

$$\widetilde{\zeta}_z := \zeta_z/(1-\eta), \quad \widetilde{\mu} := \mu/(1-\eta), \quad \chi := \frac{\varrho(\varepsilon)}{(1-\eta)2\lambda}.$$

Consequently, we get the following upper estimate:

$$\widetilde{\boldsymbol{g}}_{\mathbb{D}^2}(S(\varepsilon), z) \le \boldsymbol{m}_{\mathbb{D}}(\widetilde{\zeta}_z, \widetilde{\mu})\boldsymbol{m}_{\mathbb{D}}(\widetilde{\zeta}_z, -\widetilde{\mu})\boldsymbol{m}_{\mathbb{D}}(\widetilde{\zeta}_z, \chi)\boldsymbol{m}_{\mathbb{D}}(\widetilde{\zeta}_z, -\chi), \quad \text{or}$$
$$\widetilde{\boldsymbol{g}}_{\mathbb{D}^2}(S(\varepsilon), z) \le (1-\eta)^4\left|\frac{\zeta_z^2 - \mu^2}{(1-\eta)^2 - \zeta_z^2\overline{\mu}^2}\right| \cdot \left|\frac{\zeta_z^2 - \frac{\varrho(\varepsilon)^2}{4\lambda^2}}{(1-\eta)^2 - \zeta_z^2\frac{\varrho(\varepsilon)^2}{4\overline{\lambda}^2}}\right|$$
$$= (1-\eta)^4\left|\frac{z_2-\varepsilon}{(1-\eta)^2 - \zeta_z^2\overline{\mu}^2}\right| \cdot \left|\frac{z_2}{(1-\eta)^2 - \zeta_z^2\frac{\varrho(\varepsilon)^2}{4\overline{\lambda}^2}}\right|$$
$$\le \frac{|z_2|+\varepsilon}{1-|\mu|^2} \cdot \frac{|z_2|}{1-\frac{\varrho(\varepsilon)^2}{4|\lambda^2|}} \le \frac{2|z_2|}{1-9\varepsilon} \cdot \frac{|z_2|}{1-8\varepsilon^2} \le 8|z_2|^2. \qquad \square$$

Proof of Theorem 8.2.20. Define $\widetilde{\varrho}(\varepsilon) = \varrho(\varepsilon/\sqrt{2})$ and fix a $\widetilde{\delta} \in (0, 1/4)$. For ϱ and $\widetilde{\delta}$ choose $r_0 = r_0(\widetilde{\delta}) \in (0, 1/\sqrt{2})$ according to Lemma 8.2.21. Moreover, fix an $\eta \in (0,1)$ such that $r_0 < (1-\eta)^2 - \eta$. Now we take a point z^0 with $|z_1^0| = |z_2^0|^{3/2}$, $0 < \|z^0\| < r_0$, and $|z_2^0|^{\widetilde{\delta}} < 1/16$. Let $\varepsilon_0 = \varepsilon_0(z^0)$ be the corresponding number from Lemma 8.2.21.

Now applying Lemma 8.2.22 for $\widetilde{\varrho}$, the point $\sqrt{2}z^0$, and a $\delta < |\sqrt{2}z_2^0|^{3/2}$ leads to an $\varepsilon_1 = \varepsilon_1(\eta, \delta)$. Then, for $\varepsilon < \min\{\varepsilon_0, \varepsilon_1\}$, one obtains

$$\begin{aligned} \boldsymbol{g}_{\mathbb{B}_2}(S(\varepsilon), z^0) &\le \boldsymbol{g}_{\frac{1}{\sqrt{2}}\mathbb{D}^2}(S(\varepsilon), z^0) \le \boldsymbol{g}_{\mathbb{D}^2}(\sqrt{2}S(\varepsilon), \sqrt{2}z^0) \\ &\le 16|z_2^0|^2 < |z_2^0|^{2-\widetilde{\delta}} \le \boldsymbol{\ell}^*_{\mathbb{D}^2}(S(\varepsilon), z^0) \le \boldsymbol{\ell}^*_{\mathbb{B}_2}(S(\varepsilon), z^0), \end{aligned}$$

which completes the proof. □

In the remaining part of this section we present different examples in order to give some ideas as to what is going on with these new functions in concrete situations.

Example 8.2.23. Define

$$\mathcal{E}(p) := \Big\{(z_1, \dots, z_n) \in \mathbb{C}^n : \sum_{j=1}^{n} |z_j|^{2p_j} < 1\Big\}, \; p = (p_1, \dots, p_n) \in \mathbb{R}^n_{>0}, \; n \ge 2;$$

see § 16.4 for an extended discussion. Fix $(\nu_1, \dots, \nu_n) \in \mathbb{N}^n$. The mapping

$$\mathcal{E}(p) \ni (z_1, \dots, z_n) \overset{F}{\longmapsto} (z_1^{\nu_1}, \dots, z_n^{\nu_n}) \in \mathcal{E}((p_1/\nu_1, \dots, p_n/\nu_n))$$

is proper. Let $(a_1, \dots, a_n) \in \mathcal{E}(p)$ be such that $\det F'(a) \ne 0$ (i.e., $a_j^{\nu_j - 1} \ne 0$, $j = 1, \dots, n$) and let

$$A := F^{-1}(F(a)) = \{(\varepsilon_1 a_1, \dots, \varepsilon_n a_n) : \varepsilon_j \in \sqrt[\nu_j]{1}, \; j = 1, \dots, n\}.$$

Then, by Proposition 8.2.10,

$$\boldsymbol{g}_{\mathcal{E}(p)}(A, z) = \boldsymbol{g}_{\mathcal{E}((p_1/\nu_1, \dots, p_n/\nu_n))}(F(a), F(z)), \quad z \in \mathcal{E}(p);$$

i.e., the multi-pole pluricomplex Green function for $\mathcal{E}(p)$ is expressed by the standard one-pole pluricomplex Green function for $\mathcal{E}((p_1/\nu_1, \dots, p_n/\nu_n))$.

Notice that for some special cases $\boldsymbol{g}_{\mathcal{E}((p_1/\nu_1, \dots, p_n/\nu_n))}(F(a), F(z))$ may be effectively calculated, e.g.:

- If $\nu_j = p_j \in \mathbb{N}$, $j = 1, \dots, n$, then (cf. Corollary 2.3.5) we have

$$\begin{aligned} \boldsymbol{g}_{\mathcal{E}(p)}(A, z) &= \boldsymbol{g}_{\mathbb{B}_n}(F(a), F(z)) = \boldsymbol{m}_{\mathbb{B}_n}(F(a), F(z)) \\ &= \left(1 - \frac{(1 - \sum_{j=1}^n |a_j|^{2p_j})(1 - \sum_{j=1}^n |z_j|^{2p_j})}{|1 - \sum_{j=1}^n (z_j \bar{a}_j)^{p_j}|^2}\right)^{1/2}, \quad z \in \mathcal{E}(p); \end{aligned}$$

- Let $n = 2$, $\nu_1 = 1$, $\nu_2 = 2$, $a = (0, s)$ ($s \in (0, 1)$). Then $A = \{(0, -s), (0, s)\}$ and we get

$$\boldsymbol{g}_{\mathbb{B}_2}(\{(0, -s), (0, s)\}, (z_1, z_2)) = \boldsymbol{g}_{\mathcal{E}((1, 1/2))}((0, s^2), (z_1, z_2^2)).$$

An effective formula $\boldsymbol{m}_{\mathcal{E}((1,1/2))}((0,b),(X,Y))$ will be found in Example 16.6.1. Using this formula, we get

$$g_{\mathcal{E}((1,1/2))}((0,s^2),(z_1,z_2^2)) = \begin{cases} \frac{|z_1|}{\sqrt{1-s^2}}, & \text{if } z_2 = 0 \\ \frac{|z_2|^2}{2s^2}\frac{\sqrt{(1-s^2)v+s^2}}{1-s^2}, & \text{if } v \geq 1 \\ \frac{|z_2|^2}{2s^2}\frac{2(1-t)}{1+t}\frac{\alpha}{1-\alpha^2}, & \text{if } v < 1 \end{cases},$$

where $v := \left(\frac{2s^2|z_1|}{|z_2|^2}\right)^2$ and if $v < 1$, then

$$t := \frac{v}{2-v+2\sqrt{1-v}}, \quad \alpha := \frac{1}{2}\left((1-t)s^2 + \sqrt{(1-t)^2s^4+4t}\right);$$

cf. [156] (see also [108] for a different approach).

Example 8.2.24. Let $\mathbb{B}_2 \ni (z,w) \overset{F}{\longmapsto} (z,w^2) \in \mathcal{E}((1,1/2))$, $a := (0,0)$. Then, $\det F'(0) = 0$ and $g_{\mathbb{B}_2}(0,\cdot) \not\equiv g_{\mathcal{E}((1,1/2))}(0,F(\cdot))$.

Indeed,

$$g_{\mathbb{B}_2}((0,0),(z,w)) = \mathfrak{h}_{\mathbb{B}_2}(z,w) = \sqrt{|z|^2+|w|^2},$$

$$g_{\mathcal{E}((1,1/2))}((0,0),(z,w)) = \mathfrak{h}_{\mathcal{E}((1,1/2))}(z,w) = \frac{1}{2}\left(|w| + \sqrt{4|z|^2+|w|^2}\right),$$

where $\mathfrak{h}_D$ is the Minkowski function. In particular, for small $t > 0$, we get

$$g_{\mathbb{B}_2}((0,0),(t,t)) = t\sqrt{2}, \quad g_{\mathcal{E}((1,1/2))}((0,0),(t,t^2)) \approx t,$$

which implies the required result.

Example 8.2.25 (cf. [83]). Let

$$T := \{(z_1,z_2) \in \mathbb{D}_* \times \mathbb{D} : |z_2| < |z_1|\}$$

be the Hartogs triangle. Let $\mathfrak{p} : T \longrightarrow \mathbb{R}_+$. Consider the biholomorphism

$$\mathbb{D}_* \times \mathbb{D} \ni (z_1,z_2) \overset{F}{\longmapsto} (z_1, z_1z_2) \in T.$$

The set $\mathbb{D}^2 \setminus (\mathbb{D}_* \times \mathbb{D})$ is pluripolar. Hence, by Remark 8.2.4(m, p),

$$g_T(\mathfrak{p}, F(z)) = g_{\mathbb{D}_*\times\mathbb{D}}(\mathfrak{p}\circ F, z) = g_{\mathbb{D}^2}(\mathfrak{p}', z), \quad z \in \mathbb{D}_* \times \mathbb{D},$$

where $\mathfrak{p}' := \mathfrak{p} \circ F$ on $\mathbb{D}_* \times \mathbb{D}$ and $\mathfrak{p}' := 0$ on $\{0\} \times \mathbb{D}$. In particular,

$$g_T(a,z) = \max\{\boldsymbol{m}_{\mathbb{D}}(a_1,z_1), \boldsymbol{m}_{\mathbb{D}}(a_2/a_1, z_2/z_1)\},$$
$$a = (a_1,a_2),\ z = (z_1,z_2) \in T.$$

Example 8.2.26. For any non-empty sets $A_1, \dots, A_n \subset \mathbb{D}$ we have

$$\begin{aligned} \boldsymbol{m}_{\mathbb{D}^n}(A_1 \times \dots \times A_n, z) &= \boldsymbol{g}_{\mathbb{D}^n}(A_1 \times \dots \times A_n, z) = \max\{\boldsymbol{m}_{\mathbb{D}}(A_1, z_1), \dots, \boldsymbol{m}_{\mathbb{D}}(A_n, z_n)\} \\ &= \max\Big\{\prod_{a_j \in A_j} \boldsymbol{m}_{\mathbb{D}}(a_j, z_j) : j = 1, \dots, n\Big\}, \quad z = (z_1, \dots, z_n) \in \mathbb{D}^n. \end{aligned}$$

In particular, for any non-empty set $A \subset \mathbb{D}$ we have

$$\begin{aligned} \boldsymbol{m}_{\mathbb{D}^n}(A \times \{0\}^{n-1}, z) &= \boldsymbol{g}_{\mathbb{D}^n}(A \times \{0\}^{n-1}, z) \\ &= \max\{\boldsymbol{m}_{\mathbb{D}}(A, z_1), |z_2|, \dots, |z_n|\}, \quad z = (z_1, \dots, z_n) \in \mathbb{D}^n; \end{aligned}$$

cf. Example 8.2.27.

Indeed, by Propositions 8.2.6, 8.2.8 we may assume that $A_1, \dots, A_n$ are finite. Let

$$F_j(\lambda) := \prod_{a \in A_j} \frac{\lambda - a}{1 - \overline{a}\lambda}, \quad \lambda \in \mathbb{D}, \ j = 1, \dots, n,$$

be the corresponding Blaschke products. The mapping

$$\mathbb{D}^n \ni (z_1, \dots, z_n) \overset{F}{\longmapsto} (F_1(z_1), \dots, F_n(z_n)) \in \mathbb{D}^n$$

is proper. Moreover, $\det F'(z) = F_1'(z_1) \cdots F_n'(z_n) \neq 0$, $z \in A_1 \times \dots \times A_n$. Consequently, by Proposition 8.2.10,

$$\begin{aligned} \boldsymbol{m}_{\mathbb{D}^n}(A_1 \times \dots \times A_n, z) &= \boldsymbol{g}_{\mathbb{D}^n}(A_1 \times \dots \times A_n, z) = \boldsymbol{g}_{\mathbb{D}^n}(0, F(z)) = \max\{|F_j(z_j)| : j = 1, \dots, n\} \\ &= \max\{\boldsymbol{m}_{\mathbb{D}}(A_1, z_1), \dots, \boldsymbol{m}_{\mathbb{D}}(A_n, z_n)\}, \quad z = (z_1, \dots, z_n) \in \mathbb{D}^n. \end{aligned}$$

Example 8.2.27 (cf. [84]). Let $\mathfrak{p} : \mathbb{D}^n \longrightarrow \mathbb{R}_+$ be such that

$$|\mathfrak{p}| = \{a_1, \dots, a_N\} \subset \mathbb{D} \times \{0\}^{n-1}.$$

Put $a_j = (c_j, 0, \dots, 0)$, $k_j := \mathfrak{p}(a_j)$, $j = 1, \dots, N$, and assume that $k_1 \geq \dots \geq k_N$. Then,

$$\boldsymbol{g}_{\mathbb{D}^n}(\mathfrak{p}, z) = \prod_{j=1}^{N} u_j^{k_j - k_{j+1}}(z), \quad z \in \mathbb{D}^n,$$

where $k_{N+1} := 0$ and

$$\begin{aligned} u_j(z) :&= \max\{\boldsymbol{m}_{\mathbb{D}}(c_1, z_1) \cdots \boldsymbol{m}_{\mathbb{D}}(c_j, z_1), \ |z_2|, \dots, |z_n|\} \\ &= \max\{\boldsymbol{m}_{\mathbb{D}}(\{c_1, \dots, c_j\}, z_1), \ |z_2|, \dots, |z_n|\} \\ &= \boldsymbol{m}_{\mathbb{D}^n}(\{a_1, \dots, a_j\}, z), \quad j = 1, \dots, N. \end{aligned}$$

Moreover, if $k_1,\dots,k_N \in \mathbb{N}$, then $\boldsymbol{m}_{\mathbb{D}^n}(\mathfrak{p},\cdot) = \boldsymbol{g}_{\mathbb{D}^n}(\mathfrak{p},\cdot)$. Observe that if $k_1 = \dots = k_N = 1$, then the above formula coincides with that from Example 8.2.26.

Indeed, let $u := \prod_{j=1}^N u_j^{k_j - k_{j+1}}$. Note that $\log u$ is plurisubharmonic. Take $1 \le s \le N$ and $z = (z_1,\dots,z_n)$ in a small neighborhood of a_s. Then, for $j = s,\dots,N$, we get

$$u_j(z) \le \max\{\text{const}\,|z_1 - c_s|, |z_2|,\dots,|z_n|\} \le \text{const}\,\|z - a_s\|.$$

Consequently,

$$u(z) \le \text{const} \prod_{j=s}^N u_j^{k_j - k_{j+1}}(z) \le \prod_{j=s}^N \|z - a_s\|^{k_j - k_{j+1}} = \text{const}\,\|z - a_s\|^{k_s}.$$

Thus, $\boldsymbol{g}_{\mathbb{D}^n}(\mathfrak{p},\cdot) \ge u$. To get the opposite inequality, we first reduce the proof to the case $n = 2$. Suppose that the result holds for $n = 2$ and consider the general case $n \ge 3$. Take a point $b = (b_1,\dots,b_n) \in \mathbb{D}^n \setminus |\mathfrak{p}|$ and let $\max\{|b_2|,\dots,|b_n|\} = |b_{s_0}|$. If $b_{s_0} = 0$ (i.e., $b_2 = \dots = b_n = 0$), then consider the mapping

$$\mathbb{D} \ni \lambda \overset{F}{\longmapsto} (\lambda, 0,\dots,0) \in \mathbb{D}^n$$

and use Remark 8.2.4(m):

$$\boldsymbol{g}_{\mathbb{D}^n}(\mathfrak{p}, b) = \boldsymbol{g}_{\mathbb{D}^n}(\mathfrak{p}, F(b_1)) \le \boldsymbol{g}_{\mathbb{D}}(\mathfrak{p}\circ F, b_1) = \prod_{j=1}^N (\boldsymbol{m}_{\mathbb{D}}(c_j, b_1))^{k_j} = u(b).$$

If $b_{s_0} \ne 0$, then let $q_s := b_s / b_{s_0} \in \overline{\mathbb{D}}$, $s = 2,\dots,n$. Consider the mapping

$$\mathbb{D}^2 \ni (\lambda, \xi) \overset{F}{\longmapsto} (\lambda, q_1\xi,\dots,q_n\xi) \in \mathbb{D}^n.$$

Using Remark 8.2.4(m) and the case $n = 2$, we get

$$\boldsymbol{g}_{\mathbb{D}^n}(\mathfrak{p}, b) = \boldsymbol{g}_{\mathbb{D}^n}(\mathfrak{p}, F(b_1, b_{s_0})) \le \boldsymbol{g}_{\mathbb{D}^2}(\mathfrak{p}\circ F, (b_1, b_{s_0})) \tag{8.2.7}$$

$$= \prod_{j=1}^N (\max\{\boldsymbol{m}_{\mathbb{D}}(c_1, b_1)\cdots \boldsymbol{m}_{\mathbb{D}}(c_j, b_1), |b_{s_0}|\})^{k_j - k_{j+1}} = u(b). \tag{8.2.8}$$

Now, let $n = 2$. Observe that u is continuous on $\overline{\mathbb{D}}^2$ and $u = 1$ on $\partial(\mathbb{D}^2)$. Put

$$r_s(\lambda) := \prod_{j=1}^s \boldsymbol{m}_{\mathbb{D}}(c_j, \lambda), \quad M_s(\lambda) := \prod_{j=1}^s (\boldsymbol{m}_{\mathbb{D}}(c_j, \lambda))^{k_j}, \quad s = 1,\dots,N.$$

Note that $r_N(\lambda) < r_{N-1}(\lambda) < \cdots < r_1(\lambda) < 1$, $\lambda \in \mathbb{D} \setminus \{c_1, \dots, c_N\}$. Observe that

$$u(z_1, z_2) = \begin{cases} |z_2|^{k_1}, & \text{if } r_1(z_1) \le |z_2| < 1 \\ \left(\prod_{j=1}^{s} (\boldsymbol{m}_{\mathbb{D}}(c_j, z_1))^{k_j - k_{s+1}}\right) |z_2|^{k_{s+1}}, & \text{if } r_{s+1}(z_1) \le |z_2| \le r_s(z_1), \\ & s = 1, \dots, N-1 \\ M_N(z_1), & \text{if } |z_2| \le r_N(z_1) \end{cases}.$$

Moreover, $u(z_1, z_2) = M_s(z_1)$ if $|z_2| = r_s(z_1)$, $s = 1, \dots, N$. Using the maximum principle for psh functions and the Hadamard three circles theorem (Appendix B.4.25), we can easily check that it suffices to show that $\boldsymbol{g}_{\mathbb{D}^2}(\mathfrak{p}, (z_1, z_2)) \le M_s(z_1)$ if $|z_2| = r_s(z_1)$, $z_1 \in \mathbb{D} \setminus \{c_1, \dots, c_N\}$, $s = 1, \dots, N$. Thus, we only need to prove that for any $s \in \{1, \dots, N\}$ and any $h_j \in \operatorname{Aut}(\mathbb{D})$ with $h_j(c_j) = 0$, $j = 1, \dots s$, we have $\boldsymbol{g}_{\mathbb{D}^2}(\mathfrak{p}, (z_1, z_2)) \le M_s(z_1)$ if $z_2 = h_1(z_1) \cdots h_s(z_1)$. Consider the mapping

$$\mathbb{D} \ni \lambda \overset{F}{\longmapsto} (\lambda, h_1(\lambda) \cdots h_s(\lambda)) \in \mathbb{D}^2$$

and apply Remark 8.2.4(m) with $\mathfrak{p}_s := \mathfrak{p} \cdot \chi_{\{a_1, \dots, a_s\}}$:

$$\begin{aligned} \boldsymbol{g}_{\mathbb{D}^2}(\mathfrak{p}, (z_1, h_1(z_2) \cdots h_s(z_1))) &= \boldsymbol{g}_{\mathbb{D}^2}(\mathfrak{p}, F(z_1)) \le \boldsymbol{g}_{\mathbb{D}^2}(\mathfrak{p}_s, F(z_1)) \\ &\le \boldsymbol{m}_{\mathbb{D}}(\mathfrak{p}_s \circ F, z_1) = \prod_{j=1}^{s} (\boldsymbol{m}_{\mathbb{D}}(c_j, z_1))^{k_j} = M_s(z_1). \end{aligned}$$

We point out that in the case where $\mathfrak{p} : \mathbb{D}^n \longrightarrow \mathbb{R}_+$ is arbitrary with finite $|\mathfrak{p}|$, an effective formula for $\boldsymbol{g}_{\mathbb{D}^n}(\mathfrak{p}, \cdot)$ is not known.

Example 8.2.28 (due to W. Zwonek). Let

$$D := \mathcal{E}((1/2, 1/2)) \ni (z, w) \overset{F}{\longmapsto} (z, w^2) \in \mathcal{E}((1/2, 1/4)) =: G,$$

$a := (t, t)$, $A_t := F^{-1}(t, t) = \{(t, \sqrt{t}), (t, -\sqrt{t})\}$, $0 < t \ll 1$. Then

$$\begin{aligned} &\boldsymbol{m}_D(A_t, (0,0)) < \boldsymbol{g}_D(A_t, (0,0)) \\ &\qquad \overset{(\dagger)}{<} \min\{\boldsymbol{\ell}^*_D((t, -\sqrt{t}), (0,0)),\ \boldsymbol{\ell}^*_D((t, \sqrt{t}), (0,0))\}, \quad 0 < t \ll 1. \end{aligned}$$

Indeed, by Proposition 8.2.10, we get $\boldsymbol{g}_D(A_t, (0,0)) = \boldsymbol{g}_G((t,t), (0,0))$.

It is clear that $\boldsymbol{m}_G((t,t), (0,0)) \le \boldsymbol{m}_D(A_t, (0,0))$. Let $f \in \mathcal{O}(D, \mathbb{D})$ be such that $f|_{A_t} = 0$. Define

$$\widetilde{f}(z, w) := \tfrac{1}{2}(f(z, \sqrt{w}) + f(z, -\sqrt{w})), \quad (z, w) \in G.$$

Note that $\widetilde{f}$ is well defined, $|\widetilde{f}| < 1$, $\widetilde{f}(t,t) = 0$, $\widetilde{f}$ is continuous, and $\widetilde{f}$ is holomorphic on $G \cap \{w \ne 0\}$. In particular, $\widetilde{f}$ is holomorphic on G. Consequently, $|f(0,0)| = |\widetilde{f}(0,0)| \le \boldsymbol{m}_G((t,t), (0,0))$. Thus $\boldsymbol{m}_D(A_t, (0,0)) = \boldsymbol{m}_G((t,t), (0,0))$.

Suppose that $\boldsymbol{m}_D(A_{t_k}, (0,0)) = \boldsymbol{g}_D(A_{t_k}, (0,0))$ for a sequence $t_k \searrow 0$. Then,

$$\boldsymbol{g}_G((t_k, t_k), (0,0)) = \boldsymbol{g}_D(A_{t_k}, (0,0)) = \boldsymbol{m}_D(A_{t_k}, (0,0)) = \boldsymbol{m}_G((t_k, t_k), (0,0)), \quad k \in \mathbb{N}.$$

Thus, $\boldsymbol{m}_G((t_k, t_k), (0,0)) = \boldsymbol{g}_G((t_k, t_k), (0,0))$, $k \in \mathbb{N}$.

Consequently, using Theorem 5.1.2(d), we conclude that

$$\boldsymbol{\gamma}_G((0,0); (1,1)) = \boldsymbol{A}_G((0,0); (1,1)).$$

Hence, by Propositions 4.2.10 and 2.3.1(d), using the fact that D is the convex envelope of G, we get

$$2 = \mathfrak{h}_D(1,1) = \boldsymbol{\gamma}_G((0,0); (1,1)) = \boldsymbol{A}_G((0,0); (1,1)) = \mathfrak{h}_G(1,1) = \frac{1}{2}(3 + \sqrt{5});$$

a contradiction.

To see the inequality (†), we argue as follows. We know (cf. Theorem 5.1.2(e)) that

$$\boldsymbol{g}_D(A_t, (0,0)) = \boldsymbol{g}_G((t,t), (0,0)) \approx \boldsymbol{g}_G((0,0), (t,t)) = \mathfrak{h}_G(t,t) = \frac{t}{2}(3 + \sqrt{5})$$

for small $t > 0$. On the other hand,

$$\min\{\boldsymbol{\ell}^*_D((t, -\sqrt{t}), (0,0)), \boldsymbol{\ell}^*_D((t, \sqrt{t}), (0,0))\} = \min\{\mathfrak{h}_D(t, -\sqrt{t}), \mathfrak{h}_D(t, \sqrt{t})\} = t + \sqrt{t}.$$

It remains to observe that $\frac{t}{2}(3 + \sqrt{5}) < t + \sqrt{t}$ for small $t > 0$.

Example 8.2.29 (cf. [279]). Let $p = (p_1, \dots, p_n) \in \mathbb{R}^n_{>0}$, $\mathcal{E} := \mathcal{E}(p)$, $k \in \{1, \dots, n\}$,

$$A = A_{\mathcal{E},k} := \{(z_1, \dots, z_n) \in \mathcal{E} : z_1 \cdots z_k = 0\},$$
$$\Xi := \{(z_1, \dots, z_n) \in \mathcal{E} : p_1|z_1|^{2p_1} \le \dots \le p_k|z_k|^{2p_k}\}.$$

Our aim is to find effective formulas for $\boldsymbol{m}_{\mathcal{E}}(A_{\mathcal{E},k}, z)$ and $\boldsymbol{g}_{\mathcal{E}}(A_{\mathcal{E},k}, z)$, $z \in \Xi$. Then, the general formulas may be obtained via subdivision of $\mathcal{E}$ into $k!$ subsets of type Ξ, generated by different permutations of variables $z_1, \dots, z_k$. For $z \in \Xi$, put

$$q_s := \sum_{j=1}^{s} \frac{1}{2p_j}, \quad r_s(z) := 1 - \sum_{j=s+1}^{n} |z_j|^{2p_j} \quad (r_n(z) := 1),$$

$$c_s(z) := r_s(z)/q_s, \quad d = d(z) := \max\left\{s \in \{1, \dots, k\} : 2p_s|z_s|^{2p_s} \le c_s(z)\right\},$$

$$R_{\mathcal{E}}(A, z) := \prod_{j=1}^{d} |z_j| \left(\frac{2p_j}{c_d(z)}\right)^{\frac{1}{2p_j}} = \left(q_d^{q_d} \prod_{j=1}^{d} (2p_j)^{\frac{1}{2p_j}}\right) \frac{|z_1 \cdots z_d|}{\left(1 - \sum_{j=d+1}^{n} |z_j|^{2p_j}\right)^{q_d}}.$$

Observe that $z_{d+1} \cdots z_k \neq 0$.

Then:

(a) $\boldsymbol{g}_{\mathcal{E}}(A, z) = R_{\mathcal{E}}(A, z)$;

(b) $\boldsymbol{m}_{\mathcal{E}}(A, z) = \boldsymbol{g}_{\mathcal{E}}(A, z) = R_{\mathcal{E}}(A, z)$ if $p_j \geq 1/2$, $j = d+1, \dots, n$ (notice that this condition is empty for $d = n$);

(c) $\boldsymbol{m}_{\mathcal{E}}(A, z) = \boldsymbol{g}_{\mathcal{E}}(A, z) = R_{\mathcal{E}}(A, z)$ for $k = 1, n = 2$, $p_2 \geq 1/2$;

(d) $\boldsymbol{m}_{\mathcal{E}}(A, z) \neq \boldsymbol{g}_{\mathcal{E}}(A, z)$ if there exists a $j_0 \in \{k+1, \dots, n\}$ with $p_{j_0} < 1/2$, $0 < |z_\ell| \ll 1$, $\ell \in \{1, \dots, k, j_0\}$, $z_\ell = 0$, $\ell \in \{k+1, \dots, j_0 - 1, j_0 + 1, \dots, n\}$;

(e) $\boldsymbol{m}_{\mathcal{E}}(A, z) = \boldsymbol{g}_{\mathcal{E}}(A, z) = R_{\mathcal{E}}(A, z)$ for $k = n = 2$, $p_1 \leq p_2$, and either $p_2 \geq 1/2$ or $8p_1 + 4p_2(1 - p_2) > 1$.

It is an open question whether $\boldsymbol{m}_{\mathcal{E}}(A, z) = \boldsymbol{g}_{\mathcal{E}}(A, z) = R_{\mathcal{E}}(A, z)$ if $p_j \geq 1/2$, $j = k+1, \dots, n$ (with arbitrary n and k). **?**

Proof of (a). Step 1°. $\boldsymbol{m}_{\mathbb{D}^n}(A_{\mathbb{D}^n,k}, \zeta) = \boldsymbol{g}_{\mathbb{D}^n}(A_{\mathbb{D}^n,k}, \zeta) = |\zeta_1 \cdots \zeta_k|$, $\zeta \in \mathbb{D}^n$, where $A_{\mathbb{D}^n,k} := \{\zeta \in \mathbb{D}^n : \zeta_1 \cdots \zeta_k = 0\}$.

Indeed, it is clear that $|\zeta_1 \cdots \zeta_k| \leq \boldsymbol{m}_{\mathbb{D}^n}(A_{\mathbb{D}^n,k}, \zeta) \leq \boldsymbol{g}_{\mathbb{D}^n}(A_{\mathbb{D}^n,k}, \zeta)$. It remains to prove that $u(\zeta) := \boldsymbol{g}_{\mathbb{D}^n}(A_{\mathbb{D}^n,k}, \zeta) \leq |\zeta_1 \cdots \zeta_k|$, $\zeta \in \mathbb{D}^n$. We proceed by induction on k (with arbitrary n and log-psh function $u : \mathbb{D}^n \longrightarrow [0, 1)$ such that $u(\zeta) \leq C(a)\|\zeta - a\|$, $a \in A_{\mathbb{D}^n,k}$, $\zeta \in \mathbb{D}^n$).

For $k = 1$ the inequality follows from the Schwarz type lemma for log-sh functions $u(\cdot, \zeta_2, \dots, \zeta_n)$.

For $k > 1$, we first apply the case $k = 1$ and get $u(\zeta_1, \dots, \zeta_n) \leq |\zeta_1|$, $\zeta \in \mathbb{D}^n$. Next, we apply the inductive assumption to the functions $u(\zeta_1, \cdot)/|\zeta_1|$, $\zeta_1 \in \mathbb{D}_*$.

Step 2°. Consider the mapping

$$\mathbb{D}^d \ni (\zeta_1, \dots, \zeta_d) \overset{\iota_z}{\longmapsto} \left(\zeta_1 \left(\frac{c_d(z)}{2p_1}\right)^{\frac{1}{2p_1}}, \dots, \zeta_d \left(\frac{c_d(z)}{2p_d}\right)^{\frac{1}{2p_d}}, z_{d+1}, \dots, z_n\right) \in \mathcal{E}.$$

Using the holomorphic contractibility and Step 1°, we get $\boldsymbol{m}_{\mathcal{E}}(A, z) \leq \boldsymbol{g}_{\mathcal{E}}(A, z) \leq R_{\mathcal{E}}(A, z)$. It remains to prove that $\boldsymbol{g}_{\mathcal{E}}(A, z) \geq R_{\mathcal{E}}(A, z)$.

Step 3°. $\boldsymbol{g}_{\mathcal{E}}(A, z) \geq R_{\mathcal{E}}(A, z)$.

We may assume that $z_1 \cdots z_d \neq 0$. First, consider the case where $d = k = n$. Put

$$f(\zeta) := \prod_{j=1}^{n} \zeta_j \left(\frac{2p_j}{c_n(z)}\right)^{\frac{1}{2p_j}}, \quad \zeta \in \mathcal{E}.$$

Observe that $|f(z)| = R_{\mathcal{E}}(A, z)$, $f = 0$ on A, and $f \in \mathcal{O}(\mathcal{E}, \mathbb{D})$. Thus, $\boldsymbol{g}_{\mathcal{E}}(A, z) \geq \boldsymbol{m}_{\mathcal{E}}(A, z) \geq R_{\mathcal{E}}(A, z)$.

Now assume that $d < n$. Put $\mathcal{E}' := \mathcal{E}((p_{d+1}, \dots, p_n))$. Observe that we only need to find a log-psh function $v : \mathcal{E}' \longrightarrow [0, 1)$, $v \not\equiv 0$, such that

- $v(\zeta') \le |\zeta_j|$, $\zeta' = (\zeta_{d+1}, \dots, \zeta_n) \in \mathcal{E}'$, $j = d+1, \dots, k$ (notice that this condition is empty if $d = k$),
- the mapping $\mathcal{E}' \ni \zeta' \longmapsto v(\zeta') r_d^{q_d}(\zeta') \in \mathbb{R}_+$ attains its maximum value M at $\zeta' = (z_{d+1}, \dots, z_n)$ $(r_d(\zeta') = 1 - \sum_{j=d+1}^n |\zeta_j|^{2p_j})$.

  Indeed, suppose that such a v is already constructed. Put

$$u(\zeta) := \frac{q_d^{q_d}}{M} \left(\prod_{j=1}^d |\zeta_j| (2p_j)^{\frac{1}{2p_j}} \right) v(\zeta'), \quad \zeta = (\zeta_1, \dots, \zeta_n) = (\zeta_1, \dots, \zeta_d, \zeta').$$

Then, $\log u \in \mathcal{PSH}(\mathcal{E})$ and $u(\zeta) \le C(a)|\zeta_j| \le C(a)\|\zeta - a\|$ for any $\zeta \in \mathcal{E}$ and $a \in A$ with $a_j = 0$, where $j \in \{1, \dots, k\}$. Moreover, for $\zeta \in \mathcal{E}$ we have

$$u(\zeta) \overset{(*)}{\le} \frac{q_d^{q_d}}{M} \left(\frac{\sum_{j=1}^d |\zeta_j|^{2p_j}}{q_d} \right)^{q_d} v(\zeta') = \frac{1}{M} \left(\frac{\sum_{j=1}^d |\zeta_j|^{2p_j}}{r_d(\zeta')} \right)^{q_d} v(\zeta') r_d^{q_d}(\zeta') < 1,$$

where (*) is a direct consequence of the following elementary inequality:

$$\prod_{j=1}^d a_j^{w_j} \le \left(\frac{\sum_{j=1}^d w_j a_j}{\sum_{j=1}^d w_j} \right)^{\sum_{j=1}^d w_j}, \quad a_1, \dots, a_d \ge 0,\ w_1, \dots, w_d > 0.$$

Thus, $u : \mathcal{E} \longrightarrow [0, 1)$ and, therefore,

$$g_{\mathcal{E}}(A, z) \ge u(z) = \frac{1}{M} R_{\mathcal{E}}(A, z) v(z') r_d^{q_d}(z') = R_{\mathcal{E}}(A, z).$$

Step 4°. Construction of the function v.

We may assume that $z_{d+1}, \dots, z_n \ge 0$. For $\alpha = (\alpha_{d+1}, \dots, \alpha_n) \in \mathbb{R}_+^{n-d}$ define

$$v_\alpha(\zeta') := |\zeta_{d+1} \cdots \zeta_k| \prod_{j=d+1}^n |\zeta_j|^{\alpha_j},$$

where the first factor equals 1 if $d = k$. Obviously, $v : \mathcal{E}' \longrightarrow [0, 1)$, $\log v \in \mathcal{PSH}(\mathcal{E}')$, and $v(\zeta') \le |\zeta_j|$, $\zeta' \in \mathcal{E}'$, $j = d+1, \dots, k$. It is enough to find an α such that the function

$$\mathcal{E}' \cap \mathbb{R}_+^{n-d} \ni t' \overset{\varphi_\alpha}{\longmapsto} v_\alpha(t') r_d^{q_d}(t')$$

attains its maximum at $t' = z'$. In particular, $\frac{\partial \varphi_\alpha}{\partial t_j}(z') = 0$, $j = d+1, \dots, n$. Hence,

$$0 = 1 + \alpha_j - 2p_j q_d \frac{z_j^{2p_j}}{r_d(z')}, \quad j = d+1, \dots, k,$$

$$0 = \alpha_j - 2p_j q_d \frac{z_j^{2p_j}}{r_d(z')}, \quad j = k+1, \dots, n,$$

which gives formulas for $\alpha_{d+1}, \dots, \alpha_n$. Obviously, $\alpha_j \geq 0$, $j = k+1, \dots, n$. In the remaining cases, using the definition of the number d, we have

$$\alpha_j = \frac{2p_j q_d z_j^{2p_j} - r_d(z')}{r_d(z')} \geq 0, \quad j = d+1, \dots, k.$$

To prove that there are no other points like this, rewrite the above equations in the form

$$(1 + \alpha_j) r_d(t') = 2p_j q_d t_j^{2p_j}, \quad j = d+1, \dots, k,$$
$$\alpha_j r_d(t') = 2p_j q_d t_j^{2p_j}, \quad j = k+1, \dots, n.$$

The left sides are decreasing in any of the variables $t_{d+1}, \dots, t_n$, while the right sides are increasing. Thus, at most one common zero is allowed. □

Proof of (b). By the proof of (a), we only have to check that $\boldsymbol{m}_{\mathcal{E}}(A, z) \geq R_{\mathcal{E}}(a, z)$ in the case where $d < n$. First, observe that it is equivalent to find a function $h \in \mathcal{O}(\mathcal{E}')$, $h \not\equiv 0$, such that (**):

- $h(\zeta') = 0$ if $\zeta_{d+1} \cdots \zeta_k = 0$,
- the function $\mathcal{E}' \ni \zeta' \longmapsto |h(\zeta')| r_d^{q_d}(\zeta') \in \mathbb{R}_+$ attains its maximum M at $\zeta' = (z_{d+1}, \dots, z_n)$.

  Indeed, put

$$f(\zeta) := \frac{q_d^{q_d}}{M} \Big(\prod_{j=1}^{d} \zeta_j (2p_j)^{\frac{1}{2p_j}} \Big) h(\zeta'), \quad \zeta \in \mathcal{E}.$$

Obviously, $f(\zeta) = 0$ for $\zeta \in A$. Similar to (a), we prove that $|f| < 1$ and $|f(z)| = R_{\mathcal{E}}(A, z)$. Thus, $\boldsymbol{m}_{\mathcal{E}}(A, z) \geq |f(z)| = R_{\mathcal{E}}(A, z)$.

Conversely, suppose that $\boldsymbol{m}_{\mathcal{E}}(A, z) = R_{\mathcal{E}}(A, z)$ and let $f \in \mathcal{O}(\mathcal{E}, \mathbb{D})$, $f = 0$ on A, be such that $f(z) = \boldsymbol{m}_{\mathcal{E}}(A, z)$ (cf. Remark 8.2.4(n)). Put

$$h(\zeta') := \frac{\partial^d f}{\partial z_1 \dots \partial z_d}(0, \zeta'), \quad \zeta' \in \mathcal{E}'.$$

We have $h(\zeta') = 0$ if $\zeta_{d+1} \cdots \zeta_k = 0$. For $\zeta' \in \mathcal{E}'$, consider the mapping

$$\mathbb{D}^d \ni (\xi_1, \dots, \xi_d) \overset{\iota_{\zeta'}}{\longmapsto} \left(\xi_1 \left(\frac{c_d(\zeta')}{2p_1} \right)^{\frac{1}{2p_1}}, \dots, \xi_d \left(\frac{c_d(\zeta')}{2p_d} \right)^{\frac{1}{2p_d}}, \zeta' \right) \in \mathcal{E}.$$

Applying the Schwarz lemma to the mapping

$$\mathcal{E}' \ni \xi \longmapsto f \circ \iota_{\zeta'}(\xi) \in \mathbb{D},$$

we get $|f \circ \iota_{\zeta'}(\xi)| \le |\xi_1 \cdots \xi_n|$ and $|f \circ \iota_{z'}(\xi)| = |\xi_1 \cdots \xi_n|$. Hence,

$$|h(\zeta')| r_d^{q_d}(\zeta') \le q_d^{q_d} \left(\prod_{j=1}^{d} (2p_j)^{\frac{1}{2p_j}} \right), \quad \zeta' \in \mathcal{E}',$$

$$|h(z')| r_d^{q_d}(z') = q_d^{q_d} \left(\prod_{j=1}^{d} (2p_j)^{\frac{1}{2p_j}} \right).$$

To construct h assume that $z_{d+1}, \dots, z_n \ge 0$ and define

$$h_\alpha(\zeta') := \zeta_{d+1} \cdots \zeta_k \prod_{j=d+1}^{n} e^{\alpha_j \zeta_j},$$

where $\alpha = (\alpha_{d+1}, \dots, \alpha_n) \in \mathbb{R}_+^{n-d}$. It is enough to find an α such that the function

$$\mathcal{E}' \cap \mathbb{R}_+^{n-d} \ni t' \longmapsto h_\alpha(t') r_d^{q_d}(t')$$

attains its maximum at $t' = (z_{d+1}, \dots, z_n)$. Considering the partial derivatives results in the following equations:

$$0 = \frac{1}{z_j} + \alpha_j - 2p_j q_d \frac{z_j^{2p_j-1}}{r_d(z')}, \quad j = d+1, \dots, k,$$

$$0 = \alpha_j - 2p_j q_d \frac{z_j^{2p_j-1}}{r_d(z')}, \quad j = k+1, \dots, n.$$

We continue as in the proof of (a). □

Proof of (c). Assertion (c) follows directly from (b). □

Proof of (d). Step 1°. By the proof of (b), we already know that the equality $\boldsymbol{m}_{\mathcal{E}}(A, z) = R_{\mathcal{E}}(A, z)$ is equivalent to the existence of the function h as in (**).

Step 2°. For any $p \in (0,1)$ and $q > 0$ there exists a $c = c(p,q) \in (0,1)$ such that for any function $f \in \mathcal{O}(\mathbb{D})$, if the function

$$\mathbb{D} \ni \lambda \longmapsto |f(\lambda)|(1 - |\lambda|^p)^q$$

attains its maximum at $\lambda_0 \neq 0$, then $|\lambda_0| \geq c$.

Indeed, let

$$\varphi(t) := \frac{1}{(1 - t^p)^q}, \quad t \in [0,1).$$

Observe that there exists a $b \in (0,1)$ such that φ is strictly concave on $[0,b)$. Moreover,

$$\lim_{t \longrightarrow 0+} \frac{\varphi(t) - \varphi(0)}{t} = +\infty.$$

Consequently, there exists a $c \in (0,b)$ such that

$$\varphi(0) + \frac{b}{c}(\varphi(c) - \varphi(0)) > \varphi(b) + 2.$$

Suppose that $f \in \mathcal{O}(\mathbb{D})$ is such that the function $\mathbb{D} \ni \lambda \longmapsto |f(\lambda)|/\varphi(|\lambda|)$ attains its maximum at $\lambda_0 \neq 0$ with $|\lambda_0| < c$. We may assume that $|f(\lambda_0)| = \varphi(|\lambda_0|)$. Consider the function

$$[0,b] \ni t \overset{\psi}{\longmapsto} |f(0)| + \frac{t}{|\lambda_0|}|f(\lambda_0) - f(0)|.$$

From $\psi(0) = |f(0)| \leq \varphi(0) = 1$, $\psi(|\lambda_0|) \geq \varphi(|\lambda_0|)$, and the convexity condition we get

$$\begin{aligned}\psi(b) = |f(0)| + \frac{b}{|\lambda_0|}|f(\lambda_0) - f(0)| &\geq \varphi(0) + \frac{b}{|\lambda_0|}|\varphi(|\lambda_0|) - \varphi(0)| \\ &\geq \varphi(0) + \frac{b}{c}|\varphi(c) - \varphi(0)| > \varphi(b) + 2.\end{aligned}$$

The Schwarz lemma and the maximum principle imply that there exists a $\lambda_* \in \mathbb{D}$ with $|\lambda_*| = b$ and

$$\frac{|f(\lambda_*) - f(0)|}{|\lambda_*|} \geq \frac{|f(\lambda_0) - f(0)|}{|\lambda_0|}.$$

This means that

$$\begin{aligned}|f(\lambda_*)| &\geq |f(\lambda_*) - f(0)| - |f(0)| = |f(0)| + |f(\lambda_*) - f(0)| - 2|f(0)| \\ &\geq \psi(b) - 2|f(0)| > \varphi(b) + 2 - 2|f(0)| \geq \varphi(b) = \varphi(|\lambda_*|);\end{aligned}$$

a contradiction.

Step 3°. We may assume that $p_{k+1} < 1/2$. Assume that $0 < |z_j| < \varepsilon$, $j = 1, \dots, k+1$, $z_j = 0$, $j = k+2, \dots, n$, with $0 < \varepsilon < c(2p_{k+1}, q_k)$. Observe that $d(z) = k$ provided ε is small enough. Let h be as in Step 1°. Then, the mapping

$$\mathbb{D} \ni \lambda \longmapsto |h(\lambda, 0, \dots, 0)|(1 - |\lambda|^{2p_{k+1}})^{q_k}$$

attains its maximum at $\lambda = z_{k+1}$, which contradicts Step 2°. □

Proof of (e). Step 1°. Let $a, c > 0$, $t_0 \in (0,1)$ be such that $c \geq 1$ or $4a+2c > 1+c^2$, $t_0^c > \tau := a/(a+c)$. Then, there exist $b > 0$ and $r \geq 1$ such that

$$[0,1] \ni t \stackrel{h}{\longmapsto} \frac{t^a}{(r-t)^b}(1-t^c) \in [0,1]$$

admits its maximum at t_0.

Indeed, the condition $h'(t) = 0$ gives

$$\varphi(r,t) = a + \frac{bt}{r-t} - \frac{ct^c}{1-t^c} = 0.$$

Hence, we get a formula for r

$$r(t) = \frac{t(b - bt^c - a + at^c + t^c c)}{-a + at^c + t^c c}.$$

Observe that

$$\lim_{t \longrightarrow \tau^{1/c}} r(t) = +\infty, \qquad \lim_{t \longrightarrow 1} r(t) = 1.$$

In order to prove that $r(t) > 1$ and that $\varphi(r, \cdot)$ has only one zero, it suffices to show that $r'(t) < 0$. We have

$$r'(t) = \frac{(-(a+c)t^{2c} + (2a+c-c^2)t^c - a)b + (at^c + t^c c - a)^2}{(at^c + t^c c - a)^2}.$$

It remains to show that the coefficient $\alpha(t^c)$ next to b is negative. We have

$$\alpha(\tau) = \frac{-ac^2}{a+c} < 0, \quad \alpha(1) = -c^2 < 0,$$
$$\alpha'(u) = -2(a+c)u + 2a + c - c^2.$$

Let u_0 be the zero of $\alpha'(u)$. For $c \geq 1$ we have $u_0 \leq \tau$ and we are done. Otherwise, $u_0 \in (\tau, 1)$ and $4(a+c)\alpha(u_0) = c^2(1 + c^2 - 4a - 2c) < 0$.

Step 2°. We only need to find a function h as in (b). We may assume that $z_2 > 0$. Put $a = 2p_1$, $c = 2p_2$, $t_0 = z_2$. Let r be as in Step 1°. Putting $h(\zeta) = \frac{\zeta}{(r-\zeta)^{b/a}}$ completes the proof. □

Notice that in the case where $p_1 = \cdots = p_n = 1$, we get the following result: let $z = (z_1, \dots, z_n) \in \mathbb{B}_n$ be such that $|z_1| \le \cdots \le |z_k|$ and let

$$d := \max\Big\{s \in \{1, \dots, k\} : s|z_s|^2 + \sum_{j=s+1}^{n} |z_j|^2 \le 1\Big\}.$$

Then,

$$\boldsymbol{m}_{\mathbb{B}_n}(A_{\mathbb{B}_n,k}, z) = \boldsymbol{g}_{\mathbb{B}_n}(A_{\mathbb{B}_n,k}, z) = \left(\frac{d}{1 - \sum_{j=d+1}^{n} |z_j|^2}\right)^{\frac{d}{2}} \prod_{j=1}^{d} |z_j|.$$

8.3 Wu pseudometric

The Wu pseudometric was introduced by H. Wu in [537] (and [536]). A motivation for it has been that it combines properties of invariant metrics and regularity properties of Kähler metrics. Various properties of the Wu pseudometric have been studied in [102, 103, 300, 104, 268, 283, 286, 285].

For a domain $G \subset \mathbb{C}^n$ and a pseudometric $\eta \in \mathfrak{M}(G)$ such that $\eta(a; \cdot)$ is upper semicontinuous for every $a \in G$ (cf. § 4.3), we define the *Wu pseudometric*

$$(\mathbb{W}\eta)(a; X) := (\mathbb{W}\widehat{\eta}(a; \cdot))(X), \quad (a, X) \in G \times \mathbb{C}^n,$$

where $\widehat{\eta}$ is the Buseman pseudometric associated with η (cf. §§ 2.2, 4.3.4). Observe that $\mathbb{W}\eta \in \mathfrak{M}(G)$ (see Remark 4.3.4).

Remark 8.3.1. Concrete formulas of the Wu metric will be given later for elementary n-circled domains (see Example 10.1.12).

First we discuss a few examples showing that, in general, $\mathbb{W}\eta$ fails to have good regularity properties.

Example 8.3.2. Let $G_\varepsilon := \{(z_1, z_2) \in \mathbb{B}_2 : |z_1| < \varepsilon\}$, $0 < \varepsilon < 1/\sqrt{2}$. Recall that $\boldsymbol{\varkappa}_{\mathbb{B}_2}(0; X) = \|X\|$ and $\boldsymbol{\varkappa}_{G_\varepsilon}(0; X) = \max\{\|X\|, |X_1|/\varepsilon\}$, $X = (X_1, X_2)$. Then,

$$(\mathbb{W}\boldsymbol{\varkappa}_{G_\varepsilon})(0; (X_1, X_2)) = \sqrt{\frac{|X_1|^2}{\varepsilon^2} + \frac{|X_2|^2}{1 - \varepsilon^2}}, \quad X = (X_1, X_2) \in \mathbb{C}^2;$$

(cf. Exercise 2.12.7). In particular,

$$(\mathbb{W}\boldsymbol{\varkappa}_{\mathbb{B}_2})(0; (0, 1)) = \sqrt{2} > \frac{1}{\sqrt{1 - \varepsilon^2}} = (\mathbb{W}\boldsymbol{\varkappa}_{G_\varepsilon})(0; (0, 1)).$$

Consequently, the family $(\mathbb{W}\boldsymbol{\varkappa}_G)_G$ is not contractible even with respect to inclusions.

Example 8.3.3. There are a domain G and an upper semicontinuous metric $\eta \in \mathcal{M}(G)$ such that $\mathbb{W}\eta$ is not upper semicontinuous.

Indeed, let $\eta : \mathbb{B}_2 \times \mathbb{C}^2 \longrightarrow \mathbb{R}_+$, $\eta(z;X) := \|X\|$ for $z \neq 0$, and $\eta(0;X) := \max\{\|X\|, |X_1|/\varepsilon\}$, $X = (X_1, X_2) \in \mathbb{C}^2$ ($\varepsilon > 0$ small). Then, $(\mathbb{W}\eta)(z;X) = \sqrt{2}\|X\|$ for $z \neq 0$, and (by Example 8.3.2)

$$\{X \in \mathbb{C}^2 : (\mathbb{W}\eta)(0;X) < 1\} \not\subset \mathbb{B}(1/\sqrt{2}),$$

so $\mathbb{W}\eta$ is not upper semicontinuous.

In fact, even $\mathbb{W}\varkappa_G$ need not be upper semicontinuous, as we will see later.

Example 8.3.4. There exists a bounded domain $G \subset \mathbb{C}^2$ such that $\mathbb{W}\varkappa_G$ is not continuous (see Proposition 2 in [102], where such a continuity was claimed).

Indeed, let $D \subset \mathbb{C}^2$ be a domain such that (cf. Example 3.5.11)

- there exists a dense subset $M \subset \mathbb{C}$ such that $(M \times \mathbb{C}) \cup (\mathbb{C} \times \{0\}) \subset D$,
- $\varkappa_D(z;(0,1)) = 0$, $z \in A := M \times \mathbb{C}$,
- there exists a point $z^0 \in D \setminus A$ such that $\varkappa_D(z^0;X) \geq c\|X\|$, $X \in \mathbb{C}^2$, where $c > 0$ is a constant.

For $R > 0$, let $D_R := D \cap \mathbb{P}_2(z^0, R)$. It is known that $\varkappa_{D_R} \searrow \varkappa_D$ when $R \nearrow +\infty$. Observe that $z^0 \in D_R$ and

$$\varkappa_{D_R}(z^0;X) \geq \varkappa_D(z^0;X) \geq c\|X\|, \quad X \in \mathbb{C}^2.$$

Hence, by Proposition 2.2.10(a), $(\mathbb{W}\varkappa_{D_R})(z^0;X) \geq c\|X\|$, $X \in \mathbb{C}^2$. In particular, $(\mathbb{W}\varkappa_{D_R})(z^0;(0,1)) \geq c$.

Fix a sequence $M \ni z_k \longrightarrow z_1^0$. Note that $\{z_k\} \times (z_2^0 + \mathbb{D}(R)) \subset D_R$, which implies that $\varkappa_{D_R}((z_k, z_2^0);(0,1)) \leq 1/R$, $k = 1, 2, \dots$. In particular,

$$(\mathbb{W}\varkappa_{D_R})((z_k, z_2^0);(0,1)) \leq \sqrt{2}\varkappa_{D_R}((z_k, z_2^0);(0,1)) \leq \sqrt{2}/R, \quad k = 1, 2, \dots.$$

Now it is clear that if $R > \sqrt{2}/c$, then

$$\limsup_{k \to +\infty} (\mathbb{W}\varkappa_{D_R})((z_k, z_2^0);(0,1)) \leq \sqrt{2}/R < c \leq (\mathbb{W}\varkappa_{D_R})(z^0;(0,1)),$$

which shows that for $G := D_R$ the pseudometric $\mathbb{W}\varkappa_G$ is not continuous.

Remark 8.3.5. We point out the role played in the definition of $\mathbb{W}$ by the factor $\sqrt{m}$.

Put $\widetilde{\mathbb{W}}h := q_{s^h}$, $\widetilde{\mathbb{W}}\eta(a;X) := (\widetilde{\mathbb{W}}\hat{\eta}(a;\cdot))(X)$, $(a,X) \in G \times \mathbb{C}^n$. Let $D \subset \mathbb{C}^2$ and $D \ni z_k \longrightarrow z_0 \in D$ be such that

- $\varkappa_D(z_k;\cdot)$ is not a metric (in particular, $m(k) := \dim U(\hat{\varkappa}_D(z_k;\cdot)) \leq 1$, $k \in \mathbb{N}$),
- $\varkappa_D(z_0;\cdot)$ is a metric (take, for instance, the domain D from Example 8.3.4).

Put $G := D \times \mathbb{D} \subset \mathbb{C}^3$. Then,

$$(\widetilde{\mathbb{W}}\varkappa_G)^2((z_k,0);(0,1)) = s^{\varkappa_G((z_k,0);\cdot)}((0,1),(0,1)) = \frac{1}{m(z_k)+1} \geq \frac{1}{2}, \quad k \in \mathbb{N},$$
$$(\widetilde{\mathbb{W}}\varkappa_G)^2((z_0,0);(0,1)) = s^{\varkappa_G((z_0,0);\cdot)}((0,1),(0,1)) = \frac{1}{m(z_0)+1} = \frac{1}{3},$$

and, therefore, $\widetilde{\mathbb{W}}\varkappa_G$ is not upper semicontinuous at $((z_0,0),(0,1))$ (the example is due to W. Jarnicki).

Before continuing the discussion on the upper semicontinuity of the Wu pseudometric, we present some positive results on its behavior. For $\eta = \varkappa_G$ these properties were formulated (without proof) in [536, 537].

Proposition 8.3.6. (a) *If $\eta \in \mathfrak{M}(G)$ is a continuous metric, then, so is $\mathbb{W}\eta$ (cf. Example* 8.3.3).

(b) *If $(\delta_G)_G$ is a holomorphically contractible family of pseudometrics such that $\delta_G(a;\cdot)$ is upper semicontinuous for any $a \in G$, then:*

- *for any biholomorphic mapping $F : G \longrightarrow D$, $G, D \subset \mathbb{C}^n$, we have*

$$(\mathbb{W}\delta_D)(F(z);F'(z)(X)) = (\mathbb{W}\delta_G)(z;X), \quad (z,X) \in G \times \mathbb{C}^n;$$

- *for any holomorphic mapping $F : G \longrightarrow D$, $G \subset \mathbb{C}^{n_1}$, $D \subset \mathbb{C}^{n_2}$, we have*

$$(\mathbb{W}\delta_D)(F(z);F'(z)(X)) \leq \sqrt{n_2}(\mathbb{W}\delta_G)(z;X), \quad (z,X) \in G \times \mathbb{C}^{n_1},$$

but, for example, the family $(\mathbb{W}\varkappa_G)_G$ is not holomorphically contractible (cf. Example 8.3.2).

Proof. (a) Fix a point $z_0 \in G \subset \mathbb{C}^n$. Let $s_z := s^{\widehat{\eta}(z;\cdot)}$, $z \in G$. We are going to show that $s_z \longrightarrow s_{z_0}$ when $z \longrightarrow z_0$.

By our assumptions, there exist $r > 0$, $c > 0$ such that

$$\eta(z;X) \geq c\|X\|, \quad z \in \mathbb{B}(z_0,r) \subset G, \ X \in \mathbb{C}^n.$$

In particular, the sets

$$I_z := \{X \in \mathbb{C}^n : \widehat{\eta}(z;X) < 1\}, \quad z \in \mathbb{B}(z_0,r),$$

are contained in the ball $\mathbb{B}(C)$ with $C := 1/c$. Moreover,

$$|\widehat{\eta}(z;X) - \widehat{\eta}(z_0;X)| \leq \varphi(z)\|X\|, \quad X \in \mathbb{C}^n,$$

where $\varphi(z) \longrightarrow 0$ when $z \longrightarrow z_0$. Hence,

$$(1 + C\varphi(z))^{-1} I_z \subset I_{z_0} \subset (1 + C\varphi(z)) I_z, \quad z \in \mathbb{B}(z_0, r),$$

and, consequently,

$$\begin{aligned} I_{z_0} &\subset (1 + C\varphi(z))\mathbb{E}(s_z) = \mathbb{E}((1 + C\varphi(z))^{-2} s_z), \\ I_z &\subset (1 + C\varphi(z))\mathbb{E}(s_{z_0}) = \mathbb{E}((1 + C\varphi(z))^{-2} s_{z_0}), \quad z \in \mathbb{B}(z_0, r). \end{aligned} \tag{8.3.1}$$

Hence,

$$\begin{aligned} \operatorname{Vol}(s_{z_0}) &\le \operatorname{Vol}((1 + C\varphi(z))^{-2} s_z) = (1 + C\varphi(z))^{2n} \operatorname{Vol}(s_z), \\ \operatorname{Vol}(s_z) &\le \operatorname{Vol}((1 + C\varphi(z))^{-2} s_{z_0}) = (1 + C\varphi(z))^{2n} \operatorname{Vol}(s_{z_0}), \quad z \in \mathbb{B}(z_0, r). \end{aligned}$$

Thus $\operatorname{Vol}(s_z) \longrightarrow \operatorname{Vol}(s_{z_0})$ when $z \longrightarrow z_0$.

Take a sequence $z_\nu \longrightarrow z_0$. Since

$$|s_{z_\nu}(\boldsymbol{e}_j, \boldsymbol{e}_k)| \le \eta(z_\nu; \boldsymbol{e}_j)\eta(z_\nu; \boldsymbol{e}_k), \quad j, k = 1, \dots, n, \ \nu \in \mathbb{N},$$

we may assume that $s_{z_\nu} \longrightarrow s_*$, where s_* is a pseudo-Hermitian scalar product. We already know that $\operatorname{Vol}(s_*) = \operatorname{Vol}(s_{z_0})$. Moreover, by (8.3.1), $I_{z_0} \subset \mathbb{E}(s_*)$. Consequently, the uniqueness of s_{z_0} implies that $s_* = s_{z_0}$.

(b) Recall that the family $(\widehat{\delta}_G)_G$ is holomorphically contractible (cf. § 4.3.4).

If F is biholomorphic, then the result is obvious, because for any $z \in G$, the mapping $F'(z)$ is a $\mathbb{C}$-linear isomorphism and $\widehat{\delta}_D(F(z); F'(z)(X)) = \widehat{\delta}_G(z; X)$, $X \in \mathbb{C}^n$.

In the general case, using Proposition 2.2.10(a), we get

$$\begin{aligned} (\mathbb{W}\delta_D)(F(z); F'(z)(X)) &\le \sqrt{n_2}\widehat{\delta}_D(F(z); F'(z)(X)) \\ &\le \sqrt{n_2}\widehat{\delta}_G(z; X) \le \sqrt{n_2}(\mathbb{W}\delta_G)(z; X), \quad (z, X) \in G \times \mathbb{C}^{n_1}. \end{aligned}$$ □

? We point out that Proposition 8.3.6(a) gives the continuity of $\mathbb{W}\eta$ only in the case where η is a continuous metric. We do not know whether $\mathbb{W}\eta$ is upper semicontinuous in the case where η is a continuous pseudometric.

Now we come back to discussing the upper semicontinuity of the Wu pseudometric in more detail. The question of the upper semicontinuity of the Wu pseudometric appears in a natural way, for instance when one tries to define the associated integrated form. The results we will present here are based on [286, 285]. (It should be mentioned that in the literature it was claimed that $\mathbb{W}\varkappa_G$ is upper semicontinuous (cf. [537], Theorem 1, and [102], Proposition 2).)

In a first step we define the domain we will deal with. Let

$$u(\lambda) := 1 + \sum_{j=4}^{\infty} 2^{-j} \max\left\{\log \tfrac{|\lambda - 2^{-j}|}{2}, -2^{2j}\right\}, \quad \lambda \in \mathbb{D}.$$

Then,

- $u \in \mathcal{SH}(\mathbb{D}, (-\infty, 1))$;
- $u(0) = 1 - \frac{3}{4}\log 2 > 0$;
- $u(\lambda) > 0$ for all λ with $1/4 \le |\lambda| < 1$.

Moreover, for $R > 1$ put

$$G_R := \{z \in \mathbb{D} \times \mathbb{D}(R) : 2R|z_2|e^{u(z_1)} < 1\},$$
$$\Omega_R := (\mathbb{D} \times \mathbb{D}(1/(2R))) \cup (\mathbb{D}(1/2) \times \mathbb{D}(R)).$$

Then G_R is a bounded pseudoconvex domain satisfying

$$\mathbb{D} \times \mathbb{D}(C_R) \subset G_R \subset (\mathbb{D} \times \mathbb{D}(1/(2R))) \cup (\mathbb{D}(1/4) \times \mathbb{D}(R)) \subset \Omega_R,$$

where $C_R := \frac{1}{2eR}$. Moreover, recall that the envelope of holomorphy $\widetilde{\Omega}_R$ of Ω_R is given by

$$\widetilde{\Omega}_R = \left\{z \in \mathbb{D} \times \mathbb{D}(R) : \text{ if } |z_1| \ge 1/2, \text{ then } |z_2| < \frac{1}{2R} e^{\frac{\log|z_1|\log(2R^2)}{-\log 2}}\right\}.$$

Now let $(\alpha_D)_{D \in \mathfrak{G}}$ be a holomorphically contractible family of pseudometrics such that

- $A_D \le \alpha_D \le \varkappa_D$,
- $\alpha_D(a; \cdot)$ is upper semicontinuous for all $a \in D \in \mathfrak{G}$.

Put

$$I_R(z) := \{X \in \mathbb{C}^2 : \alpha_{G_R}(z; X) < 1\}$$

and denote by $\mathbb{E}_R(z)$ the John ellipsoid with respect to $\alpha_{G_R}(z; \cdot)$, i.e., the minimal ellipsoid containing $\widehat{I}_R(z)$, $z \in G_R$.

Lemma 8.3.7. *The following properties hold:*

(a) $I_R((z_1, 0))$ *is a Reinhardt domain,* $(z_1, 0) \in G_R$.

(b) $I_R(0) \subset \big(\mathbb{D} \times \mathbb{D}(2)\big) \cap \widetilde{\Omega}_R$.

(c) $(1, C_R) \in \partial I_R(0)$.

(d) $I_R((2^{-\nu}, 0)) \subset \widetilde{\Omega}_R$ *and* $(0, R) \in I_R((2^{-\nu}, 0))$ *for all sufficiently large* ν.

Proof. (a) Fix a point $z^0 = (z_1^0, 0) \in G_R$. Recall that $I_R(z^0)$ is balanced and that for $z \in G_R$ also $(z_1, \mu z_2) \in G_R$, $\mu \in \overline{\mathbb{D}}$. Therefore, if $X \in I_R(z^0)$ and $\theta_1, \theta_2 \in \mathbb{R}$, then

$$\alpha_{G_R}(z^0; X) = \alpha_{G_R}(z^0; e^{i\theta_1} X) \ge \alpha_{G_R}(z^0; (e^{i\theta_1} X_1, e^{i\theta_2} X_2)),$$

i.e., $I_R(z^0)$ is Reinhardt.

(b) Fix an $X \in I_R(0)$. Put $v(z) := (1/2)(|z_1|e^{u(z_1)} + |z_1|)$, $z \in G_R$. Obviously, v is log-psh, $v(z) \le e\|z\|$, and $v(z) \le (1/2)(1/(2R) + 1) < 1$. Hence,

$$1 > \alpha_{G_R}(0; X) \ge \boldsymbol{A}_{G_R}(0; X) \ge \limsup_{\mathbb{D}_* \ni \lambda \to 0} v(\lambda z)/|\lambda|$$
$$= (|X_2|e^{u(0)} + |X_1|)/2 \ge |X_2|/2,$$

i.e., $|X_2| < 2$.

Now recall that $G_R \subset \widetilde{\Omega}_R \subset \mathbb{D} \times \mathbb{C}$ is balanced pseudoconvex. So $1 > \alpha_{G_R}(0; X) \ge \boldsymbol{A}_{\widetilde{\Omega}_R}(0; X) = \mathfrak{h}_{\widetilde{\Omega}_R}(X)$ implying that $X \in \widetilde{\Omega}_R$.

(c) Put $Y := (1, C_R)$. Then, $Y \in \partial\widetilde{\Omega}_R$ and so $\alpha_{G_R}(0; Y) \ge 1$. On the other hand, note that $\varphi(\lambda) := \lambda Y \in G_R$, $\lambda \in \mathbb{D}$. Thus, $\alpha_{G_R}(0; Y) \le \boldsymbol{\varkappa}_{G_R}(0; Y) \le 1$.

(d) Let $\nu \ge 4$ and put $a_\nu := (2^{-\nu}, 0)$. Define $F_\nu(z) := (\frac{z_1 - 2^{-\nu}}{1 - z_1 2^{-\nu}}, z_2)$, $z \in \mathbb{D} \times \mathbb{C}$. Then, $F_\nu(a_\nu) = 0$. Moreover, a simple calculation shows that $F_\nu \in \mathcal{O}(G_R, \widetilde{\Omega}_R)$.

Now let $X \in I_R(a_\nu)$. Then, using the fact that $\widetilde{\Omega}_R$ is complete Reinhardt, one has

$$1 > \alpha_{G_R}(0; X) \ge \alpha_{\widetilde{\Omega}_R}(f_\nu(a_\nu); F_\nu'(a_\nu)X)$$
$$= \alpha_{\widetilde{\Omega}_R}(0; (X_1/(1 - 2^{-2\nu}), X_2) \ge \alpha_{\widetilde{\Omega}_R}(0; X) \ge \boldsymbol{A}_{\widetilde{\Omega}_R}(0; X) = \mathfrak{h}_{\widetilde{\Omega}_R}(X),$$

which means that $I_R(a_\nu) \subset \widetilde{\Omega}_R$.

For the second claim, note that $(0, R) \in \partial\widetilde{\Omega}_R$ and therefore, $\alpha_{G_R}(0; (0, R)) \ge 1$. It remains to observe that $\varphi_\nu(\lambda) := (2^{-\nu}, \lambda R)$, $\lambda \in \mathbb{D}$, is an analytic disc in G_R if ν is sufficiently large. Hence, $\alpha_{G_R}(0; (0, R)) \le \boldsymbol{\varkappa}_{G_R}(0; (0, R)) \le 1$. □

Now we are in a position to present the announced example.

Proposition 8.3.8 (cf. [286]). *Let $R \ge 9$ and let $(\alpha_D)_{D \in \mathfrak{G}}$ be a holomorphically contractible system as above. Then, neither $\mathbb{W}\alpha_{G_R}$ nor $\widetilde{\mathbb{W}}\alpha_{G_R}$ is upper semicontinuous at the point* $(0, (1, 0))$.

Proof. Fix an $R \ge 9$ and take a family $(\alpha_D)_{D \in \mathfrak{G}}$ as in the proposition. Then $I_R(0)$ satisfies the properties of Example 2.2.12 and therefore $(\sqrt{A_R}, 0) \in \mathbb{E}_R(0)$. Moreover, because of Example 2.2.13, we have $(\sqrt{A_R}, 0) \notin \mathbb{E}_R((2^{-\nu}, 0))$ for all $\nu \ge \nu_R$. Therefore,

$$\mathbb{W}\alpha_{G_R}(0; (1, 0)) = \sqrt{2}\widetilde{\mathbb{W}}\alpha_{G_R}(0; (1, 0)) < \sqrt{2/A_R}$$
$$\le \sqrt{2}\widetilde{\mathbb{W}}\alpha_{G_R}((2^{-\nu}, 0); (1, 0)) = \mathbb{W}\alpha_{G_R}((2^{-\nu}, 0); (1, 0)), \quad \nu \ge \nu_R,$$

which finishes the proof. □

Corollary 8.3.9. *Let $R \geq 9$ and assume that $(\alpha_D)_{D\in\mathfrak{G}}$ satisfies the assumption of Proposition* 8.3.8 *and, in addition, the product property, i.e.,*

$$\alpha_{D_1\times D_2}((z,w);(X,Y)) = \max\{\alpha_{D_1}(z;X), \alpha_{D_2}(w;Y))\}, \quad D_1, D_2 \in \mathfrak{G},$$

then $\mathbb{W}\alpha_{G_R\times\mathbb{D}^n}$ is not upper semicontinuous at $(0,(1,0,\dots,0))$.

Proof. Use Proposition 8.3.8 together with Proposition 2.2.10(b). □

Remark 8.3.10. Observe that in the above Corollary it was assumed that $(\alpha_D)_{D\in\mathfrak{G}}$ satisfies the product property. Moreover, we know (see Theorem 8.3.6) that for a bounded domain D all functions $\mathbb{W}\boldsymbol{\gamma}^{(k)}_{G_R}$ are continuous. In the unbounded case the domains

$$G := \{z \in \mathbb{C}^2 : |z_1|(1+|z_2|) < 1\} \text{ and } G_n := G \times \mathbb{D}^{n-2}, \quad n \geq 2,$$

play an important role (see [285]). For an arbitrary, holomorphically contractible family of pseudometrics $(\alpha_D)_{D\in\mathfrak{G}}$ the following results may be found in [285]:

- $\mathbb{W}\alpha_G$ is not upper semicontinuous.
- If $n \geq 3$, then neither $\widetilde{\mathbb{W}}\alpha_{G_n}$ nor $\mathbb{W}\alpha_{G_n}$ is upper semicontinuous.
- The pseudometric $\widetilde{\mathbb{W}}\alpha_D$ is upper semicontinuous for any domain $D \subset \mathbb{C}^2$.

Remark 8.3.11.

(a) For $p \in \mathbb{R}^n_{>0}$ put $\mathcal{E}(p) := \{z \in \mathbb{C}^n : \sum_{j=1}^n |z_j|^{2p_j} < 1\}$. The regularity of the Wu metric for $\mathcal{E}((1,m))$ was studied in [102] ($m \geq \frac{1}{2}$) and [103] ($0 < m < \frac{1}{2}$), applying formulas of the Kobayashi–Royden metric for $\mathcal{E}((1,m))$ found in [260] and in [421].

(b) In [104], the following two results are given:

- Let $G := \mathbb{B}_n \cap U$, where U is open in $\mathbb{C}^n$. Then, there exists a neighborhood V of $\partial G \cap \partial\mathbb{B}_n$ such that $\mathbb{W}\boldsymbol{\varkappa}_G = \mathbb{W}\boldsymbol{\varkappa}_{\mathbb{B}_n}$ in $V \cap G$.
- Let $p = (p_1,\dots,p_n) \in \mathbb{N}^n$, $p_j \geq 2$, $j = 1,\dots,n$. Then, any strongly pseudoconvex point $a \in \partial\mathcal{E}(p)$ has a neighborhood V such that $\mathbb{W}\boldsymbol{\varkappa}_{\mathcal{E}(p)}$ is a Kähler metric with constant negative curvature in $V \cap \mathcal{E}(p)$.

8.4 Exercises

Exercise 8.4.1. Let $G := \{z \in \mathbb{C}^2 : |z_1|^2(1-|z_2|^2) < 1\}$ and Let $(\alpha_D)_{D\in\mathfrak{G}}$ be a holomorphically contractible family of pseudometrics such that $\alpha_G(a;\cdot)$ is upper semicontinuous for all $a \in G$. Prove that $\mathbb{W}\alpha_G$ is not upper semicontinuous.

Hint. Complete the following sketch of the proof. The aim is to prove that

$$\limsup_{(0,1)\ni x_1\to 0} \mathbb{W}\alpha_G((x_1,0);(1,0)) = \sqrt{2} \limsup_{(0,1)\ni x_1\to 0} \widetilde{\mathbb{W}}\alpha_G((x_1,0);(1,0))$$
$$\geq \sqrt{2} > 1 = \mathbb{W}\alpha_G((0,0);(1,0)).$$

Step 1°. Show that the indicatrix $I(x_1)$ of $\alpha_G((x_1,0);\cdot)$ is Reinhardt, $x_1 \in (0,1)$.

Step 2°. Show that $I(x_1) \subset \{X \in \mathbb{C}^2 : (|X_1| + x_1|X_2|)/(1-x_1^2) < 1\}$ using that $\gamma_G \leq \alpha_G$ and that $F \in \mathcal{O}(G,\mathbb{D})$ for $F(z) := z_1(1+z_2)$.

Step 3°. Verify that $\widetilde{\mathbb{W}}\alpha_G((0,0);X) = \mathbb{W}\alpha_G((0,0);X) = |X_1|$.

Step 4°. Show that $\limsup_{(0,1)\ni x_1\to 0} \widetilde{\mathbb{W}}\alpha_G((x_1,0);(1,0)) \geq 1$. Assume the contrary, i.e., that $\widetilde{\mathbb{W}}\alpha_G((x_1,0);(1,0)) < t < 1$ for small x_1. Study $\Psi(\mathbb{E}) = T_{a,b}$, where $\mathbb{E}$ denotes the corresponding John ellipsoid, and observe that $at^2 > 1$ and $b > (\frac{1-x_1}{x_1})^2$. Applying $\psi(I(x_1)) \subset T_{1,x_1^{-2}} =: T$ leads to a contradiction, because it turns out that the area of $T_{a,b}$ is strictly larger that the one of T.

8.5 List of problems

Chapter 9

Contractible functions and metrics for the annulus

Summary. Let $P := \{z \in \mathbb{C} : 1/R < |z| < R\}$. The aims of this chapter are:

- to present effective formulas for $\boldsymbol{g}_P$, $\boldsymbol{A}_P$ (Proposition 9.1.2), $\boldsymbol{m}_P^{(k)}$, $\boldsymbol{\gamma}_P^{(k)}$ (Proposition 9.1.5), and $\boldsymbol{k}_P^*$, $\boldsymbol{\varkappa}_P$ (Proposition 9.1.9); Proposition 9.1.5 for $k = 1$ was proved in [77, 479] and for $k \geq 2$ in [257, 258];

- to study the relations between the above objects – Proposition 9.1.12;

- to characterize $\boldsymbol{c}_P^i$ – Proposition 9.1.13 (cf. [256]);

- to prove an analogue of the Schwarz–Pick lemma for P – Proposition 9.1.19;

- to characterize $\boldsymbol{c}$- and $\boldsymbol{\gamma}$-isometries of P – Proposition 9.1.20.

9.1 Contractible functions and metrics for the annulus

For $R > 1$ let

$$P = P(R) := \mathbb{A}(1/R, R) = \{z \in \mathbb{C} : 1/R < |z| < R\}.$$

Note that for any $0 < r_1 < r_2 < +\infty$ the annulus

$$Q := \mathbb{A}(r_1, r_2) = \{z \in \mathbb{C} : r_1 < |z| < r_2\}$$

is biholomorphic to $P(\sqrt{r_2/r_1})$.

Recall that $\boldsymbol{k}_P^* = \tanh \boldsymbol{k}_P = \tanh \boldsymbol{\ell}_P^{(k)}$ and $\boldsymbol{\varkappa}_P^{(k)} = \boldsymbol{\varkappa}_P$, cf. Remark 3.3.8(e), Corollary 3.8.9.

First, observe that P is invariant under rotations, and therefore, to calculate d_P (resp. δ_P), it is enough to describe $d_P(a, \cdot)$ (resp. $\delta_P(a; 1)$) for any a with $1/R < a < R$.

Put $q := 1/R^2$. For $1/R < a < R$ define

$$f(a, z) = \left(1 - \frac{z}{a}\right) \Pi(a, z),$$

where

$$\Pi(a, z) := \frac{\prod_{\nu=1}^{\infty}(1 - \frac{z}{a} q^{2\nu})(1 - \frac{a}{z} q^{2\nu})}{\prod_{\nu=1}^{\infty}(1 - azq^{2\nu-1})(1 - \frac{1}{az} q^{2\nu-1})}.$$

Remark 9.1.1 (cf. [115]).

(a) The function $f(a,\cdot)$ is meromorphic on $\mathbb{C}_*$. It has simple poles at $z = R^{4k-2}/a$ and simple zeros at $z = aR^{4k}$, $k \in \mathbb{Z}$. In particular, $f(a,\cdot)$ is holomorphic on $\overline{P}$ and the only zero of $f(a,\cdot)$ in $\overline{P}$ is the simple zero at $z = a$.

(b) $f(a,z)f(a,1/(R^2z)) = 1$, $f(a,z)f(a,R^2/z) = R^2/a^2$, and $f(a,\overline{z}) = \overline{f(a,z)}$. In particular,

$$|f(a,z)| = \begin{cases} 1 & \text{if } |z| = 1/R \\ R/a & \text{if } |z| = R \end{cases}.$$

Let $s = s(a) \in (0,1)$ be such that

$$a = R^{1-2s}. \tag{9.1.1}$$

Proposition 9.1.2. *For any $a \in (1/R, R)$ we have*

(a) $\boldsymbol{s}(a,z) = \boldsymbol{g}_P(a,z) = \dfrac{|f(a,z)|}{|Rz|^{s(a)}}, \quad z \in P;$

(b) $\boldsymbol{S}_P(a;1) = \boldsymbol{A}_P(a;1) = \dfrac{1}{(Ra)^{s(a)}} \cdot \dfrac{\Pi(a,a)}{a}.$

Proof. (b) is a direct consequence of (a); cf. Lemma 4.2.3. To prove (a), we only need to observe that the function $\overline{P} \ni z \longmapsto -\log(|f(a,z)|/|Rz|^{s(a)})$ is the classical Green function with pole at a (cf. Remark 4.2.8). □

Put

$$l_k(a) := \lceil ks(a) \rceil, \quad b_k(a) := R^{1-2(l_k(a)-ks(a))}, \quad k \in \mathbb{N}. \tag{9.1.2}$$

Remark 9.1.3.

(a) $l_1(a) = 1$, $b_1(a) = 1/a$.

(b) $b_k(a) \in \partial P$ iff $b_k(a) = R$ iff $ks(a) \in \mathbb{N}$.

For $\theta \in \mathbb{R}$, define

$$\boldsymbol{e}_k(z) = \boldsymbol{e}_k(a,\theta,z) := \frac{f(b_k(a), -e^{-i\theta}z)}{(Rz)^{l_k(a)}} \cdot [f(a,z)]^k, \quad z \in \overline{P},$$

where $f(R,\cdot) :\equiv 1$.

Remark 9.1.4. We have $\boldsymbol{e}_k \in \mathcal{O}(\overline{P})$, $\operatorname{ord}_a \boldsymbol{e}_k \geq k$ (Remark 9.1.1(a)), and $|\boldsymbol{e}_k| = 1$ on ∂P (Remark 9.1.1(b)). In particular,

$$\boldsymbol{m}_P^{(k)}(a,z) \geq |\boldsymbol{e}_k(z)|^{1/k}, \quad z \in P.$$

Proposition 9.1.5. *For any $a \in (1/R, R)$, we have:*

(a) $\boldsymbol{m}_P^{(k)}(a,z) = \left[\frac{f(b_k(a), -|z|)}{|Rz|^{l_k(a)}}\right]^{1/k} |f(a,z)|, \quad z \in P;$

(b) $\boldsymbol{\gamma}_P^{(k)}(a;1) = \left[\frac{f(b_k(a), -a)}{(Ra)^{l_k(a)}}\right]^{1/k} \frac{\Pi(a,a)}{a}, \quad k \in \mathbb{N}.$

Note that the right hand side of (a) is equal to $|\boldsymbol{e}_k(a, \operatorname{Arg} z, z)|^{1/k}$. Moreover, condition (b) follows from (a); cf. Proposition 4.2.20(b).

The proof of Proposition 9.1.5 will be based on the following

Lemma 9.1.6 (Robinson's fundamental lemma, cf. [453]). *Let $1/R < b < R$, $\theta \in \mathbb{R}$, $c := -e^{i\theta}b$, and let $\varphi \in \mathcal{O}(P \setminus \{c\})$. Assume that φ has at most a simple pole at c and that* $\limsup_{z\to\partial P} |\varphi(z)| \le 1$. *Then,*

(a) $|\varphi(e^{i\theta}x)| \le 1, \ 1/R < x < R;$

(b) *if* $|\varphi(e^{i\theta}x_0)| = 1$ *for some* $1/R < x_0 < R$, *then* $\varphi \equiv \text{const}$.

Proof. Without loss of generality, we may assume that $\theta = 0$ ($c = -b$).

(a) Fix $x_0 \in (1/R, R)$. We may assume that $\varphi(x_0) > 0$. Put

$$\Phi(z) := \frac{1}{2}[\varphi(z) + \overline{\varphi(\overline{z})}], \quad z \in P \setminus \{c\}.$$

Then, $\Phi \in \mathcal{O}(P \setminus \{c\})$ and Φ has at most a simple pole at c. Moreover, Φ satisfies $\limsup_{z\to\partial P} |\Phi(z)| \le 1$ and $\Phi = \operatorname{Re}\varphi$ on $(P \setminus \{c\}) \cap \mathbb{R}$. If Φ has no singularity at c, then obviously $|\Phi(x_0)| = \varphi(x_0) \le 1$. Assume now that Φ has a simple pole at c. For $\varepsilon > 0$, put

$$P_\varepsilon := \{z \in \mathbb{C} : 1/R + \varepsilon < |z| < R - \varepsilon\}, \quad M_\varepsilon := \|\varphi\|_{\partial P_\varepsilon}, \quad \Phi_\varepsilon := \Phi/M_\varepsilon;$$

note that $\limsup_{\varepsilon\to 0} M_\varepsilon \le 1$. Choose ε so small that $c, x_0 \in P_\varepsilon$. Because Φ_ε has a simple pole at c, $|\Phi_\varepsilon| \le 1$ on ∂P_ε, and $\Phi_\varepsilon((P_\varepsilon \setminus \{c\}) \cap \mathbb{R}) \subset \mathbb{R}$, we then get

$$\mathbb{R} \setminus [-1,1] \subset \Phi_\varepsilon((P_\varepsilon \setminus \{c\}) \cap \mathbb{R}_-). \tag{9.1.3}$$

Put

$$Z_\varepsilon(w) := 1 + \frac{1}{2\pi i}\int_{\partial P_\varepsilon} \frac{\Phi'_\varepsilon(\zeta)}{\Phi_\varepsilon(\zeta) - w} d\zeta, \quad |w| > 1$$

(where the boundary ∂P_ε is taken with the positive orientation). It is well known that $Z_\varepsilon(w) = \#\{z \in P_\varepsilon : \Phi_\varepsilon(z) = w\}$ (with multiplicities). It is seen that Z_ε is continuous and $Z_\varepsilon(w) \longrightarrow 1$ as $w \to \infty$. Consequently, $Z_\varepsilon \equiv 1$ and hence, (9.1.3) implies that $\Phi_\varepsilon(P_\varepsilon \cap \mathbb{R}_+) \subset [-1,1]$. In particular, $|\Phi_\varepsilon(x_0)| = \varphi(x_0)/M_\varepsilon \le 1$. Letting $\varepsilon \longrightarrow 0$ we get the required result.

(b) We may assume that $\varphi(x_0) = 1$. Let Φ be as in (a). It suffices to prove that $\Phi \equiv 1$ (since $\operatorname{Re}\varphi$ is a harmonic function on P, $\operatorname{Re}\varphi \equiv 1$, and so $\varphi \equiv 1$). If Φ has no singularity at c, then the result is trivial. Suppose that Φ has a simple pole at c. In particular, Φ is an open mapping. Let U be an open neighborhood of x_0 such that $U \subset P$, $U \cap \mathbb{R}_- = \varnothing$. Since $\Phi(U)$ is open, there exist $w_0 \in (1, +\infty)$ and $z_0 \in U$ such that $\Phi(z_0) = w_0$. Let P_ε, M_ε, and Φ_ε be as in (a). If ε is sufficiently small, we have $\Phi_\varepsilon(z_0) = w_0/M_\varepsilon > 1$. Hence, by (9.1.3) and the fact that $Z_\varepsilon \equiv 1$, we get $z_0 \in \mathbb{R}_-$; a contradiction. □

Proof of Proposition 9.1.5. In view of Remark 9.1.4, it suffices to prove the inequality $|h(z)| \le |\boldsymbol{e}_k(a, \operatorname{Arg} z, z)|$, $z \in P$, for any $h \in \mathcal{O}(P, \mathbb{D})$ with $\operatorname{ord}_a h \ge k$. Fix $z_0 = e^{i\theta}|z_0| \in P \setminus \{a\}$ and h. Put $\varphi := h/\boldsymbol{e}_k(a, \theta, \cdot)$. Then, φ satisfies all assumptions of Lemma 9.1.6, and therefore $|\varphi(z_0)| \le 1$. □

Proposition 9.1.5(a) may be extended to the multi-pole Möbius function. Let $\mathfrak{p} : P \longrightarrow \mathbb{R}_+$ be such that $|\mathfrak{p}| = \{a_1, \dots, a_N\}$. Let $k_j := \mathfrak{p}(a_j)$, $a_j = |a_j|e^{i\varphi_j}$, $|a_j| = R^{1-2s_j}$, $\ell(\mathfrak{p}) := \lceil k_1 s_1 + \dots + k_N s_N \rceil$, $b(\mathfrak{p}) := R^{1-2(\ell(\mathfrak{p}) - (k_1 s_1 + \dots + k_N s_N))}$. For $\theta \in \mathbb{R}$ define

$$\boldsymbol{e}_{\mathfrak{p},\theta}(z) := \frac{f(b(\mathfrak{p}), -e^{-i\theta} z)}{(Rz)^{\ell(\mathfrak{p})}} \prod_{j=1}^{N} (f(|a_j|, e^{-i\varphi_j} z))^{k_j}, \quad z \in \overline{P}.$$

Then, $\boldsymbol{e}_{\mathfrak{p},\theta} \in \mathcal{O}(\overline{P})$, $\operatorname{ord}_{a_j} \boldsymbol{e}_{\mathfrak{p},\theta} \ge k_j$, $j = 1, \dots, N$, and $|\boldsymbol{e}_{\mathfrak{p},\theta}| = 1$ on ∂P. Hence, $\boldsymbol{m}_P(\mathfrak{p}, z) \ge |\boldsymbol{e}_{\mathfrak{p},\theta}(z)|$, $z \in P$, $\theta \in \mathbb{R}$.

Proposition 9.1.7. *Under the above notation we have*

$$\boldsymbol{m}_P(\mathfrak{p}, z) = \frac{f(b(\mathfrak{p}), -|z|)}{|Rz|^{\ell(\mathfrak{p})}} \prod_{j=1}^{N} |f(|a_j|, e^{-i\varphi_j} z)|^{k_j} = |\boldsymbol{e}_{\mathfrak{p}, \operatorname{Arg} z}(z)|, \quad z \in P.$$

Proof. Use the same method as in the proof of Proposition 9.1.5. □

Example 9.1.8. Suppose that the analogue of Proposition 8.2.10 is true for Möbius functions. In particular,

$$\boldsymbol{m}_{P(R^2)}(1, z^2) = \boldsymbol{m}_{P(R)}(\{-1, +1\}, z), \quad z \in P(R),\ R > 1.$$

Then, using Proposition 9.1.7, we get

$$\frac{f_{R^2}(1, -|z|^2)}{R^2|z|^2} |f_{R^2}(1, z^2)| = \frac{1}{R|z|} |f_R(1, z) f_R(1, -z)|, \quad z \in P(R).$$

Consequently,

$$\frac{1}{R|z|}(1+|z|^2)\Pi_{R^2}(1,-|z|^2)|(1-z^2)\Pi_{R^2}(1,z^2)| \\ = |(1-z)\Pi_R(1,z)(1+z)\Pi_R(1,-z)|,$$

and hence,

$$\frac{1}{R|z|}(1+|z|^2)\Pi_{R^2}(1,-|z|^2)|\Pi_{R^2}(1,z^2)| = |\Pi_R(1,z)\Pi_R(1,-z)|, \quad z \in P(R);$$

a contradiction (at least for big R) (take $z = 1$ and then let $R \longrightarrow +\infty$).

According to our plans, we continue studying the Kobayashi case.

Proposition 9.1.9. *For any $a \in (1/R, R)$, we have*

(a) $\boldsymbol{k}_P^*(a,z) = \left[\dfrac{x^2+1-2x\cos(\pi(s-t))}{x^2+1-2x\cos(\pi(s+t))}\right]^{1/2}$, *$z \in P$, where $s := s(a)$, $t := s(|z|)$ (cf. (9.1.1)), $x := \exp\left(\frac{\pi \operatorname{Arg} z}{2\log R}\right)$;*

(b) $\boldsymbol{\varkappa}_P(a;1) = \dfrac{\pi}{4a(\log R)\sin(\pi s(a))}$.

Proof. Fix $a \in (1/R, R)$. The mapping

$$P \ni z \longmapsto \frac{z}{a} \in Q := \left\{z \in \mathbb{C} : \frac{1}{Ra} < |z| < \frac{R}{a}\right\}$$

is biholomorphic. In particular,

$$\boldsymbol{k}_P^*(a,z) = \boldsymbol{k}_Q^*\left(1, \frac{z}{a}\right), \ z \in P, \quad \text{and} \quad \boldsymbol{\varkappa}_P(a;1) = \frac{1}{a}\boldsymbol{\varkappa}_Q(1;1). \tag{9.1.4}$$

Let $r_1 := 1/Ra$, $r_2 := R/a$, and $S := \{w \in \mathbb{C} : \log r_1 < \operatorname{Re} w < \log r_2\}$. Define

$$\alpha := \frac{\pi i}{\log r_2/r_1}, \quad \lambda_0 := \exp\left(\frac{2\pi i \log r_1}{\log r_2/r_1}\right), \quad H(w) := \frac{\exp(\alpha w) - 1}{\exp(\alpha w) - \lambda_0}, \ w \in S. \tag{9.1.5}$$

Then, $H : S \longrightarrow \mathbb{D}$ is biholomorphic and the mapping $h := \exp \circ H^{-1} : \mathbb{D} \longrightarrow Q$ is a covering of Q with $h(0) = 1$. Consequently, by Theorem 3.3.7 and Exercise 3.9.8, we conclude that

$$\begin{aligned}\boldsymbol{k}_Q^*(1,z) &= \inf\{|\lambda| : \lambda \in h^{-1}(z)\} = \inf\{|H(w)| : w \in S,\ \exp(w) = z\} \\ &= |H(\log|z| + i \operatorname{Arg} z)|, \qquad z \in Q\end{aligned}$$

(the details are left to the reader), and

$$\varkappa_Q(1;1) = \frac{1}{|h'(0)|} = |H'(0)|.$$

Now, it suffices to use (9.1.4) and (9.1.5). □

Corollary 9.1.10. *For any $a \in (0,1)$ we have*

(a) $\boldsymbol{k}^*_{\mathbb{D}_*}(a,z) = \left[\dfrac{\theta^2 + (\log|z| - \log a)^2}{\theta^2 + (\log|z| + \log a)^2}\right]^{1/2}, \quad \theta = \operatorname{Arg} z;$

(b) $\boldsymbol{\varkappa}_{\mathbb{D}_*}(a;1) = -\dfrac{1}{2a\log a}.$

Proof. Use Propositions 9.1.9, 3.3.5(a), and Exercise 3.9.5(a), or use the covering method. □

Corollary 9.1.11.

(a) $\boldsymbol{k}^*_{\mathbb{D}_*}(a^\mu, z^\mu) = \min\{\boldsymbol{k}^*_{\mathbb{D}_*}(a, \varepsilon z) : \varepsilon \in \sqrt[\mu]{1}\}$, $a \in (0,1)$, $z \in \mathbb{D}_*$, $\mu \in \mathbb{N}$.

(b) $\boldsymbol{k}^*_{\mathbb{D}_*}(a^\mu, b^\mu) = \boldsymbol{k}^*_{\mathbb{D}_*}(a,b)$, $a, b \in (0,1)$, $\mu > 0$.

(c) $\boldsymbol{k}^*_{\mathbb{D}_*}(a,b) = \inf\{\boldsymbol{k}^*_{\mathbb{D}_*}(a, be^{i\varphi}) : \varphi \in \mathbb{R}\}$, $a, b \in (0,1)$.

Proof. Use the effective formula for $\boldsymbol{k}^*_{\mathbb{D}_*}$ from Corollary 9.1.10. □

Proposition 9.1.12. *Let $a \in (1/R, R)$. Then,*

(a) $\boldsymbol{m}^{(k)}_P(a,z) = [\boldsymbol{g}_P(b_k(a), -|z|)]^{1/k}\boldsymbol{g}_P(a,z)$, $z \in \overline{P}$,

$\boldsymbol{\gamma}^{(k)}_P(a;1) = [\boldsymbol{g}_P(b_k(a), -a)]^{1/k}\boldsymbol{A}_P(a;1)$ *with* $\boldsymbol{g}_P(R,\cdot) \equiv 1$.

In particular,

$\boldsymbol{c}^*_P(a,\cdot) < \boldsymbol{g}_P(a,\cdot)$ *in* $P \setminus \{a\}$,

$\boldsymbol{\gamma}_P(a;1) < \boldsymbol{A}_P(a;1)$.

(b) *For any $k \geq 2$*

$\boldsymbol{c}^*_P(a,\cdot) < \boldsymbol{m}^{(k)}_P(a,\cdot)$ *in* $P \setminus \{a\}$,

$\boldsymbol{\gamma}_P(a;1) < \boldsymbol{\gamma}^{(k)}_P(a;1)$.

(c) *For any $k \in \mathbb{N}$ the following statements are equivalent:*

(i) $\boldsymbol{m}^{(k)}_P(a,\cdot) \equiv \boldsymbol{g}_P(a,\cdot)$;

(ii) $\boldsymbol{m}^{(k)}_P(a,z_0) = \boldsymbol{g}_P(a,z_0)$ *for some* $z_0 \in P \setminus \{a\}$;

(iii) $\boldsymbol{\gamma}_P^{(k)}(a;1) = \boldsymbol{A}_P(a;1)$;

(iv) $ks(a) \in \mathbb{N}$ (*cf.* (9.1.1)).

(d) *For any* $k, \overline{k} \geq 2$, $k \neq \overline{k}$, *the following conditions are equivalent:*

(i) $\boldsymbol{m}_P^{(k)}(a,\cdot) \equiv \boldsymbol{m}_P^{(\overline{k})}(a,\cdot) \equiv \boldsymbol{g}_P(a,\cdot)$;

(ii) *there exist* $\theta \in \mathbb{R}$ *and a set* $I \subset (1/R, R)$ *such that* I *has an accumulation point in* $(1/R, R)$ *and* $\boldsymbol{m}_P^{(k)}(a, e^{i\theta}x) = \boldsymbol{m}_P^{(\overline{k})}(a, e^{i\theta}x)$, $x \in I$.

(e) *For any* k, $\overline{k} \geq 2$, $k \neq \overline{k}$ *and for any* $z_0 \in P \setminus \mathbb{R}_+$ *there exists* $a \in (1/R, R)$ *such that* $\boldsymbol{m}_P^{(k)}(a, z_0) = \boldsymbol{m}_P^{(\overline{k})}(a, z_0)$;

for any k, $\overline{k} \geq 2$, $k \neq \overline{k}$, *there exists* $a \in (1/R, R)$ *such that* $\boldsymbol{\gamma}_P^{(k)}(a;1) = \boldsymbol{\gamma}_P^{(\overline{k})}(a;1)$.

(f) $\boldsymbol{m}_P^{(k)} \longrightarrow \boldsymbol{g}_P$, $\boldsymbol{\gamma}_P^{(k)} \to \boldsymbol{A}_P$ *as* $k \longrightarrow +\infty$.

(g) $\boldsymbol{g}_P(a,\cdot) < \boldsymbol{k}_P^*(a,\cdot)$ *in* $P \setminus \{a\}$, $\quad \boldsymbol{A}_P(a;1) < \boldsymbol{\varkappa}_P(a;1)$.

Proof. (a) follows directly from Propositions 9.1.2 and 9.1.5.

(b) Suppose that $\boldsymbol{c}_P^*(a, z_0) = \boldsymbol{m}_P^{(k)}(a, z_0)$ for some $z_0 \in P \setminus \{a\}$ (resp. $\boldsymbol{\gamma}_P(a;1) = \boldsymbol{\gamma}_P^{(k)}(a;1)$). Let $x_0 := |z_0|$ (resp. $x_0 := a$). Define

$$\varphi(z) = \frac{\left[\frac{1}{Rz} f\left(\frac{1}{a}, -z\right)\right]^k}{\frac{1}{(Rz)^l} f(b, -z)}, \quad z \in \overline{P} \setminus \{-b\},$$

where $b := b_k(a)$, $l := l_k(a)$ (cf. (9.1.2)). Then $\varphi \in \mathcal{O}(\overline{P} \setminus \{-b\})$, φ has at most a single pole at $-b$, and $|\varphi| = 1$ on ∂P. By Proposition 9.1.5, $|\varphi(x_0)| = 1$. Hence, by Lemma 9.1.6, $\varphi \equiv$ const; a contradiction.

(c) By Remark 9.1.3(b), $\boldsymbol{g}_P(b_k(a), -x_0) = 1$ for some $x_0 \in (1/R, R)$ iff $ks(a) \in \mathbb{N}$. Thus, the result follows from (a).

(d) Suppose that (ii) is satisfied. Put

$$\psi_k(z) := \left[\frac{f(b_k(a), -e^{-i\theta}z)}{(R \cdot e^{-i\theta}z)^{l_k(a)}}\right]^{\overline{k}}, \quad \psi_{\overline{k}}(z) := \left[\frac{f(b_{\overline{k}}(a), -e^{-i\theta}z)}{(R \cdot e^{-i\theta}z)^{l_{\overline{k}}(a)}}\right]^k, \quad z \in \overline{P}.$$

By virtue of Proposition 9.1.5, $\psi_k(e^{i\theta}x) = \psi_{\overline{k}}(e^{i\theta}x)$, $x \in I$. Hence, $\psi_k \equiv \psi_{\overline{k}}$, and therefore $b_k(a) = b_{\overline{k}}(a) = R$ and $\overline{k}l_k(a) = kl_{\overline{k}}(a)$. Consequently, $ks(a)$, $\overline{k}s(a) \in \mathbb{N}$ and we can apply (c).

(e) Suppose that $\overline{k} > k$. Define

$$\varepsilon(x) := \boldsymbol{m}_P^{(k)}(x, z_0) - \boldsymbol{m}_P^{(\overline{k})}(x, z_0), \quad 1/R < x < R.$$

By (c), $\varepsilon(R^{1-2/k}) \geq 0$ and $\varepsilon(R^{1-2/\overline{k}}) \leq 0$. Hence, there is an $a \in [R^{1-2/k}, R^{1-2/\overline{k}}]$ with $\varepsilon(a) = 0$.

For the proof of the second assertion, put

$$\varepsilon(x) := \boldsymbol{\gamma}_P^{(k)}(x; 1) - \boldsymbol{\gamma}_P^{(\overline{k})}(x; 1), \quad 1/R < x < R,$$

and use the same argument.

(f) is a consequence of (a).

(g) This follows directly from the fact that P is not simply connected; cf. Proposition 4.2.22(a). □

Notice that some comparison results for $\boldsymbol{\gamma}_P$, $\boldsymbol{A}_P$, and $\boldsymbol{\varkappa}_P$ may be found in [26].

Now, we turn to the problem of characterizing $\boldsymbol{c}_P^i$.

Proposition 9.1.13. *For any $a \in (1/R, R)$ and $z \in P$, we have:*

$$\boldsymbol{c}_P^i(a, z) = \boldsymbol{c}_P(a, z) \ \textit{iff} \ z \in (1/R, R).$$

The result directly follows from the following

Lemma 9.1.14.

(a) *If $1/R < a < R$, $z_0 \in P \setminus \mathbb{R}_+$, and $\psi := \boldsymbol{e}_1(a, \operatorname{Arg} z_0, \cdot)$ (ψ is extremal for $\boldsymbol{c}_P(a, z_0)$), then*

$$|\psi'(a)| < \boldsymbol{\gamma}_P(a; 1).$$

(b) *For any $1/R < a < b < c < R$, the following equality holds*

$$\boldsymbol{c}_P(a, b) + \boldsymbol{c}_P(b, c) = \boldsymbol{c}_P(a, c). \tag{9.1.6}$$

Proof of Lemma 9.1.14. (a) By virtue of Proposition 9.1.5, it is enough to prove that

$$\left| f\left(\frac{1}{a}, -e^{-i\theta}a\right) \right| < f\left(\frac{1}{a}, -a\right), \quad 0 < \theta < 2\pi.$$

Let

$$\varphi(z) := \frac{f(\frac{1}{a}, -e^{-i\theta}z)}{f(\frac{1}{a}, -z)}, \quad z \in \overline{P} \setminus \{-1/a\}.$$

Then, $\varphi \in \mathcal{O}(\overline{P} \setminus \{-1/a\})$, φ has a simple pole at $-1/a$, $|\varphi| = 1$ on ∂P, and $\varphi \not\equiv \text{const}$. Hence, by Lemma 9.1.6, $|\varphi(a)| < 1$.

(b) First, observe that (9.1.6) is equivalent to

$$\frac{\boldsymbol{c}_P^*(a,b) + \boldsymbol{c}_P^*(b,c)}{1 + \boldsymbol{c}_P^*(a,b)\boldsymbol{c}_P^*(b,c)} = \boldsymbol{c}_P^*(a,c). \tag{9.1.7}$$

For $1/R < x < y < R$, put $\psi(x,y) := \boldsymbol{e}_1(x,0,y)$. Then (9.1.7) may be written as

$$\frac{\psi(a,b) + \psi(b,c)}{1 + \psi(a,b)\psi(b,c)} = \psi(a,c). \tag{9.1.8}$$

Now, we need some classical facts from the theory of theta functions; see [464] for details. For fixed $\tau \in \mathbb{C}$ with $\operatorname{Im}\tau > 0$ let

$$q := e^{i\pi\tau}, \quad C := \prod_{n=1}^{\infty}(1 - q^{2n}).$$

Set $w = w(z) := e^{2\pi i z}, z \in \mathbb{C}$. Then, the *theta functions* $\theta_0, \dots, \theta_3$ are defined by the formulas:

$$\begin{aligned}
\theta_0(z,q) &:= C\prod_{n=1}^{\infty}(1 - wq^{2n-1})\left(1 - \frac{1}{w}q^{2n-1}\right),\\
\theta_1(z,q) &:= -iq^{1/4}e^{i\pi z}\theta_0\left(z + \frac{\tau}{2}, q\right),\\
\theta_2(z,q) &:= \theta_1\left(z + \frac{1}{2}, q\right), \quad \theta_3(z,q) := \theta_0\left(z - \frac{1}{2}, q\right).
\end{aligned}$$

Further, put

$$\operatorname{cn}(z) := \sqrt{\frac{k'}{k}\frac{\theta_2(z,q)}{\theta_0(z,q)}}, \quad \operatorname{sn}(z) := \frac{1}{\sqrt{k}}\frac{\theta_1(z,q)}{\theta_0(z,q)}, \quad \operatorname{dn}(z) := \sqrt{k'}\frac{\theta_3(z,q)}{\theta_0(z,q)},$$

where $k := \theta_2^2(0,q)/\theta_3^2(0,q)$ and $k' := \theta_0^2(0,q)/\theta_3^2(0,q)$ are the *Jacobi moduli*; recall that $k^2 + k'^2 = 1$. The functions cn, sn, dn are called *cosinus amplitudinis, sinus amplitudinis*, and *delta amplitudinis*, respectively (with the periods $\omega_1 = 1/2$, $\omega_2 = \tau/2$). They satisfy the relations

$$\operatorname{cn}^2(z) + \operatorname{sn}^2(z) = 1, \quad k^2\operatorname{sn}^2(z) + \operatorname{dn}^2(z) = 1.$$

Moreover, the following addition-formulas are true:

$$\begin{aligned}
\operatorname{cn}(u \pm v) &= \frac{\operatorname{cn}(u)\operatorname{cn}(v) \mp \operatorname{sn}(u)\operatorname{sn}(v)\operatorname{dn}(u)\operatorname{dn}(v)}{1 - k^2\operatorname{sn}^2(u)\operatorname{sn}^2(v)},\\
\operatorname{sn}(u \pm v) &= \frac{\operatorname{sn}(u)\operatorname{cn}(v)\operatorname{dn}(v) \pm \operatorname{sn}(v)\operatorname{cn}(u)\operatorname{dn}(u)}{1 - k^2\operatorname{sn}^2(u)\operatorname{sn}^2(v)},\\
\operatorname{dn}(u \pm v) &= \frac{\operatorname{dn}(u)\operatorname{dn}(v) \mp k^2\operatorname{sn}(u)\operatorname{sn}(v)\operatorname{cn}(u)\operatorname{cn}(v)}{1 - k^2\operatorname{sn}^2(u)\operatorname{sn}^2(v)}.
\end{aligned}$$

The function ψ can be easily expressed in terms of cn, sn, and dn. Namely, if $x = e^{2\pi i t}$, $y = e^{2\pi i s}$, then

$$\psi(x, y) = -ik \frac{\operatorname{cn}(t+s)\operatorname{sn}(t-s)}{\operatorname{dn}(t-s)}. \tag{9.1.9}$$

Now, if we put

$$a =: e^{2\pi i\alpha}, \quad b =: e^{2\pi i\beta}, \quad c =: e^{2\pi i\gamma}$$

and if we introduce new variables $\xi := \alpha - \gamma, \eta := \alpha - \beta$, and $\zeta := \beta + \gamma$, then formula (9.1.8) is equivalent to the following identity:

$$\begin{aligned}\operatorname{cn}(\xi+\zeta)&\operatorname{sn}(\eta)\operatorname{dn}(\xi)\operatorname{dn}(\xi-\eta) + \operatorname{cn}(\zeta)\operatorname{sn}(\xi-\eta)\operatorname{dn}(\xi)\operatorname{dn}(\eta)\\ &- \operatorname{cn}(\eta+\zeta)\operatorname{sn}(\xi)\operatorname{dn}(\eta)\operatorname{dn}(\xi-\eta)\\ &+ k^2\operatorname{cn}(\zeta)\operatorname{cn}(\xi+\zeta)\operatorname{cn}(\eta+\zeta)\operatorname{sn}(\xi)\operatorname{sn}(\eta)\operatorname{sn}(\xi-\eta) = 0.\end{aligned} \tag{9.1.10}$$

Define

$$\begin{aligned}X_0 &:= k^2, & X_1 &:= \operatorname{cn}(\xi), & X_2 &:= \operatorname{sn}(\xi), & X_3 &:= \operatorname{dn}(\xi), & X_4 &:= \operatorname{cn}(\eta),\\ X_5 &:= \operatorname{sn}(\eta), & X_6 &:= \operatorname{dn}(\eta), & X_7 &:= \operatorname{cn}(\zeta), & X_8 &:= \operatorname{sn}(\zeta), & X_9 &:= \operatorname{dn}(\zeta).\end{aligned}$$

Using the above notation and the addition-formulas, we can write (9.1.10) in the form $R(X) = 0$ with $R := R_1 + R_2 + R_3 + R_4$, where

$$\begin{aligned}&R_1(X) := (1 - X_0X_5^2X_8^2)(X_1X_7 - X_2X_3X_8X_9)(X_3X_6 + X_0X_1X_2X_4X_5)X_3X_5,\\ &R_2(X) := (1 - X_0X_5^2X_8^2)(1 - X_0X_2^2X_8^2)(X_2X_4X_6 - X_1X_3X_5)X_3X_6X_7,\\ &R_3(X) := (1 - X_0X_2^2X_8^2)(X_5X_6X_8X_9 - X_4X_7)(X_3X_6 + X_0X_1X_2X_4X_5)X_2X_6,\\ &R_4(X) :=\\ &(X_4X_7 - X_5X_6X_8X_9)(X_1X_7 - X_2X_3X_8X_9)(X_2X_4X_6 - X_1X_3X_5)X_0X_2X_5X_7.\end{aligned}$$

Of course, the variables $X_0, \dots, X_9$ are not independent. They are connected by the relations

$$\begin{aligned}S_{2j+1}(X) &:= X_{3j+1}^2 + X_{3j+2}^2 - 1 = 0, \quad j = 0, 1, 2,\\ S_{2j}(X) &:= X_0X_{3j-1}^2 + X_{3j}^2 - 1 = 0, \quad j = 1, 2, 3.\end{aligned}$$

So, finally, the problem reduces to the following implication:
for any $X = (X_0, \dots, X_9) \in \mathbb{C}^{10}$ *if* $S_j(X) = 0$, $j = 1, \dots, 6$, *then* $R(X) = 0$.
In fact, this is true! We advise the reader to use a computer; cf. Remark 9.1.15. □

Remark 9.1.15. The polynomial R belongs to the ideal in $\mathbb{C}[X_0, \dots, X_9]$ generated by $S_1, \dots, S_6$; it is of the 16-th degree and consists of 24 monomials. If suffices to substitute all X_{3j+1}^2's by $1 - X_{3j+2}^2$, $j = 0, 1, 2$, and all X_{3j}^2's by $1 - X_0X_{3j-1}^2$, $j = 1, 2, 3$.

Remark 9.1.16. There is an alternative, more elementary, proof of Lemma 9.1.14(b), which has been presented in [324]. First, observe that for any $x \in (1/R, R)$ the function $\boldsymbol{c}_P(x,\cdot) = \tanh^{-1}\psi(x,\cdot)$ is of class $\mathcal{C}^\infty((x,R))$. Define $G_x(y) := (\boldsymbol{c}_P(x,\cdot))'(y)$, $x < y < R$. Then, using (9.1.9) and the properties of theta functions, we get, after some transformations,

$$G_x(y) = \frac{k}{2\pi y}\operatorname{sn}'(0)\operatorname{cn}(2s), \quad y = e^{2\pi i s} \in (x, R).$$

In particular, $G_a(y) = G_b(y)$, $y \in (b, R)$. Finally,

$$\boldsymbol{c}_P(a,b) + \boldsymbol{c}_P(b,c) = \int_a^b G_a(y)dy + \int_b^c G_b(y)dy = \int_a^c G_a(y)dy = \boldsymbol{c}_P(a,c).$$

Remark 9.1.17. Proposition 9.1.13 has been generalized in [259], where for arbitrary $a, b \in P$ we present a characterization of all piecewise $\mathcal{C}^1$ curves $\alpha : [0,1] \longrightarrow P$, $\alpha(0) = 1$, $\alpha(1) = b$, whose $L_{\boldsymbol{\gamma}_P}$-length is minimal in the sense that $L_{\boldsymbol{\gamma}_P}(\alpha) = \boldsymbol{c}_P^i(a,b)$. In particular, if $a \in (1/R, R)$ and $b \in P \setminus \mathbb{R}_-$, then such a curve is uniquely determined.

We move to the problems related to $\boldsymbol{c}$- and $\boldsymbol{\gamma}$-isometries of P.

Lemma 9.1.18. *For any $k \in \mathbb{N}$ and for any $z_0, w_0 \in P$, $z_0 \neq w_0$, (resp. $z_0 \in P$) the extremal function for $\boldsymbol{m}_P^{(k)}(z_0, w_0)$ (resp. $\boldsymbol{\gamma}_P^{(k)}(z_0; 1)$) is uniquely determined up to rotations.*

Proof. We may assume that $z_0 = a \in (1/R, R)$. We already know (cf. Proposition 9.1.5) that the function $\varphi_0 := \boldsymbol{e}_k(a,\theta,\cdot)$ ($w_0 = |w_0|e^{i\theta}$) (resp. $\varphi_0 := \boldsymbol{e}_k(a,0,\cdot)$) is extremal for $\boldsymbol{m}_P^{(k)}(a, w_0)$ (resp. $\boldsymbol{\gamma}_P^{(k)}(a;1)$). Let $h \in \mathcal{O}(P,\mathbb{D})$ ($\operatorname{ord}_a f \geq k$) be another extremal function for $\boldsymbol{m}_P^{(k)}(a, w_0)$ (resp. $\boldsymbol{\gamma}_P^{(k)}(a;1)$). Set $\varphi := h/\varphi_0$. Then φ satisfies the assumptions of Lemma 9.1.6 and $|\varphi(w_0)| = 1$ (resp. $|\varphi(a)| = 1$). Consequently, $h = e^{i\alpha}\varphi_0$ for some $\alpha \in \mathbb{R}$. □

Note that, in particular, the lemma says that all the extremal functions are proper.

Proposition 9.1.19 (Schwarz–Pick lemma for P). *Let $F \in \mathcal{O}(P,P)$. Then, the following conditions are equivalent:*

(i) $\boldsymbol{c}_P^*(F(z_0), F(w_0)) = \boldsymbol{c}_P^*(z_0, w_0)$ *for some* $z_0, w_0 \in P$, $z_0 \neq w_0$;

(ii) $\boldsymbol{\gamma}_P(F(z_0); F'(z_0)) = \boldsymbol{\gamma}_P(z_0; 1)$ *for some* $z_0 \in P$;

(iii) $F \in \operatorname{Aut}(P)$.

Proof. Assume that (i) (resp. (ii)) holds. The only difficulty is to show that F is proper; cf. Appendix B.1.4. Let φ_0 be an extremal function for $\boldsymbol{c}_P^*(F(z_0), F(w_0))$ (resp. for $\boldsymbol{\gamma}_P(F(z_0); F'(z_0))$. By Lemma 9.1.18, $\varphi_0 \circ F$ is proper as an extremal function for $\boldsymbol{c}_P^*(z_0, w_0)$ (resp. for $\boldsymbol{\gamma}_P(z_0; 1)$), and therefore the function F itself is proper. □

Proposition 9.1.20. *Let $F : P \longrightarrow P$. Then, the following statements are equivalent (cf. Proposition* 1.1.20*):*

(i) *F is a $\boldsymbol{c}$-isometry;*

(ii) *F is $\mathcal{C}^1$ and F is a $\boldsymbol{\gamma}$-isometry;*

(iii) $F \in \operatorname{Aut}(P) \cup \overline{\operatorname{Aut}(P)}$.

Proof. It is easily seen that (iii) $\Longrightarrow$ (ii) and (iii) $\Longrightarrow$ (i).

(i) $\Longrightarrow$ (iii): If $F : P \to P$ is a $\boldsymbol{c}$-isometry, then one can easily prove that F is continuous, injective, and proper; therefore F is a homeomorphism. Moreover, by Proposition 2.7.1(d),

$$\lim_{z \nrightarrow z_0} \left| \frac{F(z) - F(z_0)}{z - z_0} \right| = \frac{\boldsymbol{\gamma}_P(z_0; 1)}{\boldsymbol{\gamma}_P(F(z_0); 1)} \in (0, +\infty), \quad z_0 \in P.$$

Hence, by the Bohr theorem (cf. Appendix B.1.5), F is either holomorphic or anti-holomorphic. Now, we can apply Proposition 9.1.19.

(ii) $\Longrightarrow$ (iii): As in the proof of Proposition 1.1.20, we easily obtain that F belongs to $\mathcal{O}(P) \cup \overline{\mathcal{O}(P)}$, and so we can apply Proposition 9.1.19 once again. □

? **Remark 9.1.21.** We would like to emphasize that we do not know any formulas for the invariant objects on n-connected domains with $n \geq 3$.

9.2 Exercises

Exercise 9.2.1. For $\lambda \in P$ let

$$\mathcal{F}(\lambda) := \{z \in P : \exists_{F \in \mathcal{O}(P,P)} : F(1) = 1, F(z) = \lambda\}.$$

Then,

$$\mathcal{F}(\lambda) = \{\lambda, 1/\lambda\} \cup \{z \in P : \boldsymbol{c}_P^*(1, z) \geq \boldsymbol{k}_P^*(1, \lambda)\}.$$

Hint. Complete the following sketch of the proof:
Fix a $\lambda_0 \in P \setminus \{1\}$.

(a) If $z_0 \in P$ is such that $t_- := \boldsymbol{c}_P^*(1, z_0) \geq \boldsymbol{k}_P^*(1, \lambda_0) =: t_+$, then we can take

$$F(z) := \varphi\left(\frac{t_+}{t_-}\psi(z)\right), \quad z \in P,$$

where $\psi \in \mathcal{O}(P, \mathbb{D})$, $\varphi \in \mathcal{O}(\mathbb{D}, P)$, $\psi(1) = 0, \psi(z_0) = t_-$, $\varphi(0) = 1$, and $\varphi(t_+) = \lambda_0$.

(b) Let $z_0 \in P$ and $F \in \mathcal{O}(P, P)$ be such that $F(1) = 1$ and $F(z_0) = \lambda_0$. Suppose that S and H are the same as in the proof of Proposition 9.1.9 (with $r_1 := 1/R$, $r_2 := R$). Then there exists an $\widetilde{F} \in \mathcal{O}(S, S)$ such that $\widetilde{F}(0) = 0$ and $\exp \circ \widetilde{F} = F \circ \exp$. In particular, there exists a $k_0 \in \mathbb{Z}$ such that

$$\widetilde{F}(w + 2\pi i) = \widetilde{F}(w) + 2k_0\pi i, \quad w \in S.$$

By the Schwarz lemma, applied to the function $H \circ \widetilde{F} \circ H^{-1}$, we conclude that $|H(2k_0\pi i)| \leq |H(2\pi i)|$, and hence $k_0 \in \{-1, 0, 1\}$. If $k_0 = \pm 1$, then either $F(z) \equiv z$ or $F(z) \equiv 1/z$. If $k_0 = 0$, then there exists $\psi \in \mathcal{O}(P, \mathbb{D})$ such that $\psi(1) = 0$ and $H \circ \widetilde{F} = \psi \circ \exp$. Consequently, $\boldsymbol{k}_P^*(1, \lambda_0) \leq \boldsymbol{c}_P^*(1, z_0)$.

Exercise 9.2.2. Prove that if $\boldsymbol{k}_P^*(1, \lambda) \leq \boldsymbol{c}_P^*(1, z)$, then

$$\max\{\boldsymbol{c}_P^*(1, z), \boldsymbol{c}_P^*(1, \lambda)\} = \boldsymbol{c}_{P \times P}^*((1, 1), (z, \lambda)).$$

Exercise 9.2.3. Let $G_1, G_2 \subset \mathbb{C}$ be $\boldsymbol{c}$-hyperbolic domains (cf. Proposition 2.5.1). Using the methods of the proof of Proposition 9.1.19, prove that any $\boldsymbol{c}$- or $\boldsymbol{\gamma}$-isometry $F : G_1 \longrightarrow G_2$ is either holomorphic or antiholomorphic. Observe that the $\boldsymbol{c}$-hyperbolicity is essential.

Exercise 9.2.4. Let $G_j \subset \mathbb{C}$ be a bounded ν_j-connected domain which is regular with respect to the Dirichlet problem, $j = 1, 2$. Prove that any holomorphic mapping $F : G_1 \longrightarrow G_2$ with

$$\boldsymbol{c}_{G_2}^*(F(z_0'), F(z_0'')) = \boldsymbol{c}_{G_1}^*(z_0', z_0'') \quad \text{for some } z_0', z_0'' \in G_1,\ z_0' \neq z_0'',$$

is a proper k-fold ramified covering with $\nu_1 = k \cdot \nu_2$. In particular, if $\nu_1 = \nu_2$, then F has to be biholomorphic (cf. Proposition 9.1.20), and if $\nu_2 \nmid \nu_1$, then there is no such a mapping.

Hint. Use the following general result of H. Grunsky (cf. [201, 202]): if $G \subset \mathbb{C}$ is a bounded ν-connected domain regular w.r.t. the Dirichlet problem, then any holomorphic function $f : G \longrightarrow \mathbb{D}$ extremal for $\boldsymbol{c}_G^*(z', z'')$ $(z', z'' \in G,\ z' \neq z'')$ is a proper ν-fold ramified covering.

Exercise 9.2.5. Let $F(z) := \frac{1}{2}(z + \frac{1}{z})$, $z \in P$, $L := F(P)$. Prove that $F : P \longrightarrow L$ is a 2-fold ramified covering such that

$$\boldsymbol{c}_L^*(F(1), F(-1)) = \boldsymbol{c}_P^*(1, -1); \quad \text{cf. Exercise 9.2.4.}$$

Hint. For any $f \in \mathcal{O}(P, \mathbb{D})$, the function

$$\widetilde{f}(w) := \frac{1}{2}\Big[f(w + \sqrt{w^2 - 1}) + f(w - \sqrt{w^2 - 1})\Big], \quad w \in L,$$

belongs to $\mathcal{O}(L, \mathbb{D})$. □

5.6. Modify the domains P and L from Exercise 9.2.5 to get a 2ν-connected (resp. ν-connected) domain P_ν (resp. L_ν) that is regular with respect to the Dirichlet problem and such that $F : P_\nu \longrightarrow L_\nu$ is a 2-fold ramified covering with

$$\boldsymbol{c}^*_{L_\nu}(F(1), F(-1)) = \boldsymbol{c}^*_{P_\nu}(1, -1).$$

Exercise 9.2.6 (cf. [250]). Let $G := \{(z_1, z_2) \in \mathbb{C}^2 : |z_1| < 1, |z_2| < 1, |z_1 z_2| < 1/2\}$; cf. Remark 2.10.8. Take $R := 2$ and let $P := P(2)$. Define $F = (F_1, F_2) : \overline{P} \longrightarrow \mathbb{C}^2$ by the formula

$$F(\lambda) := \frac{1}{2}\left(f\left(1, \frac{1}{\lambda}\right), f(1, \lambda)\right), \quad \lambda \in \overline{P}.$$

(a) Prove that F is injective, $F(P) \subset G$, and $F(\partial P) \subset \partial G$.

(b) Using the contraction-property of $F : P \longrightarrow G$ and Proposition 9.1.5, prove that

$$\boldsymbol{c}^*_G(0, z) = 2|z_1 z_2|, \quad z = (z_1, z_2) \in \Gamma,$$

where

$$\Gamma := \{(z_1, z_2) \in \mathbb{C}^2 : \exists_{t \in (-2, -1/2)} : |z_j| = F_j(t),\ j = 1, 2\}.$$

(c) Using (b) and the maximum principle, prove that

$$\boldsymbol{c}^*_G(0, z) = 2|z_1 z_2|, \quad z = (z_1, z_2) \in \Gamma_+,$$

where

$$\Gamma_+ := \{\lambda z : z = (z_1, z_2) \in \Gamma,\ 1 \le |\lambda| < (2|z_1 z_2|)^{-1/2}\};$$

cf. Figure 2.1.

Exercise 9.2.7. Let $G := \mathbb{D}^2 \setminus (r\overline{\mathbb{D}})^2$, where $0 < r < 1$. Prove that for arbitrary $\varepsilon \in (0, 1 - r)$ we have

$$\boldsymbol{g}_G((0, t), (0, s)) \not\equiv \boldsymbol{g}_{\mathbb{D}^2}((0, t), (0, s)) \quad \text{for } r < t, s < r + \varepsilon.$$

(This example is due to S. Kołodziej.)

Hint. Fix $t \in (r, 1)$ and use the function u defined for $(z, w) \in G$ as follows:

$$u(z, w) := \begin{cases} \max\{|z|, r\boldsymbol{g}_P(t, w)\} & \text{if } |w| > r \\ |z| & \text{if } |w| \le r \end{cases},$$

where $P := \{\lambda \in \mathbb{C} : r < |\lambda| < 1\}$.

Exercise 9.2.8. Disprove that for $\widetilde{g}_P := \tanh^{-1} g_P$ the triangle inequality holds.

Exercise 9.2.9. Show that in general the following inequality $\int \gamma_P^{(s)} \leq \int \gamma_P^{(s+1)}$ does not hold.

9.3 List of problems

Chapter 10

Elementary n-circled domains III

Summary. Our aim is to continue the discussion from § 6.3 and to effectively determine $\boldsymbol{\ell}^{(k)}_{\boldsymbol{D}_\alpha}$, $\boldsymbol{k}_{\boldsymbol{D}_\alpha}$, and $\boldsymbol{\varkappa}^{(k)}_{\boldsymbol{D}_\alpha}$.

10.1 Elementary n-circled domains III

This chapter is based on [422] and [565]. Let $a \in \boldsymbol{D}_\alpha$ be as in § 6.3 with (AS) (page 238).

Proposition 10.1.1. *Assume that conditions* (AS) *are satisfied and* $q(\alpha) = n$. *Then for* $z \in \boldsymbol{D}_\alpha$, $X \in \mathbb{C}^n$, *and* $k \in \mathbb{N}$ *we have*

$$\boldsymbol{\ell}^{(k)}_{\boldsymbol{D}_\alpha}(a,z) = \boldsymbol{k}_{\boldsymbol{D}_\alpha}(a,z) = \begin{cases} \boldsymbol{k}_{\mathbb{D}_*}(a^\alpha, z^\alpha), & \text{if } \boldsymbol{D}_\alpha \text{ is of rational type} \\ \boldsymbol{k}_{\mathbb{D}_*}(a^\alpha, |z^\alpha|), & \text{if } \boldsymbol{D}_\alpha \text{ is of irrational type} \end{cases},$$

$$\boldsymbol{\varkappa}^{(k)}_{\boldsymbol{D}_\alpha}(a;X) = \boldsymbol{\varkappa}_{\mathbb{D}_*}\Big(a^\alpha; a^\alpha \sum_{j=1}^{n} \frac{\alpha_j X_j}{a_j}\Big).$$

Proof. The mapping

$$\mathbb{C}^{n-1} \times \mathbb{D}_* \ni (\lambda_1, \dots, \lambda_n) \overset{\Pi}{\longmapsto} (e^{\alpha_n\lambda_1}, \dots, e^{\alpha_n\lambda_{n-1}}, (1/\lambda_n)e^{-(\alpha_1\lambda_1 + \dots + \alpha_{n-1}\lambda_{n-1})}) \in \boldsymbol{D}_\alpha$$

is a holomorphic covering. Hence, by Theorem 3.3.7, Exercise 3.9.8, Propositions 3.7.1, 3.8.8, 3.8.7, and Corollary 3.8.9, we get

$$\begin{aligned}
\boldsymbol{\ell}^{(k)}_{\boldsymbol{D}_\alpha}(a,z) &= \inf\{\boldsymbol{\ell}^{(k)}_{\mathbb{C}^{n-1}\times\mathbb{D}_*}(\lambda',\lambda'') : \lambda'' \in \Pi^{-1}(z)\} \\
&= \inf\{\boldsymbol{\ell}^{(k)}_{\mathbb{D}_*}(\lambda'_n,\lambda''_n) : \lambda'' \in \Pi^{-1}(z)\} = \inf\{\boldsymbol{k}_{\mathbb{D}_*}(\lambda'_n,\lambda''_n) : \lambda'' \in \Pi^{-1}(z)\} \\
&= \inf\{\boldsymbol{k}_{\mathbb{C}^{n-1}\times\mathbb{D}_*}(\lambda',\lambda'') : \lambda'' \in \Pi^{-1}(z)\} = \boldsymbol{k}_{\boldsymbol{D}_\alpha}(a,z),
\end{aligned}$$

where $\lambda' \in \Pi^{-1}(a)$ is arbitrarily fixed, and

$$\boldsymbol{\varkappa}^{(k)}_{\boldsymbol{D}_\alpha}(a;X) = \boldsymbol{\varkappa}^{(k)}_{\mathbb{C}^{n-1}\times\mathbb{D}_*}(\lambda;Y) = \boldsymbol{\varkappa}^{(k)}_{\mathbb{D}_*}(\lambda_n;Y_n) = \boldsymbol{\varkappa}_{\mathbb{D}_*}(\lambda_n;Y_n),$$

where $\lambda \in \Pi^{-1}(a)$ is arbitrarily fixed and $Y := (\Pi'(\lambda))^{-1}(X)$.

We have $\Pi(\lambda) = \zeta$ iff there exist $s_1, \dots, s_{n-1} \in \mathbb{Z}$ such that

$$\lambda_j = \frac{1}{\alpha_n}(\log|\zeta_j| + i(\operatorname{Arg}\zeta_j + 2s_j\pi)), \quad j = 1, \dots, n-1,$$

$$\lambda_n = (1/\zeta_n)\exp\Big(-\frac{1}{\alpha_n}\sum_{j=1}^{n-1}\alpha_j(\log|\zeta_j| + i(\operatorname{Arg}\zeta_j + 2s_j\pi))\Big)$$

$$= |\zeta^\alpha|^{-1/\alpha_n}\exp\Big(-\frac{i}{\alpha_n}\sum_{j=1}^{n}\alpha_j(\operatorname{Arg}\zeta_j + 2s_j\pi)\Big),$$

where $s_n := 0$. Thus,

$$\boldsymbol{\ell}^{(k)}_{\boldsymbol{D}_\alpha}(a,z) = \boldsymbol{k}_{\boldsymbol{D}_\alpha}(a,z) = \inf\Big\{\boldsymbol{k}_{\mathbb{D}_*}\Big((a^\alpha)^{-1/\alpha_n},$$
$$|z^\alpha|^{-1/\alpha_n}\exp\Big(-\frac{i}{\alpha_n}\sum_{j=1}^{n}\alpha_j(\operatorname{Arg}z_j + 2s_j\pi)\Big)\Big) : s_1, \dots, s_{n-1} \in \mathbb{Z},\ s_n = 0\Big\}.$$

In the rational case ($\alpha_1, \dots, \alpha_n$ are relatively prime), using Corollary 9.1.11(a), we have

$$\boldsymbol{\ell}^{(k)}_{\boldsymbol{D}_\alpha}(a,z) = \boldsymbol{k}_{\boldsymbol{D}_\alpha}(a,z) = \inf\Big\{\boldsymbol{k}_{\mathbb{D}_*}\Big((a^\alpha)^{-1/\alpha_n},$$
$$|z^\alpha|^{-1/\alpha_n}\exp\Big(-\frac{i}{\alpha_n}\sum_{j=1}^{n}\alpha_j(\operatorname{Arg}z_j + 2s_j\pi)\Big)\Big) : s_1, \dots, s_{n-1} \in \mathbb{Z},\ s_n = 0\Big\}$$
$$= \inf\Big\{\boldsymbol{k}_{\mathbb{D}_*}\Big((a^\alpha)^{-1/\alpha_n},$$
$$|z^\alpha|^{-1/\alpha_n}\exp\Big(-\frac{i}{\alpha_n}\sum_{j=1}^{n}\alpha_j\operatorname{Arg}z_j\Big)\exp\Big(-\frac{2\pi i}{\alpha_n}\sum_{j=1}^{n-1}\alpha_j s_j\Big)\Big) : s_1, \dots, s_{n-1} \in \mathbb{Z}\Big\}$$
$$= \inf\Big\{\boldsymbol{k}_{\mathbb{D}_*}\Big((a^\alpha)^{-1/\alpha_n}, |z^\alpha|^{-1/\alpha_n}\exp\Big(-\frac{i}{\alpha_n}\sum_{j=1}^{n}\alpha_j\operatorname{Arg}z_j\Big)\varepsilon\Big) : \varepsilon \in \sqrt[-\alpha_n]{1}\Big\}$$
$$= \boldsymbol{k}_{\mathbb{D}_*}\Big(a^\alpha, |z^\alpha|\exp\Big(i\sum_{j=1}^{n}\alpha_j\operatorname{Arg}z_j\Big)\Big) = \boldsymbol{k}_{\mathbb{D}_*}(a^\alpha, z^\alpha).$$

In the irrational case, using the Kronecker theorem (Appendix B.10.1) and Corollary 9.1.11(b, c), we have

$$\begin{aligned}\boldsymbol{\ell}^{(k)}_{\boldsymbol{D}_\alpha}(a,z) &= \boldsymbol{k}_{\boldsymbol{D}_\alpha}(a,z) = \inf\Bigg\{\boldsymbol{k}_{\mathbb{D}_*}\Bigg((a^\alpha)^{-1/\alpha_n},\\ &|z^\alpha|^{-1/\alpha_n}\exp\Bigg(-\frac{i}{\alpha_n}\sum_{j=1}^{n}\alpha_j \operatorname{Arg} z_j\Bigg)\exp\Bigg(-\frac{2\pi i}{\alpha_n}\sum_{j=1}^{n-1}\alpha_j s_j\Bigg)\Bigg) : s_1,\dots,s_{n-1}\in\mathbb{Z}\Bigg\}\\ &= \inf\{\boldsymbol{k}_{\mathbb{D}_*}((a^\alpha)^{-1/\alpha_n}, |z^\alpha|^{-1/\alpha_n}e^{i\varphi}) : \varphi\in\mathbb{R}\}\\ &= \boldsymbol{k}_{\mathbb{D}_*}((a^\alpha)^{-1/\alpha_n}, |z^\alpha|^{-1/\alpha_n}) = \boldsymbol{k}_{\mathbb{D}_*}(a^\alpha, |z^\alpha|).\end{aligned}$$

Moreover, if $\Pi(\lambda) = \zeta$, then $\Pi'(\lambda)(Y) = X$ iff

$$Y_j = \frac{X_j}{\alpha_n \zeta_j}, \quad j = 1,\dots,n-1,$$

$$-\Bigg(\Bigg(\sum_{j=1}^{n-1}\frac{\alpha_j Y_j}{\zeta_n}\Bigg) + \frac{Y_n}{\zeta_n^2}\Bigg)\exp\Bigg(-\sum_{j=1}^{n-1}\alpha_j\zeta_j\Bigg) = X_n.$$

In the case $\Pi(\lambda) = a$ with $\lambda_j = (1/\alpha_n)\log a_j$, $j = 1,\dots,n-1$, $\lambda_n = (a^\alpha)^{-1/\alpha_n}$, we have

$$Y_n = -\frac{1}{\alpha_n}(a^\alpha)^{-1/\alpha_n}\sum_{j=1}^{n}\frac{\alpha_j X_j}{a_j}.$$

Consequently, using the relation $\boldsymbol{\varkappa}_{\mathbb{D}_*}(x;1) = \boldsymbol{\varkappa}_{\mathbb{D}_*}(x^\mu; \mu x^{\mu-1})$, $x\in(0,1)$, $\mu>0$ (with $x := (a^\alpha)^{-1/\alpha_n}$, $\mu := -\alpha_n$), we get

$$\boldsymbol{\varkappa}_{\boldsymbol{D}_\alpha}(a;X) = \boldsymbol{\varkappa}_{\mathbb{D}_*}(\lambda; Y_n) = \boldsymbol{\varkappa}_{\mathbb{D}_*}\Bigg(a^\alpha; a^\alpha\sum_{j=1}^{n}\frac{\alpha_j X_j}{a_j}\Bigg). \qquad \square$$

Remark 10.1.2. Observe that the proof of Proposition 10.1.1 gives a negative answer to the conjecture posed by S. Kobayashi (cf. [317], p. 48), which says that for any holomorphic covering $\Pi : \widetilde{G} \longrightarrow G$ and for any $x, y \in G$, and $\widetilde{x}\in\Pi^{-1}(x)$, there exists a $\widetilde{y}\in\Pi^{-1}(y)$ with $\boldsymbol{k}_G(x,y) = \boldsymbol{k}_{\widetilde{G}}(\widetilde{x},\widetilde{y})$.

Indeed (cf. [562]), under the assumptions of Proposition 10.1.1, suppose that $\boldsymbol{D}_\alpha$ is of irrational type, $\alpha_n = -1$, $z_j = a_j$, $j = 1,\dots,n-1$, and $|z_n| = a_n$. Then, by Proposition 10.1.1, $\boldsymbol{k}_{\boldsymbol{D}_\alpha}(a,z) = 0$. Let

$$\begin{aligned}&\mathbb{C}^{n-1}\times\mathbb{D}_* \ni (\lambda_1,\dots,\lambda_n) \overset{\Pi}{\longmapsto}\\ &\qquad (e^{-\lambda_1},\dots,e^{-\lambda_{n-1}}, (1/\lambda_n)e^{-(\alpha_1\lambda_1+\dots+\alpha_{n-1}\lambda_{n-1})}) \in \boldsymbol{D}_\alpha\end{aligned}$$

be as in the proof of Proposition 10.1.1. Fix a $\lambda' \in \Pi^{-1}(a)$. We know that

$$0 = \boldsymbol{k}_{\boldsymbol{D}_\alpha}(a,z) = \inf\{\boldsymbol{k}_{\mathbb{C}^{n-1}\times\mathbb{D}_*}(\lambda',\lambda'') : \lambda'' \in \Pi^{-1}(z)\}$$
$$= \inf\Big\{\boldsymbol{k}_{\mathbb{D}_*}\Big(a^\alpha, a^\alpha \exp\Big(-i \operatorname{Arg} z_n + 2\pi i \sum_{j=1}^{n-1} \alpha_j s_j\Big)\Big) : s_1,\dots,s_{n-1} \in \mathbb{Z}\Big\}.$$

Consequently, the infimum is attained iff there exist $s_1,\dots,s_{n-1} \in \mathbb{Z}$ such that $\frac{1}{2\pi}\operatorname{Arg} z_n + \sum_{j=1}^{n-1}\alpha_j s_j \in \mathbb{Z}$. Thus, to get a contradiction, we only need to take z_n in such a way that $\frac{1}{2\pi}\operatorname{Arg} z_n \notin \mathbb{Z}\alpha_1 + \dots + \mathbb{Z}\alpha_{n-1} + \mathbb{Z}$.

Proposition 10.1.3. *Assume that conditions* (AS) *are satisfied and $s = s(a) < n$. Then, for $z \in \boldsymbol{D}_\alpha$, $X \in \mathbb{C}^n$, and $k \in \mathbb{N}$ we have*

$$\boldsymbol{\ell}^{(k)*}_{\boldsymbol{D}_\alpha}(a,z) = \Big(\prod_{j=1}^{n} |z_j|^{\alpha_j}\Big)^{1/r}, \quad \boldsymbol{\varkappa}^{(k)}_{\boldsymbol{D}_\alpha}(a;X) = \Big(\prod_{j=1}^{s} a_j^{\alpha_j} \prod_{j=s+1}^{n} |X_j|^{\alpha_j}\Big)^{1/r},$$

where $r = r(a)$.

Proof. The proof of Proposition 6.3.3 gives

$$\Big(\prod_{j=1}^{n} |z_j|^{\alpha_j}\Big)^{1/r} = \boldsymbol{g}_{\boldsymbol{D}_\alpha}(a,z) = \boldsymbol{\ell}^{(k)*}_{\boldsymbol{D}_\alpha}(a,z), \quad z \in \boldsymbol{D}_\alpha \setminus \boldsymbol{V}_0.$$

In particular, $\boldsymbol{\ell}^*_{\boldsymbol{D}_\alpha}(a,\cdot) \in \mathcal{PSH}(\boldsymbol{D}_\alpha \setminus \boldsymbol{V}_0)$. Consequently, since $\boldsymbol{V}_0$ is pluripolar and the function $\boldsymbol{\ell}^*_{\boldsymbol{D}_\alpha}(a,\cdot)$ is upper semicontinuous, we conclude that $\boldsymbol{\ell}^*_{\boldsymbol{D}_\alpha}(a,\cdot) \in \mathcal{PSH}(\boldsymbol{D}_\alpha)$ (cf. Appendix B.4.23). Finally,

$$\Big(\prod_{j=1}^{n} |z_j|^{\alpha_j}\Big)^{1/r} = \boldsymbol{g}_{\boldsymbol{D}_\alpha}(a,z) = \boldsymbol{\ell}^{(k)*}_{\boldsymbol{D}_\alpha}(a,z), \quad z \in \boldsymbol{D}_\alpha.$$

Proposition 6.3.3 also gives

$$\Big(\prod_{j=1}^{s} a_j^{\alpha_j} \prod_{j=s+1}^{n} |X_j|^{\alpha_j}\Big)^{1/r} = \boldsymbol{A}_{\boldsymbol{D}_\alpha}(a;X) \le \boldsymbol{\varkappa}^{(k)}_{\boldsymbol{D}_\alpha}(a;X) \le \boldsymbol{\varkappa}_{\boldsymbol{D}_\alpha}(a;X), \quad X \in \mathbb{C}^n.$$

Consider first the case, where $X_{s+1}\cdots X_n \neq 0$. Put

$$\sigma := \Big(\prod_{j=1}^{s} a_j^{\alpha_j} \prod_{j=s+1}^{n} |X_j|^{\alpha_j}\Big)^{1/r} > 0.$$

Using Lemma 6.3.9(b), we find $\psi_1, \dots, \psi_n \in \mathcal{O}(\mathbb{D}, \mathbb{C}_*)$ such that

- $\psi_j(0) = a_j$, $j = 1, \dots, s$, $\psi_j(0) = X_j/\sigma$, $j = s+1, \dots, n$,
- $\sigma\psi_j'(0) = X_j$, $j = 1, \dots, s$,
- $|\psi_1|^{\alpha_1} \cdots |\psi_n|^{\alpha_n} \equiv 1$.

Define $\varphi = (\psi_1, \dots, \psi_s, \lambda\psi_{s+1}, \dots, \lambda\psi_n)$. Then, $\varphi \in \mathcal{O}(\mathbb{D}, \boldsymbol{D}_\alpha)$, $\varphi(0) = a$, and $\sigma\varphi'(0) = X$. Thus,

$$\Big(\prod_{j=1}^{s} a_j^{\alpha_j} \prod_{j=s+1}^{n} |X_j|^{\alpha_j}\Big)^{1/r} = \boldsymbol{A}_{\boldsymbol{D}_\alpha}(a; X) = \boldsymbol{\varkappa}_{\boldsymbol{D}_\alpha}^{(k)}(a; X),$$
$$X \in \mathbb{C}^n,\ X_{s+1} \cdots X_n \neq 0.$$

In particular, $\boldsymbol{\varkappa}_{\boldsymbol{D}_\alpha}(a; \cdot) \in \mathcal{PSH}(\mathbb{C}^n \setminus \{X_{s+1} \cdots X_n = 0\})$ (cf. Lemma 4.2.4). Hence, by Appendix B.4.23, $\boldsymbol{\varkappa}_{\boldsymbol{D}_\alpha}(a; \cdot) \in \mathcal{PSH}(\mathbb{C}^n)$. Finally, using the identity principle for psh functions (Appendix B.4.20), we get

$$\Big(\prod_{j=1}^{s} a_j^{\alpha_j} \prod_{j=s+1}^{n} |X_j|^{\alpha_j}\Big)^{1/r} = \boldsymbol{A}_{\boldsymbol{D}_\alpha}(a; X) = \boldsymbol{\varkappa}_{\boldsymbol{D}_\alpha}^{(k)}(a; X), \quad X \in \mathbb{C}^n. \qquad \square$$

Proposition 10.1.4. *Assume that conditions* (AS) *are satisfied, $q(\alpha) < n$, and $s(a) = n$. Then for $z \in \boldsymbol{D}_\alpha$ and $X \in \mathbb{C}^n$ we have*

$$\boldsymbol{\ell}^*_{\boldsymbol{D}_\alpha}(a, z) = \begin{cases} \min\limits_{\zeta_1 \in \sqrt[t]{a^\alpha},\ \zeta_2 \in \sqrt[t]{z^\alpha}} \{\boldsymbol{m}_{\mathbb{D}}(\zeta_1, \zeta_2)\}, & \text{if } \boldsymbol{D}_\alpha \text{ is of rational type}, z \notin V_0 \\ \boldsymbol{m}_{\mathbb{D}}(a^\alpha, |z^\alpha|), & \text{if } \boldsymbol{D}_\alpha \text{ is of irrational type}, z \notin V_0 \end{cases},$$

$$\boldsymbol{k}^*_{\boldsymbol{D}_\alpha}(a, z) = \begin{cases} \min\limits_{\zeta_1 \in \sqrt[t]{a^\alpha},\ \zeta_2 \in \sqrt[t]{z^\alpha}} \{\boldsymbol{m}_{\mathbb{D}}(\zeta_1, \zeta_2)\}, & \text{if } \boldsymbol{D}_\alpha \text{ is of rational type} \\ \boldsymbol{m}_{\mathbb{D}}(a^\alpha, |z^\alpha|), & \text{if } \boldsymbol{D}_\alpha \text{ is of irrational type} \end{cases},$$

$$\boldsymbol{\varkappa}_{\boldsymbol{D}_\alpha}(a; X) = \begin{cases} \gamma_{\mathbb{D}}\Big((a^\alpha)^{1/t}; (a^\alpha)^{1/t} \frac{1}{t} \sum\limits_{j=1}^{n} \frac{\alpha_j X_j}{a_j}\Big), & \text{if } \boldsymbol{D}_\alpha \text{ is of rational type} \\ \gamma_{\mathbb{D}}(a^\alpha; a^\alpha \sum\limits_{j=1}^{n} \frac{\alpha_j X_j}{a_j}), & \text{if } \boldsymbol{D}_\alpha \text{ is of irrational type} \end{cases},$$

where $t = t(\alpha) := \min\{\alpha_{q+1}, \dots, \alpha_n\}$ (cf. (AS)*).*

Notice that since the function $\boldsymbol{\ell}^*_{\boldsymbol{D}_\alpha}$ is symmetric, the above proposition, together with Proposition 10.1.3, in fact cover all possible configurations needed to calculate $\boldsymbol{\ell}^*_{\boldsymbol{D}_\alpha}(a, z)$ with *arbitrary* $a, z \in \boldsymbol{D}_\alpha$.

We need a few auxiliary results.

Lemma 10.1.5. *For $\mu \in \mathbb{D}_*$, define*

$$\mathbb{C}^{n-1} \ni (\zeta_1, \dots, \zeta_{n-1}) \overset{F_\mu}{\longmapsto} (e^{\alpha_n \zeta_1}, \dots, e^{\alpha_n \zeta_{n-1}}, \mu e^{-(\alpha_1 \zeta_1 + \dots + \alpha_{n-1}\zeta_{n-1})}) \in \boldsymbol{D}_\alpha,$$

$V_\mu := F_\mu(\mathbb{C}^{n-1})$. *Then,*

$$\begin{aligned} \boldsymbol{\ell}_{\boldsymbol{D}_\alpha}(a, z) &= 0, \quad a, z \in V_\mu, \\ \boldsymbol{\varkappa}_{\boldsymbol{D}_\alpha}(a; X) &= 0, \quad a \in V_\mu, \ \sum_{j=1}^{n} \frac{\alpha_j X_j}{a_j} = 0. \end{aligned}$$

Observe that $\bigcup_{\mu \in \mathbb{D}_*} V_\mu = \boldsymbol{D}_\alpha \setminus \boldsymbol{V}_0$.

Proof. The first equality follows from the holomorphic contractibility

$$\boldsymbol{\ell}_{\boldsymbol{D}_\alpha}(F_\mu(\zeta^0), F_\mu(\zeta)) \le \boldsymbol{\ell}_{\mathbb{C}^{n-1}}(\zeta^0, \zeta) = 0.$$

For the second equality, observe that

$$\boldsymbol{\varkappa}_{\boldsymbol{D}_\alpha}(F_\mu(\zeta^0); F'_\mu(\zeta^0)Y) \le \boldsymbol{\varkappa}_{\mathbb{C}^{n-1}}(\zeta^0; Y) = 0$$

and if $F_\mu(\zeta^0) = a$, then

$$F'_\mu(\zeta^0)(Y) = \left(a_1 \alpha_n Y_1, \dots, a_{n-1}\alpha_n Y_{n-1}, -a_n \sum_{j=1}^{n-1} \alpha_j Y_j\right).$$

Thus, $F'_\mu(\zeta^0)(\mathbb{C}^{n-1}) = \left\{X \in \mathbb{C}^n : \sum_{j=1}^n \frac{\alpha_j X_j}{a_j} = 0\right\}$. □

Lemma 10.1.6.

$$\begin{aligned} \boldsymbol{\ell}^*_{\boldsymbol{D}_1}(a, z) &= |z^{\mathbf{1}}|^{1/\ell(a)}, \quad a \in \boldsymbol{V}_0, \ z \in \boldsymbol{D}_1, \\ \boldsymbol{\ell}^{(k)*}_{\boldsymbol{D}_1}(a, z) &= \boldsymbol{m}_{\mathbb{D}}(a^{\mathbf{1}}, z^{\mathbf{1}}), \quad a, z \in \boldsymbol{D}_1 \setminus \boldsymbol{V}_0, \ k \in \mathbb{N}, \end{aligned}$$

where $\ell(a) = \#\{j \in \{1, \dots, n\} : a_j = 0\}$.

Proof. The case where $a \in \boldsymbol{V}_0$ follows from Proposition 10.1.3.

Since $\boldsymbol{D}_1 \ni \zeta \longmapsto \zeta^{\mathbf{1}} \in \mathbb{D}$ is well defined, we have

$$\boldsymbol{m}_{\mathbb{D}}(a^{\mathbf{1}}, z^{\mathbf{1}}) = \boldsymbol{\ell}^{(k)*}_{\mathbb{D}}(a^{\mathbf{1}}, z^{\mathbf{1}}) \le \boldsymbol{\ell}^{(k)*}_{\boldsymbol{D}_1}(a, z), \quad a, z \in \boldsymbol{D}_1.$$

Assume that $a, z \in \boldsymbol{D}_1 \setminus \boldsymbol{V}_0$. The case where $a^{\mathbf{1}} = z^{\mathbf{1}}$ follows from Lemma 10.1.5. Thus, assume that $\lambda'_0 := a^{\mathbf{1}} \neq z^{\mathbf{1}} =: \lambda''_0$. Using Lemma 6.3.9, we find a compact set $K \subset\subset \mathbb{C}_*$ and $\psi_j \in \mathcal{O}(\mathbb{D}, K)$, $j = 1, \dots, n$, such that

- $\psi_j(\lambda_0') = a_j$, $\psi_j(\lambda_0'') = z_j$, $j = 1, \dots, n-1$,
- $\psi_1 \cdots \psi_n \equiv 1$,
- K depends only on α, δ, L, provided that $a, z \in L \subset\subset \mathbb{C}_*^n$ are such that $\boldsymbol{m}_{\mathbb{D}}(a^{\mathbf{1}}, z^{\mathbf{1}}) \geq \delta > 0$.

Put

$$\varphi(\lambda) := (\psi_1(\lambda), \dots, \psi_{n-1}(\lambda), \lambda\psi_n(\lambda)), \quad \lambda \in \mathbb{D}.$$

Then, $\varphi \in \mathcal{O}(\mathbb{D}, \boldsymbol{D}_{\mathbf{1}})$, $\varphi(\lambda_0') = a$, $\varphi(\lambda_0'') = z$. Thus, $\boldsymbol{\ell}^*_{\boldsymbol{D}_{\mathbf{1}}}(a, z) \leq \boldsymbol{m}_{\mathbb{D}}(\lambda_0', \lambda_0'') = \boldsymbol{m}_{\mathbb{D}}(a^{\mathbf{1}}, z^{\mathbf{1}})$. □

Lemma 10.1.7. *Assume that $\alpha \in \mathbb{Z}_*^n$, $\alpha_1, \dots, \alpha_n$ are relatively prime. Let $a, z \in \boldsymbol{D}_\alpha \setminus V_0$, $a^\alpha \neq z^\alpha$. Assume that $\varphi = (\varphi_1, \dots, \varphi_n) : \mathbb{D} \longrightarrow \boldsymbol{D}_\alpha$ is a holomorphic disc such that $\varphi(0) = a$, $\varphi(\sigma) = z$, $\varphi_j = B_j\psi_j$, where $|B_j| \leq 1$ and $\psi_j \in \mathcal{O}(\mathbb{D}, \mathbb{C}_*)$, $j = 1, \dots, n$. Suppose that $|\psi_1|^{\alpha_1} \cdots |\psi_n|^{\alpha_n} < 1$ in $\mathbb{D}$. Then there exists a holomorphic disc $\widetilde{\varphi} = (\widetilde{\varphi}_1, \dots, \widetilde{\varphi}_n) : \mathbb{D} \longrightarrow \boldsymbol{D}_\alpha$ such that $\widetilde{\varphi}(0) = a$, $\widetilde{\varphi}(\sigma) = z$, $\widetilde{\varphi}_j = \widetilde{B}_j\widetilde{\psi}_j$, where $|\widetilde{B}_j| \leq 1$, $\widetilde{\psi}_j \in \mathcal{O}(\mathbb{D}, \mathbb{C}_*)$, $\widetilde{\psi}_j(\mathbb{D}) \subset\subset \mathbb{C}_*$, $j = 1, \dots, n$, and $\widetilde{\psi}_1^{\alpha_1} \cdots \widetilde{\psi}_n^{\alpha_n} \equiv 1$.*

Proof. Put $\gamma := |\alpha_1 \cdots \alpha_n| \in \mathbb{N}$ and consider the holomorphic mapping

$$h := (\psi_1^{\alpha_1/\gamma}, \dots, \psi_n^{\alpha_n/\gamma}) : \mathbb{D} \longrightarrow \boldsymbol{D}_{\mathbf{1}}.$$

Let $P := h(0)$, $Q := h(\sigma)$. Observe that $\boldsymbol{m}_{\mathbb{D}}(P^{\mathbf{1}}, Q^{\mathbf{1}}) \leq \boldsymbol{\ell}^*_{\boldsymbol{D}_{\mathbf{1}}}(P, Q) \leq \boldsymbol{m}_{\mathbb{D}}(0, \sigma) = \sigma$.

If $P^{\mathbf{1}} \neq Q^{\mathbf{1}}$, then we put $\widetilde{P} := P$, $\widetilde{Q} := Q$, $g :\equiv 1$.
If $P^{\mathbf{1}} = Q^{\mathbf{1}}$, then we put

$$\widetilde{P} := (P_1, \dots, P_{n-1}, P_n/g(0)), \quad \widetilde{Q} := (Q_1, \dots, Q_{n-1}, Q_n/g(\sigma)),$$

where $g \in \mathrm{Aut}(\mathbb{D})$ is chosen so that

- $\widetilde{P}, \widetilde{Q} \in \boldsymbol{D}_{\mathbf{1}}$, $\widetilde{P}^{\mathbf{1}} \neq \widetilde{Q}^{\mathbf{1}}$,
- $\boldsymbol{\ell}^*_{\boldsymbol{D}_{\mathbf{1}}}(\widetilde{P}, \widetilde{Q}) = \boldsymbol{m}_{\mathbb{D}}(\widetilde{P}^{\mathbf{1}}, \widetilde{Q}^{\mathbf{1}}) \leq \sigma$.

By Lemma 10.1.6, there exists an extremal mapping μ for $\boldsymbol{\ell}^*_{\boldsymbol{D}_{\mathbf{1}}}(\widetilde{P}, \widetilde{Q})$ of the form $\mu = (\varrho_1, \dots, \varrho_{n-1}, f\varrho_n)$ with

- $f \in \mathrm{Aut}(\mathbb{D})$,
- $\mu(0) = \widetilde{P}$,
- $\mu(R\sigma) = \widetilde{Q}, 0 < R \leq 1$,

- $\varrho_1 \cdots \varrho_n \equiv 1$,
- $\varrho_j(\mathbb{D}) \subset\subset \mathbb{C}_*$, $j = 1, \dots, n$.

Put

$$\begin{aligned}\widetilde{B}_j(\lambda) &:= B_j(\lambda), \quad \widetilde{\psi}_j(\lambda) := (\varrho_j(R\lambda))^{\gamma/\alpha_j}, \quad j = 1, \dots, n-1,\\ \widetilde{B}_n(\lambda) &:= B_n(\lambda)(g(\lambda)f(R\lambda))^{\gamma/\alpha_n}, \quad \widetilde{\psi}_n(\lambda) := (\varrho_n(R\lambda))^{\gamma/\alpha_n},\\ \widetilde{\varphi}_j &:= \widetilde{B}_j\widetilde{\psi}_j, \quad j = 1, \dots, n, \quad \widetilde{\varphi} := (\widetilde{\varphi}_1, \dots, \widetilde{\varphi}_n).\end{aligned} \tag{10.1.1}$$

Then, $\widetilde{\varphi}(0) = a$, $\widetilde{\varphi}(\sigma) = z$, $\widetilde{\psi}_j(\mathbb{D}) \subset\subset \mathbb{C}_*$, $j = 1, \dots, n$ (this is trivial if $R < 1$), and

$$\prod_{j=1}^{n} \widetilde{\varphi}_j^{\alpha_j}(\lambda) = \Big(\prod_{j=1}^{n} \widetilde{B}_j^{\alpha_j}(\lambda)\Big)(g(\lambda)f(R\lambda))^{\gamma}.$$

Hence, $\widetilde{\varphi} : \mathbb{D} \longrightarrow \boldsymbol{D}_\alpha$. □

Remark 10.1.8. We keep the notation from the proof of Lemma 10.1.7.

(a) If φ is an extremal disc for $\boldsymbol{\ell}_{\boldsymbol{D}_\alpha}(a, z)$, then so is $\widetilde{\varphi}$.

(b) Consider the special case where $\alpha_1, \dots, \alpha_q < 0, \alpha_{q+1}, \dots, \alpha_n > 0$ $(0 \le q < n)$, φ_j is bounded, and B_j is the Blaschke product for φ_j ($B_j \equiv 1$ if $\alpha_j < 0$), $j = 1, \dots, n$. Then, formulas (10.1.1) show that:

- $\widetilde{\varphi}(\mathbb{D}) \subset\subset \mathbb{C}_*^q \times \mathbb{C}^{n-q}$.
- If $R < 1$, then $\widetilde{\varphi}(\mathbb{D}) \subset\subset \boldsymbol{D}_\alpha$; in particular, in such a case, $\widetilde{\varphi}$ cannot be extremal for $\boldsymbol{\ell}_{\boldsymbol{D}_\alpha}(a, z)$ (cf. Remark 3.8.6).
- If $R = 1$, then $\widetilde{B}_j$ is the Blaschke product for $\widetilde{\varphi}_j$ (up to a multiplier $c_j \in \mathbb{T}$), $j = 1, \dots, n$.

Lemma 10.1.9. *If $\alpha \in \mathbb{Z}_*^n$, $\alpha_1, \dots, \alpha_n$ are relatively prime, $\alpha_1, \dots, \alpha_q < 0$, $\alpha_{q+1}, \dots, \alpha_n > 0$ $(0 \le q < n)$, then for arbitrary $a, z \in \boldsymbol{D}_\alpha \setminus V_0$ with $a^\alpha \neq z^\alpha$, there exists a bounded extremal disc $\varphi \in \mathcal{O}(\mathbb{D}, \boldsymbol{D}_\alpha)$, $\varphi(0) = a$, $\varphi(\sigma_0) = z$ with $\sigma_0 := \boldsymbol{\ell}^*_{\boldsymbol{D}_\alpha}(a, z)$ such that $\varphi_j = B_j\psi_j$, B_j is the Blaschke product for φ_j (up to a multiplier $c_j \in \mathbb{T}$), $\psi_j(\mathbb{D}) \subset\subset \mathbb{C}_*$, $j = 1, \dots, n$, and $\psi_1^{\alpha_1} \cdots \psi_n^{\alpha_n} \equiv 1$.*

Proof. Let $\varphi_\nu \in \mathcal{O}(\overline{\mathbb{D}}, \boldsymbol{D}_\alpha)$, $\varphi_\nu(0) = a$, $\varphi_\nu(\sigma_\nu) = z$, $\sigma_\nu \searrow \sigma_0$. Write $\varphi_{\nu,j} = B_{\nu,j}\psi_{\nu,j}$, where $B_{\nu,j}$ is the (finite) Blaschke product for $\varphi_{\nu,j}$ and $\psi_{\nu,j} \in \mathcal{O}(\mathbb{D}, \mathbb{C}_*)$, $j = 1, \dots, n$.

Put $\psi_\nu := (\psi_{\nu,1}, \dots, \psi_{\nu,n})$. Since $|\varphi_{\nu,1}^{\alpha_1} \cdots \varphi_{\nu,n}^{\alpha_n}| < 1$, we get $|\psi_{\nu,1}^{\alpha_1} \cdots \psi_{\nu,n}^{\alpha_n}| \le 1$. Thus, either $|\psi_{\nu,1}^{\alpha_1} \cdots \psi_{\nu,n}^{\alpha_n}| < 1$ or $|\psi_{\nu,1}^{\alpha_1} \cdots \psi_{\nu,n}^{\alpha_n}| \equiv 1$. In the second case, we may always assume that $\psi_{\nu,1}^{\alpha_1} \cdots \psi_{\nu,n}^{\alpha_n} \equiv 1$.

Using Lemma 10.1.7, we reduce the first case to the second one, with new functions $B_{\nu,j}$, $|B_{\nu,j}| \le 1$, which may be no longer Blaschke products.

We may assume that $B_{\nu,j} \longrightarrow B_j$ locally uniformly in $\mathbb{D}$, $j = 1, \dots, n$. Since $\psi_{\nu,1}^{\alpha_1} \cdots \psi_{\nu,n}^{\alpha_n} \equiv 1$, we get

$$a^\alpha = B_{\nu,1}^{\alpha_1}(0) \cdots B_{\nu,n}^{\alpha_n}(0) \longrightarrow B_1^{\alpha_1}(0) \cdots B_n^{\alpha_n}(0) \neq 0,$$
$$z^\alpha = B_{\nu,1}^{\alpha_1}(\sigma_\nu) \cdots B_{\nu,n}^{\alpha_n}(\sigma_\nu) \longrightarrow B_1^{\alpha_1}(\sigma) \cdots B_n^{\alpha_n}(\sigma) \neq 0.$$

Consequently,

$$\{\psi_{\nu,j}(0), \psi_{\nu,j}(\sigma_\nu) : \nu \in \mathbb{N},\ j = 1, \dots, n\} \subset L \subset\subset \mathbb{C}_*.$$

Using Lemma 6.3.9(c) (with $L_0' := \{0\}$, $\{\sigma_\nu : \nu \in \mathbb{N}\} \subset L_0'' \subset\subset \mathbb{D}$), we may modify ψ_ν so that $\psi_\nu(\mathbb{D}) \subset K^n$, $\nu \in \mathbb{N}$, where $K \subset\subset \mathbb{C}_*$ is a compact set. Thus, we may assume that also $\varphi_\nu \longrightarrow \varphi$ locally uniformly in $\mathbb{D}$. It is clear that φ is extremal for $\boldsymbol{\ell}^*_{\boldsymbol{D}_\alpha}(a, z)$. We will modify φ to get the remaining conditions.

Write $\varphi_j = B_j \psi_j$, where B_j is the Blaschke product for φ_j and $\psi_j \in \mathcal{O}(\mathbb{D}, \mathbb{C}_*)$.

In the case where $|\psi_1^{\alpha_1} \cdots \psi_n^{\alpha_n}| \equiv 1$, we apply Lemma 6.3.9(c), which easily finishes the proof.

If $|\psi_1^{\alpha_1} \cdots \psi_n^{\alpha_n}| \not\equiv 1$, then put $h_j := \psi_j^{\alpha_j/\gamma}$, $h := (h_1, \dots, h_n)$, $P := h(0)$, $Q := h(\sigma_0)$, etc. (we keep the notation from the proof of Lemma 10.1.7). We consider the following two cases:

- $R = 1$: then, by Lemma 10.1.7 and Remark 10.1.8, we may modify φ to get the required mapping.
- $R < 1$: then, by Remark 10.1.8, φ cannot be extremal – a contradiction. □

In fact, Lemma 10.1.9 may be extended to the following result:

Lemma 10.1.10. *Under the assumptions of Lemma* 10.1.9, *there exist* $\beta \in \mathbb{D}$, $c_1, \dots, c_n \in \mathbb{T}$, *and* $r_1, \dots, r_n \in \{0, 1\}$ *such that*

- $B_j(\lambda) = c_j \left(\frac{\lambda - \beta}{1 - \bar{\beta}\lambda}\right)^{r_j}$, $j = 1, \dots, n$,
- $r_1 = \dots = r_q = 0$, $r_{q+1} + \dots + r_n > 0$.

This is the most difficult part of the proof of Proposition 10.1.4. Unfortunately, at the moment we have no tools to prove it. *Lemma* 10.1.10 *will follow from Proposition* 16.3.4. Nevertheless, the proof of Proposition 10.1.4 will now be presented, assuming for a moment that Lemma 10.1.10 is proven.

Proof of Proposition 10.1.4. Step 1°. *Proof of the inequality*

$$\boldsymbol{\ell}^*_{\boldsymbol{D}_\alpha}(a,z) \le \min\left\{\boldsymbol{m}_{\mathbb{D}}(\zeta_1,\zeta_2) : \zeta_1 \in \sqrt[t]{a^\alpha},\ \zeta_2 \in \sqrt[t]{z^\alpha}\right\}$$

in the rational case.

We may assume that $t = \alpha_n$. The case where $a^\alpha = z^\alpha$ is a direct consequence of Lemma 10.1.5. Assume that $a^\alpha \ne z^\alpha$. Let ${\lambda_0'}^t = a^\alpha$, ${\lambda_0''}^t = z^\alpha$. By Lemma 6.3.9 there exist $\psi_1,\dots,\psi_{n-1} \in \mathcal{O}(\mathbb{D},\mathbb{C}_*)$ such that $\psi_j(\lambda_0') = a_j$, $\psi_j(\lambda_0'') = z_j$, $j = 1,\dots,n-1$. Put

$$\psi_n := (\psi_1^{\alpha_1}\cdots\psi_{n-1}^{\alpha_{n-1}})^{-1/\alpha_n}, \quad \varphi := (\psi_1,\dots,\psi_{n-1},\lambda\psi_n).$$

Assume that the power $(-1/\alpha_n)$ is taken so that $\psi_n(\lambda_0') = a_n/\lambda_0'$. We have $\varphi_n^{\alpha_n}(\lambda_0'') = z_n^{\alpha_n}$. Using Lemma 6.3.7(b), we get a holomorphic disc $\widetilde{\varphi} : \mathbb{D} \longrightarrow \boldsymbol{D}_\alpha$ such that $\widetilde{\varphi}(\lambda_0') = a$, $\widetilde{\varphi}(\lambda_0'') = z$. Thus $\boldsymbol{\ell}^*_{\boldsymbol{D}_\alpha}(a,z) \le \boldsymbol{m}_{\mathbb{D}}(\lambda_0',\lambda_0'')$, which implies the required inequality.

Step 2°. *Proof of the inequality*

$$\boldsymbol{\ell}^*_{\boldsymbol{D}_\alpha}(a,z) \ge \min\left\{\boldsymbol{m}_{\mathbb{D}}(\zeta_1,\zeta_2) : \zeta_1 \in \sqrt[t]{a^\alpha},\ \zeta_2 \in \sqrt[t]{z^\alpha}\right\}$$

in the rational case.

We may assume that $a^\alpha \ne z^\alpha$. Let $\varphi = (\varphi_1,\dots,\varphi_n) \in \mathcal{O}(\mathbb{D},\boldsymbol{D}_\alpha)$ be as in Lemma 10.1.10. Observe that $\varphi_1^{\alpha_1}\cdots\varphi_n^{\alpha_n} \equiv B_1^{\alpha_1}\cdots B_n^{\alpha}$. In particular, $B := (B_1,\dots,B_n) : \mathbb{D} \longrightarrow \boldsymbol{D}_\alpha$. Moreover, $B^\alpha(0) = a^\alpha$ and $B^\alpha(\sigma_0) = z^\alpha$. Let $\tau := \sum_{j=q+1}^n \alpha_j r_j \ge t$. Take a $d \in \sqrt[\tau]{c_1^{\alpha_1}\cdots c_n^{\alpha_n}}$. Then,

$$\begin{aligned}
\sigma_0 &\ge \min\left\{\sigma \in (0,1) : -d\beta \in \sqrt[\tau]{a^\alpha},\ d\frac{\sigma-\beta}{1-\overline{\beta}\sigma} \in \sqrt[\tau]{z^\alpha}\right\}\\
&\ge \min\left\{\sigma \in (0,1) : \exists_{h\in\operatorname{Aut}(\mathbb{D})} : h(0) \in \sqrt[\tau]{a^\alpha},\ h(\sigma) \in \sqrt[\tau]{z^\alpha}\right\}\\
&\ge \min\left\{\boldsymbol{m}_{\mathbb{D}}(\zeta_1,\zeta_2) : \zeta_1 \in \sqrt[\tau]{a^\alpha},\ \zeta_2 \in \sqrt[\tau]{z^\alpha}\right\}\\
&\ge \min\left\{\boldsymbol{m}_{\mathbb{D}}(\zeta_1,\zeta_2) : \zeta_1 \in \sqrt[t]{a^\alpha},\ \zeta_2 \in \sqrt[t]{z^\alpha}\right\},
\end{aligned}$$

where the last inequality follows from Proposition 1.1.19.

Step 3°. *Proof of the formula for $\boldsymbol{\ell}^*_{\boldsymbol{D}_\alpha}(a,z)$ in the irrational case.*

Fix a $z \in \boldsymbol{D}_\alpha \setminus \boldsymbol{V}_0$. Corollary 6.3.8 implies that

$$\boldsymbol{\ell}^*_{\boldsymbol{D}_\alpha}(a,z) = \boldsymbol{\ell}^*_{\boldsymbol{D}_\alpha}(a,(|z_1|,\dots,|z_n|)).$$

We may assume that $t = \alpha_n = 1$. Take $\alpha^\nu = (\alpha_1^\nu, \dots, \alpha_n^\nu) \in \mathbb{Q}^n$ such that:

- $\alpha^\nu \longrightarrow \alpha$,
- $\alpha^\nu \in \mathbb{Q}^s_{<0} \times \mathbb{Q}^{n-s}_{>0}$,
- $\min\{\alpha_{s+1}^\nu, \dots, \alpha_n^\nu\} = \alpha_n^\nu = 1$, $\nu \in \mathbb{N}$.

Let $p_\nu \in \mathbb{N}$ be such that $p_\nu \alpha^\nu \in \mathbb{Z}^n$ and the components of $p_\nu \alpha^\nu$ are relatively prime. We know that if $a, z \in \boldsymbol{D}_{\alpha^\nu}$, then

$$\begin{aligned}\boldsymbol{\ell}^*_{\boldsymbol{D}_{\alpha^\nu}}(a, (|z_1|, \dots, |z_n|)) &= \min\left\{\boldsymbol{m}_{\mathbb{D}}(\zeta_1, \zeta_2) : \zeta_1 \in \sqrt[p_\nu]{a^{p_\nu \alpha^\nu}},\ \zeta_2 \in \sqrt[p_\nu]{|z^{p_\nu \alpha^\nu}|}\right\} \\ &= \min\left\{\boldsymbol{m}_{\mathbb{D}}(a^{\alpha^\nu}, \varepsilon |z^{\alpha^\nu}|) : \varepsilon \in \sqrt[p_\nu]{1}\right\} = \boldsymbol{m}_{\mathbb{D}}\left(a^{\alpha^\nu}, |z^{\alpha^\nu}|\right).\end{aligned}$$

Observe that for every holomorphic disc $\varphi : \overline{\mathbb{D}} \longrightarrow \boldsymbol{D}_\alpha$ we have $\varphi(\mathbb{D}) \subset \boldsymbol{D}_{\alpha^\nu}$ for $\nu \gg 1$. Hence,

$$\boldsymbol{\ell}^*_{\boldsymbol{D}_\alpha}(a, z) \geq \limsup_{\nu \to +\infty} \boldsymbol{\ell}^*_{\boldsymbol{D}_{\alpha^\nu}}(a, (|z_1|, \dots, |z_n|)) \tag{10.1.2}$$

$$= \limsup_{\nu \to +\infty} \boldsymbol{m}_{\mathbb{D}}(a^{\alpha^\nu}, |z^{\alpha^\nu}|) = \boldsymbol{m}_{\mathbb{D}}(a^\alpha, |z^\alpha|). \tag{10.1.3}$$

To get the equality, first consider the case where $a^\alpha \neq |z^\alpha|$. Let $\lambda_0' := a^\alpha$, $\lambda_0'' := |z^\alpha|$. By Lemma 6.3.9, we get $\psi_1, \dots, \psi_{n-1} \in \mathcal{O}(\mathbb{D}, \mathbb{C}_*)$ such that $\psi(\lambda_0') = a_j$, $\psi_j(\lambda_0'') = |z_j|$, $j = 1, \dots, n-1$. Define $\varphi := (\psi_1, \dots, \psi_{n-1}, \lambda\psi_n)$, where $\psi_n := (\psi_1^{\alpha_1} \cdots \psi_{n-1}^{\alpha_{n-1}})^{-1}$ and the powers are chosen so that $x^{\alpha_j} > 0$ for $x > 0$, $j = 1, \dots, n-1$. Then, $\varphi(\lambda_0') = a^\alpha$, $\varphi(\lambda_0'') = |z^\alpha|$. Thus, $\boldsymbol{\ell}^*_{\boldsymbol{D}_\alpha}(a, z) \leq \boldsymbol{m}_{\mathbb{D}}(a^\alpha, |z^\alpha|)$.

The case where $a^\alpha = |z^\alpha|$ follows from Lemma 10.1.5.

Step 4°. *Proof of the formula for* $\boldsymbol{k}^*_{\boldsymbol{D}_\alpha}(a, z)$.

In the rational case, we already know that if $z \notin \boldsymbol{V}_0$, then

$$\boldsymbol{\ell}_{\boldsymbol{D}_\alpha}(a, z) = \min\{\boldsymbol{p}(\zeta_1, \zeta_2) : \zeta_1 \in \sqrt[t]{a^\alpha},\ \zeta_2 \in \sqrt[t]{z^\alpha}\} =: d(a, z).$$

Observe that the function d is well-defined for all $a, z \in \boldsymbol{D}_\alpha$ and is a continuous pseudodistance. We have $\boldsymbol{\ell}_{\boldsymbol{D}_\alpha} = d$ on the set $(\boldsymbol{D}_\alpha \setminus \boldsymbol{V}_0) \times (\boldsymbol{D}_\alpha \setminus \boldsymbol{V}_0)$. Moreover, $\boldsymbol{\ell}_{\boldsymbol{D}_\alpha}(a, z) = \boldsymbol{p}(0, |z^\alpha|^{1/r}) \geq \boldsymbol{p}(0, |z^\alpha|) = d(a, z)$ on $\boldsymbol{V}_0 \times \boldsymbol{D}_\alpha$. Hence, $\boldsymbol{\ell}_{\boldsymbol{D}_\alpha} \geq \boldsymbol{k}_{\boldsymbol{D}_\alpha} \geq d$. In particular, $\boldsymbol{k}_{\boldsymbol{D}_\alpha} = d$ on $(\boldsymbol{D}_\alpha \setminus \boldsymbol{V}_0) \times (\boldsymbol{D}_\alpha \setminus \boldsymbol{V}_0)$. Since $\boldsymbol{k}_{\boldsymbol{D}_\alpha}$ and d are continuous, we get the required formula.

In the irrational case, we argue in the same way, with

$$d(a, z) := \boldsymbol{p}(|a^\alpha|, |z^\alpha|), \quad a, z \in \boldsymbol{D}_\alpha.$$

Step 5°. *Proof of the formula for* $\boldsymbol{\varkappa}_{\boldsymbol{D}_\alpha}(a;X)$.

In the rational case, we have (cf. Theorem 4.3.3(g))

$$\begin{aligned}
\boldsymbol{\varkappa}_{\boldsymbol{D}_\alpha}(a;X) &\geq \limsup_{\lambda \to 0} \frac{1}{|\lambda|} \boldsymbol{k}_{\boldsymbol{D}_\alpha}(a, a+\lambda X) \\
&= \limsup_{\lambda \to 0} \frac{1}{|\lambda|} \min\{\boldsymbol{p}(\zeta_1,\zeta_2) : \zeta_1 \in \sqrt[t]{a^\alpha},\ \zeta_2 \in \sqrt[t]{(a+\lambda X)^\alpha}\} \\
&= \limsup_{\lambda \to 0} \frac{1}{|\lambda|} \min\{\boldsymbol{m}_{\mathbb{D}}(\zeta_1,\zeta_2) : \zeta_1 \in \sqrt[t]{a^\alpha},\ \zeta_2 \in \sqrt[t]{(a+\lambda X)^\alpha}\} \\
&= \limsup_{\lambda \to 0} \frac{1}{|\lambda|} \left| \frac{(a^\alpha)^{1/t} - \prod_{j=1}^n (a_j + \lambda X_j)^{\alpha_j/t}}{1 - (a^\alpha)^{1/t} \prod_{j=1}^n (a_j+\lambda X_j)^{\alpha_j/t}} \right| \\
&= \frac{(a^\alpha)^{1/t}}{1-(a^\alpha)^{2/t}} \limsup_{\lambda \to 0} \left| \frac{1 - \prod_{j=1}^n (1+\lambda X_j/a_j)^{\alpha_j/t}}{\lambda} \right| \\
&= \boldsymbol{\gamma}_{\mathbb{D}}\left((a^\alpha)^{1/t}; (a^\alpha)^{1/t} \frac{1}{t} \sum_{j=1}^n \frac{\alpha_j X_j}{a_j} \right).
\end{aligned}$$

In the irrational case, we have

$$\begin{aligned}
\boldsymbol{\varkappa}_{\boldsymbol{D}_\alpha}(a;X) &\geq \limsup_{\lambda \to 0} \frac{1}{|\lambda|} \boldsymbol{k}_{\boldsymbol{D}_\alpha}(a, a+\lambda X) = \limsup_{\lambda \to 0} \frac{1}{|\lambda|} \boldsymbol{p}(a^\alpha, |a+\lambda X|^\alpha) \\
&= \limsup_{\lambda \to 0} \frac{1}{|\lambda|} \boldsymbol{m}_{\mathbb{D}}(a^\alpha, |a+\lambda X|^\alpha) \\
&= \limsup_{\lambda \to 0} \frac{1}{|\lambda|} \left| \frac{a^\alpha - \prod_{j=1}^n |a_j+\lambda X_j|^{\alpha_j}}{1 - a^\alpha \prod_{j=1}^n |a_j+\lambda X_j|^{\alpha_j}} \right| \\
&= \frac{a^\alpha}{1-(a^\alpha)^2} \limsup_{\lambda \to 0} \frac{1 - \prod_{j=1}^n |1+\lambda X_j/a_j|^{\alpha_j}}{|\lambda|} \\
&= \frac{a^\alpha}{1-(a^\alpha)^2} \limsup_{\lambda \to 0} \frac{1 - \prod_{j=1}^n (1+\alpha_j \operatorname{Re}(\lambda X_j/a_j) + o(\lambda))}{|\lambda|} \\
&= \boldsymbol{\gamma}_{\mathbb{D}}\left(a^\alpha; a^\alpha \sum_{j=1}^n \frac{\alpha_j X_j}{a_j} \right).
\end{aligned}$$

To get the equalities, first consider the case where $\sum_{j=1}^n \frac{\alpha_j X_j}{a_j} = 0$. This case follows from Lemma 10.1.5.

Now assume $\sum_{j=1}^n \frac{\alpha_j X_j}{a_j} \neq 0$. To simplify notation assume that $t = \alpha_n$. Put $\lambda_0 := (a^\alpha)^{1/\alpha_n}$, $\sigma := (a^\alpha)^{1/\alpha_n} \frac{1}{\alpha_n} \sum_{j=1}^n \frac{\alpha_j X_j}{a_j}$. Using Lemma 6.3.9(b), we get $\psi_1,\dots,\psi_{n-1} \in \mathcal{O}(\mathbb{D},\mathbb{C}_*)$ such that $\psi_j(\lambda_0) = a_j, \sigma\psi_j'(\lambda_0) = X_j$, $j = 1,\dots,n-1$.

Define $\psi_n := (\psi_1^{\alpha_1} \cdots \psi_{n-1}^{\alpha_{n-1}})^{-1/\alpha_n}$, $\varphi := (\psi_1, \dots, \psi_{n-1}, \lambda\psi_n) : \mathbb{D} \longrightarrow \boldsymbol{D}_\alpha$, where the powers are taken so that $\varphi_n(\lambda_0) = a_n$. It remains to check that $\sigma\varphi_n'(\lambda_0) = X_n$. □

Proposition 10.1.11. *Assume that conditions* (AS) *are satisfied, $\boldsymbol{D}_\alpha$ is of rational type, $q(\alpha) < n$, $s(a) = n$, and $t|\alpha_j$, $j = q+1, \dots, n$. Then,*

$$\boldsymbol{\ell}_{\boldsymbol{D}_\alpha}^{(k)*}(a,z) = \boldsymbol{\ell}_{\boldsymbol{D}_\alpha}^{*}(a,z) = \min\{\boldsymbol{m}_{\mathbb{D}}(\zeta_1,\zeta_2) : \zeta_1 \in \sqrt[t]{a^\alpha},\ \zeta_2 \in \sqrt[t]{z^\alpha}\},\ z \in \boldsymbol{D}_\alpha \setminus \boldsymbol{V}_0,$$

$$\boldsymbol{\varkappa}_{\boldsymbol{D}_\alpha}^{(k)}(a;X) = \boldsymbol{\varkappa}_{\boldsymbol{D}_\alpha}(a;X) = \boldsymbol{\gamma}_{\mathbb{D}}\Big((a^\alpha)^{1/t}; (a^\alpha)^{1/t}\frac{1}{t}\sum_{j=1}^{n}\frac{\alpha_j X_j}{a_j}\Big),\quad X \in \mathbb{C}^n,\ k \in \mathbb{N}.$$

In the general case (with $k \geq 2$, $q(\alpha) < n$, and $s(a) = n$) effective formulas for $\boldsymbol{\ell}_{\boldsymbol{D}_\alpha}^{(k)}(a,\cdot)$ and $\boldsymbol{\varkappa}_{\boldsymbol{D}_\alpha}^{(k)}(a;\cdot)$ are not known.

Proof. The case $t = 1$ follows directly from Propositions 6.2.7 and 10.1.4. Assume that $t \geq 2$. Observe that

$$\boldsymbol{\ell}_{\boldsymbol{D}_1}^{(k)*}(b,w) = \boldsymbol{m}_{\mathbb{D}}(b^{\boldsymbol{1}}, w^{\boldsymbol{1}}),\quad b \in \boldsymbol{D}_{\boldsymbol{1}} \cap \mathbb{R}_{>0}^n,\ w \in \boldsymbol{D}_{\boldsymbol{1}} \setminus \boldsymbol{V}_0,$$

$$\boldsymbol{\varkappa}_{\boldsymbol{D}_1}^{(k)}(b;Y) = \boldsymbol{\gamma}_{\mathbb{D}}\Big(b^{\boldsymbol{1}}; b^{\boldsymbol{1}}\sum_{j=1}^{n}\frac{Y_j}{b_j}\Big),\quad b \in \boldsymbol{D}_{\boldsymbol{1}} \cap \mathbb{R}_{>0}^n,\ Y \in \mathbb{C}^n.$$

Let $\varphi = a + \lambda^k\psi \in \mathcal{O}(\mathbb{D}, \boldsymbol{D}_\alpha)$, $\varphi(\sigma) = z$ (resp. $\sigma\psi(0) = X$). Put

$$\widetilde{\varphi} := (\varphi_1^{\alpha_1/t}, \dots, \varphi_n^{\alpha_n/t}),$$

where the powers are taken so that $a_j^{\alpha_j/t} > 0$, $j = 1, \dots, q$. Put

$$b := \widetilde{\varphi}(0),\quad w := \widetilde{\varphi}(\sigma),\quad Y := \Big(\frac{b_1X_1}{ta_1}, \dots, \frac{b_nX_n}{ta_n}\Big).$$

Observe that $\widetilde{\varphi} = b + \lambda^k\widetilde{\psi} \in \mathcal{O}(\mathbb{D}, \boldsymbol{D}_{\boldsymbol{1}})$ and $\sigma\widetilde{\psi}(0) = Y$. Moreover, $(\widetilde{\varphi}^{\boldsymbol{1}})^t = \varphi^\alpha$. In particular, $(b^{\boldsymbol{1}})^t = a^\alpha$ and $(w^{\boldsymbol{1}})^t = z^\alpha$. Thus,

$$\begin{aligned}\sigma^k &\geq \boldsymbol{\ell}_{\boldsymbol{D}_1}^{(k)*}(b,w) = \boldsymbol{m}_{\mathbb{D}}(b^{\boldsymbol{1}}, w^{\boldsymbol{1}})\\ &\geq \inf\{\boldsymbol{m}_{\mathbb{D}}(\zeta_1,\zeta_2) : \zeta_1 \in \sqrt[t]{a^\alpha},\ \zeta_2 \in \sqrt[t]{z^\alpha}\} = \boldsymbol{\ell}_{\boldsymbol{D}_\alpha}^{*}(a,z),\end{aligned}$$

which implies that $\boldsymbol{\ell}_{\boldsymbol{D}_\alpha}^{(k)*}(a,z) = \boldsymbol{\ell}_{\boldsymbol{D}_\alpha}^{*}(a,z)$. Analogously,

$$\begin{aligned}\sigma \geq \boldsymbol{\varkappa}_{\boldsymbol{D}_1}^{(k)}(b;Y) &= \boldsymbol{\gamma}_{\mathbb{D}}\Big(b^{\boldsymbol{1}}; b^{\boldsymbol{1}}\sum_{j=1}^{n}\frac{Y_j}{b_j}\Big)\\ &= \boldsymbol{\gamma}_{\mathbb{D}}\Big((a^\alpha)^{1/t}; (a^\alpha)^{1/t}\frac{1}{t}\sum_{j=1}^{n}\frac{\alpha_j X_j}{a_j}\Big) = \boldsymbol{\varkappa}_{\boldsymbol{D}_\alpha}(a;X),\end{aligned}$$

which implies that $\boldsymbol{\varkappa}_{\boldsymbol{D}_\alpha}^{(k)}(a;X) = \boldsymbol{\varkappa}_{\boldsymbol{D}_\alpha}(a;X)$. □

Example 10.1.12. It is now relatively easy to present formulas for the Wu pseudometric in elementary n-circled domains. Let $a \in \boldsymbol{D}_\alpha$ be as in § 6.3 with (AS) (page 238) and let $\eta \in \mathfrak{M}(\boldsymbol{D}_\alpha)$, $\eta \le \boldsymbol{\varkappa}_{\boldsymbol{D}_\alpha}$, such that $\eta(z;\cdot)$ is upper semicontinuous for all $z \in \boldsymbol{D}_\alpha$. Then, we have

$$\mathbb{W}\eta(a;\cdot) = \widetilde{\mathbb{W}}\eta(a;\cdot) = \widehat{\eta}(a;\cdot).$$

Indeed, by virtue of the formulas for $\boldsymbol{\varkappa}_{\boldsymbol{D}_\alpha}$, one observes that the linear span of the zeros of $\boldsymbol{\varkappa}_{\boldsymbol{D}_\alpha}(a;\cdot)$ has codimension 0 if $s < n-1$ and 1 otherwise. So the same holds for $\eta(a;\cdot)$. Therefore, the John ellipsoids have to be built in the one-dimensional case, which immediately gives the formula.

In the case where $\eta \in \{\boldsymbol{\gamma}^{(k)}_{\boldsymbol{D}_\alpha}, \boldsymbol{A}_{\boldsymbol{D}_\alpha}, \boldsymbol{\varkappa}_{\boldsymbol{D}_\alpha}\}$, one even gets

$$\mathbb{W}\eta(a;\cdot) = \widetilde{\mathbb{W}}\eta(a;\cdot) = \begin{cases} \eta(a;\cdot), & \text{if } s \ge n-1 \\ 0, & \text{if } s < n-1 \end{cases}.$$

Indeed, using the formulas given in §§ 6.1, 6.3, 10.1 leads to $\widehat{\eta}(a;X) = \eta(a;X)$ if $s \ge n-1$ and $\widehat{\eta}(a;X) = 0$ if $s < n-1$.

In particular, one has

$$\mathbb{W}\boldsymbol{\varkappa}_{\boldsymbol{D}_{(2,1)}}(a;(1,1)) = \begin{cases} \frac{|2a_1a_2+a_1^2|}{1-|a_1^2a_2|^2}, & \text{if } a_1a_2 \ne 0 \\ |a_2|, & \text{if } a_1 = 0;\ a_s \ne 0 \end{cases}.$$

Hence, $\mathbb{W}\boldsymbol{\varkappa}_{\boldsymbol{D}_{(2,1)}}(\cdot;(1,1))$ is not continuous at each point $(0,a_2) \in \boldsymbol{D}_{(2,1)}$, $a_2 \ne 0$.

10.2 List of problems

10.1. Find effective formulas for $\boldsymbol{\ell}^{(k)}_{\boldsymbol{D}_\alpha}(a,\cdot)$ and $\boldsymbol{\varkappa}^{(k)}_{\boldsymbol{D}_\alpha}(a;\cdot)$ for the case where $k \ge 2$, $q(\alpha) < n$, and $s(a) = n$ (cf. (AS) on page 238) 356

Chapter 11

Complex geodesics. Lempert's theorem

Summary. Section 11.1 presents general properties of complex geodesics. The notion of complex geodesics is a natural generalization of the notion of extremal discs from Chapter 3. The main result of the chapter is Lempert's Theorem 11.2.1, which states that if a domain $G \subset \mathbb{C}^n$ is biholomorphic to a convex domain, then $\boldsymbol{c}_G \equiv \boldsymbol{\ell}_G$ and $\boldsymbol{\gamma}_G \equiv \boldsymbol{\varkappa}_G$. Section 11.3 collects various results related to the problem of the uniqueness of complex geodesics. Section 11.4 presents a different method of studying complex geodesics based on the Poletsky–Edigarian theorem on extremals for certain linear functionals (Theorem 11.4.5). This method may be applied to a much more general class of domains than in the Lempert Theorem (Theorem 16.3.1, Proposition 16.3.4). In § 11.5 we discuss the problem of equality in the Schwarz lemma for holomorphic mappings $F : G_q \longrightarrow G_q$, $F(0) = 0$, in a norm ball G_q (Theorems 11.5.1, 11.5.3, 11.5.4). Finally, § 11.6 presents criteria for a holomorphic mapping $F : G_1 \longrightarrow G_2$ being a $\boldsymbol{\gamma}$ or $\boldsymbol{\varkappa}$-isometry at one point, to be globally biholomorphic (Proposition 11.6.3).

Introduction. As we already observed in the previous chapters, the class $\mathcal{L}_n$ of all domains $G \subset \mathbb{C}^n$ with $\boldsymbol{c}_G \equiv \boldsymbol{\ell}_G$ and $\boldsymbol{\gamma}_G \equiv \boldsymbol{\varkappa}_G$ plays an important role. In particular, if $G \in \mathcal{L}_n$ then all holomorphically invariant functions (resp. pseudometrics) on G coincide. It is clear that $\mathcal{L}_n$ is invariant under biholomorphic mappings. Observe that $\mathcal{L}_n$ is also closed with respect to the union of an increasing sequence of domains (if $(G_s)_{s=1}^{\infty} \subset \mathcal{L}_n$ and $D_s \subset D_{s+1}$ for all s, then $\bigcup_{s=1}^{\infty} D_s \in \mathcal{L}_n$ – cf. Proposition 2.7.1(a) and Remark 3.8.2(f)). Moreover, $\mathcal{L}_n \times \mathcal{L}_m \subset \mathcal{L}_{n+m}$ (if $D \in \mathcal{L}_n$ and $G \in \mathcal{L}_m$, then $D \times G \in \mathcal{L}_{n+m}$ – cf. Proposition 3.7.1). We know that $\mathbb{B}_n$, $\mathbb{D}^n$, $\mathbb{L}_n \in \mathcal{L}_n$ or, more generally, if $G \subset \mathbb{C}^n$ is a homogeneous balanced convex domain, then $G \in \mathcal{L}_n$ (cf. Corollary 2.3.4). We also know that $\mathbb{G}_2 \in \mathcal{L}_2$ (Theorems 7.1.1, 7.1.16) and $\mathbb{E} \in \mathcal{L}_3$ (cf. Remark 7.1.23). On the other hand, $\mathbb{A}(1/R, R) \notin \mathcal{L}_1$ (cf. Proposition 4.2.22(a)) and $\mathbb{G}_n \notin \mathcal{L}_n$ for $n \geq 3$ (cf. Corollary 7.2.12). The main result of the chapter (Lempert's Theorem 11.2.1) states that every convex domain $G \subset \mathbb{C}^n$ belongs to $\mathcal{L}_n$. This result, proved by L. Lempert in [340], next was extended by him in [342] to the class of all strongly linearly convex domains with $\mathcal{C}^{\infty}$- or $\mathcal{C}^{\omega}$-boundaries. Finally, in [322] Ł. Kosiński and T. Warszawski completed some details of Lempert's proof from [342] and extended the result to the class of all strongly linearly convex domains with $\mathcal{C}^2$-boundaries (cf. Theorem A.5.5). Notice that all the above classes of domains (convex, strongly linearly convex) are characterized by "real" conditions that do not fit with the "holomorphic point of view". Recall (Remark 7.1.21(b)) that any bounded $\mathbb{C}$-convex domain with a $\mathcal{C}^2$-boundary can be exhausted by a sequence $(G_s)_{s=1}^{\infty}$ of strongly linearly convex domains with $\mathcal{C}^{\infty}$-boundaries. Thus, every bounded $\mathbb{C}$-convex domain $G \subset \mathbb{C}^n$ with $\mathcal{C}^2$-boundary belongs to the class $\mathcal{L}_n$. Therefore, one could conjecture that every bounded $\mathbb{C}$-convex domain $G \subset \mathbb{C}^n$ belongs to $\mathcal{L}_n$.

?

11.1 Complex geodesics

We start by defining a generalization of the notion of extremal discs.

Definition 11.1.1. Let $\underline{d} = (d_G)_{G\in\mathfrak{G}_0}$ be an arbitrary contractible family of functions (resp. let $\underline{\delta} = (\delta_G)_{G\in\mathfrak{G}_0}$ be an arbitrary contractible family of pseudometrics). A mapping $\varphi \in \mathcal{O}(\mathbb{D}, G)$ is said to be a *complex d_G-geodesic for* (z_0', z_0'') (resp. a *complex δ_G-geodesic for* (z_0, X_0)) if

$$\exists_{\lambda_0',\lambda_0''\in\mathbb{D}} : z_0' = \varphi(\lambda_0'),\ z_0'' = \varphi(\lambda_0''),\ d_G(z_0', z_0'') = \boldsymbol{p}(\lambda_0', \lambda_0'') \quad (11.1.1)$$

$$(\text{resp. } \exists_{\lambda_0\in\mathbb{D},\,\alpha_0\in\mathbb{C}} : z_0 = \varphi(\lambda_0),\ X_0 = \alpha_0\varphi'(\lambda_0),\ \delta_G(z_0; X_0) = \boldsymbol{\gamma}_{\mathbb{D}}(\lambda_0; \alpha_0)). \quad (11.1.2)$$

Note that if φ is a complex δ_G-geodesic for (z_0, X_0), then it is a complex δ_G-geodesic for any pair $(z_0, \lambda X_0)$, $\lambda \neq 0$.

The notion of complex d_G-geodesics extends in a natural way to $\boldsymbol{m}$-contractible families of functions. It suffices to substitute in (11.1.1) $\boldsymbol{p}(\lambda_0', \lambda_0'')$ by $\boldsymbol{m}(\lambda_0', \lambda_0'')$.

The notion of complex geodesics is a natural generalization of the notion of extremal discs (cf. Propositions 3.2.7 and 3.5.14). Recall that if $G \subset \mathbb{C}^n$ is a taut domain, then for any $z_0', z_0'' \in G$, $z_0' \neq z_0''$ (resp. for any $z_0 \in G$, $X_0 \in \mathbb{C}^n$, $X_0 \neq 0$) there exists a holomorphic mapping $\psi : \mathbb{D} \longrightarrow G$ such that

$$z_0' = \psi(0),\ z_0'' = \psi(\sigma), \text{ where } \sigma = \boldsymbol{\ell}_G^*(z_0', z_0'') \quad (11.1.3)$$

$$(\text{resp. } z_0 = \psi(0),\ X_0 = \sigma\psi'(0), \text{ where } \sigma = \boldsymbol{\varkappa}_G(z_0; X_0)); \quad (11.1.4)$$

such a mapping ψ was called an *extremal disc* for (z_0', z_0'') (resp. for (z_0, X_0)); cf. Chapter 3. Thus, extremal discs are simply complex $\boldsymbol{\ell}_G$- (resp. $\boldsymbol{\varkappa}_G$-) geodesics.

Now let $G \subset \mathbb{C}^n$ be an arbitrary domain and let $\psi \in \mathcal{O}(\mathbb{D}, G)$ be an extremal disc for (z_0', z_0'') (resp. for (z_0, X_0)). Observe that if $h \in \operatorname{Aut}(\mathbb{D})$, then the mapping $\varphi := \psi \circ h$ has the following property:

$$\exists_{\lambda_0',\lambda_0''\in\mathbb{D}} : z_0' = \varphi(\lambda_0'),\ z_0'' = \varphi(\lambda_0''),\ \boldsymbol{\ell}_G(z_0', z_0'') = \boldsymbol{p}(\lambda_0', \lambda_0'') \quad (11.1.5)$$

$$(\text{resp. } \exists_{\lambda_0\in\mathbb{D},\,\alpha_0\in\mathbb{C}} : z_0 = \varphi(\lambda_0),\ X_0 = \alpha_0\varphi'(\lambda_0),\ \boldsymbol{\varkappa}_G(z_0; X_0) = \boldsymbol{\gamma}_{\mathbb{D}}(\lambda_0; \alpha_0)); \quad (11.1.6)$$

that is, φ is an extremal disc as well. Conversely, if $\varphi : \mathbb{D} \longrightarrow G$ is a holomorphic mapping satisfying (11.1.5) (resp. (11.1.6)), then for suitable $h \in \operatorname{Aut}(\mathbb{D})$ the mapping $\psi := \varphi \circ h^{-1}$ satisfies (11.1.3) (resp. (11.1.4)).

The following simple lemma gives a relation between complex geodesics and extremal discs.

Lemma 11.1.2. *If φ is a complex d_G-geodesic for (z_0', z_0'') (resp. a complex δ_G-geodesic for (z_0, X_0)), then*

$$d_G(z_0', z_0'') = \boldsymbol{\ell}_G(z_0', z_0'') \qquad (\textit{resp. } \delta_G(z_0; X_0) = \boldsymbol{\varkappa}_G(z_0; X_0)).$$

Consequently, φ is an extremal disc for (z_0', z_0'') (resp. for (z_0, X_0)). In particular, if the pair (z_0', z_0'') admits a complex $\boldsymbol{\ell}_G$-geodesic (resp. if the pair (z_0, X_0) admits a complex $\boldsymbol{\varkappa}_G$-geodesic), e.g., if the domain G is taut, then

$$(z_0', z_0'') \text{ admits a complex } d_G\text{-geodesic iff } d_G(z_0', z_0'') = \boldsymbol{\ell}_G(z_0', z_0'')$$
$$(\text{resp. } (z_0, X_0) \text{ admits a complex } \delta_G\text{-geodesic iff } \delta_G(z_0; X_0) = \boldsymbol{\varkappa}_G(z_0; X_0)).$$

Proof. Let $\varphi(\lambda_0') = z_0'$ and $\varphi(\lambda_0'') = z_0''$ (resp. $\varphi(\lambda_0) = z_0$, $X_0 = \alpha_0\varphi'(\lambda_0)$) as in (11.1.1) (resp. (11.1.2)). Then,

$$\boldsymbol{p}(\lambda_0', \lambda_0'') = d_G(z_0', z_0'') \le \boldsymbol{\ell}_G(z_0', z_0'') \le \boldsymbol{p}(\lambda_0', \lambda_0'')$$
$$(\text{resp. } \boldsymbol{\gamma}_{\mathbb{D}}(\lambda_0; \alpha_0) = \delta_G(z_0; X_0) \le \boldsymbol{\varkappa}_G(z_0; X_0) \le \boldsymbol{\gamma}_{\mathbb{D}}(\lambda_0; \alpha_0)). \qquad \square$$

Remark 11.1.3.

(a) If $G \subset \mathbb{C}^1$ is a taut domain, z_0, z_0', $z_0'' \in G$, $z_0' \ne z_0''$, then by Proposition 4.2.20(a) the following conditions are equivalent:

(i) $\exists_{k\in\mathbb{N}} : (z_0', z_0'')$ admits a complex $\boldsymbol{m}_G^{(k)}$-geodesic;

(ii) $\exists_{k\in\mathbb{N}} : (z_0, 1)$ admits a complex $\boldsymbol{\gamma}_G^{(k)}$-geodesic;

(iii) (z_0', z_0'') admits a complex $\boldsymbol{g}_G$-geodesic;

(iv) $(z_0, 1)$ admits a complex $\boldsymbol{A}_G$-geodesic;

(v) G is biholomorphic to $\mathbb{D}$.

In particular, if $G \subset \mathbb{C}$ is a taut multi-connected domain, then there are no complex $\boldsymbol{c}_G$- or $\boldsymbol{\gamma}_G$-geodesics $\varphi : \mathbb{D} \longrightarrow G$.

(b) Let $G = G_{\mathfrak{h}} \subset \mathbb{C}^n$ be a balanced domain of holomorphy ($\mathfrak{h}$ denotes the Minkowski function of G) and let $a \in G$ be such that $\mathfrak{h}(a) > 0$. By Propositions 3.1.11 and 3.5.2, we see that the mapping $\mathbb{D} \ni \lambda \longmapsto \lambda a/\mathfrak{h}(a) \in G$ is a complex $\boldsymbol{\ell}_G$- and $\boldsymbol{\varkappa}_G$-geodesic for $(0, a)$. Consequently, by Proposition 2.3.1(b) and Lemma 11.1.2, the following conditions are equivalent:

(i) the pair $(0, a)$ admits a complex $\boldsymbol{c}_G$-geodesic;

(ii) the pair $(0, a)$ admits a complex $\boldsymbol{\gamma}_G$-geodesic;

(iii) $c^*_G(0,a) = \gamma_G(0;a) = \mathfrak{h}(a)$.

In particular, if G is not "convex at the point $a/\mathfrak{h}(a)$", i.e., if condition (v) from Proposition 2.3.1(b) is not satisfied, then there are no complex c_G- or γ_G-geodesics for $(0,a)$.

In the following we will discuss mostly the case of complex c_G- and γ_G-geodesics (from a certain point of view this is the most interesting case). More precisely, we would like to discuss domains $G \subset \mathbb{C}^n$ such that

$$\text{for } any\ z_0', z_0'' \in G,\ z_0' \neq z_0'',\ \text{the pair } (z_0', z_0'') \text{ admits a complex } c_G\text{-geodesic, and/or} \tag{11.1.7}$$

$$\text{for } any\ z_0 \in G,\ X_0 \in \mathbb{C}^n,\ X_0 \neq 0,\ \text{the pair } (z_0, X_0) \text{ admits a complex } \gamma_G\text{-geodesic.} \tag{11.1.8}$$

Recall that in the category of taut domains, condition (11.1.7) is equivalent to the equality $c_G = \ell_G$ and (11.1.8) is equivalent to the equality $\gamma_G = \varkappa_G$; cf. Lemma 11.1.2. We know that if G is biholomorphic to $\mathbb{B}_n$, $\mathbb{D}^n$, or $\mathbb{L}_n$, then (11.1.7) and (11.1.8) are fulfilled (use Remark 11.1.3(b) and the fact that $\operatorname{Aut}(G)$ acts transitively in G (cf. Remark 2.3.8)). On the other hand, by Remark 11.1.3(b), if G is a non-convex balanced domain of holomorphy, then neither (11.1.7) nor (11.1.8) is satisfied.

Proposition 11.1.4. *Let $G \subset \mathbb{C}^n$ be an arbitrary domain and let $\varphi : \mathbb{D} \longrightarrow G$ be a holomorphic mapping. Then, the following conditions are equivalent:*

(i) $\exists_{\lambda_0',\lambda_0''\in\mathbb{D},\lambda_0'\neq\lambda_0''} : c_G(\varphi(\lambda_0'),\varphi(\lambda_0'')) = p(\lambda_0',\lambda_0'')$, *i.e., φ is a complex c_G-geodesic for $(\varphi(\lambda_0'), \varphi(\lambda_0''))$;*

(ii) $\forall_{\lambda',\lambda''\in\mathbb{D}} : c_G(\varphi(\lambda'),\varphi(\lambda'')) = p(\lambda',\lambda'')$, *i.e., φ is a complex c_G-geodesic for any pair $(\varphi(\lambda'),\varphi(\lambda''))$;*

(iii) $\forall_{\lambda\in\mathbb{D}} : \gamma_G(\varphi(\lambda);\varphi'(\lambda)) = \gamma_{\mathbb{D}}(\lambda;1)$, *i.e., φ is a complex γ_G-geodesic for any pair $(\varphi(\lambda),\varphi'(\lambda))$;*

(iv) $\exists_{\lambda_0\in\mathbb{D}} : \gamma_G(\varphi(\lambda_0);\varphi'(\lambda_0)) = \gamma_{\mathbb{D}}(\lambda_0;1)$, *i.e., φ is a complex γ_G-geodesic for $(\varphi(\lambda_0),\varphi'(\lambda_0))$.*

Consequently, any complex c_G- or γ_G-geodesic φ is an embedding (injective, proper, regular mapping) of $\mathbb{D}$ into G. In particular, $\varphi(\mathbb{D})$ is a 1-dimensional complex submanifold of G.

Before we begin the proof, we would like to mention the following

Remark 11.1.5.

(a) A part of Proposition 11.1.4 remains true for more general objects like $\boldsymbol{m}^{(k)}, \boldsymbol{\gamma}^{(k)}$ or $\boldsymbol{g}$, $\boldsymbol{A}$; cf. Exercise 11.7.1.

(b) The implication (i) $\Longrightarrow$ (ii) is not true for $\boldsymbol{\ell}_G$.
In fact, let $P \subset \mathbb{C}^1$ be an annulus (as in Chapter 9). Suppose that $\varphi : \mathbb{D} \longrightarrow P$ is a "global" $\boldsymbol{k}_P$-geodesic, i.e., $\boldsymbol{k}_P(\varphi(\lambda'), \varphi(\lambda'')) = \boldsymbol{p}(\lambda', \lambda'')$, $\lambda', \lambda'' \in \mathbb{D}$. Then, φ would be injective and proper, and consequently bijective; a contradiction.

(c) Note that the implication (iv) $\Longrightarrow$ (iii) remains true for $\boldsymbol{\varkappa}_G$-geodesics in taut domains $G \subset \mathbb{C}^1$.
In fact, suppose that $\varphi : \mathbb{D} \longrightarrow G$ is a complex $\boldsymbol{\varkappa}_G$-geodesic for $(z_0, 1)$, where G is a taut domain in $\mathbb{C}$ (i.e., $\#(\mathbb{C} \setminus G) \geq 2$). We will prove that φ is the universal covering of G; then, by Exercise 3.9.8, $\boldsymbol{\varkappa}_G(\varphi(\lambda); \varphi'(\lambda)) = 1/(1 - |\lambda|^2)$ for all $\lambda \in \mathbb{D}$, that is, φ is a "global" $\boldsymbol{\varkappa}_G$-geodesic.
We may assume that $\varphi(0) = z_0, \varphi'(0) = 1/\boldsymbol{\varkappa}_G(z_0; 1)$. Let $\Pi : \mathbb{D} \longrightarrow G$ be the universal covering of G with $\Pi(0) = z_0$. Recall (Exercise 3.9.8) that $\boldsymbol{\varkappa}_G(z_0; 1) = 1/|\Pi'(0)|$. Therefore, we may assume that $\Pi'(0) = 1/\boldsymbol{\varkappa}_G(0; 1)$. Let $\widetilde{\varphi} : \mathbb{D} \longrightarrow \mathbb{D}$ denote the lifting of φ with $\widetilde{\varphi}(0) = 0$. Since $\varphi = \Pi \circ \widetilde{\varphi}$, we have $\widetilde{\varphi}'(0) = 1$. Hence, by the classical Schwarz–Pick lemma, $\widetilde{\varphi} = \mathrm{id}_{\mathbb{D}}$. Finally, $\varphi \equiv \Pi$.

Proof of Proposition 11.1.4. (i) $\Longrightarrow$ (ii). Define

$$u(\lambda) := \frac{\boldsymbol{c}_G^*(\varphi(\lambda_0'), \varphi(\lambda))}{\boldsymbol{m}(\lambda_0', \lambda)}, \quad \lambda \in \mathbb{D} \setminus \{\lambda_0'\}.$$

Then u is subharmonic, $u \leq 1$, and $u(\lambda_0'') = 1$. Hence, by the maximum principle, $u \equiv 1$, which shows that

$$\boldsymbol{c}_G(\varphi(\lambda_0'), \varphi(\lambda)) = \boldsymbol{p}(\lambda_0', \lambda), \quad \lambda \in \mathbb{D}.$$

Now, we can repeat the same argument with respect to the first variable (for fixed $\lambda^* \in \mathbb{D}$), which proves (ii).

(ii) $\Longrightarrow$ (iii). Take a $\lambda_0 \in \mathbb{D}$. Then, by Proposition 2.7.1(d), we have

$$\boldsymbol{\gamma}_{\mathbb{D}}(\lambda_0; 1) = \lim_{\lambda \nrightarrow \lambda_0} \frac{\boldsymbol{m}(\lambda_0, \lambda)}{|\lambda_0 - \lambda|} = \lim_{\lambda \nrightarrow \lambda_0} \frac{\boldsymbol{c}_G^*(\varphi(\lambda_0), \varphi(\lambda))}{|\lambda_0 - \lambda|} = \boldsymbol{\gamma}_G(\varphi(\lambda_0); \varphi'(\lambda_0)).$$

(iv) $\Longrightarrow$ (i). Put

$$u(\lambda) := \frac{\boldsymbol{c}_G^*(\varphi(\lambda_0), \varphi(\lambda))}{\boldsymbol{m}(\lambda_0, \lambda)}, \quad \lambda \in \mathbb{D} \setminus \{\lambda_0\}.$$

Then u is subharmonic and $u \le 1$. Moreover, if we put $u(\lambda_0) := \limsup_{\lambda \not\to \lambda_0} u(\lambda)$, then u is subharmonic on the whole $\mathbb{D}$ (cf. Appendix B.4.23). In light of (iv) and Proposition 2.7.1(d), we get $u(\lambda_0) = 1$, and therefore, by the maximum principle, $u \equiv 1$, which shows that

$$c_G(\varphi(\lambda_0), \varphi(\lambda)) = p(\lambda_0, \lambda), \quad \lambda \in \mathbb{D}. \qquad \square$$

Corollary 11.1.6. *Let φ, $\psi : \mathbb{D} \longrightarrow G$ be complex $\boldsymbol{c}_G$- or $\boldsymbol{\gamma}_G$-geodesics.*

(a) $\varphi(\mathbb{D}) = \psi(\mathbb{D})$ *iff there exists an* $h \in \operatorname{Aut}(\mathbb{D})$ *such that* $\psi = \varphi \circ h$, *i.e.,* $\varphi = \psi \bmod \operatorname{Aut}(\mathbb{D})$.

(b) *Assume that*

$$\#\{\lambda \in \mathbb{D} : \varphi(\lambda) = \psi(\lambda)\} \ge 2 \text{ or } \{\lambda \in \mathbb{D} : \varphi(\lambda) = \psi(\lambda),\ \varphi'(\lambda) = \psi'(\lambda)\} \ne \varnothing.$$

Then, $\varphi \equiv \psi$ *iff* $\varphi = \psi \bmod \operatorname{Aut}(\mathbb{D})$.

Proof. (a) By Proposition 11.1.4, $V := \varphi(\mathbb{D}) = \psi(\mathbb{D})$ is a submanifold of G and the mappings $\varphi, \psi : \mathbb{D} \longrightarrow V$ are biholomorphic. Hence, $h := \varphi^{-1} \circ \psi : \mathbb{D} \longrightarrow \mathbb{D}$ is an automorphism.

(b) Let $\lambda_0', \lambda_0'' \in \mathbb{D}$, $\lambda_0' \ne \lambda_0''$, be such that $\varphi(\lambda_0') = \psi(\lambda_0')$ and $\varphi(\lambda_0'') = \psi(\lambda_0'')$ (resp. $\lambda_0 \in \mathbb{D}$ be such that $\varphi(\lambda_0) = \psi(\lambda_0)$ and $\varphi'(\lambda_0) = \psi'(\lambda_0)$). Suppose that $\psi = \varphi \circ h$ for some $h \in \operatorname{Aut}(\mathbb{D})$.

In the first case, we have $\varphi(\lambda_0') = \psi(\lambda_0') = \varphi(h(\lambda_0'))$. Hence, $h(\lambda_0') = \lambda_0'$ (recall that φ is injective). Similarly, $h(\lambda_0'') = \lambda_0''$. Thus, $h \equiv \mathrm{id}_{\mathbb{D}}$.

In the second case, we have $h(\lambda_0) = \lambda_0$ and $\varphi'(\lambda_0) = \psi'(\lambda_0) = \varphi'(\lambda_0)h'(\lambda_0)$. Hence, $h'(\lambda_0) = 1$ and so $h \equiv \mathrm{id}_{\mathbb{D}}$. $\square$

Proposition 11.1.7. *Let $G \subset \mathbb{C}^n$ be a taut domain. Then, the following conditions are equivalent:*

(i) $\boldsymbol{c}_G = \boldsymbol{\ell}_G$ *and* $\boldsymbol{\gamma}_G = \boldsymbol{\varkappa}_G$ *(i.e.,* (11.1.7)+ (11.1.8)*);*

(ii) $\boldsymbol{c}_G = \boldsymbol{\ell}_G$ *(i.e., only* (11.1.7)*);*

(iii) *for any $z_0', z_0'' \in G$ with $z_0' \ne z_0''$ there exist $\varphi \in \mathcal{O}(\mathbb{D}, G)$ and $f \in \mathcal{O}(G, \mathbb{D})$ such that $z_0', z_0'' \in \varphi(\mathbb{D})$ and $f \circ \varphi = \mathrm{id}_{\mathbb{D}}$;*

(iv) *for any $z_0', z_0'' \in G$ there exist a holomorphic embedding $\varphi : \mathbb{D} \longrightarrow G$ and a holomorphic retraction $r : G \longrightarrow \varphi(\mathbb{D})$ such that $z_0', z_0'' \in \varphi(\mathbb{D})$.*

Moreover, any holomorphic mapping $\varphi : \mathbb{D} \longrightarrow G$ satisfying (iii) *or* (iv) *is a complex $\boldsymbol{c}_G$-geodesic for (z_0', z_0''). Vice versa, for any complex $\boldsymbol{c}_G$-geodesic φ for (z_0', z_0'') there exists an f (resp. r) such that* (iii) *(resp.* (iv)*) is fulfilled.*

Proof. (ii) $\Longrightarrow$ (iii). Let φ be a complex $\boldsymbol{c}_G$-geodesic for (z_0', z_0'') with $\varphi(0) = z_0'$, $\varphi(\lambda_0'') = z_0''$, $\lambda_0'' = |\lambda_0''|e^{i\theta_0}$. Take an $f \in \mathcal{O}(G, \mathbb{D})$ such that $f(z_0') = 0$ and $f(z_0'') = e^{i\theta_0}\boldsymbol{c}_G^*(z_0', z_0'')$ (f is an extremal function for $\boldsymbol{c}_G^*(z_0', z_0'')$). Then, $\psi := f \circ \varphi : \mathbb{D} \longrightarrow \mathbb{D}$ is holomorphic, $\psi(0) = 0$, and

$$\psi(\lambda_0'') = f(z_0'') = e^{i\theta_0}\boldsymbol{c}_G^*(z_0', z_0'') = e^{i\theta_0}|\lambda_0''| = \lambda_0''.$$

Hence, by the Schwarz lemma, $\psi = \mathrm{id}_{\mathbb{D}}$.

(iii) $\Longrightarrow$ (i). Take $z_0', z_0'' \in G$, $z_0' \neq z_0''$, and let φ, f be as in (iii). Suppose that $z_0' = \varphi(\lambda_0')$, $z_0'' = \varphi(\lambda_0'')$. Then,

$$\begin{aligned} \boldsymbol{p}(\lambda_0', \lambda_0'') &= \boldsymbol{p}(f(\varphi(\lambda_0')), f(\varphi(\lambda_0''))) = \boldsymbol{p}(f(z_0'), f(z_0'')) \\ &\le \boldsymbol{c}_G(z_0', z_0'') \le \boldsymbol{\ell}_G(z_0', z_0'') \le \boldsymbol{\ell}_G(\varphi(\lambda_0'), \varphi(\lambda_0'')) \le \boldsymbol{p}(\lambda_0', \lambda_0''). \end{aligned}$$

Consequently, $\boldsymbol{c}_G = \boldsymbol{\ell}_G$ (Lemma 11.1.2). In particular, (11.1.7) is true.

To prove that $\boldsymbol{\gamma}_G = \boldsymbol{\varkappa}_G$, take $z_0 \in G$ and $X_0 \in \mathbb{C}^n$, $X_0 \neq 0$. Let $\varphi_\nu : \mathbb{D} \longrightarrow G$ be a complex $\boldsymbol{c}_G$-geodesic for $(z_0, z_0 + (1/\nu)X_0)$ with $\varphi_\nu(0) = z_0$, $\varphi_\nu(\sigma_\nu) = z_0 + (1/\nu)X_0$ $(0 < \sigma_\nu < 1)$, $\nu \gg 1$. Since G is taut, we may assume that $\varphi_\nu \overset{K}{\underset{\nu\to\infty}{\Longrightarrow}} \varphi_0 \in \mathcal{O}(\mathbb{D}, G)$, $\varphi_0(0) = 0$. Then,

$$\varphi_0'(0) = \lim_{\nu\to\infty} \frac{\varphi_\nu(\sigma_\nu) - \varphi_\nu(0)}{\sigma_\nu} = \left(\lim_{\nu\to\infty} \frac{1}{\nu\sigma_\nu}\right) \cdot X_0.$$

On the other hand,

$$\lim_{\nu\to\infty} \nu\sigma_\nu = \lim_{\nu\to\infty} \frac{\boldsymbol{c}_G^*(z_0, z_0 + \frac{1}{\nu}X_0)}{\frac{1}{\nu}} = \boldsymbol{\gamma}_G(z_0; X_0).$$

Hence, $X_0 = \boldsymbol{\gamma}_G(z_0; X_0)\varphi_0'(0)$, which proves that $\boldsymbol{\gamma}_G(z_0; X_0) = \boldsymbol{\varkappa}_G(z_0; X_0)$.

(iii) $\Longrightarrow$ (iv). In view of the first part of the proof of the implication (iii) $\Longrightarrow$ (i), if φ is as in (iii), then φ is a complex $\boldsymbol{c}_G$-geodesic for (z_0', z_0''), and therefore, by Proposition 11.1.4, φ is an embedding. Now, it suffices to put $r := \varphi \circ f$.

(iv) $\Longrightarrow$ (iii). Put $f := \varphi^{-1} \circ r$. Then, f is holomorphic and $f \circ \varphi = \mathrm{id}_{\mathbb{D}}$. □

Using the same methods as in Propositions 3.1.11 and 3.5.3, one can easily derive the following

Proposition 11.1.8. *Let $G = G_{\mathfrak{h}} = \{z \in \mathbb{C}^n : \mathfrak{h}(z) < 1\}$ be a balanced domain of holomorphy ($\mathfrak{h}$ is the Minkowski function of G), let $a \in G$ be such that $\mathfrak{h}(a) > 0$, and let $\varphi : \mathbb{D} \longrightarrow G$. Then, the following conditions are equivalent:*

(i) *the mapping* φ *is a complex* $\boldsymbol{\ell}_G$*-geodesic for* $(0, a)$ *(resp.,* φ *is a complex* $\boldsymbol{\varkappa}_G$*-geodesic for* $(0, a)$*);*

(ii) *the mapping* φ *has, up to* $\operatorname{Aut}(\mathbb{D})$*, the form*

$$\varphi(\lambda) = \lambda\widetilde{\varphi}(\lambda), \quad \lambda \in \mathbb{D},$$

where $\widetilde{\varphi} \in \mathcal{O}(\mathbb{D}, \mathbb{C}^n)$ *is such that* $\mathfrak{h} \circ \widetilde{\varphi} \equiv 1$ *and* $\widetilde{\varphi}(\mathfrak{h}(a)) = a/\mathfrak{h}(a)$ *(resp.* $\widetilde{\varphi}(0) = a/\mathfrak{h}(a)$*).*

11.2 Lempert's theorem

The aim of this section is to prove the following fundamental theorem, which has already been announced several times in the previous chapters.

Theorem 11.2.1 (Lempert's Theorem, cf. [341, 342, 460]). *Let* $G \subset \mathbb{C}^n$ *be a domain biholomorphic to a convex domain. Then,*

$$\boldsymbol{c}_G = \boldsymbol{k}_G = \boldsymbol{\ell}_G \quad \text{and} \quad \boldsymbol{\gamma}_G = \boldsymbol{\varkappa}_G.$$

Proof. We may assume that the domain G itself is convex and that $n \geq 2$ (the case $n = 1$ is obvious). Let $(G_\nu)_\nu$ be a sequence of bounded convex domains such that $G_\nu \nearrow G$. Recall that $\boldsymbol{c}_{G_\nu} \searrow \boldsymbol{c}_G$ (Proposition 2.7.1(a)), $\boldsymbol{\ell}_{G_\nu} \searrow \boldsymbol{\ell}_G$ (Proposition 3.3.5(a)), $\boldsymbol{\gamma}_{G_\nu} \searrow \boldsymbol{\gamma}_G$ (Proposition 2.7.1(a)), and $\boldsymbol{\varkappa}_{G_\nu} \searrow \boldsymbol{\varkappa}_G$ (Exercise 3.9.5(a)). In particular, if the theorem is true for each G_ν, then it is true for G. Consequently, we assume that

$$G \text{ is a bounded convex domain in } \mathbb{C}^n \text{ with } 0 \in G \ (n \geq 2). \tag{11.2.1}$$

Thus, in view of Proposition 11.1.7, the Lempert Theorem is equivalent to the following claim:

(∗) *If G is as in* (11.2.1)*, then for any complex* $\boldsymbol{\ell}_G$*-geodesic* $\varphi : \mathbb{D} \longrightarrow G$ *there exists a holomorphic mapping* $f : G \longrightarrow \mathbb{D}$ *such that* $f \circ \varphi = \mathrm{id}_{\mathbb{D}}$.

Here and in what follows, "$\boldsymbol{\ell}_G$-geodesic" means "$\boldsymbol{\ell}_G$-geodesic for some (z_0', z_0'')". The proof of (∗) will be given by a (long) sequence of lemmas.

Let $\mathcal{H}^1(\mathbb{D})$ denote the first Hardy space on $\mathbb{D}$ (cf. Appendix B.9). Recall that if $\varphi \in \mathcal{H}^1(\mathbb{D})$, then for almost all $\lambda_0 \in \mathbb{T}$ the function φ has the non-tangential boundary value

$$\varphi^*(\lambda_0) = \lim_{\substack{\lambda \to \lambda_0 \\ \sphericalangle}} \varphi(\lambda);$$

here and in what follows, "for almost all (a.a)" means "for almost all (a.a) with respect to Lebesgue measure on $\mathbb{T}$". Recall that

$$z \bullet w = \langle z, \overline{w} \rangle = \sum_{j=1}^{n} z_j w_j, \quad z = (z_1, \dots, z_n), w = (w_1, \dots, w_n) \in \mathbb{C}^n.$$

Lemma 11.2.2. *Let D be an arbitrary bounded domain in $\mathbb{C}^n$. Suppose that $\varphi \in \mathcal{O}(\mathbb{D}, D)$ and $h \in \mathcal{H}^1(\mathbb{D}, \mathbb{C}^n)$ are such that*

$$\operatorname{Re}\left((z - \varphi^*(\lambda)) \bullet \frac{1}{\lambda} h^*(\lambda)\right) < 0 \text{ for all } z \in D \text{ and for almost all } \lambda \in \mathbb{T}. \tag{11.2.2}$$

Then there exists an $f \in \mathcal{O}(D, \mathbb{D})$ such that $f \circ \varphi = \mathrm{id}_{\mathbb{D}}$.

Proof of Lemma 11.2.2. For $\varepsilon \geq 0$, we define

$$\Phi_\varepsilon(z, \lambda) := (z - \varphi(\lambda)) \bullet h(\lambda) - \varepsilon\lambda, \quad z \in \mathbb{C}^n,\ \lambda \in \mathbb{D},$$
$$\Psi_\varepsilon(z, \lambda) := \frac{1}{\lambda}\Phi_\varepsilon(z, \lambda), \quad z \in \mathbb{C}^n,\ \lambda \in \mathbb{D}_*.$$

Note that $\Phi_\varepsilon = \Phi_0 - \varepsilon\lambda$ and $\Psi_\varepsilon = \Psi_0 - \varepsilon$. The mapping f we are looking for will be a holomorphic solution of the equation

$$\Phi_0(z, f(z)) = 0, \quad z \in D. \tag{11.2.3}$$

Let $a := \varphi(0)$. Observe that if we put

$$\Psi_0(a, 0) := -\varphi'(0) \bullet h(0),$$

then $\Psi_0(a, \cdot) \in \mathcal{H}^1(\mathbb{D})$. Hence, in view of (11.2.2), the maximum principle gives

$$\operatorname{Re}\Psi_0(a, \lambda) < 0, \quad \lambda \in \mathbb{D}. \tag{11.2.4}$$

In particular, $\Psi_0(a, 0) \neq 0$, and therefore

$$\frac{\partial\Phi_0}{\partial\lambda}(a, 0) = \Psi_0(a, 0) \neq 0.$$

By the implicit function theorem, there exist: an open neighborhood U of $(a, 0)$ with $U \subset D \times \mathbb{D}$, an open neighborhood V of a with $V \subset D$, and a holomorphic function $F : V \longrightarrow \mathbb{D}$ such that

$$\{(z, \lambda) \in U : \Phi_0(z, \lambda) = 0\} = \{(z, F(z)) : z \in V\}. \tag{11.2.5}$$

Obviously,

$$F(\varphi(\lambda)) = \lambda \text{ for } \lambda \text{ in a neighborhood of } 0 \in \mathbb{D}. \tag{11.2.6}$$

Suppose that $f \in \mathcal{O}(D, \mathbb{D})$ satisfies (11.2.3). Then, condition (11.2.4) implies that $f(a) = 0$. Hence, by (11.2.5), $f = F$ in a neighborhood of $a \in D$, and therefore, by (11.2.6), $f \circ \varphi = \mathrm{id}_{\mathbb{D}}$.

Thus, the problem is to find a holomorphic solution of (11.2.3). It suffices to find functions $f_\varepsilon \in \mathcal{O}(D, \mathbb{D})$, $\varepsilon > 0$, such that

$$\Phi_\varepsilon(z, f_\varepsilon(z)) = 0, \quad z \in D. \tag{11.2.7}$$

For the proof, observe that, by (11.2.4), $f_\varepsilon(a) = 0$, $\varepsilon > 0$. Hence, using Montel's argument, one can select a sequence $\varepsilon_\nu \searrow 0$ such that

$$f_{\varepsilon_\nu} \overset{K}{\underset{\nu\to\infty}{\Longrightarrow}} f \in \mathcal{O}(D, \mathbb{D}).$$

It is clear that f is the required solution of (11.2.3).

Now, we turn to the construction of f_ε, $\varepsilon > 0$. Fix an $\varepsilon > 0$. It is enough to prove that

$$\operatorname{Re} \Psi_\varepsilon(z, \lambda) < 0, \quad z \in D,\ r(\varepsilon) < |\lambda| < 1. \tag{11.2.8}$$

Suppose for a while that (11.2.8) holds. Then, for fixed $z_0 \in D$ and for $r(\varepsilon) < r < 1$, we have

$$\begin{aligned}
\#\{\lambda \in \mathbb{D} : \Phi_\varepsilon(z_0, \lambda) = 0\} &= \frac{1}{2\pi i} \int\limits_{|\lambda|=r} \frac{\frac{\partial \Phi_\varepsilon}{\partial \lambda}(z_0, \lambda)}{\Phi_\varepsilon(z_0, \lambda)} d\lambda \\
&= \frac{1}{2\pi i} \int\limits_{|\lambda|=r} \frac{d\lambda}{\lambda} + \frac{1}{2\pi i} \int\limits_{|\lambda|=r} \frac{\frac{\partial \Psi_\varepsilon}{\partial \lambda}(z_0, \lambda)}{\Psi_\varepsilon(z_0, \lambda)} d\lambda \\
&= 1 + I_0(\varepsilon, z_0, r),
\end{aligned}$$

where $I_0(\varepsilon, z_0, r)$ denotes the index at zero of the curve

$$[0, 2\pi] \ni \theta \longmapsto \Psi_\varepsilon(z_0, re^{i\theta}).$$

In view of (11.2.8), $I_0(\varepsilon, z_0, r) = 0$, $r(\varepsilon) < r < 1$. Thus, for any $z_0 \in D$ the equation $\Phi_\varepsilon(z_0, \lambda) = 0$ has exactly one solution $\lambda =: f_\varepsilon(z_0)$. Moreover, the function $D \ni z \longmapsto f_\varepsilon(z) \in \mathbb{D}$ is holomorphic, since

$$f_\varepsilon(z) = \frac{1}{2\pi i} \int\limits_{|\lambda|=r} \lambda \frac{\frac{\partial \Phi_\varepsilon}{\partial \lambda}(z, \lambda)}{\Phi_\varepsilon(z, \lambda)} d\lambda, \quad z \in D,\ r(\varepsilon) < r < 1.$$

To prove (11.2.8), it suffices to show that

$$\operatorname{Re} \Psi_0(z, \lambda) < \operatorname{Re}\left(\frac{1}{\lambda}\Phi_0(z, 0) - \lambda\overline{\Phi_0(z, 0)}\right) =: M(z, \lambda), \quad z \in D,\ \lambda \in \mathbb{D}_*. \tag{11.2.9}$$

Note that the function M is continuous (on $\mathbb{C}^n \times \mathbb{D}_*$) and $M(z, \lambda) = 0$ for $\lambda \in \mathbb{T}$. So, if (11.2.9) holds, then for any $\varepsilon > 0$ there exists $0 < r(\varepsilon) < 1$ such that $M(z, \lambda) < \varepsilon$, $z \in D$, $r(\varepsilon) < |\lambda| < 1$, which gives (11.2.8).

To prove (11.2.9), fix a $z_0 \in D$ and let $g := \Phi_0(z_0, \cdot)$, $b_0 := g(0) = \Phi_0(z_0, 0)$. We want to prove that

$$\operatorname{Re}\left(\frac{1}{\lambda} g(\lambda)\right) < \operatorname{Re}\left(\frac{1}{\lambda} b_0 - \lambda\overline{b_0}\right), \quad \lambda \in \mathbb{D}_*. \tag{11.2.10}$$

Put

$$\widetilde{g}(\lambda) := \begin{cases} \frac{g(\lambda)-b_0}{\lambda}, & \lambda \in \mathbb{D}_* \\ g'(0), & \lambda = 0 \end{cases}.$$

Then $\widetilde{g} \in \mathcal{H}^1(\mathbb{D})$, and by (11.2.2) we have

$$\operatorname{Re} \widetilde{g}^*(\lambda) < -\operatorname{Re}\left(\frac{b_0}{\lambda}\right) = -\operatorname{Re}(\overline{b_0}\lambda) \ \text{ for almost all } \lambda \in \mathbb{T}.$$

Consequently, by the maximum principle,

$$\operatorname{Re} \widetilde{g}(\lambda) < -\operatorname{Re}(\overline{b_0}\lambda), \quad \lambda \in \mathbb{D},$$

which, together with the definition of g, gives (11.2.10). □

Let q_G denote the Minkowski function of G, i.e.,

$$q_G(z) := \inf\left\{t > 0 : \frac{1}{t}z \in G\right\}, \quad z \in \mathbb{C}^n.$$

Recall that

$$G = \{z \in \mathbb{C}^n : q_G(z) < 1\}, \quad \partial G = \{z \in \mathbb{C}^n : q_G(z) = 1\}. \tag{11.2.11}$$

Let $\widehat{q}_G$ be the *dual Minkowski subnorm*, i.e.,

$$\widehat{q}_G(w) := \sup\left\{\frac{\operatorname{Re}(z \bullet w)}{q_G(z)} : z \in (\mathbb{C}^n)_*\right\} = \max\{\operatorname{Re}(z \bullet w) : z \in \partial G\}, \quad w \in \mathbb{C}^n.$$

The functions q_G and $\widehat{q}_G$ are Minkowski subnorms on $\mathbb{C}^n$. Recall that if $(F, \|\ \|_F)$ is a complex normed space, then a function $Q : F \longrightarrow \mathbb{R}_+$ is said to be a *Minkowski subnorm on* $(F, \|\ \|_F)$ if

$$Q(\varphi + \psi) \le Q(\varphi) + Q(\psi), \quad \varphi, \psi \in F, \tag{11.2.12}$$

$$Q(t\psi) = tQ(\psi), \quad t \ge 0,\ \psi \in F, \tag{11.2.13}$$

$$\frac{1}{C}\|\psi\|_F \le Q(\psi) \le C\|\psi\|_F, \quad \psi \in F, \tag{11.2.14}$$

where $C = C(Q, \|\ \|_F)$ is a positive constant. In view of (11.2.12) and (11.2.14), we get

$$|Q(\varphi) - Q(\psi)| \le C\|\varphi - \psi\|_F, \quad \varphi, \psi \in F; \tag{11.2.15}$$

in particular, any Minkowski subnorm is continuous. In the general case, the counterpart of the dual Minkowski subnorm $\widehat{q}_G$ is defined in the following way:

Let F' denote the dual of F,

$$F' := \{\mu : F \longrightarrow \mathbb{C} : \mu \text{ is } \mathbb{C}\text{-linear and continuous}\},$$

endowed with the standard sup-norm

$$\|\mu\|_{F'} := \sup\left\{\frac{|\mu(\psi)|}{\|\psi\|_F} : \psi \in F_*\right\}, \quad \mu \in F'.$$

Then, we define

$$\widehat{Q}(\mu) := \sup\left\{\frac{\operatorname{Re}\mu(\psi)}{Q(\psi)} : \psi \in F_*\right\}, \quad \mu \in F'. \tag{11.2.16}$$

One can easily prove that $\widehat{Q} : F' \longrightarrow \mathbb{R}_+$ is a Minkowski subnorm on $(F', \| \ \|_{F'})$.

Remark 11.2.3. Let $w_0 \in (\mathbb{C}^n)_*, z_0 \in \partial G$ be such that

$$\operatorname{Re}(z_0 \bullet w_0) = \widehat{q}_G(w_0).$$

Then, for any $z \in G$, we get

$$\operatorname{Re}((z - z_0) \bullet w_0) \le (q_G(z) - 1)\widehat{q}_G(w_0) < 0.$$

This means that the plane

$$\{z \in \mathbb{C}^n : \operatorname{Re}((z - z_0) \bullet w_0) = 0\}$$

is a supporting plane for G at z_0 and that the vector $\overline{w_0}$ is an outer normal vector to ∂G at z_0. In particular, if the unit outer normal vector $\nu(z_0)$ to ∂G at z_0 is uniquely determined, then $w_0 = \varrho\overline{\nu(z_0)}$ for some $\varrho > 0$.

Lemma 11.2.4. *Let G be as in* (11.2.1) *and let $\varphi : \mathbb{D} \longrightarrow G$ be a complex ℓ_G-geodesic. Then,*

$$\varphi^*(\lambda) \in \partial G \textit{ for almost all } \lambda \in \mathbb{T} \tag{11.2.17}$$

and there exists an $h \in \mathcal{H}^1(\mathbb{D}, \mathbb{C}^n), h \not\equiv 0$, such that

$$\operatorname{Re}\left(\varphi^*(\lambda) \bullet \frac{1}{\lambda}h^*(\lambda)\right) = \widehat{q}_G\left(\frac{1}{\lambda}h^*(\lambda)\right) \textit{ for almost all } \lambda \in \mathbb{T}. \tag{11.2.18}$$

Note that *Lemmas* 11.2.2 *and* 11.2.4 *imply* (∗). To see this, observe that, by the identity principle for $\mathcal{H}^1$-functions, $h^*(\lambda) \neq 0$ for almost all $\lambda \in \mathbb{T}$; cf. Appendix B.9. Then, using Remark 11.2.3, we conclude that conditions (11.2.17) and (11.2.18) ensure that the mapping h from Lemma 11.2.4 fulfills condition (11.2.2) of Lemma 11.2.2. Finally, the existence of the mapping f is a consequence of Lemma 11.2.2.

Lemma 11.2.4 will be proved via the next lemma.

Lemma 11.2.5. *Let G be as in* (11.2.1) *and let $\varphi : \mathbb{D} \longrightarrow G$ be a complex ℓ_G-geodesic for (z_0', z_0'') with $z_0' = \varphi(\lambda_0'), z_0'' = \varphi(\lambda_0'')$ $(z_0', z_0'' \in G,\ z_0' \neq z_0'')$. Then, condition* (11.2.17) *is fulfilled and there exists an $\widetilde{h} \in \mathcal{H}^1(\mathbb{D}, \mathbb{C}^n)$, $\widetilde{h} \not\equiv 0$, such that*

$$\operatorname{Re}(\varphi^*(\lambda) \bullet u^*(\lambda)) = \widehat{q}_G(u^*(\lambda)) \text{ for almost all } \lambda \in \mathbb{T},$$

where

$$u(\lambda) := \frac{\lambda}{(\lambda - \lambda_0')(\lambda - \lambda_0'')}\widetilde{h}(\lambda), \quad \lambda \in \mathbb{D} \setminus \{\lambda_0', \lambda_0''\}.$$

Proof that Lemma 11.2.5 *implies Lemma* 11.2.4. For $\zeta = re^{i\theta} \in \mathbb{D}_*$ put

$$\Phi_\zeta(\lambda) := r + \frac{1}{r} - \left(\lambda e^{-i\theta} + \frac{1}{\lambda}e^{i\theta}\right), \quad \lambda \in \mathbb{C}_*. \tag{11.2.19}$$

Then

- the only pole of Φ_ζ is the simple pole at $\lambda = 0$,
- the only zero of Φ_ζ in $\mathbb{D}$ is the simple zero at $\lambda = \zeta$,
- if $\lambda \in \mathbb{T}$, then

$$\Phi_\zeta(\lambda) = r + \frac{1}{r} - 2\operatorname{Re}(\lambda e^{-i\theta}) \geq r + \frac{1}{r} - 2 > 0.$$

Define

$$\Phi := \begin{cases} \Phi_{\lambda_0'}\Phi_{\lambda_0''} & \text{if } \lambda_0', \lambda_0'' \neq 0 \\ \Phi_{\lambda_0''} & \text{if } \lambda_0' = 0 \\ \Phi_{\lambda_0'} & \text{if } \lambda_0'' = 0 \end{cases},$$

$$h(\lambda) := \frac{\lambda^2 \Phi(\lambda)\widetilde{h}(\lambda)}{(\lambda - \lambda_0')(\lambda - \lambda_0'')}, \ \lambda \in \mathbb{D} \setminus \{0, \lambda_0', \lambda_0''\}.$$

It is clear that the function h extends holomorphically to $\mathbb{D}$, $h \not\equiv 0$, and $h \in \mathcal{H}^1(\mathbb{D}, \mathbb{C}^n)$. Moreover,

$$\frac{1}{\lambda}h^*(\lambda) = \Phi(\lambda)u^*(\lambda) \text{ for almost all } \lambda \in \mathbb{T}.$$

Using the fact that $\Phi > 0$ on $\mathbb{T}$ (and (11.2.13)), we obtain for almost all $\lambda \in \mathbb{T}$:

$$\begin{aligned} \operatorname{Re}\left(\varphi^*(\lambda) \bullet \frac{1}{\lambda}h^*(\lambda)\right) &= \Phi(\lambda)\operatorname{Re}(\varphi^*(\lambda) \bullet u^*(\lambda)) \\ &= \Phi(\lambda)\widehat{q}_G(u^*(\lambda)) = \widehat{q}_G(\tfrac{1}{\lambda}h^*(\lambda)). \end{aligned}$$

□

So far, we have shown that the proof of $(*)$ reduces to the proof of Lemma 11.2.5. Before presenting the latter, we need some auxiliary facts.

Lemma 11.2.6. *Let $G, \varphi, z_0', z_0'', \lambda_0', \lambda_0''$ be as in Lemma* 11.2.5. *Suppose that $\psi : \mathbb{D} \longrightarrow \mathbb{C}^n$ is a holomorphic mapping with $\psi(\lambda_0') = z_0'$ and $\psi(\lambda_0'') = z_0''$. Then,*

$$\sup\{q_G(\psi(\lambda)) : \lambda \in \mathbb{D}\} \geq 1.$$

Proof of Lemma 11.2.6. Suppose that $q_G(\psi(\lambda)) < r < 1$, $\lambda \in \mathbb{D}$. Then, ψ is an extremal disc for $\ell_G(z_0', z_0'')$ with $\psi(\mathbb{D}) \subset\subset G$, which contradicts Remark 3.8.6. □

Let $F := \mathcal{C}(\mathbb{T}, \mathbb{C}^n)$,

$$\|\psi\|_F := \max\{\|\psi(\lambda)\| : \lambda \in \mathbb{T}\}, \quad \psi \in F,$$
$$Q(\psi) := \max\{q_G(\psi(\lambda)) : \lambda \in \mathbb{T}\}, \quad \psi \in F.$$

It is easily seen that $Q : F \longrightarrow \mathbb{R}_+$ is a Minkowski subnorm on the Banach space $(F, \| \ \|_F)$. Recall that

$$F' = \{\mu = (\mu_1, \dots, \mu_n) : \quad \mu_1, \dots, \mu_n \text{ are complex Borel measures on } \mathbb{T}\},$$

$$\mu(\psi) := \sum_{j=1}^n \int_{\mathbb{T}} \psi_j(\lambda) d\mu_j(\lambda), \quad \mu = (\mu_1, \dots, \mu_n) \in F', \ \psi = (\psi_1, \dots, \psi_n) \in F.$$

Let $\widehat{Q} : F' \longrightarrow \mathbb{R}_+$ be the dual Minkowski subnorm (defined by (11.2.16)). Denote by $\mathfrak{m}$ the normalized Lebesgue measure on $\mathbb{T}$; in other words,

$$\int_{\mathbb{T}} v(\lambda) d\mathfrak{m}(\lambda) := \frac{1}{2\pi} \int_0^{2\pi} v(e^{i\theta}) d\theta, \quad v \in L^1(\mathbb{T}, \mathbb{C}; \mathfrak{m}).$$

Lemma 11.2.7. *Let $\mu = u\mathfrak{m} = (u_1\mathfrak{m}, \dots, u_n\mathfrak{m}) \in F'$ with $u \in L^1(\mathbb{T}, \mathbb{C}^n; \mathfrak{m})$. Then,*

$$\widehat{Q}(\mu) = \int_{\mathbb{T}} \widehat{q}_G(u(\lambda)) d\mathfrak{m}(\lambda).$$

Proof of Lemma 11.2.7. Observe that, for any $v \in L^1(\mathbb{T}, \mathbb{C}^n; \mathfrak{m})$, if we put $\nu := v\mathfrak{m}$, then by (11.2.15) we get

$$|\widehat{Q}(\nu) - \widehat{Q}(\mu)| \leq C\|\nu - \mu\|_{F'} \leq C\|v - u\|_{L^1},$$
$$\left| \int_{\mathbb{T}} (\widehat{q}_G \circ v) d\mathfrak{m} - \int_{\mathbb{T}} (\widehat{q}_G \circ u) d\mathfrak{m} \right| \leq \int_{\mathbb{T}} C\|v - u\| d\mathfrak{m} \leq C\|v - u\|_{L^1}.$$

Hence, without loss of generality, we may assume that $u \in \mathcal{C}(\mathbb{T}, \mathbb{C}^n)$, $u \not\equiv 0$.

Take a $\psi \in F_*$. Then,

$$\frac{\operatorname{Re}\mu(\psi)}{Q(\psi)} = \frac{1}{Q(\psi)}\int_{\mathbb{T}} \operatorname{Re}(\psi \bullet u)d\mathfrak{m} \le \frac{1}{Q(\psi)}\int_{\mathbb{T}} (q_G \circ \psi)(\hat{q}_G \circ u)d\mathfrak{m} \le \int_{\mathbb{T}} (\hat{q}_G \circ u)d\mathfrak{m}.$$

Hence,

$$\hat{Q}(\mu) \le \int_{\mathbb{T}} (\hat{q}_G \circ u)d\mathfrak{m}.$$

If $\int_{\mathbb{T}} (\hat{q}_G \circ u)d\mathfrak{m} = 0$, we are done. Now suppose that the integral is positive. Fix

$$0 < \varepsilon < \int_{\mathbb{T}} (\hat{q}_G \circ u)d\mathfrak{m}.$$

Observe that for each $\xi_0 \in \mathbb{T}$ there exists a $z(\xi_0) \in \partial G$ such that

$$\operatorname{Re}(z(\xi_0) \bullet u(\xi_0)) = \hat{q}_G(u(\xi_0)).$$

By continuity, there exists a neighborhood $U(\xi_0)$ of ξ_0 in $\mathbb{T}$ such that

$$\operatorname{Re}(z(\xi_0) \bullet u(\lambda)) \ge \hat{q}_G(u(\lambda)) - \varepsilon, \quad \lambda \in U(\xi_0). \tag{11.2.20}$$

Let $\xi_1, \dots, \xi_N \in \mathbb{T}$ be such that $\mathbb{T} = U(\xi_1) \cup \dots \cup U(\xi_N)$ and let $\chi_1, \dots, \chi_N$ be a continuous partition of unity for the covering $\{U(\xi_1), \dots, U(\xi_N)\}$. Define

$$\psi(\lambda) := \sum_{j=1}^{N} \chi_j(\lambda)z(\xi_j), \quad \lambda \in \mathbb{T}.$$

Then, $\psi \in F$ and by (11.2.12), (11.2.13) we have

$$q_G(\psi(\lambda)) \le \sum_{j=1}^{N} \chi_j(\lambda)q_G(z(\xi_j)) = 1, \quad \lambda \in \mathbb{T}.$$

Hence, $Q(\psi) \le 1$. Moreover, in view of (11.2.20),

$$\begin{aligned}
\operatorname{Re}\mu(\psi) &= \int_{\mathbb{T}} \operatorname{Re}(\psi \bullet u)d\mathfrak{m} = \sum_{j=1}^{N} \int_{\mathbb{T}} \chi_j(\lambda) \operatorname{Re}(z(\xi_j) \bullet u(\lambda))d\mathfrak{m}(\lambda) \\
&\ge \sum_{j=1}^{N} \int_{\mathbb{T}} \chi_j(\lambda)(\hat{q}_G(u(\lambda)) - \varepsilon)d\mathfrak{m}(\lambda) \\
&\ge \int_{\mathbb{T}} \hat{q}_G\left(\sum_{j=1}^{N} \chi_j(\lambda)u(\lambda)\right)d\mathfrak{m}(\lambda) - \varepsilon = \int_{\mathbb{T}} (\hat{q}_G \circ u)d\mathfrak{m} - \varepsilon.
\end{aligned}$$

In particular, $\psi \not\equiv 0$ and finally,

$$\hat{Q}(\mu) \geq \frac{\operatorname{Re}\mu(\psi)}{Q(\psi)} \geq \int_{\mathbb{T}} (\hat{q}_G \circ u) d\mathfrak{m} - \varepsilon.$$

With $\varepsilon \longrightarrow 0$, we get the required result. □

Let

$$A := \{\psi \in F : \exists_{\tilde{\psi} \in \mathcal{C}(\overline{\mathbb{D}},\mathbb{C}^n) \cap \mathcal{O}(\mathbb{D},\mathbb{C}^n)} : \tilde{\psi}|_{\mathbb{T}} = \psi\};$$

we will always identify ψ with $\tilde{\psi}$. Obviously, A is a closed subspace of F. Put

$$A^{\perp} := \{\mu \in F' : \mu|_A = 0\}.$$

By virtue of F. & M. Riesz theorem (cf. Appendix B.9.10), we have

$$A^{\perp} = \{\lambda h^* \mathfrak{m} : h \in \mathcal{H}^1(\mathbb{D}, \mathbb{C}^n)\}. \tag{11.2.21}$$

Proof of Lemma 11.2.5. Let $\varphi : \mathbb{D} \longrightarrow G$ be a complex $\boldsymbol{\ell}_G$-geodesic with $\varphi(\lambda_0') = z_0'$, $\varphi(\lambda_0'') = z_0''$, $\lambda_0' \neq \lambda_0''$. Put

$$V_0 := \{\psi \in A : \psi(\lambda_0') = \psi(\lambda_0'') = 0\}.$$

Observe that V_0 is a closed complex subspace of A. Fix a $\varphi_0 \in A$ such that $\varphi_0(\lambda_0') = z_0'$ and $\varphi_0(\lambda_0'') = z_0''$, e.g.,

$$\varphi_0(\lambda) := z_0' + \frac{\lambda - \lambda_0'}{\lambda_0'' - \lambda_0'}(z_0'' - z_0'), \quad \lambda \in \overline{\mathbb{D}}.$$

By Lemma 11.2.6 we have

$$Q(\varphi_0 + \psi) \geq 1, \quad \psi \in V_0. \tag{11.2.22}$$

Let us define a functional l by

$$F \supset \mathbb{R} \cdot \varphi_0 + V_0 \ni t\varphi_0 + \psi \overset{l}{\longmapsto} t \in \mathbb{R}.$$

Obviously, l is well-defined and $\mathbb{R}$-linear. Moreover, in view of (11.2.22), for any $t > 0$ we have

$$Q(t\varphi_0 + \psi) = tQ\left(\varphi_0 + \frac{1}{t}\psi\right) \geq t = l(t\varphi_0 + \psi), \quad \psi \in V_0.$$

Hence,

$$l(t\varphi_0 + \psi) \leq Q(t\varphi_0 + \psi), \quad t \in \mathbb{R},\ \psi \in V_0.$$

By the Hahn–Banach theorem, the functional l extends to an $\mathbb{R}$-linear functional $L : F \longrightarrow \mathbb{R}$ such that $L \le Q$ on F. Note that L is continuous (use (11.2.14)). Define

$$\mu(\psi) := L(\psi) - iL(i\psi), \quad \psi \in F.$$

Then, $\mu : F \longrightarrow \mathbb{C}$ is $\mathbb{C}$-linear and continuous, i.e., $\mu \in F'$. It is easily seen that $\mu|_{V_0} = 0$. Hence, by (11.2.21),

$$\mu = \frac{\lambda}{(\lambda - \lambda_0')(\lambda - \lambda_0'')}\widetilde{h}^* \mathfrak{m}$$

for some $\widetilde{h} \in \mathcal{H}^1(\mathbb{D}, \mathbb{C}^n)$. Since $\operatorname{Re}\mu(\varphi_0) = L(\varphi_0) = l(\varphi_0) = 1$, we have $\widetilde{h} \not\equiv 0$. Moreover, $\operatorname{Re}\mu = L \le Q$ on F, which shows that

$$\widehat{Q}(\mu) \le 1. \tag{11.2.23}$$

(In fact $\widehat{Q}(\mu) = 1$.) Put

$$u(\lambda) := \frac{\lambda\widetilde{h}(\lambda)}{(\lambda - \lambda_0')(\lambda - \lambda_0'')}, \quad \lambda \in \mathbb{D} \setminus \{\lambda_0', \lambda_0''\}.$$

Observe that the function $(\varphi - \varphi_0) \bullet u$ extends holomorphically to $\mathbb{D}$ as a function of the class $\lambda\mathcal{H}^1(\mathbb{D})$. Consequently,

$$\int_{\mathbb{T}} (\varphi^*(\lambda) - \varphi_0(\lambda)) \bullet u^*(\lambda) d\mathfrak{m}(\lambda) = 0,$$

and in particular,

$$\operatorname{Re}\int_{\mathbb{T}} \varphi^*(\lambda) \bullet u^*(\lambda) d\mathfrak{m}(\lambda) = \operatorname{Re}\mu(\varphi_0) = 1.$$

Finally,

$$\begin{aligned}
1 &= \int_{\mathbb{T}} \operatorname{Re}(\varphi^*(\lambda) \bullet u^*(\lambda)) d\mathfrak{m}(\lambda) \\
&\le \int_{\mathbb{T}} q_G(\varphi^*(\lambda))\widehat{q}_G(u^*(\lambda)) d\mathfrak{m}(\lambda) \text{ (use the definition of } \widehat{q}_G) \\
&\le \int_{\mathbb{T}} \widehat{q}_G(u^*(\lambda)) d\mathfrak{m}(\lambda) \text{ (use the fact that } \varphi : \mathbb{D} \longrightarrow G) \\
&= \widehat{Q}(\mu) \text{ (use Lemma 11.2.7)} \\
&\le 1 \text{ (use (11.2.23))}.
\end{aligned}$$

This shows that

$$q_G(\varphi^*(\lambda)) = 1, \text{ i.e., } \varphi^*(\lambda) \in \partial G \text{ for almost all } \lambda \in \mathbb{T} \text{ (cf. (11.2.17))},$$

and

$$\operatorname{Re}(\varphi^*(\lambda) \bullet u^*(\lambda)) = \widehat{q}_G(u^*(\lambda)) \text{ for almost all } \lambda \in \mathbb{T},$$

which finishes the proof. □

The proof of the Lempert Theorem is completed. □

In view of the Lempert Theorem (and of Proposition 11.1.4), if $G \subset \mathbb{C}^n$ is (biholomorphic to) a convex domain, then one can introduce the notion of the *complex geodesic in* G (without prefix $\boldsymbol{c}_G$, $\boldsymbol{\ell}_G$, $\boldsymbol{\gamma}_G$, $\boldsymbol{\varkappa}_G$, etc.). In this terminology, Lemma 11.2.4 reads as follows (cf. the remark after Lemma 11.2.4):

Corollary 11.2.8. *Let $G \subset \mathbb{C}^n$ be a bounded convex domain with $0 \in G$. Then, a holomorphic mapping $\varphi : \mathbb{D} \longrightarrow G$ is a complex geodesic iff*

$$\varphi^*(\lambda) \in \partial G \text{ for almost all } \lambda \in \mathbb{T}$$

and there exists an $h \in \mathcal{H}^1(\mathbb{D}, \mathbb{C}^n)$, $h \not\equiv 0$, such that

$$\operatorname{Re}\left(\varphi^*(\lambda) \bullet \frac{1}{\lambda} h^*(\lambda)\right) = \hat{q}_G\left(\frac{1}{\lambda} h^*(\lambda)\right) \text{ for almost all } \lambda \in \mathbb{T}.$$

Remark 11.2.9. The proof of Lempert's theorem (Theorem 11.2.1) is based on Lemma 11.2.2; a similar result, based on a generalized winding number formula (without proof), has been stated in [3]. In the original version (used by Lempert in [341, 342]), this lemma has been formulated and proved under the additional assumptions that $\varphi, h \in \mathcal{C}(\overline{\mathbb{D}})$ and that (11.2.2) holds for all $\lambda \in \mathbb{T}$. We found the proof without these additional conditions, and therefore the whole proof of Theorem 11.2.1 could be essentially simplified. The remaining part of the proof of Lempert's Theorem follows the ideas taken from [460] (notice that the original proof presented in [460] has some gaps). We completed and simplified the proof from [460] in a series of lemmas: Lemmas 11.2.4–11.2.7 (see also [3, 146] for a slightly modified proof). Our method of proof allowed us to skip the discussion on the regularity on $\overline{\mathbb{D}}$ of complex geodesics (this is the central problem in Lempert's proof; cf. [340, 342]; see also [458]). We will come back to this problem in Miscellanea (cf. Theorem A.5.2) and Exercises 11.7.4–11.7.12. On the other hand, Lempert's method is more general and it works also in the strongly linearly convex case; cf. Miscellanea, Theorem A.5.5. Note that in the case of convex domains $G \subset \mathbb{C}^n$, the equality $\boldsymbol{c}_G \equiv \boldsymbol{\ell}_G$ may be also proved using functional analysis methods; cf. [362].

As we already observed above, the central step in the proof of the Lempert Theorem is Lemma 11.2.5. It is worthwhile to note that this lemma may be extended to the following more general situation (cf. [147]):

For $L_j \in \mathcal{C}(\mathbb{T}, \mathbb{C}^n)$ define a $\mathbb{C}$-linear functional $\Lambda_j : L^1(\mathbb{T}, \mathbb{C}^n) \longrightarrow \mathbb{C}$,

$$\Lambda_j(h) := \frac{1}{2\pi} \int_0^{2\pi} h(e^{i\theta}) \bullet L_j(e^{i\theta}) d\theta = \frac{1}{2\pi i} \int_{\mathbb{T}} h(\zeta) \bullet L_j(\zeta) \frac{d\zeta}{\zeta}, \quad j = 1, \dots, N. \tag{11.2.24}$$

Recall that the operator $\mathcal{H}^1(\mathbb{D}, \mathbb{C}^n) \ni h \longmapsto h^* \in L^1(\mathbb{T}, \mathbb{C}^n)$ is an isometry, which permits us to identify the space $\mathcal{H}^1(\mathbb{D}, \mathbb{C}^n)$ with a closed subspace of $L^1(\mathbb{T}, \mathbb{C}^n)$

(cf. Appendix B.9.2). Consequently, the functionals $\Lambda_1, \dots, \Lambda_N$ may be defined on $\mathcal{H}^1(\mathbb{D}, \mathbb{C}^n)$ (in particular, on $\mathcal{H}^\infty(\mathbb{D}, \mathbb{C}^n)$). To simplify notation, we will write $\Lambda_j(h)$ instead of $\Lambda_j(h^*)$, $h \in \mathcal{H}^1(\mathbb{D}, \mathbb{C}^n)$.

We will always assume that $\Lambda_1, \dots, \Lambda_N$ are $\mathbb{C}$-linearly independent on the space $\mathcal{A}(\mathbb{D}, \mathbb{C}^n) := \mathcal{C}(\overline{\mathbb{D}}, \mathbb{C}^n) \cap \mathcal{O}(\mathbb{D}, \mathbb{C}^n)$.

Let $G \subset \mathbb{C}^n$ be a bounded domain and let $a_1, \dots, a_N \in \mathbb{C}$. We are interested in a characterization of the *extremal discs for* $\Lambda_1, \dots, \Lambda_N$, i.e., those holomorphic mappings $\varphi : \mathbb{D} \longrightarrow G$ for which $\Lambda_j(\varphi) = a_j$, $j = 1, \dots, N$, and there is no mapping $\psi \in \mathcal{O}(\mathbb{D}, G)$ such that $\Lambda_j(\psi) = a_j$, $j = 1, \dots, N$, and $\psi(\mathbb{D}) \subset\subset G$.

Remark 11.2.10. If $G \subset \mathbb{C}^n$ is a bounded domain, then every extremal disc for $\boldsymbol{\ell}_G^{(m)}(z_0', z_0'')$ $(z_0' \neq z_0'')$ (resp. $\boldsymbol{\varkappa}_G^{(m)}(z_0; X_0)$ $(X_0 \neq 0)$) is extremal in the sense of the above definition (with suitable $\Lambda_1, \dots, \Lambda_N$ and $a_1, \dots, a_N$).

Indeed, put $\sigma := \sqrt[m]{\boldsymbol{\ell}_G^{(m)*}(z_0', z_0'')}$ (resp. $\sigma := \boldsymbol{\varkappa}_G^{(m)}(z_0; X_0)$) and let $\varphi \in \mathcal{O}(\mathbb{D}, G)$ be such that $\varphi(0) = z_0'$, $\operatorname{ord}_0(\varphi - z_0') \geq m$, $\varphi(\sigma) = z_0''$ (resp. $\varphi(0) = z_0$, $\operatorname{ord}_0(\varphi - z_0) \geq m$, $\sigma\varphi^{(m)}(0)/m! = X_0$). Let $\boldsymbol{e}_1, \dots, \boldsymbol{e}_n$ denote the canonical basis of $\mathbb{C}^n$. Put $N := (m+1)n$ and for $j \in \{1, \dots, n\}$ define

$$L_{kn+j}(\lambda) = \frac{1}{\lambda^k}\boldsymbol{e}_j, \quad k = 0, \dots, m-1, \quad L_{mn+j}(\lambda) := \frac{\lambda}{\lambda - \sigma}\boldsymbol{e}_j,$$
$$a_j := (z_0')_j, \quad a_{kn+j} := 0, \quad k = 1, \dots, m-1, \quad a_{mn+j} := (z_0'')_j.$$

Let $\Lambda_1, \dots, \Lambda_N$ be given by (11.2.24). Using the Cauchy integral formula (cf. Appendix B.9.3), we see that a disc $\psi \in \mathcal{O}(\mathbb{D}, G)$ is extremal for $\boldsymbol{\ell}_G^{(m)}(z_0', z_0'')$ iff $\Lambda_j(\psi) = a_j$, $j = 1, \dots, N$. Suppose that there exist $\xi_1, \dots, \xi_N \in \mathbb{C}$ such that $\xi_1\Lambda_1(h) + \dots + \xi_N\Lambda_N(h) = 0$ for every $h \in \mathcal{A}(\mathbb{D}, \mathbb{C}^n)$. Then, for every $j \in \{1, \dots, n\}$ we get

$$\xi_j h(0) + \xi_{n+j}\frac{h'(0)}{1!} + \dots + \xi_{(m-1)n+j}\frac{h^{(m-1)}(0)}{(m-1)!} + \xi_{mn+j}h(\sigma) = 0$$

for all $h \in \mathcal{A}(\mathbb{D}, \mathbb{C})$, which immediately implies that $\xi_1 = \dots = \xi_N = 0$.

Now, using Remark 3.8.6, we easily conclude that φ is extremal for $\Lambda_1, \dots, \Lambda_N$. In the infinitesimal case, we take

$$L_{kn+j}(\lambda) = \frac{1}{\lambda^k}\boldsymbol{e}_j, \quad k = 0, \dots, m,$$
$$a_j := (z_0)_j, \quad a_{kn+j} := 0, \quad k = 1, \dots, m-1, \quad a_{mn+j} := (1/\sigma)(X_0)_j.$$

The details are left to the reader.

Our aim is to prove the following result:

Proposition 11.2.11. *Assume that G is a bounded convex domain with $0 \in G$. Let $\varphi \in \mathcal{O}(\mathbb{D}, G)$ be an extremal mapping for functionals $\Lambda_1, \dots, \Lambda_N$ (generated by $L_1, \dots, L_N$). Then $\varphi^*(\zeta) \in \partial G$ for a.a. $\zeta \in \mathbb{T}$ and there exist $\xi_1, \dots, \xi_N \in \mathbb{C}$ and $h \in \mathcal{H}^1(\mathbb{D}, \mathbb{C}^n)$ such that*

$$\operatorname{Re}\left(\sum_{j=1}^{N} \xi_j \Lambda_j(\varphi)\right) = 1, \quad \operatorname{Re}(\varphi^* \bullet u) = \widehat{q}_G(u) \ \text{a.e. on } \mathbb{T},$$

where $u(\lambda) := \lambda h^*(\lambda) + \sum_{j=1}^{N} \xi_j L_j(\lambda)$.

Moreover, by Remark 11.2.3, *if $\lambda \in \mathbb{T}$ is such that $\varphi^*(\lambda)$, $u(\lambda)$ exist, $\varphi^*(\lambda) \in \partial G$, and the unit outer vector $\nu(\varphi^*(\lambda))$ is uniquely determined, then $u(\lambda) = \varrho(\lambda)\overline{\nu(\varphi^*(\lambda))}$ with $\varrho(\lambda) > 0$.*

Remark 11.2.12. Observe that the above proposition generalizes Lemma 11.2.5. Indeed, in the situation of this lemma, we have (for $\lambda_0', \lambda_0'' \in \mathbb{D}$, $z_0', z_0'' \in G$):

- $N = 2n$,
- $L_j(\lambda) := \frac{\lambda}{\lambda - \lambda_0'} e_j$, $L_{n+j}(\lambda) := \frac{\lambda}{\lambda - \lambda_0''} e_j$, $j = 1, \dots, n$,
- $(a_1, \dots, a_n) := z_0'$, $(a_{n+1}, \dots, a_{2n}) := z_0''$.

Consequently, the mapping u from Proposition 11.2.11 has the form

$$u_j(\lambda) = \lambda h_j^*(\lambda) + \xi_j \frac{\lambda}{\lambda - \lambda_0'} + \xi_{n+j} \frac{\lambda}{\lambda - \lambda_0''} = \frac{\lambda}{(\lambda - \lambda_0')(\lambda - \lambda_0'')} \widetilde{h}_j^*(\lambda),$$

where $\widetilde{h}_j \in \mathcal{H}^1(\mathbb{D}, \mathbb{C})$, $j = 1, \dots, N$.

Proof of Proposition 11.2.11. The proof uses the same methods as the proof of Lemma 11.2.5. Let

$$V_0 := \{\psi \in \mathcal{A}(\mathbb{D}, \mathbb{C}^n) : \Lambda_j(\psi) = 0, \ j = 1, \dots, N\};$$

V_0 is a closed subspace of $\mathcal{A}(\mathbb{D}, \mathbb{C}^n)$. Since $\Lambda_1, \dots, \Lambda_N$ are $\mathbb{C}$-linearly independent, we find $\psi_1, \dots, \psi_n \in \mathcal{A}(\mathbb{D}, \mathbb{C}^n)$ such that $\Lambda_j(\psi_k) = \delta_{j,k}$, $j, k = 1, \dots, N$. Put $\varphi_0 := \sum_{j=1}^{N} a_j \psi_j$. Observe that

$$\Lambda_j(\varphi_0 + \psi) = a_j, \quad \psi \in V_0, \ j = 1, \dots, N.$$

Let $F := \mathcal{C}(\mathbb{T}, \mathbb{C}^n)$ and let Q be as in the proof of Lemma 11.2.5. Observe that

$$Q(\varphi_0 + \psi) \ge 1, \quad \psi \in V_0.$$

Analogously to the proof of Lemma 11.2.5, we construct a $\mathbb{C}$-linear continuous functional

$$\mu : F \longrightarrow \mathbb{C} \text{ with } \mu|_{V_0} = 0,\ \operatorname{Re}\mu = L \le Q,\ L(t\varphi_0 + \psi) := t,\ t \in \mathbb{R},\ \psi \in V_0.$$

In particular, $\hat{Q}(\mu) \le 1$. Observe that for $g \in \mathcal{A}(\mathbb{D}, \mathbb{C}^n)$ we have

$$\Lambda_k\left(g - \sum_{j=1}^{N} \Lambda_j(g)\psi_j\right) = 0, \quad k = 1, \dots, N.$$

Hence, $g - \sum_{j=1}^{N} \Lambda_j(g)\psi_j \in V_0$ and, consequently,

$$\mu - \sum_{j=1}^{N} \mu(\psi_j)\Lambda_j = 0 \text{ on } \mathcal{A}(\mathbb{D}, \mathbb{C}^n).$$

Using F. & M. Riesz theorem (cf. Appendix B.9.10), we get

$$\mu = \left(\lambda h^* + \sum_{j=1}^{N} \mu(\psi_j)L_j\right)\mathfrak{m},$$

where $h \in \mathcal{H}^1(\mathbb{D}, \mathbb{C}^n)$. Consequently, $\mu(\psi^*)$ may be defined for all $\psi \in \mathcal{H}^\infty(\mathbb{D}, \mathbb{C}^n)$ and

$$\mu = \left(\sum_{j=1}^{N} \mu(\psi_j)L_j\right)\mathfrak{m} \text{ on } \mathcal{H}^\infty(\mathbb{D}, \mathbb{C}^n).$$

Put $\xi_j := \mu(\psi_j)$, $j = 1, \dots, N$. We have

$$\begin{aligned}
\operatorname{Re}(\mu(\varphi^*)) &= \operatorname{Re}\left(\sum_{j=1}^{N} \xi_j \Lambda_j(\varphi)\right) = \operatorname{Re}\left(\sum_{j=1}^{N} \xi_j a_j\right) \\
&= \operatorname{Re}\left(\sum_{j=1}^{N} \xi_j \Lambda_j(\varphi_0)\right) = \operatorname{Re}(\mu(\varphi_0)) = \operatorname{Re}(\mu(\varphi_0)) = L(\varphi_0) = 1.
\end{aligned}$$

Put $u := \lambda h^* + \xi_1 L_1 + \dots \xi_N L_N$. We finish the proof as in Lemma 11.2.5:

$$\begin{aligned}
1 = \operatorname{Re}(\mu(\varphi^*)) &= \int_{\mathbb{T}} \operatorname{Re}(\varphi^*(\lambda) \bullet u(\lambda))d\mathfrak{m}(\lambda) \\
&\le \int_{\mathbb{T}} q_G(\varphi^*(\lambda))\hat{q}_G(u(\lambda))d\mathfrak{m}(\lambda) \le \int_{\mathbb{T}} \hat{q}_G(u(\lambda))d\mathfrak{m}(\lambda) = \hat{Q}(\mu) \le 1,
\end{aligned}$$

which gives $\varphi^*(\lambda) \in \partial G$ and $\operatorname{Re}(\varphi^*(\lambda) \bullet u(\lambda)) = \hat{q}_G(u(\lambda))$ for a.a. $\lambda \in \mathbb{T}$. □

11.3 Uniqueness of complex geodesics

We will discuss the problem of uniqueness of complex geodesics in the case of convex domains. Here, "uniqueness" means "uniqueness modulo $\operatorname{Aut}(\mathbb{D})$". If $\varphi : \mathbb{D} \longrightarrow G$ is a complex geodesic, then we identify φ with $\varphi \circ h$, where h is any automorphism of $\mathbb{D}$. Recall that complex geodesics $\varphi, \psi : \mathbb{D} \longrightarrow G$ may be identified iff $\varphi(\mathbb{D}) = \psi(\mathbb{D})$; cf. Corollary 11.1.6. We will consider the following two types of questions:

- given $z_0', z_0'' \in G$ with $z_0' \neq z_0''$, decide whether there exists exactly one (modulo $\operatorname{Aut}(\mathbb{D})$) complex $\boldsymbol{c}_G$-geodesic for (z_0', z_0'');
- given $z_0 \in G, X_0 \in (\mathbb{C}^n)_*$, decide whether there exists exactly one (modulo $\operatorname{Aut}(\mathbb{D})$) complex $\boldsymbol{\gamma}_G$-geodesic for (z_0, X_0).

First, we will prove (Proposition 11.3.2) that for convex bounded domains G the above problems are in some sense equivalent. We will need the following version of the maximum principle:

Lemma 11.3.1. *Let $G \in \mathbb{C}^n$ be a bounded convex domain and let $\varphi : \mathbb{D} \longrightarrow \mathbb{C}^n$ be a bounded holomorphic mapping. Suppose that $\varphi^*(\lambda) \in \overline{G}$ for almost all $\lambda \in \mathbb{T}$. Then, $\varphi(\mathbb{D}) \subset \overline{G}$; in particular, either $\varphi(\mathbb{D}) \subset G$ or $\varphi(\mathbb{D}) \subset \partial G$.*

Proof. One can assume that $0 \in G$. Let q_G denote the Minkowski function of G. Suppose that $z_0 := \varphi(\lambda_0) \notin \overline{G}$ for some $\lambda_0 \in \mathbb{D}$. Let $L : \mathbb{C}^n \longrightarrow \mathbb{C}$ be a $\mathbb{C}$-linear form such that $\operatorname{Re} L \leq q_G$ and $L(z_0) = q_G(z_0) > 1$. Observe that

$$\operatorname{Re}(L \circ \varphi)^*(\lambda) = \operatorname{Re} L(\varphi^*(\lambda)) \leq q_G(\varphi^*(\lambda)) \leq 1 \text{ for almost all } \lambda \in \mathbb{T}.$$

Hence, by the classical maximum principle, $\operatorname{Re} L \circ \varphi \leq 1$. In particular, $q_G(z_0) \leq 1$ and so we have a contradiction. □

Proposition 11.3.2 (cf. [191, 192]). *Let G be a bounded convex domain. Suppose that $\varphi, \psi : \mathbb{D} \longrightarrow G$ are complex $\boldsymbol{c}_G$-geodesics for (z_0', z_0''), $z_0' \neq z_0''$ (resp. $\boldsymbol{\gamma}_G$-geodesics for (z_0, X_0), $X_0 \neq 0$) such that $\varphi \not\equiv \psi \bmod \operatorname{Aut}(\mathbb{D})$. Then,*

- *for any $\lambda_0', \lambda_0'' \in \mathbb{D}$, $\lambda_0' \neq \lambda_0''$, the complex $\boldsymbol{c}_G$-geodesic for $(\varphi(\lambda_0'), \varphi(\lambda_0''))$ is not uniquely determined,*
- *for any $\lambda_0 \in \mathbb{D}$, the complex $\boldsymbol{\gamma}_G$-geodesic for $(\varphi(\lambda_0), \varphi'(\lambda_0))$ is not uniquely determined.*

Proof. For $t \in (0, 1)$, define

$$\Phi_{0,t}(\lambda) := \frac{1 + t^2 - t\lambda - t/\lambda}{(1 + t)^2}, \quad \lambda \in \mathbb{C}_*;$$

cf. (11.2.19). Then,

- $\Phi_{0,t}$ has a simple pole at $\lambda = 0$,
- the only zero of $\Phi_{0,t}$ in $\mathbb{D}$ is the simple zero at $\lambda = t$,
- $\Phi_{0,t}(\lambda) \in (0, 1]$, $\lambda \in \mathbb{T}$.

For $\xi', \xi'' \in \mathbb{D}$, $\xi' \neq \xi''$, put $\Phi_{\xi',\xi''} := \Phi_{0,t} \circ \boldsymbol{h}_{\xi',\xi''}$, where $\boldsymbol{h}_{\xi',\xi''} \in \operatorname{Aut}(\mathbb{D})$ is such that $\boldsymbol{h}_{\xi',\xi''}(\xi') = 0$ and $\boldsymbol{h}_{\xi',\xi''}(\xi'') = t \in (0, 1)$. Then,

- $\Phi_{\xi',\xi''}$ is meromorphic in a neighborhood of $\overline{\mathbb{D}}$,
- the only pole of $\Phi_{\xi',\xi''}$ is the simple pole at $\lambda = \xi'$,
- the only zero of $\Phi_{\xi',\xi''}$ is the simple zero at $\lambda = \xi''$,
- $\Phi_{\xi',\xi''}(\lambda) \in (0, 1]$, $\lambda \in \mathbb{T}$.

Moreover, we put $\Phi_{\xi,\xi} :\equiv 1$, $\xi \in \mathbb{D}$.

We may assume that

(a) $z_0' = 0 \in G$, $\varphi(0) = \psi(0) = 0$, $\varphi(\sigma) = \psi(\sigma) = z_0''$ with $\sigma \in (0, 1)$ (resp.

(b) $z_0 = 0 \in G$, $\varphi(0) = \psi(0) = 0$, $\varphi'(0) = \psi'(0) = X_0$).

Fix

(c) $\lambda_0', \lambda_0'' \in \mathbb{D}$, $\lambda_0' \neq \lambda_0''$, with $\{\lambda_0', \lambda_0''\} \neq \{0, \sigma\}$ (resp.

(d) $\lambda_0 \in \mathbb{D}_*$).

Define

$$\Phi := \begin{cases} \Phi_{0,\lambda_0'} \Phi_{\sigma,\lambda_0''}, & \text{if (a)+(c)} \\ \Phi_{0,\lambda_0} \Phi_{\sigma,\lambda_0}, & \text{if (a)+(d)} \\ \Phi_{0,\lambda_0'} \Phi_{0,\lambda_0''}, & \text{if (b)+(c)} \\ \Phi_{0,\lambda_0} \Phi_{0,\lambda_0}, & \text{if (b)+(d)} \end{cases}.$$

Then:

- Φ is meromorphic in a neighborhood of $\overline{\mathbb{D}}$,
- $\Phi(\lambda) \in (0, 1]$, $\lambda \in \mathbb{T}$,
- the function $\chi := \varphi + \Phi(\psi - \varphi)$ extends holomorphically to $\mathbb{D}$,
- $\chi^*(\lambda) \in \overline{G}$ for a.a. $\lambda \in \mathbb{T}$,
- $\chi(\lambda_0') = \varphi(\lambda_0')$ and $\chi(\lambda_0'') = \varphi(\lambda_0'')$ if (c),
- $\chi(\lambda_0) = \varphi(\lambda_0)$ and $\chi'(\lambda_0) = \varphi'(\lambda_0)$ if (d).

By Lemma 11.3.1, χ is a complex $\boldsymbol{c}_G$-geodesic for $(\varphi(\lambda_0'), \varphi(\lambda_0''))$ (resp. a $\boldsymbol{\gamma}_G$-geodesic for $(\varphi(\lambda_0), \varphi'(\lambda_0))$). It is clear that $\chi \not\equiv \varphi$. Hence, by Corollary 11.1.6(b), $\chi \neq \varphi \bmod \operatorname{Aut}(\mathbb{D})$. □

Now, we move to the case in which the complex geodesics are uniquely determined.

Proposition 11.3.3 (cf. [137]). *If G is a strictly convex bounded domain, then the complex $\boldsymbol{c}_G$- and $\boldsymbol{\gamma}_G$-geodesics are uniquely determined.*

Here, *strictly convex* means "geometrically strictly convex", i.e., if $a, b, \frac{1}{2}(a+b) \in \partial G$, then $a = b$.

Proof. Suppose that $\varphi, \psi : \mathbb{D} \longrightarrow G$ are complex $\boldsymbol{c}_G$-geodesics for (z_0', z_0'') (resp. $\boldsymbol{\gamma}_G$-geodesics for (z_0, X_0)), with $\varphi(0) = \psi(0) = z_0'$, $\varphi(\sigma) = \psi(\sigma) = z_0''$ (resp. $\varphi(0) = \psi(0) = z_0$, $\varphi'(0) = \psi'(0) = X_0$). Put

$$\chi := \frac{1}{2}(\varphi + \psi).$$

Since G is convex, $\chi(\mathbb{D}) \subset G$. Moreover, $\chi(0) = z_0'$, $\chi(\sigma) = z_0''$ (resp. $\chi(0) = z_0$, $\chi'(0) = X_0$), and therefore χ is also a complex $\boldsymbol{c}_G$-geodesic for (z_0', z_0'') (resp. $\boldsymbol{\gamma}_G$-geodesic for (z_0, X_0)). Since G is bounded, $\varphi, \psi, \chi \in \mathcal{H}^\infty(\mathbb{D}, \mathbb{C}^n)$. By Proposition 11.1.4, $\varphi^*(\lambda), \psi^*(\lambda), \chi^*(\lambda) \in \partial G$ for almost all $\lambda \in \mathbb{T}$. Now, by the strict convexity, $\varphi^* = \psi^*$ a.e. on $\mathbb{T}$ and, finally, $\varphi \equiv \psi$. □

Example 11.3.4. If G is not strictly convex, then complex geodesics need not be uniquely determined. For, let $G = \mathbb{D}^2$, $\varphi(\lambda) := (\lambda, \lambda/2)$, $\psi(\lambda) := (\lambda, \lambda^2)$, $\lambda \in \mathbb{D}$. Then, φ and ψ are both complex $\boldsymbol{c}_{\mathbb{D}^2}$-geodesics for $((0,0), (1/2, 1/4))$ but $\varphi(\mathbb{D}) \neq \psi(\mathbb{D})$, and consequently φ and ψ are not equivalent; cf. Example 8.3.8.

In the case of balanced domains of holomorphy, the uniqueness of complex geodesics may be characterized more precisely, namely, we have

Proposition 11.3.5.

(a) *Let $G = G_{\mathfrak{h}} \subset \mathbb{C}^n$ be a balanced domain of holomorphy ($\mathfrak{h}$ is the Minkowski function of G) and let $a \in G$ be such that $\mathfrak{h}(a) > 0$. Then, the following conditions are equivalent:*

 (i) *the mapping $\mathbb{D} \ni \lambda \overset{\varphi_0}{\longmapsto} \lambda a/\mathfrak{h}(a) \in G$ is the unique (modulo $\operatorname{Aut}(\mathbb{D})$) complex $\boldsymbol{\ell}_G$- or $\boldsymbol{\varkappa}_G$-geodesic for $(0, a)$;*

 (ii) *the point $b := a/\mathfrak{h}(a)$ is an "extreme point for $\overline{G}$", i.e., there is no non-constant holomorphic mapping $f : \mathbb{D} \longrightarrow \overline{G}$ with $b = f(0)$.* (∗)

(b) *(cf. [510]) Let $G = G_q \subset \mathbb{C}^n$ be a convex balanced domain (q is a seminorm) and let $a \in G$ be such that $q(a) > 0$. Then, the following conditions are equivalent:*

 (i) *the mapping $\mathbb{D} \ni \lambda \to \lambda a/q(a) \in G$ is the unique (modulo $\operatorname{Aut}(\mathbb{D})$) complex $\boldsymbol{c}_G$- or $\boldsymbol{\gamma}_G$-geodesic for $(0, a)$;*

(ii) *the point* $b := a/q(a)$ *is a "complex extreme point for* $\overline{G}$*", i.e., there is no* $\xi \in (\mathbb{C}^n)_*$ *with* $b + \mathbb{D}\xi \subset \overline{G}$.

Remark 11.3.6. Let $G \subset \mathbb{C}^n$ be an arbitrary domain. Suppose that G is strongly pseudoconvex at a point $b \in \partial G$, i.e., there exist a neighborhood U of b and a function $r \in \mathcal{C}^2(U, \mathbb{R})$ such that $G \cap U = \{r < 0\}$, $\partial G \cap U = \{r = 0\}$, $dr(z) \neq 0$, and $(\mathcal{L}r)(z; X) > 0$, $z \in U$, $X \in (\mathbb{C}^n)_*$; cf. Appendix B.7. Then, b is extreme in the sense of $(*)$. For, suppose that $f : \mathbb{D} \longrightarrow \overline{G} \cap U$ is a holomorphic mapping with $b = f(0)$. Since $r \circ f$ is a psh function, the maximum principle gives $r \circ f \equiv 0$. Consequently, $0 = (\mathcal{L}(r \circ f))(\lambda; 1) = (\mathcal{L}r)(f(\lambda); f'(\lambda))$, $\lambda \in \mathbb{D}$. Hence, $f \equiv b$.

Proof of Proposition 11.3.5. (a) (i) $\Longrightarrow$ (ii). We already know that the mapping φ_0 is both a $\boldsymbol{\ell}_G$- and $\boldsymbol{\varkappa}_G$-geodesic for $(0, a)$; cf. Remark 11.1.3(b). Suppose that $f : \mathbb{D} \longrightarrow \overline{G}$ is a holomorphic mapping with $b = f(0)$. By the maximum principle for psh functions we have $\mathfrak{h} \circ f \equiv 1$. We consider the following two cases:

1) The mapping φ_0 is the unique complex $\boldsymbol{\ell}_G$-geodesic for $(0, a)$.

Let $g \in \operatorname{Aut}(\mathbb{D})$ be such that $g(\mathfrak{h}(a)) = 0$. Define $\psi(\lambda) := \lambda f(g(\lambda))$, $\lambda \in \mathbb{D}$. Obviously, $\psi : \mathbb{D} \longrightarrow G$ and $\psi(0) = 0$, $\psi(\mathfrak{h}(a)) = \mathfrak{h}(a) f(0) = a$. Hence, ψ is a complex $\boldsymbol{\ell}_G$-geodesic for $(0, a)$. In view of the uniqueness, there exists a $\chi \in \operatorname{Aut}(\mathbb{D})$ such that $\psi \equiv \chi \cdot b$. Clearly, $\chi(0) = 0$ and $\chi(\mathfrak{h}(a)) = \mathfrak{h}(a)$. Consequently, $\chi = \mathrm{id}_{\mathbb{D}}$, and therefore $f \circ g \equiv b$, which shows that $f \equiv b$.

2) The mapping φ_0 is the unique complex $\boldsymbol{\varkappa}_G$-geodesic for $(0, a)$.

Put $\psi(\lambda) := \lambda f(\lambda)$. Then, $\psi : \mathbb{D} \longrightarrow G$, $\psi(0) = 0$, and $\psi'(0) = f(0) = b$. Hence, ψ is a complex $\boldsymbol{\varkappa}_G$-geodesic for $(0, a)$. By the uniqueness, $\psi = \chi \cdot b$ for an automorphism $\chi \in \operatorname{Aut}(\mathbb{D})$. Then, $\chi(0) = 0$ and $\chi'(0) = 1$. The end of the proof is the same as above.

(ii) $\Longrightarrow$ (i). This implication follows from Proposition 11.1.8.

(b) This is an immediate consequence of (a) and of the following lemma: □

Lemma 11.3.7 (cf. [214]). *Let* $f : \mathbb{D} \longrightarrow \mathbb{C}^n$ *be a holomorphic mapping. Then,*

$$f(0) + \eta(f(\lambda) - f(0)) \in \overline{\operatorname{conv}(f(\mathbb{D}))}$$

for all $\eta \in \mathbb{C}$ *and* $\lambda \in \mathbb{D}$ *such that* $2|\eta\lambda| \leq 1 - |\lambda|$.

In particular,

$$f(0) + \frac{1}{2}\mathbb{D} \cdot f'(0) \subset \overline{\operatorname{conv}(f(\mathbb{D}))}.$$

Consequently, if $G \subset \mathbb{C}^n$ *is a convex domain and if* $f : \mathbb{D} \longrightarrow \overline{G}$ *is a holomorphic mapping such that the point* $f(0)$ *is a complex extreme point for* $\overline{G}$*, then* $f \equiv \text{const}$.

Proof. (Cf. [199]) We may assume that $f(0) = 0$. Put $S := \overline{\operatorname{conv}(f(\mathbb{D}))}$ and suppose that there exist $\eta_0 \in \mathbb{C}_*$ and $\lambda_0 \in \mathbb{D}$ such that $2|\eta_0\lambda_0| \le 1 - |\lambda_0|$ but the point $z_0 := \eta_0 f(\lambda_0)$ is not in S. Then, by the Hahn–Banach theorem, there exists a $\mathbb{C}$-linear functional $L : \mathbb{C}^n \longrightarrow \mathbb{C}$ such that $\operatorname{Re} L \le 1$ on S and $\operatorname{Re} L(z_0) > 1$. Put $F := L \circ f : \mathbb{D} \longrightarrow \mathbb{C}$. Then, $\operatorname{Re} F \le 1$ and

$$|F(\lambda_0)| = \frac{1}{|\eta_0|}|L(z_0)| > \frac{2|\lambda_0|}{1 - |\lambda_0|},$$

which contradicts the Borel–Carathéodory lemma (cf. Exercise 2.12.4). □

Corollary 11.3.8. *Let $D \subset \mathbb{C}^n$ be a convex domain. Then,*

$$\frac{1}{2} \le \varkappa_D(z; X) d_D(z; X) \le 1, \quad z \in D,\ X \in \mathbb{C}^n,\ d_D(z; X) < +\infty,$$

where $d_D(z; X) := \sup\{r > 0 : z + \mathbb{D}(r)X \subset D\}$.

Proof. Use simply Lemma 11.3.7. □

The same inequality can be also found in [44].

The following example will show that complex $\boldsymbol{\gamma}_G$-geodesics need not be uniquely determined even for very regular strongly pseudoconvex domains $G \subset \mathbb{C}^n$; cf. Proposition 11.3.3.

Example 11.3.9 (cf. [472]). For any $k \in \mathbb{N}$ there exists a $t(k) > 0$ such that, for each $t \ge t(k)$, the domain

$$G = G_t := \{(z_1, z_2) \in \mathbb{C}^2 : |z_1|^2 + |z_2|^2 + t|z_1^{2k} - z_2^k|^2 < 1\}$$

admits at least k non-equivalent complex $\boldsymbol{\gamma}_G$-geodesics for $((0,0), (1,0))$.

Observe that for any $t \ge 0$ the domain G_t is strongly pseudoconvex with smooth $\mathcal{C}^\infty$-boundary. Obviously, $G_t \subset \mathbb{B}_2$. Moreover, the mapping

$$[0,1] \times G_t \ni (s, (z_1, z_2)) \longmapsto (sz_1, s^2 z_2) \in G_t$$

is a contraction of G to 0.

Note that for any $\alpha \in \sqrt[k]{1}$ the mapping

$$G_t \ni (z_1, z_2) \overset{T_\alpha}{\longmapsto} (z_1, \alpha z_2) \in G_t$$

is an automorphism of G_t.

We go to the proof of the example. Fix a $k \in \mathbb{N}$. We will prove that for $t \gg 0$ there exists a complex $\boldsymbol{\gamma}_{G_t}$-geodesic $\varphi : \mathbb{D} \longrightarrow G_t$ for $((0,0), (1,0))$ of the form

$$\varphi(\lambda) = (\lambda a, \lambda^2 b), \quad \lambda \in \mathbb{D},$$

with $b \neq 0$; then, the mappings $T_\alpha \circ \varphi$, $\alpha \in \sqrt[k]{1}$, are non-equivalent complex $\boldsymbol{\gamma}_{G_t}$-geodesics for $((0,0),(1,0))$.

Put

$$u_t(z_1, z_2) := |z_1|^2 + |z_2|^2 + t|z_1^{2k} - z_2^k|^2, \quad (z_1, z_2) \in \mathbb{C}^2$$

(u_t is a strictly plurisubharmonic $\mathcal{C}^\infty$-function). Let $\psi(\lambda) = (\lambda\chi_1(\lambda), \lambda^2\chi_2(\lambda))$, $\lambda \in \mathbb{D}$, be an arbitrary complex $\boldsymbol{\varkappa}_{G_t}$-geodesic for $((0,0),(1,0))$ (G_t is taut). By the maximum principle, we have $u_t(\chi_1, \chi_2) \le 1$ on $\mathbb{D}$. In particular, $u_t(\chi_1(0), \chi_2(0)) \le 1$. This shows that

$$\frac{1}{\boldsymbol{\varkappa}_{G_t}((0,0);(1,0))} = |\chi_1(0)| \le \max\{|z_1| : \exists_{z_2} : u_t(z_1, z_2) \le 1\} =: \sigma_t.$$

On the other hand, if (a,b) is such that $u_t(a,b) = \sigma_t$, then $(\lambda a, \lambda b^2) \in G_t$, $\lambda \in \mathbb{D}$. Consequently, for any $t \ge 0$, if $u_t(a,b) = \sigma_t$, then the mapping

$$\varphi(\lambda) = (\lambda a, \lambda^2 b), \quad \lambda \in \mathbb{D},$$

is a complex $\boldsymbol{\varkappa}_{G_t}$-geodesic for $((0,0),(1,0))$. Moreover, in view of the definition of σ_t, the mapping

$$G_t \ni (z_1, z_2) \overset{r}{\longmapsto} \left(z_1, \frac{b}{a^2} z_1^2\right)$$

is a holomorphic retraction of G_t onto $\varphi(\mathbb{D})$ (in fact, $r(z_1, z_2) = \varphi(z_1/a)$). Hence, φ is a complex $\boldsymbol{\gamma}_{G_t}$-geodesic for $((0,0),(1,0))$; cf. Proposition 11.1.7.

It remains to prove that for $t \gg 0$, if $u_t(a,b) = \sigma_t$, then $b \neq 0$. It suffices to take a $t > 0$ so big that

$$\max\{|z_1| : u_t(z_1, z_1^2)\} > \max\{|z_1| : u_t(z_1, 0) = 1\}$$

(note that $t(k)$ may be given by an explicit formula).

We conclude this section with an example related to Proposition 11.1.7.

Example 11.3.10 (cf. [341]). Let

$$G := \{(z_1, z_2) \in \mathbb{C}^2 : (1 + |z_1|^2)(1 + |z_2|^2) < 25\},$$

and $a := (1,1) \in G$. Then, there is no holomorphic retraction $r : G \longrightarrow S$ with $0, a \in S$ and $\dim S = 1$ (S as a holomorphic retract would be a complex submanifold of G). In particular, there are no complex $\boldsymbol{c}_G$- or $\boldsymbol{\gamma}_G$-geodesics for $(0,a)$.

For, suppose that $r : G \longrightarrow S$ is such a retraction. Note that G is balanced and strongly pseudoconvex. Moreover, G is not convex at the point $b := a/\mathfrak{h}(a) = (2,2) \in \partial G$ ($\mathfrak{h}$ denotes the Minkowski function of G). In particular, the point b is extreme (cf. the remark before the proof of Proposition 11.3.5). Hence, the mapping $\mathbb{D} \ni \lambda \overset{\varphi_0}{\longmapsto} \lambda b \in G$ is the only (modulo $\operatorname{Aut}(\mathbb{D})$) complex $\boldsymbol{\ell}_G$-geodesic for $(0,a)$;

cf. Proposition 11.3.5(a). One can easily prove that the mapping $r \circ \varphi_0$ is also a complex ℓ_G-geodesic for $(0, a)$. Hence, $r(\lambda b) = \chi(\lambda) \cdot b$, $\lambda \in \mathbb{D}$, for a suitable automorphism $\chi \in \mathrm{Aut}(\mathbb{D})$. One can see that $\chi = \mathrm{id}_{\mathbb{D}}$. Thus, $r \circ \varphi_0 = \varphi_0$. In particular, $\varphi_0(\mathbb{D}) = S$ (recall that S is one dimensional). Since $\varphi_0^{-1} \circ r : G \longrightarrow \mathbb{D}$, we conclude that $\boldsymbol{c}_G^*(0, a) = \boldsymbol{\ell}_G^*(0, a) = \mathfrak{h}(a)$. This means that G is convex in b, cf. Proposition 2.3.1; a contradiction.

11.4 Poletsky–Edigarian theorem

As we have seen, the Lempert Theorem is a powerful tool of complex analysis. Nevertheless, its usage is restricted to domains biholomorphic to convex ones. In this section, we will present a different method, which may be used in a more general category of domains. The main idea is to study extremal discs for ℓ_G using suitable systems of linear functionals (similar to Remark 11.2.10 and Proposition 11.2.11). In particular, we get an effective sufficient condition for $\varphi \in \mathcal{O}(\mathbb{D}, G)$ to be an extremal disc (Corollary 11.4.6). The whole section is based on [145] and [432].

Definition 11.4.1. Fix $N \in \mathbb{N}$, $0 < \tau < 1$, $K_1, \dots, K_N \in \mathcal{A}(\mathbb{A}(\tau, 1), \mathbb{C}^n)$, and define $\mathbb{R}$-linear functionals $\Phi_j : L^1(\mathbb{T}, \mathbb{C}^n) \longrightarrow \mathbb{R}$,

$$\Phi_j(h) = \frac{1}{2\pi} \int_0^{2\pi} \mathrm{Re}\Big(h(e^{i\theta}) \bullet K_j(e^{i\theta})\Big)\, d\theta = \mathrm{Re}\left(\frac{1}{2\pi i} \int_{\mathbb{T}} h(\zeta) \bullet K_j(\zeta) \frac{d\zeta}{\zeta}\right), \quad j = 1, \dots, N.$$

We say that $\Phi_1, \dots, \Phi_N$ are *linearly independent* if for $s \in \mathcal{H}^\infty(\mathbb{D}, \mathbb{C}_*^n)$, $g \in \mathcal{H}^\infty(\mathbb{D}, \mathbb{C}^n)$ with $g(0) = 0$, and $\xi \in \mathbb{R}^N$, if $\sum_{j=1}^N \xi_j K_j \cdot s^* = g^*$ on a subset of $\mathbb{T}$ of positive measure, then $\xi = 0$, where $z \cdot w := (z_1 w_1, \dots, z_n w_n)$.

Remark 11.4.2.

(a) $|\Phi_j(h)| \le \frac{1}{2\pi}(\max_{\mathbb{T}} \|K_j\|)\|h\|_{L^1}$, $h \in L^1(\mathbb{T}, \mathbb{C}^n)$, where

$$\|h\|_{L^1} := \int_0^{2\pi} \|h(e^{i\theta})\| d\theta.$$

In particular, the operators $\Phi_1, \dots, \Phi_N$ are continuous.

(b) Using the isometry $\mathcal{H}^1(\mathbb{D}, \mathbb{C}^n) \ni h \longmapsto h^* \in L^1(\mathbb{T}, \mathbb{C}^n)$, the functionals $\Phi_1, \dots, \Phi_N$ may be extended to the space $\mathcal{H}^1(\mathbb{D}, \mathbb{C}^n)$. To simplify notation, we will write $\Phi_j(h)$ instead of $\Phi_j(h^*)$, $h \in \mathcal{H}^1(\mathbb{D}, \mathbb{C}^n)$.

(c) If $h \in \mathcal{H}^1(\mathbb{D}, \mathbb{C}^n)$ and $h_r(\zeta) := h(r\zeta)$, $0 < r < 1$, then $h_r \longrightarrow h^*$ in $L^1(\mathbb{T}, \mathbb{C}^n)$ when $r \longrightarrow 1$ (cf. Appendix B.9.2). Consequently,

$$\lim_{r \to 1} \Phi_j(h_r) = \Phi_j(h), \quad j = 1, \dots, N.$$

(d) If $\Phi_1, \dots, \Phi_N$ are linearly independent (in the sense of the above definition), then for any $s \in \mathcal{H}^\infty(\mathbb{D}, \mathbb{C}^n_*)$ the functionals

$$\mathcal{A}(\mathbb{D}, \mathbb{C}^n) \ni h \stackrel{\Psi_{j,s}}{\longmapsto} \Phi_j(s \cdot h), \quad j = 1, \dots, N,$$

are $\mathbb{R}$-linearly independent (in the usual sense).

Indeed, suppose that $\sum_{j=1}^N \xi_j \Psi_{j,s} = 0$. Then, by F. & M. Riesz theorem (cf. Appendix B.9.10), there exists a $W \in \mathcal{H}^1(\mathbb{D}, \mathbb{C}^n)$ such that

$$\sum_{j=1}^N \xi_j K_j(\zeta) \cdot s^*(\zeta) = \zeta W^*(\zeta) \text{ for a.a. } \zeta \in \mathbb{T}.$$

Put $g(\zeta) := \zeta W(\zeta)$. Then, $g \in \mathcal{H}^\infty(\mathbb{D}, \mathbb{C}^n)$ and $g(0) = 0$. Thus, $\xi_1 = \dots = \xi_N = 0$.

Definition 11.4.3 (Problem $(\mathcal{P})$). Assume that $\Phi_1, \dots, \Phi_N$ are linearly independent. Let $G \subset\subset \mathbb{C}^n$ be a domain and let $c_1, \dots, c_N \in \mathbb{R}$. The problem $(\mathcal{P})$ is to find a $\varphi \in \mathcal{O}(\mathbb{D}, G)$ such that $\Phi_j(\varphi) = c_j$, $j = 1, \dots, N$, and there is no mapping $\psi \in \mathcal{O}(\mathbb{D}, G)$ with $\Phi_j(\psi) = c_j$, $j = 1, \dots, N$, and $\psi(\mathbb{D}) \subset\subset G$. Each such a mapping φ is called an *extremal for the problem* $(\mathcal{P})$.

We say that the problem $(\mathcal{P})$ is of type $(\mathcal{P}_m)$ if there exist $\sigma_1, \dots, \sigma_m \in \mathbb{D}$ such that for $Q(\lambda) := \prod_{k=1}^m (\lambda - \sigma_k)$, the function QK_j extends holomorphically to $\mathbb{D}$, $j = 1, \dots, N$.

Remark 11.4.4. Let $G \subset\subset \mathbb{C}^n$ be a domain. Assume that $a, b \in G$, $a \neq b$, $\sigma := \sqrt[m]{\boldsymbol{\ell}_G^{(m)*}(a, b)} > 0$ (resp. $a \in G$, $X \in (\mathbb{C}^n)_*$, $\sigma := \boldsymbol{\varkappa}_G^{(m)}(a; X) > 0$). Suppose that $\varphi \in \mathcal{O}(\mathbb{D}, G)$ is extremal for $\boldsymbol{\ell}_G^{(m)}(a, b)$ (resp. $\boldsymbol{\varkappa}_G^{(m)}(a; X)$), i.e., $\varphi(0) = a$, $\operatorname{ord}_0(\varphi - a) \geq m$, and $\varphi(\sigma) = b$ (resp. $\varphi(0) = a$, $\operatorname{ord}_0(\varphi - a) \geq m$, and $(\sigma/m!)\varphi^{(m)}(0) = X$). Then, φ is a solution of a problem of type $(\mathcal{P}_m)$.

Indeed (cf. Remark 11.2.10), let $\boldsymbol{e}_1, \dots, \boldsymbol{e}_n$ be the canonical basis of $\mathbb{C}^n$. Put $N := 2n(m+1)$ and define $K_1, \dots, K_N, c_1, \dots, c_N$ according to Table 11.1 (with $j = 1, \dots, n$, $\nu = 1, \dots, m-1$):

Table 11.1

s	$K_s(\lambda)$	c_s
j	$\boldsymbol{e}_j$	$\operatorname{Re} a_j$
$n + j$	$-i\boldsymbol{e}_j$	$\operatorname{Im} a_j$
$2n\nu + j$	$\lambda^{-\nu}\boldsymbol{e}_j$	0
$2n\nu + n + j$	$-i\lambda^{-\nu}\boldsymbol{e}_j$	0
$2nm + j$	$\lambda(\lambda - \sigma)^{-1}\boldsymbol{e}_j$	$\operatorname{Re} b_j$
$2nm + n + j$	$-i\lambda(\lambda - \sigma)^{-1}\boldsymbol{e}_j$	$\operatorname{Im} b_j$

In the infinitesimal case, we put

Table 11.2

s	$K_s(\lambda)$	c_s
j	$\boldsymbol{e}_j$	$\operatorname{Re} a_j$
$n+j$	$-i\boldsymbol{e}_j$	$\operatorname{Im} a_j$
$2n\nu+j$	$\lambda^{-\nu}\boldsymbol{e}_j$	0
$2n\nu+n+j$	$-i\lambda^{-\nu}\boldsymbol{e}_j$	0
$2nm+j$	$\lambda^{-m}\boldsymbol{e}_j$	$\operatorname{Re}(X_j)/\sigma$
$2nm+n+j$	$-i\lambda^{-m}\boldsymbol{e}_j$	$\operatorname{Im}(X_j)/\sigma$

The functions $K_1,\dots,K_N$ generate functionals $\Phi_1,\dots,\Phi_N$. Observe that for $h\in\mathcal{O}(\mathbb{D},G)$ we have $\Phi_j(h)=c_j$, $j=1,\dots,N$, iff $h(0)=a$, $\operatorname{ord}_0(h-a)\geq m$, and $h(\sigma)=b$ (resp. $(\sigma/m!)h^{(m)}(0)=X$).

Put $Q(\lambda):=\lambda^{m-1}(\lambda-\sigma)$ (resp. $Q(\lambda):=\lambda^m$). It is clear that $L_j:=QK_j$ is a polynomial, $j=1,\dots,N$.

Now we check that $\Phi_1,\dots,\Phi_N$ are linearly independent. Suppose that

$$\sum_{j=1}^{N}\xi_j K_j\cdot s^*=g^*$$

on a set $A\subset\mathbb{T}$ of positive measure, where s and g are as in Definition 11.4.3. Hence,

$$\sum_{j=1}^{N}\xi_j L_j\cdot s^*=Qg^* \text{ on } A.$$

Using the identity principle (cf. Appendix B.9.7), we get

$$\sum_{j=1}^{N}\xi_j L_j\cdot s\equiv Qg \quad\text{on } \mathbb{D}.$$

In the case of $\boldsymbol{\ell}_G^{(m)}(a,b)$, taking $\lambda=\sigma$, we obtain $\sum_{j=2nm+1}^{2n(m+1)}\xi_j L_j(\sigma)=0$, which immediately implies that $\xi_j=0$ for $j\in\{2nm+1,\dots,2n(m+1)\}$. Thus,

$$\sum_{j=1}^{2nm}\xi_j L_j\cdot s\equiv Qg.$$

Taking $\lambda=0$, we conclude that $\xi_j=0$ for $j\in\{2n(m-1)+1,\dots,2n(m+1)\}$. Consequently,

$$\sum_{j=1}^{2n(m-1)}\xi_j L_j\cdot s\equiv Qg.$$

We divide both sides by λ. Next, we put $\lambda = 0$, and we repeat the procedure. Finally, we get $\xi_j = 0$ for $j \in \{2n+1, \dots, 2n(m+1)\}$ and

$$\sum_{j=1}^{2n} \xi_j K_j \cdot s \equiv g.$$

It remains to use the assumption that $g(0) = 0$.

In the case of $\boldsymbol{\varkappa}_G^{(m)}(a; X)$, the argument is analogous.

Finally, we only need to use Remark 3.8.6.

The main result is the following theorem:

Theorem 11.4.5 (cf. [145]). *Let $G \subset\subset G_0 \subset\subset \mathbb{C}^n$ be domains and let $u \in \mathcal{PSH}(G_0) \cap \mathcal{C}(G_0)$ be such that*

$$G = \{z \in G_0 : u(z) < 0\}, \quad \partial G = \{z \in G_0 : u(z) = 0\}.$$

Suppose that $f \in \mathcal{O}(\mathbb{D}, G)$ is an extremal for $(\mathfrak{P})$ (generated by functions $K_1, \dots, K_N$). Assume that there exist $S \subset \mathbb{T}$ and $s \in \mathcal{H}^\infty(\mathbb{D}, \mathbb{C}_*^n)$ such that

(a) *$\mathbb{T} \setminus S$ is of zero measure,*

(b) *$f^*(\zeta)$, $\operatorname{grad} u(f^*(\zeta))$,*[1] *and $s^*(\zeta)$ are well defined for all $\zeta \in S$,*

(c) $\displaystyle\lim_{\mathcal{A}(\mathbb{D},\mathbb{C}^n) \ni h \to 0} \frac{\sup\{|R_h(\zeta)| : \zeta \in S\}}{\|h\|_{L^\infty}} = 0$, *where*

$R_h(\zeta) := u(f^*(\zeta) + s^*(\zeta) \cdot h(\zeta)) - u(f^*(\zeta)) - 2\operatorname{Re}(\operatorname{grad} u(f^*(\zeta)) \bullet (s^*(\zeta) \cdot h(\zeta)))$.

Then, $f^*(\zeta) \in \partial G$ for a.a. $\zeta \in \mathbb{T}$ and there exist $\varrho \in L^\infty(\mathbb{T}, \mathbb{R}_{>0})$, $g \in \mathcal{H}^\infty(\mathbb{D}, \mathbb{C}^n)$, and $(\xi_1, \dots, \xi_N) \in (\mathbb{R}^N)_*$ such that

$$\left(\sum_{k=1}^{N} \xi_k K_k(\zeta) \cdot s^*(\zeta)\right) + g^*(\zeta) = \varrho(\zeta) s^*(\zeta) \cdot \operatorname{grad} u(f^*(\zeta)) \quad \textit{for a.a. } \zeta \in \mathbb{T}.$$

Notice that if, additionally, $u \in \mathcal{C}^1(G_0)$, then one can take $s := \mathbf{1}$. Observe that the condition (c) says that $R_h(\zeta) = o(h(\zeta))$ uniformly with respect to $\zeta \in S$ and $h \in \mathcal{A}(\mathbb{D}, \mathbb{C}^n)$.

The proof will be given in § 11.4.1. Assuming for a moment that Theorem 11.4.5 is proven, we present some corollaries and applications. Directly from Theorem 11.4.5 we obtain the following useful result:

[1] That is, u has all first-order partial derivatives at each point $f^*(\zeta)$ with $\zeta \in S$.

Corollary 11.4.6 (cf. [145]). *Under the assumptions of Theorem* 11.4.5, *if f is an extremal for $(\mathcal{P}_m)$, then $f^*(\zeta) \in \partial G$ for a.a. $\zeta \in \mathbb{T}$ and there exist $\varrho \in L^\infty(\mathbb{T}, \mathbb{R}_{>0})$ and $W \in \mathcal{H}^\infty(\mathbb{D}, \mathbb{C}^n)$ such that*

$$W^*(\zeta) = Q(\zeta)\varrho(\zeta)s^*(\zeta) \cdot \operatorname{grad} u(f^*(\zeta)) \ \textit{for a.a.}\ \zeta \in \mathbb{T}$$

(and $s \equiv \mathbf{1}$ provided that $u \in \mathcal{C}^1(G_0)$).

Corollary 11.4.6 gives a tool to describe

- extremal mappings for $(\mathcal{P}_m)$ for arbitrary complex ellipsoids $\mathcal{E}(p)$ (cf. Theorem 16.3.1), and
- extremal discs for $\boldsymbol{\ell}^{(m)}_{\boldsymbol{D}_\alpha}$ and $\boldsymbol{\varkappa}^{(m)}_{\boldsymbol{D}_\alpha}$ for elementary Reinhardt domains (cf. Proposition 16.3.4).

11.4.1 Proof of Theorem 11.4.5

Note that there are two possibilities: either $u \circ f^* = 0$ a.e. on $\mathbb{T}$ or there exists an $\eta > 0$ such that the set $P_0 := \{\theta : u(f^*(e^{i\theta})) < -\eta\}$ has positive measure. In the first case, we set $P_0 := \varnothing$. Put $A_0 := [0, 2\pi) \setminus P_0$ and define $p_s : L^1(\mathbb{T}, \mathbb{C}^n) \longrightarrow \mathbb{R}$,

$$p_s(h) := \frac{1}{2\pi} \int_{A_0} \Big(\operatorname{Re}\big((s^*(e^{i\theta}) \cdot \operatorname{grad} u(f^*(e^{i\theta}))) \bullet h(e^{i\theta})\big)\Big)^+ d\theta,$$

where $x^+ := \max\{0, x\}$, $x \in \mathbb{R}$.

Remark 11.4.7.

(a) Under the assumptions of Theorem 11.4.5, there exists an $M>0$ such that

$$\|s^*(\zeta) \cdot \operatorname{grad} u(f^*(\zeta))\| \le M \quad \text{for a.a. } \zeta \in \mathbb{T}.$$

Indeed, since u_0 is continuous and $G \subset\subset G_0$, condition (c) implies that there exist $M_0, \delta > 0$ such that if $h \in \mathcal{A}(\mathbb{D}, \mathbb{C}^n)$ and $\|h\|_{L^\infty} \le \delta$, then

$$|(s^*(\zeta) \cdot \operatorname{grad} u(f^*(\zeta))) \bullet h(\zeta)| \le M_0, \quad \zeta \in S.$$

Consequently,

$$\|s^*(\zeta) \cdot \operatorname{grad} u(f^*(\zeta))\| \le M_0/\delta, \quad \zeta \in S.$$

(b) $p_s(h) \le \frac{M}{2\pi}\|h\|_{L^1}$, $h \in L^1(\mathbb{T}, \mathbb{C}^n)$. In particular, p_s is continuous.

(c) p_s is a subnorm, i.e.,

$$p_s(h+g) \le p_s(h) + p_s(g), \quad p_s(\alpha h) = \alpha p_s(g), \quad h, g \in L^1(\mathbb{T}, \mathbb{C}^n),\ \alpha \ge 0.$$

(d) If $h \in \mathcal{H}^1(\mathbb{D}, \mathbb{C}^n)$, $h_r(\zeta) := h(r\zeta)$, $0 < r < 1$, then $p_s(h_r) \longrightarrow p_s(h^*)$.

The proof of Theorem 11.4.5 is based on the following lemma.

Lemma 11.4.8. *Under the assumptions of Theorem* 11.4.5, *there exist* $T > 0$, $j_0 \in \{1, \dots, N\}$, *and* $\delta \in \{-1, 1\}$ *such that*

$$\delta\Phi_{j_0}(s \cdot h) \le T p_s(h^*), \quad h \in \Xi_{j_0} := \{h \in \mathcal{H}^1(\mathbb{D}, \mathbb{C}^n) : \forall_{\ell \ne j_0} : \Phi_\ell(s \cdot h) = 0\}.$$

Assume for a moment that Lemma 11.4.8 is proven.

Proof of Theorem 11.4.5. Let T, j_0, and δ be as in Lemma 11.4.8. Put

$$\widetilde{\Phi}(h) := \delta\Phi_{j_0}(s \cdot h), \quad h \in \Xi_{j_0}.$$

Using the Hahn–Banach theorem, we extend $\widetilde{\Phi}$ to an $\mathbb{R}$-linear functional

$$\Phi : L^1(\mathbb{T}, \mathbb{C}^n) \longrightarrow \mathbb{R}$$

with $\Phi(h) \le T p_s(h)$, $h \in L^1(\mathbb{T}, \mathbb{C}^n)$. We know that $p_s(h) \le M\|h\|_{L^1}$, so Φ is continuous on $L^1(\mathbb{T}, \mathbb{C}^n)$. Hence,

$$\Phi(h) = \frac{1}{2\pi}\int_0^{2\pi} \operatorname{Re}\left(h(e^{i\theta}) \bullet W(e^{i\theta})\right) d\theta, \quad h \in L^1(\mathbb{T}, \mathbb{C}^n).$$

where $W \in L^\infty(\mathbb{T}, \mathbb{C}^n)$. In view of the definition of the space Ξ_{j_0}, we get

$$\bigcap_{k=1}^{N} \operatorname{Ker}\left(\mathcal{H}^1(\mathbb{D}, \mathbb{C}^n) \ni h \longmapsto \Phi_k(s \cdot h)\right) \subset \operatorname{Ker}\Phi.$$

Consequently, there exists a $\xi = (\xi_1, \dots, \xi_N) \in (\mathbb{R}^N)_*$ such that

$$\Phi(h) = \sum_{k=1}^{N} \xi_k \Phi_k(s \cdot h), \quad h \in \mathcal{H}^1(\mathbb{D}, \mathbb{C}^n).$$

Define $\Psi : L^1(\mathbb{T}, \mathbb{C}) \longrightarrow \mathbb{R}$,

$$\Psi(h) := \sum_{k=1}^{N} \xi_k \Phi_k(s \cdot h) = \frac{1}{2\pi}\int_0^{2\pi} \operatorname{Re}\left((s^*(e^{i\theta}) \cdot h(e^{i\theta})) \bullet \sum_{k=1}^{N} \xi_k K_k(e^{i\theta})\right) d\theta.$$

Then, $\Phi - \Psi = 0$ on $\mathcal{H}^1(\mathbb{D}, \mathbb{C}^n)$. Hence, by the F. & M. Riesz theorem (cf. Appendix B.9.10), there exists a $g \in \mathcal{H}^\infty(\mathbb{D}, \mathbb{C}^n)$, $g(0) = 0$, such that

$$W - s^* \cdot \sum_{k=1}^{N} \xi_k K_k = g^*.$$

We have

$$\begin{aligned}\Phi(h) &= \frac{1}{2\pi}\int_0^{2\pi} \operatorname{Re}\left(\left(\sum_{k=1}^{N} \xi_k K_k(e^{i\theta}) \cdot s^*(e^{i\theta}) + g^*(e^{i\theta})\right) \bullet h(e^{i\theta})\right) d\theta \\ &\le T\frac{1}{2\pi}\int_{A_0} \left(\operatorname{Re}\left((s^*(e^{i\theta}) \cdot \operatorname{grad} u(f^*(e^{i\theta}))) \bullet h(e^{i\theta})\right)\right)^+ d\theta, \\ &\qquad h \in L^1(\mathbb{T}, \mathbb{C}^n). \quad (11.4.1)\end{aligned}$$

Let $Q_0 := \{\theta \in [0, 2\pi) : s^*(\zeta) \cdot \operatorname{grad} u(f^*(\zeta)) = 0\}$. Then, for $h \in L^1(\mathbb{T}, \mathbb{C}^n)$, if $\operatorname{Re}\left((s^*(e^{i\theta}) \cdot \operatorname{grad} u(f^*(e^{i\theta}))) \bullet h(e^{i\theta})\right) \le 0$ for all $\theta \in [0, 2\pi) \setminus (P_0 \cup Q_0)$, then $\Phi(h) \le 0$. Hence,

$$\sum_{k=1}^{N} \xi_k K_k(e^{i\theta}) \cdot s^*(e^{i\theta}) + g^*(e^{i\theta}) = 0 \text{ for a.a. } \theta \in P_0 \cup Q_0.$$

Since $\Phi_1, \dots, \Phi_N$ are linearly independent, we conclude that P_0 and Q_0 are of measure zero. In particular, $f^*(\zeta) \in \partial G$ for a.a. $\zeta \in \mathbb{T}$.

Thus there exists a full measure set $C \subset [0, 2\pi)$, for which

$$\int_C \operatorname{Re}\left(\alpha(\theta) \bullet h(e^{i\theta})\right) d\theta \le \int_C \left(\operatorname{Re}\left(\beta(\theta) \bullet h(e^{i\theta})\right)\right)^+ d\theta, \quad h \in L^1(\mathbb{T}, \mathbb{C}^n), \tag{11.4.2}$$

where

- $\alpha(\theta) := \sum_{k=1}^{N} \xi_k K_k(e^{i\theta}) \cdot s^*(e^{i\theta}) + g^*(e^{i\theta})$,
- $\beta(\theta) := T s^*(e^{i\theta}) \cdot \operatorname{grad} u(f^*(e^{i\theta}))$,
- $\alpha(\theta), \beta(\theta) \ne 0$, $\theta \in C$.

Inequality (11.4.2) implies that for all $h \in L^1(\mathbb{T}, \mathbb{C}^n)$ we have

$$\operatorname{Re}(\alpha(\theta) \bullet h(e^{i\theta})) \le (\operatorname{Re}(\beta(\theta) \bullet h(e^{i\theta})))^+, \quad \theta \in C_h \subset C,$$

where $C \setminus C_h$ is of zero measure. In particular,

$$\operatorname{Re}(\alpha(\theta) \bullet z) \le (\operatorname{Re}(\beta(\theta) \bullet z))^+, \quad \theta \in C^* \subset C,\ z \in (\mathbb{Q} + i\mathbb{Q})^n,$$

where $C \setminus C^*$ is of zero measure. Consequently,

$$\operatorname{Re}(\alpha(\theta) \bullet z) \leq (\operatorname{Re}(\beta(\theta) \bullet z))^+, \quad \theta \in C^*,\ z \in \mathbb{C}^n.$$

Fix a $\theta \in C^*$ and put $\alpha := \alpha(\theta)$, $\beta := \beta(\theta)$. We will show that $\alpha = \mu\beta$ with $0 < \mu \leq 1$.

Take a $k \in \{1, \dots, n\}$ with $\beta_k \neq 0$. We have $\operatorname{Re}(\alpha_k z_k) \leq (\operatorname{Re}(\beta_k z_k))^+$, $z_k \in \mathbb{C}$. If $z_k = \pm i\overline{\beta}_k$, then we get $\operatorname{Im}(\alpha_k \overline{\beta}_k) = 0$. Hence, $\alpha_k = \mu_k \beta_k$ with $\mu_k \in \mathbb{R}$.

Let $k \neq \ell$ be such that $\beta_k \beta_\ell \neq 0$. We have $\operatorname{Re}(\alpha_k z_+ \alpha_\ell z_\ell) \leq (\operatorname{Re}(\beta_k z_k + \beta_\ell z_\ell))^+$, $z_k, z_\ell \in \mathbb{C}$. Hence, $\operatorname{Re}(\mu_k w_k + \mu_\ell w_\ell) \leq (\operatorname{Re}(w_k + w_\ell))^+$, $w_k, w_\ell \in \mathbb{C}$. Taking $w_k = -w_\ell = w$ we get $(\mu_k - \mu_\ell)\operatorname{Re} w \leq 0$, $w \in \mathbb{C}$. Thus, $\mu_k = \mu_\ell$.

We have shown that $\alpha = \mu\beta$ with $\mu \in \mathbb{R}_*$. Hence, $\operatorname{Re}(\mu\beta \bullet z) \leq (\operatorname{Re}(\beta \bullet z))^+$, $z \in \mathbb{C}^n$, which implies that $0 < \mu \leq 1$.

Finally,

$$\sum_{k=1}^{N} \xi_k K_k(\zeta) \cdot s^*(\zeta) + g^*(\zeta) = \varrho(\zeta) s^*(\zeta) \cdot \operatorname{grad} u(f^*(\zeta)) \text{ for a.a. } \zeta \in \mathbb{T},$$

where $\varrho : \mathbb{T} \longrightarrow (0, T]$. □

Now, we will prove Lemma 11.4.8.

Proof of Lemma 11.4.8. Suppose that the lemma is not true. Then, for each $j \in \{1, \dots, N\}$ and $m \in \mathbb{N}$ there are $h^{\pm}_{j,m} \in \Xi_j$ such that

$$\pm\Phi_j(s \cdot h^{\pm}_{j,m}) > m p_s(h^{\pm}_{j,m}).$$

Observe that we may assume that $h^{\pm}_{j,m} \in \mathcal{A}(\mathbb{D}, \mathbb{C}^n)$.

Indeed, since functionals $\mathcal{A}(\mathbb{D}, \mathbb{C}^n) \ni h \longmapsto \Phi_\mu(s \cdot h)$, $\mu = 1, \dots, N$, are $\mathbb{R}$-linearly independent, there exist $g_1, \dots, g_N \in \mathcal{A}(\mathbb{D}, \mathbb{C}^n)$ such that $\Phi_\mu(s \cdot g_\nu) = \delta_{\mu,\nu}$, $\mu, \nu = 1, \dots, N$. For $h \in \{h^-_{j,m}, h^+_{j,m}\}$, define $h_r(\zeta) := h(r\zeta)$ and

$$\widetilde{h}_r(\zeta) := h_r(\zeta) - \sum_{\mu \neq j} \Phi_\mu(s \cdot h_r) g_\mu(\zeta), \quad 0 < r < 1.$$

Observe that $\widetilde{h}_r \in \mathcal{A}(\mathbb{D}, \mathbb{C}^n)$,

$$\Phi_k(s \cdot \widetilde{h}_r) = \Phi_k(s \cdot h_r) - \sum_{\mu \neq j} \Phi_\mu(s \cdot h_r)\delta_{k,\mu} = 0, \quad k \neq j.$$

Thus, $\widetilde{h}_r \in \Xi_j$. Moreover, $\Phi_\mu(s \cdot h_r) \longrightarrow \Phi_\mu(s \cdot h)$ when $r \longrightarrow 1$ (cf. Remark 11.4.2(c)). Hence, $\widetilde{h}_r \longrightarrow h - \sum_{\mu \neq j} \Phi_\mu(s \cdot h) g_\mu = h$ and $p_s(\widetilde{h}_r) \longrightarrow p_s(h)$ when $r \longrightarrow 1$ (cf. Remark 11.4.7(d)).

It is clear that we may additionally assume that $\Phi_j(s \cdot h^{\pm}_{j,m}) = \pm 1$.

For any $q = (q_1^+, q_1^-, \dots, q_N^+, q_N^-) \in \mathbb{R}_+^{2N}$, we define the function

$$f_{q,m} := f + \sum_{j=1}^{N} (q_j^+ s \cdot h_{j,m}^+ + q_j^- s \cdot h_{j,m}^-) =: f + s \cdot h_{q,m}$$

and the mapping

$$A = (A_1, \dots, A_N) : \mathbb{R}_+^{2N} \longrightarrow \mathbb{R}^N, \quad A(q) := (q_1^+ - q_1^-, \dots, q_N^+ - q_N^-).$$

Note that $\Phi_j(f_{q,m}) - \Phi_j(f) = A_j(q)$, $j = 1, \dots, N$. Moreover, $f_{q,m} \longrightarrow f$ uniformly in $\mathbb{D}$ when $q \longrightarrow 0$ (with fixed m).

Lemma 11.4.9. *Let $w \in \mathcal{SH}(\mathbb{D}, \mathbb{R}_-)$ and let Δw be the Riesz measure of w (cf. Appendix* B.5.10). *Suppose that one of the following conditions is true:*

(a) $\Delta w(\mathbb{D}(r_0)) > a > 0$ *for some* $r_0 \in (0, 1)$,

(b) $\limsup_{r \to 1} w(r\zeta) \le -a$, $\zeta \in Z$, *for a set* $Z \subset [0, 2\pi)$ *with positive measure.*

Then, $w(\zeta) \le -C(1 - |\zeta|)$, where $C > 0$ is a constant depending only on r_0, a, and Z.

Proof. It is known (cf. [218], Theorem 3.14) that

$$w(\zeta) = \frac{1}{2\pi} \int_0^{2\pi} w(re^{i\theta}) P_r(\zeta, re^{i\theta}) d\theta + \int_{\mathbb{D}(r)} \log \left| \frac{r(z - \zeta)}{r^2 - \bar{z}\zeta} \right| d\Delta w(z),$$
$$\zeta \in \mathbb{D}(r),\ 0 < r < 1,$$

where

$$P_r : \mathbb{D}(r) \times \partial\mathbb{D}(r) \longrightarrow \mathbb{R}_{>0}, \quad P_r(z, \zeta) = \frac{r^2 - |z|^2}{|\zeta - z|^2} = \operatorname{Re}\left(\frac{\zeta + z}{\zeta - z}\right)$$

is the *Poisson kernel.* We put $P := P_1$ (cf. Appendix B.5.1). Letting $r \longrightarrow 1$ and using Fatou's lemma gives

$$w(\zeta) \le \frac{1}{2\pi} \int_0^{2\pi} \vartheta(\theta) P(\zeta, e^{i\theta}) d\theta + \int_{\mathbb{D}} \log \left| \frac{z - \zeta}{1 - \bar{z}\zeta} \right| d\Delta w(z), \quad \zeta \in \mathbb{D},$$

where $\vartheta(\theta) := \limsup_{r \to 1} w(re^{i\theta})$.

- If $\Delta w(\mathbb{D}(r_0)) > a > 0$, then

$$w(\zeta) \le \int_{\mathbb{D}(r_0)} \log \left| \frac{z - \zeta}{1 - \bar{z}\zeta} \right| d\Delta w(z), \quad \zeta \in \mathbb{D}.$$

Observe that for $|z| < r_0$ we have

$$\left|\frac{z-\zeta}{1-\bar{z}\zeta}\right| \le \frac{|z|+|\zeta|}{1+|z||\zeta|} \le \frac{r_0+|\zeta|}{1+r_0|\zeta|} \le 1 - \frac{1-r_0}{2}(1-|\zeta|).$$

Hence, $\log|\frac{z-\zeta}{1-\bar{z}\zeta}| \le -\widetilde{C}(1-|\zeta|)$, $|z| < r_0$, $\zeta \in \mathbb{D}$, where $\widetilde{C} > 0$ depends only on r_0. Consequently, $w(\zeta) \le -a\widetilde{C}(1-|\zeta|)$, $\zeta \in \mathbb{D}$.

- If $\vartheta(\theta) \le -a < 0$, $\theta \in Z$, then

$$\begin{aligned} w(\zeta) &\le \frac{1}{2\pi}\int_0^{2\pi} \vartheta(e^{i\theta})P(\zeta, e^{i\theta})d\theta \\ &\le -a\frac{1}{2\pi}\int_Z P(\zeta, e^{i\theta})d\theta \le -C(1-|\zeta|), \quad \zeta \in \mathbb{D}, \end{aligned}$$

where $C > 0$ depends only on a and Z. □

Lemma 11.4.10. *There exist constants $C > 0$, $t_m > 0$, $m \in \mathbb{N}$, such that for $\|q\| < t_m$ we have:*

(a) $f_{q,m}(\mathbb{D}) \subset G_0$, *so we may define* $u_{q,m} := u \circ f_{q,m}$,

(b) $u_{q,m}(\zeta) \le v_{q,m}(\zeta) := C\log|\zeta| + \frac{1}{2\pi}\int_{A_0}\left(u^*_{q,m}(e^{i\theta})\right)^+ P(\zeta, e^{i\theta})d\theta$, $|\zeta| > 1/2$,

where $u^*_{q,m} := u(f^*_{q,m})$.

Proof. (a) It follows from the assumption that $G \subset\subset G_0$.

(b) Let $u_0 := u \circ f$. Suppose that there exists an $r_0 \in (0,1)$ such that $\Delta u_0(r_0\mathbb{D}) > a > 0$. Define

$$\widetilde{u}_{q,m}(\zeta) := u_{q,m}(\zeta) - \frac{1}{2\pi}\int_{A_0}\left(u^*_{q,m}(e^{i\theta})\right)^+ P(\zeta, e^{i\theta})d\theta, \quad \zeta \in \mathbb{D}.$$

Observe that $\Delta\widetilde{u}_{q,m} = \Delta u_{q,m}$. The continuity of u implies that if $0 < t_m \ll 1$, then $\Delta\widetilde{u}_{q,m}(r\mathbb{D}) > a/2$. Hence, from Lemma 11.4.9, we get the required result.

If $\Delta u_0(r\mathbb{D}) = 0$ for any $r \in (0,1)$ and $u^*_0(\zeta) = 0$ for a.a. $\zeta \in \mathbb{T}$, then, by the Riesz representation theorem (cf. [218], § 3.5), u_0 is harmonic in $\mathbb{D}$; a contradiction (because $u_0 \not\equiv 0$). Hence, P_0 has positive measure. The continuity of u implies that if $0 < t_m \ll 1$, then $\{\zeta : \widetilde{u}_{q,m}(\zeta) < -\tau/2\}$ has positive measure. By Lemma 11.4.9 we get the required result. □

Define $E_{q,m} := \{\zeta \in \mathbb{D} : v_{q,m}(\zeta) < 0\}$,

$$g_{q,m}(\zeta) := \zeta\exp\left(\frac{1}{2\pi C}\int_{A_0}\left(u^*_{q,m}(e^{i\theta})\right)^+ S(\zeta, e^{i\theta})d\theta\right),$$

where $S(\zeta, z) := \frac{\zeta+z}{\zeta-z}$ is the Schwarz kernel.

Remark 11.4.11. We have $C \log |g_{q,m}(\zeta)| = v_{q,m}(\zeta) \geq C \log |\zeta|$ and $E_{q,m} = g_{q,m}^{-1}(\mathbb{D})$.

Lemma 11.4.12.

(a) *$E_{q,m}$ is connected,* $0 \in E_{q,m}$,

(b) $g_{q,m} : E_{q,m} \longrightarrow \mathbb{D}$ *is biholomorphic.*

Proof. (a) Note that $E_{q,m} = \bigcup_{\delta>0}\{\zeta \in \mathbb{D} : v_{q,m}(\zeta) < -\delta\}$ and

$$\{\zeta \in \mathbb{D} : v_{q,m}(\zeta) < -\delta\} \subset \left\{\zeta \in \mathbb{D} : |\zeta| < e^{-\delta/C}\right\}.$$

Since $v_{q,m}$ is harmonic outside 0 and $v_{q,m}^*(e^{i\theta}) \geq 0$, any connected component of $\{\zeta \in \mathbb{D} : v_{q,m}(\zeta) < -\delta\}$ must contain 0.

(b) First we prove that $g_{q,m} : E_{q,m} \longrightarrow \mathbb{D}$ is proper. Indeed, let $\zeta_k \longrightarrow \zeta_0 \in \partial E_{q,m}$. If $\zeta_0 \in \mathbb{T}$, then $|g_{q,m}(\zeta_k)| \longrightarrow 1$ (because $|g_{q,m}(\zeta)| \geq |\zeta|$). If $\zeta_0 \in \mathbb{D}$, then $|g_{q,m}(\zeta_k)| \longrightarrow |g_{q,m}(\zeta_0)| = 1$.

Since $g_{q,m}'(0) \neq 0$ and $g_{q,m}^{-1}(0) = \{0\}$, the mapping $g_{q,m}$ is conformal. □

We define

$$\begin{aligned}
\widetilde{f}_{q,m}(\zeta) &= f_{q,m}(g_{q,m}^{-1}(\zeta)), \quad \widehat{f}_{q,m}(\zeta) = \widetilde{f}_{q,m}(e^{-\|q\|/m}\zeta),\\
\widetilde{A}_m(q) &= (\Phi_1(\widetilde{f}_{q,m}) - \Phi_1(f), \dots, \Phi_N(\widetilde{f}_{q,m}) - \Phi_N(f)),\\
\widehat{A}_m(q) &= (\Phi_1(\widehat{f}_{q,m}) - \Phi_1(f), \dots, \Phi_N(\widehat{f}_{q,m}) - \Phi_N(f)).
\end{aligned}$$

Observe that for $\|q\| \leq t_m$ we have $\widetilde{f}_{q,m}(\mathbb{D}) \subset G$, $\widehat{f}_{q,m}(\mathbb{D}) \subset\subset G$, and $\widetilde{A}_m(0) = \widehat{A}_m(0) = 0$.

Lemma 11.4.13. *Suppose that*

$$\Phi(h) = \frac{1}{2\pi}\int_0^{2\pi} \operatorname{Re}\left(h^*(e^{i\theta}) \bullet K(e^{i\theta})\right) d\theta, \quad h \in \mathcal{H}^1(\mathbb{D}, \mathbb{C}^n),$$

where $K \in \mathcal{A}(\mathbb{A}(\tau, 1), \mathbb{C}^n)$ *for some* $\tau \in (0, 1)$. *Let* $f \in \mathcal{H}^\infty(\mathbb{D}, \mathbb{C}^n)$, *and* $g \in \mathcal{O}(\mathbb{D}, \mathbb{D})$, $g(0) = 0$. *Then,*

$$|\Phi(f \circ g) - \Phi(f)| \leq C\|f\|_\infty \sup_{\zeta\in\mathbb{D}} |g(\tau\zeta) - \tau\zeta|,$$

where $C > 0$ *depends only on* Φ.

Proof. We have

$$\Phi(h) = \frac{1}{2\pi} \int_0^{2\pi} \operatorname{Re}\left(h(\tau e^{i\theta}) \bullet K(\tau e^{i\theta})\right) d\theta.$$

Hence,

$$|\Phi(h)| \le (\max_{\zeta \in \mathbb{T}} \|K(\tau\zeta)\|)(\max_{\zeta \in \mathbb{T}} \|h(\tau\zeta)\|). \tag{11.4.3}$$

It remains to observe that

$$\|f(g(\tau\zeta)) - f(\tau\zeta)\| \le (\sup_{\xi \in \mathbb{D}} |f'(\tau\xi)|)|g(\tau\zeta) - \tau\zeta|,$$

and $\sup_{\xi \in \mathbb{D}} |f'(\tau\xi)| \le \frac{\|f\|_\infty}{(1-\tau)^2}$. □

Lemma 11.4.14. *The mappings $\widetilde{A}_m, \widehat{A}_m : \mathbb{B}(t_m) \longrightarrow \mathbb{R}^N$ are continuous.*

Proof. If $q_k \longrightarrow q$, then $u^*_{q_k,m} \longrightarrow u^*_{q,m}$ uniformly on $\mathbb{T}$. Hence, $g_{q_k,m} \longrightarrow g_{q,m}$, $g^{-1}_{q_k,m} \longrightarrow g^{-1}_{q,m}$, $\widetilde{f}_{q_k,m} \longrightarrow \widetilde{f}_{q,m}$, and $\widehat{f}_{q_k,m} \longrightarrow \widehat{f}_{q,m}$ uniformly on compact subset of $\mathbb{D}$. Since Φ_j are continuous with respect to this convergence (cf. (11.4.3)), the proof is completed. □

Lemma 11.4.15. *For each $\varepsilon > 0$ there is an $m_0 \in \mathbb{N}$ such that for any $m \ge m_0$ there is a $\tau_m > 0$ such that $\|A(q) - \widetilde{A}_m(q)\| \le \varepsilon\|q\|$, $q \in \mathbb{B}(\tau_m)$.*

Proof. It follows from the definition of $A, \widetilde{A}_m$ that it is enough to discuss the inequality

$$|\Phi_j(\widetilde{f}_{q,m}) - \Phi_j(f_{q,m})| \le \varepsilon\|q\|.$$

By Lemma 11.4.14, it is enough to discuss the inequality

$$\sup_{\zeta \in \tau\mathbb{D}} |g^{-1}_{q,m}(\zeta) - \zeta| \le \varepsilon\|q\|.$$

Note that

$$\sup_{\zeta \in \tau\mathbb{D}} |g^{-1}_{q,m}(\zeta) - \zeta| \le \sup_{\zeta \in \tau\mathbb{D}} |g_{q,m}(\zeta) - \zeta|$$

and for small τ_m (such that $|1 - \exp \tau_m| \le 2\tau_m$)

$$\left|1 - \exp\left(\frac{1}{2\pi C} \int_{A_0} \left(u^*_{q,m}(e^{i\theta})\right)^+ S(\zeta, e^{i\theta}) d\theta\right)\right| \\ \le 2\frac{1+\tau}{1-\tau}\left(\frac{1}{2\pi C} \int_{A_0} \left(u^*_{q,m}(e^{i\theta})\right)^+ d\theta\right), \quad \zeta \in \mathbb{D}(\tau).$$

Hence, it is enough to consider

$$\int_{A_0} \left(u^*_{q,m}(e^{i\theta})\right)^+ d\theta$$
$$\le \int_{A_0} 2\left(\operatorname{Re}(\operatorname{grad} u(f^*(e^{i\theta})) \bullet s^*(e^{i\theta}) \cdot h_{q,m}(e^{i\theta}))\right)^+ d\theta + o(\|h_{q,m}\|_\infty)$$
$$= 2p_s(h_{q,m}) + o(\|h_{q,m}\|_\infty).$$

Observe that $p_s(h_{q,m}) \le \|q\| \max\{p_s(h^{\pm}_{j,m}) : j = 1, \dots, N\} \le \frac{1}{m}\|q\|$. Hence, if $m \gg 1$ and $0 < \tau_m \ll 1$, then we get the required result. □

Lemma 11.4.16. *For each $\varepsilon > 0$ there is $m_0 \in \mathbb{N}$ such that for any $m \ge m_0$ there is a $\tau_m > 0$ such that $\|\widetilde{A}_m(q) - \widehat{A}_m(q)\| \le \varepsilon\|q\|$, $q \in \mathbb{B}(\tau_m)$.*

Proof. As in Lemma 11.4.15, by Lemma 11.4.13 it is enough to prove the inequality

$$\sup_{\zeta \in \tau\mathbb{D}} |e^{-\|q\|/m}\zeta - \zeta| \le \varepsilon\|q\|.$$

But, for a small $\|q\|/m$ we have $|1 - e^{-\|q\|/m}| \le 2\frac{\|q\|}{m}$. Hence, we get the required result. □

Lemma 11.4.17. *For any continuous mapping $F : \mathbb{R}^{2N}_+ \longrightarrow \mathbb{R}^N$, if*

$$\|F(x) - A(x)\| \le \frac{1}{2\sqrt{2N}}\|x\|, \quad x \in \mathbb{B}(r) \cap \mathbb{R}^{2N}_+,$$

then there exists a $q \in \mathbb{B}_*(r) \cap \mathbb{R}^{2N}_+$ such that $F(q) = 0$.

Proof. Put $t_0 = \frac{1}{2\sqrt{2N}} \min\{1, r\}$ and let

$$\mathcal{Q} := \{(y_1, \dots, y_N) : 0 < y_j < t_0,\ j = 1, \dots, N\}$$

and

$$\mathbb{R}^N \ni (y_1, \dots, y_N) \overset{\pi}{\longmapsto} (y_1, t_0 - y_1, \dots, y_N, t_0 - y_N) \in \mathbb{R}^{2N}.$$

It easy to check that $\|\pi(y)\| \le t_0\sqrt{2N}$, $y \in \overline{\mathcal{Q}}$, and $\pi(\mathcal{Q}) \subset \mathbb{B}(r) \cap \mathbb{R}^{2N}_{>0}$. Note, that

$$\|F \circ \pi(y) - A \circ \pi(y)\| \le \frac{1}{2\sqrt{N}}\|\pi(y)\| \le \frac{t_0}{2}, \quad y \in \overline{\mathcal{Q}}.$$

Let us consider the homotopy defined by the formula $\widetilde{F}_t = tF \circ \pi + (1-t)A \circ \pi$. It is enough to show that $0 \notin \widetilde{F}_t(\partial\mathcal{Q})$.

Indeed, it then follows from the homotopical invariance of the degree of mappings [552], that $\deg(F \circ \pi, \mathcal{Q}, 0) = \deg(A \circ \pi, \mathcal{Q}, 0) \neq 0$. Thus, $0 \in F \circ \pi(\mathcal{Q})$, which will finish the proof.

To see that $0 \notin \widetilde{F}_t(\partial \mathcal{Q})$, observe that for any $y \in \partial \mathcal{Q}$ we have

$$t_0 \leq \|A \circ \pi(y)\| \leq \|\widetilde{F}_t(y)\| + t\|F \circ \pi(y) - A \circ \pi(y)\| \leq \|\widetilde{F}_t(y)\| + \frac{t_0}{2}. \qquad \square$$

Finally, we are in a position to finish the proof of Lemma 11.4.8. By Lemmas 11.4.14, 11.4.15, and 11.4.16 each operator $\widehat{A}_m$ is continuous in $\mathbb{R}^{2N}_+$ and for each $\varepsilon > 0$ there exist $m \in \mathbb{N}$ and $\tau_m > 0$ such that $\|\widehat{A}_m(q) - A(q)\| \leq \varepsilon\|q\|$, $q \in \mathbb{B}(\tau_m)$. By Lemma 11.4.17, we find a q, $0 < \|q\| \ll 1$, such that $\widehat{A}_m(q) = 0$. Hence, we have $\Phi_j(\widehat{f}_{q,m}) = a_j$, $j = 1, \dots, N$. Since $\widehat{f}_{q,m}(\mathbb{D}) \subset\subset G$, we have a contradiction with the extremality of f. $\square$

11.5 Schwarz lemma – the case of equality

Let q be a complex norm on $\mathbb{C}^n$. By

$$G_q = \{z \in \mathbb{C}^n : q(z) < 1\}$$

we denote the corresponding unit ball. Then, any holomorphic map $F : G_q \longrightarrow G_q$, $F(0) = 0$, satisfies $q(F(z)) \leq q(z)$ for all $z \in G_q$; cf. Theorem 2.1.1. Moreover, equality in a neighborhood of 0 implies that F is a linear automorphism of G_q.

Using the uniqueness of complex geodesics, J.-P. Vigué showed the following generalization of the last statement above (cf. [517]):

Theorem 11.5.1. *Let G_q be as above and suppose that every boundary point of G_q is a complex extreme point of $\overline{G}_q$. Then, any holomorphic map $F : G_q \longrightarrow G_q$ with $F(0) = 0$ and $q(F(z)) = q(z)$ for all $z \in U$, where U is a nonempty open subset of G_q, is the restriction of a linear automorphism of $\mathbb{C}^n$.*

Proof. Fix a point $z' \in U$, $z' \neq 0$. Then, $\varphi(\lambda) := \lambda z'/q(z')$, $\lambda \in \mathbb{D}$, is a complex geodesic through 0 and z'. Then, $F \circ \varphi$ is the unique complex geodesic through 0 and $F(z')$, which implies that

$$F \circ \varphi(\lambda) = \lambda F(z')/q(F(z')), \quad \lambda \in \mathbb{D}. \tag{11.5.1}$$

Now, we write F as

$$F(z) = \sum_{j=1}^{\infty} P_j(z),$$

where P_j is an n-tuple of homogeneous polynomials of degree j. So, for $\lambda \in \mathbb{D}$ we obtain

$$F(\varphi(\lambda)) = \sum_{j=1}^{\infty} P_j(\lambda z'/q(z')) = \sum_{j=1}^{\infty} P_j(z'/q(z'))\lambda^j.$$

Comparison with (11.5.1) for $j \geq 2$ yields

$$0 = P_j(z'/q(z')), \quad \text{and therefore} \quad P_j(z') = 0.$$

By the identity theorem, it follows that

$$F(z) = P_1(z) \quad \text{for all} \quad z \in G_q,$$

i.e., F is a restriction of the linear map P_1.

It remains to show that P_1 is an automorphism of $\mathbb{C}^n$. Otherwise, there exists a point $w' \in (\mathbb{C}^n)_*$ with $P_1(w') = 0$. Now, observe that the interior of

$$\{z \in \mathbb{C}^n : q(P_1(z)) = q(z)\}$$

contains a point z^0 with $q(z^0) = 1$. Since z^0 is a complex extreme point of $\overline{G}_q$, there is a sequence of complex numbers $\lambda_\nu \underset{\nu\to\infty}{\longrightarrow} 0$ with

$$q(z^0 + \lambda_\nu w') > 1 \quad \text{and} \quad q(P_1(z^0 + \lambda_\nu w')) = q(z^0 + \lambda_\nu w'), \ \nu \in \mathbb{N}.$$

Combining these pieces of information, we arrive at

$$1 = q(z^0) = q(P_1(z^0)) = q(P_1(z^0 + \lambda_\nu w')) > 1;$$

a contradiction. Hence Ker $P_1 = \{0\}$. □

Under the assumption of Theorem 11.5.1, the map F is not necessarily a linear automorphism of G_q, as the following example shows.

Example 11.5.2 (cf. [517]). Fix an $\alpha \in (0,1)$ and define on $(\mathbb{C}^2)_*$:

$$\theta(z) = \theta((z_1, z_2)) := \arccos(|z_1|/\|z\|).$$

Moreover, choose a decreasing $\mathcal{C}^\infty$-function $\chi : [0, \pi/2] \longrightarrow \mathbb{R}_+$ such that the following two properties hold:

(a) $\chi(t) = \begin{cases} 1 & \text{if } t \text{ is near } 0 \\ \alpha^2 & \text{if } t \text{ is near } \pi/2 \end{cases}$,

(b) the function $q_0(z) := \begin{cases} (\chi(\theta(z))|z_1|^2 + |z_2|^2)^{1/2} & \text{if } z \neq 0 \\ 0 & \text{if } z = 0 \end{cases}$ is a norm on $\mathbb{C}^2$ with a strictly convex unit ball G_q.

Then, the map $F(z) := (z_2, \alpha z_1)$ belongs to $\mathcal{O}(G_q, G_q)$ and $F(0) = 0$. Moreover, if $z \in G_q$ and z is near $(0, 1)$, then $q_0(F(z)) = q_0(z)$. Thus, all assumptions of Theorem 11.5.1 are fulfilled.

But $|\det F'(0)| = \alpha < 1$. So, Cartan's theorem on holomorphic mappings implies that F is not a linear automorphism of G_q; cf. § 11.6.

In the following result we sharpen the assumptions in Theorem 11.5.1 in order to be able to conclude that F is a linear automorphism of G_q.

Theorem 11.5.3 (cf. [517]). *Let q, G_q, U, and F be as in Theorem* 11.5.1. *Moreover, suppose that there exists an $a \in U \setminus \{0\}$ with $F(a) = a$. Then, F is a linear automorphism of G_q.*

Proof. According to Theorem 11.5.1, the map F is a restriction of a linear automorphism of $\mathbb{C}^n$, which we again denote by F.

Because of the linearity of F, we may assume that $q(a) = 1$ and $q(F(z)) = q(z)$ whenever $q(z - a) < r$ and $0 < r < 1$ is suitably chosen. Moreover, we recall that $q(F(z)) \le q(z)$ for all $z \in \mathbb{C}^n$. Therefore, if $q(z - a) < r$, then $q(F(z) - a) < r$.

Now, we consider a subsequence of $(F^j)_{j \in \mathbb{N}}$ that converges uniformly on compact sets to a linear map $\widehat{F}$. Obviously, $\widehat{F}(0) = 0$ and $\widehat{F}(a) = a$.

Suppose that there is a $w' \in (\mathbb{C}^n)_*$ with $\widehat{F}(w') = 0$. Since a is a complex extreme point, there exists a complex number $\lambda \neq 0$ with

$$q(a + \lambda w') > 1 \quad \text{and} \quad q(\lambda w') < r.$$

By the above construction this yields

$$q(F^j(a + \lambda w')) = q(a + \lambda w') > 1, \quad j \in \mathbb{N}. \tag{11.5.2}$$

On the other hand, we have $\widehat{F}(a + \lambda w') = \widehat{F}(a) = a$ which, because of (11.5.2), implies that

$$1 < q(a + \lambda w') = q(\widehat{F}(a)) = q(a) = 1;$$

a contradiction. □

In the next theorem, we assume that ∂G_q is a real-analytic manifold. Then, it is well-known that every boundary point of ∂G_q is a complex extreme point of $\overline{G_q}$ so that we are automatically in the situation that is discussed in Theorem 11.5.1. Nevertheless, we prefer to give a simple, direct argument.

Theorem 11.5.4. *Let q be a complex norm on $\mathbb{C}^n$ and let $F \in \mathcal{O}(G_q, G_q)$ with $F(0) = 0$. Moreover, assume that $q(F(z)) = q(z)$ for all $z \in U$, where U is a nonempty open subset of G_q. If ∂G_q is a real-analytic manifold, then F is a linear automorphism of G_q.*

Proof. Fix a $z' \in U$ with $z' \neq 0$. Then,

$$\mathbb{D} \ni \lambda \overset{\varphi}{\longmapsto} \lambda z'/q(z') \in G_q$$

is a complex geodesic through 0 and z'. So $F \circ \varphi$ is again a complex geodesic in G_q. In particular, we get

$$q(z') = \boldsymbol{\gamma}_{\mathbb{D}}(0; q(z')) = \boldsymbol{\gamma}_{G_q}(0; F'(0)z') = q(F'(0)z'). \qquad (11.5.3)$$

Now consider the map

$$\partial G_q \ni z \overset{\Phi}{\longmapsto} F'(0)z \in \mathbb{C}^n.$$

Because of (11.5.3), there is an open set V in ∂G_q, $V \neq \varnothing$, with $\Phi(V) \subset \partial G_q$. Then, the identity theorem for real-analytic functions implies that $\Phi(\partial G_q) \subset \partial G_q$, i.e., $q(F'(0)z) = q(z)$ for all $z \in \mathbb{C}^n$. Hence, $F'(0)$ is a linear automorphism of $\mathbb{C}^n$.

If we analyze the map $\Psi := F'(0)^{-1} \circ F$ we can see that $\Psi(0) = 0$, $\Psi \in \mathcal{O}(G_q, G_q)$, and $\Psi'(0) = \mathrm{id}$. Hence, Cartan's uniqueness theorem implies that $F \equiv F'(0)$, i.e., F is a linear automorphism of G_q. □

Corollary 11.5.5. *Let $F : \mathbb{B}_n \longrightarrow \mathbb{B}_n$ be a holomorphic map such that $F(0) = 0$ and $\|F(z)\| = \|z\|$ for all $z \in U$, where U is a nonempty open subset of $\mathbb{B}_n$. Then, F is a linear automorphism of $\mathbb{B}_n$.*

11.6 Criteria for biholomorphicity

Let G_1, G_2 be domains in $\mathbb{C}^n$ and let $F : G_1 \longrightarrow G_2$ be a holomorphic mapping. We say that F is a *$\boldsymbol{\gamma}$-isometry* (resp. *$\boldsymbol{\varkappa}$-isometry*) *at a point $a \in G_1$* if

$$\delta_{G_2}(F(a); F'(a)X) = \delta_{G_1}(a; X), \quad X \in \mathbb{C}^n, \qquad (11.6.1)$$

where $\delta = \boldsymbol{\gamma}$ (resp. $\delta = \boldsymbol{\varkappa}$). What is the relation between the δ-isometricity of F at a point and the global biholomorphicity of F? (Recall that any biholomorphic mapping is a δ-isometry at any point.)

In the case where $G_1 = G_2 = G$ is a bounded domain in $\mathbb{C}^n$ and $F(a) = a$, the full answer to our question is given by the following well-known theorem due to H. Cartan:

Theorem 11.6.1 (Cartan theorem, cf. [327]). *Let $G \subset \mathbb{C}^n$ be a bounded domain, let $a \in G$, and let $F : G \longrightarrow G$ be a holomorphic mapping with $F(a) = a$. Then, the following conditions are equivalent:*

(i) *if λ_0 is an eigenvalue of $F'(a)$, then $|\lambda_0| = 1$;*

(ii) *$|\det F'(a)| = 1$;*

(iii) *F is a $\boldsymbol{\gamma}$-isometry at a;*

(iv) *F is a $\boldsymbol{\varkappa}$-isometry at a;*

(v) *F is biholomorphic, i.e., $F \in \operatorname{Aut}(G)$.*

Remark 11.6.2 (cf. [516]).

(a) Let

$$G_1 := \{(z_1, z_2) \in \mathbb{C}^2 : |z_1| + |z_2| + \varepsilon|z_1 z_2| < 1\} \quad (\varepsilon > 0),$$
$$G_2 := \{(z_1, z_2) \in \mathbb{C}^2 : |z_1| + |z_2| < 1\}.$$

Then (cf. Proposition 2.3.1),

$$\boldsymbol{\gamma}_{G_1}((0,0);(X_1,X_2)) = \boldsymbol{\gamma}_{G_2}((0,0);(X_1,X_2)) = |X_1| + |X_2|, \ (X_1,X_2) \in \mathbb{C}^2.$$

In particular, $F := \mathrm{id} : G_1 \longrightarrow G_2$ is a $\boldsymbol{\gamma}$-isometry at $(0,0)$ which is not biholomorphic.

(b) Let $G_2 \subset \mathbb{C}$ be any taut domain that is not simply connected (e.g., an annulus). Let $F : \mathbb{D} \longrightarrow G_2$ be an extremal disc for $(F(0), F'(0))$, i.e., $\boldsymbol{\varkappa}_{G_2}(F(0); F'(0)) = 1$. Then, $F : \mathbb{D} \longrightarrow G_2$ is a $\boldsymbol{\varkappa}$-isometry at 0 that is not biholomorphic.

On the other hand, we have the following positive results:

Proposition 11.6.3.

(a) (Cf. [516]) Suppose that $G_1 \subset \mathbb{C}^n$ is a convex taut domain and that $G_2 \subset \mathbb{C}^n$ is an arbitrary domain. Let $F \in \mathcal{O}(G_1, G_2)$ be a $\boldsymbol{\gamma}$-isometry at a point $a \in G_1$. Then, F is biholomorphic.

(b) (Cf. [198]) Suppose that $G_1 \subset \mathbb{C}^n$ is a taut domain and that $G_2 \subset \mathbb{C}^n$ is a strictly convex bounded domain. Let $F \in \mathcal{O}(G_1, G_2)$ be a $\boldsymbol{\varkappa}$-isometry at a point $a \in G_1$. Then, F is biholomorphic.

Proof. (a) In view of Lempert's theorem (Theorem 11.2.1), $\boldsymbol{\gamma}_{G_1} = \boldsymbol{\varkappa}_{G_1}$ and $\boldsymbol{c}_{G_1} = \boldsymbol{\ell}_{G_1}$. Moreover, G_1 is $\boldsymbol{c}$-finitely compact and $\boldsymbol{\gamma}_{G_1}(a; X) > 0$ for all $X \neq 0$; cf. Chapter 13. Hence, condition (11.6.1) (with $\delta = \boldsymbol{\gamma}$) implies that $F'(a)$ is an isomorphism, and therefore $F : U_a \longrightarrow F(U_a)$ is biholomorphic for an open neighborhood $U_a \subset G_1$ of a. Since $\operatorname{top} G_1 = \operatorname{top} \boldsymbol{c}_{G_1}$, we may assume that $U_a = B_{\boldsymbol{c}_{G_1}}(a, r_0)$ for some $r_0 > 0$.

First, we will prove that

$$\boldsymbol{c}_{G_2}(F(a), F(z)) = \boldsymbol{c}_{G_1}(a, z), \quad z \in G_1. \tag{11.6.2}$$

Take a $z_0 \in G_1 \setminus \{a\}$ and let $\varphi : \mathbb{D} \longrightarrow G_1$ be a complex geodesic for (a, z_0) with $\varphi(0) = a, \varphi(\sigma) = z_0$ for some $0 < \sigma < 1$ (since G_1 is convex, the notion of complex geodesics is well-defined). By (11.6.1) and Proposition 11.1.4, we get

$$\boldsymbol{\gamma}_{G_2}(F(a); F'(a)\varphi'(0)) = \boldsymbol{\gamma}_{G_1}(\varphi(0); \varphi'(0)) = 1.$$

This means that the mapping $F \circ \varphi$ is a $\boldsymbol{\gamma}_{G_2}$-geodesic for $(F(a), F'(a)\varphi'(0))$. Again Using Proposition 11.1.4, we obtain

$$\begin{aligned} \boldsymbol{c}_{G_2}(F(a), F(z_0)) &= \boldsymbol{c}_{G_2}(F \circ \varphi(0), F \circ \varphi(\sigma)) = \boldsymbol{p}(0, \sigma) \\ &= \boldsymbol{c}_{G_1}(\varphi(0), \varphi(\sigma)) = \boldsymbol{c}_{G_1}(a, z_0), \end{aligned}$$

which finishes the proof of (11.6.2). Since G_1 is $\boldsymbol{c}$-finitely compact, condition (11.6.2) implies that F is a proper mapping. In particular, F is surjective and

$$F : B_{\boldsymbol{c}_{G_1}}(a, r_0) \longrightarrow B_{\boldsymbol{c}_{G_2}}(F(a), r_0) \text{ is biholomorphic.} \tag{11.6.3}$$

Since F is proper, there exist a proper analytic set $S \subset G_2$ and a number $N \in \mathbb{N}$ such that

$$F : G_1 \setminus F^{-1}(S) \longrightarrow G_2 \setminus S$$

is an N-fold covering (see also the proof of Proposition 19.6.7). By (11.6.3), we have $N = 1$, and therefore $F : G_1 \longrightarrow G_2$ is biholomorphic.

(b) Since G_1 is taut, $\boldsymbol{\varkappa}_{G_1}(a; X) > 0$ for all $X \neq 0$. Hence, as in (a), condition (11.6.1) (with $\delta = \boldsymbol{\varkappa}$) implies that $F : U_a \to F(U_a)$ is biholomorphic for an open neighborhood U_a of a. Let

$$\mathcal{F} := \{\varphi \in \mathcal{O}(\mathbb{D}, G_1) : \varphi(0) = a \text{ and } \boldsymbol{\varkappa}_{G_1}(a; \varphi'(0)) = 1\}, \quad M := \bigcup_{\varphi \in \mathcal{F}} \varphi(\mathbb{D}).$$

We will prove that

1° for any $\varphi \in \mathcal{F}$:

$$\boldsymbol{c}_{G_1}(\varphi(\lambda'), \varphi(\lambda'')) = \boldsymbol{c}_{G_2}(F \circ \varphi(\lambda'), F \circ \varphi(\lambda'')) = \boldsymbol{p}(\lambda', \lambda''), \ \lambda', \lambda'' \in \mathbb{D}; \tag{11.6.4}$$

2° M is closed in G_1;

3° $F : M \longrightarrow G_2$ is injective;

4° $F(M) = G_2$;

5° $(F|_M)^{-1} : G_2 \longrightarrow M$ is continuous; in particular, M is open.

Conditions 2°–5° clearly imply that $F : G_1 \longrightarrow G_2$ is biholomorphic.

Add. 1° Fix a $\varphi \in \mathcal{F}$. Since $\boldsymbol{\gamma}_{G_2} = \boldsymbol{\varkappa}_{G_2}$ (G_2 is convex), condition (11.6.1) gives

$$\boldsymbol{\gamma}_{G_2}(F(a); F'(a)\varphi'(0)) = \boldsymbol{\varkappa}_{G_1}(a; \varphi'(0)) = 1.$$

Hence by Proposition 11.1.4 we have

$$\boldsymbol{p}(\lambda', \lambda'') = \boldsymbol{c}_{G_2}(F \circ \varphi(\lambda'), F \circ \varphi(\lambda'')) \le \boldsymbol{c}_{G_1}(\varphi(\lambda'), \varphi(\lambda'')) \le \boldsymbol{p}(\lambda', \lambda''),$$

which gives (11.6.4).

Add. 2° Let $M \ni z_\nu = \varphi_\nu(\lambda_\nu) \longrightarrow z_0 \in G_1$, where $\varphi_\nu \in \mathcal{F}, \lambda_\nu \in \mathbb{D}, \nu \geq 1$. Since G_1 is taut, we may assume that $\varphi_\nu \underset{\nu\to\infty}{\overset{K}{\Longrightarrow}} \varphi_0 \in \mathcal{O}(\mathbb{D}, G_1)$. Obviously, $\varphi_0(0) = a$. Since $\varphi_\nu'(0) \to \varphi_0'(0)$, the continuity of $\boldsymbol{\varkappa}_{G_1}$ (cf. Proposition 3.5.14(b)) shows that $\varphi_0 \in \mathcal{F}$. The sequence $(\boldsymbol{c}_{G_1}(a, z_\nu))_{\nu=1}^\infty$ is bounded. Hence, by (11.6.4), the sequence $(\boldsymbol{p}(0, \lambda_\nu))_{\nu=1}^\infty$ is bounded, and consequently we may assume that $\lambda_\nu \longrightarrow \lambda_0 \in \mathbb{D}$. Finally, $z_0 = \varphi_0(\lambda_0) \in M$.

Add. 3° In view of (11.6.4), the mapping F is injective on $\varphi(\mathbb{D})$ for each $\varphi \in \mathcal{F}$. Now let $z_j = \varphi_j(\lambda_j) \in M \setminus \{a\}$, $\varphi_j \in \mathcal{F}, \lambda_j \in \mathbb{D}$, $j = 1, 2$. Assume that $z_1 \neq z_2$ and $\varphi_1(\mathbb{D}) \neq \varphi_2(\mathbb{D})$ and suppose that $F(z_1) = F(z_2) =: w_0$. By (11.6.4) the mappings $F \circ \varphi_j$, $j = 1, 2$ are complex geodesics for $(F(a), w_0)$. Since G_2 is strictly convex, $F \circ \varphi_1 = F \circ \varphi_2 \mod \operatorname{Aut}(\mathbb{D})$ (cf. Proposition 11.3.2), i.e., there exists a $\theta \in \mathbb{R}$ such that $F \circ \varphi_1(\lambda) = F \circ \varphi_2(e^{i\theta}\lambda)$, $\lambda \in \mathbb{D}$. Consequently, $\varphi_1(\lambda) = \varphi_2(e^{i\theta}\lambda)$ for λ near 0 (recall that F is biholomorphic near a) and finally, by the identity principle, $\varphi_1(\lambda) = \varphi_2(e^{i\theta}\lambda)$, $\lambda \in \mathbb{D}$. In particular, $\varphi_1(\mathbb{D}) = \varphi_2(\mathbb{D})$, which gives a contradiction.

Add. 4° Let $w_0 \in G_2 \setminus \{F(a)\}$ and let $\psi : \mathbb{D} \longrightarrow G_2$ be a complex geodesic for $(F(a), w_0)$ with $\psi(0) = F(a)$. By Proposition 11.1.4, ψ is a $\boldsymbol{\gamma}_{G_2}$-geodesic for $(F(a), \psi'(0))$. Put $X_0 := (F'(a))^{-1}\psi'(0)$ and let $\varphi \in \mathcal{F}$ be a $\boldsymbol{\varkappa}_{G_1}$-geodesic for (a, X_0). Then $F \circ \varphi$ and ψ are $\boldsymbol{\gamma}_{G_2}$-geodesic for $(F(a), \psi'(0))$ (cf. the proof of 1°). Consequently, by Proposition 11.3.2, $\psi = F \circ \varphi \mod \operatorname{Aut}(\mathbb{D})$. In particular, $w_0 \in F(\varphi(\mathbb{D})) \subset F(M)$.

Add. 5° Let $G_2 \ni w_\nu = F(z_\nu) \longrightarrow w_0 = F(z_0) \in G_2$. We have to prove that $z_\nu \longrightarrow z_0$. It suffices to show that if $z_\nu \to z_0^* \in (\mathbb{C} \cup \{\infty\})^n$, then $z_0^* = z_0$ or, equivalently (in view of continuity and injectivity of $F|_M$), that $z_0^* \in M$. Suppose that $z_\nu = \varphi_\nu(\lambda_\nu)$, where $\varphi_\nu \in \mathcal{F}$, $\lambda_\nu \in \mathbb{D}$, $\nu \geq 1$. Tautness of G_1 permits us to assume that $\varphi_\nu \underset{\nu\to\infty}{\overset{K}{\Longrightarrow}} \varphi_0 \in \mathcal{F}$ (cf. the proof of 2°). The sequence $(\boldsymbol{c}_{G_2}(F(a), w_\nu))_{\nu=1}^\infty$ is bounded. Hence, by (11.6.4), we may assume that $\lambda_\nu \longrightarrow \lambda_0 \in \mathbb{D}$. Finally, $z_0^* = \varphi_0(\lambda_0) \in M$. □

11.7 Exercises

Exercise 11.7.1. Let $\underline{d}^* := \boldsymbol{m}^{(k)}$ (resp. $\underline{d}^* := \boldsymbol{g}$) and let $\underline{\delta} := \boldsymbol{\gamma}^{(k)}$ (resp. $\underline{\delta} := \boldsymbol{A}$). Prove that for any domain $G \subset \mathbb{C}^n$ and for any point $\lambda_0 \in \mathbb{D}$ the following statements are equivalent (cf. Proposition 11.1.4):

(i) $\exists_{\lambda_0'' \in \mathbb{D}\setminus\{\lambda_0\}} : d_G^*(\varphi(\lambda_0), \varphi(\lambda_0'')) = \boldsymbol{m}(\lambda_0, \lambda_0'')$;

(ii) $\forall_{\lambda \in \mathbb{D}} : d_G^*(\varphi(\lambda_0), \varphi(\lambda)) = \boldsymbol{m}(\lambda_0, \lambda)$;

(iii) $\delta_G(\varphi(\lambda_0); \varphi'(\lambda_0)) = \boldsymbol{\gamma}_{\mathbb{D}}(\lambda_0; 1)$.

Exercise 11.7.2. Let $k \in \mathbb{N}$ and let $G \subset \mathbb{C}^n$ be a taut domain. Prove that the following conditions are equivalent (cf. Proposition 11.1.7):

(i) $\boldsymbol{m}_G^{(k)} = \boldsymbol{\ell}_G^*$ and $\boldsymbol{\gamma}_G^{(k)} = \boldsymbol{\varkappa}_G$;

(ii) $\boldsymbol{m}_G^{(k)} = \boldsymbol{\ell}_G^*$;

(iii) for any $z_0', z_0'' \in G$, $z_0' \neq z_0''$, there exist $\varphi \in \mathcal{O}(\mathbb{D}, G)$ and $f \in \mathcal{O}(G, \mathbb{D})$ such that $z_0' = \varphi(0)$, $z_0'' \in \varphi(\mathbb{D})$, $\operatorname{ord}_{z_0'} f \geq k$ and $f(\varphi(\lambda)) = \lambda^k$, $\lambda \in \mathbb{D}$.

Exercise 11.7.3. Let $k \in \mathbb{N}$ and let $G = G_{\mathfrak{h}} \subset \mathbb{C}^n$ be a balanced domain of holomorphy. Prove that for any $a \in G$ with $\mathfrak{h}(a) > 0$ the following conditions are equivalent (cf. Proposition 11.1.3(b)):

(i) the pair $(0, a)$ admits a complex $\boldsymbol{m}_G^{(k)}$-geodesic;

(ii) the pair $(0, a)$ admits a complex $\boldsymbol{\gamma}_G^{(k)}$-geodesic;

(iii) $\boldsymbol{m}_G^{(k)}(0, a) = \mathfrak{h}(a)$, cf. Exercise 4.4.4.

Exercise 11.7.4. Let $G \subset \mathbb{C}^n$ be a convex domain. Prove that there exists a sequence $G_\nu \nearrow G$ such that each G_ν is a bounded strongly convex domain with smooth real-analytic boundary.

Exercise 11.7.5 (cf. [340, 342]). Suppose that $G \subset \mathbb{C}^n$ is a convex domain such that

$$\forall_{\substack{\varphi:\mathbb{D}\longrightarrow G\\ \text{complex geodesic}}} \exists_{M_0>0}\ \exists_{\mu\in(0,1]} : \operatorname{dist}(\varphi(\lambda), \partial G) \leq M_0(1 - |\lambda|)^\mu, \quad \lambda \in \mathbb{D}, \tag{11.7.1}$$

$$\exists_{C_0>0}\ \exists_{\alpha\in(0,1]} : \boldsymbol{\varkappa}_G(z; X) \geq C_0 \frac{\|X\|}{(\operatorname{dist}(z, \partial G))^\alpha}, \quad z \in G,\ X \in \mathbb{C}^n. \tag{11.7.2}$$

Prove that any complex geodesic $\varphi : \mathbb{D} \longrightarrow G$ extends as an $(\alpha\mu)$-Hölder-continuous mapping to $\overline{\mathbb{D}}$. (Note that any bounded convex domain satisfies (11.7.1); cf. Exercise 11.7.9).

Hint. Use the Hardy–Littlewood theorem; cf. Appendix B.9.11.

Exercise 11.7.6 (cf. [138]). Let $G \subset \mathbb{C}^n$ be a bounded balanced convex domain. Prove that G satisfies (11.7.1) with $\mu = 1$.

Exercise 11.7.7 (cf. [340, 342, 460]). Let $G \subset \mathbb{C}^n$ be a bounded convex domain such that

$$\exists_{r_0>0},\ \forall_{z_0\in\partial G}\ \exists_{a_0\in G} : \|z_0 - a_0\| = r_0 \text{ and } \mathbb{B}(a_0, r_0) \subset G. \tag{11.7.3}$$

Prove that G satisfies (11.7.1) with $\mu = 1$.

Hint. Fix a complex geodesic $\varphi : \mathbb{D} \longrightarrow G$. It suffices to show that there exists a constant $M_1 > 0$ such that $\boldsymbol{c}_G(\varphi(0), \varphi(\lambda)) \leq M_1 - \frac{1}{2}\log\operatorname{dist}(\varphi(\lambda), \partial G)$, $\lambda \in \mathbb{D}$. Let $K := \{z \in G : \operatorname{dist}(z, \partial G) \geq r_0\}$. Put $M_2 := \max\{\boldsymbol{c}_G(\varphi(0), z) : z \in K\}$. Then, for $\varphi(\lambda) \in K$, we have $\boldsymbol{c}_G(\varphi(0), \varphi(\lambda)) \leq M_3 - \frac{1}{2}\log\operatorname{dist}(\varphi(\lambda), \partial G)$ with $M_3 := M_2 + \frac{1}{2}\log\operatorname{diam} G$. If $\varphi(\lambda) \notin K$, then by (11.7.3) we get $\boldsymbol{c}_G(\varphi(0), \varphi(\lambda)) \leq M_4 - \frac{1}{2}\log\operatorname{dist}(\varphi(\lambda), \partial G)$ with $M_4 := M_2 + \frac{1}{2}\log(2r_0)$.

Exercise 11.7.8. Let $G \subset \mathbb{C}^n$ be a strongly convex domain. Prove that G satisfies (11.7.3) and that

$$\exists_{R_0>0}\ \forall_{z_0\in\partial G}\ \exists_{a_0\in\mathbb{C}^n} : \|z_0 - a_0\| = R_0 \text{ and } \mathbb{B}(a_0, R_0) \supset G. \tag{11.7.4}$$

Exercise 11.7.9 (cf. [361]). Prove that any bounded convex domain $G \subset \mathbb{C}^n$ satisfies (11.7.1).

Hint. (Cf. Exercise 11.7.7) Fix a complex geodesic $\varphi : \mathbb{D} \longrightarrow G$. Let $a_0 := \varphi(0)$. It suffices to show that there exist constants $M_1 > 0$ and $0 < \mu \leq 1$ such that

$$\boldsymbol{c}_G(a_0, \varphi(\lambda)) \leq M_1 - \frac{1}{2\mu}\log\operatorname{dist}(\varphi(\lambda), \partial G), \quad \lambda \in \mathbb{D}$$

(then we can take $M_0 = \exp(2M_1\mu))$. Let $r_0 > 0$ be such that $\overline{\mathbb{B}}(a_0, 4r_0) \subset G$. Define $K := \{z \in G : \operatorname{dist}(z, \partial G) \geq r_0\}$ and $M_2 := \max\{\boldsymbol{c}_G(a_0, z) : z \in K\}$. Let $\lambda_0 \in \mathbb{D}$. If $\varphi(\lambda_0) \in K$, then for any $0 < \mu \leq 1$ we have

$$\boldsymbol{c}_G(a_0, \varphi(\lambda_0)) \leq M_3(\mu) - \frac{1}{2\mu}\log\operatorname{dist}(\varphi(\lambda_0), \partial G),$$

where $M_3(\mu) := M_2 + \frac{1}{2\mu}\log\operatorname{diam} G$. We will prove that there is a $\mu = \mu_0 \in (0, 1]$ (independent of $\varphi(\lambda_0)$) such that the above inequality is also true if $\varphi(\lambda_0) \notin K$ (then, one can take $M_1 := M_3(\mu_0)$). Suppose that $\varphi(\lambda_0) \notin K$ and let $z_0 \in \partial G$ be such that the point $\varphi(\lambda_0)$ belongs to the segment (a_0, z_0). Define

$$\Lambda = \Lambda(a_0, r_0, z_0) := \operatorname{conv}(\mathbb{B}(a_0, 2r_0) \cup \{z_0\}).$$

Let $b_0 := \frac{1}{2}(a_0 + z_0)$. Then, $\operatorname{dist}(b_0, \partial G) \geq \operatorname{dist}(b_0, \partial\Lambda) = r_0$. In particular, $b_0 \in K$. Observe that $\varphi(\lambda_0)$ belongs to the segment (b_0, z_0). For $0 < \mu \leq 1$ define

$$\psi(\lambda) = \psi_{z_0,\mu}(\lambda) := z_0 + \frac{1}{2}(1 + \lambda)^{\mu}(a_0 - z_0), \quad \lambda \in \mathbb{D},$$

where $(1 + \lambda)^{\mu} := \exp(\mu \operatorname{Log}(1 + \lambda))$ and Log denotes the principal branch of the logarithm. Observe that $\psi(0) = b_0$ and that $\psi(-t_\mu) = \varphi(\lambda_0)$, where

$$t_\mu := 1 - \left(\frac{2\|\varphi(\lambda_0) - z_0\|}{\|a_0 - z_0\|}\right)^{\frac{1}{\mu}}.$$

Moreover, $\boldsymbol{p}(0, -t_\mu) \le \frac{1}{2\mu} \log \operatorname{diam} G - \frac{1}{2\mu} \log \|\varphi(\lambda_0) - z_0\|$. Consequently, if $\varphi(\mathbb{D}) \subset G$, then

$$\begin{aligned} \boldsymbol{c}_G(a_0, \varphi(\lambda_0)) &\le \boldsymbol{c}_G(a_0, b_0) + \boldsymbol{c}_G(b_0, \varphi(\lambda_0)) \le M_2 + \boldsymbol{\ell}_G(b_0, \varphi(\lambda_0)) \\ &\le M_2 + \boldsymbol{p}(0, -t_\mu) \le M_3(\mu) - \frac{1}{2\mu} \log \operatorname{dist}(\varphi(\lambda_0), \partial G). \end{aligned}$$

It remains to observe that if

$$\mu := \min\left\{\frac{1}{2}\log_2 3,\ \frac{2}{\pi}\arctan\frac{1}{\sqrt{(\frac{\operatorname{diam} G}{2r_0})^2 - 1}}\right\},$$

then $\psi_{z_0,\mu}(\mathbb{D}) \subset \Lambda(a_0, r_0, z_0) \subset G$ for arbitrary $z_0 \in \partial G$.

Exercise 11.7.10. Let $G \subset \mathbb{C}^n$ be a convex domain. Define

$$\begin{aligned} \Delta_G(a; X) &:= \sup\{r > 0 : a + r\mathbb{D} \cdot X \subset G\}, \quad a \in G,\ X \in \mathbb{C}^n, \\ \Delta_G(\varepsilon) &:= \sup\{\Delta_G(a; X) : a \in G,\ \operatorname{dist}(a, \partial G) \le \varepsilon,\ \|X\| = 1\}, \quad \varepsilon > 0. \end{aligned}$$

Prove that

$$\boldsymbol{\varkappa}_G(a; X) \ge \frac{\|X\|}{2\Delta_G(\operatorname{dist}(a, \partial G))}, \quad a \in G,\ X \in \mathbb{C}^n.$$

In particular, if there exist $A > 0$ and $0 < \alpha \le 1$ such that

$$\Delta_G(\varepsilon) \le A\varepsilon^\alpha, \quad \varepsilon > 0, \tag{11.7.5}$$

then G satisfies (11.7.2).

Hint. Use Lemma 11.3.7 or the following argument (cf. [138]). Fix $a \in G$ and $X \in \mathbb{C}^n$ with $\|X\| = 1$. Let $\varphi : \mathbb{D} \longrightarrow G$ be any holomorphic mapping such that $\varphi(0) = a$ and $t\varphi'(0) = X$ for some $t > 0$. Put $\psi(\lambda) := \frac{1}{2\pi}\int_0^{2\pi} (1 + \cos\theta)\varphi(e^{i\theta}\lambda)\, d\theta$, $\lambda \in \mathbb{D}$. Then $\psi(\lambda) = a + \frac{\lambda}{2t}X$. Since $\frac{1}{2\pi}(1 + \cos\theta)\, d\theta$ is a probability measure on $[0, 2\pi]$, the function ψ maps $\mathbb{D}$ into G.

Exercise 11.7.11. Prove that if $G \subset \mathbb{C}^n$ is a convex domain satisfying condition (11.7.4), then condition (11.7.5) is fulfilled with $\alpha = 1/2$.

Exercise 11.7.12 (cf. [340, 342, 361, 460]). Let $G \subset \mathbb{C}^n$ be a bounded convex domain satisfying (11.7.5). Prove that for any complex geodesic $\varphi : \mathbb{D} \longrightarrow G$ there exists a $\mu \in (0, 1]$ such that φ extends to an $(\alpha\mu)$-Hölder-continuous function on $\overline{\mathbb{D}}$. In particular,

- if G is strongly convex, then one can take $\alpha = 1/2, \mu = 1$,
- if G is balanced, then one can take $\mu = 1$.

Hint. Use Exercises 11.7.5–11.7.11.

Exercise 11.7.13. Let $\mathcal{E} := \mathcal{E}(p)$ be a complex convex ellipsoid; cf. § 16.4. Prove that

$$\Delta_{\mathcal{E}}(\varepsilon) \le A(p)\varepsilon^{\alpha(p)}, \quad \varepsilon > 0, \tag{11.7.6}$$

where $A(p) > 0$ is a constant depending only on p and

$$\alpha(p) := \frac{1}{2}\frac{\min\{p_1,\dots,p_n,1\}}{\max\{p_1,\dots,p_n\}}.$$

(See Exercise 11.7.10 for the definition of the function $\Delta_{\mathcal{E}}$.) In particular, in view of Exercise 11.7.10,

$$\varkappa_{\mathcal{E}}(z;X) \ge C(p)\frac{\|X\|}{(\operatorname{dist}(z,\partial\mathcal{E}))^{\alpha(p)}}, \quad z \in \mathcal{E},\ X \in \mathbb{C}^n.$$

Hint. The proof will be based on the ideas taken from [194]. To get (11.7.6), it suffices to show that for any $z \in \mathcal{E}$, $X \in \mathbb{C}^n$ with $\|X\| = 1$, $0 < r \le 1$ such that $z + r\mathbb{D}\cdot X \subset \mathcal{E}$, the following inequality holds: $r^{2p_0^+} \le M(p)(1-u(z))^{p_0^-}$, where $p_0^- := \min\{p_1,\dots,p_n,1\}$, $p_0^+ := \max\{p_1,\dots,p_n\}$, $u(z) := \sum_{j=1}^n |z_j|^{2p_j}$, $z = (z_1,\dots,z_n) \in \mathbb{C}^n$.

Observe that this implication is a consequence of the following relation:

$$\forall_{s\ge\frac{1}{2}}\ \exists_{B(s)>0}\ \forall_{\delta\in(0,1]}\ \forall_{(\xi,\eta)\in\mathbb{D}\times\mathbb{C}}:$$

$$\text{if } \Phi(s;\xi,\eta) \le 4\delta, \text{ then } \begin{cases} |\eta|^{2s} \le B(s)\delta & \text{if } s \ge 1 \\ \min\{|\eta|^{2s}, |\eta|^2\} \le B(s)\delta & \text{if } \frac{1}{2} \le s < 1 \end{cases}, \tag{11.7.7}$$

where $\Phi(s;\xi,\eta) := |\xi+\eta|^{2s} + |\xi-\eta|^{2s} + |\xi+i\eta|^{2s} + |\xi-i\eta|^{2s} - 4|\xi|^{2s}$.

To prove (11.7.7), fix $s \ge 1/2$, $0 < \delta \le 1$, $\xi \in \mathbb{D}_*$, and $\eta \ne 0$ with $\Phi(s;\xi,\eta) \le 4\delta$. Then, $4\delta \ge \Phi(s;\xi,\eta) \ge |\eta|^{2s}\Lambda_s(|\xi/\eta|)$, where

$$\Lambda_s(r) := (r^2 + r\sqrt{2} + 1)^s + (r^2 - r\sqrt{2} + 1)^s - 2r^{2s}, \quad r \ge 0.$$

Consider two cases:

(a) $s \ge 1$: then $\lim_{r\to\infty}\Lambda_s(r) > 0$;

(b) $1/2 \le s < 1$: then

$$4\delta \ge |\eta|^{2s}\Lambda_s(|\xi/\eta|) \ge |\eta|^{2s}\Lambda_s(1/|\eta|) \ge B_1(s)\min\{|\eta|^{2s}, |\eta|^2\}.$$

Exercise 11.7.14 (cf. [192]). Let $G = G_q$ be a convex balanced domain (q is a seminorm). Suppose that $a \in G$ with $q(a) > 0$ is such that the point $b := a/q(a)$ is not a complex extreme point for $\overline{G}$ (cf. Proposition 11.3.5(b)). Prove that there exist complex $\boldsymbol{c}_G$-geodesics (resp. $\boldsymbol{\gamma}_G$-geodesics) for $(0, a)$ that cannot be continuously extended to $\overline{\mathbb{D}}$.

Hint. If $b+\mathbb{D}\cdot\xi \subset \overline{G}$ for some $\xi \in (\mathbb{C}^n)_*$, then take $\varphi(\lambda) := \lambda(b+f(\lambda)\xi)$, $\lambda \in \mathbb{D}$, where f is a suitable function from $\mathcal{O}(\mathbb{D}, \mathbb{D})$.

Exercise 11.7.15. Let $G \subset \mathbb{C}^n$ be a balanced pseudoconvex domain such that G is strongly pseudoconvex at a point $b \in \partial G$. Then, G is not biholomorphic to $\mathbb{D}^n$.

Hint. Use Corollary 3.5.8(a) and the remark before the proof of Proposition 11.3.5.

11.8 List of problems

Chapter 12

The Bergman metric

Summary. The constructions of the metrics studied so far were based on the set of bounded holomorphic functions, on special plurisubharmonic functions, or on analytic discs. The Bergman metric (Section 12.7) will be based on the Hilbert space of square integrable holomorphic functions and the associated kernel, the so-called Bergman kernel (Section 12.1). While the zeros of the Bergman kernel are discussed in Section 12.3, Sections 12.4 and 12.5 deal with the boundary growth of the kernel. Moreover, the Bergman kernel allows a characterization of L_h^2-domains of holomorphy (Section 12.6). The chapter ends with comparison results for the Bergman metric and metrics discussed before, and a modification of the Bergman distance, the so-called Skwarczyński distance.

Introduction. In the twenties of the last century, the study of square integrable functions was initiated by S. Bergman in order to solve the classification problem for domains in $\mathbb{C}^n$. Recall that the Euclidean ball and the bidisc in $\mathbb{C}^2$ (which obviously are homeomorphic) are not biholomorphically equivalent, i.e., there is no analogy for the Riemann mapping theorem in higher dimensions. These investigations had led Bergman to introduce his kernel and his metric, which became known as the Bergman kernel and the Bergman metric. In 1976, M. Skwarczyński introduced his pseudodistance, which is based on the Bergman kernel.

For a long time it was difficult to work explicitly with the Bergman metric. But with Hörmander's $\overline{\partial}$-theory an important new tool entered the theory of L^2-holomorphic functions. Deep results were found, for example, in the theory of the boundary behavior of biholomorphic or proper holomorphic mappings using the Bergman metric.

12.1 The Bergman kernel

Let M be an arbitrary set; by $\operatorname{Abb}(M, \mathbb{C})$ we denote the set of all $\mathbb{C}$-valued functions defined on M. Moreover, we suppose that a linear subspace $H \subset \operatorname{Abb}(M, \mathbb{C})$ carries a scalar product $\langle\ ,\ \rangle_H$ such that H becomes a Hilbert space. As usual, we write $\|x\|_H := \sqrt{\langle x, x\rangle_H}$, $x \in H$.

A function $K : M \times M \longrightarrow \mathbb{C}$ is called a *kernel function* (or shortly, a *kernel*) of H if

- $K(\cdot, y) \in H, \quad y \in M,$
- $f(y) = \langle f, K(\cdot, y)\rangle_H, \quad f \in H,\ y \in M.$

Note that the kernel function is uniquely defined.

Remark 12.1.1. Let M, H be as above and suppose that K is a kernel function of H. Then, the following properties of K are easily seen:

(a) $0 \le K(y,y) = \|K(\cdot,y)\|_H^2$, $y \in M$;

(b) $K(x,y) = \overline{K(y,x)}$, $x, y \in M$;

(c) $|K(x,y)|^2 \le K(x,x)K(y,y)$, $x, y \in M$;

(d) $|f(y)| \le \|f\|_H \sqrt{K(y,y)}$, $f \in H$, $y \in M$.

It is clear that (d) gives the continuity of the linear functionals $H \ni f \longmapsto f(y) \in \mathbb{C}$ $(y \in M)$.

Lemma 12.1.2. *Let M and H be as above. If we suppose that any linear functional $H \ni f \longmapsto f(y) \in \mathbb{C}$, $y \in M$, is continuous, then H carries a (unique) kernel K.*

Proof. Use the Riesz representation theorem. □

To obtain explicit kernels, a fundamental role is played by orthonormal bases. In the following we will always assume that M is a topological space with an infinite countable dense subset M' and that $H \subset \mathcal{C}(M,\mathbb{C})$. If H has a kernel function K, then H is a separable Hilbert space (take the functions $K(\cdot,y)$ when $y \in M'$). Hence, H has a countable orthonormal basis $(\varphi_j)_{j\in J}$ with $J = \{1,\dots,j_0\}$ or $J = \mathbb{N}$, and therefore K can be represented as

$$K(\cdot,y) = \sum_{j\in J} \overline{\varphi_j(y)}\varphi_j. \tag{12.1.1}$$

Observe that property (d) of Remark 12.1.1 implies that the series of functions in (12.1.1) converges uniformly on those subsets of M on which the function $x \longmapsto K(x,x)$ remains bounded. Moreover, the series is independent of the order of summation.

Remark 12.1.3. Suppose H (as above) is equipped with the kernel K. Moreover, assume that for a fixed point $y_0 \in M$ there is at least one function $f \in H$ with $f(y_0) \neq 0$. Then, the following properties show how K relates to various extremal problems:

(a) $K(y_0,y_0) > 0$;

(b) $\min\{\|f\|_H : f \in H, f(y_0) = 1\} = \|K(\cdot,y_0)/K(y_0,y_0)\|_H$;

(c) $\max\{|f(y_0)| : f \in H, \|f\|_H = 1\} = (K(\cdot,y_0)/\sqrt{K(y_0,y_0)})(y_0)$.

Moreover, the function $g := K(\cdot,y_0)/K(y_0,y_0)$ is the only function in H solving the extremal problem in (b).

So far, we discussed the kernel function in a very abstract setting. Now, we turn to the concrete situation that we are interested in during this chapter. Let $G \subset \mathbb{C}^n$ be any domain, $\varphi \in \mathcal{C}^{\downarrow}(G, \mathbb{R}_{>0})$ (i.e., φ is lower semicontinuous), and let

$$L_h^2(G,\varphi) := \left\{ f \in \mathcal{O}(G,\mathbb{C}) : \int_G |f(z)|^2 \varphi(z) d\mathcal{L}^{2n}(z) < \infty \right\}.$$

The space $L_h^2(G,\varphi)$ with the scalar product

$$\langle f, g\rangle_{L^2(G,\varphi)} = \langle f, g\rangle_{\varphi} := \int_G f(z)\overline{g(z)}\varphi(z) d\mathcal{L}^{2n}(z), \quad f, g \in L_h^2(G,\varphi),$$

is a complex Hilbert space, the *Hilbert space of all holomorphic functions on G that are square integrable with respect to the weight function* φ. In case of $\varphi_0 \equiv 1$ we write $L_h^2(G) := L_h^2(G,\varphi_0)$ and $\langle f, g\rangle = \langle f, g\rangle_{L^2(G)} = \langle f, g\rangle_{L^2(G,\varphi_0)}$ and talk simply of *the Hilbert space of square integrable functions*. Using the Cauchy integral formula, it follows that for every $z \in G$ the evaluation functional $L_h^2(G,\varphi) \ni f \longmapsto f(z)$ is continuous. Thus, by Lemma 12.1.2, $L_h^2(G,\varphi)$ carries the kernel $\boldsymbol{K}_{G,\varphi}$ which is called the *Bergman kernel with weight* φ *of* G. In case of $\varphi_0 \equiv 1$ we write $\boldsymbol{K}_G := \boldsymbol{K}_{G,\varphi_0}$ and say that $\boldsymbol{K}_G$ is the *Bergman kernel of* G. Recall that $\boldsymbol{K}_{G,\varphi}$ is uniquely defined.

$L_h^2(G)$ is a closed subspace of the Hilbert space $L^2(G)$ of all complex valued square integrable functions on G. By

$$\boldsymbol{P}_G : L^2(G) \longrightarrow L_h^2(G), \quad (\boldsymbol{P}_G f)(z) := \langle f, K(\cdot, z)\rangle_{L^2(G)},$$

we denote the orthogonal projection; $\boldsymbol{P}_G$ is called the *Bergman projection*.

Remark 12.1.4.

(a) For a bounded domain G the dimension of $L_h^2(G)$ is infinite because all polynomials belong to $L_h^2(G)$. On the other hand, observe that $L_h^2(\mathbb{C}^n) = \{0\}$. Therefore, $\boldsymbol{K}_{\mathbb{C}^n} \equiv 0$.

(b) For a plane domain G, the following properties are equivalent (cf. [484, 528]):

   (i) $\boldsymbol{K}_G(z,z) > 0$ for all $z \in G$;

   (ii) $L_h^2(G) \neq \{0\}$;

   (iii) $\dim L_h^2(G) = \infty$;

   (iv) $K(\cdot, z_1), \dots, K(\cdot, z_k)$ are linearly independent for any k pairwise different points $z_1, \dots, z_k$ in G, $k \geq 1$.

(c) In the higher dimensional case the situation changes rapidly, as the following result shows (cf. [528]): for every $k \in \mathbb{N}$ there exists an unbounded Reinhardt domain $G \subset \mathbb{C}^2$ with $\dim L_h^2(G) = k$; see Exercise 12.10.1. All of these G's are not pseudoconvex. So far, no pseudoconvex domain $G \subset \mathbb{C}^n$ with $0 < \dim L_h^2(G) < \infty$ is known.

?

(d) In the case of pseudoconvex Hartogs domains over $\mathbb{C}$ with one-dimensional fibers the following is known: For $u \in \mathcal{SH}(\mathbb{C})$ put

$$D_u := \{(z, w) \in \mathbb{C} \times \mathbb{C} : |w|e^{u(z)} < 1\}.$$

D_u is pseudoconvex. Then, the size of $L_h^2(D_u)$ is determined by the behavior of the Riesz measure $\frac{1}{2\pi}\Delta u$. To be more precise look at the following decomposition of the Riesz measure:

$$\tfrac{1}{2\pi}\Delta u = \sum_{j\geq 1} \nu(u, a_j)\delta_{a_j} + \mu,$$

where

- a_j are pairwise different points in $\mathbb{C}$,
- $\nu(u, a_j) := \lim\limits_{r\to 0} \frac{\sup\{u(a+z):|z|=r\}}{\log r}$ is the so-called *Lelong number* of u at a_j (recall that the set of points for which the Lelong number is positive is at most countable),
- δ_{a_j} is the Dirac measure at a_j,
- μ is a non-negative measure equal to zero on countable subsets of $\mathbb{C}$.

For the weights $\alpha_j := \nu(u, a_j)$ consider the condition

$$\exists_{j_1\neq j_2} : \alpha_{j_1} - \lfloor\alpha_{j_1}\rfloor, \alpha_{j_2} - \lfloor\alpha_{j_2}\rfloor, \alpha_{j_1} + \alpha_{j_2} - \lfloor\alpha_{j_1} + \alpha_{j_2}\rfloor > 0 \text{ or}$$
$$\exists_{j_1<j_2<j_3} : \alpha_{j_1} = \alpha_{j_2} = \alpha_{j_3} = 1/2. \quad (12.1.2)$$

Then, one has the following result (see [287]):

$$\dim L_h^2(D_u) = \begin{cases} \infty, & \text{if } \mu \not\equiv 0 \text{ or (12.1.2) is satisfied} \\ 0, & \text{otherwise} \end{cases}.$$

In particular, if $u = \log|f|$, where $f \in \mathcal{O}(\mathbb{C})$, then $\mu \equiv 0$ and simultaneously, since $\nu(u, a)$ is nothing other than the multiplicity of the zero of f at a, (12.1.2) is fulfilled. Hence, $L_h^2(D_u) = \{0\}$. A complete discussion of $L_h^2(D)$ for arbitrary Hartogs domain D is still not known. ?

Recall that $\boldsymbol{K}_G(z, z) > 0$ if there exists an $f \in L_h^2(G)$ with $f(z) \neq 0$. Then, according to Remark 12.1.3, we get

$$\boldsymbol{K}_G(z, z) = \sup\left\{\frac{|f(z)|^2}{\|f\|^2_{L^2(G)}} : f \in L_h^2(G), f \neq 0\right\}.$$

In particular, if $G' \subset G$, then $\boldsymbol{K}_G(z, z) \leq \boldsymbol{K}_{G'}(z, z)$ for all $z \in G'$.

Moreover, the Hartogs' theorem on separately holomorphic functions (cf. Appendix B.1.9) implies that the function $G \times \overline{G} \ni (z, w) \longmapsto \boldsymbol{K}_G(z, \overline{w})$ is holomorphic as a function of $2n$ variables. (Of course, here $\overline{G}$ is not the closure of G.) In particular, the kernel $\boldsymbol{K}_G$ is continuous on $G \times G$.

Remark 12.1.5. Here, we give another useful description of the Bergman kernel. Denote by $\varphi : \mathbb{C} \longrightarrow \mathbb{R}_+$ a rotation-invariant $\mathcal{C}^\infty$-function with $\operatorname{supp} \varphi \subset \mathbb{D}$ and $\int_{\mathbb{C}} \varphi(z) d\mathcal{L}^2(z) = 1$. Let G be a domain in $\mathbb{C}^n$ and let $w = (w_1, \dots, w_n) \in G$ with $\mathbb{P}(w, r) \subset G$ $(r > 0)$. Put

$$\psi_w(z) := (1/r^{2n})\varphi((z_1 - w_1)/r) \cdot \dots \cdot \varphi((z_n - w_n)/r), \quad z = (z_1, \dots, z_n) \in \mathbb{C}^n.$$

Then, for a holomorphic function $f : G \longrightarrow \mathbb{C}$ we obtain

$$f(w) = \int_G \psi_w(z) f(z) d\mathcal{L}^{2n}(z)$$

via the Cauchy integral formula. In particular, this formula holds for every f in $L_h^2(G)$, which implies:

$$\boldsymbol{P}_G \psi_w = \boldsymbol{K}_G(\cdot, w). \tag{12.1.3}$$

Observe that the same formula remains true if ψ_w is assumed to be invariant only under rotations with center at w.

Remark 12.1.6. The following result, which may be found in [345], describes the Bergman kernel of Hartogs domains in terms of various weighted Bergman kernels. Let $D \subset \mathbb{C}^n$ be a domain and let $\varphi \in \mathcal{C}(D, \mathbb{R}_{>0})$ be bounded. Put

$$D_{m,\varphi} := \{(z, w) \in D \times \mathbb{C}^m : \|w\| < \varphi(z)\}.$$

For $\alpha \in \mathbb{Z}_+^n$ define the number $c_\alpha \in \mathbb{R}_{>0}$ via $\int_{\mathbb{B}_m(r)} |w^{2\alpha}| d\mathcal{L}^{2m}(w) = c_\alpha r^{2|\alpha|+2m}$ and put $\varphi_\alpha := c_\alpha \varphi^{2|\alpha|+2m}$. Then, the following properties hold:

(a) if $f \in L_h^2(D_{m,\varphi})$, then $f(z, w) = \sum_{\alpha \in \mathbb{Z}_+^m} f_\alpha(z) w^\alpha$, where $f_\alpha \in L_h^2(D, \varphi_\alpha)$ and the convergence is locally uniform;

(b) $\|f\|^2_{L_h^2(D_{m,\varphi})} = \sum_{\alpha \in \mathbb{Z}_+^n} \|f_\alpha\|^2_{L_h^2(D, \varphi_\alpha)}$;

(c) $\boldsymbol{K}_{D_{m,\varphi}}((z, w), (\zeta, \omega)) = \sum_{\alpha \in \mathbb{Z}_+^n} \boldsymbol{K}_{D, \varphi_\alpha}(z, \zeta) w^\alpha \overline{\omega}^\alpha$, $(z, w), (\zeta, \omega) \in D_{m,\varphi}$.

Example 12.1.7.

(a) We mention that the functions

$$\varphi_\alpha(z) := \sqrt{\frac{(n+|\alpha|)!}{\alpha!\pi^n}} z^\alpha, \quad z \in \mathbb{B}_n,\ \alpha \in (\mathbb{Z}_+)^n,$$

form an orthonormal basis of $L_h^2(\mathbb{B}_n)$. Use induction over n to evaluate $\|z^\alpha\|_{L^2(\mathbb{B}_n)}$. Then, the Bergman kernel of $\mathbb{B}_n$ can be calculated as follows:

$$\begin{aligned} \boldsymbol{K}_{\mathbb{B}_n}(z,w) &= \sum_{\alpha\in(\mathbb{Z}_+)^n} \frac{(n+|\alpha|)!}{\alpha!\pi^n} z^\alpha \overline{w}^\alpha = \frac{1}{\pi^n}\sum_{\nu=0}^{\infty}\frac{(n+\nu)!}{\nu!}\sum_{|\alpha|=\nu}\frac{\nu!}{\alpha!}z^\alpha\overline{w}^\alpha \\ &= \frac{1}{\pi^n}\sum_{\nu=0}^{\infty}\frac{(n+\nu)!}{\nu!}\langle z,w\rangle^\nu = \frac{1}{\pi^n}\frac{d^n}{dx^n}\left(\frac{1}{1-x}\right)\bigg|_{x=\langle z,w\rangle}. \end{aligned}$$

Thus, we get

$$\boldsymbol{K}_{\mathbb{B}_n}(z,w) = \frac{n!}{\pi^n}(1-\langle z,w\rangle)^{-(n+1)}, \quad z,w \in \mathbb{B}_n. \tag{12.1.4}$$

(b) The Bergman kernel of the unit polycylinder $\mathbb{D}^n$ can be obtained via similar calculations as

$$\boldsymbol{K}_{\mathbb{D}^n}(z,w) = \prod_{j=1}^{n}\boldsymbol{K}_{\mathbb{D}}(z_j,w_j) = \frac{1}{\pi^n}\prod_{j=1}^{n}(1-z_j\overline{w}_j)^{-2}. \tag{12.1.5}$$

(c) Let $P := \{\lambda \in \mathbb{C} : r < |\lambda| < 1\}$. Then, an orthonormal basis is given by

$$\varphi_j(\lambda) = \sqrt{\frac{j+1}{\pi(1-r^{2j+2})}}\cdot\lambda^j, \quad j \in \mathbb{Z},\ j \neq -1,$$

$$\varphi_{-1}(\lambda) = \sqrt{-\frac{1}{2\pi\log r}}\cdot\lambda^{-1}.$$

Hence,

$$\boldsymbol{K}_P(z,w) = -\frac{1}{2\pi z\overline{w}\log r} + \frac{1}{\pi z\overline{w}}\sum_{\substack{j\in\mathbb{Z}\\ j\neq 0}}\frac{j(z\overline{w})^j}{1-r^{2j}}, \quad z,w \in P.$$

With the help of elliptic functions, one can rewrite the formula for the kernel as

$$\begin{aligned} \boldsymbol{K}_P(z,w) &= \frac{1}{\pi\lambda}\left(-\frac{1}{\log r^2} + \sum_{m=0}^{\infty}\left(\frac{\lambda r^{2m}}{(1-\lambda r^{2m})^2} + \frac{r^{2m+2}/\lambda}{(1-r^{2m+2}/\lambda)^2}\right)\right) \\ &=: \frac{1}{\pi\lambda}\,h(\lambda), \end{aligned}$$

where $\lambda := z\overline{w}$ with $r^2 < |\lambda| < 1$.

For h we obtain the following facts:

- h is continuous,
- $h(\lambda) \in \mathbb{R}$ if $\lambda \in \mathbb{R}$, $r^2 < |\lambda| < 1$ or if $|\lambda| = r$,
- $h(\lambda) > 0$ if $r^2 < \lambda < 1$,
- $h(-1) < -\frac{1}{\log r^2} - \frac{1}{4} < 0$, if $r < e^{-2}$ and therefore, $h(\lambda) < 0$ if $-1 < \lambda < -r^2$ and λ is near -1.

Hence, h has zeros, and therefore $\boldsymbol{K}_P$ has zeros on $P \times P$ under the above condition on r, i.e., $r < e^{-2}$. This was done in [482] (see also [483]). The general case was then solved in [457], using different methods.

(d) Let

$$\psi(t) := \begin{cases} 18, & \text{if } 0 \le t \le 1/4 \\ 1, & \text{if } 1/4 < t < 1 \end{cases}.$$

Put $\varphi(\lambda) := \psi(|\lambda|)$, $\lambda \in \mathbb{D}$. Then,

$$\boldsymbol{K}_{\mathbb{D},\varphi}(z, w) = \sum_{j=0}^{\infty} \alpha_j (z\overline{w})^j,$$

where

$$\alpha_j := \frac{1}{2\pi} \frac{16^{j+1}(2j+2)}{16^{j+1}+17}, \quad j \in \mathbb{Z}_+.$$

We have

$$\begin{aligned}(1-\xi^2)\sum_{j=0}^{\infty} \alpha_j \xi^j &= \alpha_0 + (\alpha_1 - 2\alpha_0)\xi + \sum_{j=2}^{\infty} (\alpha_j - 2\alpha_{j-1} + \alpha_{j-2})\xi^j \\ &=: L(\xi) + S(\xi).\end{aligned}$$

Note that L vanishes at $\xi_0 := -91/170$.

Assume for a moment that we knew that

$$\min\{L(\xi) : |\xi| = 1-\varepsilon\} > \max\{|S(\xi)| : |\xi| = 1-\varepsilon\}, \quad 0 < \varepsilon \ll 1. \qquad (*)$$

Then, using Rouché's theorem we can conclude that $L+S$ has a zero in $\mathbb{D}(1-\varepsilon)$, because we already know that L has one there. Hence, $\boldsymbol{K}_{\mathbb{D},\varphi}$ has zeros in $\mathbb{D}^2$ (compare the kernel given in (b)).

Finally, we will verify (*): since

$$\max\{|S(\xi)| : |\xi| = 1 - \varepsilon\} < \sum_{j=2}^{\infty} |\alpha_j - 2\alpha_{j-1} + \alpha_{j-2}|,$$

it remains to show that

$$\min\{L(\xi) : |\xi| = 1 - \varepsilon\} > \sum_{j=2}^{\infty} |\alpha_j - 2\alpha_{j-1} + \alpha_{j-2}|.$$

Using the explicit formulas for the α_j's we get for $j \geq 2$

$$\alpha_j - \alpha_{j-1} = \frac{1}{2\pi} \frac{32 \cdot 16^{2j} + 34 \cdot 16^j (15j + 16)}{(16^j + 17)(16^{j+1} + 17)},$$
$$\alpha_{j-1} - \alpha_{j-2} = \frac{1}{2\pi} \frac{2 \cdot 16^{2j-1} + 34 \cdot 16^{j-1}(15j + 1)}{(16^j + 17)(16^{j-1} + 17)}.$$

By comparing the right hand sides, we see that $\alpha_j - \alpha_{j-1} < \alpha_{j-1} - \alpha_{j-2}$ for $j \geq 3$. Therefore, $\alpha_j - 2\alpha_{j-1} + \alpha_{j-2} < 0,\ j \geq 3$. Exploiting the above telescoping sum, we see that it converges to $\alpha_1 - \alpha_0 - \lim(\alpha_j - \alpha_{j-1}) = \alpha_1 - \alpha_0 - 2$.

On the other hand, a direct calculation shows that for small ε we have

$$\min\{|L(\xi) : |\xi| = 1 - \varepsilon\} = (\alpha_1 - 3\alpha_0) - \varepsilon(\alpha_1 - 2\alpha_0).$$

Using the effective values of the α_j's, it is easy to verify that $(\alpha_1 - 3\alpha_0) - \varepsilon(\alpha_1 - 2\alpha_0) > \alpha_1 - \alpha_0 - 2$, which finishes the proof. This example is taken from [554].

Also, for more complicated domains there are explicit formulas for the Bergman kernel. We here discuss the kernel for the Thullen domains.

Example 12.1.8 (cf. [53]). For $p > 0$ put

$$D_p := \mathcal{E}((1/p, 1)) = \{z = (z_1, z_2) \in \mathbb{C}^2 : |z_1|^{2/p} + |z_2|^2 < 1\}.$$

Integration by parts shows that the functions

$$\varphi_{\mu,\nu}(z) = \left(\frac{\prod_{j=0}^{\nu+1}(p(\mu + 1) + j)}{\pi^2 p \nu!} \right)^{1/2} \cdot z_1^{\mu} z_2^{\nu}, \quad 0 \leq \nu, \mu,\ z \in D_p,$$

form an orthonormal basis of $L_h^2(D_p)$. Then, by (12.1.1) we obtain

$$\begin{aligned}
\boldsymbol{K}_{D_p}(z,w) &= \frac{1}{\pi^2 p}\sum_{\mu,\nu=0}^{\infty}\frac{(z_1\overline{w}_1)^{\mu}(z_2\overline{w}_2)^{\nu}}{\nu!}\prod_{j=0}^{\nu+1}(p(\mu+1)+j)\\
&= \frac{1}{\pi^2}\sum_{\mu=0}^{\infty}(z_1\overline{w}_1)^{\mu}(\mu+1)(p(\mu+1)+1)\\
&\quad\cdot\left(1+\sum_{\nu=1}^{\infty}\frac{(z_2\overline{w}_2)^{\nu}}{\nu!}\prod_{k=1}^{\nu}((\mu+1)p+1+k)\right)\\
&\underset{(*)}{=} \frac{1}{\pi^2(1-z_2\overline{w}_2)^{p+2}}\sum_{\mu=0}^{\infty}\left(\frac{z_1\overline{w}_1}{(1-z_2\overline{w}_2)^{p}}\right)^{\mu}\\
&\quad\cdot((\mu+1)(p+1)+p\mu(\mu+1))\\
&= \frac{1}{\pi^2(1-z_2\overline{w}_2)^{p+2}}\left(\frac{p+1}{(1-q)^2}+\frac{2pq}{(1-q)^3}\right),
\end{aligned}$$

where $q := z_1\overline{w}_1/(1-z_2\overline{w}_2)^p$, $|q|<1$. We mention that the binomial series were applied at $(*)$. Moreover, $(1-z_2\overline{w}_2)^p$ is chosen as the principal branch. Thus, we finally arrive at the following formula:

$$\boldsymbol{K}_{D_p}(z,w) = \frac{1}{\pi^2}(1-z_2\overline{w}_2)^{p-2}\frac{(p+1)(1-z_2\overline{w}_2)^p+(p-1)z_1\overline{w}_1}{((1-z_2\overline{w}_2)^p-z_1\overline{w}_1)^3}. \tag{12.1.6}$$

Very recently the following effective formula for the Bergman kernel of the complex ellipsoid $D := \mathcal{E}((2,2)) = \{z\in\mathbb{C}^2 : |z_1|^4+|z_2|^4<1\}$ was found. We just formulate it (without proof) to show how complicated the Bergman kernel may look like.

Example 12.1.9 (cf. [408]).

$$\begin{aligned}
\boldsymbol{K}_D(z,w) &= \sum_{j=1}^{2}\frac{\zeta_j(\pi+2\arcsin\zeta_j)f(\zeta_j^2,\zeta_j^2)}{\pi^3(1-\zeta_j^2)^{3/2}(1-\zeta_1^2-\zeta_2^2)^3}+\frac{8\zeta_1\zeta_2}{\pi^2(1-\zeta_1^2-\zeta_2^2)^3}\\
&\quad+\frac{2g(\zeta_1^2,\zeta_2^2)}{\pi^3(1-\zeta_1^2)(1-\zeta_2^2)(1-\zeta_1^2-\zeta_2^2)^2}, \qquad z,w\in D,
\end{aligned}$$

where $\zeta_j := z_j\overline{w}_j$, $j=1,2$, $f(x,y) := 3(1-x)^2+6y(1-x)-y^2$, and $g(x,y) := 2-x-y-(x-y)^2$.

The proof uses the standard orthogonal system $(z^{\alpha})_{\alpha\in\mathbb{Z}_+^2}$ and hypergeometric function for evaluating the infinite series, giving the Bergman kernel via the orthonormal system, which is derived from the above orthogonal one.

Next, we are interested in the behavior of the Bergman kernel under biholomorphic or even under proper holomorphic mappings.

Proposition 12.1.10. *Let $F : G \longrightarrow D$ be a biholomorphic mapping between the domains $G, D \subset \mathbb{C}^n$. Then, we have*

$$\boldsymbol{K}_D(F(z), F(w)) \det F'(z)\overline{\det F'(w)} = \boldsymbol{K}_G(z, w), \quad z, w \in G.$$

Proof. If $g \in L_h^2(D)$, then obviously $(g \circ F) \det F' \in L_h^2(G)$. Therefore, the properties of the Bergman kernel yield

$$\begin{aligned}
&\int_G (g \circ F)(w) \det F'(w)\overline{\boldsymbol{K}_G(w, z)} d\mathcal{L}^{2n}(w) \\
&\quad = (g \circ F)(z) \det F'(z) = \int_D g(\zeta)\overline{\boldsymbol{K}_D(\zeta, F(z))} \det F'(z) d\mathcal{L}^{2n}(\zeta) \\
&\quad = \int_G (g \circ F)(w)\overline{\boldsymbol{K}_D(F(w), F(z))} |\det F'(w)|^2 \det F'(z) d\mathcal{L}^{2n}(w).
\end{aligned}$$

In particular, if for $f \in L_h^2(G)$ we put $g := f \circ F^{-1} \cdot \det(F^{-1})'$, then it follows that

$$0 = \int_G f(w)(\overline{\boldsymbol{K}_G(w, z)} - \overline{\boldsymbol{K}_D(F(w), F(z)) \det F'(w)} \det F'(z)) d\mathcal{L}^{2n}(w),$$

which gives the formula in Proposition 12.1.10. □

It is clear that the domain D_2 of Example 12.1.8 is the image of $\mathbb{B}_2$ under the proper holomorphic map $F : \mathbb{B}_2 \longrightarrow D_2$, $F(z) = (z_1^2, z_2)$. Observe that Proposition 12.1.10 fails to hold for F. Nevertheless, the following more developed transformation rule for the Bergman kernels remains true also for proper holomorphic mappings.

Theorem 12.1.11 (cf. [50]). *Suppose G and D are bounded domains in $\mathbb{C}^n$ and that $F : G \longrightarrow D$ is a proper holomorphic mapping of G onto D of order m. Let $u := \det F'$ and let $\Phi_1, \dots, \Phi_m$ denote the m local inverses to F, defined locally on $D \setminus V$, where $V := \{F(z) : z \in G, u(z) = 0\}$. Moreover, let $U_k := \det \Phi_k'$. Then, the Bergman kernels transform by the rule*

$$\sum_{k=1}^{m} \boldsymbol{K}_G(z, \Phi_k(w))\overline{U_k(w)} = u(z)\boldsymbol{K}_D(F(z), w)$$

for all $z \in G$ and all $w \in D \setminus V$.

Proof. As the first step, we derive the following transformation rule for the Bergman projections:

$$\boldsymbol{P}_G(u \cdot h \circ F) = u \cdot (\boldsymbol{P}_D h) \circ F, \quad h \in L^2(D). \tag{12.1.7}$$

Obviously, (12.1.7) holds whenever $h \in L^2_h(D)$. So it is sufficient to show that $\boldsymbol{P}_G(u \cdot h \circ F) = 0$ for all $h \in L^2(D)$, h orthogonal to $L^2_h(D)$, i.e., $h \in L^2_h(D)^\perp$.

Let $g \in \mathcal{C}_0^\infty(D \setminus V)$ and $f \in L^2_h(G)$. Then,

$$\begin{aligned}&\int_G f(z)u(z)\overline{\frac{\partial g}{\partial w_j} \circ F(z)}d\mathcal{L}^{2n}(z)\\ &= \int_D \sum_{k=1}^m f \circ \Phi_k(w)\, \overline{u \circ \Phi_k(w)}|U_k(w)|^2 \overline{\frac{\partial g}{\partial w_j}(w)}d\mathcal{L}^{2n}(w), \quad 1 \le j \le n.\end{aligned}$$

Hence, integration by parts gives $\boldsymbol{P}_G(u \cdot \frac{\partial g}{\partial w_j} \circ F) = 0$, $1 \le j \le n$.

It remains to verify that

$$\hat{H} := \left\{\frac{\partial g}{\partial w_j} : g \in \mathcal{C}_0^\infty(D \setminus V),\ 1 \le j \le n\right\} \subset L^2_h(D)^\perp$$

is a dense subset of $L^2_h(D)^\perp$.

So, let $h \in L^2_h(D)^\perp \cap \hat{H}^\perp$. Then, h satisfies the Cauchy–Riemann equations on $D \setminus V$ (in the sense of distributions), since $\int_D h(w)\overline{\frac{\partial g}{\partial w_j}(w)}d\mathcal{L}^{2n}(w) = 0$ for all $g \in \mathcal{C}_0^\infty(D \setminus V)$, $1 \le j \le n$. Therefore, (cf. Appendix B.1.13), the function $h \in L^2(D)$ is holomorphic on $D \setminus V$. By Appendix B.1.14 we conclude that h extends holomorphically to the whole D, which gives $h = 0$.

We turn to the proof of the formula in Theorem 12.1.11.

Let $w \in D \setminus V$ with $w + (r\mathbb{D})^n \subset D \setminus V$ ($r > 0$) and choose the corresponding ψ as in Remark 12.1.5, i.e., $\boldsymbol{P}_D\psi = \boldsymbol{K}_D(\cdot, w)$. Applying (12.1.7), for $z \in G$ we obtain

$$\begin{aligned}u(z)\boldsymbol{K}_D(F(z), w) &= u(z)(\boldsymbol{P}_D\psi)(F(z)) = \boldsymbol{P}_G(u \cdot \psi \circ F)(z)\\ &= \int_G u(\zeta) \cdot \psi \circ F(\zeta)\overline{\boldsymbol{K}_G(\zeta, z)}d\mathcal{L}^{2n}(\zeta)\\ &= \int_D \sum_{k=1}^m \psi(\eta)\overline{U_k(\eta)\boldsymbol{K}_G(\Phi_k(\eta), z)}d\mathcal{L}^{2n}(\eta)\\ &= \sum_{k=1}^m \overline{U_k(w)\boldsymbol{K}_G(\Phi_k(w), z)}.\end{aligned}$$

□

Example 12.1.12 (cf. [209] and Exercise 12.10.3(a)). The transformation rule from Theorem 12.1.11 provides a useful tool to establish explicit formulas for the Bergman kernel. Observe, for example, that the domain

$$\mathcal{E}((1/2, 1/2)) = \{z \in \mathbb{C}^2 : |z_1| + |z_2| < 1\}$$

is the proper holomorphic image of the Thullen domain D_2 via $F(z) = (z_1, z_2^2)$ (for the definition of D_p see Example 12.1.8). Then, straightforward but tedious calculations lead to the following formula:

$$\boldsymbol{K}_{\mathcal{E}((1/2,1/2))}(z,w) = \frac{2}{\pi^2} \cdot \frac{3(1-\langle z,w\rangle)^2(1+\langle z,w\rangle) + 4z_1z_2\overline{w}_1\overline{w}_2(5-3\langle z,w\rangle)}{((1-\langle z,w\rangle)^2 - 4z_1z_2\overline{w}_1\overline{w}_2)^3}.$$

Fix points $z, w \in \mathcal{E}((1/2,1/2))$ and write, for abbreviation, $\xi_j := z_j\overline{w}_j$. Then, $\sqrt{|\xi_1|} + \sqrt{|\xi_2|} < 1$, and so $4|\xi_1\xi_2| < (1-|\xi_1|-|\xi_2|)^2$. Therefore, the numerator in the formula above allows the following estimate:

$$\begin{aligned}
3(1-\langle z,w\rangle)^2&(1+\langle z,w\rangle) + 4z_1z_2\overline{w}_1\overline{w}_2(5-3\langle z,w\rangle)\\
&= 3(1-\xi_1-\xi_2)(1-(\xi_1-\xi_2)^2) + 8\xi_1\xi_2\\
&\geq 3(1-|\xi_1|-|\xi_2|)(1-|\xi_1-\xi_2|^2) - 2(1-|\xi_1|-|\xi_2|)^2\\
&\geq 3(1-|\xi_1|-|\xi_2|)^2(1+|\xi_1-\xi_2|) - 2(1-|\xi_1|-|\xi_2|)^2\\
&> (1-|\xi_1|-|\xi_1|)^2 > 0.
\end{aligned}$$

Hence, the Bergman kernel function $\boldsymbol{K}_{\mathcal{E}((1/2,1/2))}$ has no zeros on $\mathcal{E}((1/2,1/2)) \times \mathcal{E}((1/2,1/2))$. However, there are zeros on the boundary; for instance take $((1,0), (-1,0))$.

Moreover, one has the following more general example:

Example 12.1.13. Let $D = D_p = \{z \in \mathbb{C}^2 : |z_1|^{2/p} + |z_2|^2 < 1\}$ be as in Example 12.1.8 with $p \in \mathbb{N}$. Then, there is the proper holomorphic mapping

$$F : D_p \longrightarrow G_p = \mathcal{E}((1/2,1/p)) = \{z \in \mathbb{C}^2 : |z_1| + |z_2|^{2/p} < 1\},$$
$$F(z_1,z_2) := (z_2^2, z_1).$$

Using Bell's transformation law (see Theorem 12.1.11), we obtain

$$\begin{aligned}
&\boldsymbol{K}_{G_p}((z_1^2,0),(w_1,0))2z_1\\
&\quad = \big(\boldsymbol{K}_D((z_1,0),(\sqrt{w_1},0)) - \boldsymbol{K}_D((z_1,0),(-\sqrt{w_1},0))\big)\frac{1}{2\sqrt{w_1}},
\end{aligned}$$

whenever $z_1 \in \mathbb{D}$, $w_1 \in \mathbb{D}_*$.

Now, applying Example 12.1.8 (see also [70]), it follows that

$$\boldsymbol{K}_{G_p}((z_1^2,0),(w_1^2,0))2z_1 = \frac{p+1}{2\overline{w}_1\pi^2}\big((1-z_1\overline{w}_1)^{-p-2} - (1+z_1\overline{w}_1)^{-p-2}\big).$$

Then, if $z_1 \neq 0$, the kernel function $\boldsymbol{K}_{G_p}((z_1^2,0)(w_1^2,0))$ has a zero iff $(1+x)^{p+2} = (1-x)^{p+2}$, where $x := z_1\overline{w}_1$. Observe that $\lambda \longmapsto \frac{1+\lambda}{1-\lambda}$ maps $\mathbb{D}$ biholomorphically to the right half-plane. Hence, $(\frac{1+\lambda}{1-\lambda})^{p+2} = 1$ has a non-zero solution iff $p > 2$.

We point out that also $(z_2,w_2) \longmapsto \boldsymbol{K}_{G_p}((0,z_2),(0,w_2))$ has zeros.

Other examples and methods that illustrate how to proceed may be found in [70] (see also [69]). For example, the following deflation identity can be helpful to discuss formulas for Bergman kernels.

Example 12.1.14. Fix a bounded domain $D \subset \mathbb{C}^n$, which is given as

$$D = \{z \in U : \varphi(z) < 1\},$$

where $\varphi \in \mathcal{C}(U, [0, \infty))$ for a suitable open neighborhood U of $\overline{D}$. Put

$$G_1 := \{(z, \zeta) \in D \times \mathbb{C}^1 : \varphi(z) + |\zeta|^{2/(p+q)} < 1\},$$
$$G_2 := \{(z, \zeta) \in D \times \mathbb{C}^2 : \varphi(z) + |\zeta_1|^{2/p} + |\zeta_2|^{2/q} < 1\},$$

where p, q are positive real numbers. Then, we have the following *deflation identity* (see [70])

$$\pi \boldsymbol{K}_{G_1}((z,0),(w,0)) = \frac{\pi^2 \Gamma(p+1)\Gamma(q+1)}{\Gamma(p+q+1)} \boldsymbol{K}_{G_2}((z,0,0),(w,0,0)), \quad z, w \in D. \tag{12.1.8}$$

In fact, the identity (12.1.8) holds because both sides represent the unique reproducing kernel for the Hilbert space $L_h^2(D, \pi(1-\varphi)^{p+q})$. To be more precise, fix an $h \in L_h^2(D)$. Then, h can also be thought to belong to $L_h^2(G_j)$, $j = 1, 2$. Therefore, in virtue of the reproducing property of the Bergman kernel function, we see that

$$h(z) = \int_{G_1} h(w) \boldsymbol{K}_{G_1}((z,0),(w,\zeta))\, d\mathcal{L}^{2n+2}(w,\zeta).$$

Observe that the fiber over a point $w \in D$ is a disc of radius $(1-\varphi(w))^{(p+q)/2}$. Therefore, applying the mean-value property for harmonic functions leads to

$$h(z) = \int_D h(w)(1-\varphi(w))^{p+q} \pi \boldsymbol{K}_{G_1}(z,0),(w,0))\, d\mathcal{L}^{2n}(w).$$

Hence, $\boldsymbol{K}_{G_1}(\cdot,0),(\cdot,0))$ is the reproducing kernel for $L_h^2(D, \pi(1-\varphi)^{p+q})$.

Similar reasoning leads to the same conclusion for the right side of (12.1.8), which finally proves the deflation identity.

Example 12.1.15. For example, let $G := \{z \in \mathbb{C}^2 : |z_1| + |z_2|^{1/2} < 1\}$. Recall that we know the formula for $\boldsymbol{K}_G(z,w)$ (see Example 12.1.13). Now, let $D = \mathbb{D}$ and $p = q = 2$. Applying the deflation method from above, we get

$$\pi \boldsymbol{K}_G((z,0),(w,0)) = \frac{\pi^2}{3!} \boldsymbol{K}_{G^*}((z,0,0),(w,0,0)), \quad z, w \in \mathbb{D},$$

where

$$G^* := \{z \in \mathbb{C}^3 : |z_1| + |z_2| + |z_3| < 1\}.$$

Observe that $\boldsymbol{K}_G\big((z,0),(w,0)\big) = 0$ for certain points $z, w \in \mathbb{D}$. Hence, we also have $\boldsymbol{K}_{G^*}\big((z,0,0),(w,0,0)\big) = 0$ (compare Example 12.1.12).

Example 12.1.16. Using Theorem 12.1.11, the following formula for the Bergman kernel function of $\mathbb{G}_n$ has been found in [157] (see Chapter 7 for the definition of $\mathbb{G}_n$). Recall that $\pi_n : \mathbb{D}^n \longrightarrow \mathbb{G}_n$ is a proper holomorphic mapping, where

$$\pi_n(\lambda_1,\ldots,\lambda_n) := \Big(\sum_{1\le j_1<\cdots<j_k\le n} \lambda_{j_1}\cdots\lambda_{j_k}\Big)_{k=1,\ldots,n}.$$

Then, Theorem 12.1.11 leads to

$$\boldsymbol{K}_{\mathbb{G}_n}(\pi_n(\lambda),\pi_n(\mu)) = \frac{\det\Big[\frac{1}{(1-\lambda_j\overline{\mu}_k)^2}\Big]_{1\le j,k\le n}}{\pi^n \det \pi_n'(\lambda)\overline{\det \pi_n'(\mu)}} = \frac{F_n(z,\overline{w})}{\pi^n \prod_{j,k=1}^n (1-\lambda_j\overline{\mu}_k)^2},$$
$$\lambda,\mu \in \mathbb{D}^n \setminus \mathcal{J}_n.$$

where

$$\mathcal{J}_n := \{\zeta \in \mathbb{C}^n : \det \pi_n'(\zeta) = 0\} = \{\zeta \in \mathbb{C}^n : \zeta_j = \zeta_k \text{ for some } j \ne k\}.$$

In particular,

$$\boldsymbol{K}_{\mathbb{G}_2}(\pi_2(\lambda),\pi_2(\mu)) = \boldsymbol{K}_{\mathbb{G}_2}(z,w) = \frac{F_2(z,\overline{w})}{\pi^2 \prod_{j,k=1}^2 (1-\lambda_j\overline{\mu}_k)^2}, \qquad \lambda,\mu \in \mathbb{D}^2,$$

where $(z,w) = (\pi_2(\lambda),\pi_2(\mu))$ and $F_2(z,w) := 2 - z_1w_1 + 2z_2w_2$. Moreover,

$$\boldsymbol{K}_{\mathbb{G}_3}(\pi_3(\lambda),\pi_3(\mu)) = \boldsymbol{K}_{\mathbb{G}_3}(z,w) = \frac{F_3(z,\overline{w})}{\pi^3 \prod_{j,k=1}^3 (1-\lambda_j\overline{\mu}_k)^2}, \qquad \lambda,\mu \in \mathbb{D}^3,$$

where, as above, $(z,w) = (\pi_3(\lambda),\pi_3(\mu))$ and

$$F_3(z,w) := 6-4z_1w_1-2z_2w_2+2z_1^2w_2+2z_2w_1^2-3z_1z_2w_3-3z_3w_1w_2+15z_3w_3$$
$$- z_1z_2w_1w_2 - 2z_1z_3w_1w_3 + 2z_1z_3w_2^2 + 2z_2^2w_1w_3 - 4z_2z_3w_2w_3 + 6z_3^2w_3^2.$$

Remark 12.1.17. To get the above formula Bell's transformation rule was used in [157]. There is another approach in [366]. Let S_n denote the permutation group of $\{1,\ldots,n\}$. Then, the proof here is based on an isomorphism between $L_h^2(\mathbb{G}_n)$ and

$$L^2_{h,\mathrm{anti}}(\mathbb{D}^n) := \{f \in L_h^2(\mathbb{D}^n) : f(z_\sigma) = \operatorname{sgn}\sigma f(z),\ z \in \mathbb{D}^n,\ \sigma \in S_n\},$$

where $z_\sigma := (z_{\sigma(1)},\ldots,z_{\sigma(n)})$, which finally leads to an orthonormal basis of $L_h^2(\mathbb{G}_n)$ given via so-called Schur polynomials.

It is easily seen that $\boldsymbol{K}_{\mathbb{G}_2}$ has no zeros on $\mathbb{G}_2 \times \mathbb{G}_2$ – simply use the description of $\operatorname{Aut}(\mathbb{G}_2)$ (Theorem 7.1.18) to reduce the discussion to the case $\mu_2 = 0$. What remains at the moment as a question is whether $\boldsymbol{K}_{\mathbb{G}_n}$ with $n \geq 3$ has zeros. We will come back to this kind of question soon (see Example 12.3.3).

Besides creating new explicit formulas for the Bergman kernel, Theorem 12.1.11 serves as a fundamental tool in the investigation of the boundary behavior of proper holomorphic mappings. Here we restrict ourselves to the following simple situation:

Theorem 12.1.18 (cf. [50]). *Let G_1 and G_2 be bounded complete Reinhardt domains in $\mathbb{C}^n$. Then, any biholomorphic mapping $F : G_1 \longrightarrow G_2$ extends holomorphically to an open neighborhood of $\overline{G}_1$. In particular, there exists a biholomorphic map $\widetilde{F}$ between open sets $U_j \supset \overline{G}_j$, $j = 1, 2$, with $\widetilde{F}|_{G_1} = F$.*

Proof. Let $K_j : G_j \times G_j \longrightarrow \mathbb{C}$ denote the Bergman kernel of G_j, $j = 1, 2$. Observe that the monomials z^α, $\alpha \in \mathbb{Z}_+^n$, form a complete orthogonal system for $L_h^2(G_j)$. Therefore,

$$K_j(z, \zeta) = \sum_{\alpha \in \mathbb{Z}_+^n} c_\alpha(j) z^\alpha \overline{\zeta}^\alpha, \quad z, \zeta \in G_j, \tag{12.1.9}$$

where $c_\alpha(j) := \|z^\alpha\|_{L^2(G_j)}^{-2} > 0$.

Put $u(z) := \det F'(z)$, $z \in G$, and $U(w) := \det(F^{-1})'(w)$, $w \in G_2$. Then, the transformation formula for the Bergman kernel (cf. Proposition 12.1.10) leads to

$$K_1(z, F^{-1}(w))\overline{U(w)} = u(z) K_2(F(z), w), \quad z \in G_1, \ w \in G_2. \tag{12.1.10}$$

If we differentiate (12.1.10) α-times, $\alpha \in \mathbb{Z}_+^n$, with respect to the variable $\overline{w}$, then we get

$$\frac{\partial^\alpha}{\partial \overline{w}^\alpha}\left(K_1(z, F^{-1}(\cdot))\overline{U}\right)\Big|_{w=0} = \alpha! c_\alpha(2) u(z) F^\alpha(z), \quad z \in G_1. \tag{12.1.11}$$

Choose $\varepsilon > 0$ such that $F^{-1}(0) \in \frac{1}{1+2\varepsilon} G \subset\subset G$. Then, because of (12.1.9), we have the following equalities:

$$K_1(z, \zeta) = K_1\left(z, (1+\varepsilon)\zeta \frac{1}{1+\varepsilon}\right) = K_1\left(z \frac{1}{1+\varepsilon}, (1+\varepsilon)\zeta\right), z \in G, \ \zeta \in \frac{1}{1+\varepsilon} G.$$

Thus, if $\zeta \in \frac{1}{1+\varepsilon} G$, then any derivative $\frac{\partial^\alpha}{\partial \overline{\zeta}^\alpha} K_1(z, \zeta)$, $\alpha \in \mathbb{Z}_+^n$, extends to the domain $(1+\varepsilon)G \supset \overline{G}$ as a holomorphic function in z. From (12.1.11) we now conclude that every function uF^α, $\alpha \in \mathbb{Z}_+^n$, extends as a holomorphic function g_α to a certain domain G_1' containing $\overline{G}_1$. In particular, with $\alpha = (0, \dots, 0)$ we find that there is a holomorphic function $\widetilde{u}$ on G_1' with $\widetilde{u}|_G = u$.

If $\widetilde{u}(z') \neq 0$ for a point $z' \in \partial G_1$, then $F := (F_1, \dots, F_n)$ is already holomorphic near z'. In the other case, when $\widetilde{u}(z') = 0$ ($z' \in \partial G$), we write near z':

$$\widetilde{u} = \prod_{\nu=1}^{N} p_\nu^{l_\nu} \quad \text{and} \quad g_{(0,\dots,1,\dots,0)} = \prod_{\nu=1}^{N_j} q_{j,\nu}^{k_{j,\nu}}, \quad \text{respectively,}$$

where the germs $p_{\nu,z'}$, $1 \leq \nu \leq N$, and $q_{j,\nu,z'}$, $1 \leq j \leq n$, $1 \leq \nu \leq N_j$, respectively, are irreducible elements in the UFD-ring $\mathcal{O}_{z'}$ of germs of holomorphic functions at z'. Moreover, they are pairwise not associated via units.

Now let j be fixed, $1 \leq j \leq n$. Near z' we write

$$\widetilde{u}^{m-1} g_{(0,\dots,m,\dots,0)} = \widetilde{u}^{m-1} \cdot (uF_j^m) = g^m_{(0,\dots,1,\dots,0)}, \quad m \in \mathbb{N},$$

$$\text{i.e.,} \quad \Big(\prod_{\nu=1}^{N} p_\nu^{l_\nu}\Big)^{m-1} \quad \text{divides} \quad \Big(\prod_{\nu=1}^{N_j} q_{j,\nu}^{k_{j,\nu}}\Big)^m, \quad m \in \mathbb{N}.$$

Since $p_{\nu,z'}$ are prime, it follows that $(m-1)l_\nu \leq mk_{j,\varkappa(\nu)}$ with $1 \leq \varkappa(\nu) \leq N_j$, $m \in \mathbb{N}$. Hence, $l_\nu \leq k_{j,\varkappa(\nu)}$, and so

$$\prod_{\nu=1}^{N} p_\nu^{l_\nu} \text{ divides } \prod_{\nu=1}^{N_j} q_{j,\nu}^{k_{j,\nu}},$$

i.e., $\widetilde{u}h_j = g_{(0,\dots,1,\dots,0)}$ with a suitable holomorphic function h_j near z'. Therefore, F_j is holomorphically extendable via h_j to a neighborhood of z'. Altogether, F admits a holomorphic continuation $\widetilde{F}$ to a neighborhood of $\overline{G}_1$. □

Remark. The above proof and Theorem 12.1.11 immediately show that the first part of Theorem 12.1.18 remains true even for proper holomorphic mappings (cf. Exercise 12.10.3(b)). In this form, the theorem is formulated in [50]. We mention that a similar result is true for proper mappings between circular domains that contain the origin (cf. [51]). The boundary behavior of holomorphic automorphisms of bounded Reinhardt domains was also studied by W. Kaup (cf. [290]) but his method of proof relies on the Lie theory.

We recall the form of the Bergman kernel in Example 12.1.7(b). The formula there appears as a special case of the following general result due to Bremermann (cf. [75]):

Theorem 12.1.19. *The Bergman kernel of any domain of the form $G_1 \times G_2 \subset \mathbb{C}^{n_1} \times \mathbb{C}^{n_2}$ is given by*

$$\boldsymbol{K}_{G_1\times G_2}((z_1,z_2),(w_1,w_2)) = \boldsymbol{K}_{G_1}(z_1,w_1)\cdot \boldsymbol{K}_{G_2}(z_2,w_2), \quad z_j, w_j \in G_j.$$

For the proof we will need the following preliminary result:

Lemma 12.1.20. *Suppose that $f \in L^2_h(G_1 \times G_2)$. Then, for every $z_2 \in G_2$ the function $f(\cdot, z_2)$ belongs to $L^2_h(G_1)$.*

Proof. Use, for example, the mean value inequality at z_2 for the plurisubharmonic functions $|f(z_1, \cdot)|^2$, $z_1 \in G_1$. □

Proof of the theorem. Suppose that $(\varphi_\nu)_{\nu \in \mathbb{Z}_+}$ and $(\psi_\mu)_{\mu \in \mathbb{Z}_+}$ are orthonormal bases of $L^2_h(G_1)$ and $L^2_h(G_2)$, respectively. By (12.1.1) it suffices to show that the functions

$$G_1 \times G_2 \ni (z_1, z_2) \longmapsto \varphi_\nu(z_1)\psi_\mu(z_2)$$

form an orthonormal basis of $L^2_h(G_1 \times G_2)$.

Since $(\varphi_\nu \psi_\mu)_{\nu,\mu}$ obviously form an orthonormal system in $L^2_h(G_1 \times G_2)$, it remains to prove its completeness. So, let $f \in L^2_h(G_1 \times G_2)$ be an arbitrary function with

$$\int_{G_1 \times G_2} f(z_1, z_2)\overline{\varphi_\nu(z_1)\psi_\mu(z_2)}\, d\mathcal{L}^{2n_1+2n_2}(z_1, z_2) = 0, \quad \nu, \mu \in \mathbb{Z}_+.$$

Then by the Fubini theorem we get

$$\int_{G_1} \overline{\varphi_\nu(z_1)} \int_{G_2} f(z_1, z_2)\overline{\psi_\mu(z_2)} d\mathcal{L}^{2n_2}(z_2) d\mathcal{L}^{2n_1}(z_1) = 0,$$

where $f(z_1, \cdot) \in L^2_h(G_2)$ for all $z_1 \in G_1$ according to Lemma 12.1.20.

To obtain $f = 0$ it suffices to show that the functions

$$g_\mu(z_1) := \int_{G_2} f(z_1, z_2)\overline{\psi_\mu(z_2)} d\mathcal{L}^{2n_2}(z_2)$$

are in $L^2_h(G_1)$. Choose a sequence $(D_j)_{j \in \mathbb{N}}$ of subdomains of G_2 with $D_j \subset\subset D_{j+1} \subset\subset G_2$, $\bigcup_{j=1}^\infty D_j = G_2$ and define on G_1

$$g_{\mu,j}(z_1) := \int_{D_j} f(z_1, z_2)\overline{\psi_\mu(z_2)} d\mathcal{L}^{2n_2}(z_2).$$

The functions $g_{\mu,j}$ are holomorphic on G_1 and we get

$$\begin{aligned}
&|g_\mu(z_1) - g_{\mu,j}(z_1)|^2 \\
&\quad \le \int_{G_2 \setminus D_j} |f(z_1, z_2)|^2 d\mathcal{L}^{2n_2}(z_2) \int_{G_2 \setminus D_j} |\psi_\mu(z_2)|^2 d\mathcal{L}^{2n_2}(z_2) \\
&\quad \le C(z^0) \cdot \int_{G_1 \times G_2} |f(z_1, z_2)|^2 d\mathcal{L}^{2n_1+2n_2}(z_1, z_2) \int_{G_2 \setminus D_j} |\psi_\mu(z_2)|^2 d\mathcal{L}^{2n_2}(z_2)
\end{aligned}$$

if $z_1 \in z_1^0 + (r\mathbb{D})^n \subset\subset G_1$. Here, we use the mean value inequality for the functions $|f(\cdot, z_2)|^2$, $z_2 \in G_2 \setminus D_j$. Hence, $g_{\mu,j} \underset{j\to\infty}{\overset{K}{\Longrightarrow}} g_\mu$, i.e., g_μ is holomorphic.

Moreover, the Hölder inequality implies the following estimate:

$$\begin{aligned}\int_{G_1} &|g_\mu(z_1)|^2 d\mathcal{L}^{2n_1}(z_1)\\ &\le \int_{G_1} \left[\int_{G_2} |f(z_1,z_2)|^2 d\mathcal{L}^{2n_2}(z_2) \int_{G_2} |\psi_\mu(z_2)|^2 d\mathcal{L}^{2n_2}(z_2)\right] d\mathcal{L}^{2n_1}(z_1)\\ &\le \|f\|^2_{L^2(G_1\times G_2)} \|\psi_\mu\|^2_{L^2(G_2)},\end{aligned}$$

which finally shows that $g_\mu \in L_h^2(G_1)$. □

Remark 12.1.21.

(a) Note that the use of the kernel function also gives us a direct proof of Theorem 12.1.19.

(b) Let G be a domain in $\mathbb{C}^{n_1}\times\mathbb{C}^{n_2}$ and let $\pi : \mathbb{C}^{n_1}\times\mathbb{C}^{n_2} \longrightarrow \mathbb{C}^{n_1}$ be the projection onto the first n_1 coordinates. For $f \in L_h^2(G)$, put

$$S(G, f) := \{z_1 \in \pi(G) : f(z_1,\cdot) \notin L_h^2(G \cap \pi^{-1}(z_1))\}.$$

Obviously, $S(G, f)$ has Lebesgue measure zero. According to Lemma 12.1.20, if G is of the form $G = G_1 \times G_2$, $G_j \subset \mathbb{C}^{n_j}$, then $S(G, f) = \varnothing$. In general, the set $S(G, f)$ can be a quite arbitrary set of Lebesgue measure zero; cf. [244, 245, 246]. For example, *if $S \subset \partial\mathbb{D}(r)$ is an arbitrary G_δ set, then $S = S(\mathbb{B}_2, f)$ for an existing $f \in L_h^2(\mathbb{B}_2)$.*

Using ideas analogous to the proof of Theorem 12.1.19 (cf. [369]), one can derive the following theorem (which we will need in the future):

Theorem 12.1.22. *Any holomorphic function f on a product domain $G_1 \times G_2$ can be written as a locally uniformly convergent series*

$$f(z_1, z_2) = \sum_{\nu=1}^{\infty} g_\nu(z_1) h_\nu(z_2)$$

with $g_\nu \in \mathcal{O}(G_1, \mathbb{C})$, $h_\nu \in \mathcal{O}(G_2, \mathbb{C})$.

Now, we want to describe how the Bergman kernels behave under other simple set-theoretic operations.

Theorem 12.1.23 (cf. [442, 443, 483]). *Let G be a domain in $\mathbb{C}^n$ that is the union of an increasing sequence $(G_\nu)_{\nu\in\mathbb{N}}$ of subdomains G_ν. Then $\boldsymbol{K}_G(z, w) =$* $\lim_{\nu\to\infty} \boldsymbol{K}_{G_\nu}(z, w)$, *$(z, w) \in G \times G$, and the convergence is locally uniform.*

Proof. Suppose the claim of the theorem fails to hold. Then, without loss of generality, we may assume

$$|\boldsymbol{K}_G(z_\nu, w_\nu) - \boldsymbol{K}_{G_\nu}(z_\nu, w_\nu)| \geq \varepsilon_0 > 0, \quad \text{where } z_\nu, w_\nu \in F_0 \subset\subset G. \qquad (*)$$

Now let F be an arbitrary compact subset of G. Then, $F \subset G_\nu$ if $\nu \geq \nu_0$ for a suitable ν_0 and, therefore, for $z, w \in F$ we have

$$\begin{aligned}|\boldsymbol{K}_{G_\nu}(z,w)| &\leq \sqrt{\boldsymbol{K}_{G_\nu}(z,z)}\sqrt{\boldsymbol{K}_{G_\nu}(w,w)} \\ &\leq \sqrt{\boldsymbol{K}_{G_{\nu_0}}(z,z)}\sqrt{\boldsymbol{K}_{G_{\nu_0}}(w,w)} \leq \sup\{\boldsymbol{K}_{G_{\nu_0}}(\zeta,\zeta) : \zeta \in F\} < \infty.\end{aligned}$$

Hence, the sequence $(K_\nu)_\nu$, where $K_\nu := \boldsymbol{K}_{G_\nu}$, is locally bounded on $G \times G$, which provides a subsequence $(K_{\nu_j})_j$ with $\lim_{j\to\infty} K_{\nu_j}(z,w) =: k(z,w)$. Here, the convergence is locally uniform; so the function $G \times \overline{G} \ni (z,w) \longmapsto k(z,\overline{w})$ is holomorphic.

Moreover, if D is any relatively compact subdomain of G with $G_{\nu_j} \supset D$, $j \geq j_0$, and if $w \in D$, then

$$\begin{aligned}\|k(\cdot,w)\|^2_{L^2(D)} &= \lim_{j\to\infty} \int_D |K_{\nu_j}(z,w)|^2 d\mathcal{L}^{2n}(z) \\ &\leq \liminf_{j\to\infty} \int_{G_{\nu_j}} |K_{\nu_j}(z,w)|^2 d\mathcal{L}^{2n}(z) \\ &= \lim_{j\to\infty} K_{\nu_j}(w,w) = k(w,w).\end{aligned}$$

Since this estimate is independent of D, we obtain $k(\cdot,w) \in L^2_h(G)$ for every $w \in G$.

As the last step we show that $k(\cdot,w)$ reproduces the functions of $L^2_h(G)$, i.e., $k(\cdot,w) = \boldsymbol{K}_G(\cdot,w)$.

Let $f \in L^2_h(G)$ and suppose that $w \in G_{\nu_j}$ if $j \geq j_0$. Then, for $j \geq j_0$, we get

$$\begin{aligned}f(w) &= \int_{G_{\nu_j}} f(z)\overline{K_{\nu_j}(z,w)}d\mathcal{L}^{2n}(z) \\ &= \int_G f(z)\overline{k(z,w)}d\mathcal{L}^{2n}(z) + \int_D f(z)[\overline{K_{\nu_j}(z,w)} - \overline{k(z,w)}]d\mathcal{L}^{2n}(z) \\ &\quad + \int_{G_{\nu_j}\setminus D} f(z)\overline{K_{\nu_j}(z,w)}d\mathcal{L}^{2n}(z) - \int_{G\setminus D} f(z)\overline{k(z,w)}d\mathcal{L}^{2n}(z)\end{aligned}$$

whenever $D \subset\subset G$ and $D \subset G_{\nu_j}$, $j \geq j_0$. The third summand can be estimated as follows:

$$\begin{aligned}\left|\int_{G_{\nu_j}\setminus D} f(z)\overline{K_{\nu_j}(z,w)}d\mathcal{L}^{2n}(z)\right| &\leq \|f\|_{L^2(G\setminus D)}\sqrt{K_{\nu_j}(w,w)} \\ &\leq \|f\|_{L^2(G\setminus D)}\sqrt{k(w,w)+1} \quad \text{if} \quad j \gg 1.\end{aligned}$$

If D and j are sufficiently large, then the last three summands become arbitrarily small, i.e., we obtain

$$f(w) = \int_G f(z)\overline{k(z,w)}d\mathcal{L}^{2n}(z).$$

Hence (K_{ν_j}) converges locally uniformly to $\boldsymbol{K}_G$ on $G \times G$, which contradicts assumption $(*)$. □

On the other hand, the situation for a decreasing sequence is more complicated (cf. [483]).

Theorem 12.1.24. *Let $(G_\nu)_{\nu\in\mathbb{N}}$ be a decreasing sequence of domains $G_\nu \subset \mathbb{C}^n$ with $G_\nu \supset G$, $\nu \in \mathbb{N}$. Then, $\boldsymbol{K}_G(z,w) = \lim_{\nu\to\infty} \boldsymbol{K}_{G_\nu}(z,w)$ locally uniformly on $G\times G$ if and only if for every $w \in G$, $\lim_{\nu\to\infty} \boldsymbol{K}_{G_\nu}(w,w) = \boldsymbol{K}_G(w,w)$.*

Proof. Only the sufficiency of the above condition has to be verified. Similar to the proof of Theorem 12.1.23, it follows that the sequence $(\boldsymbol{K}_{G_\nu})$ is locally bounded on $G \times G$. So, without loss of generality, we may assume that $\lim_{\nu\to\infty} \boldsymbol{K}_{G_\nu}(z,w) =: k(z,w)$ locally uniformly on $G \times G$.

As before, it turns out that

$$\|k(\cdot,w)\|^2_{L^2(G)} \leq \lim_{\nu\to\infty} \boldsymbol{K}_{G_\nu}(w,w) = k(w,w) = \boldsymbol{K}_G(w,w), \quad w \in G.$$

If $k(w,w) = 0$, then $k(\cdot,w) = 0 = \boldsymbol{K}_G(\cdot,w)$. But if $k(w,w) \neq 0$, then

$$\left\|\frac{k(\cdot,w)}{k(w,w)}\right\|_{L^2(G)} = \frac{1}{\sqrt{k(w,w)}}\frac{\|k(\cdot,w)\|_{L^2(G)}}{\sqrt{k(w,w)}} \leq \frac{1}{\sqrt{k(w,w)}} = \left\|\frac{\boldsymbol{K}_G(\cdot,w)}{\boldsymbol{K}_G(w,w)}\right\|_{L^2(G)}.$$

Therefore, by Remark 12.1.3 it follows that $k(\cdot,w) = \boldsymbol{K}_G(\cdot,w)$. □

For a more detailed discussion on how the Bergman kernels behave under intersections, see [483].

The boundary behavior of the Bergman kernel will play a fundamental role in the discussion about the Bergman completeness (cf. Chapter 13).

Theorem 12.1.25 (cf. [414]). *Let G be a bounded pseudoconvex domain in $\mathbb{C}^n$. Suppose that a point $z^* \in \partial G$ fulfills the following general* outer cone condition*:*

there exist $r \in (0,1]$, $a \geq 1$, and a sequence $(w^\nu)_{\nu\in\mathbb{N}}$ of points $w^\nu \notin G$ with $\lim_{\nu\to\infty} w^\nu = z^*$ and $G \cap \mathbb{B}(w^\nu, r\|w^\nu - z^*\|^a) = \varnothing$.

Then, for any sequence $(z^\nu)_{\nu\in\mathbb{N}} \subset G$ with $\lim_{\nu\to\infty} z^\nu = z^*$, we can find a function $f \in L^2_h(G)$, $\|f\|_{L^2(G)} = 1$, with $\sup\{|f(z^\nu)| : \nu \in \mathbb{N}\} = \infty$. Moreover, $\lim_{z\to z^*} \boldsymbol{K}_G(z,z) = \infty$.

Proof. The second assertion is a trivial consequence of the existence result. To establish this one, we need the following result due to H. Skoda (cf. [481, 413]), which we state here without proof:

Theorem* 12.1.26. *Let Ω be a pseudoconvex domain in $\mathbb{C}^n$, $\psi : \Omega \longrightarrow [-\infty, \infty)$ a psh function, $\eta > 1$, and $p \in \mathbb{N}$. Set $q := \min\{n, p-1\}$. Then, for any holomorphic functions $g_1, \dots, g_p \in \mathcal{O}(\Omega)$ and $f \in \mathcal{O}(\Omega)$ with*

$$A := \int_\Omega |f|^2 \Big(\sum_{j=1}^p |g_j|^2\Big)^{-q\eta-1} \exp(-\psi) d\mathcal{L}^{2n} < \infty,$$

there exist functions $h_1, \dots, h_p \in \mathcal{O}(\Omega)$ satisfying

(i) $f = \sum_{j=1}^p h_j g_j$,

(ii) $\displaystyle\int_\Omega \Big(\sum_{j=1}^p |h_j|^2\Big)\Big(\sum_{j=1}^p |g_j|^2\Big)^{-q\eta} \exp(-\psi) d\mathcal{L}^{2n} \le \frac{\eta}{\eta-1} A.$

We are going to prove Theorem 12.1.25 by contradiction. So, let $(z^\nu)_{\nu\in\mathbb{N}}$ be a sequence in G with $\lim_{\nu\to\infty} z^\nu = z^*$ such that every function $f \in L_h^2(G)$ is bounded along this sequence, i.e., $|f(z^\nu)| \le C(f)$, $\nu \in \mathbb{N}$. This means that the evaluation functionals $\delta_\nu : L_h^2(G) \longrightarrow \mathbb{C}$, $\delta_\nu(f) := f(z^\nu)$, are pointwise bounded. Hence, by the Banach–Steinhaus theorem, there exists a uniform bound C:

$$|f(z^\nu)| \le C\|f\|_{L^2(G)}, \quad \nu \in \mathbb{N},\ f \in L_h^2(G). \tag{12.1.12}$$

Now, we apply Skoda's theorem for the following special data: $\Omega = G$, $f \equiv 1$, $\psi \equiv 0$, $p = n$, $g_j(z) = z_j - w_j^\nu$, and $\eta > 1$ such that $2n < M := 2\eta(n-1) + 2 < 2n + 1/a$. So, we are able to find functions $h_1^\nu, \dots, h_n^\nu \in \mathcal{O}(G)$ with the following properties:

(i) $1 = \sum_{j=1}^n h_j^\nu(z)(z_j - w_j^\nu), \quad z \in G,\ \nu \in \mathbb{N}$,

(ii)
$$\int_G |h_j^\nu(z)|^2 \|z - w^\nu\|^{-2\eta(n-1)} d\mathcal{L}^{2n}(z)$$
$$\le \frac{\eta}{\eta-1} \int_G \|z - w^\nu\|^{-2\eta(n-1)-2} d\mathcal{L}^{2n}(z)$$

With $r_\nu := r\|z^* - w^\nu\|^a$ we obtain from inequality (ii) the following estimate

$$\begin{aligned}\|h_j^\nu\|^2_{L^2(G)} &\leq \hat{C}\Big(\int_{G\setminus\mathbb{B}(z^*,1/2)} \|z - w^\nu\|^{-M} d\mathcal{L}^{2n}(z) \\ &\quad + \int_{\mathbb{B}(1)\setminus\mathbb{B}(r_\nu)} \|z\|^{-M} d\mathcal{L}^{2n}(z)\Big) \\ &\leq C'(1 + r_\nu^{-M+(2n-1)+1}),\end{aligned} \tag{12.1.13}$$

and therefore $h_j^\nu \in L_h^2(G)$. Observe that $\delta := M - (2n-1) < 1 + 1/a$. Then (i), (12.1.12), and (12.1.13) lead to the following chain of inequalities:

$$\begin{aligned}1 &\leq \sum_{j=1}^n |h_j^\nu(z^\mu)|\,|z_j^\mu - w_j^\nu| \leq nC\sqrt{C'}(1 + r_\nu^{-\delta+1})\|z^\mu - w^\nu\| \\ &\leq C^*(1 + \|z^* - w^\nu\|^{a(1-\delta)})\|z^\mu - z^*\| + C^*\|z^* - w^\nu\| + C^*\|z^* - w^\nu\|^{a(1-\delta)+1} \\ &< 1 \ \text{ if } \nu \gg 1 \text{ and } \mu \gg 1,\end{aligned}$$

which gives the desired contradiction. □

Corollary 12.1.27 (cf. [415]). *Let $G \subset \mathbb{C}^n$ be a bounded pseudoconvex domain with a $\mathcal{C}^2$-boundary. Then, for any $\varepsilon \in (0,1)$, there exists a positive $\delta = \delta(\varepsilon)$ such that*

$$K_G(z,z) \geq \frac{1}{(\operatorname{dist}(z,\partial G))^{2\varepsilon}}, \quad z \in G,\ \operatorname{dist}(z,\partial G) < \delta.$$

Proof. The proof is based Theorem 12.1.26. Details are left to the reader. □

Remark 12.1.28.

(a) In fact, the above corollary remains true if the exponent 2ε is substituted just by 2 (see [182]). Its proof is based on induction and the extension theorem of Ohsawa–Takegoshi (cf. [405]).

(b) For a more precise description of the boundary behavior of K_G, the reader should consult the series of papers [130, 224, 226, 401, 417].

In this context the following localization result (see [401]) is rather useful:

Theorem 12.1.29. *Let $G \subset \mathbb{C}^n$ be a bounded pseudoconvex domain and let $z_0 \in \partial G$. Moreover, let $U_1 = U_1(z_0) \subset\subset U_2 = U_2(z_0)$ be open bounded neighborhoods of z_0. Then there exists a positive constant C such that for any connected component V of $G \cap U_2$ one has the inequality $K_V(z,z) \leq C\,K_G(z,z)$, $z \in V$.*

Proof. First, we choose a $\mathcal{C}^\infty$-function $\chi : \mathbb{C}^n \longrightarrow [0,1]$ such that χ is identical 1 in some neighborhood $\widetilde{U}_1$ of $\overline{U}_1$ and $\operatorname{supp}\chi \subset U_2$. Now fix a connected component V

and a point $a \in V$. Then there exists an $f \in L_h^2(V)$ satisfying $\|f\|_{L^2(V)} = 1$ and $|f(a)|^2 = \boldsymbol{K}_V(a,a)$. Extend f by taking its value 0 outside of V to the whole of $U_2 \cap G$. Then put $\alpha := \overline{\partial}(\chi f)$ on $U_2 \cap G$ and $\alpha = 0$ on $G \setminus U_2$. Then, α is a $\overline{\partial}$-closed $(0,1)$-form on G for which the following estimate holds:

$$\int_G \|z-a\|^{-2n}\|\alpha\|^2 d\mathcal{L}^{2n} \le C_1 \int_{V\setminus\widetilde{U}_1} \|z-a\|^{-2n}|f|^2 d\mathcal{L}^{2n} \le C_2,$$

where C_1, C_2 are independent of f and a. Applying Appendix B.7.15, one finds a $g \in \mathcal{C}^\infty(G)$ with $\overline{\partial} g = \alpha$ and $\int_G \|z-a\|^{-2n}|g|^2 d\mathcal{L}^{2n} \le C_3^2$. Put $\widehat{f} := \chi f - g$. Then, $\widehat{f} \in \mathcal{O}(G)$ and $\widehat{f}(a) = f(a)$ since the last integral is finite and so $g(a) = 0$. Moreover, we get

$$\begin{aligned}\|\widehat{f}\|_{L^2(G)} &\le \|f\|_{L^2(V)} + \|g\|_{L^2(G)} \le 1 + \sup_{z\in G}\|z-a\|^n C_3 \le 1 + C_3 \operatorname{diam} G^n \\ &=: \sqrt{C}.\end{aligned}$$

Hence, $\boldsymbol{K}_V(a,a) = |\widehat{f}(a)|^2 \le C\,\boldsymbol{K}_G(a,a)$. □

Remark 12.1.30.

(a) A similar result remains true for any plane domain D whose complement is not a polar set (see [376]).

(b) More precise information for the quotient $\boldsymbol{K}_G(z,z)/\boldsymbol{K}_V(z,z)$ if $z \longrightarrow z_0$ may be found in [231], under the hypothesis that $G \subset \mathbb{C}^n$ is a bounded pseudoconvex domain and that z_0 is a psh peak point.

Remark 12.1.31. Let D be as in Corollary 12.1.27 and $z^0 \in \partial D$ a boundary point. Then, $\lim_{z\to z^0} \boldsymbol{K}_D(z,z) = \infty$, i.e., the kernel function does not extend continuously to $(z^0, z^0) \in \partial D \times \partial D$ along the diagonal.

But if D is a smooth bounded strictly pseudoconvex domain, then $\boldsymbol{K}_D$ can be smoothly extended to $(\overline{D} \times \overline{D}) \setminus \nabla(\partial D)$, where $\nabla(\partial D) := \{(z,z) : z \in \partial D\}$ (see [294]). This result was generalized by Bell and Boas (see [49, 67]) to the following statements:

(a) Let $D \subset \mathbb{C}^n$ be a smoothly bounded pseudoconvex domain. Let $\Gamma_1, \Gamma_2 \subset \partial D$ be two open disjoint subsets of the boundary consisting of points of finite type (in the sense of d'Angelo). Then, $\boldsymbol{K}_D$ extends smoothly to $(D \cup \Gamma_1) \times (D \cup \Gamma_2)$.

(b) Let D be as in (a) and assume that D satisfies condition (R).[1] If Γ_1, Γ_2 are disjoint open subsets of ∂D and Γ_1 consists of points of finite type, then $\boldsymbol{K}_D$ extends smoothly to $(D \cup \Gamma_1) \times (D \cup \Gamma_2)$.

[1] A bounded domain is said to satisfy *condition (R)* if the Bergman projection $L^2(D) \longrightarrow L_h^2(D)$ sends $\mathcal{C}^\infty(\overline{D}) \cap L^2(D)$ to $\mathcal{C}^\infty(\overline{D}) \cap \mathcal{O}(D)$.

There was the question whether a similar extension phenomenon might be probable for any smoothly bounded pseudoconvex domain. That this is not true is shown by So-Chin Chen [100].

Theorem 12.1.32. *Let $D \subset \mathbb{C}^n$ be a smoothly bounded pseudoconvex domain, $n \geq 2$. Suppose that its boundary contains a non-trivial complex variety V. Then, $\boldsymbol{K}_D$ cannot be continuously extended to $(\overline{D} \times \overline{D}) \setminus \nabla(\partial D)$.*

Proof. Take a regular point $z^0 \in V$ and denote by $\mathfrak{n}$ the outward unit normal at z^0. Then, the smoothness assumption gives an $\varepsilon_0 > 0$ such that

$$w - \varepsilon\mathfrak{n} \in D, \quad \varepsilon \in (0, \varepsilon_0), \ w \in \partial D \cap \mathbb{B}(z^0, \varepsilon_0).$$

Moreover, we choose a holomorphic disc in V, i.e., a holomorphic embedding $\varphi : \mathbb{D} \longrightarrow V$, with $\varphi(0) = z^0$ and $\varphi(\mathbb{D}) \subset V \cap \mathbb{B}(z^0, \varepsilon_0)$.

Now assume that $\boldsymbol{K}_D \in \mathcal{C}(\overline{D} \times \overline{D} \setminus \nabla(\partial D))$. Then,

$$\sup_{|\lambda|=1/2} |\boldsymbol{K}_D(z^0, \varphi(\lambda))| < \infty.$$

Applying Theorem 12.1.25 and the maximum principle leads to

$$\begin{aligned}\sup_{|\lambda|=1/2} |\boldsymbol{K}_D(z^0, \varphi(\lambda))| &= \lim_{\varepsilon \to 0} \sup_{|\lambda|=1/2} |\boldsymbol{K}_D(z^0 - \varepsilon\mathfrak{n}, \varphi(\lambda) - \varepsilon\mathfrak{n})| \\ &\geq \lim_{\varepsilon \to 0} \boldsymbol{K}_D(z^0 - \varepsilon\mathfrak{n}, z^0 - \varepsilon\mathfrak{n}) = \infty;\end{aligned}$$

a contradiction. □

Example 12.1.33 (cf. [100]). Fix a smooth real-valued function $r : \mathbb{R} \longrightarrow \mathbb{R}$ with the following properties:

- $r(t) = 0$ if $t \leq 0$,
- $r(t) > 1$ if $t > 1$,
- $r''(t) \geq 100r'(t)$ for all t,
- $r''(t) > 0$ if $t > 0$,
- $r'(t) > 100$, if $r(t) > 1/2$.

For $s > 1$ put

$$\Omega := \Omega_s := \{z \in \mathbb{C}^2 : \varrho(z) < 0\}, \text{ where}$$
$$\varrho(z) := \varrho_s(z) := |z_1|^2 - 1 + r(|z_2|^2 - s^2).$$

Then, Ω_s is a smoothly bounded pseudoconvex domain in $\mathbb{C}^2$, it is convex and satisfies condition (R), and it is strictly pseudoconvex everywhere except on the set

$$\{z \in \mathbb{C}^2 : |z_1| = 1, \ 0 \leq |z_2| \leq s\} \subset \partial\Omega.$$

Obviously, this set contains non-trivial analytic varieties. So, Ω is an example for a domain treated in Theorem 12.1.32.

12.2 Minimal ball

Most of the domains for which an explicit formula for the Bergman kernel has been given were Reinhardt domains. Here, we describe the Bergman kernel function of a domain $\mathbb{M}_n \subset \mathbb{C}^n$ that is not biholomorphically equivalent to a Reinhardt domain if $n \geq 3$ and has almost everywhere a smooth boundary (for details see Remark 16.5.1).

For $n \geq 2$, define the *minimal ball*

$$\mathbb{M}_n := \{z \in \mathbb{C}^n : \|z\|_{\min} < 1\},$$

where

$$\begin{aligned}\|z\|_{\min} &= \left(\|z\|^2 + |z \bullet z|\right)^{1/2}\\ &= \left[\|x\|^2 + \|y\|^2 + [(\|x\|^2 - \|y\|^2)^2 + 4(x \bullet y)^2]^{1/2}\right]^{1/2},\\ &\qquad\qquad z = x + iy \in \mathbb{R}^n + i\mathbb{R}^n \cong \mathbb{C}^n.\end{aligned}$$

We mention that this definition is slightly different (by a dilatation) from the corresponding one in [269].

Remark 12.2.1 (the reader is asked to complete details).

(a) The norm $\frac{1}{\sqrt{2}}\|\ \|_{\min}$ is the *minimal complex norm* $q : \mathbb{C}^n \longrightarrow \mathbb{R}_+$ such that $q(z) \leq \|z\|$, $z \in \mathbb{C}^n$, and $q(x) = \|x\|$ for all $x \in \mathbb{R}^n$; cf. [210], see also [269], Exercise 2.1.15.

(b) $\|z\|_{\min} = \frac{1}{\sqrt{2}}\left(|z_1 - iz_2| + |z_1 + iz_2|\right), \quad z = (z_1, z_2) \in \mathbb{C}^2.$

(c) The mapping

$$\begin{aligned}\mathbb{M}_2 \ni (z_1, z_2) \longmapsto \tfrac{1}{\sqrt{2}}(z_1 - iz_2, z_1 + iz_2) \in \mathcal{E}((\tfrac{1}{2}, \tfrac{1}{2}))\\ = \{(w_1, w_2) \in \mathbb{C}^2 : |w_1| + |w_2| < 1\}\end{aligned}$$

is biholomorphic. Note that the domain on the right hand-side is a Reinhardt domain.

(d) If $n \geq 3$, then $\mathbb{M}_n$ *is never biholomorphically equivalent to a bounded Reinhardt domain.*

Indeed, assume that $F : \mathbb{M}_n \longrightarrow D$, D a bounded Reinhardt domain, is a biholomorphic mapping. Then, D is pseudoconvex. If $F(0) \neq 0$, then we may assume that $F_1(0) \neq 0$. Using an appropriate automorphism $D \ni z \overset{\varphi}{\longmapsto}$

$(e^{it}z_1, z_2, \dots, z_n) \in D$, $t \in \mathbb{R}$, we have $\varphi(F(0)) \neq F(0)$. Thus, $g := F^{-1} \circ \varphi \circ F \in \operatorname{Aut}(\mathbb{M}_n)$ and $g(0) \neq 0$; a contradiction to Theorem 16.5.12. Hence, $F(0) = 0$ and so D is a complete pseudoconvex Reinhardt domain. By the same reasoning, any $\Phi \in \operatorname{Aut}(D)$ fulfills $\Phi(0) = 0$. Then, in virtue of Sunada's work (see [492]), we know that $\operatorname{Aut}(D) = S^1 \times \cdots \times S^1$, i.e., $\operatorname{Aut}(D)$ depends on n real parameters, while $\operatorname{Aut}(\mathbb{M}_n)$ depends on $n(n-1)/2+1$ parameters (see Theorem 16.5.12). Since, by assumption, both automorphism groups are isomorphic, we get $n = n(n-1)/2 + 1$; a contradiction.

(e) For $z \in \partial\mathbb{M}_n$ we have: $\partial\mathbb{M}_n$ is smooth (real analytic smooth) at z iff $z \notin Q$, where $Q := \{z \in \mathbb{C}^n : z \bullet z = 0\}$.

(f) Smooth boundary points of $\mathbb{M}_n$ are strongly pseudoconvex.

Observe that $\mathbb{M}_n$ can be thought as a model for domains with non-smooth boundary.

For $n = 2$, a formula for its Bergman kernel function can be easily found via Example 12.1.12. The general case is contained in [400], where methods from Lie theory were used. We here follow the proof in [360], which works without results from Lie theory.

Theorem 12.2.2. *The Bergman kernel function of* $\mathbb{M} = \mathbb{M}_n$ $(n \geq 2)$ *is given by the following formula:*

$$\boldsymbol{K}_{\mathbb{M}}(z, w) = \frac{1}{n(n+1)\mathcal{L}^{2n}(\mathbb{M})} \frac{\sum_{j=0}^{\lfloor \frac{n}{2} \rfloor} \binom{n+1}{2j+1} X^{n-1-2j} Y^j (2nX - (n-2j)(X^2 - Y))}{(X^2 - Y)^{n+1}},$$

where $z, w \in \mathbb{M}$, $X = X(z, w) := 1 - \langle z, w \rangle$, *and* $Y = Y(z, w) := (z \bullet z)\overline{(w \bullet w)}$.

Proof. The main ideas of the proof are:

A) to establish a formula for the Bergman kernel function of the "cone-domain"

$$\mathfrak{M} := \{z \in \mathbb{C}^{n+1} \setminus \{0\} : \|z\| < 1, z \bullet z = 0\}$$

(observe that $\mathfrak{M}$ is an n-dimensional complex manifold),

B) to use the proper holomorphic mapping

$$\pi : \mathfrak{M} \longrightarrow \mathbb{M} \setminus \{0\}, \quad \pi(\widetilde{z}, z_{n+1}) := \widetilde{z},$$

to get a formula for the Bergman kernel function of $\mathbb{M}$.

Note that $\mathfrak{M}$ is an open subset of the n-dimensional complex manifold

$$\mathfrak{H} := \{z \in \mathbb{C}^{n+1} \setminus \{0\} : z \bullet z = 0\}$$

and that any $A \in \mathbb{O}(n+1; \mathbb{C})$ induces a biholomorphic mapping from $\mathfrak{H}$ onto $\mathfrak{H}$.

Now, we are going to present the proof in more details (*the reader who is mainly interested in the formula may skip these extremely technical details*).

Step 1°. First, we collect (without proofs) some information on $\mathfrak{M}$. It is clear that $\mathbb{O}(n+1;\mathbb{R})$ acts transitively on $\partial\mathfrak{M} \subset \mathfrak{H}$; thus, there exists a unique $\mathbb{O}(n+1;\mathbb{R})$-invariant measure μ on $\partial\mathfrak{M}$, induced by the Haar measure of $\mathbb{O}(n+1;\mathbb{R})$, with $\mu(\partial\mathfrak{M}) = 1$. We denote by $\mathfrak{P}^k(\mathfrak{M})$ the space of k-homogeneous polynomials on $\mathfrak{M}$, i.e., the restrictions of homogeneous polynomials of degree k on $\mathbb{C}^{n+1}$. It is known (see [240, 524]) that $\mathfrak{P}^k$ is spanned by a finite number of special monomials of the form $z \longmapsto \langle z, \xi\rangle^k$, where $\xi \in S^n := \partial\mathbb{B}^{\mathbb{R}}_{n+1}$, $\mathbb{B}^{\mathbb{R}}_{n+1} := \{x \in \mathbb{R}^{n+1} : \|x\| < 1\}$.

Moreover, if one puts $N(k,n) := \frac{(2k+n-1)(k+n-2)!}{k!(n-1)!}$, then one has the following identities:

$$\int_{\partial\mathfrak{M}} \langle z, \overline{w}\rangle^k \langle \zeta, w\rangle^\ell d\mu(w) = \begin{cases} \frac{\langle z,\overline{\zeta}\rangle^k}{N(k,n)}, & \text{if } k = \ell \\ 0, & \text{if } k \neq \ell \end{cases}, \tag{12.2.1}$$

where $z \in \mathfrak{M}$ and $\zeta \in \mathbb{C}^{n+1}$. Hence, if $f \in \mathfrak{P}^k(\mathfrak{M})$, then

$$f(z) = N(k,n) \int_{\partial\mathfrak{M}} f(w)\langle z, w\rangle^k d\mu(w). \tag{12.2.2}$$

Remark 12.2.3. To prove the above statements is outside of the scope of this book. Nevertheless, let us mention that these results are based on

- the knowledge on spherical harmonics (see [368, 23]):

  A function f on $S^n := \partial\mathbb{B}^{\mathbb{R}}_{n+1}$ is a *spherical harmonic of degree m* if f is the restriction of a homogeneous harmonic polynomial of degree m on $\mathbb{R}^{n+1}$. If $H_m(S^n)$ denotes the space of all spherical harmonics of degree m, then the Hilbert space $L^2(S^n)$ is equal to the orthogonal sum $\bigoplus_{m=0}^{\infty} H_m(S^n)$.

- properties of an integral transform:

  Recall that the so called Bargman-type transform is defined by

$$L^2(S^n) \ni f \longmapsto \left(\mathfrak{M} \ni z \longmapsto \hat{f}(z) := \int_{S^n} f(\xi) e^{\langle z,\xi\rangle} dS(\xi)\right),$$

  where dS is the standard surface measure on S^n. It is known (see [240, 524]), that the mapping $f \longmapsto \hat{f}$ leads to an isomorphism between $H_k(S^n)$ and $\mathfrak{P}^k(\mathfrak{M})$, i.e., properties of spherical harmonics are transformed to a certain information on $\mathfrak{M}$.

Step 2°. Observe that the n-form on $\mathbb{C}^{n+1}_*$

$$\widetilde{\alpha}(z) := \sum_{j=1}^{n+1} \frac{(-1)^{j+1}}{z_j} dz_1 \wedge \cdots \wedge \widehat{dz_j} \wedge \cdots \wedge dz_{n+1}$$

induces, by restriction, an $\mathbb{SO}(n+1;\mathbb{C})$-invariant holomorphic n-form α on the complex manifold $\mathfrak{H}$.

Indeed, fix a matrix $A \in \mathbb{SO}(n+1;\mathbb{C})$, a point $z \in \mathfrak{M}$ and put $w = Az$. Moreover, denote by $A_{j,k}$ the $n \times n$-matrix obtained from A by deleting the jth row and the kth column. Since $A \in \mathbb{SO}(n+1;\mathbb{C})$, Cramer's rule gives for the coefficients $a_{j,k}$ of the matrix A that

$$a_{j,k} = (-1)^{j+k} \det A_{j,k}. \tag{12.2.3}$$

Assume that $z_j \neq 0$. Then,

$$dz_j = -\sum_{j\neq\ell=1}^{n+1} \frac{z_\ell}{z_j} dz_\ell \quad \text{on } T_z\mathfrak{M}, \tag{12.2.4}$$

where $T_z\mathfrak{M}$ denotes the complex tangent space of $\mathfrak{M}$ at the point z. Let $A^*\alpha$ denote the pull-back of α, then

$$\begin{aligned}
(A^*\alpha)(z) &= \sum_{j=1}^{n+1} \frac{(-1)^{j-1}}{w_j} \sum_{k=1}^{n+1} \det A_{j,k} dz_1 \wedge \cdots \wedge \widehat{dz_k} \wedge \cdots \wedge dz_{n+1} \\
&= \sum_{k=1}^{n+1} (-1)^{k-1} \sum_{j=1}^{n+1} \frac{(-1)^{k+j}}{w_j} \det A_{j,k} dz_1 \wedge \cdots \wedge \widehat{dz_k} \wedge \cdots \wedge dz_{n+1} \\
&\overset{(12.2.3)}{=} \sum_{k=1}^{n+1} (-1)^{k-1} \sum_{j=1}^{n+1} \frac{a_{j,k}}{w_j} dz_1 \wedge \cdots \wedge \widehat{dz_k} \wedge \cdots \wedge dz_{n+1} \\
&\overset{(12.2.4)}{=} \sum_{k=1}^{n+1} (-1)^{k} \sum_{j=1}^{n+1} \frac{a_{j,k}}{w_j} dz_1 \wedge \cdots \wedge dz_{j-1} \wedge \Bigl(\sum_{\ell\neq j} \frac{z_\ell}{z_j} dz_\ell \Bigr) \\
&\qquad \wedge dz_{j+1} \wedge \cdots \wedge \widehat{dz_k} \wedge \cdots \wedge dz_{n+1} \\
&= \sum_{k=1}^{n+1} (-1)^{k-1} \sum_{j=1}^{n+1} (-1)^{j-k} \frac{a_{j,k}}{z_j w_j} z_k dz_1 \wedge \cdots \wedge \widehat{dz_j} \wedge \cdots \wedge dz_{n+1} \\
&= \sum_{j=1}^{n+1} \Bigl(\frac{(-1)^{j-1}}{w_j z_j} \sum_{k=1}^{n+1} a_{j,k} z_k \Bigr) dz_1 \wedge \cdots \wedge \widehat{dz_j} \wedge \cdots \wedge dz_{n+1} \\
&= \sum_{j=1}^{n+1} \frac{(-1)^{j-1}}{z_j} dz_1 \wedge \cdots \wedge \widehat{dz_j} \wedge \cdots \wedge dz_{n+1} = \alpha(z).
\end{aligned}$$

Step 3°. Next put

$$\omega(z)(V_1, \ldots, V_{2n-1}) := \alpha(z) \wedge \overline{\alpha(z)}(z, V_1, \ldots, V_{2n-1}), \quad z \in \partial\mathfrak{M},$$

where $(V_1, \ldots, V_{2n-1}) \in T_z(\partial\mathfrak{M})$. Observe that ω is a volume form on $T_z(\partial\mathfrak{M})$. Since $\alpha \wedge \overline{\alpha}$ is $\mathbb{SO}(n+1;\mathbb{C})$-invariant, ω is also $\mathbb{SO}(n+1;\mathbb{C})$-invariant. Hence, the measure on $\partial\mathfrak{M} \subset \mathfrak{H}$ induced by ω is proportional to the unique $\mathbb{SO}(n+1;\mathbb{R})$-invariant measure μ on $\partial\mathfrak{M}$ with $\mu(\partial\mathfrak{M}) = 1$. Put $\omega(\partial\mathfrak{M}) := \int_{\mathfrak{M}} \omega$.

Exploiting the definition of the form α leads to the following statement:

Lemma 12.2.4. *For any $\mathcal{C}^\infty$-function f on $\mathfrak{H}$, we have*

$$\int_{\mathfrak{H}} f(z)\alpha(z) \wedge \overline{\alpha(z)} = \omega(\partial\mathfrak{M}) \int_0^\infty t^{2n-3} \int_{\partial\mathfrak{M}} f(t\zeta)d\mu(\zeta)dt, \tag{12.2.5}$$

provided the integrals make sense.

Proof. Put $g(t,z) := tz$, where $z \in \mathbb{C}^{n+1}$ and $t \in (0,1)$, and set

$$\omega_j(z) := dz_1 \wedge \cdots \wedge \widehat{dz_j} \wedge \cdots \wedge dz_{n+1}, \quad j = 1, \ldots, n+1.$$

Then,

$$\alpha(z) \wedge \overline{\alpha}(z) = \sum_{j,k=1}^{n+1} \frac{(-1)^{j+k}}{z_j \overline{z}_k} \omega_j(z) \wedge \omega_k(\overline{z}).$$

Thus, we get

$$g^*(\alpha \wedge \overline{\alpha})(t,z) = \sum_{j,k=1}^{n+1} \frac{(-1)^{j+k}}{t^2 z_j \overline{z}_k} g^*(\omega_j)(t,z) \wedge g^*(\omega_k)(t,\overline{z}).$$

For $r, s \in \mathbb{N}$, $r < s$, calculation (use induction) leads to

$$\begin{aligned} g^*(dz_r \wedge \cdots \wedge dz_s) &= t^{s-r+1} dz_r \wedge \cdots \wedge dz_s \\ &\quad + t^{r-s} \sum_{\ell=r}^{s} (-1)^{\ell-r} z_\ell dt \wedge dz_r \wedge \cdots \wedge \widehat{dz_\ell} \wedge \cdots \wedge dz_s. \end{aligned}$$

Therefore, we get

$$g^*(\omega_j) = t^n \omega_j + t^{n-1} dt \wedge \left(\sum_{\ell=1}^{j-1} (-1)^{\ell-1} z_\ell \omega_{\ell,j} + \sum_{\ell=j+1}^{n+1} (-1)^\ell z_\ell \omega_{j,\ell} \right), \tag{12.2.6}$$

where

$$\omega_{\ell,j} := dz_1 \wedge \cdots \wedge \widehat{dz_\ell} \wedge \cdots \wedge \widehat{dz_j} \wedge \cdots \wedge dz_{n+1}.$$

In a similar way one has

$$g^*(\overline{\omega}_j) = t^n \omega_j + t^{n-1} dt \wedge \left(\sum_{\ell=1}^{j-1} (-1)^{\ell-1} \overline{z}_\ell \overline{\omega}_{\ell,j} + \sum_{\ell=j+1}^{n+1} (-1)^\ell \overline{z}_\ell \overline{\omega}_{j,\ell} \right). \tag{12.2.7}$$

Now put

$$\widetilde{\omega}_{j,k}(z) := \sum_{ell=1}^{j-1} (-1)^{\ell-1} z_\ell \omega_{\ell,j}(z) \wedge \overline{\omega_k(z)} + \sum_{\ell=j+1}^{n+1} (-1)^{\ell} z_\ell \omega_{j,\ell} \wedge \overline{\omega_k(z)}$$

$$+ \sum_{q=1}^{k-1} (-1)^{n+q-1} \overline{z}_q \omega_j(z) \wedge \overline{\omega_{q,k}(z)} + \sum_{q=k+1}^{n+1} (-1)^{n+q} \overline{z}_q \omega_j(z) \wedge \overline{\omega_{k,q}(z)}.$$

Applying (12.2.6) and (12.2.7), one computes that

$$g^*(\omega_j)(tz) \wedge g^*(\omega_k)(t\overline{z}) = t^{2n-1} dt \wedge \widetilde{\omega}_{j,k}.$$

Therefore, one obtains that

$$g^*(\alpha \wedge \overline{\alpha}) = t^{2n-3} dt \wedge \sum_{j,k=1}^{n+1} \frac{(-1)^{j+k}}{z_j \overline{z}_k} \widetilde{\omega}_{j,k}(z).$$

On the other hand, if $z \in \partial\mathfrak{M}$ and $(V_1, \dots, V_{2n-1}) \in T_z \partial\mathfrak{M}$, then

$$\omega(z)(V_1, \dots, V_{2n-1}) = \left(\sum_{j,k=1}^{n+1} \frac{(-1)j + k}{z_j \overline{z}_k} \omega_j \wedge \omega_k \right) (z, V_1, \dots, V_{2n-1}).$$

Hence, if $z = (z_1, \dots, z_n)$ and $V_0 := (z_1, \dots, z_n, \overline{z}_1, \dots, \overline{z}_n)$, then

$$\omega_j \wedge \omega_k(\overline{z})(z, V_1, \dots, V_{2n-1}) = \det(V_0, \dots, V_{2n-1}).$$

Expanding the determinant in terms of cofactors along the first column yields

$$\omega_j(z) \wedge \omega_k(\overline{z})(z, V_1, \dots, V_{2n-1}) = \widetilde{\omega}_{j,k}(z) \det(z, V_1, \dots, V_{2n-1}),$$

which shows that

$$\omega(z) = \sum_{j,k=1}^{n+1} \frac{(-1)^{j+k}}{z_j \overline{z}_k} \widetilde{\omega}_{j,k}(z) \text{ and } g^*(\alpha \wedge \overline{\alpha}) = t^{2n-3} dt \wedge \omega(z).$$

It remains to observe that g induces a $\mathcal{C}^\infty$-diffeomorphism from $(0,1) \times \partial\mathfrak{M}$ to $\mathfrak{M}$, implying that

$$\int_{\mathfrak{S}} f(z) \alpha(z) \wedge \overline{\alpha(z)} = \int_0^\infty t^{2n-3} \int_{\partial\mathfrak{M}} f(t\xi) dt \wedge \omega(\xi),$$

which implies the formula in the lemma by normalization. □

Step 4°. Next, let $f \in \mathfrak{P}^k(\mathfrak{M})$ be a homogeneous polynomial of degree k. Fix a $z \in \mathfrak{M}$. Then,

$$f(z) = C(k,n) \int_{\mathfrak{M}} \langle z, w \rangle^k f(w) \alpha(w) \wedge \overline{\alpha(w)}, \tag{12.2.8}$$

where

$$C(k,n) := \frac{2(2k+n-1)(n+k-1)!}{\omega(\partial\mathfrak{M})(n-1)!k!}.$$

In fact, using (12.2.5) and homogeneity, we have

$$\int_{\mathfrak{M}} \langle z, w \rangle^k f(w) \alpha(w) \wedge \overline{\alpha(w)} = \omega(\partial\mathfrak{M}) \int_0^1 t^{2n-3+2k} dt \int_{\partial\mathfrak{M}} \langle z, w \rangle^k f(w) d\mu(w).$$

Then, by (12.2.2) it follows that

$$f(z) = \frac{N(k,n)}{\omega(\partial\mathfrak{M}) \int_0^1 t^{2n-3+2k} dt} \int_{\mathfrak{M}} \langle z, w \rangle^k f(w) \alpha(z) \wedge \overline{\alpha(w)},$$

which is exactly (12.2.8).

Step 5°. In order to be able to continue, we need the following

Lemma 12.2.5. *Let $f \in \mathcal{O}(\mathfrak{M})$. Then, there are homogeneous polynomials f_k of degree k, $k \in \mathbb{Z}_+$, such that*

$$f(z) = \sum_{k=0}^{\infty} f_k(z), \quad z \in \mathfrak{M},$$

and the convergence is uniform on compact subsets of $\mathfrak{M}$.

The reader unfamiliar with complex analysis on complex spaces may skip the proof of this lemma. In any case, the necessary details may be found, for example, in [204, 289], and [203].

Proof. Observe that $A := \{0\} \cup \mathfrak{M}$ is an analytic subset of $\mathbb{B} = \mathbb{B}_{n+1}$. It is clear that 0 is the only singularity of A; it is a normal singularity for $n \geq 2$. Hence, A is a normal complex space and the function f extends holomorphically to a function $\tilde{f} \in \mathcal{O}(A)$. Applying Cartan's Theorem B, we find an $\hat{f} \in \mathcal{O}(\mathbb{B})$, $\hat{f}|_A = \tilde{f}$. Therefore, there are homogeneous polynomials f_k of degree k such that $\hat{f}(z) = \sum_{k=0}^{\infty} f_k(z)$, $z \in \mathbb{B}$, and the convergence is locally uniform. □

Step 6°. Denote by $L^2(\mathfrak{M})$ the space of all measurable functions on $\mathfrak{M}$ satisfying

$$\|f\|_{L^2(\mathfrak{M})} := \left(\int_{\mathfrak{M}} |f(z)|^2 \frac{\alpha(z) \wedge \overline{\alpha(z)}}{(-1)^{\frac{n(n+1)}{2}} (2i)^n} \right)^{1/2} < \infty,$$

and let $L^2_h(\mathfrak{M}) := L^2(\mathfrak{M}) \cap \mathcal{O}(\mathfrak{M})$. Then, we have the following formula for the Bergman kernel function of the space $L^2_h(\mathfrak{M})$:

Proposition 12.2.6. *The Bergman kernel function is given by*

$$\boldsymbol{K}_{\mathfrak{M}}(z,w) = \frac{2(-1)^{\frac{n(n+1)}{2}}(2i)^n}{\omega(\partial\mathfrak{M})} \left(\frac{(n-1)}{(1-\langle z,w\rangle)^n} + \frac{2n\langle z,w\rangle}{(1-\langle z,w\rangle)^{n+1}} \right), \quad z,w \in \mathfrak{M}.$$

Proof. Fix an $f \in L^2_h(\mathfrak{M})$ and a point $z \in \mathfrak{M}$. Then, applying Lemma 12.2.5 and (12.2.8), we obtain

$$\begin{aligned} f(z) = \sum_{k=0}^{\infty} f_k(z) &= \sum_{k=0}^{\infty} C(k,n) \int_{\mathfrak{M}} \langle z,w\rangle^k f_k(w)\alpha(w) \wedge \overline{\alpha(w)} \\ &= \int_{\mathfrak{M}} \boldsymbol{K}_{\mathfrak{M}}(z,w) f(w) \frac{\alpha(w) \wedge \overline{\alpha(w)}}{(-1)^{\frac{n(n+1)}{2}}(2i)^n}. \end{aligned}$$

Put $A := (-1)^{n(n+1)/2}(2i)^n$. In virtue of the orthogonality of the homogeneous polynomials of different degree we get, exploiting the last formula,

$$\begin{aligned} K_{\mathfrak{M}}(z,w) &= A \sum_{k=0}^{\infty} C(k,n)\langle z,w\rangle^k = A \sum_{k=0}^{\infty} \frac{2(n-1+2k)(n+k-1)!}{\omega(\partial\mathfrak{M})(n-1)!k!} \\ &= \frac{A}{\omega(\partial\mathfrak{M})(n-1)!} \sum_{k=0}^{\infty} \frac{(n-1+2k)(n+k-1)!}{k!} \langle z,w\rangle^k \\ &= B \left(\sum_{k=0}^{\infty} \frac{(n-1)(n+k-1)!}{k!} \langle z,w\rangle^k + 2 \sum_{k=1}^{\infty} \frac{(n+k-1)!}{(k-1)!} \langle z,w\rangle^k \right) \\ &= B \left(\frac{(n-1)(n-1)!}{(1-\langle z,w\rangle)^n} + 2 \frac{\langle z,w\rangle n!}{(1-\langle z,w\rangle)^{n+1}} \right), \end{aligned}$$

where $B := \frac{2A}{\omega(\partial\mathfrak{M})(n-1)!}$. Hence, Proposition 12.2.6 has been verified. □

To summarize, we have finished step A of the proof of Theorem 12.2.2.

Now we continue with step B.

Step 7°. Recall that $\pi : \mathfrak{M} \longrightarrow \mathbb{M} \setminus \{0\}$, $\pi(\tilde{z}, z_{n+1}) := \tilde{z}$, $z = (\tilde{z}, z_{n+1}) \in \mathfrak{M}$, is a proper map of degree 2. Let W be its branching locus and put $V := \pi(W) \cup \{0\}$. Denote the local inverses of π by φ and ψ. For $z \in \mathbb{M} \setminus V$ they are given by

$$\varphi(z) = (z, i\sqrt{z \bullet z}), \quad \psi(z) = (z, -i\sqrt{z \bullet z}).$$

Then, a calculation leads to the following description of the pull-backs of α under φ and ψ on $\mathbb{M} \setminus V$:

$$\begin{aligned} \varphi^*(\alpha) &= \frac{n+1}{i\sqrt{z \bullet z}}(-1)^n dz_1 \wedge \cdots \wedge dz_n, \\ \psi^*(\alpha) &= \frac{n+1}{-i\sqrt{z \bullet z}}(-1)^n dz_1 \wedge \cdots \wedge dz_n. \end{aligned} \tag{12.2.9}$$

Indeed, observe that the pull-back of α under (for example) φ is given by

$$\begin{aligned} \varphi^*(\alpha) &= \sum_{j=1}^{n+1} \frac{(-1)^{j-1}}{w_j} dw_1 \wedge \cdots \wedge \widehat{dw_j} \wedge \cdots \wedge dw_{n+1} \\ &= \sum_{j=1}^{n} \frac{(-1)^{j-1}}{w_j} dz_1 \wedge \cdots \wedge \widehat{dz_j} \wedge dz_n \wedge dw_{n+1} + \frac{(-1)^n}{w_{n+1}} dz_1 \wedge \cdots \wedge dz_n, \end{aligned}$$

where the coordinates in $\mathfrak{M}$ and in $\mathbb{M}$ are denoted by w_j and z_j, respectively. Now take the j-th summand of the first sum. Then:

$$\begin{aligned} &\frac{(-1)^{j-1}}{z_j} dz_1 \wedge \cdots \wedge \widehat{dz_j} \wedge \cdots \wedge dz_n \wedge d\varphi_{n+1} \\ &\qquad = \frac{(-1)^{j-1}}{z_j} dz_1 \wedge \cdots \wedge \widehat{dz_j} \wedge \cdots \wedge dz_n \wedge \left(-\sum_{k=1}^{n} \frac{z_k}{w_{n+1}} dz_k \right) \\ &\qquad = \frac{(-1)^j}{w_{n+1}} dz_1 \wedge \cdots \wedge \widehat{dz_j} \wedge \cdots \wedge dz_n \wedge dz_j \\ &\qquad = \frac{(-1)^n}{w_{n+1}} dz_1 \wedge \cdots \wedge dz_j \wedge \cdots \wedge dz_n. \end{aligned}$$

Hence, one gets

$$\varphi^*(\alpha) = (-1)^n \frac{1+n}{w_{n+1}} dz_1 \wedge \cdots \wedge dz_n = (-1)^n \frac{1+n}{i\sqrt{z \bullet z}}.$$

In a similar way, one may verify the formula for $\psi^*(\alpha)$.

Step 8°. Let $\boldsymbol{P}_{\mathfrak{M}}$ denote the Bergman projection on $\mathfrak{M}$ and $\boldsymbol{P}_{\mathbb{M}}$ the Bergman projection on $\mathbb{M}$. Then, we have the following relation:

Lemma 12.2.7. *Let $h \in L^2(\mathbb{M})$. Then,*

$$\boldsymbol{P}_{\mathfrak{M}}(\chi \cdot h \circ \pi)(z) = z_{n+1} \boldsymbol{P}_{\mathbb{M}}(h)(\pi(z)), \quad z \in \mathfrak{M},$$

where $\chi(z) := z_{n+1}$, $z \in \mathfrak{M}$.

Proof. In a first step, we will show that the lemma is true for functions $h \in L^2_h(\mathbb{M})$. So let us take such an h. Then, in virtue of (12.2.9), one has

$$\begin{aligned}\int_{\mathfrak{M}} |z_{n+1}(h \circ \pi)(z)|^2 \alpha(z) \wedge \overline{\alpha(z)} &= \int_{\mathfrak{M}\setminus W} |z_{n+1}(h \circ \pi)(z)|^2 \alpha(z) \wedge \overline{\alpha(z)} \\ &= \int_{\mathbb{M}\setminus V} |\varphi_{n+1}(w)h(w)|^2 \varphi^*(\alpha)(w) \wedge \varphi^*(\overline{\alpha})(w) \\ &\quad + \int_{\mathbb{M}\setminus V} |\psi_{n+1}(w)h(w)|^2 \psi^*(\alpha)(w) \wedge \psi^*(\overline{\alpha})(w) \\ &= 2(n+1)^2 \int_{\mathbb{M}\setminus V} |h(w)|^2 d\mathcal{L}^{2n}(w) < \infty.\end{aligned}$$

Thus, $\chi h \circ \pi \in L^2(\mathfrak{M}, \alpha \wedge \overline{\alpha})$.

Now assume that $f \in L^2(\mathfrak{M}, \alpha \wedge \overline{\alpha})$. Fix a function $g \in \mathcal{C}_0^\infty(\mathbb{M} \setminus V)$. Then,

$$\begin{aligned}\int_{\mathfrak{M}} f(z) \overline{z_{n+1} \left(\frac{\partial g}{\partial w_j} \circ \pi \right)(z)} \alpha(z) \wedge \overline{\alpha(z)} \\ = (n+1)^2 \Bigg(\int_{\mathbb{M}} \frac{(f \circ \varphi)(w)}{\varphi_{n+1}(w)} \overline{\frac{\partial g}{\partial w_j}(w)} d\mathcal{L}^{2n}(w) \\ + \int_{\mathbb{M}} \frac{(f \circ \psi)(w)}{\psi_{n+1}(w)} \overline{\frac{\partial g}{\partial w_j}(w)} d\mathcal{L}^{2n}(w) \Bigg).\end{aligned}$$

Now applying integration by parts implies that

$$\boldsymbol{P}_{\mathfrak{M}} \left(\chi \frac{\partial g}{\partial w_j} \circ \pi \right) = 0, \quad j = 1, \dots, n.$$

It remains to recall that the space $\{\frac{\partial g}{\partial w_j} : g \in \mathcal{C}_0^\infty(\mathbb{M} \setminus V)\}$ is dense in the orthogonal complement of $L^2_h(\mathbb{M})$ in $L^2(\mathbb{M})$. Therefore, the lemma holds. □

Step 9°. Now applying Lemma 12.2.7, we find a way to express the Bergman kernel function of $\mathbb{M}$ in terms of the Bergman kernel function for $\mathfrak{M}$.

Lemma 12.2.8. *Let φ and ψ be the local inverses from above. Then,*

$$z_{n+1} \boldsymbol{K}_{\mathbb{M}}(\pi(z), w) = (n+1)^2 \left(\frac{\boldsymbol{K}_{\mathfrak{M}}(z, \varphi(w))}{\overline{\varphi_{n+1}(w)}} + \frac{\boldsymbol{K}_{\mathfrak{M}}(z, \psi(w))}{\overline{\psi_{n+1}(w)}} \right),$$
$$z \in \mathfrak{M}, \ w \in \mathbb{M} \setminus V.$$

Proof. Fix a $w \in \mathbb{M} \setminus V$ and choose an $r > 0$ such that $\mathbb{P}(w,r) \subset\subset \mathbb{M} \setminus V$. In view of Remark 12.1.5, we find a $\mathcal{C}^\infty$-function $u : \mathbb{C}^n \longrightarrow [0,\infty)$, $\operatorname{supp} u \subset \mathbb{P}(w,r)$, such that

$$f(w) = \int_{\mathbb{M}} f(z)u(z)d\mathcal{L}^{2n}(z), \quad f \in \mathcal{O}(\mathbb{M}).$$

Therefore,

$$\boldsymbol{K}_{\mathbb{M}}(\cdot, w) = \boldsymbol{P}_{\mathbb{M}}(u).$$

Applying Lemma 12.2.7, it follows that

$$\begin{aligned}
z_{n+1}\boldsymbol{K}_{\mathbb{M}}(\pi(z), w) &= z_{n+1}\boldsymbol{P}_{\mathbb{M}}\big(u\big)(\pi(z)) = \boldsymbol{P}_{\mathfrak{M}}\big(\chi \cdot u \circ \pi\big)(z) \\
&= \int_{\mathfrak{M}} \zeta_{n+1} u \circ \pi(\zeta)\boldsymbol{K}_{\mathfrak{M}}(z,\zeta)\frac{\alpha(\zeta)\wedge\overline{\alpha(\zeta)}}{(-1)^{\frac{n(n+1)}{2}}(2i)^n} \\
&= (n+1)^2 \int_{\mathbb{M}\setminus V} u(\eta)\left(\frac{\boldsymbol{K}_{\mathfrak{M}}(z,\varphi(\eta))}{\overline{\varphi_{n+1}(\eta)}} + \frac{\boldsymbol{K}_{\mathfrak{M}}(z,\psi(\eta))}{\overline{\psi_{n+1}(\eta)}}\right) d\mathcal{L}^{2n}(\eta) \\
&= (n+1)^2 \left(\frac{\boldsymbol{K}_{\mathfrak{M}}(z,\varphi(w))}{\overline{\varphi_{n+1}(\eta)}} + \frac{\boldsymbol{K}_{\mathfrak{M}}(z,\psi(w))}{\overline{\psi_{n+1}(\eta)}}\right), \quad z \in \mathfrak{M}.
\end{aligned}$$

Hence, the lemma is proved. □

Step 10°. Finally, we are in a position to finish the proof of Theorem 12.2.2. According to Lemma 12.2.6, we have $\boldsymbol{K}_{\mathfrak{M}}(z,w) = Ch(\langle z,w\rangle)$, where

$$C = \frac{2(2i)^n(-1)^{\frac{n(n+1)}{2}}}{\omega(\partial\mathfrak{M})} \quad \text{and} \quad h(t) := \frac{2n}{(1-t)^{n+1}} - \frac{n+1}{(1-t)^n}, \ t \in \mathbb{C}.$$

Hence,

$$\boldsymbol{K}_{\mathfrak{M}}(\varphi(z),\varphi(w)) = Ch(x), \quad \boldsymbol{K}_{\mathfrak{M}}(\varphi(z),\psi(w)) = Ch(y), \tag{12.2.10}$$

where

$$x := \langle z,w\rangle + t, \ t := \varphi_{n+1}(z)\overline{\varphi_{n+1}(w)}, \text{ and } y := \langle z,w\rangle - t.$$

In virtue of Lemma 12.2.8, we get

$$\boldsymbol{K}_{\mathbb{M}}(z,w) = C(n+1)^2\left(\frac{h(x)-h(y)}{t}\right).$$

Using the abbreviation $r := 1 - \langle z,w\rangle$, the last expression can be written as

$$Q := \frac{h(x)-h(y)}{t} = 2n\frac{(r+t)^{n+1}-(r-t)^{n+1}}{t(r^2-t^2)^{n+1}} - (n+1)\frac{(r+t)^n-(r-t)^n}{t(r^2-t^2)^n}.$$

Then,

$$Q = \frac{2n}{(r^2-t^2)^{n+1}} 2 \sum_{k=0}^{\lfloor \frac{n}{2} \rfloor} \binom{n+1}{2k+1} r^{n-2k} t^{2k} - \frac{n+1}{(r^2-t^2)^n} 2 \sum_{k=0}^{\lfloor \frac{n-1}{2} \rfloor} \binom{n}{2k+1} r^{n-2k-1} t^{2k}.$$

Since $\binom{n}{2k+1} = \frac{n-2k}{n+1}\binom{n+1}{2k+1}$, we proceed with our calculations and get

$$Q = \frac{2}{(r^2-t^2)^{n+1}} \sum_{k=0}^{\lfloor \frac{n}{2} \rfloor} \binom{n+1}{2k+1} r^{n-1-2k} t^{2k} \left(2nr - (n-2k)(r^2-t^2)\right),$$

which immediately leads to the formula in Theorem 12.2.2. □

Remark 12.2.9. It would be nice to find a simpler proof of this result using, for example, a suitable orthonormal system from $L_h^2(\mathbb{M})$.

?

Example 12.2.10. Applying Theorem 12.2.2, the biholomorphic mapping

$$\mathbb{M}_2 \ni (z_1, z_2) \longmapsto \frac{1}{\sqrt{2}}(z_1 + iz_2, z_1 - iz_2) \in \mathcal{E}((1/2, 1/2)) =: G_2$$

leads to the well known formula of the Bergman kernel function of the domain G_2 (see Example 12.1.12):

$$K_{G_2}(z,w) = \frac{2}{\pi^2} \cdot \frac{3(1-\langle z,w\rangle)^2(1+\langle z,w\rangle) + 4z_1z_2\overline{w}_1\overline{w}_2(5-3\langle z,w\rangle)}{\left((1-\langle z,w\rangle)^2 - 4z_1z_2\overline{w}_1\overline{w}_2\right)^3}, \quad z,w \in G_2.$$

Remark 12.2.11. In [541] an explicit formula for the Bergman kernel function is given even for a more general domain Ω, which could be thought as some interpolation between the minimal balls and the Euclidean balls. Here, we only describe Ω. Fix $d \in \mathbb{N}$ and two d-tuples $m = (m_1, \dots, m_d) \in \mathbb{N}^d$ and $n = (n_1, \dots, n_d) \in \mathbb{N}^d$. Moreover, let $a = (a_1, \dots, a_d) \in [1, \infty)^d$. Then, the domain $\Omega = \Omega_{d,m,n,a}$ is given as

$$\Omega := \left\{ Z = (Z(1), \dots, Z(d)) \in \mathbb{M}(m_1 \times n_1; \mathbb{C}) \times \dots \times \mathbb{M}(m_d \times n_d; \mathbb{C}) : \sum_{j=1}^{d} \|Z(j)\|_*^{2a_j} < 1 \right\},$$

where $\mathbb{M}(p\times q;\mathbb{C})$ denotes the space of all $p\times q$-matrices with complex entities, and where

$$\|M\|_* := \left(\sum_{j=1}^{p}\left(\sum_{k=1}^{q}|z_{jk}|^2 + \left|\sum_{k=1}^{q} z_{jk}^2\right|\right)\right)^{1/2},$$
$$M = (z_{jk})_{j=1,\dots,p,\, k=1,\dots,q} \in \mathbb{M}(p\times q;\mathbb{C}).$$

Observe that for $d = 1 = a = m, n_1 = n$ the domain $\Omega_{d,m,n,a}$ is just the minimal ball $\mathbb{M} \subset \mathbb{C}^n$.

12.3 The Lu Qi-Keng problem

In this section, we collect a few observations about the zero set of the Bergman kernel function. In the following, we say that a domain $G \subset \mathbb{C}^n$ is called a *Lu Qi-Keng domain* if $\boldsymbol{K}_G(z,w) \neq 0$ for all $z, w \in G$.

It is clear that any bounded simply connected domain in $\mathbb{C}$ is a Lu Qi-Keng domain; cf. Example 12.1.7(b). On the other hand, the annulus $P := \{\lambda \in \mathbb{C} : r < |\lambda| < 1\}$, $0 < r < e^{-2}$, is not a Lu Qi-Keng domain; cf. Example 12.1.7(c).

Moreover, for plane domains the following results hold (cf. [457, 491]):

(a) Any doubly connected domain in $\mathbb{C}$ that is a Lu Qi-Keng domain is biholomorphic to the punctured disc $\mathbb{D}_*$.

(b) For any $p \geq 3$ there exists a p-connected bounded domain in $\mathbb{C}$, which is a Lu Qi-Keng domain.

(c) Any bounded domain in $\mathbb{C}$ with smooth boundary that is a Lu Qi-Keng domain is already simply connected.

For a while, it was a question (posed by Lu Qi-Keng [438]) whether the Bergman kernel function of a simply connected domain $G \subset \mathbb{C}^n$, $n \geq 2$, has no zeros. In fact, Lu Qi-Keng wrote: "But there seems to be nobody yet who has proved that for a bounded domain D in a general $\mathbb{C}^n$ the Bergman kernel $\boldsymbol{K}(z,\cdot)$ has no zero point in D, although there are many concrete examples justifying this statement." A first example of a simply connected domain of holomorphy in $\mathbb{C}^2$ that is not a Lu Qi-Keng domain was given by H. P. Boas in [66] (see also [483]). We already know (see Example 12.1.15) that there is even a convex domain, at least in $\mathbb{C}^3$, which is not Lu Qi-Keng. In fact, it turned out that the set of domains of holomorphy not being Lu Qi-Keng form a nowhere dense set in a suitable topology. For a more detailed discussion of this topic, see [68] (see also [69]). Consequently, only a few domains are Lu Qi-Keng domains. In this section, we discuss concrete examples of domains and the property to be Lu Qi-Keng.

Example 12.3.1. First, we study domains of the type

$$\Omega_{n,m} := \left\{(z,w) \in \mathbb{C}^n \times \mathbb{C}^m : \sum_{j=1}^{n} |z_j| + \sum_{k=1}^{m} |w_k|^2 < 1\right\},$$

where $n \in \mathbb{N}$ and $m \in \mathbb{Z}_+$ (cf. Example 12.1.13 for $n = m = 1$, Example 12.1.12 for $n = 2, m = 0$, and Example 12.1.15 for $n = 3, m = 0$).

In a first step, let $n = 1$ and $m \in \mathbb{Z}_+$. Then, using Bell's transformation law for the proper holomorphic mapping $F : \mathbb{B}_k \longrightarrow \Omega_{1,m}$, $F(z) := (z_1^2, z_2, \dots, z_k)$, where $k := m + 1$, we get

$$\begin{aligned}&\boldsymbol{K}_{\Omega_{1,m}}((z_1^2, z_2, \dots, z_k), (w_1^2, w_2, \dots, w_k))\\ &\quad = \frac{k!}{\pi^k 4 z_1 \overline{w}_1}\left(\frac{1}{(1 - \langle z, w\rangle)^{k+1}} - \frac{1}{(1 + z_1\overline{w}_1 - \langle \widetilde{z}, \widetilde{w}\rangle)^{k+1}}\right),\\ &\qquad\qquad z, w \in \mathbb{B}_k,\ z_1 w_1 \neq 0,\end{aligned}$$

where $\widetilde{z} := (z_2, \dots, z_k)$ and $\widetilde{w} := (w_2, \dots, w_k)$. In the case where $m + 2 > 4$, similar reasoning as in Example 12.1.13 gives $z_1, w_1 \in \mathbb{D}_*$ such that

$$\boldsymbol{K}_{\Omega_{1,m}}((z_1^2, 0, \dots, 0), (w_1^2, 0, \dots, 0)) = 0.$$

If $m + 2 \leq 4$, an easy calculation shows that $\boldsymbol{K}_{\Omega_{1,m}}$ has no zeros on $\Omega_{1,m} \times \Omega_{1,m}$. Hence, the Bergman kernel function of $\Omega_{1,m}$ has a zero iff $m + 2 > 4$.

Finally, using the above result for $n = 1$, induction over n, and the deflation method, we are led to the following result:

The Bergman kernel function of $\Omega_{n,m}$ has zeros iff $2n + m > 4$. In particular, the convex domain $\Omega_{n,0}$, $n \geq 3$, is not Lu Qi-Keng.

So far, it is not known whether there exists a convex domain in $\mathbb{C}^2$ that is not a Lu Qi-Keng domain. ?

Example 12.3.2. We also mention that the minimal ball $\mathbb{M} = \mathbb{M}_n \subset \mathbb{C}^n$, $n \geq 4$, is not a Lu Qi-Keng domain (see [419]). This result will be proved by exploiting the explicit formula given in Theorem 12.2.2. In fact, let first $n \geq 5$:

Put

$$f : \mathbb{R} \to \mathbb{R},\ f(t) := -(n+1)\arctan\frac{2t}{1-t^2} + 2\pi - \arctan\frac{2(n^2-1)t}{(n-1)^2 - (n+1)^2 t^2}.$$

Observe that $f(0) = 2\pi$ and $f(1/2) < 0$ (here we need that $n \geq 5$); so $f(t_0) = 0$ for a certain $t_0 \in (0, 1/2)$. Therefore,

$$\left(\frac{1 - it_0}{1 + it_0}\right)^{n+1} = \frac{n - 1 + it_0(n+1)}{n - 1 - it_0(n+1)}.$$

Put $z_0 := \sqrt{it_0}(1, 0, \dots, 0)$, $w_0 := \sqrt{-it_0}(0, 1, 0, \dots, 0) \in \mathbb{C}^n$. A simple calculation gives that $\|z_0\|_{\min} = \|w_0\|_{\min} = t_0 < 1/2$; thus, $z_0, w_0 \in \mathbb{M}$. Then, in virtue of Theorem 12.2.2, it follows that

$$\boldsymbol{K}_{\mathbb{M}}(z_0, w_0) = \frac{1}{n(n+1)\mathcal{L}^{2n}(\mathbb{M})} \frac{\sum_{j=0}^{\lfloor \frac{n}{2} \rfloor} \binom{n+1}{2j+1} (it_0)^{2j} (n + 2j + (n-2j)(it_0)^2)}{(1 - (it_0)^2)^{n+1}}.$$

Computing the binomial expression leads to

$$\begin{aligned} &\boldsymbol{K}_{\mathbb{M}}(z_0, w_0) \\ &= \frac{(n-1+(n+1)it_0)(1+it_0)^{n+1} - (n-1-(n+1)it_0)(1-it_0)^{n+1}}{n(n+1)\mathcal{L}^{2n}(\mathbb{M}) 2it_0 (1-(it_0)^2)^{n+1}} = 0. \end{aligned}$$

The case $n = 4$ remains. Consider the function

$$g : \mathbb{R} \longrightarrow \mathbb{R}, \ g(s) := -28s^4 + 50s^3 - 10s^2 - 15s + 5.$$

Then, $g(0) = 5$ and $g(2/5) < 0$. Therefore, there exists an $s_0 \in (0, 2/5)$ with $g(s_0) = 0$. Put

$$\begin{aligned} z_0 &:= \frac{\sqrt{s_0(i-1)}}{2}(i + \sqrt{i}, -i + \sqrt{i}, 0, 0), \\ w_0 &:= \frac{\sqrt{s_0(1-i)}}{2}(i - \sqrt{i}, -i - \sqrt{i}, 0, 0). \end{aligned}$$

Then, $\|z_0\|_{\min} = \|w_0\|_{\min} = (1 + \sqrt{2})^{s_0}/2 < 1/2$, i.e., $z_0, w_0 \in \mathbb{M}$. Moreover, using the formula in Theorem 12.2.2, calculation leads to

$$\boldsymbol{K}_{\mathbb{M}}(z_0, w_0) = \frac{g(s_0)}{5\mathcal{L}^{2n}(\mathbb{M})((1-s_0)^2 + s_0^2)^5} = 0.$$

Hence, the Bergman kernel function vanishes at the point (z_0, w_0).

? It is an open question whether the three dimensional minimal ball is a Lu Qi-Keng domain.

Using the former example, one can construct concrete strongly convex real-subalgebraic[2] domains in $\mathbb{C}^n$, $n \geq 4$, which are not Lu Qi-Keng domains and not Reinhardt domains (for details see [419]).

Example 12.3.3. Recall that the Bergman kernel function for the symmetrized bidisc $\mathbb{G}_2$ has no zeros on $\mathbb{G}_2 \times \mathbb{G}_2$. For $n \geq 3$ it turns out that $\boldsymbol{K}_{\mathbb{G}_n}$ has zeros (see [397]). The proof will be based on the formula of $\boldsymbol{K}_{\mathbb{G}_n}$ as it was given in Example 12.1.16.

[2] Here "real-subalgebraic" means that the domain is given as the sublevel set of a real polynomial.

The proof is done by induction. In fact it will be shown that for $n \geq 3$:

$(*)_n$ *there exist points* $\lambda, \mu \in \mathbb{D}^n$ *with pairwise different coordinates such that*

- $\Delta_n(\lambda, \mu) := \det\left[(1-\lambda_j\overline{\mu}_k)^{-2}\right]_{1\leq j,k\leq n} = 0$, *implying that* $\boldsymbol{K}_{\mathbb{G}_n}(\pi_n(\lambda), \pi_n(\mu)) = 0$;
- *the function* $\mathbb{D} \ni \zeta \longmapsto \Delta_n(\zeta, \lambda_2, \dots, \lambda_n, \mu)$ *is not identically zero.*

Proof of $(*)_n$ *for* $n \geq 3$. First, fix $n = 3$. Then the formula for $\boldsymbol{K}_{\mathbb{G}_3}$ in Example 12.1.16 gives for $\lambda \in \mathbb{D}^3$ and $\mu = (\mu_1, \mu_2, 0) \in \mathbb{D}^3$, $\mu_1 \neq 0$, that

$$\boldsymbol{K}_{\mathbb{G}_3}(\pi_3(\lambda), \pi_3(\mu)) = \frac{F_3(\pi_3(\lambda), \overline{\pi_3(\mu)})}{\pi^3 \prod_{j,k=1}^{3}(1-\lambda_j\overline{\mu}_k)^2}$$
$$= \frac{a(\overline{\mu}_1\lambda)(\frac{\overline{\mu}_2}{\overline{\mu}_1})^2 - b(\overline{\mu}_1\lambda)\frac{\overline{\mu}_2}{\overline{\mu}_1} + 2c(\overline{\mu}_1\lambda)}{\pi^3 \prod_{\substack{1\leq j\leq 3\\ 1\leq k\leq 2}}(1-\lambda_j\overline{\mu}_k)^2},$$

where

- $a(\nu) := 2\pi_{3,1}(\nu)\pi_{3,3}(\nu) - \pi_{3,1}(\nu)\pi_{3,2}(\nu) - 3\pi_{3,3}(\nu) + 2\pi_{3,2}(\nu)$,
- $b(\nu) := 4\pi_{3,1}(\nu) - 2\pi_{3,2}(\nu) - 2\pi_{3,1}(\nu)^2 + 4\pi_{3,3}(\nu) + \pi_{3,1}(\nu)\pi_{3,2}(\nu)$,
- $c(\nu) := 3 - 2\pi_{3,1}(\nu) + \pi_{3,2}(\nu)$, $\quad \nu \in \mathbb{C}^3$.

Put $p_t(z) := a(t)z^2 - b(t)z + 2c(t)$ for $t \in \mathbb{C}^3$ and $z \in \mathbb{C}$. If $t_0 := (e^{i\pi/6}, e^{i\pi/3}, e^{-i\pi/6})$, then calculation shows that the polynomial p_{t_0} has a zero at the point

$$z_0 := e^{-i\pi/4}\frac{6 - 3\sqrt{3} - \sqrt{40\sqrt{3} - 69}}{\sqrt{2}(3\sqrt{3} - 5)} \in \mathbb{D}.$$

Therefore, if $t \in \mathbb{D}^3$ is sufficiently near to t_0, then the polynomial p_t has a zero near z_0 in $\mathbb{D}$. Fix such a $t^* \in \mathbb{D}^3$ near t_0 with $t_1^* \neq t_2^* \neq t_3^* \neq t_1^*$, the corresponding zero z^* of p_{t^*} near z_0, and a $\mu_1^* \in \mathbb{D}$ such that $|\mu_1^*| > \max\{|t_j^*| : j = 1, 2, 3\}$. Thus, $\lambda^* := \frac{t^*}{\overline{\mu_1^*}} \in \mathbb{D}^3$ and it has pairwise different coordinates. Finally, let $\mu^* := (\mu_1^*, \overline{z^*}\mu_1^*, 0)$. Then $\boldsymbol{K}_{\mathbb{G}_3}(\pi_3(\lambda^*), \pi_3(\mu^*)) = 0$.

To complete this first induction step, let us assume that $f_3 \equiv 0$. In particular, $f_3(0) = f_3'(0) = f_3''(0) = 0$, which implies that

$$\det\begin{bmatrix} \overline{\mu_1^*}^j & \overline{\mu_2^*}^j & 0^j \\ (1-\lambda_2^*\overline{\mu_1^*})^{-2} & (1-\lambda_2^*\overline{\mu_2^*})^{-2} & 1 \\ (1-\lambda_3^*\overline{\mu_1^*})^{-2} & (1-\lambda_3^*\overline{\mu_2^*})^{-2} & 1 \end{bmatrix} = 0, \quad j = 0, 1, 2.$$

Recall that the coordinates of μ^* are pairwise different. Thus, the vectors $(\overline{\mu_1^*}^j, \overline{\mu_2^*}^j, 0^j)$, $j = 0, 1, 2$, are linearly independent which finally leads to

$$\det \begin{bmatrix} (1 - \lambda_2^* \overline{\mu_1^*})^{-2} & (1 - \lambda_2^* \overline{\mu_2^*})^{-2} \\ (1 - \lambda_3^* \overline{\mu_1^*})^{-2} & (1 - \lambda_3^* \overline{\mu_2^*})^{-2} \end{bmatrix} = 0.$$

Hence, $\boldsymbol{K}_{\mathbb{G}_2}(\pi_2(\lambda_2^*, \lambda_3^*), \pi_2(\mu_1^*, \mu_2^*)) = 0$; a contradiction. So, $(*)_3$ is verified.

Now assume that for $n \geq 3$ the condition $(*)_n$ is true with points $\lambda, \mu \in \mathbb{D}^n$. Put

$$g_{n+1}(\xi, \eta) := \Delta_{n+1}(\xi, \lambda_2, \dots, \lambda_n, \eta, \mu, \eta), \quad (\xi, \eta) \in \mathbb{D}^2.$$

Then, g_{n+1} can be written as

$$g_{n+1}(\xi, \eta) = \frac{f_n(\xi)}{(1 - |\eta|^2)^2} + h_n(\xi, \eta), \quad (\xi, \eta) \in \mathbb{D}^2,$$

where h_n is continuous on $\mathbb{D} \times \overline{\mathbb{D}}$ and holomorphic in its first coordinate. Recall that $f_n(\lambda_1) = 0$ and $f_n \not\equiv 0$. Therefore, $f_n(\xi) \neq 0$ for all $\xi \in \overline{\mathbb{D}}(\lambda_1, r) \setminus \{\lambda_1\} \subset \mathbb{D}$, where r is a sufficiently small positive number. In other words, we have $|f_n(\xi)| \geq m_1 > 0$ for all ξ with $|\xi - \lambda_1| = r$. Moreover, $|h_n| \leq m_2$ on $\partial\mathbb{D}(\lambda_1, r) \times \overline{\mathbb{D}}$. Now we choose a $\widetilde{\lambda}_{n+1} \in \mathbb{D}$ near 1 such that $\frac{m_1}{(1-|\widetilde{\lambda}_{n+1}|^2)^2} > m_2$, $\widetilde{\lambda}_{n+1} \neq \lambda_j$ and $\widetilde{\lambda}_{n+1} \neq \mu_j$ $(j = 2, \dots, n)$, and $|\widetilde{\lambda}_{n+1}| > |\lambda_1| + r$. Finally, applying Rouché's theorem leads to an isolated zero $\widetilde{\lambda}_1 \in \mathbb{D}(\lambda_1, r)$ of $g_{n+1}(\cdot, \widetilde{\lambda}_{n+1})$, which immediately gives $(*)_{n+1}$. □

Now we will discuss examples that are even Reinhardt domains. Let $n, k \in \mathbb{N}$, $m \in \mathbb{Z}_+$, and $a \in (0, 1]$. Put

$$N_{a,k}(z) := \Big(\sum_{\varepsilon_1, \dots, \varepsilon_{n+1} \in \{+1, -1\}} \alpha_{\varepsilon_1, \dots, \varepsilon_{n+1}}^{2k}(z) + a^{2k} \alpha^{2k}(z) \Big)^{\frac{1}{2k}}, \quad z \in \mathbb{C}^n \times \mathbb{C}^m,$$

where $\alpha_{\varepsilon_1, \dots, \varepsilon_{n+1}}(z) := \sum_{j=1}^n \varepsilon_j |z_j| + \varepsilon_{n+1} \sum_{j=1}^m |z_{n+j}|^2$ and $\alpha(z) := \sum_{j=1}^{n+m} |z_j|^2$. Finally, put

$$\Omega_{a,k,n,m} := \{z \in \mathbb{C}^{n+m} : N_{a,k}(z) < 1\}.$$

The following result is due to Nguyên Viêt Anh [373].

Theorem 12.3.4. *The domain $\Omega_{a,k,n,m}$ is strongly convex, real-subalgebraic, complete Reinhardt. Moreover, if $2n - m > 4$, then there is a positive integer $M = M(a, n, m)$ such that for all $k \geq M$ the domain $\Omega_{a,k,n,m}$ is not a Lu Qi-Keng domain.*

In particular, for $m = 0$ there are strongly convex algebraic complete Reinhardt domains in $\mathbb{C}^n$, $n \geq 3$, which are not Lu Qi-Keng.

What are effective values for the number $M(a,n,m)$?

To prove Theorem 12.3.4, we need the following lemma:

Lemma 12.3.5. *Let $f_j : \mathbb{R}^q \longrightarrow \mathbb{R}_+$ be a convex function, $j = 1, \dots, p$. Then, for any $k \in \mathbb{N}$ the following function*

$$\varrho(x) := \sum_{\varepsilon_1,\dots,\varepsilon_p \in \{-1,+1\}} (\varepsilon_1 f_1(x) + \cdots + \varepsilon_p f_p(x))^{2k}, \quad x \in \mathbb{R}^q,$$

is also a convex one.

Proof. Fix $x, u \in \mathbb{R}^q$. Recall that $\mathbb{R}^2 \ni (t,s) \longmapsto (t+s)^{2k} + (t-s)^{2k}$ is a convex function. Then,

$$\frac{\varrho(x)+\varrho(u)}{2} \geq \sum_{\varepsilon_1,\dots,\varepsilon_p \in \{-1,+1\}} \left(\varepsilon_1 \frac{f_1(x)+f_1(u)}{2} + \cdots + \varepsilon_p \frac{f_p(x)+f_p(u)}{2}\right)^{2k}.$$

Now recall the formula

$$\sum_{\varepsilon_1,\dots,\varepsilon_p \in \{-1,+1\}} (\varepsilon_1 b_1 + \cdots + \varepsilon_p b_p)^{2k} = 2^p \sum_{k_1+\cdots k_p = k} \frac{(2k)! b_1^{2k_1} \cdots b_p^{2k_p}}{(2k_1)! \cdots (2k_p)!}.$$

Plugging it into the first expression, we get

$$\frac{\varrho(x)+\varrho(u)}{2} \geq \sum_{k_1+\cdots+k_p=k} \frac{2^p(2k)!}{(2k_1)!\cdots(2k_p)!} \cdot \left(\frac{f_1(x)+f_1(u)}{2}\right)^{2k_1} \cdots \left(\frac{f_p(x)+f_p(u)}{2}\right)^{2k_p}.$$

By virtue of the positivity and convexity of the functions f_j, the last inequality gives $(\varrho(x)+\varrho(u))/2 \geq \varrho((x+u)/2)$, i.e., ϱ is a convex function. □

Proof of Theorem 12.3.4. Put

$$\varrho(z) := \sum_{\varepsilon_1,\dots,\varepsilon_{n+1} \in \{-1,+1\}} \alpha^{2k}_{\varepsilon_1,\dots,\varepsilon_{n+1}}(z) + a^{2k}\alpha^{2k}(z) - 1.$$

Then, ϱ is the defining function of the domain $\Omega = \Omega_{a,k,n,m}$. Using the above expansion, we see that ϱ is a polynomial with positive coefficients in $|z_1|^2, \dots, |z_n|^2$ and $\sum_{j=1}^m |z_{n+j}|^2$. Hence, Ω is an algebraic complete Reinhardt domain with a smooth boundary. Moreover, in virtue of Lemma 12.3.5, we may see that Ω is strongly convex (take simply $f_j(z) = |z_j|$, $j = 1, \dots, n$, $f_{n+1}(z) := \sum_{j=1}^m |z_{n+j}|^2$ and observe that α is strongly convex outside of the origin).

Observe that $\Omega \subset \Omega_{n,m}$, where $\Omega_{n,m}$ is the domain from Example 12.3.1, and that $N_{a,k} \le N_{a,l}$ when $l \le k$. Moreover,

$$\lim_{k\to\infty} N_{a,k}(z) = \sum_{j=1}^{n} |z_j| + \sum_{k=1}^{m} |z_{n+k}|^2, \quad z \in \Omega_{n,m}.$$

What remains is to apply Ramadanov's theorem (see Theorem 12.1.23), Example 12.3.1, and the Hurwitz theorem. □

We close this section by discussing consequences of the following result:

Theorem* 12.3.6 (cf. [160, 161, 97]). *Let $D \subset \mathbb{C}^n$ be a bounded pseudoconvex domain, $\varphi > 0$ a positive function on D, $-\log\varphi \in \mathcal{PSH}(D)$, such that $1/\varphi \in L^\infty_{loc}(D)$ fails to have a sesqui-holomorphic extension near a point $z^0 \in D$ (i.e., there is no function $f : V \times V \to \mathbb{C}$, $V \subset D$ a neighborhood of z^0, satisfying: f is holomorphic in the first coordinates and antiholomorphic in the latter, $f(z, z) = 1/\varphi(z)$ for all $z \in V$). Let $U = U(z^0) \subset D$ be a neighborhood. Then, there is an $m_U \in \mathbb{N}$ such that the Bergman kernel function $\boldsymbol{K}_{\Omega_m}((\cdot\,, 0), (\,\cdot\,, 0))$ of*

$$\Omega_m := \{(z, w) \in D \times \mathbb{C}^m : \|w\|^2 < \varphi(z)\}$$

has a zero in $U \times U$, $m \ge m_U$.

The proof is beyond the scope of this book. We only mention that it is based on a recent theorem of Ohsawa (see [404]) and the description of the Bergman kernel with weights due to E. Ligocka (see Remark 12.1.6).

We should mention that the original formulation in [97] is much stronger as the one given here. Applying Theorem 12.3.6 for certain complex ellipsoids, we obtain the following consequences:

Corollary 12.3.7 (cf. [97]).

(a) *For any real $k \ge 1$, k not an even integer, there exists an $m = m(k) \in \mathbb{N}$ such that*

$$\Omega_{k,m} := \{(z, w) \in \mathbb{D} \times \mathbb{C}^m : |z|^k + \|w\|^2 < 1\}$$

is not Lu Qi-Keng.

(b) *Let $k \in \mathbb{N}$ and put $s(k) := (2k + 1)/2$. Then there exists a natural number $m = m(k)$ such that, if*

$$\Omega := \Omega_{s(k),m} \text{ and } (z^0, w^0) := (0, 0, \dots, -1) \in \partial\Omega,$$

then Ω is convex with a $\mathcal{C}^k$-boundary and there are sequences

$$((z_j', w_j'))_j,\ ((z_j'', w_j''))_j \subset \Omega, \quad \lim_{j\to\infty} (z_j', w_j') = \lim_{j\to\infty} (z_j'', w_j'') = (z^0, w^0),$$

such that $\boldsymbol{K}_\Omega((z_j', w_j'), (z_j'', w_j'')) = 0$, $j \in \mathbb{N}$. *In particular, the set* $\{(z,w) \in \Omega \times \Omega : \boldsymbol{K}_\Omega(z,w) = 0\}$ *accumulates at* $((z^0, w^0), (z^0, w^0))$.

Proof. (a) Take $D = \mathbb{D}$ and $\varphi(z) := 1 - |z|^k$, $z \in \mathbb{D}$. Then, $-\log \varphi \in \mathcal{SH}(\mathbb{D})$ and $1/\varphi$ is not real analytic at 0. So it cannot be extended to a sesqui-holomorphic function near 0. Hence, by virtue of Theorem 12.3.6, there exists an $m = m(k) \in \mathbb{N}$ such that $\boldsymbol{K}_{\Omega_k}((\cdot\,,0),(\cdot\,,0))$ has at least one zero in $\mathbb{D} \times \mathbb{D}$.

(b) Fix a k. In virtue of part (a), there is an $m = m(k)$ such that $\boldsymbol{K}_\Omega$ has a zero at a point $((z', w'), (z'', w'')) \in \Omega \times \Omega$.

Put

$$D := \left\{\zeta \in \mathbb{C}^{m+1} : |\zeta_1|^{\frac{2k+1}{2}} + \sum_{j=2}^{m} |\zeta_j|^2 + \operatorname{Re}\zeta_{m+1} < 0\right\}.$$

Observe that

$$\Phi(\zeta) := \left(\frac{4^{\frac{2}{2k+1}}}{(\zeta_{m+1}-1)^{\frac{4}{2k+1}}}, \frac{2\zeta_2}{\zeta_{m+1}-1}, \dots, \frac{2\zeta_m}{\zeta_{m+1}-1}, \frac{\zeta_{m+1}+1}{\zeta_{m+1}-1}\right)$$

defines a biholomorphic map from D to Ω.

Moreover, for any positive ε,

$$F_\varepsilon(\zeta) := (\varepsilon^{\frac{2}{2k+1}}\zeta_1, \sqrt{\varepsilon}\zeta_2, \dots, \sqrt{\varepsilon}\zeta_m, \varepsilon\zeta_{m+1})$$

is a biholomorphic mapping from D to D. Therefore,

$$\boldsymbol{K}_\Omega(\Phi \circ F_\varepsilon \circ \Phi^{-1}(z', w'), \Phi \circ F_\varepsilon \circ \Phi^{-1}(z'', w'')) = 0, \quad \varepsilon > 0.$$

It remains to mention that

$$\lim_{\varepsilon \to 0} \Phi \circ F_\varepsilon \circ \Phi^{-1}(z', w') = \lim_{\varepsilon \to 0} \Phi \circ F_\varepsilon \circ \Phi^{-1}(z'', w'') = (z^0, w^0). \qquad \square$$

It would be interesting to find concrete numbers $m = m(k)$ in the situation of Corollary 12.3.7 **?**

So far, we saw that some of the domains $\mathcal{E}(p) \subset \mathbb{C}^n$ are not Lu Qi-Keng, and some of them are. Describe all vectors $p = (p_1, \dots, p_n)$ for which the Bergman kernel function of $\mathcal{E}(p)$ is zero–free. **?** For further open problems, see also [69]. Other examples of domains that are not Lu Qi-Keng may be found in [129, 161], and [97].

Example 12.3.8. Recently, the zeros of the Bergman kernel for domains

$$D_{n,m}(\mu) := \{(z,w) \in \mathbb{C}^n \times \mathbb{C}^m : \|w\| < e^{-\mu\|z\|^2}\}, \quad \mu > 0,\ n, m \in \mathbb{N},$$

were studied in [538]. The final result reads as follows:

For a given $n \in \mathbb{N}$ *there exists an* $m(n) \in \mathbb{N}$ *such that* $D_{n,m}(\mu)$ *is a Lu Qi-Keng domain if and only if* $m \geq m(n)$. *Moreover,* $m(n) \leq m(n+1)$ *for all* $n \in \mathbb{N}$.

Note that, in contrast to results discussed before, the number $m(n)$ here gives the exact dimension of the fiber from which on the Lu Qi-Keng property starts.

We conclude this section by a differential geometric interpretation of the zeros of the Bergman kernel, due to Lu Qi-Keng.

Remark 12.3.9. Let $D \subset \mathbb{C}^n$ be a bounded domain. Take a complete orthonormal basis $(\varphi_j)_{j\in\mathbb{N}}$ in $L_h^2(D)$. We may define the infinite dimensional projective manifold $\mathbb{CP}^\infty$ in a similar way as the finite dimensional projective space. Namely, take the sequence space ℓ_2 over $\mathbb{C}$ and identify points $\mathfrak{z} = (z_1, z_2, \dots)$ and $\mathfrak{w} = (w_1, w_2, \dots)$ from $\ell_2 \setminus \{0\}$ if and only if there is a complex number $\lambda \in \mathbb{C}_*$ such that $\lambda\mathfrak{z} = \mathfrak{w}$. Denote this quotient space by $\mathbb{CP}^\infty$. Then, as in the finite dimensional case, introducing local charts makes $\mathbb{CP}^\infty$ into an infinite dimensional manifold, into which D may be embedded by the following map:

$$D \ni z \overset{i_\varphi}{\longmapsto} [(\varphi_1(z), \varphi_2(z), \dots)] \in \mathbb{CP}^\infty.$$

Then for points $z, w \in D$ it turns out that $\boldsymbol{K}_D(z, w) = 0$ if and only if $i_\varphi(z)$ and $i_\varphi(w)$ are conjugate points in the sense of differential geometry. For details consult [439].

12.4 Bergman exhaustiveness

To simplify our notation, for further use, we will write

$$\mathfrak{K}_D(z) := \boldsymbol{K}_D(z, z), \quad z \in D.$$

In the study of the Bergman kernel, it is important to know its boundary behavior. We define

Definition 12.4.1. Let $D \subset \mathbb{C}^n$ be a domain and $z^0 \in \partial D$. We say that D is *Bergman exhaustive at* z^0 (for short, $\boldsymbol{K}$*-exhaustive*) if $\lim_{D\ni z\to z^0} \mathfrak{K}_D(z) = \infty$. Moreover, if D is $\boldsymbol{K}$-exhaustive at any of its boundary points (including ∞ if D is unbounded), then D is called $\boldsymbol{K}$*-exhaustive*.

Observe that $\mathfrak{K}_D \in \mathcal{PSH}(D)$. Therefore, any $\boldsymbol{K}$-exhaustive domain is pseudoconvex. There are a lot of general results giving sufficient condition for a pseudoconvex domain to be $\boldsymbol{K}$-exhaustive at a boundary point. Besides Theorem 12.1.25, the most general one is the following property, which relates $\boldsymbol{K}$-exhaustiveness to the boundary behavior of certain level sets of the Green function: for an arbitrary domain $D \subset \mathbb{C}^n$ and a point $a \in D$ recall the notion (see Section 5.1)

$$D(a) := D_{1/e}(a) = \{w \in D : \boldsymbol{g}_D(a, w) < 1/e\}.$$

Theorem 12.4.2. *Let D be a bounded pseudoconvex domain in $\mathbb{C}^n$ and $z_0 \in \partial D$. Assume that*

$$\lim_{z\to z_0} \mathcal{L}^{2n}(D(z)) = 0.$$

Then, D is $\boldsymbol{K}$-exhaustive at z_0.

Theorem 12.4.2 is a simple consequence of the following result (cf. [93, 230]):

Theorem 12.4.3. *For any $n \in \mathbb{N}$ there exists a positive number C_n such that for every bounded pseudoconvex domain $D \subset \mathbb{C}^n$ the following is true:*

$$\frac{|f(z)|^2}{\mathfrak{K}_D(z)} \le C_n \int_{D(z)} |f(w)|^2 d\mathcal{L}^{2n}(w), \quad f \in L_h^2(D),\ z \in D.$$

Proof. Let D be a bounded pseudoconvex domain in $\mathbb{C}^n$, $z_0 \in D$, and fix an $f \in L_h^2(D)$, $f \not\equiv 0$. Put

$$D_t := \{z \in D : \operatorname{dist}(z, \partial D) > t\}, \quad 0 < t < 1 \text{ sufficiently small, such that } D_t \neq \varnothing.$$

Moreover, let $\psi_1 \in \mathcal{C}^\infty(\mathbb{C}^n, \mathbb{R})$ be a non-negative polyradial symmetric function with $\int_{\mathbb{C}^n} \psi_1(z) d\mathcal{L}^{2n}(z) = 1$ and $\operatorname{supp} \psi_1 \subset \mathbb{B}_n$; put $\psi_t(z) := \frac{1}{t^{2n}} \psi_1(\frac{z}{t})$, $z \in \mathbb{C}^n$, $t > 0$.

On D_t, we define

$$\varphi_t(z) := 2n V_t(z) + \exp(V_t(z)) + t\|z\|^2, \quad \varphi(z) := 2n \log \boldsymbol{g}_D(z_0, \cdot) + \boldsymbol{g}_D(z_0, \cdot),$$

where $V_t := \log \boldsymbol{g}_D(z_0, \cdot) * \psi_t$. Finally, we choose a function $\chi \in \mathcal{C}^\infty(\mathbb{R}, [0,1])$ with $\chi(t) = 1$ if $t \le -2$, $\chi(t) = 0$ if $t \ge -1$, and $|\chi'| \le 2$.

We now discuss the following $\bar\partial$-closed $(0,1)$-form α_t on D_t,

$$\alpha_t := \bar\partial(\chi \circ V_t \cdot f) = \chi'(V_t) f \bar\partial V_t.$$

Observe that α_t is a smooth form, whose support is contained in the set $\{-2 \le V_t \le -1\}$. Moreover, $\varphi_t \ge -4n$ on $\operatorname{supp} \alpha_t$. Therefore, $\int_{D_t} \|\alpha_t\|^2 e^{-\varphi_t} d\mathcal{L}^{2n} < \infty$.

For the Levi form of φ_t we have the following estimate:

$$\mathcal{L}\varphi_t(z; X) \ge e^{V_t(z)} |V_t'(z) X|^2 \ge e^{-2} |V_t'(z) X|^2, \quad z \in \operatorname{supp} \alpha_t,\ X \in \mathbb{C}^n.$$

Let $Q(z)$ denote the inverse matrix of the coefficient matrix of $\mathcal{L}\varphi_t(z; \cdot)$. Then, if $z \in \operatorname{supp} \alpha_t$, we have

$$\sum_{j,k=1}^n Q_{j,k}(z) \alpha_{t,j}(z) \overline{\alpha_{t,k}(z)} \exp(-\varphi_t(z))$$
$$\le e^2 |\chi'(V_t(z))|^2 |f(z)|^2 e^{-\varphi_t(z)} \le 4e^{4n+2} |f(z)|^2.$$

Therefore, by virtue of Lemma 4.4.1 in [234], there exists a solution $u_t \in \mathcal{C}^\infty(D_t)$ of $\bar\partial u_t = \alpha_t$ with the estimates

$$\int_{D_t} |u_t|^2 e^{-\varphi_t} d\mathcal{L}^{2n} \le 4e^{4n+2} \int_{\operatorname{supp} \alpha_t} |f|^2 d\mathcal{L}^{2n}.$$

Put

$$v_t := \begin{cases} u_t e^{-\varphi_t/2} & \text{on } D_t \\ 0 & \text{on } D \setminus D_t \end{cases}.$$

Then, the family $(v_t)_t$ belongs to $L^2(D)$ and satisfies the following uniform estimate:

$$\int_D |v_t|^2 d\mathcal{L}^{2n} \le 4e^{4n+2} \int_{D(z_0)} |f|^2 d\mathcal{L}^{2n}$$

(observe that $\operatorname{supp}\alpha_t \subset \{-2 \le V_t \le -1\} \subset D(z_0)$).

By virtue of the Alaoglu–Bourbaki theorem, we may find a function $v \in L^2(D)$ satisfying

$$\int_D |v|^2 d\mathcal{L}^{2n} \le 4e^{4n+2} \int_{D(z_0)} |f|^2 d\mathcal{L}^{2n}.$$

Put $u := ve^{\varphi/2}$. Then,

$$\int_D |u|^2 d\mathcal{L}^{2n} \le e \int_D |v|^2 d\mathcal{L}^{2n} \le 4e^{4n+3} \int_{D(z_0)} |f|^2 d\mathcal{L}^{2n}. \tag{12.4.1}$$

Using distributional derivatives, we find an $\tilde{f} \in \mathcal{O}(D)$ such that

$$\tilde{f} = \chi \circ \log g_D(z_0, \cdot) - u$$

almost everywhere on D. Moreover, take a neighborhood $U \subset D$ of z_0 such that $\log g_D(z_0, \cdot) \cdot f \le -3$ on U. Then, $f - \tilde{f} = u$ almost everywhere on U. By virtue of (12.4.1), it follows that

$$\int_U |f - \tilde{f}|^2 e^{-\varphi} d\mathcal{L}^{2n} < \infty.$$

Observe that e^{φ} is not locally integrable near z_0; hence, $\tilde{f}(z_0) = f(z_0)$.

Summarizing, we have found an $\tilde{f} \in L_h^2(D)$ with $f(z_0) = \tilde{f}(z_0)$ and

$$\|\tilde{f}\|_{L^2(D)} \le (1 + 4e^{4n+3}) \int_{D(z_0)} |f|^2 d\mathcal{L}^{2n}.$$

Consequently,

$$\frac{|f(z_0)|^2}{\mathfrak{K}_D(z_0)} \le \|\tilde{f}\|_{L^2(D)}^2 \le (1 + 4e^{4n+3}) \int_{D(z_0)} |f|^2 d\mathcal{L}^{2n},$$

which finishes the proof. □

Proof of Theorem 12.4.2. By virtue of Theorem 12.4.3 there is a constant $C_n > 0$ such that

$$\frac{1}{\mathfrak{K}_D(z)} \leq C_n \int_{D(z)} d\mathcal{L}^{2n}(w) \leq C_n \mathcal{L}^{2n}(D(z)) \longrightarrow 0;$$

thus, $\mathfrak{K}_D(z) \underset{z\to z_0}{\longrightarrow} \infty$. □

Moreover, combining Theorem 12.4.2 and a result due to Błocki, we have the following (see also [402]):

Theorem 12.4.4. *For a bounded hyperconvex domain $D \subset \mathbb{C}^n$ (i.e., there is a negative $u \in \mathcal{PSH}(D)$ such that the sublevel sets $\{z \in D : u(z) < -\varepsilon\}$, $\varepsilon > 0$, are relatively compact in D), the following is true: $\mathcal{L}^{2n}(D(z)) \underset{z\to\partial D}{\longrightarrow} 0$. In particular, any hyperconvex domain is $\boldsymbol{K}$-exhaustive.*

Proof. According to [57], there is a function $u \in \mathcal{C}(\overline{D}) \cap \mathcal{PSH}(D)$ satisfying the following properties:

$$u|_{\partial D} = 0 \text{ and } (dd^c u)^n \geq \mathcal{L}^{2n}.$$

Applying [56], we get for a point $z_0 \in \partial D$:

$$\begin{aligned}\int_D (-\log g_D(z,w))^n d\mathcal{L}^{2n}(w) &\leq \lim_{k\to\infty} \int_D (-\max\{\log g_D(z,\cdot), -k\})^n (dd^c u)^n \\ &\leq n!\|u\|^{n-1}_{L^\infty(D)} |u(z)| \underset{z\to z_0}{\longrightarrow} 0,\end{aligned}$$

where the last inequality is due to Demailly (see [122]).

Finally, by virtue of Theorem 12.4.3, we get

$$\frac{1}{\mathfrak{K}_D(z)} \leq C_n \int_{D(z)} d\mathcal{L}^{2n}(w) \leq C_n \int_D (-\log g_D(z,w))^n d\mathcal{L}^{2n}(w) \underset{z\to z_0}{\longrightarrow} 0.$$

Since z_0 is arbitrary, it follows that $\mathfrak{K}_D(z) \underset{z\to\partial D}{\longrightarrow} \infty$, i.e., D is $\boldsymbol{K}$-exhaustive. □

Remark 12.4.5. (1) There is a large class of bounded pseudoconvex domains that are hyperconvex, namely

Theorem* 12.4.6 (cf. [295, 122]). *Any bounded pseudoconvex domain $D \subset \mathbb{C}^n$ with a Lipschitz boundary is hyperconvex. In particular, if D has a $\mathcal{C}^1$-boundary, then it is hyperconvex.*

(2) Hyperconvexity is even a local property.

Theorem* 12.4.7 (cf. [295]). *Suppose that D is a bounded domain in $\mathbb{C}^n$ such that every $z_0 \in \partial D$ has a neighborhood $U = U(z_0)$ for which $D \cap U$ is hyperconvex. Then D itself is hyperconvex.*

Observe that hyperconvexity is not necessary for a domain to be $\boldsymbol{K}$-exhaustive, as the following example will show.

Example 12.4.8.

(1) Put $D := \{z \in \mathbb{C}^2 : |z_1| < |z_2| < 1\}$. Then, D is $\boldsymbol{K}$-exhaustive but not hyperconvex. (For other examples of this type see also Theorems 12.4.9 and 12.4.10 and Example 12.5.16.) For D from above even more is true. Namely, there is a sequence $(z_k)_k \subset D$ tending to 0 such that $\mathcal{L}^{2n}(D(z_k)) \not\to 0$.

(2) For Reinhardt domains in $\mathbb{C}^2$ we have the following general result (due to W. Zwonek) for the pole boundary behavior of the Green function:

Theorem* 12.4.9 (cf. [568]). *Let $D \subset \mathbb{C}^2$ be a bounded pseudoconvex Reinhardt domain such that $D \cap (\mathbb{C}_* \times \{0\}) = \mathbb{D}_* \times \{0\}$. Moreover, suppose that for a $z^0 \in D$:*

$$\{v \in \mathbb{R}^2 : (\log |z_1^0|, \log |z_2^0|) + \mathbb{R}_+ v \subset \log D\} = \mathbb{R}_+(0, -1).$$

Then,

$$g_D(z, w) \underset{D \ni z \to 0}{\longrightarrow} 0, \quad w \in D \cap \mathbb{C}_*, \quad \textit{and, therefore,} \quad \mathcal{L}^{2n}(D(z)) \underset{D \ni z \to 0}{\longrightarrow} 0.$$

In particular, D is $\boldsymbol{K}$-exhaustive at the origin but not hyperconvex.

An explicit example of a Reinhardt domain satisfying the conditions of Theorem 12.4.9 is given by $D := \{z \in \mathbb{D}_* \times \mathbb{D} : |z_2| < e^{-1/|z_1|}\}$. Hence, D is $\boldsymbol{K}$-exhaustive but not hyperconvex.

For circular domains, we have the following result:

Theorem 12.4.10 (cf. [275]). *Any bounded pseudoconvex balanced domain is $\boldsymbol{K}$-exhaustive.*

Proof. Let $D = D_h = \{z \in \mathbb{C}^n : h(z) < 1\}$ be a bounded pseudoconvex balanced domain. Fix a boundary point z_0 and let M be an arbitrary positive number. Put $H := \mathbb{C} z_0$. Then, by virtue of the theorem of Ohsawa (see Appendix B.8.1), we have $\mathfrak{K}_{D \cap H}(z) \le C \mathfrak{K}_D(z)$, $z \in D \cap H$, where C is a suitable positive number. Since $D \cap H$ is a plane disc, there is an $s \in (0, 1)$ such that $M < \mathfrak{K}_{D \cap H}(s z_0)$. Using the continuity of $\mathfrak{K}_D$ leads to an open neighborhood $U = U(z_0) \subset D \setminus \{0\}$ such that $\mathfrak{K}_D(z) > M$, $z \in U$.

Now fix a $z \in U$ and define $u_z : \frac{1}{h(z)}\mathbb{D} \longrightarrow \mathbb{R}$, $u_z(\lambda) := \mathfrak{K}_D(\lambda z)$. This function is subharmonic and radial, so $u|_{[0,\frac{1}{h(z)})}$ is an increasing function. Therefore, $M < u_z(1) \le u_z(\lambda) = \mathfrak{K}_D(\lambda z)$, $1 \le |\lambda| < \frac{1}{h(z)}$. Obviously,

$$V = V_{z_0,M} := \{\lambda z : z \in U,\ \lambda \in \mathbb{C},\ |\lambda| > 1\}$$

is an open neighborhood of z_0. Since M is arbitrary, we have $\liminf_{D \ni z \to z_0} \mathfrak{K}_D(z) = \infty$, proving the theorem. □

In the case of a bounded pseudoconvex balanced domain with a continuous Minkowski function, Theorem 12.4.10 was proved in [251].

Observe that any bounded hyperconvex balanced domain is taut, and therefore its Minkowski function h is continuous. Obviously, there are a lot of bounded balanced pseudoconvex domains with a non-continuous Minkowski function. Moreover, we mention that there exists a bounded pseudoconvex balanced domain D that is not fat (i.e., $\overline{\operatorname{int} D} \neq D$); see Example 3.1.13.

Describe all bounded pseudoconvex circular domains D (i.e., $\forall_{z \in D, \theta \in \mathbb{R}} : e^{i\theta} z \in D$) that are $\boldsymbol{K}$-exhaustive. ?

Now we turn to discuss more general domains like the ones in Theorem 12.4.10, namely Hartogs domains with m-dimensional fibers (cf. Appendix B.7.7).

Lemma 12.4.11. *Let G_D be a bounded pseudoconvex Hartogs domain over $D \subset \mathbb{C}^n$ with m-dimensional balanced fibers. Then,*

(a) *there exists a $c > 0$ such that $\mathfrak{K}_D(z) \le c\mathfrak{K}_{G_D}(z, 0)$, $z \in D$;*

(b) *for any boundary point $(z_0, w_0) \in \partial G_D$ it follows that $\mathfrak{K}_{G_D}(z_0, w) \longrightarrow \infty$ when $G_D \ni (z_0, w) \longrightarrow (z_0, w_0)$.*

Proof. (a) Let $f \in L_h^2(D)$ and put $F(z, w) := f(z)$, $(z, w) \in G_D$. Observe that $G_D \subset D \times \mathbb{B}_m(R)$ for a large R, which implies that $\|F\|_{L^2(G_D)} \le c_1 \|f\|_{L^2(D)}$, where the constant c_1 may be chosen to be independent of f, which immediately implies (a).

(b) The proof is an easy consequence of Theorem 12.4.10 and Appendix B.8.1. □ *i*

Theorem 12.4.12 (cf. [275]). *Let G_D be bounded pseudoconvex Hartogs domains over D with m-dimensional balanced fibers, and let $(z_0, w_0) \in \partial G_D$. Assume that one of the following conditions is satisfied:*

(a) *$z_0 \in D$;*

(b) *$z_0 \in \partial D$ and $\lim_{D \ni z \to z_0} \mathfrak{K}_D(z) = \infty$;*

(c) *there is a neighborhood* $U = U((z_0, w_0))$ *such that*

$$U \cap G_D \subset \{(z, w) \in \mathbb{C}^n \times \mathbb{C}^m : \|w\| < \|z - z_0\|^\delta\}$$

for some $\delta > 0$.

Then, $\lim_{G_D \ni (z,w) \to (z_0,w_0)} \mathfrak{K}_{G_D}((z, w)) = \infty$. *In particular, if D is* **K***-exhaustive, then so is* G_D.

Proof. Let G_D be given via the function H. Assume that $z_0 \in D$. Take an arbitrary positive number M. Applying Lemma 12.4.11, one may find a $t = t_M \in (0, 1)$ such that $\mathfrak{K}_{G_D}((z_0, tw_0)) > M$. Now use the continuity of the kernel to find a neighborhood $U = U_1 \times U_2 \subset G_D$ of (z_0, tw_0), where $0 \notin U_2$, such that

$$\mathfrak{K}_{G_D}((z, w)) > M, \quad (z, w) \in U.$$

Now fix a point $(z, w) \in U$, observe that $H(z, w) > 0$, and put $\varrho := 1/H(z, w) > 1$. Define

$$u = u_{(z,w)} : \mathbb{D}(\varrho) \ni \lambda \longmapsto \mathfrak{K}_{G_D}((z, \lambda w)).$$

Obviously, u is a radial symmetric subharmonic function on $\mathbb{D}(\varrho)$ with $u(1) > M$. Recall that such functions are increasing on $(0, \varrho)$ (see Appendix B.4.7). Hence, $\mathfrak{K}_{G_D}((z, \lambda w)) > M$ for $\lambda \in \mathbb{A}(1, \varrho)$.

Since (z, w) was an arbitrary point in U, we have $\mathfrak{K}_{G_D} > M$ on $(U_1 \times [1, \infty)U_2) \cap G_D$. Therefore ($M$ was arbitrarily given),

$$\mathfrak{K}_{G_D}((z, w)) \underset{G_D \ni (z,w) \to (z_0,w_0)}{\longrightarrow} \infty,$$

which gives the implication for (a).

Now let us discuss the situation in (b). Take such a $z_0 \in \partial D$ and fix a positive number M. Taking into account the condition in (b) and Lemma 12.4.11(a), we find a neighborhood $U = U(z_0)$ such that $\mathfrak{K}_{G_D}((z, 0)) > M$, $z \in D \cap U$.

Fix a point $z \in D \cap U$ and a point $w \in \mathbb{C}^m$ such that $H(z, w) \in (0, 1)$. As above, define the radial symmetric subharmonic function u,

$$\mathbb{D}(1/H(z, w)) \ni \lambda \overset{u}{\longmapsto} \mathfrak{K}_{G_D}((z, \lambda w)).$$

Then $u(0) > M$ and so, as above, we get $\mathfrak{K}_{G_D}((z, w)) > M$ on $(\{z\} \times \mathbb{C}^m) \cap G_D$. Since $z \in U \cap D$ was arbitrarily chosen, it follows that $\mathfrak{K}_{G_D}((z, w)) > M$ for all $(z, w) \in G_D$, where $z \in U$. Hence,

$$\mathfrak{K}_{G_D}((z, w)) \underset{G_D \ni (z,w) \to (z_0,w_0)}{\longrightarrow} \infty.$$

What remains is to assume the situation in (c). Note that necessarily $w_0 = 0$. Moreover, we may assume that $z_0 = 0$. Now fix a point $(0, w) \notin \overline{G}_D$ with $0 < \|w\| < 1/2$. Let ε be a positive number satisfying $\varepsilon > 1$, $\varepsilon\delta > 1$, and $\delta + \varepsilon\delta > 2$. We want to apply Theorem 12.1.25, i.e., the outer cone condition at $(0, 0)$ has to be verified. Suppose that there exists a point $(\zeta, \eta) \in \mathbb{B}((0, w), \|w\|^\varepsilon/2)$. Then, $\|\zeta\| < \frac{1}{2}\|w\|^\varepsilon$ and $\|w\| - \|\eta\| \le \|\eta - w\| \le \frac{1}{2}\|w\|^\varepsilon$. Consequently, $\|w\| - \frac{1}{2}\|w\|^\varepsilon \le \|\eta\| \le \|\zeta\|^\delta < \left(\frac{1}{2}\right)^\delta \|w\|^{\delta\varepsilon}$. Therefore,

$$\frac{1}{2} < 1 - \frac{1}{2}\|w\|^{\varepsilon-1} < \left(\frac{1}{2}\right)^\delta \|w\|^{\varepsilon\delta-1} < \frac{1}{2};$$

a contradiction. So the cone condition is fulfilled and Theorem 12.1.25 applies. □

Example 12.4.13. The following example shows that Theorem 12.4.12 is far away from being optimal. Fix sequences $(a_j)_{j\in\mathbb{N}} \subset (0, 1)$ and $(n_j)_{j\in\mathbb{N}} \subset \mathbb{N}$ with $\lim_{j\to\infty} a_j = 0$ and $n_j \ge j$. Let $E_k := \mathbb{D} \setminus \{a_j : j = 1, \dots, k\}$, $u_k(\lambda) := \sum_{j=1}^k \left(\frac{a_j}{2|\lambda - a_j|}\right)^{n_j}$. Observe that $u_k(0) < 0$. Define $E_\infty := \mathbb{D} \setminus \big(\{0\} \cup \{a_j : j \in \mathbb{N}\}\big)$. Then, the sequence $(u_k)_k$ is locally bounded from above on E_∞ and globally bounded from below; moreover, it is an increasing sequence of subharmonic functions. It turns out that $u := \lim_{k\to\infty} u_k \in \mathcal{SH}(E_\infty)$ and $\lim_{(-1,0)\ni x\to 0} u(x) \le 0$. Finally, we define the following bounded pseudoconvex Hartogs domain with one-dimensional fibers:

$$G_{E_\infty} := \{(z, w) \in E_\infty \times \mathbb{C} : |w| < e^{-u(z)}\}.$$

Obviously, the point $(0, 0) \in \partial G_{E_\infty}$ does not satisfy any of the conditions in Theorem 12.4.12. Nevertheless, a correct choice of the n_j's may show that G_{E_∞} satisfies the cone condition of Theorem in [260] at $(0, 0)$. Therefore,

$$\mathfrak{K}_{G_{E_\infty}}((z, w)) \underset{G_{E_\infty}\ni(z,w)\to(0,0)}{\longrightarrow} \infty.$$

The discussion of the other boundary points, with the help of Theorem 12.4.12 and Theorem 12.1.25, even proves that G_{E_∞} is $\boldsymbol{K}$-exhaustive. Try to give a complete description of those bounded pseudoconvex Hartogs domains with m-dimensional fibers that are $\boldsymbol{K}$-exhaustive.

12.5 Bergman exhaustiveness II – plane domains

In the complex plane there is a full characterization of bounded domains being $\boldsymbol{K}$-exhaustive in terms of the potential theory (see [570, 424]). To be able to present this result we first recall a few facts from the classical plane potential theory, taken from [446].

Let $K \subset \mathbb{C}$ be compact and

$$\mathcal{PM}(K) := \{\mu : \mu \text{ a probabilistic measure of } K\}.$$

For $\mu \in \mathcal{PM}(K)$,

$$p_\mu(\lambda) := \int_K \log|\lambda - \zeta| d\mu(\zeta), \quad \lambda \in \mathbb{C},$$

is the *logarithmic potential of* μ. Recall that $p_\mu \in \mathcal{SH}(\mathbb{C})$ and that $p_\mu|_{\mathbb{C}\setminus K}$ is a harmonic function. To any such a μ one associates its *energy*

$$I(\mu) := \int_K p_\mu(\lambda) d\mu(\lambda) = \int_K \int_K \log|\lambda - \zeta| d\mu(\lambda) d\mu(\zeta).$$

A probabilistic Borel measure $\nu \in \mathcal{PM}(K)$ is called the *equilibrium measure of* K if $I(\nu) = \sup_{\mu \in \mathcal{PM}(K)} I(\mu)$. It is known that the equilibrium measure exists and is unique if K is not a polar set; we then write ν_K. Moreover, the *logarithmic capacity of any set* $M \subset \mathbb{C}$ is given by

$$\operatorname{cap} M = \operatorname{cap}(M) := \exp(\sup\{I(\mu) : K \subset M \text{ compact}, \ \mu \in \mathcal{PM}(K)\}).$$

If $M = K$ is compact and not polar, then $\operatorname{cap} K = e^{I(\nu_K)}$. Moreover, if M is any Borel set then M is polar iff $\operatorname{cap} M = 0$. In particular, $\operatorname{cap} \varnothing = 0$.

For further applications, we collect a few well-known properties of the logarithmic capacity:

- if $M_1 \subset M_2$, then $\operatorname{cap} M_1 \le \operatorname{cap} M_2$; (12.5.1)

- if $M_1 \subset M_2 \subset \dots$ are Borel sets, then $\operatorname{cap}\left(\bigcup_{j=1}^{\infty} M_j\right) = \lim_{j\to\infty} \operatorname{cap} M_j$; (12.5.2)

- if $K_1 \supset K_2 \supset \dots$ are compact sets, then $\operatorname{cap}\left(\bigcap_{k=1}^{\infty} K_k\right) = \lim_{k\to+\infty} \operatorname{cap} K_k$; (12.5.3)

- if $M = \bigcup_{j=1}^{N} M_j$, M_j Borel sets with $\operatorname{diam} M \le d$, $N \in \mathbb{N} \cup \{\infty\}$, then

$$\frac{1}{\log d - \log \operatorname{cap} M} \le \sum_{j=1}^{N} \frac{1}{\log d - \log \operatorname{cap} M_j}; \tag{12.5.4}$$

- if $M = \bigcup_{j=1}^{N} M_j$, M_j Borel sets with $\operatorname{dist}(M_j, M_k) \geq d > 0$, $k \neq j$, $N \in \mathbb{N} \cup \{\infty\}$, then $\dfrac{1}{\log^+ \frac{d}{\operatorname{cap} M}} \geq \sum_{j=1}^{N} \dfrac{1}{\log^+ \frac{d}{\operatorname{cap} M_j}}$; (12.5.5)

- **Theorem of Frostman.** *Let $K \subset \mathbb{C}$ be a non-polar compact subset and let ν_K be its equilibrium measure. Then, $p_{\nu_K} \geq \log \operatorname{cap} K$ on $\mathbb{C}$ and $p_{\nu_K} \equiv \log \operatorname{cap} K$ on $K \setminus F$, where $F \subset \partial K$ is a suitable polar $\mathcal{F}_\sigma$-set. Moreover, $p_{\nu_K}(z) = \log \operatorname{cap} K$ for $z \in \partial K$, whenever z is regular for the Dirichlet problem for the unbounded component of $\mathbb{C} \setminus K$.* (12.5.6)

- $\operatorname{cap} \mathbb{D}(z,r) = \operatorname{cap}(\partial \mathbb{D}(z,r)) = r$ and $\operatorname{cap} K = \operatorname{cap}(\partial K) \leq \operatorname{diam} K$ for any compact set $K \subset \mathbb{C}$. (12.5.7)

For a compact set in the complex plane, we introduce its Cauchy transform.

Definition 12.5.1. Let $K \subset \mathbb{C}$ be compact. The function $f_K : \mathbb{C} \setminus K \longrightarrow \mathbb{C}$,

$$f_K(z) := \begin{cases} \int_K \frac{d\nu_K(\zeta)}{z-\zeta}, & \text{if } K \text{ is not polar} \\ 0, & \text{if } K \text{ is polar} \end{cases},$$

is called *the Cauchy transform of K*. (Recall that ν_K is the equilibrium measure of K.)

Obviously, $f_K \in \mathcal{O}(\mathbb{C} \setminus K)$ and $f_K|_D \in L_h^2(D)$ for any bounded domain $D \subset \mathbb{C} \setminus K$. Then,

Lemma 12.5.2 (cf. [570]). *For a $\varrho \in (0, \frac{1}{2})$ there exist positive numbers C_1, C_2 such that for any pair of disjoint compact sets $K, L \subset \mathbb{D}(\varrho)$ and any domain $D \subset \mathbb{D}(\varrho) \setminus (K \cup L)$ the following inequalities hold:*

$$|\langle f_K, f_L \rangle_{L^2(D)}| \leq C_2 - C_1 \log \operatorname{dist}(K, L), \tag{12.5.8}$$

$$\|f_K\|_{L^2(D)}^2 \leq C_2 - C_1 \log \operatorname{cap} K. \tag{12.5.9}$$

Proof. Obviously, both inequalities are true for any constants C_j when K or L is a polar set. So we may assume that both sets are not polar.

Applying the Fubini theorem, we get the following inequality:

$$\begin{aligned} |\langle f_K, f_L \rangle_{L^2(D)}| &= \left| \int_D \int_K \frac{d\nu_K(\zeta)}{z-\zeta} \int_L \frac{d\nu_L(\eta)}{\overline{z}-\overline{\eta}} d\mathcal{L}^2(z) \right| \\ &\leq \int_K \int_L \int_{\mathbb{D}(\varrho)} \frac{1}{|z-\zeta||z-\eta|} d\mathcal{L}^2(z) d\nu_L(\eta) d\nu_K(\zeta). \end{aligned}$$

Now, we discuss the interior integral.

Take $\zeta, \eta \in \mathbb{D}(\varrho)$, $\zeta \neq \eta$. Then,

$$\int_{\mathbb{D}(\varrho)} \frac{d\mathcal{L}^2(z)}{|z-\zeta||z-\eta|} \leq \int_{\mathbb{D}} \frac{d\mathcal{L}^2(z)}{|z||z-(\zeta-\eta)|} = \int_{\mathbb{D}\left(\frac{1}{|\zeta-\eta|}\right)} \frac{d\mathcal{L}^2(z)}{|z||z-1|}$$

$$= \int_{\mathbb{D}\left(\frac{1}{2\varrho}\right)} \frac{d\mathcal{L}^2(z)}{|z||z-1|} + \int_{\mathbb{D}\left(\frac{1}{|\zeta-\eta|}\right)\setminus\mathbb{D}\left(\frac{1}{2\varrho}\right)} \frac{d\mathcal{L}^2(z)}{|z||z-1|}.$$

Observe that the first term in the last expression is finite and independent of η and ζ. For the second summand, we proceed as follows:

$$\int_{\mathbb{D}\left(\frac{1}{|\zeta-\eta|}\right)\setminus\mathbb{D}\left(\frac{1}{2\varrho}\right)} \frac{d\mathcal{L}^2(z)}{|z||z-1|} = \int_{\frac{1}{2\varrho}}^{\frac{1}{|\zeta-\eta|}} \int_0^{2\pi} \frac{dr\,d\theta}{|1-re^{i\theta}|}$$

$$= \int_{\frac{1}{2\varrho}}^{\frac{1}{|\zeta-\eta|}} \int_0^{2\pi} \left| 1 + \frac{e^{i\theta}}{r} + \frac{e^{2i\theta}}{r^2\left(1-\frac{e^{i\theta}}{r}\right)} \right| \frac{dr\,d\theta}{r}$$

$$\leq \int_{\frac{1}{2\varrho}}^{\frac{1}{|\zeta-\eta|}} \frac{C_1\,dr}{r} \leq -C_1 \log|\zeta-\eta|,$$

where C_1 is independent of the discussed ζ, η.

Consequently,

$$\int_{\mathbb{D}(\varrho)} \frac{d\mathcal{L}^2(z)}{|z-\zeta||z-\eta|} \leq C_2 - C_1 \log|\zeta-\eta|, \quad \zeta, \eta \in \mathbb{D}(\varrho),\ \zeta \neq \eta,$$

where C_1, C_2 are positive constants.

Coming back to the beginning, we obtain

$$|\langle f_K, f_L\rangle_{L^2(D)}| \leq C_2 - C_1 \int_K \int_L \log|\zeta-\eta|\,d\nu_K\,d\nu_L. \qquad \square$$

The main notion for our further discussion will be the following potential theoretic function:

Definition 12.5.3. Let $D \subset \mathbb{C}$ be a bounded domain. Put $\boldsymbol{\alpha}_D : \overline{D} \longrightarrow (-\infty, \infty]$,

$$\boldsymbol{\alpha}_D(z) := \int_0^{1/4} \frac{dr}{-r^3 \log \operatorname{cap}(\mathbb{D}(z,r) \setminus D)} = \int_0^{1/4} \frac{dr}{-r^3 \log \operatorname{cap}(\overline{\mathbb{D}}(z,r) \setminus D)}.$$

Note that the two integrals are equal because of (12.5.1), (12.5.2), and the convergence theorem of Lebesgue.

Remark 12.5.4. We denote by $\mathbb{A}_k(z)$ the annulus with center z and radii $1/2^{k+1}$, $1/2^k$, i.e.,

$$\mathbb{A}_k(z) := \mathbb{A}(z, 1/2^{k+1}, 1/2^k).$$

Then, for a bounded domain $D \subset \mathbb{C}$, there is an alternative description of $\boldsymbol{\alpha}_D$, namely,

$$\frac{1}{8}\sum_{k=3}^{\infty} \frac{2^{2k}}{-\log \operatorname{cap}(\mathbb{A}_k(z) \setminus D)} \le \boldsymbol{\alpha}_D(z) \le 8 \sum_{k=2}^{\infty} \frac{2^{2k}}{-\log \operatorname{cap}(\mathbb{A}_k(z) \setminus D)}, \quad z \in \overline{D}.$$

To get the lower estimate one only has to use the monotonicity of cap (see (12.5.1)), whereas the upper estimate is based on property (12.5.4) of cap.

Moreover, $\boldsymbol{\alpha}_D$ is semicontinuous from below on $\overline{D}$ and continuous on D; here, use properties (12.5.4) and (12.5.7) of cap and Fatou's lemma, respectively the Lebesgue theorem.

Remark 12.5.5. For a point $z_0 = x_0 + iy_0 \in \mathbb{C}$, we define the annuli with respect to the maximum norm, i.e.,

$$\widetilde{\mathbb{A}}_k(z_0) := \{z = x + iy \in \mathbb{C} : 1/2^{k+1} \le \max\{|x - x_0|, |y - y_0|\} \le 1/2^k\}, \ k \in \mathbb{N}.$$

Moreover, let $\widetilde{\mathbb{D}}(a,r) := \{z = x+iy \in \mathbb{C} : \max\{|x-\operatorname{Re} a|, |y-\operatorname{Im} a|\} < r\}$, where $a \in \mathbb{C}$ and $r > 0$. Recall that $\operatorname{cap} \overline{\widetilde{\mathbb{D}}}(a,r) = c_0 2r$, where $c_0 = 0.59\ldots$ (see [446], Table 5.1).

Then, we may define a notion similar to $\boldsymbol{\alpha}_D$, namely,

$$\widetilde{\boldsymbol{\alpha}}_D(z) := \int_0^{1/4} \frac{dr}{-r^3 \log \operatorname{cap}(\overline{\widetilde{\mathbb{D}}}(z,r) \setminus D)}, \qquad z \in D.$$

We only note that both functions $\boldsymbol{\alpha}_D$ and $\widetilde{\boldsymbol{\alpha}}_D$ are comparable and that for the new functions, inequalities like the ones in Remark 12.5.4 hold.

It turns out that, in general, the function $\boldsymbol{\alpha}_D$ is not continuous on $\overline{D}$ (see Example 12.5.6).

Example 12.5.6 (cf. [570]). For an $n \in \mathbb{N}$ put

$$M_n := \operatorname{int} \widetilde{\mathbb{A}}_n(0) \cap \left\{ \frac{j}{2^n 2^{n^3}} + i \frac{k}{2^n 2^{n^3}} : j,k \in \mathbb{Z} \right\}.$$

Then M_n has $l_n := (2^{1+n^3} - 1)^2 - (2^{n^3} + 1)^2 \le 2^{2n^3+2}$ elements. We denote them by $z_{n,k}$, $k = 1, \ldots, l_n$.

Then we define the following plane domain:

$$D := \widetilde{\mathbb{D}}(0, 1/2) \setminus \left(\bigcup_{n=2}^{\infty} \bigcup_{k=1}^{l_n} \overline{\widetilde{\mathbb{D}}}(z_{n,k}, r_n) \cup \{0\} \right).$$

Here the radii $r_n > 0$ are chosen such that

$$-\log \operatorname{cap}(\overline{\widetilde{\mathbb{D}}}(0, r_n)) = n^2 2^{2n(1+n^2)}, \quad n \geq 2.$$

Observe that the distance between two different $z_{n,k}$'s is at least $d_n := \frac{1}{2^n 2^{n^3}}$. Using the formula for cap $\overline{\widetilde{\mathbb{D}}}(0, r_n)$, we see that $r_n < \frac{1}{4 \cdot 2^{n^3} 2^n}$. Hence, the distance between different $\overline{\widetilde{\mathbb{D}}}(z_{n,k}, r_n)$'s is at least $b_n := \frac{1}{2^{1+n+n^3}}$.

After having this geometric information, we are going to estimate $\widetilde{\boldsymbol{\alpha}}_D(0)$ from above, namely,

$$\begin{aligned} \widetilde{\boldsymbol{\alpha}}_D(0) &\leq C_1 \sum_{n=2}^{\infty} \frac{2^{2n}}{-\log \operatorname{cap}(\widetilde{\mathbb{A}}_n(0) \setminus D)} \overset{(12.5.4)}{\leq} C_1 \sum_{n=2}^{\infty} \sum_{k=1}^{l_n} \frac{2^{2n}}{-\log \operatorname{cap}(\overline{\widetilde{\mathbb{D}}}(z_{n,k}, r_n))} \\ &= C_1 \sum_{n=2}^{\infty} \frac{2^{2n} 2^{2+2n^3}}{n^2 2^{2n} 2^{2n^3}} = 4C_1 \sum_{n=2}^{\infty} \frac{1}{n^2} < \infty. \end{aligned}$$

Therefore, $\boldsymbol{\alpha}_D(0) < \infty$.

To see that $\boldsymbol{\alpha}_D$ is not continuous at 0 take the points $w_n := \frac{3}{2^{n+2}} \in \widetilde{\mathbb{A}}_n(0)$, i.e., the middle points of the interval $[1/2^{n+1}, 1/2^n]$, which converge to 0. Then,

$$\begin{aligned} \widetilde{\boldsymbol{\alpha}}_D(w_n) &\geq C_2 \sum_{j=3}^{n^3-1} \frac{2^{2(n+j)}}{-\log \operatorname{cap}(\widetilde{\mathbb{A}}_{n+j}(w_n) \setminus D)} \\ &\geq C_2 \sum_{j=3}^{n^3-1} \frac{2^{2(n+j)}}{-\log \operatorname{cap}(\widetilde{\mathbb{A}}_{n+j}(w_n) \cap \widetilde{\mathbb{A}}_n(0) \setminus D)}. \end{aligned}$$

To continue with the estimate, we note that there are at least $m_{n,j} := 3 \cdot 2^{2(n^3-j)}$ of the "balls" $\overline{\widetilde{\mathbb{D}}}(z_{n,k}, r_n)$ contained in $\mathbb{A}_n(0) \cap \mathbb{A}_{n+j}(w_n)$ for $n \geq 2$ and $j \geq 3$. Denote the union of these "balls" by $B_{n,j}$. Hence, we get

$$\widetilde{\boldsymbol{\alpha}}_D(w_n) \geq C_2 \sum_{j=3}^{n^3-1} \frac{2^{2(n+j)}}{-\log \operatorname{cap} B_{n,j}} =: I(n).$$

To be able to apply (12.5.5), we have to observe that

$$2^{2(1+n+n^3)}\operatorname{cap}\overline{\widetilde{\mathbb{D}}}(0,r_n) < 1, \tag{12.5.10}$$

$$2^{2(1+n+n^3)}\operatorname{cap} B_{n,j} < 1, \tag{12.5.11}$$

if $n \geq n_0$ for a suitable $n_0 \in \mathbb{N}$.

The inequality (12.5.10) is a direct consequence of the data used so far. To obtain (12.5.11), we apply property (12.5.4) and get

$$\frac{1}{-\log\operatorname{cap} B_{n,j}} \leq \frac{m_{n,j}}{-\log\operatorname{cap}\overline{\widetilde{\mathbb{D}}}(0,r_n)} \leq \frac{3\cdot 2^{2(n^3-j)}}{n^2 2^{2n+2n^3}} = \frac{3}{n^2 2^{2j+2n}}.$$

Hence, $2^{2(1+n+n^3)}\operatorname{cap} B_{n,j} \leq \frac{2^{2(1+n+n^3)}}{\exp(n^2 2^{2(n+j)}/3)} \underset{n\to\infty}{\longrightarrow} 0$. Therefore, if $n \geq n_0$, then

$$\begin{aligned}\widetilde{\boldsymbol{\alpha}}_D(w_n) &\geq I(n) \geq C_2 \sum_{j=3}^{n^3-1} \frac{1}{2\log\frac{b_n}{\operatorname{cap} B_{n,j}}} \geq C_2 \sum_{j=3}^{n^3-1} \frac{2^{2(n+j)}\cdot 3\cdot 2^{2(n^3-j)}}{2n^2 2^{2(n+n^3)}} \\ &\geq C_2 \sum_{j=3}^{n^3-1} \frac{3}{2n^2} \geq C_2 \frac{n^3-1-2}{2n^2} \underset{n\to\infty}{\longrightarrow} \infty,\end{aligned}$$

where one has used (12.5.10) and (12.5.5).

Hence, $\lim_{D\ni z\to 0}\boldsymbol{\alpha}_D(w_n) = \lim_{D\ni z\to 0}\widetilde{\boldsymbol{\alpha}}_D(w_n) = \infty$.

Finally, we formulate the main result of this section.

Theorem 12.5.7 (cf. [570]). *Let $D \subset \mathbb{C}$ be a bounded domain, $z_0 \in \partial D$. Then, the following properties are equivalent:*

(i) *D is **K**-exhaustive at z_0 (i.e., $\lim_{D\ni z\to z_0}\mathfrak{K}_D(z) = \infty$);*

(ii) $\lim_{D\ni z\to z_0}\boldsymbol{\alpha}_D(z) = \infty$.

Proof. During the whole proof, we may assume that $D \subset \mathbb{D}(1/2)$ and $z_0 = 0 \in \partial D$.

(ii) $\Longrightarrow$ (i): Assume that (i) is not true. Then, there exists a sequence $(z_k)_{k\in\mathbb{N}} \subset D\cap\mathbb{D}(1/8)$ with $\lim_{k\to\infty} z_k = 0$ and $\sup_{k\in\mathbb{N}}\mathfrak{K}_D(z_k) =: M < \infty$.

Put $K_n^k := \mathbb{A}_n(z_k)\setminus D$, $n \geq 2$, $k \in \mathbb{N}$. Since $z_k \in D$, there is an $N_k \in \mathbb{N}$ such that $K_n^k = \varnothing$ for all $n > N_k$. Observe that necessarily $N_k \underset{k\to\infty}{\longrightarrow} \infty$.

By assumption we know that $\boldsymbol{\alpha}_D(z_k) \longrightarrow \infty$. Therefore, using Remark 12.5.4, we have

$$S_k := \sum_{n=2}^{N_k} \frac{2^{2n}}{-\log\operatorname{cap} K_n^k} \underset{k\to\infty}{\longrightarrow} \infty.$$

Put

$$K_{n,j}^k := K_n^k \cap \{z_k + re^{i\theta} : r > 0, -\pi/3 + (j-1)2\pi/3 \le \theta \le \pi/3 + (j-1)2\pi/3\}, \quad j = 1, 2, 3.$$

By virtue of property (12.5.4) for the function cap, we have

$$\frac{1}{-\log \operatorname{cap} K_n^k} \le \sum_{j=1}^{3} \frac{1}{-\log \operatorname{cap} K_{n,j}^k}.$$

Choose $j(n,k)$ such that $\operatorname{cap} K_{n,j}^k \le \operatorname{cap} K_{n,j(n,k)}^k$, $j = 1, 2, 3$, and put $\widetilde{K}_n^k := K_{n,j(n,k)}^k$. Then,

$$\frac{1}{3} S_k \le \sum_{n=2}^{N_k} \frac{2^{2n}}{-\log \operatorname{cap} \widetilde{K}_n^k} \underset{k\to\infty}{\longrightarrow} \infty.$$

Define

$$f_{n,k}(z) := \begin{cases} \int_{\widetilde{K}_n^k} \frac{d\nu_{n,k}(\zeta)}{z - e^{i\theta_{n,k}}\zeta} & \text{if } \operatorname{cap} K_n^k \ne 0 \\ 0 & \text{if } \operatorname{cap} K_n^k = 0 \end{cases}, \quad z \in \mathbb{C} \setminus \widetilde{K}_n^k,$$

where $\nu_{n,k} := \nu_{\widetilde{K}_n^k}$ and $\theta_{n,k}$ such that $\arg(z_k - e^{i\theta_{n,k}}\zeta) \in [-\pi/3, \pi/3]$ for all $\zeta \in \widetilde{K}_n^k$. Then,

$$|f_{n,k}(z_k)| \ge \operatorname{Re}\left(\int_{\widetilde{K}_n^k} \frac{d\nu_{n,k}(\zeta)}{z_k - e^{i\theta_{n,k}}\zeta}\right) \ge \widetilde{C}_3 \int_{\widetilde{K}_n^k} \frac{d\nu_{n,k}(\zeta)}{|z_k - \zeta|} \ge C_3 2^n,$$

where $\widetilde{C}_3, C_3$ are fixed positive constants.

Now the following two cases have to be discussed:

Case 1: Assume that there are a subsequence of (z_k), again denoted by (z_k), and a sequence $(n_k)_k \subset \mathbb{N}$ with $n_k \le N_k$, $k \in \mathbb{N}$, such that

$$\lim_{k\to\infty} \frac{2^{2n_k}}{-\log \operatorname{cap} \widetilde{K}_{n_k}^k} = \infty. \tag{12.5.12}$$

Put $f_k := f_{n_k,k}$. Then, by virtue of Lemma 12.5.2, we have

$$\|f_k\|_{L^2(D)}^2 \le C_2 - C_1 \log \operatorname{cap} \widetilde{K}_{n_k}^k.$$

Therefore, taking (12.5.12) into account, it follows that $\lim_{k\to\infty} \mathfrak{K}_D(z_k) = \infty$; a contradiction.

Case 2: There remains the case where we have a certain positive constant C_4 such that

$$\frac{2^{2n}}{-\log \operatorname{cap} \widetilde{K}_n^k} \leq C_4, \quad k \in \mathbb{N},\ n = 2, 3, \ldots, N_k.$$

Put $c_{k,n} := \operatorname{cap} \widetilde{K}_n^k$, $f_{k,n} := f_{\widetilde{K}_n^k}$. We are going to choose complex numbers $a_{k,n}$ with $a_{k,n} f_{k,n}(z_k) \geq 0$ such that if

$$f_k := \sum_{n=2}^{N_k} a_{k,n} f_{k,n},$$

then $(|f_k(z_k)|)_k$ is unbounded, whereas $(\|f_k\|_{L^2(D)})_k$ remains bounded by a positive constant C. In that situation we get

$$M \geq \mathfrak{K}_D(z_k) \geq \frac{|f_k(z_k)|^2}{\|f_k\|_{L^2(D)}^2} \geq \frac{1}{C^2} |f_k(z_k)|^2;$$

a contradiction.

As a first step in constructing the numbers $a_{k,n}$, we observe the following inequalities (see Lemma 12.5.2):

$$|2 \operatorname{Re}\langle f_{k,m}, f_{k,n}\rangle_{L^2(D)}| \leq \|f_{k,m}\|_{L^2(D)}^2 + \|f_{k,n}\|_{L^2(D)}^2 \leq 2C_2 - C_1 \log(c_{k,m} c_{k,n}),$$

when $|n - m| \leq 1$, and

$$\begin{aligned} &|2 \operatorname{Re}\langle f_{k,m}, f_{k,n}\rangle_{L^2(D)}| \\ &\quad \leq 2C_2 + 2C_1 \max\left\{\left|\log\left(\frac{1}{2^{m+1}} - \frac{1}{2^{m+2}}\right)\right|, \left|\log\left(\frac{1}{2^{n+1}} - \frac{1}{2^{n+2}}\right)\right|\right\} \\ &\quad \leq 2C_2 + C_5 mn, \end{aligned}$$

when $|n - m| \geq 2$.

Put $a_{k,n} := 0$ if $\operatorname{cap} \widetilde{K}_{k,n} = 0$. Then,

$$\begin{aligned} \|f_k\|_{L_h^2(D)}^2 &\leq C_6 \sum_{n=1}^{N_k} |a_{k,n}|^2 (-\log c_{k,n}) + C_6 \sum_{n,m=2,\ |n-m|\geq 2}^{N_k} |a_{k,n}||a_{k,m}| nm \\ &\leq C_7 \left(\sum_{n=2}^{N_k} |a_{k,n}|^2 (-\log c_{k,n}) + \left(\sum_{n=2}^{N_k} n |a_{k,n}| \right)^2 \right). \end{aligned}$$

Let $|a_{k,n}| := \frac{2^n}{-\log c_{k,n}} b_{k,n}$, where the numbers $b_{k,n} \geq 0$ will be fixed later. Then,

$$|f_k(z_k)| = \sum_{n=2}^{N_k} a_{k,n} f_{k,n}(z_k) \geq C_3 \sum_{n=2}^{N_k} \frac{2^n}{-\log c_{k,n}} b_{k,n} 2^n.$$

It remains to find numbers $b_{k,n}$ such that $|f_k(z_k)| \underset{k\to\infty}{\longrightarrow} \infty$, but

$$\|f_k\|^2_{L^2_h(D)} \le C_7\left(\sum_{n=2}^{N_k}\frac{2^{2n}}{-\log c_{k,n}}b^2_{k,n} + \left(\sum_{n=2}^{N_k}\frac{n}{2^n}\frac{2^{2n}}{-\log c_{k,n}}b_{k,n}\right)^2\right) \tag{12.5.13}$$

remains bounded.

Put $\nu_{k,n} := \frac{2^{2n}}{-\log c_{k,n}}$, $k \in \mathbb{N}$. Recall that $S_k = \sum_{n=2}^{N_k}\nu_{k,n} \underset{k\to\infty}{\longrightarrow} \infty$, $\nu_{k,n} \le C_4$, $k \in \mathbb{N}$, and $N_k \underset{k\to\infty}{\longrightarrow} \infty$. So, we may find sequences $(n_{k,j})_{j=0}^{q_k}$, where $n_{k,0} = 1$, $n_{k,q_k} = N_k$, and $q_k \underset{k\to\infty}{\longrightarrow} \infty$ such that

$$\nu_{k,n_{k,j}+1} + \cdots + \nu_{k,n_{k,j+1}} > 1 \text{ and } \frac{l}{2^l} < \frac{1}{j+1}, \quad j = 0,\dots,q_k-1,\ l > n_{k,j+1}.$$

Now, we define

$$b_{k,n_{k,j}+1} = \cdots = b_{k,n_{j+1}} := \frac{1}{(j+1)(\nu_{k,n_j+1} + \cdots + \nu_{k,n_{k,j+1}})}, \qquad j = 0,\dots,q_k-1.$$

With this setting, we finally obtain that $|f_k(z_k)| \underset{k\to\infty}{\longrightarrow} \infty$ and that $(\|f_k\|_{L^2(D)})_k$ remains bounded (compare (12.5.13)). Hence $\mathfrak{K}_D(z_k) \longrightarrow \infty$; a contradiction.

(i) $\Longrightarrow$ (ii): Suppose that there exists a sequence $(z_k)_k \subset D$, $z_k \underset{k\to\infty}{\longrightarrow} 0$, such that, for a suitable positive number M, $\boldsymbol{\alpha}_D(z_k) \le M$, $k \in \mathbb{N}$. Then, by virtue of Remark 12.5.4,

$$\sum_{n=2}^{\infty}\frac{2^{2n}}{-\log \operatorname{cap}(\mathbb{A}_n(z_k)\setminus D)} \le 8M.$$

In particular, if $c_{k,n} := \operatorname{cap}(\mathbb{A}_n(z_k)\setminus D)$, then $\log c_{k,n} \le -\frac{2^{2n}}{8M}$, $k,n \in \mathbb{N}$, $n \ge 2$. Therefore, we may find an $n_0 \in \mathbb{N}$ such that $\log c_{k,n} + 1 < -(n+1)\log 2 - 1$, $n > n_0$, $k \in \mathbb{N}$.

Choose a $k_0 \in \mathbb{N}$ with $|z_k| < \frac{1}{2^{n_0+1}}$ for all $k \ge k_0$. Fix a $k \ge k_0$ and let $z \in \mathbb{A}_n(z_k)$, $2 \le n < n_0$. Then,

$$\frac{1}{4} + \frac{1}{2^{n_0+1}} \ge |z - z_k| + |z_k| \ge |z| \ge |z - z_k| - |z_k| \ge \frac{1}{2^{n_0}} - \frac{1}{2^{n_0+1}} = \frac{1}{2^{n_0+1}}.$$

Now choose a domain $D' \supset D$ such that $D' \cap \mathbb{D}(\frac{1}{2^{2n_0}}) = D \cap \mathbb{D}(\frac{1}{2^{2n_0}})$ and $A_n(z_k)\setminus D' = \varnothing$, $2 \le n < n_0$, $k \ge k_0$. Applying the localization result (cf. Theorem 12.1.29) for the Bergman kernel, we conclude that $\lim_{k\to\infty}\mathfrak{K}_{D'}(z_k) = \infty$.

Now, fix a $k \geq k_0$. Recall that there is an $n_1 > 2n_0$ such that $\mathbb{D}(z_k, \frac{1}{2^{n_1}}) \subset D'$. We may exhaust D' by a sequence of domains $D'_j \subset\subset D'$ with real analytic boundaries such that

- $\displaystyle\sum_{n=2}^{\infty} \frac{2^{2n}}{-\log \operatorname{cap}(\mathbb{A}_n(z_k) \setminus D'_j)} < 8M$,
- $\partial(\mathbb{A}_2(z_k) \setminus D'_j) = \partial\mathbb{D}(z_k, 1/4)$,
- $\widetilde{K}_n := \mathbb{A}_n(z_k) \setminus D'_j$ is either empty or non polar,
- if $\widetilde{K}_n \neq \varnothing$, then any boundary point of $\widetilde{K}_n$ is a regular point with respect to the unbounded component of its complement.

So Frostman's theorem (see (12.5.6)), together with the continuity principle for logarithmic potentials (see Theorem 3.1.3 in [446]) implies that the logarithmic potential $p_n := p_{\mu_{\widetilde{K}_n}}$ is continuous on $\mathbb{C}$, if $n \geq 3$ and $\widetilde{K}_n \neq \varnothing$. Moreover, for an $n \geq 3$ with $\widetilde{K}_n = \varnothing$ put $p_n := -\infty$.

For $n \geq 3$ choose $\chi_n \in \mathcal{C}^\infty(\mathbb{R}, [0,1])$ such that $\chi_n = 0$, if $\widetilde{K}_n = \varnothing$, or

$$\chi_n(t) := \begin{cases} 1, & \text{if } t \leq \log \operatorname{cap} \widetilde{K}_n + 1/2 \\ 0, & \text{if } t \geq -(n+1)\log 2 - 1/2 \end{cases},$$

and $|\chi'_n(t)| \leq \frac{2}{-M_1 \log \operatorname{cap} \widetilde{K}_n}$, where M_1 is a suitable positive number.

Define $f_n := f_{\widetilde{K}_n}$ and $\varphi_n := \chi_n \circ p_n$ for $n \geq 3$. Note that if $\widetilde{K}_n \neq \varnothing$, then $p_n(z) \geq -(n+1)\log 2$, $z \notin A_{n-1}(z_k) \cup A_n(z_k) \cup A_{n+1}(z_k)$, and that $p_n \in \mathcal{C}^\infty(\mathbb{C} \setminus \widetilde{K}_n)$. Thus, φ_n is a smooth function with support in $A_{n-1}(z_k) \cup A_n(z_k) \cup A_{n+1}(z_k)$, satisfying $\varphi_n|_{\widetilde{K}_n} = 1$ and $\frac{\partial p_n}{\partial \bar{z}}(z) = \frac{1}{2}\overline{f}_n(z)$, $z \notin \widetilde{K}_n$.

For $n = 2$ we put $p_2(z) := \log|z|$ and take a $\chi_2 \in \mathcal{C}^\infty(\mathbb{R}, [0,1])$ such that

$$\chi_2(t) = \begin{cases} 0, & \text{if } t \leq -\log 8 \text{ or } t \geq -\log 2 \\ 1, & \text{if } t \text{ is near } -\log 4 \end{cases},$$

and $|\chi'_2| \leq \frac{2}{\log 4}$. As above, put $\varphi_2 := \chi_2 \circ p_2$ and $f_2 := 1$.

Then,

$$\left|\frac{\partial \varphi_n}{\partial \bar{z}}(z)\right| \leq \frac{|f_n(z)|}{-M_2 \log \operatorname{cap} \widetilde{K}_n}, \quad z \in \mathbb{D}(1/2) \setminus \widetilde{K}_n,\ n \geq 2.$$

Finally, we define

$$\varphi := \sup\{\varphi_n : n \geq 2\}.$$

Note that the supremum is taken over at most three functions. Then, φ is a Lipschitz function satisfying $\varphi|_{\partial D} = 1$ and $\varphi = 0$ in a neighborhood of z_k.

Now let $f \in L^2_h(D')$ be arbitrarily chosen. Then, the Cauchy formula and the Green formula lead to the following equations:

$$|f(z_k)| = \frac{1}{2\pi}\left|\int_{\partial D_j}\frac{f(\lambda)d\lambda}{\lambda - z_k}\right| = \frac{1}{2\pi}\left|\int_{\partial D_j}\frac{(f\varphi)(\lambda)d\lambda}{\lambda - z_k}\right|$$
$$= \frac{1}{\pi}\left|\int_{D_j}\frac{f(\lambda)}{\lambda}\frac{\partial\varphi}{\partial\bar\lambda}d\mathcal{L}^2(\lambda)\right|.$$

Applying various versions of the Schwarz inequalities and Lemma 12.5.2 finally gives the following inequalities:

$$|f(z_k)| \le \sum_{n=2}^{\infty} 2^n \int_{A_n(z_k)\setminus\widetilde{K}_n} |f(\lambda)|\left(\frac{|f_{n-1}(\lambda)|}{-\log\operatorname{cap}\widetilde{K}_{n-1}} + \frac{|f(\lambda)|}{-\log\operatorname{cap}\widetilde{K}_n}\right.$$
$$\left.+\frac{|f_{n+1}(\lambda)|}{-\log\operatorname{cap}\widetilde{K}_{n+1}}\right)d\mathcal{L}^2(\lambda)$$
$$\le M_4\sum_{n=2}^{\infty}\|f\|_{L^2(A_n(z_k)\setminus\widetilde{K}_n)}\left(\frac{\|f_{n-1}\|^2_{D'}}{(-\log\operatorname{cap}\widetilde{K}_{n-1})^2} + \frac{\|f_n\|^2_{D'}}{(-\log\operatorname{cap} K_n)^2}\right.$$
$$\left.+\frac{\|f_{n+1}\|^2_{D'}}{(-\log\operatorname{cap}\widetilde{K}_{n+1})^2}\right)^{1/2}$$
$$\le M_5\left(\sum_{n=2}^{\infty}\|f\|^2_{L^2_h(A_n(z_k)\setminus\widetilde{K}_n)}\right)^{1/2}\left(\sum_{n=2}^{\infty}\frac{2^{2n}}{-\log\operatorname{cap}\widetilde{K}_n}\right)^{1/2} \le \sqrt{\widetilde{M}}\|f\|_{D'}$$

for sufficiently large k, where the constant on the right hand side is independent of k and f. Therefore, $(\mathfrak{K}_{D'}(z_k))_k$ is bounded; a contradiction. □

Corollary 12.5.8 (cf. [423]). *Let $a \in \partial G$ be a regular boundary point of the bounded domain $G \subset \mathbb{C}$. Then, $\lim_{G\ni z\to a}\mathfrak{K}(z) = \infty$.*

Proof. By virtue of the former theorem and the lower semicontinuity of the function $\boldsymbol{\alpha}_G$, it suffices to verify that $\boldsymbol{\alpha}_G(a) = \infty$. This is a simple consequence of the following, so-called Wiener criterion from potential theory in the plane (see [446], Theorems 4.2.4 and 5.4.1):

a point $b \in \partial G$ is regular if and only if $$\sum_{j=0}^{\infty}\frac{1}{\log 2 - \log\operatorname{cap}(\widehat{\mathbb{A}}_j(b)\setminus D)} = \infty,$$

where $\widehat{\mathbb{A}}_j(b) := \{z \in \mathbb{C} : 1/2^{j+1} < |z-b| \le 1/2^j\}$, $j \in \mathbb{Z}_+$. Note that $\widehat{\mathbb{A}}_j(a) \subset \mathbb{A}_j(a)$ for all j. Then, applying Remark 12.5.4, it follows that $\boldsymbol{\alpha}_G(0) = \infty$. Hence, Theorem 12.5.7 implies the claim of the corollary. □

Remark. The original proof was more complicated. It was based on the following comparison result between the Bergman kernel and the Azukawa metric:

Theorem* 12.5.9 (cf. [403, 61]). *Let $G \subset \mathbb{C}$ be a bounded domain. Then,*

$$2\pi \mathfrak{K}_G(z) \geq \boldsymbol{A}_G^2(z;1), \quad z \in G.$$

N. Suita (see [490]) proved Theorem 12.5.9 for any double connected domain with no degenerated boundary component with the factor π instead of 2π. Then, Ohsawa showed in [403] the above estimate with the factor 750π instead of 2π, using the theory of the $\bar{\partial}$-equation. Theorem 12.5.9 as it is formulated was finally established in [61]. The question whether the estimate is even true with factor π was the so called *Suita conjecture*. In fact, very recently, the Suita conjecture was proved by Błocki in [63] (see also [62]).

Another application of Theorem 12.5.7 is the following higher dimensional result:

Theorem 12.5.10 (cf. [425]). *Let $D \subset \mathbb{C}^n$ be a bounded pseudoconvex domain, $u \in \mathcal{PSH}(D)$, $c \in \mathbb{R}$, D_c a connected component of $\{z \in D : u(z) < c\}$, and $a \in \partial D_c \cap D$. Then, $\mathfrak{K}_{D_c}(z) \longrightarrow \infty$ if $D_c \ni z \longrightarrow a$. In particular, any bounded balanced pseudoconvex domain is Bergman exhaustive* (*recall Theorem* 12.4.10 *for another proof*).

Proof. Choose an $r_0 < 1/4$ such that $\mathbb{B}(a, r_0) \subset D$. Now fix a $z \in D_c$ with $d = d(z) := \|z - a\| < r_0/5$. Then, $3d < r_0 - d$ and $\frac{r_0 - d}{4} \geq d$. Denote by L_z the complex line passing through z and a, and let G_z be the connected component of $D_c \cap L_z \subset \mathbb{C}$ that contains z.

Recall that, because of the Ohsawa–Takegoshi theorem (see Appendix B.8.1), we have $\mathfrak{K}_{D_c}(z) \geq C \mathfrak{K}_{G_z}(z)$, where the constant C is independent of z. Thus it remains to estimate $\mathfrak{K}_{G_z}(z)$ from below, which can be done with the help of the potential theoretic function $\boldsymbol{\alpha}_{G_z}$. Thus, we obtain

$$\mathfrak{K}_{G_z}(z) \geq C' \int_0^{1/4} \frac{dt}{-t^3 \log \operatorname{cap}(\widehat{\mathbb{D}}(z,t) \setminus G_z)} =: I(z),$$

where C' is a universal constant and $\widehat{\mathbb{D}}(z,s)$ means the disc in L_z with center z and radius s. Using the maximum principle for subharmonic functions, it follows that for any $s \in (d, r_0 - d)$ there exists a point $w \in L_z \setminus D_c$ with $\|w - z\| = s$. Then, potential theory (see [446], Exercises 5.3, 3(i)) gives that $\operatorname{cap}(\widehat{\mathbb{D}}(z,t) \setminus D_c) \geq \frac{r_0 - d}{4}$. Therefore,

$$I(z) \geq \frac{1}{-\log d} \int_{2d}^{3d} \frac{dt}{t^3} \geq \frac{5}{-72 d^2(z) \log d(z)};$$

in particular, we end up with $\mathfrak{K}_{D_c}(z) \geq C'' \frac{1}{-\|z-a\|^2 \log \|z-a\|} \longrightarrow \infty$ if $z \longrightarrow a$. □

Example 12.5.11. Let $u \in \mathcal{PSH}(\mathbb{C}^n)$ and assume such that for any $k > 0$, one has that $\lim_{\|z\|\to\infty} \|z\|^k e^{-u(z)} = 0$. Put

$$D_u := \{(z,w) \in \mathbb{C}^n \times \mathbb{C} : |w| < e^{-u(z)}\}.$$

D_u is an unbounded Hartogs domain over $\mathbb{C}^n$. Note that $D_u \ni (z,w) \longmapsto 1$ belongs to $L_h^2(D_u)$; thus, $\mathfrak{K}_{D_u} > 0$ on D_u. Moreover, using the property of u, it follows that $\mathfrak{K}_{D_u}((z,w)) \longrightarrow \infty$ if $D_u \ni (z,w) \longrightarrow \infty$. It remains to study the boundary behavior of $\mathfrak{K}_{D_u}$ at finite boundary points. So let $(z^0, w^0) \in \partial D_u$. Then, there is the biholomorphic mapping $\Phi : \mathbb{C}^n \times \mathbb{C}_* \longrightarrow \mathbb{C}^n \times \mathbb{C}_*$ defined by $(z,w) \longmapsto (wz, w)$. Put

$$G := \Phi(D_u \cap (\mathbb{C}^n \times \mathbb{C}_*)) = \{(\zeta,\eta) \in \mathbb{C}^n \times \mathbb{C}_* : \log|\eta| + u(\zeta/\eta) < 0\}.$$

Note that G is bounded and given as the sublevel set of a psh function. Thus, Theorem 12.5.10 applies and we get that $\mathfrak{K}_G((\zeta,\eta)) \longrightarrow \infty$ if $G \ni (\zeta,\eta) \longrightarrow \Phi(z^0, w^0)$. Hence, applying Proposition 12.1.10 it follows that $\mathfrak{K}_{D_u}((z,w)) \longrightarrow \infty$ when $D_u \ni (z,w) \longrightarrow (z^0,w^0)$ (recall that $\mathfrak{K}_{D_u} = \mathfrak{K}_{D_u \cap (\mathbb{C}^n \times \mathbb{C}_*)}$ on $D_u \cap (\mathbb{C}^n \times \mathbb{C}_*)$ and that $\det \Phi'(z^0, w^0) \neq 0$). Thus, D_u is an unbounded $\boldsymbol{K}$-exhaustive domain.

Remark 12.5.12. There are results similar to Theorem 12.5.7 for the so called point evaluation. To be more precise, let $z_0 \in \partial D$, where $D \subset \mathbb{C}$ is a bounded domain. Recall that $V := \{f \in L_h^2(D) : f$ is holomorphic in $D \cup \{z_0\}\}$ is dense in $L_h^2(D)$ (cf. Theorem B.8.2). Therefore, we may define the evaluation functional on V, i.e., $\Phi_{z_0} : V \longrightarrow \mathbb{C}$, $\Phi_{z_0}(f) := f(z_0)$. The point z_0 is called to be *a bounded evaluation point for* $L_h^2(D)$ if Φ extends to a continuous functional on $L_h^2(D)$. There is the following description of such points [219]:

Theorem. *Let D and z_0 be as above. Then, $\boldsymbol{\alpha}_D(z_0) = \infty$ iff z_0 is not a bounded evaluation point for $L_h^2(D)$.*

Observe, if z_0 is not a bounded evaluation point, then D is $\boldsymbol{K}$-exhaustive at z_0. Nevertheless, the converse statement is false (see Example 12.5.6).

Remark 12.5.13. For a bounded domain $D \subset \mathbb{C}$ there are notions analogous to the Bergman kernel taking derivatives into account, namely *the n-th Bergman kernel*

$$\mathfrak{K}_D^{(n)}(z) := \sup\{|f^{(n)}(z)|^2 : f \in L_h^2(D) \setminus \{0\},\ \|f\|_{L^2(D)} = 1\}, \quad n \in \mathbb{Z}_+,\ z \in D.$$

Observe that $\mathfrak{K}_D = \mathfrak{K}_D^{(0)}$. Moreover, one has the following potential theoretic function:

$$\boldsymbol{\alpha}_D^{(n)}(z) := \int_0^{1/4} \frac{dr}{-r^{2n+3}\log \operatorname{cap}(\mathbb{D}(z,r) \setminus D)}, \quad z \in \overline{D},\ n \in \mathbb{Z}_+.$$

Observe that $\boldsymbol{\alpha}_D = \boldsymbol{\alpha}_D^{(0)}$. There is the following relation between these notions (see [424]):

Theorem* 12.5.14. *Let $n \in \mathbb{Z}_+$ and $d > 1$. Then there exists a $C > 0$ such that*

- *for any domain $D \subset \mathbb{C}$ with* $\operatorname{diam} D < d$

$$C\boldsymbol{\alpha}_D^{(n)}(z) \le \mathfrak{K}_D^{(n)}(z), \quad z \in D;$$

- *for any domain $D \subset \mathbb{C}$ with $\frac{1}{d} <$* $\operatorname{diam} D < d$

$$\mathfrak{K}_D^{(n)}(z) \le C \max\{1, \boldsymbol{\alpha}_D^{(n)}(z)(\log \boldsymbol{\alpha}_D^{(n)}(z))^2\}, \quad z \in D.$$

Let $D \subset\subset \mathbb{C}$ be a domain and $z_0 \in \partial D$. Is it true that $\lim_{D \ni z \to z_0} \mathfrak{K}_D^{(n)}(z) = \infty$ implies that $\lim_{D \ni z \to z_0} \boldsymbol{\alpha}_D^{(n)}(z) = \infty$. **?**

With the help of the above theorem, there is a complete description of those Zalcman domains that are $\boldsymbol{K}$-exhaustive at all of its boundary points.

Corollary 12.5.15 (cf. [284]). *Let*

$$D := \mathbb{D} \setminus \Big(\bigcup_{k=1}^{\infty} \overline{\mathbb{D}}(x_k, r_k) \cup \{0\} \Big)$$

be a Zalcman domain,[3] *where $x_k > x_{k+1} > 0$, $\lim_{k\to\infty} x_k = 0$, and $r_k > 0$ such that $\overline{\mathbb{D}}(x_k, r_k) \subset \mathbb{D}$, $\overline{\mathbb{D}}(x_k, r_k) \cap \overline{\mathbb{D}}(x_j, r_j) = \varnothing$, $k, j \ge 1$, $k \ne j$. Assume that*

$$\exists_{\Theta_1 \in (0,1)}\ \exists_{\Theta_2 \in (\Theta_1, 1)} : \quad \Theta_1 \le \frac{x_{k+1}}{x_k} \le \Theta_2, \ k \in \mathbb{N}.$$

Then D is $\boldsymbol{K}$-exhaustive iff D is $\boldsymbol{K}$-exhaustive at 0 iff $\sum_{k=1}^{\infty} \frac{-1}{x_k^2 \log r_k} = \infty$ iff $\boldsymbol{\alpha}_D(0) = \infty$.

Observe that special cases were also treated in [402] and [93]. Moreover, we mention that the domains D in Corollary 12.5.15 are fat domains, but not all of them are $\boldsymbol{K}$-exhaustive (for another example see [275]).

Proof. First, observe that for every boundary point z_0 except the origin, we have

$$\lim_{D \ni z \to z_0} \mathfrak{K}_D(z) = \infty$$

(use Theorem 12.1.25).

[3] Observe that we here use a slightly more general notion than the one of a Zalcman type domain in Section 14.4.

Obviously, $\overline{\mathbb{D}}(x_{k+1} - r_{k+1}/2, r_{k+1}) \subset \overline{\mathbb{D}}(\delta) \setminus D$, $\delta \in (x_{k+1}, x_k)$. Then,

$$\begin{aligned}\boldsymbol{\alpha}_D(0) &\geq \sum_{k=1}^{\infty} \int_{x_{k+1}}^{x_k} \frac{dr}{-r^3 \log \operatorname{cap}(\overline{\mathbb{D}}(r) \setminus D)} \geq \sum_{k=k_0}^{\infty} \int_{x_{k+1}}^{x_k} \frac{dr}{-r^3 \log \frac{r_{k+1}}{2}} \\ &\geq \sum_{k=k_0}^{\infty} (x_k - x_{k+1}) \frac{-1}{x_k^3 \log \frac{r_{k+1}}{2}} \geq C \sum_{k=k_0}^{\infty} \frac{-1}{x_{k+1}^2 \log r_{k+1}},\end{aligned}$$

where C is a constant. Observe that for the last inequality the assumption on the centers x_k was used.

Now, the divergence of the series in the corollary implies that $\boldsymbol{\alpha}_D(0) = \infty$. By virtue of the lower semicontinuity of the function $\boldsymbol{\alpha}_D$, it follows that $\lim_{D \ni z \to 0} \boldsymbol{\alpha}_D(z) = \infty$.

On the other hand, we have

$$\begin{aligned}\boldsymbol{\alpha}_D(0) &= \left(\int_{x_1}^{1/4} + \sum_{k=1}^{\infty} \int_{x_{k+1}}^{x_k} \right) \frac{dr}{-r^3 \log \operatorname{cap}(\overline{\mathbb{D}}(r) \setminus D)} \\ &\leq C_1 + \sum_{k=1}^{\infty} \frac{x_k - x_{k+1}}{x_{k+1}^3} \sum_{j=k}^{\infty} \frac{-1}{\log r_j} \leq C_1 + C_2 \sum_{j=1}^{\infty} \frac{-1}{\log r_j} \sum_{k=1}^{j} \frac{1}{x_k^2} \\ &\leq C_1 + C_2 \sum_{j=1}^{\infty} \frac{-1}{\log r_j} \sum_{k=1}^{j} \frac{\Theta_2^{2(j-k)}}{x_j^2} \leq C_1 + C_3 \sum_{j=1}^{\infty} \frac{-1}{x_j^2 \log r_j},\end{aligned}$$

where $C_1 \geq 0$ and $C_2, C_3 > 0$ are suitable numbers. Observe that the last three inequalities follow from the assumptions on the centers x_k.

If the series in the corollary does converge, then $\boldsymbol{\alpha}_D(0) < \infty$. Moreover, directly from the definition we see that $\boldsymbol{\alpha}_D$ restricted to the interval $(-1/4, 0]$ is monotonically increasing. Hence, $\limsup_{0 > x \to 0} \boldsymbol{\alpha}_D(x) \leq \boldsymbol{\alpha}_D(0) < \infty$. So, the corollary is proved. □

Example 12.5.16. We discuss a particular case of a Zalcman domain, namely $x_k := (1/2)^k$ and $r_k := (1/2)^{kN(k)}$, where $N(k) \in \mathbb{N}$, $k \geq 2$. Then, we have

$$D \text{ is } \boldsymbol{K}\text{-exhaustive} \quad \text{iff} \quad \sum_{k=2}^{\infty} \frac{2^{2k}}{kN(k) \log 2} = \infty.$$

On the other hand, following Ohsawa [402] we have

$$D \text{ is hyperconvex} \quad \text{iff} \quad \sum_{k=2}^{\infty} 1/N(k) = \infty.$$

So we see that there are plenty of Zalcman domains that are not hyperconvex but, nevertheless are $\boldsymbol{K}$-exhaustive.

Another application of the potential theoretic description of Bergman completeness is due to X. Wang (see [525]). He used this example to show that Bergman completeness is not a quasi-conformal invariant.

Example 12.5.17. Fix $s \in (0, 1/4)$ and $t \in (0, 1/2)$. Moreover, let $\alpha_k \in (0, \pi/2)$ be defined through the relation $\sin \alpha_k = 1/e^{1/t^k}$, $k \in \mathbb{N}$. Put

- $A_{k,s,t} := \{s^k e^{i\theta} \in \mathbb{C} : -2\alpha_k \le \theta \le 2\alpha_k\}$;
- $A_{s,t} := \bigcup_{k\in\mathbb{N}} A_{k,s,t} \cup \{0\}$ and $D_{s,t} := \mathbb{D} \setminus A_{s,t}$.

Then, one has the following description of Bergman exhaustiveness:

$$D_{s,t} \text{ is } \boldsymbol{K}\text{-exhaustive at } 0 \text{ if and only if } s^2 < t.$$

Proof. Put

$$\begin{aligned} A_1 &:= \{\lambda \in \mathbb{C} : 2s^2 \le |\lambda| \le 1/4\} = \mathbb{A}(2s^2, 1/4), \\ A_k &:= \mathbb{A}(2s^{k+1}, 2s^k), \quad \mathbb{N} \ni k \ge 2. \end{aligned}$$

Write

$$\boldsymbol{\alpha}_{D_{s,t}}(0) = \left(\int_{2s^2}^{1/4} + \sum_{k=2}^{\infty} \int_{2s^{k+1}}^{2s^k}\right) \frac{d\mathcal{L}^1(r)}{-r^3 \log \operatorname{cap}(\mathbb{D}(r) \setminus D_{s,t})} =: C_1 + \sum_{k=2}^{\infty} C_k.$$

Then, one has the following estimates:

$$\begin{aligned} (1/4 - 2s^2)4^3 \frac{1}{-\log \operatorname{cap}(A_2 \setminus A_{s,t})} &\le C_1 \\ &\le (1/4 - 2s^2)(2s^2)^{-3} \sum_{k=1}^{\infty} \frac{1}{-\log \operatorname{cap}(A_k \setminus D_{s,t})} \end{aligned}$$

and

$$\begin{aligned} (2s^k)^{-2}(1-s)\frac{1}{-\log \operatorname{cap}(A_{k+1} \setminus D_{s,t})} &\le C_k \\ &\le (2s^{k+1})^{-2}(1/s - 1) \sum_{j=k}^{\infty} \frac{1}{-\log \operatorname{cap}(A_j \setminus D_{s,t})}, \quad k \ge 2. \end{aligned}$$

Using that $\operatorname{cap}(A_j \setminus D_{s,t}) = \frac{s^j}{e^{1/t^j}}$ (see [446], Table 5.1), one has

$$\frac{t^j}{1 - t \log s} \le \frac{1}{-\log \operatorname{cap}(A_j \setminus D_{s,t})} \le t^j, \quad j \in \mathbb{N}.$$

Putting everything together, one obtains

$$C(s,t)^{-1}\left(\frac{t}{s^2}\right)^k \le C_k \le C(s,t)\left(\frac{t}{s^2}\right)^k, \quad k \in \mathbb{N},$$

where $C(s,t)$ is a sufficiently large constant.

Assume that $s^2 \le t$. Then, the above estimates lead to $\boldsymbol{\alpha}_{D_{s,t}}(0) = \infty$. Taking into account that $\boldsymbol{\alpha}_{D_{s,t}}$ is lower semicontinuous gives that $\lim_{D_{s,t}\ni z\to 0} \boldsymbol{\alpha}_{D_{s,t}}(z) = \infty$, which implies, according to Theorem 12.5.7, that $D_{s,t}$ is Bergman exhaustive at the origin.

To get the other implication, let $D_{s,t}$ be Bergman exhaustive at 0. Then, Theorem 12.5.7 tells us that $\lim_{D_{s,t}\ni z\to 0} \boldsymbol{\alpha}_{D_{s,t}}(z) = \infty$. Note that for $r \in [0, 1/4)$ one has $\overline{\mathbb{D}}(x,r) \setminus D_{s,t} \subset \overline{\mathbb{D}}(r) \setminus D_{s,t}$ whenever $x \in (-1, 0)$ is sufficiently small. Therefore, $\infty = \lim_{x\nearrow 0} \boldsymbol{\alpha}_{D_{s,t}}(x) \le \boldsymbol{\alpha}_{D_{s,t}}(0)$. Now taking the previous estimates into account, we see that $s^2 \le t$. □

12.6 L_h^2-domains of holomorphy

The boundary behavior of the Bergman kernel may be used to give a complete description of L_h^2-domains of holomorphy. Recall that a domain $D \subset \mathbb{C}^n$ is an *L_h^2-domain of holomorphy* if for any pair of open sets $U_1, U_2 \subset \mathbb{C}^n$ with $\varnothing \ne U_1 \subset D \cap U_2 \ne U_2$, U_2 connected, there is an $f \in L_h^2(G)$ such that for any $F \in \mathcal{O}(U_2)$: $f|_{U_1} \ne F|_{U_1}$ (for more details see [265]). The precise result is the following:

Theorem 12.6.1 (cf. [423]). *For a bounded domain $D \subset \mathbb{C}^n$, the following conditions are equivalent:*

(i) *D is an L_h^2-domain of holomorphy;*

(ii) $\limsup_{D\ni z\to z_0} \mathfrak{K}_D(z) = \infty$ *for every boundary point $z_0 \in \partial D$;*

(iii) *for any boundary point $z_0 \in \partial D$ and for any open neighborhood $U = U(z_0)$ the set $U \setminus D$ is not pluripolar.*[4]

Proof of Theorem 12.6.1. We are mainly interested in the interaction of the Bergman kernel and the fact that D is an L_h^2-domain of holomorphy. Therefore, only the equivalence of (i) $\Longleftrightarrow$ (ii) will be verified.

The case $n = 1$ may be found in [110]. Therefore, we assume from now on that $n \ge 2$.

(ii) $\Longrightarrow$ (i): Suppose that D is not an L_h^2-domain of holomorphy. Then there exist concentric polydiscs $P \subset\subset \widetilde{P}$ satisfying $P \subset D$, $\partial P \cap \partial D \ne \varnothing$, and $\widetilde{P} \not\subset D$ such that for any function $g \in L_h^2(D)$ there exists a $\widehat{g} \in \mathcal{H}^\infty(\widetilde{P})$ with $\widehat{g}|_P = g|_P$.

[4] Recall that a set $P \subset \mathbb{C}^n$ is called to be *pluripolar* if there is a $u \in \mathcal{PSH}(\mathbb{C}^n)$, $u \not\equiv -\infty$, such that $P \subset u^{-1}(-\infty)$.

Let a be the center of P and let L be an arbitrary complex line through a. Then, $(L \cap \widetilde{P}) \setminus D =: K$ is a polar set (in L).

Indeed, suppose that K is not polar. Fix a compact non-polar subset $K' \subset K$. Then, according to Theorem 9.5 in [110], there exists a non-trivial function $f \in L_h^2(L \setminus K')$ that has no holomorphic extension to L. Since $K' \cap D = \varnothing$, Appendix B.8.1 guarantees the existence of a function $F \in L_h^2(D)$ with $F|_{L\cap D} = f|_{L\cap D}$. Hence, we find an $\widehat{F} \in \mathcal{O}(\widetilde{P})$ such that $\widehat{F}|_P = F|_P$. In particular, $\widehat{F}|_{L\cap\widetilde{P}}$ extends f to the whole of L; a contradiction.

So, $L \cap \widetilde{P} \cap D$ is connected.[5] Since L is arbitrary, $D \cap \widetilde{P}$ is connected. Therefore, for any function $g \in L_h^2(D)$ there exists a unique holomorphic extension $\widehat{g} \in \mathcal{H}^\infty(\widetilde{P})$ with $\widehat{g}|_{D\cap\widetilde{P}} = g|_{D\cap\widetilde{P}}$.

Consider the linear space

$$A := \{(g, \widehat{g}) : g \in L_h^2(D)\} \subset L_h^2(D) \times \mathcal{H}^\infty(\widetilde{P})$$

equipped with the norm $\|(g, \widehat{g})\| := \|g\|_{L^2(D)} + \|\widehat{g}\|_{\mathcal{H}^\infty(\widetilde{P})}$. Then, A is a Banach space.

Observe that the mapping $A \ni (g, \widehat{g}) \longmapsto g \in L_h^2(D)$ is a one-to-one, surjective, continuous, and linear mapping. Hence, in view of the Banach open mapping theorem, its inverse map is also continuous, i.e., there is a $C > 0$ such that

$$\|(g, \widehat{g})\| \le C\|g\|_{L^2(D)}, \quad g \in L_h^2(D).$$

In particular, $\|\widehat{g}\|_{\mathcal{H}^\infty(\widetilde{P})} \le C\|g\|_{L^2(D)}$. So, we are led to the following estimate:

$$\sup\{\mathfrak{K}_D(z) : z \in D \cap \widetilde{P}\} = \sup\left\{\frac{|g(z)|^2}{\|g\|_{L^2(D)}^2} : z \in D \cap \widetilde{P},\ 0 \not\equiv g \in L_h^2(D)\right\} \le C^2.$$

In particular, $\limsup_{z\to w} \mathfrak{K}_D(z) \le C^2$ for a point $w \in \partial P \cap \partial D \ne \varnothing$ (recall that such a point exists); a contradiction.

Before we are able to start the proof of (i) $\Longrightarrow$ (ii), we need some auxiliary results.

Lemma 12.6.2. *Let $G \subset \mathbb{C}$ be a bounded domain and let $a \in \partial G$. Assume that $\limsup_{G\ni z\to a} \mathfrak{K}_G(z) < \infty$. Then, there is a neighborhood $U = U(a)$ such that $U \setminus G$ is polar.*

Proof. Suppose that Lemma 12.6.2 is not true.

First, we claim that for any $r > 0$ the intersection $\mathbb{D}(a, r) \cap \partial G$ is not polar. Otherwise, there is an $r_0 > 0$ such that $\mathbb{D}(a, r_0) \cap \partial G$ is polar. Observe that $\mathbb{D}(a, r_0/4) \setminus G$ is not polar. Therefore, there exists a $b_0 \in \mathbb{D}(a, r_0/4) \setminus \overline{G}$. Now choose a point

[5] Recall that for a plane domain G and a relatively closed polar subset $M \subset G$, the open set $G \setminus M$ is connected – see also Appendix B.4.23.

$b \in \mathbb{D}(a, r_0/4) \cap G$ and a neighborhood $V = V(b) \subset G$. Since $\mathbb{D}(a, r_0) \cap \partial G$ is polar, there exists an $s \in (0, r_0/2)$ such that $\partial\mathbb{D}(b_0, s) \cap \partial G = \varnothing$, $b_0 + s\frac{b-b_0}{\|b-b_0\|} \in V \subset G$ (i.e., $\partial\mathbb{D}(b_0, s) \cap G \neq \varnothing$), and $\partial\mathbb{D}(b_0, s) \subset \mathbb{D}(a, r_0)$ (use, for example, [446], Exercise 5.3, 3(i)). Hence, $\partial\mathbb{D}(b_0, s) \subset G$. Therefore, for any $z \in \partial\mathbb{D}(b_0, s)$, one has $[b_0, z] \cap \partial G \neq \varnothing$. Then, by virtue of [446], Exercise 5.3, 3(ii), it follows that $\partial\mathbb{D}(b_0, s)$ is a polar set; a contradiction.

It remains to apply Kellogg's theorem (see [446], Theorem 4.2.5). Thus, there exists a sequence $(a_j)_j \subset \partial G$, $a_j \longrightarrow a$, where all the a_j's are regular boundary points. Applying Corollary 12.5.8 gives $\lim_{G \ni z \to a_j} \mathfrak{K}_G(z) = \infty$ for all j. Hence, $\limsup_{G \ni z \to a} \mathfrak{K}_G(z) = \infty$; a contradiction. □

Lemma 12.6.3. *Let $D \subset \mathbb{C}^n$, $n \geq 2$, be a domain and let $0 < r < t$. For any $z' \in \mathbb{C}^{n-1}$ define*

$$D_{z'} := \{z_n \in \mathbb{D}(t) : (z', z_n) \in D\} =: \mathbb{D}(t) \setminus K(z').$$

Assume that $K(0')$ is polar and that there is a neighborhood V of $0'$ such that for almost all $z' \in V$ the set $K(z')$ is also polar.

Then, there is a neighborhood $V' \subset V$ of $0'$ such that for any $f \in L^2_h(D)$ there exists an $\hat{f} \in \mathcal{O}(V' \times \mathbb{D}(r))$ with $f = \hat{f}$ on $D \cap (V' \times \mathbb{D}(r))$.

Proof. Since $K(0')$ is a polar set, there is an s with $r < s < t$ such that $K(0') \cap \partial\mathbb{D}(s) = \varnothing$. Therefore, we find a neighborhood $V' = V'(0') \subset V$ such that $K(z') \cap \partial\mathbb{D}(s) = \varnothing$, $z' \in V'$; i.e., $V' \times \partial\mathbb{D}(s) \subset D$. Then we may define

$$\hat{f}(z', z_n) := \frac{1}{2\pi} \int_{\partial\mathbb{D}(s)} \frac{f(z', \lambda)}{\lambda - z_n} d\lambda, \quad (z', z_n) \in V' \times \mathbb{D}(s).$$

Obviously, $\hat{f} \in \mathcal{O}(V' \times \mathbb{D}(s))$.

On the other hand, using that $f \in L^2_h(D)$, the Fubini theorem and the assumptions made in Lemma 12.6.3 give that, for almost all $z' \in V'$, the function $f(z', \cdot) \in L^2_h(\mathbb{D}(t) \setminus K(z'))$ and $K(z')$ is polar. Hence, $f(z', \cdot)$ extends to a holomorphic function on $\mathbb{D}(t)$ for almost all $z' \in V'$.[6] Applying the Cauchy integral formula, we obtain $f(z', z_n) = \hat{f}(z', z_n)$, $(z', z_n) \in V' \times \mathbb{D}(s)$, for almost all $z' \in V'$. Since this set is dense in $(V' \times \mathbb{D}(s)) \cap D$, we have reached the claim in Lemma 12.6.3. □

Now, we are able to complete the proof of Theorem 12.6.1.

(i) $\Longrightarrow$ (ii): Fix a boundary point $w \in \partial D$. First we discuss the case when $w \notin \operatorname{int}(\overline{D})$. Then, there is a sequence $(z_j)_j \subset \mathbb{C}^n$ such that $z_j \longrightarrow w$ and $z_j \notin \overline{D}$, $j \in \mathbb{N}$. By r_j we denote the largest radius such that $B_j := \mathbb{B}_n(z_j, r_j)$ does not intersect $\overline{D}$. Select $w_j \in \partial B_j \cap \partial D$. Then, $w_j \longrightarrow w$. Observe that the domain

[6] Recall that a relatively closed polar subset of a plane domain is a removable set of singularities for square-integrable holomorphic functions – see also Appendix B.1.14.

D satisfies the general outer cone condition at w_j (see Theorem 12.1.25). Therefore, $\lim_{D \ni z \to w_j} \mathfrak{K}_D(z) = \infty$. Hence, (ii) follows.

Finally, assume that $w \in \operatorname{int}(\overline{D})$. Suppose that (ii) is not true for w. Then, there are a polydisc $P := \mathbb{P}(w, \varrho) \subset\subset \overline{D}$ and a constant $C > 0$ such that

$$\mathfrak{K}_D(z) \leq C, \quad z \in D \cap P.$$

Let L be a complex line through P. Then, $(L \cap P) \setminus D$ is a polar set (in L) or it is empty. Indeed, otherwise we apply Lemma 12.6.2. Therefore,

$$\sup\{\mathfrak{K}_{D \cap L}(z) : z \in L \cap P \cap D\} = \infty.$$

Then, by virtue of Appendix B.8.1, it follows that $\sup\{\mathfrak{K}_D(z) : z \in L \cap D \cap P\} = \infty$; a contradiction.

Observe that there exists a complex line L^* passing through w and $P \cap D$. We may assume that $w = 0$ and, after a linear change of coordinates, that $P = \mathbb{D}^n$ and $L^* = \{(0, \dots, 0)\} \times \mathbb{C}$. So the assumptions of Lemma 12.6.3 are fulfilled with respect to some neighborhood $V \subset \mathbb{D}^{n-1}$ of $0' \in \mathbb{C}^{n-1}$. Therefore, we find a neighborhood $V' = V'(0') \subset V$ such that for any $f \in L_h^2(D)$ there is an $\widehat{f} \in \mathcal{O}(V' \times \mathbb{D}(1/2))$ with $f = \widehat{f}$ on $D \cap (V' \times \mathbb{D}(1/2))$, contradicting the assumption in (i). $\square$

Remark 12.6.4. In [241], the following generalization of Theorem 12.6.1 can be found:

Theorem. *Let (X, π) be a Riemann domain over $\mathbb{C}^n$ such that $\pi(X)$ is bounded. Let $(\widehat{X}, \widehat{\pi})$ be the envelope of holomorphy and $(\widetilde{X}, \widetilde{\pi})$ the $L_h^2(X)$-envelope of holomorphy of (X, π). Then, $(\widehat{X}, \widehat{\pi})$ embeds into $(\widetilde{X}, \widetilde{\pi})$ and the difference of these two sets is a pluripolar subset of $\widetilde{X}$.*

For the notions used in the former theorem the reader is asked to consult [265].

Applying Theorem 12.4.10, Theorem 12.6.1 may be used to get the following result:

Corollary 12.6.5 (cf. [262]). *Any bounded balanced domain of holomorphy is an L_h^2-domain of holomorphy.*

It is an open problem to characterize those unbounded domains of holomorphy that are L_h^2-domains of holomorphy. Even more, so far there is no description of such unbounded domains that carry a non trivial L_h^2-function.

Nevertheless, note that there are partial results in the unbounded case, for example in the class of unbounded Hartogs domains.

Remark 12.6.6 (cf. [427]). Let $D \subset \mathbb{C}$ be a domain and let $\varrho \in \mathcal{SH}(D)$ be bounded from below. Put

$$D_\varrho := \left\{ z \in D \times \mathbb{C} : |z_2| < e^{-\varrho(z_1)} \right\}$$

and let

$$S(D) := \{\zeta \in \partial D : U \setminus D \text{ is polar for some open neighborhood } U \text{ of } \zeta\}.$$

Then,

(a) If $D \neq \mathbb{C}$, then D_ϱ is an L^2_h-domain of holomorphy if and only if $\limsup_{D \ni \zeta \to a} \varrho(\zeta) = \infty$ for all points $a \in S$.

(b) If $D = \mathbb{C}$, then D_ϱ is an L^2_h-domain of holomorphy if and only if ϱ is not constant.

12.7 The Bergman pseudometric

Recall that the Bergman kernel $\boldsymbol{K}_G$ of a domain $G \subset \mathbb{C}^n$ is a $\mathcal{C}^\infty$-function on $G \times G$ and that, on the diagonal, it can be represented as

$$\mathfrak{K}(z) = \boldsymbol{K}_G(z, z) = \sup\{|f(z))|^2 : f \in L^2_h(G), \ \|f\|_{L^2(G)} = 1\}, \quad z \in G,$$

if $L^2_h(G) \neq \{0\}$. Therefore, the function $G \ni z \longmapsto \log \mathfrak{K}_G(z)$ is psh on G.

In order to have this function of class $\mathcal{C}^2$ on G, we assume for the rest of this chapter that all domains G we will deal with have the property

$$\mathfrak{K}_G(z) > 0, \quad z \in G; \tag{12.7.1}$$

in that case, we say in short that G *carries a Bergman pseudometric*. For example, all bounded domains share (12.7.1) and, moreover, whenever the coordinate functions belong to $L^2_h(G)$, then (12.7.1) is true for G.

Then $\boldsymbol{K}_G$ leads to the following positive semidefinite Hermitian form:

$$\boldsymbol{B}_G(z; X) := \sum_{\nu,\mu=1}^{n} \frac{\partial^2}{\partial z_\nu \partial \overline{z}_\mu} \log \mathfrak{K}_G(z) X_\nu \overline{X}_\mu, \quad z \in G, \ X \in \mathbb{C}^n.$$

The pseudometric

$$\boldsymbol{\beta}_G(z; X) := \sqrt{\boldsymbol{B}_G(z; X)}, \quad z \in G, \ X \in \mathbb{C}^n,$$

induced by $\boldsymbol{B}_G$ is called the *Bergman pseudometric on* G.

Observe that

(i) $\boldsymbol{\beta}_G(z; \lambda X) = |\lambda| \boldsymbol{\beta}_G(z; X), \ z \in G, \ \lambda \in \mathbb{C}, \ X \in \mathbb{C}^n,$

(ii) $\boldsymbol{\beta}_G(z; X_1 + X_2) \leq \boldsymbol{\beta}_G(z; X_1) + \boldsymbol{\beta}_G(z; X_2), \ z \in G, \ X_1, X_2 \in \mathbb{C}^n,$

(iii) $\boldsymbol{\beta}_G : G \times \mathbb{C}^n \longrightarrow \mathbb{R}_+$ is continuous.

Set

$$\boldsymbol{b}_G(z', z'') := (\textstyle\int \boldsymbol{\beta}_G)(z', z''), \quad z', z'' \in G.$$

$\boldsymbol{b}_G$ is called the *Bergman pseudodistance on G*. Obviously, $\boldsymbol{b}_G$ is continuous on $G \times G$.

Example 12.7.1.

(a)

$$\boldsymbol{\beta}_{\mathbb{B}_n}(z; X) = (\sqrt{n+1})\boldsymbol{\gamma}_{\mathbb{B}_n}(z; X), \quad z \in \mathbb{B}_n,\ X \in \mathbb{C}^n. \tag{12.7.2}$$

(b)

$$\boldsymbol{\beta}_{\mathbb{D}^n}(z; X) = \sqrt{2}\sqrt{\sum_{j=1}^{n} \boldsymbol{\gamma}_{\mathbb{D}}(z_j; X_j)^2}, \quad z \in \mathbb{D}^n,\ X \in \mathbb{C}^n. \tag{12.7.3}$$

(c)

$$\boldsymbol{\beta}_{D_p}(z; X) = \sqrt{\sum_{i,j=1}^{2} T_{i,j} X_i \overline{X}_j}, \quad z \in D_p = \mathcal{E}((1/p, 1)),\ X \in \mathbb{C}^2,$$

where

$$\begin{aligned}
T_{1,1}(z) &= (1 - |z_2|^2)^p \left(\frac{3}{C^2(z)} + \frac{p^2 - 1}{D^2(z)}\right), \\
T_{1,2}(z) &= \overline{T}_{2,1}(z) = p_1 \bar{z}_1 z_2 (1 - |z_2|^2)^{p-1} \left(\frac{3}{C^2(z)} + \frac{p^2 - 1}{D^2(z)}\right), \\
T_{2,2}(z) &= \frac{2 - p}{(1 - |z_2|^2)^2} + \frac{3p(C(z) + p|z_1|^2|z_2|^2)(1 - |z_2|^2)^{p-2}}{C^2(z)} \\
&\quad - \frac{p(p+1)(D(z) - p(p-1)|z_1|^2|z_2|^2)(1 - |z_2|^2)^{p-2}}{D^2(z)}
\end{aligned}$$

with

$$C(z) := (1 - |z_2|^2)^p - |z_1|^2 \quad \text{and} \quad D(z) := (p+1)(1 - |z_2|^2)^p + (p-1)|z_1|^2$$

(cf. Examples 12.1.7, 12.1.8 and [53]).

(d) Let $P = \{\lambda \in \mathbb{C} : 1/R < |\lambda| < R\}$ $(R > 1)$. Then there exists a positive real analytic function $\chi : (-2\log R, 2\log R) \longrightarrow \mathbb{R}_{>0}$ with the following properties:

(i) $\chi(t) = \chi(-t)$ whenever $0 \le t < 2\log R$,

(ii) $\chi'(t) > 0$ for $-2\log R < t < 0$,
$\chi'(t) = 0$ iff $t = 0$,

(iii) $0 < \lim_{t \nearrow 2\log R} \dfrac{\chi(t)}{(t - 2\log R)^2} < \infty$.

The Bergman metric for P is then given by

$$\boldsymbol{\beta}_P(\lambda; X) = \frac{|X|}{|\lambda|\sqrt{\chi(\log|\lambda|^2)}}.$$

Details of how to derive this formula can be found in [225].

In particular, we see that the general Schwarz–Pick lemma does not hold for the Bergman pseudometric: simply take $F : \mathbb{D}^2 \longrightarrow \mathbb{D}^2$, $F(z_1, z_2) := (z_1, z_1)$, and $X := (1, 0)$. Then,

$$\boldsymbol{\beta}_{\mathbb{D}^2}(F(0,0); F'(0,0)X) = 2 > \sqrt{2} = \boldsymbol{\beta}_{\mathbb{D}^2}((0,0); X).$$

On the other hand, this metric is invariant under biholomorphic mappings.

Theorem 12.7.2. *If $F : G \longrightarrow D$ is a biholomorphic mapping between the domains $G, D \subset \mathbb{C}^n$, then*

(a) $\boldsymbol{\beta}_D(F(z); F'(z)X) = \boldsymbol{\beta}_G(z; X)$, $z \in G$, $X \in \mathbb{C}^n$,

(b) $\boldsymbol{b}_D(F(z'), F(z'')) = \boldsymbol{b}_G(z', z'')$, $z', z'' \in G$.

Proof. According to Proposition 12.1.10, we get

$$\log \mathfrak{K}_D(F(z)) + \log|\det F'(z)|^2 = \log \mathfrak{K}_G(z).$$

The function $\det F'$ is a nowhere vanishing holomorphic function, and so $\log|\det F'|^2$ is pluriharmonic. Hence, the transformation rule for the Levi-form under holomorphic mappings immediately leads to (a), and (b) is a simple consequence of (a). □

Moreover, it turns out that the Bergman pseudometric is not monotone with respect to inclusions (cf. [53]).

Example 12.7.3. Put

$$G_1 := \{z \in \mathbb{C}^2 : (49/50)|z_1|^{2/3} + |z_2|^2 < 1\},$$
$$G_2 := \{z \in \mathbb{C}^2 : (49/50)|z_1|^{2/5} + (1/49)|z_2|^2 < 1\}.$$

The G_j's are Reinhardt domains and, because of

$$49(1-(49/50)x_1^{2/5}) \geq 1-(49/50)x_1^{2/3} \quad \text{whenever} \quad 0 \leq x_1 \leq (50/49)^{3/2},$$

we have $G_1 \subset G_2$.

On the other hand (cf. Example 12.1.8),

$$G_1 = F_1(D_3) \quad \text{with} \quad F_1(z_1,z_2) = ((50/49)^{3/2}z_1, z_2),$$
$$G_2 = F_2(D_5) \quad \text{with} \quad F_2(z_1,z_2) = ((50/49)^{5/2}z_1, 7z_2).$$

Hence, we obtain

$$\boldsymbol{\beta}_{G_1}^2(0;X) = \frac{7}{2}(49/50)^3|X_1|^2 + 5|X_2|^2,$$
$$\boldsymbol{\beta}_{G_2}^2(0;X) = \frac{11}{3}(49/50)^5|X_1|^2 + \frac{1}{7}|X_2|^2.$$

In particular,

$$\boldsymbol{\beta}_{G_1}(0;(1,0)) = \sqrt{\frac{7}{2}\left(\frac{49}{50}\right)^3} < \sqrt{\frac{11}{3}\left(\frac{49}{50}\right)^5} = \boldsymbol{\beta}_{G_2}(0;(1,0)).$$

According to our general remarks in Chapter 4, the system $((1/\sqrt{2})\boldsymbol{\beta}_G)_{G\in\widehat{\mathfrak{G}}}$, where

$$\widehat{\mathfrak{G}} := \{G \in \mathfrak{G} : G \text{ satisfies (12.7.1)}\},$$

is a contractible family of pseudometrics, but only with respect to biholomorphic mappings.

Remark 12.7.4. We also mention that, as a consequence of Theorem 12.1.19, we have

$$\boldsymbol{\beta}_{G_1\times G_2}((z_1,z_2);(X_1,X_2)) = \sqrt{\boldsymbol{\beta}_{G_1}^2(z_1;X_1) + \boldsymbol{\beta}_{G_2}^2(z_2;X_2)}, \tag{12.7.4}$$

whenever $z_j \in G_j \subset \mathbb{C}^{n_j} \ni X_j$.

The following representation characterizes the Bergman pseudometric also by means of a variational problem:

Theorem 12.7.5. *Suppose that G is a domain in $\mathbb{C}^n$ with property* (12.7.1). *Then,*

$$\boldsymbol{\beta}_G(z;X) = \frac{1}{\sqrt{\mathfrak{K}_G(z)}}\sup\{|f'(z)X| : f \in L_h^2(G),\ \|f\|_{L^2(G)} = 1,\ f(z) = 0\},$$
$$z \in G,\ X \in \mathbb{C}^n.$$

Proof. Fix $z^0 \in G$, $X^0 \in \mathbb{C}^n$. Then, $H'' := \{f \in L_h^2(G) : f(z^0) = 0, f'(z^0)X^0 = 0\}$ is a closed subspace of $H' := \{f \in L_h^2(G) : f(z^0) = 0\}$ whose orthogonal complement in H' is at most one-dimensional. Let $\dim H''^{\perp} = 1$. Then there exists an orthonormal basis $(\varphi_j)_{j \in \mathbb{Z}_+}$ of $L_h^2(G)$ with

$$\varphi_j(z^0) = 0, \ j \geq 1, \quad \text{and} \quad \varphi_j'(z^0)X^0 = 0, \ j \geq 2.$$

Simple calculations lead to

$$\boldsymbol{\beta}_G^2(z^0; X^0) = \frac{1}{\mathfrak{K}_G(z^0)} \sum_{j=0}^{\infty} |\varphi_j'(z^0)X^0|^2 - \frac{1}{\mathfrak{K}_G^2(z^0)} \Big| \sum_{j=0}^{\infty} \varphi_j'(z^0)X^0 \overline{\varphi_j(z^0)} \Big|^2,$$

from which we obtain

$$\boldsymbol{\beta}_G(z^0; X^0) = \frac{|\varphi_1'(z^0)X^0|}{\sqrt{\mathfrak{K}_G(z^0)}}. \tag{$*$}$$

On the other hand, any $f \in L_h^2(G)$, $f(z^0) = 0$, has the representation

$$f(z) = \sum_{j=1}^{\infty} \langle f, \varphi_j \rangle_{L^2(G)} \varphi_j(z),$$

which implies that

$$|f'(z^0)X^0| = |\langle f, \varphi_1 \rangle_{L^2(G)}| |\varphi_1'(z^0)X^0| \leq \|f\|_{L^2(G)} |\varphi_1'(z^0)X^0|.$$

This, together with $(*)$, yields the claimed formula. In the case where $H' = H''$, we only mention that φ_1 does not occur. □

Corollary 12.7.6. *Let G be as in Theorem* 12.7.5, $z \in G$. *Then,* $\boldsymbol{\beta}_G(z; \cdot)$ *is positive definite if and only if for any* $X \in \mathbb{C}^n$, $X \neq 0$, *there exists an* $f \in L_h^2(G)$ *with* $f'(z)X \neq 0$.

Example 12.7.7. To describe the Bergman metric for the symmetrized bidisc, it suffices to know it for the points $(s, 0)$, $s \in [0, 1)$ (use simply $\operatorname{Aut}(\mathbb{G}_2)$). Then, one has the following result (see [503]), which is based on the description of the Bergman metric as in Theorem 12.7.5:

$$\boldsymbol{\beta}_{\mathbb{G}_2}((s,0); X) = \tfrac{1}{\pi} \left(B_1|X_1|^2 + B_2|X_2|^2 - 2B_3 \operatorname{Re}(X_1 \overline{X}_2) \right)^{1/2}, \quad X \in \mathbb{C}^2,$$

where $B_1 := \frac{2(3-2s^2)}{(1-s^2)^2(2-s^2)^2}$, $B_2 := \frac{2(2s^4-6s^2+5)}{(1-s^2)^2(2-s^2)^2}$, and $B_3 := \frac{2s(2-s^2)}{(1-s^2)^2}$.

Because of the coordinate functions, the Bergman pseudometric for a bounded domain is indeed a metric. A domain G for which (12.7.1) holds and for which $\boldsymbol{\beta}_G$ is a metric is called *$\boldsymbol{\beta}$-hyperbolic.*

Remark 12.7.8. Let M_G denote the numerator of the formula in Theorem 12.7.5, i.e.,

$$M_G(z;X) := \sup\{|f'(z)X| : f \in L_h^2(G),\ \|f\|_{L^2(G)} = 1,\ f(z) = 0\}.$$

The description of $\boldsymbol{\beta}_G$ in Theorem 12.7.5 has the advantage that $(M_G)_{G\in\widehat{\mathfrak{G}}}$ is monotonic in the following sense:

$$\text{if } G \subset D, \text{ then } M_D(z;X) \le M_G(z;X), \quad z \in G,\ X \in \mathbb{C}^n.$$

Moreover, the system $(M_G)_{G\in\widehat{\mathfrak{G}}}$ satisfies the following transformation rule:

$$\begin{aligned}&\text{if } F : G \longrightarrow D \text{ is biholomorphic, then}\\ &\qquad M_D(F(z); F'(z)X)|\det F'(z)| = M_G(z;X), \quad z \in G \subset \mathbb{C}^n \ni X.\end{aligned}$$

We have already observed that the Bergman metric is not holomorphically contractible, even for inclusions (see Example 12.7.3). Even more can happen, as the following example will show. Namely, there exist two plane domains $D_2 \subset D_1$ such that $\boldsymbol{\beta}_{D_1} \le c\boldsymbol{\beta}_{D_2}$ on D_1 is false for every positive constant c.

Example 12.7.9 (cf. [424]). First, we mention that for a bounded plane domain D we have $M_D^2(z;1) \le \mathfrak{K}_D^{(1)}(z)$, $z \in D$. Hence, using Theorem 12.5.14, it follows that

$$\frac{M_D^2(z;1)}{C\max\{1, \alpha_D(z)\log^2\alpha_D(z)\}} \le \boldsymbol{\beta}_D^2(z;1) \le C\frac{\max\{1, \boldsymbol{\alpha}_D^{(1)}(z)\log^2\boldsymbol{\alpha}_D^{(1)}(z)\}}{\boldsymbol{\alpha}_D(z)},$$
$$z \in D. \quad (12.7.5)$$

For the construction of the desired domains $D_1 \subset D_2 \subset\subset \mathbb{C}$ we choose points $x_n < 1/2^n$, near $1/2^n$, $n \ge 2$. Moreover, put

- $r_n := \exp(-2^{2(2n+\ell)}) < R_n := \exp(-2^{2(2n+\ell)}/n^3), \quad n \ge 2;$
- $K_{n,\ell} := \overline{\mathbb{D}}(R_n), \quad \ell = 1,\dots,n,\ n \ge 2;$
- $K_{n,\ell} := \overline{\mathbb{D}}(r_n), \quad \ell = n+1,\dots,2n,\ n \ge 2;$
- $L_{n,\ell} := \overline{\mathbb{D}}(r_n), \quad \ell = 1,\dots,2n,\ n \ge 2.$

Finally, we define the compact sets $L_n := \bigcup_{\ell=1}^{2n} L_{n,\ell} \subset K_n := \bigcup_{\ell=1}^{2n} K_{n,\ell}$.

Now, choose a sequence $(n_k)_{k\in\mathbb{N}} \subset \mathbb{N}$ with $n_k > k$ and $3n_k < n_{k+1} + 1$, positive numbers r_k, and points $x_{n_k,\ell}$, $\ell = 1,\dots,2n_k$, such that

- $K_{n_k} \subset \mathbb{D}(1/4)$;
- $x_{n_k,\ell} \in L_{n_k,\ell} \subset K_{n_k,\ell} \subset \operatorname{int} \mathbb{A}_{2n_k+\ell}(x_{n_k}), \quad \ell = 1,\dots,2n_k, k \in \mathbb{N};$ (12.7.6)
- $K_{n_j} \subset \operatorname{int} \mathbb{A}_{n_j}(x_{n_k}), \quad j = 1,\dots,k-1, \quad k \in \mathbb{N};$ (12.7.7)
- $K_{n_j} \subset \mathbb{D}(r_k) \cap \mathbb{A}_{n_k}(x_{n_k}), \quad j > k.$ (12.7.8)

Finally, we introduce those domains we are interested in,

$$D_1 := \mathbb{D} \setminus \left(\bigcup_{k=1}^{\infty} K_{n_k} \cup \{0\} \right) \subset D_2 := \mathbb{D} \setminus \left(\bigcup_{k=1}^{\infty} L_{n_k} \cup \{0\} \right).$$

We intend to prove that

$$\frac{\boldsymbol{\beta}_{D_2}^2(x_{n_k};1)}{\boldsymbol{\beta}_{D_1}^2(x_{n_k};1)} \underset{k\to\infty}{\longrightarrow} \infty.$$

First, we present an estimate of $\boldsymbol{\beta}_{D_1}^2(x_{n_k};1)$ from above:
Recall that

$$\boldsymbol{\alpha}_{D_1}(z) \geq \tfrac{1}{8} \sum_{j=3}^{\infty} \frac{2^{2j}}{-\log\big(\operatorname{cap}(\mathbb{A}_j(z) \setminus D_1)\big)} =: \hat{\boldsymbol{\alpha}}_{D_1}(z), \quad z \in D_1.$$

Therefore, using (12.5.7), (12.7.6) and summing only over $j = 2n_k + \ell$, $\ell = 1,\dots,2n_k$, gives

$$8\hat{\boldsymbol{\alpha}}_{D_1}(x_{n_k}) \geq \sum_{j=\ell}^{n_k} \frac{2^{2(2n_k+\ell)} n_k^3}{2^{2(2n_k+\ell)}} + \sum_{\ell=n_k+1}^{2n_k} \frac{2^{2(2n_k+\ell)}}{2^{2(2n_k+\ell)}} = n_k^4 + n_k.$$

Moreover, recall that

$$\boldsymbol{\alpha}_{D_1}^{(1)}(x_{n_k}) \leq 2^5 \sum_{s=2}^{\infty} \frac{2^{4s}}{-\log\big(\operatorname{cap}(\mathbb{A}_s(x_{n_k}) \setminus D)\big)} =: \hat{\boldsymbol{\alpha}}_{D_1}^{(1)}(x_{n_k}).$$

Hence, summing first over $j = 2n_k + \ell$, $\ell = 1, \dots, 2n_k$, and over $s = n_j$, $j = 1, \dots, k-1$, and applying (12.5.4), (12.7.8), and (12.7.7) we get

$$\begin{aligned}
\tfrac{1}{32}\widehat{\boldsymbol{\alpha}}_{D_1}(x_{n_k}) &\le \sum_{\ell=1}^{2n_k} \frac{2^{4(2n_k+\ell)}}{-\log\big(\operatorname{cap}(\mathbb{A}_{2n_k+\ell}(x_{n_k}) \setminus D)\big)} \\
&\quad + \sum_{j=1}^{k-1} \frac{2^{4n_j}}{-\log\big(\operatorname{cap}(\mathbb{A}_{n_j}(x_{n_k}) \setminus D)\big)} + \text{the remaining sum} \\
&\le \sum_{\ell=1}^{n_k} \frac{2^{4(2n_k+\ell)} n_k^3}{2^{2(2n_k+\ell)}} + \sum_{\ell=n_k+1}^{2n_k} \frac{2^{4(2n_k+\ell)}}{2^{2(2n_k+\ell)}} \\
&\quad + \sum_{j=1}^{k-1} 2^{4n_j} \left(\sum_{\ell=1}^{n_j} \frac{n_j^3}{2^{2(2n_j+\ell)}} + \sum_{j=n_j+1}^{2n_j} \frac{1}{2^{2(2n_j+\ell)}} \right) + 1 \\
&\le \sum_{\ell=1}^{2n_k} 2^{2(2n_k+\ell)} n_k^3 + \sum_{j=1}^{k-1} n_j^3 \left(\sum_{\ell=1}^{2n_j} \frac{1}{2^{2\ell}} \right) + 1 \le C_1 2^{8n_k}.
\end{aligned}$$

Taking into account that $(0,\infty) \ni t \longmapsto t(\log t)^2$ is monotonically increasing we obtain $\frac{1}{\boldsymbol{\beta}^2_{D_1}(x_{n_k};1)} \ge C_2 \frac{n_k + n_k^4}{2^{8n_k} n_k^2}$.

What remains is a lower estimate of $\boldsymbol{\beta}^2_{D_2}(x_{n_k};1) = \frac{M^2_{D_2}(x_{n_k};1)}{\mathfrak{K}_{D_2}(x_{n_k})}$. Similar to the above, we get

$$\begin{aligned}
&\sum_{j=1}^{\infty} \frac{2^{2j}}{-\log \operatorname{cap}(\mathbb{A}_j(x_{n_k}) \setminus D_2)} \\
&\qquad \le \sum_{\ell=1}^{2n_k} \frac{2^{2(2n_k+\ell)}}{2^{2(2n_k+\ell)}} + \sum_{j=1}^{k-1} 2^{2n_j} \left(\sum_{\ell=1}^{2n_j} \frac{1}{2^{2(2n_j+\ell)}} \right) + 1 \le 2n_k + 2.
\end{aligned}$$

To find a lower estimate of $M_{D_2}(x_{n_k};1)$, we define

$$f_k(\lambda) := \frac{1}{\lambda - x_{n_k,2n_k}} \text{ and } g_k(\lambda) := f_k(\lambda) - f_k(x_{n_k}), \quad \lambda \in D_2.$$

Then, $g_k(x_{n_k}) = 0$ and by virtue of the definition of D_2, we see that

$$\|g_k\|^2_{L^2(D_2)} \le C_3\big(\|f_k\|^2_{L^2(D_2)} + |f_k(x_{n_k})|^2\big) \le C_4 2^{8n_k} \text{ and } |g_k'(x_{n_k})|^2 \ge C_5 2^{16n_k}.$$

Hence, $M^2_{D_2}(x_{n_k};1) \geq C_6 2^{8n_k}$. Putting all this information together, we end up with

$$\frac{\boldsymbol{\beta}^2_{D_2}(x_{n_k};1)}{\boldsymbol{\beta}^2_{D_1}(x_{n_k};1)} \geq C_7 \frac{\frac{2^{8n_k}}{n_k(\log n_k)^2}}{\frac{2^{8n_k} n_k^2}{n_k^4}} \underset{k\to\infty}{\longrightarrow} \infty,$$

which proves the desired property.

12.8 Comparison and localization

The representation of $\boldsymbol{\beta}_G$ in Theorem 12.7.5 reminds us of the definition of the Carathéodory–Reiffen pseudometric. With this in mind, we obtain the following comparison (cf. [78, 205, 206, 350]):

Theorem 12.8.1. *If the domain $G \subset \mathbb{C}^n$ satisfies* (12.7.1), *then*

(a) $\boldsymbol{\gamma}_G(z;X) \leq \boldsymbol{\beta}_G(z;X)$ *for all $z \in G$, $X \in \mathbb{C}^n$,*

(b) $\boldsymbol{c}_G(z',z'') \leq \boldsymbol{c}^i_G(z',z'') \leq \boldsymbol{b}_G(z',z'')$, $z',z'' \in G$.

Proof. Fix $z_0 \in G$, $X \in \mathbb{C}^n$, with $\boldsymbol{\gamma}_G(z_0;X) > 0$ and choose $f \in \mathcal{O}(G,\mathbb{D})$ to be extremal in the sense of the Carathéodory–Reiffen pseudometric, i.e., $f(z_0) = 0$ and $|f'(z_0)X| = \boldsymbol{\gamma}_G(z_0;X)$.

Then, the function $g : G \longrightarrow \mathbb{C}$, $g(z) := f(z)\boldsymbol{K}_G(z,z_0)/\sqrt{\mathfrak{K}_G(z_0)}$, belongs to $L^2_h(G)$ with $g(z_0) = 0$ and $\|g\|_{L^2(G)} < 1$. Hence, by Theorem 12.7.5, we get

$$\boldsymbol{\beta}_G(z_0;X) \geq \frac{1}{\sqrt{\mathfrak{K}_G(z_0)}} \frac{|g'(z_0)X|}{\|g\|_{L^2(G)}} \geq |f'(z_0)X| = \boldsymbol{\gamma}_G(z_0;X).$$

To obtain (b), use Remark 2.7.5. □

Remark 12.8.2. In the case where $\boldsymbol{\gamma}_G(z;X) > 0$, we even have $\boldsymbol{\gamma}_G(z;X) < \boldsymbol{\beta}_G(z;X)$.

On the other hand, there is, in general, no chance to compare the Bergman and the Kobayashi pseudometrics; cf. [127, 128].

Theorem 12.8.3. *There exist: a bounded pseudoconvex domain $G \subset \mathbb{C}^3$ with real-analytic smooth boundary, a sequence $(z_\nu)_{\nu\in\mathbb{N}} \subset G$ with $\lim_{\nu\to\infty} z_\nu =: z^* \in \partial G$, and a vector $X \in \mathbb{C}^3$ such that*

$$\frac{\boldsymbol{\beta}_G(z_\nu;X)}{\boldsymbol{\varkappa}_G(z_\nu;X)} \underset{\nu\to\infty}{\longrightarrow} \infty.$$

Remark 12.8.4. In Chapter 13 we will see that there also exists a bounded balanced pseudoconvex domain in $\mathbb{C}^3$ whose Minkowski function is continuous such that the sequence $(\boldsymbol{\beta}_G(z_\nu;X_\nu)/\boldsymbol{\varkappa}_G(z_\nu;X_\nu))_{\nu\in\mathbb{N}}$ is unbounded for suitable $(z_\nu)_{\nu\in\mathbb{N}} \subset G$ and $(X_\nu)_{\nu\in\mathbb{N}} \subset \mathbb{C}^3$.

Proof. With $q(z) := \operatorname{Re} z_1 + |z_1|^2 + |z_2|^{12} + |z_3|^{12} + |z_2|^4|z_3|^2 + |z_2|^2|z_3|^6$, $z \in \mathbb{C}^3$, we put $G := \{z \in \mathbb{C}^3 : q(z) < 0\}$. Then, it is easy to see that G is a bounded pseudoconvex domain with real-analytic smooth boundary. Moreover, $G \subset 2\mathbb{B}_3$.

We put $z(t) := (-t, 0, 0) \in G$, $0 < t < 1/2$, and we define $g_t : \mathbb{D}((\frac{t}{2})^{1/12}) \longrightarrow G$ by $g_t(\lambda) := (-t, 0, \lambda)$. Observe that $g_t(0) = z(t)$ and $g_t'(0) = (0, 0, 1) =: X$, which implies that

$$\varkappa_G(z(t); X) = \varkappa_G(g_t(0); g_t'(0)) \le \varkappa_{\mathbb{D}((t/2)^{1/12})}(0; 1) = \left(\frac{2}{t}\right)^{1/12}.$$

Hence, we obtain

$$\frac{\boldsymbol{\beta}_G(z(t); X)}{\varkappa_G(z(t); X)} \ge \boldsymbol{\beta}_G(z(t); X)\left(\frac{t}{2}\right)^{1/12}.$$

Bearing in mind the representation of $\boldsymbol{\beta}_G$ in Theorem 12.7.5, we estimate the Bergman kernel $\mathfrak{K}_G(z(t))$ from above. We mention that the polycylinder

$$P_\varepsilon(t) := \mathbb{D}(-t, \varepsilon t) \times \mathbb{D}((\varepsilon t)^{1/5}) \times \mathbb{D}((\varepsilon t)^{1/10})$$

is contained in G if $0 < \varepsilon \ll 1$. So, we get

$$\mathfrak{K}_G(z(t)) \le \mathfrak{K}_{P_\varepsilon(t)}(z(t)) = \pi^{-3}(\varepsilon t)^{-2}(\varepsilon t)^{-2/5}(\varepsilon t)^{-2/10} = \pi^{-3}(\varepsilon t)^{-13/5}.$$

The last step in the proof consists of a lower estimate for $M_G(z(t); X)$, $0 < t < 1/2$. Put $\widetilde{G} := \{z \in \mathbb{C}^3 : \operatorname{Re} z_1 + |z_2|^4|z_3|^2 + |z_2|^2|z_3|^6 < 0\}$. Obviously, $G \subset (2\mathbb{B}_3) \cap \widetilde{G} =: G'$ and G' is connected. Then, according to Remark 12.7.8, we have $M_G(z(t); X) \ge M_{G'}(z(t); X)$.

Now, for $0 < t < 1/2$, we define

$$f_t(z) := 4t^2 z_3/(t - z_1)^2, \quad z = (z_1, z_2, z_3) \in G'.$$

Then f_t is holomorphic on G' and, moreover, we obtain the following estimate for the $L^2(G')$-norm of f_t:

$$\int_{G'} |f_t(z)|^2 d\mathcal{L}^6(z) \le 16t^4 \int_{2\mathbb{B}_2} |z_3|^2 \int_{-\infty}^{-r(z_2,z_3)} \int_{\mathbb{R}} \frac{dy_1}{((t-x_1)^2 + y_1^2)^2} dx_1 d\mathcal{L}^4(z_2, z_3),$$

where $r(z_2, z_3) = |z_2|^4|z_3|^2 + |z_2|^2|z_3|^6$. The substitution $v(t - x_1) = y_1$ then yields

$$\begin{aligned}
\|f_t\|^2_{L^2(G')} &\le 16t^4 \int_{2\mathbb{B}_2} |z_3|^2 \int_{-\infty}^{-r(z_2,z_3)} \frac{1}{(t - x_1)^3} dx_1 \, d\mathcal{L}^4(z_2, z_3) \cdot \int_{\mathbb{R}} \frac{dv}{(1 + v^2)^2} \\
&\le 16t^4 \int_{|z_3|<2} |z_3|^2 \int_{|z_2|<2} \frac{1}{(t + r(z_2, z_3))^2} d\mathcal{L}^2(z_2) d\mathcal{L}^2(z_3) \\
&= 16t^4 4\pi^2 \int_0^2 s^2 \int_0^2 (t + r^4 s^2 + r^2 s^6)^{-2} r dr \, s ds.
\end{aligned}$$

Finally, the substitutions $r = t^{1/5}\varrho$ and $\sigma \cdot t^{1/10} = s$ lead to

$$\begin{aligned}\|f_t\|^2_{L^2(G')} &\le C_1 t^4 \int_0^{2t^{-1/10}} \int_0^{2t^{-1/5}} \frac{\sigma^3 \varrho t^{8/10}}{t^2(1+\varrho^2\sigma^2(\varrho^2+\sigma^4))^2} d\varrho d\sigma \\ &\le C_1 \int_0^\infty \int_0^\infty \frac{\sigma^3\varrho}{(1+\varrho^2\sigma^2(\varrho^2+\sigma^4))^2} d\varrho d\sigma \cdot t^{14/5} =: C_2 t^{14/5}\end{aligned}$$

with positive constants C_1, C_2, which are independent of t. Because of $f_t(z(t)) = 0$, $f_t'(z(t))X = 1$, we find that

$$M_G(z(t); X) \ge C_2^{-1/2} \cdot t^{-7/5},$$

and therefore

$$\boldsymbol{\beta}_G(z(t); X)/\boldsymbol{\varkappa}_G(z(t); X) \ge C t^{1/12-1/10} \underset{t\to 0+}{\longrightarrow} \infty. \qquad \square$$

Remark 12.8.5. The above example shows that, in general, the Bergman metric is not majorized by a multiple of the Kobayashi–Royden metric. On the other hand, for the punctured disc $\mathbb{D}_*$ we have

$$\frac{\boldsymbol{\varkappa}_{\mathbb{D}_*}(t;1)}{\boldsymbol{\beta}_{\mathbb{D}_*}(t;1)} = \frac{\boldsymbol{\varkappa}_{\mathbb{D}_*}(t;1)}{\boldsymbol{\beta}_{\mathbb{D}}(t;1)} = \frac{1-t^2}{2\sqrt{2}t|\log t|} \underset{t\to 0+}{\longrightarrow} \infty.$$

Thus, in general, there is no relation between these two metrics. Nevertheless, if we restrict our considerations to good classes of domains, e.g., the strongly pseudoconvex domains, we will find positive results; cf. the formula for the ball.

We conclude this section with a general localization result for the Bergman metric; cf. [128, 401].

Theorem 12.8.6. *Suppose that G is a bounded domain of holomorphy and let $z_0 \in \partial G$. Then for neighborhoods $U_1 = U_1(z_0) \subset\subset U_2 = U_2(z_0)$ of z_0 there exists a positive constant C such that for all $z \in V \cap U_1$, $X \in \mathbb{C}^n$, where V denotes any connected component of $G \cap U_2$, we have*

(i) $(1/C)M_V(z;X) \le M_G(z;X) \le M_V(z;X)$,

(ii) $(1/C)\boldsymbol{\beta}_V(z;X) \le \boldsymbol{\beta}_G(z;X) \le C\boldsymbol{\beta}_V(z;X)$.

Proof. Obviously (ii) is an immediate consequence of (i) and Theorem 12.1.29. So, we restrict ourselves to the proof of (i).

Remark 12.7.8 gives the second inequality in (i); so, only the first one has to be verified.

Choose intermediate neighborhoods $U_1 \subset\subset U_3 := U_3(z_0) \subset\subset U_4 := U_4(z_0) \subset\subset U_2$ and a cut-off function $\chi \in \mathcal{C}_0^\infty(U_4)$, $0 \le \chi \le 1$, with $\chi|_{U_3} \equiv 1$.

Now fix $z' \in V \cap U_1$, $X \in \mathbb{C}^n$, and $f \in L_h^2(V)$ with $\|f\|_{L^2(V)} = 1$, $f(z') = 0$. Moreover, put $\varphi(z) := 2(n+2)\log\|z - z'\|$ and define the $(0,1)$-form

$$\alpha := \overline{\partial}(f \cdot \chi) = f \sum_{j=1}^{n} \frac{\partial \chi}{\partial \overline{z}_j} d\overline{z}_j$$

(by trivial extension) as a $\overline{\partial}$-closed $\mathcal{C}^\infty$-form on the whole G.

Because of

$$\int_G \|\alpha\|^2(z)\exp(-\varphi(z))d\mathcal{L}^{2n}(z)$$
$$= \int_V |f(z)|^2 \sum_{j=1}^{n} \left|\frac{\partial \chi}{\partial \overline{z}_j}(z)\right|^2 \exp(-\varphi(z))d\mathcal{L}^{2n}(z) \leq C_1 < \infty$$

(C_1 independent of f and V), the form α belongs to $L^2_{(0,1)}(G, \exp(-\varphi))$, the Hilbert space of $(0,1)$-forms that are square-integrable with respect to the weight function $\exp(-\varphi)$. Observe that φ is plurisubharmonic.

By Hörmander's theory (cf. Appendix B.7) there exists a $\mathcal{C}^\infty$-function $g : G \longrightarrow \mathbb{C}$ with $\overline{\partial} g = \alpha$ and

$$\int_G |g(z)|^2(1 + \|z\|^2)^{-2}\exp(-\varphi(z))d\mathcal{L}^{2n}(z) \leq C_1.$$

Because of $\exp(-\varphi(z))(1+\|z\|^2)^{-2} \geq C_2 > 0$ on G, it follows that $g \in L^2(G)$ with $\|g\|_{L^2(G)} \leq \sqrt{C_1/C_2} =: C_3$.

Put $h := \chi f - g$; then, h is a $\mathcal{C}^\infty$-function on G, which satisfies the Cauchy–Riemann equations $\overline{\partial} h \equiv 0$, i.e., h is holomorphic. Moreover, we obtain

$$\|h\|_{L^2(G)} \leq \|f\|_{L^2(V)} + \|g\|_{L^2(G)} \leq 1 + C_3 =: C.$$

According to the choice of χ, we have $\overline{\partial}(\chi f) \equiv 0$ on $U_3 \cap G$, which implies that g is also holomorphic on $U_3 \cap G$ and satisfies

$$\int_G |g(z)|^2 \frac{1}{\|z - z'\|^{2n+4}} \frac{1}{(1 + \|z\|^2)^2} d\mathcal{L}^{2n}(z) < \infty.$$

Now, the existence of this singular integral means that

$$g(z') = 0 = \frac{\partial g}{\partial z_j}(z'), \quad 1 \leq j \leq n,$$

and therefore

$$h(z') = 0, \quad |f'(z')X| = |h'(z')X| \leq \|h\|_{L^2(G)} \left|\frac{h'(z')}{\|h\|_{L^2(G)}} X\right| \leq C M_G(z'; X).$$

We may assume that $h \not\equiv 0$. Since f is arbitrary, it follows that $M_V(z'; X) \leq C M_G(z'; X)$. □

12.9 The Skwarczyński pseudometric

Let $(H, \langle\ ,\ \rangle_H)$ be an arbitrary Hilbert space. By $\mathbb{P}H$, we denote the associated projective space, i.e., $\mathbb{P}H := H \setminus \{0\}/_\sim$, where $x \sim y$ iff $x = \lambda y$ for a suitable $\lambda \in \mathbb{C}_*$, $x, y \in H \setminus \{0\}$. Then, $\mathbb{P}H$ is a complete metric space with respect to the distance

$$d_H([x],[y]) := \operatorname{dist}([x] \cap S_H, [y] \cap S_H) = \left(2 - 2\frac{|\langle x, y\rangle_H|}{\|x\|_H \|y\|_H}\right)^{1/2}, \quad [x],[y] \in \mathbb{P}H, \tag{12.9.1}$$

where $S_H := \{\xi \in H : \|\xi\|_H = 1\}$.

This general observation was used by M. Skwarczyński (cf. [483]) to introduce another pseudodistance on domains in $\mathbb{C}^n$. Let G be a domain in $\mathbb{C}^n$ with (12.7.1), i.e., $\mathfrak{K}_G(z) > 0$ for all $z \in G$. The following map

$$\tau : G \longrightarrow \mathbb{P}(L^2_h(G)), \quad \tau(z) := [\boldsymbol{K}_G(\cdot, z)],$$

enables us to introduce the following continuous pseudodistance on $G \times G$:

$$\varrho_G(z', z'') := \frac{1}{\sqrt{2}} d_{L^2_h(G)}(\tau(z'), \tau(z'')) = \left(1 - \frac{|\boldsymbol{K}_G(z', z'')|}{\sqrt{\mathfrak{K}_G(z')}\sqrt{\mathfrak{K}_G(z'')}}\right)^{1/2};$$

ϱ_G is called the *Skwarczyński pseudodistance*.

Observe that the following conditions are equivalent:

(i) τ is injective;

(ii) for each two distinct points $z', z'' \in G$ the functions $\boldsymbol{K}_G(\cdot, z')$, $\boldsymbol{K}_G(\cdot, z'')$ are linearly independent;

(iii) ϱ_G is a distance.

This is, for example, the case when G is bounded.

Remark 12.9.1.

(a) $\varrho_{\mathbb{D}}(z', z'') = \boldsymbol{m}(z', z'')$, $z', z'' \in \mathbb{D}$.

(b) By Proposition 12.1.10, we conclude that if $F : G \longrightarrow D$ is biholomorphic, then $\varrho_D(F(z'), F(z'')) = \varrho_G(z', z'')$, $z', z'' \in G$.

(c) $\varrho_{\mathbb{D}\times\mathbb{D}}((0,0), (\lambda, \lambda)) = (1 - (1 - |\lambda|^2)^2)^{1/2} > (1 - (1 - |\lambda|^2))^{1/2} = \varrho_{\mathbb{D}}(0, \lambda)$, $0 < |\lambda| < 1$, i.e., the Skwarczyński pseudodistance fails to fulfill the general Schwarz–Pick lemma.

Nevertheless, according to Chapter 4, the system $(\varrho_G)_{G \in \widehat{\mathfrak{G}}}$ is an $\boldsymbol{m}$-contractible family of pseudodistances with respect to all biholomorphic mappings.

Proposition 12.9.2. *For any domain G, we have* $\operatorname{top}\varrho_G = \operatorname{top} G$ *if $L_h^2(G)$ contains the coordinate functions.*

Proof. Obviously, $\operatorname{top}\varrho_G \subset \operatorname{top} G$. Now fix $z' \in G$ and a sequence $(z_j)_{j\in\mathbb{N}} \subset G$ with $\lim_{j\to\infty} \varrho_G(z', z_j) = 0$. Because of (12.9.1), we may assume that

$$\lim_{j\to\infty} \left\| \frac{e^{i\theta_j} \boldsymbol{K}_G(\cdot, z_j)}{\sqrt{\mathfrak{K}_G(z_j)}} - \frac{\boldsymbol{K}_G(\cdot, z')}{\sqrt{\mathfrak{K}_G(z')}} \right\|_{L^2(G)} = 0,$$

i.e., $C_j \boldsymbol{K}_G(\cdot, z_j) \overset{L^2(G)}{\longrightarrow} \boldsymbol{K}_G(\cdot, z')$ with $C_j := e^{i\theta_j} \sqrt{\mathfrak{K}_G(z')}/\sqrt{\mathfrak{K}_G(z_j)}$.
Since $1 \in L_h^2(G)$, we get

$$\lim_{j\to\infty} \overline{C}_j = \lim_{j\to\infty} \langle 1, C_j \boldsymbol{K}_G(\cdot, z_j)\rangle_{L^2(G)} = \langle 1, \boldsymbol{K}_G(\cdot, z')\rangle_{L^2(G)} = 1.$$

If we denote by π_k the k-th coordinate function, then we obtain

$$\begin{aligned}\lim_{j\to\infty} \pi_k(z_j) &= \lim_{j\to\infty} \frac{1}{\overline{C}_j} \langle \pi_k, C_j \boldsymbol{K}_G(\cdot, z_j)\rangle_{L^2(G)} \\ &= \lim_{j\to\infty} \langle \pi_k, C_j \boldsymbol{K}_G(\cdot, z_j)\rangle_{L^2(G)} = \pi_k(z'), \quad \text{i.e.,} \quad \lim_{j\to\infty} z_j = z'.\end{aligned}$$

Hence, the two topologies coincide. □

Recall that $(\varrho_G)_{G\in\widehat{\mathfrak{G}}}$ is not an $\boldsymbol{m}$-contractible family of pseudodistances in the strong sense. So it is not clear how to compare ϱ_G and $\boldsymbol{c}_G^*$. To provide an answer, we need the following estimate of the Skwarczyński pseudodistance by an expression similar to that of Theorem 12.7.5 (cf. [483]):

Theorem 12.9.3. *Suppose that $G \subset \mathbb{C}^n$ satisfies* (12.7.1). *Then,*

$$\varrho_G(z', z'') \le \widetilde{M}_G(z', z'')/\sqrt{\mathfrak{K}_G(z')} \le \sqrt{2}\varrho_G(z', z''), \quad z', z'' \in G,$$

where $\widetilde{M}_G(z', z'') := \sup\{|f(z')| : f \in L_h^2(G),\ \|f\|_{L^2(G)} = 1,\ f(z'') = 0\}$.

Proof. Write $\boldsymbol{K}_G(\cdot, z') = \alpha \boldsymbol{K}_G(\cdot, z'') + g$ with $g \in [\mathbb{C}\boldsymbol{K}_G(\cdot, z'')]^\perp$ and $\alpha \in \mathbb{C}$. Then,

$$g(z'') = \langle g, \boldsymbol{K}_G(\cdot, z'')\rangle_{L^2(G)} = 0 \quad \text{and} \quad \alpha = \boldsymbol{K}_G(z'', z')/\mathfrak{K}_G(z'').$$

Therefore, by definition, if $\|g\|_{L^2(G)} \ne 0$, then

$$\widetilde{M}_G(z', z'') \ge \frac{|g(z')|}{\|g\|_{L^2(G)}} = \frac{|\langle g, \boldsymbol{K}_G(\cdot, z')\rangle_{L^2(G)}|}{\|g\|_{L^2(G)}} = \|g\|_{L^2(G)}.$$

On the other hand, if $f \in L^2_h(G)$, $\|f\|_{L^2(G)} = 1$, and $f(z'') = 0$, then

$$|f(z')| = |\langle f, \boldsymbol{K}_G(\cdot, z')\rangle_{L^2(G)}| = |\langle f, \alpha \boldsymbol{K}_G(\cdot, z'') + g\rangle_{L^2(G)}| \le \|g\|_{L^2(G)}.$$

Hence, with $\|g\|_{L^2(G)} = \widetilde{M}_G(z', z'')$, we find that

$$\begin{aligned}\frac{\widetilde{M}_G(z', z'')}{\sqrt{\mathfrak{K}_G(z')}} &= \frac{1}{\sqrt{\mathfrak{K}_G(z')}}\sqrt{\mathfrak{K}_G(z') - \frac{|\boldsymbol{K}_G(z', z'')|^2}{\mathfrak{K}_G^2(z'')}\mathfrak{K}_G(z'')} \\ &= \varrho_G(z', z'')\sqrt{1 + \frac{|\boldsymbol{K}_G(z', z'')|}{\sqrt{\mathfrak{K}_G(z')}\sqrt{\mathfrak{K}_G(z'')}}}.\end{aligned}$$

By virtue of the Schwarz inequality, the claimed inequality follows. □

Theorem 12.9.4 (cf. [79]). *Let G satisfy* (12.7.1). *Then,* $\boldsymbol{c}^*_G \le \sqrt{2}\varrho_G$.

Proof. Use Theorem 12.9.3 and proceed as in the proof of Theorem 12.8.1. □

Remark 12.9.5.

(a) The proof in [79] is different from the one presented here. It is based on the inequality

$$\frac{|\boldsymbol{K}_G(z', z'')|^2}{\mathfrak{K}_G(z')\mathfrak{K}_G(z'')} \le \frac{(1 - |f(z')|^2)(1 - |f(z'')|^2)}{|1 - f(z')\overline{f(z'')}|^2}, \quad f \in \mathcal{O}(G, \mathbb{D}),\ z', z'' \in G,$$

and gives the more precise inequality $(2 - \varrho_G^2)\varrho_G^2 \ge \boldsymbol{c}^{*2}_G$.

(b) In Chapter 13, we will construct a bounded pseudoconvex balanced domain in $\mathbb{C}^3$ whose Bergman distance and then, in particular, its Skwarczyński distance, is not majorized by any multiple of its Kobayashi distance (cf. Theorem 12.9.6).

We conclude this section by calculating the inner distance associated to ϱ_G (cf. [358]).

Theorem 12.9.6. *Let $G \in \widehat{\mathfrak{G}}$. Then,* $\varrho^i_G = (1/\sqrt{2})\boldsymbol{b}_G$.

Proof. According to Lemma 6.1.5, we only have to prove that ϱ_G is a $\mathcal{C}^1$-pseudodistance with $\mathcal{D}\varrho_G = (1/\sqrt{2})\boldsymbol{\beta}_G$. So, fix an $a \in G$ and take $(z'_\nu)_{\nu\in\mathbb{N}}, (z''_\nu)_{\nu\in\mathbb{N}} \subset G$ with $\lim_{\nu\to\infty} z'_\nu = \lim_{\nu\to\infty} z''_\nu = a$, $z'_\nu \ne z''_\nu$ for all ν and $\lim_{\nu\to\infty}(z'_\nu - z''_\nu)/\|z'_\nu - z_\nu\| =: X \in \mathbb{C}^n$. Then we write

$$\varrho_G(z'_\nu, z''_\nu) = \left(\frac{\Phi(z'_\nu, z''_\nu)}{\sqrt{\mathfrak{K}_G(z'_\nu)}\sqrt{\mathfrak{K}_G(z''_\nu)}(\sqrt{\mathfrak{K}_G(z'_\nu)}\sqrt{\mathfrak{K}_G(z''_\nu)} + |\boldsymbol{K}_G(z'_\nu, z''_\nu)|)}\right)^{\frac{1}{2}}.$$

Here, $\Phi(z, w) := \mathfrak{K}_G(z)\mathfrak{K}_G(w) - \boldsymbol{K}_G(z, w)\boldsymbol{K}_G(w, z)$, $z, w \in G$, is a $\mathcal{C}^\infty$-function with $\Phi \ge 0$ and $\Phi(z, z) = 0$, $z \in G$. Therefore, the Taylor formula for $\Phi(\cdot, z''_\nu)$ up to second order and the holomorphicity properties of the Bergman kernel lead to

$$\lim_{\nu\to\infty} \frac{\varrho_G(z'_\nu, z''_\nu)}{\|z'_\nu - z''_\nu\|}$$

$$= \frac{1}{2\mathfrak{K}_G^2(a)} \left(\sum_{\nu,\mu=1}^{n} \left(\frac{\partial^2 \boldsymbol{K}_G}{\partial z_\nu \partial \overline{z}_\mu}(a,a)\mathfrak{K}_G(a) - \frac{\partial \boldsymbol{K}_G}{\partial z_\nu}(a,a)\frac{\partial \boldsymbol{K}_G}{\partial \overline{z}_\mu}(a,a) \right) X_\nu \overline{X}_\mu \right)^{1/2}$$

$$= (1/2\boldsymbol{B}_G(a;X))^{1/2}. \qquad \square$$

Corollary 12.9.7.

(a) *For* $G \in \widehat{\mathfrak{G}}$, *we have* $\boldsymbol{c}_G^* \le \sqrt{2}\varrho_G \le \sqrt{2}\varrho_G^i = \boldsymbol{b}_G \ge \boldsymbol{c}_G$.

(b) *If G is bounded, then* $\operatorname{top} G = \operatorname{top} \boldsymbol{b}_G = \operatorname{top} \varrho_G^i$.

Remark 12.9.8. In a recent paper [22], distance functions for general reproducing kernel Hilbert spaces, i.e., for triples (H, X, K), where H is a Hilbert space of functions on the set X and $K : X \times X \to \mathbb{C}$ is such that for any $x \in X$ one has that $K(\cdot, x) \in H$ and $f(x) = \langle f, K(\cdot, x)\rangle_H$, $f \in H$, are studied. Note that $(L_h^2(G), G, \boldsymbol{K})$ is such a triple.

12.10 Exercises

Exercise 12.10.1 (cf. [528]). Put $\Omega_k := \Omega \cup \widetilde{\Omega}_{4k}$, $k \in \mathbb{N}$, where

$$\begin{aligned}\Omega := &\{(z,w) \in \mathbb{C}^2 : |z| < 2e, |w| < 2e\}\\ &\cup \{(z,w) \in \mathbb{C}^2 : |z| > e, |w| < 1/(|z| \log |z|)\}\\ &\cup \{(z,w) \in \mathbb{C}^2 : |w| > e, |z| < 1/(|w| \log |w|)\}\end{aligned}$$

and

$$\widetilde{\Omega}_m := \left\{(z,w) \in \mathbb{C}^2 : |z| > 1, |w| > 1, ||z| - |w|| < \frac{1}{(|z| + |w|)^m}\right\}, \quad m \in \mathbb{N}.$$

Prove that $\dim L_h^2(\Omega_k) = k$, or more that $L_h^2(\Omega_k) = \operatorname{span}\{1, zw, \dots, (zw)^{k-1}\}$.

Hint. Verify first that a monomial $z^p w^q$ belongs to $L_h^2(\Omega)$ iff $p = q$. Then, show that $z^p w^p \in L_h^2(\Omega_k)$ iff $p < k$.

Exercise 12.10.2. Let $D = \mathbb{R} + i(a,b)$, where $a < b$. Verify that

$$\boldsymbol{K}_D(z,w) = \frac{\pi}{4(b-a)^2} \cdot \frac{1}{\cosh(\frac{\pi}{2(b-a)})(z - \overline{w} - i(a+b))}.$$

Hint. Use a correct biholomorphic map.

Exercise 12.10.3.

(a) Prove the formula in Example 12.1.12.

(b) Prove the first part of Theorem 12.1.18 only under the assumption that F is a proper holomorphic mapping.

Hint. Use Theorem 12.1.11 instead of Proposition 12.1.10.

Exercise 12.10.4. Use Remark 12.7.4 to prove that

$$\boldsymbol{b}_{G_1\times G_2}((z_1',z_2'),(z_1'',z_2''))=\sqrt{\boldsymbol{b}_{G_1}^2(z_1',z_1'')+\boldsymbol{b}_{G_2}^2(z_2',z_2'')},\quad z_j',z_j''\in G_j\subset\subset\mathbb{C}^{n_j}.$$

Exercise 12.10.5. Let

$$D(k):=\{z=(z_1,\dots,z_n)\in\mathbb{C}^n:|z_1|^{2k}+\sum_{j=2}^{n}|z_j|^2<1\},\ k>0.$$

Prove:

$$\boldsymbol{K}_{D(k)}(z,w)=k\pi^{-n}g(z,w)^{-n-1/k}\sum_{j=2}^{n+1}b_j g(z,w)^{j/k}((1-g(z,w))^{1/k}-z_1\overline{w}_1)^{-j},$$

where $g(z,w):=1-\sum_{j=2}^{n}z_j\overline{w}_j$ (cf. [119]).
Hint. Use an orthonormal basis.

Exercise 12.10.6. Let

$$D:=D(p_1,\dots,p_n):=\{z\in\mathbb{C}^n:|z_1|^{2/p_1}+\dots+|z_n|^{2/p_n}<1\},\quad p_j\in\mathbb{N}.$$

Prove that

$$\boldsymbol{K}_D(z,w)=\frac{1}{\pi^n}\frac{1}{p_1\cdots p_n}\frac{\partial^n}{\partial x_1\dots\partial x_n}\sum_{j_1=1}^{p_1}\cdots\sum_{j_n=1}^{p_n}\frac{1}{1-y_{j_1,1}-\dots-y_{j_n,n}}\bigg|_{\substack{x_\nu=z_\nu\overline{w}_\nu\\ 1\le\nu\le n}},$$

where $y_{j,i}=x_i^{1/p_i}\varepsilon_{j,i}$, $1\le j\le p_i$, $1\le i\le n$; here $\varepsilon_{j,i}$ are all the p_i-th roots of unity (cf. [Zin]).

Exercise 12.10.7. Let $D(k)$ be as in Exercise 12.10.5. Prove (cf. [539]) that $\boldsymbol{\beta}_{D(k)}\le\delta_{D(k)}$ on $D(k)$, where

$$\delta_{D(k)}(z;X):=\sqrt{\sum_{i,j=1}^{n}\frac{\partial^2\log g(z)}{\partial z_i\partial\overline{z}_j}X_i\overline{X}_j},$$

$$g(z):=(1-\widetilde{g}(z))^{-\lambda}(1-\|\widetilde{z}\|^2)^{-(nk+1)/k},$$

$$\widetilde{g}(z):=|z_1|^2/(1-\|\widetilde{z}\|^2)^{1/k},\quad\lambda\gg1,\ z=(z_1,\widetilde{z})\in D(k),\ X\in\mathbb{C}^n.$$

Exercise 12.10.8. Let G be a bounded domain in $\mathbb{C}^n$ with a transitive group of automorphisms. Prove (cf. [350]) that there exists a positive constant $k = k(G)$ such that for any bounded domain $D \subset \mathbb{C}^n$ and any $F \in \mathcal{O}(D, G)$ the following inequality is true:

$$\boldsymbol{\beta}_G(F(z); F'(z)X) \le k\boldsymbol{\beta}_D(z; X), \quad z \in D,\ X \subset \mathbb{C}^n.$$

Exercise 12.10.9. Let $G \subset \mathbb{C}$ be a bounded domain with smooth $\mathcal{C}^\infty$-boundary. Denote by $\mathfrak{g}_G(a, \cdot)$ the classical Green function of G with pole at $a \in G$; see Appendix B.5. Recall from the classical potential theory that $\mathfrak{g}_G$ is a $\mathcal{C}^\infty$-function of both variables on $\overline{G} \times \overline{G} \setminus \{(z, z) : z \in \overline{G}\}$.

Prove that

$$-\frac{2}{\pi}\frac{\partial^2 \mathfrak{g}_G}{\partial \overline{w} \partial z}(z, w) = \boldsymbol{K}_G(z, w), \quad z, w \in G,\ z \ne w.$$

Hint. Use Stokes' formula to show that the function

$$-\frac{2}{\pi}\frac{\partial^2 \mathfrak{g}_G}{\partial \overline{w} \partial z}(z, \cdot)$$

has the reproducing property (with respect to the L^2-scalar product) for all $h \in \mathcal{C}^\infty(\overline{G}) \cap \mathcal{O}(G)$. Finally, observe that the space $\mathcal{C}^\infty(\overline{G}) \cap \mathcal{O}(G)$ is dense in $L^2_h(G)$.

12.11 List of problems

Chapter 13

Hyperbolicity

Summary. As we have seen, the Kobayashi hyperbolicity implies that the topology induced by $\boldsymbol{k}_D$ leads to the standard one, but such a result fails for Carathéodory hyperbolicity. While in Section 13.1 general notions of hyperbolicity are discussed, localization properties of hyperbolicity will be studied in Section 13.2. Most of the general investigations on hyperbolicity have been initiated by S. Kobayashi [317]. Particular cases were treated by K. Azukawa [24], T. J. Barth [35], N. Sibony [474], and J. Siciak (private communication). The remaining sections contain the description of hyperbolicity for concrete classes of domains such as Reinhardt domains (Section 13.3), Hartogs domains (Section 13.4), and tube domains (Section 13.6).

13.1 Global hyperbolicity

Let $G \subset \mathbb{C}^n$ be an arbitrarily given domain. We will consider the following "hyperbolicity"-conditions and discuss some of their relations:

(H_1) G is bounded;

(H_2) G is biholomorphic to a bounded domain;

(H_3) $\operatorname{top} G = \operatorname{top} \boldsymbol{c}_G$;

(H_4) G is $\boldsymbol{c}$-hyperbolic;

(H_5) G is $\boldsymbol{k}$-hyperbolic;

(H_6) G does not contain non-trivial entire curves, i.e., every holomorphic map $f : \mathbb{C} \longrightarrow G$ is constant;

(H_7) for any $f : \mathbb{C} \longrightarrow \mathbb{C}^n$, $f(\lambda) = a + \lambda v$, $v \neq 0$, the image $f(\mathbb{C})$ does not lie in G, i.e., G does not contain any affine complex line;

(H_8) no complex line through 0 stays inside G.

Then, obviously, (H_1) $\Longrightarrow$ (H_2) $\Longrightarrow$ (H_3) $\Longrightarrow$ (H_4) $\Longrightarrow$ (H_5) $\Longrightarrow$ (H_6) $\Longrightarrow$ (H_7) $\Longrightarrow$ (H_8).

A domain G satisfying the property (H_6) is sometimes called *Brody hyperbolic*.

We will see that for domains of a more restrictive shape, some of the converse implications are true.

Theorem 13.1.1 (cf. [318]). *Any $\boldsymbol{k}$-hyperbolic balanced domain G in $\mathbb{C}^n$ is bounded.*

Proof. Suppose that $\mathbb{B}(R) \subset\subset G$ $(R > 0)$ and assume that G is not bounded. Then, G contains a sequence $(z_\nu)_{\nu\in\mathbb{N}}$ of points z_ν with $R < \|z_\nu\| \underset{\nu\to\infty}{\longrightarrow} \infty$. If we put $\varphi_\nu(\zeta) := \zeta z_\nu$, $\zeta \in \mathbb{D}$, then we get

$$\boldsymbol{k}_G(0, Rz_\nu/\|z_\nu\|) \le \boldsymbol{p}(0, R/\|z_\nu\|) \underset{\nu\to\infty}{\longrightarrow} 0.$$

Since $(Rz_\nu/\|z_\nu\|)_{\nu\in\mathbb{N}}$ has an accumulation point a with $\|a\| = R$, it follows that $\boldsymbol{k}_G(0,a) = 0$, which contradicts the assumption of hyperbolicity. □

By Theorem 13.1.1, we see that there is no difference between $\boldsymbol{k}$- and $\boldsymbol{c}$-hyperbolicity in the class of general balanced domains. For balanced pseudoconvex domains in $\mathbb{C}^2$, the following characterization of the Brody hyperbolicity is due to J. Siciak:

Theorem 13.1.2. *Let $G = \{z \in \mathbb{C}^2 : \mathfrak{h}(z) < 1\}$ be a balanced pseudoconvex domain. Then, the following properties are equivalent:*

(i) *G is Brody hyperbolic;*

(ii) *G does not contain complex lines through 0;*

(iii) *$\mathfrak{h}(z) > 0$ if $z \neq 0$.*

Proof. The implications (i) $\Longrightarrow$ (ii) $\Longrightarrow$ (iii) are trivial and they remain true even in arbitrary dimension.

Now, we assume that (iii) holds. Suppose that there exists an entire curve $f = (f_1, f_2) : \mathbb{C} \longrightarrow G$ with, for example, f_1 not identically constant.

Put $u := \mathfrak{h} \circ f$. Then, u is a subharmonic function on $\mathbb{C}$ with $u < 1$, which implies that $u \equiv C$ for a certain constant $C \in [0,1)$; cf. Appendix B.4.27. If $C = 0$, then our assumption leads to $f \equiv 0$, contradicting our choice of f_1. Thus, we may assume that $0 < C < 1$. In particular, $f_2(\lambda) \neq 0$ if $f_1(\lambda) = 0$.

Put $v := -\log C + \log \mathfrak{h}(1,\cdot)$ on $\mathbb{C}$. Then, v is a subharmonic function satisfying the following estimate:

$$-\infty < v(\lambda) \le C_1 + \log(1 + |\lambda|), \quad \lambda \in \mathbb{C}. \tag{13.1.1}$$

Observe that for $\lambda \in \mathbb{C} \setminus f_1^{-1}(0)$ we have

$$\log|f_1(\lambda)| = -v(m(\lambda)), \tag{13.1.2}$$

where $m := f_2/f_1 \in \mathcal{O}(\mathbb{C} \setminus f_1^{-1}(0))$.

Because of (13.1.2), m is not identically constant. Applying Picard's little theorem (Theorem 1.2.5), it follows that there are points $\zeta_1, \zeta_2 \in \mathbb{C}$ such that $D := \mathbb{C} \setminus \{\zeta_1, \zeta_2\} \subset m(\mathbb{C} \setminus f_1^{-1}(0))$.

Now, we study v on D. Fix a $\zeta_0 \in D$. Using the local invertibility of m, we conclude via (13.1.2) that v is a harmonic function in a punctured neighborhood of ζ_0. Moreover, v is locally bounded, which implies that v is harmonic in a full neighborhood of ζ_0. Hence, we have v harmonic on D with the estimate (13.1.1).

On the other hand, we note that v cannot be harmonic on the whole $\mathbb{C}$. Otherwise $v = \operatorname{Re} g$, $g \in \mathcal{O}(\mathbb{C})$, and therefore we find from (13.1.1) that

$$|e^{g(\lambda)}| = e^{v(\lambda)} \le e^{C_1}(1 + |\lambda|) \quad \text{for } \lambda \in \mathbb{C}.$$

So Liouville's theorem implies that the entire function e^g is of the form $\lambda \longmapsto a + b\lambda$, which obviously leads to e^g and also v being identically constant. By (13.1.2), this contradicts the assumption that f_1 is not identically constant.

Thus, we may assume that v is not harmonic in a neighborhood of ζ_1. Using the Riesz decomposition theorem, v can be written as

$$v(\lambda) = v_0(\lambda) + \frac{1}{2\pi}\int_{|\lambda-\zeta_1|<r} \log|\lambda - \zeta| d\mu(\lambda), \quad |\lambda - \zeta_1| < r \ll 1,$$

where μ is a non-negative measure given by Δv and v_0 is harmonic in $\mathbb{B}(\zeta_1, r)$; cf. Appendix B.5.10. Because of the harmonicity of v outside ζ_1, it follows that μ is concentrated on $\{\zeta_1\}$, hence

$$v(\lambda) = v_0(\lambda) + \sigma \log|\lambda - \zeta_1|, \quad |\lambda - \zeta_1| < r \ll 1$$

with $\sigma \ge 0$. Recall that a generalized function whose support consists of a single point can be represented as a linear combination of derivatives of the Dirac delta distribution. Now, $\mu \ge 0$ allows us to exclude derivatives in that combination.

Because of the definition of v, the constant σ has to be zero. Hence, v is harmonic near ζ_1; a contradiction. □

Example 13.1.3 (cf. [24]). We start with the following subharmonic function $\varphi : \mathbb{C} \longrightarrow [-\infty, \infty)$,

$$\varphi(\lambda) := \max\left\{\log|\lambda|, \sum_{k\ge 2} \frac{1}{k^2} \log\left|\lambda - \frac{1}{k}\right|\right\}, \quad \lambda \in \mathbb{C},$$

and on $\mathbb{C}^2$, we define

$$\mathfrak{h}(z) := \begin{cases} |z_2| \exp\varphi(z_1/z_2) & \text{if} \quad z_2 \ne 0 \\ |z_1| & \text{if} \quad z_2 = 0 \end{cases}.$$

Observe, that $\mathfrak{h}$ is plurisubharmonic on $\mathbb{C} \times \mathbb{C}_*$ and upper semicontinuous on $\mathbb{C} \times \mathbb{C}$. By Appendix B.4.23, it follows that $\mathfrak{h}$ is plurisubharmonic on $\mathbb{C}^2$.

Therefore, $\mathfrak{h}$ defines a balanced pseudoconvex domain $G := \{z \in \mathbb{C}^2 : \mathfrak{h}(z) < 1\}$ in $\mathbb{C}^2$, which is not bounded. For example $(1/2, k/2) \in G$, $k \geq 2$. Hence, G is not $\boldsymbol{k}$-hyperbolic.

On the other hand, Theorem 13.1.2 leads to the Brody hyperbolicity of G (see Exercise 13.7.3 for a slightly more general situation).

This example shows that the property "Brody hyperbolic" is strictly weaker than $\boldsymbol{k}$-hyperbolicity, even in the class of balanced pseudoconvex domains; see also Exercise 13.7.1.

Remark 13.1.4. If a balanced domain G has a continuous Minkowski function $\mathfrak{h}$, then the condition (H_8) implies that G is bounded. Observe that, in particular, all complete Reinhardt domains belong to this special class of balanced domains (cf. Remark 2.2.1(r)).

Corollary 13.1.5. *For a complete Reinhardt domain G, the properties (H_1)–(H_8) are equivalent.*

Proposition 13.1.6. *For a balanced pseudoconvex domain the conditions (H_7) and (H_8) are equivalent.*

Proof. Without loss of generality we may assume that $n \geq 2$. Let $f(\lambda) = a + \lambda b$, $a, b \in \mathbb{C}^n$, $b \neq 0$, be an affine complex line inside G. If a, b are linearly dependent, then the line $f(\mathbb{C})$ contains the origin.

Now suppose that a, b are linearly independent. Since $\lambda a + \mu b = \lambda(a + \frac{\mu}{\lambda} b) \in G$ if $0 < |\lambda| \leq 1$, we obtain $\mathbb{D}_* \cdot a + \mathbb{C} \cdot b \subset G$. Moreover, since $0 \in G$ we have $\varepsilon\mathbb{D} \cdot a + \varepsilon\mathbb{D} \cdot b \subset G$ for a suitable $\varepsilon > 0$. By hypothesis, G is a domain of holomorphy. Hence, the Kontinuitätssatz implies that $\mathbb{C} \cdot b$ belongs to G. □

In the class of convex domains containing the origin, the following result due to T. J. Barth (cf. [35]) is true:

Theorem 13.1.7. *Any convex domain $G \subset \mathbb{C}^n$, $0 \in G$, which satisfies condition (H_8) is biholomorphically equivalent to a bounded domain.*

Proof. Let $b_1 := (1, 0, \dots, 0)$. By assumption, there exist a complex number $\alpha_1 \neq 0$ and a unit vector $a_1 \in \mathbb{C}^n$ such that the following is true:

$$\alpha_1 b_1 \in \partial G \quad \text{and} \quad G \subset \{z \in \mathbb{C}^n : \operatorname{Re}\langle z - \alpha_1 b_1, a_1 \rangle < 0\}.$$

Now, if $n \geq 2$, choose a unit vector b_2 with $\langle b_2, a_1 \rangle = 0$. Then, the same argument leads to a number $\alpha_2 \neq 0$ and a unit vector a_2 with

$$\alpha_2 b_2 \in \partial G \quad \text{and} \quad G \subset \{z \in \mathbb{C}^n : \operatorname{Re}\langle z - \alpha_2 b_2, a_2 \rangle < 0\}.$$

Obviously, the vectors a_1, a_2 are linearly independent. Continuing this construction, we obtain linearly independent unit vectors $a_1, \dots, a_n$ such that

$$G \subset \bigcap_{j=1}^{n} \{z \in \mathbb{C}^n : \operatorname{Re}\langle z - \alpha_j b_j, a_j \rangle < 0\}$$

with suitable numbers α_j and vectors b_j.

Therefore, the mapping

$$\pi : \mathbb{C}^n \longrightarrow \mathbb{C}^n, \quad \pi(z) := (\langle z - \alpha_1 b_1, a_1 \rangle, \dots, \langle z - \alpha_n b_n, a_n \rangle)$$

gives a biholomorphic map from G onto a domain $D \subset \{z \in \mathbb{C}^n : \operatorname{Re} z_j < 0,\ 1 \leq j \leq n\}$. The latter is biholomorphically equivalent to a bounded domain. □

Remark 13.1.8. In [74], one may find an even longer list of equivalent properties for a convex domain to be $\boldsymbol{k}$-hyperbolic. Similar results for $\mathbb{C}$-convex domains are given in [392].

Remark 13.1.9. There are examples of domains, most of them easy to obtain, which satisfy the condition (H_{j+1}) from the beginning, but not (H_j), $1 \leq j < 7$ (cf. [17] and Theorem 2.6.3). Here we only mention that $\mathbb{C} \setminus \{0, 1\}$ is $\boldsymbol{k}$-hyperbolic but not $\boldsymbol{c}$-hyperbolic. This result follows from the well-known fact that the unit disc is the universal cover of $\mathbb{C} \setminus \{0, 1\}$, and the following theorem:

Theorem 13.1.10. *Let $\Pi : \widetilde{G} \longrightarrow G$ be a holomorphic covering. Then we have: $\widetilde{G}$ is $\boldsymbol{k}$-hyperbolic iff G is $\boldsymbol{k}$-hyperbolic.*

Proof. Since for any $\boldsymbol{k}$-hyperbolic domain the $\boldsymbol{k}$-topology and the $\| \ \|$-topology coincide, the implication "$\Longrightarrow$" is an immediate consequence of Theorem 3.3.7.

Now let us assume that G is $\boldsymbol{k}$-hyperbolic. Of course, it suffices to show that $\boldsymbol{k}_{\widetilde{G}}(\widetilde{x}, \widetilde{y}) > 0$ whenever $\widetilde{x}, \widetilde{y} \in \widetilde{G}$ are different points with $\Pi(\widetilde{x}) = \Pi(\widetilde{y}) =: x$. Since any $\| \ \|$-open subset of G is also $\boldsymbol{k}_G$-open, there exist a $\boldsymbol{k}_G$-ball $B_{\boldsymbol{k}_G}(x, r) =: U$ and open disjoint neighborhoods $\widetilde{U}_1 := \widetilde{U}_1(\widetilde{x})$ and $\widetilde{U}_2 := \widetilde{U}_2(\widetilde{y})$ such that $\Pi|_{\widetilde{U}_j} : \widetilde{U}_j \longrightarrow U$ is biholomorphic. Thus, any continuous curve α in $\widetilde{G}$ connecting $\widetilde{x}$ and $\widetilde{y}$ has the $\boldsymbol{k}_{\widetilde{G}}$-length greater or equal than $2r$, which, by Proposition 3.3.1, implies that $\boldsymbol{k}_{\widetilde{G}}(\widetilde{x}, \widetilde{y}) \geq 2r$. Hence, $\widetilde{G}$ is $\boldsymbol{k}$-hyperbolic. □

A similar result is due to A. Eastwood (see [144]).

Theorem 13.1.11. *Let $G_j \subset \mathbb{C}^{n_j}$, $j = 1, 2$, be domains and let $F \in \mathcal{O}(G_1, G_2)$. Assume that G_2 is $\boldsymbol{k}$-hyperbolic and that there exists an open covering $(U_\alpha)_{\alpha \in A}$ of G_2 such that each connected component of $F^{-1}(U_\alpha)$ is a $\boldsymbol{k}$-hyperbolic domain, $\alpha \in A$. Then, G_1 is $\boldsymbol{k}$-hyperbolic.*

Proof. Since G_2 is $\boldsymbol{k}$-hyperbolic it suffices to discuss different points $z', z'' \in G_1$ with $F(z') = F(z'') =: w^0 \in G_2$. Choose an $\alpha' \in A$ such that $w^0 \in U_{\alpha'}$. Again applying the $\boldsymbol{k}$-hyperbolicity of G_2, we know that the standard topology coincides with the one induced by $\boldsymbol{k}_{G_2}$. So there exists a Kobayashi ball $B_{\boldsymbol{k}_{G_2}}(w^0, 2s) \subset U_{\alpha'}$. Denote by V the connected component of $F^{-1}(U_{\alpha'})$ that contains z'.

Put $\beta = \frac{e^{2s}-1}{e^{2s}+1}$ and recall that there is a universal constant $\widetilde{\beta} > 1$ such that $\tanh^{-1}(t/\beta) \le \widetilde{\beta} \tanh^{-1}(t)$ whenever $0 \le t \le \frac{e^{s/2}-1}{e^{s/2}+1} < \beta$.

Now assume that $\boldsymbol{k}_{G_1}(z', z'') = 0$ and let $\varepsilon \in (0, s/4)$ be, at the moment, an arbitrary number. Then, we choose a sequence of points $z' = z_0, z_1, \dots, z_N = z''$ in G_1 such that $\sum_{j=0}^{N-1} \boldsymbol{\ell}_{G_1}(z_j, z_{j+1}) < \varepsilon$. Note that this sequence depends on ε. In the case that there is some j' with $F(z_{j'}) \notin B_{\boldsymbol{k}_{G_2}}(w^0, \varepsilon)$ (note that $j' < N$), one is lead to

$$\begin{aligned}\varepsilon &> \sum_{j=0}^{N-1} \boldsymbol{\ell}_{G_1}(z_j, z_{j+1}) \ge \sum_{j=j'}^{N-1} \boldsymbol{\ell}_{G_1}(z_j, z_{j+1}) \\ &\ge \sum_{j=j'}^{N-1} \boldsymbol{\ell}_{G_2}(F(z_j), F(z_{j+1}) \ge \boldsymbol{k}_{G_2}(F(z_{j'}), F(z'')) \ge \varepsilon;\end{aligned}$$

a contradiction.

Hence, we have $F(z_j) \in B_{\boldsymbol{k}_{G_2}}(w^0, \varepsilon)$, $j = 0, \dots, N$. Now we can choose analytic discs $\varphi_j \in \mathcal{O}(\mathbb{D}, G_1)$ with $\varphi(0) = z_j$, $\varphi_j(\sigma_j) = z_{j+1}$, $0 < \sigma_j$, and

$$\sum_{j=0}^{N-1} \boldsymbol{k}_{\mathbb{D}}(0, \sigma_j) = \sum_{j=0}^{N-1} \tanh^{-1}(\sigma_j) < \varepsilon < s/4.$$

Observe that $\sigma_j < \frac{e^{2\varepsilon}-1}{e^{2\varepsilon}+1} < \frac{e^{s/2}-1}{e^{s/2}+1}$.

Finally, define $\gamma : \mathbb{D} \longrightarrow \mathbb{D}(\beta) = B_{\boldsymbol{k}_{\mathbb{D}}}(s) \subset \mathbb{D}$ by $\gamma(\lambda) := \beta\lambda$. Put $\widetilde{\varphi}_j := \varphi_j \circ \gamma : \mathbb{D} \longrightarrow G_1$. Then,

$$\begin{aligned}\boldsymbol{k}_{G_2}(F \circ \widetilde{\varphi}_j(\lambda), F(z'')) &\le \boldsymbol{k}_{G_2}(F \circ \widetilde{\varphi}_j(\lambda), F \circ \widetilde{\varphi}_j(0)) + \boldsymbol{k}_{G_2}(F(z_j), F(z'')) \\ &\le \boldsymbol{k}_{\mathbb{D}}(\gamma(\lambda), 0) + \varepsilon < 2s,\end{aligned}$$

which shows that $\widetilde{\varphi}_j(\mathbb{D}) \subset F^{-1}(U_{\alpha'})$, i.e., the discs $\widetilde{\varphi}_j$ can be treated as discs inside of $F^{-1}(U_{\alpha'})$. Moreover, $\widetilde{\varphi}_j(\sigma_j/\beta) = \widetilde{\varphi}_{j+1}(0)$ for $j = 0, \dots, N-1$, which gives that $z'' \in V$ and that the discs may be seen as discs in V. Then, by assumption, $\boldsymbol{k}_V(z', z'') =: r > 0$.

Now we specify the ε to be a number with the additional property that $\varepsilon < r/\widetilde{\beta}$. Taking the corresponding sequence from above we get

$$r \le \sum_{j=0}^{N-1} \boldsymbol{\ell}_V(z_j, z_{j+1}) \le \sum_{j=0}^{N-1} \tanh^{-1}(\sigma_j/\beta) \le \widetilde{\beta} \sum_{j=1}^{N-1} \tanh^{-1}(\sigma_j) < \widetilde{\beta}\varepsilon < r;$$

a contradiction. □

As a simple application of the above theorem, we present a class of $\boldsymbol{k}$-hyperbolic tube domains.

Corollary 13.1.12. *Let $G \subset \mathbb{R}^2$ be a domain sitting inside of the "rectangle" $(a,b)\times\mathbb{R}$, where $(a,b) \neq \mathbb{R}$. Assume that $(a,b) = \bigcup_{j\in\mathbb{N}}(\alpha_j,\beta_j)$, where $\alpha_j,\beta_j \in \mathbb{R}$, such that*

$$G_j := G \cap ((\alpha_j,\beta_j)\times\mathbb{R})$$

is connected and contained in $(\alpha_j,\beta_j)\times(\gamma_j,\delta_j) =: R_j$, where $(\gamma_j,\delta_j) \neq \mathbb{R}$. Then, the tube domain $T_G := G + i\mathbb{R}^2$ is $\boldsymbol{k}$-hyperbolic.

Proof. We may assume that $b < \infty$. Then, the strip domain $(a,b) + i\mathbb{R}$ is $\boldsymbol{k}$-hyperbolic (it is simply biholomorphic to $\mathbb{D}$). Moreover, each of the convex tube domains $R_j + i\mathbb{R}^2$ is $\boldsymbol{k}$-hyperbolic (use Theorem 13.1.7). Then, also the smaller domain $G_j + i\mathbb{R}^2$ is $\boldsymbol{k}$-hyperbolic, $j \in \mathbb{N}$. To be able to apply the former theorem, it suffices to take the simple holomorphic map $F : T_G \longrightarrow (a,b) + i\mathbb{R}$ defined by $F(z) := z_1$. □

So far, we have discussed the notion of hyperbolicity on the level of the Carathéodory or Kobayashi pseudodistance. Now, we ask whether it is possible to express the property "hyperbolic" in terms of the associated metrics (cf. also Chapters 2, 3, and 4).

13.2 Local hyperbolicity

Let G be any domain in $\mathbb{C}^n$ and let $\delta = \delta_G : G \times \mathbb{C}^n \longrightarrow [0,\infty)$ be an arbitrary pseudometric on G. We say that G is *δ-hyperbolic* if for any $z_0 \in G$ there exist a neighborhood $U = U(z_0) \subset G$ and a positive real number C such that $\delta_G(z;X) = \delta(z;X) \geq C\|X\|$, $z \in U$, $X \in \mathbb{C}^n$ (cf. § 4.1).

Remark 13.2.1.

(a) Any bounded domain is $\boldsymbol{\gamma}$-hyperbolic, and therefore $\boldsymbol{S}$-hyperbolic, $\boldsymbol{A}$-hyperbolic, and $\boldsymbol{\varkappa}$-hyperbolic.

(b) Because of Proposition 2.5.1, the notions of "$\boldsymbol{\gamma}$-hyperbolic" and "$\boldsymbol{c}$-hyperbolic" coincide for plane domains. But nothing is known about their relations in higher dimensions. ?

On the other hand, we have the following complete description of Kobayashi hyperbolicity.

Theorem 13.2.2. *For a domain G in $\mathbb{C}^n$ the following properties are equivalent:*

(i) *G is $\boldsymbol{k}$-hyperbolic;*

(ii) $\operatorname{top} G = \operatorname{top} \boldsymbol{k}_G$;

(iii) *for any domain $G' \subset \mathbb{C}^m$, any $w' \in G'$, any $z' \in G$, and any neighborhood $U = U(z') \subset G$, there exist neighborhoods $V = V(w') \subset G'$ and $\widetilde{U} = \widetilde{U}(z') \subset U$ such that if $f \in \mathcal{O}(G', G)$ with $f(w') \in \widetilde{U}$, then $f(V) \subset U$;*

(iv) *condition* (iii) *is true for $G' = \mathbb{D}$ and $w' = 0 \in \mathbb{D}$;*

(v) *G is $\boldsymbol{\varkappa}$-hyperbolic;*

(vi) *for any $z' \in G$ there exists a Kobayashi-ball around z' with finite radius r, which is a bounded subset of $\mathbb{C}^n$;*

(vii) *any point $z' \in G$ has a neighborhood $U = U(z') \subset G$ such that, for $z, w \in U$, $\boldsymbol{k}_G(z, w) \geq M \|z - w\|$, where M is a suitable positive constant.*

Proof. Because of Proposition 5.3.1, the proof is left as an exercise. □

Another description of the $\boldsymbol{k}$-hyperbolicity of a domain G in terms of topological properties of the embedding $\mathcal{O}(\mathbb{D}, G) \longrightarrow C(\mathbb{D}, G^*)$, where G^* denotes the one-point compactification of G, is given in [4].

Remark 13.2.3. Observe that there exists a domain G in $\mathbb{C}^2$ that is not $\boldsymbol{k}$-hyperbolic and thus not $\boldsymbol{\varkappa}$-hyperbolic, but such that $\boldsymbol{\varkappa}_G(z; X) > 0$ for every $z \in G$ and $X \in \mathbb{C}^n \setminus \{0\}$ (cf. Remark 3.5.12). Therefore, in general, we have to distinguish carefully between the $\underline{\delta}$-hyperbolicity and the pointwise $\underline{\delta}$-hyperbolicity.

The property (iv) of Theorem 13.2.2 can be used to verify the following sufficient criterion for hyperbolicity:

Corollary 13.2.4. *Any taut domain G in $\mathbb{C}^n$ is $\boldsymbol{k}$-hyperbolic.*

Proof. Suppose the contrary, which means that condition (iv) is violated. This implies the existence of a point $z' \in G$, a neighborhood $U = U(z')$, a sequence $\lambda_\nu \underset{\nu\to\infty}{\longrightarrow} 0$ in $\mathbb{D}$, and a sequence $(\varphi_\nu) \subset \mathcal{O}(\mathbb{D}, G)$ with $\varphi_\nu(0) \underset{\nu\to\infty}{\longrightarrow} z'$ but $\varphi_\nu(\lambda_\nu) \notin U$. For taut domains, such a combination is impossible. □

Remark 13.2.5. In the case of an unbounded domain $G \subset \mathbb{C}^n$, a similar argument shows that the following properties are equivalent (see Proposition 3.1 in [380]):

(i) G is $\boldsymbol{k}$-hyperbolic;

(ii) $\liminf\limits_{G \ni z \to \infty,\, G \ni w \to b} \boldsymbol{\ell}_G(z, w) > 0$ for any $b \in G$.

In general, it seems rather difficult to calculate the Kobayashi–Royden metric explicitly. On the other hand, plurisubharmonic functions are very flexible. So the Sibony metric may serve as a tool to find a large class of $\boldsymbol{S}$-hyperbolic, and therefore also $\boldsymbol{k}$-hyperbolic domains in $\mathbb{C}^n$.

Theorem 13.2.6 (cf. [474]). *Let G be a domain in $\mathbb{C}^n$. Suppose that there is a bounded function $u \in \mathcal{PSH}(G)$, which is $\mathcal{C}^2$ and strictly plurisubharmonic near a point $z_0 \in G$. Then, $\boldsymbol{S}_G(z;X) \geq C\|X\|$, $z \in V$, $X \in \mathbb{C}^n$, for a suitable $C > 0$ and a suitably chosen neighborhood $V = V(z_0) \subset G$.*

Proof. Without loss of generality we may assume that

- $-1 < u < 0$ on G,
- u is strongly plurisubharmonic and $\mathcal{C}^2$ on $U := \mathbb{B}(z_0, 3R) \subset\subset G$,
- $(\mathcal{L}u)(z;X) \geq \alpha\|X\|^2$, $z \in U$, where $\alpha > 0$.

To prove the required inequality, we are going to construct a function of the Sibony class $\mathcal{S}_G(z')$, $z' \in V := \mathbb{B}(z_0, R)$ (cf. Remark 4.2.1). Put

$$v_t(z) := \chi\left(\frac{\|z - z'\|^2}{R^2}\right)\exp(tu(z)), \quad z \in G,\ t > 0,$$

where $\chi : \mathbb{R} \longrightarrow [0,1]$ denotes an increasing $\mathcal{C}^\infty$-function with $\chi(t) = t$ if $0 \leq t \leq 1/2$, and $\chi(t) = 1$ if $t \geq (3/4)^2$. Obviously, v_t is of class $\mathcal{C}^2$ near z', $v_t(z') = 0$, and $0 \leq v_t < 1$.

Now, we have to show that v_t is a log-psh function on G whenever $t \gg 1$. First, observe that

$$\log v_t = tu \quad \text{on} \quad G \setminus \overline{\mathbb{B}(z', (3/4)R)},$$
$$\log v_t = \log\frac{\|\cdot - z'\|^2}{R^2} + tu \quad \text{on} \quad \mathbb{B}(z', R/\sqrt{2}).$$

Moreover, for $1/2 < \|z - z'\|/R < 1$ we easily obtain

$$(\mathcal{L}\log v_t)(z;X) \geq (-m + \alpha t)\|X\|^2 \geq \|X\|^2 \quad \text{if } t \geq t_0 = (m+1)/\alpha.$$

We point out that t_0 can be chosen independently of z'.

Hence, for any $z' \in \mathbb{B}(z_0, R)$, we have constructed the function $\sqrt{v_{t_0}}$ belonging to $\mathcal{S}_G(z')$. Therefore,

$$\begin{aligned}\boldsymbol{S}_G(z';X) \geq \left((\mathcal{L}v_{t_0})(z';X)\right)^{1/2} &= \|X\|(1/R)\exp(t_1 u(z')) \\ &\geq \|X\|(1/R)\exp(-t_1) \quad \text{with} \quad t_1 := t_0/2. \quad \square\end{aligned}$$

Corollary 13.2.7. *Any connected component of the set $\{z \in \mathbb{C}^n : \psi(z) < 0\}$, where ψ is a strictly psh $\mathcal{C}^2$-function on $\mathbb{C}^n$, is $\boldsymbol{S}$-hyperbolic, and so also $\boldsymbol{k}$-hyperbolic.*

Remark 13.2.8. Suppose that we are in the situation of Theorem 13.2.6, except for the $\mathcal{C}^2$-condition. Then, as in the proof above, we may define the function v_{t_0} for $z' \in B(z_0, R)$. Similar to the above, the reader may verify that v_{t_0} is a log-psh function on G with $0 \leq v_{t_0} < 1$ and $v_{t_0}(z') = 0$.

Now let $X \in \mathbb{C}^n$ and $\varphi \in \mathcal{O}(\mathbb{D}, G)$ with $\varphi(0) = z'$ and $\sigma\varphi'(0) = X$ ($\sigma > 0$, $X \neq 0$). Then, $v := v_{t_0} \circ \varphi$ is log-subharmonic on $\mathbb{D}$ with $v(0) = 0$, $0 \le v < 1$, and $v(\lambda)/|\lambda|^2$ is bounded near the origin. Therefore, the extension theorem for subharmonic functions and the maximum principle lead to

$$1 \ge \limsup_{\lambda \to 0} v(\lambda)/|\lambda|^2 = \frac{\|X\|^2}{\sigma^2 R^2} \exp(t_0 u(z')),$$

i.e., $\boldsymbol{\varkappa}_G(z'; X) \ge \|X\|(1/R)\exp(-t_0/2)$.

It is unclear whether this inequality still holds for $\boldsymbol{S}_G$. ?

Remark 13.2.9. Let G denote the unbounded component of

$$\{z = (z_1, z_2) \in \mathbb{C}^2 : \operatorname{Re}(z_1^3 + z_2^3 + z_1) + |z_1|^2 + |z_2|^2 < 0\}$$

(cf. Corollary 13.2.7). In [494] it is shown that G is $\boldsymbol{\gamma}$-hyperbolic. Using a similar argument, one can also prove that G is even $\boldsymbol{c}$-hyperbolic. We do not know of any example of a domain like the one in Corollary 13.2.7 that is neither $\boldsymbol{\gamma}$- nor $\boldsymbol{c}$-hyperbolic. ?

Under the hypothesis that G is a $\boldsymbol{k}$-hyperbolic domain, the following localization result was formulated by H. Royden: (cf. [459], also [197]).

Proposition 13.2.10. *Suppose that G is a $\boldsymbol{k}$-hyperbolic domain in $\mathbb{C}^n$ and let $U \subset G$ be any subdomain. Then,*

$$\boldsymbol{\varkappa}_U(z; X) \le \coth \boldsymbol{\ell}_G(z, G \setminus U)\boldsymbol{\varkappa}_G(z; X) \le \coth \boldsymbol{k}_G(z, G \setminus U)\boldsymbol{\varkappa}_G(z; X), \quad z \in U,\ X \in \mathbb{C}^n,$$

where $\boldsymbol{\ell}_G(z, G \setminus U) := \inf\{\boldsymbol{\ell}_G(z, w) : w \in G \setminus U\}$.

Proof. Recall that $\coth \boldsymbol{p}(0, t) = 1/t$ for $0 < t < 1$. Therefore, we may write

$$\coth \boldsymbol{\ell}_G(z, w) = \sup\{1/r > 1 : \exists_{\varphi \in \mathcal{O}(\mathbb{D}, G)} : \varphi(0) = z,\ \varphi(r) = w\}, \quad z, w \in G,\ z \neq w.$$

Now fix a $z_0 \in U$ and an $X \in \mathbb{C}^n$ and let $\varphi \in \mathcal{O}(\mathbb{D}, G)$ be an arbitrary analytic disc with $\varphi(0) = z_0$, $\alpha\varphi'(0) = X$ with $\alpha > 0$. Moreover, choose an $s > 0$ with $1/s > \coth \boldsymbol{\ell}_G(z_0, G \setminus U)$. Observe that $\boldsymbol{\ell}_G(z_0, G \setminus U) > 0$ since $\operatorname{top} \boldsymbol{k}_G = \operatorname{top} G$ by hyperbolicity. Hence, every $\psi \in \mathcal{O}(\mathbb{D}, G)$ with $\psi(0) = z_0$ maps $s\mathbb{D}$ into U. In particular, if we put $\widetilde{\varphi}(\lambda) := \varphi(s\lambda)$, $\lambda \in \mathbb{D}$, then $\widetilde{\varphi} \in \mathcal{O}(\mathbb{D}, U)$ with $\widetilde{\varphi}(0) = z_0$, $(\alpha/s)\widetilde{\varphi}'(0) = X$, i.e., $\alpha/s \ge \boldsymbol{\varkappa}_U(z_0; X)$. Since the choice of α and s was arbitrary, the claimed inequality follows. □

In a forthcoming chapter, we will discuss more details of the localization of the Kobayashi–Royden metric. We will then exploit Proposition 13.2.10

We conclude this section with a generalization of the big Picard theorem (Theorem 1.2.7). Recall that this theorem implies the following:

Any holomorphic function $f : \mathbb{D}_* \longrightarrow \mathbb{C}$ omitting at least two complex values extends as a holomorphic or a meromorphic function to the whole unit disc.

The following generalization of this formulation of the Picard theorem is given by M. H. Kwack (cf. [332]).

Theorem 13.2.11. *Let $f : \mathbb{D}_* \longrightarrow G$ be a holomorphic map, where G is a $\boldsymbol{k}$-hyperbolic domain in $\mathbb{C}^n$. Assume that for a sequence $(\lambda_k)_{k=1}^{\infty} \subset \mathbb{D}_*$ with $\lambda_k \longrightarrow 0$, the sequence $(f(\lambda_k))_{k=1}^{\infty}$ converges to a point $z_0 \in G$. Then, f extends to a holomorphic map $\tilde{f} : \mathbb{D} \longrightarrow G$.*

Proof. Without loss of generality, we may assume that the sequence $r_k := |\lambda_k|$ is strictly decreasing and that $z_0 = 0 \in G$. We are going to prove that the function $\tilde{f}$ given by $\tilde{f}(\lambda) := f(\lambda)$, $\lambda \in \mathbb{D}_*$, $\tilde{f}(0) := 0$ is continuous on $\mathbb{D}$. Fix an $\varepsilon \in (0, \operatorname{dist}(0, \partial G))$. Since G is $\boldsymbol{k}$-hyperbolic, we have $B_{\boldsymbol{k}_G}(0, \delta_\varepsilon) \subset \mathbb{B}(\varepsilon)$ with δ_ε being a suitable positive number. By assumption, there is a $k_0 \in \mathbb{N}$ such that for $k \geq k_0$ we have $f(\lambda_k) \in B_{\boldsymbol{k}_G}(0, \delta_\varepsilon/2)$. Applying Corollary 9.1.10 for $\lambda \in \mathbb{D}_*, |\lambda| = r_k$, we obtain

$$\begin{aligned} \boldsymbol{k}_G(0, f(\lambda)) &\leq \boldsymbol{k}_G(0, f(\lambda_k)) + \boldsymbol{k}_G(f(\lambda_k), f(\lambda)) \\ &\leq \boldsymbol{k}_G(0, f(\lambda_k)) + \boldsymbol{k}_{\mathbb{D}_*}(\lambda_k, \lambda) < \frac{\delta_\varepsilon}{2} + \frac{\pi}{-\log r_k} < \delta_\varepsilon \ \text{ if } k \geq k_1 \geq k_0. \end{aligned}$$

Therefore, $f(\partial\mathbb{D}(r_k)) \subset \mathbb{B}(\varepsilon)$ if $k \geq k_1$.

Now, it suffices to show that if $r_{k+1} < |\lambda| < r_k$, $k \gg 1$, then $f(\lambda) \in \mathbb{B}(\varepsilon)$. We proceed by supposing the contrary. Then we may assume that there are numbers $r_{k+1} < b_{k+1} \leq a_k < r_k < b_k$, $k \geq 1$, such that

$$\begin{gathered} f(\{\lambda \in \mathbb{D} : a_k < |\lambda| < b_k\}) \subset \mathbb{B}(\varepsilon), \\ f(\partial\mathbb{D}(a_k)) \cap \partial\mathbb{B}(\varepsilon) \neq \varnothing, \quad f(\partial\mathbb{D}(b_k)) \cap \partial\mathbb{B}(\varepsilon) \neq \varnothing. \end{gathered}$$

Similar to the above, we may also assume that there are points $z', z'' \in \partial\mathbb{B}(\varepsilon)$ with

$$\lim_{k\to\infty} f(\partial\mathbb{D}(a_k)) = z', \quad \lim_{k\to\infty} f(\partial\mathbb{D}(b_k)) = z''$$

(otherwise, take an appropriate subsequence). Choose a holomorphic function $g : \mathbb{C}^n \longrightarrow \mathbb{C}$ with $g(0) = 0$ and $g(z')g(z'') \neq 0$. Consequently, we have

$$\begin{aligned} &g \circ f(\partial\mathbb{D}(a_k)) \underset{k\to\infty}{\longrightarrow} g(z') \neq 0, \\ &g \circ f(\partial\mathbb{D}(b_k)) \underset{k\to\infty}{\longrightarrow} g(z'') \neq 0, \quad g \circ f(\lambda_k) \underset{k\to\infty}{\longrightarrow} 0. \end{aligned}$$

Let B' (resp. B'') be a small disc around $g(z')$ (resp. $g(z'')$) such that $0 \notin \overline{B}'$ (resp. $0 \notin \overline{B}''$). Then, for sufficiently large k, we obtain

$$g \circ f(\partial\mathbb{D}(a_k)) \subset B' \subset \overline{B}' \not\ni g \circ f(\lambda_k), \quad g \circ f(\partial\mathbb{D}(b_k)) \subset B'' \subset \overline{B}'' \not\ni g \circ f(\lambda_k).$$

Application of the Cauchy theorem leads to

$$0 = \int_{g\circ f(\partial\mathbb{D}(c_k))} \frac{d\lambda}{\lambda - g \circ f(\lambda_k)} = \int_{\partial\mathbb{D}(c_k)} \frac{(g \circ f)'(\lambda)d\lambda}{g \circ f(\lambda) - g \circ f(\lambda_k)},$$

where $c = a$ or $c = b$. From here, the principle of the argument implies that the function $g \circ f - g \circ f(\lambda_k)$ is without zeros in the annulus $\{\lambda \in \mathbb{C} : a_k < |\lambda| < b_k\}$; a contradiction. □

13.3 Hyperbolicity for Reinhardt domains

Before we discuss the different notions of hyperbolicity in the case of pseudoconvex Reinhardt domains, let us recall (for the convenience of the reader) some notations which will be frequently used:

- $V_j := \{z \in \mathbb{C}^n : z_j = 0\}, \quad j = 1, \dots, n;$
- $D_{\alpha,C} = \{z \in \mathbb{C}^n(\alpha) : |z_1|^{\alpha_1} \cdots |z_n|^{\alpha_n} < e^C\}$, where $\alpha \in (\mathbb{R}^n)_*$ and $C \in \mathbb{R}$.

Moreover, for a matrix $A = (A^j_k)_{j=1,\dots,n,\, k=1,\dots,n} \in \mathbb{M}(n \times n; \mathbb{Z})$, we denote by A^j its j-th row. Put

$$\Phi_A : \mathbb{C}^n_* \to \mathbb{C}^n_*, \quad \Phi(z) := (z^{A^1}, \dots, z^{A^n}).$$

Theorem 13.3.1 (cf. [564]). *Let G be a pseudoconvex Reinhardt domain in $\mathbb{C}^n$. Then, the following properties are equivalent:*

(i) *G is $\boldsymbol{c}$-hyperbolic;*

(ii) *G is $\boldsymbol{\ell}$-hyperbolic (i.e., $\boldsymbol{\ell}_G(z,w) > 0$ for all $(z,w) \in G \times G$, $z \neq w$);*

(iii) *G is Brody hyperbolic (i.e., $\mathcal{O}(\mathbb{C}, G) = \mathbb{C}$);*

(iii') *$\log G$ [1] contains no affine lines, and either $V_j \cap G = \varnothing$ or $V_j \cap G$ (treated as a domain in $\mathbb{C}^{n-1}$) is $\boldsymbol{c}$-hyperbolic, $j = 1, \dots, n$;*

[1] $\log G := \{x \in \mathbb{R}^n : (e^{x_1}, \dots, e^{x_n}) \in G\}$.

(iv) *there exist* $A = (A_k^j)_{j=1,\dots,n,\,k=1,\dots,n} \in \mathbb{M}(n\times n;\mathbb{Z})$, rank $A = n$, *and a vector* $C = (C_1,\dots,C_n) \in \mathbb{R}^n$ *such that*

- $G \subset \boldsymbol{D}_{A,C} := \boldsymbol{D}_{A^1,C_1} \cap \cdots \cap \boldsymbol{D}_{A^n,C_n}$,
- *either* $\boldsymbol{V}_j \cap G = \varnothing$ *or* $\boldsymbol{V}_j \cap G$ *is* $\boldsymbol{c}$*-hyperbolic as a domain in* $\mathbb{C}^{n-1}$, $j = 1,\dots,n$;

(iv') *there exist* $A \in \mathbb{M}(n\times n;\mathbb{Z})$, $|\det A| = 1$, *and a vector* $C \in \mathbb{R}^n$ *such that*

- $G \subset \boldsymbol{D}_{A,C}$ (cf. (iv)),
- *either* $\boldsymbol{V}_j \cap G = \varnothing$ *or* $\boldsymbol{V}_j \cap G$ *is* $\boldsymbol{c}$*-hyperbolic as a domain in* $\mathbb{C}^{n-1}$, $j = 1,\dots,n$;

(v) G *is algebraically equivalent to a bounded domain (i.e., there exists a matrix* $A \in \mathbb{M}(n\times n;\mathbb{Z})$ *such that* Φ_A *is defined on* G *and gives a biholomorphic mapping from* G *to the bounded domain* $\Phi_A(G)$*);*

(vi) G *is* $\boldsymbol{k}$*-complete, (i.e.,* G *is* $\boldsymbol{k}$*-hyperbolic and every* $\boldsymbol{k}_G$*-Cauchy sequence in* G *converges (in the standard topology) to a point in* G*).*

In the following a domain of the type $\boldsymbol{D}_{A,C}$ (cf. (iv) in Theorem 13.3.1) will be shortly called a *quasi-elementary Reinhardt domain.*

Remark 13.3.2. So far, we did not discuss $\boldsymbol{k}$-complete domains; nevertheless, because of the above result it seems quite natural to include the statement (vi) in Theorem 13.3.1. A detailed discussion on completeness with respect to holomorphically contractible distances will follow in the next chapter.

To prove Theorem 13.3.1, we need the following lemmas:

Lemma 13.3.3 (cf. [564]). *Let* $\boldsymbol{D}_{A,C}$ *be as in Theorem* 13.3.1(iv). *Then,*

(a) *there exist a matrix* $\widetilde{A} \in \mathbb{M}(n\times n;\mathbb{Z})$, $|\det \widetilde{A}| = 1$, *and a vector* $\widetilde{C} \in \mathbb{R}^n$ *such that* $\boldsymbol{D}_{A,C} \subset \boldsymbol{D}_{\widetilde{A},\widetilde{C}}$;

(b) $\boldsymbol{c}_{\boldsymbol{D}_{A,C}}(z,w) > 0$ *for any points* $z,w \in \boldsymbol{D}_{A,C} \cap \mathbb{C}_*^n$, $z \neq w$.

Proof. Fix a matrix A and a vector C as in Theorem 13.3.1(iv).

Step 1°. To prove (a) it suffices to construct, for an arbitrary quasi-elementary Reinhardt domain $G := \boldsymbol{D}_{B,D}$ with $|\det B| > 1$, a new quasi-elementary Reinhardt domain $\widetilde{G} := \boldsymbol{D}_{\widetilde{B},\widetilde{D}}$ such that $G \subset \widetilde{G}$ and $|\det \widetilde{B}| < |\det B|$.

Recall that

- $\mathcal{S}(G) := \{\alpha \in \mathbb{Z}^n : z^\alpha \in \mathcal{H}^\infty(G)\}$,
- $\mathcal{B}(G) := \mathcal{S}(G) \setminus (\mathcal{S}(G) + \mathcal{S}(G))$.

It is known (cf. Lemma 2.10.6) that

$$\mathcal{S} := \mathcal{S}(G) = \mathbb{Z}^n \cap (\mathbb{Q}_+ B^1 + \cdots + \mathbb{Q}_+ B^n),$$
$$\mathcal{B} := \mathcal{B}(G) \subset \mathbb{Z}^n \cap (\mathbb{Q} \cap [0,1) B^1 + \cdots + \mathbb{Q} \cap [0,1) B^n) \cup \{B^1, \dots, B^n\}.$$

Claim: $\mathcal{B} \not\subset \{B^1, \dots, B^n\}$.
Assume the contrary, i.e., $\mathcal{B} \subset \{B^1, \dots, B^n\}$. Define

$$r(B) := \min\{r \in \mathbb{N} : \text{ if } x \in \mathbb{Q}^n,\ xB \in \mathbb{Z}^n, \text{ then } rx \in \mathbb{Z}^n\}.$$

Observe that $B^{-1}B \in \mathbb{M}(n \times n; \mathbb{Z})$, i.e., all the rows of B^{-1} are special vectors in the definition of the number $r(B)$. So $r(B)B^{-1} \in \mathbb{M}(n \times n; \mathbb{Z})$, from which $r(B)^n = \det(r(B)B^{-1}B) = \det(r(B)B^{-1})\det(B)$ follows. Therefore, if $r(B) = 1$, then $|\det B| = 1$, which gives a contradiction.

So it remains to prove that $r(B) = 1$.

Take an arbitrary rational vector $x \in \mathbb{Q}^n$ with $xB \in \mathbb{Z}^n$. We have to show that $x \in \mathbb{Z}^n$. In fact, we write $xB = uB + \nu B$, where $u = (u_1, \dots, u_n)$, $u_j := x_j - \lfloor x_j \rfloor \geq 0$ and $\nu = (\nu_1, \dots, \nu_n)$, $\nu_j := \lfloor x_j \rfloor \in \mathbb{Z}$, $j = 1, \dots, n$ (here $\lfloor x \rfloor$ denotes the largest integer smaller or equal x). Obviously, $uB \in \mathbb{Z}^n$. Applying the above description of $\mathcal{S}$, it follows that $uB = u_1 B^1 + \cdots + u_n B^n \in \mathcal{S}$ (recall that B^j is the j-th row of B). By virtue of the assumption, uB is an entire linear combination of the vectors $B^1, \dots, B^n$; in particular, $u \in \mathbb{Z}^n$ (recall that the B^j's are linearly independent). Hence, $x = u + \nu \in \mathbb{Z}^n$, i.e., $r(B) = 1$. So, the claim is verified.

Therefore, there is a $\beta \in \mathcal{B} \setminus \{B^1, \dots, B^n\}$ such that $\beta = t_1 B^1 + \cdots + t_n B^n$ with $t_j \in [0,1)$ and one of the t_j's is positive. We may assume that $t_1 > 0$. We denote by $\widetilde{B}$ that matrix whose rows $\widetilde{B}^j$ are given by $\widetilde{B}^1 := \beta$, $\widetilde{B}^j := B^j$, $j = 2, \dots, n$. Moreover, with $\widetilde{D}_1 := \sum_{j=1}^n t_j D_j$ and $\widetilde{D}_j := D_j$, $j = 2, \dots, n$, we put

$$\widetilde{G} := \boldsymbol{D}_{\widetilde{B},\widetilde{D}}, \text{ where } \widetilde{D} := (\widetilde{D}_1, \dots, \widetilde{D}_n).$$

Then, $|\det \widetilde{B}| = t_1 |\det B| < |\det B|$ and $G \subset \widetilde{G}$. Hence, (a) is verified.

Step 2°. Recall that for a matrix $A \in \mathbb{M}(n \times n; \mathbb{Z})$ the mapping

$$\Phi_A : \mathbb{C}_*^n \longrightarrow \mathbb{C}_*^n, \quad \Phi_A(z) := (z^{A^1}, \dots, z^{A^n}),\ z \in \mathbb{C}_*^n,$$

is proper iff $\det A \neq 0$, and that in this case its multiplicity is given by $|\det A|$. In particular, the mapping $\Phi_{\widetilde{A}}$, where $\widetilde{A}$ is taken from (a), is a biholomorphic mapping from $\mathbb{C}_*^n$ to itself.

Now fix two different points $z, w \in \boldsymbol{D}_{A,C} \cap \mathbb{C}_*^n$. Then,

$$\boldsymbol{c}_{\boldsymbol{D}_{A,C}}(z,w) \geq \boldsymbol{c}_{\boldsymbol{D}_{\widetilde{A},\widetilde{C}}}(z,w) = \boldsymbol{c}_{\boldsymbol{D}_{\widetilde{A},\widetilde{C}} \cap \mathbb{C}_*^n}(z,w) = \boldsymbol{c}_{\mathbb{D}^n}(\Psi(z), \Psi(w)) > 0,$$

where $\Psi(z) := (\Phi_1(z)/e^{\widetilde{C}_1}, \dots, \Phi_n(z)/e^{\widetilde{C}_n})$ and $\Phi_{\widetilde{A}} =: (\Phi_1, \dots, \Phi_n)$. □

The next lemma is Lemma 1.4.12 in [269].

Lemma 13.3.4. *Let $\Omega \subset \mathbb{R}^n$ be a convex domain containing no straight lines. Then there are linearly independent vectors $A^1, \dots, A^n \in \mathbb{Z}^n$ and a vector $C \in \mathbb{R}^n$ such that*

$$\Omega \subset \{x \in \mathbb{R}^n : \langle x, A^j \rangle < C_j,\ j = 1, \dots, n\}.$$

Proof. See [269]. □

After these preparations, we turn our attention to the proof of Theorem 13.3.1.

Proof of Theorem 13.3.1. First, observe that the implications (i) $\Longrightarrow$ (ii) $\Longrightarrow$ (iii) are obvious and that (iv) $\Longrightarrow$ (iv') is true due to Lemma 13.3.3.

The remaining proof uses induction on the dimension n. Obviously, the theorem is true in the case where $n = 1$. Now, let $n \geq 2$.

(iii) $\Longrightarrow$ (iii'): The first condition is an obvious consequence of (iii). The second one follows from the induction process.

(iii') $\Longrightarrow$ (iv): Note that the second condition in (iv) follows from applying the theorem in the case where $n - 1$. From (iii') we see that $\log G$ does not contain straight lines. Therefore, we immediately get (iv) from Lemma 13.3.4.

(iv') $\Longrightarrow$ (i): Take $z, w \in G$, $z \neq w$.

Case 1°: If both points belong to $\mathbb{C}_*^n$, then, by virtue of Lemma 13.3.3, we have

$$\boldsymbol{c}_G(z, w) \geq \boldsymbol{c}_{\boldsymbol{D}_{A,C}}(z, w) > 0.$$

Case 2°: Let $z \in \mathbb{C}_*^n$, $w \notin \mathbb{C}_*^n$. Without loss of generality, we may assume that $w = (w_1, \dots, w_k, 0, \dots, 0)$ with $w_1 \cdots w_k \neq 0$. Then, $k < n$ and $A_s^j \geq 0$, $j = 1, \dots, n$, $s = k + 1, \dots, n$. Since $\operatorname{rank} A = n$ we find a $j \in \{1, \dots, n\}$ and an $r \in \{k + 1, \dots, n\}$ such that $A_r^j > 0$. Thus, $w^{A^j} = 0 \neq z^{A^j}$. Therefore,

$$\boldsymbol{c}_G(z, w) \geq \boldsymbol{c}_{\boldsymbol{D}_{A^j, C_j}}(z, w) \geq c_D(z^{A^j}, w^{A^j}) > 0, \text{ where } D := e^{C_j}\mathbb{D}.$$

Case 3°: Let $z, w \notin \mathbb{C}_*^n$. We may assume that $z_1 = 0$ and $z_2 \neq w_2$. Consequently, $\pi_{2,\dots,n}(G)$ [2] is $\boldsymbol{c}$-hyperbolic and $\pi_{2,\dots,n}(z) \neq \pi_{2,\dots,n}(w)$. Therefore,

$$\boldsymbol{c}_G(z, w) \geq \boldsymbol{c}_{\pi_{2,\dots,n}(G)}(\pi_{2,\dots,n}(z), \pi_{2,\dots,n}(w)) > 0.$$

Hence, G is $\boldsymbol{c}$-hyperbolic.

[2] $\pi_{i_1,\dots,i_k}(z_1, \dots, z_n) := (z_{i_1}, \dots, z_{i_k})$.

(iv') $\Longrightarrow$ (v): By (iv') we know that there is a matrix $A \in \mathbb{M}(n \times n; \mathbb{Z})$, $|\det A| = 1$, and a vector $C \in \mathbb{R}^n$ with $G \subset \boldsymbol{D}_{A,C}$. Moreover, the mapping $\Phi_A : \mathbb{C}_*^n \longrightarrow \mathbb{C}_*^n$, $\Phi_A(z) := (z^{A^1}, \dots, z^{A^n})$ is biholomorphic.

Therefore, if the domain G is contained in $\mathbb{C}_*^n$, then $\Phi_A : G \longrightarrow \Phi_A(G)$ is a biholomorphic mapping and $\Phi_A(G)$ is bounded.

The remaining case is proved by induction:

Obviously, the case $n = 1$ is clear. So, we may assume that $n \geq 2$ and, without loss of generality, that $V_n \cap G \neq \varnothing$.

Claim: It suffices to prove (v) under the additional assumption that

$$V_n \cap G \neq \varnothing \text{ and } \pi_j(G) \text{ is bounded, } j = 1 \dots, n-1. \tag{13.3.1}$$

In fact, put $\widetilde{G} := G \cap V_n$. By assumption, $\widetilde{G}$ is a $\boldsymbol{c}$-hyperbolic pseudoconvex Reinhardt domain in $\mathbb{C}^{n-1}$. By the induction hypothesis there exists a matrix $\widetilde{A} \in \mathbb{M}((n-1) \times (n-1); \mathbb{Z})$ such that $\Phi_{\widetilde{A}}$ is defined on $\widetilde{G}$, $\Phi_{\widetilde{A}}(\widetilde{G})$ is bounded, and $\Phi_{\widetilde{A}} : \widetilde{G} \longrightarrow \Phi_{\widetilde{A}}(\widetilde{G})$ is biholomorphic. Put

$$B := \begin{bmatrix} \widetilde{A} & 0 \\ 0 & 1 \end{bmatrix} \in \mathbb{M}(n \times n; \mathbb{Z}).$$

Then Φ_B satisfies condition (13.3.1), and so the claim has been verified.

For the remaining part of the proof of (v), we may now assume that (13.3.1) is fulfilled. Without loss of generality, assume that

$$V_j \cap G \neq \varnothing,\ j = 1, \dots, k, \quad V_j \cap G = \varnothing,\ j = k+1, \dots, n-1.$$

Put $\widetilde{G} := V_1 \cap \dots \cap V_k \cap G$. Then $\widetilde{G}$ is a (non-empty) $\boldsymbol{c}$-hyperbolic pseudoconvex Reinhardt domain. Then there is $\alpha = (0, \dots, 0, \alpha_{k+1}, \dots, \alpha_n) \in \mathcal{S}(\widetilde{G})$, $\alpha_n \neq 0$. The fact that $\widetilde{G} \cap V_n \neq \varnothing$ implies $\alpha_n > 0$. Moreover, by virtue of (13.3.1), it is clear that $\boldsymbol{e}_j := (0, \dots, 0, 1, 0, \dots, 0) \in \mathcal{S}(\widetilde{G})$ (the number 1 in $\boldsymbol{e}_j$ is at the j-th place), $j = k+1, \dots, n-1$. Thus,

$$\widetilde{\alpha} := \frac{1}{\alpha_n}\alpha + \sum_{j=k+1}^{n-1} \left(\left\lfloor \frac{\alpha_j}{\alpha_n} \right\rfloor + 1 - \frac{\alpha_j}{\alpha_n} \right) \boldsymbol{e}_j \in \mathcal{S}(\widetilde{G}) \subset \mathcal{S}(G).$$

Define

$$A := \begin{bmatrix} & & \mathbb{I}_{n-1} & & & & 0 \\ 0 & \dots & 0 & \widetilde{\alpha}_{k+1} & \dots & \widetilde{\alpha}_{n-1} & 1 \end{bmatrix}.$$

Then, A fulfills all the required properties. Hence, condition (v) is proved.

(v) $\Longrightarrow$ (vi): By assumption, we may assume that G is a bounded pseudoconvex Reinhardt domain. Fix a point $w \in G$. To verify that G is $\boldsymbol{k}$-complete, we only have to disprove the existence of a sequence $(z^j)_{j \in \mathbb{N}} \subset G$ such that $(\boldsymbol{k}_G(w, z^j))_{j \in \mathbb{N}}$ is bounded, but $z^j \underset{j \to \infty}{\longrightarrow} z^0 \in \partial G$.

Case $z^0 \in \mathbb{C}_*^n$: We may assume that $z^0 = (1, \dots, 1)$. It is clear that there is an $\alpha \in \mathbb{R}^n$, $\alpha \neq 0$, such that $G \subset \boldsymbol{D}_\alpha$, where $\boldsymbol{D}_\alpha$ denotes the elementary Reinhardt domain for α. Moreover, we may assume that $\alpha_j \neq 0$, $j = 1, \dots, k$, and $\alpha_{k+1} = \dots \alpha_n = 0$, where $k \geq 1$. So we get

$$\begin{aligned} \boldsymbol{k}_G(w, z^j) &\geq \boldsymbol{k}_{\boldsymbol{D}_\alpha}(w, z^j) \\ &= \max\{\boldsymbol{k}_{\boldsymbol{D}_{\widetilde{\alpha}}}(\widetilde{w}, \widetilde{z}^j),\ \boldsymbol{k}_{\mathbb{C}^{n-k}}((w_{k+1}, \dots, w_n), (z^j_{k+1}, \dots, z^j_n))\} \\ &= \boldsymbol{k}_{\boldsymbol{D}_{\widetilde{\alpha}}}(\widetilde{w}, \widetilde{z}^j), \end{aligned} \tag{13.3.2}$$

where $\widetilde{\alpha} := (\alpha_1, \dots, \alpha_k)$, $\widetilde{w} := (w_1, \dots, w_k)$, $\widetilde{z}^j := (z^j_1, \dots, z^j_k)$.

Observe that the sequence $(\widetilde{z}^j)_{j \in \mathbb{N}}$ converges to the boundary point $\widetilde{z}^0$ of $\boldsymbol{D}_{\widetilde{\alpha}}$. Then, applying Proposition 10.1.1, we see that the sequence in (13.3.2) tends to infinity.

Case $z^0 \notin \mathbb{C}_*^n$: Let us assume that $z^0_j \neq 0$ for $j = 1, \dots, k$ with a suitable k, $0 \leq k < n$, and $z^0_{k+1} = \dots = z^0_n = 0$. We have to discuss two subcases:

(a) There is an $s \in \{k+1, \dots, n\}$ such that $G \cap V_s = \varnothing$.
Then,

$$\boldsymbol{k}_G(w, z^j) \geq \boldsymbol{k}_{\pi_s(G)}(w_s, z^j_s).$$

Here, $\pi_s(G)$ is a plane Reinhardt domain not containing the origin, but $0 \in \partial\pi_s(G)$. Therefore, the right side tends to infinity.

(b) All intersections $G \cap \boldsymbol{V}_j$, $j = k+1, \dots, n$, are non-empty.
Obviously, $k > 0$, otherwise $z^0 = 0 \in G$; contradiction. Then,

$$\boldsymbol{k}_G(w, z^j) \geq \boldsymbol{k}_{\widetilde{G}}((w_1, \dots, w_k), (z^j_1, \dots, z^j_k)),$$

where $\widetilde{G} := \pi_{1,\dots,k}(G)$. Since $(z^0_1, \dots, z^0_k) \in \partial\widetilde{G}$ and $\widetilde{G}$ is a Reinhardt domain of the first case, the right side again tends to infinity.

Hence, the Kobayashi completeness of G has been verified.

What remains is to mention that (vi) trivially implies (iv). $\square$

Remark 13.3.5. Observe that Theorem 13.3.1 shows that *all notions of hyperbolicity coincide in the class of pseudoconvex Reinhardt domains*. That is why we will often speak only of *hyperbolic pseudoconvex Reinhardt domains*.

Remark 13.3.6. The pseudoconvex Reinhardt domain

$$D := \{z \in \mathbb{C}^3 : \max\{|z_1 z_2|, |z_1 z_3|, |z_2|, |z_3|\} < 1\}$$

is not $\boldsymbol{k}$-hyperbolic, since $\mathbb{C} \times \{0\} \times \{0\} \subset D$; in particular, D is not $\boldsymbol{c}$-hyperbolic. Let $\widetilde{D} := D \setminus (\mathbb{C} \times \{0\} \times \{0\})$. Then, $\widetilde{D}$ is $\boldsymbol{c}$-hyperbolic (the functions $z_1 z_2$, $z_1 z_3$, z_2, and z_3 separate the points of $\widetilde{D}$). Observe that D is the envelope of holomorphy of $\widetilde{D}$, i.e., $D = \mathcal{E}(\widetilde{D})$. Hence, in general, $\boldsymbol{c}$-hyperbolicity of a Reinhardt domain and its envelope of holomorphy may be different.

But in the two-dimensional case, there is the following positive result [136]:

Theorem 13.3.7. *Let $G \subset \mathbb{C}^2$ be a $\boldsymbol{c}$-hyperbolic Reinhardt domain. Then, its envelope of holomorphy $\mathcal{E}(G)$ is $\boldsymbol{c}$-hyperbolic.*

Proof. Recall that the envelope of holomorphy $\mathcal{E}(D)$ of a Reinhardt domain $D \subset \mathbb{C}_*^n$ always exists and satisfies the following properties:

- $\mathcal{E}(D) \subset \mathbb{C}_*^n$,
- $\log \mathcal{E}(D) = \operatorname{conv}(\log D)$;

see, for example, [269].

Put $G_* := G \cap \mathbb{C}_*^2$; G_* is a Reinhardt domain. Assume that $\log \mathcal{E}(G_*)$ contains an affine line ℓ. Fix a point $x_0 \in \log G \setminus \ell$. Denote by ℓ' the line passing through x_0 which is parallel to ℓ. Then, $\ell' \subset \log \mathcal{E}(G_*)$. Let

$$\ell' = \{(a_1 t + b_1, a_2 t + b_2) : t \in \mathbb{R}\},$$

where $a_1, a_2, b_1, b_2 \in \mathbb{R}$ and $a_1^2 + a_2^2 \neq 0$. Hence,

$$A := \{(e^{a_1 \lambda + b_1}, e^{a_2 \lambda + b_2}) : \lambda \in \mathbb{C}\} \subset \mathcal{E}(G_*).$$

Using Liouville's theorem and the fact that G is $\boldsymbol{c}$-hyperbolic, we get $A \cap G = \varnothing$ or $\ell' \cap \log G = \varnothing$; a contradiction.

Now assume that $\log \mathcal{E}(G)$ contains an affine line. As in the previous step, this leads to a non-trivial entire map $\varphi : \mathbb{C} \longrightarrow \mathcal{E}(G) \cap \mathbb{C}_*^n$. Recall that $\mathcal{E}(G_*) = \mathcal{E}(G) \cap \mathbb{C}_*^n$ (see Theorem 2.5.9 in [265]). Hence, $\mathcal{E}(G_*)$ contains an affine line; a contradiction.

Without loss of generality, assume finally that $\mathcal{E}(G) \cap V_2 \neq \varnothing$. Denote this intersection by $G' \subset \mathbb{C}$. Suppose that G' is not $\boldsymbol{c}$-hyperbolic. Then, either $G' = \mathbb{C}$ or $G' = \mathbb{C}_*$. Therefore, either $A_1 := \mathbb{C} \times \{0\} \subset \mathcal{E}(G)$ or $A_2 := \mathbb{C}_* \times \{0\} \subset \mathcal{E}(G)$. By virtue of the $\boldsymbol{c}$-hyperbolicity of G, we conclude that $A_1 \cap G = \varnothing$ or that $A_2 \cap G = \varnothing$. Therefore, $G \cap V_2 = \varnothing$; a contradiction. Thus, Theorem 13.3.1 implies that $\mathcal{E}(G)$ is $\boldsymbol{c}$-hyperbolic. □

We conclude this section with the following result, which will be useful later:

Proposition 13.3.8 (cf. [566]). *Let $G \subset \mathbb{C}^n$ be a hyperbolic pseudoconvex Reinhardt domain. Then the following conditions are equivalent:*

(i) *G is algebraically equivalent to an unbounded Reinhardt domain;*

(ii) *G is algebraically equivalent to a bounded Reinhardt domain D, for which there is a j_0, $1 \le j_0 \le n$, such that $\overline{D} \cap V_{j_0} \neq \varnothing$, but $D \cap V_{j_0} = \varnothing$.*

Proof. (i) $\Longrightarrow$ (ii): We may assume that G is an unbounded hyperbolic pseudoconvex Reinhardt domain. By virtue of Theorem 13.3.1, there are a bounded Reinhardt domain D and a biholomorphic mapping $\Phi_A : D \longrightarrow G$ (here we use the notation from Theorem 13.3.1). Suppose that D satisfies the following property:

$$\text{if } \overline{D} \cap V_j \neq \varnothing, \text{ then } D \cap V_j \neq \varnothing, \quad j = 1, \dots, n.$$

Without loss of generality, we may assume that there is a $k \in \{0, 1, \dots, n\}$ such that

$$\overline{D} \cap V_j \neq \varnothing,\ j = 1, \dots, k, \quad \overline{D} \cap V_j = \varnothing,\ j = k+1, \dots, n. \tag{13.3.3}$$

Now, let $A = (A_j^r)_{r=1,\dots,n,\ j=1,\dots,n} \in \mathbb{M}(n \times n; \mathbb{Z})$. Then, $A_j^r \ge 0$, $j = 1, \dots, k$, $r = 1, \dots, n$. Moreover, using (13.3.3) and the fact that D is bounded, gives a positive M such that

$$|z_j| \ge M, \quad z \in D,\ k+1 \le j \le n.$$

Hence, $\sup\{|z^{A^r}| : z \in D\} < \infty$, $r = 1, \dots, n$, which implies that G is bounded; a contradiction.

(ii) $\Longrightarrow$ (i): Observe that the mapping

$$D \ni z \longmapsto (z_1, \dots, z_{j_0-1}, 1/z_{j_0}, z_{j_0+1}, \dots, z_n)$$

maps D biholomorphically onto an unbounded pseudoconvex Reinhardt domain $\widetilde{D}$. Thus, D is algebraically equivalent to $\widetilde{D}$ and so is G. □

13.4 Hyperbolicities for balanced domains

Recall that any balanced domain is $\boldsymbol{k}$-hyperbolic if and only if it is bounded (cf. Theorem 13.1.1). Moreover, there exists an unbounded pseudoconvex balanced domain $G \subset \mathbb{C}^2$ that is Brody hyperbolic (see Example 13.1.3). Recall that G was given as

$$G := \{z \in \mathbb{C}^2 : \mathfrak{h}(z) < 1\}, \text{ where } \mathfrak{h}(z) := \begin{cases} |z_2| e^{\varphi\left(\frac{z_1}{z_2}\right)} & \text{if } z_2 \neq 0 \\ |z_1| & \text{if } z_2 = 0 \end{cases},$$

and

$$\varphi(\lambda) := \max\left\{\log|\lambda|, \sum_{j=2}^{\infty} \frac{1}{k^2} \log\left|\lambda - \frac{1}{k}\right|\right\}, \quad \lambda \in \mathbb{C}.$$

S.-H. Park [409] has shown that G is almost $\boldsymbol{\ell}$*-hyperbolic*, i.e., $\boldsymbol{\ell}_G(z, w) > 0$, whenever $z_1 \neq w_1$ or ($z_1 = w_1 \neq 0$ and $z_2 \neq w_2$) (see Exercise 13.7.4 for some more details). It is still unclear what happens to $\boldsymbol{\ell}_G((0, z_2), (0, w_2))$.

?

Nevertheless, there is the following result (cf. [409]; see also [410]):

Proposition 13.4.1. *For any $n \geq 3$ there exists a pseudoconvex balanced domain $G \subset \mathbb{C}^n$ such that*

- *G is Brody hyperbolic,*
- *G is not $\boldsymbol{\ell}$-hyperbolic.*

Proof. Obviously, it suffices to construct such an example G in $\mathbb{C}^3$. Then, in the general case, $G \times \mathbb{D}^{n-3}$ will do the job in $\mathbb{C}^n$.

So, let $n = 3$. Put $r_j := e^j$, $s_j := 1/(r_j^2 + r_j)$, $t_j := \sqrt{j}/s_j$, $\varepsilon_j := 2^{-j-1}$, and $\eta_j := t_j s_j$, $j \in \mathbb{N}$. Then,

$$\sum_{j=1}^{\infty} \varepsilon_j = 1/2, \quad \sum_{j=1}^{\infty} \varepsilon_j \log \frac{1}{\eta_j} \geq \sum_{j=1}^{\infty} \varepsilon_j \log \frac{1}{t_j} > -\infty.$$

For $j \in \mathbb{N}$ define

$$Q_j(z) := z_1 z_2 - s_j(z_3 - z_2)(z_3 - 2z_2), \quad z = (z_1, z_2, z_3) \in \mathbb{C}^3.$$

Put

$$G := \{z \in \mathbb{C}^3 : \mathfrak{h}(z) < 1\} \text{ with } \mathfrak{h}(z) := \max\{|z_1|, |z_2|/2, h_0(z)\},$$

where

$$h_0(z) := \prod_{j=1}^{\infty} \left|\frac{Q_j(z)}{\eta_j}\right|^{\varepsilon_j} = \exp\left(\sum_{j=1}^{\infty} \varepsilon_j \log \frac{|Q_j(z)|}{\eta_j}\right).$$

We claim that G is a pseudoconvex balanced domain that is Brody hyperbolic, but not $\boldsymbol{\ell}$-hyperbolic.

Step 1°. $\mathfrak{h}$ is absolutely homogeneous and positive definite.

It suffices to discuss h_0. Fix $z \in \mathbb{C}^3$ and $\lambda \in \mathbb{C}$. Then,

$$\begin{aligned} \sum_{j=1}^{\infty} \varepsilon_j \log \frac{|Q_j(\lambda z)|}{\eta_j} &= \sum_{j=1}^{\infty} \varepsilon_j \log|\lambda|^2 + \sum_{j=1}^{\infty} \varepsilon_j \log \frac{|Q_j(z)|}{\eta_j} \\ &= \log|\lambda| + \sum_{j=1}^{\infty} \varepsilon_j \log \frac{|Q_j(z)|}{\eta_j}. \end{aligned}$$

Hence, $h_0(\lambda z) = |\lambda| h_0(z)$.

Now assume that $\mathfrak{h}(z) = 0$. Then, $z_1 = z_2 = 0 = h_0(z)$, which implies that

$$-\infty = \sum_{j=1}^{\infty} \varepsilon_j \log \frac{|Q_j(0,0,z_3)|}{\eta_j} = \sum_{j=1}^{\infty} \varepsilon_j \log \frac{1}{t_j} + \frac{1}{2} \log |z_3|^2,$$

from which we obtain that $z_3 = 0$. Hence, $\mathfrak{h}$ is positively defined.

Step 2°. $h_0 \in \mathcal{PSH}(\mathbb{C}^3)$ (in particular, G is pseudoconvex).

Fix a positive R and let $z \in (R\mathbb{D})^3$. Then, $|Q_j(z)| \le (1+6)R^2$. Recall that $\eta_j \longrightarrow \infty$. Therefore, there is a j_R such that

$$|Q_j(z)|/\eta_j < 1, \quad z \in (R\mathbb{D})^3, \ j \ge j_R.$$

So it follows that $h_0 \in \mathcal{PSH}((R\mathbb{D})^3)$ for arbitrary R. Hence, $h_0 \in \mathcal{PSH}(\mathbb{C}^3)$.

Step 3°. G is not $\boldsymbol{\ell}$-hyperbolic.

Let

$$\varphi_j \in \mathcal{O}(\mathbb{C}, \mathbb{C}^3), \quad \varphi_j(\lambda) := (s_j \lambda(\lambda - 1), 1, \lambda + 1), \quad j \in \mathbb{N}.$$

Observe that $Q_j \circ \varphi_j = 0$ on $\mathbb{C}$, $j \in \mathbb{N}$. Therefore, $\varphi_j(\lambda) \in G$ if $|\lambda| < r_j$. In particular,

$$\boldsymbol{\ell}_G\big((0,1,1),(0,1,2)\big) = \boldsymbol{\ell}_G\big(\varphi_j(0), \varphi_j(1)\big) \le \boldsymbol{k}_{\mathbb{D}}(0, 1/r_j) \underset{j \to \infty}{\longrightarrow} 0,$$

meaning that G is not $\boldsymbol{\ell}$-hyperbolic.

Step 4°. G is Brody hyperbolic.

Let $f = (f_1, f_2, f_3) \in \mathcal{O}(\mathbb{C}, G)$. By virtue of the form of G, f_j is bounded and so $f_j \equiv: a_j$, $j = 1, 2$. Suppose that f_3 is not constant. Then, by virtue of Picard's theorem (Theorem 1.2.5), we have $\mathbb{C} \setminus \{w\} \subset f_3(\mathbb{C})$ for a suitable $w \in \mathbb{C}$. Hence, $\mathfrak{h}(a_1, a_2, \cdot) < 1$ on $\mathbb{C} \setminus \{w\}$. Using Liouville's theorem for subharmonic functions (Appendix B.4.27), we conclude that $h_0(a_1, a_2, \cdot) \equiv \text{const}$. Note that $h_0(a_1, a_2, \lambda) = 0$ if $Q_j(a_1, a_2, \lambda) = 0$ for at least one j. Therefore, $h_0(a_1, a_2, \cdot) \equiv 0$.

To get a contradiction, we discuss different cases of a_1, a_2.

Case $a_2 = 0$: then, $Q_j(a_1, 0, \lambda) = -s_j \lambda^2$, $j \in \mathbb{N}$. Therefore,

$$\log h_0(a_1, 0, 1) = \sum_{j=1}^{\infty} \varepsilon_j \log \frac{|Q_j(a_1, 0, 1)|}{\eta_j} = \sum_{j=1}^{\infty} \varepsilon_j \log \frac{1}{t_j} > -\infty;$$

contradiction.

Case $a_2 \neq 0$, $a_1 = 0$: then, $Q_j(0, a_2, 0) = -2s_j a_2^2$, $j \in \mathbb{N}$. Therefore,

$$\log h_0(0, a_2, 0) = \sum_{j=1}^{\infty} \varepsilon_j \log \frac{1}{t_j} + \log(2|a_2|^2) \sum_{j=1}^{\infty} \varepsilon_j > -\infty;$$

a contradiction.

Case $a_1 a_2 \neq 0$: then,

$$\log h_0(a_1, a_2, a_2) = \sum_{j=1}^{\infty} \varepsilon_j \log \frac{|a_1 a_2|}{\eta_j} > -\infty;$$

contradiction.

Hence, G is Brody hyperbolic. □

Remark 13.4.2. It remains an open question whether such an example does exist in $\mathbb{C}^2$. ?

13.5 Hyperbolicities for Hartogs type domains

For Hartogs type domains we have the following hyperbolicity criterion (cf. [496, 380], see also [495]):

Theorem 13.5.1. *Let $D = D_H \subset G \times \mathbb{C}^m$ be a Hartogs domain over $G \subset \mathbb{C}^n$ with m-dimensional balanced fibers (cf. Appendix B.7.7). Then, the following properties are equivalent:*

(i) *D is $\boldsymbol{k}$-hyperbolic,*

(ii) *G is $\boldsymbol{k}$-hyperbolic and, for any compact set $K \subset G$, the function $\log H$ is bounded from below on $K \times \partial\mathbb{B}_m$.*

Proof. (i) $\Longrightarrow$ (ii): If D is $\boldsymbol{k}$-hyperbolic, then $\boldsymbol{k}_G(z', z'') \geq \boldsymbol{k}_D((z', 0), (z'', 0)) > 0$ for all $z', z'' \in G$, $z' \neq z''$. Hence, G is $\boldsymbol{k}$-hyperbolic.

Now assume that there are two sequences $(z^j)_{j \in \mathbb{N}} \subset G$, $\lim z^j =: z^0 \in G$, $(w^j)_{j \in \mathbb{N}} \subset \partial\mathbb{B}_m$, $\lim w^j = w^0 \in \partial\mathbb{B}_m$ such that $\lim_{j \to \infty} H(z^j, w^j) = 0$. We may assume that $(z^j, w^j) \in D$, $j \in \mathbb{N}$. Then, $\varphi_j \in \mathcal{O}(\mathbb{C}, G \times \mathbb{C}^m)$, $\varphi_j(\lambda) := (z^j, \lambda w^j)$, maps $R_j \mathbb{D}$ into D for a suitable sequence $(R_j)_{j \in \mathbb{N}}$ with $R_j \underset{j \to \infty}{\longrightarrow} \infty$. Therefore,

$$\boldsymbol{k}_D((z^j, 0), (z^j, w^j)) = \boldsymbol{k}_D(\varphi_j(0), \varphi_j(1)) \leq \boldsymbol{k}_{\mathbb{D}}(0, 1/R_j) \longrightarrow 0;$$

hence, $\boldsymbol{k}_D((z^0, 0), (z^0, w^0)) = 0$; contradiction.

(ii) $\Longrightarrow$ (i): Suppose that D is not $\boldsymbol{k}$-hyperbolic. In particular, D is unbounded. Applying Remark 13.2.5, we find a point $b = (b', b'') \in D \subset \mathbb{C}^n \times \mathbb{C}^m$ and sequences $(\varphi_j)_{j \in \mathbb{N}} \subset \mathcal{O}(\mathbb{D}, D)$ $(\varphi_j = (\varphi_{j,1}, \varphi_{j,2}))$ and $(\sigma_j)_{j \in \mathbb{N}} \subset [0, 1)$ such that

- $\varphi_j(0) = (z'_j, w'_j) \longrightarrow b$,
- $\varphi_j(\sigma_j) = (z''_j, w''_j) \longrightarrow \infty$,
- $\sigma_j \longrightarrow 0$.

Let now $\mathbb{B}_n(b', r) \subset\subset G$. Assume there is a subsequence $(z''_{j_k})_k \subset (z''_j)_j$ with $\|z''_{j_k} - b'\| > r$, $k \in \mathbb{N}$. Since G is $\boldsymbol{k}$-hyperbolic, we have for large k

$$\begin{aligned}0 < \inf_{\zeta \in \partial\mathbb{B}_n(b',r)} \boldsymbol{k}_G(\zeta, b') \le \boldsymbol{k}_G(b', z''_{j_k}) &\le \boldsymbol{k}_G(z'_{j_k}, z''_{j_k}) + \boldsymbol{k}_G(b', z'_{j_k}) \\ &\le \boldsymbol{\ell}_G(z'_{j_k}, z''_{j_k}) + \boldsymbol{k}_G(b', z'_{j_k}) \longrightarrow 0;\end{aligned}$$

a contradiction. Therefore, all points z''_j belong to a compact subset K of G. Thus, $\varphi_j(\sigma_j) = w''_j \longrightarrow \infty$. Hence, we get

$$\begin{aligned}1 \ge H(\varphi_j(\sigma_j)) &= \|\varphi_{j,2}(\sigma_j)\| H\left(\varphi_{j,1}(\sigma_j), \frac{\varphi_{j,2}(\sigma_j)}{\|\varphi_{j,1}(\sigma_j)\|}\right) \\ &\ge \|\varphi_{j,2}(\sigma_j)\| \inf_{K \times \partial\mathbb{B}_m} H \longrightarrow \infty;\end{aligned}$$

a contradiction. □

Remark 13.5.2. In a recent paper [497] the relative notion of $\boldsymbol{k}$-hyperbolicity modulo an analytic set has been discussed also for Hartogs domains. Let $\Omega \subset \mathbb{C}^n$ be a domain and $A \subset \Omega$ an analytic set. Recall that then Ω is said to be *$\boldsymbol{k}$-hyperbolic modulo A* if $\boldsymbol{k}_\Omega(z', z'') > 0$ for all pairs of different points $z', z'' \in \Omega$ unless both points belong to A (see [316]). Then, the following result similar to Theorem 13.5.1 is given: *Let G, H, and D be as in Theorem* 13.5.1 *and let $A \subset G$ be an analytic subset. Put $\widetilde{A} := A \times \mathbb{C}^m$. Then, the following properties are equivalent:*

(i) *D_H is $\boldsymbol{k}$-hyperbolic modulo $\widetilde{A} \cap D_H$,*

(ii) *G is $\boldsymbol{k}$-hyperbolic modulo A and if $G \setminus A \ni z_k \longrightarrow z^* \in G \setminus A$ and $\mathbb{C}^m \ni w_k \longrightarrow w^* \neq 0$, then* $\liminf H(z_k, w_k) \neq 0$.

Remark 13.5.3. In Remark 13.3.6 we mentioned that, if a Reinhardt domain in $\mathbb{C}^2$ is $\boldsymbol{c}$-hyperbolic, then its envelope of holomorphy is also $\boldsymbol{c}$-hyperbolic. In the class of Hartogs domains and the case of $\boldsymbol{k}$-hyperbolicity, such a conclusion is false even in dimension 2 (cf. [136]).

Let $u : [0, 1) \longrightarrow (-\infty, 0)$ be a continuous function satisfying $\lim_{t \nearrow 1} \varphi(t) = -\infty$. Put $u(z_1) := \varphi(|z_1|)$. Then, the domain

$$D := \left\{z \in \mathbb{D} \times \mathbb{C} : |z_2| < e^{-u(z_1)}\right\}$$

is $\boldsymbol{k}$-hyperbolic (see Theorem 13.5.1). Recall that

$$\mathcal{E}(D) = \left\{z \in \mathbb{D} \times \mathbb{C} : |z_2| < e^{-\hat{u}(z_1)}\right\},$$

where $\hat{u}$ is the largest subharmonic minorant of u. By virtue of the maximum principle for subharmonic function, it is clear that $\hat{u} \equiv -\infty$. Therefore, $\mathcal{E}(D) = \mathbb{D} \times \mathbb{C}$,which is not $\boldsymbol{k}$-hyperbolic.

Remark 13.5.4. So far, we discussed hyperbolicity. We close this part by a remark on the opposite situation. There is the following result due to E. Fornæss and N. Sibony [173]: Let $D \subset \mathbb{C}^2$ be a domain that can be monotonically exhausted by domains D_j, where each of the D_j is biholomorphically equivalent to $\mathbb{B}_2$. If $\boldsymbol{\varkappa}_D \not\equiv 0$, then D is biholomorphically equivalent either to $\mathbb{B}_2$ or to $\mathbb{D} \times \mathbb{C}$. Observe that $\mathbb{B}_2(j) \nearrow \mathbb{C}^2$, $\boldsymbol{\varkappa}_{\mathbb{C}^2} \equiv 0$, but, obviously, $\mathbb{C}^2$ is neither biholomorphic to $\mathbb{B}_2$ nor to $\mathbb{D} \times \mathbb{C}$.

It turns out that there is a domain $D \subset \mathbb{C}^n$, $n \geq 2$, $D_j \nearrow D$, each D_j is biholomorphically equivalent to $\mathbb{B}_n$, such that

- $\boldsymbol{\varkappa}_D \equiv 0$,
- $\exists_{u \in \mathcal{PSH}(\mathbb{C}^n)} : D = \{z \in \mathbb{C}^n : u(z) < 0\}$ and $u|_D \not\equiv \text{const}$; in particular, D is not biholomorphic to $\mathbb{C}^n$.

Domains of that type are called *short* $\mathbb{C}^n$*'s* (see [167]). The domain D is obtained in the following way: Let $d \in \mathbb{N}$, $d \geq 2$, and $\eta > 0$. Denote by $\operatorname{Aut}_{d,\eta}$ the set of all polynomial automorphisms Φ of $\mathbb{C}^n$ of the form

$$\Phi(z) = \Phi(z_1, \dots, z_n) = (z_1^d + P_1(z), P_2(z), \dots, P_n(z)),$$

where $\deg P_j \leq d - 1$, $j = 1, \dots, n-1$, and where each coefficient of the polynomials P_j has a modulus of at most η. Choosing sufficiently good sequences $a_j \searrow 0$, $a_j \in (0,1)$, and $F_j \in \operatorname{Aut}_{d,\eta_j}$, where $\eta_j := a_j^{d^j}$, $j \in \mathbb{N}$, gives D as

$$D = \{z \in \mathbb{C}^n : \lim_{k \to \infty} F_k \circ \dots \circ F_1(z) = 0\}.$$

13.6 Hyperbolicities for tube domains

Recall that a *tube domain with basis* G is a domain of the form $T_G = G + i\mathbb{R}^n$, where $G \subset \mathbb{R}^n$ is a domain. It is well-known (theorem of Bochner, see [265]) that the envelope of holomorphy of T_G is given by $T_{\operatorname{conv} G}$. The discussion of hyperbolicity for tube domains in this section is mainly based on [348] and [239].

Proposition 13.6.1. *Let $G \subset \mathbb{R}^n$ be a given domain. Then, the following properties are equivalent:*

(i) $T_{\operatorname{conv} G}$ *is biholomorphically equivalent to a bounded domain in* $\mathbb{C}^n$*;*

(ii) T_G *is biholomorphic to a bounded domain in* $\mathbb{C}^n$*;*

(iii) $\operatorname{conv} G$ *does not contain any affine real line;*

(iv) $T_{\operatorname{conv} G}$ *is* $\boldsymbol{c}$*-hyperbolic;*

(v) $T_{\operatorname{conv} G}$ *is* $\boldsymbol{k}$*-hyperbolic.*

Proof. By virtue of Theorem 13.1.7, we know that (ii) $\Longleftarrow$ (i) $\Longleftrightarrow$ (iii) $\Longleftrightarrow$ (iv) $\Longleftrightarrow$ (v). Now assume that (ii) holds, but (iii) fails. Then, there is a non trivial real line L in $\operatorname{conv} G$. Using convexity, any line L' through a point $a \in G$ parallel to L is contained in $\operatorname{conv} G$. Let $L' = a + \mathbb{R}b$ be such a line with $b \in \mathbb{R}^n$, $b \neq 0$. Then, the complex line $\widehat{L}' := a + \mathbb{C}b$ lies inside of $T_{\operatorname{conv} G}$.

Now assume that the biholomorphic mapping in (ii) is given by an $F \in \mathcal{O}(T_G, \mathbb{C}^n)$, where $F(T_G)$ is bounded. Applying the Bochner result gives that F can be holomorphically extended to a bounded holomorphic mapping $\widetilde{F} : T_{\operatorname{conv} G} \longrightarrow \mathbb{C}^n$. Then, by the Liouville result, $\widetilde{F}$ is constant on the complex line $\widehat{L}'$, which contradicts the fact that $\widetilde{F} = F$ is injective near the point $a \in G$; a contradiction. □

For tube domains one has the following reformulation of $\boldsymbol{k}$-hyperbolicity of tube domains in terms of harmonic mappings:

Lemma 13.6.2. *For a tube domain $T_G \subset \mathbb{C}^n$, the following properties are equivalent:*

(a) T_G *is* $\boldsymbol{k}$*-hyperbolic;*

(b) G *is 2-hyperbolic, i.e., for every $x^0 \in G$ there exist an open neighborhood $V = V(x^0) \subset G$ and a positive number M such*

$$\|\operatorname{grad} h(0)\| := \max\{\|\operatorname{grad} h_j(0)\| : j = 1, \dots, n\} \le M$$

for all harmonic mappings $h = (h_1, \dots, h_n) : \mathbb{D} \longrightarrow G$ with $h(0) \in V$.

Proof. The proof is mainly based on the equivalence of $\boldsymbol{k}$- and $\boldsymbol{\varkappa}$-hyperbolicity. Details are left to the reader. □

For a differentiable function $f : D \longrightarrow \mathbb{R}$, $D \subset \mathbb{C}$ a domain, we introduce the following notion of the "derivative"

$$\widetilde{f}(z) := \|\operatorname{grad} f(z)\| / \cosh(f(z)), \quad z \in D,$$

which is needed in the following discussions.

Theorem 13.6.3 (Reparametrization lemma). *Let $(f_j)_{j\in\mathbb{N}} \subset \mathcal{C}^1(\mathbb{D},\mathbb{R})$ be a given sequence and let $\mathbb{D} \ni p_j \longrightarrow p \in \mathbb{D}$ be such that $\widetilde{f}_j(p_j) \longrightarrow \infty$. Then there exist positive numbers $\alpha_j \longrightarrow 0$ and points $\mathbb{D} \ni q_j \longrightarrow p$ such that, if*

$$g_j(z) := f_j(\alpha_j z + q_j), \quad z \in D_j := (\mathbb{D} - q_j)/\alpha_j,$$

then $\widetilde{g}_j(0) = 1$, $j \in \mathbb{N}$, and for any compact set $K \subset \mathbb{C}$ there are positive numbers $\varepsilon_j \longrightarrow 0$ with $\widetilde{g}_j|_K \le 1 + \varepsilon_j$ for sufficiently large j's.

To get the claimed reparametrization, the following elementary lemma is needed.

Lemma 13.6.4. *Let $\varphi : \mathbb{D} \longrightarrow \mathbb{R}_+$ be bounded on all Carathéodory balls $B_{c_\mathbb{D}}(a,r)$ ($a \in \mathbb{D}$, $r > 0$). Let $\tau > 1, \varepsilon > 0$, and $p \in \mathbb{D}$ with $\varphi(p) > 0$. Then there exists a point $q \in \mathbb{D}$ satisfying*

(a) $c_\mathbb{D}(q,p) \le \frac{\tau}{\varepsilon\varphi(p)(\tau-1)}$,

(b) $\varphi(q) \ge \varphi(p)$,

(c) $\varphi(z) \le \tau\varphi(q)$ *for all* $z \in \mathbb{D}$ *with* $c_\mathbb{D}(z,q) \le \frac{1}{\varepsilon\varphi(q)}$.

Proof. Put $q_0 = p$. If the condition (c) for q_0 is fulfilled, set $q := q_0$. Obviously, then, (a) and (b) hold.

Assume that (c) does not hold for q_0. Then choose a point $q_1 \in B_{c_\mathbb{D}}\left(q_0, \frac{1}{\varepsilon\varphi(q_0)}\right)$ with $\varphi(q_1) > \tau\varphi(q_0) = \tau\varphi(p)$. In case q_1 satisfies (c), put $q := q_1$. Obviously, (a) and (b) are true for q.

Otherwise choose a point $q_2 \in \mathbb{D}$ with $c_\mathbb{D}(q_2,q_1) < \frac{1}{\varepsilon\varphi(q_1)}$ and $\varphi(q_2) > \tau\varphi(q_1) > \tau^2\varphi(p)$. Then,

$$c_\mathbb{D}(q_2,p) \le c_\mathbb{D}(q_2,q_1) + c_\mathbb{D}(q_1,p) \le \frac{1}{\varepsilon\varphi(p)}\left(1 + \frac{1}{\tau}\right) \le \frac{\tau}{\varepsilon\varphi(p)(\tau-1)}.$$

If (c) is true for q_2, put $q := q_2$. Obviously then, (a) and (b) are satisfied. Otherwise, fix a point $q_3 \in \mathbb{D}$ with $c_\mathbb{D}(q_3,q_2) < \frac{1}{\varepsilon\varphi(q_2)}$ and $\varphi(q_3) > \tau\varphi(q_2) > \tau^3\varphi(p)$. Then,

$$c_\mathbb{D}(q_3,p) \le c_\mathbb{D}(q_3,q_2) + c_\mathbb{D}(q_2,p) < \frac{1}{\varepsilon\varphi(p)}\left(1 + \frac{1}{\tau} + \frac{1}{\tau^2}\right) \le \frac{\tau}{\varepsilon\varphi(p)(\tau-1)}.$$

Suppose we can repeat this construction. This will lead to points

$$q_j \in B_{c_\mathbb{D}}\left(p, \tfrac{\tau}{\varepsilon\varphi(p)(\tau-1)}\right) \subset\subset \mathbb{D}$$

satisfying $\varphi(q_j) > \tau^j\varphi(p) \longrightarrow \infty$, which contradicts the local boundedness assumption for φ. Hence, there is a last point q_{j_0} in this construction. Put $q := q_{j_0}$. Then, q will satisfy (a), (b), and (c). □

Proof of Theorem 13.6.3. Put $\varphi_j := \widetilde{f}_j$ on $\mathbb{D}$, $\varepsilon_j := \varphi_j(p_j)^{-1/3} \longrightarrow 0$, and $\tau_j := 1 + \varepsilon_j$. Applying Lemma 13.6.4 gives points $q_j \in \mathbb{D}$ with the following properties:

- $\boldsymbol{c}_{\mathbb{D}}(q_j, p_j) \le \frac{\tau_j}{\varepsilon_j \varphi_j(p_j)(\tau_j - 1)}$,
- $\varphi_j(q_j) \ge \varphi_j(p_j)$,
- $\varphi_j(z) \le \tau_j \varphi_j(q_j)$, if $\boldsymbol{c}_{\mathbb{D}}(z, q_j) < \frac{1}{\varepsilon_j \varphi_j(q_j)}$.

Then, $\boldsymbol{c}_{\mathbb{D}}(q_j, p) \le \boldsymbol{c}_{\mathbb{D}}(q_j, p_j) + \boldsymbol{c}_{\mathbb{D}}(p_j, p) \le (1 + \varepsilon_j)\varepsilon_j + \boldsymbol{c}_{\mathbb{D}}(p_j, p) \longrightarrow 0$; in particular, $q_j \longrightarrow p$. Put $0 < \alpha_j := \frac{1}{\widetilde{f}_j(q_j)} \longrightarrow 0$ and define on $D_j := (\mathbb{D} - q_j)/\alpha_j$ the following new function $g_j(z) := f_j(\alpha_j z + q_j)$. Then, $\widetilde{g}_j(0) = \alpha_j \widetilde{f}_j(q_j) = 1$.

Moreover, put $s := (1 - |p|)/4$. Fix an $R > 0$. Then there exists a j'_R such that $\alpha_j z + q_j \in \mathbb{D}(q_j, s) \subset \mathbb{D}(p, 2s) \subset\subset \mathbb{D}$ whenever $z \in \mathbb{D}(R)$ and $j \ge j'_R$. Consequently, $\boldsymbol{c}_{\mathbb{D}}(\alpha_j z + q_j, q_j) \le \tanh^{-1}(\alpha_j R/s)$ and the last number is majorized by $\varphi_j(p_j)^2 \le \frac{1}{\varepsilon_j \varphi(q_j)}$ if $j \ge j_R \ge j'_R$. Therefore, $\widetilde{g}_j(z) = \alpha_j \varphi_j(\alpha_j z + q_j) \le \tau_j \alpha_j \varphi_j(q_j) = 1 + \varepsilon_j$ if $z \in \mathbb{D}(R)$ and $j \ge j_R$. $\square$

Theorem 13.6.5. *Assume the situation of Theorem* 13.6.3 *with* $p = p_j = 0$ *and, in addition, that all functions* f_j *are harmonic on* $\mathbb{D}$. *Let* (g_j) *be the sequence after renormalization. Then there exist a subsequence* $(g_{j_k})_k$ *of* $(g_j)_j$ *and a non-trivial affine linear function* $g : G \longrightarrow \mathbb{R}$ *such that* $g_{j_k} \underset{k\to\infty}{\overset{K}{\Longrightarrow}} g$.

Proof. Recall that $\|\operatorname{grad} g_j\|$ remains locally bounded. Therefore, the family $(g_j)_j$ is equicontinuous on all discs $\mathbb{D}(R)$.

Assume that for any point $a \in \mathbb{C}$ the sequence $g_j(a)$ is bounded. Then, by Ascoli's theorem, there exists a subsequence $(g_{j_k})_k$ of $(g_j)_j$ that converges locally uniformly to a function $g : \mathbb{C} \longrightarrow \mathbb{R}$. Obviously, g is harmonic and because of $\widetilde{g}_j(0) = 1$ it is not identically constant.

Now assume that there exists a point $a \in G$ such that $\sup\{g_j(a)\} = \infty$. Without loss of generality we may assume that $g_j(a) \longrightarrow \infty$. Then, using the Harnack inequality, we see that $g_j \underset{j\to\infty}{\overset{K}{\Longrightarrow}} \infty$; in particular, $g_j(0) \longrightarrow \infty$. Taking large j's then, again by Harnack inequality, we see that the sequence $(g_j/g_j(0))_j$ is bounded on a disc $\mathbb{D}(r)$. Hence there exists a subsequence (which we call again by g_j) that converges locally uniformly on $\mathbb{D}(r)$ to a harmonic function h. Therefore, $\|\operatorname{grad} g_j\|/g_j(0)$ is bounded. On the other side we have

$$1 = \widetilde{g}_j(0) = \frac{\|\operatorname{grad} g_j(0)\|}{g_j(0)} \frac{g_j(0)}{\cosh g_j(0)} \longrightarrow 0;$$

a contradiction.

The case that there exists a point $a \in \mathbb{C}$ such that $\inf\{g_j(a)\} = -\infty$ can be handled in the same way, which finally shows that a subsequence $(g_{j_k})_k$ of (g_j) converges locally uniformly to a harmonic function.

It remains to verify that the limit function g is affine linear. Observe that $\widetilde{g} \leq 1$ and that $\Delta \log(\cosh) = \widetilde{g}$ on the whole complex plane. Therefore, $v := |z|^2 - \log(\cosh) - 2 \in \mathcal{SH}(\mathbb{C})$ and so, using the mean value inequality for subharmonic functions, one has

$$\frac{1}{2\pi}\int_0^{2\pi} \log(\cosh(g(re^{it})))dt \leq r^2 + 2.$$

Finally recall that $|t| - \log(\cosh t) \leq \log 2$ for real t. Hence,

$$\frac{1}{2\pi}\int_0^{2\pi} |g(r^{it})|dt \leq r^2 + 2 + \log 2, \quad r > 0.$$

Now let $g = \sum_{k=0}^{\infty} p_k$ be the representation of g by harmonic homogeneous real valued polynomials p_k of degree k. Using their orthogonality, it follows that

$$\begin{aligned} r^{2k}\int_0^{2\pi} p_k^2(e^{it})dt &= \frac{1}{2\pi}\int_0^{2\pi} p_k^2(re^{it})dt = \int_0^{2\pi} (gp_k)(re^{it})dt \\ &\leq \frac{1}{2\pi} r^k \|p_k\|_{\mathbb{T}} \int_0^{2\pi} |g(re^{it})|dt \leq r^k \|p_k\|_{\mathbb{T}}(r^2 + 2 + \log 2), \end{aligned}$$

which implies that $p_k = 0$ for all $k \geq 3$ (let $r \longrightarrow \infty$). Therefore, g is a harmonic polynomial of degree at most two, i.e., $g(z) - g(0) = L(z) + zQz^t$, where L is linear and Q is a real square matrix generating the second order harmonic monomial. Assume that $Q \neq 0$. Then fix a vector $v \in \mathbb{C} \equiv \mathbb{R}^2$ with $vQv^t = 0$. Then,

$$g(z + jv) - g(0) = j\left(L(v) + 2zQv^t\right) + zQz^t, \quad z \in \mathbb{C}^2.$$

Observe that the term in parentheses can take on any real value such that there is no subsequence $(g_{j_k})_k$ of $(g_j)_j$, where $g_j := g(\cdot + jv)$, such that g_{j_k} converges locally uniformly on $\mathbb{C}$ either to a harmonic function or to $+\infty$ or to $-\infty$. On the other hand, we know that $\widetilde{g}_j \leq 1$, $j \in \mathbb{N}$, meaning that our sequence is locally equicontinuous, which implies via the Ascoli theorem that such a subsequence has to exist; a contradiction. Hence, $Q = 0$. □

In a next step, we will study tube domains $T_G \subset \mathbb{C}^2$ for which conv G does contain non trivial real lines, i.e., G is sitting in a halfplane. To discuss hyperbolicity, we may assume that $G \subset \mathbb{R} \times \mathbb{R}_{>0}$.

Theorem 13.6.6. *Let $G \subset \mathbb{R} \times \mathbb{R}_{>0}$ be a domain. Then, the following properties hold:*

(a) *if T_G is $\boldsymbol{k}$-hyperbolic, then for every $a \in G$ there is no real sequence $\mathbb{R} \ni b_j \longrightarrow a_2$ such that $[-j, j] \times \{b_j\} \subset G$, $j \in \mathbb{N}$;*

(b) *if T_G is not $\boldsymbol{k}$-hyperbolic, then for every $a \in G$ and $k \in \mathbb{N}$ there exists a real analytic curve $[-k, k] \ni t \longmapsto (t, \gamma(t))$ such that $|\gamma(t) - a_2| \le 1/k$, $t \in [-k, k]$.*

Remark 13.6.7.

(a) In [348] it is even claimed that non-$\boldsymbol{k}$-hyperbolicity of such a tube domain T_G implies the linear property stated in (a). The argument given there seems not to work. But, according to our knowledge, there is so far no counterexample.

(b) Using the property given in Theorem 13.6.6(b) it is an easy exercise to present a tube domain $T_G \subset \mathbb{C}^2$ with $\operatorname{conv} G = \mathbb{R} \times (1, 2)$, which is $\boldsymbol{k}$-hyperbolic, but its envelope of holomorphy is not.

Proof of Theorem 13.6.6. (a) Suppose the contrary. Then there exists a sequence $\mathbb{R} \ni b_j \to a_2$ with $[-j, j] \times \{b_j\} \subset G$, $j \in \mathbb{N}$. Put $\varphi_j(\lambda) := (a_1 + \lambda j/2, b_j)$, $\lambda \in \mathbb{D}$. Then, $\varphi_j \in \mathcal{O}(\mathbb{D}, G)$ for large j. Hence,

$$\tfrac{j}{2}\varkappa_{T_G}((a_1, b_j); (1, 0)) = \varkappa_{T_G}((a_1, b_j); \varphi_j'(0)) \le \varkappa_{\mathbb{D}}(0; 1) = 1;$$

a contradiction with Theorem 13.2.2.

(b) Assume that T_G is not $\boldsymbol{k}$-hyperbolic. Then, using again Theorem 13.2.2, we find a point $a \in T_G$, a sequence $(b_j)_j \subset T_G$ with $b_j \longrightarrow a$, and unit vectors $X_j \in \mathbb{C}^n$ such that $\varkappa_{T_G}(b_j; X_j) < \frac{1}{2j}$. Therefore, there are $\varphi_j \in \mathcal{O}(\mathbb{D}, T_G)$ such that $\alpha_j \varphi_j'(0) = X_j$ with $0 \le \alpha_j < 1/j$. In particular, $\|\varphi_j'(0)\| \ge j \longrightarrow \infty$. Let $a = a' + ia''$, $b_j = b_j' + ib_j'' \in \mathbb{R}^2 + i\mathbb{R}^2$ and $\varphi_j = h_j + ig_j$. Then, $h_j(0) = b_j' \longrightarrow a' \in G$; in particular, $h_{j,2}(0) = b_{j,2}' \longrightarrow a_2' \in \mathbb{R}$. Moreover,

$$\|\operatorname{grad} h_{j,1}(0)\|^2 + \|\operatorname{grad} h_{j,2}(0)\|^2 \longrightarrow \infty. \tag{*}$$

Recall that $h_{j,2} : G \longrightarrow \mathbb{R}_+$ for all j. Since the family of harmonic functions with non-negative values is a normal one, we may extract a subsequence $(h_{j_k,2})_k$ which converges locally uniformly to a harmonic function h_2 on $\mathbb{D}$. By virtue of (*) it follows that $\|\operatorname{grad} h_{j,1}(0)\| \longrightarrow \infty$.

Now applying Theorem 13.6.5, one can find positive numbers $\alpha_k \longrightarrow 0$ and points $\mathbb{D} \ni p_k \longrightarrow 0$ such that there is a locally uniformly convergent subsequence $(u_{k_\ell,1})_\ell$ of $(u_{k,1})_k$, where $u_{k,1}(z) := h_{j_k,1}(\alpha_k z + p_k)$, which converges to a non-trivial linear function h_1 on $\mathbb{C}$. Put $u_{k,2} := h_{j_k,2}(\alpha_k z + p_k)$. Then, $u_{k_\ell,2} \underset{\ell\to\infty}{\overset{K}{\Longrightarrow}} a_2'$ on $\mathbb{C}$.

Let $h(x) = \alpha_0 + \alpha_1 x_1 + \alpha_2 x_2$. We may assume that $\alpha_1 > 0$. Fix a number $k \in \mathbb{N}$: then, there are points $x_k^{\pm} \in \mathbb{R}$, $x_k^- < x_k^+$, such that $h(x_k^{\pm}) = k + 1$. Now choose a large j such that $|u_{j,1}(x_k^{\pm})| < k$. Then, $u_{j,1}$ is strictly increasing if j is sufficiently large. Put $\gamma_k(t) := u_{j,2}(u_{j,1}^{-1}(t))$, $t \in [-k, k]$. Then we may assume, having chosen a large j, that $|\gamma_k(t) - a_2| < 1/k$ for all $t \in [-k, k]$. □

Turning to the case of base-domains $G \subset \mathbb{R}^2$ with $\operatorname{conv} G = \mathbb{R}^2$ we will see that there is a necessary condition for T_G to be not $\boldsymbol{k}$-hyperbolic.

Theorem 13.6.8. *Let $G \subset \mathbb{R}^2$ be such that T_G is not $\boldsymbol{k}$-hyperbolic. Then at least one of the following properties holds:*

(a) *the closure of G contains a non-trivial affine real line;*

(b) *for any $t \in \mathbb{R}$ there is a sequence of points $(x_k)_k \subset G$ with $x_{k,1} \longrightarrow t$ and $x_{k,2} \longrightarrow \infty$ (or $x_{k,2} \longrightarrow -\infty$) or $x_{k,2} \longrightarrow t$ and $x_{k,1} \longrightarrow \infty$ (or $x_{k,1} \longrightarrow -\infty$).*

Proof. By the assumption, there exist a point $p \in G$ and a sequence of harmonic mappings $h_j : \mathbb{D} \longrightarrow G$ with $h_j(0) \longrightarrow p$ and $\|\operatorname{grad} h_j(0)\| \longrightarrow \infty$. Put $\widetilde{h}_j(0) = \widetilde{h}_{j,1}(0) + \widetilde{h}_{j,2}(0)$, where $h_j = (h_{j,1}, h_{j,2})$. Following the proof of Theorem 13.6.5, this leads to harmonic mappings $g_j(z) = h_j(\alpha_j + z q_j)$, where $0 < \alpha_j \longrightarrow 0$ and $G \ni q_j \longrightarrow p$, such that there is a subsequence $(g_{j_k})_k$ which converges on $\mathbb{C}$ locally uniformly to a mapping (g_1, g_2), where at least one of the components is a non-trivial linear function and the other one is identically $\pm\infty$, or again non-trivial linear. With this information in hand, (a) or (b) is an immediate consequence. □

Corollary 13.6.9. *There is a tube domain $T_G \subset \mathbb{C}^2$ with $\operatorname{conv}(T_G) = \mathbb{C}^2$ such that T_G is $\boldsymbol{k}$-hyperbolic.*

Proof. As an example, simply take a sufficiently small open neighborhood of

$$\mathbb{R}_+(0,1) \cup \mathbb{R}_+(-1,-1) \cup \mathbb{R}_+(1,-1)$$

and apply Theorem 13.6.8. Obviously, (a) and (b) of this theorem are not satisfied, implying that T_G has to be $\boldsymbol{k}$-hyperbolic. □

In [239], different classes of domains $G \subset \mathbb{R}^2$ with $\operatorname{conv} G = \mathbb{R}^2$ are discussed, which all are $\boldsymbol{k}$-hyperbolic. For example, we have the following result:

Theorem 13.6.10. *Let*

$$A_{\alpha,s,t} := \{x \in \mathbb{R}_{>0} \times \mathbb{R} : x_2 > t x_1^{\alpha}\} \cup \{x \in \mathbb{R}_{<0} \times \mathbb{R} : x_2 > s(-x_1)^{\alpha}\} \cup (\{0\} \times \mathbb{R}_{>0}),$$

where $\alpha > 0$, $\alpha \neq 1$, $s < 0$, *and* $t > 0$;

$$B_{s,t} := \{x \in \mathbb{R}_{>0} \times \mathbb{R} : x_2 > x_1 \log(tx_1)\} \cup \{x \in \mathbb{R}_{<0} \times \mathbb{R} : x_2 > x_1 \log(-sx_1)\} \cup (\{0\} \times \mathbb{R}_{>0}),$$

where $s, t > 0$. *Then, the associated tube domain* $T_{A_{\alpha,s,t}}$, $T_{B_{s,t}}$ *is* $\boldsymbol{k}$*-hyperbolic.*

Proof. Let us start with a simple geometric observation with respect to the domain involved in the theorem. For short, we write $G = A_{\alpha,s,t}$. Fix a real c and put

$$G_c^{\pm} := G \cap \{x \in \mathbb{R}^2 : \pm(x_1 - c) > 0\}.$$

Obviously, $G_c^{\pm}$ is a domain. Moreover, let $I_c^{\pm m} := \{x_1 \in \mathbb{R} : 0 < \pm(x_1 - c) < m\}$, where $m \in \mathbb{N}$. Observe that we have that $G_c^{\pm} \cap \{x \in \mathbb{R}^2 : x_1 \in I_c^{\pm m}\}$ is contained in a rectangle $R_c^{\pm m}$ over the interval $I_c^{\pm m}$, which remains bounded into the negative x_2-direction. Hence, applying Corollary 13.1.12 implies that the tube domain $T_{G_c^{\pm}}$ is $\boldsymbol{k}$-hyperbolic.

Assume now that T_G is not $\boldsymbol{k}$-hyperbolic. Then, in virtue of Lemma 13.6.2, there exist a point $a \in G$ and a sequence of harmonic mappings $h_k = (h_{k,1}, h_{k,2}) : \mathbb{D} \longrightarrow G$ such that $h_k(0) \longrightarrow a$ and $\| \operatorname{grad} h_k \| \longrightarrow \infty$.

Fix a $c \in \mathbb{R} \setminus \{a_1\}$. Then, either $c < a_1$ or $c > a_1$. Hence, either $h_{k,1}(0) > c$ or $h_{k,1}(0) < c$ for all $k \geq k_0$. Assume now that $c \notin h_{k,1}(\mathbb{D})$ for infinitely many $k \in K \subset \mathbb{N}$, $k \geq k_0$. Then, $h_{k,1}(\mathbb{D}) \subset (c, \infty)$ or $h_{k,1}(\mathbb{D}) \subset (-\infty, c)$ for $k \in K$. Therefore, the harmonic mapping h_k sends $\mathbb{D}$ to G_c^+ (respectively, to G_c^-), $k \in K$. And so Lemma 13.6.2 gives that one of the tube domains $T_{G_c^{\pm}}$ is not $\boldsymbol{k}$-hyperbolic; a contradiction. Summarizing, what we have shown is that any real number $c \neq a_1$ is contained in $h_{k,1}(\mathbb{D})$ for almost all k's.

Again, let $c \neq a_1$ be a real number. Put $L_k(c) := \{\lambda \in \mathbb{D} : h_{k,1}(\lambda) = c\}$. Then we claim that

$$\lim_{k \to +\infty} \operatorname{dist}(0, L_k(c)) = 0. \tag{13.6.1}$$

Suppose the contrary. Then there are a subsequence $(L_{k_j})_j$ of $(L_k)_k$ and a positive number $r \in (0, 1)$ such that $\operatorname{dist}(0, L_{k_j}(c)) \geq r, \in \mathbb{N}$. Therefore, either $h_{k_j}(\mathbb{D}(r)) \subset \{x \in G : x_1 < c\}$ or $h_{k_j}(\mathbb{D}(r)) \subset \{x \in G : x_1 > c\}$. Put $\widetilde{h}_j(\lambda) := h_{k_j}(r\lambda)$. Obviously, $\widetilde{h}_j$ are harmonic mappings with $\widetilde{h}_j(0) \longrightarrow a$ and $\| \operatorname{grad} \widetilde{h}_j(0) \| \longrightarrow \infty$. So, we are back in the situation we have discussed just before. Hence, we again end up with a contradiction.

The remaining argument in the proof is based on information on the harmonic measure. Choose $R, p, q > 0$, and $\varepsilon \in (0, 1)$, such that

- $(q/p)^{\alpha-1} > |s|/t$,
- $a_1 - pR < h_{k,1}(0) < a_1 + qR$ for $k \geq k_0$,

- $|h_{k,1}(0)| < 1 - \varepsilon, k \geq k_0$,
- $a_1 - pR$ and $a_1 + qR$ are not critical values of $h_{k,1}$, $k \geq k_0$ (use the Sard lemma).

Put

$$\widetilde{\Omega}_k := \{\lambda \in \mathbb{D} : a_1 - pR < h_{k,1}(\lambda) < a_1 + qR;\ |\lambda| < 1 - \varepsilon\}.$$

Note that $0 \in \widetilde{\Omega}_k$, $k \geq k_0$. From now on, we assume that $k \geq k_0$. Denote by Ω_k that connected component of $\widetilde{\Omega}_k$ that contains $h_{k,1}(0)$. Applying the maximum principle for harmonic functions implies that Ω_k is simply connected. Moreover, $\partial\Omega_k$ is locally connected and has no cut points. Therefore, according to the Carathéodory theorem (see [437], Theorem 2.6), Ω_k is a simply connected Jordan domain.

The boundary of $\partial\Omega_k$ is given by three parts, namely $\Gamma_k \subset L_k(a_1 - pR)$, $\Gamma'_k \subset L_k(a_1 + qR)$, and $\gamma_k = \partial\Omega_k \cap \mathbb{D}(1 - \varepsilon)$. Then, Theorem IV, 6.2 in [188] (see page 149 there) leads to

$$\omega(0, \gamma_k, \Omega_k) \leq \frac{8}{\pi} \exp\left(-\pi \int_{\operatorname{dist}(0,\partial\Omega_k)}^{1-\varepsilon} \frac{dr}{2\pi r}\right) = \frac{8}{\pi}\sqrt{\frac{\operatorname{dist}(0, \partial\Omega_k)}{1 - \varepsilon}},$$

where $\omega_k(0, E, \Omega_k)$ denotes the harmonic measure for $E \subset \partial\Omega_k$ in 0. Thus, by (13.6.1), we get $\omega(0, \gamma_k, \Omega_k) \longrightarrow 0$.

Using the Poisson integral formula, we have

$$\begin{aligned} h_{k,1}(0) &= \int_{\Gamma_k} h_{k,1} d\omega(0, \cdot, \Omega_k) + \int_{\Gamma'_k} h_{k,1} d\omega(0, \cdot, \Omega_k) + \int_{\gamma_k} h_{k,1} d\omega(0, \cdot, \Omega_k) \\ &= (a_1 - pR)\mu_k + (a_1 + qR)\mu'_k + \int_{\gamma_k} h_{k,1} d\omega(0, \cdot, \Omega_k), \end{aligned}$$

where $\mu_k := \int_{\Gamma_k} d\omega(0, \cdot, \Omega_k)$ and $\mu'_k := \int_{\Gamma'_k} d\omega(0, \cdot, \Omega_k)$. Finally, recall that $h_{k,1}$ is uniformly bounded on γ_k, which implies that the last summand above tends to zero if $k \longrightarrow \infty$. So, we end up with the following two limits:

(a) $(a_1 - pR)\mu_k + (a_1 + qR)\mu'_k \underset{k\to\infty}{\longrightarrow} a_1$,

(b) $\mu_k + \mu'_k \underset{k\to\infty}{\longrightarrow} 1$,

from which one derives that $\mu_k \longrightarrow \frac{q}{p+q}$ and $\mu'_k \longrightarrow \frac{p}{p+q}$.

Coming back to the definition of the domain G, we see that

(a) $h_{k,1}(\lambda) = a_1 - pR < 0$ for $\lambda \in \Gamma_k$, thus $h_{k,2}(\lambda) > s(-(a_1 - pR))^\alpha$;

(b) $h_{k,1}(\lambda) = a_1 + qR > 0$ for $\lambda \in \Gamma'_k$, thus $h_{k,2}(\lambda) > t(a_1 + qR)^\alpha$.

Therefore, applying the Poisson integral formula for $h_{k,2}$, implies that

$$\begin{aligned} h_{k,2}(0) &= \int_{\Gamma_k} h_{k,2} d\omega(0,\cdot,\Omega_k) + \int_{\Gamma_k'} h_{k,2} d\omega(0,\cdot,\Omega_k) + \int_{\gamma_k} h_{k,2} d\omega(0,\cdot,\Omega_k) \\ &\geq s(pR-a_1)^\alpha \mu_k + t(a_1+qR)^\alpha \mu_k' + s(pR-a_1)^\alpha \omega(0,\gamma_k,\Omega_k) \\ &\longrightarrow \frac{R^\alpha}{p+q}\left(sq\left(p-\frac{a_1}{R}\right)^\alpha + tp\,(a_1+qR)^\alpha\right), \end{aligned}$$

where the last expression tends to infinity when $R \longrightarrow \infty$ (recall that $(q/p)^{\alpha-1} > |s|/t$ according our choice at the beginning). But $h_{k,2}(0)$ remains bounded; a contradiction. □

Remark 13.6.11. It should be mentioned that other tube domains in $\mathbb{C}^2$, whose envelopes of holomorphy are the whole $\mathbb{C}^2$, are also treated in [239] by similar methods; the interested reader may consult the paper itself. Nevertheless, according to our knowledge there is no complete description for such tube domains to be $\boldsymbol{k}$-hyperbolic.

?

13.7 Exercises

Exercise 13.7.1. Define

$$G := \{(z_1,z_2) \in \mathbb{C}^2 : |z_1| < 1, |z_1 z_2| < 1\} \setminus F,$$

where $F := \{(z_1,z_2) \in \mathbb{C}^2 : z_1 = 0, |z_2| \leq 1\}$. Prove that G is Brody hyperbolic but not $\boldsymbol{k}$-hyperbolic. (A similar example is due to D. Eisenman and L. Taylor; cf. [317], page 130.)

Exercise 13.7.2. Prove that any domain G in $\mathbb{C}^n$ admitting a bounded psh exhaustion function u (i.e., $u : G \longrightarrow [0,1)$ with $\{z \in G : u(z) \leq \alpha\} \subset\subset G$ for every $0 \leq \alpha < 1$) is $\boldsymbol{k}$-hyperbolic (cf. [474]).

Exercise 13.7.3. Let $g : \mathbb{C}^{n-1} \longrightarrow \mathbb{R}_+$, $n \geq 2$, be upper semicontinuous such that $\lim_{\|z'\|\to\infty} g(z')/\|z'\|$ exists and is finite. Call it ℓ. Put $\mathfrak{h} : \mathbb{C}^{n-1} \times \mathbb{C} \longrightarrow \mathbb{R}_+$,

$$\mathfrak{h}(z) = \mathfrak{h}(z',z_n) := \begin{cases} |z_n| g(z'/z_n), & \text{if } z_n \neq 0 \\ \ell\|z'\|, & \text{if } z_n = 0 \end{cases}.$$

(a) Prove that $\mathfrak{h}$ is upper semicontinuous and absolutely homogeneous.

(b) Assume, in addition, that $\mathfrak{h} \in \mathcal{PSH}(\mathbb{C}^n)$ and that there exists a $C > 0$ such that $\mathfrak{h}(z',1) \geq C\|z'\|$, $z' \in \mathbb{C}^{n-1}$. Prove that $G := \{z \in \mathbb{C}^n : \mathfrak{h}(z) < 1\}$ is Brody hyperbolic if $\{0'\} \times \mathbb{C} \subsetneq G$.

Exercise 13.7.4. Prove that the pseudoconvex balanced domain from Example 13.1.3 satisfies

$$\boldsymbol{\ell}_G(z,w) > 0, \text{ if } z_1 \neq w_1 \text{ or } z_1 = w_1 \neq 0,\ z_2 \neq w_2.$$

Hint. (1) Observe that $\boldsymbol{\ell}_G(z,w) > 0$, if $z_1 \neq w_1$.

(2) Assume now that $\boldsymbol{\ell}_G((a,z_2),(a,w_2)) = 0$, where $a \neq 0$ and $z_2 \neq w_2$. Using the theorems of Montel and Liouville, construct a sequence $(\varphi_j)_{j\in\mathbb{N}} \subset \mathcal{O}(\mathbb{D}(s_j), G)$ ($s_j \nearrow \infty$) such that for all $j \in \mathbb{N}$

- $\varphi_j(0) = (f_j, g_j)(0) = (a, z_2)$,
- $\varphi_j(1) = (f_j, g_j)(1) = (a, w_2)$,
- $\varphi_j(\mathbb{D}(s_j)) \subset \mathbb{D}(a,\varepsilon) \times \mathbb{C}$ for a small positive ε.

(3) Exploit the shape of G over the disc $\mathbb{D}(a,\varepsilon)$ to prove that

$$(a,z_2),(a,w_2) \in \bigcup_{j=1}^{\infty} \varphi_j(\mathbb{D}(s_j)) \subset \Omega \subset \mathbb{D}(a,\varepsilon) \times \mathbb{C},$$

where Ω is a bounded domain.

(4) Finally, again apply the Montel and Liouville theorems, to end up with a contradiction.

13.8 List of problems

Chapter 14

Completeness

Summary. Completeness with respect to various families of holomorphically contractible distances will be discussed. While in Section 14.1 completeness is studied for general continuous inner distances, completeness with respect to the Carathéodory and Kobayashi distances will be investigated in Sections 14.2 and 14.5, respectively. In the class of Reinhardt domains, a full characterization for Carathéodory completeness in geometric terms is given in Section 14.3. Finally, completeness with respect to the integrated higher Carathéodory–Reiffen metrics is studied for Reinhardt and Zalcman domains (cf. Sections 14.3, 14.4).

14.1 Completeness – general discussion

Let G be an arbitrary domain in $\mathbb{C}^n$ equipped with a continuous distance d_G, e.g., $d_G = \boldsymbol{c}_G$ when G is $\boldsymbol{c}$-hyperbolic. In general, the d_G-topology may be different from $\operatorname{top} G$. Therefore, we have to distinguish very carefully between different notions of completeness.

We say that G is *weakly d_G-complete* if the metric space (G, d_G) is complete in the usual sense. Since we are mainly interested in the interplay between $\operatorname{top} d_G$ and $\operatorname{top} G$, the notion of "d_G-complete" seems to be more convenient for our purpose. A domain G is called *d_G-complete* if any d_G-Cauchy sequence $(z_\nu)_{\nu\in\mathbb{N}} \subset G$ converges to a point $z_0 \in G$ with respect to $\operatorname{top} G$, i.e., $\|z_\nu - z_0\| \underset{\nu\to\infty}{\longrightarrow} 0$.

Moreover, there is also another important notion which has been borrowed from differential geometry. Namely, we call the domain G *d_G-finitely compact* if all d_G-balls are relatively compact (with respect to $\operatorname{top} G$) inside G. Of course, here we may also formulate the "weaker" property *weakly d_G-finitely compact*.

If we deal with the Carathéodory, Bergman, or Kobayashi distance, we will always omit, for simplicity, the subscript G, e.g., we just say that G is $\boldsymbol{c}$-, $\boldsymbol{b}$-, or $\boldsymbol{k}$-complete etc.

Remark 14.1.1.

(a) For any d_G-finitely compact domain G, the topologies $\operatorname{top} G$ and $\operatorname{top} d_G$ coincide.

(b) Moreover, the following implications (i) $\Longrightarrow$ (ii) $\Longrightarrow$ (iii) are obvious, where

  (i) G is d_G-finitely compact;

  (ii) G is d_G-complete;

(iii) G is weakly d_G-complete.

(c) Already here we would like to point out that for the Carathéodory distance the question whether (iii) implies (i) is still unsolved; cf. Exercises 14.6.2 for the case of complex spaces.

As is known from differential geometry, a theorem of H. Hopf asserts the equivalence of Cauchy-completeness and finite compactness. This result was generalized by W. Rinow (cf. [452]) and S. Cohn-Vossen (cf. [107]) to the situation we are interested in.

Theorem 14.1.2. *Let d_G denote a continuous inner distance on a domain $G \subset \mathbb{C}^n$. Then,* $\operatorname{top} G = \operatorname{top} d_G$ *and, moreover, the following properties are equivalent:*

(i) *G is d_G-finitely compact;*

(ii) *G is weakly d_G-finitely compact;*

(iii) *G is d_G-complete;*

(iv) *G is weakly d_G-complete;*

(v) *any half-segment $\alpha : [0, b) \longrightarrow G$ (i.e., any $\| \ \|$-continuous curve $\alpha : [0, b) \longrightarrow G$ with $d_G(\alpha(t'), \alpha(t'')) = t'' - t'$, $0 \le t' < t'' < b < \infty$) has a continuous extension $\widetilde{\alpha} : [0, b] \longrightarrow G$.*

To prepare the necessary tools for the proof of the main implication (v) $\Longrightarrow$ (i), we first derive the following two lemmas:

Lemma 14.1.3. *Under the assumptions of Theorem* 14.1.2, *the d_G-closure of a d_G-ball can be described as*

$$\overline{B_{d_G}(a, r)}^{d_G} = \overline{B_{d_G}(a, r)} \cap G = B_{d_G}(a, r) \cup S_{d_G}(a, r), \quad a \in G,\ r > 0,$$

where $S_{d_G}(a, r) := \{z \in G : d_G(a, z) = r\}$.

Proof. The only thing we have to verify is that each point in $S_{d_G}(a, r)$ belongs to $\overline{B_{d_G}(a, r)}^{d_G}$. So, we fix a $z_0 \in S_{d_G}(a, r)$. Because of $d^i_G(a, z_0) = d_G(a, z_0) = r$ there exists a sequence of $\| \ \|$-rectifiable curves $\alpha_\nu : [0, 1] \longrightarrow G$ with $\alpha_\nu(0) = a$, $\alpha_\nu(1) = z_0$, and $L_{d_G}(\alpha_\nu) \underset{\nu\to\infty}{\searrow} r$.

Recall that $d_G(a, \alpha_\nu(\cdot))$ is continuous on $[0, 1]$. Thus, for suitable $t_\nu \in (0, 1)$ we conclude that $d_G(a, \alpha_\nu(t_\nu)) = \frac{\nu-1}{\nu} r$, i.e., $\alpha_\nu(t_\nu) \in B_{d_G}(a, r)$. On the other hand, the relations

$$\frac{\nu - 1}{\nu} r + d_G(\alpha_\nu(t_\nu), z_0) \le L_{d_G}(\alpha_\nu) \underset{\nu\to\infty}{\searrow} r$$

imply that $\lim_{\nu\to\infty} d_G(\alpha_\nu(t_\nu), z_0) = 0$. Hence, $z_0 \in \overline{B_{d_G}(a, r)}^{d_G}$. □

Remark 14.1.4. The reader should recall that for the equality in Lemma 14.1.3 it is essential that d_G is inner; cf. Example 2.7.10.

Lemma 14.1.5. *Let (G, d_G) be as in Theorem 14.1.2. Suppose that $\overline{B_{d_G}(a,r)}^{d_G}$ is a compact subset of G. Then, for any point $b \in \overline{B_{d_G}(a,r)}^{d_G}$, $a \neq b$, there exists a d_G-geodesic α (i.e., α is a curve $\alpha : [0,1] \longrightarrow G$ with $\alpha(0) = a$, $\alpha(1) = b$, and $L_{d_G}(\alpha) = d_G(a,b)$).*

Proof. If $d_G(a,b) < r$, we choose a sequence of curves $\alpha_\nu : [0,1] \longrightarrow G$ connecting a and b with $r > L_{d_G}(\alpha_\nu) \searrow d_G(a,b)$, i.e., $\alpha_\nu([0,1]) \subset B_{d_G}(a,r)$. Reparametrization with respect to the d_G-arc-length and the use of the fact that $L_{d_G}(\alpha_\nu) \le r < \infty$, $\nu \ge 1$, enable us to assume that the family (α_ν) is equicontinuous. Thus, the Arzela–Ascoli theorem leads to a subsequence $(\alpha_{\nu_\mu})_{\mu\in\mathbb{N}}$ that converges uniformly to a curve $\alpha : [0,1] \longrightarrow G$ with $\alpha(0) = a$, $\alpha(1) = b$, and $d_G(a,b) \le L_{d_G}(\alpha) \le \liminf_{\mu\to\infty} L_{d_G}(\alpha_{\nu_\mu}) = d_G(a,b)$, i.e., α is a d_G-geodesic in G from a to b.

In the case where $d_G(a,b) = r$, we approximate b by points $b_\nu \in B_{d_G}(a,r)$. We choose d_G-geodesics α_ν in $B_{d_G}(a,r)$ from a to b_ν such that the family (α_ν) is again equicontinuous. Then, the same argument as above finishes the proof. □

Proof of Theorem 14.1.2. Obviously, the only non-trivial implication is (v) $\Longrightarrow$ (i). To establish it we argue by contradiction, i.e., we suppose (i) to be false. Hence, there exists a d_G-ball $B_{d_G}(a,R)$ whose d_G-closure is not compact.

Put $r_0 := \sup\{r > 0 : \overline{B_{d_G}(a,r)}^{d_G}$ is compact$\}$. Because of $\operatorname{top} G = \operatorname{top} d_G$ we obtain $0 < r_0 \le R$. We claim that

(*) $\overline{B_{d_G}(a,r_0)}^{d_G}$ is not compact.

For otherwise, by a simple compactness argument, we would be able to find a finite covering

$$\overline{B_{d_G}(a,r_0)}^{d_G} \subset \bigcup_{j=1}^{N} B_{d_G}(z_j,r_j) =: U,$$

where $z_j \in \overline{B_{d_G}(a,r_0)}^{d_G}$ and where $\overline{B_{d_G}(z_j,r_j)}^{d_G}$ are compact subsets of G. In particular, $U \subset\subset G$.

Suppose that

(**) there are points $w_\nu \in G \setminus U$, $\nu \in \mathbb{N}$, with $d_G(a,w_\nu) < r_0 + 1/\nu$.

Then we will find curves $\alpha_\nu : [0,1] \longrightarrow G$ with $\alpha_\nu(0) = a$, $\alpha_\nu(1) = w_\nu$, and $L_{d_G}(\alpha_\nu) < r_0 + 1/\nu$ and real numbers $0 < t_\nu < 1$ with $d_G(a,\alpha_\nu(t_\nu)) = r_0$, i.e., $\alpha_\nu(t_\nu) \in S_{d_G}(a,r_0)$. Without loss of generality, we may assume that $\alpha_\nu(t_\nu) \underset{\nu\to\infty}{\longrightarrow} z' \in S_{d_G}(a,r_0)$ and $z' \in B_{d_G}(z_{j_0},r_{j_0})$. Thus $\alpha_\nu(t_\nu) \in B_{d_G}(z_{j_0},r_{j_0})$ for almost all ν. Combining all our information gathered so far, we obtain

$$r_0 + 1/\nu \ge L_{d_G}(\alpha_\nu) \ge r_0 + d_G(\alpha(t_\nu), w_\nu),$$

i.e., $\lim_{\nu\to\infty} d_G(\alpha_\nu(t_\nu), w_\nu) = 0$ and $d_G(w_\nu, z_{j_0}) \le d_G(w_\nu, \alpha_\nu(t_\nu)) + d_G(\alpha_\nu(t_\nu), z_{j_0}) < r_{j_0}$, $\nu \gg 1$. Thus, $w_\nu \in U$ if $\nu \gg 1$, which contradicts (**).

Hence, we can conclude that a larger ball $B_{d_G}(a, r_0 + \varepsilon)$ is contained in U, which contradicts the definition of r_0.

Summarizing,

(***) we have found a ball $B_{d_G}(a, r_0)$ whose d_G-closure contains a sequence $(z_\nu)_\nu$ without an accumulation point in G.

By Lemma 14.1.3 we may assume that all of the z_ν's belong to $B_{d_G}(a, r_0)$ and that $d_G(a, z_\nu) =: r_\nu \nearrow r_0$.

Now applying Lemma 14.1.5, we find d_G-geodesics $\alpha_\nu : [0, r_\nu] \longrightarrow B_{d_G}(a, r_0)$ for a and z_ν, which are parametrized by the arc-length, i.e., the α_ν are segments in the sense of (v). A successive exploitation of the Arzela–Ascoli argument then leads to a chain of subsequences

$$(\alpha_{j,\nu})_{\nu\ge j} \subset (\alpha_{j-1,\nu})_{\nu\ge j-1} \subset \cdots \subset (\alpha_{1,\nu} = \alpha_\nu)_{\nu\ge 1}$$

such that $\alpha_{j,\nu}|_{[0,r_j]} \underset{\nu\to\infty}{\overset{K}{\Longrightarrow}} \beta_j$. Then, $\beta_j : [0, r_j] \longrightarrow G$ is a continuous curve with $\beta_j(0) = a$ and $\beta_{j+1}|_{[0,r_j]} = \beta_j$.

Set $\beta : [0, r_0) \longrightarrow G$ as $\beta(t) := \beta_j(t)$ if $0 \le t \le r_j$. Then, β is a segment in G and according to the property (v) it can be continuously extended to $\widetilde{\beta} : [0, r_0] \longrightarrow G$. Put $z^* := \widetilde{\beta}(r_0) \in G$. Observe that for $\mu \ge j$ we have

$$\begin{aligned} d_G(z^*, z_{j,\mu}) &= d_G(z^*, \alpha_{j,\mu}(r_{j,\mu})) \\ &\le d_G(z^*, \beta_j(r_j)) + d_G(\beta_j(r_j), \alpha_{j,\mu}(r_j)) + d_G(\alpha_{j,\mu}(r_j), \alpha_{j,\mu}(r_{j,\mu})) \\ &\le d_G(z^*, \widetilde{\beta}(r_j)) + d_G(\beta_j(r_j), \alpha_{j,\mu}(r_j)) + (r_0 - r_j). \end{aligned}$$

Because of $\alpha_{j,\mu}(r_j) \underset{\mu\to\infty}{\longrightarrow} \beta_j(r_j)$ we see that z^* is an accumulation point of the sequence $(z_\nu)_{\nu\in\mathbb{N}}$ contradicting (***). □

We conclude this section by introducing the completeness concepts for families of invariant distances. Let $\underline{d} = (d_G)_G$ be an arbitrary holomorphically contractible family of pseudodistances, e.g., $\underline{d} = \boldsymbol{c}$ or $\underline{d} = \boldsymbol{k}$. A domain G is called *$\underline{d}$-complete* if d_G is a distance and any d_G-Cauchy sequence in G does converge to a point in G (with respect to top G); similarly, G is called *$\underline{d}$-finitely compact* if d_G is a distance and if any d_G-ball $B_{d_G}(a, r)$, $r \in \mathbb{R}_+$, is relatively compact in G (with respect to top G).

Since the Kobayashi distance is inner, we immediately obtain

Corollary 14.1.6. *For any $\boldsymbol{k}$-hyperbolic domain $G \subset \mathbb{C}^n$ all the properties* (i)–(v) *in Theorem* 14.1.2 *are equivalent for $d_G = \boldsymbol{k}_G$. In particular, $\boldsymbol{k}$-completeness is equivalent to $\boldsymbol{k}$-finite compactness.*

14.2 Carathéodory completeness

In this section, we study relations between the notions of Carathéodory completeness and well-known properties of domains of holomorphy. Moreover, we present a sufficiently large class of $\boldsymbol{c}$-finitely compact domains.

Theorem 14.2.1 (cf. [416]). *For a $\boldsymbol{c}$-hyperbolic domain $G \subset \mathbb{C}^n$ the following statements are equivalent:*

(i) *G is $\boldsymbol{c}$-finitely compact;*

(ii) *for any $z_0 \in G$ and for any sequence $(z_\nu)_{\nu\in\mathbb{N}} \subset G$ without accumulation points* (*with respect to* top G) *in G there exists an $f \in \mathcal{O}(G,\mathbb{D})$ with $f(z_0) = 0$ and* $\sup_{\nu\in\mathbb{N}} |f(z_\nu)| = 1$.

Proof. It suffices to establish the implication (i) $\Longrightarrow$ (ii). Because of the fact that $(z_\nu)_\nu$ does not accumulate in G we may assume, without loss of generality, that $\boldsymbol{c}^*_G(z_0, z_\nu) > 1-1/2^{2\nu}$, $\nu \in \mathbb{N}$. Thus, we are able to choose functions $f_\nu \in \mathcal{O}(G,\mathbb{D})$ with $f_\nu(z_0) = 0$ and $f_\nu(z_\nu) > 1 - 1/2^{2\nu}$. Put $g_\nu(z) := (1 + f_\nu(z))/(1 - f_\nu(z))$, $z \in G$. Then, $g_\nu \in \mathcal{O}(G, H^+)$, $g_\nu(z_0) = 1$, where H^+ denotes the right half plane. Now, we define $g(z) := \sum_{\nu=1}^{\infty}(1/2^\nu)g_\nu(z)$, $z \in G$. Observe that on any compact subset $K \subset G$ the functions f_ν are uniformly bounded by a number $R = R(K) < 1$. With $|g_\nu(z)| \le 2/(1 - R)$, $z \in K$, it follows that $g \in \mathcal{O}(G, H^+)$. Moreover, we obtain

$$|g(z_\nu)| \ge |\operatorname{Re} g(z_\nu)| \ge (1/2^\nu)g_\nu(z_\nu) \ge 2^\nu \underset{\nu\to\infty}{\longrightarrow} \infty.$$

Hence, with $f(z) := (g(z) - 1)/(g(z) + 1)$, $z \in G$, we get a function $f \in \mathcal{O}(G,\mathbb{D})$ with $f(z_0) = 0$ and

$$|f(z_\nu)| \ge \frac{|g(z_\nu)| - 1}{|g(z_\nu)| + 1}\frac{2^\nu - 1}{2^\nu + 1} \underset{\nu\to\infty}{\longrightarrow} 1. \qquad \square$$

Moreover, in the class of $\mathbb{C}$-convex domains the following is true:

Proposition 14.2.2 (cf. [390]). *Any bounded $\mathbb{C}$-convex domain $G \subset \mathbb{C}^n$ is $\boldsymbol{c}$-finitely compact* (*e.g.,* $\mathbb{G}_2$).

Proof. Assuming the contrary, one may find a point $a \in G$, a sequence $(b_j)_j \subset G$ with $b_j \longrightarrow b \in \partial G$, and an $M \in \mathbb{R}_+$ such that $\boldsymbol{c}_G(a, b_j) \le M$, $j \in \mathbb{N}$. Since G is $\mathbb{C}$-convex, there is a complex hyperplane H through b not intersecting G. Without loss of generality, we may assume that $b = 0$ and that H is given as $H = \{z \in \mathbb{C}^n : z_1 = 0\}$. Then, by virtue of Proposition 2.3.6 in [20], one has that

$$G' := \{z_1 \in \mathbb{C} : \exists_{\tilde{z}\in\mathbb{C}^{n-1}} : (z_1, \tilde{z}) \in G\}$$

is a simply connected domain with $0 \in \partial G'$, and therefore biholomorphic to $\mathbb{D}$. Since $b_{j,1} \longrightarrow b_1 = 0 \in \partial G'$, there exists an $f \in \mathcal{O}(G', \mathbb{D})$ with $f(a_1) = 0$ and $|f(b_{j,1})| \longrightarrow 1$ as $j \longrightarrow \infty$ (eventually, one has to take a subsequence). Hence,

$$M \geq \boldsymbol{c}_G(a, b_j) \geq \boldsymbol{c}_{G'}(a_1, b_{j,1}) \longrightarrow \infty;$$

a contradiction. □

Remark 14.2.3. Even more is true (see [323]), namely: for any bounded $\mathbb{C}$-convex domain $G \subset \mathbb{C}^n$ and any boundary point $a \in \partial G$ there exists an $f \in \mathcal{O}(G, \mathbb{D})$ with $\lim_{G \ni z \to a} f(z) = 1$.

Example 14.2.4.

(a) Any bounded strongly pseudoconvex domain in $\mathbb{C}^n$ and any bounded pseudoconvex domain in $\mathbb{C}^2$ with real analytic boundary is $\boldsymbol{c}$-finitely compact. Use the existence of peak-functions (cf. [39, 171, 173]).

(b) Any bounded convex domain is $\boldsymbol{c}$-finitely compact. Again argue by means of peak-functions obtained via convexity.

(c) A convex domain $G \subset \mathbb{C}^n$ is $\boldsymbol{c}$-finitely compact iff G is $\boldsymbol{k}$-hyperbolic (cf. Theorem 13.1.7).

(d) $\mathbb{G}_n$ is $\boldsymbol{c}$-finitely compact; use Corollary 7.2.7 (see also the following result).

Proposition 14.2.5 (cf. [323]). *For bounded domains $D, G \subset \mathbb{C}^n$ let $F : D \longrightarrow G$ be a proper holomorphic mapping. Then, the following properties are equivalent:*

(i) *G is $\boldsymbol{c}$-finitely compact;*

(ii) *D is $\boldsymbol{c}$-finitely compact.*

Proof. (i) $\Longrightarrow$ (ii): Fix a point $a \in D$ and let $R > 0$. By assumption, we know that $B := B_{\boldsymbol{c}_G}(F(a), R) \subset\subset G$. Then, since F is proper, $F^{-1}(B)$ is a relatively compact subset of D. Note that $\boldsymbol{c}_G(F(a), F(z)) \leq \boldsymbol{c}_D(a, z)$; therefore, $B_{\boldsymbol{c}_D}(a, R) \subset F^{-1}(B)$ and so is relatively compact in D. Hence, D is $\boldsymbol{c}$-finitely compact.

(ii) $\Longrightarrow$ (i): Fix a $b \in G$ and assume that there is a sequence $(w_j)_j \subset G$ with $w_j \longrightarrow w^* \in \partial G$ such that $\boldsymbol{c}^*_G(b, w_j) \leq R < 1$ for a certain R. Without loss of generality, one may assume that all the w_j's are not critical values of F. Because of the properness of F, one may assume that there are points $a \in D$, $z_j \in D$ such that $F(a) = b$, $F(z_j) = w_j$, $j \in \mathbb{N}$, and $z_j \longrightarrow z^* \in \partial D$ (eventually one has to choose a subsequence). Then, Theorem 14.2.1 provides an $f \in \mathcal{O}(D, \mathbb{D})$ with $f(a) = 0$ and $|f(z_j)| \longrightarrow 1$ (again after taking a correctly chosen subsequence). Then, outside of the analytic set A of critical values of F one may define the following holomorphic mapping:

$$\tilde{g} := \pi_n \circ (f \times \cdots \times f) \circ F^{-1} : G \setminus A \longrightarrow \mathbb{G}_n.$$

Then, $\tilde{g}$ extends to a holomorphic mapping $g : G \longrightarrow \mathbb{G}_n$ (use Riemann's Theorem on removable singularities). Taking into account the boundary behavior of f, one sees that (after again choosing a correct subsequence) that $g(w_j) \longrightarrow a \in \partial\mathbb{G}_n$. Now, one applies Corollary 7.2.7 to get a function $h \in \mathcal{O}(\mathbb{G}_n, \mathbb{D})$ such that $|h(g(w_j))| \longrightarrow 1$; a contradiction. □

This result may be used to reprove that $\mathbb{G}_n$ is $\boldsymbol{c}$-finitely compact, although not $\mathbb{C}$-convex for $n \geq 3$.

Proposition 14.2.6. *Any $\boldsymbol{c}$-complete domain G in $\mathbb{C}^n$ is an $\mathcal{H}^\infty$-domain of holomorphy. In particular, G is pseudoconvex.*

Proof. Suppose the contrary. Then there exist $z_0 \in G$ and $0 < r < R$ with the following properties:

$\mathbb{B}(z_0, r) \subset G$, but $\mathbb{B}(z_0, R) \not\subset G$ and for any $f \in \mathcal{H}^\infty(G)$ the restriction $f|_{\mathbb{B}(z_0,r)}$ extends to a holomorphic function $\tilde{f}$ on $\mathbb{B}(z_0, R)$.

Let $z' \in \mathbb{B}(z_0, R) \setminus G$ and choose the point z^* on the segment $\overline{z_0 z'}$ with $\overline{z_0 z^*} \subset G \cup \{z^*\}$ but $z^* \in \partial G$. Then if $z_\nu := z_0 + (1 - 1/(2\nu))(z^* - z_0)$, $\nu \in \mathbb{N}$, we get (cf. Remark 2.1.4)

$$\boldsymbol{c}_G^*(z_\nu, z_\mu) \leq \boldsymbol{c}_{\mathbb{B}(z_0,R)}^*(z_\nu, z_\mu) \underset{\mu\to\infty}{\longrightarrow} 0,$$

i.e., $(z_\nu)_{\nu\in\mathbb{N}}$ is a $\boldsymbol{c}_G$-Cauchy sequence with $\lim_{\nu\to\infty} z_\nu = z^* \in \partial G$, which contradicts the hypothesis. □

Corollary 14.2.7. *Any $\boldsymbol{c}$-finitely compact domain G in $\mathbb{C}^n$ is necessarily $\mathcal{H}^\infty(G)$-convex and it is an $\mathcal{H}^\infty$-domain of holomorphy.*

Remark 14.2.8.

?

(a) It is not known whether the property "$\boldsymbol{c}$-complete" implies that G is $\mathcal{H}^\infty(G)$-convex.

(b) Observe that although the notions "$\mathcal{H}^\infty$-domain of holomorphy" and "$\mathcal{H}^\infty(G)$-convex" coincide for bounded plane domains (cf. [15]) they are not comparable in higher dimensions (cf. [471]).

(c) In Theorem 14.5.7 we will construct a bounded pseudoconvex balanced domain G in $\mathbb{C}^3$ with a continuous Minkowski function that is not $\boldsymbol{c}$-complete, although it is an $\mathcal{H}^\infty$-convex $\mathcal{H}^\infty$-domain of holomorphy. Other examples of domains sharing these properties were given by P. R. Ahern & R. Schneider (cf. [15]) and N. Sibony (cf. [471]).

(d) It is well known that any bounded pseudoconvex domain with smooth $\mathcal{C}^\infty$-boundary is $\mathcal{H}^\infty$-convex and is an $\mathcal{H}^\infty$-domain of holomorphy. Despite many efforts, no example has been constructed of a domain of this type not being $\boldsymbol{c}$-finitely compact.

(e) For a fairly complete discussion of $\mathcal{H}^\infty$-domains of holomorphy and $\mathcal{H}^\infty$-convex ones see [265], § 4.1.

The only case in which the relation between "$\boldsymbol{c}$-complete" and "$\boldsymbol{c}$-finitely compact" is well understood is the plane case, which was studied by M. A. Selby (cf. [468]) and by N. Sibony (cf. [471]).

Theorem 14.2.9. *Let $G \subset \mathbb{C}$ be $\boldsymbol{c}$-hyperbolic. Then G is $\boldsymbol{c}$-complete iff G is $\boldsymbol{c}$-finitely compact.*

Proof. It remains to verify the implication "$\Longrightarrow$". Unfortunately, so far, the only available proof requires a detour into the theory of function algebras. For the convenience of the reader, we will quote all the results we need but without going into proofs.

The first result, due to A. Browder (cf. [76], also Th. 26.12 in [489]), presents a metric condition for a point $z_0 \in K$, K a compact subset of $\mathbb{C}$, to be a peak point for the uniform algebra $R(K)$. Here, $R(K)$ denotes the subalgebra of those $f \in C(K)$ that can be uniformly approximated by rational functions with their poles outside K.

Lemma 14.2.10. *Let $K \subset \mathbb{C}$ be compact. Suppose that $z_0 \in K$ is not a peak point for $R(K)$. Then,*

$$\lim_{\delta \to 0} \frac{\mathcal{L}^2(P_\varepsilon(z_0) \cap \mathbb{D}(z_0, \delta))}{\pi \delta^2} = 1,$$

where $P_\varepsilon(z_0) := \{z \in K : \text{ if } f \in R(K),\ \|f\|_K \le 1, \text{ then } |f(z) - f(z_0)| < \varepsilon\}$.

For the domain G in $\mathbb{C}$, we denote by Spec $=$ Spec $\mathcal{H}^\infty(G)$ the set of all non-zero continuous algebra homomorphisms $\varphi : \mathcal{H}^\infty(G) \longrightarrow \mathbb{C}$. Spec is called the *spectrum of* $\mathcal{H}^\infty(G)$ and, endowed with the Gelfand topology, it is a compact topological space.

A $\varphi \in$ Spec is said to belong to the fiber over z, $z \in \partial G$, if for any $f \in \mathcal{H}^\infty(G)$ that extends holomorphically through z we have $\varphi(f) = f(z)$. This fiber is called Spec_z.

In order to prove "$\Longrightarrow$" it suffices to see that for any boundary point $z \in \partial G$ the fiber Spec_z is a peak set for $\widehat{\mathcal{H}^\infty}(G)$, the algebra of the Gelfand transforms $\widehat{f}$, $f \in \mathcal{H}^\infty(G)$.

For the future let us assume that

Spec_ζ *is not a peak set for $\widehat{\mathcal{H}^\infty}(G)$ with a fixed $\zeta \in \partial G$.* $(*)$

To continue, we need the following theorem of M. S. Melnikov (cf. VIII.4.5 in [185]):

Lemma 14.2.11. *Let $K \subset \mathbb{C}$ be compact and let $z_0 \in K$. Fix $0 < a < 1$. Then, z_0 is a peak point for $R(K)$ iff $\sum_{j=1}^{\infty} a^j \gamma(E_j \setminus K) = \infty$, where*

$$E_j := \{z \in \mathbb{C} : a^{j+1} \le |z - z_0| \le a^j\},$$
$$\gamma(A) := \sup\{|f'(\infty)| : f \in \mathcal{O}(\overline{\mathbb{C}} \setminus L),\ L \subset A \text{ compact},\ |f| \le 1,\ f(\infty) = 0\}.$$

We call $\gamma(A)$ the *analytic capacity of A*.

Moreover (cf. [186]), we have

Lemma 14.2.12. *Let G be a $\mathbf{c}$-hyperbolic domain in $\mathbb{C}$, $z_0 \in \partial G$, $0 < a < 1$. Then,*

$$\operatorname{Spec}_{z_0} \text{ is a peak set for } \widehat{\mathcal{H}^\infty}(G) \quad \text{iff} \quad \sum_{j=1}^{\infty} a^j \gamma(E_j \setminus G) = \infty.$$

Thus, if we replace z_0 by ζ, then $(*)$ gives $\sum_{j=1}^{\infty} a^j \gamma(E_j \setminus G) < \infty$. Now, we choose a compact set $K \subset (\mathbb{D}(\zeta, a) \cap G) \cup \{\zeta\}$ with $\sum_{j=1}^{\infty} a^j \gamma(E_j \setminus K) < \infty$. So ζ is not a peak point for $R(K)$.

Even more, using the characterization of peak points for uniform algebras in terms of representing measures (cf. II.3 in [185]), we find a positive measure σ on $K \setminus \{\zeta\}$ with $f(\zeta) = \int f(x) d\sigma(x)$, $f \in R(K)$.

By means of the localization operator, the following approximation result can be derived (cf. VIII. 10.8 in [185]).

Lemma 14.2.13. *For any $f \in \mathcal{H}^\infty(G)$ there exists a sequence $(f_j)_{j \in \mathbb{N}} \subset \mathcal{H}^\infty(G)$ with $\|f_j\|_G \le 17 \|f\|_G$ such that $f_j \longrightarrow f$ uniformly on $G \cap (\mathbb{C} \setminus \mathbb{D}(\zeta, \varepsilon))$, $\varepsilon > 0$, and, moreover, f_j extends holomorphically to a neighborhood of ζ.*

Therefore, $(*)$ together with Lemma 14.2.13 give an element $\Phi_0 \in \operatorname{Spec}_\zeta$, which is represented by σ.

According to Lemma 14.2.10 we can choose points $z_\nu \in K \setminus \{\zeta\}$, $\lim_{\nu\to\infty} z_\nu = \zeta$, such that $|f(\zeta) - f(z_\nu)| < 1/\nu$ for all $f \in R(K)$ with $\|f\|_K \le 1$. In particular, if $f \in \mathcal{O}(G, \mathbb{D})$ extends holomorphically to ζ with $f(\zeta) = 0$, then $|f(z_\nu)| < 1/\nu$.

Now let $f \in \mathcal{O}(G, \mathbb{D})$ with $\widehat{f}(\Phi_0) = 0$. Then, Lemma 14.2.13 delivers a sequence of functions $f_j \in \mathcal{O}(G, 17\mathbb{D})$ such that f_j extends through ζ and $\lim_{j\to\infty} f_j(z_\nu) = f(z_\nu)$, $\nu \in \mathbb{N}$. Because of the properties of Φ_0 we also obtain $\Phi_0(f_j) = f_j(\zeta) \underset{j\to\infty}{\longrightarrow} \Phi_0(f) = 0$. Put $g_j := (f_j - f_j(\zeta))/(17(1 + |f_j(\zeta)|))$. Then, $g_j \in \mathcal{O}(G, \mathbb{D})$, g_j extends to ζ, and $g_j(\zeta) = 0$. Therefore, $|g_j(z_\nu)| < 1/\nu$, which implies

$$|f(z_\nu)| = \lim_{j\to\infty} |f_j(z_\nu)| \le \limsup_{j\to\infty} 17(1 + |f_j(\zeta)|)\, |g_j(z_\nu)| + \lim_{j\to\infty} |f_j(\zeta)| \le \frac{17}{\nu}.$$

To estimate $c_G^*(z_\nu, z_\mu)$, choose $f \in \mathcal{O}(G, \mathbb{D})$ with $f(z_\nu) = 0$ and define $g := (f - \widehat{f}(\Phi_0))/2$. Then, $|g(z_\mu)| \le 17/\mu$ or

$$|f(z_\mu)| \le 2|g(z_\mu)| + 2|g(z_\nu)| \underset{\nu,\mu\to\infty}{\longrightarrow} 0.$$

Thus we have found that the sequence $(z_\nu)_{\nu\in\mathbb{N}}$ is a c-Cauchy sequence, which contradicts G being c-complete. □

Remark 14.2.14.

(a) We emphasize that Theorem 14.2.9 fails to hold for one dimensional complex spaces (cf. [273]). We provide some hints to this counterexample in Exercise 14.6.2. In the case of domains, it is still an open problem whether Theorem 14.2.9 remains true in higher dimensions.

(b) Obviously, any c-finitely compact domain is c^i-finitely compact. In [48], a complex space X of finite dimension is given such that c_X and c_X^i define the topology of X and X is c^i-complete, but is not c-complete. It seems to be still open whether such an example exists in the category of domains.

We end this section by presenting an example that shows that c-finite compactness is not a "local property" (cf. [144]).

Example 14.2.15. We study the following example of N. Sibony (cf. [471]):

$$G := \{(z_1, z_2) \in \mathbb{D} \times \mathbb{C} : |z_2| < \exp(-u(z_1))\},$$

where $u : \mathbb{D} \longrightarrow [-\infty, \infty)$ denotes the following psh function on $\mathbb{D}$

$$u(\lambda) := \exp\left(\sum_{j=1}^{\infty} \alpha_j \log \frac{|\lambda - a_j|}{2}\right).$$

Here, $(a_j)_{j\in\mathbb{N}}$ is a discrete sequence of points in $\mathbb{D}$ such that every boundary point of $\mathbb{D}$ is a non-tangential limit of a subsequence and α_j are positive numbers with $\sum_{j=1}^{\infty} \alpha_j \log \frac{|a_j|}{2} > -\infty$ (cf. Example 2.7.10(6°)).

Using Hartogs' series and the Fatou theorem, it is easy to see that any bounded holomorphic function on G extends holomorphically to the bidisc $\mathbb{D}^2$, i.e., G is not c-complete (cf. Remark 2.1.3).

On the other hand, let $z' \in \partial G$ be fixed. In the case where $|z_1'| = 1$ or $|z_2'| = 1$ it is clear that there is no c_G-Cauchy sequence in G converging to z'. In the remaining case $|z_1'| < 1$ and $\operatorname{dist}(z_1', \{a_j : j \ge 1\}) =: \alpha > 0$ put $U := \mathbb{D}(z_1', \alpha/2) \times \mathbb{C}$ and $V := G \cap U$. Observe that V is connected. Now, for any j we choose $\log \frac{z_1 - a_j}{2}$ to be holomorphic on $\mathbb{D}(z_1', \alpha/2)$. If we define $g(z_1) = e^{i\theta} \prod_{j=1}^{\infty} (\frac{z_1 - a_j}{2})^{\alpha_j}$ with $g(z_1') \in [0, \infty)$, we obtain the following holomorphic function on V:

$$f(z_1, z_2) = z_2 \exp g(z_1) \quad \text{with} \quad f(z') = 1 \quad \text{and} \quad f \in \mathcal{O}(V, \mathbb{D}).$$

Again we conclude that there is no c_V-Cauchy sequence in V tending to z'.

14.3 c-completeness for Reinhardt domains

In this section, Carathéodory completeness for Reinhardt domains will we discussed. Recall that a domain $G \subset \mathbb{C}^n$ is said to be $\boldsymbol{c}$*-complete* (respectively, $\boldsymbol{c}$-finitely compact) if $\boldsymbol{c}_G$ is a distance and if any $\boldsymbol{c}_G$-Cauchy sequence converges to a point in G (in top G) (respectively, if $\boldsymbol{c}_G$ is a distance and if any $\boldsymbol{c}_G$-ball with a finite radius is a relatively compact subset of G). Moreover, recall that any $\boldsymbol{c}$-complete domain G is pseudoconvex.

Theorem 14.3.1. *Let $G \subset \mathbb{C}^n$ be a pseudoconvex Reinhardt domain. Then the following conditions are equivalent:*

(i) *G is $\boldsymbol{c}$-finitely compact;*

(ii) *G is $\boldsymbol{c}$-complete;*

(iii) *there is no boundary sequence $(z_\nu)_{\nu\in\mathbb{N}} \subset G$ with $\sum_{\nu=1}^{\infty} \boldsymbol{g}_G(z_\nu, z_{\nu+1}) < \infty$;*

(iv) *G is bounded and fulfills the following so-called* Fu-condition:

$$\text{if} \quad \overline{G} \cap V_j \neq \varnothing, \text{ then } G \cap V_j \neq \varnothing, \tag{14.3.1}$$

where, as usual, $V_j = \{z \in \mathbb{C}^n : z_j = 0\}$.

This result is due to P. Pflug ([416]), S. Fu ([181]), and W. Zwonek ([566], see also [565]).

For the proof of Theorem 14.3.1, we shall need the following three lemmas:

Lemma 14.3.2 (cf. [566]). *Let $G \subset \mathbb{C}^n_*$ be a pseudoconvex Reinhardt domain. Then $\boldsymbol{k}_G = \boldsymbol{\ell}_G$. In particular, the Lempert function $\boldsymbol{\ell}_G$ is continuous on $G \times G$.*

Proof. Observe that $\widehat{G} := \log G$ is a convex domain in $\mathbb{R}^n$ and that the mapping

$$T := \widehat{G} + i\mathbb{R}^n \ni z \overset{\Phi}{\longmapsto} (e^{z_1}, \dots, e^{z_n}) \in G$$

is a holomorphic covering. Therefore, for $z, w \in G$ we have

$$\begin{aligned} \boldsymbol{\ell}_G(z,w) &= \inf\{\boldsymbol{\ell}_T(\widetilde{z}, \widetilde{w}) : \widetilde{z}, \widetilde{w} \in T \text{ with } \Phi(\widetilde{z}) = z,\ \Phi(\widetilde{w}) = w\} \\ &= \inf\{\boldsymbol{k}_T(\widetilde{z}, \widetilde{w}) : \widetilde{z}, \widetilde{w} \in T \text{ with } \Phi(\widetilde{z}) = z,\ \Phi(\widetilde{w}) = w\} = \boldsymbol{k}_G(z,w). \end{aligned}$$

Here, we have used Theorem 11.2.1 of Lempert. □

Lemma 14.3.3. *Let $\Omega \subset \mathbb{R}^n$ be an unbounded convex domain that is contained in $\{x \in \mathbb{R}^n : x_j < R,\ j = 1, \dots, n\}$ for a certain number R. Then, for any point $a \in \Omega$ there exist a vector $v \in \mathbb{R}^n_- \setminus \{0\}$ and a neighborhood $V = V(a) \subset \Omega$ such that $V + \mathbb{R}_+ v \subset \Omega$.*

Proof. Take without loss of generality the point $a = 0$. Then, the continuity of the Minkowski function h of Ω and the assumptions on Ω lead to a vector v on the unit sphere with $h(v) = 0$. Obviously, $v \in \mathbb{R}^n_- \setminus \{0\}$ and $\mathbb{R}_+ v \subset \Omega$. Finally, using the convexity of Ω, we see that for any open ball $V \subset \Omega$ with center a the following inclusion holds: $V + \mathbb{R}_+ v \subset \Omega$. □

Now we are in the position to proceed with the proof of Theorem 14.3.1.

Proof of Theorem 14.3.1. Observe that the following two implications (i) $\Longrightarrow$ (ii) $\Longrightarrow$ (iii) are obvious.

(iii) $\Longrightarrow$ (iv): Suppose that this implication is false. Then, by virtue of Proposition 13.3.8, we may assume that G is bounded and does not fulfill the Fu-condition (14.3.1). Moreover, without loss of generality, we only have to deal with the following situation:

$$\overline{G} \cap V_j \neq \varnothing, \text{ but } G \cap V_j = \varnothing, \quad j = 1, \dots, k,$$
$$\overline{G} \cap V_j = \varnothing, \quad j = k+1, \dots, n, \ 1 \le k \le n.$$

In fact, if $G \cap V_j \neq \varnothing$, then one can go to the intersection of G with those coordinate axes.

Hence, $G \subset \mathbb{C}^n_*$. We may also assume that $(1, \dots, 1) \in G$.

Observe that $\log G$ is convex, bounded into all positive directions, unbounded into the first k negative directions, and bounded in the remaining negative directions. Thus, by virtue of Lemma 14.3.3 we find a small ball $V = V(0) \subset \log G$ with center 0 and a vector $v \in \mathbb{R}^n_- \setminus \{0\}$ such that $V + \mathbb{R}_+ v \subset \log G$. It is clear that $v_j = 0$, $j = k+1, \dots, n$. Without loss of generality, we may assume that $v_j < 0$, $j = 1, \dots, \ell$, where $\ell \le k$, $v_1 = -1$, and $v_{\ell+1} = \dots v_n = 0$. Hence,

$$(e^{x_1} e^{-t}, e^{x_2} e^{t v_2}, \dots, e^{x_n} e^{t v_n}) \in G, \quad t > 0, \ x \in V.$$

Then, with $\alpha := -v$, we have an $\varepsilon > 0$ such that

$$(e^{\lambda}, \mu_2 e^{\lambda \alpha_2}, \dots, \mu_\ell e^{\lambda \alpha_\ell}, 1, \dots, 1) \in G, \quad \lambda \in H, \ e^{-\varepsilon} < |\mu_j| < e^{\varepsilon}, \ j = 2, \dots, \ell.$$

Put

$$A := \{(\mu_2, \dots, \mu_\ell) \in \mathbb{C}^{\ell-1} : e^{-\varepsilon} < |\mu_j| < e^{\varepsilon}, \ j = 2, \dots, \ell\},$$
$$H_R := \{\lambda \in \mathbb{C} : \operatorname{Re} \lambda < R\}, \quad R \ge 0,$$
$$\Phi : \mathbb{C} \times A \longrightarrow \mathbb{C}^\ell, \quad \Phi(\lambda, \mu) := (e^{\lambda}, \mu_2 e^{\lambda \alpha_2}, \dots, \mu_\ell e^{\lambda \alpha_\ell}).$$

It is clear that $\Phi(H_R \times A) =: D_R \subset \mathbb{C}^\ell_*$, $R \ge 0$, is a pseudoconvex Reinhardt domain with $D_R \underset{R \to \infty}{\nearrow} D_\infty := \Phi(\mathbb{C} \times A) \subset \mathbb{C}^\ell_*$. Therefore,

$$\ell_{D_\infty}(\Phi(-1, 1 \dots, 1), \Phi(\lambda, 1, \dots, 1)) \le \ell_{\mathbb{C}}(-1, \lambda) = 0, \quad \lambda \in \mathbb{C}.$$

By virtue of Lemma 14.3.3, $\boldsymbol{\ell}_{D_\infty}(\Phi(-1,1,\dots,1),z) = 0$ for all $z \in D_\infty \cap M$, where $M := \overline{\Phi(\mathbb{C} \times \{(1,\dots,1)\})}$.

Observe that $D_0 \times \{(1,\dots,1)\} \subset G$, but $(0,\dots,0,1,\dots,1) \notin G$. Now choose positive numbers a_j, $j \in \mathbb{N}$, with $\sum_{j=1}^{\infty} a_j < \infty$. It suffices to find points $z^j \in D_0$, $j \in \mathbb{N}$, with $\lim_{j\to\infty} z_1^j = 0$ such that

$$\boldsymbol{g}_G((z^j,1,\dots,1),(z^{j+1},1,\dots,1)) \le \boldsymbol{g}_{D_0}(z^j,z^{j+1}) \le a_j, \quad j \in \mathbb{N}.$$

Applying the fact that $\boldsymbol{\ell}_{D_R}$ is continuous on $G_R \times G_R$, the theorem of Dini, and that

$$\lim_{R\to\infty} \boldsymbol{\ell}_{D_R}(\Phi(-1,1,\dots,1),z) = \boldsymbol{\ell}_{D_\infty}(\Phi(-1,1,\dots,1),z) = 0,$$
$$z \in D_\infty \cap \overline{\Phi(\mathbb{C} \times \{(1,\dots,1)\})},\ e^{-2} < |z_1| < e^{-1},$$

we conclude that this convergence is a uniform one.

Hence we have a sequence $(R_j)_{j\in\mathbb{N}}$, $\lim_{j\to\infty} R_j = \infty$, such that

$$\boldsymbol{\ell}^*_{D_{R_j}}(\Phi(-1,1,\dots,1),\Phi(\lambda,1,\dots,1)) < a_j, \quad -2 \le \operatorname{Re}\lambda \le -1.$$

Observe that the mapping

$$\psi_R : D_0 \longrightarrow D_R, \quad \psi(z) := (e^R z_1, z_2 e^{\alpha_2 R}, \dots, z_\ell e^{\alpha_\ell R}),$$

is biholomorphic. Therefore,

$$\boldsymbol{\ell}^*_{D_0}(\Phi(-1-R_j,1\dots,1),\Phi(\lambda,1,\dots,1)) < a_j, \quad -2-R_j \le \operatorname{Re}\lambda \le -1-R_j.$$

Define

$$u_j(\lambda) := \log \boldsymbol{g}_{D_0}(\Phi(-1-R_j,1,\dots,1),\Phi(\lambda,1,\dots,1)), \quad \lambda \in H_0.$$

Observe that $u_j \in \mathcal{SH}(H_0)$. By virtue of Appendix B.4.8, it follows that $u_j(\lambda) < \log a_j$ whenever $\operatorname{Re}\lambda \le -1-R_j$. Therefore, we may take $z^j := \Phi(-1-R_j,1,\dots,1)$ as the desired point-sequence.

(iv) $\Longrightarrow$ (i): Without loss of generality we may assume that

(a) $G \subset \mathbb{D}^n$,

(b) $G \cap V_j \neq \varnothing$ for $j = 1,\dots,m$,

(c) $\overline{G} \cap V_j = \varnothing$ for $j = m+1,\dots,n$, where $m \in \{0,1,\dots,n\}$.

Because of (c) one may find a positive ε_0 such that, if $z \in G$, then $|z_j| \ge \varepsilon_0$ for all $j = m+1,\dots,n$.

Now suppose that (i) does not hold. Then there exists a $\boldsymbol{c}_G$-Cauchy sequence $(z^\nu)_{\nu\in\mathbb{N}} \subset G$ with $\lim_{\nu\to\infty} z^\nu = z^0 = (z_1^0,\dots,z_n^0) \in \partial G$.

Step 1°. Assume that $z_1^0 \cdots z_n^0 \neq 0$. Then the real point $x^0 := (\log|z_1^0|, \dots, \log|z_n^0|)$ becomes a boundary point of the convex domain $\log G$. Its convexity implies the existence of a linear functional $L : \mathbb{R}^n \longrightarrow \mathbb{R}$ satisfying $L(x) = \sum_{j=1}^n \xi_j x_j < L(x^0) =: C$ on $\log G$.

Because of (b), we obtain $\xi_j \geq 0$, $j = 1, \dots, m$. Without loss of generality we may assume that $\xi_1, \dots, \xi_\ell$ are positive while the ξ_j's vanish for $j = \ell+1, \dots, m$ for a suitable $\mathbb{Z} \ni \ell \in \{0, \dots, m\}$.[1] Fix an $N \in \mathbb{N}$. Then applying the Dirichlet pigeon hole principle (cf. Appendix B.10.3), we choose natural numbers $\beta_{N,1}, \dots, \beta_{N,n}, q_N \in \mathbb{Z}$ with $\operatorname{sgn} \beta_{N,j} = \operatorname{sgn} \xi_j$ for $j = 1, \dots, n$ such that $q_N > N$ and

$$\left| \xi_\nu - \frac{\beta_{N,\nu}}{q_N} \right| \leq \frac{1}{N q_N}, \quad 1 \leq \nu \leq n.$$

Put

$$f_N(z) := e^{-C q_N} z_1^{\beta_{N,1}} \dots z_n^{\beta_{N,n}}.$$

Then, by virtue of (b), $f_N \in \mathcal{O}(G) \cap \mathcal{C}(\overline{G})$. Moreover, let $g_N := f_N / \|f_N\|_G$; thus $g_N \in \mathcal{O}(G, \mathbb{D})$.

Set $\delta := \min\{|z_j^0|/2 : j = 1, \dots, n\}$. If $z \in G \cap \mathbb{P}(z^0, \delta)$, then we establish the following inequality:

$$\begin{aligned}
\log|f_N(z)| &= -C q_N + \sum_{\nu=1}^n \beta_{N,\nu} \log|z_\nu| \\
&= -q_N\big(C - L(\log|z_1|, \dots, \log|z_n|)\big) + q_N \sum_{\nu=1}^n \left(\frac{\beta_{N,\nu}}{q_N} - \xi_\nu \right) \log|z_\nu| \\
&\geq -q_N\big(C - L(\log|z_1|, \dots, \log|z_n|)\big) - n|\log\delta|/N,
\end{aligned}$$

i.e.,

$$|f_N(z)| \geq \exp\big(-q_N\big(C - L(\log|z_1|, \dots, \log|z_n|)\big) - n|\log\delta|/N\big).$$

Now, we study points $z = (z_1, \dots, z_n) \in G \cap \mathbb{C}_*^n =: G^*$. Observe first that there exists a positive $\varepsilon_1 \leq \delta$ such that if $z \in G^*$ and $0 < |z_{j_0}| \leq \varepsilon_1$ for at least one $j_0 \in \{1, \dots, \ell\}$, then $-C q_N + \sum_{\nu=1}^n \beta_{N,\nu} \log|z_\nu| \leq 0$. Indeed, otherwise we would have a sequence $(z^s)_{s \in \mathbb{N}} \subset G^*$ such that, without loss of generality, $|z_1^s| < 1/s$ and,

[1] Note that $\ell = 0$ is possible. Then, some part of the following argument has to be slightly modified.

since the $\beta_{N,j}$'s are positive for $j = 1, \dots, \ell$,

$$\begin{aligned}
Cq_N &\le \beta_{N,1} \log |z_1^s| + \sum_{j=m+1}^{n} \beta_{N,j} \log |z_j^s| \\
&\le \beta_{N,1} \log(1/s) + \sum_{j=m+1}^{n} |\beta_{N,j}| |\log |z_j^s|| \\
&\le \beta_{N,1} \log(1/s) + \sum_{j=m+1}^{n} |\beta_{N,j}| |\log \varepsilon_0| \underset{s\to\infty}{\longrightarrow} -\infty;
\end{aligned}$$

a contradiction.

Therefore, it remains to investigate points $z \in G^*$ with $|z_j| \ge \varepsilon_1$, $1 \le j \le \ell$. For such points we have

$$\begin{aligned}
\log |f_N(z)| &= -q_N\big(C - L(\log |z_1|, \dots, \log |z_n|)\big) + q_N \sum_{\nu=1}^{\ell} \left(\frac{\beta_{N,\nu}}{q_N} - \xi_\nu\right) \log |z_\nu| \\
&\quad + q_N \sum_{\nu=m+1}^{n} \left(\frac{\beta_{N,\nu}}{q_N} - \xi_\nu\right) \log |z_\nu| \\
&\le (\ell |\log \varepsilon_1| + (n-m)|\log \varepsilon_0|)/N.
\end{aligned}$$

Therefore,

$$|f_N(z)| \le \exp(M/N), \quad z \in G^*,$$

where $M := \ell |\log \varepsilon_1| + (n-m)|\log \varepsilon_0|$. But then the same estimate holds for all $z \in G$.

This implies for $\nu \gg 1$ (i.e., for $z^\nu \in G \cap \mathbb{P}(z^0, \delta)$) the following estimate

$$|g_N(z^\nu)| \ge \exp\big(-q_N(C - L(\log |z_1^\nu|, \dots, \log |z_n^\nu|)) - (n|\log \delta| + M)/N\big).$$

If $\ell \ge 1$, then we may fix a point $a = (0, a_2, \dots, a_n) \in G$. Thus, $g_N(a) = 0$. Choosing N first and then taking sufficiently large ν, we obtain holomorphic functions $g_N \in \mathcal{O}(G, \mathbb{D})$, $g_N(0) = 0$, with values almost one at z^ν. Hence, $\boldsymbol{c}_G(a, z^\nu) \longrightarrow \infty$; a contradiction.

Let us now assume that $\ell = 0$. Then fix a point $a \in G \cap \mathbb{P}(z^0, \delta)$ and observe that

$$\begin{aligned}
|g_N(a)| &\le \exp\big(-q_N(C - L\big(\log |a_1|, \dots, \log |a_n|)) - (n|\log \delta| + M)/N\big) \\
&\le \exp\left(-2\alpha + (n|\log \delta| + M)/N\right) \le \exp(-\alpha) < 1 \text{ for large } N,
\end{aligned}$$

where $2\alpha := C - L\big(\log |a_1|, \dots, \log |a_n|) > 0$. Then, choosing N and ν as above, it follows that $\boldsymbol{c}_G(a, z^\nu) \longrightarrow \infty$; a contradiction.

Step 2°. It remains to consider the case that $z^0 \in V_1 \cup \dots \cup V_m$. Without loss of generality we may assume that then $z_1^0 = \dots = z_k^0 = 0$, and the next $m - k$ coordinates are different from 0. Projecting $\mathbb{C}^n$ onto the space $\mathbb{C}^k$ by $(z_1, \dots, z_n) \overset{\pi}{\longmapsto} (z_1, \dots, z_k)$, we obtain a new Reinhardt domain $\pi(G) =: G'$ sharing all properties of G. Then, the result in Step 1° together with $\boldsymbol{c}_{G'}(\pi(a), \pi(z^\nu)) \le \boldsymbol{c}_G(a, z^\nu)$ finishes the proof. □

Remark 14.3.4. Even more is true, namely *any hyperbolic Reinhardt domain is* $\boldsymbol{k}$*-complete*. The proof is similar to the former one and can be found in [181].

Remark 14.3.5. Let $G \subset \mathbb{C}^2$ be a bounded pseudoconvex Reinhardt domain, $a \in G$, and $z^0 \in \partial G \cap \mathbb{C}_*^2$. Then $\boldsymbol{c}_G(a, z) \underset{z \to z^0}{\longrightarrow} \infty$ (see [565]). So, that part of ∂G not lying on a coordinate axis is $\boldsymbol{c}_G$-infinitely far away from any point of G.

We point out that this phenomenon does not hold in higher dimensions.

Example 14.3.6 (cf. [565]). Let $\alpha > 0$ be an irrational number. Put

$$G := \{z \in \mathbb{C}^3 : |z_1||z_2|^\alpha |z_3|^{\alpha+1} < 1,\ |z_2||z_3| < 1,\ |z_3| < 1\}.$$

Then G is a pseudoconvex Reinhardt domain. Fixing points $z^0 \in G \cap \mathbb{C}_*^3$ and $w \in \mathbb{C}_*^3$ with $|w_1||w_2|^\alpha |w_3|^{\alpha+1} = 1$, $|w_2|^{-1}|w_3|^2 < 1$, and $|w_3| < 1$, we get

$$\limsup_{G \ni z \to w} \boldsymbol{c}_G(z^0, z) < \infty.$$

Moreover, the biholomorphic map

$$\Phi : G \cap \mathbb{C}_*^3 \longrightarrow \mathbb{C}_*^3, \quad \Phi(z) := (z_1 z_2^{\lfloor \alpha \rfloor + 1} z_3^{\lfloor \alpha \rfloor + 3}, z_2 z_3^2, z_3),$$

has as its image a bounded pseudoconvex Reinhardt domain G^* contained in

$$\{z \in \mathbb{C}_*^3 : |z_2| < 1,\ |z_3| < 1,\ |z_1||z_2|^{\alpha - \lfloor \alpha \rfloor - 1} |z_3|^{\lfloor \alpha \rfloor - \alpha} < 1\}.$$

In the class of Reinhardt domains, we have the following characterization of hyperconvexity (cf. [566] and [83]):

Theorem 14.3.7. *Let $G \subset \mathbb{C}^n$ be a pseudoconvex Reinhardt domain. Then the following conditions are equivalent:*

(i) *G is hyperconvex;*

(ii) *G is bounded and fulfills the Fu-condition.*

Proof. The direction (ii) $\Longrightarrow$ (i) follows directly from Theorem 14.3.1(i).

To prove the converse, suppose that G does not satisfy the conditions in (ii). According to Proposition 13.3.8, we may assume that G is bounded and does not satisfy the Fu condition. Hence, without loss of generality, we may assume that $G = D_0$ (compare the proof of Theorem 14.3.1), i.e.,

$$G := \{(\zeta, \mu_2\zeta^{\alpha_2}, \dots, \mu_n\zeta^{\alpha_n}) \in \mathbb{C}^n : \zeta \in \mathbb{D}_*,\ \mu_j \in \mathbb{C},\ e^{-\varepsilon_0} < |\mu_j| < e^{\varepsilon_0},\ j = 2, \dots, n\},$$

where $\varepsilon_0 > 0$, $\alpha_j > 0$, $j = 2, \dots, n$.

Let $u \in \mathcal{PSH}(G) \cap \mathcal{C}(G)$, $u < 0$, be such that $\{z \in G : u(z) < -\varepsilon\} \subset\subset G$ for any $\varepsilon > 0$. Define

$$v(z) := \sup\{u(z_1 e^{i\theta_1}, \dots, z_n e^{i\theta_n}) : \theta_j \in \mathbb{R}\}.$$

Obviously, v is an exhausting function of G with $v(z) = v(|z_1|, \dots, |z_n|)$. Therefore, the function

$$\mathbb{D}_* \ni \lambda \longmapsto v(|\lambda|, |\lambda|^{\alpha_2}, \dots, |\lambda|^{\alpha_n})$$

is subharmonic and bounded from above by 0. Hence, it can be continued as a function $v^* \in \mathcal{SH}(\mathbb{D})$. Then, by virtue of the hyperconvexity of G, it follows that $v^*(0) = 0$ implying that $v = 0$; a contradiction. □

Recall that the Carathéodory distance is not inner. So, in general, we have $\boldsymbol{c}_G \le \boldsymbol{c}^i_G$, $\boldsymbol{c}_G \ne \boldsymbol{c}^i_G$ (cf. § 1.2.1). Moreover, it is known that $\boldsymbol{c}^i_G = \int\boldsymbol{\gamma}_G = \int\boldsymbol{\gamma}^{(1)}_G \le \int\boldsymbol{\gamma}^{(k)}_G$, $k \in \mathbb{N}$. Thus,

$$\boldsymbol{c}\text{–complete} \Longrightarrow \boldsymbol{c}^i\text{-complete} \Longrightarrow \int\boldsymbol{\gamma}^{(k)}\text{-complete},\ k \in \mathbb{N}.$$

For Reinhardt domains, there is even the following result (see [569, 545]):

Theorem 14.3.8. *Let $G \subset \mathbb{C}^n$ be a bounded pseudoconvex Reinhardt domain. Then the following properties are equivalent:*

(i) *G is $\boldsymbol{c}$-complete;*

(ii) *G is $\boldsymbol{c}^i$-complete;*

(iii) *G is $\int\boldsymbol{\gamma}^{(k)}$-complete, $k \in \mathbb{N}$;*

(iv) *there is a $k \in \mathbb{N}$ such that G is $\int\boldsymbol{\gamma}^{(k)}$-complete.*

In order to be able to prove Theorem 14.3.8, we first recall a fact on multi-dimensional Vandermonde's determinants (for example, see [476]), namely,

Let $X_s := (s, \dots, s) \in \mathbb{C}^n$, $s \in \mathbb{N}$,[2] and $N_k := \#\{\alpha \in \mathbb{Z}_+^n : |\alpha| \le k\}$, $k \in \mathbb{N}$. Then,

$$\det\left([X_s^\alpha]_{1\le s\le N_k,\, |\alpha|\le k}\right) \neq 0. \tag{14.3.2}$$

Using this information, we get the following

Lemma 14.3.9. *Let $P(z) = \sum_{1\le|\beta|\le k} b_\beta z^\beta$, $z \in \mathbb{C}^n$, be a polynomial in $\mathbb{C}^n$. Then there are numbers $(N_j)_{1\le j\le k} \subset \mathbb{N}$, $(c_{j,s})_{1\le j\le k,\, 1\le s\le N_j} \subset \mathbb{C}$, and vectors $(X_{j,s})_{1\le j\le k,\, 1\le s\le N_j} \subset \mathbb{C}^n$ such that*

$$P(z) = \sum_{j=1}^{k}\sum_{s=1}^{N_j} c_{j,s} \sum_{|\beta|=j} \frac{j!}{\beta!} p_\beta(z) X_{j,s}^\beta, \quad z = (z_1, \dots, z_n) \in \mathbb{C}^n,$$

where

$$p_\beta(z) := \prod_{j=1}^{n} p_{\beta,j}(z), \quad p_{\beta,j}(z) := z_j(z_j - 1)\cdots(z_j - \beta_j + 1).$$

Proof. The proof is by induction on $k \in \mathbb{N}$. Obviously, the case $k = 1$ is true. So we may assume that Lemma 14.3.9 holds for a $k \in \mathbb{N}$. Now take a polynomial $P(z) = \sum_{1\le|\beta|\le k+1} b_\beta z^\beta$, $z \in \mathbb{C}^n$, and write

$$P(z) = \sum_{1\le|\beta|\le k} b_\beta z^\beta + \sum_{|\beta|=k+1} b_\beta (z^\beta - p_\beta(z)) + \sum_{|\beta|=k+1} b_\beta p_\beta(z), \quad z \in \mathbb{C}^n.$$

Observe that the first two terms are of degree less than or equal to k. The third one may be written as

$$\sum_{|\beta|=k+1} b_\beta p_\beta(z) = \sum_{|\beta|=k+1} \frac{(k+1)!}{\beta!} p_\beta(z) \frac{\beta! b_\beta}{(k+1)!}.$$

Using (14.3.2), we find $(c_s)_{1\le s\le N_{k+1}} \subset \mathbb{C}$ and $(X_s)_{1\le s\le N_{k+1}} \subset \mathbb{C}^n$ such that

$$\frac{\beta! b_\beta}{(k+1)!} = \sum_{s=1}^{N_{k+1}} c_s X_s^\beta, \quad |\beta| = k+1.$$

Hence, Lemma 14.3.9 has been proved. □

Proof of Theorem 14.3.8. It remains to prove (iv) $\Longrightarrow$ (i).

We may assume that $n \ge 2$. Suppose that G does not fulfill the Fu-condition, but is $\int\gamma^{(k)}$-complete for a suitable k. Then we may assume that there exists an $\ell \in \mathbb{N}$ such that

[2] Notice that here X_s is a vector and not the s-th coordinate of a vector.

- $(1,\dots,1) \in G$;
- $\overline{G} \cap V_j \neq \varnothing$, but $G \cap V_j = \varnothing$, $\quad j = 1,\dots,\ell$;
- $\overline{G} \cap V_j = \varnothing$, $\quad j = \ell+1,\dots,n$.

In particular, we have $G \subset \mathbb{C}_*^n$. Indeed, in general we have natural numbers $1 \leq \ell \leq m \leq n$ such that

- $\overline{G} \cap V_j \neq \varnothing$, but $G \cap V_j = \varnothing$, $\quad j = 1,\dots,\ell$;
- $\overline{G} \cap V_j = \varnothing$, $\quad j = \ell+1,\dots,m$;
- $G \cap V_j \neq \varnothing$, $\quad j = m+1,\dots,n$.

Put $\widetilde{G} := \{\widetilde{z} \in \mathbb{C}^m : (\widetilde{z}, 0) \in G\}$. Then, $\widetilde{G}$ is a bounded pseudoconvex Reinhardt domains, which does not fulfill the Fu-condition. Moreover, note that $\int \boldsymbol{\gamma}_{\widetilde{G}}^{(k)}(\widetilde{z}, \widetilde{w}) \geq \int \boldsymbol{\gamma}_{G}^{(k)}((\widetilde{z}, 0), (\widetilde{w}, 0))$. Thus, if $(\widetilde{z}_j)_j \subset \widetilde{G}$ is a $\int \boldsymbol{\gamma}_{\widetilde{G}}^{(k)}$-Cauchy sequence, then $((\widetilde{z}_j, 0))_j \subset G$ is a $\int \boldsymbol{\gamma}_{G}^{(k)}$-Cauchy sequence. Thus, it converges to a point $(\widetilde{z}, 0) \in G$, i.e., $\widetilde{z}_j \longrightarrow \widetilde{z} \in \widetilde{G}$. Hence we may deal with the case $G = \widetilde{G}$.

In particular, we have $G \subset \mathbb{C}_*^n$; thus $\log G$ is unbounded and convex and it contains the origin. Again applying Lemma 14.3.3, there are a neighborhood $V = V(0) \subset \mathbb{R}^n$ and a vector $v \in \mathbb{R}_+^n$, $v \neq 0$, such $V - \mathbb{R}_+ v \subset G$. By virtue of the geometry of G one concludes that there is an $\mathbb{N} \ni m \leq \ell$ such that $v_j > 0$, $j = 1,\dots,m$, and $v_{m+1} = \dots v_n = 0$. Rescaling allows us to assume that $v_1 = 1$. Finally, choose a small positive ε such that $(-\varepsilon,\varepsilon)^n \subset V$. Put

$$D := \{z \in \mathbb{C}_*^n : (\log|z_1|,\dots,\log|z_n|) \in \{0\} \times (-\varepsilon,\varepsilon)^{n-1} - \mathbb{R}_+ v\}.$$

Then, D is a bounded pseudoconvex Reinhardt domain sitting inside of G. Put

$$\chi : (0,1) \longrightarrow G, \quad \chi(t) := (t^{v_1},\dots,t^{v_n}).$$

Then, χ gives a curve inside of D with endpoint $\chi(0) \in \partial G$. It remains to show that $L_{\gamma_D^{(k)}}(\chi|_{(0,1/2]}) < \infty$ to get a $\int \boldsymbol{\gamma}_D^{(k)}$-Cauchy sequence, and so a $\int \boldsymbol{\gamma}_G^{(k)}$-Cauchy sequence not converging in G, which would lead to the desired contradiction.

For a fixed $t \in (0,1)$, we are going to estimate $\boldsymbol{\gamma}_G^{(k)}(\chi(t); \chi'(t))$.

Fix an $f \in \mathcal{O}(D,\mathbb{D})$, $\operatorname{ord}_{\chi(t)} f \geq k$. Then, using Laurent expansion, we get

$$f(z) = \sum a_\alpha z^\alpha, \text{ where } \quad a_\alpha = \frac{1}{(2\pi i)^n} \int_{|\zeta_1|=r_1,\dots,|\zeta_n|=r_n} \frac{f(\zeta)d\zeta_1\dots d\zeta_n}{\zeta^{\alpha+1}}$$

is independent of $r = (r_1,\dots,r_n) \in D \cap \mathbb{R}_{>0}^n$. Note that

$$|a_\alpha| \leq \frac{1}{r^\alpha} \text{ for any } r \in D. \tag{14.3.3}$$

From (14.3.3), it follows that $a_\alpha = 0$ if $\langle \alpha, v\rangle < 0$ (recall that the monomial z^α, $\alpha \in \mathbb{Z}^n$, is bounded on D if and only if $\langle \alpha, v\rangle \geq 0$). Therefore,

$$f(z) = \sum_{\alpha\in\mathbb{Z}^n:\ \langle\alpha,v\rangle\geq 0} a_\alpha z^\alpha, \quad z \in D.$$

Taking $r_1 < 1$ in (14.3.3) arbitrarily large and r_j arbitrarily close to δ (or to δ^{-1}), $j = 2, \dots, n$, then

$$|a_\alpha| \leq \delta^{|\alpha_2|+\dots+|\alpha_n|}.$$

Let $s = 0, \dots, k$ and $X = (X_1, \dots, X_n) \in \mathbb{C}^n$. Then, taking derivatives, we have

$$\frac{1}{s!} f^{(s)}(\chi(t))(X_1 t^{v_1-\frac{k}{s}}, \dots, X_n t^{v_n-\frac{k}{s}}) = \sum_{\langle\alpha,v\rangle\geq 0} \Big(a_\alpha \sum_{|\beta|=s} \frac{1}{\beta!} p_\beta(\alpha) X^\beta t^{\langle\alpha,v\rangle-k}\Big),$$

where $p_\beta(\alpha) := p_{\beta_1}(\alpha_1)\cdots p_{\beta_n}(\alpha_n)$ and $p_{\beta_j}(\alpha_j) := \alpha_j(\alpha_j-1)\cdots(\alpha_j-\beta_j+1)$, $j = 1, \dots, n$.

Since, by assumption, $\operatorname{ord}_{\chi(t)} f \geq k$, it follows that

$$\sum_{\langle\alpha,v\rangle\geq 0} a_\alpha t^{\langle\alpha,v\rangle-k} \Big(\sum_{|\beta|=s} \frac{1}{\beta!} p_\beta(\alpha) X^\beta\Big) = 0, \quad 0 \leq s < k,\ X \in \mathbb{C}^n. \tag{14.3.4}$$

Moreover,

$$\begin{aligned}\frac{1}{k!} f^{(k)}(\chi(t))(\chi'(t)) &= \frac{1}{k!} \sum_{\langle\alpha,v\rangle\geq 0} a_\alpha \langle\alpha, v\rangle^k t^{\langle\alpha,v\rangle-k} \\ &+ \frac{1}{k!} \sum_{\langle\alpha,v\rangle\geq 0} a_\alpha t^{\langle\alpha,v\rangle-k} \sum_{|\beta|=k} \frac{k!}{\beta!} (p_\beta(\alpha) - \alpha^\beta) v^\beta. \end{aligned} \tag{14.3.5}$$

Applying Lemma 14.3.9 and (14.3.4) shows that the second term in (14.3.5) vanishes. Hence, we have

$$\frac{1}{k!} f^{(k)}(\chi(t))(\chi'(t)) = \frac{1}{k!} \sum_{\langle\alpha,v\rangle\geq 0} a_\alpha \langle\alpha, v\rangle^k t^{\langle\alpha,v\rangle-k}.$$

By virtue of the above estimate, we obtain the following inequality:

$$\gamma_D^{(k)}(\chi(t)); \chi'(t)) \leq \frac{1}{\sqrt[k]{k!}} \sum_{\alpha\in\mathbb{Z}^n:\langle\alpha,v\rangle>0} \delta^{(|\alpha_2|+\dots+|\alpha_n|)/k} \langle\alpha, v\rangle t^{(\langle\alpha,v\rangle/k)-1}, \ t \in (0,1).$$

What remains is to show that $L_{\gamma_D^{(k)}}(\chi|_{(0,1/2]})$ is finite, which would give the desired contradiction.

So, we have the following estimate:

$$
\begin{aligned}
L_{\gamma_D^{(k)}}(\chi|_{(0,1/2]}) &\le \int_0^{1/2} \gamma_D^{(k)}(\chi(t);\chi'(t))dt \\
&\le \frac{1}{\sqrt[k]{k!}} \sum_{\langle\alpha,v\rangle\ge 0} \delta^{(|\alpha_2|+\dots+|\alpha_n|)/k} \int_0^{1/2} \langle\alpha, v\rangle t^{(\langle\alpha,v\rangle/k)-1} dt \\
&= \frac{1}{\sqrt[k]{k!}} \sum_{\langle\alpha,v\rangle>0} \delta^{(|\alpha_2|+\dots+|\alpha_n|)/k} \frac{k}{2^{\langle\alpha,v\rangle/k}} \\
&= \frac{k}{\sqrt[k]{k!}} \sum_{\alpha_2,\dots,\alpha_n\in\mathbb{Z}} \delta^{(|\alpha_2|+\dots+|\alpha_n|)/k} \sum_{\alpha_1\in\mathbb{Z}:\alpha_1>-\langle\alpha',v'\rangle} \frac{1}{2^{\langle\alpha,v\rangle/k}} \\
&\le \frac{k}{\sqrt[k]{k!}} \frac{1}{2^{1/k}-1} \sum_{\alpha_2,\dots,\alpha_n\in\mathbb{Z}} \delta^{(|\alpha_2|+\dots+|\alpha_n|)/k} \frac{1}{2^{(\langle\alpha',v'\rangle+\lfloor-\langle\alpha',v'\rangle\rfloor)/k}},
\end{aligned}
$$

where $\alpha' := (\alpha_2,\dots,\alpha_n)$, $v' := (v_2,\dots,v_n)$. Obviously, the last number is finite, which finishes the proof. □

Remark 14.3.10. Observe that in the case where $v \in \mathbb{Q}^n$ the above proof may be essentially simplified. Namely, the punctured unit disc can then be embedded into D. So the fact that D is not $\int \gamma^{(k)}$-complete follows immediately from the non-completeness of $\mathbb{D}_*$.

Remark 14.3.11. Let $G \subset \mathbb{C}^n$ be an arbitrary domain and $A \subset G$ finite. In generalization of the notion of $\boldsymbol{c}_G$-finite compactness we say that G is *$\boldsymbol{m}_G(A,\cdot)$-finitely compact* if for any $R > 0$ the set $\{z \in G : \boldsymbol{m}_G(A,z) < R\}$ is relatively compact in G. Obviously, any $\boldsymbol{m}_G(A,\cdot)$-finitely compact domain is $\boldsymbol{c}_G$-finitely compact. Is there a geometrical characterization for $\boldsymbol{m}_G(A,\cdot)$-finite compactness in the class of all pseudoconvex Reinhardt domains as in Theorem 14.3.1?

?

14.4 $\int \gamma^{(k)}$-completeness for Zalcman domains

First, let us introduce the class of domains we will study in this section. Take two sequences $(a_j)_{j\in\mathbb{N}}$ and $(r_j)_{j\in\mathbb{N}}$ of positive real numbers such that:

- $2r_j < a_j$, $j \in \mathbb{N}$,
- $a_j \underset{j\to\infty}{\searrow} 0$,
- $\overline{\mathbb{D}}(a_j, r_j) \subset \mathbb{D}$, $\overline{\mathbb{D}}(a_j, r_j) \cap \overline{\mathbb{D}}(a_k, r_k) = \varnothing$, $j \neq k$.

Then, $G := \mathbb{D}_* \setminus \bigcup_{j=1}^{\infty} \overline{\mathbb{D}}(a_j, r_j)$ is called a *Zalcman type domain*. Obviously, such a G is $\boldsymbol{c}$-hyperbolic.

The main result here is the following one due to P. Zapałowski (see [544] and [546]):

Theorem 14.4.1. *For any $k \in \mathbb{N}$ there exists a Zalcman type domain G which is $\int\gamma^{(\ell)}$-complete, but not $\int\gamma^{(m)}$-complete, whenever $m \le k < \ell$.*

Remark 14.4.2.

(a) It seems to be an open problem whether for different $k, l \in \mathbb{N}$, $k < l$, there exists a Zalcman type domain G, which is $\int\gamma^{(k)}$-complete, but not $\int\gamma^{(l)}$-complete. Note that for $l = sk$, $s \in \mathbb{N}$, such a phenomenon is impossible, because of $\int\gamma^{(k)} \le \int\gamma^{(sk)}$. **?**

(b) According to our knowledge, there is also no characterization of $\int\gamma^{(k)}$-complete Zalcman type domains. **?**

Before giving the proof of Theorem 14.4.1, we mention the following sufficient condition for a Zalcman type domain to be not $\int\gamma^{(k)}$-complete.

Proposition 14.4.3. *Let $G \subset \mathbb{C}$ be a Zalcman type domain (as above) and let $k \in \mathbb{N}$, $\alpha \in (0,1)$, and $c > 0$. Assume that*

$$\gamma_G^{(k)}(t;1) \le c|t|^{-\alpha}, \quad t \in (-1,0). \tag{14.4.1}$$

Then G is not $\int\gamma^{(\ell)}$-complete for any $\ell \ge k$.

Proof. Fix an $\ell \in \mathbb{N}$, $\ell \ge k$, and a point $t \in (-1,0)$. Take an $f \in \mathcal{O}(G,\mathbb{D})$ with $f(t) = f'(t) = \cdots = f^{(\ell-1)}(t) = 0$ such that $(\gamma_G^{(\ell)}(t;1))^\ell = \frac{1}{\ell!}|f^{(\ell)}(t)|$ and define

$$g(z) = \begin{cases} \frac{f(z)}{(z-t)^{\ell-k}}, & \text{if } z \ne t \\ 0, & \text{if } z = t \end{cases}.$$

Then g is holomorphic on G with

$$g^{(m)}(t) = 0, \quad m = 0,\dots,k-1, \text{ and } g^{(k)}(t) = \frac{k!}{\ell!} f^{(\ell)}(t).$$

Moreover, by virtue of the maximum principle, we have

$$\|g\|_G \le \operatorname{dist}(t,\partial G)^{-(\ell-k)}.$$

Therefore, with $h := g \operatorname{dist}(t,\partial G)^{\ell-k} \in \mathcal{O}(G,\mathbb{D})$ we obtain

$$\begin{aligned}\left(\gamma_G^{(k)}(t;1)\right)^k \ge \frac{1}{k!}|h^{(k)}(t)| &= \frac{\operatorname{dist}(t,\partial G)^{\ell-k}}{\ell!}|f^{(\ell)}(t)| \\ &= \operatorname{dist}(t,\partial G)^{(\ell-k)}\left(\gamma_G^{(\ell)}(t;1)\right)^\ell. \end{aligned} \tag{14.4.2}$$

Finally, from the assumed inequality (14.4.1), the following estimate follows:

$$\boldsymbol{\gamma}_G^{(\ell)}(t;1) \le \frac{c^{k/\ell}|t|^{-\alpha k/\ell}}{|t|^{(\ell-k)/\ell}} = c'|t|^{-\alpha'},$$

where $c' := c^{k/\ell}$ and $\alpha' := (\alpha k + (\ell - k))/\ell < 1$. Then, integrating along the segment $(-1/2, 0)$ shows that G is not $\int \boldsymbol{\gamma}^{(\ell)}$-complete. □

Consequently, to find examples as claimed in Theorem 14.4.1, we should try to deal with situations where the boundary behavior of $\boldsymbol{\gamma}_G^{(k)}$ is of the type

$$\boldsymbol{\gamma}_G^{(k)}(\cdot;1) \le c \operatorname{dist}(\cdot, \partial G)^{-1} |\log \operatorname{dist}(\cdot, \partial G)|^{-\alpha},$$

with some $\alpha > 1$, $c > 0$.

Lemma 14.4.4. *Let $G \subset \mathbb{C}$ be a Zalcman type domain and $k \in \mathbb{N}$. Then there exists a $C > 0$ such that*

$$|f^{(k)}(z)| \le C\left(1 + \sum_{j=1}^{\infty} \frac{r_j}{(a_j - z)^{k+1}}\right), \quad z \in (-\tfrac{1}{2}, 0), \ f \in \mathcal{O}(G, \mathbb{D}).$$

Proof. Choose numbers $\widetilde{a}_j \in (0, a_j)$ and $\widetilde{r}_j \in (\widetilde{a}_j, 1)$ such that

$$\overline{\mathbb{D}}(a_s, r_s) \subset \mathbb{D}(\widetilde{a}_j, \widetilde{r}_j), \ s > j, \text{ and } \overline{\mathbb{D}}(\widetilde{a}_j, \widetilde{r}_j) \cap \overline{\mathbb{D}}(a_j, r_j) = \varnothing.$$

Put

$$G_j := \mathbb{D} \setminus \left(\overline{\mathbb{D}}(\widetilde{a}_j, \widetilde{r}_j) \cup \bigcup_{s=1}^{j} \overline{\mathbb{D}}(a_s, r_s)\right).$$

Obviously, G_j is a $(j+2)$-connected domain with $G_j \subset G$, $j \in \mathbb{N}$. Then, for sufficiently small positive ε_j's (we may assume that $\varepsilon_j \underset{j\to\infty}{\longrightarrow} 0$), we have

$$G_{j,\varepsilon_j} := \mathbb{D}(1 - \varepsilon_j) \setminus \left(\overline{\mathbb{D}}(\widetilde{a}_j, \widetilde{r}_j + \varepsilon_j) \cup \bigcup_{s=1}^{j} \overline{\mathbb{D}}(a_s, r_s + \varepsilon_j)\right) \subset\subset G_j.$$

By virtue of the Cauchy integral, it follows that

$$\begin{aligned} f^{(k)}(z) = \frac{k!}{2\pi i} \int_{|\zeta|=1-\varepsilon_j} \frac{f(\zeta)}{(\zeta - z)^{k+1}} d\zeta - \frac{k!}{2\pi i} \int_{|\zeta - \widetilde{a}_j| = \widetilde{r}_j + \varepsilon_j} \frac{f(\zeta)}{(\zeta - z)^{k+1}} d\zeta \\ - \sum_{s=1}^{j} \frac{k!}{2\pi i} \int_{|\zeta - a_s| = r_s + \varepsilon_j} \frac{f(\zeta)}{(\zeta - z)^{k+1}} d\zeta, \quad z \in G_{j,\varepsilon_j}, \ f \in \mathcal{O}(G, \mathbb{D}). \end{aligned}$$

Now fix a $z \in (-1/2, 0)$. Then, $z \in (-1/2, \widetilde{a}_j - \widetilde{r}_j - \varepsilon_j - \sqrt[2(k+1)]{\widetilde{r}_j + \varepsilon_j})$ and $z < -\varepsilon_j$ for all sufficiently large j. Hence, we obtain

$$
\begin{aligned}
|f^{(k)}(z)| &\le \frac{k!}{2\pi}\int_0^{2\pi}\frac{1-\varepsilon_j}{|(1-\varepsilon_j)e^{it}-z|^{k+1}}dt \\
&\quad + \frac{k!}{2\pi}\int_0^{2\pi}\frac{\widetilde{r}_j+\varepsilon_j}{|(\widetilde{r}_j+\varepsilon_j)e^{it}+\widetilde{a}_j-z|^{k+1}}dt \\
&\quad + \sum_{s=1}^{j}\frac{k!}{2\pi}\int_0^{2\pi}\frac{r_s+\varepsilon_j}{|(r_s+\varepsilon_j)e^{it}+a_s-z|^{k+1}}dt \\
&\le k!\Bigg(\frac{1-\varepsilon_j}{(1/2-\varepsilon_j)^{k+1}}+\frac{\widetilde{r}_j+\varepsilon_j}{(\sqrt[2(k+1)]{\widetilde{r}_j+\varepsilon_j})^{k+1}} \\
&\quad + \sum_{s=1}^{j}\frac{r_s+\varepsilon_j}{(1/2(a_s-z-\varepsilon_j))^{k+1}}\Bigg).
\end{aligned}
$$

Observe that here the assumption $2r_j < a_j$, $j \in \mathbb{N}$, is used to estimate the third term.

Since $\varepsilon_j \longrightarrow 0$, we finally receive the following inequality:

$$
|f^{(k)}(z)| \le k!\left(2^{k+1}+\sqrt{\widetilde{r}_j}+2^{k+1}\sum_{s=1}^{j}\frac{r_s}{(a_s-z)^{k+1}}\right).
$$

Recall that $\widetilde{r}_j \underset{j\to\infty}{\longrightarrow} 0$. Therefore, letting $j \longrightarrow \infty$, we finally have

$$
|f^{(k)}(z)| \le k!2^{k+1}\left(1+\sum_{s=1}^{\infty}\frac{r_s}{(a_s-z)^{k+1}}\right). \qquad \square
$$

Lemma 14.4.5. *For every $k \in \mathbb{N}$ there exist a $\widetilde{k} \in \mathbb{N}$, $\widetilde{k} \ge 2$, and a Zalcman type domain G such that*

(a) $\displaystyle\limsup_{(-1,0)\ni z\to 0} (\int \gamma_G^{(m)})(-1/2^{\widetilde{k}-1}, z) < \infty, \quad 1 \le m \le k,$

(b) $\displaystyle\lim_{G\ni z\to 0} (\int \gamma_G^{(\ell)})(w, z) = \infty, \quad w \in G,\ k < \ell.$

Observe that Lemma 14.4.5 immediately implies Theorem 14.4.1.

Proof of Theorem 14.4.1. Fix a $k \in \mathbb{N}$ and take the corresponding Zalcman type domain G from Lemma 14.4.5. Let $m \in \mathbb{N}$, $1 \le m \le k$. As a direct consequence of (a) and the fact that the $\int \gamma^{(m)}$-completeness is equivalent to the $\int \gamma^{(m)}$-finite compactness (recall Theorem 14.1.2), we see that G is not $\int \gamma^{(m)}$-complete.

It remains to see that G is $\int\gamma^{(\ell)}$-complete whenever $\ell > k$. So let us fix such an ℓ, a point $a \in G$, and a boundary point $z^0 \in \partial G$. We have to show that $\lim_{z\to z^0}(\int\gamma_G^{(\ell)})(a,z) = \infty$.

Case 1°: If $z^0 = 0$, then, using (b), we are done.

Case 2°: If $|z^0| = 1$, it follows that

$$\lim_{z\to z^0}(\int\gamma_G^{(\ell)})(a,z) \geq \lim_{z\to z^0}(\int\gamma_{\mathbb{D}}^{(\ell)})(a,z) = \lim_{z\to z^0} c_{\mathbb{D}}(a,z) = \infty.$$

Case 3°: If $z^0 \in \partial\mathbb{B}(a_j, r_j)$ for some j, then

$$\begin{aligned}\lim_{z\to z^0}(\int\gamma_G^{(\ell)})(a,z) &\geq \lim_{z\to z^0} c_G(a,z) \geq \lim_{z\to z^0} c_{\mathbb{C}\setminus\overline{\mathbb{D}}(a_j,r_j)}(a,z)\\ &= \lim_{z\to z^0} c_{\mathbb{D}_*}\left(\frac{r_j}{a-a_j}, \frac{r_j}{z-a_j}\right) = \lim_{z\to z^0} c_{\mathbb{D}}\left(\frac{r_j}{a-a_j}, \frac{r_j}{z-a_j}\right) = \infty,\end{aligned}$$

since $|r_j/(z-a_j)| \longrightarrow 1$ as $z \to z^0$. □

What remains is the proof of Lemma 14.4.5.

Proof of Lemma 14.4.5. Fix a $k \in \mathbb{N}$ and put $a_j := 2^{-j}$ and $r_{k,j} := 2^{-j} j^{-k-1}$, $j \in \mathbb{N}$. Since $\lim_{s\to\infty}\left(\frac{s}{s-1}\right)^2 \frac{1}{\sqrt[k]{2}} = \frac{1}{\sqrt[k]{2}} < 1$, we may choose a $\widetilde{k} \in \mathbb{N}$ such that $\left(\frac{s}{s-1}\right)^2 \frac{1}{\sqrt[k]{2}} < 1$ for all $s \geq \widetilde{k}$. Put

$$G = G_k := \mathbb{D}_* \setminus \bigcup_{j\geq\widetilde{k}} \overline{\mathbb{D}}(a_j, r_{k,j}).$$

Obviously, G is a Zalcman type domain.

To prove (a) it suffices to verify the following inequality:

$$\exists_{c=c(k)>0} : \gamma_G^{(m)}(z;1) \leq \frac{c}{-z(-\log(-z))^{\frac{k+1}{m}}}, \quad z \in \left[-\tfrac{1}{2^{\widetilde{k}-1}}, 0\right),\ m \leq k. \tag{14.4.3}$$

In fact, let $z \in \left[-\frac{1}{2^{\widetilde{k}-1}}, 0\right)$. Then there exist a unique $N \in \mathbb{N}$, $N \geq \widetilde{k}$, and a $b \in (1, 2]$ such that $z = -b/2^N$. Therefore,

$$\sum_{j=\widetilde{k}}^{N} \frac{r_{k,j}}{(a_j - z)^{m+1}} \leq \sum_{j=\widetilde{k}}^{N} \frac{r_{k,j}}{a_j^{m+1}} = \sum_{j=\widetilde{k}}^{N} \frac{2^{jm}}{j^{k+1}} \leq \frac{2^{Nm}}{N^{k+1}} \sum_{j=0}^{\infty} \delta^j \leq \frac{1}{1-\delta} \frac{2^{Nm}}{N^{k+1}}, \tag{14.4.4}$$

$$\begin{aligned}\sum_{j=N}^{\infty} \frac{r_{k,j}}{(a_j - z)^{m+1}} &\leq \sum_{j=N}^{\infty} \frac{r_{k,j}}{(-z)^{m+1}} = \sum_{j=N}^{\infty} \frac{2^{N(m+1)}}{2^j j^{k+1} b^{m+1}} \\ &\leq \frac{2^{N(m+1)}}{2^N N^{k+1}} \sum_{j=0}^{\infty} \frac{1}{2^j} \leq \frac{2^{Nm+1}}{N^{k+1}}. \end{aligned} \tag{14.4.5}$$

(The second inequality in (14.4.4) easily follows from the observation that there is a positive $\delta < 1$ such that $\frac{2^{(s-1)m}}{(s-1)^{k+1}} \leq \delta \frac{2^{sm}}{s^{k+1}}$, $s \geq \widetilde{k}$, $m \leq k$.)

We put $\widehat{c} := \frac{1}{1-\delta}$. Using (14.4.4) and (14.4.5), we get

$$\sum_{j=\widetilde{k}}^{\infty} \frac{r_{k,j}}{(a_j - z)^{m+1}} \leq \frac{(\widehat{c} + 2)2^k (\log 2)^{k+1} 2^{Nm}}{b^m (\log(2^m/b))^{k+1}} =: \frac{C_1}{(-z)^m (-\log(-z))^{k+1}}.$$

By virtue of Lemma 14.4.4, we obtain

$$|f^{(m)}(z)| \leq C \left(1 + \frac{C_1}{(-z)^m (-\log(-z))^{k+1}}\right) \leq \frac{2CC_1}{(-z)^m (-\log(-z))^{k+1}}, \quad f \in \mathcal{O}(G, \mathbb{D}),$$

which finally proves (14.4.3). (Note that we may take $C = k!2^{k+1} \geq m!2^{m+1}$, $m \leq k$; thus, the constant $C = C(k)$ from Lemma 14.4.4 works for all m, $m \leq k$.)

To prove (b), we claim that

$$\forall_{\ell > k} \, \exists_{c = c(k,\ell) > 0} : \gamma_G^{(\ell)}(z; 1) \geq \frac{c}{|z| \log(1/|z|)}, \quad |z| < \frac{1}{2^{\ell-2}}, \; z \in G. \tag{14.4.6}$$

Assume for a while that (14.4.6) is correct. Fix an $\ell \in \mathbb{N}$, $\ell > k$, and a point $w \in G$. Take a $z \in G$, $|z| < \frac{1}{2^{\widetilde{k}-2}}$, and a $\mathcal{C}^1$-curve $\alpha : [0, 1] \longrightarrow G$ connecting z with w. Then we have

$$\begin{aligned}\int_0^1 \gamma_G^{(\ell)}(\alpha(t); \alpha'(t)) dt &\geq c \int_0^{t_\alpha} \frac{|\alpha'(t)| dt}{|\alpha(t)| \log(1/|\alpha(t)|)} \geq c \int_0^{t_\alpha} \frac{\frac{d}{dt}|\alpha(t)| dt}{|\alpha(t)| \log(1/|\alpha(t)|)} \\ &\geq c \int_0^{t_\alpha} \frac{d}{dt}\bigl(-\log\log(1/|\alpha(t)|)\bigr) dt \\ &= c \left(\log\log \frac{1}{|z|} - \log\log 2^{\widetilde{k}-2}\right),\end{aligned}$$

where $t_\alpha := \sup\{t \in [0, 1] : |\alpha(\tau)| < \frac{1}{2^{\widetilde{k}-2}}, \; 0 \leq \tau \leq t\}$.

Since the curve α was an arbitrary one connecting z and w in G, it follows that

$$\int \gamma_G^{(\ell)}(w,z) \geq c\left(\log\log\frac{1}{|z|} - \log\log 2^{\widetilde{k}-2}\right) \underset{z\to 0}{\longrightarrow} \infty,$$

Hence, (b) is verified.

What remains is the proof of (14.4.6). Fix a $z \in G_k \cap \mathbb{D}(\frac{1}{2^{\widetilde{k}-2}})$. Then we have to find an $f \in \mathcal{O}(G,\mathbb{D})$ satisfying the following conditions:

- $f(z) = f'(z) = \cdots = f^{(\ell-1)}(z) = 0$,
- $|f^{(\ell)}(z)| \geq \dfrac{c}{(|z|\log(1/|z|))^\ell}$, where c is independent of z.

Again, we write z as $z = be^{i\theta}/2^N$ with $N \in \mathbb{N}$, $b \in (1,2]$, and $\theta \in [0,2\pi)$. Observe that $N \geq \widetilde{k} - 1$. Put

$$g(\lambda) := \sum_{j=0}^{\ell-1} \alpha_{b,\theta,j}(2^{-N-j-1} - \lambda)^{-1} + 2^{N+1}\beta_{b,\theta}, \quad \lambda \in G, \tag{14.4.7}$$

where $\alpha_{b,\theta,0} := 1$ and $\alpha_{b,\theta,1},\dots,\alpha_{b,\theta,\ell-1},\beta_{b,\theta} \in \mathbb{C}$ depend only on b and θ such that (obviously, $g \in \mathcal{O}(G)$) $g(z) = \cdots = g^{(\ell-1)}(z) = 0$.

We proceed under the assumption that we already have chosen g as in (14.4.7). Then it follows that

$$\begin{aligned}
|g^{(\ell)}(z)| &= \left|\sum_{j=0}^{\ell-1} \alpha_{b,\theta,j}\,\ell!\left(\frac{1}{2^{N+j+1}} - \frac{be^{i\theta}}{2^N}\right)^{-\ell-1}\right| \\
&\geq \left|\sum_{j=0}^{\ell-1} \alpha_{b,\theta,j}\left(\frac{2^{N+j+1}}{1-2^{j+1}be^{i\theta}}\right)^{\ell+1}\right| = 2^{(N+1)(\ell+1)}|B_{\ell,b,\theta}|,
\end{aligned}$$

where $B_{\ell,b,\theta} := \sum_{j=0}^{\ell-1}\alpha_{b,\theta,j}(\frac{2^j}{1-2^{j+1}be^{i\theta}})^{\ell+1}$.

Moreover, let us assume that $|B_{\ell,b,\theta}| \geq B_\ell > 0$, where B_ℓ is independent of b and θ. Then,

$$\begin{aligned}
\|g\|_G &\leq \sum_{j=0}^{\ell-1}\frac{|\alpha_{b,\theta,j}|}{r_{k,N+j+1}} + 2^{N+1}|\beta_{b,\theta}| \leq \alpha\frac{\ell+1}{r_{k,N+\ell}} \\
&= \alpha(\ell+1)2^{N+\ell}(N+\ell)^{k+1} \leq c2^N(N-1)^{k+1} \leq c2^N(N-1)^\ell,
\end{aligned}$$

where $\alpha := \max\{|\alpha_{b,\theta,j}|, |\beta_{b,\theta}|\}$ and c depends only on k and ℓ.

Put $f := g/\|g\|_G \in \mathcal{O}(G,\mathbb{D})$. Then, the following estimate is true:

$$|f^{(\ell)}(z)| \geq \frac{2^{\ell+1}B_\ell 2^{N\ell}}{c(N-1)^\ell} \geq \widetilde{c}_1\left(\frac{2^N\log 2}{b(\log 2^N - \log b)}\right)^\ell \geq \frac{\widetilde{c}}{(|z|\log(1/|z|))^\ell},$$

where $\widetilde{c}_1$ and $\widetilde{c}$ are constants that only depend on k.

In order to finish the proof of Lemma 14.4.4 we need the following lemma:

Lemma 14.4.6. *For an $\ell \in \mathbb{N}$ there are positive numbers α and B_ℓ such that for every $z = be^{i\theta}/2^N$, where $b \in [1,2)$, $\theta \in [0,2\pi]$, and $N \geq \ell - 1$, there exist complex numbers $\alpha_{b,\theta,j}$, $j = 1,\dots,\ell-1$, and $\beta_{b,\theta}$ such that*

- $\max\{|\alpha_{b,\theta,j}| : j = 1,\dots,\ell-1, |\beta_{b,\theta}|, b \text{ and } \theta \text{ as above}\} \leq \alpha$,
- $\min\{|B_{\ell,b,\theta}| : b \in [1,2],\ \theta \in [0,2\pi]\} \geq B_\ell$,
- $g(z) = g'(z) = \cdots = g^{(\ell-1)}(z) = 0$ *(for g see* (14.4.7)).

Proof. Let g be a function as in (14.4.7) with unknown numbers $\alpha_{b,\theta,j}$. Then, the condition $g'(z) = \cdots = g^{(\ell-1)}(z) = 0$ gives the following system of $\ell-1$ equations:

$$\sum_{j=0}^{\ell-1} s!\left(\frac{2^{N+j+1}}{1-2^{j+1}be^{i\theta}}\right)^{s+1} \alpha_{b,\theta,j} = 0, \quad s = 1,\dots,\ell-1,$$

which is equivalent to

$$\sum_{j=1}^{\ell-1} s!\left(\frac{2^{j}}{1-2^{j+1}be^{i\theta}}\right)^{s+1} \alpha_{b,\theta,j} = -\left(\frac{1}{1-2be^{i\theta}}\right)^{s+1}, \quad s = 1,\dots,\ell-1. \tag{14.4.8}$$

To further simplify discussions, we put

$$A_{b,\theta,j} := \frac{2^j}{1-2^{j+1}be^{i\theta}}, \quad j = 0,\dots,\ell-1.$$

Observe that $|A_{b,\theta,j}| \in [1/8,1]$ and that $A_{b,\theta,\mu} \neq A_{b,\theta,\nu}$ for $\mu \neq \nu$. Now we can rewrite the system of equations (14.4.8) in the following form:

$$\sum_{j=1}^{\ell-1} A_{b,\theta,j}^{s+1}\alpha_{b,\theta,j} = -A_{b,\theta,0}^{s+1}, \quad s = 1,\dots,\ell-1.$$

From here, we conclude that

$$\left|\det\left[A_{b,\theta,j}^{s+1}\right]_{j,s=1,\dots,\ell-1}\right| = \left|\prod_{j=1}^{\ell-1} A_{b,\theta,j}\right|^2 \prod_{1\leq\mu<\nu\leq k} |A_{b,\theta,\mu} - A_{b,\theta,\nu}| \geq \varepsilon > 0,$$

where ε is independent of b and θ. Hence, the claimed choice of the $\alpha_{b,\theta,j}$, $j = 1,\dots,\ell-1$, is always possible. Next, we put

$$\beta_{b,\theta} := -\sum_{j=0}^{\ell-1} A_{b,\theta,j}\alpha_{b,\theta,j}.$$

For the upper estimate, observe that

$$|\beta_{b,\theta}| \leq \sum_{j=0}^{\ell-1} |A_{b,\theta,j}\alpha_{b,\theta,j}| \leq \ell \max\{|\alpha_{b,\theta,j}| : j = 0, \dots, \ell - 1\}.$$

Therefore, it suffices to estimate the $\alpha_{b,\theta,j}$'s.

Recall that $|A_{b,\theta,j}|^{s+1} \in [2^{-3\ell}, 1]$, $j = 0, \dots, \ell - 1$, $s = 1, \dots, \ell - 1$, $b \in [1, 2]$, and $\theta \in [0, 2\pi]$. Applying Cramer's formula and the continuity of det-function, we see there is a number $\widetilde{\alpha} > 0$ such that all $|\alpha_{b,\theta,j}| \leq \widetilde{\alpha}$.

Finally, the lower estimate remains. Since $|B_{\ell,b,\theta}|$ is continuous with respect to (b, θ) it suffices to show that $B_{\ell,b,\theta} \neq 0$ or, equivalently,

$$\sum_{j=1}^{\ell-1} A_{b,\theta,j}^{\ell+1}\alpha_{b,\theta,j} \neq -A_{b,\theta,0}^{\ell+1}.$$

Suppose that this is false. Then, the $\alpha_{b,\theta,j}$'s fulfill the following ℓ equations:

$$\sum_{j=1}^{\ell-1} A_{b,\theta,j}^{s+1}\alpha_{b,\theta,j} = -A_{b,\theta,0}^{s+1}, \quad s = 1, \dots, \ell,$$

implying that $A_{b,\theta,0}/A_{b,\theta,j} = 1$, $j = 1, \dots, \ell - 1$. But this is impossible. Thus the lower estimate has also been proved. □ □

We conclude this section by discussing the boundary behavior of the $\gamma^{(k)}$'s for bounded plane domains, which should be compared with its consequences in Proposition 14.4.3.

Proposition 14.4.7. *Let $G \subset \mathbb{C}$ be a bounded domain. Then, the following properties are equivalent:*

(i) *there exist $k_0 \in \mathbb{N}$, $\alpha_0 \in [0, 1)$, and $c_0 > 0$ such that*

$$\gamma_G^{(k_0)}(z; 1) \leq \frac{c_0}{\big(\operatorname{dist}(z, \partial G)\big)^{\alpha_0}}, \quad z \in G;$$

(ii) *for any $k \in \mathbb{N}$ there exist $\alpha \in [0, 1)$ and $c > 0$ such that*

$$\gamma_G^{(k)}(z; 1) \leq \frac{c}{\big(\operatorname{dist}(z, \partial G)\big)^{\alpha}}, \quad z \in G.$$

The proof will be based on the following lemma:

Lemma 14.4.8. *Let $G \subset \mathbb{C}$ be a bounded plane domain and let $k, \ell \in \mathbb{N}$ with $k \le \ell$. Then,*

$$\left(\gamma_G^{(\ell)}(z;1)\right)^{\ell/k} \le \frac{\gamma_G^{(k)}(z;1)}{\left(\operatorname{dist}(z;\partial G)\right)^{(\ell-k)/k}}, \quad z \in G.$$

Proof. Fix natural numbers $k, \ell \in \mathbb{N}$ with $k \le \ell$ and a point $a \in G$. Let $f \in \mathcal{O}(G, \mathbb{D})$ be a competitor for the definition of $\gamma_G^{(k)}(a;1)$; in particular, one has $f(a) = f'(a) = \dots = f^{(\ell-1)}(a) = 0$. Put

$$g(z) := \begin{cases} \frac{f(z)}{(z-a)^{\ell-k}}, & \text{if } a \ne z \in G \\ 0, & \text{if } z = a \end{cases}.$$

Then $g \in \mathcal{O}(G)$, $g^{(m)}(a) = 0$ for all $m = 0, \dots, k-1$, $g^{(k)}(a) = k! f^{(\ell)}(a)/\ell!$, and $\|g\|_G \le 1/\left(\operatorname{dist}(a, \partial G)\right)^{\ell-k}$. Put $h := g/\left(\operatorname{dist}(\cdot, \partial G)\right)^{\ell-k}$. Then $h \in \mathcal{O}(G, \mathbb{D})$ and $\operatorname{ord}_a h \ge k$. Therefore,

$$\gamma_G^{(k)}(a;1) \ge |h^{(k)}(a)|/k! = \left(\frac{\left(\operatorname{dist}(a;\partial G)\right)^{\ell-k}}{k!}|f^{(\ell)}(0)|\right)^{1/k}.$$

Since f was arbitrarily chosen, our claimed inequality is an immediate consequence. □

Proof of Proposition 14.4.7. Obviously it suffices to verify the implication (i) $\Longrightarrow$ (ii). So, choose k_0, α_0, and c_0 as in (i). Since $\gamma_G^{(1)} \le \gamma_G^{(k_0)}$, we may assume that $k_0 = 1$. Now take a $k \in \mathbb{N}$, $k > 1$. Then, Lemma 14.4.8 implies that

$$\gamma_G^{(k)}(z;1) \le \frac{c_0^{1/k}}{\left(\operatorname{dist}(z, \partial G)\right)^{(\alpha_0+k-1)/k}}, \quad z \in G.$$

It remains to mention that the exponent $(\alpha_0 + k - 1)/k \in [0, 1)$. □

14.5 Kobayashi completeness

Since $\boldsymbol{c} \le \boldsymbol{k}$, every $\boldsymbol{c}$-complete domain is $\boldsymbol{k}$-complete. So, section 14.2 provides a lot of examples of $\boldsymbol{k}$-complete domains. On the other hand, the following necessary condition shows that there are many domains which are not $\boldsymbol{k}$-complete:

Proposition 14.5.1. *Any $\boldsymbol{k}$-complete domain is taut.*

Proof. Let $(\varphi_j)_{j\in\mathbb{N}}$ be an arbitrary sequence of $\varphi_j \in \mathcal{O}(\mathbb{D}, G)$. Assume that $(\varphi_j)_{j\in\mathbb{N}}$ is not uniformly divergent. This implies that there are compact sets $K \subset \mathbb{D}$ and $L \subset G$ such that, without loss of generality, $\varphi_j(\lambda_j) \in L$ with $\lambda_j \in K$. Fix $z^* \in L$ and let $0 < r < 1$ with $K \subset \mathbb{D}(r)$. Then for $\lambda \in \mathbb{D}(r)$ we obtain

$$\begin{aligned} \boldsymbol{k}_G(\varphi_j(\lambda), z^*) &\le \boldsymbol{k}_G(\varphi_j(\lambda), \varphi_j(\lambda_j)) + \boldsymbol{k}_G(\varphi_j(\lambda_j), z^*) \\ &\le \boldsymbol{p}(\lambda, \lambda_j) + \sup\{\boldsymbol{k}_G(z, z^*) : z \in L\} =: C. \end{aligned}$$

Hence, $\bigcup_{j\in\mathbb{N}} \varphi_j(\mathbb{D}(r)) \subset B_{\boldsymbol{k}_G}(z^*, C+1) \subset\subset G$. Therefore, Montel's theorem guarantees the existence of a subsequence $(\varphi_{j_\nu})_\nu \subset (\varphi_j)_j$ that converges locally uniformly to a map in $\mathcal{O}(\mathbb{D}, G)$. □

Corollary 14.5.2. *Any $\boldsymbol{k}$-complete domain is a domain of holomorphy.*

Remark 14.5.3.

(a) For a while, there was the question whether tautness can imply $\boldsymbol{k}$-completeness. The first negative example was found by J.-P. Rosay (cf. [455]). Later in this section, we will present two other examples.

(b) Because of Remark 3.2.3(c) and Theorem 13.1.1, any $\boldsymbol{k}$-complete balanced domain is bounded and it admits a continuous Minkowski function.

There is a simple example of a domain that is not $\boldsymbol{c}$-complete, but which is $\boldsymbol{k}$-complete, namely the punctured disc $\mathbb{D}_*$. This observation is a direct consequence of the next result due to S. Kobayashi (cf. [311, 317]).

Theorem 14.5.4. *If $\Pi : \widetilde{G} \longrightarrow G$ denotes a holomorphic covering between domains in $\mathbb{C}^n$, then the following statements are equivalent:*

(i) *$\widetilde{G}$ is $\boldsymbol{k}$-complete;*

(ii) *G is $\boldsymbol{k}$-complete.*

Proof. (i) $\Longrightarrow$ (ii): According to Theorem 13.1.10, G is $\boldsymbol{k}$-hyperbolic. Fix a ball $B_{\boldsymbol{k}_G}(z_0, r)$ in G. By Theorem 3.3.7, it is clear that $B_{\boldsymbol{k}_G}(z_0, r) \subset \Pi(B_{\boldsymbol{k}_{\widetilde{G}}}(\widetilde{z}_0, r))$, where $\widetilde{z}_0$ is a point in $\widetilde{G}$ with $\Pi(\widetilde{z}_0) = z_0$. Then, the assumption and Corollary 14.1.6 imply that $B_{\boldsymbol{k}_{\widetilde{G}}}(\widetilde{z}_0, r) \subset\subset \widetilde{G}$, and therefore $B_{\boldsymbol{k}_G}(z_0, r) \subset\subset G$.

(ii) $\Longrightarrow$ (i): As above, $\widetilde{G}$ is $\boldsymbol{k}$-hyperbolic. Fix a $\boldsymbol{k}_{\widetilde{G}}$-Cauchy sequence $(\widetilde{z}_\nu)_{\nu\in\mathbb{N}} \subset \widetilde{G}$. Then obviously $(\Pi(\widetilde{z}_\nu))_{\nu\in\mathbb{N}}$ is a $\boldsymbol{k}_G$-Cauchy sequence. By assumption, this sequence converges to a point $z_0 \in G$. Using again Theorem 3.3.7 it is easy to construct a subsequence $(z_{\nu_\mu})_{\mu\in\mathbb{N}}$ of $(z_\nu)_\nu$ and points $\widetilde{z}_{0,\mu} \in \widetilde{G}$ with $\Pi(\widetilde{z}_{0,\mu}) = z_0$ and $\boldsymbol{k}_{\widetilde{G}}(\widetilde{z}_{\nu_\mu}, \widetilde{z}_{0,\mu}) < 1/\mu$. Thus, $\boldsymbol{k}_{\widetilde{G}}(\widetilde{z}_{0,\mu}, \widetilde{z}_{0,\lambda}) \underset{\lambda,\mu\to\infty}{\longrightarrow} 0$. On the other hand, there exist a neighborhood $B_{\boldsymbol{k}_G}(z_0, r)$ of z_0 and neighborhoods U_μ of $\widetilde{z}_{0,\mu}$ such that $\Pi|_{U_\mu} : U_\mu \longrightarrow B_{\boldsymbol{k}_G}(z_0, r)$ is biholomorphic, $\mu \in \mathbb{N}$. Put $V_\mu := \Pi^{-1}(B_{\boldsymbol{k}_G}(z_0, r/2)) \cap U_\mu$;

then $\boldsymbol{k}_{\widetilde{G}}(\widetilde{z}_{0,\mu}, \partial V_\mu) \ge r/2$. This observation, together with $\boldsymbol{k}_{\widetilde{G}} = \boldsymbol{k}^i_{\widetilde{G}}$, shows that for a sufficiently large μ_0 we obtain $\widetilde{z}_{0,\mu_0} = \widetilde{z}_{0,\mu}$, $\mu \ge \mu_0$. Put $\widetilde{z}_0 := \widetilde{z}_{0,\mu_0}$. Then, a standard argument leads to $\lim_{\nu\to\infty} \boldsymbol{k}_{\widetilde{G}}(\widetilde{z}_\nu, \widetilde{z}_0) = 0$. □

Example. $\mathbb{C} \setminus \{0, 1\}$ is a $\boldsymbol{k}$-complete domain, but it is not even $\boldsymbol{c}$-hyperbolic.

We will see that the property of $\boldsymbol{k}$-completeness is a local one (cf. [144]), in contrast to Example 14.2.15.

Theorem 14.5.5. *Let G be a bounded domain in $\mathbb{C}^n$. Suppose that any boundary point $z_0 \in \partial G$ permits a bounded neighborhood $U = U(z_0)$ such that any connected component of $G \cap U$ is $\boldsymbol{k}$-complete. Then, G itself is $\boldsymbol{k}$-complete.*

Proof. Assume that there exists a $\boldsymbol{k}_G$-Cauchy sequence $(z_\nu)_{\nu\in\mathbb{N}} \subset G$ with $\lim_{\nu\to\infty} z_\nu = z_0 \in \partial G$. Choose $R > 0$ so large that $G \cup U(z_0) \subset \mathbb{B}(R) =: V$. As a consequence of the equality of the topologies, $U(z_0)$ contains a $\boldsymbol{k}_V$-ball $B_{\boldsymbol{k}_V}(z_0, 2s)$ $(s > 0)$. Then, for $0 < \varepsilon < s/3$, take $\nu_\varepsilon \in \mathbb{N}$ such that $\boldsymbol{k}_G(z_\nu, z_\mu) < \varepsilon$ and $z_\nu \in B_{\boldsymbol{k}_V}(z_0, s/3)$ whenever $\nu, \mu \ge \nu_\varepsilon$. Now fix such ν, μ. According to the definition of $\boldsymbol{k}_G$, we find analytic discs $\varphi_j \in \mathcal{O}(\mathbb{D}, G)$, $1 \le j \le k$, and points $a_j \in \mathbb{D}$, $1 \le j \le k$, with the following properties:

$$\varphi_1(0) = z_\nu,\ \varphi_j(a_j) = \varphi_{j+1}(0),\ 1 \le j < k,\ \varphi_k(a_k) = z_\mu,\ \text{ and } \sum_{j=1}^{k} \boldsymbol{p}(0, a_j) < \varepsilon.$$

Observe that, in particular, $\boldsymbol{p}(0, a_j) < \varepsilon < s/3$.

For $\lambda \in B_{\boldsymbol{p}}(0, s) \subset \mathbb{D}$, we obtain the following inequalities:

$$\begin{aligned} \boldsymbol{k}_V(\varphi_j(\lambda), z_0) &\le \boldsymbol{k}_V(\varphi_j(\lambda), \varphi_j(0)) + \boldsymbol{k}_V(\varphi_j(0), z_\mu) + \boldsymbol{k}_V(z_\mu, z_0) \\ &< \boldsymbol{p}(\lambda, 0) + \sum_{\nu=j}^{k} \boldsymbol{k}_V(\varphi_\nu(0), \varphi_\nu(a_\nu)) + s/3 \\ &< s + \sum_{\nu=1}^{k} \boldsymbol{p}(0, a_\nu) + s/3 < 2s, \end{aligned}$$

i.e., $\varphi_j(B_{\boldsymbol{p}}(0, s)) \subset U(z_0)$.

If we denote by $\gamma : \mathbb{D} \longrightarrow B_{\boldsymbol{p}}(0, s)$ the biholomorphic dilatation, then we have analytic discs $\widetilde{\varphi}_j := \varphi_j \circ \gamma \in \mathcal{O}(\mathbb{D}, U \cap G)$ with $\widetilde{\varphi}_1(0) = z_\nu$, $\widetilde{\varphi}_j(\gamma^{-1}(a_j)) = \widetilde{\varphi}_{j+1}(0)$, $1 \le j < k$, $\widetilde{\varphi}_k(\gamma^{-1}(a_k)) = z_\mu$, and $\sum_{j=1}^k \boldsymbol{p}(0, \gamma^{-1}(a_j)) \le c \sum_{j=1}^k \boldsymbol{p}(0, a_j) < c\varepsilon$, where c is independent of ε. In particular, the above reasoning also shows that the sequence $(z_\nu)_{\nu\gg1}$ belongs to a connected component U' of $U(z_0) \cap G$. So, we obtain a $\boldsymbol{k}_{U'}$-Cauchy sequence with $\lim_{\nu\to\infty} z_\nu = z_0 \notin U'$, which contradicts the assumption that U' is $\boldsymbol{k}$-complete. □

Example 14.5.6.

(a) Applying Theorem 14.5.5 to Example 14.2.15 shows that the Sibony domain is $\boldsymbol{k}$-complete.

(b) The above theorem also provides a simple argument to show that any strongly pseudoconvex domain is $\boldsymbol{k}$-complete, using only the existence of local peak functions.

By Theorem 14.3.1 we know that any bounded pseudoconvex Reinhardt domain containing 0 is $\boldsymbol{k}$-complete. Moreover, tautness, i.e., the continuity of its Minkowski function, is necessary for a balanced domain to be $\boldsymbol{k}$-complete. Nevertheless, the following result shows that tautness, even in the case of a balanced domain, does not imply $\boldsymbol{k}$-completeness.

Theorem 14.5.7 (cf. [255]). *There exists a bounded balanced pseudoconvex domain* $G = G_{\mathfrak{h}} = \{z \in \mathbb{C}^n : \mathfrak{h}(z) < 1\}$ $(n \geq 3)$ *with continuous Minkowski function* $\mathfrak{h}$ *that is not* $\boldsymbol{k}$*-complete.*

Proof. It suffices to construct an example of G in $\mathbb{C}^3$, since taking $G \times \mathbb{D}^{n-3}$ if $n > 3$ leads to an example in $\mathbb{C}^n$.

Our construction will be based on a series of reduction steps.

1st reduction. It suffices to construct a continuous log-psh function $g : \mathbb{C}^2 \longrightarrow \mathbb{R}_+$ satisfying the following properties:

(a) $\lim_{\|z\| \to \infty} g(z)/\|z\| = l \in \mathbb{R}$,

(b) the set $G := \{z \in \mathbb{C}^2 : g(z) < 1\}$ is bounded and it has a connected component G', which is not $\boldsymbol{k}$-complete.

Assume that g, l, G', and G are given. We define a new function $h_0 : \mathbb{C}^2 \times \mathbb{C} \longrightarrow [0, \infty)$ by putting

$$h_0(z, z_3) := \begin{cases} |z_3| g(z_1/z_3, z_2/z_3) & \text{if } \quad z_3 \neq 0 \\ l \|z\| & \text{if } \quad z_3 = 0 \end{cases}.$$

Observe that h_0 is a continuous homogeneous function on $\mathbb{C}^3$ that is log-psh on $\mathbb{C}^2 \times \mathbb{C}_*$. By Appendix B.4.23, we conclude that h_0 is even log-psh on the whole $\mathbb{C}^3$. By (b) we know that $G \subset \mathbb{B}(r_0)$ for some positive r_0. Define $\mathfrak{h} : \mathbb{C}^3 = \mathbb{C}^2 \times \mathbb{C} \longrightarrow \mathbb{R}_+$ by

$$\mathfrak{h}(z, z_3) = \max \left\{ h_0(z, z_3), \sqrt{\frac{\|z\|^2 + |z_3|^2}{1 + r_0^2}} \right\}.$$

Then $\mathfrak{h}$ is a continuous log-plurisubharmonic homogeneous function with $\mathfrak{h}^{-1}(0) = \{0\}$. Thus, the domain

$$\widetilde{G} := \{(z, z_3) \in \mathbb{C}^2 \times \mathbb{C} : \mathfrak{h}(z, z_3) < 1\}$$

is a bounded pseudoconvex balanced domain and $G' \times \{1\}$ is a connected component of $\widetilde{G} \cap (\mathbb{C}^2 \times \{1\}) = G \times \{1\}$. Therefore, $\widetilde{G}$ is not $\boldsymbol{k}$-complete.

2nd reduction. It suffices to find a connected set $X \subset \mathbb{B}(r_0) \subset \mathbb{C}^2$ for some $r_0 > 0$ and a sequence $(z_j)_{j \in \mathbb{N}} \subset X$, $\lim_{j\to\infty} z_j = 0$, with the following properties:

(c) there are holomorphic maps $\varphi_j : \mathbb{D} \longrightarrow \mathbb{C}^n$, such that

$$\varphi_j(\mathbb{D}) \subset X,\ \varphi_j(\lambda_j') = z_j,\ \varphi_j(\lambda_j'') = z_{j+1}\ (\lambda_j', \lambda_j'' \in \mathbb{D}),$$

$$\text{and} \sum_{j=1}^{\infty} \boldsymbol{p}(\lambda_j', \lambda_j'') =: A < \infty;$$

(d) there exists $\varphi : \mathbb{C}^2 \longrightarrow \mathbb{R}_+$, continuous and log-psh, such that

(d)$_1$ $\varphi(0) = 1, \varphi|_X < 1$,

(d)$_2$ for suitable positive C, α, R, we have the inequality $\varphi(z) \le C\|z\|^\alpha$ for $\|z\| \ge R$.

Drawing on these assumptions, we will construct a function g as in the first reduction step. Taking $\varphi^{1/2\alpha}$ instead of φ we may assume that $\alpha = 1/2$. Put

$$g(z) := \max\{\varphi(z), \|z\|/(2r_0)\}, \quad z \in \mathbb{C}^n.$$

Obviously, g is continuous and log-psh on $\mathbb{C}^n$ with $\lim_{\|z\|\to\infty} g(z)/\|z\| = 1/(2r_0)$. Moreover, $X \subset G := \{z \in \mathbb{C}^n : g(z) < 1\} \subset \mathbb{B}(2r_0)$ and the connected component G' of G with $X \subset G'$ is not $\boldsymbol{k}$-complete because of $\boldsymbol{k}_{G'}(z_1, z_j) \le \sum_{\nu=1}^{j} \boldsymbol{p}(\lambda_\nu', \lambda_\nu'') \le A < \infty$ and $z_j \longrightarrow 0 \in \partial G'$.

What remains is to produce the data of the second reduction step.

Let $c_j := 1/2^{j+1}$ and $a_j := 1/2^{2(j+1)}$; put

$$X_j := \{(z, w) \in \mathbb{C}^2 : w = z(a_{j+1} + a_j) - a_j a_{j+1}, |z| \le c_j\}, \quad j \in \mathbb{N}. \tag{14.5.1}$$

Then, X_j is a connected compact set with $X_j \subset \mathbb{B}_2$ and $X_j \underset{j\to\infty}{\longrightarrow} \{0\}$. Moreover, for $z_j := (a_j, a_j^2)$ we obtain $z_j, z_{j+1} \in X_j$. Let $\varphi_j : \mathbb{D} \longrightarrow \mathbb{C}^2$ be a holomorphic map defined by $\varphi_j(\lambda) := (c_j\lambda, c_j\lambda(a_{j+1} + a_j) - a_j a_{j+1})$. Then, $\varphi_j(\mathbb{D}) \subset X_j$, $\varphi_j(\lambda_j') = z_j$, and $\varphi_j(\lambda_j'') = z_{j+1}$, where $\lambda_j' := a_j/c_j$, $\lambda_j'' := a_{j+1}/c_j \in \mathbb{D}$.

Direct calculations lead to $\sum_{j=1}^{\infty} \boldsymbol{p}(\lambda_j', \lambda_j'') < \infty$. If we put $X := \bigcup_{j=1}^{\infty} X_j$ we have found all we wanted in (c).

To construct the function φ as in (d) put

$$P_j(z, w) := w - z(a_j + a_{j+1}) + a_j a_{j+1}$$

and observe that

$$|P_j(\xi)| \le 3\|\xi\|, \quad \|\xi\| \ge 1, \quad \xi = (z, w) \in \mathbb{C}^2.$$

Next, we choose sequences $(r_j)_j$ and $(\varepsilon_j)_j$ of positive numbers satisfying the following properties:

$$r_j \geq 3, \quad r_j \geq 2\max\{|P_j(\xi)| : \|\xi\| \leq j\}, \quad \log 3/4 = \sum_{j=1}^{\infty} \varepsilon_j \log(a_j a_{j+1}/r_j).$$

In particular, $\log(4/3) \geq \sum_{j=1}^{\infty} \varepsilon_j =: \alpha$.

Using these sequences, we get the following psh function $\psi : \mathbb{C}^2 \longrightarrow [-\infty, \infty)$:

$$\psi(\xi) := \sum_{j=1}^{\infty} \varepsilon_j \log \frac{|P_j(\xi)|}{r_j}, \qquad \xi \in \mathbb{C}^2.$$

Note that $\psi(0) = \log 3/4$, $\psi|_X = -\infty$, $\psi < 0$ on $\overline{\mathbb{B}}_2$, and $\psi(\xi) \leq \alpha \log \|\xi\|$ if $\|\xi\| \geq 1$.

Let $\Phi : \mathbb{C}^2 \longrightarrow \mathbb{R}_+$ be a $\mathcal{C}_0^\infty$ function on $\mathbb{C}^2$ with support in $\mathbb{B}_2$ and satisfying $\Phi(z, w) = \Phi(|z|, |w|)$ and $\int_{\mathbb{C}^2} \Phi(\xi) d\lambda(\xi) = 1$. For $\varepsilon > 0$, we put

$$\psi_\varepsilon(\xi) := \int_{\mathbb{C}^2} \psi(\xi - \varepsilon\eta)\Phi(\eta) d\lambda(\eta), \quad \xi \in \mathbb{C}^2.$$

Thus, we obtain a sequence of psh $\mathcal{C}^\infty$-functions ψ_ε on $\mathbb{C}^2$ with $\psi_\varepsilon \underset{\varepsilon \searrow 0}{\searrow} \psi$. In particular, $\psi_\varepsilon(0) \geq \log 3/4$ and $\psi_\varepsilon(\xi) < 0$ if $\|\xi\| \leq 1$, $0 < \varepsilon \leq \varepsilon_0 < 1$. Moreover, the definition of ψ_ε leads to

$$\psi_\varepsilon(\xi) \leq \alpha \log(2\|\xi\|) \quad \text{if} \quad 0 < \varepsilon \leq 1 \quad \text{and} \quad \|\xi\| \geq 2.$$

Setting

$$\widetilde{\varphi}_\varepsilon := \exp \psi_\varepsilon / \exp \psi_\varepsilon(0), \quad 0 < \varepsilon \leq \varepsilon_0, \tag{14.5.2}$$

we receive a family of positive log-psh $\mathcal{C}^\infty$-functions on $\mathbb{C}^2$ having the following properties:

$$\widetilde{\varphi}_\varepsilon(0) = 1, \ \widetilde{\varphi}_\varepsilon(\xi) < 4/3 \quad \text{if} \quad \|\xi\| \leq 1, \qquad \text{and} \quad \widetilde{\varphi}_\varepsilon(\xi) \leq C\|\xi\|^\alpha \quad \text{if} \quad \|\xi\| \geq 2,$$

where C is a suitable positive constant.

Now, we follow Bishop's construction of peak-functions. Put $U_1 := \mathbb{B}(1/2)$ and choose $\varepsilon_1 \in (0, \varepsilon_0)$ such that $\varphi_1 := \widetilde{\varphi}_{\varepsilon_1}$ is less than $1/3$ on $X \setminus U_1$.

Suppose we have already constructed neighborhoods $U_1 \supset U_2 \supset \cdots \supset U_k$ of 0 and numbers $\varepsilon_1 > \cdots > \varepsilon_k$ such that, with $\varphi_j := \widetilde{\varphi}_{\varepsilon_j}$, we have

$$\varphi_j < 1/3 \quad \text{on} \quad X \setminus U_j, \quad 1 \leq j \leq k,$$
$$\varphi_\nu < 1 + 2^{-j}/3 \quad \text{on} \quad U_j, \quad 1 < j \leq k, \ 1 \leq \nu < j.$$

Define

$$U_{k+1} := \{\xi \in U_k : \varphi_j(\xi) < 1 + 2^{-(k+1)}/3, \quad 1 \le j \le k\}$$

and choose ε_{k+1}, $0 < \varepsilon_{k+1} < \varepsilon_k$, such that the function $\varphi_{k+1} := \widetilde{\varphi}_{\varepsilon_{k+1}}$ is below $1/3$ on $X \setminus U_{k+1}$. Given the sequence $(\varphi_j)_{j\in\mathbb{N}}$, we put

$$\varphi := \sum_{j=1}^{\infty} (1/2)^j \varphi_j.$$

This series is locally uniformly convergent. Therefore, the function φ is continuous log-psh on $\mathbb{C}^2$ with $\varphi(0) = 1$ and $\varphi(\xi) \le C\|\xi\|^\alpha$ if $\|\xi\| \ge 2$. Moreover, for $\xi \in X$ we have:

$$\text{if } \xi \in X \setminus U_1, \quad \text{then} \quad \varphi(\xi) < 1;$$
$$\text{if } \xi \in X \cap (U_k \setminus U_{k+1}), \quad \text{then} \quad \varphi(\xi) \le 1 - 2^{-(2k-1)}/3 < 1.$$

Thus, the function φ satisfies all the properties needed in (d). □

Remark 14.5.8.

(a) It would be very interesting to know whether such an example could be also constructed in $\mathbb{C}^2$.We emphasize that the method used above does not work in the two-dimensional case. **?**

(b) So far it is totally unclear how to characterize the $\boldsymbol{k}$-completeness (or the $\boldsymbol{c}$-completeness) of a bounded pseudoconvex balanced domain via the properties of its Minkowski function. **?**

(c) We would like to point out that the example of Theorem 14.5.7 is a taut $\mathcal{H}^\infty$-convex $\mathcal{H}^\infty$-domain of holomorphy; cf. [478].

Up to now it is an open problem whether every bounded pseudoconvex domain with $\mathcal{C}^\infty$-smooth boundary is $\boldsymbol{k}$-complete.The strongest result in the negative direction is the following unpublished one due to N. Sibony, cf. [473]: We mention that a part of the construction of Theorem 14.5.7 was based on this work. **?**

Theorem 14.5.9. *There exists a pseudoconvex non $\boldsymbol{k}$-complete domain $G \subset\subset \mathbb{B}_2$, given as a connected component of $\{z \in \mathbb{B}(3) : u(z) < 1\}$, where u is a continuous psh function in $\mathbb{B}(3)$, $\mathcal{C}^\infty$ outside zero with* $\operatorname{grad} u(z) \ne 0$ *if $z \ne 0$, and $u(0) = 1$.*

Proof. Following the construction of the proof of Theorem 14.5.7, we may assume (adding suitable multiples of $\|z\|^2$ to $\widetilde{\varphi}_\varepsilon$, cf. (14.5.2)) that all the functions $\widetilde{\varphi}_\varepsilon$ are strictly psh on $\mathbb{C}^2$. Thus, the function $\varphi = (1/2)\varphi_1 + \sum_{j=2}^\infty (1/2^j)\varphi_j$ is strictly psh and, even more, $\widehat{\varphi} := \varphi - A\|\ \|^2$ is a continuous strictly psh function on $\mathbb{C}^2$, where A is a suitable positive number. By the approximation theorem of R. Richberg (cf. Appendix B.4.32), a $\mathcal{C}^\infty$-strictly psh function $u : \mathbb{C}^2 \setminus \{0\} \longrightarrow \mathbb{R}$ can be found with $\widehat{\varphi} \le u \le \varphi - (A/2)\| \; \|^2$. Then, putting $u(0) = 1$ the function u extends to a continuous psh function on the whole $\mathbb{C}^2$ satisfying

$$u(0) = 1 \quad \text{and} \quad u(z) \le 1 - (A/2)\|z\|^2, \quad z \in X.$$

Let ϱ be any $\mathcal{C}^\infty$-psh function on $\mathbb{B}(3)$ with $\varrho \equiv 0$ on $\mathbb{B}_2$, $\varrho(z) \ge 2$ if $\|z\| \ge 2$, and $\lim_{\|z\| \to 3} \varrho(z) = \infty$.

Define

$$u_\varepsilon(z) := u(z) + \varepsilon\|z\|^2 + \varrho(z), \quad z \in \mathbb{B}(3),\ 0 < \varepsilon < A/2$$

Observe that

$$X \subset \{z \in \mathbb{B}(3) : u_\varepsilon(z) < 1\} =: \widetilde{G}_\varepsilon \subset\subset \mathbb{B}(3) \quad \text{and} \quad 0 \in \partial\widetilde{G}_\varepsilon.$$

Using Sard's lemma for the function $(1 - \varrho(z) - u(z))/\|z\|^2$ on $\mathbb{B}(3) \setminus \{0\}$, we are able to find a small ε_0, $0 < \varepsilon_0 < A/2$, such that the function u_{ε_0} is regular at every $z \in \mathbb{B}(3) \setminus \{0\}$ with $u_{\varepsilon_0}(z) = 1$. Hence, the connected component $G := G_{\varepsilon_0}$ of $\widetilde{G}_{\varepsilon_0}$ that contains X is the domain whose existence we claimed. □

14.6 Exercises

Exercise 14.6.1. Let $G_1 \subset \mathbb{C}^2$ be the domain introduced in Example 14.2.15. Put $G = \{z = (z', z_3) \in \mathbb{C}^3 : z' \in G_1, |z_3| < \operatorname{dist}(z', \partial G_1)\}$. Prove that G is an $\mathcal{H}^\infty(G)$-domain of holomorphy that is not $\boldsymbol{c}$-complete (cf. [471]).

Exercise 14.6.2. Complete the details of the following construction of a connected complex space X, which satisfies: (a) $\operatorname{top} X = \operatorname{top} \boldsymbol{c}_X$, (b) X is $\boldsymbol{c}$-complete but (c) X is not $\boldsymbol{c}$-finitely compact (cf. [273]).

Sketch of the construction. Write $\lambda_{j,k} := (1 - 1/(j+1)) \exp(2\pi i k/(j+1)) \in \mathbb{D}$, $j \ge 1$ and $0 \le k \le j$, and put $D_j := \mathbb{D}$, $j \in \mathbb{Z}_+$. X is defined by identifying the points $\lambda_{j,k} \in D_j$ $(j \ge 1)$ with $\lambda_{j,k} \in D_0$ in $\bigcup_{j=0}^\infty D_j$. Then, a holomorphic function f on X is a collection $(f_j)_{j \in \mathbb{Z}_+}$ of functions $f_j \in \mathcal{O}(\mathbb{D})$ with $f_0(\lambda_{j,k}) = f_j(\lambda_{j,k})$, $0 \le k \le j$, $j \ge 1$. Hence, the collection $(g_j)_{j \in \mathbb{Z}_+}$ with $g_j(\lambda) = \prod_{k=0}^j \frac{\lambda - \lambda_{j,k}}{1 - \overline{\lambda}_{j,k}\lambda}$ represents a holomorphic function g on X. Moreover, there is a subsequence $g_{j_\nu} \overset{K}{\underset{\nu\to\infty}{\Longrightarrow}} g$ with $g \in \mathcal{O}(\mathbb{D})$ and $|g(0)| = 1/e$. These facts lead to a proof of the equality $\operatorname{top} X = \operatorname{top} \boldsymbol{c}_X$ and to the $\boldsymbol{c}$-completeness of X.

To see that X is not $\boldsymbol{c}$-finitely compact, observe that any $(f_j)_{j \in \mathbb{Z}_+} \in \mathcal{O}(X)$ with $f_0(0) = 0$ yields holomorphic functions $h_j \in \mathcal{O}(\mathbb{D})$ with $h_j(\lambda) := (1/2)(f_j(\lambda) - f_0(\lambda))$, $j \ge 1$. Then, $h_j(\lambda_{j,k}) = 0$ for $0 \le k \le j$. Therefore, by the Schwarz lemma, it follows that $|f_j(0)| \le 2(1 - 1/(j+1))^{j+1} \underset{j\to\infty}{\longrightarrow} 2/e$, which can be read as $0_j \in B_{\boldsymbol{c}_X}(0_0, 2)$, $j \gg 1$. Here, 0_j denotes the origin of D_j in X.

Exercise 14.6.3. Let $(a_j)_{j\in\mathbb{N}}$ be a sequence of points in $\mathbb{D}$, $a_j \neq a_k$ if $j \neq k$, such that every boundary point of $\mathbb{D}$ is the non-tangential limit of a subsequence of (a_j). Choose sequences $(n_j)_{j\in\mathbb{N}}$ and $(m_j)_{j\in\mathbb{N}}$ of natural numbers satisfying

$$\sum_{j=1}^{\infty}(1/n_j)\log\frac{|a_j|}{2} > -\infty, \quad m_j \geq n_j,$$

$\mathbb{B}(a_j, 3\exp(-jm_j)) \cap \mathbb{B}(a_k, 3\exp(-km_k)) = \varnothing$ if $k \neq j$, and $\mathbb{B}(a_j, 3\exp(-jm_j)) \subset \mathbb{D}$ for all $j \in \mathbb{N}$. Then consider the following domain in $\mathbb{C}^2$:

$$G := \{z = (z_1, z_2) \in \mathbb{C}^2 : |z_1| < 1,\ |z_2| \cdot \exp u(z_1) < 1\},$$

where

$$u(\lambda) := \sum_{j=1}^{\infty}(1/n_j)\max\{\log(|z - a_j|/2), -jm_j\}.$$

Prove that u is continuous psh, that the pseudoconvex domain G is $\boldsymbol{k}$-complete, and that any bounded holomorphic function f on G depends only of the first variable, i.e., $f(z_1, z_2) = g(z_1)$ with $g \in \mathcal{O}(\mathbb{D})$; cf. [474].

Exercise 14.6.4. Let G be a $\boldsymbol{k}$-complete domain in $\mathbb{C}^n$. Prove that for $f \in \mathcal{H}^\infty(G)$ the domain $G' := \{z \in G : f(z) \neq 0\}$ is again $\boldsymbol{k}$-complete.

Exercise 14.6.5. Let $G_j \subset \mathbb{C}^{n_j}$ be a domain, $j = 1, 2$. Assume that G_2 is $\boldsymbol{k}$-complete. Let $A \subset G_1$ be relatively closed with $H^{2n_1-2}(A) = 0$. Then every holomorphic mapping $F : G_1 \setminus A \longrightarrow G_2$ extends to a holomorphic map $\widetilde{F} : G_1 \longrightarrow G_2$.

Hint. Use Appendix B.1.11.

14.7 List of problems

Chapter 15

Bergman completeness

Summary. Conditions for bounded domains in $\mathbb{C}^n$ are studied, which imply their Bergman completeness, i.e., the completeness with respect to the Bergman distance, which was introduced in Section 12.7. As was already observed by S. Kobayashi (see [309, 310]) this property is strongly related to the boundary behavior of the Bergman kernel (recall Section 12.4), which leads to a lot of results. For example, any hyperconvex bounded domain is Bergman complete (see Theorem 15.1.5). Moreover, characterizations for Zalcman domains (Theorem 15.1.14) and for Reinhardt domains (Section 15.2) are given to be Bergman complete.

15.1 Bergman completeness

Let $D \subset \mathbb{C}^n$ be a domain satisfying inequality (12.7.1), i.e.,

$$\mathfrak{K}_D(z) > 0, \quad z \in D. \tag{15.1.1}$$

In that case we will simply say that D *allows a Bergman pseudometric*. Then, such a D is called ***b**-complete* if $\boldsymbol{b}_D$ is, in fact, a metric and any $\boldsymbol{b}_D$-Cauchy sequence converges to a point in D in the standard topology.

In this section, we will restrict ourselves mostly to bounded domains in $\mathbb{C}^n$. Thus, (almost) all domains under consideration allow a Bergman pseudometric and they are automatically $\boldsymbol{\varrho}$- and $\boldsymbol{b}$-hyperbolic. Moreover, since the Bergman distance is inner, all completeness notions with respect to $\boldsymbol{b}$ coincide. Therefore, we will only speak of $\boldsymbol{b}$-completeness. Observe that Theorem 12.9.4 implies that any bounded $\boldsymbol{\varrho}$-complete domain is also $\boldsymbol{b}$-complete. The inverse implication is still an open problem. ?

Again we obtain the necessity of the pseudoconvexity.

Theorem 15.1.1 (cf. [75]). *Any bounded **b**-complete domain G is pseudoconvex.*

Proof. Suppose the contrary. Then, there exist polydiscs $\Delta := \mathbb{P}(z_0, r) \subset G$ and $\Delta' := \mathbb{P}(z_0, R) \not\subset G$, $R > r$, such that for every $f \in \mathcal{O}(G)$ the restriction $f|_\Delta$ extends holomorphically to Δ'. In particular, by Hartogs' theorem, there exists a function $F : \Delta' \times \Delta' \longrightarrow \mathbb{C}$ with the following properties:

(a) $F|_{\Delta \times \Delta} = \boldsymbol{K}_G|_{\Delta \times \Delta}$,

(b) $\Delta' \times \widehat{\Delta}' \ni (z, w) \longmapsto F(z, \overline{w})$ is holomorphic, where $\widehat{\Delta}' := \{w \in \mathbb{C}^n : \overline{w} \in \Delta'\}$.

By hypothesis, there exists $z' \in \Delta' \cap \partial G$ such that the segment $[z_0, z') \subset G \cap \Delta'$. Since $\log \mathfrak{K}_G(z) = \log F(z, z)$ near $[z_0, z')$, it follows that any sequence $(z_j)_{j \in \mathbb{N}} \subset [z_0, z')$ with $\lim_{j\to\infty} z_j = z'$ is a $\boldsymbol{b}_G$-Cauchy sequence; a contradiction. □

Example 15.1.2. The converse of Theorem 15.1.1 fails to hold. For example, consider the pseudoconvex domain

$$G := \{z \in \mathbb{C}^2 : |z_1| < |z_2| < 1\}.$$

Recalling that G is biholomorphic to $\mathbb{D} \times \mathbb{D}_*$, we obtain

$$\mathfrak{K}_G(z) = \frac{|z_2|^2}{\pi^2(|z_2|^2 - |z_1|^2)^2(1 - |z_2|^2)^2}.$$

Then, a direct calculation shows that $\boldsymbol{b}_G((0, 1/\nu), (0, 1/\mu)) \underset{\nu,\mu\to\infty}{\longrightarrow} 0$, i.e., $((0, 1/\nu))_{\nu=2}^{\infty}$ is a $\boldsymbol{b}_G$-Cauchy sequence with $\lim_{\nu\to\infty}(0, 1/\nu) = (0, 0) \in \partial G$. Observe that $\mathcal{H}^\infty(G)$ is not dense in $L_h^2(G)$, e.g., the $L_h^2(G)$-function $1/z_2$ cannot be approximated (see [269]).

There is the following localization, which will be helpful in concrete situation:

Lemma 15.1.3. *Let G be a bounded pseudoconvex domain in $\mathbb{C}^n$ with $z' \in \partial G$. Suppose that there is a neighborhood $U = U(z')$ with the following properties:*

(i) *$U \cap G$ is connected,*

(ii) *any sequence $(z_\nu)_{\nu\in\mathbb{N}} \subset U \cap G$ with $\lim_{\nu\to\infty} z_\nu = z'$ is not a $\varrho_{G\cap U}$-Cauchy sequence.*

Then, there is no ϱ_G-Cauchy sequence converging to z'.

Proof. Let $U' = U'(z') \subset\subset U$. Then, an argument similar the one used in the proof of Theorem 12.8.6, but simpler, leads to the following inequality:

$$\widetilde{M}_{G\cap U}(z', z'') \le C\widetilde{M}_G(z', z''), \quad z', z'' \in U' \cap G,$$

where

$$\widetilde{M}_{G\cap U}(z', z'') := \sup\{|f(z')| : f \in L_h^2(G \cap U),\ \|f\|_{L^2(G\cap U)} = 1,\ f(z'') = 0\}.$$

Then, applying Theorem 12.9.3, we obtain

$$\varrho_{G\cap U}(z', z'') \le \frac{\widetilde{M}_{G\cap U}(z', z'')}{\sqrt{\mathfrak{K}_{G\cap U}(z')}} \le \frac{C\widetilde{M}_G(z', z'')}{\sqrt{\mathfrak{K}_G(z')}} \le \sqrt{2}C\varrho_G(z', z''),$$

$$z', z'' \in U' \cap G. \qquad \square$$

Observe that the above lemma immediately leads to the conclusion that the Sibony domain in Example 14.2.15 is $\boldsymbol{\varrho}$-complete.

The main tool for proving $\boldsymbol{b}$-completeness is the following so-called Kobayashi criterion:

Theorem 15.1.4. *Let $D \subset \mathbb{C}^n$ be a bounded pseudoconvex domain.*

(a) *Assume that*

$$\limsup_{z \to \partial D} \frac{|f(z)|}{\sqrt{\mathfrak{K}_D(z)}} < \|f\|^2_{L^2_h(D)}, \quad f \in L^2_h(D) \setminus \{0\}.$$

Then, D is $\boldsymbol{b}$-complete.

(b) *Let $H \subset L^2_h(D)$ be a dense subset. Moreover, assume that for any sequence $(z_j)_j \in D$, $z_j \longrightarrow z_0 \in \partial D$, and any $g \in H$, there is a subsequence $(z_{j_k})_k$ such that*

$$\lim_{k \to \infty} \frac{|g(z_{j_k})|}{\sqrt{\mathfrak{K}_D(z_{j_k})}} = 0.$$

Then, D is $\boldsymbol{\varrho}$-complete and thus also $\boldsymbol{b}$-complete.

Proof. (a) Suppose the contrary. Then, there exists a $\boldsymbol{\varrho}_D$-Cauchy sequence $(z_\nu)_{\nu \in \mathbb{N}} \subset D$ such that $\lim_{\nu \to \infty} z_\nu = z' \in \partial D$.

Then, by the definition of $\boldsymbol{\varrho}_D$ and the completeness of $L^2_h(D)$, one finds a function $f \in L^2_h(D)$, $\|f\|_{L^2_h(D)} = 1$, such that

$$\frac{\boldsymbol{K}_D(\cdot, z_\nu)}{\sqrt{\mathfrak{K}_D(z_\nu)}} e^{i\theta_\nu} \longrightarrow f \quad \text{in} \quad L^2_h(D),$$

where θ_ν are suitable real numbers. Hence, we obtain

$$\frac{|f(z_\nu)|}{\sqrt{\mathfrak{K}_D(z_\nu)}} = \frac{|\langle f, \boldsymbol{K}_D(\cdot, z_\nu) e^{i\theta_\nu} \rangle_{L^2(D)}|}{\sqrt{\mathfrak{K}_D(z_\nu)}} \longrightarrow \|f\|^2_{L^2(D)} = 1;$$

a contradiction.

(b) Suppose again that D is not $\boldsymbol{b}$-complete and choose f and θ_j as above. Moreover, take a $g \in H$ with $\|g - f\|_{L^2_h(D)} < 1/2$. Then, by virtue of our assumption, there is a subsequence $(z_{j_k})_k$ such that

$$1 \underset{k \to \infty}{\longleftarrow} \frac{|f(z_{j_k})|}{\sqrt{\mathfrak{K}_D(z_{j_k})}} \le \|f - g\|_{L^2_h(D)} + \frac{|g(z_{j_k})|}{\sqrt{\mathfrak{K}_D(z_{j_k})}} \underset{k \to \infty}{\longrightarrow} \|f - g\|_{L^2_h(D)} < \frac{1}{2};$$

a contradiction. □

Now, we list some immediate consequences of Theorem 15.1.4:

Theorem 15.1.5 (cf. [64, 230]). *Assume that $D \subset \mathbb{C}^n$ is a bounded pseudoconvex domain such that*

$$\lim_{D \ni z \to \partial D} \mathcal{L}^{2n}(D(z)) = 0.$$

Then, D is $\boldsymbol{\varrho}$-complete and so also $\boldsymbol{b}$-complete. In particular, if D is hyperconvex, then it is $\boldsymbol{b}$-complete.

Proof. Use Theorems 12.4.3 and 12.4.4. □

Recall that any bounded pseudoconvex domain with a $\mathcal{C}^1$-smooth boundary is hyperconvex (cf. Remark 3.2.3(b)). Therefore,

Corollary 15.1.6. *If $D \subset \mathbb{C}^n$ is a bounded pseudoconvex domain with a $\mathcal{C}^1$-smooth boundary, then D is $\boldsymbol{b}$-complete.*

Remark 15.1.7. Because of the former corollary it would be interesting to understand how fast the Bergman distance blows up at the boundary. The most general result at the moment seems to be the following one (see [60]), which is an improvement of a former result in [132]: *Let $D \subset \mathbb{C}^n$ be a bounded pseudoconvex domain with a $\mathcal{C}^2$-smooth boundary. Given a point $a \in D$, then there exists a positive constant c such that*

$$\boldsymbol{b}_D(a,z) \geq c \frac{-\log(\operatorname{dist}(z,\partial D))}{\log\big(-\log(\operatorname{dist}(z,\partial D))\big)}, \qquad z \in D, \textit{ near to } \partial D.$$

Recall that bounded balanced pseudoconvex domains are not necessarily hyperconvex (see Remark 3.2.3(a)). Nevertheless, Theorem 12.4.10 gives (see [262] and [251] in case that $\mathfrak{h}_D$ is continuous)

Proposition 15.1.8. *Any bounded balanced pseudoconvex domain $D \subset \mathbb{C}^n$ is $\boldsymbol{b}$-complete.*

Proof. Recall that any $f \in \mathcal{O}(D)$ can be written as a series $\sum_{k=1}^{\infty} Q_k$, where Q_k are homogeneous polynomials, and that this convergence is an L_h^2-convergence. Therefore, the bounded holomorphic functions on D are dense in $L_h^2(D)$. Then, Theorem 15.1.4(b) finishes the proof. □

Recall that $\boldsymbol{c} \leq \boldsymbol{b}$. Thus, any $\boldsymbol{c}$-complete bounded domain is $\boldsymbol{b}$-complete (use Corollary 12.9.7). Along with Theorem 14.5.7, we obtain the

Corollary 15.1.9.

(a) *For any bounded pseudoconvex balanced domain $G \subset \mathbb{C}^n$ with non continuous Minkowski function, any comparison of the type $\boldsymbol{b}_G \le C\boldsymbol{k}_G$ fails to hold.*

(b) *There exists a bounded pseudoconvex balanced domain G with continuous Minkowski function in $\mathbb{C}^3$ for which any comparison of the type $\boldsymbol{b}_G \le C\boldsymbol{k}_G$ fails to hold.*

Remark 15.1.10. Nevertheless, if $D \subset \mathbb{C}^n$ is a $\mathbb{C}$-convex domain that does not contain any complex line, then $\boldsymbol{k}_D$ and $\boldsymbol{b}_D$ are comparable, i.e.,

$$\boldsymbol{k}_D \le 4\boldsymbol{b}_D \le c_n \boldsymbol{c}_D^i \le c_n \boldsymbol{k}_D,$$

where the constant c_n depends only on the dimension n. These inequalities are a direct consequence of the inequalities between the corresponding infinitesimal versions, $\boldsymbol{\varkappa}_D \le 4\boldsymbol{\beta}_D \le c_n\boldsymbol{\gamma}_D$ (see [391]).

In the planar case we have

Corollary 15.1.11. *Any bounded $\boldsymbol{K}$-exhaustive domain $D \subset \mathbb{C}$ is $\boldsymbol{b}$-complete.*

Proof. Again use Theorem 15.1.4 and Appendix B.8.2. □

There are also sufficient conditions for Hartogs domains with m-dimensional fibers to be $\boldsymbol{b}$-complete (see [275]).

Theorem 15.1.12. *Let $D \subset \mathbb{C}^n$ be a domain and let G_D be bounded pseudoconvex Hartogs domain with m-dimensional balanced fibers (cf. Appendix* B.7.7*). Assume that D is $\boldsymbol{K}$-exhaustive, that $\mathcal{H}^\infty(D)$ is dense in $L_h^2(D)$, and that there is an $\varepsilon > 0$ such that $D \times \mathbb{P}_m(\varepsilon) \subset G_D$.*

Then, G_D is $\boldsymbol{b}$-complete.

Proof. According to Theorems 15.1.4 and 12.4.12 it suffices to show that $\mathcal{H}^\infty(G_D)$ is dense in $L_h^2(G_D)$. So, take an $F \in L_h^2(G_D)$. Then,

$$F(z,w) = \sum_{k=0}^{\infty} F_k(z,w) = \sum_{k=0}^{\infty} \sum_{\alpha\in\mathbb{Z}_+^m, |\alpha|=k} f_\alpha(z) w^\alpha, \quad (z,w) \in G_D,$$

where the convergence is locally uniform. Because of the orthogonality of the functions g_α ($g_\alpha(w) := w^\alpha$) it follows that the functions $G_N := \sum_{k=0}^N F_k$ converge to F in $L_h^2(G_D)$. Therefore, it suffices to approximate $f_\alpha g_\alpha$ by bounded holomorphic functions. Using the theorem of Fubini together with the last assumption shows that each $f_\alpha \in L_h^2(D)$. Hence, there are bounded holomorphic functions $h_{\alpha,k} \in \mathcal{H}^\infty(D)$, $k \in \mathbb{N}$, such that $h_{\alpha,k} \longrightarrow f_\alpha$ in $L_h^2(D)$ if $k \longrightarrow \infty$. Consequently, $h_{\alpha,k} g_\alpha$ tends to $f_\alpha g_\alpha$ in $L_h^2(G_D)$, $\alpha \in \mathbb{Z}_+^m$. □

Remark 15.1.13. Note that the condition $D \times \mathbb{P}_m(\varepsilon) \subset G_D$ is essential. For instance, $\mathcal{H}^\infty(\mathbb{D}_*) = \mathcal{H}^\infty(\mathbb{D})|_{\mathbb{D}_*}$ is dense in $L^2_h(\mathbb{D}_*) = L^2_h(\mathbb{D})|_{\mathbb{D}_*}$, but $\mathcal{H}^\infty(G_{\mathbb{D}_*})$ is not dense in the Hartogs triangle $G_{\mathbb{D}_*} := \{(z,w) \in \mathbb{D}_* \times \mathbb{C} : |w| < |z|\}$.

We mention without proof that G_D is also $\boldsymbol{b}$-complete if the base domain D is $\boldsymbol{c}^i$-complete (see [275]). For further results on $\boldsymbol{b}$-complete Hartogs domains see also [96]. ? Is there a complete characterization of such domains that are $\boldsymbol{b}$-complete?

Moreover, the following result may be found in [95]:

Theorem. *Let $u \in \mathcal{PSH}(\mathbb{C}^n)$, $u \not\equiv -\infty$, and $h \in \mathcal{O}(\mathbb{C}^m)$, $h \not\equiv 0$, be such that $u \in \mathcal{C}(\mathbb{C}^n \setminus u^{-1}(-\infty))$. Let $r > 0$ and assume that*

$$\Omega := \{(z', z'') \in \mathbb{B}_n(r) \times \mathbb{B}_m(r) \subset \mathbb{C}^n \times \mathbb{C}^m : u(z') + e^{\frac{1}{|h(z'')|}} < 1\}$$

is a domain. Then, Ω is $\boldsymbol{b}$-complete.

If, in addition, there is a point $(z_0', z_0'') \in \mathbb{B}_n(r) \times \mathbb{B}_m(r)$ with $u(z_0') = -\infty$, $h(z_0'') = 0$, then Ω is not hyperconvex.

Observe that the boundary behavior of the level sets of the Green function implies both $\boldsymbol{K}$-exhaustiveness and $\boldsymbol{b}$-completeness. We already saw that there exist $\boldsymbol{K}$-exhaustive domains that are not $\boldsymbol{b}$-complete. It was a long-standing question whether a $\boldsymbol{b}$-complete domain is automatically $\boldsymbol{K}$-exhaustive. The first counterexample to that question was given by W. Zwonek [568] (see also [570]). The following Theorem 15.1.14 (see [284]) gives even a large variety of domains that are $\boldsymbol{b}$-complete but not $\boldsymbol{K}$-exhaustive.

Theorem 15.1.14. *Let $D \subset \mathbb{C}$ be a Zalcman domain as in Corollary* 12.5.15. *Then,*

$$D \text{ is } \boldsymbol{b}\text{-complete} \quad \text{iff} \quad \sum_{k=1}^{\infty} \frac{1}{x_k \sqrt{-\log r_k}} = \infty.$$

Proof. The proof "$\Longrightarrow$" is similar to the one of Theorem 12.5.7; so it is omitted here.

Proof of "$\Longleftarrow$": Suppose that D is not $\boldsymbol{b}$-complete. Then, D is not $\boldsymbol{b}$-exhaustive. Therefore, in view of Corollary 12.5.15, we have

$$\sum_{k=1}^{\infty} \frac{1}{x_k \sqrt{-\log r_k}} = \infty \quad \text{and} \quad \lim_{j \to \infty} \frac{1}{-x_j^2 \log r_j} = 0. \tag{15.1.2}$$

Moreover, there is a $\boldsymbol{b}_D$-Cauchy sequence $(z_k)_k \subset D$ with $\lim_{k\to\infty} z_k = 0$. We may even assume that $\boldsymbol{b}_D(z_k, z_{k+1}) < \frac{1}{2^k}$. So there exist $\mathcal{C}^1$-curves $\gamma_k : [0,1] \longrightarrow D$ such that $L_{\boldsymbol{\beta}_D}(\gamma_k) < 1/2^k$. Gluing all these curves together, we obtain a piecewise $\mathcal{C}^1$-curve $\gamma : [0,1) \to D$ with a finite $\boldsymbol{\beta}_D$-length.

We claim that the Bergman kernel remains bounded along γ. In fact, if not, then there is a sequence $(w_k)_k \subset \gamma([0,1))$ such that

$$\lim_{k\to\infty} \mathfrak{K}_D(w_k) = \infty, \quad \lim_{k\to\infty} w_k = 0.$$

Obviously, the sequence $(w_k)_k$ is again a $\boldsymbol{b}_D$-Cauchy sequence. As in the proof of Theorem 15.1.4, there exist an $f \in L_h^2(D)$ and a subsequence $(w_{k_j})_j$ such that

$$\lim_{j\to\infty} \frac{|f(w_{k_j})|^2}{\mathfrak{K}_D(w_{k_j})} = 1.$$

Applying Theorem B.8.2, we find a $g \in L_h^2(D)$, locally bounded near 0, such that $\|f-g\|_{L_h^2(D)} < 1/2$. Therefore,

$$0 \underset{j\to\infty}{\longleftarrow} \frac{|g(w_{k_j})|}{\sqrt{\mathfrak{K}_D(w_{k_j})}} \geq \frac{|f(w_{k_j})|}{\sqrt{\mathfrak{K}_D(w_{k_j})}} - \|f-g\|_{L_h^2(D)} \geq \frac{|f(w_{k_j})|}{\sqrt{\mathfrak{K}_D(w_{k_j})}} - \frac{1}{2} \underset{j\to\infty}{\longrightarrow} \frac{1}{2};$$

a contradiction. Hence, there is a positive C such that $\mathfrak{K}_D(\gamma(t)) \leq C$, $t \in [0,1)$.

To be able to continue, we need the following lemma:

Lemma 15.1.15. *Let D be a domain as above, satisfying* (15.1.2), *and let $\gamma : [0,1) \longrightarrow D$ be a piecewise $\mathcal{C}^1$-curve with $\lim_{t\to 1}\gamma(t) = 0$. Then,*

$$\lim_{\tau\to 1} \int_0^\tau \sqrt{M_D(\gamma(t);\gamma'(t))}dt = \infty.$$

Proof. We may assume that $|\gamma(0)| > x_1$ and that $x_1\sqrt{-\log r_1} < x_j\sqrt{-\log r_j}$, $j \geq j_0$ for a suitable j_0 (use (15.1.2)). Now, fix an $N \in \mathbb{N}$, $N \geq j_0$, and let $z_N \in D$ be an arbitrary point with $x_{N+2} \leq |z_N| \leq x_{N+1}$.

We define

$$f := f_{\overline{\mathbb{D}}(x_1,r_1)} - \frac{x_N - z_N}{x_1 - z_N} f_{\overline{\mathbb{D}}(x_N,r_N)},$$

where f_K denotes the Cauchy transform of K. Or, more explicitly, we have

$$f(z) = \frac{1}{x_1 - z} - \frac{x_N - z_N}{x_1 - z_N}\frac{1}{x_N - z}, \quad z \in D.$$

Therefore, we see that $f(z_N) = 0$ and $f'(z_N) = \frac{x_N - x_1}{(x_1-z_N)^2(x_N-z_N)}$. What remains is to estimate the $L_h^2(D)$-norm of the function f. Applying the relation between x_n and x_{n+1}, we get

$$\begin{aligned}\|f\|_{L_h^2(D)} &\leq \|f_{\overline{\mathbb{D}}(x_1,r_1)}\|_{L_h^2(D)} + \frac{|x_N - z_N|}{|x_1 - z_N|}\|f_{\overline{\mathbb{D}}(x_N,r_N)}\|_{L_h^2(D)} \\ &\leq C_2 \frac{|x_N - z_N|}{|x_1 - z_N|}\sqrt{-\log r_N},\end{aligned}$$

where C_1, C_2 are positive constants, independent of N and z_N.

Therefore, if $x_{N+2} \le |z| \le x_{N+1}$ then,

$$\sqrt{M_D(z;X)} \ge |X| \frac{|x_1 - x_N|}{C_2|x_1 - \gamma(t)||x_n - \gamma(t)|^2\sqrt{-\log r_N}} \ge |X| \frac{C_3}{x_N^2\sqrt{-\log r_N}},$$

where $C_3 > 0$ is a constant (use again that $x_{k+1} \le \Theta_2 x_k$ for all k). Finally, we obtain

$$\begin{aligned}\lim_{\tau\to 1}\int_0^\tau \sqrt{M_D(\gamma(t);\gamma'(t))}dt &\ge \sum_{N=j_0}^{\infty} C_3 \frac{x_{n+1} - x_{N+2}}{x_N^2\sqrt{-\log r_N}} \\ &\ge \sum_{N=j_0}^{\infty} C_3 \frac{\Theta_1 x_N - \Theta_2^2 x_N}{x_N^2\sqrt{-\log r_N}} \\ &\ge C_4 \sum_{N=j_0}^{\infty} \frac{1}{x_N\sqrt{-\log r_N}} = \infty,\end{aligned}$$

where $C_4 > 0$. Hence, the proof of this lemma is complete. □

Now, applying Lemma 15.1.15 leads to the following contradiction:

$$\infty > \lim_{\tau\to 1}\int_0^\tau \boldsymbol{\beta}_D(\gamma(t);\gamma'(t))dt \ge \frac{1}{C}\int_0^\tau \sqrt{M_D(\gamma(t);\gamma'(t))}dt = \infty. \qquad \square$$

The boundary behavior of the Bergman metric on a Zalcman domain is partially described in the following result the proof of which is based on methods from the proof of Theorem 15.1.14.

Theorem 15.1.16 (cf. [284]). *Let D be a domain as in Theorem* 15.1.14.

(a) *If* $\sum_{k=1}^{\infty} \frac{1}{x_k^2\sqrt{-\log r_k}} < \infty$, *then* $\limsup_{(-1,0)\ni t\to 0} \boldsymbol{\beta}_D(t;1) < \infty$.

(b) *If* $\limsup_{(-1,0)\ni t\to 0} \boldsymbol{\beta}_D(t;1) < \infty$, *then* $\limsup_{k\to\infty} \frac{1}{x_k^2\sqrt{-\log r_k}} < \infty$.

? It seems to be open how to characterize those Zalcman domains that are *$\boldsymbol{\beta}$-exhaustive*, i.e., $\lim_{z\to\partial D} \boldsymbol{\beta}_D(z;1) = \infty$.

The $\boldsymbol{b}$-completeness means, heuristically, that boundary points are infinitely far away from inner points, so one might think that for a $\boldsymbol{b}$-complete domain, the Bergman metric $\boldsymbol{\beta}_D$ becomes infinite at the boundary. The following example shows that this is not true.

Example 15.1.17. There exists a $\boldsymbol{b}$-complete bounded domain D in the plane that is not $\boldsymbol{\beta}_D$-exhaustive, i.e., there is a boundary sequence $(w_k)_k \subset D$ such that the sequence $(\boldsymbol{\beta}_D(w_k;1))_{k\in\mathbb{N}}$ is bounded (cf. [424]). To be more precise:

Put

$$x_n := \frac{1}{2^{n+1}} + \frac{1}{2^{n+2}}, \quad z_{n,j} := \exp\left(i\frac{2\pi j}{2^{4n}}\right), \qquad n \in \mathbb{N},\ j = 0, \dots, 2^{4n} - 1.$$

Moreover, let $r_n := \exp(-C_1 2^{9n})$, $n \in \mathbb{N}$, where $C_1 > 0$ is chosen such that

- the discs $\mathbb{B}(z_{n,j}, r_n) \subset \mathbb{C}$, $n \in \mathbb{N}$, $j = 0, \dots, 2^{4n} - 1$, are pairwise disjoint,
- $\overline{\mathbb{B}}(z_{n,j}, r_n) \subset A_n(0)$, $n \in \mathbb{N}$, $j = 0, \dots, 2^{4n} - 1$.

Then, there is a sequence $(n_k)_k \subset \mathbb{N}$ such that the domain

$$D := \mathbb{D} \setminus \left(\bigcup_{k=1}^{\infty} \bigcup_{j=0}^{2^{4n_k}-1} \overline{\mathbb{D}}(z_{n_k,j}, r_n) \right)$$

is a domain satisfying the above desired properties.

At the end of this section, we add a few remarks for unbounded domains.

Theorem 15.1.18 (cf. [92]). *Let $D \subset \mathbb{C}^n$ be a pseudoconvex domain (not necessarily bounded). Assume for any point $w \in D$ that there exists an $r > 0$ such that*

$$D_w(r) := \{z \in D : \log \boldsymbol{g}_D(w, z) < -r\} \subset\subset D.^1$$

Then, D allows a Bergman metric and $\boldsymbol{b}_D$ is a distance.

We will not present the full proof of this result. We only mention that for any $z \in D$ and any non zero vector $X \in \mathbb{C}^n$, one has to construct functions $f, g \in L_h^2(D)$ such that

$$g(z) \neq 0 \quad \text{and} \quad f'(z)X \neq 0,$$

which is done, as always, by solving a $\bar{\partial}$-problem.

In particular, D allows a Bergman distance, if one of the following conditions is true (see [92]):

(a) there exists a bounded strongly psh function on D;

(b) D is hyperconvex, i.e., there is a negative psh function u on D such that all the level sets $\{z \in D : u(z) < -r\}$, $r > 0$, are relatively compact subsets of D;

(c) $D \subset \mathbb{C}$ and its complement $\mathbb{C} \setminus D$ is not a polar set. (15.1.3)

Moreover, the following result due to N. Nikolov (private communication) is also a consequence of Theorem 15.1.18.

Corollary 15.1.19. *Let $D \subset \mathbb{C}^n$ be an unbounded domain. Assume that there are $R > 0$ and $\psi \in \mathcal{PSH}(D \setminus \overline{\mathbb{B}(R)})$ such that*

- $\psi < 0$ *on* $D \setminus \overline{\mathbb{B}(R)}$,
- $\lim_{z\to\infty} \psi(z) = 0$,
- $\limsup_{z\to a} \psi(z) < 0$, $a \in (\partial D) \setminus \overline{\mathbb{B}(R)}$.

Then, D has the Bergman metric.

Proof. Fix a $z_0 \in D$ and choose positive numbers $R_3 > R_2 > R_1 > R$ such that $\|z_0\| < R_1$ and

$$2 \inf_{D\setminus\mathbb{B}(R_2)} \psi \geq \sup_{D\cap\partial\mathbb{B}(R_1)} \psi =: c < 0.$$

Moreover, put

$$\begin{aligned} d &:= \inf_{w\in D\cap\partial\mathbb{B}(R_1)} \log \boldsymbol{g}_{\mathbb{B}(R_3)}(z_0, w) > -\infty, \\ u(w) &:= 2\psi(w)(d/c) - d, \quad w \in D \setminus \overline{\mathbb{B}(R)}. \end{aligned}$$

Observe that

$$\begin{aligned} u(w) &\leq d \leq \log \boldsymbol{g}_{\mathbb{B}(R_3)}(z_0, w), \quad w \in D \cap \partial\mathbb{B}(R_1), \\ u(w) &\geq 0 \geq \log \boldsymbol{g}_{\mathbb{B}(R_3)}(z_0, w), \quad w \in D \cap \partial\mathbb{B}(R_2). \end{aligned}$$

Hence, the function

$$v(w) := \begin{cases} \log \boldsymbol{g}_{\mathbb{B}(R_3)}(z_0, w), & w \in D \cap \mathbb{B}(R_1) \\ \max\{\log \boldsymbol{g}_{\mathbb{B}(R_3)}(z_0, w),\ u(w)\}, & w \in D \cap (\mathbb{B}(R_2) \setminus \mathbb{B}(R_1)) \\ u(w), & w \in D \setminus \mathbb{B}(R_2) \end{cases}$$

is psh on D with logarithmic pole at z_0. Therefore, $v + d \leq \log \boldsymbol{g}_D(z_0, \cdot)$ on D. Since $v = u \geq 0$ on $D \setminus \mathbb{B}(R_2)$, we have $\log \boldsymbol{g}_D(z_0, \cdot) \geq d$ on $D \setminus \mathbb{B}(R_2)$. And if $w \in D \cap \mathbb{B}(R_2)$, then $\log \boldsymbol{g}_D(z_0, w) \geq \log \boldsymbol{g}_{\mathbb{B}(R_3)}(z_0, w) + d \geq \log \boldsymbol{g}_{\mathbb{B}(z_0, R_3+\|z_0\|)}(z_0, w) + d$.

Let $\mathbb{B}(z_0, s) \subset\subset D \cap \mathbb{B}(R_2)$, then there is a $d_1 < 0$ such that $\log \boldsymbol{g}_D(z_0, w) \geq d + d_1$, $w \in D \cap \mathbb{B}(R_2) \setminus \mathbb{B}(z_0, s)$. Therefore, $D_{z_0}(-d - d_1) \subset\subset D$ and, since z_0 was arbitrarily chosen, Theorem 15.1.18 finishes the proof. □

Due to N. Nikolov [376], $\boldsymbol{b}$-completeness for plane domains (also unbounded ones) is a local property.

Theorem 15.1.20. *Let $D \subset \mathbb{C}$ be a domain such that $\mathbb{C} \setminus D$ is not a polar set. Assume that D is locally $\boldsymbol{b}$-complete.*[2] *Then, D is $\boldsymbol{b}$-complete.*

The proof of Theorem 15.1.20 is based on the following lemma:

Lemma 15.1.21. *Let $D \subset \mathbb{C}$ be a domain such that $\mathbb{C} \setminus D$ is not polar. Moreover, let $a \in \partial D$ and $U = U(a)$ be an open neighborhood of a. Then, there exist a neighborhood $V = V(a) \subset U$ and a constant $C > 0$ such that*

$$C\boldsymbol{\beta}_{\widehat{U}}(z;1) \le \boldsymbol{\beta}_D(z;1), \quad z \in V \cap D,$$

where $\widehat{U}$ denotes that connected component of $D \cap U$ with $z \in \widehat{U}$.

Proof. Since $\mathbb{C} \setminus D$ is not polar, there is an $r_0 > 0$ such that $\mathbb{C} \setminus \big(D \cup \mathbb{D}(a, r_0)\big)$ is not polar. Hence, $\log \boldsymbol{g}_{D \cup \mathbb{D}(a,r_0)}$ is harmonic on $\big(D \cup \mathbb{D}(a, r_0)\big) \setminus \{a\}$. Fix an $r_1 \in (0, r_0)$ and define $D_1 := D \cup \mathbb{D}(a, r_1)$. Applying that $\boldsymbol{g}_{D_1}(a, z) \ge \boldsymbol{g}_{D \cup \mathbb{D}(a,r_0)}(a, z)$, $z \in D_1$, we have

$$\inf\{\log \boldsymbol{g}_{D_1}(a, z) - |z - a|^2 : z \in \partial\mathbb{D}(a, r_1) \cap D\} =: m > -\infty.$$

Put

$$u(z) := \begin{cases} \max\{|z - a|^2 + m, \log \boldsymbol{g}_{D_1}(a, z)\}, & \text{if } z \in D \cap \mathbb{D}(a, r_1) \\ \log \boldsymbol{g}_{D_1}(a, z), & \text{if } z \in D \setminus \mathbb{D}(a, r_1) \end{cases}.$$

Observe that $|z - a|^2 + m \le \boldsymbol{g}_{D_1}(a, z)$, $z \in D \cap \mathbb{D}(a, r_1)$. Therefore, $0 \ge u \in \mathcal{SH}(D_1)$ and $u(z) = |z - a|^2 + m$, $z \in \mathbb{D}(a, r_2)$, for a sufficiently small $r_2 < r_1$.

Choose numbers $0 < r_4 < r_3 < r_2$ and a $\mathcal{C}^\infty$ cut-off function χ such that $\chi \equiv 1$ on $\mathbb{D}(a, r_4)$ and $\chi \equiv 0$ outside of $\mathbb{D}(a, r_3)$.

Fix a point $z_0 \in D \cap \mathbb{D}(a, r_4)$ and let $\widehat{U}$ be the connected component of $D \cap U$ with $z_0 \in \widehat{U}$. Take an $f \in L^2_h(\widehat{U})$ with $f(z_0) = 0$. Put

$$\alpha(z) := \begin{cases} \bar\partial(\chi f)(z), & \text{if } z \in \widehat{U} \\ 0, & \text{if } z \in D \setminus \widehat{U} \end{cases}.$$

Then, α is a $\bar\partial$-closed $\mathcal{C}^\infty_{(0,1)}$-form on D satisfying the inequality

$$\int_D |\alpha(z)|^2 e^{-6\log \boldsymbol{g}_D(z_0,z) - u(z)} d\mathcal{L}^2(z) \le \widetilde{C} \int_{\widehat{U}} |f(z)|^2 d\mathcal{L}^2(z) < \infty,$$

[2] Observe that here the point ∞ is counted as a boundary point of D; so we also assume that there is a compact set $K \subset \mathbb{C}$ such that any (non-empty) connected component of $D \setminus K$ is $\boldsymbol{b}$-complete.

where $\widetilde{C} > 0$ is independent of f and z_0. Observe that the subharmonic weight function is strictly subharmonic near the support of α. Therefore, using Hörmander's L^2-theory,[3] we get a function $h \in \mathcal{C}^\infty(D)$ with $\bar{\partial} h = \alpha$ on D such that

$$\|h\|^2_{L^2(D)} \leq \int_D |h(z)|^2 e^{-6\log g_D(z_0,z)-u(z)} d\mathcal{L}^2(z) \leq C'\|f\|^2_{L^2_h(\widehat{U})},$$

where C' is a positive number, which is independent of f and z_0. Moreover, since the second integral is finite, it follows that $h(z_0) = h'(z_0) = 0$. Hence, the function

$$\widehat{f}(z) := \begin{cases} (\chi f)(z) - h(z), & \text{if } z \in \widehat{U} \\ -h(z), & \text{if } z \in D \setminus \widehat{U} \end{cases}$$

is holomorphic on D satisfying $\widehat{f}(z_0) = 0$, $\widehat{f}'(z_0) = f'(z_0)$, and $\|\widehat{f}\|_{L^2_h(D)} \leq C\|f\|_{L^2_h(\widehat{U})}$, where $C > 0$ is independent of f and z_0. Therefore, by virtue of Theorem 12.7.5, we get $\widetilde{C}\boldsymbol{\beta}_D(z_0; 1) \geq \boldsymbol{\beta}_{\widehat{U}}(z_0; 1)$. Since z_0 was arbitrary, the lemma is proved. □

Now, we turn to the proof of Theorem 15.1.20.

Proof of Theorem 15.1.20. First of all, let us mention that, by virtue of (15.1.3), D has a Bergman metric.

Now suppose that D is not $\boldsymbol{b}$-complete. Then, there is a $\boldsymbol{b}_D$-Cauchy sequence $(z_j)_j \subset D$ with $z_j \longrightarrow a \in \partial D$ or $z_j \longrightarrow \infty$. The second case can be reduced to the first one using the biholomorphic transformation $z \longmapsto \frac{1}{z-c}$, where $c \notin D$. So, we only have to deal with the first case.

By the assumption, there is an open neighborhood $U = U(a)$ such that any connected component of $U \cap D$ is $\boldsymbol{b}$-complete. Fix a positive r_1 such that $\mathbb{D}(a, r_1) \subset\subset U$. Applying Lemma 15.1.21, we may find positive numbers $r_2 < r_1$ and C such that $C\boldsymbol{\beta}_D(z; 1) \leq \min\{\boldsymbol{\beta}_{\widehat{U}}(z; 1), \boldsymbol{\beta}_{\widetilde{U}}(z; 1)\}$, $z \in D \cap \mathbb{D}(a, r_2)$, where $\widehat{U}$ and $\widetilde{U}$ denote the connected components of $D \cap U$ and $D \cap \mathbb{D}(a, r_1)$, respectively, with $z \in \widehat{U} \cap \widetilde{U}$. Choose an $r_3 \in (0, r_2)$. Put

$$d := \inf\{\boldsymbol{b}_{\widetilde{U}}(z, w) : z \in \partial\mathbb{D}(a, r_3) \cap D,\ w \in \partial\mathbb{D}(a, r_2) \cap D,\ z, w \in \widetilde{U},\ \widetilde{U} \text{ a connected component of } D \cap \mathbb{D}(a, r_1)\}.$$

[3] Here we use the following form of Hörmander's result.

Theorem. *Let $D \subset \mathbb{C}^n$ be a pseudoconvex domain, $\varphi \in \mathcal{PSH}(D)$, and $\alpha \in \mathcal{C}^\infty_{(0,1)}(D)$. Assume that $\bar{\partial}\alpha = 0$ and that on an open set $U \subset D$, $\operatorname{supp}\alpha \subset U$, the function φ can be written as $\varphi = \psi + \chi$, $\psi, \chi \in \mathcal{PSH}(U)$, such that $\mathcal{L}\psi(z; X) \geq C\|X\|^2$, $z \in U$, $X \in \mathbb{C}^n$. Then, there exists an $h \in \mathcal{C}^\infty(D)$, $\bar{\partial} h = \alpha$, such that $\int_D |h|^2 e^{-\varphi} d\mathcal{L}^{2n} \leq C' \int_D |\alpha|^2 e^{-\varphi} d\mathcal{L}^{2n}$, where $C' > 0$ depends only on C.*

By virtue of the inequality $\boldsymbol{c} \le \boldsymbol{b}$, it follows that $d > 0$. So, we may take an index $k_0 \in \mathbb{N}$ such that $\boldsymbol{b}_D(z_k, z_\ell) < \frac{Cd}{2}$ and $z_k \in \mathbb{D}(a, r_3)$, $k, \ell \ge k_0$.

Fix such k, ℓ with $z_k \ne z_\ell$. Then, there is a $\mathcal{C}^1$-curve $\alpha_{k,\ell} : [0,1] \longrightarrow D$ such that

$$2\boldsymbol{b}_D(z_k, z_\ell) > \int_0^1 \boldsymbol{\beta}_D\big(\alpha_{k,\ell}(t); \alpha'_{k,\ell}(t)\big)dt.$$

Suppose this curve is not lying in $\mathbb{D}(a, r_1)$. Then, there are numbers $0 < s_1 < s_2 < 1$ such that $\alpha_{k,\ell}(s_1) \in \partial\mathbb{D}(a, r_3)$, $\alpha_{k,\ell}(s_2) \in \partial\mathbb{D}(a, r_2)$, and $\alpha_{k,\ell}([s_1, s_2]) \subset \mathbb{D}(a, r_1)$. Hence,

$$2\boldsymbol{b}_D(z_k, z_\ell) > \int_{s_1}^{s_s} \boldsymbol{\beta}_D\big(\alpha_{k,\ell}(t); \alpha'_{k,\ell}(t)\big)dt \ge C\boldsymbol{b}_{\widetilde{U}}(z_k, z_\ell) \ge Cd,$$

where $\widetilde{U}$ is the connected component of $D \cap \mathbb{D}(a, r_1)$ containing this part of the curve; a contradiction.

Hence, we obtain for $k, \ell \ge k_0$ that $C\boldsymbol{b}_{\widetilde{U}_{k,\ell}}(z_k, z_\ell) \le \boldsymbol{b}_D(z_k, z_\ell)$, where $\widetilde{U}_{k,\ell}$ denotes the connected component of $U \cap D$ that contains the curve $\alpha_{k,\ell}$. Hence, $(z_j)_j$ is even a $\boldsymbol{b}_{U_{k,\ell}}$-Cauchy sequence; a contradiction. □

Remark 15.1.22. Let $D \subset \mathbb{C}$ be an unbounded $\boldsymbol{b}$-complete domain. Due to N. Nikolov (private communication), the following inverse statement to that of Theorem 15.1.20 is true: For any open disc $U \subset \mathbb{C}$, any connected component of $U \cap D$ (resp. $D \setminus \overline{U}$) is also $\boldsymbol{b}$-complete.

15.2 Reinhardt domains and b-completeness

In the class of pseudoconvex Reinhardt domains, there is a complete geometric characterization of $\boldsymbol{b}$-complete domains (see [563, 565]).

Let $D \subset \mathbb{C}^n$ be a pseudoconvex Reinhardt domain. Then, $\Omega := \Omega_D := \log D$ is a convex domain in $\mathbb{R}^n$. Let us fix a point $a \in \Omega$. Put

$$\mathfrak{C}(\Omega, a) := \{v \in \mathbb{R}^n : a + \mathbb{R}_+ v \subset \Omega\}.$$

It is easy to see that $\mathfrak{C}(\Omega, a)$ is a closed convex cone with vertex at 0, i.e., $tx \in \mathfrak{C}(\Omega, a)$ for all $x \in \mathfrak{C}(\Omega, a)$ and $t \in \mathbb{R}_+$. Moreover, this cone is independent of the point a, i.e., $\mathfrak{C}(\Omega, a) = \mathfrak{C}(\Omega, b)$, $b \in \Omega$. So we will write shortly $\mathfrak{C}(\Omega) := \mathfrak{C}(\Omega, a)$. Observe that $\mathfrak{C}(\Omega) = \{0\}$ iff $\Omega \subset\subset \mathbb{R}^n$.

We now define

$$\widetilde{\mathfrak{C}}(D) := \{v \in \mathfrak{C}(\Omega_D) : \overline{\exp(a + \mathbb{R}_+ v)} \subset D\}, \quad \mathfrak{C}'(D) := \mathfrak{C}(\Omega_D) \setminus \widetilde{\mathfrak{C}}(D).$$

Observe that the definition of $\widetilde{\mathfrak{C}}(D)$ and $\mathfrak{C}'(D)$ is independent of the point a.

With the help of this geometric notions, there is the following complete description of those bounded Reinhardt domains that are Bergman complete:

Theorem 15.2.1 (cf. [563]). *Let $D \subset \mathbb{C}^n$ be a bounded pseudoconvex Reinhardt domain. Then, the following conditions are equivalent:*

(i) *D is $\boldsymbol{b}$-complete;*

(ii) $\mathfrak{C}'(D) \cap \mathbb{Q}^n = \varnothing$.

Before presenting the proof we will discuss examples, to illustrate how to use Theorem 15.2.1.

Example 15.2.2. Put

$$D_1 := \{z \in \mathbb{C}^2 : |z_1|^2/2 < |z_2| < 2|z_1|^2,\ |z_1| < 2\}.$$

Obviously, D_1 is a bounded pseudoconvex domain, which contains the point $(1,1)$. It turns out that $\mathfrak{C}'(D_1) = \mathbb{R}_{>0}(-1,-2)$; so it contains the rational vector $(-1,-2)$.

Using the map

$$\Phi : \mathbb{C}_*^2 \longrightarrow \mathbb{C}_*^2, \quad \Phi(z) := (z_1^3 z_2^{-1}, z_1^{-1} z_2), \quad z = (z_1, z_2),$$

we see that D_1 is biholomorphic to

$$\widetilde{D}_1 := \{z \in \mathbb{C}_*^2 : 1/2 < |z_2| < 2,\ |z_1 z_2| < 2\}.$$

It may be directly seen that $\widetilde{D}_1$ and, therefore also D_1, is not $\boldsymbol{b}$-complete.

On the other hand, let

$$D_2 := \{z \in \mathbb{C}^2 : \tfrac{1}{2}|z_1|^{\sqrt{2}} < |z_2| < 2|z_1|^{\sqrt{2}},\ |z_1| < 2\}.$$

Again, D_2 is a bounded pseudoconvex Reinhardt domain; now, a simple calculation gives $\mathfrak{C}'(D_2) = \mathbb{R}_{>0}(-1,-\sqrt{2})$, i.e., $\mathfrak{C}'(D_2)$ does not contain any rational vector. Hence, Theorem 15.2.1 tells us the D_2 is $\boldsymbol{b}$-complete. Recall that D_2 is not hyperconvex.

The next example can be found in [230]. Let

$$D := \{z \in \mathbb{C}^2 : |z_2|^2 < \exp(-1/|z_1|^2),\ |z_1| < 1\}.$$

Again, D is a bounded pseudoconvex Reinhardt domain. Here, we have $\mathfrak{C}(D) = \widetilde{\mathfrak{C}}(D) = \{0\} \times \mathbb{R}_-$ and $\mathfrak{C}'(D) = \varnothing$. So, Theorem 15.2.1 gives that D is $\boldsymbol{b}$-complete (in [230], a direct proof of this fact is presented). Again, observe that D is not hyperconvex.

For the proof of Theorem 15.2.1 we need the following lemma:

Lemma 15.2.3. *Let $C \subset \mathbb{R}^n$ be a convex closed cone with $C \cap \mathbb{Q}^n = \{0\}$. Assume that C contains no straight lines. Then, for any positive δ and any vector $v \in C \setminus \{0\}$ there is a $\beta \in \mathbb{Z}^n$ such that*

$$\langle \beta, v\rangle > 0 \quad \textit{and} \quad \langle \beta, w\rangle < \delta,\ w \in C,\ \|w\| = 1.$$

Since this lemma is based on geometric number theory, we will omit its proof. For more details, we refer to [563].

Proof of Theorem 15.2.1. In a first step, we are going to verify (i) $\Longrightarrow$ (ii):

Suppose that (ii) does not hold, i.e., there is a non-trivial vector $v \in \mathfrak{C}'(D) \cap \mathbb{Q}^n$. We may assume that $0 \in \log D$, $v = (v_1, \dots, v_n) \in \mathbb{Z}_-^n$, and that $v_1, \dots, v_n$ are relatively prime. It suffices to see that the Bergman length $L_{\boldsymbol{\beta}_D}$ of the curve $(0, 1] \overset{\gamma}{\longmapsto} (t^{-v_1}, \dots, t^{-v_n}) \in D$ is finite.

In fact, put $\varphi(\lambda) := (\lambda^{-v_1}, \dots, \lambda^{-v_n})$, $\lambda \in \mathbb{D}_*$. Then, $\varphi \in \mathcal{O}(\mathbb{D}_*, D)$. Let $u(\lambda) := \mathfrak{K}_D(\varphi(\lambda))$, $\lambda \in \mathbb{D}_*$. To continue we need a part of the following lemma (see [565]), the proof of which will be given later:

Lemma 15.2.4. *Let $D \subset \mathbb{C}^n$ be a pseudoconvex Reinhardt domain, $\alpha \in \mathbb{Z}^n$, and $p \in (0, \infty)$. Then, the following properties hold:*

(a) *the monomial z^α belongs to $L_h^p(D)$ iff $\langle \frac{p}{2}\alpha + \mathbf{1}, v\rangle < 0$ for any $v \in \mathfrak{C}(D) \setminus \{0\}$;*

(b) *if $\langle \alpha, v\rangle < 0$ for any $v \in \mathfrak{C}(D) \setminus \{0\}$, then $z^\alpha \in \mathcal{H}^\infty(D)$;*

(c) *if $z^\alpha \in \mathcal{H}^\infty(D)$, then $\langle \alpha, v\rangle \le 0$ for any $v \in \mathfrak{C}(D)$.*

By virtue of Lemma 15.2.4(a) ($p = 2$) it follows that

$$u(\lambda) = \sum_{\alpha \in \mathbb{Z}^n:\, \langle \alpha + \mathbf{1}, v\rangle < 0} a_\alpha |\lambda|^{-2\langle \alpha, v\rangle} = \sum_{j=j_0}^{\infty} b_j |\lambda|^{2j},$$

where $b_{j_0} \neq 0$. Therefore,

$$\boldsymbol{\beta}_D(\varphi(\lambda); \varphi'(\lambda)) = \frac{\partial^2 \log u(\lambda)}{\partial\lambda \partial\overline{\lambda}} = \frac{\partial^2}{\partial\lambda\partial\overline{\lambda}}\Bigl(\log \sum_{j=j_0}^{\infty} b_j |\lambda|^{2j-2j_0}\Bigr).^4$$

Obviously, the last expression remains bounded along $(0, 1)$, which gives the desired contradiction.

Now we turn to the proof of (ii) $\Longrightarrow$ (i).

We start with the following observation: put $\mathfrak{E} := \operatorname{span}\{z^\alpha : z^\alpha \in L_h^2(D)\}$. Then, $\mathfrak{E}$ is a dense subspace of $L_h^2(D)$. By virtue of Theorem 15.1.4, we only have to show that for any point $z^0 \in \partial D$ and for any sequence $(z^j)_j \subset D$ with $\lim_{j\to\infty} z^j = z_0$, we find a subsequence $(z^{j_k})_k$ such that

$$\frac{|f(z^{j_k})|}{\sqrt{\mathfrak{K}_D(z^{j_k})}} \underset{k\to\infty}{\longrightarrow} 0, \quad f \in \mathfrak{E}. \tag{15.2.1}$$

First, we discuss the case where z^0 satisfies the following property: if $z_j^0 = 0$ then $V_j \cap D \neq \varnothing$, $j = 1, \dots, n$, where $V_j := \{z \in \mathbb{C}^n : z_j = 0\}$.

Fix an $\alpha \in \mathbb{Z}^n$ such that $z^\alpha \in L_h^2(D)$. Then, $\alpha_j \geq 0$ for all j with $z_j^0 = 0$. So it suffices to verify that $\mathfrak{K}_D(z) \underset{z \to z^0}{\longrightarrow} \infty$. Without loss of generality, we may assume that $z_j^0 = 0$, $j = 1, \dots, s$, and $z_j^0 \neq 0$, $j = s+1, \dots, n$. Obviously, $s < n$. Then, there is an $R > 0$ such that $D \subset \mathbb{B}_s(0, R) \times \widetilde{\pi}_s(D)$, where $\widetilde{\pi}_s := \pi_{s+1,\dots,n}$ denotes the projection of $\mathbb{C}^n$ onto $\mathbb{C}^{n-s}$ if $s \geq 1$ or the identity if $s = 0$. Then, $\widetilde{\pi}_s(D)$ is a bounded pseudoconvex Reinhardt domain with $\widetilde{\pi}_s(z^0) \in \partial\widetilde{\pi}_s(D)$, where all coordinates of $\widetilde{\pi}_s(z^0)$ are different from zero. Hence, $\widetilde{\pi}_s(D)$ satisfies the general outer cone condition at $\widetilde{\pi}_s(z^0)$. By virtue of Theorem 12.1.25, it follows that $\lim_{z'' \to \widetilde{\pi}_s(z^0)} \mathfrak{K}_{\widetilde{\pi}_s(D)}(z'') = \infty$. Using the monotonicity and the product formula of the Bergman kernel, we finally get

$$\mathfrak{K}_D(z) \geq \mathfrak{K}_{\mathbb{B}_s(0,R)}(z')\mathfrak{K}_{\widetilde{\pi}_s(D)}(z'') \underset{D \ni z = (z', z'') \to z^0}{\longrightarrow} \infty.$$

In the remaining part of the proof, we assume that there is at least one j such that $z_j^0 = 0$, but $V_j \cap D = \varnothing$.

We may assume that $D \cap V_j \neq \varnothing$, $j = 1, \dots, k$, $D \cap V_j = \varnothing$, $j = k+1, \dots, n$,[5] $z_{k+1}^0 = 0$, and $\mathbf{1} \in D$.

Let $v \in \big(\mathbb{Q}^n \cap \mathfrak{C}(D)\big) \setminus \{0\}$. By assumption we know that $v \in \widetilde{\mathfrak{C}}(D)$. Therefore, $\lim_{t \to \infty} \exp(tv) = w \in D$. So, if $v_j < 0$ then $w_j = 0$, and if $v_j = 0$ then $w_j = 1$. In particular, if there is a $v \in \mathfrak{C}(D) \cap \mathbb{Q}^n$, $v_j < 0$, then $j \leq k$.

Observe that $\mathbb{R}_-^k \times \{0\}^{n-k} \subset \mathfrak{C}(D)$. Now, we claim that for any $v \in \mathfrak{C}(D) \setminus \big(\mathbb{R}^k \times \{0\}^{n-k}\big)$ we have that $v \notin \mathbb{R}^k \times \mathbb{Q}^{n-k}$.

Indeed, suppose that $v \in \mathbb{R}^k \times \mathbb{Q}^{n-k}$. So $v_j < 0$ for some $j > k$. Then, we may choose a suitable vector $w \in \mathbb{R}_-^k \times \{0\}^{n-k} \subset \mathfrak{C}(D)$ such that $\widetilde{v} := v + w \in \mathfrak{C}(D) \cap \mathbb{Q}^n$ and $\widetilde{v}_j < 0$. Hence, $j \leq k$; a contradiction.

Put $\pi : \mathbb{R}^n \longrightarrow \mathbb{R}^n$, $\pi(x) := (0, \dots, 0, x_{k+1}, \dots, x_n)$, where $x = (x_1, \dots, x_n)$. Then, $\pi\big(\mathfrak{C}(D)\big)$ is a closed convex cone in $\{0\}^k \times \mathbb{R}_-^k$. By virtue of the above property, we conclude that

$$\pi\big(\mathfrak{C}(D)\big) \cap \big(\{0\}^k \times \mathbb{Q}^{n-k}\big) = \{0\}.$$

Recall that $z_{k+1}^0 = 0$. Now, let $z^j \in D \cap \mathbb{C}_*^n$ be a sequence tending to z^0.[6] Put $x^j := (\log|z_1^j|, \dots, \log|z_n^j|) \in \mathbb{R}^n$. Obviously, $\|x^j\| \longrightarrow \infty$. Moreover, without loss of generality, we may assume that the sequence $(x^j/\|x^j\|)_j$ converges to a vector $\widetilde{v} \in \mathfrak{C}(D)$.

Fix an $\alpha \in \mathbb{Z}^n$ such that $z^\alpha \in L_h^2(D)$. Then, using Lemma 15.2.4, we conclude that

$$\inf\{-\langle \alpha + \mathbf{1}, w\rangle : w \in \mathfrak{C}(D),\ \|w\| = 1\} =: \delta_0 > 0.$$

[5] Then, necessarily, $k < n$.

[6] Observe that it suffices to prove (15.2.1) for sequences in $\mathbb{C}_*^n$.

Two cases have to be discussed.

Case 1°. $\widetilde{v}_j < 0$ for some $j > k$.

Applying Lemma 15.2.3 for $C = \pi(\mathfrak{C}(D))$, $v = \pi(\widetilde{v})$, and δ_0, we get the existence of a $\beta \in \{0\}^k \times \mathbb{Z}^{n-k}$ such that

$$\langle \beta, \widetilde{v} \rangle = \langle \beta, \pi(\widetilde{v}) \rangle > 0,$$
$$\langle \beta, w \rangle = \|\pi(w)\| \Big\langle \beta, \frac{\pi(w)}{\|\pi(w)\|} \Big\rangle < \delta, \ w \in \mathfrak{C}(D), \ \pi(w) \neq 0.$$

Observe that $\langle \beta, w \rangle = 0$ if $\pi(w) = 0$.

Then, $z^{\alpha+\beta} \in L^2_h(D)$ (use Lemma 15.2.4) and

$$\frac{|(z^j)^\alpha|}{\sqrt{\mathfrak{K}_D(z^j)}} \le \|z^{\alpha+\beta}\|_{L^2_h(D)} \frac{|(z^j)^\alpha|}{|(z^j)^{\alpha+\beta}|} = \|z^{\alpha+\beta}\|_{L^2_h(D)} |(z^j)^{-\beta}| \underset{j\to\infty}{\longrightarrow} 0.$$

Hence, the assumption of Theorem 15.1.4 is satisfied.

Case 2°. $\widetilde{v}_{k+1} = \cdots = \widetilde{v}_n = 0$.

Recall that $\|\pi(x^j)\| \longrightarrow \infty$. So, we may assume that

$$\frac{\pi(x^j)}{\|\pi(x^j)\|} \longrightarrow \widetilde{w} = (0, \ldots, 0, \widetilde{w}_{k+1}, \ldots, \widetilde{w}_n).$$

If $\widetilde{w} \in \pi(\mathfrak{C}(D))$, then, by virtue of Lemma 15.2.3, there is a $\beta \in \{0\}^k \times \mathbb{Z}^{n-k}$ such that $\langle \beta, \widetilde{w} \rangle > 0$ and $\langle \beta, w \rangle < \delta_0$, $w \in \mathfrak{C}(D) \setminus \{0\}$.

If $\widetilde{w} \notin \pi(\mathfrak{C}(D))$, let $\widetilde{C}$ be the smallest convex closed cone containing $\pi(\mathfrak{C}(D))$ and $-\widetilde{w}$. Then, $\widetilde{C} \subset \{0\}^k \times \mathbb{R}^{n-k}$ and $\widetilde{w} \notin \widetilde{C}$. Therefore,

$$\{\widetilde{\beta} \in \{0\}^k \times \mathbb{R}^{n-k} : \langle \widetilde{\beta}, u \rangle < 0, u \in \widetilde{C} \setminus \{0\}\}$$

is a non-empty convex open cone (see [522], § 25). So, it contains a $\beta \in \{0\}^k \times \mathbb{Z}^{n-k}$. Thus, $\langle \beta, -\widetilde{w} \rangle < 0$ and $\langle \beta, w \rangle = \|\pi(w)\| \langle \beta, \frac{\pi(w)}{\|\pi(w)\|} \rangle < 0 < \delta_0$, $w \in \mathfrak{C}(D)$, $\pi(w) \neq 0$.

Now we are able to complete the proof as in Case 1° using the β we had just constructed. Namely, we conclude that $z^{\alpha+\beta} \in L^2_h(D)$ and

$$\frac{|(z^j)^\alpha|}{\sqrt{\mathfrak{K}_D(z^j)}} \le \|z^{\alpha+\beta}\|_{L^2_h(D)} |(z^j)^{-\beta}| = \|z^{\alpha+\beta}\|_{L^2_h(D)} \prod_{\nu=k+1}^{n} |(z^j_\nu)^{-\beta_\nu}| \underset{j\to\infty}{\longrightarrow} 0.$$

Hence, Theorem 15.1.4 may be applied. ☐

Finally, we will prove that part of Lemma 15.2.4 that is used during the proof of Theorem 15.2.1.

Proof of Lemma 15.2.4. We restrict ourselves to prove only the following statement (the other ones in Lemma 15.2.4 may be taken as exercises) :

if D is as in Theorem 15.2.1 (in particular, D is bounded)

$$\text{and if } \langle \alpha + \mathbf{1}, v\rangle < 0,\ v \in \mathfrak{C}(D) \setminus \{0\}, \text{ then } z^\alpha \in L_h^2(D). \quad (\dagger)$$

Assume that $\mathbf{1} \in D$. In the case where $\mathfrak{C}(D) = \{0\}$, then ($\dagger$) is obvious. So let us assume that $\mathfrak{C}(D) \neq \{0\}$. Then, there is a $\delta_0 < 0$ such that $\langle \alpha + \mathbf{1}, v\rangle < \delta_0$, $v \in \mathfrak{C}(D)$, $\|v\| = 1$.

We claim that for any $\varepsilon > 0$ there is a cone T such that $\log D \setminus T$ is bounded and $\|w - v\| < \varepsilon$, $v \in T$, $w \in \mathfrak{C}(D)$, $\|v\| = \|w\| = 1$.

Indeed, fix an $\varepsilon > 0$ and let h be the Minkowski function of the convex set $\log D$. Observe that h is continuous and $h^{-1}(0) = \mathfrak{C}(D)$. Therefore, there is a $\delta > 0$ such that

$$\{w \in \mathbb{R}^n : h(w) \le \delta,\ \|w\| = 1\} \subset \{w \in \mathbb{R}^n : \|w\| = 1,\ \exists_{v \in \mathfrak{C}(D)} : \|v\| = 1,\ \|v - w\| < \varepsilon\}.$$

Set T as the smallest cone containing $\{w \in \mathbb{R}^n : h(w) \le \delta,\ \|w\| = 1\}$. Then, $\log D \setminus T$ is bounded; otherwise there would exist an unbounded sequence $x^j \in \log D \setminus T$ such that $h(x^j) < 1$. Therefore, $h(\frac{x^j}{\|x^j\|}) < \frac{1}{\|x^j\|}$, i.e., $x^j \in T$ for large j; a contradiction.

Now observe that $\langle \alpha + \mathbf{1}, v\rangle \le \frac{\delta_0}{2}\|v\|$, $v \in T$, and

$$\int_D |z^\alpha|^2 d\mathcal{L}^{2n}(z) < \infty \text{ iff } \int_{\log D} e^{2\langle \alpha+\mathbf{1}, x\rangle} d\mathcal{L}^n(x) < \infty$$
$$\text{iff } \int_T e^{2\langle \alpha+\mathbf{1}, x\rangle} d\mathcal{L}^n(x) < \infty.$$

It remains to estimate the last integral. We get

$$\int_T e^{2\langle \alpha+\mathbf{1}, x\rangle} d\mathcal{L}^n(x) \le \int_T e^{\delta_0\|x\|} d\mathcal{L}^n(x) < \int_{\mathbb{R}^n} e^{\delta_0\|x\|} d\mathcal{L}^n(x) < \infty.$$

Hence, the monomial $z^\alpha \in L_h^2(D)$. □

We conclude this section with a short discussion on unbounded pseudoconvex (Reinhardt) domains.

Example 15.2.5. Put

$$D_k := \{z \in \mathbb{C}^2 : |z_2| < 1,\ |z_2 z_1^k| < 1\}.$$

Note that if k is sufficiently large, then D_k has a Bergman metric. Put $D_k' := D_k \cap$

$\mathbb{C}_*^2$. Let $F \in \operatorname{Aut}(\mathbb{C}_*^2)$ be given by $F(z) := (z_1^{k+1}, z_2 z_1^k)$. Then,

$$F(D_k') = \{w \in \mathbb{C}_*^2 : |w_2| < 1, |w_2|^{k+1} < |w_1|^k\} =: G_k'.$$

Now we discuss the following sequence of points $(s/2, 2^k/s^{k+1}) \in D_k$ for large $s \in \mathbb{N}$. Obviously, this sequence does not converge inside of D_k. But its F-image $((1/2, 1/s))_s$ converges to the point $(1/2, 0) \in G_k$, where

$$G_k := \{w \in \mathbb{C}^2 : |w_2| < 1,\ |w_2|^{k+1} < |w_1|^k\}.$$

Since $\boldsymbol{b}_{G_k'} = \boldsymbol{b}_{G_k}|_{G_k' \times G_k'}$, it is clear that this sequence is a $\boldsymbol{b}_{G_k'}$-Cauchy sequence. Thus, the starting sequence is a $\boldsymbol{b}_{D_k'}$-Cauchy sequence and so it is also a $\boldsymbol{b}_{D_k}$-Cauchy sequence. Hence, D_k is not $\boldsymbol{b}$-complete.

As a positive result, even for Hartogs domains, we mention the following one without giving its proof. It is mainly based on the methods which were discussed before.

Theorem (cf. [425]). *Let $\varrho \in \mathcal{PSH}(\mathbb{C}^n)$ be bounded from below. If we also assume that $\lim_{\|z\|\to\infty} \|z\|^k e^{-\varrho(z)} = 0$ for all $k > 0$, then the following Hartogs domain*

$$\{(z, w) \in \mathbb{C}^n \times \mathbb{C} : |w| < e^{-\varrho(z)}\}$$

*admits a Bergman metric and is **K**-exhaustive and Bergman complete.*

Observe that the extra condition in the former theorem is not true for Example 15.2.5.

Remark 15.2.6. To our knowledge, there is so far no complete description for $\boldsymbol{b}$-complete unbounded Reinhardt domains.

15.3 List of problems

Chapter 16

Complex geodesics – effective examples

Summary. As an application and illustration of results from Chapter 11, we present characterizations of complex geodesics in the classical complex balls: the unit Euclidean ball, polydisc, Lie ball, and minimal ball – §§ 16.1, 16.5. We also give a detailed description of complex geodesics in complex ellipsoids – §§ 16.2, 16.3. Biholomorphisms between complex ellipsoids are characterized in § 16.4 (Theorem 16.4.1). Section 16.6 contains effective formulas for the Kobayashi–Royden metric in certain complex ellipsoids. Finally, in Sections 16.7 and 16.8 we discuss complex geodesics in the symmetrized bidisc and tetrablock.

16.1 Complex geodesics in the classical unit balls

Example 16.1.1. The case of the unit Euclidean ball $\mathbb{B}_n$.

Note that the group $\operatorname{Aut}(\mathbb{B}_n)$ acts transitively on $\mathbb{B}_n$ and that any boundary point $a \in \partial\mathbb{B}_n$ is a complex extreme point for $\overline{\mathbb{B}}_n$ ($\mathbb{B}_n$ is strictly convex).

All complex geodesics $\varphi : \mathbb{D} \longrightarrow \mathbb{B}_n$ are of the form

$$\varphi(\lambda) = \lambda a, \quad \lambda \in \mathbb{D} \quad (\operatorname{mod} \operatorname{Aut}(\mathbb{D}) \text{ and } \operatorname{mod} \operatorname{Aut}(\mathbb{B}_n)),$$

where $a \in \partial\mathbb{B}_n$ (this means that any complex geodesic $\varphi : \mathbb{D} \longrightarrow \mathbb{B}_n$ is of the form $\mathbb{D} \ni \lambda \to F(\chi(\lambda)a) \in G$, where $F \in \operatorname{Aut}(\mathbb{B}_n)$, $\chi \in \operatorname{Aut}(\mathbb{D})$, and $a \in \partial\mathbb{B}_n$). Moreover, by Proposition 11.3.3 the complex geodesics in $\mathbb{B}_n$ are uniquely determined (modulo $\operatorname{Aut}(\mathbb{D})$).

Example 16.1.2. The case of the unit polydisc $\mathbb{D}^n$.

$\operatorname{Aut}(\mathbb{D}^n)$ acts transitively. A point $a = (a_1, \dots, a_n) \in \partial(\mathbb{D}^n)$ is a complex extreme point for $\overline{\mathbb{D}}^n$ iff $|a_1| = \dots = |a_n|$ (i.e., iff $a \in \mathbb{T}^n$).

All complex geodesics $\varphi : \mathbb{D} \longrightarrow \mathbb{D}^n$ are of the form

$$\varphi(\lambda) = \lambda\widetilde{\varphi}(\lambda), \quad \lambda \in \mathbb{D} \quad (\operatorname{mod} \operatorname{Aut}(\mathbb{D}) \text{ and } \operatorname{mod} \operatorname{Aut}(\mathbb{D}^n)),$$

where $\widetilde{\varphi} : \mathbb{D} \longrightarrow \partial(\mathbb{D}^n)$ is an arbitrary holomorphic mapping. In particular, the complex geodesics in $\mathbb{D}^n$ are not uniquely determined.

Example 16.1.3. The case of the unit Lie ball $\mathbb{L}_n := \{z \in \mathbb{C}^n : L_n(z) < 1\}$, where

$$\begin{aligned} L_n(z) :&= \left(\frac{\|z\|^2 + |z \bullet z|}{2}\right)^{1/2} + \left(\frac{\|z\|^2 - |z \bullet z|}{2}\right)^{1/2} \\ &= [\|z\|^2 + (\|z\|^4 - |z \bullet z|^2)^{1/2}]^{1/2} \\ &= \left[\|x\|^2 + \|y\|^2 + 2[\|x\|^2\|y\|^2 - (x \bullet y)^2]^{1/2}\right]^{1/2}, \\ &\qquad z = x + iy \in \mathbb{R}^n + i\mathbb{R}^n \cong \mathbb{C}^n; \end{aligned}$$

cf. Remark 2.3.8. The norm L_n is the maximal complex norm $q : \mathbb{C}^n \longrightarrow \mathbb{R}_+$ such that $q(x) = \|x\|$ for all $x \in \mathbb{R}^n \cong \mathbb{R}^n + i0$. L_n is called the *Lie norm* in $\mathbb{C}^n$; the ball $\mathbb{L}_n$ is called the *Lie ball* in $\mathbb{C}^n$. Observe that

$$L_2(z_1, z_2) = \max\{|z_1 - iz_2|, |z_1 + iz_2|\},$$

and consequently the mapping

$$\mathbb{L}_2 \ni (z_1, z_2) \longmapsto (z_1 - iz_2, z_1 + iz_2) \in \mathbb{D}^2 \tag{16.1.1}$$

is biholomorphic.

The group $\mathrm{Aut}(\mathbb{L}_n)$ acts transitively on $\mathbb{L}_n$; in fact, $\mathbb{L}_n$ is one of the classical Cartan domains. Moreover,

$$\mathrm{Aut}_0(\mathbb{L}_n) := \{F \in \mathrm{Aut}(\mathbb{L}_n) : F(0) = 0\} = \{e^{i\theta} A : \theta \in \mathbb{R},\ A \in \mathbb{O}(n)\}$$

(cf. [237]), where $\mathbb{O}(n) :=$ the group of all orthogonal operators $A : \mathbb{R}^n \longrightarrow \mathbb{R}^n$ acting on $\mathbb{C}^n$ according to the formula

$$\mathbb{C}^n \ni x + iy \longmapsto A(x) + iA(y) \in \mathbb{C}^n.$$

We will prove that a point $a = x_0 + iy_0 \in \partial\mathbb{L}_n$ is a complex extreme point for $\overline{\mathbb{L}}_n$ iff the vectors x_0 and y_0 are $\mathbb{R}$-linearly dependent in $\mathbb{R}^n$ (see also Exercise 16.9.1).

If $n = 2$, then the above statement directly follows from (16.1.1). The general case may be reduced to the case of $n = 2$ using the following three remarks:

- the result is invariant under the action of $\mathbb{O}(n)$,
- for any $a \in \mathbb{C}^n$ there exists an $A \in \mathbb{O}(n)$ such that $A(a) \in \mathbb{C}^2 \times \{0\}$,
- $\mathbb{L}_n \cap (\mathbb{C}^2 \times \{0\}) = \mathbb{L}_2 \times \{0\}$.

All complex geodesics $\varphi : \mathbb{D} \longrightarrow \mathbb{L}_n$ are of the form (cf. [2])

$$\varphi(\lambda) = \lambda\left(\frac{\psi_1(\lambda)+\psi_2(\lambda)}{2}, \frac{\psi_1(\lambda)-\psi_2(\lambda)}{2i}, 0,\dots,0\right), \quad \lambda \in \mathbb{D}$$
$$(\text{mod Aut}(\mathbb{L}_n) \text{ and mod Aut}(\mathbb{D})), \quad (16.1.2)$$

where $(\psi_1, \psi_2) \in \mathcal{O}(\mathbb{D}, \partial(\mathbb{D}^2))$. In particular, the geodesics are not uniquely determined.

To prove (16.1.2), observe that the case $n = 2$ follows from (16.1.1). The general case may be reduced to $n = 2$ in the following way:

let $\varphi : \mathbb{D} \longrightarrow \mathbb{L}_n$ be a complex geodesic with $\varphi(0) = 0$. Then, we have

$$L_n(\varphi(\lambda)) = |\lambda|, \quad \lambda \in \mathbb{D}.$$

Fix $0 < \sigma < 1$ and choose $A \in \mathbb{O}(n)$ such that $A(\varphi(\sigma)) \in \mathbb{C}^2 \times \{0\}$. Put $\chi = (\chi_1,\dots,\chi_n) := A \circ \varphi$.

Note that

$$L_k(z) \le L_{k+l}(z,w), \quad z \in \mathbb{C}^k,\ w \in \mathbb{C}^l, \tag{16.1.3}$$

with the equality only for $w = 0$.

In particular, $(\chi_1, \chi_2) : \mathbb{D} \longrightarrow \mathbb{L}_2$ and $L_2(\chi_1(\sigma), \chi_2(\sigma)) = L_n(\chi(\sigma)) = \sigma$. This shows that (χ_1, χ_2) is a complex geodesic in $\mathbb{L}_2$. Consequently,

$$|\lambda| = L_2(\chi_1(\lambda), \chi_2(\lambda)) \le L_n(\chi(\lambda)) = |\lambda|, \quad \lambda \in \mathbb{D},$$

which by (16.1.3) implies that $\chi_3 \equiv \dots \equiv \chi_n \equiv 0$. The proof of (16.1.2) is completed.

16.2 Geodesics in convex complex ellipsoids

For $p = (p_1,\dots,p_n)$ with $p_1,\dots,p_n > 0$, $n \ge 2$, define the *complex ellipsoid* (cf. Examples 8.2.23, 8.2.29, 12.1.8)

$$\mathcal{E}(p) := \{(z_1,\dots,z_n) \in \mathbb{C}^n : \sum_{j=1}^n |z_j|^{2p_j} < 1\}.$$

Note that $\mathcal{E}(p) \subset \mathbb{D}^n$ is a balanced n-circled pseudoconvex domain and that $\mathbb{B}_n = \mathcal{E}((1,\dots,1))$.

Remark 16.2.1.

(a) $\mathcal{E}(p)$ is convex iff $p_1,\dots,p_n \ge 1/2$.

(b) $\mathcal{E}(p)$ is geometrically strictly convex if and only if $p_1,\dots,p_n \ge 1/2$ and $\#\{j : p_j = 1/2\} \le 1$.

(c) $\partial\mathcal{E}(p)$ is C^ω-smooth and strongly pseudoconvex at all points z belonging to $(\partial\mathcal{E}(p)) \cap (\mathbb{C}_*)^n$.

(d) If $p_1,\dots,p_n > 1/2$, then $\mathcal{E}(p)$ is strongly convex at all points z belonging to $(\partial\mathcal{E}(p)) \cap (\mathbb{C}_*)^n$.

Recall that a bounded domain $D \subset \mathbb{R}^N$ is called *strongly convex at a boundary point* $a \in \partial D$ if there exist a neighborhood U of a and a $\mathcal{C}^2$-function $r : U \longrightarrow \mathbb{R}$ such that $U \cap D = \{x \in U : r(x) < 0\}, U \cap \partial D = \{x \in U : r(x) = 0\}$, $\operatorname{grad} r \neq 0$ on U, and $\mathcal{H}r(x;\xi) > 0$, $x \in U$, $\xi \in (\mathbb{R}^N)_*$, where

$$\mathcal{H}r(x;\xi) := \sum_{j,k=1}^{N} \frac{\partial^2 r}{\partial x_j \partial x_k}(x)\xi_j\xi_k.$$

(e) $\partial\mathcal{E}(p)$ is $\mathcal{C}^1$-smooth iff $p_1,\dots,p_n > 1/2$.

(f) $\partial\mathcal{E}(p)$ is $\mathcal{C}^2$-smooth iff $p_1,\dots,p_n \geq 1$.

(g) For $p_1,\dots,p_n \geq 1$, the following conditions are equivalent:

(i) $\mathcal{E}(p)$ is strongly convex;

(ii) $\mathcal{E}(p)$ is strongly pseudoconvex;

(iii) $\mathcal{E}(p) = \mathbb{B}_n$ (i.e., $p_1 = \cdots = p_n = 1$).

(h) If $p_1,\dots,p_n \geq 1/2$, then any boundary point of $\mathcal{E}(p)$ is a complex extreme point for $\overline{\mathcal{E}(p)}$.

Throughout this section we will assume that $\mathcal{E}(p)$ is convex, i.e., $p_1,\dots,p_n \geq 1/2$ (cf. Remark 16.2.1(a)).

Our aim is to characterize all complex geodesics $\varphi : \mathbb{D} \longrightarrow \mathcal{E}(p)$. Observe that if $\varphi = (\varphi_1,\dots,\varphi_n) : \mathbb{D} \longrightarrow \mathcal{E}(p)$ is a complex geodesic with $\varphi_n \equiv 0$, then the mapping $\widetilde{\varphi} := (\varphi_1,\dots,\varphi_{n-1}) : \mathbb{D} \longrightarrow \mathcal{E}(\widetilde{p})$, $\widetilde{p} := (p_1,\dots,p_{n-1})$, is a "lower dimensional" complex geodesic. For we have

$$\boldsymbol{c}_{\mathbb{D}}(\lambda',\lambda'') = \boldsymbol{c}_{\mathcal{E}(p)}(\varphi(\lambda'),\varphi(\lambda'')) \leq \boldsymbol{c}_{\mathcal{E}(\widetilde{p})}(\widetilde{\varphi}(\lambda'),\widetilde{\varphi}(\lambda'')) \leq \boldsymbol{c}_{\mathbb{D}}(\lambda',\lambda''), \quad \lambda',\lambda'' \in \mathbb{D}.$$

Hence, it suffices to describe only those complex geodesics $\varphi : \mathbb{D} \longrightarrow \mathcal{E}(p)$ for which

$$\varphi_j \not\equiv 0, \quad j = 1,\dots,n. \tag{16.2.1}$$

Moreover, after a suitable permutation of variables, we may assume that for some $0 \leq s \leq n$

$$\varphi_1,\dots,\varphi_s \text{ have zeros in } \mathbb{D} \text{ and } \varphi_{s+1},\dots,\varphi_n \text{ are without zeros in } \mathbb{D}. \tag{16.2.2}$$

Proposition 16.2.2 (cf. [274]). *A mapping $\varphi = (\varphi_1, \dots, \varphi_n) : \mathbb{D} \longrightarrow \mathbb{C}^n$ with* (16.2.1) *and* (16.2.2) *is a complex geodesic in $\mathcal{E}(p)$ iff*

$$\varphi_j(\lambda) = \begin{cases} a_j \frac{\lambda - \alpha_j}{1 - \bar{\alpha}_j \lambda} \left(\frac{1 - \bar{\alpha}_j \lambda}{1 - \bar{\alpha}_0 \lambda} \right)^{1/p_j}, & j = 1, \dots, s \\ a_j \left(\frac{1 - \bar{\alpha}_j \lambda}{1 - \bar{\alpha}_0 \lambda} \right)^{1/p_j}, & j = s+1, \dots, n \end{cases}, \tag{16.2.3}$$

where

$$a_1, \dots, a_n \in \mathbb{C}_*, \tag{16.2.4}$$

$$\alpha_0, \dots, \alpha_s \in \mathbb{D}, \ \alpha_{s+1}, \dots, \alpha_n \in \overline{\mathbb{D}}, \tag{16.2.5}$$

$$\alpha_0 = \sum_{j=1}^{n} |a_j|^{2p_j} \alpha_j, \tag{16.2.6}$$

$$1 + |\alpha_0|^2 = \sum_{j=1}^{n} |a_j|^{2p_j} (1 + |\alpha_j|^2), \tag{16.2.7}$$

$$\text{the case } s = 0, \ \alpha_0 = \alpha_1 = \dots = \alpha_n \ \text{ is excluded,} \tag{16.2.8}$$

$$\text{the branches of the powers are such that } 1^{1/p_j} = 1, \ j = 1, \dots, n. \tag{16.2.9}$$

Moreover, the complex geodesics in $\mathcal{E}(p)$ are unique modulo $\operatorname{Aut}(\mathbb{D})$.

Remark 16.2.3.

(a) In the case where $p_1 = \dots = p_n > 1/2$ the problem of characterizing the complex geodesics in $\mathcal{E}(p)$ has been studied in [432] and [138].

(b) The case where $p_1 = \dots = p_n = 1/2$ has been investigated in [193] and [138].

(c) The case where $n = 2$, $p_1 = 1$, $p_2 \geq 1/2$ has been discussed in [55]; cf. Example 16.6.1.

Note that condition (16.2.8) says that $\varphi \not\equiv \text{const}$. Condition (16.2.9) is of technical character, since we prefer to have a one-to-one correspondence between φ and the parameters $a_j, \alpha_j, \ j = 1, \dots, n$.

Corollary 16.2.4. *Let $\varphi : \mathbb{D} \longrightarrow \mathcal{E}(p)$ be a complex geodesic.*

(a) *If $p_1, \dots, p_n \in \{\frac{1}{2}, 1\}$, then φ extends holomorphically to a neighborhood of $\overline{\mathbb{D}}$.*

(b) *If $t := \max\{p_1, \dots, p_n\} > 1$, then φ extends to an $(1/t)$-Hölder continuous mapping on $\overline{\mathbb{D}}$ ($\varphi \in \mathcal{C}^{1/t}(\overline{\mathbb{D}})$).*

(c) *If $u := \max\{p_j : p_j \neq 1\} < 1$, then φ extends to a $\mathcal{C}^1$-mapping on $\overline{\mathbb{D}}$ whose first order partial derivatives are $(1/u - 1)$-Hölder continuous ($\varphi \in \mathcal{C}^{1,1/u-1}(\overline{\mathbb{D}})$).*

Remark 16.2.5. Proposition 16.2.2 gives a tool for finding effective formulas for $\tau := \varkappa_{\mathcal{E}(p)}(z_0; X_0)$. Since the case where $z_{0,j_0} = X_{0,j_0} = 0$ (for some $1 \le j_0 \le n$) may be reduced to the lower dimensional one, we can assume that for each $j \in \{1, \dots, n\}$ either $z_{0,j} \neq 0$ or $X_{0,j} \neq 0$. This condition assures that the complex geodesic $\varphi : \mathbb{D} \longrightarrow \mathcal{E}(p)$ with $\varphi(0) = z_0$, $\tau\varphi'(0) = X_0$ satisfies (16.2.1) (note that by Proposition 16.2.2, φ is uniquely determined). Thus, to calculate τ it is enough to find a permutation σ of $(1, \dots, n)$, a number $0 \le s \le n$, and a_j, α_j, $j = 1, \dots, n$, with (16.2.4–16.2.9) such that

$$z_{0,\sigma(j)} = \varphi_j(0) = \begin{cases} -a_j\alpha_j, & j = 1, \dots, s \\ a_j, & j = s+1, \dots, n \end{cases},$$

$$\frac{1}{\tau} X_{0,\sigma(j)} = \varphi_j'(0) = \begin{cases} a_j(1 - |\alpha_j|^2 - \frac{\alpha_j}{p_j}(\overline{\alpha}_0 - \overline{\alpha}_j)), & j = 1, \dots, s \\ \frac{a_j}{p_j}(\overline{\alpha}_0 - \overline{\alpha}_j), & j = s+1, \dots, n \end{cases}.$$

There are situations where the above problem has an explicit solution (cf. §§ 16.6.1, 16.6.2, and Exercise 16.9.4), and consequently $\tau = \varkappa(z_0; X_0)$ may be described by an effective formula.

Proof of Proposition 16.2.2. First, observe that if $z_0 \in (\partial\mathcal{E}(p)) \cap (\mathbb{C}_*)^n$, then the unit outer normal vector $\nu(z_0) = (\nu_1(z_0), \dots, \nu_n(z_0))$ to $\partial\mathcal{E}(p)$ at z_0 is uniquely determined and

$$\nu_j(z_0) = \widetilde{\varrho}(z_0) p_j |z_{0j}|^{2(p_j-1)} z_{0j}, \quad j = 1, \dots, n,$$

where $\widetilde{\varrho}(z_0) > 0$. If $\varphi = (\varphi_1, \dots, \varphi_n) : \mathbb{D} \longrightarrow \mathbb{C}^n$ is a bounded holomorphic mapping with (16.2.1), then by the identity principle $\varphi^*(\lambda) \in (\mathbb{C}_*)^n$ for almost all $\lambda \in \mathbb{T}$. Combining these facts and using Remark 11.2.3, Corollary 11.2.8, Lemma 11.3.1, and Remark 16.2.1(h), we get the following criterion:

Corollary 16.2.6. *Let $\varphi = (\varphi_1, \dots, \varphi_n) : \mathbb{D} \longrightarrow \mathbb{C}^n$ be a non-constant bounded holomorphic mapping with* (16.2.1). *Then, φ is a complex geodesic in $\mathcal{E}(p)$ iff*

$$\sum_{j=1}^{n} |\varphi_j^*|^{2p_j} = 1 \quad \text{a.e. on } \mathbb{T} \tag{16.2.10}$$

and there exist functions $h \in \mathcal{H}^1(\mathbb{D}, \mathbb{C}^n)$ and $\varrho : \mathbb{T} \longrightarrow \mathbb{R}_{>0}$ such that

$$\frac{1}{\lambda} h_j^* = \varrho p_j |\varphi_j^*|^{2(p_j-1)} \overline{\varphi_j^*} \quad \text{a.e. on } \mathbb{T}, \quad j = 1, \dots, n. \tag{16.2.11}$$

The proof of Proposition 16.2.2 will be divided into four steps.

Step 1°: *Any mapping of the form* (16.2.3) *with* (16.2.4–16.2.9) *is a complex geodesic in* $\mathcal{E}(p)$.

Suppose that φ is given by (16.2.3) with (16.2.4–16.2.9). To verify that φ is a complex geodesic, we apply Corollary 16.2.6($\Longleftarrow$). Obviously, φ is continuous on $\overline{\mathbb{D}}$ and non-constant. If $\lambda \in \mathbb{T}$, then by (16.2.6, 16.2.7) we get

$$\sum_{j=1}^{n} |\varphi_j(\lambda)|^{2p_j} = \sum_{j=1}^{n} |a_j|^{2p_j} \left|\frac{1-\overline{\alpha}_j\lambda}{1-\overline{\alpha}_0\lambda}\right|^2$$
$$= \frac{\sum_{j=1}^{n} |a_j|^{2p_j}(1+|\alpha_j|^2) - 2\operatorname{Re}(\overline{\lambda}\sum_{j=1}^{n}|a_j|^{2p_j}\alpha_j)}{1+|\alpha_0|^2 - 2\operatorname{Re}(\overline{\lambda}\alpha_0)} = 1,$$

which gives (16.2.10). Define

$$h_j(\lambda) := p_j|a_j|^{2(p_j-1)}\overline{a}_j \begin{cases} \dfrac{(1-\overline{\alpha}_j\lambda)^2}{\left(\frac{1-\overline{\alpha}_j\lambda}{1-\overline{\alpha}_0\lambda}\right)^{\frac{1}{p_j}}}, & j=1,\dots,s \\ \dfrac{(\lambda-\alpha_j)(1-\overline{\alpha}_j\lambda)}{\left(\frac{1-\overline{\alpha}_j\lambda}{1-\overline{\alpha}_0\lambda}\right)^{\frac{1}{p_j}}}, & j=s+1,\dots,n \end{cases}, \quad \lambda \in \mathbb{D},$$
$$\varrho(\lambda) := |1-\overline{\alpha}_0\lambda|^2, \quad \lambda \in \mathbb{T}$$

(the branches of the powers are the same as in (16.2.3)). One can easily prove that $h_j \in \mathcal{H}^\infty(\mathbb{D})$, $j = 1,\dots,n$. Direct calculations give (16.2.11).

The proof of Step 1° is completed.

Step 2°: *Any complex geodesic* $\varphi : \mathbb{D} \longrightarrow \mathcal{E}((p_0,\dots,p_0))$ *with* (16.2.1) *and* (16.2.2) *is of the form* (16.2.3) *with* (16.2.4–16.2.9).

Let $p_1 = \dots = p_n = p_0$ and let $\varphi : \mathbb{D} \longrightarrow \mathcal{E}(p)$ be a complex geodesic with (16.2.1) and (16.2.2). We are going to apply Corollary 16.2.6($\Longrightarrow$). Let h and ϱ be as in this corollary. In view of (16.2.11) we have

$$\frac{1}{\lambda}\varphi_j^* h_j^* \in \mathbb{R}_{>0} \text{ a.e. on } \mathbb{T}, \quad j=1,\dots,n. \tag{16.2.12}$$

Now, we need an auxiliary lemma (at the moment only with $m = 1$).

Lemma 16.2.7 (cf. [193] for $m = 1$ and [145] for $m \ge 2$). *Let $f \in \mathcal{H}^1(\mathbb{D})$ be such that*

$$\frac{f^*(\lambda)}{\prod_{k=1}^{m}(\lambda-\sigma_k)} \in \mathbb{R}_{>0} \textit{ for a.a. } \lambda \in \mathbb{T},$$

where $\sigma_k \in \mathbb{C}$, $k = 1,\dots,m$. Then there exist $r > 0$ and $\alpha_k \in \overline{\mathbb{D}}$, $k = 1,\dots,m$, such that

$$f(\lambda) = r\frac{\prod_{k=1}^{m}(\lambda-\alpha_k)(1-\overline{\alpha}_k\lambda)}{\prod_{k=1}^{m}(1-\overline{\sigma}_k\lambda)}, \quad \lambda \in \mathbb{D}.$$

In particular (for $m = 1$), if $f \in \mathcal{H}^1(\mathbb{D})$ is such that

$$\frac{1}{\lambda} f^*(\lambda) \in \mathbb{R}_{>0} \text{ for a.a. } \lambda \in \mathbb{T},$$

then there exist $r > 0$ and $\alpha \in \overline{\mathbb{D}}$ such that

$$f(\lambda) = r(\lambda - \alpha)(1 - \overline{\alpha}\lambda), \quad \lambda \in \mathbb{D}.$$

Proof. Put $\widetilde{f}(\lambda) = f(\lambda)\prod_{k=1}^m (1 - \overline{\sigma}_k\lambda)$. Then, $\widetilde{f} \in \mathcal{H}^1(\mathbb{D})$ and

$$\frac{1}{\lambda^m}\widetilde{f}^*(\lambda) = \frac{f^*(\lambda)}{\prod_{k=1}^m(\lambda - \sigma_k)} \prod_{k=1}^m |\lambda - \sigma_k|^2 \in \mathbb{R}_{>0} \text{ for a.a. } \lambda \in \mathbb{T}.$$

Hence, it is enough to prove the lemma for $\sigma_k = 0$, $k = 1, \dots, m$.

Let $f(\lambda) = \sum_{k=0}^\infty a_k\lambda^k$, $\lambda \in \mathbb{D}$. Observe that

$$a_m = \frac{1}{2\pi i}\int_{\mathbb{T}} \frac{f^*(\zeta)}{\zeta^{m+1}} d\zeta = \frac{1}{2\pi}\int_0^{2\pi} \frac{f^*(e^{i\theta})}{e^{im\theta}} d\theta > 0.$$

Let

$$P(\lambda) = \sum_{k=0}^m a_k\lambda^k + \sum_{k=0}^{m-1} \overline{a}_k\lambda^{2m-k}.$$

If $\lambda \in \mathbb{T}$, then

$$\begin{aligned}\frac{P(\lambda)}{\lambda^m} - \frac{\overline{P(\lambda)}}{\overline{\lambda}^m} &= \sum_{k=0}^m a_k\lambda^{k-m} + \sum_{k=0}^{m-1} \overline{a}_k\lambda^{m-k} - \sum_{k=0}^m \overline{a}_k\lambda^{m-k} - \sum_{k=0}^{m-1} a_k\lambda^{k-m} \\ &= a_m - \overline{a}_m = 0.\end{aligned}$$

Put

$$\psi(\lambda) := \begin{cases} \frac{f(\lambda) - P(\lambda)}{\lambda^m}, & \text{if } \lambda \in \mathbb{D}_* \\ 0, & \text{if } \lambda = 0 \end{cases}.$$

Then, $\psi \in \mathcal{H}^1(\mathbb{D})$ and $\psi^*(\lambda) \in \mathbb{R}$ for a.a. $\lambda \in \mathbb{T}$. Hence, $\psi \equiv 0$, i.e., $f \equiv P$.

Suppose that $\lambda_0 = e^{i\theta_0}$ is such that $P(\lambda_0) = 0$. Let $T(\theta) := \frac{P(e^{i\theta})}{e^{i\theta m}}$. Then T is $\mathbb{R}$-analytic, $T \geq 0$, and $T(\theta_0) = 0$. Consequently, $T(\theta) = (\theta - \theta_0)^{2\ell}\widetilde{T}(\theta)$. This means that $P(\lambda) = ((\lambda - \lambda_0)(1 - \overline{\lambda}_0\lambda))^\ell \widetilde{P}(\lambda)$. Observe that $\frac{\widetilde{P}(\lambda)}{\lambda^{m-\ell}} > 0$ for a.a. $\lambda \in \mathbb{T}$.

Indeed,

$$\frac{\widetilde{P}(e^{i\theta})}{e^{i(m-\ell)\theta}} = \left|\frac{\theta - \theta_0}{e^{i\theta} - e^{i\theta_0}}\right|^{2\ell} \widetilde{T}(\theta) > 0 \text{ for a.a. } \theta \in \mathbb{R}.$$

Thus, after a finite number of steps in which we replace P by $\widetilde{P}$, we reach the situation where P has no zeros on $\mathbb{T}$ – notice that in each step we change P and m.

Suppose that $P(0) = 0$. Write $P(\lambda) = \lambda^\ell \widetilde{P}(\lambda)$. Observe that $\frac{\widetilde{P}(\lambda)}{\lambda^{m-\ell}} = \frac{P(\lambda)}{\lambda^m} > 0$ for a.a. $\lambda \in \mathbb{T}$. Thus, we may assume that P has no zeros on $\mathbb{T}$ and $P(0) \neq 0$.

Suppose that $P(\lambda_0) = 0$ ($\lambda_0 \notin \mathbb{T}$, $\lambda_0 \neq 0$). We have

$$\overline{P(1/\overline{\lambda})} = \sum_{k=0}^{m} \overline{a}_k \lambda^{-k} + \sum_{k=0}^{m-1} a_k \lambda^{k-2m} = \frac{P(\lambda)}{\lambda^{2m}}.$$

Hence, $P(1/\overline{\lambda}_0) = 0$. Thus, $P(\lambda) = (\lambda - \lambda_0)(1 - \overline{\lambda}_0 \lambda)\widetilde{P}(\lambda)$ with $\lambda_0 \in \mathbb{D}_*$. Observe that $\frac{\widetilde{P}(\lambda)}{\lambda^{m-1}} > 0$ for a.a. $\lambda \in \mathbb{T}$.

Indeed,

$$\frac{\widetilde{P}(\lambda)}{\lambda^{m-1}} = \frac{P(\lambda)}{\lambda^m} \frac{1}{|\lambda - \lambda_0|^2} > 0 \text{ for a.a. } \lambda \in \mathbb{T}.$$

Thus, after a finite number of steps, we get the required result. □

In view of the lemma, condition (16.2.12) implies that there exist $r_0, \dots, r_n > 0$, $\alpha_0, \dots, \alpha_n \in \overline{\mathbb{D}}$ such that:

$$\varphi_j(\lambda) h_j(\lambda) = r_j(\lambda - \alpha_j)(1 - \overline{\alpha}_j \lambda), \quad \lambda \in \mathbb{D},\ j = 1, \dots, n, \tag{16.2.13}$$

$$\varphi(\lambda) \bullet h(\lambda) = r_0(\lambda - \alpha_0)(1 - \overline{\alpha}_0 \lambda), \quad \lambda \in \mathbb{D}. \tag{16.2.14}$$

Replacing h by $\frac{1}{r_0} h$ (and ϱ by $\frac{1}{r_0}\varrho$), we can always assume that $r_0 = 1$. Note that by (16.2.2) we have

$$\alpha_1, \dots, \alpha_s \in \mathbb{D}. \tag{16.2.15}$$

Moreover, (16.2.13) and (16.2.14) give

$$\alpha_0 = \sum_{j=1}^{n} r_j \alpha_j \quad \text{and} \quad 1 + |\alpha_0|^2 = \sum_{j=1}^{n} r_j (1 + |\alpha_j|^2). \tag{16.2.16}$$

On the other hand, conditions (16.2.10), (16.2.11), and (16.2.14) imply that for almost all $\lambda \in \mathbb{T}$

$$\varrho(\lambda) p_0 = \varrho(\lambda) p_0 \sum_{j=1}^{n} |\varphi_j^*(\lambda)|^{2p_0} = \frac{1}{\lambda} \varphi^*(\lambda) \bullet h^*(\lambda) = |1 - \overline{\alpha}_0 \lambda|^2.$$

Hence, by (16.2.11) we conclude that

$$|h_j^*(\lambda)| = |1 - \overline{\alpha}_0 \lambda|^2 |\varphi_j^*(\lambda)|^{2p_0 - 1} \quad \text{for almost all } \lambda \in \mathbb{T}, \quad j = 1, \dots, n.$$

In particular,

$$\text{the functions } h_1, \dots, h_n \text{ are bounded,} \tag{16.2.17}$$

and by (16.2.13) we have

$$|\varphi_j^*(\lambda)||1-\overline{\alpha}_0\lambda|^{\frac{1}{p_0}} = r_j^{\frac{1}{2p_0}}|1-\overline{\alpha}_j\lambda|^{\frac{1}{p_0}} \quad \text{for almost all } \lambda \in \mathbb{T}, \quad j = 1,\dots,n. \tag{16.2.18}$$

Note that the functions $(1-\overline{\alpha}_j\lambda)^{1/p_0}$, $j = 1,\dots,n$, are outer. Consequently, by (16.2.18), the decomposition theorem for $\mathcal{H}^1$-functions (cf. Appendix B.9.6) shows

$$\varphi_j(\lambda)(1-\overline{\alpha}_0\lambda)^{1/p_0} = a_j B_j(\lambda) S_j(\lambda)(1-\overline{\alpha}_j\lambda)^{1/p_0}, \quad \lambda \in \mathbb{D}, \tag{16.2.19}$$

where

$$|a_j| = r_j^{\frac{1}{2p_0}}, \quad j = 1,\dots,n, \tag{16.2.20}$$

$$B_j(\lambda) := \begin{cases} \frac{\lambda-\alpha_j}{1-\overline{\alpha}_j\lambda}, & j = 1,\dots,s \\ 1, & j = s+1,\dots,n \end{cases} \quad \text{(cf. (16.2.2), (16.2.13), (16.2.15)),} \tag{16.2.21}$$

$$S_j(\lambda) := \exp\Bigl(-\int_{-\pi}^{\pi} \frac{e^{i\theta}+\lambda}{e^{i\theta}-\lambda} d\sigma_j(\theta)\Bigr);$$

σ_j is a singular non-negative Borel measure, $j = 1,\dots,n$, and the branches of the powers are chosen such that $1^{1/p_0} = 1$.

It remains to prove that $\sigma_j = 0$, $j = 1,\dots,n$. If we suppose for a moment that all the measures vanish, then

- conditions (16.2.19) and (16.2.21) imply (16.2.3);
- conditions (16.2.16) and (16.2.20) imply (16.2.6), (16.2.7);
- $\alpha_0 \in \mathbb{D}$, otherwise $s = 0$ and $\alpha_0 = \alpha_1 = \dots = \alpha_n \in \mathbb{T}$, which would imply that $\varphi \equiv \text{const}$.

We come back to the proof of $\sigma_j = 0$, $j = 1,\dots,n$. First, observe that by (16.2.13), (16.2.17), and (16.2.19) there exists an $\varepsilon > 0$ such that

$$|S_j(\lambda)| \geq \varepsilon|\lambda-\alpha_j||1-\overline{\alpha}_j\lambda|^{2-\frac{1}{p_0}}|1-\overline{\alpha}_0\lambda|^{\frac{1}{p_0}}, \quad \lambda \in \mathbb{D},\ j = 1,\dots,n.$$

On the other hand (cf. Appendix B.9.6),

$$S_j^*(\lambda) = 0 \quad \text{for } \sigma_j\text{-almost all } \lambda \in \mathbb{T}, \quad j = 1,\dots,n.$$

Combining the two conditions above and using the fact that for any $\beta \in \mathbb{R}$, $b > 0$ the function

$$\mathbb{D} \ni \lambda \longmapsto |\lambda-1|^{\beta} \exp\left(b\frac{1-|\lambda|^2}{|\lambda-1|^2}\right)$$

is unbounded, we easily conclude that $\sigma_j = 0$ for all $j = 1,\dots,n$.

The proof of Step 2° is finished.

Step 3°. *Any complex geodesic* $\varphi : \mathbb{D} \longrightarrow \mathcal{E}(p)$ *with* (16.2.1) *and* (16.2.2) *is of the form* (16.2.3) *with* (16.2.4–16.2.9).

Note that Steps 1° and 3° give the proof of the first part of Proposition 16.2.2.

Fix $p = (p_1, \dots, p_n)$, let $p_0 := \max\{p_1, \dots, p_n\}$, and let $\varphi : \mathbb{D} \longrightarrow \mathcal{E}(p)$ be an arbitrary complex geodesic with (16.2.1) and (16.2.2). Let h, ϱ correspond to φ as in Corollary 16.2.6. Write

$$\varphi_j = B_j \psi_j, \quad j = 1, \dots, n,$$

where B_j is the Blaschke product for φ_j and ψ_j is nowhere vanishing (define $B_j :\equiv 1$ for $j = s+1, \dots, n$). Put

$$\widetilde{\varphi}_j := B_j \psi_j^{\frac{p_j}{p_0}}, \quad \widetilde{h}_j := \frac{p_0}{p_j} h_j \frac{\varphi_j}{\widetilde{\varphi}_j}. \tag{16.2.22}$$

One can easily prove that by (16.2.10) we have

$$\sum_{j=1}^{n} |\widetilde{\varphi}_j^*|^{2p_0} = \sum_{j=1}^{n} |\varphi_j^*|^{2p_j} = 1 \text{ a.e. on } \mathbb{T},$$

and that, by (16.2.11),

$$\frac{1}{\lambda} \widetilde{h}_j^* = \varrho p_0 |\widetilde{\varphi}_j^*|^{2(p_0-1)} \overline{\widetilde{\varphi}_j^*} \text{ a.e. on } \mathbb{T}, \quad j = 1, \dots, n.$$

Moreover, $\widetilde{h}_j \in \mathcal{H}^1(\mathbb{D})$; here, it is important that $p_j \le p_0$, $j = 1, \dots, n$. Thus, by Corollary 16.2.6, the mapping $\widetilde{\varphi}$ is a complex geodesic in $\mathcal{E}((p_0, \dots, p_0))$ with (16.2.1) and (16.2.2). Consequently, by Step 2°, the geodesic $\widetilde{\varphi}$ is of form (16.2.3) with (16.2.4–16.2.9). Finally, using relation (16.2.22), we conclude that the same is true for the geodesic φ.

The proof of Step 3° is finished.

We move to the last part of the proof of Proposition 16.2.2.

Step 4°. *Proof of the uniqueness of complex geodesics.* The case where $\mathcal{E}(p)$ is geometrically strictly convex (cf. Remark 16.2.1(b)) follows directly from Proposition 11.3.3. In the general case, we proceed as follows:

By Proposition 11.3.2 it suffices to prove the uniqueness of the $\varkappa_{\mathcal{E}(p)}$-geodesics. Let $\varphi, \psi : \mathbb{D} \longrightarrow \mathcal{E}(p)$ be two complex geodesics with

$$\varphi(0) = \psi(0) \quad \text{and} \quad \varphi'(0) = \psi'(0). \tag{16.2.23}$$

So far, Proposition 16.2.2 shows that $\varphi_j \equiv 0$ iff $\psi_j \equiv 0$. Thus, without loss of generality, we may assume that φ and ψ satisfy (16.2.1). Moreover, we assume that

φ fulfills condition (16.2.2). Put $I_0 := \{j : \psi_j \text{ has a zero in } \mathbb{D}\}$. Now observe that $\chi := \frac{1}{2}(\varphi + \psi)$ is again a complex geodesic in $\mathcal{E}(p)$ ($\mathcal{E}(p)$ is convex, cf. the proof of Proposition 11.3.3). In particular,

$$\chi(\lambda),\ \varphi(\lambda),\ \psi(\lambda) \in \partial\mathcal{E}(p) \quad \text{for all } \lambda \in \mathbb{T}.$$

Therefore,

$$\begin{aligned} \arg\varphi_j(\lambda) = \arg\psi_j(\lambda) &= \arg\chi_j(\lambda) \ \text{ if } \varphi_j(\lambda)\psi_j(\lambda) \neq 0,\ \lambda \in \mathbb{T}, \\ |\varphi_j| = |\psi_j| &= |\chi_j| \quad \text{on } \mathbb{T} \ \text{ if } p_j > 1/2. \end{aligned} \tag{16.2.24}$$

Consequently, if $p_j > 1/2$, then $\varphi_j = \psi_j$ on $\mathbb{T}$ and so $\varphi_j \equiv \psi_j$ on $\mathbb{D}$.

It remains to discuss j with $p_j = 1/2$. Fix such a j. First, note that by (16.2.24) we have

$$\varphi_j\overline{\psi}_j = \psi_j\overline{\varphi}_j \quad \text{on} \quad \mathbb{T}. \tag{16.2.25}$$

There are four cases.

(a) $1 \le j \le s$ and $j \in I_0$. Then,

$$\varphi_j(\lambda) = a_j\frac{(\lambda - \alpha_j)(1 - \overline{\alpha}_j\lambda)}{(1 - \overline{\alpha}_0\lambda)^2} \quad \text{and} \quad \psi_j(\lambda) = b_j\frac{(\lambda - \beta_j)(1 - \overline{\beta}_j\lambda)}{(1 - \overline{\beta}_0\lambda)^2}.$$

(b) $1 \le j \le s$ and $j \notin I_0$. Then,

$$\varphi_j(\lambda) = a_j\frac{(\lambda - \alpha_j)(1 - \overline{\alpha}_j\lambda)}{(1 - \overline{\alpha}_0\lambda)^2} \quad \text{and} \quad \psi_j(\lambda) = b_j\left(\frac{1 - \overline{\beta}_j\lambda}{1 - \overline{\beta}_0\lambda}\right)^2.$$

(c) $s + 1 \le j \le n$ and $j \in I_0$. This case is symmetric to (b).

(d) $s + 1 \le j \le n$ and $j \notin I_0$. Then,

$$\psi_j(\lambda) = a_j\left(\frac{1 - \overline{\alpha}_j\lambda}{1 - \overline{\alpha}_0\lambda}\right)^2 \quad \text{and} \quad \psi_j(\lambda) = b_j\left(\frac{1 - \overline{\beta}_j\lambda}{1 - \overline{\beta}_0\lambda}\right)^2.$$

Add. (a) Using (11.2.21) one gets

$$\frac{a_j\overline{b}_j}{(1 - \overline{\alpha}_0\lambda)^2(\lambda - \beta_0)^2} \equiv \frac{\overline{a}_j b_j}{(1 - \overline{\beta}_0\lambda)^2(\lambda - \alpha_0)^2}.$$

So we obtain $\alpha_0 = \beta_0$. Moreover, because of (16.2.23) (cf. Remark 16.2.5), we get

$$\alpha_j a_j = \beta_j b_j \quad \text{and} \quad a_j(1 + |\alpha_j|^2) = b_j(1 + |\beta_j|^2),$$

which directly implies that $\alpha_j = \beta_j$ and $a_j = b_j$.

Add. (b) By (16.2.25), we have

$$\frac{\overline{b}_j a_j}{(1-\overline{\alpha}_0\lambda)^2}\left(\frac{\lambda-\beta_j}{\lambda-\beta_0}\right)^2 \equiv \frac{\overline{a}_j b_j}{(\lambda-\alpha_0)^2}\left(\frac{1-\overline{\beta}_j\lambda}{1-\overline{\beta}_0\lambda}\right)^2.$$

So we conclude that $\alpha_0 = \beta_0$, and therefore

$$\overline{b}_j a_j(\lambda-\beta_j)^2 \equiv \overline{a}_j b_j(1-\overline{\beta}_j\lambda)^2.$$

Hence, we have $|\beta_j| = 1$. Again using (16.2.23) we get

$$-\alpha_j a_j = b_j \quad \text{and} \quad a_j(1+|\alpha_j|^2) = -2b_j\overline{\beta}_j,$$

which implies that $|\alpha_j| = 1$; a contradiction.

Add. (c) The argument is the same as (b).

Add. (d) Directly from (16.2.23), using the forms of φ_j and ψ_j, we conclude that $a_j = b_j$. Then, in view of (16.2.25), we obtain $\alpha_0 = \beta_0$ and next, using once again (16.2.23), we prove that $\alpha_j = \beta_j$; the details are left to the reader. □

The proof of Proposition 16.2.2 is finished. □

16.3 Extremal discs in arbitrary complex ellipsoids

We are going to discuss the case of arbitrary (not necessarily convex) complex ellipsoids. The main result is the following theorem:

Theorem 16.3.1. *Let $\varphi : \mathbb{D} \longrightarrow \mathcal{E}(p)$ be an extremal for $(\mathcal{P}_m)$ (cf. Definition 11.4.3) such that $\varphi_j \not\equiv 0$, $j = 1,\dots,n$. Then,*

$$\varphi_j(\lambda) = a_j \prod_{k=1}^{m}\left(\frac{\lambda-\alpha_{k,j}}{1-\overline{\alpha}_{k,j}\lambda}\right)^{r_{k,j}}\left(\frac{1-\overline{\alpha}_{k,j}\lambda}{1-\overline{\alpha}_{k,0}\lambda}\right)^{1/p_j}, \quad j = 1,\dots,n,$$

where

$$a_1,\dots,a_n \in \mathbb{C}_*, \quad \alpha_{k,j} \in \overline{\mathbb{D}}, \quad r_{k,j} \in \{0,1\}, \quad r_{k,j} = 1 \Longrightarrow \alpha_{k,j} \in \mathbb{D},$$

$$\sum_{j=1}^{n}|a_j|^{2p_j}\prod_{k=1}^{m}(\zeta-\alpha_{k,j})(1-\overline{\alpha}_{k,j}\zeta) = \prod_{k=1}^{m}(\zeta-\alpha_{k,0})(1-\overline{\alpha}_{k,0}\zeta), \quad \zeta \in \mathbb{D},$$

the case $r_{j,k} = 0$, $j = 1,\dots,n$, $k = 1,\dots,m$, is excluded,

the branches of the powers are such that $1^{1/p_j} = 1$, $j = 1,\dots,n$.

In particular, if φ is extremal for $\boldsymbol{\ell}^{(m)*}_{\mathcal{E}(p)}(a,b)$ $(a \neq b)$ or for $\boldsymbol{\varkappa}^{(m)}_{\mathcal{E}(p)}(a;X)$ $(X \neq 0)$, then φ must be of the above form (cf. Remark 11.4.4).

Remark 16.3.2. Theorem 16.3.1 generalizes the "only if" implication in Proposition 16.2.2. Recall that in the case where $\mathcal{E}(p)$ is convex, any mapping described in Theorem 16.3.1 with $m = 1$ is a complex geodesic in $\mathcal{E}(p)$. This is no longer true if $\mathcal{E}(p)$ is not convex – cf. Example 16.6.4.

Lemma 16.3.3. *Let S_1, S_2 be singular inner functions and let $S_1 S_2 \equiv 1$. Then, $S_1 \equiv S_2 \equiv 1$.*

Proof. Suppose that

$$S_j(\lambda) = \exp\left(-\int_0^{2\pi} \frac{e^{i\theta} + \lambda}{e^{i\theta} - \lambda} d\mu_j(\theta)\right), \quad j = 1, 2,$$

where μ_1, μ_2 are non-negative Borel measures, singular with respect to the Lebesgue measure (cf. Appendix B.9.6). Then, $S_1 S_2 \equiv 1$ is equivalent to $\mu_1 + \mu_2 = 0$. Since $\mu_j \geq 0$, $j = 1, 2$, we get $\mu_1 = \mu_2 = 0$. □

Proof of Theorem 16.3.1. In the situation of Theorem 11.4.5, let $G_0 := \mathbb{C}^n$, $u(z) := -1 + \sum_{j=1}^n |z_j|^{2p_j}$, $G := \mathcal{E}(p)$. Observe that

$$\frac{\partial u}{\partial z_j} = p_j \frac{|z_j|^{2p_j}}{z_j}, \quad j = 1, \dots, n.$$

We know that $\varphi_j = B_j S_j F_j$, where B_j is a Blaschke product, S_j is a singular inner function, and F_j is an outer function.

We will show that the assumptions (a, b, c) of Theorem 11.4.5 are satisfied with $s := (F_1, \dots, F_n)$. We know that $\varphi_j \not\equiv 0$, $j = 1, \dots, n$. Hence, $\operatorname{grad} u(\varphi^*(\zeta))$ exists for a.a. $\zeta \in \mathbb{T}$. Thus, (a, b) are satisfied. The main problem is to check (c). Take an $h = (h_1, \dots, h_n) \in \mathcal{A}(\mathbb{D}, \mathbb{C}^n)$. Then, for a.a. $\zeta \in \mathbb{T}$ we have

$$\begin{aligned} R_h(\zeta) &= \sum_{j=1}^n \Bigg(|\varphi_j^*(\zeta) + F_j^*(\zeta) h_j(\zeta)|^{2p_j} - |\varphi_j^*(\zeta)|^{2p_j} \\ &\qquad - 2 \operatorname{Re}\left(p_j \frac{|\varphi_j^*(\zeta)|^{2p_j}}{\varphi_j^*(\zeta)} F_j^*(\zeta) h_j(\zeta) \right) \Bigg) \\ &= \sum_{j=1}^n |\varphi_j^*(\zeta)|^{2p_j} \left(\left| 1 + \frac{F_j^*(\zeta)}{\varphi_j^*(\zeta)} h_j(\zeta) \right|^{2p_j} - 1 - 2p_j \operatorname{Re}\left(\frac{F_j^*(\zeta)}{\varphi_j^*(\zeta)} h_j(\zeta) \right) \right). \end{aligned}$$

We have $|\varphi_j^*(\zeta)/F_j^*(\zeta)| = 1$ for a.a. $\zeta \in \mathbb{T}$. Now, in order to get (c), we only need to use the following elementary equality:

$$\lim_{\delta \to 0} \frac{\sup\{||1 + z|^\alpha - 1 - \alpha \operatorname{Re} z| : |z| \leq \delta\}}{\delta} = 0, \quad \alpha > 0.$$

Thus, by Corollary 11.4.6, there exist $g \in \mathcal{H}^\infty(\mathbb{D}, \mathbb{C}^n)$ and $\varrho \in L^\infty(\mathbb{T}, \mathbb{R}_{>0})$ such that

$$Q(\zeta)\varrho(\zeta)F_j^*(\zeta)\frac{|\varphi_j^*(\zeta)|^{2p_j}}{\varphi_j^*(\zeta)} = g_j^*(\zeta) \text{ for a.a. } \zeta \in \mathbb{T},\ j = 1, \dots, n,$$

where $Q(\zeta) = \prod_{k=1}^m (\zeta - \sigma_k)$ is the polynomial from the definition of $(\mathcal{P}_m)$. It is equivalent to

$$Q(\zeta)\varrho(\zeta)|F_j^*(\zeta)|^{2p_j} = B_j^*(\zeta)S_j^*(\zeta)g_j^*(\zeta) \text{ for a.a. } \zeta \in \mathbb{T},\ j = 1, \dots, n.$$

By Lemma 16.2.7, there exist $r_j > 0$ and $\alpha_{k,j} \in \overline{\mathbb{D}}$ such that

$$B_j^*(\zeta)S_j^*(\zeta)g_j^*(\zeta) = r_j \frac{\prod_{k=1}^m (\zeta - \alpha_{k,j})(1 - \overline{\alpha}_{k,j}\zeta)}{\prod_{k=1}^m (1 - \overline{\sigma}_k\zeta)} \tag{16.3.1}$$

and there exist $r_0 > 0$ and $\alpha_{k,0} \in \overline{\mathbb{D}}$ such that

$$Q(\zeta)\varrho(\zeta) = \sum_{j=1}^n B_j^*(\zeta)S_j^*(\zeta)g_j^*(\zeta) = r_0 \frac{\prod_{k=1}^m (\zeta - \alpha_{k,0})(1 - \overline{\alpha}_{k,0}\zeta)}{\prod_{k=1}^m (1 - \overline{\sigma}_k\zeta)}. \tag{16.3.2}$$

We have

$$r_0 \prod_{k=1}^m (\zeta - \alpha_{k,0})(1 - \overline{\alpha}_{k,0}\zeta)|F_j(\zeta)|^{2p_j} = r_j \prod_{k=1}^m (\zeta - \alpha_{k,j})(1 - \overline{\alpha}_{k,j}\zeta). \tag{16.3.3}$$

Hence,

$$F_j(\zeta) = a_j \prod_{k=1}^m \left(\frac{1 - \overline{\alpha}_{k,j}\zeta}{1 - \overline{\alpha}_{k,0}\zeta}\right)^{1/p_j}, \tag{16.3.4}$$

where $a_j \in \mathbb{C}_*$. From (16.3.4) it follows that

$$B_j(\zeta) = \prod_{k=1}^m \left(\frac{\zeta - \alpha_{k,j}}{1 - \overline{\alpha}_{k,j}\zeta}\right)^{r_{k,j}}, \text{ where } r_{k,j} \in \{0, 1\}.$$

Hence

$$S_j(\zeta)g_j(\zeta) = r_j \frac{\prod_{k=1}^m (\zeta - \alpha_{k,j})^{1-r_{k,j}}(1 - \overline{\alpha}_{k,j}\zeta)^{1+r_{k,j}}}{\prod_{k=1}^m (1 - \overline{\sigma}_k\zeta)}.$$

Since the right hand side is an outer function, we get from Lemma 16.3.3 that $S_j \equiv 1$, $j = 1, \dots, n$.

From (16.3.3) and (16.3.4) we see that $|a_j|^{2p_j} = \frac{r_j}{r_0}$ and from (16.3.1) and (16.3.2) it follows that $\sum_{j=1}^n |a_j|^{2p_j} \prod_{k=1}^m (\zeta - \alpha_{k,j})(1 - \overline{\alpha}_{k,j}\zeta) = \prod_{k=1}^m (\zeta - \alpha_{k,0})(1 - \overline{\alpha}_{k,0}\zeta)$, $\zeta \in \mathbb{D}$. So, we get the required result. □

Observe that Corollary 11.4.6 may also be used to characterize extremal discs in more complicated situations. For example:

Proposition 16.3.4. *Let* $\alpha = (\alpha_1, \dots, \alpha_n) \in \mathbb{R}^s_{<0} \times \mathbb{R}^{n-s}_{>0}$ *with* $0 \le s \le n-1$. *Take* $a, b \in \boldsymbol{D}_\alpha$ *with* $\sigma := \sqrt[m]{\boldsymbol{\ell}^{(m)*}_{\boldsymbol{D}_\alpha}(a,b)} > 0$ *(resp.* $a \in \boldsymbol{D}_\alpha$, $X \in \mathbb{C}^n$ *with* $\sigma := \boldsymbol{\varkappa}^{(m)}_{\boldsymbol{D}_\alpha}(a;X) > 0$*). Suppose that* $\varphi \in \mathcal{O}(\mathbb{D}, G)$ *is an extremal for* $\boldsymbol{\ell}^{(m)*}_{\boldsymbol{D}_\alpha}(a,b)$ *(resp.* $\boldsymbol{\varkappa}^{(m)}_{\boldsymbol{D}_\alpha}(a;X)$*) such that* $\varphi(\mathbb{D}) \subset\subset \mathbb{C}^s_* \times \mathbb{C}^{n-s}$ *and the function* $\varphi_{s+1}\cdots\varphi_n$ *has zeros. Then,*

$$\varphi_j = B_j \psi_j, \quad j = 1, \dots, n,$$

where

- $\psi_j \in \mathcal{O}(\mathbb{D}, \mathbb{C}_*)$, $j = 1, \dots, n$,
- $B_j = 1$, $j = 1, \dots, s$,

$$B_j(\lambda) = \left(\frac{\lambda - \beta_1}{1 - \overline{\beta}_1 \lambda}\right)^{r_{j,1}} \cdots \left(\frac{\lambda - \beta_\mu}{1 - \overline{\beta}_\mu \lambda}\right)^{r_{j,\mu}}, \quad j = s+1, \dots, n,$$

- $\mu \in \{1, \dots, m\}$,
- $\beta_1, \dots, \beta_\mu \in \mathbb{D}$ *are pairwise different,*
- $r_{j,k} \in \{0, \dots, r_j\}$, $j = s+1, \dots, n$, $k = 1, \dots, \mu$,
- $r_1, \dots, r_\mu \in \mathbb{N}$ *are such that* $r_1 + \dots + r_\mu \le m$,
- $\sum_{j=s+1}^n r_{j,k} > 0$, $k = 1, \dots, \mu$.

Notice that the above proposition implies Lemma 10.1.10

Proof of Proposition 16.3.4. Suppose that $\varphi(\mathbb{D}) \subset\subset \mathbb{A}_s(1/R, R) \times \mathbb{P}_{n-s}(R) =: P$ and let $G \subset \boldsymbol{D}_\alpha \cap P$, $G \subset\subset P$, be a $\mathcal{C}^1$-smooth bounded pseudoconvex Reinhardt domain such that $\varphi(\mathbb{D}) \subset G$ and $(\partial \boldsymbol{D}_\alpha) \cap P = \{z \in P : |z^\alpha| = 1\} \subset \partial G$. Let $u \in \mathcal{PSH}(G_0) \cap \mathcal{C}^1(G_0)$ be a defining function of G, where $G_0 \subset P$ is an open neighborhood of $\overline{G}$ and $u(z) = |z^\alpha| - 1$ in an open neighborhood of $U \subset P$ of $(\partial \boldsymbol{D}_\alpha) \cap P$. Note that $\operatorname{grad} u(z) = |z^\alpha|(\alpha_1/z_1, \dots, \alpha_n/z_n)$ in U.

Observe that φ is extremal for $\boldsymbol{\ell}^{(m)*}_G(a,b)$ (resp. $\boldsymbol{\varkappa}^{(m)}_G(a;X)$). Indeed, $\sigma^m = \boldsymbol{\ell}^{(m)*}_{\boldsymbol{D}_\alpha}(a,b) \le \boldsymbol{\ell}^{(m)*}_G(a,b) \le \sigma^m$ (resp. $\sigma = \boldsymbol{\varkappa}^{(m)}_{\boldsymbol{D}_\alpha}(a;X) \le \boldsymbol{\varkappa}^{(m)}_G(a;X) \le \sigma$).

By Corollary 11.4.6, $|(\varphi_1^*)^{\alpha_1} \cdots (\varphi_n^*)^{\alpha_n}| = 1$ for a.a. $\lambda \in \mathbb{T}$ and there exist functions $\varrho \in L^\infty(\mathbb{T}, \mathbb{R}_{>0})$, $h \in \mathcal{H}^\infty(\mathbb{D}, \mathbb{C}^n)$ such that

$$h_j^*(\lambda)\varphi_j^*(\lambda) = \alpha_j Q(\lambda)\varrho(\lambda) \text{ for a.a. } \lambda \in \mathbb{T}, \; j = 1, \dots, n,$$

where $Q(\lambda) = \prod_{k=1}^{m}(\lambda - \sigma_k)$ is as in Remark 11.4.4. In particular, $\frac{1}{\alpha_j} h_j \varphi_j \equiv \frac{1}{\alpha_k} h_k \varphi_k$, $j \neq k$. Hence, by Lemma 16.2.7 we get

$$h_j(\lambda)\varphi_j(\lambda) = \tau_j \frac{\prod_{k=1}^{m}(\lambda - \beta_k)(1 - \overline{\beta}_k \lambda)}{\prod_{k=1}^{m}(1 - \overline{\sigma}_k \lambda)}, \quad \lambda \in \mathbb{D},\ j = 1, \dots, n,$$

where $\tau_1, \dots, \tau_n > 0$, $\beta_1, \dots, \beta_m \in \overline{\mathbb{D}}$, and $\tau_j/\alpha_j = \tau_k/\alpha_k$ for $j \neq k$. Observe that $\varphi_j^{-1}(0) \subset \{\beta_1, \dots, \beta_m\}$, $j = s+1, \dots, n$. Let $\{\beta_1, \dots \beta_\mu\} := \bigcup_{j=s+1}^{m} \varphi_j^{-1}(0)$ $(\mu \leq m)$, $\beta_j \neq \beta_k$, $j, k = 1, \dots, \mu$, $j \neq k$. Put $r_{j,k} := \operatorname{ord}_{\beta_k} \varphi_j$, $r_k := \max\{r_{j,k} : j = s+1, \dots, n\}$, $k = 1, \dots, \mu$. Then, $r_k \geq 1$, $r_1 + \dots + r_\mu \leq m$, and $r_{s+1,k} + \dots + r_{n,k} \geq 1$, $k = 1, \dots, \mu$. □

16.4 Biholomorphisms of complex ellipsoids

Fix $p = (p_1, \dots, p_n), q = (q_1, \dots, q_n) \in (\mathbb{R}_{>0})^n$, $n \geq 2$, and consider the complex ellipsoids $\mathcal{E}(p)$ and $\mathcal{E}(q)$; cf. § 16.6. It is natural to ask when the ellipsoids $\mathcal{E}(p)$ and $\mathcal{E}(q)$ are biholomorphic.

Theorem 16.4.1. *The ellipsoids $\mathcal{E}(p), \mathcal{E}(q)$ are biholomorphic iff $p = q$ up to a permutation.*

Remark 16.4.2.

(a) It was N. Kritikos [330] who already in 1927 studied the group of automorphisms of $\mathcal{E}((1/2, 1/2))$ by exploiting complex geodesics (he himself used the notion *metrische Ebene*); cf. Corollary 16.4.6.

(b) The biholomorphic equivalence problem for all complex ellipsoids was studied first by I. Naruki (cf. [372]) and then by M. Ise (cf. [242]); the equivalence problem for general bounded Reinhardt domains containing the origin is due to T. Sunada (cf. [492]).

(c) Similar investigations for proper holomorphic mappings were done by M. Landucci (cf. [334]) and S. M. Webster (cf. [527]).

(d) More generally, we have the following result:

If two bounded balanced pseudoconvex domains G_1, $G_2 \subset \mathbb{C}^n$ are biholomorphic, then there exists a biholomorphic mapping $F : G_1 \longrightarrow G_2$ with $F(0) = 0$.

Then, by the results of Chapter 3, it easily follows that G_1, G_2 are bilinearly equivalent. This result is due to W. Kaup & H. Upmeier (cf. [291]) and also to W. Kaup & J.-P. Vigué (cf. [292]). Their proofs are based on Lie theory and so are beyond the scope of this book. There is another proof, given by K.-T. Kim (cf. [299]); he exploits the fact that maximal compact subgroups of a connected Lie group are conjugate.

Our proof of Theorem 16.4.1 is more consistent with complex analysis. It is based on the fact that any biholomorphic mapping between bounded complete Reinhardt domains extends to a biholomorphism of neighborhoods of the closures (cf. Theorem 12.1.18).

First observe that, without loss of generality, we may assume that

$$p_1 = \cdots = p_k = 1, \; p_{k+1}, \dots, p_n \neq 1, \tag{16.4.1}$$

$$q_1 = \cdots = q_l = 1, \; q_{l+1}, \dots, q_n \neq 1, \tag{16.4.2}$$

$$0 \le l \le k \le n, \; l \le n-1. \tag{16.4.3}$$

(In the case where $k = l = n$, we simply have $\mathcal{E}(p) = \mathcal{E}(q) = \mathbb{B}_n$.)

Lemma 16.4.3. *Assume that p satisfies* (16.4.1) *and let $\Psi \in \operatorname{Aut}(\mathbb{B}_k)$, $a' := \Psi^{-1}(0)$, $\theta_{k+1}, \dots, \theta_n \in \mathbb{R}$. Define*

$$\Phi(z_1, \dots, z_n) = \Phi(z', z_{k+1}, \dots, z_n)$$
$$= \left(\Psi(z'), e^{i\theta_{k+1}} z_{k+1} \left(\frac{1 - \|a'\|^2}{(1 - \langle z', a' \rangle)^2} \right)^{\frac{1}{2p_{k+1}}}, \dots, e^{i\theta_n} z_n \left(\frac{1 - \|a'\|^2}{(1 - \langle z', a' \rangle)^2} \right)^{\frac{1}{2p_n}} \right).$$

Then, $\Phi \in \operatorname{Aut}(\mathcal{E}(p))$.

Proof. The cases $k = 0$ or $k = n$ are obvious. Assume $1 \le k \le n-1$. Clearly, Φ is holomorphic and injective in a neighborhood of $\overline{\mathcal{E}(p)}$. Observe that by Corollary 2.3.5 we have

$$\|\Psi(z')\| = \boldsymbol{c}^*_{\mathbb{B}_k}(0, \Psi(z')) = \boldsymbol{c}^*_{\mathbb{B}_k}(a', z') = \left(1 - \frac{(1 - \|a'\|^2)(1 - \|z'\|^2)}{|1 - \langle z', a' \rangle|^2}\right)^{\frac{1}{2}}, \quad z' \in \mathbb{B}_k.$$

Hence, for $z = (z', z_{k+1}, \dots, z_n) \in \overline{\mathcal{E}(p)}$ we obtain

$$\sum_{j=1}^n |\Phi_j(z)|^{2p_j} = 1 - \frac{1 - \|a'\|^2}{|1 - \langle z', a' \rangle|^2} \left(1 - \sum_{j=1}^n |z_j|^{2p_j}\right),$$

which finishes the proof. □

Lemma 16.4.4. *Let $p, q \in (\mathbb{R}_{>0})^n$ satisfy* (16.4.1–16.4.3) *and let $L : \mathbb{C}^n \longrightarrow \mathbb{C}^n$ be a $\mathbb{C}$-linear isomorphism such that $L(\mathcal{E}(q)) = \mathcal{E}(p)$. Then, $p = q$ up to a permutation σ of $(k+1, \dots, n)$ (in particular, $k = l$) and the mapping L is of the form*

$$L(w_1, \dots, w_n) = L(w', w_{k+1}, \dots, w_n) = (U(w'), e^{i\theta_{k+1}} w_{\sigma(k+1)}, \dots, e^{i\theta_n} w_{\sigma(n)}),$$

where $U : \mathbb{C}^{n-k} \longrightarrow \mathbb{C}^{n-k}$ is a unitary isomorphism and $\theta_{k+1}, \dots, \theta_n \in \mathbb{R}$.

Proof. Recall (cf. Remark 16.2.1) that $\partial\mathcal{E}(q)$ is strongly pseudoconvex at a point $a \in \partial\mathcal{E}(q)$ iff

$$a \notin H_1 := \{(w_1,\dots,w_n) \in \mathbb{C}^n : w_{l+1}\cdots w_n = 0\}.$$

Moreover, $\partial\mathcal{E}(q)$ is strongly pseudoconvex at a iff $\partial\mathcal{E}(p)$ is strongly pseudoconvex at $L(a)$. Hence,

$$L(H_1) = H_2 := \{(z_1,\dots,z_n) \in \mathbb{C}^n : z_{k+1}\cdots z_n = 0\}.$$

Now, since L is an isomorphism, we easily conclude that $k = l$ and that there exists a permutation σ of $(k+1,\dots,n)$ such that $L_{\sigma(j)}(\{w_j = 0\}) = \{0\}$. We can assume that $\sigma = \mathrm{id}$. Thus,

$$L(w', w_{k+1},\dots,w_n) = (U(w') + M(w_{k+1},\dots,w_n), c_{k+1}w_{k+1},\dots,c_n w_n),$$

where $U : \mathbb{C}^k \longrightarrow \mathbb{C}^k$ and $M : \mathbb{C}^{n-k} \longrightarrow \mathbb{C}^k$ are $\mathbb{C}$-linear and $c_{k+1},\dots,c_n \in \mathbb{C}_*$. The case $k = 0$ is elementary. Assume $k \ge 1$. It is clear that U is an isomorphism and that $U(\mathbb{B}_k) = \mathbb{B}_k$ (take $w_{k+1} = \dots = w_n = 0$). Hence, U is unitary. It remains to prove that $M \equiv 0$. Since $L(\partial\mathcal{E}(q)) = \partial\mathcal{E}(p)$, we get

$$\|U(w')+M(w_{k+1},\dots,w_n)\|^2 + \sum_{j=k+1}^{n} |c_j w_j|^{2p_j} = 1,\ (w', w_{k+1},\dots,w_n) \in \partial\mathcal{E}(q).$$

This implies that

$$-\sum_{j=k+1}^{n} |w_j|^{2q_j} + \|M(w_{k+1},\dots,w_n)\|^2 + \sum_{j=k+1}^{n} |c_j w_j|^{2p_j} = 0$$

$$\text{for all } (w_{k+1},\dots,w_n) \text{ with } \sum_{j=k+1}^{n} |w_j|^{2q_j} \le 1.$$

Taking $w_{k+1} = \dots = w_{j-1} = w_{j+1} = \dots = w_n = 0$, we have

$$-x^{q_j} + \|M(0,\dots,0,\overset{j}{1},0,\dots,0)\|^2 x + |c_j|^{2p_j} x^{p_j} = 0,\ 0 \le x \le 1, j = k+1,\dots,n.$$

Hence, we get $p_j = q_j, |c_j| = 1$, and $M(0,\dots,0,\overset{j}{1},0,\dots,0) = 0$ for $j = k + 1,\dots,n$. □

Proof of Theorem 16.4.1 Recall that if $\Phi : \mathcal{E}(q) \longrightarrow \mathcal{E}(p)$ is a biholomorphic mapping with $\Phi(0) = 0$, then $L := \Phi'(0)$ is a $\mathbb{C}$-linear isomorphism with $L(\mathcal{E}(q)) = \mathcal{E}(p)$. Thus, in view of Lemmas 16.4.3 and 16.4.4, it remains to prove the following fact:

Lemma 16.4.5. *Assume that $\Phi : \mathcal{E}(q) \longrightarrow \mathcal{E}(p)$ is a biholomorphic mapping with $\Phi(0) = (0, \dots, 0, b_{k+1}, \dots, b_n)$. If p, q satisfy (16.4.1–16.4.3), then $b_{k+1} = \dots = b_n = 0$.*

Proof. Fix ν with $l + 1 \le \nu \le n$ and let $H_\nu := \mathcal{E}(q) \cap \{z \in \mathbb{C}^n : z_\nu = 0\}$. Because of the fact that Φ extends biholomorphically to $\overline{\mathcal{E}(q)}$, we have $\Phi_{k+1} \cdot \dots \cdot \Phi_n = 0$ on H_ν; cf. the proof of Lemma 16.4.4. Then, $\Phi(H_\nu) \subset \mathcal{E}(p) \cap \{w \in \mathbb{C}^n : w_{j(\nu)} = 0\}$ with $l + 1 \le j(\nu) \le n$. In particular, $\Phi_{j(\nu)}(0) = b_{j(\nu)} = 0$. Since $\nu \longmapsto j(\nu)$ is bijective, we have $b_{k+1} = \dots = b_n = 0$. □ □

Corollary 16.4.6. *If $p_j \ne 1$ for all $j = 1, \dots, n$, then any $\Phi \in \operatorname{Aut}(\mathcal{E}(p))$ has the origin as a fixed point.*

Observe that the main tool in the proof of Theorem 16.4.1 was Theorem 12.1.18 on the extendability of biholomorphic mappings. It seems interesting to find a proof of Theorem 16.4.1 (at least in the convex case) using only the complex geodesics from Proposition 16.2.2. ?

16.5 Complex geodesics in the minimal ball

Recall (cf. § 12.2) that for $n \ge 2$ the minimal ball is the domain

$$\mathbb{M}_n := \{z \in \mathbb{C}^n : \|z\|_{\min} < 1\},$$

where

$$\begin{aligned}\|z\|_{\min} &= \left(\|z\|^2 + |z \bullet z|\right)^{1/2} \\ &= \left[\|x\|^2 + \|y\|^2 + [(\|x\|^2 - \|y\|^2)^2 + 4(x \bullet y)^2]^{1/2}\right]^{1/2}, \\ &\qquad\qquad z = x + iy \in \mathbb{R}^n + i\mathbb{R}^n \cong \mathbb{C}^n.\end{aligned}$$

Remark 16.5.1 (the reader is asked to complete details, cf. Remark 12.2.1).

(a) Since $\mathbb{M}_n$ is convex, Lempert's theorem applies. Therefore, $\boldsymbol{c}_{\mathbb{M}_n} \equiv \boldsymbol{\ell}_{\mathbb{M}_n}$, $\boldsymbol{\gamma}_{\mathbb{M}_n} \equiv \boldsymbol{\varkappa}_{\mathbb{M}_n}$, and we have existence of complex geodesics.

(b) Any point $a \in \partial\mathbb{M}_n$ is a complex extreme point for $\overline{\mathbb{M}_n}$ (to recall the definition of a complex extreme point see Proposition 11.3.5).

The case $n = 2$ follows from Remark 12.2.1(c), the general case may be proved by the same methods as in Example 16.1.3.

(c) In view of (b), all complex geodesics $\varphi : \mathbb{D} \longrightarrow \mathbb{M}_n$ with $0 \in \varphi(\mathbb{D})$ are uniquely determined and they are of the form

$$\varphi(\lambda) = \lambda a, \quad \lambda \in \mathbb{D} \quad (\operatorname{mod} \operatorname{Aut}(\mathbb{D})),$$

where $a \in \partial\mathbb{M}_n$; cf. Proposition 11.3.5(b).

(d) Since $\|z'\|_{\min} \le \|(z', z_n)\|_{\min}$, $(z', z_n) \in \mathbb{C}^{n-1} \times \mathbb{C}$, using the holomorphic contractibility with respect to the mappings

$$\mathbb{M}_{n-1} \ni z' \longmapsto (z', 0) \in \mathbb{M}_n, \quad \mathbb{M}_n \ni (z', z_n) \longmapsto z' \in \mathbb{M}_{n-1},$$

we get

$$\boldsymbol{c}_{\mathbb{M}_n}((z', 0), (w', 0)) = \boldsymbol{c}_{\mathbb{M}_{n-1}}(z', w'), \; \boldsymbol{\gamma}_{\mathbb{M}_n}((z', 0); (X', 0)) = \boldsymbol{\gamma}_{\mathbb{M}_{n-1}}(z'; X').$$

Consequently, if $\psi = (\psi_1, \dots, \psi_{n-1}, 0) : \mathbb{D} \longrightarrow \mathbb{M}_n$ is a holomorphic mapping, then ψ is a complex geodesic iff $(\psi_1, \dots, \psi_{n-1}) : \mathbb{D} \longrightarrow \mathbb{M}_{n-1}$ is a complex geodesic. In particular, similar to the case of complex convex ellipsoids, we may always restrict our considerations to the case where $\varphi_j \not\equiv 0$, $j = 1, \dots, n$.

(e) $\operatorname{Aut}_0(\mathbb{M}_n) = \{e^{i\theta} A : \theta \in \mathbb{R},\ A \in \mathbb{O}(n)\}$.

Indeed (cf. [298]), it is clear that $e^{i\theta} A \in \operatorname{Aut}_0(\mathbb{M}_n)$. If $\Phi \in \operatorname{Aut}_0(\mathbb{M}_n)$, then by Cartan's theorem (cf. [269], Proposition 2.1.9), Φ must be $\mathbb{C}$-linear and $\|\Phi(z)\|_{\min} = \|z\|_{\min}$, $z \in \mathbb{C}^n$. Hence, we conclude that $\Phi(Q) = Q$, where $Q := \{z \in \mathbb{C}^n : z \bullet z = 0\}$. Observe that to prove that $\Phi = e^{i\theta} A$ for some $A \in \mathbb{O}(n)$, we only need to show that $S := \Phi^t \Phi = e^{2i\theta} \mathbb{I}_n$, where $\mathbb{I}_n$ denotes the unit matrix. We have $z \bullet z = 0 \Longrightarrow z^t S z = 0$. Taking $z = \boldsymbol{e}_j \pm i \boldsymbol{e}_k$ with $j \neq k$, we easily get the conclusion.

(f) $\mathbb{M}_n$ is not biholomorphic to $\mathbb{B}_n$.

(g) In fact, we have

$$\operatorname{Aut}(\mathbb{M}_n) = \operatorname{Aut}_0(\mathbb{M}_n) = \{e^{i\theta} A : \theta \in \mathbb{R},\ A \in \mathbb{O}(n)\};$$

cf. [298]. A direct proof using complex geodesics will be given in Theorem 16.5.12. In particular, the group $\operatorname{Aut}(\mathbb{M}_n)$ does not act transitively in $\mathbb{M}_n$.

Our aim is to characterize all complex geodesics $\varphi = (\varphi_1, \dots, \varphi_n) : \mathbb{D} \longrightarrow \mathbb{M}_n$. The remaining part of this section is based on [420].

Theorem 16.5.2. *Complex geodesics in* $\mathbb{M}$ *are uniquely determined* (mod $\operatorname{Aut}(\mathbb{D})$). *Every complex geodesic* $\varphi = (\varphi_1, \dots, \varphi_n) : \mathbb{D} \longrightarrow \mathbb{M}$ *with* $\varphi_j \not\equiv 0$, $j = 1, \dots, n$, *is of the form*

$$\varphi_j(\lambda) = a_j \left(\frac{\lambda - \alpha_{j,1}}{1 - \overline{\alpha}_{j,1}\lambda} \right)^{r_{j,1}} \left(\frac{\lambda - \alpha_{j,2}}{1 - \overline{\alpha}_{j,2}\lambda} \right)^{r_{j,2}} \frac{(1 - \overline{\alpha}_{j,1}\lambda)(1 - \overline{\alpha}_{j,2}\lambda)}{(1 - \overline{\alpha}\lambda)^2}, \quad (16.5.1)$$

where $a_j \in \mathbb{C}_*$, $\alpha \in \mathbb{D}$, $r_{j,s} \in \{0, 1\}$, $\alpha_{j,s} \in \overline{\mathbb{D}}$ *are such that* $(r_{j,s} = 1 \Longrightarrow \alpha_{j,s} \in \mathbb{D})$, $s = 1, 2$, *and* $r_{j,1} + r_{j,2} \le 1$, $j = 1, \dots, n$.

Moreover,

$$\varphi \bullet \varphi = b\left(\frac{\lambda-\beta_1}{1-\overline{\beta}_1\lambda}\right)^{k_1}\left(\frac{\lambda-\beta_2}{1-\overline{\beta}_2\lambda}\right)^{k_2}\frac{(1-\overline{\beta}_1\lambda)^2(1-\overline{\beta}_2\lambda)^2}{(1-\overline{\alpha}\lambda)^4}, \tag{16.5.2}$$

where $b \in \mathbb{C}$, $k_s \in \{0,1\}$, $\beta_s \in \overline{\mathbb{D}}$ *are such that* ($k_s = 1 \Longrightarrow \beta_s \in \mathbb{D}$), $s = 1, 2$.

For the proof, we need a few auxiliary results. Let

$$\widehat{q}_{\mathbb{M}}(w) := \max\{\operatorname{Re}(z \bullet w) : z \in \partial\mathbb{M}\}, \quad w \in \mathbb{C}^n;$$

cf. § 11.2. Fix a complex geodesic $\varphi = (\varphi_1,\dots,\varphi_n) : \mathbb{D} \longrightarrow \mathbb{M}$ with $\varphi_j \not\equiv 0$, $j = 1,\dots,n$. By Corollary 11.2.8 we get $\varphi^*(\lambda) \in \partial\mathbb{M}$ for a.a. $\lambda \in \mathbb{T}$, and

$$\operatorname{Re}\left(\varphi^*(\lambda) \bullet \frac{1}{\lambda}h^*(\lambda)\right) = \widehat{q}_{\mathbb{M}}\left(\frac{1}{\lambda}h^*(\lambda)\right) \text{ for a.a. } \lambda \in \mathbb{T},$$

where $h \in \mathcal{H}^1(\mathbb{D},\mathbb{C}^n)$.

Lemma 16.5.3. *Let* $w_0 \in (\mathbb{C}^n)_*$, $z_0 \in \partial\mathbb{M}$ *be such that*

$$\operatorname{Re}(z_0 \bullet w_0) = \widehat{q}_{\mathbb{M}}(w_0).$$

Then, there are numbers $\varrho > 0$ *and* $\eta \in \overline{\mathbb{D}}(\varrho)$ *such that*

$$w_0 = \begin{cases} \varrho\left(\overline{z}_0 + \frac{\overline{z_0 \bullet z_0}}{|z_0 \bullet z_0|}z_0\right), & \text{if } z_0 \bullet z_0 \neq 0 \\ \varrho\overline{z}_0 + \eta z_0, & \text{if } z_0 \bullet z_0 = 0 \end{cases}. \tag{16.5.3}$$

Proof. By Remark 11.2.3 we get

$$\operatorname{Re}((z - z_0) \bullet w_0) < 0, \quad z \in \mathbb{M}. \tag{16.5.4}$$

If $z_0 \bullet z_0 \neq 0$, then z_0 is a smooth boundary point of $\mathbb{M}$. Consequently (cf. Remark 11.2.3), $\overline{w}_0$ is the outer normal vector to $\partial\mathbb{M}$ at z_0, i.e., $w_0 = \varrho\left(\overline{z}_0 + \frac{\overline{z_0 \bullet z_0}}{|z_0 \bullet z_0|}z_0\right)$ for some $\varrho > 0$.

Now assume that $z_0 \bullet z_0 = 0$. Then, $\overline{w}_0 \perp T_{z_0}^{\mathbb{R}}V$, where

$$\begin{aligned} V :&= \{z \in \partial\mathbb{B}_n : z \bullet z = 0\} \\ &= \{x + iy \in \mathbb{R}^n + i\mathbb{R}^n : x \bullet x + y \bullet y = 1,\ x \bullet x - y \bullet y = 0,\ x \bullet y = 0\}; \end{aligned}$$

note that V is a $(2n-3)$-dimensional real analytic manifold. Thus,

$$\overline{w}_0 = \varrho z_0 + \nu\overline{z}_0 + \mu i\overline{z}_0 = (\nu + i\mu)\overline{z}_0 + \varrho z_0 =: \eta\overline{z}_0 + \varrho z_0$$

for some $\varrho, \nu, \mu \in \mathbb{R}$. By (16.5.4) we get

$$\operatorname{Re}(\eta(z \bullet z_0)) + \varrho\operatorname{Re}(z \bullet \overline{z}_0) < \varrho, \quad z \in \mathbb{M}.$$

Setting $z = \lambda\overline{z}_0$ with $\lambda \in \mathbb{D}$, we get $\operatorname{Re}(\eta\lambda) < \varrho$ for all $\lambda \in \mathbb{D}$. Consequently, $|\eta| \le \varrho$ and $\varrho > 0$. □

Lemma 16.5.4. *Let $\varphi = (\varphi_1, \dots, \varphi_n) : \mathbb{D} \longrightarrow \mathbb{M}$ be a complex geodesic such that $\varphi \bullet \varphi \not\equiv 0$ and $\varphi_j \not\equiv 0$ with $j = 1, \dots, n$. Then, φ is of the form* (16.5.1) *with* (16.5.2).

Proof. Using Lemma 16.5.3, we conclude that

$$\frac{1}{\lambda} h_j^*(\lambda) = \varrho(\lambda) \left(\overline{\varphi}_j^*(\lambda) + \frac{\overline{\varphi}^*(\lambda) \bullet \overline{\varphi}^*(\lambda)}{|\varphi^*(\lambda) \bullet \varphi^*(\lambda)|} \varphi_j^*(\lambda) \right) \text{ for a.a. } \lambda \in \mathbb{T}.$$

where $h = (h_1, \dots, h_n) \in \mathcal{H}^1(\mathbb{D}, \mathbb{C}^n)$ and $\varrho : \mathbb{T} \longrightarrow \mathbb{R}_{>0}$. Since

$$\frac{1}{\lambda} \varphi^*(\lambda) \bullet h^*(\lambda) = \varrho(\lambda) > 0 \text{ for a.a. } \lambda \in \mathbb{T},$$

by Lemma 16.2.7 (with $m = 1$), there are $r > 0$ and $\alpha \in \overline{\mathbb{D}}$ such that

$$(\varphi \bullet h)(\lambda) = r(\lambda - \alpha)(1 - \overline{\alpha}\lambda), \quad \lambda \in \mathbb{D}.$$

We may assume that $r = 1$. Thus,

$$\varrho(\lambda) = |1 - \lambda\overline{\alpha}|^2 \text{ for a.a. } \lambda \in \mathbb{T}.$$

In particular, $h \in \mathcal{H}^\infty(\mathbb{D}, \mathbb{C}^n)$. Observe that

$$\frac{1}{\lambda^2}(h^*(\lambda) \bullet h^*(\lambda))(\varphi^*(\lambda) \bullet \varphi^*(\lambda)) = 2|1 - \lambda\overline{\alpha}|^4 |\varphi^*(\lambda) \bullet \varphi^*(\lambda)| > 0 \quad \text{for a.a. } \lambda \in \mathbb{T}. \tag{16.5.5}$$

Hence, by Lemma 16.2.7 (with $m = 2$) there exist $a > 0$, $\beta_1, \beta_2 \in \overline{\mathbb{D}}$ such that

$$\begin{aligned}(h(\lambda) \bullet h(\lambda))(\varphi(\lambda) \bullet \varphi(\lambda)) &= a(\lambda - \beta_1)(1 - \overline{\beta}_1\lambda)(\lambda - \beta_2)(1 - \overline{\beta}_2\lambda) \\ &= \prod_{j \in J} \frac{\lambda - \beta_j}{1 - \overline{\beta}_j\lambda} a \prod_{j \in J} (1 - \overline{\beta}_j\lambda)^2 \prod_{j \notin J} (-\beta_j)(1 - \overline{\beta}_j\lambda)^2, \quad \lambda \in \mathbb{D}, \end{aligned} \tag{16.5.6}$$

where $J := \{j \in \{1, 2\} : \beta_j \in \mathbb{D}\}$. Let $h \bullet h = B_1 S_1 F_1$, $\varphi \bullet \varphi = B_2 S_2 F_2$, where B_i is the Blaschke product, S_i is the singular inner functions, and F_i is the singular outer function. Using (16.5.6) and Lemma 16.3.3, we conclude that $S_1 \equiv S_2 \equiv 1$. By (16.5.5) we get

$$\begin{aligned} F_1(\lambda) &= \exp\left(\frac{1}{2\pi} \int_0^{2\pi} \frac{e^{i\theta} + \lambda}{e^{i\theta} - \lambda} \log |h^*(e^{i\theta}) \bullet h^*(e^{i\theta})| d\theta \right) \\ &= \exp\left(\frac{1}{2\pi} \int_0^{2\pi} \frac{e^{i\theta} + \lambda}{e^{i\theta} - \lambda} \log 2|1 - e^{i\theta}\overline{\alpha}|^4 d\theta \right) = 2(1 - \lambda\overline{\alpha})^4, \quad \lambda \in \mathbb{D}. \end{aligned}$$

Thus,

$$\begin{aligned}B(\lambda)(\varphi(\lambda)\bullet\varphi(\lambda)) &= \frac{h(\lambda)\bullet h(\lambda)}{F_1(\lambda)}(\varphi(\lambda)\bullet\varphi(\lambda))\\ &= \prod_{j\in J}\frac{\lambda-\beta_j}{1-\overline{\beta}_j\lambda}\\ &\quad\cdot\frac{a\prod_{j\in J}(1-\overline{\beta}_j\lambda)^2\prod_{j\notin J}(-\beta_j)(1-\overline{\beta}_j\lambda)^2}{(1-\lambda\overline{\alpha})^4}, \quad \lambda\in\mathbb{D},\end{aligned}$$

which, in view of (16.5.6), implies that (16.5.2) must be fulfilled.

Define

$$g_j(\lambda) := 2(\lambda-\alpha)(1-\overline{\alpha}\lambda)h_j(\lambda) - (h(\lambda)\bullet h(\lambda))\varphi_j(\lambda), \quad \lambda\in\mathbb{D}.$$

Observe that

$$\frac{g_j^*(\lambda)\varphi_j^*(\lambda)}{4(\lambda-\alpha)^2(1-\overline{\alpha}\lambda)^2} = \frac{1}{2}|\varphi_j^*(\lambda)|^2 > 0 \text{ for a.a. } \lambda\in\mathbb{T}. \tag{16.5.7}$$

First assume that $\alpha\in\mathbb{D}$. Then, using Lemma 16.2.7 with $m = 2$ and $f = \frac{g_j\varphi_j}{4(1-\overline{\alpha}\lambda)^2}$ (note that $f\in\mathcal{H}^\infty(\mathbb{D})$), we get $r_j > 0$, $\alpha_{j,1},\alpha_{j,2}\in\overline{\mathbb{D}}$ such that

$$\begin{aligned}g_j(\lambda)\varphi_j(\lambda) &= r_j\prod_{s=1}^{2}(\lambda-\alpha_{j,s})(1-\overline{\alpha}_{j,s}\lambda)\\ &= \prod_{s\in J_j}\frac{\lambda-\alpha_{j,s}}{1-\overline{\alpha}_{j,s}\lambda}r_j\prod_{s\in J_j}(1-\overline{\alpha}_{j,s}\lambda)^2\\ &\quad\cdot\prod_{s\notin J_j}(-\alpha_{j,s})(1-\overline{\alpha}_{j,s}\lambda)^2, \quad \lambda\in\mathbb{D},\end{aligned} \tag{16.5.8}$$

where $J_j := \{s\in\{1,2\} : \alpha_{j,s}\in\mathbb{D}\}$. Write $g_j = B_{j,1}S_{j,1}F_{j,1}$, $\varphi_j = B_{j,2}S_{j,2}F_{j,2}$, where $B_{j,s}$ is the Blaschke product, $S_{j,s}$ – the singular inner function, and $F_{j,s}$ – the singular outer function. Using (16.5.8) and Lemma 16.3.3, we get $S_{j,s}\equiv 1$. Moreover, by (16.5.7), we have

$$\begin{aligned}|F_{j,2}^*(\lambda)|^2 = |\varphi_j^*(\lambda)|^2 &= 2\left|\frac{g_j^*(\lambda)\varphi_j^*(\lambda)}{4(\lambda-\alpha)^2(1-\overline{\alpha}\lambda)^2}\right|\\ &= 2\left|\frac{r_j\prod_{s=1}^2(\lambda-\alpha_{j,s})(1-\overline{\alpha}_{j,s}\lambda)}{4(\lambda-\alpha)^2(1-\overline{\alpha}\lambda)^2}\right|\\ &= \frac{r_j}{2}\left|\prod_{s=1}^{2}\frac{1-\overline{\alpha}_{j,s}\lambda}{1-\overline{\alpha}\lambda}\right|^2 \text{ for a.a. } \lambda\in\mathbb{T}.\end{aligned}$$

Thus,

$$F_{j,2}(\lambda) = \sqrt{\frac{r_j}{2}} \prod_{s=1}^{2} \frac{1-\overline{\alpha}_{j,s}\lambda}{1-\overline{\alpha}\lambda}, \quad \lambda \in \mathbb{D},$$

which implies (16.5.1).

It remains to exclude the situation where $\alpha \in \mathbb{T}$. Suppose that $\alpha \in \mathbb{T}$. Then, using (16.5.7) and Lemma 16.2.7 with $m = 4$ and $f = \frac{g_j \varphi_j}{4\overline{\alpha}^2}$, we get $r_j > 0$, $\alpha_{j,s} \in \overline{\mathbb{D}}$ such that

$$g_j(\lambda)\varphi_j(\lambda) = 4r_j\alpha^2 \frac{\prod_{s=1}^{4}(\lambda-\alpha_{j,s})(1-\overline{\alpha}_{j,s}\lambda)}{(\lambda-\alpha)^4}, \quad \lambda \in \mathbb{D}. \tag{16.5.9}$$

Since the left hand side of (16.5.9) is bounded, we conclude that $\alpha_{j,s} = \alpha$, $j = 1,\dots,n$, $s = 1,\dots,4$. Now, by (16.5.7) and (16.5.9), we get

$$|\varphi_j^*(\lambda)|^2 = 2\left|\frac{g_j^*(\lambda)\varphi_j^*(\lambda)}{4(\lambda-\alpha)^2(1-\overline{\alpha}\lambda)^2}\right| = \frac{2r_j}{|\lambda-\alpha|^4} \text{ for a.a. } \lambda \in \mathbb{T},$$

which implies that φ_j is unbounded; a contradiction. □

Lemma 16.5.5. *Let $\varphi = (\varphi_1,\dots,\varphi_n) : \mathbb{D} \longrightarrow \mathbb{M}$ be a complex geodesic such that $\varphi \bullet \varphi \equiv 0$ and $\varphi_j \not\equiv 0$ with $j = 1,\dots,n$. Then, φ is of the form* (16.5.1).

Proof. By Lemma 16.5.3 we get $h \in \mathcal{H}^1(\mathbb{D},\mathbb{C}^n)$, $\varrho : \mathbb{T} \longrightarrow \mathbb{R}_{>0}$, and $\eta : \mathbb{T} \longrightarrow \mathbb{C}$ such that $|\eta| \le \varrho$ and

$$\frac{1}{\lambda}h_j^*(\lambda) = \varrho(\lambda)\overline{\varphi}_j^*(\lambda) + \eta(\lambda)\varphi_j^*(\lambda) \text{ for a.a. } \lambda \in \mathbb{T}. \tag{16.5.10}$$

Since $\frac{1}{\lambda}\varphi^*(\lambda) \bullet h^*(\lambda) = \varrho(\lambda) > 0$ for a.a. $\lambda \in \mathbb{T}$, applying Lemma 16.2.7 (with $m = 1$) we get $r > 0$ and $\alpha \in \overline{\mathbb{D}}$ such that

$$(\varphi \bullet h)(\lambda) = r(\lambda-\alpha)(1-\overline{\alpha}\lambda), \quad \lambda \in \mathbb{D}. \tag{16.5.11}$$

We may assume that $r = 1$. Hence, $\varrho(\lambda) = |1-\lambda\overline{\alpha}|^2$ for a.a. $\lambda \in \mathbb{T}$. Observe that

$$\frac{1}{\lambda^2}h^*(\lambda) \bullet h^*(\lambda) = 2\varrho(\lambda)\eta(\lambda), \quad \|h^*(\lambda)\|^2 = \varrho^2(\lambda) + |\eta^2(\lambda)| \text{ for a.a. } \lambda \in \mathbb{T}. \tag{16.5.12}$$

First consider the case where $\eta = 0$ on a set of positive measure on $\mathbb{T}$. Then, by (16.5.12), $h \bullet h \equiv 0$. Moreover, by (16.5.10),

$$\frac{1}{\lambda}h_j^*(\lambda) = \varrho(\lambda)\overline{\varphi}_j^*(\lambda) \text{ for a.a. } \lambda \in \mathbb{T}.$$

This implies that $\varphi : \mathbb{D} \longrightarrow \mathbb{B}_n$ is a complex geodesic and we may therefore apply Proposition 16.2.2.

Now assume that $\eta(\lambda) \neq 0$ for a.a. $\lambda \in \mathbb{T}$. Put

$$g(\lambda) := 2h_j^*(\lambda)(h^* \bullet \varphi^*)(\lambda) - (h^* \bullet h^*)(\lambda)\varphi_j^*(\lambda), \quad \lambda \in \mathbb{D}.$$

Using (16.5.10) and (16.5.12), we get

$$\frac{g_j^*(\lambda)\varphi_j^*(\lambda)}{\lambda^2} = 2\varrho^2(\lambda)|\varphi_j^*(\lambda)|^2 > 0 \text{ for a.a. } \lambda \in \mathbb{T}. \tag{16.5.13}$$

Hence, by Lemma 16.2.7 (with $m = 2$) we have

$$g_j(\lambda)\varphi_j(\lambda) = r_j \prod_{s=1}^{2} (\lambda - \alpha_{j,s})(1 - \overline{\alpha}_{j,s}\lambda), \quad \lambda \in \mathbb{D}, \tag{16.5.14}$$

where $r_j > 0$ and $\alpha_{j,s} \in \overline{\mathbb{D}}$. Let $B_{j,s}, S_{j,s}, F_{j,s}$ be as in the proof of Lemma 16.5.4 ($S_{j,s} \equiv 1$). Then, by (16.5.13) and (16.5.14), we get

$$\begin{aligned} |F_{j,2}^*(\lambda)|^2 = |\varphi_j^*(\lambda)|^2 &= \frac{|g_j^*(\lambda)\varphi_j^*(\lambda)|}{2\varrho^2(\lambda)} = \frac{|r_j \prod_{s=1}^2 (\lambda - \alpha_{j,s})(1 - \overline{\alpha}_{j,s}\lambda)|}{2|1 - \overline{\alpha}\lambda|^4} \\ &= \frac{r_j}{2}\left|\prod_{s=1}^{2} \frac{1 - \overline{\alpha}_{j,s}\lambda}{1 - \overline{\alpha}\lambda}\right|^2 \text{ for a.a. } \lambda \in \mathbb{T}, \end{aligned}$$

which, similar to Lemma 16.5.4, finishes the proof. □

Lemma 16.5.6. *Let $\varphi, \psi : \mathbb{D} \to \mathbb{M}$ be complex geodesics with $\varphi(0) = \psi(0)$ and $\varphi'(0) = \psi'(0)$. Write*

$$\varphi(\lambda) = \frac{P(\lambda)}{(1 - \overline{\alpha}\lambda)^2}, \quad \psi(\lambda) = \frac{Q(\lambda)}{(1 - \overline{\beta}\lambda)^2},$$

where $\alpha, \beta \in \mathbb{D}$ and $P, Q \in \mathcal{P}_2(\mathbb{C}, \mathbb{C}^n)$ ($P, Q : \mathbb{C} \longrightarrow \mathbb{C}^n$ are polynomial mappings of degree ≤ 2); cf. Lemmas 16.5.4, 16.5.5*). Then, $\alpha = \beta$.*

Proof. Suppose that $\alpha \neq \beta$. Since $\chi := (1/2)(\varphi + \psi)$ is also a complex geodesic, we may write

$$\chi(\lambda) = \frac{R(\lambda)}{(1 - \overline{\gamma}\lambda)^2},$$

where $\gamma \in \mathbb{D}$ and $R \in \mathcal{P}_2(\mathbb{C}, \mathbb{C}^n)$. Thus,

$$P(\lambda)(1 - \overline{\beta}\lambda)^2(1 - \overline{\gamma}\lambda)^2 + Q(\lambda)(1 - \overline{\alpha}\lambda)^2(1 - \overline{\gamma}\lambda)^2 = 2R(\lambda)(1 - \overline{\alpha}\lambda)^2(1 - \overline{\beta}\lambda)^2.$$

First, consider the case where $\gamma = 0$. We may assume that $\alpha \neq 0$. Then,

$$P(\lambda)(1-\overline{\beta}\lambda)^2 + Q(\lambda)(1-\overline{\alpha}\lambda)^2 = 2R(\lambda)(1-\overline{\alpha}\lambda)^2(1-\overline{\beta}\lambda)^2,$$

which implies that $(1-\overline{\alpha}\lambda)^2$ divides P. Consequently, φ is constant; a contradiction.

Now assume that $\gamma \neq 0$. If $\alpha \neq \gamma$ and $\beta \neq \gamma$, then $(1-\overline{\gamma}\lambda)^2$ divides R. Hence, $\chi \equiv$ const; a contradiction. Therefore, we may assume that $\alpha = \gamma$. Then,

$$P(\lambda)(1-\overline{\beta}\lambda)^2 + Q(\lambda)(1-\overline{\alpha}\lambda)^2 = 2R(\lambda)(1-\overline{\beta}\lambda)^2,$$

which implies that $(1-\overline{\beta}\lambda)^2$ divides Q; a contradiction. □

Lemma 16.5.7. *Complex geodesics in $\mathbb{M}$ are uniquely determined.*

Proof. By Proposition 11.3.2 it suffices to check that if $\varphi, \psi : \mathbb{D} \longrightarrow \mathbb{M}$ are complex geodesics with

$$\varphi(0) = \psi(0) \text{ and } \varphi'(0) = \psi'(0), \tag{16.5.15}$$

then $\varphi \equiv \psi$. By Lemma 16.5.6 we know that

$$\varphi(\lambda) = \frac{P(\lambda)}{(1-\overline{\alpha}\lambda)^2}, \quad \psi(\lambda) = \frac{Q(\lambda)}{(1-\overline{\alpha}\lambda)^2},$$

where $\alpha \in \mathbb{D}$ and $P, Q \in \mathcal{P}_2(\mathbb{C}, \mathbb{C}^n)$. Conditions (16.5.15) imply that $P(\lambda) - Q(\lambda) = C\lambda^2$ for some $C \in \mathbb{C}^n$. Put $Z := \psi - \varphi$ and suppose that $Z \not\equiv 0$. Consequently, by the identity principle, $Z(\lambda) \neq 0$ for $\lambda \in \mathbb{T}$ except for a finite number of points. Write $P(\lambda) = a_0 + a_1\lambda + a_2\lambda^2$, where $a_0, a_1, a_2 \in \mathbb{C}^n$. We are going to prove that

$$0 = a_0 \bullet a_0 = a_0 \bullet a_1 = a_1 \bullet a_1 + 2a_0 \bullet a_2 = a_1 \bullet a_2. \tag{16.5.16}$$

Suppose for a moment that (16.5.16) is true. Then,

$$\varphi(\lambda) \bullet \varphi(\lambda) = \frac{(a_2 \bullet a_2)\lambda^4}{(1-\overline{\alpha}\lambda)^4}, \quad \lambda \in \mathbb{D}.$$

Since $\varphi \circ \varphi$ satisfies (16.5.2), we conclude that $\varphi \circ \varphi \equiv 0$. The same argument gives $\psi \bullet \psi \equiv 0$ and $\frac{\varphi+\psi}{2} \bullet \frac{\varphi+\psi}{2} \equiv 0$ ($\frac{\varphi+\psi}{2}$ is also a complex geodesic). Therefore, $\varphi \bullet \psi \equiv 0$, which implies that $Z \bullet Z \equiv 0$; a contradiction.

We move to the proof of (16.5.16). We know that for every $t \in [0,1]$ the mapping $\chi_t := \varphi + tZ : \mathbb{D} \longrightarrow \mathbb{M}$ is also a complex geodesic. Thus,

$$(1 - \|\varphi + tZ\|^2)^2 - |(\varphi + tZ) \bullet (\varphi + tZ)|^2 = 0 \text{ on } \mathbb{T} \text{ for } t \in [0,1].$$

Expand the left hand side with respect to the powers of t. Since the coefficient near t^4 must vanish, we have $\|Z\|^2 = |Z \bullet Z|$ on $\mathbb{T}$. Observe that

$$Z(\lambda) \bullet Z(\lambda) = \frac{(C \bullet C)\lambda^4}{(1-\overline{\alpha}\lambda)^4}.$$

In particular, $C \bullet C \neq 0$.

Since the coefficient near t^2 must also vanish, we get

$$2\left(|\varphi \bullet Z|^2 - (\operatorname{Re}(\varphi \bullet Z))^2\right) + \left(|\varphi \bullet \varphi|\|Z\|^2 + \operatorname{Re}((\overline{\varphi} \bullet \overline{\varphi})(Z \bullet Z))\right) = 0 \text{ on } \mathbb{T}.$$

Observe that both terms are non-negative. Hence, $|\varphi \bullet \varphi|\|Z\|^2 + \operatorname{Re}((\overline{\varphi} \bullet \overline{\varphi})(Z \bullet Z)) = 0$ on $\mathbb{T}$. Consequently, $|\varphi \bullet \varphi||Z \bullet Z| + \operatorname{Re}((\overline{\varphi} \bullet \overline{\varphi})(Z \bullet Z)) = 0$ on $\mathbb{T}$, which implies that $(\overline{\varphi} \bullet \overline{\varphi})(Z \bullet Z) \in \mathbb{R}$ on $\mathbb{T}$. Observe that for $\lambda \in \mathbb{T}$ we have

$$\begin{aligned}(\overline{\varphi}(\lambda) \bullet \overline{\varphi}(\lambda))(Z(\lambda) \bullet Z(\lambda)) &= \frac{(\overline{P}(\lambda) \bullet \overline{P}(\lambda))(C \bullet C)\lambda^4}{|1-\overline{\alpha}\lambda|^4} \\ &= \frac{(P^*(\lambda) \bullet P^*(\lambda))(C \bullet C)}{|1-\overline{\alpha}\lambda|^4},\end{aligned}$$

where $P^*(\lambda) := \overline{a}_2 + \overline{a}_1\lambda + \overline{a}_0\lambda^2$. Thus, the complex polynomial $(P^* \bullet P^*)(C \bullet C)$ is real on $\mathbb{T}$. Consequently $P^* \bullet P^* \equiv \text{const}$, which immediately gives (16.5.16). □

Lemma 16.5.8. *Let $\varphi = (\varphi_1, \dots, \varphi_n) : \mathbb{D} \longrightarrow \mathbb{M}_n$ be a complex geodesic such that $\varphi_n \not\equiv 0$. Then, φ_n has at most one zero in $\mathbb{D}$.*

Proof. Write $\varphi = (\psi, \varphi_n)$. Suppose that $\varphi_n(\lambda_0') = \varphi_n(\lambda_0'') = 0$ for some $\lambda_0', \lambda_0'' \in \mathbb{D}$, $\lambda_0' \neq \lambda_0''$ (resp. $\varphi_n(\lambda_0) = \varphi_n'(\lambda_0) = 0$ for some $\lambda_0 \in \mathbb{D}$). Observe that $\psi(\lambda_0') \neq \psi(\lambda_0'')$ (resp. $\psi'(\lambda_0) \neq 0$). Using Lempert's theorem, we find a complex $\boldsymbol{c}$- (resp. $\boldsymbol{\gamma}$-) geodesic $\varrho : \mathbb{D} \longrightarrow \mathbb{M}_{n-1}$ for $(\psi(\lambda_0'), \psi(\lambda_0''))$ (resp. $(\psi(\lambda_0), \psi'(\lambda_0))$). Then, the mapping $\widetilde{\varphi} := (\varrho, 0) : \mathbb{D} \longrightarrow \mathbb{M}_n$ is a complex $\boldsymbol{c}$- (resp. $\boldsymbol{\gamma}$-) geodesic with $\widetilde{\varphi}(\lambda_0') = \varphi(\lambda_0')$ and $\widetilde{\varphi}(\lambda_0'') = \varphi(\lambda_0'')$ (resp. $\widetilde{\varphi}(\lambda_0) = \varphi(\lambda_0)$ and $\widetilde{\varphi}'(\lambda_0) = \varphi'(\lambda_0)$). Since $\varphi_n \not\equiv 0$, we get a contradiction with Lemma 16.5.7. □

Proof of Theorem 16.5.2. The proof follows directly from the previous lemmas. □

It is natural to ask when the converse to Theorem 16.5.2 is true. Let $\mathfrak{F}$ stand for the family of all mappings $\varphi : \mathbb{D} \longrightarrow \mathbb{C}^n$ of the form (16.5.1) with (16.5.2) and $\varphi_j \not\equiv 0$, $j = 1, \dots, n$. We may assume that

- if φ_j has exactly one zero, then $r_{j,1} = 1$, $r_{j,2} = 0$, $j = 1, \dots, n$,
- if $\varphi \bullet \varphi$ has exactly one zero, then $k_1 = 1$, $k_2 = 0$.

Put $s_j := \overline{\alpha}_{j,1} + \overline{\alpha}_{j,2}$, $p_j := \overline{\alpha}_{j,1}\overline{\alpha}_{j,2}$, $j = 1, \dots, n$, $s := \overline{\beta}_1 + \overline{\beta}_2$, $p := \overline{\beta}_1\overline{\beta}_2$.

Directly from the relation $\|\varphi\|^2 + |\varphi \bullet \varphi| = 1$ on $\mathbb{T}$, one gets the following result:

Lemma 16.5.9. *Let $\varphi \in \mathfrak{F}$. Then, $\varphi(\lambda) \in \partial\mathbb{M}$ for all $\lambda \in \mathbb{T}$ iff*

$$\sum_{j=1}^{n} |a_j|^2 p_j + |b|p = \overline{\alpha}^2, \tag{16.5.17}$$

$$\sum_{j=1}^{n} |a_j|^2(1 + |s_j|^2 + |p_j|^2) + |b|(1 + |s|^2 + |p|^2) = 1 + 4|\alpha|^2 + |\alpha|^4, \tag{16.5.18}$$

$$\sum_{j=1}^{n} |a_j|^2(s_j + \overline{s}_j p_j) + |b|(s + \overline{s}p) = 2\overline{\alpha}(1 + |\alpha|^2). \tag{16.5.19}$$

The following effective criterion has been proved in [420].

Proposition 16.5.10. *Let $\varphi \in \mathfrak{F}$.*

(a) *If $\varphi \bullet \varphi \not\equiv 0$, then φ is a complex geodesic iff* (16.5.17, 16.5.18, 16.5.19) *are satisfied and there exists an $\ell \in \{0, 1, 2\}$ such that*

$$\overline{a}_j(\alpha - \alpha_{j,1})(\alpha - \alpha_{j,2}) + a_j N_\ell(\alpha)\frac{(1 - \overline{\alpha}_{j,1}\alpha)(1 - \overline{\alpha}_{j,2}\alpha)}{(1 - |\alpha|^2)^2} = 0 \text{ if } r_{j,1} = r_{j,2} = 0, \tag{16.5.20}$$

$$\overline{a}_j(1 - \overline{\alpha}_{j,1}\alpha)(\alpha - \alpha_{j,2}) + a_j N_\ell(\alpha)\frac{(\alpha - \alpha_{j,1})(1 - \overline{\alpha}_{j,2}\alpha)}{(1 - |\alpha|^2)^2} = 0 \text{ if } r_{j,1} = 1, r_{j,2} = 0, \tag{16.5.21}$$

where

$$N_0(\lambda) := \frac{\overline{b}}{|b|}\frac{(1 - \overline{\alpha}\lambda)^2(\lambda - \beta_1)(\lambda - \beta_2)}{(1 - \overline{\beta}_1\lambda)(1 - \overline{\beta}_2\lambda)} \text{ if } k_1 = k_2 = 0,$$
$$N_1(\lambda) := \frac{\overline{b}}{|b|}\frac{(1 - \overline{\alpha}\lambda)^2(\lambda - \beta_2)}{1 - \overline{\beta}_2\lambda} \text{ if } k_1 = 1,\ k_2 = 0,$$
$$N_2(\lambda) := \frac{\overline{b}}{|b|}(1 - \overline{\alpha}\lambda)^2 \text{ if } k_1 = k_2 = 1.$$

(b) *If $\varphi \bullet \varphi \equiv 0$, then φ is a complex geodesic iff there exists a $\gamma \in \overline{\mathbb{D}}$ such that $P_j^*(\alpha) = \gamma P_j(\alpha)$, where $P_j(\lambda) := (1 - \overline{\alpha}\lambda)^2\varphi_j(\lambda)$ ($P_j \in \mathcal{P}_2(\mathbb{C}, \mathbb{C})$) and $P_j^* \in \mathcal{P}_2(\mathbb{C}, \mathbb{C})$ is such that $P_j^*(\lambda) = (\lambda - \alpha)^2\overline{\varphi}_j(\lambda)$, $\lambda \in \mathbb{T}$, $j = 1, \dots, n$.*

Proof. (a) Assume that φ is a complex geodesic. Then, by Theorem 16.5.2, $\varphi \in \mathfrak{F}$. Moreover, by Lemma 16.5.9, φ satisfies (16.5.17, 16.5.18, 16.5.19). The proof of Lemma 16.5.4 implies that there exist $h \in \mathcal{H}^1(\mathbb{D}, \mathbb{C}^n)$ and $\alpha \in \mathbb{D}$ such that

$$\frac{1}{\lambda} h_j^*(\lambda) = |1 - \lambda\overline{\alpha}|^2 \left(\overline{\varphi_j^*}(\lambda) + \frac{\overline{\varphi}^*(\lambda) \bullet \overline{\varphi}^*(\lambda)}{|\varphi^*(\lambda) \bullet \varphi^*(\lambda)|} \varphi_j^*(\lambda) \right) \text{ for a.a. } \lambda \in \mathbb{T}. \quad (16.5.22)$$

Direct calculations give

$$\frac{\overline{\varphi}^*(\lambda) \bullet \overline{\varphi}^*(\lambda)}{|\varphi^*(\lambda) \bullet \varphi^*(\lambda)|} = \frac{1}{(\lambda - \alpha)^2} \begin{cases} N_0(\lambda), & \text{if } k_1 = k_2 = 0 \\ N_1(\lambda), & \text{if } k_1 = 1, k_2 = 0 \\ N_2(\lambda), & \text{if } k_1 = k_2 = 1 \end{cases} \text{ for a.a. } \lambda \in \mathbb{T}.$$

In the case where $k_1 = k_2 = r_{j,1} = r_{j,2} = 0$ we have

$$\begin{aligned}
(\lambda - \alpha) h_j^*(\lambda) &= \lambda(\lambda - \alpha)|1 - \lambda\overline{\alpha}|^2 \left(\overline{\varphi_j^*}(\lambda) + \frac{N_0(\lambda)}{(\lambda - \alpha)^2} \varphi_j^*(\lambda) \right) \\
&= (\lambda - \alpha)^2 (1 - \lambda\overline{\alpha}) \left(\overline{\varphi_j^*}(\lambda) + \frac{N_0(\lambda)}{(\lambda - \alpha)^2} \varphi_j^*(\lambda) \right) \\
&= (1 - \lambda\overline{\alpha}) \\
&\quad \cdot \left(\overline{a}_j (\lambda - \alpha_{j,1})(\lambda - \alpha_{j,2}) + a_j N_0(\lambda) \frac{(1 - \overline{\alpha}_{j,1}\lambda)(1 - \overline{\alpha}_{j,2}\lambda)}{(1 - \overline{\alpha}\lambda)^2} \right)
\end{aligned}$$

for a.a. $\lambda \in \mathbb{T}$. This implies that

$$\begin{aligned}
&(\lambda - \alpha) h_j(\lambda) \\
&= (1 - \lambda\overline{\alpha}) \left(\overline{a}_j (\lambda - \alpha_{j,1})(\lambda - \alpha_{j,2}) + a_j N_0(\lambda) \frac{(1 - \overline{\alpha}_{j,1}\lambda)(1 - \overline{\alpha}_{j,2}\lambda)}{(1 - \overline{\alpha}\lambda)^2} \right), \quad \lambda \in \mathbb{D}.
\end{aligned}$$

Putting $\lambda = \alpha$ gives (16.5.20). The remaining cases can be proved similarly.

Now suppose that $\varphi \in \mathfrak{F}$ satisfies (16.5.17–16.5.21). We consider the case where $k_1 = k_2 = r_{j,1} = r_{j,2} = 0$ (the other cases can be checked in a similar way). Define

$$h_j(\lambda) := \frac{1 - \lambda\overline{\alpha}}{\lambda - \alpha} \left(\overline{a}_j (\lambda - \alpha_{j,1})(\lambda - \alpha_{j,2}) + a_j N_0(\lambda) \frac{(1 - \overline{\alpha}_{j,1}\lambda)(1 - \overline{\alpha}_{j,2}\lambda)}{(1 - \overline{\alpha}\lambda)^2} \right), \quad \lambda \in \mathbb{D}.$$

Then, $h_j \in \mathcal{H}^1(\mathbb{D})$ and (16.5.22) is satisfied. It remains to use Corollary 11.2.8.

(b) Define

$$P_j(\lambda) := (1 - \overline{\alpha}\lambda)^2 \varphi_j(\lambda) = \sum_{k=0}^{2} d_{j,k} \lambda^k, \quad P_j^*(\lambda) := \sum_{k=0}^{2} \overline{d}_{j,k} \lambda^{2-k}.$$

Observe that $(\lambda-\alpha)^2\overline{\varphi}_j(\lambda) = P_j^*(\lambda)$, $\lambda \in \mathbb{T}$.

First, assume that φ is a complex geodesic. Then, by (16.5.10), there exists an $h \in \mathcal{H}^1(\mathbb{D}, \mathbb{C}^n)$ such that

$$h_j^*(\lambda) = \frac{1-\overline{\alpha}\lambda}{\lambda-\alpha}\left(P_j^*(\lambda) + \lambda(\lambda-\alpha)\eta(\lambda)\frac{P_j(\lambda)}{(1-\overline{\alpha}\lambda)^3}\right), \quad \lambda \in \mathbb{T}.$$

Recall (cf. (16.5.12)) that $h(\lambda) \bullet h(\lambda) = 2\varrho(\lambda)\eta(\lambda)\lambda^2$, $\lambda \in \mathbb{T}$. Therefore,

$$h_j^*(\lambda) = \frac{1-\overline{\alpha}\lambda}{\lambda-\alpha}\left(P_j^*(\lambda) + \frac{h(\lambda)\bullet h(\lambda)}{2(1-\overline{\alpha}\lambda)^4}P_j(\lambda)\right), \quad \lambda \in \mathbb{T}.$$

Hence, we get

$$P_j^*(\alpha) = -\frac{h(\alpha)\bullet h(\alpha)}{2(1-|\alpha|^2)^4}P_j(\alpha).$$

It remains to prove that $\gamma := -\frac{h(\alpha)\bullet h(\alpha)}{2(1-|\alpha|^2)^4} \in \mathbb{T}$. By (16.5.12) we have $|h \bullet h| \le 2\varrho|\eta| \le 2\varrho^2$ a.e. on $\mathbb{T}$. Hence, by the maximum principle (applied to the function $\lambda \longmapsto \frac{h(\lambda)\bullet h(\lambda)}{(1-\overline{\alpha}\lambda)^4}$) we get $|P_j^*(\alpha)| \le |P_j(\alpha)|$.

Now assume all the conditions in (b) are satisfied. Put

$$h_j(\lambda) := \frac{1-\overline{\alpha}\lambda}{\lambda-\alpha}\Big(P_j^*(\lambda) - \gamma P_j(\lambda)\Big), \quad \lambda \in \mathbb{D},\ j = 1,\dots,n.$$

Then, $h_j \in \mathcal{H}^1(\mathbb{D})$ and

$$\frac{1}{\lambda}h_j(\lambda) = |1-\overline{\alpha}\lambda|^2\overline{\varphi}_j(\lambda) - \gamma\frac{(1-\overline{\alpha}\lambda)^3}{\lambda(\lambda-\alpha)}\varphi_j(\lambda), \quad \lambda \in \mathbb{T}.$$

Setting $\eta(\lambda) := -\gamma\frac{(1-\overline{\alpha}\lambda)^3}{\lambda(\lambda-\alpha)}$, $\lambda \in \mathbb{T}$, we see that $|\eta(\lambda)| \le |1-\overline{\alpha}\lambda|^2 =: \varrho(\lambda)$, $\lambda \in \mathbb{T}$. Hence, by Corollary 11.2.8, φ is a complex geodesic. □

Remark 16.5.11. Observe that a geodesic φ for $\mathbb{M}$ with $\varphi \bullet \varphi \equiv 0$ is always a proper map to $\mathbb{B}_n$. By virtue of the proof of Proposition 16.5.10, it is also a geodesic for $\mathbb{B}_n$ iff γ can be chosen to be 0.

Theorem 16.5.12 (cf. [559]).

$$\operatorname{Aut}(\mathbb{M}_n) = \operatorname{Aut}_0(\mathbb{M}_n) = \{e^{i\theta}A : \theta \in \mathbb{R},\ A \in \mathbb{O}(n)\}.$$

The proof needs a few lemmas.

Lemma 16.5.13. *If $\varphi : \mathbb{D} \longrightarrow \mathbb{M}_n$ is a complex geodesic, $\varphi(0) = z_0' = x_0' + iy_0'$, $\varphi(\sigma) = z_0'' = x_0'' + iy_0''$, $\sigma \in (0,1)$, then $\varphi(\mathbb{D}) \subset \mathbb{C}x_0' + \mathbb{C}y_0' + \mathbb{C}x_0'' + \mathbb{C}y_0'' =: L_{\mathbb{C}}$.*

Proof. Let $A \in \mathbb{O}(n)$ be the orthogonal symmetry with respect to the subspace $L_{\mathbb{R}} := \mathbb{R}x_0' + \mathbb{R}y_0' + \mathbb{R}x_0'' + \mathbb{R}y_0''$. In particular, $\{x \in \mathbb{R}^n : A(x) = x\} = L_{\mathbb{R}}$. Then, $\psi := A \circ \varphi : \mathbb{D} \longrightarrow \mathbb{M}_n$ is also a complex geodesic with $\psi(0) = \varphi(0)$ and $\psi(\sigma) = \varphi(\sigma)$. We know that complex geodesics in $\mathbb{M}_n$ are uniquely determined (Lemma 16.5.7). Thus, $A \circ \varphi \equiv \varphi$ and, therefore, $\varphi(\mathbb{D}) \subset \{z \in \mathbb{C}^n : A(z) = z\} = L_{\mathbb{C}}$. □

Lemma 16.5.14. *Let $G = G_{\mathfrak{h}} \subset \mathbb{C}^n$ be a bounded balanced convex domain such that every point $a \in \partial G$ is complex extreme for $\overline{G}$. Let $\Phi \in \operatorname{Aut}(G)$. Then,*

$$G \cap (\mathbb{C} \cdot \Phi(0)) \subset \boldsymbol{O}(G, 0) := \{\Psi(0) : \Psi \in \operatorname{Aut}(G)\}.$$

Proof. We may assume that $w := \Phi(0) \neq 0$. For $\xi, \eta \in \mathbb{T}$ define

$$\Phi_{\xi,\eta} := \xi\Phi(\eta\Phi^{-1}) \in \operatorname{Aut}(G).$$

Our aim is to show that for every $z \in G \cap (\mathbb{C} \cdot w)$ there exist $N \in \mathbb{N}$ and $\xi_j, \eta_j \in \mathbb{T}$, $j = 1, \dots, N$, such that $z = \Phi_{\xi_1,\eta_1} \circ \cdots \circ \Phi_{\xi_N,\eta_N}(0)$.

Let $\widetilde{w} := \Phi^{-1}(0)$. Observe that the mapping $\mathbb{D} \ni \lambda \longmapsto \Phi(\frac{\lambda}{\mathfrak{h}(\widetilde{w})}\widetilde{w})$ is a complex geodesic for $(0, w)$. Hence, by Proposition 11.3.5(b),

$$\Phi\left(\frac{\lambda}{\mathfrak{h}(\widetilde{w})}\widetilde{w}\right) = \frac{h(\lambda)}{\mathfrak{h}(w)}w, \quad \lambda \in \mathbb{D},$$

where $h \in \operatorname{Aut}(\mathbb{D})$. Put

$$h_{\xi,\eta} := \xi h(\eta h^{-1}) \in \operatorname{Aut}(\mathbb{D}), \quad \xi, \eta \in \mathbb{T},$$
$$C_1 := \{h_{\xi,\eta}(0) : \xi, \eta \in \mathbb{T}\}, \quad C_{j+1} := \{h_{\xi,\eta}(\lambda) : \xi, \eta \in \mathbb{T},\ \lambda \in C_j\}.$$

We have

$$\Phi_{\xi,\eta}\left(\frac{\lambda}{\mathfrak{h}(w)}w\right) = \frac{h_{\xi,\eta}(\lambda)}{\mathfrak{h}(w)}w, \quad \lambda \in \mathbb{D}.$$

Thus, it suffices to show that $\bigcup_{j=1}^{\infty} C_j = \mathbb{D}$. We will prove that, in fact, $C_j = \overline{\mathbb{D}}(\varrho_j)$ with $\varrho_j \nearrow 1$.

Observe that $h = \tau \boldsymbol{h}_t$, where $\tau \in \mathbb{T}$ and $t := \mathfrak{h}(\widetilde{w}) \in (0, 1)$. Take an $r \in [0, 1)$ and let

$$K_r := \{h_{\xi,\eta}(\lambda) : \xi, \eta \in \mathbb{T},\ \lambda \in \overline{\mathbb{D}}(r)\}, \quad \varrho(r) := \frac{r + \frac{2t}{1+t^2}}{1 + r\frac{2t}{1+t^2}}.$$

We are going to prove that $K_r = \overline{\mathbb{D}}(\varrho(r))$ (then we take $\varrho_j := \underbrace{\varrho \circ \cdots \circ \varrho}_{j\times}(0) \nearrow 1$).

In fact, it suffices to show that

$$\Delta_r := \{|h_{\xi,\eta}(\lambda)| : \xi, \eta \in \mathbb{T},\ \lambda \in \overline{\mathbb{D}}(r)\} = [0, \varrho(r)].$$

Observe that $\Delta_r = \{|\boldsymbol{h}_t(\eta\boldsymbol{h}_{-t}(\lambda))| : \eta \in \mathbb{T},\ \lambda \in \overline{\mathbb{D}}(r)\}$. In the case where $r = 0$ we get

$$\Delta_0 = \{|\boldsymbol{h}_t(\eta t)| : \eta \in \mathbb{T}\} = \left\{t\left|\frac{\eta - 1}{1 - t^2\eta}\right| : \eta \in \mathbb{T}\right\} = \left[0, \frac{2t}{1+t^2}\right] = [0, \varrho(0)].$$

In the case where $r \in (0,1)$ we have $\Delta_r := \{|\boldsymbol{h}_{-a(\eta)}(\lambda)| : \eta \in \mathbb{T},\ \lambda \in \overline{\mathbb{D}}(r)\}$, where $a(\eta) = t\frac{\eta-1}{1-t^2\eta}$. We already know that $\{|a(\eta)| : \eta \in \mathbb{T}\} = [0, \varrho(0)]$. Using Exercise 1.3.2, we get

$$\begin{aligned}\Delta_r &= \left\{|w| : w \in \overline{\mathbb{D}}\left(\frac{a(\eta)(1-r^2)}{1-r^2|a(\eta)|^2}, \frac{r(1-|a(\eta)|^2)}{1-r^2|a(\eta)|^2}\right),\ \eta \in \mathbb{T}\right\} \\ &= \left[0, \sup\left\{\frac{r+\mu}{1+r\mu},\ \mu \in [0, \varrho(0)]\right\}\right] = [0, \varrho(r)].\end{aligned}$$

□

Lemma 16.5.15. $\boldsymbol{O}(\mathbb{M}_n, 0) \subset Q$, *where* $Q := \{z \in \mathbb{C}^n : z \bullet z = 0\}$.

Proof. Let $\Phi \in \operatorname{Aut}(\mathbb{M}_n)$. Suppose that $z := \Phi(0) \notin Q$. By Lemma 16.5.14, there exists a sequence $(\Psi_j)_{j=1}^{\infty} \in \operatorname{Aut}(\mathbb{M}_n)$ such that $\Psi_j(0) \in \mathbb{M}_n \cap (\mathbb{C}\cdot z)$ and $\Psi_j(0) \longrightarrow \widetilde{z} \in \partial\mathbb{M}_n$. Observe that $\widetilde{z}$ is a point of strong pseudoconvexity of $\partial\mathbb{M}_n$. Hence, in view of the Rosay–Wong theorem (cf. [454, 532]), $\mathbb{M}_n$ must be biholomorphic to $\mathbb{B}_n$; a contradiction (cf. Remark 16.5.1(f)). □

Proof of Theorem 16.5.12. By Remark 16.5.1(e), we have to prove that for every $\Phi \in \operatorname{Aut}(\mathbb{M}_n)$ we have $\Phi(0) = 0$. Take a $\Phi \in \operatorname{Aut}(\mathbb{M}_n)$ and suppose that $z := \Phi(0) \neq 0$. We already know that $z \in Q$ (Lemma 16.5.14). In particular, $z \neq \overline{z}$. Put $\Psi(z) := \overline{\Phi(\overline{z})}$, $z \in \mathbb{M}_n$. Then, $\Psi \in \operatorname{Aut}(\mathbb{M}_n)$. Put $w := \Phi^{-1} \circ \Psi(0)$. By Lemmas 16.5.14, 16.5.15 we get $L := \Phi(\mathbb{M}_n \cap (\mathbb{C}\cdot w)) \subset \boldsymbol{O}(\mathbb{M}_n, 0) \subset Q$. On the other hand, by Lemma 16.5.13, $L \subset \mathbb{C}z + \mathbb{C}\overline{z}$. Thus, $L \subset Q \cap (\mathbb{C}z + \mathbb{C}\overline{z}) \subset (\mathbb{C}z) \cup (\mathbb{C}\overline{z})$. Recall that L is a one-dimensional complex submanifold of $\mathbb{M}_n$ with $z, \overline{z} \in L$ and $z \neq \overline{z}$. Consequently, the identity principle gives a contradiction. □

16.6 Effective formula for the Kobayashi–Royden metric in certain complex ellipsoids

16.6.1 Formula for $\boldsymbol{\varkappa}_{\mathcal{E}((1,m))}$

Fix an $m > 0$, put $\mathcal{E} := \mathcal{E}((1,m))$, and fix $(a,b) \in \mathcal{E}$, $(X,Y) \in (\mathbb{C}^2)_*$. We are going to find an effective formula for $\tau := \boldsymbol{\varkappa}_{\mathcal{E}}((a,b);(X,Y))$. We already know that if $a = b = 0$, then

$$\tau = \boldsymbol{\varkappa}_{\mathcal{E}}((0,0);(X,Y)) = q_{\mathcal{E}}(X,Y),$$

where $q_{\mathcal{E}}$ denotes the Minkowski function of $\mathcal{E}$. Observe that the number $u := 1/q_{\mathcal{E}}(X, Y)$ is the only positive solution of the equation

$$u^2|X|^2 + u^{2m}|Y|^{2m} = 1. \tag{16.6.1}$$

Assume that $(a, b) \neq (0, 0)$. One can easily prove that for any $\theta \in \mathbb{R}$ the mapping

$$\mathcal{E} \ni (z_1, z_2) \longmapsto \left(\frac{z_1 - a}{1 - \bar{a} z_1}, e^{i\theta} \frac{(1 - |a|^2)^{\frac{1}{2m}} z_2}{(1 - \bar{a} z_1)^{\frac{1}{m}}} \right) \in \mathcal{E}$$

is an automorphism of $\mathcal{E}$. Thus, without loss of generality, we may assume that $a = 0$ and $0 < b < 1$. If $X = 0$, then

$$\tau = \varkappa_{\mathcal{E}}((0, b); (0, Y)) = \varkappa_{\mathbb{D}}(b; Y) = \frac{|Y|}{1 - b^2},$$

and therefore we may also assume that $X \neq 0$.

Let $\varphi = (\varphi_1, \varphi_2) : \mathbb{D} \longrightarrow \mathcal{E}$ be an extremal disc with

$$\varphi(0) = (0, b), \quad \tau\varphi'(0) = (X, Y), \quad \tau = \varkappa_{\mathcal{E}}((0, b); (X, Y)) > 0.$$

Note that φ satisfies (16.2.1) and (16.2.2) (with $s = 1$ or $s = 2$). Hence, by Theorem 16.3.1,

$$\varphi(\lambda) = \left(a_1 \frac{\lambda}{1 - \bar{\alpha}_0 \lambda}, a_2 \left(\frac{1 - \bar{\alpha}_2 \lambda}{1 - \bar{\alpha}_0 \lambda} \right)^{\frac{1}{m}} \right) \quad \text{(for } s = 1)$$

or

$$\varphi(\lambda) = \left(a_1 \frac{\lambda}{1 - \bar{\alpha}_0 \lambda}, a_2 \frac{\lambda - \alpha_2}{1 - \bar{\alpha}_2 \lambda} \left(\frac{1 - \bar{\alpha}_2 \lambda}{1 - \bar{\alpha}_0 \lambda} \right)^{\frac{1}{m}} \right) \quad \text{(for } s = 2),$$

where a_1, a_2, α_0, and α_2 satisfy (16.2.4–16.2.9) (with $\alpha_1 = 0$). In particular,

$$\alpha_0,\ \alpha_2 \in \overline{\mathbb{D}}, \quad \alpha_0 = |a_2|^{2m} \alpha_2, \quad 1 + |\alpha_0|^2 = |a_1|^2 + |a_2|^{2m}(1 + |\alpha_2|^2).$$

Recall that in the convex case ($m \geq 1/2$) the complex geodesics are uniquely determined (Proposition 16.2.2). Thus, in the convex case, for given b, X, and Y, the extremal disc φ is either of the first form ($s = 1$) or of the second one ($s = 2$). In the non-convex case ($0 < m < 1/2$) we have to consider both situations and then choose the one for which the corresponding τ is smaller.

In the case where $s = 1$ we have (cf. Remark 16.2.5)

$$\alpha_2 \in \overline{\mathbb{D}}, \quad a_1 = \frac{X}{\tau}, \quad a_2 = b, \quad \alpha_0 = b^{2m} \alpha_2, \quad \frac{Y}{\tau} = -\frac{b}{m}(1 - b^{2m}) \bar{\alpha}_2,$$
$$1 + b^{4m} |\alpha_2|^2 = \frac{|X|^2}{\tau^2} + b^{2m}(1 + |\alpha_2|^2).$$

Direct calculations show that the above conditions are fulfilled iff

$$m|Y| \le b|X| \text{ and } \tau = \begin{cases} \frac{m|Y|}{b}\frac{\sqrt{(1-b^{2m})v+b^{2m}}}{1-b^{2m}}, & \text{if } Y \neq 0 \\ \frac{|X|}{\sqrt{1-b^{2m}}}, & \text{if } Y = 0 \end{cases},$$

where

$$v := \left(\frac{b|X|}{m|Y|}\right)^2. \tag{16.6.2}$$

In the case where $s = 2$ we have:

$$\alpha_2 \in \mathbb{D}_*, \quad a_1 = \frac{X}{\tau}, \quad a_2 = -b/\alpha_2, \quad \alpha_0 = \frac{b^{2m}}{|\alpha_2|^{2m}}\alpha_2, \tag{16.6.3}$$

$$\frac{Y}{\tau} = -\frac{b}{m\alpha_2}\left(m(1-|\alpha_2|^2) + |\alpha_2|^2 - \frac{b^{2m}}{|\alpha_2|^{2m-2}}\right), \tag{16.6.4}$$

$$1 + \frac{b^{4m}}{|\alpha_2|^{4m-2}} = \frac{|X|^2}{\tau^2} + \frac{b^{2m}}{|\alpha_2|^{2m}}(1+|\alpha_2|^2). \tag{16.6.5}$$

Observe that (16.6.5) implies that the case $|\alpha_0| = 1$ is excluded, i.e., $|\alpha_2|^{2m-1} > b^{2m}$. Note that this inequality is automatically satisfied if $0 < m \le 1/2$.

Assume that $Y \neq 0$. Direct calculations show that conditions (16.6.3–16.6.5) are fulfilled iff there exists an $\alpha \in (0,1)$ $(\alpha := |\alpha_2|)$ such that $\alpha^{2m-1} > b^{2m}$,

$$v((m-1)\alpha^{2m} - m\alpha^{2m-2} + b^{2m})^2 - (\alpha^{4m-2} - b^{2m}\alpha^{2m} - b^{2m}\alpha^{2m-2} + b^{4m}) = 0, \tag{16.6.6}$$

$$\tau = \frac{m|Y|}{b}\frac{\alpha^{2m-1}}{|(m-1)\alpha^{2m} - m\alpha^{2m-2} + b^{2m}|}, \tag{16.6.7}$$

First consider the convex case ($m \ge 1/2$). We already know that in this case we have $0 < v < 1$. It is easy to check that (16.6.6) may be written in the form

$$(\alpha^{2m} - t\alpha^{2m-2} - (1-t)b^{2m}) \cdot \left((m-1)^2 v\alpha^{2m} - \frac{m^2 v}{t}\alpha^{2m-2} + \frac{(1-v)b^{2m}}{1-t}\right) = 0, \tag{16.6.8}$$

where

$$t = T(v) := \frac{2m^2 v}{1 + 2m(m-1)v + \sqrt{1+4m(m-1)v}}. \tag{16.6.9}$$

Note that t is well defined and $0 < t < 1$. Moreover,

$$(m-1)^2 vt^2 - (1 + 2m(m-1)v)t + m^2 v = 0, \quad v = \frac{t}{(t(1-m)+m)^2}. \tag{16.6.10}$$

The equation

$$f(\alpha) := \alpha^{2m} - t\alpha^{2m-2} - (1-t)b^{2m} = 0 \tag{16.6.11}$$

has exactly one solution $\alpha \in (0,1)$, which satisfies the inequality $\alpha^{2m-1} > b^{2m}$.

Indeed, putting $u := 1/\alpha$, we get

$$u^2 t + u^{2m}(1-t)b^{2m} = 1;$$

cf. (16.6.1). Since $(t^{1/2}, b(1-t)^{1/(2m)}) \in \mathcal{E}$, we conclude that

$$\alpha = q_{\mathcal{E}}(t^{1/2}, b(1-t)^{1/(2m)}).$$

One can easily check that $\alpha^{2m-1} > b^{2m}$.

Observe that, using (16.6.11), we have

$$\tau = \frac{m|Y|}{b}\frac{1-t}{m(1-t)+t}\frac{\alpha}{1-\alpha^2}. \tag{16.6.12}$$

Finally, we obtain

Example 16.6.1 (cf. [55]). If $m \geq 1/2$, $0 < b < 1$, $X, Y \in \mathbb{C}$, then

$$\varkappa_{\mathcal{E}}((0,b);(X,Y)) = \begin{cases} \frac{|X|}{\sqrt{1-b^{2m}}}, & \text{if } Y = 0 \\ \frac{m|Y|}{b}\frac{\sqrt{(1-b^{2m})v+b^{2m}}}{1-b^{2m}}, & \text{if } v \geq 1 \\ \frac{m|Y|}{b}\frac{1-t}{m(1-t)+t}\frac{\alpha}{1-\alpha^2}, & \text{if } v < 1 \end{cases},$$

where t is given by (16.6.9), v – by (16.6.2), and $\alpha := q_{\mathcal{E}}(t^{1/2}, b(1-t)^{1/(2m)})$.

Remark 16.6.2. The effective formulas from Example 16.6.1 may be used to determine the regularity of $\varkappa_{\mathcal{E}}$ on $\mathcal{E} \times (\mathbb{C}^2)_*$. For example:

- If $m = 1/2$, then $\varkappa_{\mathcal{E}} \notin \mathcal{C}^1(\mathcal{E} \times (\mathbb{C}^2)_*)$.

  Indeed, suppose that $\varkappa_{\mathcal{E}} \in \mathcal{C}^1(\mathcal{E} \times (\mathbb{C}^2)_*)$ and consider the function

  $$(0,1) \ni b \overset{F}{\longmapsto} \varkappa_{\mathcal{E}}((0,b);(1,2b)) = \frac{1}{1-b}.$$

  We have

  $$\begin{aligned} 1 = \lim_{b\to 0} F'(b) &= \lim_{b\to 0}\left(\frac{\partial \varkappa_{\mathcal{E}}}{\partial b}((0,b);(1,2b)) + 2\frac{\partial \varkappa_{\mathcal{E}}}{\partial Y}((0,b);(1,2b))\right) \\ &= \frac{\partial \varkappa_{\mathcal{E}}}{\partial b}((0,0);(1,0)) + 2\frac{\partial \varkappa_{\mathcal{E}}}{\partial Y}((0,0);(1,0)). \end{aligned}$$

  On the other hand, $\varkappa_{\mathcal{E}}((0,b);(1,0)) = \frac{1}{\sqrt{1-b}}$ (which gives $\frac{\partial \varkappa_{\mathcal{E}}}{\partial b}((0,0);(1,0)) = 1/2$) and $\varkappa_{\mathcal{E}}((0,0);(1,Y)) = \frac{1}{2}(Y + \sqrt{Y^2+4})$ (which gives $\frac{\partial \varkappa_{\mathcal{E}}}{\partial Y}((0,0);(1,0)) = 1/2$) – a contradiction.

- (Cf. [55]) If $m > 1/2$, then $\varkappa_{\mathcal{E}} \in \mathcal{C}^1(\mathcal{E} \times (\mathbb{C}^2)_*)$.
- (Cf. [354]) If $m \geq 1$, then $\varkappa_{\mathcal{E}} \in \mathcal{C}^2(\mathcal{E} \times (\mathbb{C}^2)_*)$.
- (Cf. [354]) If $m \geq 3/2$, then $\varkappa_{\mathcal{E}}$ is piecewise $\mathcal{C}^3$ on $\mathcal{E} \times (\mathbb{C}^2)_*$, but $\varkappa_{\mathcal{E}} \notin \mathcal{C}^3(\mathcal{E} \times (\mathbb{C}^2)_*)$.

We move to the non-convex case ($0 < m < 1/2$) with $Y \neq 0$. To simplify notation, put $m^* := 1 - m$.

First observe that if $v > 0$ satisfies (16.6.8) with some $\alpha \in (0,1)$, then $v \leq \frac{1}{4mm^*}$. Indeed, in view of (16.6.6), it suffices to check that

$$\begin{aligned}((m-1)\alpha^{2m} &- m\alpha^{2m-2} + b^{2m})^2 \\ &- 4m(1-m)(\alpha^{4m-2} - b^{2m}\alpha^{2m} - b^{2m}\alpha^{2m-2} + b^{4m}) \\ &\qquad = ((m-1)\alpha^{2m} + m\alpha^{2m-2} + (1-2m)b^{2m})^2 \geq 0.\end{aligned}$$

The function T given in (16.6.9) is well defined for $v \leq \frac{1}{4mm^*}$,

$$T : \left(0, \frac{1}{4mm^*}\right] \longrightarrow \left(0, \frac{m}{m^*}\right]$$

is strictly increasing,

$$T(0+) = 0, \quad T(1) = \left(\frac{m}{m^*}\right)^2, \quad T\left(\frac{1}{4mm^*}\right) = \frac{m}{m^*} < 1. \tag{16.6.13}$$

Recall that the inequality $\alpha^{2m-1} > b^{2m}$ is automatically fulfilled. As before, for any $v \leq \frac{1}{4mm^*}$, we get one solution $\alpha_1 := q_{\mathcal{E}}(t^{1/2}, b(1-t)^{1/(2m)})$ of (16.6.11). Let τ_1 be defined by (16.6.12) with $\alpha = \alpha_1$.

Consider the second equation

$$(m-1)^2 v\alpha^{2m} - \frac{m^2 v}{t}\alpha^{2m-2} + \frac{(1-v)b^{2m}}{1-t} = 0.$$

Using (16.6.10) we easily check that it is equivalent to

$$g(\alpha) := m^{*2}t\alpha^{2m} - m^2\alpha^{2m-2} - (m^{*2}t - m^2)b^{2m} = 0. \tag{16.6.14}$$

Observe that the function $g : \mathbb{R}_{>0} \longrightarrow \mathbb{R}$ is strictly increasing, $g(0+) = -\infty$, and

$$g(1) = (m^{*2}t - m^2)(1 - b^{2m}) \begin{cases} \leq 0, & \text{if } v \leq 1 \\ > 0, & \text{if } v > 1 \end{cases}.$$

Consequently,

- if $v \le 1$, then there is no $\alpha \in (0,1)$ with $g(\alpha) = 0$,
- if $v > 1$, then there is exactly one $\alpha_2 \in (0,1)$ with $g(\alpha_2) = 0$.

In the case where $v > 1$, putting $u := 1/\alpha_2$, we get

$$u^2 \frac{m^2}{m^{*2}t} + u^{2m}\left(1 - \frac{m^2}{m^{*2}t}\right) b^{2m} = 1$$

and

$$\alpha_2 = q_\mathcal{E}\left(\frac{m}{m^*\sqrt{t}}, b\left(1 - \frac{m^2}{m^{*2}t}\right)^{1/(2m)}\right).$$

Let τ_2 be defined via (16.6.7) with $\alpha = \alpha_2$.

Consider $\alpha_1, \tau_1, \alpha_2, \tau_2$ as functions of $t \in [(\frac{m}{m^*})^2, \frac{m}{m^*}]$. Note that $\alpha_1(\frac{m}{m^*}) = \alpha_2(\frac{m}{m^*})$. We are going to show that $\tau_1 < \tau_2$ in $[(\frac{m}{m^*})^2, \frac{m}{m^*})$. We have

$$\tau_1 = \frac{m|Y|}{b} \frac{1-t}{b(m^*t + m)} \frac{\alpha_1}{1 - \alpha_1^2},$$

$$\tau_2 = \frac{m|Y|}{b} \frac{m^{*2}t - m^2}{mm^*(m^*t + m)} \frac{\alpha_2}{1 - \alpha_2^2}.$$

Observe that $\tau_1(\frac{m}{m^*}) = \tau_2(\frac{m}{m^*})$. From (16.6.11) and (16.6.14), we get

$$\alpha_1'(t) = \frac{1}{2} \frac{\alpha_1(1 - \alpha_1^2)}{(1-t)(m\alpha_1^2 + m^*t)},$$

$$\alpha_2'(t) = -\frac{1}{2} \frac{mm^*\alpha_2(1 - \alpha_2^2)}{(m^{*2}t - m^2)(m^*t\alpha_2^2 + m)}.$$

Hence, after some calculations, we get

$$\tau_1'(t) - \tau_2(t) = \frac{m}{2b} \frac{m - m^*t}{(m + m^*t)^2} \frac{(\alpha_1\alpha_2 - 1)(m^*t\alpha_2 - m\alpha_1)}{(m\alpha_1^2 + m^*t)(m^*t\alpha_2^2 + m)}.$$

Moreover, $m^*t\alpha_2 - m\alpha_1 < 0$ in $[(\frac{m}{m^*})^2, \frac{m}{m^*})$.

Indeed, since $\alpha_1(\frac{m}{m^*}) = \alpha_2(\frac{m}{m^*})$, it suffices to check that $(\frac{\alpha_1}{t\alpha_2})' < 0$. After some calculations, we get

$$\left(\frac{\alpha_1}{t\alpha_2}\right)' = \frac{t\alpha_1'\alpha_2 - t\alpha_1\alpha_2' - \alpha_1\alpha_2}{(t\alpha_2)^2}$$
$$= -\frac{b^{2m}(m + m^*t)(\alpha_1^{2m-2}(m\alpha_1^2 + m^*t) + \alpha_2^{2m-2}(m^*t\alpha_2^2 + m))}{2(t\alpha_2)^2(\alpha_1\alpha_2)^{2m-3}(m\alpha_1^2 + m^*t)(m^*t\alpha_2^2 + m)} < 0.$$

Consequently, $\tau_1' - \tau_2' > 0$ in $[(\frac{m}{m^*})^2, \frac{m}{m^*})$, which implies that $\tau_1 - \tau_2 < \tau_1(\frac{m}{m^*}) - \tau_2(\frac{m}{m^*}) = 0$ in $[(\frac{m}{m^*})^2, \frac{m}{m^*})$.

In the case where $Y = 0$, we get

$$\varkappa_{\mathcal{E}}((0,b);(X,0)) = \lim_{\mathbb{C}_* \ni Y \to 0} \varkappa_{\mathcal{E}}((0,b);(X,Y)) = \frac{|X|}{\sqrt{1-b^{2m}}}.$$

Finally, we get

Example 16.6.3 (cf. [421]). If $0 < m < 1/2$, $0 < b < 1$, $X, Y \in \mathbb{C}$, then

$$\varkappa_{\mathcal{E}}((0,b);(X,Y))$$
$$= \begin{cases} \frac{\sqrt{(1-b^{2m})|X|^2+m^2b^{2m-2}|Y|^2}}{1-b^{2m}}, & \text{if } v > \frac{1}{4m(1-m)} \\ \min\left\{\frac{m|Y|}{b}\frac{\sqrt{(1-b^{2m})v+b^{2m}}}{1-b^{2m}}, \frac{m|Y|}{b}\frac{\alpha}{1-\alpha^2}\frac{1-t}{m(1-t)+t}\right\}, & \text{if } 1 \le v \le \frac{1}{4m(1-m)} \\ \frac{m|Y|}{b}\frac{\alpha}{1-\alpha^2}\frac{1-t}{m(1-t)+t}, & \text{if } v < 1 \end{cases},$$

where v is given by (16.6.2), t – by (16.6.9), and $\alpha = q_{\mathcal{E}}(t^{1/2}, b(1-t)^{1/(2m)})$.

Example 16.6.4 (cf. [421]). If $0 < m < 1/2$, then there are mappings given by Theorem 16.3.1 that are not extremal for $\varkappa_{\mathcal{E}((1,m))}$.

We keep the notation from the above discussion of $\varkappa_{\mathcal{E}}((0,b);(X,Y))$. Take $0 < b < 1$ and $X > 0$, $Y = 1$ such that $bX = mY$, i.e., $v = 1$. In particular, $t = (\frac{m}{1-m})^2$ (cf. (16.6.13)). Note that if $\varphi : \mathbb{D} \longrightarrow \mathbb{C}^2$ as in Theorem 16.3.1 with $\varphi(0) = (0,b)$, $\sigma\varphi'(0) = (X,1)$, and $s \in \{1,2\}$, then $\varphi(\mathbb{D}) \subset \mathcal{E}$ (cf. Step 1° of the proof of Corollary 16.2.6).

Our aim is to choose $b \in (0,1)$ so that the following three conditions (a, b, c) are satisfied:

(a) There exists a mapping $\varphi = (\varphi_1, \varphi_2) : \mathbb{D} \longrightarrow \mathbb{C}^2$ as in Theorem 16.3.1 with $\varphi(0) = (0,b)$, $\sigma_0\varphi'(0) = (X,1)$, $s = 1$, and

$$\tau_0 := |\sigma_0| = \frac{m|Y|}{b}\frac{\sqrt{(1-b^{2m})v+b^{2m}}}{1-b^{2m}} = \frac{m}{b}\frac{1}{1-b^{2m}}.$$

(b) There exists a mapping $\psi = (\psi_1, \psi_2) : \mathbb{D} \longrightarrow \mathbb{C}^2$ as in Theorem 16.3.1 with $\psi(0) = (0,b)$, $\sigma_1\psi'(0) = (X,1)$, $s = 2$, and

$$\tau_1 := |\sigma_1| = \frac{m|Y|}{b}\frac{\alpha}{1-\alpha^2}\frac{1-t}{m(1-t)+t} = \frac{m}{b}\frac{\alpha}{1-\alpha^2}\frac{1-2m}{m(1-m)},$$

where $\alpha \in (0,1)$ is a unique solution of the equation (cf. (16.6.11))

$$(1-m)^2\alpha^{2m} - m^2\alpha^{2m-2} - (1-2m)b^{2m} = 0. \tag{16.6.15}$$

(c) $\tau_1 < \tau_0$. In particular, φ *is not extremal for* $\varkappa_{\mathcal{E}}((0,b);(X,1))$.

Add. (a) Define

$$\varphi(\lambda) = \left(a_1 \frac{\lambda}{1-\overline{\alpha}_0\lambda}, a_2 \left(\frac{1-\overline{\alpha}_2\lambda}{1-\overline{\alpha}_0\lambda}\right)^{\frac{1}{m}}\right),$$

with

$$a_1 := 1-b^{2m}, \quad a_2 := b, \quad \alpha_0 := -b^{2m}, \quad \alpha_2 := -1, \quad \sigma_0 := \frac{X}{1-b^{2m}}.$$

Add. (b) Define

$$\psi(\lambda) = \left(a_1 \frac{\lambda}{1-\overline{\alpha}_0\lambda}, a_2 \frac{\lambda-\alpha_2}{1-\overline{\alpha}_2\lambda}\left(\frac{1-\overline{\alpha}_2\lambda}{1-\overline{\alpha}_0\lambda}\right)^{\frac{1}{m}}\right)$$

with

$$a_1 := X/\sigma_1, \quad a_2 := -b/\alpha, \quad \alpha_0 := b^{2m}/\alpha^{2m-1},$$
$$\alpha_2 := \alpha, \quad -\sigma_1 := \frac{m}{b}\frac{\alpha}{1-\alpha^2}\frac{1-2m}{m(1-m)},$$

where $\alpha \in (0,1)$ satisfies (16.6.15).

Add. (c) The inequality $\tau_1 < \tau_0$ means that

$$f(\alpha) := -(1-m)^2\alpha^{2m+1} + m^2\alpha^{2m-1}$$
$$+ m(1-m)\alpha^2 + (1-2m)\alpha - m(1-m) < 0.$$

We have

$$f(1) = f'(1) = f''(1) = 0, \quad f'''(1) = 2m(1-m)(2m-1)^2 > 0.$$

Thus, if we take an $\alpha < 1$, $\alpha \approx 1$, then $f(\alpha) < 0$, and if we define b via (16.6.15), then $b \in (0,1)$ and we get the required situation.

16.6.2 Formula for $\varkappa_{\mathcal{E}((\frac{1}{2},\frac{1}{2}))}$

Let $\Delta_2 := \mathcal{E}((\frac{1}{2},\frac{1}{2}))$. Fix $b = (b_1,b_2) \in \Delta_2 \cap (\mathbb{R}_{>0})^2$, $X = (X_1,X_2) \in (\mathbb{R}^2)_*$. Using the same methods as in Example 16.6.1, one can prove the following formulas. The details are left to the reader.

Define

$$d := 1 - b_1 - b_2,\ c_0 := 1 + b_1 + b_2,\ c_1 := 1 + b_1 - b_2,\ c_2 := 1 - b_1 + b_2,$$
$$L_1(X) := (1 - b_2)X_1 + b_1 X_2, \quad L_2(X) := b_2 X_1 + (1 - b_1)X_2,$$

$$A := dc_1 + i2db_2,\ B := 2db_1 + idc_2,$$
$$C := -2b_1c_2 + idc_2,\ D := -dc_1 + i2b_2c_1,$$
$$E := -2d\sqrt{b_1} + i\frac{2db_2}{1 - \sqrt{b_1}},\ F := -\frac{2db_1}{1 - \sqrt{b_2}} + i2d\sqrt{b_2},$$

$$C_{\text{lin}}^{(1)} := \{z \in \mathbb{C}_* : \operatorname{Arg} A < \operatorname{Arg} z < \operatorname{Arg} B \text{ or } \operatorname{Arg}(-A) < \operatorname{Arg} z < \operatorname{Arg}(-B)\},$$
$$C_{\text{lin}}^{(2)} := \{z \in \mathbb{C}_* : \operatorname{Arg} C < \operatorname{Arg} z < \operatorname{Arg} D \text{ or } \operatorname{Arg}(-C) < \operatorname{Arg} z < \operatorname{Arg}(-D)\},$$
$$C_{\text{par}}^{(1)} := \{z \in \mathbb{C}_* : \operatorname{Arg}(-D) \le \operatorname{Arg} z \le \operatorname{Arg} A \text{ or } \operatorname{Arg} D \le \operatorname{Arg} z \text{ or } \operatorname{Arg} z \le \operatorname{Arg}(-A)\},$$
$$C_{\text{par}}^{(2)} := \{z \in \mathbb{C}_* : \operatorname{Arg}(B) \le \operatorname{Arg} z \le \operatorname{Arg} C \text{ or } \operatorname{Arg}(-B) \le \operatorname{Arg} z \le \operatorname{Arg}(-C)\},$$
$$C_{\text{par}}^{(3)} := \{z \in \mathbb{C}_* : \operatorname{Arg}(-F) < \operatorname{Arg} z \le \operatorname{Arg} A \text{ or } \operatorname{Arg} F < \operatorname{Arg} z \text{ or } \operatorname{Arg} z \le \operatorname{Arg}(-A)\},$$
$$C_{\text{par}}^{(4)} := \{z \in \mathbb{C}_* : \operatorname{Arg} B \le \operatorname{Arg} z < \operatorname{Arg} D \text{ or } \operatorname{Arg}(-B) \le \operatorname{Arg} z < \operatorname{Arg}(-E)\},$$
$$C_{\text{eli}} := \{z \in \mathbb{C}_* : \operatorname{Arg}(D) < \operatorname{Arg} z \le \operatorname{Arg} F \text{ or } \operatorname{Arg}(-E) \le \operatorname{Arg} z \le \operatorname{Arg}(-F)\}.$$

Let $\tau := \boldsymbol{\varkappa}_{\mathcal{E}((\frac{1}{2},\frac{1}{2}))}(b; X)$. There are two cases:

(a) $\sqrt{b_1} + \sqrt{b_2} < 1$ (cf. Figure 16.1); then,

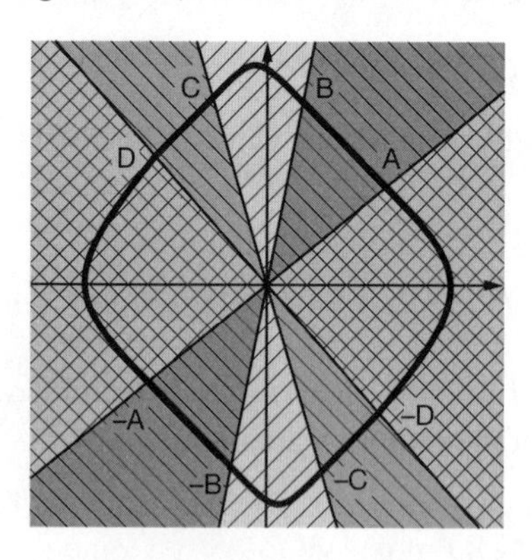

Figure 16.1. The indicatrix of $\boldsymbol{\varkappa}_{\Delta_2}(b; \cdot)$.

$$\tau = \begin{cases} \frac{1}{dc_0}\left(|X_1| + |X_2|\right), & \text{if } X \in C_{\text{lin}}^{(1)} \\ \frac{1}{c_1c_2}\left(|X_1| + |X_2|\right), & \text{if } X \in C_{\text{lin}}^{(2)} \\ \frac{1}{2db_2c_1}\left(|L_1(X)| + \sqrt{|L_1(X)|^2 + \frac{1}{b_2}dc_1|X_2|^2}\right), & \text{if } X \in C_{\text{par}}^{(1)} \\ \frac{1}{2db_1c_2}\left(|L_2(X)| + \sqrt{|L_2(X)|^2 + \frac{1}{b_1}dc_2|X_1|^2}\right), & \text{if } X \in C_{\text{par}}^{(2)} \end{cases}.$$

(b) $\sqrt{b_1} + \sqrt{b_2} \ge 1$ (cf. Figure 16.2); then,

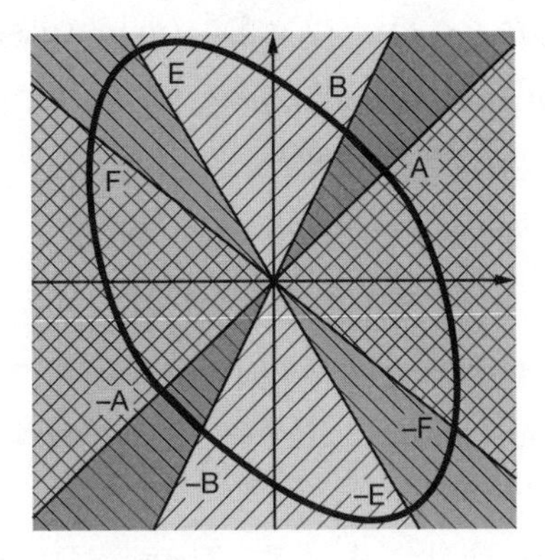

Figure 16.2. The indicatrix of $\boldsymbol{\varkappa}_{\Delta_2}(b;\cdot)$.

$$\tau = \begin{cases} \frac{1}{dc_0}\left(|X_1| + |X_2|\right), & \text{if } X \in C_{\text{lin}}^{(1)} \\ \frac{1}{2db_2c_1}\left(|L_1(X)| + \sqrt{|L_1(X)|^2 + \frac{1}{b_2}dc_1|X_2|^2}\right), & \text{if } X \in C_{\text{par}}^{(3)} \\ \frac{1}{2db_1c_2}\left(|L_2(X)| + \sqrt{|L_2(X)|^2 + \frac{1}{b_1}dc_2|X_1|^2}\right), & \text{if } X \in C_{\text{par}}^{(4)} \\ \sqrt{\frac{1}{4d^3b_1b_2}\left(b_2|L_1(X)|^2 + b_1|L_2(X)|^2 - b_1b_2|X_1 + X_2|^2\right)}, & \text{if } X \in C_{\text{eli}} \end{cases}.$$

16.7 Complex geodesics in the symmetrized bidisc

We keep the notation from § 7.1:

$$\begin{aligned} &\pi : \mathbb{C}^2 \longrightarrow \mathbb{C}^2, \quad \pi(\lambda_1, \lambda_2) := (\lambda_1 + \lambda_2, \lambda_1\lambda_2), \\ &\mathbb{G}_2 := \pi(\mathbb{D}^2) = \{(\lambda_1 + \lambda_2, \lambda_1\lambda_2) : \lambda_1, \lambda_2 \in \mathbb{D}\}, \quad \sigma_2 := \pi(\mathbb{T}^2) \subset \partial\mathbb{G}_2, \\ &\Delta_2 := \{(\lambda, \lambda) : \lambda \in \mathbb{D}\}, \quad \Sigma_2 := \pi(\Delta_2) = \{(2\lambda, \lambda^2) : \lambda \in \mathbb{D}\}, \\ &\boldsymbol{F}_a(s, p) := \frac{2ap - s}{2 - as}, \quad a \in \overline{\mathbb{D}},\ (s, p) \in (\mathbb{C} \setminus \{2/a\}) \times \mathbb{C}. \end{aligned}$$

Also recall (Theorem 7.1.1) that $\boldsymbol{c}_{\mathbb{G}_2} \equiv \boldsymbol{\ell}_{\mathbb{G}_2}$ and $\boldsymbol{\gamma}_{\mathbb{G}_2} \equiv \boldsymbol{\varkappa}_{\mathbb{G}_2}$, so the notion of a complex geodesic in $\mathbb{G}_2$ is well defined. Our aim is to find the full characterization of all complex geodesics $\varphi : \mathbb{D} \longrightarrow \mathbb{G}_2$ up to $\operatorname{Aut}(\mathbb{D})$ and $\operatorname{Aut}(\mathbb{G}_2)$, i.e., we identify φ with $H_g \circ \varphi \circ h$, where $g, h \in \operatorname{Aut}(\mathbb{D})$.

Theorem 16.7.1 (cf. [426]). *Let $\varphi = (S, P) : \mathbb{D} \longrightarrow \mathbb{G}_2$. Then,*

(a) *If $\#(\varphi(\mathbb{D}) \cap \Sigma_2) \ge 2$, then φ is a complex geodesic iff*

$$\varphi(\lambda) = (-2\lambda, \lambda^2), \quad \lambda \in \mathbb{D} \quad (\operatorname{mod} \operatorname{Aut}(\mathbb{D}));$$

in particular, if φ is such a complex geodesic, then $\varphi(\mathbb{D}) = \Sigma_2$.

(b) *If* $\#(\varphi(\mathbb{D}) \cap \Sigma_2) = 1$, *then* φ *is a complex geodesic iff*

$$\varphi(\lambda) = \pi(B(\sqrt{\lambda}), B(-\sqrt{\lambda})), \quad \lambda \in \mathbb{D} \quad (\mathrm{mod}\,\mathrm{Aut}(\mathbb{D}) \text{ and } \mathrm{mod}\,\mathrm{Aut}(\mathbb{G}_2)),$$

where B *is a Blaschke product of order* ≤ 2 *with* $B(0) = 0$ *and, moreover, the case* $B(\lambda) = \tau\lambda^2$ $(\tau \in \mathbb{T})$ *is excluded.*

(c) *If* $\varphi(\mathbb{D}) \cap \Sigma_2 = \varnothing$, *then* φ *is a complex geodesic iff*

$$\varphi(\lambda) = \pi(h_1(\lambda), h_2(\lambda)), \quad \lambda \in \mathbb{D},$$

where $h_1, h_2 \in \mathrm{Aut}(\mathbb{D})$ *are such that* $h_1 - h_2$ *has no zero in* $\mathbb{D}$.

In particular, any complex geodesic $\varphi : \mathbb{D} \longrightarrow \mathbb{G}_2$ *extends holomorphically to* $\overline{\mathbb{D}}$ *and* $\varphi(\mathbb{T}) \subset \sigma_2$ *(notice that* σ_2 *is a very thin part of* $\partial\mathbb{G}_2$*).*

Proof. (a) Let $\varphi(\lambda) := (-2\lambda, \lambda^2)$, $\lambda \in \mathbb{D}$. Then, $\varphi(\mathbb{D}) = \Sigma_2$. Consequently, by Lemma 7.1.13, φ is a complex geodesic.

Now, let $\varphi : \mathbb{D} \longrightarrow \mathbb{G}_2$ be a complex geodesic with $\varphi(\xi) = (2\mu, \mu^2) \in \Sigma_2$ $(\xi, \mu \in \mathbb{D})$. Taking $\varphi \circ \boldsymbol{h}_{-\xi}$ instead of φ, we may assume that $\xi = 0$. Taking $H_{\boldsymbol{h}_\mu} \circ \varphi$ instead of φ, we may assume that $\mu = 0$, i.e., $\varphi(0) = (0, 0)$. By Theorem 7.1.1, there exists an $\alpha \in \mathbb{T}$ such that $\boldsymbol{F}_\alpha \circ \varphi = \frac{2\alpha P - S}{2 - \alpha S} =: h \in \mathrm{Aut}(\mathbb{D})$. Taking $R_\alpha \circ \varphi$ instead of φ, we may assume that $\alpha = 1$. Observe that h must be a rotation and, therefore, we may also assume that $h = \mathrm{id}$, i.e., $\frac{2P - S}{2 - S} = \mathrm{id}$. Thus, $S(\lambda) = 2\frac{P(\lambda) - \lambda}{1 - \lambda}$, $\lambda \in \mathbb{D}$.

Let $\eta \in \mathbb{D}_*$ be such that $\varphi(\eta) \in \Sigma_2$. Then, $S^2(\eta) = 4P(\eta)$, i.e., $\left(\frac{P(\eta) - \eta}{1 - \eta}\right)^2 = P(\eta)$. Hence, $P(\eta) = \eta^2$ and so $S(\eta) = -2\eta$. The Schwarz lemma implies that $S(\lambda) = -2\lambda$, $\lambda \in \mathbb{D}$, and, finally, $P(\lambda) = \lambda^2$, $\lambda \in \mathbb{D}$.

(b) Let $\varphi(\lambda) := \pi(B(\sqrt{\lambda}), B(-\sqrt{\lambda}))$, $\lambda \in \mathbb{D}$, where B is a Blaschke product of order ≤ 2 with $B(0) = 0$.

In the case $B(\lambda) = \tau\lambda$, $\lambda \in \mathbb{D}$ $(\tau \in \mathbb{T})$, we get $\varphi(\lambda) = (0, -\tau^2\lambda)$, $\lambda \in \mathbb{D}$. Consequently, $\boldsymbol{F}_a \circ \varphi \in \mathrm{Aut}(\mathbb{D})$ for any $a \in \mathbb{T}$ (cf. Step 1° of the proof of Theorem 7.1.1).

In the case where $B(\lambda) = \tau\lambda\boldsymbol{h}_b(\lambda)$, $\lambda \in \mathbb{D}$ $(\tau \in \mathbb{T}, b \in \mathbb{D}_*)$, we get $\varphi(\mathbb{T}) \subset \sigma_2$ and

$$\varphi(\lambda) = \left(2\tau\lambda\frac{1 - |b|^2}{1 - \overline{b}^2\lambda}, \tau^2\lambda\frac{\lambda - b^2}{1 - \overline{b}^2\lambda}\right), \quad \lambda \in \mathbb{D}.$$

To apply Lemma 7.1.13, we only need to observe that $\varphi(\xi) = (2\mu, \mu^2)$ for some $\xi, \mu \in \mathbb{T}$. Take $\xi := b/\overline{b}$. Then, $\varphi(\xi) = (2\tau\xi, \tau^2\xi^2)$.

Now, let $\varphi : \mathbb{D} \longrightarrow \mathbb{G}_2$ be a complex geodesic with $\#(\varphi(\mathbb{D}) \cap \Sigma_2) = 1$. Then, as in (a), we may assume that $\varphi(0) = (0, 0)$, $\frac{2P - S}{2 - S} = \mathrm{id}$. Observe that $\Delta(\lambda) := S^2(\lambda) - 4P(\lambda) \neq 0$ for $\lambda \in \mathbb{D}_*$. Write $\Delta(\lambda) = \lambda^k\widetilde{\Delta}(\lambda)$, where $\widetilde{\Delta}(\lambda) \neq 0$, $\lambda \in \mathbb{D}$. Define

$$B(\lambda) := \frac{1}{2}\left(S(\lambda^2) + \lambda^k\sqrt{\widetilde{\Delta}(\lambda^2)}\right), \quad \lambda \in \mathbb{D}.$$

Then,

$$S^2(\lambda^2) - 4P(\lambda^2) = \Delta(\lambda^2) = \lambda^{2k}\widetilde{\Delta}(\lambda^2) = 4B^2(\lambda) - 4B(\lambda)S(\lambda^2) + S^2(\lambda^2), \quad \lambda \in \mathbb{D},$$

which implies that

$$B(\lambda)S(\lambda^2) - B^2(\lambda) = P(\lambda^2) = B(-\lambda)S(\lambda^2) - B^2(-\lambda)$$

and, consequently,

$$(B(\lambda) - B(-\lambda))(S(\lambda^2) - (B(\lambda) + B(-\lambda))) = 0, \quad \lambda \in \mathbb{D}.$$

We have the following two cases:

(i) $S(\lambda^2) = B(\lambda) + B(-\lambda)$, $\lambda \in \mathbb{D}$: then,

$$\begin{aligned} P(\lambda^2) &= B(\lambda)S(\lambda^2) - B^2(\lambda) \\ &= B(\lambda)(B(\lambda) + B(-\lambda)) - B^2(\lambda) = B(\lambda)B(-\lambda), \quad \lambda \in \mathbb{D}. \end{aligned}$$

Hence, $\varphi(\lambda) = \pi(B(\sqrt{\lambda}), B(-\sqrt{\lambda}))$, $\lambda \in \mathbb{D}$.

Fix a $t_0 \in (0,1)$. Let $(s_0, p_0) := \varphi(t_0^2) = \pi(B(t_0), B(-t_0)) = \pi(\lambda_1^0, \lambda_2^0)$.

Suppose that there exists a function $f \in \mathcal{O}(\mathbb{D}, \mathbb{D})$ such that $f(0) = 0$, $f(t_0) = \lambda_1^0$, $f(-t_0) = \lambda_2^0$, and $f(\mathbb{D}) \subset\subset \mathbb{D}$. Put $\psi := \pi(f, f) : \mathbb{D} \longrightarrow \mathbb{G}_2$ and observe that $\psi(0) = (0,0)$ and $\psi(t_0^2) = (s_0, p_0)$. Hence, ψ would be a complex geodesic with $\psi(\mathbb{D}) \subset\subset \mathbb{G}_2$ – a contradiction.

Thus, the function B solves an extremal problem of 2-type in the sense of Definition 11.4.3 and therefore, by Theorem 16.3.1, B must be a Blaschke product of order ≤ 2. Observe that if $B(\lambda) = \tau\lambda^2$ ($\tau \in \mathbb{T}$), then $\varphi(\lambda) = (2\tau\lambda, \tau^2\lambda^2)$, which contradicts the condition $\#(\varphi(\mathbb{D}) \cap \Sigma_2) = 1$.

(ii) $B(\lambda) = B(-\lambda)$, $\lambda \in \mathbb{D}$: then, there exists a function $B_1 \in \mathcal{O}(\mathbb{D}, \mathbb{D})$ such that

$$B(\lambda) = B_1(\lambda^2) = \frac{1}{2}\left(S(\lambda^2) + \lambda^k\sqrt{\widetilde{\Delta}(\lambda^2)}\right), \quad \lambda \in \mathbb{D}.$$

Using the same argument, we reduce the proof to the case where there exists a $B_2 \in \mathcal{O}(\mathbb{D}, \mathbb{D})$ such that

$$B_2(\lambda^2) = \frac{1}{2}\left(S(\lambda^2) - \lambda^k\sqrt{\widetilde{\Delta}(\lambda^2)}\right), \quad \lambda \in \mathbb{D}.$$

Hence, $\varphi = \pi(B_1, B_2)$. Since $\frac{2P-S}{2-S} = \mathrm{id}$, we get

$$2B_1(\lambda)B_2(\lambda) - (B_1(\lambda) + B_2(\lambda)) = \lambda(2 - (B_1(\lambda) + B_2(\lambda))), \quad \lambda \in \mathbb{D},$$

which gives $-(B_1'(0)+B_2'(0))=2$. Consequently, by the Schwarz lemma, $B_1(\lambda)=B_2(\lambda)=-\lambda$, and finally $\varphi(\lambda)=(-2\lambda,\lambda^2)$, $\lambda\in\mathbb{D}$. Thus, $\Delta\equiv 0$; contradiction.

(c) Let $\varphi:=\pi(h_1,h_2)$, where $h_1,h_2\in\operatorname{Aut}(\mathbb{D})$ are such that h_1-h_2 has no zero in $\mathbb{D}$. Observe that φ satisfies (7.1.1) and $\varphi(\mathbb{T})\subset\sigma_2$. To use Lemma 7.1.13 we only need to check that $\varphi(\xi)=(2\eta,\eta^2)$ for some $\xi,\eta\in\mathbb{T}$, i.e., $h_1(\xi)=h_2(\xi)$ for some $\xi\in\mathbb{T}$. Let $h_j=\tau_j\boldsymbol{h}_{a_j}$ ($\tau_j\in\mathbb{T}$, $a_j\in\mathbb{D}$), $j=1,2$. Then, we have to find a root $z=\xi$ of the equation

$$\tau_1(z-a_1)(1-\bar{a}_2z)-\tau_2(z-a_2)(1-\bar{a}_1z)=A_2z^2+A_1z+A_0=0$$

with $|\xi|=1$. We have $A_2=-\tau_1\bar{a}_2+\tau_2\bar{a}_1$, $A_0=-\tau_1a_1+\tau_2a_2$. Observe that $|A_2|=|A_0|$. Since the equation $h_1-h_2=0$ has no roots in $\mathbb{D}$, we get $A_2\neq 0$. Let z_1,z_2 be the roots of the above equation. We have $|z_1|,|z_2|\geq 1$ and $|z_1z_2|=|A_0/A_2|=1$. Thus, $|z_1|=|z_2|=1$.

Now, let $\varphi:\mathbb{D}\longrightarrow\mathbb{G}_2$ be a complex geodesic with $\varphi(\mathbb{D})\cap\Sigma_2=\varnothing$. Then, there exists a holomorphic mapping $\psi:\mathbb{D}\longrightarrow\mathbb{D}^2$ with $\pi\circ\psi=\varphi$. Consequently, ψ must be a complex geodesic ($\boldsymbol{m}_{\mathbb{D}}(\lambda',\lambda'')=\boldsymbol{c}^*_{\mathbb{G}_2}(\varphi(\lambda'),\varphi(\lambda''))\leq\boldsymbol{c}^*_{\mathbb{D}^2}(\psi(\lambda'),\psi(\lambda''))\leq\boldsymbol{m}_{\mathbb{D}}(\lambda',\lambda'')$). Hence, $\psi=(h_1,h_2)$, where $h_1,h_2\in\mathcal{O}(\mathbb{D},\mathbb{D})$ and at least one of h_1 and h_2 is an automorphism. Assume that $h_1\in\operatorname{Aut}(\mathbb{D})$.

Fix a $t_0\in(0,1)$ and suppose that $\boldsymbol{m}_{\mathbb{D}}(h_2(0),h_2(t_0))<t_0$. Let

$$\varrho:=\boldsymbol{m}_{\mathbb{D}}(h_2(0),h_2(t_0))/t_0\in(0,1).$$

There exists a $g\in\operatorname{Aut}(\mathbb{D})$ such that $g(0)=h_2(0)$, $g(\varrho t_0)=h_2(t_0)$. Put $f(\lambda):=g(\varrho\lambda)$, $\lambda\in\overline{\mathbb{D}}$. Then, $f(0)=h_2(0)$, $f(t_0)=h_2(t_0)$, and $f(\mathbb{D})\subset\subset\mathbb{D}$. Put $\chi=(\chi_1,\chi_2):=\pi(h_1,f)$. Then, $\chi(0)=\varphi(0)$ and $\chi(t_0)=\varphi(t_0)$. Thus, χ is also a complex geodesic. Notice that by the Rouché theorem the function h_1-f has a zero in $\mathbb{D}$. Hence, $\chi(\mathbb{D})\cap\Sigma_2\neq\varnothing$. In particular, in view of (a) and (b), $\chi(\mathbb{T})\subset\sigma_2$. On the other hand $\chi_2=h_1f$; contradiction.

Consequently, $\boldsymbol{m}_{\mathbb{D}}(h_2(0),h_2(t_0))=t_0$, and, therefore, $h_2\in\operatorname{Aut}(\mathbb{D})$. □

Remark 16.7.2 (the reader is asked to complete the details). The proof of Theorem 16.7.1 shows that all complex geodesics $\varphi=(S,P):\mathbb{D}\longrightarrow\mathbb{G}_2$ are of the following four types (mod $\operatorname{Aut}(\mathbb{D})$):

(i) $\varphi(\lambda)=(-2\lambda,\lambda^2)$, $\lambda\in\mathbb{D}$.

(ii) $$\begin{aligned}\varphi(\lambda)&=(\tau\boldsymbol{h}_a(\sqrt{\lambda})+\tau\boldsymbol{h}_a(-\sqrt{\lambda}),\tau^2\boldsymbol{h}_a(\sqrt{\lambda})\boldsymbol{h}_a(-\sqrt{\lambda}))\\&=\left(2\tau\frac{\bar{a}\lambda-a}{1-\bar{a}^2\lambda},-\tau^2\frac{\lambda-a^2}{1-\bar{a}^2\lambda}\right)=(\xi h(\lambda)+\bar{\xi},h(\lambda)),\quad\lambda\in\mathbb{D},\end{aligned}$$

where

$$\tau\in\mathbb{T},\quad a\in\mathbb{D},\quad \xi=\xi(a,\tau):=-\frac{2\tau\bar{a}}{1+|a|^2},\quad h:=-\tau^2\boldsymbol{h}_{a^2}.$$

Observe that $\{\xi(a,\tau) : a \in \mathbb{D},\ \tau \in \mathbb{T}\} = \mathbb{D}$. Consequently, $\varphi(\lambda) = (\xi\lambda + \overline{\xi}, \lambda)$, $\lambda \in \mathbb{D}$, where $\xi \in \mathbb{D}$.

(iii)
$$\begin{aligned}\varphi(\lambda) &= (\tau \boldsymbol{h}_{a_1}(\sqrt{\lambda})\boldsymbol{h}_{a_2}(\sqrt{\lambda}) + \tau \boldsymbol{h}_{a_1}(-\sqrt{\lambda})\boldsymbol{h}_{a_2}(-\sqrt{\lambda}),\\ &\qquad \tau^2 \boldsymbol{h}_{a_1}(\sqrt{\lambda})\boldsymbol{h}_{a_2}(\sqrt{\lambda})\boldsymbol{h}_{a_1}(-\sqrt{\lambda})\boldsymbol{h}_{a_2}(-\sqrt{\lambda})),\\ &= \left(2\tau\frac{\overline{a}_1\overline{a}_2\lambda^2 + (1-|a_1+a_2|^2+|a_1a_2|^2)\lambda + a_1a_2}{(1-\overline{a}_1^2\lambda)(1-\overline{a}_2^2\lambda)},\right.\\ &\qquad \left.\tau^2\frac{\lambda - a_1^2}{1-\overline{a}_1^2\lambda}\cdot\frac{\lambda-a_2^2}{1-\overline{a}_2^2\lambda}\right),\end{aligned}$$

where $\tau \in \mathbb{T}$, $a_1, a_2 \in \mathbb{D}$, and the case $a_1 + a_2 = 0$ is excluded.

(iv) $\varphi(\lambda) = (h_1(\lambda)+h_2(\lambda), h_1(\lambda)h_2(\lambda))$, $\lambda \in \mathbb{D}$, where $h_1, h_2 \in \operatorname{Aut}(\mathbb{D})$ are such that $h_1(\lambda) + h_2(\lambda) \neq 0$, $\lambda \in \mathbb{D}$.

Observe that, except for (ii), P is a Blaschke product of order 2.

Different descriptions of complex geodesics $\varphi : \mathbb{D} \longrightarrow \mathbb{G}_2$ have been presented in [113] and [13].

Proposition 16.7.3 (cf. [13]). *Let $\varphi = (S, P) : \mathbb{D} \longrightarrow \mathbb{G}_2$. Then, φ is a complex geodesic* (mod Aut($\mathbb{D}$)) *iff either*

(a) $\varphi(\lambda) = (\beta\lambda + \overline{\beta}, \lambda)$, $\lambda \in \mathbb{D}$, *where* $\beta \in \mathbb{D}$, *or*

(b)
$$P(\lambda) := \frac{(4 - 2\omega_0 s_0 + \omega_0^2 p_0)\lambda^2 + 2(s_0 - \omega_0 p_0)\lambda + p_0}{4 - 2\overline{\omega}_0\overline{s}_0 + \overline{\omega}_0^2\overline{p}_0 + 2(\overline{s}_0 - \overline{\omega}_0\overline{p}_0)\lambda + \overline{p}_0\lambda^2}, \tag{16.7.1}$$
$$S(\lambda) := 2\frac{\omega_0 P(\lambda) - \lambda}{1 - \omega_0\lambda},$$

where $(s_0, p_0) \in \mathbb{G}_2$, $\omega_0 \in \mathbb{T}$ *are such that*

$$\operatorname{Re}(2\omega_0 s_0 - \omega_0^2 p_0) = \max\{\operatorname{Re}(2\omega s_0 - \omega^2 p_0) : \omega \in \mathbb{T}\}. \tag{16.7.2}$$

Proof. It is clear that (a) coincides with type (ii) from Remark 16.7.2. We are going to prove that (b) coincides with the union of (i), (iii), and (iv). First we prove that each mapping of the form (b) is a complex geodesic. Consider the following three cases:

- $(s_0, p_0) = \pi(\lambda_1^0, \lambda_2^0)$ with $\lambda_1^0, \lambda_2^0 \in \mathbb{D}$, $\lambda_1^0 \neq \lambda_2^0$. Let $\omega_0 \in \mathbb{T}$ be such that

$$\boldsymbol{c}^*_{\mathbb{G}_2}((\lambda_1^0, 0), (\lambda_2^0, 0)) = \boldsymbol{m}_{\mathbb{D}}(\boldsymbol{F}_{\omega_0}(\lambda_1^0, 0), \boldsymbol{F}_{\omega_0}(\lambda_1^0, 0)) \tag{16.7.3}$$

(cf. Theorem 7.1.1) and let $\varphi = (S, P) : \mathbb{D} \longrightarrow \mathbb{G}_2$ be a complex $\boldsymbol{c}_{\mathbb{G}_2}$-geodesic for $(\lambda_1^0, 0)$, $(\lambda_2^0, 0)$ with $\varphi(\mu_j) = (\lambda_j^0, 0)$, $j = 1, 2$. Put $h := \boldsymbol{F}_{\omega_0} \circ \varphi$. Then, (16.7.3) implies that $h \in \operatorname{Aut}(\mathbb{D})$. Substituting φ by $\varphi \circ h^{-1}$, we may assume that $h = \mathrm{id}$. We have $P(\mu_j) = 0$, $j = 1, 2$. In particular, P must be a Blaschke product of order 2. Moreover, $\boldsymbol{F}_{\omega_0} \circ \varphi = \mathrm{id}$, i.e.,

$$S(\lambda) = 2\frac{\omega_0 P(\lambda) - \lambda}{1 - \omega_0 \lambda}, \quad \lambda \in \mathbb{D}.$$

Since the function S is bounded, we conclude that $P(\overline{\omega}_0) = \overline{\omega}_0^2$. Hence,

$$P = \frac{\overline{\omega}_0^2}{\boldsymbol{h}_{\mu_1}(\overline{\omega}_0)\boldsymbol{h}_{\mu_2}(\overline{\omega}_0)} \boldsymbol{h}_{\mu_1} \boldsymbol{h}_{\mu_2}.$$

We have $\mu_j = \boldsymbol{F}_{\omega_0} \circ \varphi(\mu_j) = \boldsymbol{F}_{\omega_0}(\lambda_j^0, 0) = \frac{-\lambda_j^0}{2 - \omega_0 \lambda_j^0}$, $j = 1, 2$. Direct calculations give formula (16.7.1). It remains to check that ω_0 satisfies (16.7.2). For, we only need to observe that

$$\begin{aligned} \boldsymbol{m}_{\mathbb{D}}(\boldsymbol{F}_\omega(\lambda_1^0, 0), \boldsymbol{F}_\omega(\lambda_1^0, 0)) &= \boldsymbol{m}_{\mathbb{D}}\left(\frac{-\lambda_1^0}{2 - \omega\lambda_1^0}, \frac{-\lambda_2^0}{2 - \omega\lambda_2^0}\right) \\ &= \frac{|\lambda_1^0 - \lambda_2^0|}{\sqrt{2 - \operatorname{Re}(2\omega s_0 - \omega^2 p_0)}}. \end{aligned}$$

- $(s_0, p_0) = \pi(\lambda_0, \lambda_0)$. Let $\omega_0 \in \mathbb{T}$ be such that

$$\sigma_0 := \boldsymbol{\gamma}_{\mathbb{G}_2}((\lambda_0, 0); (1, 0)) = \boldsymbol{\gamma}_{\mathbb{D}}(\boldsymbol{F}_{\omega_0}(\lambda_0, 0); \boldsymbol{F}'_{\omega_0}(\lambda_0, 0)(1, 0))$$

(cf. Theorem 7.1.16) and let $\varphi = (S, P) : \mathbb{D} \longrightarrow \mathbb{G}_2$ be a complex $\boldsymbol{\gamma}_{\mathbb{G}_2}$-geodesic for $((\lambda_0, 0), (1, 0))$ with $\varphi(\mu_0) = (\lambda_0, 0)$, $\sigma_0 \varphi'(\mu_0) = (1, 0)$. Then, $\boldsymbol{F}_{\omega_0} \circ \varphi \in \operatorname{Aut}(\mathbb{D})$ and we may assume that $\boldsymbol{F}_{\omega_0} \circ \varphi = \mathrm{id}$. We have $P(\mu_0) = P'(\mu_0) = 0$ and $P(\overline{\omega}_0) = \overline{\omega}_0^2$. Hence, $P = \frac{\overline{\omega}_0^2}{\boldsymbol{h}_{\mu_0}^2(\overline{\omega}_0)} \boldsymbol{h}_{\mu_0}^2$. Moreover, $\mu_0 = \frac{-\lambda_0}{2 - \omega_0 \lambda_0}$. Thus, after some calculations, we get (16.7.1). We have to check that ω_0 satisfies (16.7.2). We have

$$\boldsymbol{\gamma}_{\mathbb{D}}(\boldsymbol{F}_\omega(\lambda_0, 0); \boldsymbol{F}'_{\omega_0}(\lambda_0, 0)(1, 0)) = \frac{1}{2(1 - \operatorname{Re}(\omega_0 \lambda_0))}.$$

Hence, $\omega_0 = |\lambda_0|/\lambda_0$. On the other hand,

$$\operatorname{Re}(2\omega s_0 - \omega^2 p_0) = \operatorname{Re}(4\omega\lambda_0 - \omega^2\lambda_0^2) = 4 - \operatorname{Re}((2 - \omega\lambda_0)^2).$$

It remains to observe that the term $\operatorname{Re}((2 - \omega\lambda_0)^2)$ is minimal for $\omega_0 = |\lambda_0|/\lambda_0$.

- $(s_0, p_0) = (0, 0)$. Then, ω_0 is arbitrary and $\varphi(\lambda) = (-2\lambda, \lambda^2)$, $\lambda \in \mathbb{D}$, i.e., we get the former type (i).

Now we prove that each complex geodesic $\varphi = (S, P)$ of type (i), (iii), or (iv) may be written in the form (b) (mod $\operatorname{Aut}(\mathbb{D})$). Since the case (i) is obvious, we may assume that φ is of the type (iii) or (iv). In view of the above constructions, it suffices to show that either φ is a complex $\boldsymbol{c}_{\mathbb{G}_2}$-geodesic for some points $(\lambda_1^0, 0)$, $(\lambda_2^0, 0)$ with $\lambda_1^0, \lambda_2^0 \in \mathbb{D}$, $\lambda_1^0 \neq \lambda_2^0$, or φ is a complex $\boldsymbol{\gamma}_{\mathbb{G}_2}$-geodesic for $((\lambda_0, 0), (1, 0))$ with $\lambda_0 \in \mathbb{D}_*$. Consider the following two cases:

- φ is of the form (iii). Let $\mu_j := a_j^2$, $j = 1, 2$. First consider the case where $\mu_1 \neq \mu_2$. Put

$$\lambda_j^0 := S(\mu_j) = \tau \boldsymbol{h}_{a_1}(-a_j)\boldsymbol{h}_{a_2}(-a_j) \in \mathbb{D}, \quad j = 1, 2.$$

One can easily check that $\lambda_1^0 \neq \lambda_2^0$, which finishes the proof.

Now suppose that $\mu_1 = \mu_2 =: \mu_0$, i.e., $a_1 = a_2 =: a \neq 0$ (recall that the case $a_1 + a_2 = 0$ is excluded). Then, $\lambda_0 := S(\mu_0) = \tau \boldsymbol{h}_a^2(-a) \in \mathbb{D}_*$, so we are done.

- φ is of the form (iv). We may assume that $h_1 = \mathrm{id}$, $h_2 = \tau \boldsymbol{h}_b$ with $\tau \in \mathbb{T}$, $b \in \mathbb{D}_*$. Let $\mu_1 := 0$, $\mu_2 := b$, $\lambda_1^0 := S(\mu_1) = -\tau b$, $\lambda_2^0 := S(\mu_2) = b$. If $\tau \neq -1$, then $\lambda_1^0 \neq \lambda_2^0$ and we are done. The case where $\tau = -1$ does not occur, because the equation $h_1(\lambda) - h_2(\lambda) = \lambda + \frac{\lambda - b}{1 - \bar{b}\lambda} = 0$ has a root in $\mathbb{D}$ (namely $\frac{1 - \sqrt{1 - |b|^2}}{\bar{b}}$). □

Proposition 16.7.4 (cf. [13]). *The complex $\boldsymbol{c}_{\mathbb{G}_2}$- and $\boldsymbol{\gamma}_{\mathbb{G}_2}$-geodesics are uniquely determined* (mod $\operatorname{Aut}(\mathbb{D})$).

The uniqueness of the complex geodesics $\varphi : \mathbb{D} \longrightarrow \mathbb{G}_2$ passing through $(0, 0)$ has been proved in [11] and, using different methods, in [426].

Proof. Let $\varphi_j = (S_j, P_j) : \mathbb{D} \longrightarrow \mathbb{G}_2$ be a complex $\boldsymbol{c}_{\mathbb{G}_2}$-geodesic with $\varphi_j(\mu_{j,k}) = z_k = (s_k, p_k)$, $k = 1, 2$, $z_1 \neq z_2$, $j = 1, 2$. Let $\omega_0 \in \mathbb{T}$ be such that $\boldsymbol{m}_{\mathbb{D}}(\mu_{j,1}, \mu_{j,2}) = \boldsymbol{c}^*_{\mathbb{G}_2}(z_1, z_2) = \boldsymbol{m}_{\mathbb{D}}(\boldsymbol{F}_{\omega_0}(z_1), \boldsymbol{F}_{\omega_0}(z_2))$. Taking $\varphi_2 \circ h$ instead of φ_2 with suitable $h \in \operatorname{Aut}(\mathbb{D})$, we may assume that $\mu_{2,k} = \mu_{1,k} =: \mu_k$, $k = 1, 2$. Then, $h_j := \boldsymbol{F}_{\omega_0} \circ \varphi_j \in \operatorname{Aut}(\mathbb{D})$ and $h_j(\mu_k) = \boldsymbol{F}_{\omega_0}(z_k)$, $j = 1, 2$. Thus, $h_1 \equiv h_2$ and we may assume that $h_1 = h_2 = \mathrm{id}$. We have $P_1(\overline{\omega}_0) = \overline{\omega}_0^2 = P_2(\overline{\omega}_0)$ and $P_1(\mu_k) = p_k = P_2(\mu_k)$, $k = 1, 2$. Recall that P_j is a Blaschke product of order $d_j \in \{1, 2\}$, $j = 1, 2$. Let

$$Q_j := \frac{\boldsymbol{h}_{p_1} \circ P_j}{\boldsymbol{h}_{\mu_1}}, \quad Q_j(\mu_1) := \frac{\boldsymbol{h}'_{p_1}(p_1) P'_j(\mu_1)}{\boldsymbol{h}'_{\mu_1}(\mu_1)}, \quad j = 1, 2.$$

Then, Q_j is a Blaschke product of order $d_k - 1$ (order 0 means that Q_j is a constant from $\mathbb{T}$), $Q_1(\overline{\omega}_0) = Q_2(\overline{\omega}_0) \in \mathbb{T}$, and $Q_1(\mu_2) = Q_2(\mu_2)$. Consequently, $d_1 = d_2$ and $P_1 \equiv P_2$. Finally, since $\boldsymbol{F}_{\omega_0} \circ \varphi_j = \mathrm{id}$, we also get $S_1 \equiv S_2$.

Now, let $\varphi_j = (S_j, P_j) : \mathbb{D} \longrightarrow \mathbb{G}_2$ be a complex $\boldsymbol{\gamma}_{\mathbb{G}_2}$-geodesic with $\varphi_j(\mu_j) = z_0 = (s_0, p_0)$, $\alpha_j \varphi'_j(\mu_j) = Z_0 = (X_0, Y_0) \neq (0, 0)$, and $\boldsymbol{\gamma}_{\mathbb{D}}(\mu_j; \alpha_j) = \boldsymbol{\gamma}_{\mathbb{G}_2}(z_0; Z_0)$, $j = 1, 2$. We may assume that $\mu_1 = \mu_2 = 0$ and $\alpha_1 = \alpha_2 = 1$. Let $\omega_0 \in \mathbb{T}$ be such that $1 = \boldsymbol{\gamma}_{\mathbb{D}}(0;1) = \boldsymbol{\gamma}_{\mathbb{G}_2}(z_0;Z_0) = \boldsymbol{\gamma}_{\mathbb{D}}(\boldsymbol{F}_{\omega_0}(z_0);\boldsymbol{F}'_{\omega_0}(z_0)(Z_0))$. Then, $h_j := \boldsymbol{F}_{\omega_0} \circ \varphi_j \in \operatorname{Aut}(\mathbb{D})$, $h_j(0) = \boldsymbol{F}_{\omega_0}(z_0)$, and $h'_j(0) = \boldsymbol{F}'_{\omega_0}(z_0)(Z_0)$, $j = 1,2$. Thus, $h_1 \equiv h_2$ and we may assume that $h_1 = h_2 = \mathrm{id}$. We have $P_1(\overline{\omega}_0) = \overline{\omega}_0^2 = P_2(\overline{\omega}_0)$, $P_1(0) = p_0 = P_2(0)$, and $P'_1(0) = Y_0 = P'_2(0)$, $k = 1,2$. Recall that φ_j is also a complex $\boldsymbol{c}_{\mathbb{G}_2}$-geodesic (Proposition 11.1.4). In particular, P_j is a Blaschke product of order $d_j \in \{1,2\}$, $j = 1,2$. Let

$$Q_j := \frac{\boldsymbol{h}_{p_0} \circ P_j}{\boldsymbol{h}_0}, \quad Q_j(0) := \boldsymbol{h}'_{p_0}(p_0)Y_0, \quad j = 1,2.$$

Then, Q_j is a Blaschke product of order $d_k - 1$, $Q_1(\overline{\omega}_0) = Q_2(\overline{\omega}_0) \in \mathbb{T}$ and $Q_1(0) = Q_2(0)$. Consequently, $d_1 = d_2$, $P_1 \equiv P_2$, and $S_1 \equiv S_2$. □

Remark 16.7.5. For $n \geq 3$ the problem of characterization of all $\boldsymbol{c}$- (resp. $\boldsymbol{\gamma}$-, $\boldsymbol{\ell}$-, $\boldsymbol{\varkappa}$-) geodesics $\varphi : \mathbb{D} \longrightarrow \mathbb{G}_n$ remains open.

16.8 Complex geodesics in the tetrablock

Recall (cf. Remark 7.1.23) that the *tetrablock* is the domain

$$\begin{aligned}\mathbb{E} &= \{(z_1,z_2,z_3) \in \mathbb{C}^3 : \forall_{\zeta_1,\zeta_2 \in \overline{\mathbb{D}}} : 1 - z_1\zeta_1 - z_2\zeta_2 + z_3\zeta_1\zeta_2 \neq 0\} \\ &= \{(z_1,z_2,z_3) \in \mathbb{C}^3 : |z_2 - \overline{z}_1 z_3| + |z_1 z_2 - z_3| + |z_1|^2 < 1\} \\ &= \{(z_1,z_2,z_3) \in \mathbb{C}^3 : |z_1 - \overline{z}_2 z_3| + |z_1 z_2 - z_3| + |z_2|^2 < 1\};\end{aligned}$$

cf. [7]. Recall (Remark 7.1.23) that $\boldsymbol{c}_{\mathbb{E}} \equiv \boldsymbol{\ell}_{\mathbb{E}}$ and $\boldsymbol{\gamma}_{\mathbb{E}} \equiv \boldsymbol{\varkappa}_{\mathbb{E}}$, which implies that the notion of a complex geodesic in $\mathbb{E}$ is well-defined.

Theorem* 16.8.1.

(a) (*Cf.* [540])

$$\mathcal{T} := \{\Phi(0) : \Phi \in \operatorname{Aut}(\mathbb{E})\} = \{(\lambda,\mu,\lambda\mu) : \lambda,\mu \in \mathbb{D}\}.$$

(b) (*Cf.* [154]) *For every* $z^0 \in \mathbb{E} \setminus \mathcal{T}$ *there exists a* $\Psi \in \operatorname{Aut}(\mathbb{E})$ *such that* $\Psi(z^0) = (0,0,-\beta^2)$.

Theorem* 16.8.2.

(a) (*Cf.* [158]) *Every complex geodesic* $\varphi : \mathbb{D} \longrightarrow \mathbb{E}$ *with* $\varphi(0) = 0$ *is of the form (up to a permutation of the first two variables).*

$$\varphi(\lambda) = \left(\omega_1 \frac{h(\lambda) + C}{1 + C}, \omega_2\lambda\frac{1 + Ch(\lambda)}{1 + C}, \omega_1\omega_2\lambda h(\lambda)\right), \quad \lambda \in \mathbb{D},$$

where $h : \mathbb{D} \longrightarrow \overline{\mathbb{D}}$ *is a holomorphic function with* $h(0) = -C \in [-1,0]$ *and* $\omega_1,\omega_2 \in \mathbb{T}$. *Conversely, every mapping of the above form is a complex geodesic.*

(b) (*Cf.* [154]) *Every complex geodesic* $\varphi : \mathbb{D} \longrightarrow \mathbb{E}$ *with* $\varphi(\mathbb{D}) \cap \mathcal{T} = \varnothing$ *and* $\varphi(0) = (0, 0, -\beta^2)$ *is of the form*

$$\varphi = \tfrac{1}{\Delta}((1-\beta^2)A, (1-\beta^2)C, AC - (B+\beta)^2),$$

where

$$Z \in \mathcal{O}(\mathbb{D}, \mathbb{D}), \quad Z(0) = 0,$$
$$A(\lambda) := a^2\lambda + b^2 Z(\lambda), \quad B(\lambda) := ac\lambda + bdZ(\lambda), \quad C(\lambda) = c^2 + d^2 Z(\lambda),$$
$$a, b, c, d \in \overline{\mathbb{D}}, \quad |a|^2 + |b|^2 = |c|^2 + |d|^2 = 1, \quad a\bar{c} + b\bar{d} = 0,$$
$$\Delta := (1 + \beta B)^2 - \beta^2 AC.$$

Notice that in view of Theorem 16.8.1, the above theorem describes in fact all the complex geodesics $\varphi : \mathbb{D} \longrightarrow \mathbb{E}$.

Proposition 16.8.3 (cf. [154], see also [7]). *There is no uniqueness of complex geodesics (even passing through* 0).

Proof. Let $\psi \in \mathcal{O}(\mathbb{D}, \mathbb{D})$ be such that $\psi(0) = 0$ and $\psi(1/2) = 1/4$ (e.g., $\psi(\lambda) := \lambda^2$ or $\psi(\lambda) := (1/2)\lambda$). Define

$$\varphi : \mathbb{D} \longrightarrow \mathbb{E}, \quad \varphi(\lambda) := (\lambda, \psi(\lambda), \lambda\psi(\lambda)).$$

Using the contractibility with respect to the mappings

$$\mathbb{D}^2 \ni (\lambda, \mu) \longmapsto (\lambda, \mu, \lambda\mu) \in \mathbb{E}, \quad \mathbb{E} \ni (\lambda, \mu, \eta) \longmapsto (\lambda, \mu) \in \mathbb{D}^2,$$

we get

$$\begin{aligned} \boldsymbol{m}_{\mathbb{E}}((\lambda_1, \mu_1, \lambda_1\mu_1), (\lambda_2, \mu_2, \lambda_2\mu_2)) &= \boldsymbol{m}_{\mathbb{D}^2}((\lambda_1, \mu_1), (\lambda_2, \mu_2)) \\ &= \max\{\boldsymbol{m}_{\mathbb{D}}(\lambda_1, \lambda_2), \boldsymbol{m}_{\mathbb{D}}(\mu_1, \mu_2)\}. \end{aligned}$$

Hence,

$$\boldsymbol{m}_{\mathbb{E}}((\lambda', \psi(\lambda'), \lambda'\psi(\lambda')), (\lambda'', \psi(\lambda''), \lambda''\psi(\lambda''))) = \boldsymbol{m}_{\mathbb{D}}(\lambda', \lambda''), \quad \lambda', \lambda'' \in \mathbb{D},$$

which shows that φ is a complex geodesic with $\varphi(0) = 0$. □

16.9 Exercises

Exercise 16.9.1.

(a) (Cf. [277]) Prove that the ball

$$\mathbb{H}_n := \{z \in \mathbb{C}^n : H_n(z) < 1\},$$

where

$$\begin{aligned} H_n(z) :&= \left(\|z\|^2 + (\|z\|^4 - \mu^2(z))^{1/2}\right)^{1/2} \\ &= \begin{cases} \sqrt{2}\|z\|, & \text{if } \sqrt{2}\|z\|_\infty \le \|z\| \\ \|z\|_\infty + \sqrt{\|z\|^2 - \|z\|_\infty^2}, & \text{if } \sqrt{2}\|z\|_\infty > \|z\| \end{cases}, \\ \mu(z) :&= \max\{0, 2\|z\|_\infty - \|z\|^2\}, \end{aligned}$$

is the maximal Reinhardt domain contained in the Lie ball $\mathbb{L}_n$.

Hint. Note that

$$H_n(z_1, \dots, z_n) = \max\{L_n(\xi_1 z_1, \dots, \xi_n z_n) : \xi_1, \dots, \xi_n \in \mathbb{T}\}.$$

(b) Observe that $H_2(z_1, z_2) = |z_1| + |z_2|$.

(c) Prove that every boundary point of $\partial\mathbb{H}_n$ is a complex extremal point for $\overline{\mathbb{H}}_n$.

Hint. Suppose that $H_n(a + \lambda v) \le 1$, $\lambda \in \mathbb{D}$, for $a \in \partial\mathbb{H}_n$ and $v \in \mathbb{C}^n$. We are going to prove that $v = 0$. By the maximum principle for psh functions, we get $H_n(a + \lambda v) = 1$, $\lambda \in \mathbb{D}$. The case where $\sqrt{2}\|a + \lambda_0 v\|_\infty < \|a + \lambda_0 v\|$ for some $\lambda_0 \in \mathbb{D}$ reduces to the Euclidean ball. Thus, we may assume that

$$\|a + \lambda v\|_\infty + \sqrt{\|a + \lambda v\|^2 - \|a + \lambda v\|_\infty^2} = 1, \quad \lambda \in \mathbb{D}.$$

Let $A_j := \{\lambda \in \mathbb{D} : |a_j + \lambda v_j| = \|a + \lambda v\|_\infty\}$, $j = 1, \dots, n$. Let j_0 be such that $U := \operatorname{int} A_{j_0} \neq \varnothing$. We may assume that $j_0 = n$. Write $z = (z', z_n) \in \mathbb{C}^{n-1} \times \mathbb{C}$. Then, $|a_n + \lambda v_n| + \|a' + \lambda v'\| = 1$, $\lambda \in U$. Since the function $\lambda \longmapsto |a_n + \lambda v_n| + \|a' + \lambda v'\|$ is real analytic on $\mathbb{C} \setminus M$, where M is finite, the identity principle gives $|a_n + \lambda v_n| + \|a' + \lambda v'\| = 1$, $\lambda \in \mathbb{C}$. Now, letting $\lambda \longrightarrow \infty$ gives $v = 0$.

(d) Using the above method try to find a criterion for a boundary point of a ball $B \subset \mathbb{C}^n$ to be a complex extremal point for $\overline{B}$.

Exercise 16.9.2. Let $\mathcal{E}(p), \mathcal{E}(q)$ be convex complex ellipsoids, $p = (p_1, \dots, p_n)$, $p_1, \dots, p_n \ge 1/2$, $q = (q_1, \dots, q_n)$, $q_1, \dots, q_n \ge 1/2$. Let $\varphi = (\varphi, \dots, \varphi) : \mathbb{D} \longrightarrow \mathcal{E}(p)$ be a complex geodesic. Write $\varphi_j = B_j \cdot \psi_j$, where B_j is the Blaschke product for φ_j (we put $B_j :\equiv 1$ if $0 \notin \varphi_j(\mathbb{D})$ or $\varphi_j \equiv 0$), $j = 1, \dots, n$; cf. Step 3° in the proof of Proposition 16.2.2. Put $\widetilde{\varphi}_j := B_j \psi_j^{p_j/q_j}$, $j = 1, \dots, n$, $\widetilde{\varphi} := (\widetilde{\varphi}_1, \dots, \widetilde{\varphi}_n)$. Prove that $\widetilde{\varphi} : \mathbb{D} \longrightarrow \mathcal{E}(q)$ is a complex geodesic.

Hint. Use Proposition 16.2.2.

Exercise 16.9.3. Let $\mathcal{E}(p)$ be a convex complex ellipsoid. For $b \in \mathcal{E}(p) \cap (\mathbb{C}_*)^n$ denote by $\mathcal{Z}_{\mathcal{E}(p)}(b)$ the set of all $X \in (\mathbb{C}^n)_*$ such that the pair (b, X) admits a complex $\boldsymbol{\gamma}_{\mathcal{E}(p)}$-geodesic $\varphi = (\varphi_1, \dots, \varphi_n) : \mathbb{D} \longrightarrow \mathcal{E}(p)$ with $0 \notin \varphi_j(\mathbb{D})$, $j = 1, \dots, n$. Prove that

$$\mathcal{Z}_{\mathbb{B}_n}(b) = \{X = (X_1, \dots, X_n) \in (\mathbb{C}^n)_* : \|X_k b - b_k X\| \geq |X_k|,\ k = 1, \dots, n\},$$
$$b = (b_1, \dots, b_n) \in \mathbb{B}_n \cap (\mathbb{C}_*)^n.$$

Hint. If $\varphi : \mathbb{D} \longrightarrow \mathbb{B}_n$ is a complex geodesic, then $\varphi(\mathbb{D}) = \mathbb{B}_n \cap L$, where L is a one-dimensional affine subspace of $\mathbb{C}^n$.

Exercise 16.9.4. Let $\mathcal{E}(p)$ be a convex complex ellipsoid, $p = (p_1, \dots, p_n)$, $p_1, \dots, p_n \geq 1/2$. Fix $b = (b_1, \dots, b_n) \in \mathcal{E}(p) \cap (\mathbb{C}_*)^n$ and put $b_j^* := b_j^{p_j}$ (the power is arbitrarily fixed), $j = 1, \dots, n$. Let $b^* := (b_1^*, \dots, b_n^*)$. Note that $b^* \in \mathbb{B}_n$. For $X = (X_1, \dots, X_n) \in \mathbb{C}^n$, write $X_j^* := p_j (b_j^*/b_j) X_j$, $j = 1, \dots, n$, $X^* := (X_1^*, \dots, X_n^*)$. Prove that (cf. Exercise 16.9.3)

(a) $$\mathcal{Z}_{\mathcal{E}(p)}(b) = \{X \in \mathbb{C}^n : X^* \in \mathcal{Z}_{\mathbb{B}_n}(b^*)\} = \{X = (X_1, \dots, X_n) \in (\mathbb{C}^n)_* :$$
$$\sum_{j=1}^n |p_k b_k^{p_k - 1} X_k b_j^{p_j} - b_k^{p_k} p_j b_j^{p_j - 1} X_j|^2 \geq p_k^2 |b_k|^{2p_k - 2} |X_k|^2,$$
$$k = 1, \dots, n\}, \text{ and}$$

(b) $$\boldsymbol{\gamma}_{\mathcal{E}(p)}(b; X) = \boldsymbol{\gamma}_{\mathbb{B}_n}(b^*; X^*)$$
$$= \left(\frac{\sum_{j=1}^n p_j^2 |b_j|^{2p_j - 2} |X_j|^2}{1 - \sum_{j=1}^n |b_j|^{2p_j}} + \frac{|\sum_{j=1}^n p_j |b_j|^{2p_j - 2} b_j \overline{X}_j|^2}{(1 - \sum_{j=1}^n |b_j|^{2p_j})^2} \right)^{1/2},$$
$$X \in \mathcal{Z}_{\mathcal{E}(p)}(b).$$

Hint. Use Exercises 16.9.2 and 16.9.3.

Exercise 16.9.5. Prove that $\mathbb{E}$ is starlike with center at 0.

16.10 List of problems

Chapter 17

Analytic discs method

Summary. Section 17.1 contains a few properties of the relative extremal function (which will also be used § 18.5). Next, we present properties of the Poisson, Green, Lelong, and Lempert functionals (§§ 17.2, 17.3, 17.4). The main results are Theorems 17.3.4 and 17.4.3, which will be applied in §§ 18.5, 18.6.

Introduction. From some general point of view, the invariant objects we have studied so far may be collected into the following three groups:

(a) objects related to certain extremal problems concerning holomorphic mappings $f : G \longrightarrow \mathbb{D}$, e.g., $\boldsymbol{m}_G^{(k)}(a,z)$, $\boldsymbol{m}_G(\mathfrak{p},z)$, $\boldsymbol{\gamma}_G^{(k)}(a;X)$;

(b) objects related to certain extremal problems concerning log-psh functions $u : G \longrightarrow [0,1)$, e.g., $\boldsymbol{g}_G(a,z)$, $\boldsymbol{g}_G(\mathfrak{p},z)$, $\boldsymbol{s}_G(a,z)$, $\boldsymbol{A}_G(a;X)$, $\boldsymbol{S}_G(a;X)$;

(c) objects related to certain extremal problems concerning analytic discs $\varphi : \mathbb{D} \longrightarrow G$, e.g., $\boldsymbol{\ell}_G^{(k)*}(a,z)$, $\boldsymbol{h}_G^*(a,z)$, $\boldsymbol{\varkappa}_G^{(k)}(a;X)$, $\boldsymbol{\eta}_G(a;X)$.

At the end of the eighties, E. A. Poletsky invented and partially developed a general method that reduces, in some sense, problems of type (b) to (c). This method found various important applications, mainly due to A. Edigarian (cf. [151] and the references given there) and E. A. Poletsky (cf. [435, 436, 155]) – see for instance §§ 18.5, 18.6. In the present chapter, we are inspired by the exposition of the analytic disc theory presented in [337] and [151].

We point out that the disc method is one of the important tools of modern complex analysis.

17.1 Relative extremal function

Definition 17.1.1. Let $G \subset \mathbb{C}^n$ be a domain. For any function $\mathfrak{p} : G \longrightarrow [-\infty,\infty)$ let

$$\mathcal{P}_{\mathfrak{p}}(G) := \{u \in \mathcal{PSH}(G) : u \le \mathfrak{p}\},\ \boldsymbol{h}_{\mathfrak{p},G}(z) := \sup\{u(z) : u \in \mathcal{P}_{\mathfrak{p}}(G)\},\ z \in G.$$

The function $\boldsymbol{h}_{\mathfrak{p},G}$ is called the *generalized relative extremal function with weights* $\mathfrak{p}$.

For $A \subset G$ set

$$\boldsymbol{h}_{A,G} := \boldsymbol{h}_{\chi_{G \setminus A},G}.$$

The function $\boldsymbol{h}_{A,G}$ is called the *relative extremal function of A in G*.

Let $\boldsymbol{h}^*_{A,G}$ denote the upper semicontinuous regularization of $\boldsymbol{h}_{A,G}$.

Observe that if $\mathfrak{p}$ is upper semicontinuous, then $\boldsymbol{h}_{\mathfrak{p},G} \in \mathcal{P}_{\mathfrak{p}}(G)$. Notice that if A is open, then $\boldsymbol{h}_{A,G} = \boldsymbol{h}^*_{A,G}$ (the reader is advised to consult [270], Ch. 3, for properties of $\boldsymbol{h}_{A,G}$).

Some authors prefer the following different definition of the relative extremal function (see e.g., [305], § 4.5): $\omega_{A,G} := \boldsymbol{h}_{A,G} - 1$.

Remark 17.1.2. Let $F : G \longrightarrow D$ be a holomorphic mapping and let $A \subset G$, $B \subset D$ be such that $F(A) \subset B$. Then, $\boldsymbol{h}_{B,D} \circ F \le \boldsymbol{h}_{A,G}$.

Proposition 17.1.3. *Let $D, G \subset \mathbb{C}^n$ be domains, $F \in \mathcal{O}(G, D)$, and $B \subset D$. Put $A := F^{-1}(B)$.*

(a) *If F is proper, then $\boldsymbol{h}_{B,D} \circ F = \boldsymbol{h}_{A,G}$.*

(b) *If F is a holomorphic covering (cf. Appendix* B.1.6) *and B is open, then $\boldsymbol{h}_{B,D} \circ F = \boldsymbol{h}_{A,G}$.*

Proof. Let $v \in \mathcal{PSH}(G)$, $v \le 1$, $v \le 0$ on A.

(a) Define

$$u(w) := \max\{v(z) : z \in F^{-1}(w)\}, \quad w \in D.$$

Then, $u \in \mathcal{PSH}(D)$ (cf. Appendix B.4.33), $u \le 1$, and $u \le 0$ on B. Thus, $u \le \boldsymbol{h}_{B,D}$. Consequently, $v(z) \le u(F(z)) \le \boldsymbol{h}_{B,D}(F(z))$, $z \in G$. Hence, $\boldsymbol{h}_{B,D} \circ F \ge \boldsymbol{h}_{A,G}$. Now, it remains to apply Remark 17.1.2.

(b) Define

$$u_0(w) := \sup\{v(z) : z \in F^{-1}(w)\}, \quad w \in D, \quad u := u_0^*.$$

Then, $u \in \mathcal{PSH}(D)$ (cf. Appendix B.4.16), $u \le 1$, and $u \le 0$ on B. Thus, $u \le \boldsymbol{h}_{B,D}$. Consequently, $v(z) \le u_0(F(z)) \le u(F(z)) \le \boldsymbol{h}_{B,D}(F(z))$, $z \in G$. Hence, $\boldsymbol{h}_{B,D} \circ F \ge \boldsymbol{h}_{A,G}$. As above, Remark 17.1.2 finishes the proof. □

Proposition 17.1.4 (cf. [150]). *Let $G \subset \mathbb{C}^n$ be a domain, let $\mathfrak{p} : G \longrightarrow \mathbb{R}_+$ be such that the set $\#|\mathfrak{p}|$ is finite. Fix an $R > 0$ so small that*

- *$\mathbb{P}(a, R^{1/\mathfrak{p}(a)}) \subset\subset G$ for any $a \in |\mathfrak{p}|$,*
- *$\mathbb{P}(a, R^{1/\mathfrak{p}(a)}) \cap \mathbb{P}(b, R^{1/\mathfrak{p}(b)}) = \varnothing$ for any $a, b \in |\mathfrak{p}|$, $a \ne b$.*

Let $A_r := \bigcup_{a \in |\mathfrak{p}|} \mathbb{P}(a, r^{1/\mathfrak{p}(a)})$, $0 < r < R$. Then,

$$\left(\log \frac{R}{r}\right)(\boldsymbol{h}_{A_r,G} - 1) \searrow \log \boldsymbol{g}_G(\mathfrak{p}, \cdot) \text{ when } r \searrow 0.$$

Proof. Let

$$v_r := \left(\log \frac{R}{r}\right)(\boldsymbol{h}_{A_r,G} - 1), \quad 0 < r < R.$$

Step 1°. $v_{r_1} \le v_{r_2}$ *for* $0 < r_1 < r_2$.

Indeed, fix $0 < r_1 < r_2 < R$ and define

$$u := \frac{v_{r_1}}{\log \frac{R}{r_2}} = \frac{\left(\log \frac{R}{r_1}\right)(\boldsymbol{h}_{A_{r_1},G} - 1)}{\log \frac{R}{r_2}}.$$

Then, $u \in \mathcal{PSH}(G)$ and $u \le 0$. It suffices to show that $u \le -1$ on A_{r_2}. Fix an $a \in |\mathfrak{p}|$. Let $k := \mathfrak{p}(a)$. Take a $z \in \mathbb{P}(a, r_2^{1/k})$. Then, (cf. [305], Lemma 4.5.8; see also Exercise 17.5.1),

$$u(z) \le \frac{\left(\log \frac{R}{r_1}\right)\left(\boldsymbol{h}_{\mathbb{P}(a,r_1^{1/k}),\, \mathbb{P}(a,R^{1/k})}(z) - 1\right)}{\log \frac{R}{r_2}}$$
$$= \frac{\log \frac{R}{r_1}}{\log \frac{R}{r_2}} \left(\frac{\log^+ \frac{\|z-a\|_\infty}{r_1^{1/k}}}{\log \frac{R^{1/k}}{r_1^{1/k}}} - 1 \right) \le -1.$$

Let

$$v := \lim_{r \to 0+} v_r = \lim_{r \to 0+} \left(\log \frac{R}{r}\right)(\boldsymbol{h}_{A_r,G} - 1).$$

Note that $v \in \mathcal{PSH}(G)$.

Step 2°. $v_r \ge \log \boldsymbol{g}_G(\mathfrak{p}, \cdot)$, $0 < r < R$. *In particular,* $v \ge \log \boldsymbol{g}_G(\mathfrak{p}, \cdot)$.

Indeed, fix $0 < r < R$ and let

$$u_r := \frac{\log \boldsymbol{g}_G(\mathfrak{p}, \cdot)}{\log \frac{R}{r}}.$$

Then, $u_r \in \mathcal{PSH}(G)$ and $u_r \le 0$. Fix $a \in |\mathfrak{p}|$ and $z \in \mathbb{P}(a, r^{1/k})$ ($k := \mathfrak{p}(a)$). Then,

$$u_r(z) \le \frac{k \log \boldsymbol{g}_{\mathbb{P}(a,R^{1/k})}(a, z)}{\log \frac{R}{r}} = \frac{k \log \frac{\|z-a\|_\infty}{R^{1/k}}}{\log \frac{R}{r}} \le -1.$$

Thus, $u_r \le \boldsymbol{h}_{A_r,G} - 1$.

Step 3°. $v \le \log \boldsymbol{g}_G(\mathfrak{p}, \cdot)$.

Indeed, it suffices to check the growth of v near every point $a \in |\mathfrak{p}|$. Fix an $a \in |\mathfrak{p}|$ and let $z \in \mathbb{P}(a, R^{1/k})$, $z \ne a$ ($k := \mathfrak{p}(a)$). Let $0 < r < \|z - a\|_\infty^k$. Then,

$$\begin{aligned}
v(z) - k \log \|z - a\| &\le \left(\log \frac{R}{r}\right) (\boldsymbol{h}_{A_r, G}(z) - 1) - k \log \|z - a\|_\infty \\
&\le \left(\log \frac{R}{r}\right) \left(\boldsymbol{h}_{\mathbb{P}(a, r^{1/k}),\, \mathbb{P}(a, R^{1/k})}(z) - 1\right) - k \log \|z - a\|_\infty \\
&= \left(\log \frac{R}{r}\right) \left(\frac{\log^+ \frac{\|z-a\|_\infty}{r^{1/k}}}{\log \frac{R^{1/k}}{r^{1/k}}} - 1\right) - k \log \|z - a\|_\infty = -\log R. \qquad \square
\end{aligned}$$

17.2 Disc functionals

Definition 17.2.1. Let $G \subset \mathbb{C}^n$ be a domain. By a *disc functional* (on G) we mean any function

$$\Xi : \mathcal{O}(\overline{\mathbb{D}}, G) \longrightarrow \overline{\mathbb{R}}.$$

The *envelope* of a disc functional $\Xi : \mathcal{O}(\overline{\mathbb{D}}, G) \longrightarrow \overline{\mathbb{R}}$ is the function $\mathcal{E}_\Xi : G \longrightarrow \overline{\mathbb{R}}$ defined by the formula

$$\mathcal{E}_\Xi(z) := \inf\{\Xi(\varphi) : \varphi \in \mathcal{O}(\overline{\mathbb{D}}, G),\ \varphi(0) = z\}, \quad z \in G.$$

Definition 17.2.2. The following four types of disc functionals play an important role in complex analysis:

- *Poisson functional:*

$$\Xi_{\text{Poi}}^{\mathfrak{p}}(\varphi) := \frac{1}{2\pi} \int_0^{2\pi} \mathfrak{p} \circ \varphi(e^{i\theta}) d\theta, \quad \varphi \in \mathcal{O}(\overline{\mathbb{D}}, G),\ \mathfrak{p} : G \longrightarrow [-\infty, +\infty),$$

  where $\mathfrak{p}$ is upper semicontinuous, $\mathfrak{p} \not\equiv -\infty$ (in fact, $\Xi_{\text{Poi}}^{\mathfrak{p}}$ may be defined for more general functions $\mathfrak{p}$ – see [151]).

- *Green functional:*

$$\Xi_{\text{Gre}}^{\mathfrak{p}}(\varphi) := \sum_{\lambda \in \mathbb{D}_*} \mathfrak{p}(\varphi(\lambda)) \log |\lambda|, \quad \varphi \in \mathcal{O}(\overline{\mathbb{D}}, G),$$ [1]

  where $\mathfrak{p} : G \longrightarrow \mathbb{R}_+,\ \mathfrak{p} \not\equiv 0$.

[1] If $f : A \longrightarrow [-\infty, 0]$, then $\sum_{\lambda \in A} f(\lambda) := \inf_{\substack{B \subset A \\ \#B < +\infty}} \sum_{\lambda \in B} f(\lambda)$.

- *Lelong functional:*

$$\Xi^{\mathfrak{p}}_{\mathrm{Lel}}(\varphi) := \sum_{\lambda\in\mathbb{D}_*} \mathfrak{p}(\varphi(\lambda))\,\mathrm{ord}_\lambda(\varphi-\varphi(\lambda))\log|\lambda|, \quad \varphi\in\mathcal{O}(\overline{\mathbb{D}},G),\ \mathfrak{p}:G\longrightarrow\mathbb{R}_+,$$

where $\mathfrak{p}:G\longrightarrow\mathbb{R}_+,\ \mathfrak{p}\not\equiv 0$.

- *Lempert functional:*

$$\Xi^{\mathfrak{p}}_{\mathrm{Lem}}(\varphi) := \inf\{\mathfrak{p}(\varphi(\lambda))\log|\lambda| : \lambda\in\mathbb{D}_*\}, \quad \varphi\in\mathcal{O}(\overline{\mathbb{D}},G),$$

where $\mathfrak{p}:G\longrightarrow\mathbb{R}_+,\ \mathfrak{p}\not\equiv 0$.

Put $\mathcal{E}^{\mathfrak{p}}_{\mathrm{Poi}} := \mathcal{E}_{\Xi^{\mathfrak{p}}_{\mathrm{Poi}}}$, $\mathcal{E}^{\mathfrak{p}}_{\mathrm{Gre}} := \mathcal{E}_{\Xi^{\mathfrak{p}}_{\mathrm{Gre}}}$, $\mathcal{E}^{\mathfrak{p}}_{\mathrm{Lel}} := \mathcal{E}_{\Xi^{\mathfrak{p}}_{\mathrm{Lel}}}$, $\mathcal{E}^{\mathfrak{p}}_{\mathrm{Lem}} := \mathcal{E}_{\Xi^{\mathfrak{p}}_{\mathrm{Lem}}}$.

Remark 17.2.3. The reader is asked to complete the details.

(a) $\Xi^{\mathfrak{p}}_{\mathrm{Lel}} \le \Xi^{\mathfrak{p}}_{\mathrm{Gre}} \le \Xi^{\mathfrak{p}}_{\mathrm{Lem}}$ and, consequently, $\mathcal{E}^{\mathfrak{p}}_{\mathrm{Lel}} \le \mathcal{E}^{\mathfrak{p}}_{\mathrm{Gre}} \le \mathcal{E}^{\mathfrak{p}}_{\mathrm{Lem}}$.

(b) Let $\Xi\in\{\Xi^{\mathfrak{p}}_{\mathrm{Gre}}, \Xi^{\mathfrak{p}}_{\mathrm{Lel}}, \Xi^{\mathfrak{p}}_{\mathrm{Lem}}\}$. Then, $\Xi(\varphi)$ is well defined for $\varphi\in\mathcal{O}(\mathbb{D},G)$. Moreover, $\mathcal{E}_\Xi(z) = \inf\{\Xi(\varphi) : \varphi\in\mathcal{O}(\mathbb{D},G),\ \varphi(0)=z\}$, $z\in G$.

Indeed, for $\varphi\in\mathcal{O}(\mathbb{D},G)$ let $\varphi_r(\lambda) := \varphi(r\lambda)$, $|\lambda|<1/r$, $0<r<1$. Then, $\varphi_r\in\mathcal{O}(\overline{\mathbb{D}},G)$, $\varphi_r(0)=\varphi(0)$, and $\Xi(\varphi) = \inf_{0<r<1}\Xi(\varphi_r)$.

(c) If $F:G\longrightarrow D$ is holomorphic and $\mathfrak{q}:D\longrightarrow[-\infty,+\infty)$ (or $\mathfrak{q}:D\longrightarrow\mathbb{R}_+$), then $\Xi^{\mathfrak{q}}\circ F_* = \Xi^{\mathfrak{q}\circ F}$, $\Xi\in\{\Xi_{\mathrm{Poi}}, \Xi_{\mathrm{Gre}}, \Xi_{\mathrm{Lem}}\}$, and $\Xi^{\mathfrak{q}}_{\mathrm{Lel}}\circ F_* \le \Xi^{\mathfrak{q}\circ F}_{\mathrm{Lel}}$, where $F_*:\mathcal{O}(\overline{\mathbb{D}},G)\longrightarrow\mathcal{O}(\overline{\mathbb{D}},D)$, $F_*(\varphi) := F\circ\varphi$.

(d) Let $F:G\longrightarrow D$ be holomorphic and let $\Xi:\mathcal{O}(\overline{\mathbb{D}},D)\longrightarrow\overline{\mathbb{R}}$ be a disc functional on D. Then, the mapping $\Xi\circ F_*$ is a disc functional on G and $\mathcal{E}_\Xi\circ F_* \le \mathcal{E}_{\Xi\circ F}$. If, moreover, F is a covering, then $\mathcal{E}_\Xi\circ F_* = \mathcal{E}_{\Xi\circ F}$.

Indeed, we only need to observe that if F is a covering, then for any disc $\psi\in\mathcal{O}(\overline{\mathbb{D}},D)$ with $\psi(0)=F(z)$ there exists a $\varphi\in\mathcal{O}(\overline{\mathbb{D}},G)$ such that $\varphi(0)=z$ and $F\circ\varphi=\psi$.

(e) $\mathcal{E}^{\mathfrak{p}}_{\mathrm{Lem}}(z) = \inf\{\mathfrak{p}(a)\log\boldsymbol{\ell}^*_G(a,z) : a\in|\mathfrak{p}|\}$, $z\in G$; in particular, the function $G\ni z\longmapsto\mathcal{E}^{\mathfrak{p}}_{\mathrm{Lem}}(z)$ is upper semicontinuous.

(f) If $\mathbb{B}(a,r)\subset G$, then for any $z\in\mathbb{B}(a,r)$ we get

$$\begin{aligned}\mathcal{E}^{\mathfrak{p}}_{\mathrm{Lem}}(z) &\le \mathfrak{p}(a)\log\boldsymbol{\ell}^*_G(a,z) \le \mathfrak{p}(a)\log\boldsymbol{\ell}^*_{\mathbb{B}(a,r)}(a,z) = \mathfrak{p}(a)\log\frac{\|z-a\|}{r}\\ &= \mathfrak{p}(a)\log\|z-a\| - \mathfrak{p}(a)\log r.\end{aligned}$$

(g) In the case where $\mathfrak{p} = \chi_{\{a\}}$ we have:

$$\begin{aligned}\Xi^{\mathfrak{p}}_{\mathrm{Poi}}(\varphi) &= \mathfrak{m}_{\mathbb{T}}(\varphi^{-1}(a) \cap \mathbb{T}),\\ \Xi^{\mathfrak{p}}_{\mathrm{Gre}}(\varphi) &= \sum_{\lambda\in\varphi^{-1}(a)\cap\mathbb{D}_*} \log|\lambda|,\\ \Xi^{\mathfrak{p}}_{\mathrm{Lel}}(\varphi) &= \sum_{\lambda\in\varphi^{-1}(a)\cap\mathbb{D}_*} \mathrm{ord}_\lambda(\varphi - a)\log|\lambda|,\\ \Xi^{\mathfrak{p}}_{\mathrm{Lem}}(\varphi) &= \inf\{\log|\lambda| : \lambda \in \varphi^{-1}(a)\cap\mathbb{D}_*\},\end{aligned}$$

where $\mathfrak{m}_{\mathbb{T}}$ denotes the normalized Lebesgue measure on $\mathbb{T}$, $\mathfrak{m}_{\mathbb{T}}(\mathbb{T}) = 1$.

17.3 Poisson functional

Remark 17.3.1.

(a) $\mathcal{E}^{\mathfrak{p}}_{\mathrm{Poi}} \le \mathfrak{p}$.

(b) If $\mathfrak{p}_k \searrow \mathfrak{p}$, then $\Xi^{\mathfrak{p}_k}_{\mathrm{Poi}} \searrow \Xi^{\mathfrak{p}}_{\mathrm{Poi}}$ and $\mathcal{E}^{\mathfrak{p}_k}_{\mathrm{Poi}} \searrow \mathcal{E}^{\mathfrak{p}}_{\mathrm{Poi}}$.

Proposition 17.3.2. *For any upper semicontinuous function $\mathfrak{p} : G \longrightarrow [-\infty,\infty)$ we have $\boldsymbol{h}_{\mathfrak{p},G} \le \mathcal{E}^{\mathfrak{p}}_{\mathrm{Poi}}$. Consequently, if $\mathcal{E}^{\mathfrak{p}}_{\mathrm{Poi}} \in \mathcal{PSH}(G)$, then $\mathcal{E}^{\mathfrak{p}}_{\mathrm{Poi}} \in \mathcal{P}_{\mathfrak{p}}(G)$ and $\boldsymbol{h}_{\mathfrak{p},G} \equiv \mathcal{E}^{\mathfrak{p}}_{\mathrm{Poi}}$.*

Proof. For $u \in \mathcal{P}_{\mathfrak{p}}(G)$ and $\varphi \in \mathcal{O}(\overline{\mathbb{D}}, G)$ we have

$$u(\varphi(0)) \le \frac{1}{2\pi}\int_0^{2\pi} u\circ\varphi(e^{i\theta})d\theta \le \frac{1}{2\pi}\int_0^{2\pi} \mathfrak{p}\circ\varphi(e^{i\theta})d\theta = \Xi^{\mathfrak{p}}_{\mathrm{Poi}}(\varphi). \qquad \square$$

Lemma 17.3.3. *The function $\mathcal{E}^{\mathfrak{p}}_{\mathrm{Poi}}$ is upper semicontinuous on G.*

Proof. Fix a $z_0 \in G$ and suppose that $\mathcal{E}^{\mathfrak{p}}_{\mathrm{Poi}}(z_0) < A$. Then, there exists a $\varphi_0 \in \mathcal{O}(\overline{\mathbb{D}}, G)$ such that $\varphi_0(0) = z_0$ and $\Xi^{\mathfrak{p}}_{\mathrm{Poi}}(\varphi_0) < A$. Take $0 < r < \mathrm{dist}(\varphi_0(\overline{\mathbb{D}}), \partial G)$. Then, for $z \in \mathbb{B}(z_0, r)$ we get

$$\mathcal{E}^{\mathfrak{p}}_{\mathrm{Poi}}(z) \le \Xi^{\mathfrak{p}}_{\mathrm{Poi}}(\varphi_0 + z - z_0) = \frac{1}{2\pi}\int_0^{2\pi} \mathfrak{p}\circ(\varphi_0(e^{i\theta}) + z - z_0)d\theta.$$

Consequently, there exists $0 < \delta < r$ such that $\mathcal{E}^{\mathfrak{p}}_{\mathrm{Poi}}(z) < A$, $z \in \mathbb{B}(z_0, \delta)$. $\square$

Theorem 17.3.4. *For any upper semicontinuous function $\mathfrak{p} : G \longrightarrow [-\infty,\infty)$ we have $\mathcal{E}^{\mathfrak{p}}_{\mathrm{Poi}} \in \mathcal{PSH}(G)$. Consequently, by Proposition* 17.3.2,

$$\begin{aligned}\inf\left\{\frac{1}{2\pi}\int_0^{2\pi} \mathfrak{p}\circ\varphi(e^{i\theta})d\theta : \varphi \in \mathcal{O}(\overline{\mathbb{D}}, G),\ \varphi(0) = z\right\} &= \mathcal{E}^{\mathfrak{p}}_{\mathrm{Poi}}(z)\\ &= \boldsymbol{h}_{\mathfrak{p},G}(z) = \sup\{u(z) \in \mathcal{PSH}(G) : u \le \mathfrak{p}\}, \quad z \in G.\end{aligned}$$

In particular, if $U \subset G$ is open, then

$$\begin{aligned} \boldsymbol{h}_{U,G}(z) &= \boldsymbol{h}_{\chi_{G\setminus U},G}(z) = \mathcal{E}_{\Xi_{\mathrm{Poi}}^{\chi_{G\setminus U}}}(z) \\ &= \inf\left\{\frac{1}{2\pi}\int_0^{2\pi} \chi_{G\setminus U} \circ \varphi(e^{i\theta})d\theta : \varphi \in \mathcal{O}(\overline{\mathbb{D}}, G),\ \varphi(0) = z\right\} \\ &= 1 - \sup\{\mathfrak{m}_{\mathbb{T}}(\varphi^{-1}(U) \cap \mathbb{T}) : \varphi \in \mathcal{O}(\overline{\mathbb{D}}, G),\ \varphi(0) = z\}, \quad z \in G. \end{aligned}$$

Taking $\mathcal{E}^{\mathfrak{p}}_{\mathrm{Lem}}$ instead of $\mathfrak{p}$ and using Remark 17.2.3(e), *we conclude that*

$$\mathcal{E}_{\Xi_{\mathrm{Poi}}^{\mathcal{E}^{\mathfrak{p}}_{\mathrm{Lem}}}} \in \mathcal{PSH}(G).$$

Notice that the above result translates the extremal problem for psh functions (from the definition of $\boldsymbol{h}_{\mathfrak{p},G}$) to an extremal problem for analytic discs.

A class of complex manifolds G for which the above result is true was presented in [337]. The case where G is an arbitrary complex manifold was proved in [456, 152].

Proof. By Remark 17.3.1, we may assume that $\mathfrak{p} : G \longrightarrow \mathbb{R}$ is continuous. Let $u_0 := \mathcal{E}^{\mathfrak{p}}_{\mathrm{Poi}}$. By Lemma 17.3.3, we only need to show that

$$u_0(\varphi(0)) \le \frac{1}{2\pi}\int_0^{2\pi} u_0 \circ \varphi(e^{i\theta})d\theta, \quad \varphi \in \mathcal{O}(\overline{\mathbb{D}}, G).$$

Fix a $\varphi_0 \in \mathcal{O}(\overline{\mathbb{D}}, G)$. It suffices to prove that for any $\varepsilon > 0$ there exists a $\widetilde{\varphi} \in \mathcal{O}(\overline{\mathbb{D}}, G)$ such that $\widetilde{\varphi}(0) = \varphi_0(0)$ and

$$\Xi^{\mathfrak{p}}_{\mathrm{Poi}}(\widetilde{\varphi}) = \frac{1}{2\pi}\int_0^{2\pi} \mathfrak{p} \circ \widetilde{\varphi}(e^{i\theta}) \le \frac{1}{2\pi}\int_0^{2\pi} u_0 \circ \varphi_0(e^{i\theta})d\theta + \varepsilon. \tag{17.3.1}$$

Fix an $\varepsilon > 0$. The proof will be divided into four steps (Lemmas 17.3.5, 17.3.6, 17.3.7, 17.3.8):

Lemma 17.3.5. *There exist $r > 1$ and $\Phi \in \mathcal{C}^\infty(\mathbb{D}(r) \times \mathbb{T}, G)$ such that:*

(i) $\Phi(\cdot, \xi) \in \mathcal{O}(\overline{\mathbb{D}}, G), \quad \xi \in \mathbb{T}$,

(ii) $\Phi(0, \xi) = \varphi_0(\xi), \quad \xi \in \mathbb{T}$,

(iii)
$$\int_0^{2\pi} \Xi^{\mathfrak{p}}_{\mathrm{Poi}}(\Phi(\cdot, e^{it}))dt \le \int_0^{2\pi} u_0(\varphi_0(e^{it}))dt + \varepsilon. \tag{17.3.2}$$

Proof. Since u_0 is upper semicontinuous (Lemma 17.3.3), there exists a $v \in \mathcal{C}(G, \mathbb{R})$ with $v \geq u_0$ such that

$$\int_0^{2\pi} v(\varphi_0(e^{it}))dt \leq \int_0^{2\pi} u_0(\varphi_0(e^{it}))dt + \varepsilon/2.$$

For any $\xi_0 \in \mathbb{T}$ there exist $\varphi \in \mathcal{O}(\overline{\mathbb{D}}, G)$, $0 < \delta < \operatorname{dist}(\varphi(\overline{\mathbb{D}}), \partial G)$, an open arc $I \subset \mathbb{T}$, and $r > 1$ such that

- $\xi_0 \in I$, $\varphi(0) = \varphi_0(\xi_0)$,
- $\Xi^{\mathfrak{p}}_{\mathrm{Poi}}(\varphi + z - \varphi_0(\xi_0)) < v(z) + \varepsilon/4, \quad z \in \mathbb{B}(\varphi_0(\xi_0), \delta)$,
- $\varphi_0(\xi) \in \mathbb{B}(\varphi_0(\xi_0), \delta), \quad \xi \in I$,
- $\Phi_0(\mathbb{D}(r) \times I) \subset\subset G$, where $\Phi_0(\lambda, \xi) := \varphi(\lambda) + \varphi_0(\xi) - \varphi_0(\xi_0)$.

By a compactness argument we find a finite covering $\mathbb{T} = \bigcup_{\nu=1}^{N_0} I_\nu$, $r > 1$, and functions $\Phi_\nu \in \mathcal{C}^\infty(\mathbb{D}(r) \times I_\nu, G)$, $\nu = 1, \dots, N_0$, such that

- $\Phi_\nu(\cdot, \xi) \in \mathcal{O}(\overline{\mathbb{D}}, G), \quad \xi \in I_\nu$,
- $\Phi_\nu(0, \xi) = \varphi_0(\xi), \quad \xi \in I_\nu$,
- $\Phi_\nu(\mathbb{D}(r) \times I_\nu) \subset\subset G$,
- $\Xi^{\mathfrak{p}}_{\mathrm{Poi}}(\Phi_\nu(\cdot, \xi)) < v(\varphi_0(\xi)) + \varepsilon/4, \quad \xi \in I_\nu, \nu = 1, \dots, N_0$.

Let K be the closure of the set $\varphi_0(\mathbb{T}) \cup \bigcup_{\nu=1}^{N_0} \Phi_\nu(\mathbb{D}(r) \times I_\nu)$ and let $C > 0$ be such that $C > \max\{v(z) : z \in K\}$. There exist disjoint closed arcs $J_\nu \subset I_\nu$, $\nu \in A \subset \{1, \dots, N_0\}$, such that

$$\mathfrak{m}_{\mathbb{T}}(\mathbb{T} \setminus \bigcup_{\nu \in A} J_\nu) < \varepsilon/(8C).$$

We may assume that $A = \{1, \dots, N\}$ for some $N \leq N_0$. Fix open disjoint arcs K_ν with $J_\nu \subset K_\nu \subset I_\nu$, $\nu = 1, \dots, N$, and let $\varrho \in \mathcal{C}^\infty(\mathbb{T}, [0, 1])$ be such that $\varrho = 1$ on $\bigcup_{\nu=1}^N J_\nu$ and $\operatorname{supp} \varrho \subset \bigcup_{\nu=1}^N K_\nu$. Now we define $\Phi : \mathbb{D}(r) \times \mathbb{T} \longrightarrow G$ by the formula

$$\Phi(\lambda, \xi) := \begin{cases} \Phi_\nu(\varrho(\xi)\lambda, \xi), & (\lambda, \xi) \in \mathbb{D}(r) \times K_\nu \\ \varphi_0(\xi), & (\lambda, \xi) \in \mathbb{D}(r) \times (\mathbb{T} \setminus \bigcup_{\nu=1}^N K_\nu) \end{cases}.$$

It is clear that Φ is well-defined, $\Phi \in \mathcal{C}^\infty(\mathbb{D}(r) \times \mathbb{T}, G)$, $\Phi(\mathbb{D}(r) \times \mathbb{T}) \subset K$, and Φ satisfies (i) and (ii). It remains to check (iii). Let $\widetilde{J}_\nu := \{t \in [0, 2\pi) : e^{it} \in J_\nu\}$,

$\nu = 1, \dots, N$. We have

$$\begin{aligned}\int_0^{2\pi} \Xi^{\mathfrak{p}}_{\mathrm{Poi}}(\Phi(\cdot, e^{it}))dt &\le \sum_{\nu=1}^{N} \int_{\tilde{J}_\nu} \Xi^{\mathfrak{p}}_{\mathrm{Poi}}(\Phi_\nu(\cdot, e^{it}))dt + \varepsilon/8 \\ &\le \sum_{\nu=1}^{N} \int_{\tilde{J}_\nu} v(\varphi_0(e^{it}))dt + 3\varepsilon/8 \\ &\le \int_0^{2\pi} v(\varphi_0(e^{it}))dt + \varepsilon/2 \\ &\le \int_0^{2\pi} u_0(\varphi_0(e^{it}))dt + \varepsilon. \qquad \square\end{aligned}$$

Lemma 17.3.6. *Let $r > 1$ be as in Lemma 17.3.5. There exists an $s \in (1, r)$ such that for any $j \ge 1$ there exist an open annulus $A_j \supset \mathbb{T}$ and $\Phi_j \in \mathcal{O}(\mathbb{D}(s) \times A_j, G)$ with*

(i) *$\Phi_j \longrightarrow \Phi$ uniformly on $\mathbb{D}(s) \times \mathbb{T}$,*

(ii) *there exist $1 < s_j < s$ and $k_j \in \mathbb{N}$, $k_j \ge j$, such that the mapping*

$$(\lambda, \xi) \longrightarrow \Phi_j(\lambda \xi^{k_j}, \xi)$$

extends to a $\Psi_j \in \mathcal{O}(\mathbb{D}(s_j) \times \mathbb{D}(s_j), G)$,

(iii) *$\Psi_j(0, \xi) = \varphi_0(\xi)$, $\xi \in \mathbb{D}(s_j)$.*

Proof. Let

$$\Phi_j(\lambda, \xi) := \varphi_0(\xi) + \sum_{k=-j}^{j} \left(\frac{1}{2\pi} \int_0^{2\pi} (\Phi(\lambda, e^{i\theta}) - \varphi_0(e^{i\theta}))e^{-ik\theta} d\theta \right) \xi^k,$$
$$(\lambda, \xi) \in \mathbb{D}(r) \times \mathbb{D}_*(r);$$

observe that the second term is the j-th partial sum of the Fourier series of the function $\xi \longrightarrow \Phi(\lambda, \xi) - \varphi_0(\xi)$; Φ_j is holomorphic and $\Phi_j(0, \xi) = \varphi_0(\xi)$, $\xi \in \mathbb{D}_*(r)$. Moreover, for any $1 < t < r$, $\Phi_j \longrightarrow \Phi$ uniformly on $\mathbb{D}(t) \times \mathbb{T}$. Indeed, it follows directly from Fourier series theory that $\Phi_j(\lambda, \cdot) \longrightarrow \Phi(\lambda, \cdot)$ uniformly on $\mathbb{T}$ for any $\lambda \in \mathbb{D}(r)$. Thus, we only need to show that the series

$$\sum_{k=-\infty}^{\infty} \left(\frac{1}{2\pi} \int_0^{2\pi} (\Phi(\lambda, e^{i\theta}) - \varphi_0(e^{i\theta}))e^{-ik\theta} d\theta \right) \xi^k$$

converges uniformly on $\mathbb{D}(t) \times \mathbb{T}$. Using integration by parts, we obtain

$$\Big| \int_0^{2\pi} (\Phi(\lambda, e^{i\theta}) - \varphi_0(e^{i\theta})) e^{-ik\theta} d\theta \xi^k \Big|$$
$$\leq \frac{1}{k^2} \sup_{\lambda \in \mathbb{D}(t),\, \theta \in [0,2\pi)} \Big| \frac{\partial^2}{\partial \theta^2} (\Phi(\lambda, e^{i\theta}) - \varphi_0(e^{i\theta})) \Big|, \quad k \in \mathbb{Z}_*,\ (\lambda, \xi) \in \mathbb{D}(t) \times \mathbb{T},$$

which implies the required convergence.

Fix $1 < s < r$. It follows that $\Phi_j(\mathbb{D}(s) \times \mathbb{T}) \subset\subset G$ for $j \geq j_0$. Hence, one can find an open annulus $A_j \supset \mathbb{T}$ such that $\Phi_j(\mathbb{D}(s) \times A_j) \subset G$, $j \geq j_0$.

For any $\xi \in \mathbb{D}_*(r)$ the mapping $\Phi_j(\cdot, \xi) - \varphi_0(\xi)$ has a zero at $\lambda = 0$. For any $\lambda \in \mathbb{D}(r)$ the mapping $\Phi_j(\lambda, \cdot) - \varphi_0$ has a pole of order $\leq j$ at $\xi = 0$. Consequently, for any $k \geq j$ the mapping $(\lambda, \xi) \longmapsto \Phi_j(\lambda \xi^k, \xi)$ extends holomorphically to a $\Psi_{j,k} \in \mathcal{O}(\mathbb{D}(r) \times \mathbb{D}(r), \mathbb{C}^n)$. It remains to find $s_j \in (1, s)$ and $k_j \geq j$ such that $\Psi_{j,k_j}(\mathbb{D}(s_j) \times \mathbb{D}(s_j)) \subset G$. Recall that $\Phi_j(0, \cdot) = \varphi_0$. Hence, there exists a $\delta_j > 0$ such that $\Phi_j(\mathbb{D}(\delta_j) \times \mathbb{D}(s)) \subset G$. Since $\Phi_j(\mathbb{D}(s) \times A_j) \subset G$, $j \geq j_0$, we can find $0 < \varepsilon_j < 1$ such that $\Phi_j(\mathbb{D}(s) \times \mathbb{A}(1 - \varepsilon_j, 1 + \varepsilon_j)) \subset G$, $j \geq j_0$. Now, let $k_j \geq j$ be so big that $(1 - \varepsilon_j)^{k_j} < \delta_j$. Then, $\Phi_j(\lambda \xi^{k_j}, \xi) \in G$, $(\lambda, \xi) \in \overline{\mathbb{D}} \times \overline{\mathbb{D}}$, $j \geq j_0$. □

Lemma 17.3.7. *There exist $s \in (1, r)$ and $\Psi \in \mathcal{O}(\mathbb{D}(s) \times \mathbb{D}(s), G)$ such that*

(i) $\Psi(0, \xi) = \varphi_0(\xi), \quad \xi \in \mathbb{D}(s),$

(ii)

$$\int_0^{2\pi} \Xi^{\mathfrak{p}}_{\mathrm{Poi}}(\Psi(\cdot, e^{it}))dt \leq \int_0^{2\pi} \Xi^{\mathfrak{p}}_{\mathrm{Poi}}(\Phi(\cdot, e^{it}))dt + \varepsilon. \tag{17.3.3}$$

Proof. Let Φ_j, Ψ_j be as in Lemma 17.3.6. Then, for $j \geq j(\varepsilon)$ we have

$$\begin{aligned}
\int_0^{2\pi} \Xi^{\mathfrak{p}}_{\mathrm{Poi}}(\Psi_j(\cdot, e^{it}))dt &= \int_0^{2\pi} \left(\frac{1}{2\pi} \int_0^{2\pi} \mathfrak{p}(\Phi_j(e^{i(\theta + k_j t)}, e^{it}))d\theta \right) dt \\
&= \frac{1}{2\pi} \int_0^{2\pi} \int_0^{2\pi} \mathfrak{p}(\Phi_j(e^{i\theta}, e^{it}))d\theta dt \\
&\leq \frac{1}{2\pi} \int_0^{2\pi} \int_0^{2\pi} \mathfrak{p}(\Phi(e^{i\theta}, e^{it}))d\theta dt + \varepsilon \\
&= \int_0^{2\pi} \Xi^{\mathfrak{p}}_{\mathrm{Poi}}(\Phi(\cdot, e^{it}))dt + \varepsilon.
\end{aligned}$$ □

Lemma 17.3.8. *Let $s > 1$ be as in Lemma 17.3.7. There exists a $\theta_0 \in \mathbb{R}$ such that if we put*

$$\widetilde{\varphi}(\lambda) = \varphi_{\theta_0}(\lambda) := \Psi(e^{i\theta_0}\lambda, \lambda), \quad \lambda \in \mathbb{D}(s),$$

then

$$\Xi^{\mathfrak{p}}_{\mathrm{Poi}}(\widetilde{\varphi}) \le \frac{1}{2\pi}\int_0^{2\pi} \Xi^{\mathfrak{p}}_{\mathrm{Poi}}(\Psi(\cdot, e^{it}))dt. \tag{17.3.4}$$

Proof. We have

$$\int_0^{2\pi}\int_0^{2\pi} \mathfrak{p}(\Psi(e^{i\theta}, e^{it}))d\theta dt = \int_0^{2\pi}\int_0^{2\pi} \mathfrak{p}(\Psi(e^{i\theta}e^{it}, e^{it}))dt d\theta.$$

Consequently, there exists a $\theta_0 \in \mathbb{R}$ such that

$$\begin{aligned}\Xi^{\mathfrak{p}}_{\mathrm{Poi}}(\varphi_{\theta_0}) &= \frac{1}{2\pi}\int_0^{2\pi} \mathfrak{p}(\Psi(e^{i\theta_0}e^{it}, e^{it}))dt \le \frac{1}{(2\pi)^2}\int_0^{2\pi}\int_0^{2\pi} \mathfrak{p}(\Psi(e^{i\theta}e^{it}, e^{it}))dt d\theta \\ &= \frac{1}{(2\pi)^2}\int_0^{2\pi}\int_0^{2\pi} \mathfrak{p}(\Psi(e^{i\theta}, e^{it}))d\theta dt = \frac{1}{2\pi}\int_0^{2\pi} \Xi^{\mathfrak{p}}_{\mathrm{Poi}}(\Psi(\cdot, e^{it}))dt.\end{aligned}$$

□

Now, using (17.3.4), (17.3.3), and (17.3.2) gives (17.3.1). □

17.4 Green, Lelong, and Lempert functionals

For any function $\mathfrak{p} : G \longrightarrow \mathbb{R}_+$ let

$$\begin{aligned}\mathcal{G}_{\mathfrak{p}}(G) := \{&u \in \mathcal{PSH}(G) : \\ &u \le 0,\ \forall_{a\in|\mathfrak{p}|}\ \exists_{C(u,a)\in\mathbb{R}}\ \forall_{z\in G} : u(z) \le \mathfrak{p}(a)\log\|z-a\| + C(a)\}.\end{aligned}$$

Observe that $\log \boldsymbol{g}_G(\mathfrak{p}, z) = \sup\{u(z) : u \in \mathcal{G}_{\mathfrak{p}}(G)\}$, $z \in G$ (cf. Definition 8.2.1). Obviously, $\mathcal{G}_{\chi_{\{a\}}}(G) = \log \mathcal{K}_G(a)$ (cf. § 4.2).

Proposition 17.4.1. $\log \boldsymbol{g}_G(\mathfrak{p}, \cdot) \le \mathcal{E}^{\mathfrak{p}}_{\mathrm{Lel}}$. *Consequently,*

(a) *for* $\Xi \in \{\Xi^{\mathfrak{p}}_{\mathrm{Gre}}, \Xi^{\mathfrak{p}}_{\mathrm{Lel}}, \Xi^{\mathfrak{p}}_{\mathrm{Lem}}\}$ *if* $\mathcal{E}_{\Xi} \in \mathcal{PSH}(G)$, *then* $\mathcal{E}_{\Xi} \in \mathcal{G}_{\mathfrak{p}}(G)$ *and*

$$\log \boldsymbol{g}_G(\mathfrak{p}, \cdot) \equiv \mathcal{E}_{\Xi},$$

(b) $\mathcal{E}_{\Xi_{\mathrm{Poi}}^{\mathcal{E}^{\mathfrak{p}}_{\mathrm{Lem}}}} \le \log \boldsymbol{g}_G(\mathfrak{p}, \cdot) \le \mathcal{E}^{\mathfrak{p}}_{\mathrm{Lel}}$.

Proof. (a) See the proof of Lemma 8.2.19(b).

(b) By Theorem 17.3.4, we get $\mathcal{E}_{\Xi_{\mathrm{Poi}}^{\mathcal{E}^{\mathfrak{p}}_{\mathrm{Lem}}}} \in \mathcal{PSH}(G)$. Hence, using Remark 17.2.3(f), we conclude that $\mathcal{E}_{\Xi_{\mathrm{Poi}}^{\mathcal{E}^{\mathfrak{p}}_{\mathrm{Lem}}}} \in \mathcal{G}_{\mathfrak{p}}(G)$. □

Proposition 17.4.2. $\mathcal{E}^{\mathfrak{p}}_{\text{Gre}} = \mathcal{E}^{\mathfrak{p}}_{\text{Lel}}$.

Proof. We have to prove that

$$\begin{aligned} L(z) &:= \inf\Big\{\sum_{\lambda\in\mathbb{D}_*} \mathfrak{p}(\varphi(\lambda))\operatorname{ord}_\lambda(\varphi-\varphi(\lambda))\log|\lambda| : \varphi\in\mathcal{O}(\overline{\mathbb{D}},G),\ \varphi(0)=z\Big\} \\ &= \inf\Big\{\sum_{\lambda\in\mathbb{D}_*} \mathfrak{p}(\varphi(\lambda))\log|\lambda| : \varphi\in\mathcal{O}(\overline{\mathbb{D}},G),\ \varphi(0)=z\Big\} =: R(z),\ z\in G. \end{aligned}$$

The inequality "$L \le R$" is obvious. Fix a $z\in G$ and an arbitrary constant $C > L(z)$. We want to show that $C \ge R(z)$.

Since $C > L(z)$, there exist $\varphi\in\mathcal{O}(\overline{\mathbb{D}},G)$, $\varphi(0)=z$, and a finite set $B\subset\mathbb{D}_*\cap\varphi^{-1}(|\mathfrak{p}|)$ such that

$$\sum_{\lambda\in B}\mathfrak{p}(\varphi(\lambda))\operatorname{ord}_\lambda(\varphi-\varphi(\lambda))\log|\lambda| < C.$$

Write $B=\{b_1,\dots,b_N\}$, $a_j := \varphi(b_j)$, $r(j) := \operatorname{ord}_{b_j}(\varphi-a_j)$, $j=1,\dots,N$. Consider the family of all systems $\mathfrak{c}$ of pairwise different points $c_{j,k}\in\mathbb{D}$, $j=1,\dots,N$, $k=1,\dots,r(j)$, such that $c_{j,1}\cdots c_{j,r(j)} = b_j^{r(j)}$, $j=1,\dots,N$. Define polynomials

$$Q_{\mathfrak{c},\mu,\nu}(\lambda) := \prod_{\substack{j=1,\dots,N\\ k=1,\dots,r(j)\\ (j,k)\neq(\mu,\nu)}} (\lambda-c_{j,k}),\quad \nu=1,\dots,r(\mu),$$

$$P_{\mathfrak{c},\mu}(\lambda) := \sum_{\nu=1}^{r(\mu)} \frac{Q_{\mathfrak{c},\mu,\nu}(\lambda)}{Q_{\mathfrak{c},\mu,\nu}(c_{\mu,\nu})},\quad \mu=1,\dots,N,\ \lambda\in\mathbb{C}.$$

Observe that

- $\deg P_{\mathfrak{c},j} \le r(1)+\dots+r(N)-1$,
- $P_{\mathfrak{c},j}(c_{\mu,\nu}) = 0$ if $\mu\neq j$ and $P_{\mathfrak{c},j}(c_{j,\nu}) = 1$,
- $P_{\mathfrak{c},1}+\dots+P_{\mathfrak{c},N}\equiv 1$.

Define

$$\varphi_{\mathfrak{c}}(\lambda) := \sum_{\mu=1}^{N} P_{\mathfrak{c},\mu}(\lambda)\left(\frac{\varphi(\lambda)-a_\mu}{(\lambda-b_\mu)^{r(\mu)}}(\lambda-c_{\mu,1})\cdots(\lambda-c_{\mu,r(\mu)})+a_\mu\right).$$

Observe that $\varphi_{\mathfrak{c}}\in\mathcal{O}(\overline{\mathbb{D}},\mathbb{C}^n)$, $\varphi_{\mathfrak{c}}(0)=\varphi(0)=z$, and $\varphi_{\mathfrak{c}}(c_{j,k}) = a_j$ for all $j=1,\dots,N$, $k=1,\dots,r(j)$. Moreover,

$$\sum_{j=1}^{N}\sum_{k=1}^{r(j)}\mathfrak{p}(a_j)\log|c_{j,k}| = \sum_{j=1}^{N}\mathfrak{p}(a_j)r(j)\log|b_j| < C.$$

It remains to observe that $\varphi_c(\overline{\mathbb{D}}) \subset G$ provided that $c_{j,k} \approx b_j$, $j = 1,\dots,N$, $k = 1,\dots,r(j)$. □

Theorem 17.4.3. *If* $|\mathfrak{p}|$ *is finite, then* $\mathcal{E}_{\Xi_{\mathrm{Poi}}^{\mathcal{E}^{\mathfrak{p}}_{\mathrm{Lem}}}} = \mathcal{E}^{\mathfrak{p}}_{\mathrm{Lel}}$. *Consequently, by Propositions* 17.4.1(b) *and* 17.4.2, *we have*

$$\mathcal{E}_{\Xi_{\mathrm{Poi}}^{\mathcal{E}^{\mathfrak{p}}_{\mathrm{Lem}}}} = \log \boldsymbol{g}_G(\mathfrak{p},\cdot) = \mathcal{E}^{\mathfrak{p}}_{\mathrm{Lel}} = \mathcal{E}^{\mathfrak{p}}_{\mathrm{Gre}}.$$

Moreover, by Proposition 8.2.6, *for arbitrary* $\mathfrak{p} : G \longrightarrow \mathbb{R}_+$ *we get the* Poletsky formula

$$\log \boldsymbol{g}_G(\mathfrak{p},\cdot) = \mathcal{E}^{\mathfrak{p}}_{\mathrm{Lel}} = \mathcal{E}^{\mathfrak{p}}_{\mathrm{Gre}}.$$

The Poletsky formula and the main ideas of the proof are due to E. A. Poletsky, cf. [434, 435, 436]. The first complete proof was given by A. Edigarian in [148]. We follow the exposition of A. Edigarian.

Proof. By Proposition 17.4.1(b), we only need to show that $\mathcal{E}^{\mathfrak{p}}_{\mathrm{Lel}} \le \mathcal{E}_{\Xi_{\mathrm{Poi}}^{\mathcal{E}^{\mathfrak{p}}_{\mathrm{Lem}}}}$. Fix $\varphi_0 \in \mathcal{O}(\overline{\mathbb{D}}, G)$ and $\varepsilon > 0$. It suffices to find a $\widetilde{\varphi} \in \mathcal{O}(\overline{\mathbb{D}}, G)$ such that $\widetilde{\varphi}(0) = \varphi_0(0)$ and

$$\Xi^{\mathfrak{p}}_{\mathrm{Lel}}(\widetilde{\varphi}) \le \frac{1}{2\pi}\int_0^{2\pi} \mathcal{E}^{\mathfrak{p}}_{\mathrm{Lem}}(\varphi_0(e^{it}))dt + \varepsilon.$$

The existence of $\widetilde{\varphi}$ will be a consequence of a sequence of lemmas (Lemmas 17.4.4–17.4.8):

Lemma 17.4.4. *There exist*

- $1 < s < r$,
- $\Phi \in \mathcal{C}^\infty(\mathbb{D}(r) \times \mathbb{T}, G)$,
- $N \in \mathbb{N}$,
- $a_1,\dots,a_N \in |\mathfrak{p}|$,
- $\sigma_1,\dots,\sigma_N \in \mathcal{C}^\infty(\mathbb{T}, \mathbb{C}_*)$,
- *disjoint closed arcs* $J_1,\dots,J_N \subset \mathbb{T}$

such that:

(i) $\Phi(\cdot,\xi) \in \mathcal{O}(\overline{\mathbb{D}}, G)$, $\Phi(0,\xi) = \varphi_0(\xi)$, $\xi \in \mathbb{T}$,

(ii) *if* $|\sigma_\nu(\xi)| < s$, *then* $|\sigma_\mu(\xi)| > s$, $\mu \neq \nu$, *and* $\Phi(\sigma_\nu(\xi),\xi) = a_\nu$,

(iii) $|\sigma_\nu(\xi)| < 1$, $\xi \in J_\nu$, $\nu = 1,\dots,N$, $\mathfrak{m}_{\mathbb{T}}(\mathbb{T} \setminus \bigcup_{\nu=1}^N J_\nu) < \varepsilon$,

(iv) $\sigma_\nu(\xi) \neq \sigma_\mu(\xi)$, $\xi \in \mathbb{T}$, $\nu \neq \mu$,

(v) $2\pi N \max\limits_{\nu=1,\dots,N} \{\mathfrak{p}(a_\nu) \max_{\mathbb{T}} \log|\sigma_\nu|\} < \varepsilon/2$,

(vi) $\sum_{\nu=1}^{N} \mathfrak{p}(a_\nu) \int_0^{2\pi} \log|\sigma_\nu(e^{it})|dt \leq \int_0^{2\pi} \mathcal{E}^{\mathfrak{p}}_{\mathrm{Lem}}(\varphi_0(e^{it}))dt + \varepsilon$.

Proof. Let $u_0 := \mathcal{E}^{\mathfrak{p}}_{\mathrm{Lem}}$. Since u_0 is upper semicontinuous, there exists a $v \in \mathcal{C}(G, \mathbb{R})$ with $v \geq u_0$ such that

$$\int_0^{2\pi} v(\varphi_0(e^{it}))dt \leq \int_0^{2\pi} u_0(\varphi_0(e^{it}))dt + \varepsilon/2.$$

For any $\xi_0 \in \mathbb{T}$ there exist $\varphi \in \mathcal{O}(\overline{\mathbb{D}}, G)$, $\lambda_0 \in \mathbb{D}_*$, $\delta > 0$, an open arc $I \subset \mathbb{T}$, and $r > 1$ such that (the reader is asked to complete details)

- $\xi_0 \in I$, $\varphi(0) = \varphi_0(\xi_0)$, $\varphi(\lambda_0) =: a \in |\mathfrak{p}|$,
- $\mathfrak{p}(a) \log|\lambda_0| < v(z) + \varepsilon/8$, $\quad z \in \mathbb{B}(\varphi_0(\xi_0), \delta) \subset G$,
- $\varphi(\lambda) + (1 - \lambda/\lambda_0)(z - \varphi_0(\xi_0)) \in G$, $\quad (\lambda, z) \in \mathbb{D}(r) \times \mathbb{B}(\varphi_0(\xi_0), \delta)$,
- $\varphi_0(\xi) \in \mathbb{B}(\varphi_0(\xi_0), \delta)$, $\quad \xi \in I$,
- $\Phi_0(\mathbb{D}(r) \times I) \subset\subset G$, where $\Phi_0(\lambda, \xi) := \varphi(\lambda) + (1 - \lambda/\lambda_0)(\varphi_0(\xi) - \varphi_0(\xi_0))$.

By a compactness argument we find a covering $\mathbb{T} = \bigcup_{\nu=1}^{N_0} I_\nu$, a new $r > 1$, functions $\Phi_\nu \in \mathcal{C}^\infty(\mathbb{D}(r) \times I_\nu, G)$, $\nu = 1, \dots, N_0$, and points $\lambda_1, \dots, \lambda_{N_0} \in \mathbb{D}_*$ such that

- $\Phi_\nu(\cdot, \xi) \in \mathcal{O}(\overline{\mathbb{D}}, G)$, $\quad \xi \in I_\nu$,
- $\Phi_\nu(0, \xi) = \varphi_0(\xi)$, $\quad \xi \in I_\nu$,
- $\Phi_\nu(\lambda_\nu, \xi) =: a_\nu \in |\mathfrak{p}|$, $\quad \xi \in I_\nu$,
- $\Phi_\nu(\mathbb{D}(r) \times I_\nu) \subset\subset G$,
- $\mathfrak{p}(a_\nu) \log|\lambda_\nu| < v(\varphi_0(\xi)) + \varepsilon/8$, $\quad \xi \in I_\nu$, $\nu = 1, \dots, N_0$.

Replacing Φ_ν by the function $(\lambda, \xi) \longmapsto \Phi_\nu(e^{i\theta_\nu}\lambda, \xi)$ with a suitable $\theta_\nu \approx 0$, we may assume that the points $\lambda_1, \dots, \lambda_{N_0}$ have different arguments.

Fix $1 < s < s_0 < r$ with $2\pi N_0 \max\limits_{\nu=1,\dots,N_0} \mathfrak{p}(a_\nu) \log s_0 < \varepsilon/8$. Let K be the closure of the set

$$\varphi_0(\mathbb{T}) \cup \bigcup_{\nu=1}^{N_0} \Phi_\nu(\mathbb{D}(r) \times I_\nu)$$

and let $C > 0$ be such that

$$C > 2\pi N_0 \max\{\mathfrak{p}(a_\nu)|\log|\lambda_\nu|| : \nu = 1, \dots, N_0\} + \max\{v(z) : z \in K\}.$$

There exist disjoint closed arcs $J_\nu \subset I_\nu$, $\nu \in A \subset \{1, \dots, N_0\}$, such that

$$\mathfrak{m}_{\mathbb{T}}\Big(\mathbb{T} \setminus \bigcup_{\nu \in A} J_\nu\Big) < \frac{\varepsilon}{8C}.$$

We may assume that $A = \{1, \dots, N\}$ for some $N \le N_0$. Fix open disjoint arcs K_ν with $J_\nu \subset K_\nu \subset I_\nu$, $\nu = 1, \dots, N$, and let $\varrho \in \mathcal{C}^\infty(\mathbb{T}, [0, 1])$ be such that $\varrho = 1$ on $\bigcup_{\nu=1}^N J_\nu$ and $\operatorname{supp} \varrho \subset \bigcup_{\nu=1}^N K_\nu$. We define $\Phi : \mathbb{D}(r) \times \mathbb{T} \longrightarrow G$ by the formula

$$\Phi(\lambda, \xi) := \begin{cases} \Phi_\nu(\varrho(\xi)\lambda, \xi), & (\lambda, \xi) \in \mathbb{D}(r) \times K_\nu \\ \varphi_0(\xi), & (\lambda, \xi) \in \mathbb{D}(r) \times (\mathbb{T} \setminus \bigcup_{\nu=1}^N K_\nu) \end{cases}.$$

It is clear that Φ is well-defined, $\Phi \in \mathcal{C}^\infty(\mathbb{D}(r) \times \mathbb{T}, G)$, and Φ satisfies (i).

Let $K_\nu = \{e^{i\theta} : \theta \in (\alpha_\nu, \beta_\nu)\}$, $J_\nu = \{e^{i\theta} : \theta \in [\gamma_\nu, \delta_\nu]\}$ with $\alpha_\nu < \gamma_\nu < \delta_\nu < \beta_\nu$. We may assume that ϱ increases on (α_ν, γ_ν) and decreases on (δ_ν, β_ν).

Then, the set $J'_\nu := \{\xi \in K_\nu : |\lambda_\nu|/\varrho(\xi) \le s\}$ is a closed arc with $J_\nu \subset J'_\nu \subset K_\nu$. Take a $\sigma_\nu \in \mathcal{C}^\infty(\mathbb{T}, \mathbb{R}_{>0}\lambda_\nu)$ with

- $\sigma_\nu(\xi) = \lambda_\nu/\varrho(\xi), \quad \xi \in J'_\nu$,
- $s < |\sigma_\nu(\xi)| < s_0, \quad \xi \in K_\nu \setminus J'_\nu$,
- $|\sigma_\nu(\xi)| = s_0, \quad \xi \in \mathbb{T} \setminus K_\nu$.

Then, (ii), (iii), (iv), and (v) are satisfied. It remains to check (vi). Let $\widetilde{J}_\nu := \{\theta \in [0, 2\pi) : e^{i\theta} \in J_\nu\}$, $\nu = 1, \dots, N$. We have

$$\begin{aligned}
&\sum_{\nu=1}^N \mathfrak{p}(a_\nu) \int_0^{2\pi} \log|\sigma_\nu(e^{it})| dt \\
&\qquad \le \sum_{\nu=1}^N \mathfrak{p}(a_\nu) \int_{\widetilde{J}_\nu} \log|\lambda_\nu| dt + \varepsilon/8 \le \sum_{\nu=1}^N \int_{\widetilde{J}_\nu} v(\varphi_0(e^{it})) dt + \varepsilon/4 \\
&\qquad \le \int_0^{2\pi} v(\varphi_0(e^{it})) dt + \varepsilon/2 \le \int_0^{2\pi} \mathcal{E}^{\mathfrak{p}}_{\mathrm{Lem}}(\varphi_0(e^{it})) dt + \varepsilon. \qquad \Box
\end{aligned}$$

Lemma 17.4.5. *There exists a $j_0 \in \mathbb{N}$ such that for any $j \ge j_0$ there exist $1 < s_j < s$, $\Psi_j \in \mathcal{O}(\mathbb{D}(s_j) \times \mathbb{D}(s_j), G)$, and $\tau_{\nu,j} \in \mathcal{O}(\mathbb{A}(1/s_j, s_j))$, $\nu = 1, \dots, N$, such that*

(i) $\Psi_j(0, \xi) = \varphi_0(\xi), \quad \xi \in \mathbb{D}(s_j)$,

(ii) $|\tau_{\nu,j}| \longrightarrow |\sigma_\nu|$ *uniformly on* $\mathbb{T}$,

(iii) $\Psi_j(\tau_{\nu,j}(\xi), \xi) = a_\nu, \quad \xi \in \mathbb{A}(1/s_j, s_j)$ *with* $|\tau_{\nu,j}(\xi)| < s_j$,

(iv) $|\tau_{\nu,j}(\xi)| < 1, \quad \xi \in J_\nu, \ \nu = 1, \dots, N$.

Proof. Recall that for any $\xi \in \mathbb{T}$ the numbers $0, \sigma_1(\xi), \dots, \sigma_N(\xi)$ are pairwise different. Let $P : \mathbb{C} \times \mathbb{T} \longrightarrow \mathbb{C}$ be defined by the formula

$$P(\lambda, \xi) := \varphi_0(\xi) \prod_{\ell=1}^{N} \frac{\lambda - \sigma_\ell(\xi)}{-\sigma_\ell(\xi)} + \sum_{\mu=1}^{N} \frac{\lambda a_\mu}{\sigma_\mu(\xi)} \prod_{\substack{\ell=1 \\ \ell \neq \mu}}^{N} \frac{\lambda - \sigma_\ell(\xi)}{\sigma_\mu(\xi) - \sigma_\ell(\xi)};$$

observe that $P(\cdot, \xi)$ is the Lagrange interpolation polynomial with $P(0, \xi) = \varphi_0(\xi)$, $P(\sigma_\nu(\xi), \xi) = a_\nu$, $\nu = 1, \dots, N$. We will prove that there exists a function $\Phi_0 \in \mathcal{C}^\infty(\mathbb{D}(s) \times \partial U)$ such that

$$\Phi(\lambda, \xi) = P(\lambda, \xi) + (\lambda - \sigma_1(\xi)) \cdots (\lambda - \sigma_N(\xi)) \Phi_0(\lambda, \xi), \quad (\lambda, \xi) \in \mathbb{D}(s) \times \mathbb{T}.$$

Indeed, the only problem is to check that Φ_0 is $\mathcal{C}^\infty$ near a point $(\sigma_\nu(\xi_0), \xi_0)$ with $|\sigma_\nu(\xi_0)| < s$. Then, $|\sigma_\mu(\xi_0)| > s$ for $\mu \neq \nu$, and there exists a neighborhood V of ξ_0 such that $|\sigma_\nu(\xi)| < s$, $|\sigma_\mu(\xi)| > s$, $\mu \neq \nu$, for $\xi \in V$. Observe that

$$\begin{aligned} \Phi(\lambda, \xi) &= a_\nu + (\lambda - \sigma_\nu(\xi)) \widehat{\Phi}(\lambda, \xi), \\ P(\lambda, \xi) &= a_\nu + (\lambda - \sigma_\nu(\xi)) \widehat{P}(\lambda, \xi), \quad (\lambda, \xi) \in \mathbb{D}(s) \times V, \end{aligned}$$

where $\widehat{\Phi}$ and $\widehat{P}$ are $\mathcal{C}^\infty$ mappings. Hence,

$$\Phi_0(\lambda, \xi) = (\widehat{\Phi}(\lambda, \xi) - \widehat{P}(\lambda, \xi)) \prod_{\mu \neq \nu} \frac{1}{\lambda - \sigma_\mu(\xi)}, \quad (\lambda, \xi) \in \mathbb{D}(s) \times V,$$

and, consequently, $\Phi_0 \in \mathcal{C}^\infty(\mathbb{D}(s) \times V)$. Notice that $\Phi_0(0, \cdot) \equiv 0$.

Let $\Phi_{0,j}$ and $\sigma_{\nu,j}$ be the j-th partial sums of the Fourier series of Φ_0 and σ_ν, respectively, i.e.,

$$\begin{aligned} \Phi_{0,j}(\lambda, \xi) : &= \sum_{k=-j}^{j} \left(\frac{1}{2\pi} \int_0^{2\pi} \Phi_0(\lambda, e^{it}) e^{-ikt} dt \right) \xi^k, \\ \sigma_{\nu,j}(\xi) : &= \sum_{k=-j}^{j} \left(\frac{1}{2\pi} \int_0^{2\pi} \sigma_\nu(e^{it}) e^{-ikt} dt \right) \xi^k, \quad (\lambda, \xi) \in \mathbb{D}(s) \times \mathbb{C}_*; \end{aligned}$$

cf. the proof of Lemma 17.3.6. One can easily show that $\Phi_{0,j} \longrightarrow \Phi_0$ and $\sigma_{\nu,j} \longrightarrow \sigma_\nu$ uniformly on $\mathbb{D}(t) \times \mathbb{T}$ for any $1 < t < s$. Define

$$\begin{aligned} P_j(\lambda, \xi) &:= \varphi_0(\xi) \prod_{\ell=1}^{N} \frac{\lambda - \sigma_{\ell,j}(\xi)}{-\sigma_{\ell,j}(\xi)} + \sum_{\mu=1}^{N} \frac{\lambda a_\mu}{\sigma_{\mu,j}(\xi)} \prod_{\substack{\ell=1 \\ \ell \neq \mu}}^{N} \frac{\lambda - \sigma_{\ell,j}(\xi)}{\sigma_{\mu,j}(\xi) - \sigma_{\ell,j}(\xi)}, \\ \Phi_j(\lambda, \xi) &:= P_j(\lambda, \xi) + (\lambda - \sigma_{1,j}(\xi)) \cdots (\lambda - \sigma_{N,j}(\xi)) \Phi_{0,j}(\lambda, \xi). \end{aligned}$$

Then,

- $\Phi_{0,j} \in \mathcal{O}(\mathbb{D}(s) \times \mathbb{C}_*)$,
- for any $\lambda \in \mathbb{D}(s)$ the function $\Phi_{0,j}(\lambda, \cdot)$ has a pole of order $\le j$ at $\xi = 0$,
- for any $\xi \in \mathbb{C}_*$ the function $\Phi_{0,j}(\cdot, \xi)$ has a zero at $\lambda = 0$,
- $\Phi_j \longrightarrow \Phi$ uniformly on $\mathbb{D}(t) \times \mathbb{T}$, $1 < t < s$,
- Φ_j is holomorphic on $\mathbb{D}(t) \times \mathbb{T}$, $1 < t < s$, $j \gg 1$,
- for any $\lambda \in \mathbb{D}(s)$ the function $\Phi_j(\lambda, \cdot)$ has a pole of order $\le j$ at $\xi = 0$, $j \gg 1$.

Suppose that $j \gg 1$ is such that $\sigma_{\nu,j}(\xi) \neq 0$, $\xi \in \mathbb{T}$. In particular, the set $Z_{\nu,j} := \mathbb{D}_* \cap \sigma_{\nu,j}^{-1}(0)$ is finite. Observe that $\Phi_j \in \mathcal{O}(\mathbb{D}(s) \times (\overline{\mathbb{D}}_* \setminus Z_j))$, where $Z_j := \bigcup_{\nu=1}^{N} Z_{\nu,j}$. Put $B_j := B_{1,j} \cdots B_{N,j}$, where $B_{\nu,j}$ denotes the Blaschke product for $Z_{\nu,j}$ with the zeros counted with multiplicities.[2] For every $\xi \in \mathbb{C}_* \setminus Z_j$ with $|\sigma_{\nu,j}(\xi)| < s$, we get $\Phi_j(\sigma_{\nu,j}(\xi), \xi) = a_\nu$. For any $k \ge j$

- the mapping $\Phi_{j,k}(\lambda, \xi) := \Phi_j(\lambda \xi^k B_j(\xi), \xi)$ is holomorphic on $\overline{\mathbb{D}} \times \overline{\mathbb{D}}$ and
- the mapping $\sigma_{\nu,j,k}(\xi) := \sigma_{\nu,j}(\xi)/(\xi^k B_j(\xi))$ is meromorphic in $\mathbb{C}_*$ and zero-free holomorphic in $\mathbb{D}_*$.

Moreover, $\Phi_{j,k}(\sigma_{\nu,j,k}(\xi), \xi) = a_\nu$ for all $\xi \in \mathbb{D}_*(s)$ such that $|\sigma_{\nu,j,k}(\xi)\xi^k B_j(\xi)| < s$. Using the same method as in the proof of Lemma 17.3.6, we get the required result with

$$\begin{aligned} \Psi_j(\lambda, \xi) &:= \Phi_{j,k_j}(\lambda, \xi) = \Phi_j(\lambda \xi^{k_j} B_j(\xi), \xi), \\ \tau_{\nu,j}(\xi) &:= \sigma_{\nu,j,k_j}(\xi) = \sigma_{\nu,j}(\xi)/(\xi^{k_j} B_j(\xi)), \end{aligned}$$

where $k_j \ge j$ is sufficiently big (and $1 < s_j < s$, $s_j \approx 1$). □

Taking a $j \gg 1$ in Lemma 17.4.5 gives the following result.

Lemma 17.4.6. *There exist* $1 < t < s$, $\Psi \in \mathcal{O}(\mathbb{D}(t) \times \mathbb{D}(t), G)$, $\tau_\nu \in \mathcal{O}(\mathbb{A}(1/t, t), \mathbb{C}_*)$, $\nu = 1, \dots, N$, *such that*

(i) $\Psi(0, \xi) = \varphi_0(\xi)$, $\xi \in \mathbb{D}(t)$,

(ii) $|\tau_\nu(\xi)| < 1$, $\xi \in J_\nu$,

(iii) $\Psi(\tau_\nu(\xi), \xi) = a_\nu$, $\xi \in \mathbb{D}_*(t)$ *with* $|\tau_\nu(\xi)| < t$, $\nu = 1, \dots, N$,

(iv) $2\pi N \max_{\nu=1,\dots,N} \{\mathfrak{p}(a_\nu) \max_{\mathbb{T}} \log |\tau_\nu|\} < \varepsilon/2$,

(v) $\sum_{\nu=1}^{N} \mathfrak{p}(a_\nu) \int_0^{2\pi} \log |\tau_\nu(e^{it})| dt \le \sum_{\nu=1}^{N} \mathfrak{p}(a_\nu) \int_0^{2\pi} \log |\sigma_\nu(e^{it})| dt + \varepsilon$.

[2] That is, the function $\sigma_{\nu,j}/B_{\nu,j}$ extends to a zero-free holomorphic function on $\mathbb{D}_*$.

Lemma 17.4.7. *There exist $\eta_0 \in \mathbb{T}$, k, $c > 0$, and $0 < \varrho < 1$ such that the functions*

$$f(\xi) := \Psi(\eta_0 \xi^k, \xi), \quad F(\lambda, \eta) := \eta \frac{\varrho\lambda + e^{-c/k}}{1 + e^{-c/k}\varrho\lambda}$$

satisfy

$$\int_0^{2\pi} \Xi_{\mathrm{Lel}}^{\mathfrak{p}}(f(F(\cdot, e^{it})))dt \le \sum_{\nu=1}^{N} \mathfrak{p}(a_\nu) \int_0^{2\pi} \log|\tau_\nu(e^{it})|dt + \varepsilon.$$

Proof. Since $\tau_\nu(\xi) \neq 0$, $\xi \in \mathbb{A}(1/t, t)$, there exists $c \gg 1$ such that

$$\log\left|\frac{\eta e^{-c} - \tau_\nu(\xi)}{1 - \overline{\tau_\nu(\xi)}\eta e^{-c}}\right| < \log|\tau_\nu(\xi)| + \varepsilon/(2M), \quad \eta \in \overline{\mathbb{D}},\ \xi \in \mathbb{T},\ \nu = 1, \dots, N, \tag{17.4.1}$$

where $M := \sum_{\nu=1}^{N} \mathfrak{p}(a_\nu)$. Let

$$\psi(\lambda) := \exp\left(c\,\frac{\lambda - 1}{\lambda + 1}\right), \quad \lambda \in \mathbb{C} \setminus \{-1\};$$

observe that $\psi(\mathbb{D}) = \mathbb{D}_*$, $\psi(\mathbb{T} \setminus \{-1\}) = \mathbb{T}$.

Define

$$\varphi_\nu(\lambda; \eta, \xi) := \frac{\eta\psi(\lambda) - \tau_\nu(\xi)}{1 - \overline{\tau_\nu(\xi)}\eta\psi(\lambda)}, \quad (\lambda, \eta, \xi) \in (\mathbb{C} \setminus \{-1\}) \times \mathbb{T} \times J_\nu;$$

we have $|\varphi_\nu(\lambda; \eta, \xi)| = 1$, $(\lambda, \eta, \xi) \in (\mathbb{T} \setminus \{-1\}) \times \mathbb{T} \times J_\nu$. Moreover, $\varphi_\nu(t; \eta, \xi) \longrightarrow -\tau_\nu(\xi)$ when $t \longrightarrow -1^-$. Thus, $\varphi_\nu(\cdot; \eta, \xi)$ is an inner function with non-zero radial limits. Consequently, by (B.9.8), $\varphi_\nu(\cdot; \eta, \xi)$ is a Blaschke product. Moreover, since $\psi'(\lambda) \neq 0$, the zeros of $\varphi_\nu(\cdot; \eta, \xi)$ are simple and, by the implicit mapping theorem, for any point $(\lambda_0, \eta_0, \xi_0)$ with $\varphi_\nu(\lambda_0; \eta_0, \xi_0) = 0$, there exists a holomorphic function $h = h_{\lambda_0,\eta_0,\xi_0}$ defined in a neighborhood V_0 of (η_0, ξ_0) such that $h(\eta_0, \xi_0) = \lambda_0$ and $\varphi_\nu(h(\eta, \xi); \eta, \xi) = 0$, $(\eta, \xi) \in V_0$. Observe that

$$h(\eta, \xi) = \frac{1}{2\pi i} \int_{\partial\mathbb{D}(\lambda_0, r)} \frac{\lambda\eta\psi'(\lambda)}{\eta\psi(\lambda) - \tau_\nu(\xi)} d\lambda,$$

where $\mathbb{D}(\lambda_0, r)$ is so small that $\lambda = \lambda_0$ is the only zero of $\varphi_\nu(\cdot; \eta_0, \xi_0)$ in $\overline{\mathbb{D}}(\lambda_0, r)$. Let $(\lambda_{\nu,\ell})_{\ell=1}^{\infty}$ be the zeros of $\varphi_\nu(\cdot; \eta_0, \xi_0)$ in $\mathbb{D}_*$. Since $\varphi_\nu(\cdot; \eta_0, \xi_0)$ is a Blaschke product, we get

$$|\varphi_\nu(0; \eta_0, \xi_0)| = \left|\frac{\eta e^{-c} - \tau_\nu(\xi)}{1 - \overline{\tau_\nu(\xi)}\eta e^{-c}}\right| = \prod_{\ell=1}^{\infty} |\lambda_{\nu,\ell}|.$$

Hence, using (17.4.1), we conclude that there exist $L \in \mathbb{N}$ and $\varrho > 1$ such that

$$\sum_{\ell=1}^{L} \log(|\lambda_{\nu,\ell}|/\varrho) < \log|\tau_\nu(\xi_0)| + \varepsilon/(2M).$$

Consequently,

$$\sum_{\ell=1}^{L} \log(|h_{\lambda_{\nu,\ell},\eta_0,\xi_0}(\eta,\xi)|/\varrho) < \log|\tau_\nu(\xi)| + \varepsilon/(2M)$$

for (η, ξ) in a neighborhood of (η_0, ξ_0).

Using a compactness argument, we see that there exist $L \in \mathbb{N}$ and $\varrho > 1$ such that for any point $(\eta, \xi) \in \mathbb{T} \times J_\nu$ there exist $\lambda_{\nu,1}(\eta,\xi), \dots, \lambda_{\nu,L}(\eta,\xi)$ such that

$$\varphi_\nu(\lambda_{\nu,\ell}(\eta,\xi); \eta, \xi) = 0, \quad \ell = 1, \dots, L,$$

and

$$\sum_{\ell=1}^{L} \log(|\lambda_{\nu,\ell}(\eta,\xi)|/\varrho) < \log|\tau_\nu(\xi)| + \varepsilon/(2M).$$

Let

$$\psi_k(\lambda) := \frac{\lambda + e^{-c/k}}{1 + e^{-c/k}\lambda} = 1 + (1 - e^{-c/k})\frac{\lambda - 1}{1 + e^{-c/k}\lambda}, \quad \lambda \in \mathbb{C} \setminus \{-e^{c/k}\}.$$

Observe that $\psi_k \longrightarrow 1$ locally uniformly in $\mathbb{D}$ and

$$\lim_{k\to+\infty} k \operatorname{Log} \psi_k(\lambda) = \lim_{k\to+\infty} k(1 - e^{-c/k})\frac{\lambda - 1}{1 + e^{-c/k}\lambda} = c\frac{\lambda - 1}{\lambda + 1}$$

locally uniformly in $\mathbb{D}$. Consequently, $\psi_k^k \longrightarrow \psi$ locally uniformly in $\mathbb{D}$.

Fix a $1 < t_0 < 1/\varrho$ and let V_ν be a neighborhood of J_ν such that $|\tau_\nu(\xi)| < 1$, $\xi \in V_\nu$. Let $k_0 \in \mathbb{N}$ be such that $\xi\psi_k(\varrho\lambda) \in V_\nu$, $(\lambda, \xi) \in \mathbb{D}(t_0) \times J_\nu$, $k \ge k_0$. Hence, by (iii) of Lemma 17.4.6, we get

$$\Psi(\tau_\nu(\xi\psi_k(\varrho\lambda)), \xi\psi_k(\varrho\lambda)) = a_\nu, \quad (\lambda, \xi) \in \mathbb{D}(t_0) \times J_\nu,\ k \ge k_0.$$

Recall that

$$\eta\psi_k^k(\varrho\lambda) - \tau_\nu(\xi\psi_k(\varrho\lambda)) \longrightarrow \eta\psi(\varrho\lambda) - \tau_\nu(\xi)$$

uniformly with respect to $(\lambda, \eta, \xi) \in \mathbb{D}(t_0) \times \mathbb{T} \times J_\nu$. Hence, by the Hurwitz theorem, for $k \gg 1$, there are zeros $\lambda_{\nu,\ell,k}(\eta,\xi)$ of the function $\lambda \longmapsto \eta\psi_k^k(\varrho\lambda) - \tau_\nu(\xi\psi_k(\varrho\lambda))$, which are so close to $\lambda_{\nu,\ell}(\eta,\xi)$ that

$$\sum_{\ell=1}^{L} \log|\lambda_{\nu,\ell,k}(\eta,\xi)| < \log|\tau_\nu(\xi)| + \varepsilon/(2M), \quad (\eta,\xi) \in \mathbb{T} \times J_\nu.$$

Observe that

$$\Psi(\eta\psi_k^k(\varrho\lambda_{\nu,\ell,k}(\eta,\xi)), \xi\psi_k(\varrho\lambda_{\nu,\ell,k}(\eta,\xi))) = a_\nu, \quad (\eta,\xi) \in \mathbb{T} \times J_\nu,\ k \gg 1.$$

Hence,

$$\Xi_{\text{Lel}}^{\mathfrak{p}}(\lambda \longmapsto \Psi(\eta\psi_k^k(\varrho\lambda), \xi\psi_k(\varrho\lambda))) < \sum_{\nu=1}^{N} \mathfrak{p}(a_\nu)\log|\tau_\nu(\xi)| + \varepsilon/2,$$
$$(\eta,\xi) \in Q := \bigcup_{\nu=1}^{N}(\mathbb{T} \times J_\nu).$$

Consider the diffeomorphism $H : \mathbb{T}^2 \longrightarrow \mathbb{T}^2$ given by $H(\eta,\xi) := (\eta\xi^{-k}, \xi)$. Let $S := H(Q)$. Then, $\mathfrak{m}_{\mathbb{T}}^2(S) = \mathfrak{m}_{\mathbb{T}}^2(Q) \geq 1 - \varepsilon$ (because the modulus of the Jacobian of H is equal to 1). Consequently, there exists an $\eta_0 \in \mathbb{T}$ such that $\mathfrak{m}_{\mathbb{T}}(R) \geq 1 - \varepsilon$, where $R := \{\xi \in \mathbb{T} : (\eta_0, \xi) \in S\}$. We have

$$\Xi_{\text{Lel}}^{\mathfrak{p}}(\lambda \longmapsto \Psi(\eta_0(\xi\psi_k(\varrho\lambda))^k, \xi\psi_k(\varrho\lambda))) \leq \sum_{\nu=1}^{N} \mathfrak{p}(a_\nu)\log|\tau_\nu(\xi)| + \varepsilon/2, \quad \xi \in R.$$

Finally, by Lemma 17.4.6, we conclude that

$$\int_0^{2\pi} \Xi_{\text{Lel}}^{\mathfrak{p}}(\lambda \longmapsto \Psi(\eta_0(e^{it}\psi_k(\varrho\lambda))^k, e^{it}\psi_k(\varrho\lambda)))dt$$
$$\leq \sum_{\nu=1}^{N} \mathfrak{p}(a_\nu) \int_0^{2\pi} \log|\tau_\nu(e^{it})|dt + \varepsilon,$$

which directly implies the required result. □

Lemma 17.4.8. *There exists a $\theta_0 \in \mathbb{R}$ such that the mapping*

$$\varphi(\xi) := f(F(e^{i\theta_0}\xi, \xi))$$

satisfies

$$\Xi_{\text{Lel}}^{\mathfrak{p}}(\varphi) \leq \frac{1}{2\pi}\int_0^{2\pi} \Xi_{\text{Lel}}^{\mathfrak{p}}(f(F(\cdot, e^{it})))dt. \tag{17.4.2}$$

Proof. First we will prove that for any $\varphi \in \mathcal{O}(\overline{\mathbb{D}}, G)$ we have

$$\Xi_{\text{Lel}}^{\mathfrak{p}}(\varphi) = \int_{\mathbb{D}} (\log|\lambda|)\Delta v_\varphi(\lambda) d\mathcal{L}^2(\lambda), \tag{17.4.3}$$

where

$$v_\varphi(\lambda) := \frac{1}{2\pi} \sum_{b \in B_\varphi} \mathfrak{p}(\varphi(b)) \operatorname{ord}_b(\varphi - \varphi(b)) \log \boldsymbol{m}_{\mathbb{D}}(b,\lambda), \quad \lambda \in \mathbb{D}(r),$$

$$B_\varphi := \{b \in \mathbb{D}_* : \mathfrak{p}(\varphi(b)) > 0\} = \mathbb{D}_* \cap \varphi^{-1}(|\mathfrak{p}|)$$

(for some $r > 1$). Observe that B_φ is discrete and $v_\varphi \in \mathcal{SH}(\mathbb{D}(r))$. To prove (17.4.3), we use the Riesz representation formula (cf. Appendix B.5.10):

$$\begin{aligned}\int_{\mathbb{D}} (\log|\lambda|) \Delta v_\varphi(\lambda) d\mathcal{L}^2(\lambda) &= 2\pi v_\varphi(0) - \int_0^{2\pi} v_\varphi(e^{i\theta}) d\theta \\ &= \sum_{b \in B_\varphi} \mathfrak{p}(\varphi(b)) \operatorname{ord}_b(\varphi - \varphi(b)) \log|b| = \Xi^{\mathfrak{p}}_{\mathrm{Lel}}(\varphi).\end{aligned}$$

Next, we are going to show that for any function $h \in \mathcal{O}(\overline{\mathbb{D}})$ with $h(\overline{\mathbb{D}}) \subset \overline{\mathbb{D}}$ we have

$$\Delta v_{\varphi \circ h} = \Delta(v_\varphi \circ h) \text{ in } \mathbb{D}, \tag{17.4.4}$$

with $\Delta_{-\infty} := 0$. If $\varphi \equiv$ const or $h \equiv$ const or $h(\mathbb{D}) \cap B_\varphi = \varnothing$, then (17.4.4) is obviously true. Assume that $\varphi \not\equiv$ const and $h \not\equiv$ const and $h(\mathbb{D}) \cap B_\varphi \neq \varnothing$. It is clear that $v_{\varphi\circ h}$ and $v_\varphi \circ h$ are harmonic on $\mathbb{D} \setminus h^{-1}(B_\varphi)$ and, consequently, $\Delta v_{\varphi\circ h} = \Delta(v_\varphi \circ h) = 0$ on $\mathbb{D} \setminus h^{-1}(B_\varphi)$.

Take $b \in B_\varphi$ and $c \in h^{-1}(b)$. Write $h(\lambda) = (\lambda - c)^m g(\lambda)$, where $g \in \mathcal{O}(\overline{\mathbb{D}})$ and $g(c) \neq 0$. Then,

$$v_\varphi \circ h(\lambda) = \frac{1}{2\pi} \mathfrak{p}(\varphi(b)) \operatorname{ord}_b(\varphi - \varphi(b)) m \log|\lambda - c| + u(\lambda),$$

where u is harmonic near c. Thus,

$$\begin{aligned}\Delta(v_\varphi \circ h) &= \mathfrak{p}(\varphi(b)) \operatorname{ord}_b(\varphi - \varphi(b)) m \delta_c \\ &= \mathfrak{p}(\varphi(h(c))) \operatorname{ord}_c(\varphi \circ h - \varphi \circ h(c)) \delta_c = \Delta v_{\varphi\circ h}\end{aligned}$$

in a neighborhood of c.

Applying (17.4.4) to $\varphi := f$ and $h := F(\cdot, \xi)$, we get

$$\begin{aligned}\Xi^{\mathfrak{p}}_{\mathrm{Lel}}(f(F(\cdot,\xi))) &= \int_{\mathbb{D}} (\log|\lambda|) \Delta v_{f\circ F(\cdot,\xi)}(\lambda) d\mathcal{L}^2(\lambda) \\ &= \int_{\mathbb{D}} (\log|\lambda|) \Delta_\lambda (v_f \circ F(\lambda,\xi)) d\mathcal{L}^2(\lambda).\end{aligned} \tag{17.4.5}$$

Now we need the following auxiliary result:

Lemma 17.4.9. *Let $w \in \mathcal{PSH}(\mathbb{D}(r) \times \mathbb{D}(r))$ $(r > 1)$ and let $w_\theta(\xi) := w(e^{i\theta}\xi, \xi)$. Then, there exists a $\theta_0 \in \mathbb{R}$ such that*

$$\int_{\mathbb{D}} (\log|\lambda|) \Delta w_{\theta_0}(\lambda) d\mathcal{L}^2(\lambda) \le \frac{1}{2\pi} \int_0^{2\pi} \left(\int_{\mathbb{D}} (\log|\lambda|) \Delta_\lambda w(\lambda, e^{i\theta}) d\mathcal{L}^2(\lambda) \right) d\theta.$$

Proof. The Riesz representation formula gives

$$w(0,0) = w_\theta(0) = \frac{1}{2\pi} \int_{\mathbb{D}} (\log|\lambda|) \Delta w_\theta(\lambda) d\mathcal{L}^2(\lambda) + \frac{1}{2\pi} \int_0^{2\pi} w(e^{i(\theta+t)}, e^{it}) dt.$$

Hence,

$$\begin{aligned} 2\pi w(0,0) &= \frac{1}{2\pi} \int_0^{2\pi} \left(\int_{\mathbb{D}} (\log|\lambda|) \Delta w_\theta(\lambda) d\mathcal{L}^2(\lambda) \right) d\theta \\ &\quad + \frac{1}{2\pi} \int_0^{2\pi} \left(\int_0^{2\pi} w(e^{i(\theta+t)}, e^{it}) dt \right) d\theta. \end{aligned}$$

On the other hand, using the Riesz representation formula for the functions $w(0,\cdot)$ and $w(\cdot, e^{i\theta})$, we get

$$\begin{aligned} 2\pi w(0,0) &= \int_{\mathbb{D}} (\log|\lambda|) \Delta_\lambda w(0,\lambda) d\mathcal{L}^2(\lambda) \\ &\quad + \int_0^{2\pi} w(0, e^{i\theta}) d\theta \le \int_0^{2\pi} w(0, e^{i\theta}) d\theta \\ &= \int_0^{2\pi} \left(\frac{1}{2\pi} \int_{\mathbb{D}} (\log|\lambda|) \Delta_\lambda w(\lambda, e^{i\theta}) d\mathcal{L}^2(\lambda) \right. \\ &\quad \left. + \frac{1}{2\pi} \int_0^{2\pi} w(e^{it}, e^{i\theta}) dt \right) d\theta. \end{aligned}$$

Consequently,

$$\int_0^{2\pi} \left(\int_{\mathbb{D}} (\log|\lambda|) \Delta w_\theta(\lambda) d\mathcal{L}^2(\lambda) \right) d\theta \le \int_0^{2\pi} \left(\int_{\mathbb{D}} (\log|\lambda|) \Delta_\lambda w(0,\lambda) d\mathcal{L}^2(\lambda) \right) d\theta,$$

which implies the required result. □

Applying Lemma 17.4.9 to (17.4.5) gives (17.4.2). □

This finishes the proof of Theorem 17.4.3. □

17.5 Exercises

Exercise 17.5.1 (cf. [270], Proposition 3.3.1). Let $q : \mathbb{C}^n \longrightarrow \mathbb{R}_+$ be a complex norm. Put $B(r) := \{z \in \mathbb{C}^n : q(z) < r\}$. Prove that for $0 < r < R$ we have

$$\boldsymbol{h}_{B(r),B(R)}(z) = \max\left\{0, \frac{\log \frac{q(z)}{r}}{\log \frac{R}{r}}\right\}, \quad z \in B(R).$$

Hint. The inequality "$\geq$" is obvious. Moreover, both sides vanish on $B(r)$. Thus, by the Oka theorem (cf. Appendix B.4.26), they vanish on $\overline{B}(r)$. Let Φ stand for the right hand side. Take an $a \in B(R) \setminus \overline{B}(r)$ and let

$$A := \mathbb{A}(r/q(a), R/q(a)) \ni \lambda \overset{v}{\longmapsto} \boldsymbol{h}_{B(r),B(R)}(\lambda a) - \Phi(\lambda a).$$

Observe that v is subharmonic and $v \geq 0$. Moreover, $\limsup_{\lambda \to \partial A} v(\lambda) \leq 0$. Thus, by the maximum principle (cf. Appendix B.4.6), $v \leq 0$. In particular, $v(1) = \boldsymbol{h}_{B(r),B(R)}(a) - \Phi(a) \leq 0$.

Chapter 18

Product property

Summary. The aim of this chapter is to discuss the product properties (18.1.1) and (18.1.2) for all holomorphically contractible families of functions and pseudometrics that have been defined so far.

Introduction. The product property we are going to discuss in this chapter may be intuitively formulated as follows: Suppose we are given a contractible family of functions $\underline{d}$ (resp. a contractible family of pseudometrics $\underline{\delta}$). We are interested in characterizing those domains G_1, G_2 for which $d_{G_1\times G_2}$ (resp. $\delta_{G_1\times G_2}$) is equal to the maximum of the corresponding functions (resp. pseudometrics) on G_1 and G_2.

18.1 Product property – general theory

Definition 18.1.1. Let $\underline{d} = (d_G)_{G\in\mathfrak{G}_0}$ be a contractible (or an $\boldsymbol{m}$-contractible) family of functions. We say that $\underline{d}$ has the *product property on* $G_1 \times G_2$ *at* $(z_1', z_2'), (z_1'', z_2'')$ $(G_j \in \mathfrak{G}_0,\ z_j', z_j'' \in G_j,\ j = 1, 2)$ if

$$d_{G_1\times G_2}((z_1', z_2'), (z_1'', z_2'')) = \max\{d_{G_1}(z_1', z_1''), d_{G_2}(z_2', z_2'')\}. \tag{18.1.1}$$

If (18.1.1) is satisfied for all $z_j', z_j'' \in G_j$, $j = 1, 2$, then we say that $\underline{d}$ has the *product property on* $G_1 \times G_2$. If $\underline{d}$ has the product property on $G_1 \times G_2$ for all $G_1, G_2 \in \mathfrak{G}_0$, then we simply say that $\underline{d}$ has the *product property*.

We already know that $\boldsymbol{\ell}^{(m)}$, $\boldsymbol{\ell}^{(m)*}$, $\boldsymbol{k}$, and $\boldsymbol{k}^*$ have the product property (cf. Proposition 3.8.7 and Theorem 3.7.2) and that $\boldsymbol{c}^{(*)}$ has the product property on $G \times \mathbb{D}$ for arbitrary G (cf. Example 2.7.10(2°)). Moreover, we know that $\boldsymbol{h}^*$ does not have the product property (cf. Remark 8.1.2(k)).

Notice that the product properties for $\underline{d}$ and $\underline{d}^*$ (cf. Remark 4.1.3) are equivalent, and therefore, to verify the product property we can always pass from $\underline{d}$ to $\underline{d}^*$, or vice-versa.

Observe that the product properties defined above are different from the "product properties" discussed for the Bergman kernel (cf. Theorem 12.1.19) and the Bergman pseudometric (cf. Remark 12.7.4).

Definition 18.1.2. In the case where $\underline{\delta} = (\delta_G)_{G\in\mathfrak{G}_0}$ is a contractible family of pseudometrics, the *product property of* $\underline{\delta}$ *on* $G_1 \times G_2$ *at* (z_1^0, z_2^0), (X_1^0, X_2^0) ($G_j \in \mathfrak{G}_0, G_j \subset \mathbb{C}^{n_j}$, $z_j^0 \in G_j$, $X_j^0 \in \mathbb{C}^{n_j}$, $j = 1, 2$) has the form

$$\delta_{G_1\times G_2}((z_1^0, z_2^0); (X_1^0, X_2^0)) = \max\{\delta_{G_1}(z_1^0; X_1^0), \delta_{G_2}(z_2^0; X_2^0)\}. \qquad (18.1.2)$$

Similarly as above, we introduce the notion of the *product property of* $\underline{\delta}$ on $G_1 \times G_2$, etc.

Recall that $\hat{\boldsymbol{\varkappa}}$ and $\boldsymbol{\varkappa}^{(m)}$ have the product property; cf. Propositions 3.7.1 and 3.8.7. Moreover, we know that $\boldsymbol{\eta}$ does not have the product property (cf. Remark 8.1.2(k)).

Note that the product properties may also be studied for more general objects, e.g., for $\mathcal{D}\underline{d}$; see Proposition 18.1.5(b).

It is clear that the product properties are very useful in all cases where we want to determine $d_{G_1\times G_2}$ (or $\delta_{G_1\times G_2}$) given formulas for d_{G_1} and d_{G_2} (resp. for δ_{G_1} and δ_{G_2}) separately. This problem frequently appears when we construct examples.

For some special families $\underline{d}$ or $\underline{\delta}$ (like $\boldsymbol{m}^{(k)}$ or $\boldsymbol{\gamma}^{(k)}$) one can also study more general "product properties"; see Exercise 18.8.2.

We would like to point out right now that, in general, for given G_1 and G_2 there exists a large set of (z_1', z_2'), (z_1'', z_2'') (resp. (z_1^0, z_2^0), (X_1^0, X_2^0)) at which the product property is fulfilled for *any* $\underline{d}$ (resp. $\underline{\delta}$); see Remark 18.1.3(b, c). Nevertheless, the fact that for given G_1, G_2, and $\underline{d}$ (or $\underline{\delta}$) the product property holds for $\underline{d}$ (or $\underline{\delta}$) on the *whole* $G_1 \times G_2$ is rather exceptional; see Examples 18.1.4 and 18.2.5.

Remark 18.1.3.

(a) Since the natural projections $G_1 \times G_2 \longrightarrow G_j$, $j = 1, 2$, are contractions, the inequalities "$\geq$" in (18.1.1) and (18.1.2) are always true.

(b) Suppose that for fixed $(z_1', z_2'), (z_1'', z_2'') \in G_1 \times G_2$ there exists a holomorphic mapping $F : G_1 \longrightarrow G_2$ with $F(z_1') = z_2'$ and $F(z_1'') = z_2''$. Then, using the map $G_1 \ni z_1 \longmapsto (z_1, F(z_1)) \in G_1 \times G_2$, we easily conclude that in (18.1.1) the inequality "$\leq$" is also satisfied. Consequently, in this case, the product property holds for *any* $\underline{d}$ at $(z_1', z_2'), (z_1'', z_2'') \in G_1 \times G_2$.

It is clear that such a mapping F exists if $z_2' = z_2''$. In particular, (18.1.1) is fulfilled if $z_1' = z_1''$ or $z_2' = z_2''$.

If G_2 is taut, then one can easily construct such a mapping F for all points for which we have $\boldsymbol{\ell}_{G_2}(z_2', z_2'') \leq \boldsymbol{c}_{G_1}(z_1', z_1'')$; cf. Exercise 9.2.2. For let $f \in \mathcal{O}(G_1, \mathbb{D})$, $\varphi \in \mathcal{O}(\mathbb{D}, G_2)$ be such that $f(z_1') = 0$, $f(z_1'') = \sigma_1 := \boldsymbol{c}^*_{G_1}(z_1', z_1'')$ and $\varphi(0) = z_2'$, $\varphi(\sigma_2) = z_2''$, where $\sigma_2 := \boldsymbol{\ell}^*_{G_2}(z_2', z_2'')$ (cf. Proposition 3.2.7). Then, we can take $F := \varphi(\frac{\sigma_2}{\sigma_1} f)$.

Observe that for an arbitrary domain G_2, if $\boldsymbol{\ell}_{G_2}(z_2', z_2'') < \boldsymbol{c}_{G_1}(z_1', z_1'')$, then the mapping F may also be constructed. In fact, we take an extremal f as above and a $\varphi \in \mathcal{O}(\mathbb{D}, G_2)$ such that $\varphi(0) = z_2'$ and $\varphi_2(\sigma_2) = z_2''$ with $\sigma_2 \le \sigma_1$. Then, we define F as above.

(c) The discussion in (b) can be easily repeated in the case of pseudometrics. If $F : G_1 \longrightarrow G_2$ is holomorphic and $F(z_1^0) = z_2^0$, $F'(z_1^0)X_1^0 = X_2^0$, then (18.1.2) is fulfilled for *any* $\underline{\delta}$. In particular, the product property is satisfied if $X_1^0 = 0$ or $X_2^0 = 0$. If G_2 is taut, then such an F exists in all cases where $\varkappa_{G_2}(z_2^0; X_2^0) \le \gamma_{G_1}(z_1^0; X_1^0)$. Moreover, such an F exists in all cases where $\varkappa_{G_2}(z_2^0; X_2^0) < \gamma_{G_1}(z_1^0; X_1^0)$ (with an arbitrary domain G_2).

(d) If $d_{G_1 \times G_2}$ is a pseudodistance, then (a) implies

$$\begin{aligned}\max\{d_{G_1}(z_1', z_1''), d_{G_2}(z_2', z_2'')\} &\le d_{G_1\times G_2}((z_1', z_2'), (z_1'', z_2''))\\ &\le d_{G_1}(z_1', z_1'') + d_{G_2}(z_2', z_2'').\end{aligned}$$

Similarly, if $\delta_{G_1\times G_2}((z_1^0, z_2^0); \cdot)$ is a seminorm, then in view of (a) we get

$$\begin{aligned}\max\{\delta_{G_1}(z_1^0; X_1^0), \delta_{G_2}(z_2^0; X_2^0)\} &\le \delta_{G_1\times G_2}((z_1^0, z_2^0); (X_1^0, X_2^0))\\ &\le \delta_{G_1}(z_1^0; X_1^0) + \delta_{G_2}(z_2^0; X_2^0).\end{aligned}$$

(e) If $\boldsymbol{c}_{G_j} \equiv \boldsymbol{\ell}_{G_j}$ and $\boldsymbol{\gamma}_{G_j} \equiv \varkappa_{G_j}$, $j = 1, 2$, then obviously the product property is fulfilled on the whole $G_1 \times G_2$ for *any* $\underline{d}$ and $\underline{\delta}$.

Observe that the above condition is satisfied if $G_j = \bigcup_{i\in I_j} G_{j,i}$, where $G_{j,i}$ is a subdomain of G_j with $\boldsymbol{c}_{G_{j,i}} \equiv \boldsymbol{\ell}_{G_{j,i}}$, $\boldsymbol{\gamma}_{G_{j,i}} \equiv \varkappa_{G_{j,i}}$, and for any compact $K \subset\subset G_j$ there exists an $i_0 \in I_j$ with $K \subset G_{j,i_0}$ (cf. Exercise 4.4.1).

Notice that the above situation holds if each $G_{j,i}$ is biholomorphic to

- a convex domain – Lempert Theorem 11.2.1, or even to
- a strongly linearly convex domain – Theorem A.5.5.

Example 18.1.4. If $\boldsymbol{c}_{G_0} \not\equiv \boldsymbol{\ell}_{G_0}$ (e.g., $G_0 := P$ = the annulus), then there exists a contractible family of functions $\underline{d}$ such that $\underline{d}$ does not have the product property on $G_0 \times \mathbb{D}$; cf. Remark 18.1.3(e). We will now prove a slightly more general result.

Let $\underline{d}^{(0)} = (d_G^{(0)})_{G\in\mathfrak{G}_0}$ and $\underline{d}^{(1)} = (d_G^{(1)})_{G\in\mathfrak{G}_0}$ be contractible families of functions. Put

$$d_G^{(t)} := (1-t)d_G^{(0)} + t d_G^{(1)}, \quad \underline{d}^{(t)} := (d_G^{(t)})_{G\in\mathfrak{G}_0}, \quad 0 < t < 1.$$

It is clear that $\underline{d}^{(t)}$ is also a contractible family. Suppose that $d_{G_0}^{(0)} \not\equiv d_{G_0}^{(1)}$ for some $G_0 \in \mathfrak{G}_0$. Then, $\underline{d}^{(t)}$ does not have the product property on $G_0\times\mathbb{D}$ for any $0 < t < 1$.

For let $z_0', z_0'' \in G_0$ be such that $M_0 := d_{G_0}^{(0)}(z_0', z_0'') \ne d_{G_0}^{(1)}(z_0', z_0'') =: M_1$. Assume that $M_0 < M_1$. Fix $0 < t < 1$ and let $\sigma(t) \in (0, 1)$ be such that $p(0, \sigma(t)) \in (M_0, M_1)$ and

$$t < \frac{p(0, \sigma(t)) - M_0}{M_1 - M_0}.$$

Then, using Remark 18.1.3(a), we get

$$d^{(t)}_{G_0\times\mathbb{D}}((z_0',0),(z_0'',\sigma(t))) > \boldsymbol{p}(0,\sigma(t)) = \max\{d^{(t)}_{G_0}(z_0',z_0''), d^{(t)}_{\mathbb{D}}(0,\sigma(t))\}.$$

Notice that similar examples can be easily constructed (using the same idea) on the level of $\boldsymbol{m}$-contractible families of functions, ($\boldsymbol{m}$-) contractible families of pseudodistances, and contractible families of pseudometrics; the details are left to the reader. Note that one could start the above construction with the annulus $G_0 = P$ and

$$\underline{d}^{(0)},\ \underline{d}^{(1)} \in \{\boldsymbol{m}^{(k)}, \boldsymbol{g}, \boldsymbol{\ell}^*\},\quad \underline{d}^{(0)} \neq \underline{d}^{(1)},$$
$$\underline{\delta}^{(0)},\ \underline{\delta}^{(1)} \in \{\boldsymbol{\gamma}^{(k)}, \boldsymbol{A}, \boldsymbol{\varkappa}\},\quad \underline{\delta}^{(0)} \neq \underline{\delta}^{(1)}.$$

Now, we would like to select basic operations under which the product properties are invariant.

Proposition 18.1.5. *Let $\underline{d} = (d_G)_{G\in\mathfrak{G}_0}$ be a contractible family of pseudodistances such that $\underline{d}$ has the product property on $G_1 \times G_2$ ($G_1, G_2 \in \mathfrak{G}_0$). Then (cf. § 4.3),*

(a) *$\underline{d}^i$ has the product property on $G_1 \times G_2$;*

(b) *$\mathfrak{D}\underline{d}$ has the product property on $G_1 \times G_2$.*

Let $\underline{\delta} = (\delta_G)_{G\in\mathfrak{G}_0}$ be a contractible family of pseudometrics such that $\underline{\delta}$ has the product property on $G_1 \times G_2$. Then,

(c) *$\widehat{\underline{\delta}}$ has the product property on $G_1 \times G_2$;*

(d) *$\int\underline{\delta}$ has the product property on $G_1 \times G_2$ provided that δ_{G_1} and δ_{G_2} are upper semicontinuous;*

(e) *$\widetilde{\underline{\delta}}$ has the product property on $G_1 \times G_2$.*

In particular,

[*the product property for* $\boldsymbol{\varkappa}^{(k)}$] $\Longrightarrow$ [*the product property for* $\widetilde{\boldsymbol{\varkappa}}^{(k)}$, $\widehat{\boldsymbol{\varkappa}}$, *and* $\boldsymbol{k}$],

[*the product property for* $\boldsymbol{c}$] $\Longrightarrow$ [*the product property for* $\boldsymbol{c}^i$ *and* $\boldsymbol{\gamma}$].

Proof. (a) Let us first make a remark on the L_{d_G}-length. Fix a $G \in \mathfrak{G}_0$ and let $\alpha : [0,1] \longrightarrow G$ be an arbitrary continuous curve with $\ell := L_{d_G}(\alpha) < +\infty$ (cf. § 4.3). Put

$$\psi_0(u) := L_{d_G}(\alpha|_{[0,u]}),\quad 0 \le u \le 1.$$

Then, $\psi_0 : [0,1] \longrightarrow [0,\ell]$ is a continuous and increasing function (which, in general, need not be strictly increasing). Define $\psi_\varepsilon(u) := \psi_0(u) + \varepsilon u$, $0 \le u \le 1$, $\varepsilon > 0$. It is clear that $\psi_\varepsilon : [0,1] \longrightarrow [0,\ell+\varepsilon]$ is a strictly increasing bijection. Define $\alpha_\varepsilon(t) := \alpha(\psi_\varepsilon^{-1}(t\cdot(\ell+\varepsilon)))$, $0 \le t \le 1$, $\varepsilon > 0$. Then,

$$L_{d_G}(\alpha_\varepsilon|_{[t',t'']}) \le (\ell+\varepsilon)(t''-t'),\quad 0 \le t' < t'' \le 1.$$

Now, we go to the main proof of (a). Fix $z_j', z_j'' \in G_j$, $j = 1, 2$, and $\varepsilon > 0$. Let $\alpha_j : [0, 1] \longrightarrow G_j$ be a $\|\ \|$-rectifiable curve such that $\alpha_j(0) = z_j'$, $\alpha_j(1) = z_j''$, and

$$L_{d_{G_j}}(\alpha_j) - d^i_{G_j}(z_j', z_j'') \le \varepsilon, \quad j = 1, 2.$$

In view of the above remark, we can assume that

$$L_{d_{G_j}}(\alpha_j|_{[t',t'']}) \le (\ell_j + \varepsilon)(t'' - t'), \quad 0 \le t' < t'' \le 1,$$

where $\ell_j := L_{d_{G_j}}(\alpha_j)$, $j = 1, 2$. Put $\widetilde{\alpha}(t) := (\alpha_1(t), \alpha_2(t))$, $0 \le t \le 1$. Obviously, $\widetilde{\alpha} : [0, 1] \longrightarrow G_1 \times G_2$ is $\|\ \|$-rectifiable. Moreover, for $0 = t_0 < \cdots < t_N = 1$ we have

$$\begin{aligned}\sum_{j=1}^{N} d_{G_1\times G_2}(\widetilde{\alpha}(t_{j-1}), \widetilde{\alpha}(t_j)) &= \sum_{j=1}^{N} \max\{d_{G_1}(\alpha_1(t_{j-1}), \alpha_1(t_j)), d_{G_2}(\alpha_2(t_{j-1}), \alpha_2(t_j))\} \\ &\le \sum_{j=1}^{N} \max\{(\ell_1 + \varepsilon)(t_j - t_{j-1}), (\ell_2 + \varepsilon)(t_j - t_{j-1})\} \\ &\le \max\{\ell_1, \ell_2\} + \varepsilon \le \max\{d^i_{G_1}(z_1', z_1''), d^i_{G_2}(z_2', z_2'')\} + 2\varepsilon.\end{aligned}$$

Consequently,

$$d^i_{G_1\times G_2}((z_1', z_2'), (z_1'', z_2'')) \le L_{d_{G_1\times G_2}}(\widetilde{\alpha}) \le \max\{d^i_{G_1}(z_1', z_1''), d^i_{G_2}(z_2', z_2'')\} + 2\varepsilon.$$

Letting $\varepsilon \longrightarrow 0$ we get the required result.

(b) Use the definition of $\mathfrak{D}\underline{d}$.

(c) Use standard functional analysis argument.

(d) Use the same method as in the proof of Theorem 3.7.2.

(e) Use Remark 2.2.7. □

Remark 18.1.6. By Lemma 4.2.3 and Proposition 4.2.20(b) we get

(a) if $\boldsymbol{g}$ has the product property on $G_1 \times G_2$, then so has $\boldsymbol{A}$;

(b) if $\boldsymbol{m}^{(k)}$ has the product property on $G_1 \times G_2$, then so has $\boldsymbol{\gamma}^{(k)}$.

Now, after all the above general remarks, we would like to discuss more concrete situations. Namely, we are going to prove that

- the Carathéodory pseudodistance has the product property – Theorem 18.2.1;
- the higher-order Möbius functions $\boldsymbol{m}^{(k)}$, $k \ge 2$, do not have the product property – Example 18.2.5 (see also Exercise 18.8.2);

- the generalized (multi-pole) Möbius function has the product property under certain additional assumptions – Theorem 18.3.2;
- the Green function has the product property – Theorems 18.4.1, 18.6.1;
- the relative extremal function has the product property – Theorem 18.5.1;
- the generalized (multi-pole) Green function has the product property under certain additional assumptions – Theorem 18.6.1 and Remark 18.6.2;
- the generalized Lempert function has the product property under certain additional assumptions – Theorem 18.7.1.

18.2 Product property for the Möbius functions

Theorem 18.2.1 (cf. [252]). *The pseudodistance $\boldsymbol{c}^{(*)}$ has the product property on $G_1 \times G_2$ for arbitrary domains G_1, G_2. In particular, $\boldsymbol{c}^i$ and $\boldsymbol{\gamma}$ have the product property.*

Remark 18.2.2. In fact, the above theorem will be a direct consequence of Theorem 18.3.2 the proof of which will be a modification of the original proof of Theorem 18.2.1. Consequently, at the moment, we postpone the proof of Theorem 18.2.1.

The product property for the Carathéodory pseudodistance (with G_1 and G_2 being connected complex analytic spaces) was first stated without proof by Kobayashi in [314] (in fact, as it turned out later, no proof did exist at that time). The first proof, for the case where G_1, G_2 are countable at infinity connected complex manifolds, was given in [252]. In his monograph [316], § 4.9, Kobayashi stated once again the product property for the Carathéodory pseudodistance for arbitrary connected complex analytic spaces and he gave a proof based on [252]. Unfortunately, the methods from [252] apply only for those countable at infinity connected complex spaces G_1, G_2 for which the space $\mathcal{O}(G_1) \otimes \mathcal{O}(G_2)$ is dense in $\mathcal{O}(G_1 \times G_2)$ in the topology of locally uniform convergence. The final step was made in [264], where (23 years after Kobayashi stated the result) we proved that *the product property holds for arbitrary connected complex spaces*.

We will apply Theorem 18.2.1 in order to present another proof of Theorem 2.9.1 (cf. [412]). Observe that K. Peters worked with the Kobayashi–Buseman metric instead of the Carathéodory–Reiffen metric.

Theorem 18.2.3. *Let $G_j, \widetilde{G}_j \subset \mathbb{C}^{n_j}$ be bounded domains, $j = 1, 2$. Let $\varphi = (\varphi_1, \varphi_2) : G_1 \times G_2 \longrightarrow \widetilde{G}_1 \times \widetilde{G}_2$ be a biholomorphic mapping. Assume that there is an open non-empty subset U of $G_1 \times G_2$ such that for every $(z, w) \in U$ we have*

$$\varphi_1'(z, w)(\mathbb{C}^{n_1} \times \{0\}) = \mathbb{C}^{n_1}.$$

Then, φ is of the form $\varphi(z,w) = (g_1(z), g_2(w)), z \in G_1, w \in G_2$, where $g_j : G_j \longrightarrow \widetilde{G}_j$, $j = 1, 2$, are biholomorphic mappings.

In the case where $G_j = \widetilde{G}_j$ for $j = 1, 2$, Theorem 18.2.3 immediately leads to a new proof of Theorem 2.9.1.

The proof of Theorem 18.2.3 is based on the following general lemma from linear algebra:

Lemma 18.2.4. *Let V, $\widetilde{V}$ and W, $\widetilde{W}$ be n- and m-dimensional normed real vector spaces, respectively. Assume that $V \times W$ and $\widetilde{V} \times \widetilde{W}$ are endowed with the corresponding maximum norms. If $\varphi = (\varphi_1, \varphi_2) : V \times W \longrightarrow \widetilde{V} \times \widetilde{W}$ is a linear surjective isometry with $\varphi_1(V \times \{0\}) = \widetilde{V}$, then we have*

$$\varphi_2(V \times \{0\}) = \{0\},\ \varphi_2(\{0\} \times W) = W,\ \varphi_1(\{0\} \times W) = \{0\},$$
$$\|\varphi_2(0,w)\| = \|w\|,\ \|\varphi_1(v,0)\| = \|v\|, \quad v \in V,\ w \in W.$$

In particular, $f : V \longrightarrow \widetilde{V}$ given by $f(v) := \varphi_1(v,0), v \in V$, and $g : W \longrightarrow \widetilde{W}$ given by $g(w) = \varphi_2(0,w), w \in W$, are linear surjective isometries satisfying $\varphi(v,w) = (f(v), g(w))$ on $V \times W$.

We postpone the proof of Lemma 18.2.4 and continue the proof of Theorem 18.2.3.

Proof of Theorem 18.2.3. Without loss of generality, we may assume that there are points $a \in G_1$, $b \in G_2$, and a number $r > 0$ such that $U = \mathbb{B}(a,r) \times \mathbb{B}(b,r)$. Then, fixing (z,w) in U, we see that

$$\varphi'(z,w) : \mathbb{C}^{n_1} \times \mathbb{C}^{n_2} \longrightarrow \mathbb{C}^{n_1} \times \mathbb{C}^{n_2}$$

is a $\mathbb{C}$-linear isomorphism. Now, we endow the "left" $\mathbb{C}^{n_1}, \mathbb{C}^{n_2}$, and $\mathbb{C}^{n_1} \times \mathbb{C}^{n_2}$ with the norms $\gamma_{G_1}(z;\cdot), \gamma_{G_2}(w;\cdot)$, and $\gamma_{G_1 \times G_2}((z,w);\cdot)$, respectively, and the "right" $\mathbb{C}^{n_1}, \mathbb{C}^{n_2}$, and $\mathbb{C}^{n_1} \times \mathbb{C}^{n_2}$ with the norms $\gamma_{\widetilde{G}_1}(\varphi_1(z,w);\cdot), \gamma_{\widetilde{G}_2}(\varphi_2(z,w);\cdot)$, and $\gamma_{\widetilde{G}_1 \times \widetilde{G}_2}(\varphi(z,w);\cdot)$, respectively. Then, $\varphi'(z,w)$ becomes a linear surjective isometry. By virtue of Theorem 18.2.1, Lemma 18.2.4 applies and we obtain

$$\varphi_2'(z,w)(\mathbb{C}^{n_1} \times \{0\}) = \{0\} \text{ and } \varphi_1'(z,w)(\{0\} \times \mathbb{C}^{n_2}) = \{0\}.$$

Thus, we have

$$\frac{\partial \varphi_2}{\partial z_j}(z,w) = 0 \ \text{ for } 1 \le j \le n_1, \qquad \frac{\partial \varphi_1}{\partial w_k}(z,w) = 0 \ \text{ for } 1 \le k \le n_2.$$

Since $(z,w) \in U$ was arbitrary, it follows on U that

$$\varphi_2(z,w) = \varphi_2(a,w) \text{ and } \varphi_1(z,w) = \varphi_1(z,b).$$

The identity theorem then implies that for every $(z, w) \in G_1 \times G_2$

$$\varphi_2(z, w) = \varphi_2 \circ i_a(w), \text{ where } i_a(w) := (a, w),$$
$$\varphi_1(z, w) = \varphi_1 \circ j_b(z), \text{ where } j_b(z) := (z, b).$$

Finally, it is clear that $\varphi_2 \circ i_a : G_2 \longrightarrow \widetilde{G}_2$ and $\varphi_1 \circ j_b : G_1 \longrightarrow \widetilde{G}_1$ are biholomorphic mappings, which completes the proof. □

Proof of Lemma 18.2.4. First, we introduce the following temporary notation for a normed real vector space $(T, \| \ \|)$:

$$B_T := \{x \in T : \|x\| < 1\}, \ \overline{B}_T := \{x \in T : \|x\| \le 1\}, \ S_T := \{x \in T : \|x\| = 1\}.$$

Moreover, if L is an arbitrary linear subspace of T, we write

$$N_T(L) := \{x \in S_T : \|x - y\| \ge 1 \text{ for all } y \in L\}.$$

Then, the following inclusions are easy to obtain:

$$\begin{aligned}\overline{B}_{\widetilde{V}} \times N_{\widetilde{W}}(\varphi_2(V \times \{0\})) &\subset N_{\widetilde{V} \times \widetilde{W}}(\varphi(V \times \{0\})) \\ &\subset (B_{\widetilde{V}} \times N_{\widetilde{W}}(\varphi_2(V \times \{0\}))) \cup (N_{\widetilde{V}}(\varphi_1(V \times \{0\})) \times B_{\widetilde{W}}) \cup (S_{\widetilde{V}} \times S_{\widetilde{W}}).\end{aligned}$$

Since $\varphi_1(V \times \{0\}) = \widetilde{V}$, it is clear that $N_{\widetilde{V}}(\varphi_1(V \times \{0\})) = \varnothing$.

Let us assume that there is a point $(\widetilde{v}, \widetilde{w})$ in $N_{\widetilde{V} \times \widetilde{W}}(\varphi(V \times \{0\}))$ that does not belong to the closed set $\overline{B}_{\widetilde{V}} \times N_{\widetilde{W}}(\varphi_2(V \times \{0\}))$. Then, we find an open subset $\widetilde{U}$ of $N_{\widetilde{V} \times \widetilde{W}}(\varphi(V \times \{0\}))$ with $(\widetilde{v}, \widetilde{w}) \in \widetilde{U}$ and $\widetilde{U} \subset S_{\widetilde{V}} \times S_{\widetilde{W}}$. By hypothesis, φ is a linear isometry from $V \times W$ onto $\widetilde{V} \times \widetilde{W}$. Therefore, $U := \varphi^{-1}(\widetilde{U})$ is an open subset of $N_{V \times W}(V \times \{0\}) = \overline{B}_V \times S_W$. Thus, $\varphi|_U$ is a topological map from U onto $\widetilde{U}$, although the dimensions of $S_{\widetilde{V}} \times S_{\widetilde{W}}$ and $\overline{B}_V \times S_W$ are different; a contradiction.

We have therefore obtained

$$\begin{aligned}\overline{B}_{\widetilde{V}} \times N_{\widetilde{W}}(\varphi_2(V \times \{0\})) &= N_{\widetilde{V} \times \widetilde{W}}(\varphi(V \times \{0\})) \\ &= \varphi(N_{V \times W}(V \times \{0\})) = \varphi(\overline{B}_V \times S_W).\end{aligned}$$

Define $g : W \longrightarrow \widetilde{W}$ by $g(w) := \varphi_2(0, w)$ for $w \in W$. Then,

$$g(S_W) \subset \varphi_2(\overline{B}_V \times S_W) = N_{\widetilde{W}}(\varphi_2(V \times \{0\})) \subset S_{\widetilde{W}}, \tag{18.2.1}$$

i.e., g is a linear map preserving the norms, and therefore g is injective. Because of $\dim W = \dim \widetilde{W}$, it follows that g is a surjective linear isometry from W onto $\widetilde{W}$. So, from (18.2.1) we deduce that $N_{\widetilde{W}}(\varphi_2(V \times \{0\})) = S_{\widetilde{W}}$, and so $\varphi_2(V \times \{0\}) = \{0\}$.

On the other hand, the fact that $\varphi_2(\{0\} \times W) = g(W) = \widetilde{W}$ implies, with the same argument as before, that $f : V \longrightarrow \widetilde{V}$ defined by $f(v) := \varphi_1(v, 0)$ is a surjective linear isometry and $\varphi_1(\{0\} \times W) = \{0\}$. □

We move to the case of higher order Möbius functions.

Example 18.2.5. For any $k \geq 2$ the family $\boldsymbol{\gamma}^{(k)}$ does not have the product property on $P \times \mathbb{D}$, where $P = P(R) = \mathbb{A}(1/R, R)$ is an annulus with sufficiently big radius $R > 1$. Consequently, for any $k \geq 2$ the family $\boldsymbol{m}^{(k)}$ does not have the product property on $P \times \mathbb{D}$ (cf. Remark 18.1.6(b)).

For we fix a $k \geq 2$ and we will prove that there exists an $R(k) > 1$ such that for any $R \geq R(k)$ we have

$$\boldsymbol{\gamma}^{(k)}_{P\times\mathbb{D}}((a,0);(1,Y)) > \max\{\boldsymbol{\gamma}^{(k)}_P(a;1), \boldsymbol{\gamma}^{(k)}_{\mathbb{D}}(0;Y)\}, \tag{18.2.2}$$

where $P := P(R)$, $a = a(R,k) := R^{\frac{k-1}{k+1}}$, $Y = Y(R,k) := \boldsymbol{\gamma}^{(k)}_P(a;1)$. Obviously, the right hand side of (18.2.2) is equal to Y. By Proposition 9.1.5(b), we have

$$\boldsymbol{\gamma}^{(k)}_P(a;1) = \left[\frac{1}{Ra} f(a,-a)\right]^{1/k} \frac{\Pi(a,a)}{a} =: \psi_+(R,k).$$

For $R > 2$ define

$$h(\lambda,\xi) := \alpha_1 h_1(\lambda)\xi^{k-1} + \alpha_k h_k(\lambda),$$

where

$$\alpha_1 := \frac{2}{2 + R^{2/(k+1)}}, \quad \alpha_k := \frac{R^{2/(k+1)}}{2 + R^{2/(k+1)}},$$
$$h_1(\lambda) := \frac{1}{R\lambda} f(a,\lambda) f\left(\frac{2}{R}, -\lambda\right), \quad h_k(\lambda) := \frac{1}{R\lambda}[f(a,\lambda)]^k f\left(\frac{R}{2}, -\lambda\right).$$

Observe that $\operatorname{ord}_{(a,0)} h \geq k$ and $\alpha_1|h_1| + \alpha_k|h_k| = 1$ on ∂P. This implies that

$$\begin{aligned}
&\boldsymbol{\gamma}^{(k)}_{P\times\mathbb{D}}((a,0);(1,Y)) \geq (\alpha_1|h_1'(a)|Y^{k-1} + \alpha_k|(h_k)_{(k)}(a)|)^{1/k} \\
&= \left[\alpha_1 \frac{1}{Ra}\frac{\Pi(a,a)}{a} f\left(\frac{2}{R}, -a\right) Y^{k-1} + \alpha_k \frac{1}{Ra}\left(\frac{\Pi(a,a)}{a}\right)^k f\left(\frac{R}{2}, -a\right)\right]^{1/k} \\
&=: \psi_-(R,k).
\end{aligned}$$

Then, direct calculations show that

$$\lim_{R\to\infty} \frac{\psi_-(R,k)}{\psi_+(R,k)} = \left(\frac{2^{\frac{k-1}{k}} + 1}{2}\right)^{1/k} > 1,$$

which gives (18.2.2).

18.3 Product property for the generalized Möbius function

In the context of Theorem 18.2.1 it is natural to ask whether the generalized Möbius function has the product property, i.e., whether for arbitrary domains $G_j \subset \mathbb{C}^{n_j}$, $j = 1,2$, and sets $\varnothing \neq A_j \subset G_j$, $j = 1,2$, we have

$$\boldsymbol{m}_{G_1\times G_2}(A_1 \times A_2, (z_1, z_2)) = \max\{\boldsymbol{m}_{G_1}(A_1, z_1), \boldsymbol{m}_{G_2}(A_2, z_2)\}, \quad (z_1, z_2) \in G_1 \times G_2; \quad \text{(P)}$$

cf. Definition 8.2.1. The product property from Theorem 18.2.1 is just the case where $\#A_1 = \#A_2 = 1$. We do not know whether the above *generalized product property* (P) is true. The positive answer will be given only in the case where $\min\{\#A_1, \#A_2\} = 1$ – cf. Theorem 18.3.2. The main idea of the proof of Theorem 18.3.2 is based in the following proposition:

Proposition 18.3.1. *Assume that for any $N \in \mathbb{N}$, the system $(\boldsymbol{m}_G)_G$ has the following* special product property*:*

$$|\Psi(z_1, z_2)| \le (\sup_{G_1\times G_2} |\Psi|) \max\{\boldsymbol{m}_{G_1}(A_1, z_1), \boldsymbol{m}_{G_2}(A_2, z_2)\}, \quad (z_1, z_2) \in G_1 \times G_2, \quad (\text{P}_0)$$

where

- $\Psi(z_1, z_2) = \Psi_N(z_1, z_2) := z_1 \bullet z_2$, $(z_1, z_2) \in \mathbb{C}^N \times \mathbb{C}^N$,
- *$G_1, G_2 \subset \mathbb{C}^N$ are balls with respect to $\mathbb{C}$-norms,*
- *$\varnothing \neq A_1 \subset G_1$, $\varnothing \neq A_2 \subset G_2$ are finite sets with $A_1 \times A_2 \subset \Psi^{-1}(0)$.*

Then, the system $(\boldsymbol{m}_G)_G$ has the product property (P) *in the full generality.*

Moreover, if (P_0) *holds with $\#A_2 = 1$ (and arbitrary other elements), then* (P) *holds with $\#A_2 = 1$.*

We point out once again that we do not know whether condition (P_0) holds in the full generality. Later, in Theorem 18.3.2, we will show that (P_0) is true if $\min\{\#A_1, \#A_2\} = 1$. Observe that in the case where $\#A_1 = \#A_2 = 1$, condition (P_0) means that

$$|\Psi(z_1, z_2)| \le (\sup_{G_1\times G_2} |\Psi|) \max\{\boldsymbol{m}_{G_1}(a_1, z_1), \boldsymbol{m}_{G_2}(a_2, z_2)\}, \quad (a_1, a_2), (z_1, z_2) \in G_1 \times G_2,\ \Psi(a, b) = 0. \quad (*)$$

Using the Lempert Theorem and the product property for $\boldsymbol{\ell}$ we easily see that (*) holds true. Nevertheless, we do not know how to prove (*) directly.

Proof of Proposition 18.3.1. We have to prove that for arbitrary domains $G_j \subset \mathbb{C}^{n_j}$, $j = 1, 2$, sets $\varnothing \neq A_j \subset G_j$, $j = 1, 2$, $(z_1^0, z_2^0) \in G_1 \times G_2$, and $F \in \mathcal{O}(G_1 \times G_2, \mathbb{D})$ with $A_1 \times A_2 \subset F^{-1}(0)$, we have

$$|F(z_1^0, z_2^0)| \le \max\{\boldsymbol{m}_{G_1}(A_1, z_1^0), \boldsymbol{m}_{G_2}(A_2, z_2^0)\}.$$

By Proposition 8.2.8, we may assume that A_1, A_2 are finite. Let $(G_{j,\nu})_{\nu=1}^{\infty}$ be a sequence of relatively compact subdomains of G_j such that $A_j \cup \{z_j^0\} \subset G_{j,\nu} \nearrow G$, $j = 1, 2$. By Proposition 8.2.5, it suffices to show that

$$|F(z_1^0, z_2^0)| \le \max\{\boldsymbol{m}_{G_{1,\nu}}(A_1, z_1^0), \boldsymbol{m}_{G_{2,\nu}}(A_2, z_2^0)\}, \quad \nu \in \mathbb{N}.$$

Fix a $\nu_0 \in \mathbb{N}$ and let $G_j' := G_{j,\nu_0}$, $j = 1, 2$. The function F may be approximated locally uniformly in $G_1 \times G_2$ by functions of the form

$$F_s(z_1, z_2) = \sum_{\mu=1}^{N_s} f_{1,s,\mu}(z_1) f_{2,s,\mu}(z_2), \quad (z_1, z_2) \in G_1 \times G_2, \tag{18.3.1}$$

where $f_{j,s,\mu} \in \mathcal{O}(G_j)$, $j = 1, 2$, $s \in \mathbb{N}$, $\mu = 1, \dots, N_s$ (cf. Theorem 12.1.22 and Appendix B.1.15). In particular, $F_s \longrightarrow 0$ uniformly on $A_1 \times A_2$. Using the Lagrange interpolation formula, we find polynomials $P_s : \mathbb{C}^{n_1} \times \mathbb{C}^{n_2} \longrightarrow \mathbb{C}$ such that $P_s = F_s$ on $A_1 \times A_2$ and $P_s \longrightarrow 0$ locally uniformly in $\mathbb{C}^{n_1} \times \mathbb{C}^{n_2}$. The functions $\widehat{F}_s := F_s - P_s$, $s \in \mathbb{N}$, also have the form (18.3.1) and $\widehat{F}_s \longrightarrow F$ locally uniformly in $G_1 \times G_2$. Hence, without loss of generality, we may assume that $F_s = 0$ on $A_1 \times A_2$, $s \in \mathbb{N}$. Moreover, we may assume that $\|F_s\|_{G_1' \times G_2'} < 1$, $s \in \mathbb{N}$, i.e., $F_s(G_1' \times G_2') \subset \mathbb{D}$, $s \in \mathbb{N}$. It is enough to prove that

$$|F_s(z_1^0, z_2^0)| \le \max\{\boldsymbol{m}_{G_1'}(A_1, z_1^0), \boldsymbol{m}_{G_2'}(A_2, z_2^0)\}, \quad s \ge 1.$$

Fix an $s = s_0 \in \mathbb{N}$ and let $N := N_{s_0}$, $f_{j,\mu} := f_{j,s_0,\mu}$, $j = 1, 2$, $\mu = 1, \dots, N$. Let $f_j := (f_{j,1}, \dots, f_{j,N}) : G_j \longrightarrow \mathbb{C}^N$, $j = 1, 2$. Put

$$K_1 := \{\xi \in \mathbb{C}^N : |\xi_\mu| \le \|f_{1,\mu}\|_{G_1'},\ \mu = 1, \dots, N,\ |\Psi(\xi, f_2(z_2))| \le 1,\ z_2 \in G_2'\};$$

K_1 is a balanced convex compact set with $f_1(G_1') \subset K_1$. Let

$$K_2 := \{\eta \in \mathbb{C}^N : |\eta_\mu| \le \|f_{2,\mu}\|_{G_2'},\ \mu = 1, \dots, N,\ |\Psi(\xi, \eta)| \le 1,\ \xi \in K_1\};$$

K_2 is a balanced convex compact set with $f_2(G_2') \subset K_2$. Let $(W_{j,\sigma})_{\sigma=1}^{\infty}$ be a sequence of absolutely convex bounded domains in $\mathbb{C}^N$ such that $W_{j,\sigma+1} \subset\subset W_{j,\sigma}$ and $W_{j,\sigma} \searrow K_j$, $j = 1, 2$. Put $M_\sigma := \|\Psi\|_{W_{1,\sigma} \times W_{2,\sigma}}$, $\sigma \in \mathbb{N}$. By (P_0) and by the holomorphic contractibility applied to the mappings $f_j : G_j' \longrightarrow W_{j,\sigma}$, $j = 1, 2$, we have

$$\begin{aligned}|F_{s_0}(z_1^0, z_2^0)| &= |\Psi(f_1(z_1^0), f_2(z_2^0))| \\ &\le M_\sigma \max\{\boldsymbol{m}_{W_{1,\sigma}}(f_1(A_1), f_1(z_1^0)), \boldsymbol{m}_{W_{2,\sigma}}(f_2(A_2), f_2(z_2^0))\} \\ &\le M_\sigma \max\{\boldsymbol{m}_{G_1'}(f_1^{-1}(f_1(A_1)), z_1^0), \boldsymbol{m}_{G_2'}(f_2^{-1}(f_2(A_2)), z_2^0)\} \\ &\le M_\sigma \max\{\boldsymbol{m}_{G_1'}(A_1, z_1^0), \boldsymbol{m}_{G_2'}(A_2, z_2^0)\}.\end{aligned}$$

Letting $\sigma \longrightarrow +\infty$, we get the required result. □

Theorem 18.3.2 (cf. [247], see also [135]). *The system $(\boldsymbol{m}_G)_G$ has the product property* (P) *whenever* $\min\{\#A_1, \#A_2\} = 1$.

Proof. By Proposition 18.3.1, it suffices to check (P) in the case where G_2 is a bounded convex domain, A_1 is finite, and $A_2 = \{b\}$. Fix $(z_1^0, z_2^0) \in G_1 \times G_2$. Let $\varphi : \mathbb{D} \longrightarrow G_2$ be a complex geodesic with $\varphi(0) = b$ and $\varphi(\boldsymbol{m}_{G_2}(b, z_2^0)) = z_2^0$. Consider the mapping $F : G_1 \times \mathbb{D} \longrightarrow G_1 \times G_2$, $F(z_1, \lambda) := (z_1, \varphi(\lambda))$. Then,

$$\boldsymbol{m}_{G_1 \times G_2}(A_1 \times \{b\}, (z_1^0, z_2^0)) \le \boldsymbol{m}_{G_1 \times \mathbb{D}}(A_1 \times \{0\}, (z_1^0, \boldsymbol{m}_{G_2}(b, z_2^0))).$$

Consequently, it suffices to show that

$$\boldsymbol{m}_{G_1 \times \mathbb{D}}(A_1 \times \{0\}, (z_1^0, \lambda)) \le \max\{\boldsymbol{m}_{G_1}(A_1, z_1^0), |\lambda|\}, \quad \lambda \in \mathbb{D}. \tag{18.3.2}$$

The case where $\boldsymbol{m}_{G_1}(A_1, z_1^0) = 0$ is elementary: if $f \in \mathcal{O}(G \times \mathbb{D}, \mathbb{D})$ and $f = 0$ on $A_1 \times \{0\}$, then (by the Schwarz lemma) $|f(z_1^0, \lambda)| \le |\lambda|$, $\lambda \in \mathbb{D}$.

Thus, we may assume that $r := \boldsymbol{m}_{G_1}(A_1, z_1^0) > 0$. First observe that it suffices to prove (18.3.2) only on the circle $|\lambda| = r$. Indeed, in the disc $\mathbb{D}(r)$ we apply the maximum principle to the sh function $\lambda \longmapsto \boldsymbol{m}_{G_1 \times \mathbb{D}}(A_1 \times \{0\}, (z_1^0, \lambda))$. In the annulus $\mathbb{A}(r, 1)$, we apply the maximum principle to the sh function $\lambda \longmapsto \frac{1}{|\lambda|}\boldsymbol{m}_{G_1 \times \mathbb{D}}(A_1 \times \{0\}, (z_2^0, \lambda))$.

Now fix a $\lambda_0 \in \mathbb{D}$ with $|\lambda_0| = r$. Let f be an extremal function for $\boldsymbol{m}_{G_1}(A_1, z_1^0)$ with $f = 0$ on A_1 and $f(z_1^0) = \lambda_0$. Consider $F : G_1 \longrightarrow G_1 \times \mathbb{D}$, $F(z) := (z, f(z))$. Then,

$$\boldsymbol{m}_{G_1}(A_1 \times \{0\}, (z_1^0, \lambda_0)) \le \boldsymbol{m}_{G_1}(A_1, z_1^0) = \max\{\boldsymbol{m}_{G_1}(A_1, z_1^0), |\lambda_0|\},$$

which completes the proof. □

18.4 Product property for the Green function

Now, we turn to the product property for the complex Green function.

Theorem 18.4.1. *If G_1, G_2 are domains of holomorphy, then $\boldsymbol{g}$ has the product property on $G_1 \times G_2$. Consequently, $\boldsymbol{A}$ has the product property in the class of all domains of holomorphy (cf. Remark* 18.1.6(a)). *In particular, if $G_1, \dots, G_n \subset \mathbb{C}^1$, then*

$$\boldsymbol{g}_{G_1\times\dots\times G_n}((z_1',\dots,z_n'),(z_1'',\dots,z_n'')) = \max\{\boldsymbol{g}_{G_1}(z_1',z_1''),\dots,\boldsymbol{g}_{G_n}(z_n',z_n'')\},$$
$$z_j',z_j'' \in G_j,\ j=1,\dots,n,$$

$$\boldsymbol{A}_{G_1\times\dots\times G_n}((z_1,\dots,z_n);(X_1,\dots,X_n)) = \max\{\boldsymbol{A}_{G_1}(z_1;X_1),\dots,\boldsymbol{A}_{G_n}(z_n;X_n)\},$$
$$z_j \in G_j,\ X_j \in \mathbb{C},\ j=1,\dots,n.$$

We will see in Theorem 18.6.1 *that in fact, using the Poletsky disc method, one can eliminate the assumption that G_1, G_2 are domains of holomorphy, i.e., the product property holds for $\boldsymbol{g}$ and $\boldsymbol{A}$ on arbitrary domains G_1, G_2.*

Proof. The proof will be based on the methods of the Monge–Ampère operator (cf. Appendix B.6). Recall that any domain of holomorphy may be exhausted by an increasing sequence of bounded hyperconvex domains. Hence, in view of Proposition 4.2.10(a), it suffices to prove the product property on products $G := G_1 \times G_2$ of bounded hyperconvex domains $G_j \subset \mathbb{C}^{n_j}$, $j = 1,2$. Fix $a = (a_1,a_2)$ and $b = (b_1,b_2) \in G$. Let $n := n_1 + n_2$. We want to show that

$$\boldsymbol{g}_G(a,b) = \max\{\boldsymbol{g}_{G_1}(a_1,b_1), \boldsymbol{g}_{G_2}(a_2,b_2)\}.$$

Recall that the inequality "$\geq$" is always true (cf. Remark 18.1.3(a)). By Remark 18.1.3(b), we can assume that $a_j \neq b_j$, $j = 1,2$. Fix an $\varepsilon > 0$ and define

$$u_-(z) := (1+\varepsilon)\log \boldsymbol{g}_G(a,z),$$
$$u_+(z) := \log\max\{\boldsymbol{g}_{G_1}(a_1,z_1), \boldsymbol{g}_{G_2}(a_2,z_2)\}, \quad z=(z_1,z_2)\in G.$$

It suffices to show that $u_-(b) \leq u_+(b)$ (then, letting $\varepsilon \longrightarrow 0$ we get the required inequality). We are going to apply the domination principle of E. Bedford and B. A. Taylor (cf. Appendix B.6.1). For $r > 0$, define $G_r := G \setminus \overline{\mathbb{B}}(a,r)$. We will prove that for sufficiently small r the functions u_- and u_+ satisfy all the necessary assumptions of this principle on G_r. It is clear that $u_-, u_+ \in \mathcal{PSH}(G_r) \cap L^\infty(G_r)$; cf. the proof of Proposition 4.2.10(i). Moreover, by Proposition 4.2.10(h), for any $\zeta \in \partial G$ we obtain

$$\lim_{z\to\zeta} u_-(z) = \lim_{z\to\zeta} u_+(z) = 0.$$

Fix an $R > 0$ such that $\mathbb{B}(a, R) \subset G$. Then, for $z = (z_1, z_2) \in \mathbb{B}(a, R)$ we have

$$u_+(z) - u_-(z) \geq \log\left(\max\left\{\frac{\|z_1 - a_1\|}{\operatorname{diam} G_1}, \frac{\|z_2 - a_2\|}{\operatorname{diam} G_2}\right\}\right) - (1+\varepsilon)\log\frac{\|z-a\|}{R}$$
$$\geq \log M - \varepsilon \log\frac{\|z-a\|}{R},$$

where $M > 0$ is a constant independent of z. Consequently, there exists an $r_0 = r_0(\varepsilon) \in (0, R)$ such that for any $0 < r \leq r_0$ and for any $\zeta \in \partial\mathbb{B}(a, r)$ we have

$$\liminf_{z\to\zeta}\,(u_+(z) - u_-(z)) \geq 0.$$

Thus, if r is sufficiently small, then

$$\liminf_{z\to\zeta}\,(u_+(z) - u_-(z)) \geq 0, \quad \zeta \in \partial G_r.$$

By Proposition 4.2.10(i), $(dd^c u_-)^n = 0$ in $G \setminus \{a\}$. In view of Appendix B.6.3 (and Proposition 4.2.10(i)), we have

$$(dd^c u_+)^n = 0 \text{ in } (G_1 \setminus \{a_1\}) \times (G_2 \setminus \{a_2\}).$$

It remains to show that $(dd^c u_+)^n = 0$ in an open neighborhood of the set

$$((G_1 \setminus \{a_1\}) \times \{a_2\}) \cup (\{a_1\} \times (G_2 \setminus \{a_2\})).$$

Fix, for instance, a point $z^0 = (a_1, z_2^0)$ with $z_2^0 \neq a_2$. Then, $u_+(z) = \log g_{G_2}(a_2, z_2)$ for $z = (z_1, z_2)$ in an open neighborhood of z^0, and therefore (by Proposition 4.2.10(i)) $(dd^c u_+)^n = 0$ in a neighborhood of z^0. □

Remark 18.4.2. Theorem 18.4.1 has been generalized in [261] to the case where one of the domains G_1, G_2 is a domain of holomorphy. Note that the proof presented in [261] was based on methods independent of those used for Theorem 18.4.1.

18.5 Product property for the relative extremal function

Theorem 18.5.1 (Product property; cf. [505, 155, 151], Theorem 4.1). *Let $G_j \subset \mathbb{C}^{n_j}$ be a domain, $A_j \subset G_j$, $j = 1, 2$. Assume that A_1, A_2 are open or A_1, A_2 are compact. Then,*

$$\boldsymbol{h}_{A_1\times A_2,\, G_1\times G_2}(z_1, z_2) = \max\{\boldsymbol{h}_{A_1,G_1}(z_1), \boldsymbol{h}_{A_2,G_2}(z_2)\}, \quad (z_1, z_2) \in G_1 \times G_2.$$

Moreover, if G_j is biholomorphic to a bounded domain, $j = 1, 2$, then for arbitrary subsets $A_j \subset G_j$, $j = 1, 2$, we have

$$\boldsymbol{h}^*_{A_1\times A_2,\, G_1\times G_2}(z_1, z_2) = \max\{\boldsymbol{h}^*_{A_1,G_1}(z_1), \boldsymbol{h}^*_{A_2,G_2}(z_2)\}, \quad (z_1, z_2) \in G_1 \times G_2.$$

Observe that the inequality "$\geq$" is elementary and that it holds for arbitrary $A_j \subset G_j$, $j = 1, 2$. We need the following auxiliary results:

Lemma 18.5.2. *Let $A \subset \mathbb{D}$ be a compact polar set and let $\Pi : \mathbb{D} \longrightarrow \mathbb{D} \setminus A$ be a universal covering. Then, Π is an inner function. Moreover, if $0 \notin A$, then Π is a Blaschke product.*

Proof. Obviously, $\Pi^*(\zeta) \in A \cup \mathbb{T}$ for each $\zeta \in \mathbb{T}$ such that $\Pi^*(\zeta)$ exists. Hence, by Appendix B.9.9, we conclude that $\Pi^*(\zeta) \in \mathbb{T}$ for almost all $\zeta \in \mathbb{T}$ and, consequently, Π is an inner function. Now, if $0 \notin A$, then Appendix B.9.8 implies that Π is a Blaschke product. □

Remark 18.5.3. Let B be a finite Blaschke product and let $\varphi \in \mathcal{O}(\mathbb{D}, \mathbb{D})$. Then, φ is an inner function iff $B \circ \varphi$ is inner.

Lemma 18.5.4 (Löwner theorem, cf. [151]). *Let $\varphi \in \mathcal{O}(\mathbb{D}, \mathbb{D})$ be an inner function such that $\varphi(0) = 0$. Then, for any open set $I \subset \mathbb{T}$ we have $\mathfrak{m}_{\mathbb{T}}((\varphi^*)^{-1}(I)) = \mathfrak{m}_{\mathbb{T}}(I)$.*[1]

Proof. We may assume that I is an arc. Put $J := (\varphi^*)^{-1}(I)$ (observe that J is measurable). Consider the following holomorphic functions:

$$u_I(z) := \frac{1}{2\pi}\int_0^{2\pi} P(z, e^{i\theta})\chi_I(e^{i\theta})d\theta,$$

$$u_J(z) := \frac{1}{2\pi}\int_0^{2\pi} P(z, e^{i\theta})\chi_J(e^{i\theta})d\theta, \quad z \in \mathbb{D},$$

$$u := u_I \circ \varphi - u_J,$$

where $P(z, \zeta)$ is the Poisson kernel and χ_S denotes the characteristic function of S. Let A denote the set of all $\zeta \in \mathbb{T}$ such that

- $u_J^*(\zeta)$ does not exist or
- $u_J^*(\zeta)$ exists but $u_J^*(\zeta) \neq \chi_I(\zeta)$ or
- $\varphi^*(\zeta)$ does not exist or
- $\varphi^*(\zeta)$ exists and $\varphi^*(\zeta) \in \partial_{\mathbb{T}} I$ (here, $\partial_{\mathbb{T}} I$ denotes the boundary of I in $\mathbb{T}$).

Note that A is of zero measure (use Appendix B.9.9). Observe that $u^*(\zeta) = 0$ on $J \setminus A$. Moreover, $u^*(\zeta) \leq 0$ on $(\mathbb{T} \setminus J) \setminus A$. Thus, $u^* \leq 0$ almost everywhere on $\mathbb{T}$ and hence $u \leq 0$. In particular, $u(0) = \mathfrak{m}_{\mathbb{T}}(I) - \mathfrak{m}_{\mathbb{T}}(J) \leq 0$.

Using the same argument to the arc $\mathbb{T} \setminus I$ shows that $\mathfrak{m}_{\mathbb{T}}(\mathbb{T} \setminus I) \leq \mathfrak{m}_{\mathbb{T}}(\mathbb{T} \setminus J)$, which finishes the proof. □

[1] Recall that $\mathfrak{m}_{\mathbb{T}}$ denotes the normalized Lebesgue measure on $\mathbb{T}$.

Lemma 18.5.5 (cf. [151]). *Let* $(I_j)_{j=1}^k \subset \mathbb{T}$ *be a family of disjoint open arcs, let* $I := \bigcup_{j=1}^k I_j$, *and let* $\alpha := 2\pi\mathfrak{m}_{\mathbb{T}}(I)$. *Then, for every* $\varepsilon > 0$ *there exists a finite Blaschke product* B *such that*

- $B(0) = 0$,
- $B'(z) \neq 0$ *for* $z \in B^{-1}(0)$,
- $B^{-1}(J_\varepsilon) \subset I$, *where* $J_\varepsilon = \{e^{i\theta} : 0 < \theta < \alpha - \varepsilon\}$.

Proof. We may assume that $\alpha < 1$. Let $I_j = \{e^{i\theta} : \theta_{j,1} < \theta < \theta_{j,2}\}$, $j = 1, \dots, k$, $J_0 := \{e^{i\theta} : 0 < \theta < \alpha\}$. For each $j \in \{1, \dots, k\}$, consider the homography

$$g_j(z) := e^{(-i/2)(\theta_{j,2}-\theta_{j,1})} \frac{z - e^{i\theta_{j,2}}}{z - e^{i\theta_{j,1}}}.$$

Then,

- $g_j(\mathbb{D}) = \{w \in \mathbb{C} : \operatorname{Im} w > 0\}$,
- $g_j(0) = e^{(i/2)(\theta_{j,2}-\theta_{j,1})}$,
- $g_j(I_j) = \mathbb{R}_{<0}$,
- $g_j(\mathbb{T} \setminus \overline{I}_j) = \mathbb{R}_{>0}$.

Put $h_j := (-i/\pi) \operatorname{Log} g_j$. Observe that h_j extends homeomorphically to a mapping $\overline{\mathbb{D}} \setminus \{e^{i\theta_{j,1}}, e^{i\theta_{j,2}}\} \longrightarrow \overline{\mathbb{S}}$, where $\mathbb{S} := \{w \in \mathbb{C} : 0 < \operatorname{Re} w < 1\}$. We have,

- $h_j(\mathbb{D}) = \mathbb{S}$,
- $h_j(0) = \frac{\theta_{j,2}-\theta_{j,1}}{2\pi}$,
- $h_j(I_j) = \{\operatorname{Re} w = 1\}$,
- $h_j(\mathbb{T} \setminus \overline{I}_j) = \{\operatorname{Re} w = 0\}$.

In particular,

$$\operatorname{Re} h_j(z) = \frac{1}{2\pi} \int_0^{2\pi} P(z, e^{i\theta}) \chi_{I_j}(e^{i\theta}) d\theta.$$

Let $h := h_1 + \dots + h_k$. Observe that

$$\operatorname{Re} h(z) = \frac{1}{2\pi} \int_0^{2\pi} P(z, e^{i\theta}) \chi_I(e^{i\theta}) d\theta.$$

In particular, $h : \mathbb{D} \longrightarrow \mathbb{S}$. Moreover,

- h extends to a continuous mapping $\overline{\mathbb{D}} \setminus \partial_{\mathbb{T}}(I) \longrightarrow \overline{\mathbb{S}}$,
- $h(0) = \alpha/2$,
- $h(I) = \{\operatorname{Re} w = 1\}$,
- $h(\mathbb{T} \setminus \overline{I}) \subset \{\operatorname{Re} w = 0\}$.

Let

$$F(w) := \frac{e^{\pi i w} - e^{\alpha i/2}}{e^{\pi i w} - e^{-\alpha i/2}}.$$

Then,

- $F : \mathbb{S} \longrightarrow \mathbb{D}$ is biholomorphic and F extends homeomorphically to $\overline{\mathbb{S}} \longrightarrow \overline{\mathbb{D}} \setminus \{1, e^{i\alpha}\}$,
- $F(\alpha/2) = 0$,
- $F(\{\operatorname{Re} w = 1\}) = J_0$,
- $F(\{\operatorname{Re} w = 0\}) = \mathbb{T} \setminus \overline{J}_0$.

Put

$$B_0(z) := F(h(z)) = \frac{\prod_{j=1}^k (z - e^{i\theta_{j,2}}) - e^{i\alpha} \prod_{j=1}^k (z - e^{i\theta_{j,1}})}{\prod_{j=1}^k (z - e^{i\theta_{j,2}}) - \prod_{j=1}^k (z - e^{i\theta_{j,1}})}.$$

Thus,

- $B_0 \in \mathcal{O}(\mathbb{D}, \mathbb{D})$, B_0 extends continuously to $\overline{\mathbb{D}} \setminus \partial_{\mathbb{T}} I \longrightarrow \overline{\mathbb{D}} \setminus \{1, e^{i\alpha}\}$,
- $B_0(0) = 0$,
- $B_0(I) = J_0$,
- $B_0(\mathbb{T} \setminus \overline{I}) \subset \mathbb{T} \setminus \overline{J}_0$.

Observe that, in fact, B_0 must be a finite Blaschke product (cf. e.g., [187], Theorem 2.4). Write

$$B_0(z) = e^{i\tau} \prod_{j=1}^{N} \left(\frac{z - a_j}{1 - \overline{a}_j z} \right)^{m_j}.$$

Take a closed arc $\widetilde{J}_0 \subset J_0$ such that $2\pi \mathfrak{m}_{\mathbb{T}}(\widetilde{J}_0) \ge \alpha - \varepsilon$. Then, for different points $a_{j,1}, \dots, a_{j,m_j}$, sufficiently close to a_j, such that $a_j \in \{a_{j,1}, \dots, a_{j,m_j}\}$, if

$$\widetilde{B}_0(z) = e^{i\tau} \prod_{j=1}^{N} \prod_{\ell=1}^{m_j} \left(\frac{z - a_{j,\ell}}{1 - \overline{a}_{j,\ell} z} \right),$$

then $\widetilde{B}_0(\mathbb{T} \setminus I) \subset \mathbb{T} \setminus \widetilde{J}_0$. Finally, we put $B(z) := \widetilde{B}_0(e^{i\theta_0} z)$ (with suitable θ_0). $\square$

Proposition 18.5.6 (cf. [58]). *Let $G \subset \mathbb{C}^n$ be biholomorphic to a bounded domain, $A \subset G$. Put*

$$\Delta(\varepsilon) := \{z \in G : \boldsymbol{h}^*_{A,G}(z) < \varepsilon\}, \quad 0 < \varepsilon < 1.$$

Then,

$$\frac{\boldsymbol{h}^*_{A,G} - \varepsilon}{1-\varepsilon} \le \boldsymbol{h}^*_{\Delta(\varepsilon),G} \le \boldsymbol{h}^*_{A,G},$$

Consequently, $\boldsymbol{h}^*_{\Delta(\varepsilon),G} \nearrow \boldsymbol{h}^*_{A,G}$ as $\varepsilon \searrow 0$.

? We do not know whether the result is true for an arbitrary domain $G \subset \mathbb{C}^n$.

Proof. It is clear that $\boldsymbol{h}_{\Delta(\varepsilon),G} \ge \frac{\boldsymbol{h}^*_{A,G} - \varepsilon}{1-\varepsilon}$. Put

$$P := \{z \in A : 0 = \boldsymbol{h}_{A,G}(z) < \boldsymbol{h}^*_{A,G}(z)\}.$$

Then, P is pluripolar and, hence, $\boldsymbol{h}^*_{A \setminus P,G} \equiv \boldsymbol{h}^*_{A,G}$ (cf. [305], Theorem 4.7.6). Observe that $A \setminus P \subset \Delta(\varepsilon)$. Thus, $\boldsymbol{h}_{\Delta(\varepsilon),G} \le \boldsymbol{h}^*_{A \setminus P,G} \equiv \boldsymbol{h}^*_{A,G}$ on $\Delta(\varepsilon)$. □

Lemma 18.5.7. *If $G_j \subset \mathbb{C}^{n_j}$ is a domain, $A_j \subset G_j$, $j = 1, 2$, then*

$$\begin{aligned}\boldsymbol{h}_{A_1 \times A_2,\, G_1 \times G_2}(z_1, z_2) &\le 1 - (1 - \boldsymbol{h}_{A_1,G_1}(z_1))(1 - \boldsymbol{h}_{A_2,G_2}(z_2)) \\ &= \boldsymbol{h}_{A_1,G_1}(z_1) + \boldsymbol{h}_{A_2,G_2}(z_2) - \boldsymbol{h}_{A_1,G_1}(z_1)\boldsymbol{h}_{A_2,G_2}(z_2), \quad (z_1, z_2) \in G_1 \times G_2.\end{aligned}$$

Proof. Fix a $u \in \mathcal{PSH}(G_1 \times G_2)$ with $u \le 1$, $u|_{A_1 \times A_2} \le 0$. For $(a_1, a_2) \in G_1 \times G_2$ with $\boldsymbol{h}_{A_j,G_j}(a_j) < 1$, $j = 1, 2$, define

$$v_{a_1} := \frac{u(a_1, \cdot) - \boldsymbol{h}_{A_1,G_1}(a_1)}{1 - \boldsymbol{h}_{A_1,G_1}(a_1)}, \quad v^{a_2} := \frac{u(\cdot, a_2) - \boldsymbol{h}_{A_2,G_2}(a_2)}{1 - \boldsymbol{h}_{A_2,G_2}(a_2)}.$$

Observe that

$$\begin{aligned}\boldsymbol{h}_{A_1 \times A_2,\, G_1 \times G_2}(a_1, a_2) &\le 1 - (1 - \boldsymbol{h}_{A_1,G_1}(a_1))(1 - \boldsymbol{h}_{A_2,G_2}(a_2)) \\ &\iff v_{a_1}(a_2) \le \boldsymbol{h}_{A_2,G_2}(a_2) \iff v^{a_2}(a_1) \le \boldsymbol{h}_{A_1,G_1}(a_1).\end{aligned}$$

It is clear that $v_{a_1} \in \mathcal{PSH}(G_2)$, $v_{a_1} \le 1$, $v^{a_2} \in \mathcal{PSH}(G_1)$, $v^{a_2} \le 1$. If $a_2 \in A_2$, then $v^{a_2} \le 0$ on A_1. Thus, if $a_2 \in A_2$, then $v^{a_2} \le \boldsymbol{h}_{A_1,G_1}$.

Fix $(z_1^0, z_2^0) \in G_1 \times G_2$. The inequality in the lemma is trivial if $\boldsymbol{h}_{A_1,G_1}(z_1^0) = 1$ or $\boldsymbol{h}_{A_2,G_2}(z_2^0) = 1$. Assume that $h_{A_j,G_j}(z_j^0) < 1$, $j = 1, 2$.

Then, $v_{z_1^0}(a_2) \le \boldsymbol{h}_{A_2,G_2}(a_2)$, $a_2 \in A_2$. Hence, $v_{z_1^0} \le 0$ on A_2, which gives $v_{z_1^0}(z_2^0) \le \boldsymbol{h}_{A_2,G_2}(z_2^0)$. □

Proposition 18.5.8 (cf. [343]). *Let $G \subset \mathbb{C}^n$ be a domain and let $A \subset G$. Then,*

$$\boldsymbol{h}_{A,G} = \sup\{\boldsymbol{h}_{U,G} : A \subset U \in \operatorname{top} G\}.$$

In particular, if A is compact, then for any neighborhood basis $(U_k)_{k=1}^{\infty}$ of A with $G \supset U_{k+1} \subset U_k$, we have

$$\boldsymbol{h}_{A,G} = \lim_{k\to\infty} \boldsymbol{h}_{U_k,G}.$$

Proof. The inequality "$\geq$" is obvious. Let $u \in \mathcal{PSH}(G)$, $u \leq 1$, $u \leq 0$ on A. Fix $0 < \varepsilon < 1$ and define

$$U_\varepsilon := \{z \in G : u < \varepsilon\}.$$

Then, $\frac{u-\varepsilon}{1-\varepsilon} \leq \boldsymbol{h}_{U_\varepsilon,G}$. Consequently,

$$u \leq \varepsilon + (1-\varepsilon)\sup\{\boldsymbol{h}_{U,G} : A \subset U \in \operatorname{top} G\}.$$

Taking $\varepsilon \longrightarrow 0$, we get the required result. □

Proof of Theorem 18.5.1. First assume that A_1, A_2 are open. Put $u_j = \chi_{G_j \setminus A_j}$, $j = 1, 2$. Let $(z_1^0, z_2^0) \in G_1 \times G_2$ be fixed and let $\beta \in (0,1)$ be such that

$$\max\{\boldsymbol{h}_{A_1,G_1}(z_1^0), \boldsymbol{h}_{A_2,G_2}(z_2^0)\} < \beta.$$

By Theorem 17.3.4, there is a $\varphi_j \in \mathcal{O}(\overline{\mathbb{D}}, G_j)$ such that $\varphi_j(0) = z_j$ and

$$\frac{1}{2\pi}\int_0^{2\pi} u_j(\varphi_j(e^{i\theta}))d\theta < \beta, \quad j = 1, 2.$$

Note that $\varphi_1^{-1}(A_1) \cap \mathbb{T}$ is an open set in $\mathbb{T}$. So, we may choose a finite set of disjoint open arcs $I_1^1, \dots, I_m^1 \subset \varphi_1^{-1}(A_1) \cap \mathbb{T}$ such that $\mathfrak{m}_{\mathbb{T}}(I^1) > 1 - \beta$, where $I^1 = \bigcup_{j=1}^m I_j^1$. Similarly, we choose $I_1^2, \dots, I_k^2$ with $I^2 = \bigcup_{j=1}^k I_j^2$. By Lemma 18.5.5, we find Blaschke products B_1, B_2 and a closed arc $I \subset \mathbb{T}$ with $\mathfrak{m}_{\mathbb{T}}(I) > 1 - \beta$ such that $B_j^{-1}(I) \subset I^j$, $j = 1, 2$.

Let $A := \bigcup_{j=1}^2 \{B_j(z) : B_j'(z) = 0\}$. Note that $0 \notin A$. Let $\Pi : \mathbb{D} \longrightarrow \mathbb{D} \setminus A$ be the holomorphic universal covering with $\Pi(0) = 0$. Observe that Π is inner (Lemma 18.5.2). If $\widetilde{I} = \Pi^{-1}(I)$, then, according to Lemma 18.5.4, $\mathfrak{m}_{\mathbb{T}}(\widetilde{I}) = \mathfrak{m}_{\mathbb{T}}(I)$. There are liftings $\psi_1, \psi_2 : \mathbb{D} \longrightarrow \mathbb{D}$ of Π such that $\Pi = B_1 \circ \psi_1 = B_2 \circ \psi_2$ and $\psi_1(0) = \psi_2(0) = 0$. By Remark 18.5.3, ψ_1, ψ_2 are inner. Moreover, $\psi_j^*(\widetilde{I}) \subset I^j$. Put $\widetilde{\varphi}_j = \varphi_j \circ \psi_j$, $j = 1, 2$. Then,

$$\begin{aligned}
&\frac{1}{2\pi}\int_0^{2\pi} \chi_{G_1\times G_2 \setminus A_1 \times A_2}(\widetilde{\varphi}_1(e^{i\theta}), \widetilde{\varphi}_2(e^{i\theta}))d\theta \\
&\quad = \frac{1}{2\pi}\int_0^{2\pi} \max\{u_1(\widetilde{\varphi}_1(e^{i\theta})), u_2(\widetilde{\varphi}_2(e^{i\theta}))\}d\theta \leq 1 - \mathfrak{m}_{\mathbb{T}}(\widetilde{I}) < \beta.
\end{aligned}$$

By Fatou's theorem, the same inequality holds if we replace $\widetilde{\varphi}_j(z)$, $j = 1,2$, with $\widetilde{\varphi}_j(rz)$, where $r < 1$ is sufficiently close to 1. Hence, $\boldsymbol{h}_{A_1\times A_2,\,G_1\times G_2}(z_1^0, z_2^0) < \beta$. Since β was arbitrary, we get the required result.

The case where A_1, A_2 are compact follows from Proposition 18.5.8.

We move to the second part of the theorem. Let

$$\Delta_j(\varepsilon) := \{z_j \in G_j : \boldsymbol{h}^*_{A_j,G_j}(z_j) < \varepsilon\}, \quad j = 1,2,$$
$$\Delta(\varepsilon) := \{(z_1,z_2) \in G_1 \times G_2 : \boldsymbol{h}^*_{A_1\times A_2,\,G_1\times G_2}(z_1,z_2) \le 1 - (1-\varepsilon)^2\}.$$

By Lemma 18.5.7 we get $\boldsymbol{h}_{\Delta_1(\varepsilon)\times\Delta_2(\varepsilon),\,G_1\times G_2} < 1-(1-\varepsilon)^2$ on $\Delta_1(\varepsilon)\times\Delta_2(\varepsilon)$. Hence, $\boldsymbol{h}^*_{\Delta_1(\varepsilon)\times\Delta_2(\varepsilon),\,G_1\times G_2} \le 1-(1-\varepsilon)^2$ on $\Delta_1(\varepsilon)\times\Delta_2(\varepsilon)$, which implies that $\Delta_1(\varepsilon)\times\Delta_2(\varepsilon) \subset \Delta(\varepsilon)$. Now, by the first part of the theorem, we have

$$\begin{aligned}\max\{\boldsymbol{h}_{\Delta_1(\varepsilon),G_1}(z_1), \boldsymbol{h}_{\Delta_2(\varepsilon),G_2}(z_2)\} &= \boldsymbol{h}_{\Delta_1(\varepsilon)\times\Delta_2(\varepsilon),\,G_1\times G_2}(z_1,z_2)\\ &= \boldsymbol{h}^*_{\Delta_1(\varepsilon)\times\Delta_2(\varepsilon),\,G_1\times G_2}(z_1,z_2) \ge \boldsymbol{h}^*_{\Delta(\varepsilon),\,G_1\times G_2}(z_1,z_2).\end{aligned}$$

To finish the proof, it remains to observe that by Proposition 18.5.6 we get $\boldsymbol{h}_{\Delta_j(\varepsilon),G_j} \nearrow \boldsymbol{h}^*_{A_j,G_j}$, $j = 1,2$, and $\boldsymbol{h}^*_{\Delta(\varepsilon),G_1\times G_2} \nearrow \boldsymbol{h}^*_{A_1\times A_2,\,G_1\times G_2}$ when $\varepsilon \searrow 0$. $\square$

Remark 18.5.9. Using the analytic discs method, F. Lárusson, P. Lassere, and R. Sigurdsson proved in [336] the following result:

Let $G \subset \mathbb{C}^n$ be a convex domain and let $A \subset G$ be an open or compact convex set. Then, for any $\alpha \in [0,1)$, the level set $\{z \in G : \boldsymbol{h}_{A,G}(z) < \alpha\}$ is convex.

18.6 Product property for the generalized Green function

Theorem 18.6.1 (cf. [149, 150]). *For any domains $G_1 \subset \mathbb{C}^{n_1}$, $G_2 \subset G^{n_2}$ and for any sets $A_j \subset G_j$, $j = 1,2$, the pluricomplex Green function with many poles has the* product property

$$\boldsymbol{g}_{G_1\times G_2}(A_1\times A_2, (z_1,z_2)) = \max\{\boldsymbol{g}_{G_1}(A_1,z_1), \boldsymbol{g}_{G_2}(A_2,z_2)\}, \qquad (z_1,z_2) \in G_1\times G_2.$$

In particular, the one-pole Green function has the product property

$$\boldsymbol{g}_{G_1\times G_2}((a_1,a_2),(z_1,z_2)) = \max\{\boldsymbol{g}_{G_1}(a_1,z_1), \boldsymbol{g}_{G_2}(a_2,z_2)\}, \qquad (a_1,a_2),\ (z_1,z_2) \in G_1\times G_2. \quad (18.6.1)$$

*Consequently, **A** has the product property (cf. Remark* 18.1.6(a)).

Proof. Using Proposition 8.2.6, we may assume that A_1, A_2 are finite. Let $R > 0$ be such that

- $\mathbb{P}(a_j, R) \subset\subset G_j$ for any $a_j \in A_j$,
- $\mathbb{P}(a_j, R) \cap \mathbb{P}(b_j, R) = \varnothing$ for any $a_j, b_j \in A_j, a_j \neq b_j, j = 1, 2$.

Put

$$B_r^j := \bigcup_{a_j \in A_j} \mathbb{P}(a_j, r), \quad j = 1, 2,$$

$$B_r := B_r^1 \times B_r^2 = \bigcup_{a \in A_1 \times A_2} \mathbb{P}(a, r), \quad 0 < r < R.$$

Now, we use Proposition 17.1.4 and Theorem 18.5.1. We have,

$$\begin{aligned}
\log \boldsymbol{g}_{G_1\times G_2}(A_1 \times A_2, (z_1, z_2)) &= \lim_{r\to 0}\left(\log\frac{R}{r}\right)(\boldsymbol{h}_{B_r, G_1\times G_2}(z_1, z_2) - 1) \\
&= \lim_{r\to 0}\left(\log\frac{R}{r}\right)(\boldsymbol{h}_{B_r^1\times B_r^2, G_1\times G_2}(z_1, z_2) - 1) \\
&= \lim_{r\to 0}\left(\log\frac{R}{r}\right)(\max\{\boldsymbol{h}_{B_r^1, G_1}(z_1), \boldsymbol{h}_{B_r^2, G_2}(z_2)\} - 1) \\
&= \max\left\{\lim_{r\to 0}\left(\log\frac{R}{r}\right)(\boldsymbol{h}_{B_r^1, G_1}(z_1) - 1), \lim_{r\to 0}\left(\log\frac{R}{r}\right)(\boldsymbol{h}_{B_r^2, G_2}(z_2) - 1)\right\} \\
&= \max\{\log \boldsymbol{g}_{A_1, G_1}(z_1), \log \boldsymbol{g}_{A_2, G_2}(z_2)\}.
\end{aligned}$$

□

Remark 18.6.2. One could try to generalize the above product property to arbitrary pole functions $\mathfrak{p}_j : G_j \longrightarrow \mathbb{R}_+$ with $\max_{G_j} \mathfrak{p}_j = 1$, $j = 1, 2$. For instance, one could conjecture that

$$\boldsymbol{g}_{G_1\times G_2}(\mathfrak{p}, (z_1, z_2)) = \max\{\boldsymbol{g}_{G_1}(\mathfrak{p}_1, z_1), \boldsymbol{g}_{G_2}(\mathfrak{p}_2, z_2)\}, \quad (z_1, z_2) \in G_1 \times G_2,$$

where $\mathfrak{p}(a_1, a_2) := \min\{\mathfrak{p}_1(a_1), \mathfrak{p}_2(a_2)\}$. Unfortunately, such a formula is false.

Take for instance $G_1 = G_2 = \mathbb{D}$, $\mathfrak{p}_1 := \chi_{\{0\}} + \frac{1}{2}\chi_{\{c\}}$, $\mathfrak{p}_2 := \chi_{\{0\}}$, where $0 < c < 1$. Observe that $\mathfrak{p} = \chi_{\{(0,0)\}} + \frac{1}{2}\chi_{\{(c,0)\}}$. Hence, by Example 8.2.27, we get

$$\boldsymbol{g}_{\mathbb{D}^2}(\mathfrak{p}, (z_1, z_2)) = \left(\max\{|z_1|, |z_2|\} \max\{|z_1|\boldsymbol{m}(z_1, c), |z_2|\}\right)^{1/2}.$$

In particular, if $z_1 = c$, $z_2 = c^2$, then $\boldsymbol{g}_{\mathbb{D}^2}(\mathfrak{p}, (c, c^2)) = c^{3/2}$. On the other hand, $\max\{\boldsymbol{g}_{\mathbb{D}}(\mathfrak{p}_1, c), \boldsymbol{g}_{\mathbb{D}}(\mathfrak{p}_2, c^2)\} = c^2$.

18.7 Product property for the generalized Lempert function

Theorem 18.7.1 (cf. [396]). *Let $G_j \subset \mathbb{C}^{n_j}$ be a domain, $j = 1, 2$, and let $a_1, b_1 \in G_1$, $b_2 \in G_2$ be fixed. Then, the following conditions are equivalent:*

(i) *for any set $\varnothing \neq A_2 \subset G_2$ the following* product property *holds*

$$\boldsymbol{\ell}^*_{G_1 \times G_2}(\{a_1\} \times A_2, (b_1, b_2)) = \max\{\boldsymbol{\ell}^*_{G_1}(a_1, b_1), \boldsymbol{\ell}^*_{G_2}(A_2, b_2)\};$$

(ii) $\boldsymbol{\ell}^*_{G_1}(a_1, b_1) = \boldsymbol{g}_{G_1}(a_1, b_1)$.

Proof. First observe that

- by Theorem 8.2.11(b) we easily conclude that the product property in (i) holds for arbitrary A_2 iff it holds for finite A_2;
- the inequality "$\geq$" in (i) follows directly from the definition of the generalized Lempert function;
- if $a_1 = b_1$, then the equality in (i) is true;
- the inequality "$\leq$" in (ii) is always satisfied – cf. Remark 8.2.14(a);
- $\boldsymbol{g}_{G_1}(a_1, b_1) = \inf\{\prod_{\mu \in \mathbb{D}_* \cap \varphi_1^{-1}(a_1)} |\mu| : \varphi_1 \in \mathcal{O}(\mathbb{D}, G_1) : \varphi_1(0) = b_1\}$ – cf. Theorem 17.4.3.

(i) $\Longrightarrow$ (ii): We may assume that $a_1 \neq b_1$. Take an $\varepsilon > 0$ and let $\varphi_1 \in \mathcal{O}(\mathbb{D}, G_1)$ and $\mu_1, \dots, \mu_N \in \mathbb{D}_*$, pairwise different, such that $\varphi_1(0) = b_1$, $\varphi_1(\mu_j) = a_1$, $j = 1, \dots, N$, and $\prod_{j=1}^N |\mu_j| \leq \boldsymbol{g}_{G_1}(a_1, b_1) + \varepsilon$. Take an arbitrary $\varphi_2 \in \mathcal{O}(\mathbb{D}, G_2)$ such that $\varphi_2(0) = b_2$ and $\varphi_2(\mu_j) \neq \varphi_2(\mu_k)$ for $j \neq k$. Put $A_2 := \{\varphi_2(\mu_j) : j = 1, \dots, N\}$. Then, using (i), we get

$$\boldsymbol{\ell}^*_{G_1}(a_1, b_1) \leq \boldsymbol{\ell}^*_{G_1 \times G_2}(\{a_1\} \times A_2, (b_1, b_2)) \leq \prod_{j=1}^N |\mu_j| \leq \boldsymbol{g}_{G_1}(a_1, b_1) + \varepsilon.$$

(ii) $\Longrightarrow$ (i): We may assume that $A_2 = \{a_{2,1}, \dots, a_{2,N}\}$ is finite and $a_1 \neq b_1$. Let $1 > \alpha > \max\{\boldsymbol{\ell}^*_{G_1}(a_1, b_1), \boldsymbol{\ell}^*_{G_2}(A_2, b_2)\}$. Then, there exist $\varphi_1 \in \mathcal{O}(\mathbb{D}, G_1)$, $\mu_1 \in \mathbb{D}$, $\varphi_2 \in \mathcal{O}(\mathbb{D}, G_2)$, $\mu_{2,1}, \dots, \mu_{2,N} \in \mathbb{D}$ such that $\varphi_1(0) = b_1$, $\varphi_1(\mu_1) = a_1$, $\varphi_2(0) = b_2$, $\varphi_2(\mu_{2,j}) = a_{2,j}$, $j = 1, \dots, N$, and $\max\{|\mu_1|, \prod_{j=1}^N |\mu_{2,j}|\} < \alpha$. By Lemma 18.7.2 (see below) there exist $f \in \mathcal{O}(\mathbb{D}, \mathbb{D})$ and $\eta_1, \dots, \eta_N \in \mathbb{D}$ such that $f(0) = 0$, $f(\eta_j) = \mu_{2,j}$, $j = 1, \dots, N$, and $\prod_{j=1}^N |\eta_j| = \alpha$. Define

$$B(\lambda) := \prod_{j=1}^N \frac{\overline{\eta}_j}{|\eta_j|} \frac{\eta_j - \lambda}{1 - \overline{\eta}_j \lambda}, \quad \psi(\lambda) := \left(\varphi_1\left(\frac{\mu_1}{\alpha} \frac{\alpha - B(\lambda)}{1 - \alpha B(\lambda)}\right), \varphi_2(f(\lambda))\right), \quad \lambda \in \mathbb{D}.$$

Then,

$$B \in \mathcal{O}(\mathbb{D},\mathbb{D}), \quad B(\eta_j) = 0, \quad B(0) = \prod_{j=1}^{N} |\eta_j| = \alpha,$$

$$\psi \in \mathcal{O}(\mathbb{D}, G_1 \times G_2),\ \psi(0) = (b_1, b_2),\ \psi(\eta_j) = (\varphi_1(\mu_1), \varphi_2(\mu_{2,j})) = (a_1, a_{2,j}).$$

Thus, $\boldsymbol{\ell}^*_{G_1\times G_2}(\{a_1\} \times A_2, (b_1, b_2)) \le \prod_{j=1}^N |\eta_j| = \alpha$. □

Lemma 18.7.2. *Let $\mu_1, \dots, \mu_N \in \mathbb{D}$ be pairwise different, $\prod_{j=1}^N |\mu_j| < \alpha < 1$. Then, there exist $f \in \mathcal{O}(\mathbb{D},\mathbb{D})$ and $\eta_1, \dots, \eta_N \in \mathbb{D}$ such that $f(0) = 0$, $f(\eta_j) = \mu_j$, $j = 1, \dots, N$, and $\prod_{j=1}^N |\eta_j| = \alpha$.*

Proof. Put $\beta := \prod_{j=1}^N |\mu_j|$. Define

$$f_a(\lambda) := \lambda \frac{a - \lambda}{1 - a\lambda}, \quad \lambda \in \mathbb{D},\ a \in [0, 1).$$

First consider the case where $\mu_1, \dots, \mu_N \in \mathbb{D}_*$. For each j, consider the equation $f_a(\lambda) = \mu_j$. It has two roots $\eta_j^-(a)$, $\eta_j^+(a) \in \mathbb{D}_*$ (counted with multiplicity). We may assume that $|\eta_j^-(a)| \le |\eta_j^+(a)|$, $j = 1, \dots, N$. Observe that the functions $[0, 1) \ni a \longmapsto |\eta_j^{\mp}(a)|$ may be given by effective formulas – in particular, they are continuous. We have $|\eta_j^-(a)| \le \sqrt{|\mu_j|} \le |\eta_j^+(a)|$ and $|\eta_j^-(0)| = |\eta_j^+(0)| = \sqrt{|\mu_j|}$. Moreover, $|\eta_j^-(a)| \longrightarrow |\mu_j|$ and $|\eta_j^+(a)| \longrightarrow 1$ when $a \longrightarrow 1$. Let $g^{\mp}(a) := \prod_{j=1}^N |\eta_j^{\mp}(a)|$. Then, $g^{\mp}(0) = \sqrt{\beta}$, $g^-(a) \longrightarrow \beta$ and $g^+(a) \longrightarrow 1$ when $a \longrightarrow 1$. Thus, if $\alpha \le \sqrt{\beta}$, then there exists an a such that $g^-(a) = \alpha$ and if $\alpha \ge \sqrt{\beta}$, then there exists an a such that $g^+(a) = \alpha$. It remains to put $f := f_a$.

Now, suppose that some of the numbers $\mu_1, \dots, \mu_N$ are zero. Take an $a \approx 1$ and $\mu_1', \dots, \mu_N' \in \mathbb{D}_*$ such that $f_a(\mu_j') = \mu_j$, $j = 1, \dots, N$, and $\prod_{j=1}^N |\mu_j'| < \alpha$. Using the first part of the proof, we find a $g \in \mathcal{O}(\mathbb{D},\mathbb{D})$ and $\eta_1, \dots, \eta_N \in \mathbb{D}$ such that $g(0) = 0$, $g(\eta_j) = \mu_j'$, $j = 1, \dots, N$, and $\prod_{j=1}^N |\eta_j| = \alpha$. Now we only need to set $f := f_a \circ g$. □

18.8 Exercises

Exercise 18.8.1. Let $G_k := \{(z_1, z_2) \in \mathbb{C}^2 : |z_1 z_2^{k-2}| < 1\}$, $k \in \mathbb{N}$, $k \ge 3$. Using Proposition 2.10.5, prove that $\boldsymbol{m}^{(k)}$ and $\boldsymbol{\gamma}^{(k)}$ do not have the product property on $G_k \times \mathbb{D}$.

Exercise 18.8.2 (Generalized product properties for $\boldsymbol{m}^{(k)}$ and $\boldsymbol{\gamma}^{(k)}$). We say that $\boldsymbol{m}^{(k)}$ has the *generalized product property on* $G_1 \times G_2$ if for any $z'_j, z''_j \in G_j$, $j = 1, 2$,

$$\boldsymbol{m}^{(k)}_{G_1\times G_2}((z'_1, z'_2), (z''_1, z''_2)) = \max\{[(\boldsymbol{m}^{(\ell)}_{G_1}(z'_1, z''_1))^\ell (\boldsymbol{m}^{(k-\ell)}_{G_2}(z'_2, z''_2))^{k-\ell}]^{1/k} : \ell = 0, \dots, k\}, \quad (18.8.1)$$

where $\boldsymbol{m}^{(0)} :\equiv 1$. Similarly, we say that $\boldsymbol{\gamma}^{(k)}$ has the *generalized product property on* $G_1 \times G_2$ if, whenever $z_j \in G_j \subset \mathbb{C}^{n_j} \ni X_j$, $j = 1, 2$, then

$$\boldsymbol{\gamma}^{(k)}_{G_1\times G_2}((z_1, z_2); (X_1, X_2)) = \max\{[(\boldsymbol{\gamma}^{(\ell)}_{G_1}(z_1; X_1))^\ell (\boldsymbol{\gamma}^{(k-\ell)}_{G_2}(z_2; X_2))^{k-\ell}]^{1/k} : \ell = 0, \dots, k\}, \quad (18.8.2)$$

where $\boldsymbol{\gamma}^{(0)} :\equiv 1$.

(a) Prove that (18.8.2) is a consequence of (18.8.1).

(b) Prove that in (18.8.1) and (18.8.2) the inequalities "$\geq$" are always satisfied.

(c) Observe that for $k = 1, 2$, the generalized product properties coincide with the standard ones.

(d) Note that the standard product property implies the generalized one.

(e) Let $G_j \subset \mathbb{C}^{n_j}$ be an n_j-circled domain with $0 \in G_j$, $j = 1, 2$. Assume that G_1 and G_2 satisfy the cone condition. Prove (using Proposition 2.10.5) that the generalized product properties are satisfied for $\boldsymbol{m}^{(k)}$ (resp. for $\boldsymbol{\gamma}^{(k)}$) if $(z'_1, z'_2) = (0, 0)$ (resp. if $(z_1, z_2) = (0, 0)$).

(f) Prove that in the situation described in Example 18.2.5, the generalized product property does not hold.

Hint to (f). If $R \gg 1$, then $\max\{[(\boldsymbol{\gamma}^{(\ell)}_P(a; 1))^\ell Y^{k-\ell}]^{1/k} : \ell = 0, \dots, k\} = Y$.

? (g) Try to find the correct form of the product property for $\boldsymbol{m}^{(k)}$ and $\boldsymbol{\gamma}^{(k)}$.

18.9 List of problems

Chapter 19

Comparison on pseudoconvex domains

Summary. § 19.2 of this chapter, which is mainly based on work of M. Abate [1] and F. Forstnerič & J.-P. Rosay [176], studies the boundary behavior of the Carathéodory and the Kobayashi distances, respectively, on strongly pseudoconvex domains. As an application, the boundary behavior of proper holomorphic mappings between strongly pseudoconvex domains is discussed (see Theorem 19.2.10). § 19.3 deals with localization results for $\boldsymbol{\gamma}$, $\boldsymbol{\varkappa}$, and $\boldsymbol{k}$. In § 19.4, the precise boundary behavior of the invariant pseudometrics is described in terms of the directional vector X, i.e., in terms of its tangential and its normal components. It is based on work of I. Graham [197], G. Aladro [19], and D. Ma [351, 352], and [353]. The analogous results for the Bergman metric are due to L. Hörmander [234] and K. Diederich [124, 125]. Moreover, it turns out that the Carathéodory and the Kobayashi distance are almost equal for strongly pseudoconvex domains (§ 19.5). The chapter concludes with the characterization of the unit ball via its automorphism group; this kind of results was initiated by B. Wong [532] and J.-P. Rosay [454]. The general formulation given here is due to E. B. Lin and B. Wong [346].

Introduction. From Chapter 11 we know that the equalities $\boldsymbol{\gamma}_G = \boldsymbol{\varkappa}_G$ and $\boldsymbol{c}_G = \boldsymbol{k}_G$ hold for any convex domain G. Moreover, on the unit ball $\mathbb{B}_n$ the Bergman metric and distance coincide (up to a constant) with $\boldsymbol{\gamma}_{\mathbb{B}_n}$ and $\boldsymbol{c}_{\mathbb{B}_n}$, respectively. Since it is well known that strongly pseudoconvex domains share a lot of properties with $\mathbb{B}_n$, one may expect that all the above objects behave very similarly on strongly pseudoconvex domains, at least near the boundary. The discussion of this problem is exactly the content of this chapter.

Most of the results in this chapter are based on the existence of precise solutions of the $\overline{\partial}$-equation on strongly pseudoconvex domains; cf. for instance, [223, 265], and [445] for detailed information. The results on the boundary behavior of the Carathéodory and Kobayashi distances are mainly taken from the work of M. Abate [1] and F. Forstnerič & J.-P. Rosay [176]. They lead to a weak form of the beautiful extension theorem of C. Fefferman [165]; see also [52]. More general results are also true for proper holomorphic mappings. The book of K. Diederich and I. Lieb [131] may serve as a source of further information; see also [166].

The boundary behavior of the Carathéodory–Reiffen (resp. the Kobayashi–Royden) metric was studied by I. Graham [197], G. Aladro [19], and by D. Ma [351, 352, 353]. In Chapter 19, we try to follow the estimates given by D. Ma. The analogous results for the Bergman metric are due to L. Hörmander [234] and K. Diederich [124, 125].

In the case of domains that are of finite type, the boundary behavior of the metrics is studied by D. Catlin [88]; see also [228].

The characterization of the unit ball by its automorphism group was initiated by B. Wong [532] and J.-P. Rosay [454]. The general formulation given here is due to E. B. Lin and B. Wong [346]. Much effort was made to generalize this result by substituting the unit ball by complex ellipsoids as model domains; e.g., see [319].

19.1 Strongly pseudoconvex domains

Let G be a bounded domain in $\mathbb{C}^n$. We recall that G is *strongly pseudoconvex*, if there exist a neighborhood U of ∂G and a $\mathcal{C}^2$-function $r : U \longrightarrow \mathbb{R}$ satisfying

(i) $G \cap U = \{z \in U : r(z) < 0\}$, (19.1.1)

(ii) $(\mathbb{C}^n \setminus \overline{G}) \cap U = \{z \in U : r(z) > 0\}$, (19.1.2)

(iii) $dr(z) \neq 0$ for every $z \in \partial G$, (19.1.3)

(iv') $(\mathcal{L}r)(z; X) > 0$ for all $z \in \partial G,\ X \in (\mathbb{C}^n)_*$ with $\sum_{j=1}^{n} \frac{\partial r}{\partial z_j}(z) X_j = 0$.

Observe that if G, U, and r are as above satisfying (19.1.1) and $dr(z) \neq 0$ for all $z \in U$, then r automatically satisfies property (19.1.2).

Under these assumptions it is well known that then the signed boundary distance gives a new $\mathcal{C}^2$-defining function for ∂G; cf. [196, 327, 329]. Therefore, one can choose new U and r such that, in addition to (i), (ii), (iii), the following conditions are also satisfied:

(iv) $(\mathcal{L}r)(z; X) > 0$ for all $z \in U,\ X \in (\mathbb{C}^n)_*$; (19.1.4)

(v) $\|\operatorname{grad} r(z)\| = 1,\ z \in \partial G$, where $\operatorname{grad} r(z) = \left(\frac{\partial r}{\partial \overline{z}_1}(z), \dots, \frac{\partial r}{\partial \overline{z}_n}(z)\right)$; (19.1.5)

(vi) for any $z \in G \cap U$ there exists exactly one point $\pi(z) \in \partial G$ such that

$$\operatorname{dist}(z, \partial G) = \|z - \pi(z)\|. \qquad (19.1.6)$$

We say that a strongly pseudoconvex domain is given by a pair (U, r) if (i)–(vi) are satisfied.

Most of the theorems in this chapter are based on the existence of good solutions of $\overline{\partial}$-equations. For the convenience of the reader, we formulate the result we need in such a form that is used in what follows. Of course, the proof of that deep result is beyond the scope of this book and so is omitted. We refer the reader to, for example, the books [223, 445], and [265], where other references may also be found.

Theorem 19.1.1. *Any strongly pseudoconvex domain $G \subset \mathbb{C}^n$ admits a positive constant C such that if $\alpha = \sum_{j=1}^n \alpha_j d\overline{z}_j$ is a $\overline{\partial}$-closed $(0,1)$-form of class $\mathcal{C}^\infty$ on G with $\|\alpha\|_G := \sum_{j=1}^n \|\alpha_j\|_G < \infty$, then there exists a $\mathcal{C}^\infty$-function f on G with $\overline{\partial} f = \alpha$ and $\|f\|_G \le C\|\alpha\|_G$.*

We know that any strongly pseudoconvex domain is locally biholomorphic to a strictly convex one; cf. [445]. Thus, it is clear that for any boundary point $\zeta \in \partial G$ one can find a *local peak function* f, i.e., $f \in \mathcal{O}(V \cap \overline{G})$ for $V = V(\zeta)$ sufficiently small with $f(\zeta) = 1$ and $|f(z)| < 1$ whenever $z \in \overline{G} \cap V \setminus \{\zeta\}$. In fact, even more is true, as the following result shows.

Theorem 19.1.2. *Let G be a strongly pseudoconvex domain in $\mathbb{C}^n$. Then, for any sufficiently small positive η_1 we can find $0 < \eta_2 < \eta_1$, constants d_1, d_2, and a domain $\widetilde{G} \supset \overline{G}$ such that there exist functions $h(\cdot;\zeta) \in \mathcal{O}(\widetilde{G})$, $\zeta \in \partial G$, that satisfy the following inequalities:*

(i) $h(\zeta;\zeta) = 1$ *and* $|h(z;\zeta)| < 1$, $z \in \overline{G} \setminus \{\zeta\}$,

(ii) $|1 - h(z;\zeta)| \le d_1\|z - \zeta\|$, $z \in \widetilde{G} \cap \mathbb{B}(\zeta, \eta_2)$,

(iii) $|h(z;\zeta)| \le d_2 < 1$, $z \in \overline{G}, \|z - \zeta\| \ge \eta_1$.

Proof. Because of the strong pseudoconvexity of G, we find a neighborhood $U = U(\partial G)$ and a $\mathcal{C}^2$-function r on U such that properties (19.1.1–19.1.6) are satisfied. For a point $\zeta \in \partial G$ put

$$P(z;\zeta) := -\sum_{j=1}^n \frac{\partial r}{\partial z_j}(\zeta)(z_j - \zeta_j) - \frac{1}{2}\sum_{i,j=1}^n \frac{\partial^2 r}{\partial z_i \partial z_j}(\zeta)(z_i - \zeta_i)(z_j - \zeta_j);$$

$P(\cdot;\zeta)$ is called the *Levi polynomial of r at ζ*.

We choose a sufficiently small $\varepsilon_1 > 0$ such that $U' := \bigcup_{\zeta\in\partial G} \mathbb{B}(\zeta, \varepsilon_1) \subset\subset U$. Then, the assumptions on r yield $(\mathcal{L}r)(z;X) \ge C_1\|X\|^2$, $z \in U'$, $X \in \mathbb{C}^n$, where $C_1 < 1$ is a suitable positive constant. With $C_2 := C_1/2$, the Taylor formula then leads to the following inequality:

$$\begin{aligned} r(z) &= r(\zeta) - 2\operatorname{Re} P(z;\zeta) + (\mathcal{L}r)(\zeta; z - \zeta) + o(\|z - \zeta\|^2) \\ &\ge -2\operatorname{Re} P(z;\zeta) + C_2\|z - \zeta\|^2 \quad \text{if} \quad \|z - \zeta\| < \varepsilon_2 < \varepsilon_1,\ \zeta \in \partial G, \end{aligned}$$

where ε_2 is independent of ζ. Thus, on $\mathbb{B}(\zeta, \varepsilon_2)$ we have

$$2\operatorname{Re} P(z;\zeta) \ge C_2\|z - \zeta\|^2 - r(z). \tag{19.1.7}$$

Now fix any positive number $\eta_1 < \varepsilon_2$ and choose a $\mathcal{C}^\infty$-function $\widehat{\chi} : \mathbb{R} \longrightarrow [0,1]$ with $\widehat{\chi}(t) = 1$ for $t \le \eta_1/2$ and $\widehat{\chi}(t) = 0$ for $t \ge \eta_1$. Putting $\chi(z;\zeta) := \widehat{\chi}(\|z - \zeta\|)$, we obtain a $\mathcal{C}^\infty$-function on $\mathbb{C}^n \times \mathbb{C}^n$. Let

$$\varphi(z;\zeta) := \chi(z;\zeta)P(z;\zeta) + (1 - \chi(z;\zeta))\|z - \zeta\|^2, \quad \zeta \in \partial G,\ z \in \mathbb{C}^n.$$

If $\|z-\zeta\| \le \eta_1/2$, then $\varphi(z;\zeta) = P(z;\zeta)$. In particular, $\varphi(\cdot;\zeta)$ is holomorphic on the ball $\mathbb{B}(\zeta,\eta_1/2)$. Moreover, (19.1.7) and $C_2 < 1$ yield the inequality

$$2\operatorname{Re}\varphi(z;\zeta) \ge C_2\eta_1^2/8 > 0 \quad \text{if} \quad \|z-\zeta\| \ge \eta_1/2 \text{ and } r(z) < C_2\eta_1^2/8. \quad (19.1.8)$$

We choose $\eta < C_2\eta_1^2/8$ such that $\widetilde{G} := G \cup \{z \in U' : r(z) < \eta\} \subset\subset G \cup U'$ is a strongly pseudoconvex domain containing $\overline{G}$. Again we point out that η can be taken independent of ζ.

Summarizing, $\varphi(\cdot;\zeta)$ is a $\mathcal{C}^\infty$-function on $\mathbb{C}^n$, which does not vanish on $\widetilde{G} \setminus \mathbb{B}(\zeta,\eta_1/2)$ and which is holomorphic on $\mathbb{B}(\zeta,\eta_1/2)$. Therefore, $\overline{\partial}(1/\varphi(\cdot;\zeta))$ defines a $\overline{\partial}$-closed form $\alpha(\cdot;\zeta) = \sum_{j=1}^n \alpha_j(\cdot;\zeta)dz_j$ of class $\mathcal{C}^\infty$ on $\widetilde{G}$. Because of

$$\alpha_j(z;\zeta) = \begin{cases} 0 & \text{if } z \in \widetilde{G} \cap B(\zeta,\eta_1/2) \\ -\frac{\partial\varphi}{\partial\overline{z}_j}(z;\zeta)/\varphi^2(z;\zeta) & \text{if } z \in \widetilde{G} \setminus \mathbb{B}(\zeta,\eta_1/2) \end{cases}$$

and (19.1.8), it follows that $\|\alpha_j(\cdot;\zeta)\|_{\widetilde{G}} \le C_3$, where the constant C_3 is independent of ζ.

Hence, by Theorem 19.1.1 there exist $\mathcal{C}^\infty$-functions $v(\cdot;\zeta)$ on $\widetilde{G}$ with $\overline{\partial}v(\cdot;\zeta) = \alpha(\cdot;\zeta)$ and $\|v(\cdot;\zeta)\|_{\widetilde{G}} \le C_4$, where C_4 does not depend on ζ.

Put $f(\cdot;\zeta) := 1/\varphi(\cdot;\zeta) + C_4 - v(\cdot;\zeta)$ on $\widetilde{G} \setminus Z(\zeta)$, where $Z(\zeta) := \{z \in \widetilde{G} : \varphi(z;\zeta)=0\}$. Then, $f(\cdot;\zeta)$ belongs to $\mathcal{O}(\widetilde{G}\setminus Z(\zeta))$ and because of (19.1.7) and (19.1.8) we have $\operatorname{Re} f(\cdot;\zeta) > 0$ on $(\widetilde{G}\setminus\mathbb{B}(\zeta,\eta_1/2)) \cup (\overline{G}\setminus\{\zeta\}) =: \widehat{G}$. Therefore, the function $h(\cdot;\zeta) := \exp(-g(\cdot;\zeta))$ with $g(\cdot;\zeta) := 1/f(\cdot;\zeta)$ is holomorphic on $\widehat{G}$ and its values on $\overline{G} \setminus \{\zeta\}$ lie inside the unit disc. If $z \in \widetilde{G} \cap \mathbb{B}(\zeta,\eta_1/2)$, $z \notin Z(\zeta)$, then

$$g(z;\zeta) = \frac{P(z;\zeta)}{1 - P(z;\zeta)(v(z;\zeta) - C_4)}.$$

But, since $g(\cdot;\zeta)$ remains bounded near $Z(\zeta)$, it extends holomorphically through $Z(\zeta) \cap \widetilde{G} \cap B(\zeta,\eta_1/2)$, and therefore $h(\cdot;\zeta) \in \mathcal{O}(\widetilde{G})$. Observe that there is a positive C_5 such that

$$|P(z;\zeta)| \le C_5\|z-\zeta\|, \quad \zeta \in \partial G,\ z \in \widetilde{G}.$$

Thus, if $0 < \eta_2 < \min\{\eta_1/2, 1/(4C_4C_5)\}$, then it follows that

$$|g(z;\zeta)| \le \frac{C_5\|z-\zeta\|}{1 - 2C_4C_5\|z-\zeta\|} < C_6\|z-\zeta\|, \quad z \in \widetilde{G},\ \|z-\zeta\| < \eta_2,$$

where $C_6 := 2C_5$.

Now, we choose C_7 in such a way that $|e^\lambda - 1| \le C_7|\lambda|$ for $|\lambda| \le C_6\eta_2$. Then, we obtain

$$|1 - h(z;\zeta)| \le C_7|g(z;\zeta)| \le C_6C_7\|z-\zeta\| =: d_1\|z-\zeta\|, \quad z \in \widetilde{G},\ \|z-\zeta\| < \eta_2.$$

In particular, this shows that $h(\zeta;\zeta) = 1$.

What remains is the upper estimate for $|h(\cdot;\zeta)|$ away from the boundary point ζ. Take $z \in \overline{G}$, $\|z-\zeta\| \ge \eta_1$. Then, we get

$$\begin{aligned}\operatorname{Re} g(z;\zeta) &= \|z-\zeta\|^2 \frac{1+\|z-\zeta\|^2(C_4-\operatorname{Re} v(z;\zeta))}{|1-\|z-\zeta\|^2(v(z;\zeta)-C_4)|^2} \\ &\ge \eta_1^2/(1+(\operatorname{diam} G)^2\cdot 2C_4)^2 =: C_8.\end{aligned}$$

Hence, we have $|h(z;\zeta)| \le e^{-C_8} =: d_2 < 1$. □

Before presenting a result on the approximation of bounded holomorphic functions, we recall the following *stability property* of strongly pseudoconvex domains: if K is an arbitrary compact subset of the boundary of a strongly pseudoconvex domain, then there are arbitrarily small, strongly pseudoconvex enlargements $\widehat{G}$ with $K \subset \widehat{G}$ and $\partial G = \partial\widehat{G}$ away from K. Observe that for the case where $K = \partial G$ this property has been just exploited in the proof of Theorem 19.1.2. In the general form this property will be used in the proof of the following result on approximation, which has been found by I. Graham (cf. [197]):

Theorem 19.1.3. *Let G be a strongly pseudoconvex domain in $\mathbb{C}^n$. Then, for sufficiently small R the set $G \cap \mathbb{B}(\zeta,R)$ is connected whenever $\zeta \in \partial G$ and there exists a $\varrho = \varrho(R) < R$ such that the following property holds:*

Given $\varepsilon > 0$, there exists a number $L = L(\varepsilon,R) > 0$ such that for any $\zeta \in \partial G$, $f \in \mathcal{H}^\infty(G \cap B(\zeta,R))$, and $w \in G \cap \mathbb{B}(\zeta,\varrho)$ there is an $\hat{f} \in \mathcal{H}^\infty(G)$ satisfying

(i) $D^\alpha \hat{f}(w) = D^\alpha f(w)$, $\alpha \in (\mathbb{Z}_+)^n$, $|\alpha| \le 1$,

(ii) $\|\hat{f}\|_G \le L\|f\|_{G\cap\mathbb{B}(\zeta,R)}$,

(iii) $\|\hat{f}-f\|_{G\cap\mathbb{B}(\zeta,\varrho)} < \varepsilon$.

Proof. According to Theorem 19.1.2 we choose the data $\eta_2 < \eta_1$, $d_1, d_2 < 1$, $\widetilde{G} \supset \overline{G}$, and $h(\cdot;\zeta) \in \mathcal{O}(\widetilde{G})$, $\zeta \in \partial G$, so that $G \cap \mathbb{B}(\zeta,R)$ with $R := 2\eta_1$ is connected for all $\zeta \in \partial G$. Introducing $(h(\cdot;\zeta)+3)/4$, we can also require that $|h(z;\zeta)| \ge 1/2$, $z \in \overline{G}$, $\zeta \in \partial G$.

Fix $d_3 \in (d_2,1)$ and then take $0 < \eta \le \eta_2$ such that $\mathbb{B}(\zeta,\eta) \subset \widetilde{G}$ and $|h(z;\zeta)| \ge d_3$, $\|z-\zeta\| \le \eta$, $\zeta \in \partial G$. With $\varrho := \min\{\eta, \eta_1/5\}$ there is a finite number of boundary points $\zeta_1,\dots,\zeta_N$ such that $\partial G \subset \bigcup_{j=1}^N \mathbb{B}(\zeta_j,\varrho)$. Then, we choose strongly pseudoconvex domains $G_j \supset G$ by modifying G near ζ_j so that

$$\overline{G} \cap \overline{\mathbb{B}(\zeta_j,2\varrho)} \subset\subset G_j \subset \widetilde{G} \quad\text{and}\quad G\setminus\mathbb{B}(\zeta_j,4\varrho) = G_j\setminus\mathbb{B}(\zeta_j,4\varrho).$$

Now let $\zeta_0 \in \partial G$, $f \in \mathcal{H}^\infty(G\cap\mathbb{B}(\zeta_0,R))$, and $w \in G\cap\mathbb{B}(\zeta_0,\varrho)$. Then, we find ζ_{j_0} with $\zeta_0 \in \mathbb{B}(\zeta_{j_0},\varrho)$. For simplicity, we take $\zeta_{j_0} = \zeta_1$.

We denote by χ a cut-off function with $0 \le \chi \le 1$, $\chi \equiv 1$ on $\mathbb{B}(\zeta_0,6\eta_1/5)$, and $\chi \equiv 0$ outside $\mathbb{B}(\zeta_0,9\eta_1/5)$. Then, put $\alpha := (\bar\partial\chi)f$ on $G\cap\mathbb{B}(\zeta_0,2\eta_1)$ and $\alpha := 0$ on

$G \setminus \mathbb{B}(\zeta_0, 2\eta_1)$. Observe that $\alpha \equiv 0$ on $G \cap \mathbb{B}(\zeta_0, 6\eta_1/5)$ and on $G \setminus \mathbb{B}(\zeta_0, 9\eta_1/5)$. Hence, by trivial extension, α can be thought of as a $\overline{\partial}$-closed $(0,1)$-form on G_1 of class $\mathcal{C}^\infty$.

Instead of solving the equation $\overline{\partial} u = \alpha$, we will deal with the equations $\overline{\partial} v_k = h(\cdot; \zeta_0)^k \alpha$, $k \in \mathbb{N}$, where the right hand side is again a $\overline{\partial}$-closed $(0,1)$-form of class $\mathcal{C}^\infty$ on G_1. Applying Theorem 19.1.1 gives a solution $v_k \in C^\infty(G_1)$ with

$$\|v_k\|_{G_1} \le C_1 \|h(\cdot; \zeta_0)^k\|_{\operatorname{supp}\alpha} \|\alpha\|_{G_1} \le C_1 C d_2^k \|f\|_{G \cap \mathbb{B}(\zeta_0, R)},$$

where the constant C depends only on χ, and so only on η_1. Then, the functions $f_k := \chi f - h(\cdot; \zeta_0)^{-k} v_k$ and $h(\cdot; \zeta_0)^{-k} v_k$ are holomorphic on G and on $G_1 \cap \mathbb{B}(\zeta_0, \eta)$, respectively, with

$$\|h(\cdot; \zeta_0)^{-k} v_k\|_{G_1 \cap \mathbb{B}(\zeta_0, \eta)} \le C_1 C (d_2/d_3)^k \|f\|_{G \cap \mathbb{B}(\zeta_0, R)}.$$

Now, if $z \in \overline{G} \cap \mathbb{B}(\zeta_0, \varrho)$, then $\|z - \zeta_1\| \le \|z - \zeta_0\| + \|\zeta_0 - \zeta_1\| < 2\varrho$, i.e., $\overline{G} \cap \overline{\mathbb{B}(\zeta_0, \varrho)} \subset \overline{G} \cap \mathbb{B}(\zeta_1, 2\varrho) \subset\subset G_1$.

Hence, there is an $L_1 < 1$, independent of $w \in G \cap \mathbb{B}(\zeta_0, \varrho)$, such that the Cauchy inequalities imply that

$$\left|\frac{\partial f_k}{\partial z_j}(w) - \frac{\partial f}{\partial z_j}(w)\right| \le \frac{C_1 C}{L_1} \left(\frac{d_2}{d_3}\right)^k \|f\|_{G \cap \mathbb{B}(\zeta_0, R)}.$$

Let $\varepsilon < 1$. We fix the exponent k so large that $\frac{C_1 C}{L_1} \left(\frac{d_2}{d_3}\right)^k < \widetilde{\varepsilon}$ with

$$\widetilde{\varepsilon} := \varepsilon \left(2(1 + n(\eta + \operatorname{diam} G))(1 + \|f\|_{G \cap \mathbb{B}(\zeta_0, R)})\right)^{-1}.$$

So we obtain

$$\max\left\{\|f_k - f\|_{G_1 \cap \mathbb{B}(\zeta_0, \eta)}, \left|\frac{\partial f_k}{\partial z_j}(w) - \frac{\partial f}{\partial z_j}(w)\right|\right\} < \widetilde{\varepsilon} \|f\|_{G \cap \mathbb{B}(\zeta_0, R)}.$$

In the next step, we define on G a new holomorphic function $\widehat{f}$ by

$$\widehat{f}(z) := f_k(z) + f(w) - f_k(w) + \sum_{j=1}^n \left(\frac{\partial f}{\partial z_j}(w) - \frac{\partial f_k}{\partial z_j}(w)\right)(z_j - w_j).$$

We note that $\widehat{f}(w) = f(w)$ and $\frac{\partial \widehat{f}}{\partial z_j}(w) = \frac{\partial f}{\partial z_j}(w)$. Moreover, since $\varrho \le \eta$, it follows that

$$\|\widehat{f} - f\|_{G \cap \mathbb{B}(\zeta_0, \varrho)} \le \varepsilon,$$
$$\|\widehat{f}\|_G \le \|f_k\|_G + \|f\|_{G \cap \mathbb{B}(\zeta_0, R)} \le (2 + 2^k C_1 C d_2^k) \|f\|_{G \cap \mathbb{B}(\zeta_0, R)}.$$

With $L := \max\{2 + 2^k C_j C d_2^k : 1 \le j \le N\}$, the proof is complete. □

19.2 The boundary behavior of the Carathéodory and the Kobayashi distances

Looking at the explicit formula of the Carathéodory distance for the unit ball, one easily sees that $c_{\mathbb{B}_n}(0,\cdot)$ behaves like $-(1/2)(\log \operatorname{dist}(\cdot, \partial\mathbb{B}_n))$ near the boundary. It will turn out that the same boundary behavior remains true in any strongly pseudoconvex domain (cf. [1, 163, 523]).

Theorem 19.2.1. *Let K be a compact subset of a strongly pseudoconvex domain $G \subset \mathbb{C}^n$. Then, there exists a constant $C > 0$ such that*

$$c_G(z_0, z) \geq -(1/2)\log \operatorname{dist}(z, \partial G) - C, \quad z_0 \in K,\ z \in G.$$

Proof. Because of $\mathcal{C}^2$-smoothness of ∂G we can find a positive ε_0 such that for any point $z \in G$ with $\operatorname{dist}(z, \partial G) < \varepsilon_0$ there is exactly one point $\zeta(z) \in \partial G$ satisfying

$$z = \zeta(z) - \operatorname{dist}(z, \partial G) \cdot \nu(\zeta(z)). \tag{19.2.1}$$

Here, $\nu(\zeta(z))$ denotes the unit outer normal to ∂G at the point $\zeta(z)$. Now, we choose positive real numbers $\eta_2 < \eta_1$, d_1, d_2, and the function h according to Theorem 19.1.2 with $2\eta_1 < \varepsilon_0$ and $\eta_1 < \operatorname{dist}(K, \partial G)$.

Fix $z_0 \in K$. Now let $z \in G$ with $\operatorname{dist}(z, \partial G) < \eta_3 := \min\{\eta_2, (1-d_2)/d_1\}$. Then, with the aid of the point $\zeta(z) \in \partial G$, we get the following estimate:

$$\begin{aligned} c_G^*(z_0, z) &\geq \frac{|h(z;\zeta(z)) - h(z_0;\zeta(z))|}{|1 - \overline{h(z_0;\zeta(z))}h(z;\zeta(z))|} \geq \frac{|h(z_0;\zeta(z)) - 1| - |h(z;\zeta(z)) - 1|}{|h(z_0;\zeta(z)) - 1| + |h(z;\zeta(z)) - 1|} \\ &\geq \frac{1 - d_2 - |h(z;\zeta(z)) - 1|}{1 - d_2 + |h(z;\zeta(z)) - 1|} \geq \frac{1 - d_2 - d_1\|z - \zeta(z)\|}{1 - d_2 + d_1\|z - \zeta(z)\|} > 0. \end{aligned}$$

Using (19.2.1), it follows that

$$c_G(z_0, z) \geq \frac{1}{2}\log\frac{1 - d_2}{d_1\|z - \zeta(z)\|} = \frac{1}{2}\log\frac{1 - d_2}{d_1} - \frac{1}{2}\log \operatorname{dist}(z, \partial G).$$

On the other hand, if $z \in G$ with $\operatorname{dist}(z, \partial G) \geq \eta_3$, then

$$c_G(z_0, z) \geq 0 \geq (-1/2)\log \operatorname{dist}(z, \partial G) - C_1$$

with a suitable $C_1 > 0$. □

Analyzing the proof of Theorem 19.2.1 immediately gives the following local version of Theorem 19.2.1:

Theorem 19.2.2. *Let G be as in Theorem 19.2.1 and let $\varepsilon > 0$. Then, there exist positive numbers $\varrho_2 < \varrho_1 < \varepsilon$, and C such that for any $\zeta \in \partial G$ the following inequality is true:*

$$c_G(z, w) \geq -(1/2)\log \operatorname{dist}(z, \partial G) - C, \quad z \in G \cap \mathbb{B}(\zeta, \varrho_2),\ w \in G \setminus \mathbb{B}(\zeta, \varrho_1).$$

Remark 19.2.3.

(a) The proof of Theorem 19.2.1 shows a strong relation between the lower estimate of the Carathéodory distance and the existence of good peak functions.

(b) Moreover, if G is as above and if ζ', ζ'' are different boundary points of G, then even the following inequality has been claimed in [163, 523]:

$$c_G(z', z'') \geq -(1/2)\log \operatorname{dist}(z', \partial G) - (1/2)\log \operatorname{dist}(z'', \partial G) - C,$$

whenever $z', z'' \in G$, z' is near ζ', and z'' is near ζ''.

We will not use this result. Later on, we will show a similar inequality for the Kobayashi distance that is much easier to obtain.

As we have already mentioned, the lower estimate for the Carathéodory distance depends on deep results of complex analysis whereas the following upper estimate relies only on the smoothness of the boundary:

Proposition 19.2.4. *For a bounded domain G in $\mathbb{C}^n$ with smooth $\mathcal{C}^2$-boundary and a compact subset K of G there is a constant C such that*

$$\boldsymbol{k}_G(z_0, z) \leq -(1/2)\log \operatorname{dist}(z, \partial G) + C, \quad z_0 \in K,\ z \in G.$$

Proof. If we choose $\varepsilon_0 > 0$ sufficiently small, then any point $z \in G$ such that $\operatorname{dist}(z, \partial G) < \varepsilon_0$ lies inside the ball $\mathbb{B}(z', \varepsilon_0) \subset G$ with $z' := \zeta(z) - \varepsilon_0 \nu(\zeta(z))$, $\zeta(z) \in \partial G$, and, moreover, $z = \zeta(z) - \operatorname{dist}(z, \partial G)\nu(\zeta(z))$. Applying the triangle inequality, we obtain

$$\begin{aligned}\boldsymbol{k}_G(z, z_0) &\leq \boldsymbol{k}_G(z, z') + \boldsymbol{k}_G(z', z_0) \leq \boldsymbol{k}_{\mathbb{B}(z',\varepsilon_0)}(z, z') + C_1\\ &= \frac{1}{2}\log\frac{\varepsilon_0 + \|z - z'\|}{\varepsilon_0 - \|z - z'\|} + C_1 \leq -\frac{1}{2}\log \operatorname{dist}(z, \partial G) + \frac{1}{2}\log(2\varepsilon_0) + C_1,\end{aligned}$$

where $C_1 := \sup\{\boldsymbol{k}_G(w, \omega) : \omega \in K \text{ and } w \in G \text{ with } \operatorname{dist}(w, \partial G) \geq \varepsilon_0\}$.

The remaining case, namely $z \in G$ with $\operatorname{dist}(z, \partial G) \geq \varepsilon_0$, can be handled as in the proof of Theorem 19.2.1. ☐

Combining the previous two theorems yields the following result:

Theorem 19.2.5 (cf. [1]). *For a strongly pseudoconvex domain $G \subset \mathbb{C}^n$ and $z_0 \in G$ we have*

$$\lim_{z\to\partial G}\frac{c_G(z_0, z)}{-\log \operatorname{dist}(z, \partial G)} = \lim_{z\to\partial G}\frac{\boldsymbol{k}_G(z_0, z)}{-\log \operatorname{dist}(z, \partial G)} = \frac{1}{2},$$

where the limits are locally uniform in the first variable z_0.

Moreover, we get a precise description of the boundary behavior of complex geodesics in strongly pseudoconvex domains (cf. Exercise 11.7.9).

Corollary 19.2.6. *Let $\varphi : \mathbb{D} \longrightarrow G$ be a complex $\boldsymbol{c}_G$-geodesic in a strongly pseudoconvex domain G. Then, there exist $k_1, k_2 > 0$ (depending only on $\varphi(0)$) such that*

$$k_1 \operatorname{dist}(\varphi(\lambda), \partial G) \le 1 - |\lambda| \le k_2 \operatorname{dist}(\varphi(\lambda), \partial G), \quad \lambda \in \mathbb{D}.$$

Proof. For a compact set $K \subset G$ with $\varphi(0) \in K$, using Theorem 19.2.1 and Proposition 19.2.4, we obtain

$$\begin{aligned}-C_1 - (1/2) \log \operatorname{dist}(\varphi(\lambda), \partial G) &\le \boldsymbol{c}_G(\varphi(0), \varphi(\lambda)) = p(0, \lambda) \\ &= \frac{1}{2} \log \frac{1 + |\lambda|}{1 - |\lambda|} \le C_2 - \frac{1}{2} \log \operatorname{dist}(\varphi(\lambda), \partial G)\end{aligned}$$

with appropriate constants C_1, C_2 that depend only on K. Thus, it follows that

$$\exp(-2C_1)/\operatorname{dist}(\varphi(\lambda), \partial G) \le \frac{1 + |\lambda|}{1 - |\lambda|} \le \exp(2C_2)/\operatorname{dist}(\varphi(\lambda), \partial G),$$

and so

$$\exp(-2C_2) \operatorname{dist}(\varphi(\lambda), \partial G) \le 1 - |\lambda| \le 2 \exp(2C_1) \operatorname{dist}(\varphi(\lambda), \partial G). \qquad \square$$

Corollary 19.2.6 will be used at the end of this section to show that complex $\boldsymbol{c}_G$-geodesics in strongly pseudoconvex domains are continuous up to the boundary.

Now, we turn to the lower estimate of the Kobayashi distance between two points near two different boundary points; cf. Remark 19.2.3(b).

Proposition 19.2.7. *Let ζ', ζ'' be two different boundary points of a strongly pseudoconvex domain $G \subset \mathbb{C}^n$. Then, for a suitable constant C, we have*

$$\boldsymbol{k}_G(z', z'') \ge -(1/2) \log \operatorname{dist}(z', \partial G) - (1/2) \log \operatorname{dist}(z'', \partial G) - C,$$

whenever $z', z'' \in G$, z' is near ζ' and z'' is near ζ''.

Proof. Theorem 19.2.2 implies that there are disjoint neighborhoods $U' = U'(\zeta')$ and $U'' = U''(\zeta'')$ such that for suitable smaller $V' = V'(\zeta') \subset\subset U'$ and $V'' = V''(\zeta'') \subset\subset U''$ and for an appropriate constant C the following inequalities are true:

$$\boldsymbol{k}_G(z', G \setminus U') \ge -(1/2) \log \operatorname{dist}(z', \partial G) - C, \quad z' \in V', \tag{19.2.2}$$

$$\boldsymbol{k}_G(z'', G \setminus U'') \ge -(1/2) \log \operatorname{dist}(z'', \partial G) - C, \quad z'' \in V''. \tag{19.2.3}$$

Now fix $z' \in V'$ and $z'' \in V''$ and choose $\varepsilon > 0$. Then, we can find a $\mathcal{C}^1$-curve $\alpha : [0, 1] \longrightarrow G$, $\alpha(0) = z'$, $\alpha(1) = z''$, such that

$$\boldsymbol{k}_G(z', z'') + \varepsilon > \int_0^1 \boldsymbol{\varkappa}_G(\alpha(t); \alpha'(t)) dt. \tag{19.2.4}$$

With $0 < t_1 < t_2 < 1$ such that $\alpha([0,t_1)) \subset U'$, $\alpha(t_1) \in \partial U'$, and $\alpha((t_2,1]) \subset U''$, $\alpha(t_2) \in \partial U''$, conditions (19.2.4) and (19.2.2, 19.2.3) imply that

$$\begin{aligned} \boldsymbol{k}_G(z',z'') + \varepsilon &\geq \boldsymbol{k}_G(z',\alpha(t_1)) + \boldsymbol{k}_G(\alpha(t_2),z'') \\ &\geq -(1/2)\log \operatorname{dist}(z',\partial G) - (1/2)\log\operatorname{dist}(z'',\partial G) - 2C. \end{aligned}$$

Since ε is arbitrary, the proposition is verified. □

Remark 19.2.8. A version of Proposition 19.2.7 with less restrictive conditions may be found in a paper of F. Forstnerič and J.-P. Rosay (cf. [176]). They only assume that G is a bounded domain whose boundary is $\mathcal{C}^2$ and which is strongly pseudoconvex in neighborhoods of the points ζ', ζ''.

In the case where two points, whose $\boldsymbol{k}$-distance is measured, converge to the same boundary point, the following upper estimate is also due to F. Forstnerič and J.-P. Rosay. They dealt with bounded domains with $\mathcal{C}^{1+\varepsilon}$-boundary. Here, we will assume that the boundary is of class $\mathcal{C}^2$.

Proposition 19.2.9. *Let G be a bounded domain in $\mathbb{C}^n$ with smooth $\mathcal{C}^2$-boundary and let $\zeta_0 \in \partial G$. Then, there exist a neighborhood $U = U(\zeta_0)$ and a constant $C > 0$ such that*

$$\boldsymbol{k}_G(z_1,z_2) \leq -\frac{1}{2}\sum_{j=1}^{2}\log\operatorname{dist}(z_j,\partial G) + \frac{1}{2}\sum_{j=1}^{2}\log(\operatorname{dist}(z_j,\partial G) + \|z_1 - z_2\|) + C$$

for z_1, $z_2 \in G \cap U$.

Proof. Since the boundary of G is of class $\mathcal{C}^2$, there exists an $R \ll 1$ such that

(i) $\|\nu(\zeta) - \nu(\zeta_0)\| < 1/8$, $\zeta \in \partial G \cap \mathbb{B}(\zeta_0, 2R)$;

(ii) $z - \delta\nu(\zeta) \in G$ and $\operatorname{dist}(z - \delta\nu(\zeta), \partial G) > 3\delta/4$, whenever $z \in G \cap U$, $U := \mathbb{B}(\zeta_0, R)$, $\zeta \in \partial G \cap \mathbb{B}(\zeta_0, 8R)$ and $\delta \leq 2R$;

(iii) $\overline{\mathbb{B}(\zeta - 4R\nu(\zeta), 4R)} \subset G \cup \{\zeta\}$, $\zeta \in \partial G$;

here, $\nu(\zeta)$ denotes, as before, the unit outer normal to ∂G at ζ.

Now fix two points z_1, $z_2 \in G \cap U$ and choose the uniquely determined points $\zeta_j \in \partial G$ with $\|z_j - \zeta_j\| = \operatorname{dist}(z_j, \partial G)$. Recall that $z_j = \zeta_j - \|z_j - \zeta_j\|\nu(\zeta_j)$. Moreover, we obtain that

$$\|\zeta_j - \zeta_0\| \leq \|\zeta_j - z_j\| + \|z_j - \zeta_0\| \leq 2\|z_j - \zeta_0\| < 2R,$$

i.e., $\zeta_j \in \partial G \cap \mathbb{B}(\zeta_0, 2R)$. Therefore, the points $w_j := z_j - \|z_1 - z_2\|\nu(\zeta_j)$ lie in G and $\operatorname{dist}(w_j, \partial G) \geq (3/4)\|z_1 - z_2\|$, because of (ii) above. So the triangle inequality leads to the following upper estimate:

$$\begin{aligned}
\boldsymbol{k}_G(z_1,z_2) &\le \boldsymbol{k}_G(w_1,w_2)+\sum_{j=1}^{2}\boldsymbol{k}_G(w_j,z_j) \le \boldsymbol{k}_G(w_1,w_2)+\sum_{j=1}^{2}\boldsymbol{k}_{\mathbb{B}(w_j,\varrho_j)}(w_j,z_j)\\
&= \boldsymbol{k}_G(w_1,w_2)+\sum_{j=1}^{2}\frac{1}{2}\log\frac{\varrho_j+\|z_j-w_j\|}{\varrho_j-\|z_j-w_j\|}\\
&\le \boldsymbol{k}_G(w_1,w_2)+\frac{1}{2}\sum_{j=1}^{2}\log\frac{2\varrho_j}{\operatorname{dist}(z_j,\partial G)},
\end{aligned}$$

where $\varrho_j := \|w_j-\zeta_j\| < 4R$.

What remains is to estimate the term $\boldsymbol{k}_G(w_1,w_2)$. First, observe that we have $\|w_1-w_2\| \le (5/4)\|z_1-z_2\|$; here we use (i) above. Then, we consider the analytic curve $\varphi : \mathbb{C} \longrightarrow \mathbb{C}^n$ defined by $\varphi(\lambda) := w_1+\lambda(w_2-w_1)$, $\lambda\in\mathbb{C}$. If $|\lambda|<3/5$ (resp. $|\lambda-1|<3/5$), we obtain

$$\begin{aligned}
&\|\varphi(\lambda)-w_1\| < (3/5)(5/4)\|z_1-z_2\| \le \operatorname{dist}(w_1,\partial G)\\
(\text{resp. } &\|\varphi(\lambda)-w_2\| = |1-\lambda|\,\|w_1-w_2\| < (3/4)\|z_1-z_2\| \le \operatorname{dist}(w_2,\partial G)),
\end{aligned}$$

i.e., if $D := \mathbb{B}(3/5)\cup\mathbb{B}(1,3/5)\subset\mathbb{C}$, then $\varphi|_D$ is a holomorphic map into G. So we finally conclude that

$$\begin{aligned}
\boldsymbol{k}_G(z_1,z_2) \le \boldsymbol{k}_D(0,1)+\log 2-\frac{1}{2}\sum_{j=1}^{2}\log\operatorname{dist}(z_j,\partial G)\\
+\frac{1}{2}\sum_{j=1}^{2}\log(\operatorname{dist}(z_j,\partial G)+\|z_1-z_2\|). \qquad \square
\end{aligned}$$

Now, we intend to apply Theorem 19.2.1 and Proposition 19.2.9 to prove a theorem on the boundary behavior of proper holomorphic mappings between strongly pseudoconvex domains. For a more general treatment and an extensive bibliography, the reader should consult the book of K. Diederich and I. Lieb [131]; see also [175].

For the convenience of the reader, we collect those properties of proper holomorphic mappings that we will need. Details can be found in Rudin's book [462], pages 300–305 (see also Appendix B.2).

Let $F : G_1 \longrightarrow G_2$ be proper and holomorphic (G_j, $j=1,2$, are domains in $\mathbb{C}^n$). Put $M := \{z\in G_1 : \det F'(z)=0\}$. Then, $M \subsetneq G_1$, F is an open mapping, and the set of *critical values* $F(M)$ of F is a proper analytic subset of G_2. Moreover, $F|_{G_1\setminus F^{-1}(F(M))} : G_1\setminus F^{-1}(F(M)) \longrightarrow G_2\setminus F(M)$ is an unbranched proper holomorphic covering of finite order.

After this short summary, we can formulate the result we want to discuss; also cf. Exercise 19.7.5.

Theorem 19.2.10. *Let G_1 and G_2 be two strongly pseudoconvex domains in $\mathbb{C}^n$. Then, any proper holomorphic map $F : G_1 \longrightarrow G_2$ extends to a continuous map from $\overline{G}_1$ into $\overline{G}_2$.*

Proof. We assume that G_1 is given by a pair (U_1, r_1) satisfying conditions (19.1.1–19.1.6). Now let U_1' be an open set with $\partial G \subset U_1' \subset\subset U_1$. Put $K_2 := F(G_1 \setminus U_1')$ and define

$$r_2(w) := \sup\{r_1(z) : z \in G \text{ and } F(z) = w\}, \quad w \in G_2 \setminus K_2.$$

Since F is proper and holomorphic, it turns out that r_2 is negative and continuous on $G_2 \setminus K_2$ and psh on $(G_2 \setminus K_2) \setminus F(M)$. So, r_2 is a psh function on $G_2 \setminus K_2$ (cf. Appendix B.4.23) with $r_2(w) \underset{w \to \partial G_2}{\longrightarrow} 0$.

Since G_2 has a $\mathcal{C}^2$-boundary, there exist positive numbers ε_0, δ_0, and C_0 with the following properties: for any point $w \in G_2$ with $\operatorname{dist}(w, \partial G_2) < \varepsilon_0$ one can find points $w_0 \in G_2$ and $\zeta(w) \in \partial G_2$ satisfying

(i) $\mathbb{B}(w_0, \varepsilon_0) \subset G_2 \setminus K_2$,

(ii) $\overline{\mathbb{B}(w_0, \varepsilon_0)} \cap \partial G_2 = \{\zeta(w)\}$,

(iii) $\|w - \zeta(w)\| = \operatorname{dist}(w, \partial G_2)$,

(iv) $\mathbb{B}(\zeta(w) - 2(w_0 - \zeta(w)), \delta_0) =: B_w \subset G_2 \setminus K_2$,

(v) $r_2 \le -C_0$ on B_w.

Now, we are going to prove a fact known as the Hopf lemma. Fix w, w_0, and $\zeta(w)$ as above. Then, the function $u : \overline{\mathbb{D}} \longrightarrow \mathbb{R}$ defined as

$$u(\lambda) := \begin{cases} r_2(w_0 + \lambda(\zeta(w) - w_0)) & \text{if } \lambda \neq 1 \\ 0 & \text{if } \lambda = 1 \end{cases}$$

is continuous and its restriction to $\mathbb{D}$ is subharmonic. If $h : \overline{\mathbb{D}} \longrightarrow (-\infty, 0]$ is the solution of the Dirichlet problem with u as boundary values, then Harnack's inequality implies that

$$u(\lambda) \le h(\lambda) \le h(0)\frac{1 - |\lambda|}{1 + |\lambda|} \le h(0)(1 - |\lambda|), \quad \lambda \in \mathbb{D}.$$

In particular, we obtain

$$r_2(w) = u\left(\frac{\varepsilon_0 - \operatorname{dist}(w, \partial G_2)}{\varepsilon_0}\right) \le \frac{h(0)}{\varepsilon_0} \operatorname{dist}(w, \partial G_2).$$

Because of (iv) and (v), the number $C_1 := h(0) = \frac{1}{2\pi}\int_0^{2\pi} u(e^{i\theta})d\theta$ is negative, i.e.,

$$r_2(w) \le C_1 \operatorname{dist}(w, \partial G_2)/\varepsilon_0 = -C_2 \operatorname{dist}(w, \partial G_2) \quad \text{with} \quad C_2 := -C_1/\varepsilon_0 > 0.$$

Hence, for $z \in G_1$ sufficiently near ∂G_1, we obtain $C_2 \operatorname{dist}(F(z), \partial G_2) \le |r_2(F(z))| \le |r_1(z)| \le C_3 \operatorname{dist}(z, \partial G)$, where C_3 denotes a suitable positive constant.

With a correctly chosen positive C, we finally get the following inequality:

$$\operatorname{dist}(F(z), \partial G_2) \le C \operatorname{dist}(z, \partial G_1), \quad z \in G_1. \tag{19.2.5}$$

The last step is to prove that F extends continuously to $\overline{G}_1$. Here, Propositions 19.2.7 and 19.2.9 will do the main job. Obviously, it suffices to prove that for any sequence $(z_\nu)_{\nu\in\mathbb{N}} \subset G_1$ with $\lim_{\nu\to\infty} z_\nu = z_0 \in \partial G_1$ the image sequence $(F(z_\nu))_{\nu\in\mathbb{N}}$ is also convergent. Observe that $\operatorname{dist}(F(z_\nu), \partial G_2) \underset{\nu\to\infty}{\longrightarrow} 0$ since F is proper.

Now, we argue via a contradiction. Suppose that there are two different accumulation points $w^{(1)}$, $w^{(2)}$ of $(F(z_\nu))_{\nu\in\mathbb{N}}$, i.e., there exist two subsequences $(z_\nu^{(j)})_{\nu\in\mathbb{N}} \subset (z_\nu)_{\nu\in\mathbb{N}}$ with $\lim_{\nu\to\infty} F(z_\nu^{(j)}) = w^{(j)} \in \partial G_2$, $j = 1, 2$. Then, Propositions 19.2.7 and 19.2.9 imply the following chain of inequalities:

$$\begin{aligned}
-\sum_{j=1}^{2} \frac{1}{2} &\log \operatorname{dist}(F(z_\nu^{(j)}), \partial G_2) - C_1 \\
&\le \boldsymbol{k}_{G_2}(F(z_\nu^{(1)}), F(z_\nu^{(2)})) \le \boldsymbol{k}_{G_1}(z_\nu^{(1)}, z_\nu^{(2)}) \\
&\le -\sum_{j=1}^{2} \frac{1}{2} \log \operatorname{dist}(z_\nu^{(j)}, \partial G_1) \\
&\quad + \sum_{j=1}^{2} \frac{1}{2} \log(\operatorname{dist}(z_\nu^{(j)}, \partial G_1) + \|z_\nu^{(1)} - z_\nu^{(2)}\|) + C_2,
\end{aligned}$$

where C_1, C_2 are suitable positive constants.

With the aid of (19.2.5), we can easily continue this estimate, and so we end up with

$$-C_3 \le \sum_{j=1}^{2} \frac{1}{2} \log(\operatorname{dist}(z_\nu^{(j)}, \partial G_1) + \|z_\nu^{(1)} - z_\nu^{(2)}\|) + C_2, \quad \nu \in \mathbb{N},$$

which is obviously impossible. □

As a simple consequence, we get the following weak version of a deep result of C. Fefferman (cf. [165], see also [523]).

Corollary 19.2.11. *Any biholomorphic mapping between two strongly pseudoconvex domains in $\mathbb{C}^n$ extends to a homeomorphism between their closures.*

As promised after Corollary 19.2.6, we conclude this section with a general regularity result for complex $\boldsymbol{c}_G$-geodesics in strongly pseudoconvex domains.

Theorem 19.2.12. *Let G be a strongly pseudoconvex domain in $\mathbb{C}^n$. Then, any complex $\boldsymbol{c}_G$-geodesic $\varphi : \mathbb{D} \longrightarrow G$ is $1/2$-Hölder continuous, and therefore it extends continuously to $\overline{\mathbb{D}}$.*

Before we start the proof, we establish the following lemma on good exhaustions of strongly pseudoconvex domains (cf. [174]):

Lemma 19.2.13. *Any strongly pseudoconvex domain $G \subset \mathbb{C}^n$ admits a strictly psh $\mathcal{C}^2$-function $\varrho : V \longrightarrow \mathbb{R}$ defined on an open set $V \supset \overline{G}$ such that*

(i) $G = \{z \in V : \varrho(z) < 0\}$,

(ii) $d\varrho(z) \neq 0$ *for* $z \in \partial G$.

Proof. Assume that G is given by a pair (U, r) satisfying (19.1.1–19.1.6). Now choose an open set $U' \subset\subset U$ with $\partial G \subset U'$ and put $5\varepsilon := \sup\{r(z) : z \in G \cap \partial U'\} < 0$.

Moreover, let $\chi_1 : \mathbb{R} \longrightarrow \mathbb{R}$ be an increasing convex $\mathcal{C}^\infty$-function with the following properties:

$$\chi_1(t) = t \text{ if } t \geq 3\varepsilon, \quad \chi_1(t) = 3.5\varepsilon \text{ if } t \leq 4\varepsilon.$$

Then, the function $\chi_1 \circ r$ can be regarded as a $\mathcal{C}^2$-function on $V := G \cup U'$ (by setting $\chi_1 \circ r \equiv 3.5\varepsilon$ on $G \setminus U'$), which is psh and, in addition, strictly psh on $\{z \in U' : r(z) > 3\varepsilon\}$.

Finally, we choose a $\mathcal{C}^\infty$-function χ_2 on $\mathbb{C}^n$ with

$$\chi_2(z) = 1 \text{ if } z \in G \setminus U' \text{ or if } z \in G \cap U' \text{ and } r(z) < 2\varepsilon, \tag{19.2.6}$$

$$\chi_2(z) = 0 \text{ if } z \notin G \text{ or if } z \in G \cap U' \text{ and } r(z) > \varepsilon. \tag{19.2.7}$$

Obviously, for a suitably chosen positive constant c the function $r_1(z) := c\chi_2(z)\|z\|^2 + \chi_1 \circ r(z)$ fulfills all the requirements of Lemma 19.2.13. □

Proof of Theorem 19.2.12. According to Corollary 19.2.6, it suffices to show that there is a positive C such that

$$\boldsymbol{\varkappa}_G(z; X) \geq C\|X\|/\sqrt{\operatorname{dist}(z, \partial G)}, \quad z \in G,\ X \in \mathbb{C}^n. \tag{19.2.8}$$

Namely, from (19.2.8) and Corollary 19.2.6 it follows that

$$\frac{1}{1 - |\lambda|} \geq \boldsymbol{\varkappa}_{\mathbb{D}}(\lambda; 1) = \boldsymbol{\varkappa}_G(\varphi(\lambda); \varphi'(\lambda)) \geq \frac{C_1\|\varphi'(\lambda)\|}{(1 - |\lambda|)^{1/2}},$$

and therefore

$$\|\varphi'(\lambda)\| \leq \frac{1}{C_1(1 - |\lambda|)^{1/2}}, \quad \lambda \in \mathbb{D}.$$

Hence, the Hardy–Littlewood theorem (cf. Appendix B.9.11) implies the claim of Theorem 19.2.12.

To verify the inequality (19.2.8), we may assume that $r : V \longrightarrow \mathbb{R}$ is the function from Lemma 19.2.13. Then, there exists a positive α_1 such that

$$(\mathcal{L}r)(z;X) \geq \alpha_1 \|X\|^2, \quad z \in G,\ X \in \mathbb{C}^n.$$

We mention that it suffices to prove (19.2.8) for points $z \in G$ with sufficiently small boundary distance such that $-r(z) \leq \alpha_2 \operatorname{dist}(z, \partial G)$ with $\alpha_2 > 0$.

Now fix such a $z_0 \in G$ and an $X_0 \in (\mathbb{C}^n)_*$ and let $\psi \in \mathcal{O}(\overline{\mathbb{D}}, G)$ with $\psi(0) = z_0$ and $\sigma\psi'(0) = X_0$ for $\sigma > 0$. Let $\zeta_0 \in \partial G$ with $\|\zeta_0 - z_0\| = \operatorname{dist}(z_0, \partial G)$.

Then, $\widetilde{r}(z) := r(z) - \alpha_3\|z - \zeta_0\|^2$ is a psh function on G, where $\alpha_3 := \alpha_1/2$. Therefore, we obtain

$$\widetilde{r}(z_0) = \widetilde{r} \circ \psi(0) \leq -\frac{\alpha_3}{2\pi} \int_0^{2\pi} \|\psi(e^{i\theta}) - \zeta_0\|^2 d\theta.$$

Applying the Cauchy integral formula it follows that

$$\begin{aligned}\|\psi'(0)\| &= \left(\sum_{j=1}^{n} \left|\frac{1}{2\pi i}\int_{|\lambda|=1} \frac{\psi_j(\lambda) - \zeta_{0,j}}{\lambda^2} d\lambda\right|^2\right)^{1/2} \\ &\leq \left(\frac{1}{2\pi}\int_0^{2\pi} \|\psi(e^{i\theta}) - \zeta_0\|^2 d\theta\right)^{1/2} \leq (1 + \alpha_2/\alpha_3)^{1/2}\sqrt{\operatorname{dist}(z_0, \partial G)}.\end{aligned}$$

Hence, we end up with (19.2.8). □

Remark 19.2.14. The main argument in the proof of Theorem 19.2.12 uses the fact that G admits a good strictly psh exhaustion function. Such functions also exist for any $\mathcal{C}^\infty$-smooth bounded pseudoconvex domain whose boundary is B-regular (for a precise definition see [475]). To be more concrete, for any $0 < \eta < 1$ there exists a defining function r on $U = U(\overline{G})$ of ∂G such that the function $r_1(z) := -(-r(z))^\eta$, $z \in G$, is a negative psh exhaustion function of G with $(\mathcal{L}r)(z;X) \geq C\|X\|^2$, $z \in G$, $X \in \mathbb{C}^n$, where $C > 0$. Modifying the argument in the proof of Theorem 19.2.12, we find that any complex c_G-geodesics $\varphi : \mathbb{D} \longrightarrow G$ is $\eta/2$-Hölder continuous on G. In particular, this result holds if the boundary of ∂G is real-analytic.

19.3 Localization

In this section, we study how to estimate our metrics on strongly pseudoconvex domains dealing only with local information. Both results we will present were found by I. Graham (cf. [197]).

Theorem 19.3.1. *Let G be a strongly pseudoconvex domain in $\mathbb{C}^n$. Then, for a sufficiently small R the set $G \cap \mathbb{B}(\zeta, R)$ is connected and we have*

$$\lim_{\substack{z \to \zeta \\ z \in G \cap \mathbb{B}(\zeta, R)}} \frac{\boldsymbol{\gamma}_{G \cap \mathbb{B}(\zeta,R)}(z; X)}{\boldsymbol{\gamma}_G(z; X)} = 1$$

whenever $\zeta \in \partial G$, $X \in (\mathbb{C}^n)_*$. The convergence is uniform in $\zeta \in \partial G$ and $X \in (\mathbb{C}^n)_*$.

Proof. According to Theorem 19.1.3 we can choose R, sufficiently small, and $\varrho = \varrho(R) < R$. We recall that for $z \in G \cap \mathbb{B}(\zeta, R)$, $\zeta \in \partial G$, the inequality $\boldsymbol{\gamma}_G(z; \cdot) \le \boldsymbol{\gamma}_{G \cap \mathbb{B}(\zeta,R)}(z; \cdot)$ always holds.

Now fix $\varepsilon > 0$ and put $\varepsilon' := \varepsilon/(2+\varepsilon)$. For this ε' we choose $L = L(\varepsilon', R)$ via Theorem 19.1.3. Moreover, for $\zeta \in \partial G$ let $h(\cdot; \zeta)$ denote the peak function of Theorem 19.1.2 with the data $\eta_2 < \eta_1 \le \varrho$, d_2, d_1, and $\widetilde{G}$. With a suitable $k \in \mathbb{N}$, we find that $d_2^k L < 1$ and

$$|h(z; \zeta)^k| \ge (1 - d_1 \|z - \zeta\|)^k > 1 - \varepsilon' \quad \text{if } \|z - \zeta\| \le \eta(\varepsilon) \le \eta_2.$$

After these preparations, we fix $z_0 \in G \cap \mathbb{B}(\zeta, \eta(\varepsilon))$ and $X \in (\mathbb{C}^n)_*$. Then, we find an $f \in \mathcal{O}(G \cap \mathbb{B}(\zeta, R), \mathbb{D})$ with

$$f(z_0) = 0 \quad \text{and} \quad \boldsymbol{\gamma}_{G \cap \mathbb{B}(\zeta,R)}(z_0; X) = \left| \sum_{j=1}^n \frac{\partial f}{\partial z_j}(z_0) X_j \right| = |f'(z_0) X|.$$

According to what we said in Theorem 19.1.3, there exists an $\widehat{f} \in \mathcal{H}^\infty(G)$ with the following properties:

$$\widehat{f}(z_0) = 0,\ \boldsymbol{\gamma}_{G \cap \mathbb{B}(\zeta,R)}(z_0; X) = |\widehat{f}'(z_0) X|,\ \|\widehat{f}\|_G \le L, \text{ and } \|f - \widehat{f}\|_{G \cap \mathbb{B}(\zeta,\varrho)} < \varepsilon'.$$

Putting $\widetilde{f} := h(\cdot; \zeta)^k \cdot \widehat{f}$ gives a holomorphic function with

$$\widetilde{f}(z_0) = 0,\ |\widetilde{f}'(z_0) X| = |h(z_0; \zeta)|^k \boldsymbol{\gamma}_{G \cap \mathbb{B}(\zeta,R)}(z_0; X) \ge (1 - \varepsilon') \boldsymbol{\gamma}_{G \cap \mathbb{B}(\zeta,R)}(z_0; X).$$

Moreover, we get

$$\|\widetilde{f}\|_{G \setminus \mathbb{B}(\zeta,\eta_1)} \le d_2^k L < 1 \quad \text{and} \quad \|\widetilde{f}\|_{G \cap \mathbb{B}(\zeta,\eta_1)} \le 1 + \varepsilon'.$$

Thus, if $z_0 \in \mathbb{B}(\zeta, \eta(\varepsilon))$, then

$$\boldsymbol{\gamma}_{G \cap \mathbb{B}(\zeta,R)}(z_0; X) \le \frac{1+\varepsilon'}{1-\varepsilon'} \left| \sum_{j=1}^n \frac{\partial(\widetilde{f}/(1+\varepsilon'))}{\partial z_j}(z_0) X_j \right| \le (1+\varepsilon) \boldsymbol{\gamma}_G(z_0; X). \qquad \square$$

The analogous result for the Kobayashi–Royden metric is also due to I. Graham (cf. [197]) and will be a consequence of Proposition 13.2.10 and Theorem 19.1.2.

Theorem 19.3.2. *Let G be a strongly pseudoconvex domain and let $U = U(\zeta_0)$ be a neighborhood of a boundary point $\zeta_0 \in \partial G$ such that $G \cap U$ is connected. Then,*

$$\lim_{\substack{z\to\zeta_0 \\ z\in G\cap U}} \frac{\varkappa_{G\cap U}(z;X)}{\varkappa_G(z;X)} = 1\,,$$

where the convergence is uniform in $X \in (\mathbb{C}^n)_*$.

Proof. First, recall from Proposition 13.2.10 that

$$\varkappa_{G\cap U}(z;X) \le \coth \boldsymbol{k}_G(z, G\setminus U)\varkappa_G(z;X), \quad z\in G\cap U,\ X\in\mathbb{C}^n.$$

Therefore, we only have to prove a lower estimate of $\boldsymbol{k}_G(z; G\setminus U)$ for z that is sufficiently near ζ_0. According to Theorem 19.1.2 we can find a peak function h for ζ_0. With $r := \inf\{|1-h(w)| : w\in \overline{G}\setminus U\} > 0$ we see that $V(s) := \{z\in\overline{G} : |1-h(z)| < s\}\subset U$ if $0<s<r$. Obviously, $V(s)\supset G\cap V_s$ for an appropriate $V_s = V_s(\zeta_0)$. Following the calculation of the proof of Theorem 19.2.1, we obtain for $z\in G\cap V_s$ and $w\in G\setminus U$ that

$$\boldsymbol{k}_G(z,w)\ge \boldsymbol{c}_G(z,w)\ge (1/2)\log(r/s) \underset{s\to 0}{\longrightarrow} \infty.$$

Hence, for any $\varepsilon > 0$ we have

$$\varkappa_{G\cap U}(z;X)\le (1+\varepsilon)\varkappa_G(z;X) \quad \text{if} \quad z\in G\cap V_s,\ s\ll 1. \qquad \square$$

Observe that the claim of Theorem 19.3.2 remains true if G is only assumed to be a bounded domain that admits a global peak function for $\zeta_0\in\partial G$.

As a consequence of the previous two results, it turns out that the Kobayashi–Royden and the Carathéodory–Reiffen metrics have the same asymptotic boundary behavior on strongly pseudoconvex domains.

Theorem 19.3.3. *For any strongly pseudoconvex domain G we have*

$$\lim_{z\to\partial G}\frac{\varkappa_G(z;X)}{\boldsymbol{\gamma}_G(z;X)} = 1 \quad \textit{uniformly for } X\in(\mathbb{C}^n)_*.$$

Proof. Suppose the contrary. Then, there exist $\varepsilon_0 > 0$ and sequences $(z_\nu)_{\nu\in\mathbb{N}}\subset G$, $(X_\nu)_{\nu\in\mathbb{N}}\subset(\mathbb{C}^n)_*$ with $\lim_{\nu\to\infty} z_\nu = \zeta_0\in\partial G$ and

$$\varkappa_G(z_\nu;X_\nu)\ge(1+\varepsilon_0)\boldsymbol{\gamma}_G(z_\nu;X_\nu). \qquad (*)$$

Because of the strong pseudoconvexity, Lempert's Theorem gives a neighborhood $U = U(\zeta_0)$ such that $\boldsymbol{\gamma}_{G\cap U} = \boldsymbol{\varkappa}_{G\cap U}$. Therefore, by Theorems 19.3.1 and 19.3.2, it follows that

$$\lim_{\nu\to\infty} \frac{\boldsymbol{\varkappa}_G(z_\nu; X_\nu)}{\boldsymbol{\gamma}_G(z_\nu; X_\nu)} = \lim_{\nu\to\infty} \frac{\boldsymbol{\varkappa}_G(z_\nu; X_\nu)}{\boldsymbol{\varkappa}_{G\cap U}(z_\nu; X_\nu)} \frac{\boldsymbol{\gamma}_{G\cap U}(z_\nu; X_\nu)}{\boldsymbol{\gamma}_G(z_\nu; X_\nu)} = 1,$$

which contradicts $(*)$. □

So far, we have discussed localization theorems for the Carathéodory–Reiffen and the Kobayashi–Royden metric, respectively. Similar results are also true for the corresponding distances. At the moment, we can only prove a theorem for the Kobayashi distance. The result for the Carathéodory distance will follow from a more general comparison result.

Theorem 19.3.4 (cf. [507]). *Let $U = U(\zeta_0)$ be a neighborhood of a boundary point ζ_0 of a strongly pseudoconvex domain $G \subset \mathbb{C}^n$. Suppose that $U \cap G$ is connected. Then, for any sequences $(z_\nu)_{\nu\in\mathbb{N}}, (w_\nu)_{\nu\in\mathbb{N}} \subset G \cap U$ with $\lim_{\nu\to\infty} z_\nu = \lim_{\nu\to\infty} w_\nu = \zeta_0$, $z_\nu \neq w_\nu$, we have*

$$\lim_{\nu\to\infty} \frac{\boldsymbol{k}_{G\cap U}(z_\nu, w_\nu)}{\boldsymbol{k}_G(z_\nu, w_\nu)} = 1.$$

The proof of this theorem relies on the following lemma.

Lemma 19.3.5. *Let G and ζ_0 be as in Theorem 19.3.4. Moreover, assume that there are sequences $(z_\nu)_{\nu\in\mathbb{N}}, (w_\nu)_{\nu\in\mathbb{N}}, (t_\nu)_{\nu\in\mathbb{N}}$ of points z_ν, w_ν, t_ν in G, respectively, and a positive constant C with*

$$\lim_{\nu\to\infty} z_\nu = \lim_{\nu\to\infty} w_\nu = \zeta_0 \quad \textit{and}$$
$$\boldsymbol{k}_G(z_\nu, t_\nu) + \boldsymbol{k}_G(t_\nu, w_\nu) \le \boldsymbol{k}_G(z_\nu, w_\nu) + C, \ \nu \in \mathbb{N}.$$

Then, $\lim_{\nu\to\infty} t_\nu = \zeta_0$.

Proof. Suppose the contrary. Then, there is a subsequence of $(t_\nu)_\nu$, which we again denote by $(t_\nu)_\nu$ with $\lim_{\nu\to\infty} t_\nu = t_0 \neq \zeta_0$.

Case 1°. If $t_0 \in G$, then Theorem 19.2.1 implies that

$$\begin{aligned} &-(1/2)\log \operatorname{dist}(z_\nu, \partial G) - (1/2)\log \operatorname{dist}(w_\nu, \partial G) - C_1 \\ &\qquad \le \boldsymbol{k}_G(z_\nu, t_0) + \boldsymbol{k}_G(w_\nu, t_0) \le \boldsymbol{k}_G(z_\nu, w_\nu) + 2\boldsymbol{k}_G(t_\nu, t_0) + C, \end{aligned}$$

which contradicts the upper estimate in Proposition 19.2.9.

Case 2°. If $t_0 \in \partial G$, then by Proposition 19.2.7 we have

$$-(1/2)\log \operatorname{dist}(z_\nu, \partial G) - (1/2)\log \operatorname{dist}(w_\nu, \partial G) - \log \operatorname{dist}(t_\nu, \partial G) - C_1$$
$$\leq \boldsymbol{k}_G(z_\nu, t_\nu) + \boldsymbol{k}_G(t_\nu, w_\nu) \leq \boldsymbol{k}_G(z_\nu, w_\nu) + C,$$

which again does not fit with Proposition 19.2.9. □

Proof of Theorem 19.3.4. Applying the localization result of Theorem 19.3.2, we find for given $\varepsilon > 0$ a neighborhood $V = V(\zeta_0) \subset U$ such that

$$\boldsymbol{\varkappa}_{G\cap U}(z; X) \leq (1+\varepsilon)\boldsymbol{\varkappa}_G(z; X), \quad z \in V \cap G,\ X \in \mathbb{C}^n.$$

Moreover, there are $\mathcal{C}^1$-curves $\alpha_\nu : [0,1] \longrightarrow G$ connecting z_ν and w_ν such that

$$\int_0^1 \boldsymbol{\varkappa}_G(\alpha_\nu(t); \alpha_\nu'(t))dt \leq (1+\varepsilon_\nu)\boldsymbol{k}_G(z_\nu, w_\nu)$$
$$\text{with } 0 < \varepsilon_\nu < \min\{\varepsilon, 1/\boldsymbol{k}_G(z_\nu, w_\nu)\}.$$

Now, we claim that if $\nu \gg 1$, then $\alpha_\nu([0,1]) \subset V$. For otherwise there is a sequence $(\tau_j)_{j\in\mathbb{N}} \subset (0,1)$ with $\alpha_{\nu_j}(\tau_j) =: t_j \notin V$. On the other hand, we obtain

$$\boldsymbol{k}_G(z_{\nu_j}, t_j) + \boldsymbol{k}_G(t_j, w_{\nu_j}) \leq \int_0^1 \boldsymbol{\varkappa}_G(\alpha_{\nu_j}(t); \alpha_{\nu_j}'(t))dt \leq \boldsymbol{k}_G(z_{\nu_j}, w_{\nu_j}) + 1.$$

Hence, by Lemma 19.3.5 it follows that $\lim_{j\to\infty} t_j = \zeta_0$, i.e., $t_j \in V$ for large j; a contradiction.

Therefore, we conclude that for $\nu \gg 1$ we have

$$\boldsymbol{k}_{G\cap U}(z_\nu, w_\nu) \leq \int_0^1 \boldsymbol{\varkappa}_{G\cap U}(\alpha_\nu(t); \alpha_\nu'(t))dt$$
$$\leq (1+\varepsilon)\int_0^1 \boldsymbol{\varkappa}_G(\alpha_\nu(t); \alpha_\nu'(t))dt \leq (1+\varepsilon)^2\boldsymbol{k}_G(z_\nu, w_\nu). \quad \square$$

Finally, we formulate a localization result for the Bergman metric, more precise than the one of Theorem 12.8.6 (cf. [124, 234]). It can easily be obtained by modifying the $\overline{\partial}$-problem in the proof of Theorem 12.8.6 with the aid of an appropriate peak function (cf. Theorem 19.1.2); details are left to the reader.

Theorem 19.3.6. *Let G be a strongly pseudoconvex domain and let $R > 0$ be such that for any $\zeta_0 \in \partial G$ the intersection $G \cap \mathbb{B}(\zeta_0, R)$ is connected. Then, for every $\varepsilon > 0$ there exists a $\delta = \delta_\varepsilon \in (0, R)$ such that*

(a) $M_G(z;X) \le M_{G\cap\mathbb{B}(\zeta_0,R)}(z;X) \le (1+\varepsilon)M_G(z;X)$,

(b) $\mathfrak{K}_G(z) \le \mathfrak{K}_{G\cap\mathbb{B}(\zeta_0,R)}(z) \le (1+\varepsilon)\mathfrak{K}_G(z)$,

(c) $(1+\varepsilon)^{-1}\boldsymbol{\beta}_{G\cap\mathbb{B}(\zeta_0,R)}(z;X) \le \boldsymbol{\beta}_G(z;X) \le \sqrt{(1+\varepsilon)}\boldsymbol{\beta}_{G\cap\mathbb{B}(\zeta_0 R)}(z;X)$,

whenever $z \in G \cap \mathbb{B}(\zeta_0,\delta)$, $X \in \mathbb{C}^n$. Moreover, if R is sufficiently small, then $\delta = \delta_\varepsilon$ can be chosen independently of the boundary point ζ_0.

19.4 Boundary behavior of the Carathéodory–Reiffen and the Kobayashi–Royden metrics

Trying to understand the boundary growth of various metrics on a strongly pseudoconvex domain, the following observation may be important: strongly pseudoconvex domains look locally (up to biholomorphisms) like strictly convex domains, and therefore, up to localization, metrics should "coincide" near the boundary. The precise formulation of this asymptotical equality will be the main goal of this section.

It was I. Graham who studied the boundary behavior of $\boldsymbol{\varkappa}_G$ and $\boldsymbol{\gamma}_G$ on strongly pseudoconvex domains (cf. [197], also [222]). He obtained an asymptotic estimate of the length of normal and tangential vectors separately. In [19], G. Aladro obtained an estimate for the length of a general vector without specifying the asymptotic constant. He needed the boundary to be of class $\mathbb{C}^n$. The theorem we are going to present here is in the spirit of the one found by D. Ma (cf. [351, 352]), which is more precise than Graham's and Aladro's results. N. Sibony (cf. [474]) also established estimates of the Kobayashi–Royden metric on domains that carry a "good" psh function. Moreover, estimates of these metrics near the boundary of pseudoconvex domains of finite type in $\mathbb{C}^2$ were found by D. Catlin [88]. Next, J. McNeal [359] extended this investigation to smoothly bounded pseudoconvex domains G of finite type in $\mathbb{C}^n$ under the additional hypothesis that G is "decoupled near $z_0 \in \partial G$", i.e., up to a biholomorphic change of coordinates $w = w(z)$ near z_0, $w(z_0) = 0$, the domain G is locally given as $\{w \in \mathbb{C}^n : r(w) = 2\operatorname{Re} w_1 + \sum_{j=2}^n r_j(w_j) < 0\}$ with r_j smooth, subharmonic but not harmonic, and such that $r_j(0) = dr_j(0) = 0$.

To prepare the proof of the main theorem, we have to introduce the notion of an analytic ellipsoid. Let

$$H(z,w) := \sum_{i,j=1}^n a_{i,j} z_i \overline{w}_j, \quad z = (z_1,\dots,z_n),\ w = (w_1,\dots,w_n) \in \mathbb{C}^n,$$

be a positive definite Hermitian form. Then, the domain

$$\mathcal{E} = \mathcal{E}(H) := \{z = (z_1,\dots,z_n) \in \mathbb{C}^n : \Phi_{\mathcal{E}}(z) := -2\operatorname{Re} z_1 + H(z,z) < 0\} \tag{19.4.1}$$

is called an *analytic ellipsoid*. We point out that the Levi form of a strictly psh function can be taken as an example of such an H.

Since H is given by a positive definite Hermitian matrix $A = (a_{i,j})_{1\le i,j\le n}$, there exists a unitary $(n-1)\times(n-1)$-matrix $\widetilde{S}$ such that if we put

$$S := \begin{pmatrix} 1 & 0 \\ 0 & \widetilde{S} \end{pmatrix},$$

then the matrix $B = (b_{i,j})_{1\le i,j\le n} := S^t \cdot A \cdot \overline{S}$ has the following properties:

$$b_{j,j} > 0 \text{ if } j \ge 2, \quad b_{i,j} = 0 \text{ if } i,j \ge 2,\ i \ne j.$$

Setting

$$T := \begin{pmatrix} 1 & 0 & \dots & 0 \\ -b_{1,2}/b_{2,2} & 1 & \dots & 0 \\ \vdots & \vdots & \ddots & \vdots \\ -b_{1,n}/b_{n,n} & 0 & \dots & 1 \end{pmatrix},$$

we find that $C = (c_{i,j})_{1\le i,j\le n} := T^t \cdot B \cdot \overline{T}$ is a diagonal matrix with $c_{j,j} > 0$, $j \ge 1$. Therefore,

$$\begin{aligned} \mathcal{E}' := (S\cdot T)^{-1}(\mathcal{E}) &= \Big\{ z \in \mathbb{C}^n : -z_1 - \overline{z}_1 + \sum_{j=1}^n c_{j,j} z_j \overline{z}_j < 0 \Big\} \\ &= \Big\{ z \in \mathbb{C}^n : c_{1,1}|z_1 - 1/c_{1,1}|^2 + \sum_{j=2}^n c_{j,j} z_j \overline{z}_j < 1/c_{1,1} \Big\}. \end{aligned}$$

Using the transformation

$$\Phi(z) := (z_1/\sqrt{c_{1,1}}, \dots, z_n/\sqrt{c_{n,n}}) + (1/c_{1,1}, 0, \dots, 0)$$

we get

$$\mathcal{E}'' := \Phi^{-1}(\mathcal{E}') = \Big\{ z \in \mathbb{C}^n : \sum_{j=1}^n |z_j|^2 < 1/c_{1,1} \Big\}.$$

Combining all these maps, we have the explicit biholomorphic mapping $F : \mathcal{E} \longrightarrow \mathbb{B}_n$ with

$$F(z) := \sqrt{c_{11}}\,\Phi^{-1}(((S\cdot T)^{-1} z^t)^t).$$

Using F and the formula for $\boldsymbol{\varkappa}_{\mathbb{B}_n}$, the next lemma becomes an easy exercise.

Lemma 19.4.1. *Let* $\mathcal{E} = \mathcal{E}(H)$ *and* $\Phi_{\mathcal{E}}$ *be as in* (19.4.1). *Then,*

$$\boldsymbol{\varkappa}_{\mathcal{E}}(z;X) = \boldsymbol{\gamma}_{\mathcal{E}}(z;X) = \left(\frac{H(X,X)}{-\Phi_{\mathcal{E}}(z)} + \left| \frac{H(X,z) - X_1}{-\Phi_{\mathcal{E}}(z)} \right|^2 \right)^{1/2}, \quad z \in \mathcal{E},\ X \in \mathbb{C}^n.$$

In the following, analytic ellipsoids will serve as local comparison domains in the process of estimating metrics on strongly pseudoconvex domains. The main result is the following (cf. [351, 352]):

Theorem 19.4.2. *Let G be a strongly pseudoconvex domain in $\mathbb{C}^n$ that is given by (U, r) (cf. (19.1.1–19.1.6)). Then, for every $\varepsilon \in (0,1)$ there exists a $\delta = \delta(\varepsilon) > 0$ such that the following properties hold:*

(i) *for every $z \in G$ with $\operatorname{dist}(z, \partial G) < \delta$ there is a unique $\zeta =: \pi(z) \in \partial G$ with $\|z - \zeta\| = \operatorname{dist}(z, \partial G)$,*

(ii) *for every such $z \in G$ and every $X \in \mathbb{C}^n$ the following inequalities are true*

$$(1-\varepsilon)\left(\frac{(\mathcal{L}r)(\pi(z); X_{(t)})}{2\operatorname{dist}(z,\partial G)} + \frac{\|X_{(n)}\|^2}{4\operatorname{dist}(z,\partial G)^2}\right)^{1/2} \le \boldsymbol{\gamma}_G(z;X) \le \boldsymbol{\varkappa}_G(z;X)$$

$$\le (1+\varepsilon)\left(\frac{(\mathcal{L}r)(\pi(z); X_{(t)})}{2\operatorname{dist}(z,\partial G)} + \frac{\|X_{(n)}\|^2}{4\operatorname{dist}(z,\partial G)^2}\right)^{1/2},$$

where X is split into its normal and tangential components $X_{(n)}$ and $X_{(t)}$ at the point $\pi(z)$, i.e., $X = X_{(n)} + X_{(t)}$, $\langle X_{(n)}, X_{(t)}\rangle = 0$, and

$$\sum_{j=1}^{n} (\partial r/\partial z_j)(\pi(z))(X_{(t)})_j = 0.$$

The following technical lemma will be used in the proof.

Lemma 19.4.3. *Let $\mathcal{E} = \mathcal{E}(H)$ be as in (19.4.1) with $H(z,z) = \sum_{i,j=1}^n a_{i,j} z_i \overline{z}_j \ge \alpha\|z\|^2$, $z \in \mathbb{C}^n$, where $0 < \alpha < 1$. For $0 \le \lambda < \alpha$ let $\mathcal{E}_\lambda^\pm$ denote the analytic ellipsoid*

$$\mathcal{E}_\lambda^\pm := \{z \in \mathbb{C}^n : -2\operatorname{Re} z_1 + H(z,z) \pm \lambda\|z\|^2 < 0\}.$$

Let $a \in \mathbb{C}$ and assume that $|a| \le M$ and $|a_{i,j}| \le M$, $1 \le i, j \le n$, for fixed $M \ge 1$. Finally, for $t > 0$ put $z(t) := (t - at^2, 0, \dots, 0) \in \mathbb{C}^n$. Then,

$$z(t) \in \mathcal{E}_\lambda^+ \subset \mathcal{E}_\lambda^- \text{ if } 0 \le \lambda < \alpha \text{ and } 0 < t < \varrho_1 := 1/(10M^3).$$

Moreover, for every $\varepsilon > 0$ there exists a positive constant $\varrho_0 < \min(\alpha, \varrho_1)$, depending only on α, M, and ε, such that

$$(1-\varepsilon)\left(\frac{H(X,X)}{2t} + \frac{|X_1|^2}{4t^2}\right)^{1/2} \le \boldsymbol{\varkappa}_{\mathcal{E}_\lambda^-}(z(t);X) \le \boldsymbol{\varkappa}_{\mathcal{E}_\lambda^+}(z(t);X)$$

$$\le (1+\varepsilon)\left(\frac{H(X,X)}{2t} + \frac{|X_1|^2}{4t^2}\right)^{1/2},$$

whenever $0 \le \lambda \le \varrho_0$, $0 < t \le \varrho_0$, and $X \in \mathbb{C}^n$.

Proof. First, a simple calculation shows that for $0 < t < \varrho_1$ the following inequality is true:

$$-2t(1+5M^3t) \leq \Phi_{\mathcal{E}_\lambda^-}(z(t)) \leq \Phi_{\mathcal{E}_\lambda^+}(z(t)) \leq -2t(1-5M^3t)\,;$$

in particular, $z(t) \in \mathcal{E}_\lambda^+$.

For the remaining inequality, we will only prove, for example, the lower estimate. So, if $0 < t < \varrho_1$, Lemma 19.4.1 leads to

$$\left(\varkappa_{\mathcal{E}_\lambda^-}(z(t);X)\right)^2 = \frac{H(X,X)-\lambda\|X\|^2}{-\Phi_{\mathcal{E}_\lambda^-}(z(t))} + \left|\frac{H(X,z(t))-\lambda X_1 t(1-\overline{a}t)-X_1}{-\Phi_{\mathcal{E}_\lambda^-}(z(t))}\right|^2$$

$$\geq \frac{1-\lambda/\alpha}{(1+5M^3t)^2}\left(\frac{H(X,X)}{2t}+\frac{|X_1|^2}{4t^2}-\frac{C}{4t^2}\left(\|X\|^2t^2+|X_1|^2t+\|X\|\,|X_1|t\right)\right),$$

where the positive constant C can be chosen in such a way that it depends only on M.

If $|X_1| \leq \|X\|\sqrt{t}$, then it follows that

$$\left(\varkappa_{\mathcal{E}_\lambda^-}(z(t);X)\right)^2 \geq \frac{1-\lambda/\alpha}{(1+5M^3t)^2}\left(\frac{|X_1|^2}{4t^2}+\frac{H(X,X)}{2t}\left(1-\frac{3C}{2\alpha}\sqrt{t}\right)\right),$$

whereas, for $|X_1| > \|X\|\sqrt{t}$ we obtain

$$\left(\varkappa_{\mathcal{E}_\lambda^-}(z(t);X)\right)^2 \geq \frac{1-\lambda/\alpha}{(1+5M^3t)^2}\left(\frac{H(X,X)}{2t}+\frac{|X_1|^2}{4t^2}(1-3C\sqrt{t})\right).$$

Therefore, we have the following lower estimate:

$$\varkappa_{\mathcal{E}_\lambda^-}(z(t);X) \geq \frac{(1-\lambda/\alpha)^{1/2}\left(1-\frac{3C}{2\alpha}\sqrt{t}\right)^{1/2}(1-3C\sqrt{t})^{1/2}}{1+5M^3t}\left(\frac{H(X,X)}{2t}+\frac{|X_1|^2}{4t^2}\right)^{1/2},$$

whenever $0 < \sqrt{t} < \alpha/(3C)$. Now, it is clear how to choose the constant ϱ_0 in Lemma 19.4.3. □

Proof of Theorem 19.4.2. First, we choose a $\delta_1 < 1$ sufficiently small to have the following situation:

(a) $(\partial G)_{\delta_1} := \{z \in \mathbb{C}^n : \operatorname{dist}(z,\partial G) < \delta_1\} \subset\subset U$;

(b) if $z \in (\partial G)_{\delta_1}$, then $\operatorname{dist}(z,\partial G) = \|z-\pi(z)\|$, where $\pi(z) \in \partial G$ is uniquely determined;

(c) $(\mathcal{L}r)(z;X) \geq \alpha\|X\|^2$ and $\alpha \leq \|dr(z)\| \leq 1/\alpha$ for a suitable $\alpha \in (0,1)$ whenever $z \in (\partial G)_{\delta_1}$, $X \in \mathbb{C}^n$.

Moreover, for any $\gamma > 0$ we can find $\delta(\gamma) < \min\{\delta_1, \alpha^2\}$ such that $|D^\beta r(z) - D^\beta r(\pi(z))| < \gamma$ for all $z \in (\partial G)_{\delta(\gamma)}$ and for all multi-indices $\beta \in (\mathbb{Z}_+)^{2n}$, $|\beta| \le 2$. Here we use the fact that r is a $\mathcal{C}^2$-function.

Now fix $\varepsilon > 0$ and let ζ_0 be an arbitrary point of ∂G. We point out that further construction will depend on ζ_0. Nevertheless, we will omit the index ζ_0 to keep the notation simple. On the other hand, all constants in the estimates will be chosen independently of the specific ζ_0.

For ζ_0, we take a unitary matrix $A = A_{\zeta_0}$ transforming the vector $\operatorname{grad} r(\zeta_0)$ into the vector $(-1, 0, \dots, 0)$, i.e., $(-1, 0, \dots, 0) = (A \operatorname{grad} r(\zeta_0)^t)^t$. Thus, $T : \mathbb{C}^n \longrightarrow \mathbb{C}^n$, $T(z) := (A(z - \zeta_0)^t)^t$, biholomorphically maps G onto the strongly pseudoconvex domain $\widetilde{G} := T(G)$, which is given by the pair $\widetilde{U} := T(U)$ and $\widetilde{r} := r \circ T^{-1}$. Note that $\operatorname{grad} \widetilde{r}(0) = (-1, 0, \dots, 0)$ and that $(\partial \widetilde{G})_{\delta_1} \subset\subset \widetilde{U}$. Therefore, we obtain the following Taylor expansion of $\widetilde{r}$ on $\mathbb{B}(\delta_1)$:

$$\widetilde{r}(z) = -2 \operatorname{Re}\Bigl(z_1 - \frac{1}{2} \sum_{i,j=1}^{n} c_{i,j} z_i z_j \Bigr) + (\mathcal{L}\widetilde{r})(0; z) + \beta(z)\|z\|^2, \tag{19.4.2}$$

where for any $\gamma > 0$ we have $|\beta(z)| \le C_1 \gamma$, provided that $\|z\| < \delta(\gamma) < \delta_1$. Here, the constant $C_1 > 1$ can be chosen independently of ζ_0. Moreover, we have $(\mathcal{L}\widetilde{r})(0; a) = (\mathcal{L} r)(\zeta_0; (A^{-1} a^t)^t) \ge \alpha \|a\|^2$, $c_{i,j} = c_{j,i}$, and $|c_{i,j}| \le C_2$; C_2 is again independent of ζ_0.

Let $F : \mathbb{C}^n \longrightarrow \mathbb{C}^n$ be the map defined by

$$F(z_1, \dots, z_n) := \Bigl(z_1 - \frac{1}{2} \sum_{i,j=1}^{n} c_{i,j} z_i z_j, z_2, \dots, z_n \Bigr). \tag{19.4.3}$$

We will show that for every λ with $0 < \lambda < \alpha$ and for all sufficiently small (depending on λ) positive $\delta \le \delta_1/2$ the following inclusions are true:

$$\mathcal{E}_\lambda^+ \cap \mathbb{B}(\delta) \subset F(\widetilde{G} \cap \mathbb{B}(2\delta)) \subset \mathcal{E}_\lambda^-, \tag{19.4.4}$$

where $\mathcal{E}_\lambda^\pm := \{w \in \mathbb{C}^n : -2 \operatorname{Re} w_1 + (\mathcal{L}\widetilde{r})(0; w) \pm \lambda \|w\|^2 < 0\}$.

Moreover, for $0 < t \le \delta/2$ we claim that

$$F(t, 0, \dots, 0) = \Bigl(t - \frac{1}{2} c_{1,1} t^2, 0, \dots, 0 \Bigr) \in \mathcal{E}_\lambda^+ \cap \mathbb{B}(\delta). \tag{19.4.5}$$

Observe that (19.4.5) is an immediate consequence of Lemma 19.4.3. (Note that the coefficients $a_{i,j}$ of $(\mathcal{L}\widetilde{r})(0; z)$ and $c_{i,j}$ are uniformly bounded by a constant $M \ge 1$, which is independent of ζ_0.)

To prove (19.4.4), we choose a positive $R_0 < \delta_1$, independent of ζ_0, such that

$$\|F(z') - z' - F(z'') + z''\| \le \|z' - z''\|/2, \quad z', z'' \in \mathbb{B}(R_0). \tag{19.4.6}$$

Therefore, $F := F|_{\mathbb{B}(R_0)}$ biholomorphically maps $\mathbb{B}(R_0)$ onto an open set V containing $\mathbb{B}(R_0/2)$; compare a proof of the inverse mapping theorem.

Now suppose that $0 < \lambda < \alpha$ is fixed and take a point $w \in \mathcal{E}_\lambda^+ \cap \mathbb{B}(R_0/2)$. Put $z := F^{-1}(w) \in \mathbb{B}(R_0)$. Then, since $\|z\| \le 2\|w\|$ (cf. (19.4.6)) and since $(\mathcal{L}\widetilde{r})(0; z) - (\mathcal{L}\widetilde{r})(0; F(z)) = O(\|z\|^3)$, we obtain

$$\begin{aligned}\widetilde{r}(z) &= -2\operatorname{Re} w_1 + (\mathcal{L}\widetilde{r})(0; z) + \beta(z)\|z\|^2 \\ &< (\mathcal{L}\widetilde{r})(0; z) - (\mathcal{L}\widetilde{r})(0; F(z)) + (4\beta(F^{-1}(w)) - \lambda)\|w\|^2 \\ &\le (C_3\|w\| + 4\beta(F^{-1}(w)) - \lambda)\|w\|^2 < 0\end{aligned}$$

provided that $\|w\| < \delta \le R_0/2$, i.e., $z \in \widetilde{G} \cap \mathbb{B}(2\delta)$.

The second inclusion of (19.4.4) will follow similarly. Let $z \in \widetilde{G} \cap \mathbb{B}(R_0)$ and put $w := F(z)$. Then,

$$\begin{aligned}-2\operatorname{Re} w_1 + (\mathcal{L}\widetilde{r})(0; w) - \lambda\|w\|^2 &\le \widetilde{r}(z) + (\mathcal{L}\widetilde{r})(0; w) - (\mathcal{L}\widetilde{r})(0; z) \\ &\quad - \lambda\|w\|^2 - \beta(z)\|z\|^2 \\ &\le (C_3\|z\| - \beta(z) - \lambda/4)\|z\|^2 < 0\end{aligned}$$

provided that $\|z\| < 2\delta$ and δ sufficiently small.

Hence, the inclusions (19.4.4) are proved. So $\mathcal{E}_\lambda^+$ and $\mathcal{E}_\lambda^-$ can, locally, serve as an inner and an outer comparison domain for $\widetilde{G}$, respectively.

Now, we are able to apply Lemma 19.4.3 with $H(z, z) = (\mathcal{L}\widetilde{r})(0; z)$. We fix $\lambda = \varrho_0 < \alpha$ and we choose a positive $\delta < \min\{R_0/2, \varrho_0\}$ such that (19.4.4) and (19.4.5) hold and that for all $0 < t \le \delta$ the point $z(t) := (t, 0, \dots, 0)$ belongs to $\widetilde{G}$ with $\operatorname{dist}(z(t), \partial\widetilde{G}) = t$. Of course, δ can be chosen independently of ζ_0. Then, using Proposition 13.2.10, for $X \in \mathbb{C}^n$ and $0 < t < \delta/2$ we obtain the following upper estimate (note that $F(z(t)) \in \mathcal{E}_\lambda^+ \cap \mathbb{B}(\delta)$):

$$\begin{aligned}\boldsymbol{\varkappa}_{\widetilde{G}}(z(t); X) &\le \boldsymbol{\varkappa}_{\mathcal{E}_\lambda^+ \cap \mathbb{B}(\delta)}(F(z(t)); F'(z(t))X) \\ &\le \coth \boldsymbol{k}_{\mathcal{E}_\lambda^+}(F(z(t)), \mathcal{E}_\lambda^+ \setminus \mathbb{B}(\delta)) \cdot \boldsymbol{\varkappa}_{\mathcal{E}_\lambda^+}(F(z(t)); F'(z(t))X) \\ &\le (1+\varepsilon) \coth \boldsymbol{k}_{\mathcal{E}_\lambda^+}(F(z(t)), \mathcal{E}_\lambda^+ \setminus \mathbb{B}(\delta)) \\ &\quad \cdot \left(\frac{(\mathcal{L}\widetilde{r})(0; F'(z(t))X)}{2t} + \frac{|F_1'(z(t))X|^2}{4t^2} \right)^{1/2}.\end{aligned} \tag{19.4.7}$$

On the other hand, if we assume that δ is sufficiently small, then we can apply Theorem 19.3.1. So we get

$$\boldsymbol{\gamma}_{\widetilde{G}}(z(t); X) \ge (1-\varepsilon)\boldsymbol{\gamma}_{\widetilde{G} \cap \mathbb{B}(\delta)}(z(t); X)$$

provided that t is sufficiently small, say $t < \delta'_\varepsilon < \delta/2$. Then (19.4.4) and Lemma 19.4.3 give

$$\begin{aligned}\boldsymbol{\gamma}_{\tilde{G}}(z(t); X) &\geq (1-\varepsilon)\boldsymbol{\gamma}_{\mathcal{E}_\lambda^-}(F(z(t)); F'(z(t))X)\\ &\geq (1-\varepsilon)^2 \left(\frac{(\mathcal{L}\tilde{r})(0; F'(z(t))X)}{2t} + \frac{|F_1'(z(t))X|^2}{4t^2}\right)^{1/2}. \qquad (19.4.8)\end{aligned}$$

To replace $F'(z(t))X$ and $F_1'(z(t))X$ by $X_{(t)} := (0, X_2, \dots, X_n)$ and $X_{(n)} := (X_1, 0, \dots, 0)$, respectively, we make the following observation:

there are positive constants C_4, C_5 such that the following two inequalities are true:

$$\begin{aligned}|(\mathcal{L}\tilde{r})(0; F'(z(t))X) - (\mathcal{L}\tilde{r})(0; X_{(t)})| &\leq C_4(|X_1|^2/\sqrt{t} + \sqrt{t}\|X_{(t)}\|^2),\\ \big||F_1'(z(t))X|^2 - |X_1|^2\big| &\leq C_5(\sqrt{t}|X_1|^2 + t^{3/2}\|X_{(t)}\|^2).\end{aligned}$$

Thus, since $(\mathcal{L}\tilde{r})(0; X_{(t)}) \geq \alpha\|X_{(t)}\|^2$, we derive

$$(1 + O(\sqrt{t}))\left(\frac{(\mathcal{L}\tilde{r})(0; X_{(t)})}{2t} + \frac{|X_1|^2}{4t^2}\right) = \frac{(\mathcal{L}\tilde{r})(0; F'(z(t))X)}{2t} + \frac{|F_1'(z(t))X|^2}{4t^2}. \qquad (19.4.9)$$

Hence (19.4.7), (19.4.8), and (19.4.9) imply that

$$\boldsymbol{\gamma}_{\tilde{G}}(z(t); X) \geq (1-\varepsilon)^3 \left(\frac{(\mathcal{L}\tilde{r})(0; X_{(t)})}{2t} + \frac{|X_1|^2}{4t^2}\right)^{1/2}$$

and

$$\begin{aligned}&\boldsymbol{\varkappa}_{\tilde{G}}(z(t); X)\\ &\quad\leq (1+\varepsilon)^2 \left(\frac{(\mathcal{L}\tilde{r})(0; X_{(t)})}{2t} + \frac{|X_1|^2}{4t^2}\right)^{1/2} \cdot \coth \boldsymbol{k}_{\mathcal{E}_\lambda^+}(F(z(t)), \mathcal{E}_\lambda^+ \setminus \mathbb{B}(\delta))\end{aligned}$$

if $0 < t < \delta'_\varepsilon \leq 1/2\delta$.

It remains to show that $\coth \boldsymbol{k}_{\mathcal{E}_\lambda^+}(F(z(t)), \mathcal{E}_\lambda^+ \setminus \mathbb{B}(\delta)) < 1 + \varepsilon$ for small t. Therefore, we consider the function $g(z) := \exp(-2z_1)$ on $\mathcal{E}_\lambda^+$, which peaks at 0 because $g(0) = 1$ and $2\operatorname{Re} z_1 > (\mathcal{L}\tilde{r})(0; z) + \lambda\|z\|^2 \geq \alpha\|z\|^2$, $z \in \mathcal{E}_\lambda^+$. As in the proof of Theorem 19.2.1, we obtain

$$\begin{aligned}\boldsymbol{k}_{\mathcal{E}_\lambda^+}(F(z(t)), \mathcal{E}_\lambda^+ \setminus \mathbb{B}(\delta)) &\geq \boldsymbol{c}_{\mathcal{E}_\lambda^+}(F(z(t)), \mathcal{E}_\lambda^+ \setminus \mathbb{B}(\delta))\\ &\geq \frac{1}{2}\log\frac{1-d_2}{d_1\|z(t)\|} \geq \frac{1}{2}\log\frac{1-d_2}{d_1(1+M)t}\end{aligned}$$

with $d_2 = d_2(\delta) < 1$, $d_1 > 0$, and t sufficiently small.

Hence, we have verified that

$$\varkappa_{\tilde{G}}(z(t); X) \leq (1+\varepsilon)^3 \left(\frac{(\mathcal{L}\tilde{r})(0; X_{(t)})}{2t} + \frac{|X_1|^2}{4t^2} \right)^{1/2}$$

provided that $0 < t < \delta_\varepsilon \leq \delta'_\varepsilon \leq 1/2\delta$. So, the proof is complete for any $z \in G$ with $\operatorname{dist}(z, \partial G) < \delta_\varepsilon$. □

Remark 19.4.4. Under the additional hypothesis that G has a $\mathcal{C}^3$-boundary D. Ma (cf. [352, 353]) obtained stronger results, in which he even specified the order of the asymptotical convergence. Moreover, in the case of the Kobayashi–Royden metric he has the following precise result:

Let G be a bounded domain in $\mathbb{C}^n$, not necessarily pseudoconvex. Assume that M_0 is a relatively open subset of ∂G, M_0 is a $\mathcal{C}^3$ strongly pseudoconvex hypersurface, and G is on the pseudoconvex side of M_0. Let M be a compact subset of M_0. For $\delta > 0$ let $Q_\delta := \{z \in \mathbb{C}^n : \operatorname{dist}(z, M) < \delta\}$ and $G_\delta := G \cap Q_\delta$. Moreover, assume that for a positive δ_0 we have

(a) $\partial G \cap Q_{\delta_0}$ is relatively compact in M_0;

(b) there is a strictly psh function $\varphi \in \mathcal{C}^3(Q_{\delta_0})$ with $G_{\delta_0} := \{z \in Q_{\delta_0} : \varphi(z) < 0\}$ and $\|\operatorname{grad} \varphi(z)\| = 1$ whenever $z \in Q_{\delta_0} \cap M$.

Let $\delta < \delta_0$ be a positive number such that for each $z \in G_\delta$ there is a unique point $\pi(z) \in M_0 \cap Q_{\delta_0}$ with $\operatorname{dist}(z, \partial G) = \|\pi(z) - z\|$ and such that $M_1 := \pi(G_\delta) \subset\subset M_0 \cap Q_{\delta_0}$.

Then, there exists a positive $C = C(G, M, \delta)$ such that for each $z \in G_\delta$ the Kobayashi–Royden metric satisfies the following estimate:

$$\begin{aligned} \exp(-C\sqrt{u(z)}) \left(\frac{(\mathcal{L}\varphi)(\pi(z); X_{(t)})}{2u(z)} + \frac{\|X_{(n)}\|^2}{4u^2(z)} \right)^{1/2} &\leq \varkappa_G(z; X) \\ &\leq \exp(C\sqrt{u(z)}) \left(\frac{(\mathcal{L}\varphi)(\pi(z); X_{(t)})}{2u(z)} + \frac{\|X_{(n)}\|^2}{4u^2(z)} \right)^{1/2}, \end{aligned}$$

where $u(z) := \operatorname{dist}(z, \partial G)$ and $X = X_{(t)} + X_{(n)} \in \mathbb{C}^n$ as in Theorem 19.4.2.

There is an example (cf. Exercise 19.7.1) that shows that in the above estimate the factors $\exp(\pm C\sqrt{u(z)})$ cannot be improved to $\exp(\pm C(u(z))^{(1/2)+\varepsilon})$.

It should be mentioned that the proof of this precise result cannot be used as an approach to find the optimal estimate for the Carathéodory–Reiffen metric.

From Theorem 19.4.2, we immediately derive Graham's formulation of the asymptotic behavior.

Corollary 19.4.5. *Let G and r be as in Theorem* 19.4.2 *and fix $\zeta_0 \in \partial G$. Then,*

$$\lim_{z\to\zeta_0} \delta_G(z;X)\operatorname{dist}(z,\partial G) = (1/2)\|X_{(n)}\|, \quad X \in \mathbb{C}^n,$$

where $X_{(n)}$ denotes the normal component of X at ζ_0 and where δ_G belongs to a holomorphically contractible family of pseudometrics $\underline{\delta}$. If $X_{(n)} = 0$, i.e., X is a complex tangent vector to ∂G at ζ_0, then

$$\lim_{z\to\zeta_0} \delta_G(z;X)^2\operatorname{dist}(z,\partial G) = (1/2)(\mathcal{L}r)(\zeta_0;X).$$

We point out that the proof of Theorem 19.4.2 is based on the holomorphic contractibility. Nevertheless, it turns out that the estimates there remain true if we substitute $\boldsymbol{\gamma}_G$ and $\boldsymbol{\varkappa}_G$ with the Bergman metric $\boldsymbol{\beta}_G$. This result, and much more information about the boundary behavior of derivatives of the Bergman kernel, were obtained by K. Diederich (cf. [124, 125]).

Theorem 19.4.6. *The inequalities of Theorem* 19.4.2 *remain true if $\boldsymbol{\gamma}_G$ and $\boldsymbol{\varkappa}_G$ there are replaced by the Bergman metric divided by $\sqrt{n+1}$.*

Before we go into the necessary modifications of the proof of Theorem 19.4.2, we have to present the formula of the Bergman kernel for analytic ellipsoids.

Lemma 19.4.7. *Let $\mathcal{E} = \mathcal{E}(H)$ be as in* (19.4.1) *with $H(z,z) = \sum_{i,j=1}^n a_{i,j} z_i \overline{z}_j$. Then, the Bergman kernel of $\mathcal{E}$ is given by*

$$\mathfrak{K}_{\mathcal{E}}(z) = \frac{n!}{\pi^n}\frac{\det(a_{i,j})_{i,j=2}^n}{(-\Phi_{\mathcal{E}}(z))^{n+1}}, \quad z \in \mathcal{E},$$

where $\Phi_{\mathcal{E}}(z) = -2\operatorname{Re} z_1 + H(z,z)$.

Proof. Use a biholomorphic map from $\mathcal{E}$ to $\mathbb{B}_n$; cf. the proof of Lemma 19.4.1. □

Proof of Theorem 19.4.6. Here, we will use the same notations as in the proof of Theorem 19.4.2. We will discuss only those steps of the proof that are now different.

Fix $\varepsilon > 0$ and choose $\lambda = \varrho_0$ and $\delta_\varepsilon < \delta$ as in the proof of Theorem 19.4.2. Moreover, we may assume that δ is chosen so small that the estimate of Theorem 19.3.6 holds uniformly with respect to the boundary points.

Now let $z_0 \in G$ with $\operatorname{dist}(z_0,\partial G) =: t < \delta_\varepsilon < \delta$, $\zeta_0 \in \partial G$ with $\|\zeta_0 - z_0\| = t$, and $Y \in \mathbb{C}^n$. As above, we put $z(t) := (t,0,\dots,0)$ and $\widetilde{G} := T(G)$.

Then, we begin the upper estimate of M_G (see Remark 12.7.8 for the definition):

$$\begin{aligned} M_G(z_0;Y) &= M_{\widetilde{G}}(z(t);X) \le M_{\widetilde{G}\cap\mathbb{B}(2\delta)}(z(t);X)\\ &= M_{F(\widetilde{G}\cap\mathbb{B}(2\delta))}(F(z(t));F'(z(t))X)\cdot|\det F'(z(t))|\\ &\le M_{\mathcal{E}_\lambda^+\cap\mathbb{B}(\delta)}(F(z(t));F'(z(t))X)\cdot|\det F'(z(t))|, \end{aligned}$$

where $X := AY$. Here, we have used the transformation rule for M_G (cf. Remark 12.7.8) and its monotonicity in G.

Moreover, if δ_ε is sufficiently small, by Theorem 19.3.6 we obtain

$$\begin{aligned}\mathfrak{K}_G(z_0) &= \mathfrak{K}_{\widetilde{G}}(z(t)) \geq (1-\varepsilon)\mathfrak{K}_{\widetilde{G}\cap\mathbb{B}(2\delta)}(z(t))\\ &\geq (1-\varepsilon)\mathfrak{K}_{F(\widetilde{G}\cap\mathbb{B}(2\delta))}(F(z(t)))\cdot|\det F'(z(t))|^2\\ &\geq (1-\varepsilon)\mathfrak{K}_{\mathcal{E}_\lambda^-}(F(z(t)))\cdot|\det F'(z(t))|^2.\end{aligned}$$

Hence, the above inequalities yield (cf. Theorem 12.7.5)

$$\boldsymbol{\beta}_G(z_0;Y) \leq \frac{M_{\mathcal{E}_\lambda^+\cap\mathbb{B}(\delta)}(F(z(t));F'(z(t))X)}{(1-\varepsilon)^{1/2}\sqrt{\mathfrak{K}_{\mathcal{E}_\lambda^-}(F(z(t)))}}. \tag{19.4.10}$$

To increase the numerator of (19.4.10), we apply Theorem 19.3.6 again and we find that

$$M_{\mathcal{E}_\lambda^+\cap\mathbb{B}(\delta)}(F(z(t));F'(z(t))X) \leq (1+\varepsilon)M_{\mathcal{E}_\lambda^+}(F(z(t));F'(z(t))X)$$

whenever $0 < t < \delta_\varepsilon$ and δ_ε sufficiently small.

On the other hand, the explicit formula for the Bergman kernel gives that

$$\mathfrak{K}_{\mathcal{E}_\lambda^-}(F(z(t))) \geq (1-\varepsilon)\mathfrak{K}_{\mathcal{E}_\lambda^+}(F(z(t)))$$

provided that $\lambda = \varrho_0$ is sufficiently small.

So, the final inequality looks like this:

$$\begin{aligned}\boldsymbol{\beta}_G(z_0;X) &\leq \frac{1+\varepsilon}{1-\varepsilon}\boldsymbol{\beta}_{\mathcal{E}_\lambda^+}(F(z(t));F'(z(t))X)\\ &= \frac{1+\varepsilon}{1-\varepsilon}\sqrt{(n+1)}\boldsymbol{\varkappa}_{\mathcal{E}_\lambda^+}(F(z(t));F'(z(t))X);\end{aligned}$$

cf. Example 12.7.1. Now, we can turn to the proof of Theorem 19.4.2 and continue, starting with (19.4.7).

We now move to the lower estimate. First, we again apply Theorem 19.3.6 to get

$$\begin{aligned}M_G(z_0;Y) &= M_{\widetilde{G}}(z(t);X) \geq (1-\varepsilon)M_{\widetilde{G}\cap\mathbb{B}(2\delta)}(z(t);X)\\ &\geq (1-\varepsilon)M_{F(\widetilde{G}\cap\mathbb{B}(2\delta))}(F(z(t));F'(z(t))X)\cdot|\det F'(z(t))|\\ &\geq (1-\varepsilon)M_{\mathcal{E}_\lambda^-}(F(z(t));F'(z(t))X)\cdot|\det F'(z(t))|\end{aligned}$$

if δ_ε is sufficiently small. Moreover, we have the following chain of inequalities for the Bergman kernel (cf. Theorem 19.3.6):

$$\begin{aligned}\mathfrak{K}_G(z_0) &= \mathfrak{K}_{\widetilde{G}}(z(t)) \le \mathfrak{K}_{\widetilde{G}\cap\mathbb{B}(2\delta)}(z(t))\\ &= \mathfrak{K}_{F(\widetilde{G}\cap\mathbb{B}(2\delta))}(F(z(t)))\cdot|\det F'(z(t))|^2\\ &\le \mathfrak{K}_{\mathcal{E}_\lambda^+\cap\mathbb{B}(\delta)}(F(z(t)))\cdot|\det F'(z(t))|^2\\ &\le (1+\varepsilon)\mathfrak{K}_{\mathcal{E}_\lambda^+}(F(z(t)))\cdot|\det F'(z(t))|^2\end{aligned}$$

if δ_ε is sufficiently small.

Combining the last two inequalities, we can conclude that

$$\boldsymbol{\beta}_G(z_0;Y) \ge \frac{(1-\varepsilon)}{(1+\varepsilon)^{1/2}}\frac{M_{\mathcal{E}_\lambda^-}(F(z(t));F'(z(t))X)}{\sqrt{\mathfrak{K}_{\mathcal{E}_\lambda^+}(F(z(t)))}}.$$

The last step then is similar to the one in the upper estimate, and therefore is left to the reader. □

19.5 A comparison of distances

Recall that Theorem 19.2.5 has taught us that the quotient $\boldsymbol{c}_G(z_0,z)/\boldsymbol{k}_G(z_0,z)$ tends to one if z approaches the boundary. Moreover, the convergence is locally uniform in the first variable z_0. Some more work leads to the fact that the above convergence is even uniform in the first variable. More precisely, we have

Theorem 19.5.1 (cf. [507]). *Let G be a strongly pseudoconvex domain in $\mathbb{C}^n$. Then, for every $\varepsilon > 0$ there exists a compact set $K = K(\varepsilon) \subset G$ such that*

$$\boldsymbol{c}_G(z',z'') \le \boldsymbol{k}_G(z',z'') \le (1+\varepsilon)\boldsymbol{c}_G(z',z''),$$

whenever $z' \in G$ and $z'' \in G\setminus K$.

The proof of Theorem 19.5.1 needs a very deep result of J. E. Fornæss (cf. [166]), which we state here in a form appropriate to our purposes.

Theorem 19.5.2. *Let G be a strongly pseudoconvex domain in $\mathbb{C}^n$.*

(a) *For any $\zeta_0 \in \partial G$ there exist a domain $G' \supset \overline{G}$, a neighborhood $U = U(\zeta_0) \subset G'$, a convex domain $D \subset\subset \mathbb{C}^n$ with $\mathcal{C}^2$-boundary, and a holomorphic mapping $\Phi : G' \longrightarrow \mathbb{C}^n$ which satisfy the following properties:*

 (i) $\Phi(G) \subset D$,

 (ii) $\Phi(U\setminus\overline{G}) \subset \mathbb{C}^n\setminus\overline{D}$,

(iii) $\Phi^{-1}(\Phi(U)) = U$,

(iv) $\Phi|_U$ *is injective.*

(b) *There exist a domain* $\widetilde{G} \supset \overline{G}$, *a bounded strictly convex domain* $\widetilde{D} \subset \mathbb{C}^N$, *and a holomorphic map* $\psi : \widetilde{G} \longrightarrow \mathbb{C}^N$ *such that*

(i) ψ *is biholomorphic onto a closed submanifold of* $\mathbb{C}^N$,

(ii) $\psi(G) \subset \widetilde{D}$ *and* $\psi(\widetilde{G} \setminus \overline{G}) \subset \mathbb{C}^N \setminus \overline{\widetilde{D}}$,

(iii) $\psi(\widetilde{G})$ *intersects* $\partial\widetilde{D}$ *transversally.*

Proof of Theorem 19.5.1. Let us suppose that the claim of Theorem 19.5.1 does not hold. Then, we find $\varepsilon_0 > 0$ and sequences $(z'_\nu)_{\nu\in\mathbb{N}}, (z''_\nu)_{\nu\in\mathbb{N}} \subset G$ with $z''_\nu \underset{\nu\to\infty}{\longrightarrow} \partial G$ such that

$$(1+\varepsilon_0)\boldsymbol{c}_G(z'_\nu, z''_\nu) < \boldsymbol{k}_G(z'_\nu, z''_\nu).$$

In particular, we have $z'_\nu \neq z''_\nu$. We may assume that $\lim_{\nu\to\infty} z''_\nu =: z'' \in \partial G$ and $\lim_{\nu\to\infty} z'_\nu =: z' \in \overline{G}$ exist.

If $z' \in G$, then Theorem 19.2.1 and Proposition 19.2.4 imply that

$$(1+\varepsilon_0) \le \frac{\boldsymbol{k}_G(z'_\nu, z''_\nu)}{\boldsymbol{c}_G(z'_\nu, z''_\nu)} \le \frac{-(1/2)\log \operatorname{dist}(z''_\nu, \partial G) + C_1}{-(1/2)\log \operatorname{dist}(z''_\nu, \partial G) - C_2} \underset{\nu\to\infty}{\longrightarrow} 1,$$

which gives a contradiction.

If $z' \in \partial G$ with $z' \neq z''$, we are led via Propositions 19.2.4, 19.2.7, and Theorem 19.5.2(b) to the following chain of inequalities:

$$\begin{aligned}(1+\varepsilon_0) &\le \frac{\boldsymbol{k}_G(z'_\nu, z''_\nu)}{\boldsymbol{c}_G(z'_\nu, z''_\nu)} \le \frac{\boldsymbol{k}_G(z'_\nu, z''_\nu)}{\boldsymbol{c}_{\widetilde{D}}(\psi(z'_\nu), \psi(z''_\nu))} = \frac{\boldsymbol{k}_G(z'_\nu, z''_\nu)}{\boldsymbol{k}_{\widetilde{D}}(\psi(z'_\nu), \psi(z''_\nu))} \\ &\le \frac{-(1/2)\log \operatorname{dist}(z'_\nu, \partial G) - (1/2)\log \operatorname{dist}(z''_\nu, \partial G) + C_1}{-(1/2)\log \operatorname{dist}(\psi(z'_\nu), \partial\widetilde{D}) - (1/2)\log \operatorname{dist}(\psi(z''_\nu), \partial\widetilde{D}) - C_2}.\end{aligned}$$

Employing $\psi(\partial G) \subset \partial\widetilde{D}$ gives

$$\operatorname{dist}(\psi(w_\nu), \partial\widetilde{D}) \le C_3 \operatorname{dist}(w_\nu, \partial G) \quad \text{for } \nu \gg 1,$$

where $w_\nu = z'_\nu$ or $w_\nu = z''_\nu$. Inserting this in the upper inequality again leads to a contradiction.

It remains to consider the case that $z' = z''$. According to Theorem 19.5.2(a), we choose G', $U = U(\zeta_0)$ with $\zeta_0 := z' = z''$, and Φ. Put $V := \Phi(U)$; V is an open neighborhood of $\Phi(\zeta_0) \in \partial D$ and $\Phi : U \longrightarrow V$ is biholomorphic. Now choose a ball V' around $\Phi(\zeta_0)$ with $V' \subset V$ and put $U' := \Phi^{-1}(V')$. Then, $U' = U'(\zeta_0) \subset U$.

Moreover, $D \cap V'$ is connected, and therefore $G \cap U'$ is connected, too. So, for $\nu \gg 1$ we obtain

$$(1+\varepsilon_0) \le \frac{\boldsymbol{k}_G(z'_\nu, z''_\nu)}{\boldsymbol{c}_G(z'_\nu, z''_\nu)} \le \frac{\boldsymbol{k}_G(z'_\nu, z''_\nu)}{\boldsymbol{k}_{G\cap U'}(z'_\nu, z''_\nu)} \cdot \frac{\boldsymbol{k}_{D\cap V'}(\Phi(z'_\nu), \Phi(z''_\nu))}{\boldsymbol{k}_D(\Phi(z'_\nu), \Phi(z''_\nu))} \underset{\nu\to\infty}{\longrightarrow} 1$$

because of Theorem 19.3.4; a contradiction. Therefore, Theorem 19.5.1 is completely verified. □

Corollary 19.5.3. *Let G be a strongly pseudoconvex domain in $\mathbb{C}^n$ and let $U = U(\zeta_0)$ be an open neighborhood of $\zeta_0 \in \partial G$ such that $G \cap U$ is connected. Then,*

$$\lim_{\substack{z',z''\to\zeta_0 \\ z'\neq z''}} \frac{\boldsymbol{c}_{G\cap U}(z', z'')}{\boldsymbol{c}_G(z', z'')} = 1.$$

Proof. Suppose the contrary. Then, there are sequences $(z'_\nu)_{\nu\in\mathbb{N}}, (z''_\nu)_{\nu\in\mathbb{N}} \subset G \cap U$, $z'_\nu \neq z''_\nu$, with $\boldsymbol{c}_{G\cap U}(z'_\nu, z''_\nu) > (1+\varepsilon_0)\boldsymbol{c}_G(z'_\nu, z''_\nu)$. Hence, by Theorem 19.5.1, $\boldsymbol{c}_G(z'_\nu, z''_\nu)(1+\varepsilon_0)^{1/2} \ge \boldsymbol{k}_G(z'_\nu, z''_\nu)$ provided that $\nu \gg 1$. Thus, it follows that

$$\boldsymbol{k}_{G\cap U}(z'_\nu, z''_\nu) \ge (1+\varepsilon_0)^{1/2}\boldsymbol{k}_G(z'_\nu, z''_\nu),$$

which contradicts Theorem 19.3.4. □

? **Remark 19.5.4.** It would be interesting to find a direct proof of Corollary 19.5.3.

19.6 Characterization of the unit ball by its automorphism group

The aim of this section is to characterize the unit Euclidean ball in $\mathbb{C}^n$ by its automorphism group. An even stronger result dealing with unbranched proper holomorphic mappings will be presented. Roughly speaking, we will show that if there are sufficiently many unbranched proper mappings between two domains in $\mathbb{C}^n$, then both domains are biholomorphically equivalent to the unit Euclidean ball.

Theorem 19.6.1 (cf. [346]). *Let G and D be two bounded domains in $\mathbb{C}^n$, where G has a smooth $\mathcal{C}^1$-boundary, and let $q \in G$. Assume that there exists a sequence of unbranched proper maps $F_j : G \longrightarrow D$, $j \in \mathbb{N}$, with $\lim_{j\to\infty} F_j(q) = w_0 \in \partial D$, such that D has a strongly pseudoconvex boundary near w_0. Then, G and D are biholomorphically equivalent to the unit ball $\mathbb{B}_n$.*

Before we go into the proof, we should mention two consequences, which were already announced in the title of this section (cf. [211, 454, 532]):

Corollary 19.6.2.

(a) *Any strongly pseudoconvex domain $G \subset \mathbb{C}^n$ is biholomorphically equivalent to $\mathbb{B}_n$ iff* $\mathrm{Aut}(G)$ *is not compact.*

(b) *A bounded domain with smooth $\mathcal{C}^2$-boundary is biholomorphically equivalent to $\mathbb{B}_n$ iff the group* $\mathrm{Aut}(G)$ *acts transitively on G.*

Proof. One has only to recall that any bounded domain with smooth $\mathcal{C}^2$-boundary admits at least one strongly pseudoconvex boundary point. Then, setting $G = D$, Theorem 19.6.1 applies. □

Remark 19.6.3. Because of Corollary 19.6.2, the only (up to a biholomorphism) strongly pseudoconvex domain with a non compact automorphism group is the unit Euclidean ball. In the case of bounded pseudoconvex domains of finite type, there are analogous results by E. Bedford and S. Pinchuk (cf. [41, 42, 43]). For example, a pseudoconvex domain $G \subset\subset \mathbb{C}^2$ of finite type, for which $\mathrm{Aut}(G)$ is not compact, is biholomorphically equivalent to a Thullen domain $\{z \in \mathbb{C}^2 : |z_1|^2 + |z_2|^{2k} < 1\}$ with $k \in \mathbb{N}$.

Now, we are going to prepare the proof of Theorem 19.6.1 giving a series of various lemmas.

Lemma 19.6.4. *Under the assumptions of Theorem* 19.6.1 *there exists a subsequence $(F_{j_\nu})_{\nu\in\mathbb{N}} \subset (F_j)_{j\in\mathbb{N}}$ with $F_{j_\nu} \overset{K}{\underset{\nu\to\infty}{\Longrightarrow}} F \in \mathcal{O}(G, \mathbb{C}^n)$, where $F(z) = w_0$ for every $z \in G$.*

Proof. Since D is bounded, there is a subsequence $(F_{j_\nu}) \subset (F_j)$ such that $F_{j_\nu} \overset{K}{\underset{\nu\to\infty}{\Longrightarrow}} F \in \mathcal{O}(G, \overline{D})$. In particular, we have $F(q) = w_0$.

By hypothesis, the point w_0 is a strongly pseudoconvex boundary point of D. Therefore, there exist a neighborhood $V = V(w_0)$ and a function $f \in \mathcal{O}(V, \mathbb{C})$ with $f(w_0) = 1$, $|f(w)| < 1$ whenever $w \in \overline{D} \cap V \setminus \{w_0\}$. If $U = U(q)$ is so small that $F(U) \subset V$, then we obtain a holomorphic function $f \circ (F|_U)$ on U which attains its maximum at q. Thus, $f \circ (F|_U) \equiv 1$, which implies that $F(z) = w_0$ for every $z \in U$. Finally, the identity theorem yields $F \equiv w_0$. □

From now on, we will denote this subsequence again by $(F_j)_{j\in\mathbb{N}}$. (19.6.1)

Remark 19.6.5. Observe that this lemma shows that all information about G is already hidden in the shape of D near w_0. So, using the local convexity of D at w_0, it is easy to deduce that G is a simply connected domain. Since we will not use this fact, we omit the details of the proof.

The proof of the next auxiliary result could be based on a certain fixed point theorem from differential geometry, which is due to E. Cartan; cf. [220], Théorème 13.5. One only has to observe that the sectional curvature of the Bergman metric of $\mathbb{B}_n$ is non-positive. Nevertheless, we here give a proof that does not depend on the concept of curvature.

Lemma 19.6.6. *Any unbranched proper holomorphic mapping $\pi : \mathbb{B}_n \longrightarrow D$ to a domain $D \subset \mathbb{C}^n$ is biholomorphic.*

Proof. Since π is proper, it is surjective. So, it remains for us to show its injectivity. Let us suppose the contrary, i.e., the existence of an integer $N \geq 2$ such that $\#\pi^{-1}(w) = N$ for every $w \in D$.

Now fix $w^* \in D$ and let $\pi^{-1}(w^*) = \{z_1, \dots, z_N\} \subset \mathbb{B}_n$. Then, obviously,

$$r_0 := \inf\left\{ r > 0 : \bigcap_{j=1}^{N} \overline{B_{\boldsymbol{k}_{\mathbb{B}_n}}(z_j, r)} \neq \varnothing \right\}$$

is a positive number. We put $K := \bigcap_{j=1}^{N} \overline{B_{\boldsymbol{k}_{\mathbb{B}_n}}(z_j, r_0)}$. Observe that K is a non-empty convex compact subset of $\mathbb{B}_n$.

Let $\mathfrak{g} \subset \operatorname{Aut}(\mathbb{B}_n)$ denote the finite group of covering transformations for $\pi : \mathbb{B}_n \longrightarrow D$. Then, for $g \in \mathfrak{g}$ and $z \in K$ we get

$$\boldsymbol{k}_{\mathbb{B}_n}(z_j, g(z)) = \boldsymbol{k}_{\mathbb{B}_n}(g^{-1}(z_j), z) = \boldsymbol{k}_{\mathbb{B}_n}(z_{l(j)}, z) \leq r_0,$$

i.e., the set K is invariant under the action of $\mathfrak{g}$. So, K consists of at least N points with $N \geq 2$. Let us take two of them, say w', $w'' \in K$ with $w' \neq w''$. Since K is convex, the segment $\overline{w'w''}$ belongs to K.

Because of the minimality of r_0 and the convexity of $\boldsymbol{k}_{\mathbb{B}_n}(z_j, \cdot)$, we conclude that there is at least one index j_0, say $j_0 = 1$, such that $\boldsymbol{k}_{\mathbb{B}_n}(z_1, w) = r_0$ for every $w \in \overline{w'w''}$.

Now, if we look at the formula for $\boldsymbol{k}_{\mathbb{B}_n} = \tanh^{-1}(\boldsymbol{c}^*_{\mathbb{B}_n})$, we immediately see (using the identity theorem for real analytic functions) that $\boldsymbol{k}_{\mathbb{B}_n}(z_1, w) = r_0$ as long as w lies on the line through w' and w'' and in $\mathbb{B}_n$. But this contradicts the completeness of $\mathbb{B}_n$. □

By virtue of Lemma 19.6.6 it remains to establish that G is biholomorphic to $\mathbb{B}_n$. The idea is to apply the following classification result, which is due to C. M. Stanton; cf. [488].

Proposition 19.6.7. *Let G be a bounded pseudoconvex domain in $\mathbb{C}^n$ with a smooth $\mathcal{C}^1$-boundary. Moreover, assume that $\boldsymbol{\gamma}_G \equiv \boldsymbol{\varkappa}_G \equiv \frac{1}{\sqrt{n+1}} \boldsymbol{\beta}_G$. Then, G is biholomorphically equivalent to $\mathbb{B}_n$.*

Proof. By Corollary 15.1.6, G is complete with respect to the Bergman distance. So G is a complete Hermitian manifold the distance of which is given by $\boldsymbol{k}_G = \boldsymbol{c}_G = \frac{1}{\sqrt{n+1}} \boldsymbol{b}_G$.

Let us fix a point $z_0 \in G$ and put

$$\widetilde{\mathbb{B}}_n := \left\{ v \in \mathbb{C}^n : N(v) := \frac{1}{\sqrt{n+1}} \boldsymbol{\beta}_G(z_0; v) < 1 \right\}.$$

Observe that $\widetilde{\mathbb{B}}_n$ is given by a Hermitian scalar product, so it is obviously biholomorphic to the Euclidean ball $\mathbb{B}_n$.

In the next step, we will use some well-known facts from differential geometry. Namely, for any vector $v \in \mathbb{C}^n$ there exists a uniquely determined $\mathcal{C}^\infty$-curve $\alpha_v : \mathbb{R} \longrightarrow G$ with $\alpha_v(0) = z_0$ and $\alpha_v'(0) = v$, which is a geodesic with respect to the Hermitian structure of G and such that $\alpha_{\tau v}(t) = \alpha_v(\tau t)$, $t, \tau \in \mathbb{R}$. Moreover, there is a small positive ε_0 such that the mapping

$$\Phi : \{v \in \mathbb{C}^n : N(v) < \varepsilon_0\} \longrightarrow G \quad \text{with} \quad \Phi(v) := \exp_{z_0}(v) := \alpha_v'(1)$$

is a $\mathcal{C}^\infty$-diffeomorphism onto a neighborhood of z_0.

Now, we are able to introduce the map $F_0 : \widetilde{\mathbb{B}}_n \longrightarrow G$ that we will be interested in. We put $F_0(v) := \Phi(\tanh^{-1}(N(v))\frac{v}{N(v)})$, $v \in \widetilde{\mathbb{B}}_n$. Note that $v \longmapsto \tanh^{-1}(N(v))/N(v)$ extends to $\widetilde{\mathbb{B}}_n$ as a $\mathcal{C}^\infty$-function. Hence, F_0 is a $\mathcal{C}^\infty$-map near the origin. To see that F_0 is also holomorphic, we want to apply Forelli's theorem (see [462], Theorem 4.4.5). Thus, we have to verify that F_0 is slicewise holomorphic.

For if we fix $a \in \mathbb{C}^n$ with $N(a) = 1$, then there exists an extremal analytic disc $\varphi \in \mathcal{O}(\mathbb{D}, G)$ for $\boldsymbol{\varkappa}_G$ (recall that G is taut) with $\varphi(0) = z_0$, $\varphi'(0) = a$. In particular, φ is a complex $\boldsymbol{c}_G$-geodesic. Then, if $\lambda \in \mathbb{D}_*$, we study the $\mathcal{C}^\infty$-curve $\widetilde{\alpha}$ over $\mathbb{R}$ defined by

$$\widetilde{\alpha}(t) := \varphi(\tanh(t|\lambda|)\lambda/|\lambda|), \quad t \in \mathbb{R}.$$

It is clear that $\widetilde{\alpha}(0) = z_0$ and $\widetilde{\alpha}'(0) = \lambda a$. In addition, for positive τ we have

$$\boldsymbol{k}_G(z_0, \widetilde{\alpha}(\tau)) = p(0, \tanh(\tau|\lambda|)\lambda/|\lambda|) = \tau|\lambda| = \int_0^\tau \boldsymbol{\varkappa}_G(\widetilde{\alpha}(t); \widetilde{\alpha}'(t))dt.$$

Hence, $\widetilde{\alpha}$ is the shortest curve. In particular, it is the geodesic (starting at z_0 into direction of λa), and so we obtain

$$F_0(\lambda a) = \exp_{z_0}\left(\frac{\tanh^{-1}(|\lambda|)}{|\lambda|}\lambda a\right) = \widetilde{\alpha}\left(\frac{\tanh^{-1}(|\lambda|)}{|\lambda|}\right) = \varphi(\lambda).$$

Since λ is arbitrary in $\mathbb{D}_*$, we find that the map $E \ni \lambda \longmapsto F_0(\lambda a)$ is holomorphic. Hence, Forelli's theorem implies that F_0 is holomorphic.

The considerations above have also led to the following equality:

$$\boldsymbol{k}_G(z_0, F_0(a)) = \tanh^{-1}(\boldsymbol{\varkappa}_G(z_0; a)), \quad a \in \widetilde{\mathbb{B}}_n. \tag{19.6.2}$$

Therefore, F_0 is a proper holomorphic mapping from $\widetilde{\mathbb{B}}_n$ to G. So it is surjective.

According to the general properties of proper holomorphic maps, we know that if $M := \{v \in \widetilde{\mathbb{B}}_n : \det F_0'(v) = 0\}$, then

$$F : \widetilde{\mathbb{B}}_n \setminus F_0^{-1}(F_0(M)) \longrightarrow G \setminus F_0(M)$$

is a proper covering map and that $F_0(M)$ is a proper analytic subset of G. On the other hand, (19.6.2) shows that F_0 maps $\boldsymbol{\varkappa}_G(z_0;\cdot)$-balls into $\boldsymbol{k}_G(z_0,\cdot)$-balls. Thus, generically, the fibres of F_0 near z_0 consist of one element only, which implies that F_0 is globally injective. □

Remark 19.6.8. The above proposition remains true under slightly weaker assumptions; cf. [488]. In this context we should also mention the work by M. Abate and G. Patrizio [6].

The next steps toward the proof of Theorem 19.6.1 consist of successively establishing the assumptions of Proposition 19.6.7.

Lemma 19.6.9. *Let* G, D, q, w_0, *and* F_j *be as in Theorem* 19.6.1 *and* (19.6.1). *Then,* G *is a pseudoconvex domain.*

Proof. Obviously, it suffices to exhaust G by a sequence of pseudoconvex domains.

By assumption, the domain D is strongly pseudoconvex near w_0. Therefore, we find a neighborhood $V = V(w_0)$ of w_0 such that $\widetilde{D} := D \cap V$ is biholomorphic to a convex domain.

Now, we are going to construct sequences $(\widetilde{D}_k)_{k\in\mathbb{N}}$ and $(G_k)_{k\in\mathbb{N}}$ of subdomains $\widetilde{D}_k$ of $\widetilde{D}$ and G_k of G, respectively, and a subsequence $(F_{j_k})_{k\in\mathbb{N}}$ of $(F_j)_{j\in\mathbb{N}}$ such that

$$F_{j_k}(q) \in \widetilde{D}_k \subset\subset \widetilde{D}_{k+1}, \; \widetilde{D}_k \text{ is biholomorphic to a convex domain, } \bigcup_{k=1}^{\infty} \widetilde{D}_k = \widetilde{D}; \tag{19.6.3}$$

$$q \in G_k \subset G_{k+1}, \; \bigcup_{k=1}^{\infty} G_k = G, \text{ and } F_{j_k}|_{G_k} : G_k \longrightarrow \widetilde{D}_k \text{ is biholomorphic.} \tag{19.6.4}$$

We only indicate the first step of this construction. Without loss of generality, we may assume that $F_1(q) \in \widetilde{D}$. Then, we fix an arbitrary subdomain $\widetilde{D}_1 \subset\subset \widetilde{D}$ with $F_1(q) \in \widetilde{D}_1$ such that $\widetilde{D}_1$ is biholomorphic to a convex domain. We denote by $\widetilde{G}_1$ the connected component of $F_1^{-1}(\widetilde{D}_1)$ that contains the point q. Obviously, G_1 is a relatively compact subdomain of G. Moreover, since $\widetilde{D}_1$ is simply connected and since F_1 is a covering map, $F_1|_{G_1}$ is a biholomorphic map between G_1 and $\widetilde{D}_1$; in particular, G_1 is pseudoconvex.

For $\nu \in \mathbb{N}$ and for a sufficiently small positive number a we put

$$G_\nu^* := \{z \in G : \operatorname{dist}(z, \partial G) > a/\nu\} \subset\subset G, \tag{19.6.5}$$

$$D_\nu^* := \{w \in \widetilde{D} : \operatorname{dist}(w, \partial \widetilde{D}) > a/\nu\} \subset\subset \widetilde{D}. \tag{19.6.6}$$

Then, we fix an index $j_2 > j_1 := 1$ such that $F_{j_2}(G_1 \cup G_1^*) \subset\subset \widetilde{D}$. After that, we choose a domain $\widetilde{D}_2 \subset\subset \widetilde{D}$ in such a way that $\widetilde{D}_1 \cup D_1^* \cup F_{j_2}(G_1 \cup G_1^*) \subset\subset \widetilde{D}_2$ and that $\widetilde{D}_2$ is biholomorphic to a convex domain. Denote by G_2 the connected component of $F_{j_2}^{-1}(\widetilde{D}_2)$ containing q. Then, G_2 is a relatively compact subdomain of G with $G_1 \subset G_2$. Moreover, $F_{j_2}|_{G_2} \longrightarrow \widetilde{D}_2$ is biholomorphic and G_2 is pseudoconvex.

The continuation of this procedure then results in sequences the existence of which was postulated. □

Now, we turn to the comparison between the Kobayashi–Royden and the Carathéodory–Reiffen metrics on $G \times \mathbb{C}^n$.

Lemma 19.6.10. *Under the assumptions of Theorem* 19.6.1, *we have* $\boldsymbol{\gamma}_G = \boldsymbol{\varkappa}_G$ *on* $G \times \mathbb{C}^n$.

Proof. We will use the sequences $(\widetilde{D}_k)_{k\in\mathbb{N}}$, $(G_k)_{k\in\mathbb{N}}$, and $(F_{j_k})_{k\in\mathbb{N}}$ from the proof of Lemma 19.6.9; cf. (19.6.3) and (19.6.4). If $(z, X) \in G \times \mathbb{C}^n$, then, applying Lempert's Theorem, we conclude that

$$\begin{aligned}
\boldsymbol{\gamma}_G(z;X) &= \lim_{k\to\infty} \boldsymbol{\gamma}_{G_k}(z;X) = \lim_{k\to\infty} \boldsymbol{\gamma}_{\widetilde{D}_k}(F_{j_k}(z); F'_{j_k}(z)X) \\
&= \lim_{k\to\infty} \boldsymbol{\varkappa}_{\widetilde{D}_k}(F_{j_k}(z); F'_{j_k}(z)X) \\
&= \lim_{k\to\infty} \boldsymbol{\varkappa}_{G_k}(z;X) \geq \boldsymbol{\varkappa}_G(z;X) \geq \boldsymbol{\gamma}_G(z;X),
\end{aligned}$$

i.e., $\boldsymbol{\gamma}_G = \boldsymbol{\varkappa}_G$ on $G \times \mathbb{C}^n$. □

We recall that the Bergman metric is not holomorphically contractible. Nevertheless, the following statement is true:

Lemma 19.6.11. *If* $G \subset\subset \mathbb{C}^n$ *is the union of an increasing sequence of subdomains* G_k, *then* $\boldsymbol{\beta}_G = \lim_{k\to\infty} \boldsymbol{\beta}_{G_k}$.

Proof. We already know (cf. Theorem 12.1.23) that $\lim_{k\to\infty} \boldsymbol{K}_{G_k} = \boldsymbol{K}_G$. It remains to prove that $\lim_{k\to\infty} M_{G_k} = M_G$; cf. Theorem 12.7.5 and Remark 12.7.8. But this convergence is a simple consequence of Montel's argument. The details are left to the reader. □

Finally, we are in a position to complete the proof of Theorem 19.6.1.

Proof of Theorem 19.6.1. First, we recall that G is a pseudoconvex domain with smooth $\mathcal{C}^1$-boundary for which $\boldsymbol{\gamma}_G = \boldsymbol{\varkappa}_G$ holds. To apply Proposition 19.6.7, it suffices to verify that $\boldsymbol{\varkappa}_G = \boldsymbol{\beta}_G/\sqrt{n+1}$ on $G \times \mathbb{C}^n$.

Now fix $z \in G$ and $X \in (\mathbb{C}^n)_*$. Moreover, choose a strongly pseudoconvex domain $\approx{D} \subset \widetilde{D}$ in such a way that $D \cap U \subset \approx{D}$, where $U = U(w_0) \subset\subset V$ is a sufficiently small neighborhood of w_0 and such that $\approx{D}$ is biholomorphically equivalent to a convex domain. Here, $\widetilde{D}$ and V are taken from the proof of Lemma 19.6.9.

We denote by G_ν^* the domains of (19.6.5) and by $\widetilde{D}_\nu^*$ the domains corresponding to (19.6.6), now defined with respect to $\approx{D}$ instead of $\widetilde{D}$.

We are going to slightly modify the construction of the sequences $(G_k)_{k\in\mathbb{N}}$, $(\widetilde{D}_k)_{k\in\mathbb{N}}$, and $(F_{j_k})_{k\in\mathbb{N}}$; cf. Lemma 19.6.9.

Assume that we have constructed subdomains $\widetilde{G}_1 \subset \cdots \subset \widetilde{G}_k \subset\subset G$, $\approx{D}_1 \subset \cdots \subset \approx{D}_k \subset\subset \approx{D}$, and mappings $(F_{j_\nu})_{\nu=1}^k$ satisfying

(a) $z \in G_\nu^* \subset \widetilde{G}_{\nu+1}$, $\widetilde{D}_\nu^* \subset \approx{D}_{\nu+1}$, $\quad 1 \le \nu \le k-1$;

(b) $F_{j_\nu}|_{\widetilde{G}_\nu} : \widetilde{G}_\nu \longrightarrow \approx{D}_\nu$ is biholomorphic;

(c) $1 - \dfrac{1}{2\nu} \le \dfrac{\delta_{\approx{D}}(F_{j_\nu}(z); F'_{j_\nu}(z)X)}{\delta_{\approx{D}_\nu}(F_{j_\nu}(z); F'_{j_\nu}(z)X)} \le 1 + \dfrac{1}{2\nu}$, $\quad 1 \le \nu \le k$,

where $\delta = \boldsymbol{\varkappa}, \boldsymbol{\gamma}$, or $\boldsymbol{\beta}$.

Next, we choose an index $j_{k+1} > j_k$ such that $F_{j_{k+1}}(\widetilde{G}_k \cup G_k^*) \subset\subset D \cap U$. Moreover, we take a domain $\approx{D}_{k+1} \subset\subset \approx{D}$ with $\approx{D}_k \subset\subset \approx{D}_{k+1}$ in such a way that $\approx{D}_{k+1} \supset \widetilde{D}_k^* \cup F_{j_{k+1}}(\widetilde{G}_k \cup G_k^*)$, that $\approx{D}_{k+1}$ is biholomorphic to a convex domain, and (c) becomes true for $\nu = k+1$. The further construction proceeds as before.

Then, we have the following chain of equalities:

$$\begin{aligned}
\frac{\boldsymbol{\beta}_G(z;X)}{\boldsymbol{\gamma}_G(z;X)} &= \lim_{k\to\infty} \frac{\boldsymbol{\beta}_{\widetilde{G}_k}(z;X)}{\boldsymbol{\gamma}_{\widetilde{G}_k}(z;X)} = \lim_{k\to\infty} \frac{\boldsymbol{\beta}_{\approx{D}_k}(F_{j_k}(z); F'_{j_k}(z)X)}{\boldsymbol{\gamma}_{\approx{D}_k}(F_{j_k}(z); F'_{j_k}(z)X)} \\
&= \lim_{k\to\infty} \frac{\boldsymbol{\beta}_{\approx{D}_k}(F_{j_k}(z); F'_{j_k}(z)X)}{\boldsymbol{\beta}_{\approx{D}}(F_{j_k}(z); F'_{j_k}(z)X)} \lim_{k\to\infty} \frac{\boldsymbol{\gamma}_{\approx{D}}(F_{j_k}(z); F'_{j_k}(z)X)}{\boldsymbol{\gamma}_{\approx{D}_k}(F_{j_k}(z); F'_{j_k}(z)X)} \\
&\quad \times \lim_{k\to\infty} \frac{\boldsymbol{\beta}_{\approx{D}}(F_{j_k}(z); F'_{j_k}(z)X)}{\boldsymbol{\gamma}_{\approx{D}}(F_{j_k}(z); F'_{j_k}(z)X)} \\
&= \lim_{k\to\infty} \frac{\boldsymbol{\beta}_{\approx{D}}(F_{j_k}(z); F'_{j_k}(z)X)}{\boldsymbol{\gamma}_{\approx{D}}(F_{j_k}(z); F'_{j_k}(z)X)} \underset{(*)}{=} \sqrt{n+1},
\end{aligned}$$

where $(*)$ is a consequence of Theorems 10.4.2 and 10.4.6. Since z and X are arbitrary, we find, making use of Lemma 19.6.10, that $\boldsymbol{\gamma}_G = \boldsymbol{\varkappa}_G = \boldsymbol{\beta}_G/\sqrt{n+1}$. $\square$

Remark 19.6.12. According to a result due to S. Pinchuk (cf. [431]) it is known that any proper holomorphic mapping between strongly pseudoconvex domains in $\mathbb{C}^n$ is unbranched. Therefore, the formulation of Theorem 19.6.1 becomes very simple if G and D are assumed to be strongly pseudoconvex domains.

19.7 Exercises

The following exercises deal with domains that are not necessarily strongly pseudoconvex. So, they may serve as a kind of introduction to the next chapter.

Exercise 19.7.1 (cf. [353]). Define

$$G := \{z \in \mathbb{C}^2 : -2\operatorname{Re}(z_1 + z_1 z_2/2) + |z_1 + z_1 z_2/2|^2 + |z_2|^2 < 0\}.$$

(a) Using $(z_1, z_2) \longmapsto (z_1 - 1 + z_1 z_2/2, z_2)$, prove that G is biholomorphic to $\mathbb{B}_2$.

(b) For $z(t) := (t, 0)$ and $X(t) := (\sqrt{t}, 1)$, where t is a small positive number, calculate the Kobayashi–Royden metric as

$$\varkappa_G(z(t); X(t)) = \frac{\sqrt{3t + t^{3/2} - 3t^2/4}}{2t - t^2}.$$

(c) Compare this formula with the estimate stated in Remark 19.4.4.

Exercise 19.7.2 (cf. [325]). For $3/4 \le t < 1$, $m := 1/(2 - 2t)$, put

$$G_t := \{z \in \mathbb{C}^2 : 1 < |z_1|^2 + |z_2|^m < 4\}.$$

Note that G_t is a domain with smooth $\mathcal{C}^2$-boundary. Prove that there exists a $C > 1$ such that for points $z(\delta) := (-1 - \delta, 0) \in G_t$ and $X := (1, 0)$ (δ small) the following inequalities are true:

$$(1/C)\operatorname{dist}(z(\delta), \partial G_t)^{-t} \le \varkappa_{G_t}(z(\delta); X) \le C\operatorname{dist}(z(\delta), \partial G_t)^{-t}. \qquad (*)$$

Compare $(*)$ with Theorems 19.4.2 and 20.1.2.

Hint. Use the analytic disc $\varphi \in \mathcal{O}(\mathbb{D}, G_t)$, $\varphi(\lambda) := (-1 - \delta + (\delta^t/10)\lambda, \lambda^2)$, for the estimate from above. Observe that for any $\varphi = (\varphi_1, \varphi_2) \in \mathcal{O}(\mathbb{D}, G_t)$, $\varphi(0) = z(\delta), \sigma\varphi'(0) = X$ ($\sigma > 0$), the function $g(\lambda) := \varphi_1(\delta^{1-t}\lambda/\sqrt{2})$, $\lambda \in \mathbb{D}$, has its values in the annulus $\{\lambda \in \mathbb{C} : 1 - \delta < |\lambda| < 4\}$.

Exercise 19.7.3. Let G be any bounded domain in $\mathbb{C}^2$ with smooth $\mathcal{C}^2$-boundary. Show that there are positive numbers ε_0 and C such that whenever $z \in \partial G$ and $t \in (0, \varepsilon_0)$, then $z - t\nu(z) \in G$ and $\varkappa_G(z - t\nu(z); \nu(z)) \ge Ct^{-3/4}$, where, as usual, $\nu(z)$ denotes the unit outer normal of G at z.

Hint. Use the shells $\mathbb{B}(z + \varepsilon_0\nu(z), R) \setminus \overline{\mathbb{B}}(z + \varepsilon_0\nu(z), \varepsilon_0) \supset G$ as comparison domains.

Exercise 19.7.4 (cf. [444]). Let

$$G := \{z \in \mathbb{C}^3 : |z_1|^2 + |z_2|^2 + |z_3|^4 < 1\}$$

and put $X := (0,1,0)$, $X^* := (0,0,1)$. Prove that $\boldsymbol{\gamma}_G(z;X) \sim \operatorname{dist}(z,\partial G)^{-1/2}$ and $\boldsymbol{\gamma}_G(z;X^*) \sim \operatorname{dist}(z,\partial G)^{-1/4}$ if $z \longrightarrow (1,0,0)$ along the inner normal to ∂G at $(1,0,0)$.

Exercise 19.7.5 (cf. [131]). Let $F : G_1 \longrightarrow G_2$ be a proper holomorphic mapping between bounded domains with smooth $\mathcal{C}^2$-boundaries in $\mathbb{C}^n$. For $\underline{\delta} = \boldsymbol{\gamma}$ or $\underline{\delta} = \boldsymbol{\varkappa}$ assume that

(a) there exist $C, \varepsilon > 0$ such that $\operatorname{dist}(F(z), \partial G_2) \le C \operatorname{dist}(z, \partial G_1)^{\varepsilon}$, $z \in G_1$;

(b) there is $\eta > 0$ with $\delta_{G_2}(w;X) \ge C\|X\| \operatorname{dist}(w,\partial G_2)^{-\eta}$, $w \in G_2, X \in \mathbb{C}^n$.

Prove that F extends continuously to $\overline{G}_1$.

19.8 List of problems

Chapter 20

Boundary behavior of invariant functions and metrics on general domains

Summary. While in the previous chapter the domains were mainly assumed to be strongly pseudoconvex, we now study the boundary behavior of invariant functions and metrics on more general domains, which may be even non-pseudoconvex. In § 20.1, the boundary behavior of various invariant metrics at a non-pseudoconvex boundary point in normal direction is given on arbitrary domains with a smooth boundary (Theorem 20.1.1), showing that, in general, all these metrics are different. The following § 20.2, which is based on [169] and [428], treats the boundary behavior of the Kobayashi–Royden metric in the normal direction on pseudoconvex domains. For domains with a good boundary (i.e., a $\mathcal{C}^{1+\varepsilon}$-smooth one) a general upper estimate for the Lempert function is presented in § 20.3.

20.1 Boundary behavior of pseudometrics for non pseudoconvex domains

So far, we mainly discussed the boundary behavior of invariant pseudodistances for strongly pseudoconvex domains. Now, we turn our interest to non-pseudoconvex domains. The main goal here is to prove the following result (see [170] and also [134]):

Theorem 20.1.1. *Let $D \subset \mathbb{C}^n$, $n > 1$, be a bounded domain with $\mathcal{C}^2$-boundary. Fix a boundary point $z_0 \in \partial D$ at which D is not pseudoconvex (i.e., there is a defining function r for D and a vector $X \in \mathbb{C}^n$, $\sum_{j=1}^n r'_{z_j}(z_0)X_j = 0$, such that $(\mathcal{L}r)(z_0; X) < 0$). Denote by z_δ the point on the inner normal at z_0 at distance δ to z_0 and by ν the outer unit normal at ∂D in z_0. Then, there are positive numbers δ_0 and $C_k > 1$ $(k = 0, 1, \dots)$ such that*

(a) $1/C_0 \le \boldsymbol{\gamma}_D(z_\delta; \nu) \le C_0, \quad 0 < \delta \le \delta_0;$

(b) $(1/C_k)\delta^{-\left(1-\frac{k}{2(k+1)}\right)} \le \boldsymbol{\varkappa}_D^{(k)}(z_\delta; \nu) \le C_k\delta^{-\left(1-\frac{k}{2(k+1)}\right)}, \quad 0 < \delta \le \delta_0,\ k \in \mathbb{N};$

(c) $(1/C_0)\delta^{-1/2} \le \boldsymbol{S}_D(z_\delta; \nu) \le \boldsymbol{A}_D(z_\delta; \nu) \le \widetilde{\boldsymbol{\varkappa}}_D(z_\delta; \nu) \le C_0\delta^{-1/2}, \quad 0 < \delta \le \delta_0$.

The proof of this result is mainly based on a localization result and the following proposition:

Proposition 20.1.2. *Let $n \in \mathbb{N}$, $n \geq 2$, $s \in (0,1)$, $m = (m_1, \dots, m_n)$ with $2 = m_1 \leq m_2 \leq m_3 \leq \dots \leq m_n$, and*

$$D_m = D_m(s) := \left\{ z \in \mathbb{C}^n : s^2 < \sum_{j=1}^{n} |z_j|^{m_j} < 1 \right\}.$$

Let $z_0 := (s, 0, \dots, 0)$. Put $z_\delta := (s + \delta, 0, \dots, 0) \in D_m$ and $v := (1, 0, \dots, 0)$. Then, there exist positive constants $C_k > 1$ ($k = 0, 1, \dots$) and $\delta_0 < \min\{\frac{s}{2}, \frac{1-s}{2}\}$ such that

(a) $(1/C_k)\delta^{-\left(1-\frac{k}{(k+1)m_2}\right)} \leq \varkappa^{(k)}_{D_m}(z_\delta; v) \leq C_0\delta^{-\left(1-\frac{k}{(k+1)m_2}\right)}$, $0 < \delta \leq \delta_0$, $k \in \mathbb{N}$,

(b) $(1/C_0)\delta^{-\left(1-\frac{1}{m_2}\right)} \leq \boldsymbol{S}_{D_m}(z_\delta; v) \leq C_0\delta^{-\left(1-\frac{1}{m_2}\right)}$, $0 < \delta \leq \delta_0$.

The result in Proposition 20.1.2 for the Kobayashi–Royden metric is due to S. Krantz (see [325]), while the inequality for the Sibony metric has been proved by J. E. Fornæss and L. Lee (see [170]). In particular, this proposition shows that the Kobayashi–Royden metric and the Sibony metric are not equal for the ring domain $D_{(2,2)}$. For another proof of this fact, see [168].

Moreover, let us emphasize that the boundary point in Proposition 20.1.2 does not satisfy the properties in Theorem 20.1.1.

Proof. Step 1°: *The upper estimate for $\varkappa^{(k)}_{D_m}(z_\delta; v)$.*

Fix a $k \in \mathbb{N}$. By the simple embedding

$$D_{(2,m_2)}(s) \ni (z_1, z_2) \longmapsto (z_1, z_2, 0, \dots, 0) \in D_m(s),$$

it is enough to find the upper estimate for the corresponding two-dimensional case.

For that case, it suffices to observe that, if $\delta \in (0, (1-s)/2)$ and $c > \frac{2(s+3)}{1-s} > 1$, then $\varphi_k \in \mathcal{O}(\mathbb{D}, D_{(2,m_2)}(s))$ (see below), where

$$\varphi_k(\lambda) := \left(s + \delta + \frac{\delta^{1-\frac{k}{(k+1)m_2}}}{c_k}\lambda^k, c_k^{-1/m_2}\lambda^{k+1} \right),$$

and $\varphi_k(0) = (s + \delta, 0)$, $\varphi_k^{(j)}(0) = (0,0)$, $j = 1, \dots, k-1$, and

$$\frac{1}{k!}\varphi_k^{(k)}(0) = \frac{\delta^{1-\frac{k}{(k+1)m_2}}}{c_k}(1,0).$$

Let us give some details on how to show that φ_k maps $\mathbb{D}$ into $D_{(2,m_2)}(s)$: fix a $\lambda \in \mathbb{D}$ and a $\delta \le \frac{1-s}{2}$. Moreover, put for a moment $\alpha = \alpha_k := 1 - \frac{k}{(k+1)m_2}$. Then,

$$\begin{aligned}\left|s + \delta + \frac{\delta^\alpha}{c}\lambda^k\right|^2 + \frac{1}{c}|\lambda|^{(k+1)m_2} &< \left(s + \delta + \frac{1}{c}\right)^2 + \frac{1}{c} \le \left(\frac{s+1}{2} + \frac{1}{c}\right)^2 + \frac{1}{c} \\ &\le \frac{s+1}{2} + \frac{1}{c}\left(s + 1 + \frac{1}{c} + 1\right) \\ &\le \frac{1+s}{2} + \frac{1-s}{2(s+3)}(s+3) = 1\end{aligned}$$

because of the choice of the constant c.

What remains is the proof of the lower estimate

$$s^2 < \left|s + \delta + \frac{\delta^\alpha}{c}\lambda^k\right|^2 + \frac{1}{c}|\lambda|^{(k+1)m_2} =: R.$$

Calculation leads to

$$\begin{aligned}R &= s^2 + \left|\delta + \frac{\delta^\alpha}{c}\lambda^k\right|^2 + 2s\,\mathrm{Re}\left(\delta + \frac{\delta^\alpha}{c}\lambda^k\right) + \frac{1}{c}|\lambda|^{(k+1)m_2} \\ &\ge s^2 + 2s\delta - 2s\frac{\delta^\alpha}{c}|\lambda|^k + \frac{1}{c}|\lambda|^{(k+1)m_2} \\ &= s^2 + 2s\delta + |\lambda|^k\left(\frac{1}{c}|\lambda|^{(k+1)m_2-k} - 2s\frac{\delta^\alpha}{c}\right).\end{aligned}$$

Thus, we have to verify that the function $h(t) := 2s\delta + \frac{t^k}{c}\left(t^{m_2(k+1)-k} - 2s\delta^\alpha\right)$ is positive on $[0,1)$. Put

$$g(t) := t^{(k+1)m_2} - 2s\delta^\alpha t^k, \quad t \in [0,1).$$

Then, $g(0) = 0$ and $g'(t) = (k+1)m_2 t^{m_2(k+1)-1} - 2ks\delta^\alpha t^{k-1}$.

Therefore, $g'(t_0) = 0$ if and only if $t_0 = 0$ $(k \ge 1)$ or $t_0 = \left(\frac{2ks\delta^\alpha}{(k+1)m_2}\right)^{1/(m_2(k+1)-k)}$. Note that $t_0 \in (0,1)$. So it remains to estimate

$$0 \overset{!}{<} 2s\delta + \frac{t_0^k}{c}\left(t_0^{m_2(k+1)-k} - 2s\delta^\alpha\right) = 2s\delta - 2s\delta^\alpha t_0^k \tfrac{1}{c}\alpha,$$

or, equivalently,

$$c \overset{!}{>} \frac{1}{\delta} t_0^k \delta^\alpha \alpha = t_0^k \delta^{-k/(k+1)m_2} \alpha$$
$$= \left(\frac{2ks\delta^\alpha}{(k+1)m_2}\right)^{k/((k+1)m_2-k)} \delta^{-k/((k+1)m_2)} \alpha$$
$$= \alpha \left(\frac{2ks}{(k+1)m_2}\right)^{\frac{k}{(k+1)m_2-k}},$$

which is obviously true, since $c > 1$.

Step 2°: *The upper estimate for* $\boldsymbol{S}_{D_m}(z_\delta; v)$. Repeat that the constant c in Step 1° is independent of k. Hence, the upper estimate for the Sibony metric now follows directly from Proposition 4.2.11 and (4.2.5).

Step 3°: *The lower estimate for* $\varkappa_{D_m}^{(k)}(z_\delta; v)$.

Fix a $k \in \mathbb{N}$. Take an arbitrary analytic disc $\varphi \in \mathcal{O}(\mathbb{D}, D_m)$ with $\varphi(0) = z_\delta$, $\varphi^{(j)}(0) = (0, \dots, 0)$ if $1 \le j < k$, and $\frac{\alpha}{k!}\varphi^{(k)}(0) = v$. We may assume that $\alpha > 0$. Then, using the maximum principle, we see that $|\varphi_j(\lambda)| \le |\lambda|^{k+1}$ on $\mathbb{D}$, $j = 2, \dots, n$. Therefore, if $\lambda \in \mathbb{D}$, then

$$s^2 < \sum_{j=1}^n |\varphi_j(\lambda)|^{m_j} \le |\varphi_1(\lambda)|^2 + \sum_{j=2}^n |\lambda|^{(k+1)m_j} \le |\varphi_1(\lambda)|^2 + (n-1)|\lambda|^{(k+1)m_2}.$$

In particular, if $|\lambda| < \left(\frac{s^2\delta}{n-1}\right)^{1/((k+1)m_2)}$, $\delta < \delta_0 := s(1-s)/4$, then $s(1-\delta) < |\varphi_1(\lambda)| < 1$. Therefore, the holomorphic map

$$\psi(\lambda) := \varphi_1\left(\left(\frac{s^2\delta}{n-1}\right)^{1/((k+1)m_2)} \lambda\right)$$

sends $\mathbb{D}$ into $\mathbb{A}(s(1-\delta), 1)$ with $\psi(0) = s + \delta$, $\psi^{(j)}(0) = 0$ for $j = 1, \dots, k-1$, and $\frac{1}{k!}\psi^{(k)}(0) = \frac{1}{\alpha}\left(\frac{s^2\delta}{n-1}\right)^{k/((k+1)m_2)}$. Hence,

$$\alpha \left(\frac{n-1}{s^2\delta}\right)^{k/((k+1)m_2)} \ge \varkappa_{\mathbb{A}(s(1-\delta),1)}^{(k)}(s+\delta; 1) \overset{(*)}{=} \varkappa_{\mathbb{A}(s(1-\delta),1)}(s+\delta; 1);$$

for $(*)$ see Corollary 3.8.9. Put $R := 1/\sqrt{s(1-\delta)}$. Then, using the holomorphic mapping

$$\mathbb{A}(s(1-\delta), 1) \ni \lambda \longmapsto R\lambda \in \mathbb{A}(1/R, R)$$

gives

$$\alpha \ge \left(\frac{s^2\delta}{n-1}\right)^{k/((k+1)m_2)} R\varkappa_{\mathbb{A}(1/R,R)}(R(s+\delta); 1).$$

Applying Proposition 9.1.9, we get

$$\varkappa_{\mathbb{A}(1/R,R)}(R(s+\delta);1) = \frac{\pi}{4R(s+\delta)\log R\sin(\pi s(a))},$$

where $s(a) := \frac{1}{2}\left(1 - \frac{\log a}{\log R}\right) = \frac{\log(s+\delta)}{\log(s(1-\delta))} \geq 1/2$ (see 9.1.1). So we may increase the sinus by the following term $\pi\left(1 - \frac{\log(s+\delta)}{\log(s(1-\delta))}\right)$. Hence, we get the following lower estimate of α:

$$\begin{aligned}\alpha &\geq \frac{1}{2}\left(\frac{s^2\delta}{n-1}\right)^{k/((k+1)m_2)} \frac{1}{\log(s+\delta) - \log(s(1-\delta))} \\ &\geq \frac{1}{2}\left(\frac{s^2\delta}{n-1}\right)^{k/((k+1)m_2)} \frac{s(1-\delta)}{\delta(1-s)} \geq \frac{s}{2}\left(\frac{s^2}{n-1}\right)^{k/((k+1)m_2)} \frac{1}{\delta^{1-\frac{k}{(k+1)m_2}}}.\end{aligned}$$

Since φ was an arbitrary competitor for the Kobayashi–Royden metric, the claim is verified.

Step 4°: *The lower estimate for* $\boldsymbol{S}_{D_m}(z_\delta;\nu)$ if $0 \leq \delta < \min\{s/2, (1-s)/2\}$.

Recall that we are looking for a good competitor for $\boldsymbol{S}_{D_m}(z_\delta;\nu)$, i.e., for one with a good $\frac{\partial^2}{\partial z_1 \partial \overline{z}_1}$-derivative at the point z_δ. With this in mind, put

$$f(z) = \delta^{\frac{2}{m_2}}\left|\frac{z_1 - s - \delta}{z_1 - s + \delta}\right|^2, \quad z \in \Omega := \{w \in \mathbb{C}^2 : w_1 \neq s - \delta\}.$$

Then, $(\mathcal{L}\widetilde{f})(z_\delta;\nu) = \frac{1}{4\delta^{2-\frac{2}{m_2}}}$, where $z = (z_1, \widetilde{z})$ and $\widetilde{f}(z) := f(z) + \|\widetilde{z}\|^2$.

It remains to modify $\widetilde{f}$ to get a log-plurisubharmonic function on the whole of D_m. We put on D_m

$$\widetilde{u}(z) := \begin{cases} \max\{\log(\widetilde{f}(z)), \log(L\|\widetilde{z}\|^{2+\varepsilon})\} - L', & \text{if } \|\widetilde{z}\| < c^{\frac{2}{m_2}}\delta^{\frac{1}{m_2}} \\ \log(L\|\widetilde{z}\|^{2+\varepsilon}) - L', & \text{if } \|\widetilde{z}\| \geq c^{\frac{2}{m_2}}\delta^{\frac{1}{m_2}} \end{cases},$$

where $c, L, L', \varepsilon, \delta$ are positive numbers such that

(a) $\delta < c^2 = 11s/19$,

(b) $\log(5/L) < (2/m_2)\log(c^2/2)$,

(c) $\varepsilon < m_2 \frac{\log\frac{5}{L} - \frac{2}{m_2}\log\frac{c^2}{2}}{\log\delta + \log\frac{c^2}{2}}$,

(d) $L' > \max\{\log L, \log 5\}$.

Observe that, using the fact that the l^2-norm is majorized by the Euclidean one, it is easy to see that $\widetilde{u}$ is well-defined.

First, we show that $\widetilde{u} \in \mathcal{PSH}(D_m)$ using Appendix B.4.18.

Note that $|z_1 - s - \delta|^2 \le |z_1 - s + \delta|^2$ if $\operatorname{Re} z_1 \ge s$. Therefore,

$$f(z) \le \delta^{2/m_2}, \quad \text{if } z \in D_m,\ \operatorname{Re} z_1 \ge s.$$

It remains to estimate f for those $z = (z_1, \widetilde{z}) \in D_m$, $z_1 = x_1 + iy_1$, with $x_1 < s$ and $\|\widetilde{z}\|^{m_2} < c^2\delta$:

Again applying the above fact on the norms, it follows that $\alpha := \sqrt{s^2 - c^2\delta} < |z_1| < 1$. Moreover, note that $0 < s - \delta < \alpha$.

In a first step, assume that, in addition, $-1 < x_1 \le -\alpha$ or $\alpha \le x_1 < s$. Then,

$$f(z) = \delta^{\frac{2}{m_2}} \frac{(x_1 - s - \delta)^2 + y_1^2}{(x_1 - s + \delta)^2 + y_1^2} =: \delta^{\frac{2}{m_2}} \frac{A + y_1^2}{B + y_1^2} \le \delta^{\frac{2}{m_2}} \frac{A}{B},$$

since $A := (x_1 - s - \delta)^2 \ge (x_1 - s + \delta)^2 =: B > 0$.

Hence, we have

$$\begin{aligned} f(z) &\le \delta^{\frac{2}{m_2}} \frac{(x_1 - s - \delta)^2}{(x_1 - s + \delta)^2} \\ &= \delta^{\frac{2}{m_2}} \begin{cases} \frac{(-x_1+s+\delta)^2}{(-x_1+s-\delta)^2}, & \text{if } -1 < x_1 \le -\alpha \\ \frac{(-x_1+s+\delta)^2}{(x_1-s+\delta)^2}, & \text{if } \alpha \le x_1 < s \end{cases}. \end{aligned}$$

Then, it is easy to verify that both quotients without squares are below 2. Hence, we get $f(z) \le 4\delta^{2/m_2}$.

For the remaining case, let z be as above, now with $|x_1| < \alpha$. Then, the function g,

$$g(x_1, t) := \frac{(x_1 - s - \delta)^2 + s^2 - c^2\delta - x_1^2 + t}{(x_1 - s + \delta)^2 + s^2 - c^2\delta - x_1^2 + t} = \frac{\widetilde{A} + t}{\widetilde{B} + t}, \quad t \ge 0,$$

where $\widetilde{A} \ge \widetilde{B} > 0$, is decreasing. Recall that $y_1^2 > s^2 - c^2\delta - x_1^2$. Therefore, $f(z) \le \delta^{\frac{2}{m_2}} g(x_1, 0)$. Simple calculation then gives that $g(x_1, 0) \le 4$. Indeed, note that the function $g(\cdot, 0)$ is increasing if we assume, in addition, that $\delta < c^2$. Then, we have $g(x_1, 0) \le g(\alpha, 0)$, and this term was already estimated before by the number 4. Hence, $f(z) \le 4\delta^{2/m_2}$.

To summarize, for $\delta < c^2$ we have the following inequality:

$$f(z) + \|\widetilde{z}\|^2 \le 4\delta^{2/m_2} + (c^2\delta)^{2/m_2} \le 5\delta^{2/m_2}, \quad z \in D_m,\ \|\widetilde{z}\| < (c^2\delta)^{m_2}.$$

Now let $z \in D_m$ with $\|\widetilde{z}\| > (c^2\delta/2)^{m_2}$. Using the fact that

$$5\delta^{2/m_2} \le L(c^2\delta/2)^{(2+\varepsilon)/m_2} \le L\|\widetilde{z}\|^{2+\varepsilon},$$

it follows that the function $\widetilde{u}$ is plurisubharmonic on D_m. Moreover, the choice of L' implies that $\widetilde{u} < 0$ on D_m. So we conclude that the function $u := e^{\widetilde{u}}$ is log-psh on D_m, $0 \le u < 1$, and $u(z) = e^{-L'}\widetilde{f}(z)$ near the point z_δ; therefore, u is near the point z_δ of class $\mathcal{C}^2$. Finally, we end up with $\boldsymbol{S}_{D_m}(z_\delta; v) \ge e^{-L'/2}\frac{1}{\delta^{1-\frac{1}{m_2}}}$. □

Remark 20.1.3. With respect to the Carathéodory–Reiffen metric, we only mention that $\boldsymbol{\varkappa}_{D_m} = \boldsymbol{\gamma}_{\mathbb{B}_n}$, since $\mathcal{H}^\infty(D_m) = \mathcal{H}^\infty(\mathbb{B}_n)$.

Proof of Theorem 20.1.1. In order to prove the lower estimates it suffices to observe that there are balls $\mathbb{B}(z^*, r)$ and $\mathbb{B}(z^*, R)$ such that $z_0 \in \partial\mathbb{B}(z^*, r)$ and $D \subset \mathbb{B}(z^*, R)\setminus \overline{\mathbb{B}}(z^*, r)$. These balls exist, since D is bounded and has a $\mathcal{C}^2$-smooth boundary. Then, one has only to apply Proposition 20.1.2.

Now we turn to discuss the upper estimate. We may assume that $z_0 = 0 \in \mathbb{C}^n$ and that there are a neighborhood

$$U := (-s_1, s_1) \times \{(y_1, \widetilde{z}) \in \mathbb{R} \times \mathbb{C}^{n-1} : y_1^2 + \|\widetilde{z}\|^2 < s_2^2\} =: (-s_1, s_1) \times U_2$$

and a function $r \in \mathcal{C}^2(U_2)$ with $r(0) = 0$, $\frac{\partial r}{\partial x_j}(0, \widetilde{0}) = \frac{\partial r}{\partial y_k}(0, \widetilde{0}) = 0$ $(1 \le k \le n,$ $2 \le j \le n)$ and

$$\begin{aligned} D \cap U = {} & \{(x_1 + iy_1, \widetilde{z}) \in U : x_1 < r(y_1, \widetilde{z})\} \\ & \supset \{(-s_1/2 + iy_1, \widetilde{z}) : y_1^2 + \|\widetilde{z}\|^2 < s_2^2\}. \end{aligned}$$

Then, $z_\delta = (-\delta, 0, \dots, 0)$, $0 < \delta < s_1$, and $v = (1, 0, \dots, 0)$. Moreover, we may assume (after a rotation if necessary) that $\frac{\partial^2 r}{\partial z_2 \partial \overline{z}_2}(0) < 0$.

Put

$$\widetilde{D} := \{(z_1, z_2) \in \mathbb{C}^2 : (z_1, z_2, 0, \dots, 0) \in D \cap U\},$$

$\widetilde{z}_\delta := (-\delta, 0) \in \mathbb{C}^2$, and $\widetilde{v} := (1, 0)$. Applying holomorphic contractibility, we have

$$\mu_{\widetilde{D}}(\widetilde{z}_\delta; \widetilde{v}) \ge \mu_D(z_\delta; v),$$

where μ stands for one of the invariant pseudometrics under discussion. Let

$$\widetilde{r}(z) := \operatorname{Re} z_1 - r(y_1, z_2, 0, \dots, 0)$$

on the cylinder $\widetilde{U} := (-s_1, s_1) \times \{(y_1, z_2) : y_1^2 + |z_2|^2 < s_2^2\}$. Then, we may write on $\widetilde{U}$

$$\widetilde{r}(z) = \operatorname{Re}\Big(z_1 + \sum_{j,k=1}^{2} a_{j,k} z_j z_k\Big) + \sum_{j,k=1}^{2} b_{j,k} z_j \overline{z}_k + o(\|z\|^2\Big)$$

with $b_{2,2} < 0$. Applying the holomorphic mapping

$$\Phi(z) = \left(z_1 + \sum_{j,k=1}^{2} a_{j,k} z_j z_k, z_2\right), \quad z \in \mathbb{C}^2,$$

we observe that Φ maps a connected neighborhood $V \subset \widetilde{U}$ of 0 biholomorphically onto a domain W with $\Phi(0) = 0 \in W$, $\Phi(\widetilde{z}_\delta) = (\delta + a_{1,1}\delta^2, 0) =: (\widehat{\delta}, 0) =: \widehat{z}_\delta$, and $\Phi'(0)\widetilde{v} = \widetilde{v}$. Then,

$$\widetilde{G} := \{w \in W : \varrho(w) := r(\Phi^{-1}(w)) < 0\} = \Phi(V \cap \widetilde{D}).$$

Hence, $\mu_{\widetilde{D}}(\widetilde{z}_\delta; \widetilde{v}) \le \mu_{\widetilde{G}}(\widehat{z}_\delta; \widetilde{v})$.

Rewriting ϱ, we get

$$\varrho(w) = \operatorname{Re} w_1 + \sum_{j,k=1}^{2} c_{j,k} w_j \overline{w}_k + o(\|w\|^2), \quad w \in W,$$

where $c_{2,2} < 0$. Then, the following estimate is true for w in a small ball $W' \subset W$ with center 0:

$$\varrho(w) \le \operatorname{Re} w_1 + c_1|w_1|^2 - c_2|w_2|^2 \text{ with } c_1, c_2 > 0.$$

(To get this estimate, use that $2\alpha\beta \le m\alpha^2 + 1/m\beta^2$ for positive numbers α, β, and a big m.)

Set

$$G := \{w \in W' : \operatorname{Re} w_1 + c_1|w_1|^2 - c_2|w_2|^2 < 0\} \subset \widetilde{G}.$$

Hence, we have $\mu_D(z_\delta; v) \le \mu_G(\widehat{z}_\delta; \widetilde{v})$, if δ is sufficiently small.

To summarize: we have reduced our problem of proving the upper estimates for the case where the domain is given by G.

With respect to the Kobayashi–Royden metric (and the Sibony metric, too), we only have to follow the ideas of Step 1° and Step 2° in the proof of Proposition 20.1.2, respectively. Therefore, details are left to the reader as an exercise.

To find an upper estimate for the remaining metrics, it suffices to establish such an estimate for $\widetilde{\varkappa}_D$ (see subsection 4.3.5).

Choose an $s \in (0,1)$ with $\mathbb{P}(2s) \subset W'$ and a positive $c < 1/2$ such that $(-\widehat{\delta} + c\mathbb{D}) \times c\mathbb{D} \subset \mathbb{P}(2s)$ for any $0 < \widehat{\delta} < \min\{s, c_2/c_1\}$. Now fix a vector $X \in \mathbb{C}^2$ with $|X_1| \le \sqrt{\widehat{\delta}}$ and $|X_2| = 1/\sqrt{c_2}$. Then, the analytic disc $\varphi(\lambda) := (-\delta + \lambda c X_1, \lambda c X_2)$ belongs to $\mathcal{O}(\mathbb{D}, W')$ if $0 < \delta < s$. Moreover, we have $\varphi(0) = (-\delta, 0)$ and $\varphi'(0) = cX$. To get that $\varkappa_G((-\widehat{\delta}, 0); cX) \le 1$, it suffices to prove that the image of φ belongs to G. Indeed, let us first assume that $|\lambda| \le \sqrt{\widehat{\delta}}$. Then,

$$\operatorname{Re}\varphi_1(\lambda) + c_1|\varphi_1(\lambda)|^2 - c_2|\varphi_2(\lambda)|^2 \le -\widehat{\delta}(1 - c - c_1\widehat{\delta}(1 + 2c + c^2)) < 0$$

if $\widehat{\delta}$ is sufficiently small.

In the remaining step, we assume that $\sqrt{\widehat{\delta}} < |\lambda| < 1$. Then,

$$\begin{aligned}\operatorname{Re}\varphi_1(\lambda)+&c_1|\varphi_1(\lambda)|^2 - c_2|\varphi_2(\lambda)|^2 \\ &\leq -\widehat{\delta} + c\sqrt{\widehat{\delta}}|\lambda| + c_1(\widehat{\delta}^2 + 2c\widehat{\delta}\sqrt{\widehat{\delta}}|\lambda| + c^2\widehat{\delta}|\lambda|^2) - c^2|\lambda|^2 \\ &= -\widehat{\delta} + c_1\widehat{\delta}^2 + c\sqrt{\widehat{\delta}}(1+2c_1\widehat{\delta})|\lambda| + c^2(\widehat{\delta}c_1 - 1)|\lambda|^2 =: A(\lambda).\end{aligned}$$

Observe that $\widehat{\delta}c_1 - 1 < 0$ for sufficiently small $\widehat{\delta}$, which we will assume from now on. The maximal value of the last term on the interval $(\sqrt{\widehat{\delta}}, 1)$ is taken at the point $t_0 := \frac{\sqrt{\widehat{\delta}}(1+2c_1\widehat{\delta})}{2c(1-\widehat{\delta}c_1)} \in (\sqrt{\widehat{\delta}}, 1)$, if $\widehat{\delta}$ is sufficiently small. Hence, for such small $\widehat{\delta}$, we have

$$A(\lambda) < \widehat{\delta}(-1 + c_1\widehat{\delta}) - \frac{(1+2c_1\widehat{\delta})^2}{4(1-\widehat{\delta}c_1)} < 0$$

on $\mathbb{A}(\sqrt{\widehat{\delta}}, 1)$ (the last "$<$" is true if $\widehat{\delta}$ is, in addition, suitable small); hence, $\varphi \in \mathcal{O}(\mathbb{D}, G)$. In particular, we have $\boldsymbol{\varkappa}_G(\widehat{z}_\delta; X) < 1$ if

$$X \in \left\{ Y \in \mathbb{C}^2 : |Y_1| < \frac{c\sqrt{\widehat{\delta}}}{2},\ |Y_2| = \frac{c}{2\sqrt{c_2}} \right\}.$$

In a similar way one can prove that $\boldsymbol{\varkappa}_G(\widehat{z}_\delta; (0; X_2)) < 1$ if $|X_2| < \frac{c}{2\sqrt{c_2}}$ (use $\varphi(\lambda) := (-\widehat{\delta}, cX_2\lambda/\sqrt{c_2})$).

Using the Kontinuitätssatz, it follows that for all $X \in \mathbb{C}^2$ with $|X_1| < c\sqrt{\widehat{\delta}}/2$ and $|X_2| < \frac{c}{2\sqrt{c_2}}$ one gets $\widetilde{\boldsymbol{\varkappa}}_D(\widehat{z}_\delta; X) < 1$. In particular, this implies (for sufficiently small $\widehat{\delta}$) that $\widetilde{\boldsymbol{\varkappa}}_G(\widehat{z}_\delta; \widetilde{\nu}) \leq \frac{2}{c\sqrt{\widehat{\delta}}}$.

Hence, taking all the previous information together, we end up with the following inequality:

$$\boldsymbol{S}_D(z_\delta; \nu) \leq \boldsymbol{A}_D(z_\delta; \nu) \leq \widetilde{\boldsymbol{\varkappa}}_D(z_\delta; \nu) \leq \frac{2}{c\sqrt{\delta - a_{1,1}\delta^2}} \leq \frac{4}{c\sqrt{\delta}}.$$

It remains to discuss the Carathéodory–Reiffen metric. We choose a polydisc $\mathbb{P}_2(0, \varepsilon) \subset\subset W'$. Then, $\{0\} \times \overline{\mathbb{D}}_*(\varepsilon) \subset\subset G$. Now we may take an $\varepsilon' < \varepsilon$ such that $\mathbb{D}(\varepsilon') \times \partial\mathbb{D}(\varepsilon) \subset\subset G$.

After this geometric preparation, let $f \in \mathcal{O}(G, \mathbb{D})$ and put on $\widetilde{G} := \mathbb{D}(\varepsilon') \times \mathbb{D}(\varepsilon)$,

$$\widetilde{f}(z) := \frac{1}{2\pi i} \int_{\partial\mathbb{D}(\varepsilon)} \frac{f(z_1, \zeta_2)}{\zeta_2 - z_2} d\zeta_2.$$

Obviously, $\widetilde{f} \in \mathcal{O}(\widetilde{G}, \mathbb{D})$. Note that for small $t_1 < t_2 < 0$ with $-\varepsilon' < t_1$ we have $[t_1, t_2] \times \overline{\mathbb{D}}(\varepsilon) \subset G$ and so, for a small positive δ we get that $\widehat{G} := ([t_1, t_2] +$

$i[-\delta,\delta]) \times \mathbb{D}(\varepsilon) \subset\subset \widetilde{G}$. Hence, $\widetilde{f} = f$ on $\widehat{G}$. It remains to prove that $G \cap \widetilde{G}$ is connected in order to see that $\widetilde{f} = f$ on $\widetilde{G}$. Hence, f extends to a holomorphic function $\widetilde{f} \in \mathcal{O}(\widetilde{G}, \mathbb{D})$, which finally implies the estimate also for this metric. □

Remark 20.1.4. In fact, even the following is true for the boundary behavior of the Kobayashi–Royden metric (see [183], for the case where ∂D is of class $\mathcal{C}^\infty$, see [180]):

- If $D = \{z \in \mathbb{C}^n : r(z) < 0\}$ is a bounded domain with a $\mathcal{C}^{1,1}$-smooth boundary (r is a defining function of D), then there exists a $C > 0$ such that

$$\varkappa_D(z;X) \geq C \frac{|\sum_{j=1}^n \frac{\partial r}{\partial z_j}(z)X_j|}{|r(z)|^{1/2}}, \quad z \in D,\ X \in \mathbb{C}^n.$$

- Let $D \subset \mathbb{C}^2$ be a bounded domain that is locally defined near the origin by $r(z) := \operatorname{Re} z_2 - |z_1|^2 < 0$. Then,

$$\varkappa_D((-\delta,0);(\delta^{-1/2},1)) \leq C \frac{1}{\delta^{1/2}}, \quad 0 < \delta << 1,$$

  showing that the exponent $1/2$ in the former estimate is sharp.

- Let D be as in the first point. If there exist constants $C > 0$ and $\alpha > 1/2$ such that

$$\varkappa_D(z;X) \geq C \frac{|\sum_{j=1}^n \frac{\partial r}{\partial z_j}(z)X_j|}{|r(z)|^{\alpha}}, \quad z \in D,\ X \in \mathbb{C}^n,$$

  then D is necessarily pseudoconvex.

- Conversely, if D is a bounded pseudoconvex domain with a $\mathcal{C}^3$-smooth boundary, then there exists a positive constant C such that

$$\varkappa_D(z;X) \geq C \frac{|\sum_{j=1}^n \frac{\partial r}{\partial z_j}(z)X_j|}{|r(z)|^{2/3}}, \quad z \in D,\ X \in \mathbb{C}^n.$$

- In case that D is given by a $\mathcal{C}^4$-smooth psh defining function on a neighborhood U of ∂D, then the asymptotic lower estimate may even be improved by an exponent $7/8$ instead of $2/3$ as in the former case (see [428]). Note that not all smooth bounded pseudoconvex domains are locally the sublevel sets of a psh function.

The proofs of these results are based on the method of comparison domains. As we have seen (see Proposition 20.1.2), the exponent $2/3$ (even if we only discuss the estimates in the normal direction) cannot replaced by 1.

In the case of plane domains, we have the following result on the boundary behavior of the Kobayashi–Royden metric:

Remark 20.1.5. Put $D := \mathbb{C}\setminus\{0,1\}$. We already know that $\varkappa_D(\lambda;1) \le \varkappa_{\mathbb{D}_*}(\lambda;1) = -\frac{1}{2|\lambda|\log|\lambda|}$, $\lambda \in \mathbb{D}_*$.

Let $p : \mathbb{H}^+ \longrightarrow D$, $\mathbb{H}^+$ the upper halfplane, be a universal covering and $\mathbb{H}^+ \ni \widetilde{\lambda}$ such that $p(\widetilde{\lambda}) = \lambda \in \mathbb{D}_*$. Moreover, take a biholomorphic mapping $m : \mathbb{H}^+ \longrightarrow \mathbb{D}$ with $m(\widetilde{\lambda}) = 0$. Then, $\varkappa_D(\lambda;1) = \left|\frac{m'(\widetilde{\lambda})}{p'(\widetilde{\lambda})}\right|$.

Then, using the effective formulas for p (as elliptic modular function) and m, one can prove (see [288]) that $\varkappa_D$ even fulfills a similar lower estimate, namely there exist positive numbers C and δ_0 such that

$$\varkappa_D(\lambda;1) \ge -C\frac{1}{|\lambda|\log|\lambda|}; \quad 0 < |\lambda| < \delta_0.$$

20.2 Boundary behavior of $\varkappa$ on pseudoconvex domains in normal direction

Let $D \subset \mathbb{C}^n$ be a bounded pseudoconvex domain with a $\mathcal{C}^2$-smooth boundary. Recall that it is not known whether D is $\boldsymbol{k}$-complete. To get a positive answer, it would suffice to know the following boundary behavior of the Kobayashi–Royden metric:

$$\varkappa_D(z;X) \ge C\|X\|/\operatorname{dist}(z,\partial D), \quad z \in D \text{ with } \operatorname{dist}(z,\partial D) < \delta_0,\ X \in \mathbb{C}^n, \tag{20.2.1}$$

where C and δ_0 are appropriate positive constants. Under this assumption, it is easy to conclude that D is $\boldsymbol{k}$-complete. Indeed, assume this is not true. Then, there exists a $\boldsymbol{k}_D$-Cauchy sequence $(z_j)_{j=1}^\infty \subset D$ with $z_j \longrightarrow a \in \partial D$. Thus, $\boldsymbol{k}_D(b,z_j) \le M$, $j \in \mathbb{N}$, where $b \in D$ is arbitrarily fixed. In particular, there are $\mathcal{C}^1$-curves $\alpha_j : [0,1] \longrightarrow D$ connecting b with z_j such that

$$\int_0^1 \varkappa_D(\alpha_j(t);\alpha_j'(t))dt \le M+1, \quad j \in \mathbb{N}.$$

Let r be the signed boundary distance function on a small strip $b \notin U = U(\partial D)$ such that $\|\operatorname{grad} r\| \le c$ and $\operatorname{dist}(\cdot,\partial D) \le \delta_0$ on U. Moreover, assume that U contains all points with $\operatorname{dist}(z,\partial D) \le \varepsilon$ for some positive ε. Now choose t_j such that $\alpha_j([t_j,1]) \subset$

U and $-r(\alpha_j(t_j)) = \varepsilon$, $j \in \mathbb{N}$. Then, we have

$$\begin{aligned} M+1 &\geq \int_{t_j}^{1} \varkappa_D(\alpha_j(t);\alpha_j'(t))dt \geq C\int_{t_j}^{1} \frac{\|\alpha_j'(t)\|}{-r(\alpha_j(t))}dt \\ &\geq \frac{C}{2c}\int_{t_j}^{1} |(\log(-r\circ\alpha_j))'(t)|dt \\ &= \frac{C}{2c}\left(-\log(-r(z_j)) + \log(-r(\alpha_j(t_j)))\right) \longrightarrow \infty; \end{aligned}$$

a contradiction. Therefore, it is an interesting question whether (20.2.1) holds for any bounded pseudoconvex domain with a smooth boundary.

Recall that we already know that if $D \subset \mathbb{C}^n$ is an arbitrary bounded domain with a $\mathcal{C}^2$-smooth boundary, then even $\varkappa_D(z_\delta;\nu) \geq C/\delta$ for some $C > 0$, is, in general, not true for non pseudoconvex domains, where $z_0 \in \partial D$, $z_\delta := z_0 - \delta\nu$, and ν denotes the outer unit normal vector at ∂D in z_0.

In this paragraph we will show that the lower estimate is not even correct for smooth bounded pseudoconvex domains and for the normal direction. Therefore, to prove $\boldsymbol{k}$-completeness we need another idea.

Theorem 20.2.1 (cf. [428]). *For any positive integer k there exist a $\mathcal{C}^k$-smooth bounded pseudoconvex domain $D \subset \mathbb{C}^3$, a boundary point $z_0 \in \partial D$, positive numbers ε and C, and a decreasing sequence $(\delta_j)_j$ of positive numbers with $\delta_j \longrightarrow 0$ such that*

$$\varkappa_D(z_{\delta_j};\nu) \leq C\frac{1}{\delta_j^{1-\varepsilon}}, \quad j \in \mathbb{N},$$

where $z_\delta := z_0 - \delta\nu$ and ν is the outer normal unit vector at ∂D in z_0.

Remark 20.2.2. Our proof is based on the one given in [169], where a weaker version has been shown. The argument used there goes back to an idea which can be found in [326].

Proof. Before we start with the technical details of the proof, we would like to sketch the main ideas of the proof. The main one consists in constructing a function $\widetilde{\varrho} \in \mathcal{C}^k(\mathbb{B}_2(M_0)) \cap \mathcal{PSH}(\mathbb{B}_2(M_0))$ ($M_0 \gg 1$), which satisfies the following condition:

$$\widetilde{\varrho}(\lambda^3,\lambda^2) \leq 2\delta_n - \frac{a_n\delta_n}{r_n}\operatorname{Re}\lambda, \quad \lambda \in \mathbb{D}(r_n),$$

where δ_n, r_n, and a_n ($n \in \mathbb{N}$) are correctly chosen positive numbers with $\delta_n \searrow 0$, $r_n \searrow 0$, and $a_n \nearrow \infty$. Then, put

$$\widetilde{\Omega}' := \{(z,w) \in \mathbb{C}^2 \times \mathbb{C} : \operatorname{Re} w + \widetilde{\varrho}(z) < 0\} \cap \mathbb{B}_3(2).$$

Looking at the following analytic discs:

$$\varphi_n(\lambda) := (r_n^3\lambda^3, r_n^2\lambda^2, -2\delta_n + a_n\delta_n\lambda), \quad \lambda \in \mathbb{D},$$

it is easy to see that $\varphi_n \in \mathcal{O}(\mathbb{D}, \widetilde{\Omega}')$, $\varphi(0) = -2\delta_n$, and $\varphi_n'(0) = (0, 0, a_n\delta_n)$ (for large n). Therefore, we get that

$$\varkappa_{\widetilde{\Omega}'}((0,0,-2\delta_n);(0,0,1)) \le \frac{1}{\delta_n a_n} = \frac{2}{2\delta_n a_n} \le \frac{C}{(2\delta_n)^{1-\varepsilon}}, \quad n \gg 1,$$

where the last inequality requires us to work with the correct numbers a_n, δ_n and r_n. Here, $z_0 = (0,0,0)$ and $\nu = (0,0,1)$ is the normal unit vector at $\partial\widetilde{\Omega}'$ in z_0. Finally, it remains "to round" $\widetilde{\Omega}'$ in order to end up with the claimed domain $\widetilde{\Omega}$.

In order to obtain $\widetilde{\varrho}$ as above, we first need a function $\varrho \in \mathcal{C}^k(\mathbb{C}) \cap \mathcal{PSH}(\mathbb{C})$ with the following property:

$$\varrho(\lambda) < 2\delta_n - \frac{a_n\delta_n}{r_n}\operatorname{Re}\lambda, \quad |\lambda| < r_n,\ n \gg 1.$$

Then, $\widetilde{\varrho}$ will be a suitable extension of the function $N \ni (\lambda^3, \lambda^2) \longmapsto \varrho(\lambda)$, $\lambda \in \mathbb{C}$, to the whole $\mathbb{C}^2$, where N is the Neile parabola from section 2.11.

The proof will be done in several steps.

A) *Construction of the function ϱ.*

Step 1°. Fix real numbers $a > 1$ and $0 < \varepsilon < 1/(9k+5)$. Note that $\varepsilon < 2/5$. Put

$$r_n := a^{-4\cdot 3^n\varepsilon} < 1, \quad \delta_n := a^{-3^n}, \quad a_n := a^{3^n\varepsilon},$$
$$A_n := 1 + \frac{3}{8} + \frac{a_n}{r_n} - \frac{\log r_n}{4\log a_n}, \quad n \in \mathbb{N}.$$

Then, by virtue of these definitions, it follows that $16 < a_n \nearrow \infty$, $r_{n+1} \le \frac{r_n^2}{a_{n+1}}$, $2r_{n-1}^2 \le 1$, and $\delta_n \le \frac{\delta_{n-1}}{A_n 2^n}$ for $n \ge n_0 \gg 1$. Choose n_0 so that $4\log a_{n_0} > 1$ and $2 < e^{3^{n_0}\varepsilon}$.

Step 2°. Let us have a look at the following functions:

$$g_n(t) := \frac{1}{8} - t + \frac{\log t}{4\log a_n}, \quad t \in (0,1], \quad n \ge n_0.$$

Then, for $t_n := \frac{1}{4\log a_n} \in (0,1)$ we get $g_n(t_n) > 0$, $g_n' > 0$ on $(0, t_n)$, and $g_n' < 0$ on $(t_n, 1]$. Finally, let b_n denote the zero of g_n in $(0, t_n)$. Also note that $g_n < 0$ on $(0, a_n^{-1/2})$, which implies that $2r_n < a_n^{-1/2} \le b_n$, $n \ge n_0$.

For $n \ge n_0$ (all n's we are dealing with are now supposed to be greater than or equal to n_0, where the n_0 will be chosen more precisely during the proof) we put

$$u_n(\lambda) := \frac{1}{8} - \operatorname{Re}\lambda + \frac{\log|\lambda|}{4\log a_n}, \quad \lambda \in \mathbb{C}.$$

Obviously, $u_n \in \mathcal{SH}(\mathbb{C})$ and is harmonic outside zero.

Finally, put $\varepsilon_n := r_n/2$. Then, the following global estimate for all $|\lambda| > r_n$ and $w \in \mathbb{D}$ is true:

$$u_n(\lambda - \varepsilon_n w) - u_n(\lambda) \le \varepsilon_n + \frac{\log(1 + \varepsilon_n/r_n)}{4\log a_n} = \varepsilon_n + \frac{\log(3/2)}{4\log a_n}. \tag{20.2.2}$$

Step 3°. Put

$$R_n(\lambda) := \begin{cases} \max\{u_n(\lambda), 0\}, & \text{if } \operatorname{Re}\lambda < b_n \\ u_n(\lambda), & \text{if } \operatorname{Re}\lambda \ge b_n \end{cases}, \quad \lambda \in \mathbb{C}.$$

If $\operatorname{Re}\lambda < b_n$ and $\lambda \longrightarrow b_n + iy$, then $u_n(\lambda) \longrightarrow \frac{1}{8} - b_n + \frac{\log|b_n + iy|}{4\log a_n} \ge g_n(b_n) = 0$. Therefore, using Appendix B.4.18, we conclude that $R_n \in \mathcal{SH}(\mathbb{C})$ and is continuous on $\mathbb{C}$.

Moreover, observe that $R_n(\lambda) \ge 0$, if $\operatorname{Re}\lambda < b_n$.

Step 4°. Estimates for R_n.

(a) Let $M > 1$ be an arbitrary large number. Then, there is a number $C_M > 2M$ such that for all $\lambda \in \mathbb{D}(2Ma_n/r_n)$ with $R_n(\lambda) = u_n(\lambda)$ the following inequality is true:

$$R_n(\lambda) \le \frac{1}{8} + \frac{2Ma_n}{r_n} + \frac{\log(2M) + \log(a_n/r_n)}{4\log a_n} < C_M \frac{a_n}{r_n}.$$

For the lower estimate, using the minimum principle for harmonic functions we have

$$u_n(\lambda) \ge \frac{1}{8} - 2M\frac{a_n}{r_n} \ge -C_M\frac{a_n}{r_n} \quad \text{on } \mathbb{D}(2Ma_n/r_n) \cap \{\lambda \in \mathbb{C} : \operatorname{Re}\lambda \ge b_n\}.$$

Hence,

$$|R_n| \le C_M \frac{a_n}{r_n}, \quad \lambda \in \mathbb{D}(2Ma_n/r_n).$$

(b) Moreover, if $|\lambda| < 2r_n$, then

$$u_n(\lambda) \le \frac{1}{8} + 2\frac{r_{n-1}^2}{a_n} + \frac{-\log a_n + \log(2r_{n-1}^2)}{4\log a_n} \le -\frac{1}{8} + \frac{2}{a_n} < 0,$$

because of the choice of n_0. Therefore, $R_n = 0$ on $\mathbb{D}(2r_n)$ (recall that $2r_n \le b_n$).

(c) If $|\lambda| < a_n$, then $u_n(\lambda) \le 3/8 - \operatorname{Re}\lambda$ and if, in addition, $\operatorname{Re}\lambda < b_n$, then the right side is positive for large n_0. Thus,

$$R_n(\lambda) \le \frac{3}{8} - \operatorname{Re}\lambda, \quad \lambda \in \mathbb{D}(a_n). \tag{20.2.3}$$

Step 5°. Note that R_n is a subharmonic function. To get a smooth subharmonic function that is larger than or equal to R_n and almost equal to R_n one has to use convolution. So take a non-constant function $\chi \in \mathcal{C}^\infty(\mathbb{C}, \mathbb{R}_+)$ with $\chi(\lambda) = \chi(|\lambda|)$, $\lambda \in \mathbb{C}$, $\chi = 0$ outside of the unit disc $\mathbb{D}$, and $\int_{\mathbb{C}} \chi(w) d\mathcal{L}^2(w) = 1$. Put

$$\widetilde{R}_n(\lambda) := \int_{\mathbb{C}} R_n(\lambda - \varepsilon_n w)\chi(w) d\mathcal{L}^2(w), \quad \lambda \in \mathbb{C},$$

where, as above, $0 < \varepsilon_n = r_n/2 < 1/2$. Then, $R_n \le \widetilde{R}_n \in \mathcal{SH}(\mathbb{C}) \cap \mathcal{C}^\infty(\mathbb{C})$.

Step 6°. Now, we study estimates for the derivatives of $\widetilde{R}_n$. Obviously, we have

$$\widetilde{R}_n^{(\ell)}(\lambda) = \int_{\mathbb{C}} R_n(\lambda - \varepsilon_n w)\chi^{(\ell)}(w)(-\varepsilon_n)^{-\ell} d\mathcal{L}^2(w).$$

Therefore, using the estimates from Step 4° we get for $\lambda \in \mathbb{D}(Ma_n/r_n)$

$$|\widetilde{R}_n^{(\ell)}(\lambda)| \le C_M C' \frac{a_n}{r_n \varepsilon_n^\ell}, \quad 0 \le \ell \le k,$$

where $C' := \max\{\int_{\mathbb{C}} |\chi^{(\ell)}(w)| d\mathcal{L}^2(w) : 0 \le \ell \le k\}$.

Put $\varrho_n(\lambda) := \widetilde{R}_n(\frac{a_n}{r_n}\lambda)$, $\lambda \in \mathbb{C}$. Then, $\varrho_n \in \mathcal{SH}(\mathbb{C}) \cap \mathcal{C}^\infty(\mathbb{C})$. Moreover, its derivatives allow the following estimates:

$$|\delta_n \varrho_n^{(\ell)}(\lambda)| \le 2^\ell C_M C' a^{3^n(\varepsilon(5+9\ell)-1)} \le 2^k C_M C' a^{3^n(\varepsilon(5+9k)-1))}, \quad \lambda \in \mathbb{D}(M). \tag{20.2.4}$$

Recall that M could be taken as any large positive number. Therefore, if

$$\varrho(\lambda) := \sum_{n=n_0}^{\infty} \delta_n \varrho_n(\lambda), \quad \lambda \in \mathbb{C},$$

then $\varrho \in \mathcal{SH}(\mathbb{C}) \cap \mathcal{C}^k(\mathbb{C})$ (recall the choice of ε).

We also note that if $j < n$ and $|\lambda| < r_n$, then $|\frac{a_j}{r_j}\lambda| \le \frac{a_j r_n}{r_j} < r_j$. Thus,

$$\varrho_j(\lambda) = \widetilde{R}_j\left(\frac{a_j}{r_j}\lambda\right) = 0, \quad |\lambda| < r_n. \tag{20.2.5}$$

Step 7°. A global estimate for $\widetilde{R}_n - R_n$. Note that if $|\lambda| < r_n$, then $\widetilde{R}_n(\lambda) = R_n(\lambda) = 0$. So it suffices to assume that $|\lambda| \ge r_n$. Then,

$$0 \le \widetilde{R}_n(\lambda) - R_n(\lambda) \le \int_{\mathbb{C}} (R_n(\lambda - \varepsilon_n w) - u_n(\lambda))\, \chi(w) d\mathcal{L}^2(w).$$

Now fix $|\lambda| \geq r_n$ and $w \in \mathbb{D}$. Then, either $R_n(\lambda - \varepsilon_n w) = u_n(\lambda - \varepsilon_n w)$ or $R_n(\lambda - \varepsilon_n w) = 0$. Assume the first case. Then, we get (use (20.2.2))

$$R_n(\lambda - \varepsilon_n w) - u_n(\lambda) \leq \frac{1}{2} r_n + \frac{\log(3/2)}{4 \log a_n} < 1,$$

provided that n_0 is sufficiently large. In the remaining case, we know that necessarily $\operatorname{Re}(\lambda - \varepsilon_n w) \leq b_n$ and therefore we have

$$R_n(\lambda - \varepsilon_n w) - R_n(\lambda) \leq -u_n(\lambda) \leq -\frac{1}{8} + b_n + \varepsilon_n + 1 < 1,$$

if n_0 is sufficiently large. Hence, $0 \leq \widetilde{R}_n - R_n < 1$ for all $n \geq n_0$.

Step 8°. Estimates for ϱ on $\mathbb{D}(r_n)$, $n \geq n_0$. Fix a $\lambda \in \mathbb{D}(r_n)$. Then,

$$\varrho_n(\lambda) = \widetilde{R}_n\left(\frac{a_n}{r_n}\lambda\right) \leq 1 + R_n\left(\frac{a_n}{r_n}\lambda\right).$$

Now observe that the argument of R_n lies in $\mathbb{D}(a_n)$. Thus, using (20.2.3), we see that $\varrho_n(\lambda) \leq 1 + \frac{3}{8} - \frac{a_n}{r_n} \operatorname{Re} \lambda$ and therefore, $\delta_n \varrho_n(\lambda) \leq 2\delta_n - \frac{a_n \delta_n}{r_n} \operatorname{Re} \lambda$.

If $j \in \mathbb{N}$ is such that $n_0 \leq j < n$, then $\varrho_j(\lambda) = 0$ (see (20.2.5)). Therefore,

$$\varrho(\lambda) = \sum_{j=n_0}^{\infty} \delta_j \varrho_j(\lambda) \leq \left(1 + \frac{3}{8}\right) \delta_n - \frac{a_n \delta_n}{r_n} \operatorname{Re} \lambda + \sum_{j=n+1}^{\infty} \delta_j \varrho_j(\lambda).$$

It remains to estimate the remaining summands. Fix a $j > n$. Then,

$$\begin{aligned}
\varrho_j(\lambda) = \widetilde{R}_j\left(\frac{a_j}{r_j}\lambda\right) &\leq 1 + R_j\left(\frac{a_j}{r_j}\lambda\right) \leq 1 + \left|u_j\left(\frac{a_j}{r_j}\lambda\right)\right| \\
\leq 1 + \frac{1}{8} + \frac{a_j}{r_j} + \frac{\log a_j + \log(r_n/r_j)}{4 \log a_j} &\leq 1 + \frac{3}{8} + \frac{a_j}{r_j} + \frac{\log(1/r_j)}{4 \log a_j} = A_j.
\end{aligned}$$

By our assumption, we conclude

$$\delta_j \varrho_j(\lambda) \leq \delta_j A_j \leq \frac{\delta_{j-1}}{2^j} \leq \frac{\delta_n}{2^j}.$$

Thus, $\sum_{j=n+1}^{\infty} \delta_j \varrho_j(\lambda) \leq \delta_n \sum_{j=2}^{\infty} 2^{-j} = \frac{\delta_n}{2}$. Hence, $\varrho(\lambda) < 2\delta_n - \frac{a_n \delta_n}{r_n} \operatorname{Re} \lambda$.

B) *Summary and an interim result.*

So far, we have found a function $\varrho \in \mathcal{SH}(\mathbb{C}) \cap \mathcal{C}^k(\mathbb{C})$ satisfying the following estimate:

$$\varrho(\lambda) < 2\delta_n - \frac{a_n \delta_n}{r_n} \operatorname{Re} \lambda, \quad |\lambda| < r_n, \ n \geq n_0. \tag{20.2.6}$$

Put

$$\Omega' := \{(z,w) \in \mathbb{C} \times \mathbb{C} : \operatorname{Re} w + \varrho(z) < 0\} \cap \mathbb{B}_2(2)$$

and denote the connected component of Ω' that contains the normal segment $\{0\} \times \{w \in \mathbb{C} : w = \operatorname{Re} w \in (-2,0)\}$ by Ω. Then, Ω is a bounded pseudoconvex domain with a piecewise $\mathcal{C}^k$-smooth boundary. Now we introduce the following analytic discs:

$$\varphi_n(\lambda) := (r_n\lambda, -2\delta_n + a_n\delta_n\lambda), \quad \lambda \in \mathbb{D},\ n \geq n_0.$$

Note that $\varphi_n \in \mathcal{O}(\mathbb{D}, \Omega)$ (n large), $\varphi_n(0) = (0, -2\delta_n)$, and $\varphi_n'(0) = (r_n, a_n\delta_n) = a_n\delta_n(\frac{r_n}{a_n\delta_n}, 1) =: a_n\delta_n X_n$. Therefore,

$$\varkappa_\Omega(z_{2\delta_n}; X_n) = \varkappa_\Omega((0, -2\delta_n); X_n) \leq \frac{2}{a_n 2\delta_n} \leq C(2\delta_n)^{-(1-\varepsilon)}, \quad n \gg 1.$$

We should emphasize that if $n \longrightarrow \infty$ the vectors X_n become more and more tangential at $\partial\Omega$ in $(0,0)$. Exactly this situation was treated in [326].

Now we continue the proof of Theorem 20.2.1(a).

C) *Construction of the function* $\widetilde{\varrho}$.

Step 9°. Let us first recall some facts from section 2.11. Let $N = N_{2,3} = \{z \in \mathbb{C}^2 : z_1^2 = z_2^3\}$ be the Neile parabola. We already know that N has a global bijective holomorphic parametrization

$$\mathbb{C} \ni \lambda \overset{p=p_{2,3}}{\longmapsto} (\lambda^3, \lambda^2)$$

satisfying the following property: the mapping $q = q_{2,3} := p^{-1}$ is holomorphic on $N^* := N \setminus \{(0,0)\}$, continuous on N, and it is given by $q(z) = z_1 z_2^{-1}$, $z \in N^*$, $q(0,0) = 0$.

Put $B_n := \mathbb{B}_2(r_{n+1}^3)$, $B_n' := \mathbb{B}_2((3/4)r_{n+1}^2)$, and $\widetilde{\varrho}_n(z) := \varrho_n \circ q(z)$, $z \in N$. Note that if $z \in N \cap B_n$, then $|q(z)| < r_{n+1}$, and thus $\widetilde{\varrho}_n(z) = 0$ (see (20.2.5)). Defining $\widehat{\varrho}_n := 0$ on B_n gives that $\widehat{\varrho}_n|_{N\cap B_n} = \widetilde{\varrho}|_{N\cap B_n}$; i.e., $\widetilde{\varrho}_n$ is extended to the whole ball B_n. It remains to extend $\widetilde{\varrho}_n$ to the whole of $\mathbb{C}^2$.

Observe that N^* consists of the two disjoint parts

$$V_1 := \{z \in N^* : z_1 = r^{3/2}e^{i3\theta/2},\ z_2 = re^{i\theta},\ \theta \in [0,2\pi)\},$$
$$V_2 := \{z \in N^* : z_1 = r^{3/2}e^{i(3\theta/2+\pi)},\ z_2 = re^{i\theta},\ \theta \in [0,2\pi)\}.$$

Then, we may take a small open neighborhood U_n of $V_1 \cup V_2$ such that the projection $\pi : U_n \setminus \overline{B_n'} \longrightarrow N$, $\pi(z) := (z_2^{3/2}, z_2) \in V_1 \cup V_2$ is well-defined (with a properly chosen branch of the power). Finally, we define

$$\widehat{\varrho}_n(z) := \widetilde{\varrho}_n \circ \pi(z), \quad z \in U_n \setminus \overline{B_n'}.$$

Note that $\widehat{\varrho}_n(z) = \varrho_n(\pm\sqrt{z_2})$. Thus, $\widehat{\varrho}_n = 0$ on $(U_n \setminus \overline{B'_n}) \cap B_n$. Hence, we have a well defined $\mathcal{C}^\infty$-function $\widehat{\varrho}_n$ on $B_n \cup (U_n \setminus \overline{B'_n})$ that extends the original function $\widetilde{\varrho}_n$.

Then, using the Faà di Bruno formula, we see that

$$|D^\alpha \widehat{\varrho}_n(z)| \le C_k F_k(z)\frac{1}{r_{n+1}^{3\ell}} \le C_k F_k(z)\frac{1}{r_n^{9\ell}}, \quad z \in U_n \setminus \overline{B}'_n,\ \alpha \in \mathbb{N}_0^2,\ |\alpha| = \ell \le k, \tag{20.2.7}$$

where C_k is a universal constant and $F_k(z) := \max\{|\varrho_n^{(\ell)}(\pm\sqrt{z_2})| : 1 \le \ell \le k\}$.

Finally, take a $\mathcal{C}^\infty$-function $\chi : \mathbb{R} \longrightarrow [0,1]$ that is equal to 1 on $[0, 1/2]$ and equal to 0 on $[1, \infty)$. Then, we may introduce the following globally defined $\mathcal{C}^\infty$-function p_n by

$$p_n(z) := \begin{cases} 0, & \text{if } z \in B_n \\ \widehat{\varrho}_n(z), & \text{if } z \in U_n \setminus B_n,\ \|z - \pi(z)\| < d_n^2/2 \\ \widehat{\varrho}_n(z)\chi\left(\frac{\|z-\pi(z)\|^2}{d_n^2}\right), & \text{if } z \in U_n,\ d_n^2/2 \le \|z - \pi(z)\|^2 \le d_n^2 \\ 0, & \text{if } z \notin U_n \cup B_n \end{cases},$$

where $d_n = c r_{n+1}$ and c is chosen such that $\{z \in \mathbb{C}^2 \setminus \overline{B}'_n : \|z - \pi(z)\| < d_n\} \subset U_n$ for all $n \ge n_0$.

First observe that $p_n|_N = \widetilde{\varrho}_n$ and $p_n \in \mathcal{C}^\infty(\mathbb{C}^2)$, i.e., p_n is a $\mathcal{C}^\infty$-extension of $\widetilde{\varrho}_n$ to the whole $\mathbb{C}^2$. Now fix $M_0 := 4$. Then, using the formula for p_n, we have a constant $D_{M_0} > 0$ such that

$$|D^\alpha p_n(z)| \le C'_k D_{M_0} F_k(z) r_n^{-27\ell}, \quad z \in \mathbb{B}_2(M_0),\ |\alpha| = \ell \le k,$$

with a new constant C'_k. Taking (20.2.4) into account we finally obtain the following inequality:

$$\delta_n |D^\alpha p_n(z)| \le 2^k C'_k D_{M_0} C_{M_0} C' a^{3^n(\varepsilon(5+9k)-1)} r_n^{-27k}, \quad |\alpha| \le k,\ z \in \mathbb{B}_2(M_0).$$

Hence,

$$\delta_n |D^\alpha p_n(z)| \le C a^{3^n(\varepsilon(5+120k)-1)}, \quad |\alpha| \le k,\ z \in \mathbb{B}_2(M_0),$$

where C is a universal constant. From now on *we even assume that* $\varepsilon < 1/(5+120k)$. Hence, $\widetilde{p} := \sum_{n=n_0}^\infty \delta_n p_n$ gives a $\mathcal{C}^k$-function on $\mathbb{B}_2(M_0)$.

Moreover, using only the first two derivatives, we get $\mathcal{L}p_n(z; X) \ge -C_n\|X\|^2$, where C_n is a multiple of r_n^{-60}, i.e., $C_n = s r_n^{-60}$, where the s may be chosen to be independent of n.

Step 10°. A psh modification of p_n. Put

$$h(z) := e^{\|z\|^2}|z_1^2 - z_2^3|^2, \quad z \in \mathbb{C}^2.$$

Our aim now is to define a new function of the form $\widetilde{p}_n := p_n + K_n h$ that is a psh $\mathcal{C}^\infty$-extension of $\widetilde{\varrho}_n$ to the whole of $\mathbb{C}^2$. Observe that

$$\mathcal{L}h(z;X) \geq |z_1^2 - z_2^3|^2\|X\|^2, \quad z \in \mathbb{C}^2,\ X \in \mathbb{C}^2;$$

use the fact that $h(z) = e^{\|z\|^2+\log|z_1^2-z_2^3|^2} =: e^g(z)$ and

$$\mathcal{L}h(z;X) = \mathcal{L}g(z;X)e^{g(z)} + \Big|\sum_{j=1}^{2}\frac{\partial g}{\partial z_j}(z)X_j\Big|^2.$$

Note that $|z_1^2 - z_2^3|^2 = \|z - (z_2^{3/2}, z_2)\|^2\|(z - (-z_2^{3/2}, z_2)\|^2$. Therefore, there exists a positive constant D such that

$$\mathcal{L}h(z;X) \geq Dd_n^2/2\|X\|^2, \quad d^2 \geq \|z - \pi(z)\|^2 \geq d_n^2/2,\ z \notin B_n',\ X \in \mathbb{C}^2. \tag{20.2.8}$$

Now, take $K_n := \frac{2C_n}{Dd_n^2}$. Then, $\widetilde{p}_n \in \mathcal{C}^\infty(\mathbb{C}^2) \cap \mathcal{PSH}(\mathbb{B}_2(M_0))$ and $\widetilde{p}_n|_N = \widetilde{\varrho}_n$.

Note that $\delta_n K_n = 2\frac{\delta_n C_n}{Dd_n^2} = \frac{2s}{D}a^{3^n(4\cdot 69\varepsilon-1)}$. From now on we also assume that $\varepsilon < 1/(4\cdot 69)$. Then, $\sum_{n=n_0}^{\infty}\delta_n K_n < \infty$.

Step 11°. Construction of $\widetilde{\varrho}$. Finally, we define on $\mathbb{B}_2(M_0)$

$$\widetilde{\varrho}(z) := \sum_{n\geq n_0}^{\infty}\delta_n\widetilde{p}_n(z) = \widetilde{p}(z) + \sum_{n=n_0}^{\infty}\delta_n K_n h(z).$$

It remains to note that this function belongs to $\mathcal{C}^k(\mathbb{B}_2(M_0)) \cap \mathcal{PSH}(\mathbb{B}_2(M_0))$.

Step 12°. The domain $\widetilde{\Omega}$. Put

$$\widetilde{\Omega}' := \{(z,w) \in \mathbb{C}^2 \times \mathbb{C} : r(z,w) := \operatorname{Re} w + \widetilde{\varrho}(z) < 0\} \cap \mathbb{B}_3(2).$$

Recall that if $z = (\lambda^3, \lambda^2) \in N$ with $|\lambda| < r_n$, then $\widetilde{\varrho}(z) = \varrho(\lambda) < 2\delta_n - \frac{a_n\delta_n}{r_n}\operatorname{Re}\lambda$.

Now put

$$\varphi_n(\lambda) := (r_n^3\lambda^3, r_n^2\lambda^2, -2\delta_n + a_n\delta_n\lambda), \quad \lambda \in \mathbb{D}.$$

Then,

$$\begin{aligned} r(\varphi_n(\lambda)) &= -\delta_n + a_n\delta_n\operatorname{Re}\lambda + \widetilde{\varrho}(r_n^3\lambda^3, r_n^2\lambda^2) \\ &< -2\delta_n + a_n\delta_n\operatorname{Re}\lambda + 2\delta_n - a_n\delta_n\operatorname{Re}\lambda = 0; \end{aligned}$$

thus φ_n is an analytic disc inside of $\widetilde{\Omega}'$ (for large n) satisfying the following properties: $\varphi_n(0) = (0, 0, -2\delta_n)$ and $\varphi_n'(0) = a_n\delta_n(0, 0, 1)$. Denote the connected component of $\widetilde{\Omega}'$, that contains the normal segment $\{(0,0)\} \times \{w \in \mathbb{C} : w = \operatorname{Re} w \in (-2, 0)\}$, by $\widetilde{\Omega}$. Hence, $\varkappa_{\widetilde{\Omega}}(z_{2\delta_n}; \nu) \le \frac{1}{a_n\delta_n}$ (n large), where $\nu = (0, 0, 1)$ and $z_\delta := (0, 0, -\delta)$.

D) *Smoothing of* $\widetilde{\Omega}$.

Step 13°. Smoothing of $\widetilde{\Omega}$.

Take $M_0 = 10$ and take a convex $\mathcal{C}^\infty$-function $\theta : \mathbb{R} \to \mathbb{R}_+$ such that $\theta = 0$ on $(-\infty, 4]$, θ is strongly increasing on $(4, \infty)$, and $\theta \ge \sup\{|\operatorname{Re} w + \widetilde{\varrho}(z)| : (z, w) \in \mathbb{B}_3(10)\}$ on $[25, \infty)$. Put

$$u_t(z, w) := \operatorname{Re} w + \widetilde{\varrho}(z) + t\theta(|w|^2 + \|z\|^2) =: h(z, w) + tg(z, w), \qquad (z, w) \in \mathbb{C}^3,\ t \in \mathbb{R}_+.$$

Then, $u_t \in \mathcal{PSH}(\mathbb{B}_3(5)) \cap \mathcal{C}^k(\mathbb{B}_3(5))$. From now on, we may assume that $k \ge 6$. Then, using Sard's lemma, there is a $t_0 > 1$, near 1, such that $\operatorname{grad}(\frac{h}{g})(z, w) \ne 0$, whenever $\frac{h(z,w)}{g(z,w)} = -t_0$, $(z, w) \in \mathbb{B}_3(5) \setminus \overline{\mathbb{B}}_3(2)$. Hence, $\operatorname{grad}(h + t_0 g)(z, w) \ne 0$ if $(h + t_0 g)(z, w) = 0$ and $(z, w) \in \mathbb{B}_3(5) \setminus \overline{\mathbb{B}}_3(2)$. Now we define

$$\widehat{\Omega} := \text{the connected component of } \{(z, w) \in \mathbb{B}_3(5) : u_{t_0}(z, w) < 0\} \text{ containing } (0, 0, w),\ -2 < w = \operatorname{Re} w < 0.$$

Then, $\widehat{\Omega}$ is a pseudoconvex domain with a $\mathcal{C}^k$-smooth boundary satisfying $\widetilde{\Omega} \subset \widehat{\Omega} \subset \mathbb{B}_3(5)$. Hence,

$$\varkappa_{\widehat{\Omega}}(z_{2\delta_n}; \nu) \le \varkappa_{\widetilde{\Omega}}(z_{2\delta_n}; \nu) \le \frac{1}{a_n\delta_n}, \quad n \gg 1,$$

which ends the proof. □

Corollary 20.2.3. *For any $k \in \mathbb{N}$ there exist a $\mathcal{C}^k$-smooth bounded pseudoconvex domain $D \subset \mathbb{C}^3$ and a boundary point $z_0 \in \partial D$ such that for any $C > 0$ the following inequality fails to hold:*

$$\varkappa_D(z_\delta; \nu) \ge C\frac{1}{\delta^{1-\varepsilon}}, \quad 0 < \delta \ll 1.$$

? It seems to be an open question whether there exists an example similar to Theorem 20.2.1 in dimension $n = 2$?

Remark 20.2.4. In the case of bounded pseudoconvex domains with a $\mathcal{C}^\infty$-smooth boundary, we even have (see [428]):

for any $\alpha > 0$ there exist a $\mathcal{C}^\infty$-smooth bounded pseudoconvex domain $D \subset \mathbb{C}^3$, a boundary point $z_0 \in \partial D$, a positive number C, and a decreasing sequence $(\delta_j)_j$ of positive numbers with $\delta_j \longrightarrow 0$ such that

$$\varkappa_D(z_{\delta_j}; \nu) \le C \frac{1}{\delta_j(-\log \delta_j)^\alpha}, \quad j \in \mathbb{N},$$

where z_δ and ν is as in Theorem 20.2.1.

We conclude this section with the following remark on the blowing up of the Kobayashi–Royden metric for plane domains (see [326]):

Remark 20.2.5. Let $D \subset \mathbb{C}$ be a bounded domain with a $\mathcal{C}^{1+\varepsilon}$-smooth boundary. Then, there exists a constant $C > 1$ such that

$$\frac{1}{C \operatorname{dist}(\lambda, \partial D)} \le \varkappa_D(\lambda; X) \le \frac{C}{\operatorname{dist}(\lambda, \partial D)}, \quad \lambda \in D,\ X \in \mathbb{C}.$$

20.3 An upper boundary estimate for the Lempert function

We start this section with a discussion on the boundary behavior of γ and $\varkappa$ when one approaches a $\mathcal{C}^1$-smooth boundary point. This result will be used to present an example of a $\mathcal{C}^1$-smooth plane domain for which the boundary behavior of the Lempert function differs from the one for domains with a $\mathcal{C}^{1+\varepsilon}$-smooth boundary (see Example 20.3.3).

Lemma 20.3.1 (cf. [375, 248, 379]). *Let a_0 be a $\mathcal{C}^1$-smooth boundary point of a bounded plane domain D. Then,*

$$\lim_{D \ni z \to a_0} \gamma_D(z; 1) \operatorname{dist}(z, \partial D) = \lim_{D \ni z \to a_0} \varkappa_D(z; 1) \operatorname{dist}(z, \partial D) = \tfrac{1}{2}.$$

Proof. Let us start by observing that we only have to prove that

(a) $\limsup_{D \ni z \to a_0} \varkappa_D(z; 1) \operatorname{dist}(z, \partial D) \le 1/2$,

(b) $1/2 \le \liminf_{D \ni z \to a_0} \gamma_D(z; 1) \operatorname{dist}(z, \partial D)$.

As a general preparation, without loss of generality, we may assume that $a_0 = 0$ and that there is a rectangle $Q := (-r_1', r_1') + i(-r_2, r_2) \subset \mathbb{C}$, a $\mathcal{C}^1$-function $s : (-r_1', r_1') \longrightarrow (-r_2, r_2)$, and an $r_1 < r_1'$ such that

- $s(0) = 0$ and $s'(0) = 0$,
- $|s'| < 1/2$ on $[-r_1, r_1]$,
- for any j there exists a positive η_j such that $|s'(\xi') - s'(\xi'')| < 3/(4j)$ whenever $|\xi' - \xi''| < \eta_j$ and $\xi', \xi'' \in [-r_1, r_1]$,
- $D \cap Q = \{z = x + iy \in Q : y < s(x)\}$.

Moreover, we will use the abbreviation $d(z) := \operatorname{dist}(z, \partial D)$, $z \in D$.

To (a): The idea of the proof here is to embed symmetric triangles around the inner normal at boundary points of ∂D, which approach a_0 with the additional property that the angle at a_0 tends to π, and then to use $\varkappa$ of the right half-plane.

Now let us turn to the technical details of the proof. First, take positive numbers R_j with $R_j \le \frac{\min\{r_1, r_2, \eta_j\}}{4(j+1)}$ and radii δ_j with $2\delta_j < \frac{\min\{r_1, r_2\}}{2}$ and $2j\delta_j < R_j$, $j \in \mathbb{N}$. Finally, take a sequence $(a_j) \subset D \cap Q$ such that $|\operatorname{Re} a_j| < r_1$, $a_j \in \mathbb{D}(\delta_j)$, $j \in \mathbb{N}$, and

$$\lim_{j\to\infty} \varkappa_D(a_j; 1) d(a_j) = \limsup_{D \ni z \to 0} \varkappa_D(z; 1) d(a_j).$$

Then, for all j, we find a point $\hat{a}_j = (\xi_j, s(\xi_j)) \in \mathbb{D}(2\delta_j) \cap \partial D$ with $|a_j - \hat{a}_j| = d(a_j) < \delta_j$. Put

$$\Delta_j := \{w = u + iv \in \mathbb{C} : 0 < u < R_j,\ |v| < ju\}, \quad j \in \mathbb{N}.$$

Then, we study the holomorphic mapping $\psi_j : \Delta_j \longrightarrow \mathbb{C}$ by $\psi_j(w) := \hat{a}_j + B_j(s'(\xi_j) - i) \cdot (u + iv)$, where $B_j := \frac{1}{1 + (s'(\xi_j))^2}$. We claim that $\psi_j(\Delta_j) \subset D$. Indeed,

- $|\operatorname{Re} \psi_j(w)| = |\xi_j + B_j(s'(\xi_j)u + v)| \le 2\delta_j + R_j(1 + j) \le 2\delta_j + r_1/2 < r_1$;
- $|\operatorname{Im} \psi_j(w)| = |s(\xi_j) + B_j(vs'(\xi_j) - u)| \le 2\delta_j + R_j(j+1) < r_2/2 + r_2/2 = r_2$ (hence, $\psi_j(\Delta_j) \subset Q$);
- it remains to verify that $s(\xi_j) + B_j(vs'(\xi_j) - u) < s\big(\xi_j + B_j(s'(\xi_j)u + v)\big)$. Using Taylor expansion, we only have to show that $vs'(\xi_j) - u < (s'(\xi_j)u + v)s'(\widehat{\xi}_j)$, where $\widehat{\xi}_j \in [\xi_j, \xi_j + B_j(s'(\xi_j)u + v)]$. Note that $\max\{|\xi_j|, |\widehat{\xi}_j|\} < r_1$ and $|\xi_j - \widehat{\xi}_j| \le 2R_j(1 + j) < \eta_j$, which means that $|s'(\widehat{\xi}_j) - s'(\xi_j)| < \frac{3}{4j}$. Then, $v(s'(\xi_j) - s'(\widehat{\xi}_j)) \le \frac{3u}{4} < u(1 - |s'(\xi_j)s'(\widehat{\xi}_j)|) \le u(1 + s'(\xi_j)s'(\widehat{\xi}_j))$ implying the wanted inequality. Hence, $\psi_j(\Delta_j) \subset D$.

Take the inverse of ψ_j, i.e., $\varphi_j(z) = -\frac{\hat{a}_j - z}{B_j(s'(\xi_j) - i)}$. Then, $\varphi_j(a_j)$ lies on the positive real axis and, moreover, we have that

$$v_j := \varphi_j(a_j) = \operatorname{Re} \varphi_j(a_j) = \sqrt{1 + (s'(\xi_j))^2}\, d(a_j) \le 2\delta_j < R_j$$

(after eventually taking a new subsequence). Hence, we obtain

$$\begin{aligned}\varkappa_D(a_j;1) &\le \varkappa_{\Delta_j}(v_j;\varphi_j'(a_j)) = \varkappa_{\Delta_j}(v_j;1)\sqrt{1+(s'(\xi_j))^2}\\ &= \varkappa_{\frac{1}{v_j}\Delta_j}(1;1/v_j)\sqrt{1+(s'(\xi_j))^2} = \frac{1}{d(a_j)}\varkappa_{\widetilde{\Delta}_j}(1;1),\end{aligned}$$

where $\widetilde{\Delta}_j := \{z \in \mathbb{C} : 0 < u < R_j/v_j,\ |v| < ju\}$. Note that $\widetilde{\Delta}_j \nearrow \mathbb{H}^+$ (after taking a subsequence), where $\mathbb{H}^+$ denotes the right halfplane. Hence,

$$\lim_{j\to\infty} \varkappa_D(a_j;1)d(a_j) \le \varkappa_{\mathbb{H}^+}(1;1) = 1/2,$$

where the last equality is left to the reader as an exercise.

To (b): As in (a), we may assume that $a_0 = 0$ and $D \cap Q = \{x+iy \in Q : y < s(x)\}$, where $Q := (-r_1', r_1') + i(-r_2, r_2)$, $s : (-r_1', r_1') \longrightarrow (-r_2, r_2)$ is a $\mathcal{C}^1$-function with $s(0) = s'(0) = 0$. Let $r_0 > 0$ be such that $\mathbb{D}(3r_0) \subset Q$. For $a \in \mathbb{D}(r_0)$ let $\widehat{a} = (\xi_a, s(\xi_a)) \in \partial D$ be such that $|a - \widehat{a}| = d(a) := \operatorname{dist}(a, \partial D)$. Note that $\widehat{a} \longrightarrow 0$ when $a \longrightarrow 0$. Define $\Phi_a(z) := (s'(\xi_a) - i)(\widehat{a} - z)$, $z \in \mathbb{C}$. Observe that $\Phi_a(a) = d(a)\sqrt{(s'(\xi_a))^2 + 1}$ and $|\Phi_a'(a)| = \sqrt{(s'(\xi_a))^2 + 1}$. Put $E_r := \{w \in \mathbb{C} : \operatorname{Re} w + r|w| > 0\}$, $F_r := \{w \in \mathbb{C} : |w| > r\}$. We are going to prove that

(*) for $0 < \varepsilon \ll 1$ there exist $\delta(\varepsilon) > 0$, $\eta(\varepsilon) \in (0, r_0)$ such that $\Phi_a(D) \subset E_\varepsilon \cup F_{\delta(\varepsilon)}$ for $a \in \mathbb{D}(\eta(\varepsilon))$.

Suppose for a moment that we have (*). Then,

$$\begin{aligned}\boldsymbol{\gamma}_D(a;1) &\ge \boldsymbol{\gamma}_{E_\varepsilon \cup F_{\delta(\varepsilon)}}(\Phi_a(a);\Phi_a'(a)) = \boldsymbol{\gamma}_{E_\varepsilon \cup F_{\delta(\varepsilon)/\Phi_a(a)}}(1;1)\frac{|\Phi_a'(a)|}{\Phi_a(a)}\\ &= \boldsymbol{\gamma}_{E_\varepsilon \cup F_{\delta(\varepsilon)/\Phi_a(a)}}(1;1)\frac{1}{d(a)}, \quad a \in \mathbb{D}(\eta(\varepsilon)),\ 0 < \varepsilon \ll 1.\end{aligned}$$

Note that $E_\varepsilon \cup F_{\delta(\varepsilon)/\Phi_a(a)} \underset{a\to 0}{\longrightarrow} E_\varepsilon$ and $E_\varepsilon \underset{\varepsilon\searrow 0}{\searrow} \overline{\mathbb{H}}^+ \setminus \{0\}$. Hence, by Exercise 3.9.5(c),

$$\liminf_{a\to 0} \boldsymbol{\gamma}_D(a;1)d(a) \ge \lim_{\varepsilon\to 0} \boldsymbol{\gamma}_{E_\varepsilon}(1;1) = \boldsymbol{\gamma}_{\mathbb{H}^+}(1;1) = \frac{1}{2},$$

which finishes the proof.

We move to the proof of (*). First, observe that

$$\lim_{\xi,x\to 0,\ \xi\neq x} \frac{s'(\xi)(\xi - x) + s(\xi) - s(x)}{|\xi - x|} + \varepsilon\sqrt{(s'(\xi))^2 + 1} = \varepsilon.$$

Consequently, there exists an $\eta_0 \in (0, r_0)$ such that for $a \in \mathbb{D}(\eta_0)$ and $z = x + iy \in D \cap \mathbb{D}(\eta_0)$ we get

$$\begin{aligned}
&\operatorname{Re} \Phi_a(z) + \varepsilon|\Phi_a(z)| \\
&= s'(\xi_a)(\xi_a - x) + s(\xi_a) - y + \varepsilon\sqrt{(s'(\xi_a))^2 + 1}\sqrt{(\xi_a - x)^2 + (s(\xi_a) - y)^2} \\
&> s'(\xi_a)(\xi_a - x) + s(\xi_a) - s(x) + \varepsilon\sqrt{(s'(\xi_a))^2 + 1}\,|\xi_a - x| \geq \frac{\varepsilon}{2}|\xi_a - x|.
\end{aligned}$$

Let $0 < \eta(\varepsilon) < \eta_0$ be so small that $|\Phi_a(z) - \Phi_0(z)| < \eta_0/2$ for $a \in \mathbb{D}(\eta(\varepsilon))$ and $z \in \mathbb{D}(r_0)$. Then,

$$|\Phi_a(z)| > |\Phi_0(z)| - \eta_0/2 = |z| - \eta_0/2 \geq \eta_0/2 =: \delta(\varepsilon), \qquad a \in \mathbb{D}(\eta(\varepsilon)),\ z \in D \setminus \mathbb{D}(\eta_0).$$

Finally, $\Phi_a(D) \subset E_\varepsilon \cup F_{\delta(\varepsilon)}$, $a \in \mathbb{D}(\eta(\varepsilon))$. □

Remark 20.3.2.

(a) In the previous lemma, the smoothness condition is essential. To see this, the reader may discuss the boundary behavior for a plane rectangle when the point is approaching a corner.

(b) In higher dimensions, there is the following analogous result (see [375]): let $D \subset \mathbb{C}^n$ be a bounded domain with a $\mathcal{C}^1$-smooth convex boundary point a. Assume that ∂D does not contain any germ of a complex line through a. Then, e.g., $\lim_{D \ni z \to a} \varkappa_D(z; \nu) = 1/2$, where ν is the outer unit normal vector in a at ∂D.

Moreover, Exercise 20.4.1 shows that the assumption of the non-existence of a non-trivial complex line through a is essential.

Example 20.3.3.

(a) Put

$$\mathbb{D} \ni \lambda \overset{f}{\longmapsto} 2\lambda + (1 - \lambda)\log(1 - \lambda), \quad \lambda \in \mathbb{D}.$$

Note that f is an injective holomorphic mapping and its image $D := f(\mathbb{D})$ is a simply connected bounded domain with a $\mathcal{C}^1$-smooth boundary (see [437], page 46) with $2 \in \partial D$. Then,

$$\boldsymbol{\ell}^*_D(0, w) = \boldsymbol{\ell}^*_{\mathbb{D}}(0, g(w)) = |g(w)|, \quad w \in D,$$

where $g := f^{-1}$. Moreover, by virtue of Lemma 20.3.1 we have

$$\lim_{D \ni w \to 2} |g'(w)| \frac{\operatorname{dist}(w, \partial D)}{\operatorname{dist}(g(w), \partial\mathbb{D})} = 1.$$

Therefore, one obtains

$$\lim_{\mathbb{R}\ni u\nearrow 2} \frac{1-\boldsymbol{\ell}_D^*(0,u)}{\operatorname{dist}(u,\partial D)} = \lim_{\mathbb{R}\ni u\nearrow 2} |g'(u)| = 0.$$

In particular, this limit behavior proves that there is no general estimate of the following type $\boldsymbol{\ell}_D^*(z,w) \le 1 - c\operatorname{dist}(z,\partial D)\operatorname{dist}(w,\partial D)$, $z,w \in D$, with a suitable positive constant c.

(b) (see [378]) On the other hand, for a convex domain $D \subset \mathbb{C}^n$, $D \ne \mathbb{C}^n$, we have the following universal upper estimate:

$$\boldsymbol{\ell}_D(z,w) \le \frac{\|z-w\|}{d_D(z)-d_D(w)} \log\frac{d_D(z)}{d_D(w)}, \qquad z,w \in D,\ z \ne w,$$

where $d_D(\zeta) := \operatorname{dist}(\zeta,\partial D)$ for $\zeta \in D$.

In particular, if D is, in addition, bounded, then for any compact $K \subset D$ there exist positive constants c_1, c_2 such that

$$\boldsymbol{\ell}_D(z,w) \le c_1 + c_2 \log\frac{1}{d_D(w)}, \qquad z \in K,\ w \in D.$$

The proof is left to the reader. (*Hint*: Use

$$C_{z,w} := \operatorname{conv}\big((z + \mathbb{C}(w-z)) \cap (\mathbb{B}(z,d_D(z)) \cup \mathbb{B}(w,d_D(w)))\big)$$

and the equality of $\boldsymbol{\ell}_{C_{z,w}}$ and $\boldsymbol{k}_{C_{z,w}}$.)

Now we deal with bounded domains with a $\mathcal{C}^{1+\varepsilon}$-smooth boundary. In that case, there is a uniform upper estimate for the Lempert function.

Theorem 20.3.4 (cf. [387]). *Let $D \subset \mathbb{C}^n$ be a bounded domain with a $\mathcal{C}^{1+\varepsilon}$-smooth boundary. Then, there exists a positive constant c such that*

$$\boldsymbol{\ell}_D^*(z,w) \le 1 - c\operatorname{dist}(z,\partial D)\operatorname{dist}(w,\partial D), \qquad z,w \in D.$$

Corollary 20.3.5. *Let D be as in Theorem* 20.3.4*. Then, there exists a constant c such that*

$$\boldsymbol{\ell}_D(z,w) \le -(1/2)\log\operatorname{dist}(z,\partial D) - (1/2)\log\operatorname{dist}(w,\partial D) + c, \qquad z,w \in D.$$

In particular, similar estimates for $\boldsymbol{k}_D$ (see the former chapter) are now a direct consequence of the above result.

A main tool in the proof of the former theorem is the following lemma:

Lemma 20.3.6. *Let $D \subset \mathbb{C}^n$ be a bounded domain with a $\mathcal{C}^1$-smooth boundary and let $a, b \in \partial D$. Then, there exists a polynomial mapping $\varphi : \mathbb{C} \longrightarrow \mathbb{C}^n$ such that*

$$\varphi\big((-1,1)\big) \subset D,\ \varphi(-1) = a,\ \varphi(1) = b,\ \varphi'(-1) = -\nu_a,\ \varphi'(1) = \nu_b,$$

where ν_ζ, $\zeta \in \partial D$, denotes an outer normal vector (not necessarily a unit vector) in ζ at ∂D (i.e., $\nu_\zeta/\|\nu_\zeta\|$ is the outer normal unit vector in ζ).

Proof. Let $U = U(\partial D)$ and $r : U \longrightarrow \mathbb{R}$ be a $\mathcal{C}^1$-smooth defining function of D with $1 \le \|\overline{\partial} r(z)\| \le M$, $z \in U$, where $\overline{\partial} r(z) = \big(\frac{\partial r}{\partial \overline{z}_1}(z), \dots, \frac{\partial r}{\partial \overline{z}_n}(z)\big)$. Moreover, put $\widetilde{\nu}_a = -2\overline{\partial} r(a)$ and $\nu_b := 2\overline{\partial} r(b)$. Note that $\widetilde{\nu}_a$ (respectively ν_b) is a vector parallel to the inner normal unit vector in a (respectively, the outer normal unit vector in b) at ∂D. Take a $\mathcal{C}^2$-smooth curve $\widetilde{\varphi} : [-1,1] \longrightarrow \mathbb{C}^n$ such that

$$\widetilde{\varphi}\big((-1,1)\big) \subset D,\ \widetilde{\varphi}(-1) = a,\ \widetilde{\varphi}(1) = b,\ \widetilde{\varphi}'(-1) = \widetilde{\nu}_a = -\nu_a,\ \widetilde{\varphi}'(1) = \nu_b,$$

Obviously, such a curve exists. Then, we are going to approximate this curve by polynomial curves, i.e., we choose polynomial mappings $\varphi_j : \mathbb{C} \longrightarrow \mathbb{C}^n$, $j \in \mathbb{N}$, such that

- $\max\{\|\varphi_j^{(k)}(t) - \widetilde{\varphi}^{(k)}(t)\| : k = 0, 1\} < \frac{1}{j}$ for all $t \in [-1,1]$,
- $\widetilde{\varphi}$ and φ_j coincide at ± 1 up to order 1,

(use Weierstrass approximation). We will now show that for a large j such a φ_j satisfies all properties claimed in the lemma.

First, choose positive numbers $j_1 \in \mathbb{N}$ and ε_1 such that $\widetilde{\varphi}([1-\delta_1, 1]) \subset U$ and $\varphi_j([1-\delta_1, 1]) \subset U$, $j \ge j_1$. Now fix a $t \le \delta_1$ and a $j \ge j_1$. Thus,

$$(r \circ \varphi_j)(1-t) = -2\int_{1-t}^{1} \operatorname{Re}\langle \overline{\partial} r(\varphi_j(s)), \varphi_j'(s)\rangle ds =: -\int_{1-t}^{1} A_j(s) ds.$$

Then, the integrand allows the following lower estimate:

$$\begin{aligned}
A_j(s) &\ge \|\nu_b\|^2 + \operatorname{Re}\langle 2\overline{\partial} r(\varphi_j(s)) - \nu_b, \nu_b\rangle + 2\operatorname{Re}\langle \overline{\partial} r(\varphi_j(s)), \varphi_j'(s) - \nu_b\rangle \\
&\ge \|\nu_b\|^2 - \|\nu_b\| \cdot \|2\overline{\partial} r(\varphi_j(s)) - \nu_b\| - 2\|\overline{\partial} r(\varphi_j(s))\| \cdot \|\varphi_j'(s) - \nu_b\| \\
&\ge 1 - M\|2\overline{\partial} r(\varphi_j(s)) - \nu_b\| - 2M\|\varphi_j'(s) - \nu_b\| \\
&\ge 1 - 2M\Big(\|2\overline{\partial} r(\varphi_j(s)) - 2\overline{\partial} r(\widetilde{\varphi}(s))\| + \|2\overline{\partial} r(\widetilde{\varphi}(s)) - 2\overline{\partial} r(\widetilde{\varphi}(1))\| \\
&\quad + \|2\overline{\partial} r(\widetilde{\varphi}(1)) - \nu_b\| + \|\varphi_j'(s) - \widetilde{\varphi}'(s)\| + \|\widetilde{\varphi}'(s) - \widetilde{\varphi}'(1)\| + \|\widetilde{\varphi}'(1) - \nu_b\|\Big) \\
&\ge 1 - 1/2, \quad t \in (0, \delta_1) \text{ and } j \ge j_1,
\end{aligned}$$

after we have substituted the former δ_1 (respectively, j_0) by a smaller (respectively larger) one. Thus, we obtain that $A_j(s) > 0$. Hence, $r(\varphi_j(1-t)) < 0$ whenever $t \le \delta_1$ and $j \ge j_1$.

In an analogous way, we may find a $\delta_2 \in (0, 1-\delta_1)$ and a $j_2 \in \mathbb{N}$ such that

$$r(\varphi_j(t)) < 0, \quad 0 < t < \delta_2, \ j \ge j_2.$$

Therefore, we get $\varphi_j((0,\delta_2) \cup (1-\delta_1, 1)) \subset D$, when $j \ge \max\{j_1, j_2\}$. Note that $\widetilde{\varphi}([\delta_2, 1-\delta_1])$ belongs to D and therefore, if j is sufficiently large, then $\varphi_j([\delta_2, 1-\delta_1]) \subset D$, which completes the proof. □

Remark 20.3.7. Observe that, in fact, it suffices to only assume that ∂D is $\mathcal{C}^1$-smooth near a and b.

The following geometric fact will be used in the estimate of invariant metrics that follows:

Lemma 20.3.8. *Let $D \subset \mathbb{C}$ with $0 \in \partial D$ be given as $D = \{z \in \mathbb{C} : r(z) < 0\}$, where $r \in \mathcal{C}^1(\mathbb{C})$, $2\frac{\partial r}{\partial \bar z}(0) = \frac{\partial r}{\partial x}(0) \le -\delta < 0$, and $\max\{|\frac{\partial r}{\partial x}(z)|, |\frac{\partial r}{\partial y}(z)|\} \le M|z|^\delta$ for all $z \in \mathbb{D}(R)$. Then, there exists an $\varepsilon = \varepsilon(\delta, M, R)$ such that*

$$D^i := \{z \in \mathbb{D}(2\varepsilon) : x > |y|^{1+\varepsilon}\} \subset D \text{ and } (-\overline{D^i}) \cap D = \varnothing.$$

Proof. Use Taylor expansion. □

Lemma 20.3.9. *Let $G_j := \{z \in \mathbb{C} : r_j(z) < 0\}$ be a simply connected bounded plane domain with $r_j \in \mathcal{C}^1(\mathbb{C})$, $j \in \mathbb{N}$. Assume that G is a bounded domain such that $G_j \longrightarrow G$ in the sense that for any open sets K, L with $K \subset\subset G \subset\subset L$ there exist a $j_0 \in \mathbb{N}$ with $K \subset G_j \subset L$, $j \ge j_0$. Moreover, let U be an open neighborhood of ∂G and M and δ positive numbers such that*

- $|\frac{\partial r_j}{\partial z}(z)| \ge \delta > 0$, $j \in \mathbb{N}$,
- $|\frac{\partial r_j}{\partial z}(z') - \frac{\partial r_j}{\partial z}(z'')| \le M|z'-z''|^\delta$, $z', z'' \in U$, $j \in \mathbb{N}$.

Then, there exists a positive constant C and a natural number $j_0 \in \mathbb{N}$ and a positive ε such that for any sequence of biholomorphic mapping $f_j : \mathbb{D} \longrightarrow G_j$, $j \ge j_0$, with $\operatorname{dist}(f_j(0), \partial G_j) \ge \varepsilon$ the following uniform estimate $|f_j'| \le C$ holds on $\mathbb{D}$.

Proof. To simplify our notation, we write, during the proof, $d_D(w) = \operatorname{dist}(w, \partial D)$ for any domain D and any point $w \in D$.

Let us start with the following observation: it suffices to find a j_0 and positive constants c_1 and c_2 such that

(a) $d_{G_j}(f_j(z))\varkappa_{G_j}(f_j(z); 1) \ge c_1$,

(b) $\dfrac{1-\boldsymbol{\ell}^*_{G_j}(f_j(0), f_j(z))}{d_{G_j}(f_j(z))} \ge c_2$,

for $j \ge j_0$, $z \in \mathbb{D}$.

Indeed, fix a $z \in \mathbb{D}$ and a $j \geq j_0$. Then, $1 - |z| = 1 - \boldsymbol{\ell}^*_{\mathbb{D}}(0, z) = 1 - \boldsymbol{\ell}^*_{G_j}(f_j(0), f_j(z))$. Then, the second estimate gives $\frac{d_{\mathbb{D}}(z)}{d_{G_j}(f_j(z))} \geq c_2$. Moreover,

$$\begin{aligned}\frac{1}{d_{\mathbb{D}}(z)} \geq \frac{1}{1-|z|^2} &= \boldsymbol{\varkappa}_{\mathbb{D}}(z; 1) = \boldsymbol{\varkappa}_{G_j}(f_j(z); f_j'(z)) \\ &= \frac{|f_j'(z)|}{d_{G_j}(f_j(z))} \boldsymbol{\varkappa}_{G_j}(f_j(z); 1) d_{G_j}(f_j(z)) \geq \frac{|f_j'(z)|}{d_{G_j}(f_j(z))} c_1.\end{aligned}$$

Therefore,

$$|f_j'(z)| = \frac{|f_j'(z)| d_{\mathbb{D}}(z)}{d_{G_j}(f_j(z))} \cdot \frac{d_{G_j}(f_j(z))}{d_{\mathbb{D}}(z)} \leq \frac{1}{c_1 c_2}.$$

So it remains to verify the above estimates. We start with a general observation for all the domain G_j, we are dealing with. Take a neighborhood $V \subset\subset U$ of ∂G. Put $K = G \setminus \overline{V}$ and $L := G \cup V$. Then, by assumption there exists an index j_0 such that $K \subset G_j \subset L$ whenever $j \geq j_0$. In particular, all the corresponding boundaries lie in U. Moreover, one may find a universal positive R such that $\mathbb{D}(z, R) \subset U$ for all $z \in \partial G_j$, $j \geq j_0$.

Now take such a J and a boundary point $a \in \partial G_j$. Put $\Phi_{j,a}(z) := (z - a)e^{i\theta_{j,a}}$, where the angle $\theta_{j,a}$ is chosen such that $\frac{\partial r_j}{\partial z}(a)e^{i\theta_{j,a}} < 0$. Put $G_{j,a} := \Phi_{j,a}(G_j)$. By virtue of the assumption, Lemma 20.3.8 implies that there is a positive $\varepsilon < 1$ such that

$$A := \{z \in \mathbb{D}(2\varepsilon) : x > |y|^{1+\varepsilon}\} \subset G_{j,a}, \ G_{j,a} \subset B, \quad a \in \partial G_j, \ j \geq j_0.$$

where $B := \mathbb{C} \setminus (-\overline{A})$. By rounding the two corners of A we may assume that A is the inner of a Jordan curve of class $\mathcal{C}^{1+\delta}$. By the Riemann mapping theorem, we may find a biholomorphic mapping $\psi : B \cup \{\infty\} \longrightarrow \mathbb{D}$ with $\psi(\infty) = 0$. Moreover, note that $G_j \subset \mathbb{D}(R)$, $j \geq j_0$, for a suitable R. Hence, $|\psi(w)| \geq r$ whenever $w \in G_j$, $j \geq j_0$.

Now we can start the verification of (a) and (b).

(a) Fix $j \geq j_0$ and a point $w \in G_j$ with $d_{G_j}(w) > \varepsilon$. Then, choose an $a \in \partial G_j$ with $|w - a| = d_{G_j}(w)$. Put $w_{j,a} := \Phi_{j,a}(w)$. Then,

$$\begin{aligned}\boldsymbol{\varkappa}_{G_j}(w; 1) = \boldsymbol{\varkappa}_{G_{j,a}}(w_{j,a}; 1) &\geq \boldsymbol{\varkappa}_B(w_{j,a}; 1) \geq \boldsymbol{\varkappa}_{\mathbb{D}}(\psi(w_{j,a}); \psi'(w_{j,a})) \\ &= \frac{|\psi'(w_{j,a})|}{1 - |\psi(w_{j,a})|^2} \geq \frac{|\psi'(w_{j,a})|}{2d_{\mathbb{D}}(\psi(w_{j,a}))}.\end{aligned}$$

Put $(\psi|_B)^{-1} =: \varphi$, i.e., $\varphi : \mathbb{D}^* \longrightarrow B$ is a biholomorphic mapping. Now define the mapping

$$\mathbb{D} \ni \lambda \overset{g}{\longmapsto} \frac{\varphi(\psi(w_a) + r\lambda d_{\mathbb{D}}(w_a)) - \varphi(\psi(w_a))}{\varphi'(\psi(w_a)) r d_{\mathbb{D}}(\psi(w_a))}.$$

Obviously, $g \in \mathcal{O}(\mathbb{D})$ and we have $g(0) = 0$ and $g'(0) = 1$. Thus, using Koebe's theorem one gets $\mathbb{D}(w_a, (1/4)|\varphi'(\psi(w_a))r|d_{\mathbb{D}}(\psi(w_a))) \subset B$; in particular,

$$d_B(w_a) \geq (1/4)|\varphi'(\psi(w_a))r|d_{\mathbb{D}}(\psi(w_a)).$$

Hence,

$$\varkappa_{G_j}(w;1) \geq \frac{r|\psi'(w_a)| \cdot |\varphi'(\psi(w_a))|}{8d_B(w_a)} = \frac{r}{8d_B(w_a)} = \frac{r}{8d_{G_j}(w)},$$

which gives the first estimate with $c_1 := r/8$.

(b) Again fix a $j \geq j_0$ and a $z \in G_j$. Let $a \in \partial G_j$ as above. Using the triangle equation for $\boldsymbol{\ell}_{G_j}$, we get $\boldsymbol{\ell}_{G_j}(z,w) \leq \boldsymbol{\ell}_{G_{j,a}}(z_{j,a},\varepsilon) + \boldsymbol{\ell}_{G_{j,a}}(\varepsilon, w_a)$ (G_j is simply bounded connected), where $w_a := \Phi_{j,a}(w)$. Thus,

$$(1 - \boldsymbol{\ell}^*_{G_{j,a}}(z_a,\varepsilon))(1 - \boldsymbol{\ell}^*_{G_{j,a}}(\varepsilon, w_a)) \leq 4(1 - \boldsymbol{\ell}_{G_j}(z,w)). \tag{20.3.1}$$

To continue, observe that there are domains $G' \subset\subset G'' \subset\subset G$ and an index $j_1 \geq j_0$ such that the following is true for any $j \geq j_1$:

- $G'' \subset G_j$,
- if $w \in G_j$ with $d_{G_j}(w) \geq \varepsilon$, then $w \in G'$,
- if $z \in G_j$ and $a \in \partial G_j$ with $d_{G_j}(z) = |z - a|$, then $\Phi_{j,a}(\varepsilon) \in G'$.

Now we assume that our $j \geq j_1$. Using the former information implies that in (20.3.1) we can continue with $1 - \boldsymbol{\ell}^*_{G_{j,a}}(\varepsilon, w_a) \geq 1 - \boldsymbol{\ell}^*_{G''}(\Phi^{-1}_{j,a}(\varepsilon), w) \geq 1 - \sup \boldsymbol{\ell}_{G''}(\zeta_1,\zeta_2) : \zeta_k \in G'\} \geq d_1 > 0$ for a suitable positive d_1. Thus,

$$1 - \boldsymbol{\ell}^*_{G_j}(z,w) \geq (1/4)d_1(1 - \boldsymbol{\ell}^*_{G_{j,a}}(z_{j,a},\varepsilon)).$$

In case that $d_{G_j}(z) \geq \varepsilon$ we have, as before, that $1 - \boldsymbol{\ell}^*_{G_j}(z,w) \geq (1/4)d_1^2 \geq d_2 d_{G_j}(z)$, where d_2 is a suitable positive number independent of j and z. So, let us assume that $d_{G_j}(z) < \varepsilon$. Then, $z_{j,a} \in A$ and

$$1 - \boldsymbol{\ell}^*_{G_{j,a}}(z_{j,a},\varepsilon) \geq 1 - \boldsymbol{\ell}^*_A(z_{j,a},\varepsilon) = 1 - |\psi(z_{j,a})| = d_{\mathbb{D}}(\psi(z_{j,a})),$$

where $\psi : A \longrightarrow \mathbb{D}$ is a conformal mapping sending ε to 0. Using Koebe's theorem and the Kellogg–Warschawski theorem (i.e., $|\psi'| > M > 0$ on A), one concludes that the last term can be estimated from below by $(1/4)Md_A(z_{j,a})$; thus, $1 - \boldsymbol{\ell}^*_{G_j}(z,w) \geq (1/16)Md_1 d_A(z_{j,a})$. The last step in this estimate uses the fact that a part of the cone $\{z \in \mathbb{C} : |y| < x\}$ near the origin is contained in A. Thus, there exists a positive constant d_3 such that $d_A(z_{j,a}) \geq d_3|z_{j,a}| \geq d_3 d_{G_j}(z)$. Hence, (b) has been verified. □

Proof of Theorem 20.3.4. Before starting with the proof, we should emphasize that because of the lack of a triangle inequality for the Lempert function, there is no chance to prove the above result via localization as in the discussion on the Kobayashi distance. Nevertheless, the idea of how to prove this result seems to be clear: join two points (possibly on the boundary of D) by a suitable real analytic curve and perturb a holomorphic extension of this curve in order to cover neighborhoods of these two points.

By virtue of a usual compactness argument it suffices to verify only the following two statements:

(a) for any two points $a, b \in \partial D$ there exist neighborhoods $U_{a,b}$ and $V_{a,b}$ of a and b, respectively, and a positive constant $c_{a,b}$ such that

$$\ell_D^*(z,w) \le 1 - c_{a,b} d(z) d(w), \quad z \in U_{a,b},\ w \in V_{a,b},$$

where we always write $d(\xi) = \operatorname{dist}(\xi, \partial D)$, $\xi \in D$.

(b) for any point $a \in \partial D$ and for any compact set $K \subset D$ there exist a neighborhood $U_{a,K}$ of a and a positive constant $c_{a,K}$ such that

$$\ell_D^*(z,w) \le 1 - c_{a,K} d(z) d(w), \quad z \in U_{a,K},\ w \in K.$$

Let us repeat that, by assumption, there are an open neighborhood U of ∂D and a $\mathcal{C}^{1+\varepsilon}$-function $r : \mathbb{C}^n \longrightarrow \mathbb{R}$ satisfying the following properties:

- $D = \{z \in \mathbb{C}^n : r(z) < 0\}$, $\mathbb{C}^n \setminus \overline{D} = \{z \in \mathbb{C}^n : r(z) > 0\}$;
- $\|\partial r\| \ge 1$ on U, where $\partial r(z) := \big(\frac{\partial r}{\partial z_1}(z), \dots, \frac{\partial r}{\partial z_n}(z)\big)$.

Proof of (a): Fix points $a, b \in \partial D$. Then, by virtue of Lemma 20.3.6, we find a polynomial mapping $\varphi : \mathbb{C} \to \mathbb{C}^n$, which satisfies the following properties:

$$\varphi\big((-1,1)\big) \subset D,\ \varphi(-1) = a,\ \varphi(1) = b,\ \varphi'(-1) = -2\overline{\partial} r(a),\ \varphi'(1) = 2\overline{\partial} r(b).$$

Moreover, take orthonormal bases $E_2, \dots, E_n \in \mathbb{C}^n$ of $T_a^{\mathbb{C}} \partial D$ and $D_2, \dots, D_n \in \mathbb{C}^n$ of $T_b^{\mathbb{C}} \partial D$. Define $\Phi : \mathbb{C} \times \mathbb{C}^{n-1} \times \mathbb{C} \times \mathbb{C}^{n-1} \longrightarrow \mathbb{C}^n \times \mathbb{C}^n$ via

$$\Phi(\zeta_1, \xi, \zeta_2, \eta) := (\varphi_{\xi,\eta}(\zeta_1), \varphi_{\xi,\eta}(\zeta_2)),$$

where

$$\varphi_{\xi,\eta}(\zeta) := \varphi(\zeta) + \Big(\frac{\zeta - 1}{2}\Big)^2 \sum_{j=2}^n \xi_j E_j + \Big(\frac{\zeta + 1}{2}\Big)^2 \sum_{j=2}^n \eta_j D_j,$$

$\xi = (\xi_2, \dots, \xi_n) \in \mathbb{C}^{n-1}$, $\eta = (\eta_2, \dots, \eta_n) \in \mathbb{C}^{n-1}$, and $\zeta \in \mathbb{C}$. Note that $\Phi(-1,0,1,0) = (a,b)$ and $\det \Phi'(-1,0,1,0) \ne 0$. Thus, there exist open neighborhoods $V = V_1(-1) \times V_2(0) \times V_3(1) \times V_4(0)$ and $W = W(a,b)$ such that

$\Phi|_V : V \longrightarrow W$ is a biholomorphic map. Set $\Psi := (\Phi|_V)^{-1}$. Write $\Psi(z, w) = (\zeta^-(z, w), \xi(z, w), \zeta^+(z, w), \eta(z, w))$ for $(z, w) \in W$.

For $(z, w) \in W$ define the holomorphic mapping $\psi_{z,w} := \varphi_{\xi(z,w),\eta(z,w)} : \mathbb{C} \longrightarrow \mathbb{C}^n$. Note that $\psi_{a,b} = \varphi$. Put $G_{z,w} := \psi_{z,w}^{-1}(D)$. Then, $G_{z,w}$ is an open subset of $\mathbb{C}^n$. Moreover, let $r_{z,w} := r \circ \psi_{z,w}$. So $r_{z,w} \in \mathcal{C}^{1+\varepsilon}(\mathbb{C})$.

Then, we have the following properties for a point $\zeta \in \mathbb{C}$:

- $\zeta \in G_{z,w} \iff \psi_{z,w}(\zeta) \in D \iff r_{z,w}(\zeta) < 0$,
- $\zeta \in \partial G_{z,w} \Longrightarrow \psi_{z,w}(\zeta) \in \partial D \iff r_{z,w}(\zeta) = 0$,
- $r_{z,w}(\zeta) > 0 \Longrightarrow \zeta \notin \overline{G}_{z,w}$,
- $(-1, 1) \subset G_{a,b}$ and $\pm 1 \in \partial G_{a,b}$.

Recall that $r_{a,b}(\pm 1) = 0$, $r_{a,b} < 0$ on $(-1, 1)$, $2\overline{\partial} r_{a,b}(-1) = -4\|\overline{\partial} r(a)\|^2 =: -t_1 < 0$, and $2\overline{\partial} r_{a,b}(1) = 4\|\overline{\partial} r(b)\|^2 =: t_2 > 0$. In particular, $\frac{\partial r_{a,b}}{\partial s}(-1) = -t_1 < 0$, $\frac{\partial r_{a,b}}{\partial s}(1) = t_2 > 0$, and $\frac{\partial r_{a,b}}{\partial t}(\pm 1) = 0$, where we will always write $\zeta = s + it$.

Therefore, by virtue of the implicit function theorem, there exist positive $\delta_1 < 1/2$, δ_2, a neighborhood $W' = W'(a, b) \subset W$, and a $\mathcal{C}^{+\varepsilon}1$-function $\tau^\pm : (-\delta_2, \delta_2) \times W' \longrightarrow (\pm 1 - \delta_1, \pm 1 + \delta_1) =: I^\pm$ such that

- $(\pm 1) r_{z,w}(\cdot + it)$ is strictly monotonically increasing on $(\pm 1 - \delta_1, \pm 1 + \delta_1)$, $t \in (-\delta_2, \delta_2)$,
- $\tau^\pm(0, z, w) = \pm 1$,
- $r_{z,w}(\tau^\pm(t, z, w) + it) = 0$ for all $(z, w) \in W'$.

Hence, if $(z, w) \in W'$, then

$$M_{z,w}^\pm := \{\zeta = s + it \in I^\pm + i(-\delta_2, \delta_2) : (\pm 1)s < (\pm 1)\tau^\pm(t, z, w)\} \subset G_{z,w}.$$

Now we may choose a positive $\delta_2' < \delta_2$ and a perhaps smaller neighborhood of (a, b) again denoted by W' satisfying that the strip $S := [-1 + \delta_1/2, 1 - \delta_1/2] + i(-\delta_2', \delta_2') \subset G_{z,w}$ for all $(z, w) \in W'$. For $(z, w) \in W'$, put

$$\widehat{G}_{z,w} := S \cup \left(M_{z,w}^\pm \cap \{\zeta = s + it \in \mathbb{C} : |t| < \delta_2'\}\right), \quad (z, w) \in W'.$$

Note that $\widehat{G}_{z,w} = \{\zeta = s + it \in \mathbb{C} : |t| < \delta_2',\ \tau_{z,w}^-(t) < s < \tau_{z,w}^+(t)\}$, where $\tau_{z,w}^\pm := \tau^\pm(\cdot, z, w)$. Obviously, $\widehat{G}_{z,w}$ is a simply connected bounded domain, which has a $\mathcal{C}^{1+\varepsilon}$-smooth boundary except at the four boundary points

$$\tau_{z,w}^+(\delta_2') + i\delta_2',\ \tau_{z,w}^+(-\delta_2') - i\delta_2',\ \tau_{z,w}^-(\delta_2') + i\delta_2',\ \tau_{z,w}^-(-\delta_2') - i\delta_2'.$$

Moreover, we have $\widehat{G}_{z,w} \longrightarrow \widehat{G}_{a,b}$ in the sense of Lemma 20.3.9.

Fix a δ_2'' with $3\delta_2'/4 < \delta_2'' < \delta_2'$. Then, choose a $\mathcal{C}^\infty$ function $\chi : \mathbb{R} \longrightarrow [0,1]$ satisfying $\chi|_{[-3\delta_2'/4, 3\delta_2'/4]} \equiv 1$ and $\chi = 0$ on $(-\infty, -\delta_2''] \cup [\delta_2'', \infty)$. For $(z,w) \in W'$, define the following new simply connected domain:

$$\widetilde{G}_{z,w} := \{\zeta \in \mathbb{R} + i(-\delta_2', \delta_2') : \\ \chi(t)\tau_{z,w}^-(t) + (1-\chi(t))r < s < \chi(t)\tau_{z,w}^+(t) + (1-\chi(t))r\},$$

where $\zeta = s + it$. Note that all these new domains coincide in the rectangle $\{\zeta = s + it \in \mathbb{C} : |s| < r,\ |y| < \delta_2'\}$, have a $\mathcal{C}^{1+\varepsilon}$-boundary outside of the four corners $\pm r \pm i\delta_2'$, and are contained in $G_{z,w}$. Near these points, the domains are bounded by two orthogonal lines. In a last step, we may round these corners by inscribing four small discs, which touch the lines in an orthogonal way. What we get are $\mathcal{C}^{1+\varepsilon}$-smooth simply connected domains $G_{z,w}^*$, which coincide with the starting domains $\widehat{G}_{z,w}$ in the strip $\{\zeta = s + it : |t| < 3\delta_2'/4\}$ and for which we have that $G_{z,w}^* \underset{(z,w)\to(a,b)}{\longrightarrow} G_{a,b}^*$ in the sense of Lemma 20.3.9.

So we may assume that $G_{z,w}^* = \{\zeta \in \mathbb{C} : r_{z,w}^* < 0\}$ with suitable $\mathcal{C}^1$-functions $r_{z,w}^*$ satisfying the assumptions of Lemma 20.3.9 (observe that the δ in the Lemma is now the ε).

Now recall that if $(z,w) \in D \times D$ is sufficiently near (a,b), i.e., $(z,w) \in W'' = W''(a,b) \subset W'$, then $\zeta^\pm(z,w) \in \mathbb{D}(\pm 1, \delta/2)$ and therefore, $\zeta^\pm(z,w) \in \widetilde{G}_{z,w}$. For these z, w take a biholomorphic mapping $\eta_{z,w} : \mathbb{D} \longrightarrow G_{z,w}^*$ with $\eta_{z,w}(0) = 0$ and $\eta_{z,w}(p_{z,w}^\pm) = \zeta^\pm(z,w)$, where $p_{z,w}^\pm \in \mathbb{D}$ is different from zero. Because of the good boundary of $G_{z,w}^*$, the conformal mapping $\eta_{z,w}$ extends to a diffeomorphism between the closures of the corresponding domains. Put $q_{z,w}^\pm := p_{z,w}^\pm / |p_{z,w}^\pm|$. We claim that there exists a smaller neighborhood of (a,b), again denoted by W'', such that $|p_{z,w}^\pm - q_{z,w}^\pm| < \delta/2$, $(z,w) \in W'' \cap D$. Indeed, suppose this is not true. Then, one can find a sequence $(z_j, w_j) \in D$ with $(z_j, w_j) \longrightarrow (a,b)$ such that with $p_j := p_{z_j,w_j}^+$ and $q_j := q_{z_j,w_j}^+$ one has that $|p_j - q_j| \geq \delta/2$, i.e., $|p_j| < 1 - \delta/2 =: R < 1$. Then, we may assume that $p_j \longrightarrow p'$ and $\eta_j \longrightarrow \eta \in \mathcal{O}(\mathbb{D}, \mathbb{C})$ (use Montel). Note that $\eta(0) = 0$ and $\eta(p') = 1$ (use that $\eta_j(p_j) = \zeta^+(z_j, w_j) + \longrightarrow 1$) meaning that η is not identically constant. Thus, $\mathbb{D}(1,r) \subset \eta(\mathbb{D})$ and hence, $\mathbb{D}(1, r/2) \subset \eta_j(\mathbb{D})$ for large j; a contradiction. The situation near the point -1 is similar.

Then, using Lemma 20.3.9 and the former claim, one obtains

$$|\eta_{z,w}(q_{z,w}^\pm) - (\pm 1)| \leq C|q_{z,w}^\pm - p_{z,w}^\pm| + \delta/2 < \delta;$$

thus, $\eta_{z,w}(q_{z,w}^\pm) \in \partial G_{z,w}$. Finally, put $\theta_{z,w} = \psi_{z,w} \circ \eta_{z,w} : \overline{\mathbb{D}} \longrightarrow \overline{D}$. Note that $\theta_{z,w}$ is a holomorphic disc into D and that $\theta_{z,w}(q_{z,w}^\pm) \in \partial D$. Fix neighborhoods $W''(a)$ and $W''(b)$ such that $W''(a) \times W''(b) \subset W''$. Then, we conclude that

$$d_D(z) \leq |\theta_{z,w}(q_{z,w}^-) - \theta_{z,w}(p_{z,w}^+)| \leq C' d_{\mathbb{D}}(p_{z,w}^-), \quad z \in W''(a) \cap D,$$

and similarly, $d_D(w) \le C' d_{\mathbb{D}}(p^+_{z,w})$ for $w \in W''(b) \cap D$. Hence, we end up with

$$1 - \ell^*_D(z,w) \ge 1 - \left| \frac{p^+_{z,w} - p^-_{z,w}}{1 - p^+_{z,w}\overline{p^-_{z,w}}} \right| \ge \frac{d_{\mathbb{D}}(p^+_{z,w}) d_{\mathbb{D}}(p^-_{z,w})}{4} \ge \frac{d_D(z) d_D(w)}{4C'^2},$$

which ends the proof of (a).

The proof of (b) is left to the reader. □

Remark 20.3.10.

(a) For an arbitrary convex bounded domain $D \subset \mathbb{C}^n$, the following lower/upper estimates of the Kobayashi distance are known (see [361]): for a $z_0 \in D$ there exist positive constants C_1, C_2, and $\alpha \ge 1$ such that

$$C_1 - \tfrac{1}{2}\log(\operatorname{dist}(\cdot, \partial D)) \le \boldsymbol{k}_D(z_0, \cdot) \le C_2 - \frac{\alpha}{2}\log(\operatorname{dist}(\cdot, \partial D))$$

holds on D. In general, α has to be larger than 1. Indeed, let $D_\alpha := f_\alpha(\mathbb{D})$, $\alpha > 1$, where $f_\alpha(\lambda) := (1+\lambda)^{1/\alpha}$, $\lambda \in \mathbb{D}$. Fix $z_0 := 1 \in D_\alpha$. Looking at $\boldsymbol{k}_{D_\alpha}(1, f_\alpha(-t))$, $0 < t \nearrow +1$, leads to such an example.

(b) In case of a bounded $\mathbb{C}$-convex domain $D \subset \mathbb{C}^n$ one even has the following lower estimate for the Carathéodory distance (see [378]):

$$\boldsymbol{c}_D(z,w) \ge \frac{1}{4}\log\left(\frac{\operatorname{dist}(z,\partial D)}{\operatorname{dist}(w,\partial D)}\right), \quad z, w \in D.$$

In particular, for any compact set $K \subset D$ there exists a positive constant c such that $\boldsymbol{c}_D(z,w) \ge c - \log(\operatorname{dist}(w,\partial D))$, $\quad z \in K,\ w \in D$.

Note that the factor $1/4$ is optimal, as the domain $D := \mathbb{C} \setminus [0, +\infty)$ may show.

(c) Recently, the boundary behavior of $\boldsymbol{k}_D$ for Reinhardt domains has been studied in [526].

Let $D \subset \mathbb{C}^n$ be an arbitrary pseudoconvex Reinhardt domain and $\zeta \in \partial D$. Then, for an $a \in D$ there exists a positive constant c such that

(i) $\boldsymbol{k}_D(a,z) \le -\log\operatorname{dist}(z,\partial D) + c, \quad z \in D,\ z$ near ζ;

(ii) if $\zeta \in \mathbb{C}^n_*$, then we even have

$$\boldsymbol{k}_D(a,z) \le -(1/2)\log\operatorname{dist}(z,\partial D) + c, \quad z \in D,\ z \text{ near } \zeta;$$

(iii) if $\zeta \in (\partial D) \cap \operatorname{int}\overline{D}$, then

$$\boldsymbol{k}_D(a,z) \le (1/2)\log(-\log\operatorname{dist}(z,\partial D)) + c, \quad z \in D,\ z \text{ near } \zeta.$$

Moreover, in the case where D has, in addition, a $\mathcal{C}^1$-smooth boundary, then

$$\boldsymbol{k}_D(a,z) \ge (1/2)\log(-\log\operatorname{dist}(z,\partial D)) + c,$$

when $z \in D$ tends non-tangentially to ζ.

20.4 Exercises

Exercise 20.4.1 (cf. [375]). Put

$$D := \mathbb{D}^2 \setminus \{(z_1, z_2) \in \mathbb{D}^2 : \operatorname{Re} z_1 \le 0,\ |z_2| \le 1/4\}.$$

Show that $\limsup_{D \ni z \to 0} \varkappa_D(z; (-1, 0)) \operatorname{dist}(z, \partial D) \le 3/8$.

Exercise 20.4.2. In the proof of Theorem 20.3.4, we had to deal with the boundary regularity of conformal mappings from $\mathbb{D}$ onto a simply connected domain with a good boundary. Therefore, this exercise makes sense in this place. Let $D := \mathbb{D} \cap \{z \in \mathbb{C} : \operatorname{Im} z > 0\}$. Prove that

$$\varphi(z) = \frac{1 + i\left(\frac{i(1-z)}{1+z}\right)^{1/2}}{1 - i\left(\frac{i(1-z)}{1+z}\right)^{1/2}}, \quad z \in \mathbb{D},$$

is a biholomorphic mapping from $\mathbb{D}$ onto D. Decide whether φ has a $\mathcal{C}^1$-extension at the points ± 1.

20.5 List of problems

Appendix A
Miscellanea

As we have already mentioned in the Preface, we here collect various topics that belong to the theory, but which are somehow outside the main scope of the book. We report (without proofs) on the following topics:

A.1: Carathéodory balls,

A.2: Lie structure of $\mathrm{Aut}(G)$,

A.3: symmetrized ellipsoids,

A.4: holomorphic curvature,

A.5: fixed points of holomorphic mappings and boundary regularity of complex geodesics,

A.6: criteria for biholomorphicity,

A.7: Carathéodory and Kobayashi isometries,

A.8: boundary behavior of contractible metrics in weakly pseudoconvex domains,

A.9: a spectral ball.

A.1 Carathéodory balls

Consider the following general problem: given a bounded convex balanced domain $G \subset \mathbb{C}^n$ $(n \geq 2)$ with Minkowski function $\mathfrak{h}_G$,[1] find conditions on $a, b \in G$ and $r, R \in (0,1)$ under which the Carathéodory ball

$$B_{\boldsymbol{m}_G}(a,r) := \{z \in G : \boldsymbol{c}_G^*(a,z) < r\}\ {}^{2}$$

coincides with the norm ball $B_{\mathfrak{h}_G}(b,R) := \{z \in \mathbb{C}^n : \mathfrak{h}_G(z-b) < R\}$. Since $\boldsymbol{m}_G(0,\cdot) = \mathfrak{h}_G(\cdot)$, we always have

$$B_{\boldsymbol{m}_G}(0,r) = B_{\mathfrak{h}_G}(0,r), \quad r \in (0,1).$$

---

[1] Notice that under our assumptions $\mathfrak{h}_G$ is a complex norm.

[2] Recall that in the case of the unit disc we have

$$B_{\boldsymbol{m}_{\mathbb{D}}}(a,r) = \mathbb{B}\left(\frac{a(1-r^2)}{1-r^2|a|^2}, \frac{r(1-|a|^2)}{1-r^2|a|^2}\right), \quad a \in \mathbb{D},\ r \in (0,1).$$

In the case where

$$G = \mathcal{E}(p,\alpha) := \Big\{(z_1,\dots,z_n) \in \mathbb{C}^n : 2\alpha|z_1|^{p_1}|z_2|^{p_2} + \sum_{j=1}^{n} |z_j|^{2p_j} < 1\Big\},$$
$$p = (p_1,\dots,p_n) \in \mathbb{R}^n_{>0},\ \alpha \ge 0,$$

the problem was studied in the following cases:

$n = 2, \alpha = 0, p_1 = p_2 = 1$	[466]
$\alpha = 0, p_1 = \dots = p_n = 1$	[486, 556]
$\alpha = 0, 1 < p_1 = \dots = p_n \notin \mathbb{N}$	[467]
$\alpha = 0$	[560, 565]
the general case	[521]

Observe that $\mathcal{E}(p,0) = \mathcal{E}(p)$. The methods introduced by W. Zwonek and developed by B. Visintin are based on the complex geodesic (see Chapter 11). The most general result is the following one:

Theorem A.1.1 (cf. [521]). *Assume that $\alpha \ge 0$ and $p \in \mathbb{R}^n_{>0}$ are such that $\mathcal{E}(p,\alpha)$ is convex. Then, the following conditions are equivalent:*

(i) *there exist $a, b \in \mathcal{E}(p,\alpha)$, $a \ne 0$, $r, R \in (0,1)$ such that*

$$B_{\boldsymbol{m}_{\mathcal{E}(p,\alpha)}}(a,r) = B_{\mathfrak{h}_{\mathcal{E}(p,\alpha)}}(b,R);$$

(ii) *$\alpha = 0$, $\{j \in \{1,\dots,n\} : a_j \ne 0\} = \{j_0\}$, $p_{j_0} = 1$, and $p_j = 1/2$ for all $j \ne j_0$.*

A.2 The automorphism group of bounded domains

Let G be a bounded domain in $\mathbb{C}^n$. Then the Carathéodory distance $\boldsymbol{c}_G$ defines the standard topology of G. If the automorphism group $\operatorname{Aut}(G)$ of G is provided with the compact-open topology, then $\operatorname{Aut}(G)$ carries the structure of a real Lie group. This result is due to H. Cartan (cf. [86]).

Here, we discuss a method of proof that uses invariant distances (cf. [317]). We define

$$\operatorname{Iso}_{\boldsymbol{c}}(G) := \{\Phi : G \longrightarrow G : \Phi \text{ is a bijective } \boldsymbol{c}_G\text{-isometry}\}.$$

Then, according to a result of van Dantzig and van der Waerden (cf. [506]), $\operatorname{Iso}_{\boldsymbol{c}}(G)$ provided with the compact-open topology is a locally compact group.

Moreover, $\operatorname{Aut}(G)$ is an effective transformation group of G, i.e.,

$$\Phi : \operatorname{Aut}(G) \times G \longrightarrow G,\ \Phi(g,z) := g(z), \quad g \in \operatorname{Aut}(G),\ z \in G$$

is continuous with

(i) $\Phi(g_2, \Phi(g_1, z)) = \Phi(g_2 \circ g_1, z), \quad g_j \in \operatorname{Aut}(G), z \in G;$

(ii) $\Phi(\operatorname{id}_G, z) = z, \quad z \in G;$

(iii) id_G is the only group element satisfying (ii).

Obviously, $\Phi(g,\cdot)$ describes a $\mathcal{C}^2$-transformation of G. Therefore, we can apply the following theorem of S. Bochner and D. Montgomery (cf. [71, 367]):

Theorem A.2.1. *Let $\mathcal{G}$ be a locally compact effective transformation group of a connected $\mathcal{C}^1$-manifold M (i.e., $\Phi : \mathcal{G} \times M \longrightarrow M$ as above) and let each transformation $\Phi(g,\cdot) : M \longrightarrow M$ be of class $\mathcal{C}^1$. Then, $\mathcal{G}$ is a real Lie group.*

Summarizing, we obtain

Theorem A.2.2. *The automorphism group $\operatorname{Aut}(G)$ of any bounded domain G in $\mathbb{C}^n$ is a real Lie group.*

We only mention that the Lie-structure of $\operatorname{Aut}(G)$ is often used to study the biholomorphical equivalence problem for classes of domains in $\mathbb{C}^n$.

A.3 Symmetrized ellipsoids

For $p > 0$, $n \geq 2$ define

$$\mathbb{B}_{p,n} := \{(z_1, \dots, z_n) \in \mathbb{C}^n : |z_1|^{2p} + \dots + |z_n|^{2p} < 1\}.$$

The set $\mathbb{E}_{p,n} := \pi_n(\mathbb{B}_{p,n})$ is called the *symmetrized (p,n)-ellipsoid*, where $\pi_n : \mathbb{C}^n \longrightarrow \mathbb{C}^n$ has been defined in § 7.2. Moreover, put

$$\Delta_{p,n} := \{(\lambda, \dots, \lambda) \in \mathbb{C}^n : |\lambda| < n^{-1/(2p)}\}, \quad \Sigma_{p,n} := \pi_n(\Delta_{p,n}).$$

The mapping $\pi_n : \mathbb{B}_{p,n} \longrightarrow \mathbb{E}_{p,n}$ is proper (with multiplicity $n!$) and $\pi_n : \mathbb{B}_{p,n} \setminus \Delta_{p,n} \longrightarrow \mathbb{E}_{p,n} \setminus \Sigma_{p,n}$ is a holomorphic covering.

The geometry of the domains $\mathbb{E}_{p,n}$ is only partially understood – the following particular properties have been proved in [548] and [550]:

Proposition A.3.1.

(a) *If $p > 1/2$, $n \geq 3$, and $\log_{n(n-1)} n < p$, then $\mathbb{E}_{p,n}$ is not $\mathbb{C}$-convex. In particular, $\mathbb{E}_{p,n}$ is not $\mathbb{C}$-convex for any $p > \log_6 3$ and $n \geq 3$.*

(b) *For any* $p \in (0, 1/2) \cup (1, +\infty)$, *the domain* $\mathbb{E}_{p,n}$ *is not convex.*

(c) *For any* $p > \log_3 \frac{3}{2}$ *and* $n \geq 3$, *the domain* $\mathbb{E}_{p,n}$ *is not convex.*

(d) *The domains* $\mathbb{E}_{1,2}$ *and* $\mathbb{E}_{1/2,2}$ *are convex.*

(e) *The domain* $\mathbb{E}_{3/2,2}$ *is starlike with respect to the origin.*

(f) *If* $\mathbb{E}_{p,2}$ *is starlike with respect to the origin, then so is* $\mathbb{E}_{\frac{p}{2},2}$.

(g) *For* $p > 3/2$ *the domain* $\mathbb{E}_{p,n}$ *is not starlike with respect to the origin.*

(h) *For* $p > 1$ *the domain* $\mathbb{E}_{p,2}$ *cannot be exhausted by domains biholomorphic to convex domains.*

(i) $\mathbb{E}_{1,2}$ *is the Lu Qi-Keng domain.*

Proper holomorphic mappings $\mathbb{E}_{p,n} \longrightarrow \mathbb{E}_{q,n}$ have been completely characterized in [549]. We need the following two auxiliary definitions:

For every $s \in \mathbb{N}$ we denote by P_s a unique polynomial mapping $\mathbb{C}^n \longrightarrow \mathbb{C}^n$ such that $\pi_n(z_1^s, \dots, z_n^s) = P_s(\pi_n(z_1, \dots, z_n))$. Moreover, for any $L = (L_1, \dots, L_n)$, with $L_j(z) = a \sum_{k=1}^n z_k + b z_j + c$, $j = 1, \dots, n$ $(a, b, c \in \mathbb{C})$, we denote by S_L a unique polynomial mapping $\mathbb{C}^n \longrightarrow \mathbb{C}^n$ such that

$$\pi_n \circ L = S_L \circ \pi_n. \qquad (\dagger)$$

Theorem A.3.2.

(a) *Proper mappings* $\mathbb{E}_{p,n} \longrightarrow \mathbb{E}_{q,n}$ *exist iff* $p/q \in \mathbb{N}$.

(b) *If* $s := p/q \in \mathbb{N}$, *then the only proper mappings* $F : \mathbb{E}_{p,n} \longrightarrow \mathbb{E}_{q,n}$ *are of the form*

(1) *if* $p \neq 1$, *or* $q \notin \{1/2m : m \in \mathbb{N}\}$, *or* $n \neq 2$, *then* $F = P_s \circ \Phi_{p,n}$, *where* $\Phi_{p,n} \in \operatorname{Aut}(\mathbb{E}_{p,n})$;

(2) *if* $p = 1$, $q = 1/(2m)$, *and* $n = 2$, *then* $F = P_{2m} \circ \Phi_{1,2}$ *or* $F = P_m \circ \Phi_{1,2} \circ P_2 \circ \Phi_{1/2,2}$, *where* $\Phi_{1,2} \in \operatorname{Aut}(\mathbb{E}_{1,2})$, $\Phi_{1/2,2} \in \operatorname{Aut}(\mathbb{E}_{1/2,2})$.

(c) *Any proper mapping* $\mathbb{E}_{p,n} \longrightarrow \mathbb{E}_{p,n}$ *is an automorphism.*

(d) *The only automorphisms* $F : \mathbb{E}_{p,n} \longrightarrow \mathbb{E}_{p,n}$ *are of the form*

(1) *if* $p \neq 1$ *and* $(p, n) \neq (1/2, 2)$, *then there exists a* $\zeta \in \mathbb{T}$ *such that* $F(z) = \Psi_\zeta(z) := (\zeta z_1, \dots, \zeta^n z_n)$;

(2) *if* $p = 1$, *then there exist* $\zeta_1, \zeta_2 \in \mathbb{T}$, *and* $a_0 \in \mathbb{R}$ *with* $a_0^2 < 1/n$ *such that*

$$F(z) = \left(\frac{S_1(z)}{n(1 - a_0 z_1)}, \dots, \frac{S_n(z)}{n^n (1 - a_0 z_1)^n} \right),$$

where $(S_1, \dots, S_n) := S_L$ *is defined by* (†) *with*

$$L_j(z) := \zeta_1\Big(\sum_{k=1}^{n} z_k - na_0\Big) + \zeta_2\sqrt{1-na_0^2}\Big(\sum_{k=1}^{n} z_k - nz_j\Big),$$

$$j = 1, \dots, n;$$

(3) *if* $(p, n) = (1/2, 2)$, *then there exists a* $\zeta \in \mathbb{T}$ *such that* $F = \Psi_\zeta$ *or* $F(z) = (\zeta z_1, \zeta^2(\frac{1}{4}z_1^2 - z_2))$.

Remark A.3.3. In [551] there are other results on proper holomorphic mappings between complex ellipsoids and generalized Hartogs domains.

A.4 Holomorphic curvature

A thorough study of the Ahlfors–Schwarz lemma in Chapter 1 leads us to the conclusion that this result can be thought of as one from differential geometry, where the notion of curvature is mainly involved. To be able to deal with metrics that are only upper semicontinuous, we first introduce the notion of the generalized (lower) Laplacian. Let $G \subset \mathbb{C}$ be an open set and let $u : G \longrightarrow \mathbb{R}$ be an upper semicontinuous function. Then,

$$(\Delta u)(\lambda) := 4 \liminf_{r\to 0} \frac{1}{r^2}\left(\frac{1}{2\pi}\int_0^{2\pi} u(\lambda + re^{i\theta})d\theta - u(\lambda)\right) \in [-\infty, \infty]$$

is called the *generalized Laplacian* of u.

If u is of class $\mathcal{C}^2$, then Δu coincides with the standard Laplace operator. Moreover, if a general u takes a local maximum at $\lambda_0 \in G$, then, as usual, $(\Delta u)(\lambda_0) \le 0$.

By means of this generalized Laplacian, we introduce the notion of holomorphic curvature (cf. [493, 533]). Namely, let G be a domain in $\mathbb{C}^n$ and let $\delta_G : G \times \mathbb{C}^n \longrightarrow [0, \infty)$ denote an upper semicontinuous pseudometric on G. Then the *holomorphic curvature of* δ_G *at* $(z_0; X) \in G \times (\mathbb{C}^n)_*$ with $\delta_G(z_0; X) > 0$ is defined as the number

$$\begin{aligned}&\text{h-curv}(z_0; X; \delta_G)\\ &:= \sup\left\{\frac{(\Delta \log \delta_G^2(\varphi;\varphi'))(0)}{-2\delta_G^2(z_0; X)} : r > 0,\ \varphi \in \mathcal{O}(r\mathbb{D}, G),\ \varphi(0) = z_0,\ \varphi'(0) = X\right\}.\end{aligned}$$

Obviously, if $\delta_G(z_0; X) > 0$, then

$$\text{h-curv}(z_0; X; \delta_G) = \text{h-curv}(z_0; X/\|X\|e^{i\theta}; \delta_G), \quad \theta \in \mathbb{R},$$

i.e., the holomorphic curvature only depends on the complex direction of X. Moreover, the holomorphic curvature is a biholomorphic invariant when it is considered with respect to holomorphically contractible metrics:

For the Carathéodory–Reiffen and the Kobayashi–Royden metrics the following estimates are true (cf. [493, 533]).

Theorem A.4.1.

(a) *For any $\boldsymbol{\gamma}$-hyperbolic domain $G \subset \mathbb{C}^n$ we have the following inequality:* $\text{h-curv}(\cdot;\cdot;\boldsymbol{\gamma}_G) \le -4$ *on* $G \times (\mathbb{C}^n)_*$.

(b) *If G denotes a $\boldsymbol{\varkappa}$-hyperbolic domain, then* $\text{h-curv}(\cdot;\cdot;\boldsymbol{\varkappa}_G) \ge -4$ *on* $G \times (\mathbb{C}^n)_*$.

As a consequence of the Lempert Theorem we obtain the following corollary:

Corollary A.4.2. *For a bounded convex domain G in $\mathbb{C}^n$, the holomorphic curvature with respect to $\boldsymbol{\gamma}_G$ and $\boldsymbol{\varkappa}_G$ is identically equal to -4.*

Now, we rewrite the Ahlfors–Schwarz lemma of Chapter 1 in terms of holomorphic curvature.

Theorem A.4.3. *Let G be a domain in $\mathbb{C}^n$ and let δ_G be an upper semicontinuous metric on G (recall that $\delta_G(z;\lambda X) = |\lambda|\delta_G(z;X)$ and $\delta_G(z;X) > 0$ if $X \neq 0$). If we assume that* $\text{h-curv}(\cdot;\cdot;\delta_G) \le -c^2 < 0$ *on $G \times (\mathbb{C}^n)_*$, then the following inequality holds:*

$$\delta_G(z;X) \le (2/c)\boldsymbol{\varkappa}_G(z;X), \quad z \in G,\ X \in (\mathbb{C}^n)_*.$$

A proof of this theorem can be found in, for example, [137].

Theorem A.4.3 gives a tool to, at least theoretically, find comparison results for general metrics and the Kobayashi–Royden metric. Since it seems to be very difficult to estimate the holomorphic curvature in general, we restrict ourselves to the case of Hermitian metrics, in particular, to the Bergman metric.

Lemma A.4.4. *Let $G \subset \mathbb{C}^n$ and let $\delta_G(z;X) = [\sum_{i,j=1}^n g_{i,j}(z)X_i\overline{X}_j]^{1/2}$ be a Hermitian metric on G with $g_{i,j} \in \mathcal{C}^2(G)$ ($z \in G$, $X \in \mathbb{C}^n$). Then the holomorphic curvature of δ_G at $(z_0, X) \in G \times (\mathbb{C}^n)_*$ coincides with the holomorphic sectional curvature at z_0 in the direction of X, i.e.,*

$$\text{h-curv}(z_0;X;\delta_G) = -2\sum_{i,j,k,l=1}^n R_{i,j,k,l}(z)X_i\overline{X}_jX_k\overline{X}_l/\delta_G^4(z_0;X) =: 2S(z_0;X;\delta_G), \tag{A.4.1}$$

where

$$R_{i,j,k,l}(z) := \frac{\partial g_{i,j}}{\partial z_k\partial\overline{z}_l}(z) - \sum_{\alpha,\beta=1}^n \frac{\partial g_{i,\alpha}}{\partial z_k}(z)g^{\alpha,\beta}(z)\frac{\partial g_{\beta,j}}{\partial\overline{z}_l}(z)$$

with $(g^{\alpha,\beta}(z))(g_{i,j}(z))$ being the unit matrix.

Formula (A.4.1) can be found in [533] and [535]; for a more explicit calculation, compare [5].

In the case of Thullen domains (complex ellipsoids) K. Azukawa and M. Suzuki (cf. [25, 31]) found upper and lower estimates for the holomorphic sectional curvature of the Bergman metric, and therefore for the holomorphic curvature.

Theorem A.4.5. *Let* $D_p := \mathcal{E}(1, 1/p) = \{z \in \mathbb{C}^2 : |z_1|^2 + |z_2|^{2/p} < 1\}$, $p > 0$, *and* $D_0 := \{z \in \mathbb{C}^2 : |z_1| < 1, |z_2| < 1\}$. *Then the following inequalities are true:*

(a) *for* $0 \le p \le 1$*:*

$$-\frac{1 + 4p + p^2}{(1 + 2p)^2} \le \frac{1}{2} \cdot \text{h-curv}(z; X; \boldsymbol{\beta}_G) \le -\frac{2(2 + 11p + 15p^2 + 8p^3)}{(2 + p)(1 + 3p)(4 + 5p)};$$

(b) *for* $p > 1$*:* $-\dfrac{2(2 + 11p + 15p^2 + 8p^3)}{(2 + p)(1 + 3p)(4 + 5p)} \le \dfrac{1}{2} \cdot \text{h-curv}(z; X; \boldsymbol{\beta}_G) \le -\dfrac{2}{2 + p}$.

Remark A.4.6. The above theorem shows that for all D_p, $p \ge 0$, the holomorphic curvature of the Bergman metric is bounded from above by a negative constant. But, in general, such an estimate does not remain true for all complex ellipsoids. For example we have $\lim_{p\to 0} \sup_{X\ne 0} \text{h-curv}(0; X; \boldsymbol{\beta}_{\mathcal{E}(p,p)}) = 4$, where $\mathcal{E}(p, p) = \{z \in \mathbb{C}^2 : |z_1|^{2p} + |z_2|^{2p} < 1\}$ (cf. [25]).

Applying Theorems A.4.3 and A.4.5, we obtain a comparison between the Bergman metric and the Kobayashi–Royden metric on Thullen domains D_p. As Remark A.4.6 shows, Theorem A.4.3 cannot be used to get comparison for general complex ellipsoids. Nevertheless, using the localization for the Bergman and the Kobayashi–Royden metric, one is led to the following result (cf. [209]):

Theorem A.4.7. *For* $p > 0$, $q > 0$ *there exists a positive constant* $C_{p,q}$ *such that* $\boldsymbol{\beta}_{\mathcal{E}(p,q)}(z; X) \le C_{p,q} \boldsymbol{\varkappa}_{\mathcal{E}(p,q)}(z; X)$, $z \in \mathcal{E}(p, q)$, $X \in \mathbb{C}^2$.

We do not know whether such a comparison is true for all pseudoconvex balanced Reinhardt domain in $\mathbb{C}^2$. Moreover, it seems to be open whether Theorem A.4.7 remains true in higher dimensions. ?

Evaluating Theorem A.4.5 in the case where $p = 1$, i.e., $D_1 = \mathbb{B}_2$, gives $\text{h-curv}(\cdot; \cdot; \boldsymbol{\beta}_{\mathbb{B}_2}) = -4/3$. In general, it is easily seen that

$$\text{h-curv}(\cdot; \cdot; \boldsymbol{\beta}_{\mathbb{B}_n}) = -4/(n + 1).$$

The following result of P. Klembeck (cf. [302]) again illustrates the affinity of strongly pseudoconvex domains to the ball:

Theorem A.4.8. *Let $G \subset \mathbb{C}^n$ be a strongly pseudoconvex domain with $\mathcal{C}^\infty$-boundary. Then, near the boundary of G, the holomorphic curvature of the Bergman metric approaches the constant value $-4/(n+1)$ of the holomorphic curvature of the ball $\mathbb{B}_n$.*

The proof of Theorem A.4.8 relies on the asymptotic formula for the Bergman kernel, due to C. Fefferman (cf. [165]).

It is known that for a bounded domain $G \subset \mathbb{C}^n$ one can express $S(\cdot;\cdot;\boldsymbol{\beta}_G)$ in terms of L_h^2-functions. In fact, the following formula

$$S(z;X;\boldsymbol{\beta}_G) = 2 - \frac{J_G^{(0)}(z;X)J_G^{(2)}(z;X)}{\big(J_G^{(1)}(z;X)\big)^2}$$

holds, where

$$J_G^{(0)}(z;X) := \mathfrak{K}_G(z),$$

$$J_G^{(1)}(z;X) := \sup\Big\{\Big|\sum_{j=1}^n \frac{\partial f}{\partial z_j}(z)X_j\Big|^2 : f \in L_h^2(G),\ f(z)=0,\ \|f\|_{L^2(G)} \le 1\Big\},$$

$$J_G^{(2)}(z;X) := \sup\Big\{\Big|\sum_{j,k=1}^n \frac{\partial^2 f}{\partial z_j \partial z_k}(z)X_jX_k\Big|^2 : f \in L_h^2(G),\ f(z) = \frac{\partial f}{\partial z_j}(z) = 0,\ j=1,\dots,n,\ \|f\|_{L^2(g)} \le 1\Big\}.$$

Recently, the asymptotic behavior of the holomorphic sectional curvatures for the Bergman metric on annuli $\mathbb{A}(r,1)$ was studied in [571] (see also [139]). In fact, it turns out that

$$\lim_{r\to 0+} S(r^\alpha;1;\boldsymbol{\beta}_{\mathbb{A}(r,1)}) = \begin{cases} -\infty, & \text{if } \alpha \in (1/3,2/3) \\ 2, & \text{if } \alpha \in (0,1/3] \cup [2/3,1) \end{cases}.$$

These results are used to construct a bounded planar domain D of Zalcman type such that

$$-\infty = \inf\{S(z;1;\boldsymbol{\beta}_D) : z \in D\} \quad \text{and} \quad \sup\{S(z;1;\boldsymbol{\beta}_D) : z \in D\} = 2.$$

A.5 Complex geodesics

Lempert's theorem (Theorem 11.2.1) is a powerful tool of complex analysis on convex domains. There are various applications of this result. For example, we have

Theorem A.5.1 (cf. [515]). *Let $G \subset \mathbb{C}^n$ be a bounded convex domain and let M be an analytic subset of G. Then, the following conditions are equivalent:*

(i) *there is a holomorphic mapping* $F : G \longrightarrow G$ *such that* $M = \{z \in G : F(z) = z\}$;

(ii) *there exists a holomorphic retraction* $r : G \longrightarrow M$.

The proof of Theorem 11.2.1 that we present in the book is based on the ideas taken from [460]. Our proof is simpler and more elementary than the original proof by Lempert (cf. [340, 341]). On the other hand, Lempert's results on regularity of complex geodesics are deeper, and his methods may also be applied to more general situations. For instance, we have

Theorem A.5.2 (cf. [340]). *Let* $G \subset \mathbb{C}^n$ *be a strongly convex domain with* $\mathcal{C}^k$*-boundary, where* $3 \le k \le \omega$. *Then,*

(a) *any complex geodesic* $\varphi : \mathbb{D} \longrightarrow G$ *extends to a* $\mathcal{C}^{k-2}$*-mapping on* $\overline{\mathbb{D}}$;

(b) *if* h *is a mapping as in* (11.2.18), *then* h *extends to a* $\mathcal{C}^{k-2}$*-mapping on* $\overline{\mathbb{D}}$;

(c) *if* $k \ge 6$, *then* $\boldsymbol{k}_G$ *is a* $\mathcal{C}^{k-4}$*-function on* $G \times G \setminus$ diagonal.

Recently, the above result has been generalized in the following way:

Theorem A.5.3 (cf. [458]). *Let* $G \subset \mathbb{C}^n$ *be a convex domain with* $\mathcal{C}^k$*-boundary* ($k \in \mathbb{Z}_+$). *Then,*

(a) *any complex geodesic* $\varphi : \mathbb{D} \longrightarrow G$ *extends to a* $\mathcal{C}^k$*-mapping on* $\overline{\mathbb{D}}$;

(b) *if* $k \ge 1$, *then the mapping* h *from* (11.2.18) *is uniquely determined.*

In [342] L. Lempert generalized his results from [340, 341] to the case of strongly linearly convex domains (where our methods do not work); cf. p. 260.

Example A.5.4. Any strongly convex domain is obviously strongly linearly convex, but not vice versa. For we put

$$G := \{z = (z_1, \dots, z_n) \in \mathbb{C}^n : \|z\|^2 + (\mathrm{Re}(z_1^2))^2 < 1\}.$$

Then, G is a strongly linearly convex domain (with $\mathcal{C}^\omega$-boundary) but not convex. We do not know whether G is biholomorphic to a convex domain. **?**

Theorem A.5.5 (cf. [342, 322]). *Let* $G \subset \mathbb{C}^n$ *be a strongly linearly convex domain with* $\mathcal{C}^k$*-boundary, where* $k \in \mathbb{N}_2 \cup \{\infty, \omega\}$. *Then,*

(a) $\boldsymbol{c}_G = \boldsymbol{k}_G = \boldsymbol{\ell}_G$, $\boldsymbol{\gamma}_G = \boldsymbol{\varkappa}_G$,

(b) *complex* $\boldsymbol{c}_G$*-geodesics are uniquely determined* mod Aut($\mathbb{D}$),

(c) *complex* $\boldsymbol{\gamma}_G$*-geodesics are uniquely determined* mod Aut($\mathbb{D}$),

(d) *any complex geodesic extends to a* $\mathcal{C}^{k-3/2}(\overline{\mathbb{D}})$*-mapping,*

(e) *if* $k \geq 3$*, then any complex geodesic extends to a* $\mathcal{C}^{k-1-\varepsilon}(\overline{\mathbb{D}})$*-mapping* ($\varepsilon > 0$).

Notice that the case $k \in \{\infty, \omega\}$ has been discussed in [342] and the general case has recently been solved in [322]. We would like to propose the reader to study this paper, because it seems to be the first one with complete detailed proofs.

Recalling Remark 7.1.21(b), one has the following consequence:

Corollary A.5.6 (cf. [243]). *If* $G \subset \mathbb{C}^n$ *is a bounded* $\mathbb{C}$*-convex domain with a smooth* $\mathcal{C}^2$*-boundary, then* $\boldsymbol{c}_G = \boldsymbol{\ell}_G$ *and* $\boldsymbol{\gamma}_G = \boldsymbol{\varkappa}_G$.

New approaches to complex geodesics were proposed by J. Agler ([9]) and Z. Słodkowski ([485]). For instance, using the methods of dilatation theory Agler proved the following generalization of Theorem 11.2.1:

Theorem A.5.7 (cf. [9]). *Let* $G \subset \mathbb{C}^n$ *be a bounded domain and let* z_0', z_0'' *be points in* G, $z_0' \neq z_0''$. *Then there exist a holomorphic mapping* $\varphi : \mathbb{D} \longrightarrow \operatorname{conv} G$ *and points* $\lambda_0', \lambda_0'' \in \mathbb{D}$ *such that* $z_0' = \varphi(\lambda_0')$, $z_0'' = \varphi(\lambda_0'')$*, and* $\boldsymbol{c}_G(z_0', z_0'') = \boldsymbol{p}(\lambda_0', \lambda_0'')$.

In the meantime, C. H. Chang, M. C. Hu, and H.-P. Lee studied complex geodesics for points in the closure of a domain. Let G be a taut domain in $\mathbb{C}^n$ such that $\boldsymbol{c}_G = \boldsymbol{\ell}_G$ (and $\boldsymbol{\gamma}_G = \boldsymbol{\varkappa}_G$; cf. Proposition 11.1.7). Let $z_0', z_0'' \in \overline{G}, z_0' \neq z_0''$. We say that a complex geodesic $\varphi : \mathbb{D} \longrightarrow G$ is a *complex geodesic for* (z_0', z_0'') if φ extends to a continuous mapping on $\overline{\mathbb{D}}$ with $z_0', z_0'' \in \varphi(\overline{\mathbb{D}})$. Similarly, if $z_0 \in \partial G$ and $X_0 \in (\mathbb{C}^n)_*$, then we say that a complex geodesic $\varphi : \mathbb{D} \longrightarrow G$ is a *complex geodesic for* (z_0, X_0) if φ extends to a $\mathcal{C}^1$-mapping on $\overline{\mathbb{D}}$ such that there exist $\lambda_0 \in \mathbb{T}$ and $\alpha_0 \in \mathbb{C}$ with $z_0 = \varphi(\lambda_0)$ and $\alpha_0\varphi'(\lambda_0) = X_0$.

Theorem A.5.8. *Let* $G \subset \mathbb{C}^n$ *be a strongly linearly convex domain with* $\mathcal{C}^3$*-boundary. Then,*

(a) (Cf. [90]) *For any points* $z_0', z_0'' \in \overline{G}$ *with* $z_0' \neq z_0''$ *there exists exactly one (modulo* $\operatorname{Aut}(\mathbb{D})$*) complex geodesic for* (z_0', z_0'').

(b) (Cf. [90, 238]) *For any* $z_0 \in \partial G$ *and for any* $X_0 \in \mathbb{C}^n \setminus T_{z_0}^{\mathbb{C}}(\partial G)$ *there exists exactly one (modulo* $\operatorname{Aut}(\mathbb{D})$*) complex geodesic for* (z_0, X_0).

If G is not strongly linearly convex, then the above theorem is no longer true (even if G is strictly convex).

Theorem A.5.9 (cf. [91]). *Let* $\mathcal{E} = \mathcal{E}(p)$ *be a complex ellipsoid in* $\mathbb{C}^n$ *with* $p = (p_1, \dots, p_n) \in \mathbb{N}^n$. *Suppose that* $\mathcal{E}$ *is not strongly convex (i.e.,* $p \neq (1, \dots, 1)$*; cf. Remark* 16.2.1*). Fix* $z_0' := (1, 0, \dots, 0) \in \partial\mathcal{E}$.

(a) *Let* $z_0'' = (\xi_1, \dots, \xi_n) \in \partial\mathcal{E} \setminus \{z_0'\}$. *Suppose that either* $p_1 = 1$ *or* $\xi_1 \in \{0\} \cup \mathbb{T}$. *Then there exists a complex geodesic for* (z_0', z_0'').

(b) *There exists a function* $M : [0, 1) \longrightarrow \mathbb{R}_{>0}$ *such that if* $z_0'' = (\xi_1, \dots, \xi_n) \in \mathcal{E}$ *and* $\sum_{j=2}^n |\xi_j|^{2p_j} < M(|\xi_1|)$, *then* (z_0', z_0'') *admits a complex geodesic.*

(c) *If* $p_1 > 1$, *then there exists a point* $z_0'' \in \partial\mathcal{E}$ *(resp.* $z_0'' \in \mathcal{E}$*) such that* (z_0', z_0'') *admits at least two non-equivalent complex geodesics.*

A.6 Criteria for biholomorphicity

Criteria for biholomorphicity (cf. § 11.6) were studied by several authors. There are two main streams of problems:

1° We are given two domains $G_1, G_2 \subset \mathbb{C}^n$ and a holomorphic mapping $F : G_1 \longrightarrow G_2$ such that F is a δ-isometry at a point $a \in G_1$, where $\delta \in \{\boldsymbol{\gamma}, \boldsymbol{\varkappa}\}$. We would like to decide under which conditions F is biholomorphic; cf. Proposition 11.6.3.

2° We are given a domain $G \subset \mathbb{C}^n$ and a point $a \in G$ such that $\boldsymbol{\gamma}_G(a;\cdot) = \boldsymbol{\varkappa}_G(a;\cdot)$. We would like to decide whether G is biholomorphic to the indicatrix

$$\mathcal{I}_G(a) := \{X \in \mathbb{C}^n : \boldsymbol{\varkappa}_G(a; X) < 1\};$$

cf. Proposition 19.6.7.

Note that $\mathcal{I}_G(a)$ is biholomorphic to $\mathbb{B}_n$ iff the mapping $\mathbb{C}^n \ni X \longmapsto \boldsymbol{\varkappa}_G^2(a; X) \in \mathbb{R}_+$ is a Hermitian form.

In direction 1° we mention, for instance, the following results:

Theorem A.6.1. *Let* $F : G_1 \longrightarrow G_2$ *be a holomorphic mapping and let* $a \in G_1$ *be such that* $\delta_{G_2}(F(a); F'(a)X) = \delta_{G_1}(a; X)$, $X \in \mathbb{C}^n$. *Then,* F *is biholomorphic in each of the following cases:*

(a) (Cf. [46, 47]) G_1 *is bounded* $\boldsymbol{c}$*-complete,* G_2 *is the unit ball with respect to a* $\mathbb{C}$*-norm,* $F(a) = 0$*, and* $\delta = \boldsymbol{\varkappa}$.

(b) (Cf. [47, 518]) G_1 *is* $\boldsymbol{c}^i$*-complete,* G_2 *is the unit ball with respect to a* $\mathbb{C}$*-norm,* $F(a) = 0$*, and* $\delta = \boldsymbol{\varkappa}$.

(c) (Cf. [518]) G_1 *is bounded convex,* G_2 *is bounded, and* $\delta = \boldsymbol{\gamma}$.

(d) (Cf. [518]) G_1 *is taut,* G_2 *is bounded strictly convex, and* $\delta = \boldsymbol{\varkappa}$.

(e) (Cf. [520]) $G_1 = G_2 = G \subset \mathbb{C}$ *is a bounded domain for which there exists an* $\eta > 0$ *such that for any piecewise* $\mathcal{C}^1$*-loop* $\alpha : [0, 1] \longrightarrow G$ *that is not homotopic to a point, we have* $\operatorname{length}(\alpha) \geq \eta$ *(e.g.,* G *is a non-degenerated annulus), and* $\delta = \boldsymbol{\varkappa}$.

Theorem A.6.2 (cf. [520]). *Let $F : G_1 \longrightarrow G_2$ be a holomorphic mapping and let $a \in G_1$ be such that $\varkappa_{G_2}(F(a); F'(a)X) = \varkappa_{G_1}(a; X)$, $X \in \mathbb{C}^n$. Then F is a covering in each of the following cases:*

(a) *G_1 is taut and the universal covering of G_2 is biholomorphic to a bounded strictly convex domain.*

(b) *The universal coverings of G_1 and G_2 are biholomorphic to bounded convex domains.*

Problems of type 2° are more difficult than those of 1°. We would like to mention the following results (in chronological order):

Theorem A.6.3 (cf. [487]). *Let $G \subset \mathbb{C}^n$ be a $\boldsymbol{c}$-finitely compact domain. Suppose that $\gamma_G(a;\cdot) = \varkappa_G(a;\cdot)$ for a point $a \in G$ and $\{X \in \mathbb{C}^n : \gamma_G(a; X) < 1\} = \mathbb{D}^n$. Then the domain G is biholomorphic to $\mathbb{D}^n$.*

Theorem A.6.4 (cf. [488]). *Let $G \subset \mathbb{C}^n$ be a complete hyperbolic domain. Suppose that $\gamma_G(a;\cdot) = \varkappa_G(a;\cdot)$ for a point $a \in G$ and suppose that γ_G or $\varkappa_G$ is a $\mathcal{C}^\infty$ Hermitian metric on $G \times \mathbb{C}^n$. Then G is biholomorphic to $\mathbb{B}_n$.*

Theorem A.6.5 (cf. [411]). *Let $G \subset \mathbb{C}^n$ be a strongly convex domain with $\mathcal{C}^\infty$-boundary. Then the following conditions are equivalent:*

(i) *G is biholomorphic to $\mathbb{B}_n$;*

(ii) *there exists an $a \in G$ such that the function $G \ni z \longmapsto \boldsymbol{k}_G^2(a, z)$ is of class $\mathcal{C}^\infty$;*

(iii) *there exist $a \in G$ and its neighborhood U such that $\varkappa_G$ is a $\mathcal{C}^\infty$ Hermitian metric on $U \times \mathbb{C}^n$.*

We say that $\mathcal{R}$ is a *real ellipsoid* in $\mathbb{C}^n$ if, after a $\mathbb{C}$-linear change of coordinates, $\mathcal{R}$ may be written as

$$\mathcal{R} = \Big\{(z_1, \dots, z_n) \in \mathbb{C}^n : \sum_{j=1}^{n} (|z_j|^2 + \lambda_j \operatorname{Re}(z_j^2)) < 1\Big\},$$

where $0 \le \lambda_j < 1,\ j = 1, \dots, n$.

Theorem A.6.6 (cf. [293]). *Let $\mathcal{R} \subset \mathbb{C}^n$ be a real ellipsoid. Then, $\mathcal{R}$ is biholomorphic to $\mathbb{B}_n$ iff $\varkappa_{\mathcal{R}}(0;\cdot)$ is a Hermitian form.*

Notice that a simpler proof of Theorem A.6.6 is given in [561].

Theorem A.6.7 (cf. [6]). *Let $G \subset \mathbb{C}^n$ be a taut domain. Suppose that $a \in G$ is such that,*

(a) $\gamma_G(a;\cdot) = \varkappa_G(a;\cdot)$,

(b) *there exists a neighborhood U of a such that $\varkappa_G$ is a $\mathcal{C}^\infty$-function on $U \times (\mathbb{C}^n \setminus \{0\})$,*

(c) *the indicatrix $\mathfrak{I}_G(a)$ is strongly pseudoconvex,*

(d) *there exist $0 < r_1 < r_2$ and a biholomorphic mapping $\Phi : B_{k_G}(a, r_1) \longrightarrow B_{k_G}(a, r_2)$ such that $\Phi(a) = a$.*

Then G is biholomorphic to $\mathfrak{I}_G(a)$.

Theorem A.6.8 (cf. [6]). *Let $G \subset \mathbb{C}^n$ be a taut domain. Then G is biholomorphic to $\mathbb{B}_n$ iff there exists a point $a \in G$ such that $\gamma_G(a;\cdot) = \varkappa_G(a;\cdot)$ and $\varkappa_G$ is a $\mathcal{C}^\infty$ Hermitian metric.*

Theorem A.6.9 (cf. [501]). *Let $G \subset \mathbb{C}^n$ be a strongly pseudoconvex domain with simply connected boundary. Then the following implications hold.*

(a) *If ∂G is $\mathcal{C}^\infty$-smooth and $\varkappa_G$ is a Hermitian metric in a neighborhood of ∂G, then G is biholomorphic to $\mathbb{B}_n$.*

(b) *If ∂G is $\mathcal{C}^2$ smooth and $\varkappa_G$ is a $\mathcal{C}^\infty$-Hermitian metric in a neighborhood of ∂G, then G is biholomorphic to $\mathbb{B}_n$.*

(c) *If ∂G is smooth real analytic and $\varkappa_G$ is a Hermitian metric in a neighborhood of a boundary point, then G is biholomorphic to $\mathbb{B}_n$.*

A.7 Isometries

Theorem A.7.1 (cf. [469]). *Let $F : G_1 \longrightarrow G_2$ be a $\mathcal{C}^1$-Kobayashi (resp. inner Carathéodory) isometry, where G_1, G_2 are bounded strongly pseudoconvex domains in $\mathbb{C}^n$ (with $\mathcal{C}^3$-boundaries in the inner Carathéodory case). Suppose that F extends to a $\mathcal{C}^1$-mapping on $\overline{G}_1$. Then:*

- *$F|_{\partial G_1} : \partial G_1 \longrightarrow \partial G_2$ is a CR or anti-CR-diffeomorphism,*
- *G_1 and G_2 are biholomorphic or anti-biholomorphic.*

Theorem A.7.2 (cf. [470]). *Let G be a bounded strongly convex domain in $\mathbb{C}^n$ with $\mathcal{C}^6$-boundary. Then, the group of Kobayashi isometries of G is compact, unless G is biholomorphic or anti-biholomorphic to $\mathbb{B}_n$. In particular, when G is equivalent to $\mathbb{B}_n$, each isometry of G is biholomorphic or anti-biholomorphic.*

Theorem A.7.3 (cf. [301]). *Let G be a bounded strongly pseudoconvex domain in $\mathbb{C}^n$ such that its Kobayashi isometry group is non-compact. Then,*

- *there exists a Kobayashi isometry $F : G \longrightarrow \mathbb{B}_n$;*
- *if $\partial G \in \mathcal{C}^{2,\varepsilon}$, then G is biholomorphic to $\mathbb{B}_n$;*
- *if $\partial G \in \mathcal{C}^{2,\varepsilon}$ and $F : G \longrightarrow \mathbb{B}_n$ is a Kobayashi isometry, then F is either biholomorphic or anti-biholomorphic.*

Theorem A.7.4 (cf. [190]). *Let $G_j \subset \mathbb{C}^{n_j}$ be a bounded strictly convex domain with $\mathcal{C}^3$-boundary, $j = 1, 2$. Then, every Kobayashi isometry $F : G_1 \longrightarrow G_2$ is either biholomorphic or anti-biholomorphic.*

A.8 Boundary behavior of contractible metrics on weakly pseudoconvex domains

In Chapter 19 we discussed estimates from below and above for the Bergman, Carathéodory, and Kobayashi metrics near the boundary of strongly pseudoconvex domains. For various purposes it is worthwhile to know similar estimates also on pseudoconvex domains that are not strongly pseudoconvex. Since the investigation of the boundary behavior of various invariant objects is still an active field of research, we point out that we are not able to mention all existing results.

In dimension two the following result is due to E. Bedford and J. E. Fornæss:

Theorem A.8.1 (cf. [40]). *Let G be a bounded pseudoconvex domain in $\mathbb{C}^2$ with smooth real analytic boundary. Then there exist $C > 0$ and $0 < \varepsilon < 1$ such that*

$$\gamma_G(z; X) \geq C\|X\| \operatorname{dist}(z, \partial G)^{-\varepsilon}, \quad z \in G,\ X \in \mathbb{C}^2.$$

The necessary information for this inequality is taken from the construction of peak functions (cf. [39]). It is still an open question whether the estimate in Theorem A.8.1 remains true in higher dimensions.

Remark A.8.2. Note that such an inequality needs a special boundary behavior of the domain G. To be more precise, let $G \subset \mathbb{C}^n$ be a bounded domain with a $\mathcal{C}^\infty$-smooth boundary. Let $a \in \partial G$ and let r be a defining function near a. Recall the *d'Angelo variety type of* a, i.e.,

$$\Delta_1(a; G) := \sup\left\{\frac{\operatorname{ord}_0(r \circ \varphi)}{\operatorname{ord}_0(\varphi)} : \varphi \in \mathcal{O}(\mathbb{D}, \mathbb{C}^n),\ \varphi(0) = a,\ \varphi \not\equiv a\right\},$$

where ord_0 is the order of vanishing of φ (resp. $r \circ \varphi$) at 0 (for more details see [120]). Then,

Theorem A.8.3 (cf. [542]). *Let G and a be as before. If there exists a number $m \geq 1$ such that for any $X \in (T_a^{\mathbb{C}}(\partial D))_*$ one can find a constant $C(X)$ with*

$$\gamma_G(z^j; X) \geq C(X)\,\mathrm{dist}(z^j, \partial G)^{-1/m} \text{ for a non-tangential sequence } G \ni z^j \longrightarrow a,$$

then a is of finite variety type with $\Delta_1(a; G) \leq m$.

If G has a real analytic boundary, then G is of *finite type*, i.e., $\sup\{\Delta_1(a; G) : a \in \partial G\} < \infty$.

In the case where, in addition, G is convex, the following result due to M. Range is true even in arbitrary dimensions:

Theorem A.8.4 (cf. [444]). *Let G be a bounded convex domain in $\mathbb{C}^n$ with smooth real analytic boundary. Assume that G is given by $G = \{z \in \mathbb{C}^n : r(z) < 0\}$ with a defining function r. Then there are positive numbers C and $\varepsilon < 1$ such that*

$$\gamma_G(z; X) \geq C\left(\frac{\|X\|}{\mathrm{dist}(z, \partial G)^{\varepsilon}} + \frac{|\sum_{\nu=1}^{n}(\partial r/\partial z_\nu)(z)X_\nu|}{\mathrm{dist}(z, \partial G)}\right), \quad z \in G,\ X \in \mathbb{C}^n.$$

We only mention that the exponent ε in Theorem A.8.4 is somehow related to the order of contact certain supporting analytic hypersurfaces have with ∂G.

Similar estimates were already mentioned in the text for arbitrary convex domains (see Corollary 11.3.8). Moreover, there is the following generalization:

Theorem A.8.5 (cf. [391]). *Let $G \subset \mathbb{C}^n$ be a $\mathbb{C}$-convex domain not containing any complex line. Then,*

$$1/4 \leq \gamma_G(z; X) d_G(z; X) \leq \varkappa_G(z; X) d_G(z; X) \leq 1, \quad z \in G,\ X \in (\mathbb{C}^n)_*,$$

where $d_G(z; X) := \sup\{r > 0 : z + \mathbb{D}(r)X \subset G\}$.

If G is convex, then the lower estimate is even true with the factor $1/2$ (see [44]). Moreover, we should mention that the factor $1/4$ is the optimal one; indeed, take the domain $G := f(\mathbb{C} \setminus [1/4, +\infty)) \subset \mathbb{C}$, where f denotes the Koebe function, i.e., $f(\lambda) := \frac{\lambda}{(1+\lambda)^2}$.

Remark A.8.6. Let G be as in the former theorem. Assume that, in addition, a is a smooth boundary point of G. Fix an $X \in (\mathbb{C}^n)_*$ and define $l_{a,X}$ to be the order of contact of the line $a + \mathbb{C}X$ and ∂G at the point a. If $l_{a,X} < \infty$, then there are a constant $c > 1$ and a neighborhood $U = U(a)$ such that

$$\tfrac{1}{c}\,\mathrm{dist}(z, \partial G) \leq d_G(z; X)^{1/l_{a,X}} \leq c\,\mathrm{dist}(z, \partial G), \quad z \in G \cap U \cap \nu_a,$$

where ν_a is the inner normal at ∂D in a. From here the estimate in Theorem A.8.5 may be reformulated in terms of $\mathrm{dist}(\cdot, \partial G)$. For a similar result for the convex case, see [338].

For arbitrary dimensions a lower estimate for the Kobayashi–Royden metric has been found by K. Diederich and J. E. Fornæss.

Theorem A.8.7 (cf. [126]). *If G is a pseudoconvex domain in $\mathbb{C}^n$ with smooth real analytic boundary, then for suitable $C > 0$ and $\varepsilon > 0$ the following inequality is true:*

$$\varkappa_G(z; X) \geq C \|X\| \operatorname{dist}(z, \partial G)^{-\varepsilon}, \quad z \in G, \ X \in \mathbb{C}^n.$$

Details can also be found in [131]. As a consequence (cf. Exercise 19.7.5) one can prove that any proper holomorphic mapping $F : G_1 \longrightarrow G_2$ is Hölder continuous on G_1, where $G_1 \subset\subset \mathbb{C}^n$ is a pseudoconvex domain with smooth $\mathcal{C}^2$-boundary and $G_2 \subset\subset \mathbb{C}^n$ is pseudoconvex with smooth real analytic boundary.

In [542] a slightly more general variety type, called the *k-type*, is introduced, which is connected with the boundary behavior of the k-th Kobayashi–Royden metric. Let $G \subset \mathbb{C}^n$ be a bounded domain, $a \in \partial G$ a $\mathcal{C}^\infty$-smooth boundary point, and r a local defining function of G near a. Then the k-order type of G in a, $k \in \mathbb{N}$, is defined as

$$\Delta_1(a, k) = \Delta_1(a, k; G) := \sup\{\Delta_1(a, X, k; G) : X \in (T_a^{\mathbb{C}}(\partial G))_*\},$$

where

$$\Delta_1(a, X, k; G) := \tfrac{1}{k} \sup\{\operatorname{ord}_0(r \circ \varphi) : \\ \varphi \in \mathcal{O}(\mathbb{D}, \mathbb{C}^n),\ \varphi(0) = a,\ \operatorname{ord}_0 \varphi = k,\ \varphi^{(k)}(0) = X\}.$$

Note that this definition is independent of r. With this notion in mind, one has the following result:

Theorem A.8.8 (cf. [542]). *Let G and a be as before and $k \in \mathbb{N}$. Then,*

$$\Delta_1(a, k) \leq \sup\left\{\limsup_{t \to 0+} \frac{-\log \operatorname{dist}(a - t\nu_a, \partial D)}{\log \varkappa_G^{(k)}(a - t\nu_a; X)} : X \in (T_a^{\mathbb{C}}(\partial G))_*\right\},$$

where ν_a is the unit outer normal vector at ∂G in a.

Note that the right side in this inequality measures how fast the k-th Kobayashi–Royden metric blows up near the point a.

Even more is true, namely: *If $G \subset \mathbb{C}^n$ is a bounded domain with a $\mathcal{C}^\infty$-smooth boundary point a and if there are constants $k, m \in \mathbb{N}$, a neighborhood U of a, a tangential vector $X \in T_a^{\mathbb{C}}(\partial G)$, and a non-tangential sequence $(z^j)_j \subset G$ with $z^j \longrightarrow a$ such that $\varkappa_{G \cap U}^{(k)}(z^j; X) \geq C(k, X) \operatorname{dist}(z^j, \partial G)^{-1/m}$, $j \in \mathbb{N}$, for some positive constant $C(k, X)$, then $\Delta_1(a, X, k; G) \leq m$.* Hence, a necessary condition for a good lower estimate of the Kobayashi metrics seems to be that a finite-type condition is fulfilled.

Now, we come back to $\mathbb{C}^2$. We assume that a bounded pseudoconvex domain $G \subset \mathbb{C}^2$ is given by $G = \{z \in U \cap G : r(z) < 0\}$, where $U = U(\partial G)$ and $r \in \mathcal{C}^\infty(U, \mathbb{R})$ with $dr(z) \neq 0$, $z \in U$. A boundary point ζ_0 of G is said to be of *regular type* $m \in \mathbb{N} \cup \{+\infty\}$ if

$$m = \Delta_1^r(\zeta_0; G) := \sup\{\operatorname{ord}_0 r \circ \varphi : \varphi \in \mathcal{O}(\mathbb{D}, \mathbb{C}^n),\ \varphi(0) = \zeta_0,\ \varphi'(0) \neq 0\}.$$

Recall that for a boundary point of a smooth domain $D \subset \mathbb{C}^2$, one knows that both forms of types coincide, i.e., $\Delta_1(a; D) = \Delta_1^r(a; D)$. For more information on the variety (resp. regular) type of points see [65, 120], and [320].

For simplicity, we will assume that $\zeta_0 = 0$ and $r_{z_2}(0) \neq 0$. (∗)

Then for z near ζ_0 we put

$$L_1(z) := \left(1, -\frac{r_{z_1}(z)}{r_{z_2}(z)}\right), \qquad L_2(z) := (0, 1).$$

Obviously, $L_1(z)$ and $L_2(z)$ form a basis of $\mathbb{C}^2$. Moreover, for $j, k \in \mathbb{N}$ we set

$$(\mathcal{L}_{j,k})(z) := \left(\frac{\partial}{\partial \bar{z}_1} - \frac{r_{\bar{z}_1}(z)}{r_{\bar{z}_2}(z)} \frac{\partial}{\partial \bar{z}_2}\right)^{j-1} \left(\frac{\partial}{\partial z_1} - \frac{r_{z_1}(z)}{r_{z_2}(z)} \frac{\partial}{\partial z_2}\right)^{k-1} (\mathcal{L}r)(z; L_1(z)).$$

Observe that the assumption on the type implies that

$$\begin{aligned} &(\mathcal{L}_{j,k})(0) = 0 \quad \text{if } j + k < m, \\ &(\mathcal{L}_{j_0,k_0})(0) \neq 0 \quad \text{for at least one pair } (j_0, k_0) \text{ with } j_0 + k_0 = m. \end{aligned}$$

By means of the $\mathcal{L}_{j,k}$'s, for $l \in \mathbb{N}$ and z near 0 we define

$$\mathcal{C}_l(z) := \max\{|(\mathcal{L}_{j,k})(z)| : j + k = l\}.$$

If now X is an arbitrary vector in $\mathbb{C}^2$ and if z is near 0, we have the following unique representation of X: $X =: X_1(z)L_1(z) + X_2(z)L_2(z)$. With this notion in mind, we finally define

$$M_m(z; X) := |X_2(z)||r(z)|^{-1} + |X_1(z)| \sum_{l=2}^{m} |\mathcal{C}_l(z)/r(z)|^{1/l}.$$

After these preparations, we can describe the size of the metrics in the following "small constant–large constant" sense. These estimates have been found by D. Catlin.

Theorem A.8.9 (cf. [88]). *Let $G \subset \mathbb{C}^2$ be a pseudoconvex domain with $\mathcal{C}^\infty$-boundary. Assume that ζ_0 is a boundary point of G of finite type m (we may suppose that (∗) is satisfied). Then there exist a neighborhood $U = U(\zeta_0)$ and positive constants c and C such that for all $z \in G \cap U$ and all $X = X_1(z)L_1(z) + X_2(z)L_2(z) \in \mathbb{C}^2$ we have*

$$c M_m(z; X) \le \delta_G(z; X) \le C M_m(z; X),$$

where $\delta_G = \boldsymbol{\gamma}_G$, $\delta_G = \boldsymbol{\varkappa}_G$, or $\delta_G = \boldsymbol{\beta}_G$.

Similar results in higher dimension, but under additional assumptions on the Levi form, can be found in, for example, [106, 229].

We mention that if G is a bounded convex domain with smooth $\mathcal{C}^\infty$-boundary in $\mathbb{C}^n$, n arbitrary, and if $\zeta_0 \in \partial G$ is a boundary point of a suitable finite linear type, then a similar "small constant–large constant" estimate has been established by J.-H. Chen [99]. But, in general, the boundary behavior of contractible metrics on weakly pseudoconvex domains in $\mathbb{C}^n, n > 2$, seems to be unknown except for special cases. We conclude this section with a result of G. Herbort in this direction.

Let G be a domain in $\mathbb{C}^n, n \geq 2$, which is given by

$$G := \{z = (z_1, \widetilde{z}) \in \mathbb{C}^n = \mathbb{C}^1 \times \mathbb{C}^{n-1} : \operatorname{Re} z_1 + P(\widetilde{z}) < 0\},$$

where P is a real-valued psh polynomial on $\mathbb{C}^{n-1}$ without pluriharmonic terms. We say that G is of *homogeneous finite diagonal type* if

(a) there exist $m_j \in \mathbb{N}, 2 \leq j \leq n$, such that

$$P(t^{1/2m_2} z_2, \ldots, t^{1/2m_n} z_n) = tP(\widetilde{z}), \quad \widetilde{z} = (z_2, \ldots, z_n) \in \mathbb{C}^{n-1},\ t > 0;$$

(b) $P(\widetilde{z}) - 2s \sum_{j=2}^{n} |z_j|^{2m_j}$ is psh on $\mathbb{C}^{n-1}$ for a suitable $s > 0$.

To be able to formulate the final result, the following auxiliary functions are needed:

$$A_{l,j}(z) := \max\left\{ \left| \frac{\partial^l r}{\partial z_j^\nu \partial \overline{z}_j^\mu}(z) \right| : \nu, \mu \in \mathbb{N},\ \nu + \mu = l \right\},\ 2 \leq j \leq n,\ 2 \leq l \leq 2m_j;$$

$$\mathcal{C}_j(z) := \sum_{l=2}^{2m_j} \left(\frac{A_{l,j}}{-r(z)} \right)^{1/l}, \quad 2 \leq j \leq n;$$

$$M_G^*(z; X) := \left| \sum_{j=1}^{n} \frac{\partial r}{\partial z_j}(z) X_j \right| / |r(z)| + \sum_{j=2}^{n} \mathcal{C}_j(z) |X_j|, \quad z \in G,\ X \in \mathbb{C}^n.$$

Under the additional hypothesis that any term of P contains at most two of the variables $z_2, \ldots, z_n$, the following comparison between M_G^*, γ_G, and $\varkappa_G$ is true:

Theorem A.8.10 (cf. [228, 227]). *Suppose that*

$$G := \{z \in \mathbb{C}^n : \operatorname{Re} z_1 + P(\widetilde{z}) < 0\}$$

is a domain of homogeneous finite diagonal type and suppose that P has the form

$$P(z_2, \ldots, z_n) = \sum_{j=2}^{n} P_j(z_j) + \sum_{2 \leq j < k \leq n} P_{j,k}(z_j, z_k),$$

where P_j, $P_{j,k}$ are real-valued polynomials and where for $j < k$ we have

$$P_{j,k}(z_j, z_k) = \sum_{\alpha+\beta\ge 1, \gamma+\delta\ge 1} c_{\alpha,\beta,\gamma,\delta} z_j^{\alpha} \bar{z}_j^{\beta} z_k^{\gamma} \bar{z}_k^{\delta}.$$

Then there exist positive constants c and C such that

$$c M_G^*(z; X) \le \delta_G(z; X) \le C M_G^*(z; X), \quad z \in G,\ X \in \mathbb{C}^n,$$

where $\delta_G = \boldsymbol{\gamma}_G$, $\delta_G = \boldsymbol{\varkappa}_G$, or $\delta_G = \boldsymbol{\beta}_G$.

A particular case of Theorem A.8.10 is contained in [359].

In the flavor of Theorem A.8.9 there are also estimates for the invariant distances. For example (see [232]):

Theorem A.8.11. *Let $D := \{z \in \mathbb{C}^2 : r(z) < 0\}$ be a $\mathcal{C}^\infty$-smooth bounded pseudoconvex domain such that all boundary points are of finite type. Then there exists a positive constant C such that*

$$C\varphi \le \boldsymbol{c}_D^i \le \boldsymbol{k}_D \le (1/C)\varphi \quad \text{and} \quad C\varphi \le \boldsymbol{b}_D \le (1/C)\varphi,$$

where the function φ is given effectively in terms of the defining function r.

A.9 Spectral ball

For $n \ge 2$, the *spectral ball* Ω_n is defined as follows:

$$\Omega_n := \{A \in \mathbb{M}(n \times n; \mathbb{C}) : r(A) < 1\},$$

where

$$r(A) := \max\{|\lambda| : \lambda \text{ is an eigenvalue of } A\}$$

is the *spectral radius* of A (cf. Example 5.3.8); Ω_n may be considered as a domain in $\mathbb{C}^{n^2}$. Note that

- Ω_n is balanced.
- If $A = [a_{j,k}]$, $a_{n,1} := 1$, $a_{j,k} := 0$ otherwise, then $r(\mu A) = 0$, $\mu \in \mathbb{C}$. In particular, Ω_n contains complex lines.

Define $\sigma_n : \mathbb{M}(n \times n; \mathbb{C}) \longrightarrow \mathbb{C}^n$, $\sigma_n(A) := \pi_n(\lambda_1, \dots, \lambda_n)$, where $\{\lambda_1, \dots, \lambda_n\}$ denotes the set of all eigenvalues of A (with multiplicities) and $\pi_n : \mathbb{C}^n \longrightarrow \mathbb{C}^n$ has been defined in § 7.2. One can easily prove that σ_n is holomorphic. Obviously, $\sigma_n(\Omega_n) = \mathbb{G}_n$. Thus, $\Omega_n = \{A \in \mathbb{M}(n \times n; \mathbb{C}) : h \circ \sigma_n(A) < 1\}$, where h is the $(1, \dots, n)$-Minkowski function of $\mathbb{G}_n$ (cf. § 7.2). In particular, Ω_n is pseudoconvex.

Proposition A.9.1 (cf. [379]).

$$\begin{aligned} \boldsymbol{m}_{\Omega_n}(A_1, A_2) &= \boldsymbol{m}_{\mathbb{G}_n}(\sigma_n(A_1), \sigma_n(A_2)), \quad A_1, A_2 \in \Omega_n, \\ \boldsymbol{\gamma}_{\Omega_n}(A; X) &= \boldsymbol{\gamma}_{\mathbb{G}_n}(\sigma_n(A); \sigma_n'(A)(X)), \quad A \in \Omega_n,\ X \in \mathbb{M}(n \times n; \mathbb{C}). \end{aligned}$$

We say that a matrix $A \in \Omega_n$ is *cyclic* ($A \in \mathcal{C}_n$) if there exists a $v \in \mathbb{C}^n$ such that the vectors $v, Av, \dots, A^{n-1}v$ are $\mathbb{C}$-linearly independent. One can prove that $\mathcal{C}_n$ is dense in Ω_n (cf. [236]).

Given different points $\lambda_1, \dots, \lambda_m \in \mathbb{D}$ and arbitrary matrices $A_1, \dots, A_m \in \Omega_n$, we may consider the following *spectral Nevanlinna–Pick problem*:

Find an $F \in \mathcal{O}(\mathbb{D}, \Omega_n)$ *such that* $F(\lambda_j) = A_j$, $j = 1, \dots, m$.

In the case of cyclic matrices, the above problem may be translated to an interpolation problem for $\mathbb{G}_n$.

Theorem A.9.2 (cf. [114, 10]). *Let* $f \in \mathcal{O}(\mathbb{D}, \mathbb{G}_n)$, *let* $\lambda_1, \dots, \lambda_m \in \mathbb{D}$ *be pairwise different, and let* $A_1, \dots, A_m \in \mathcal{C}_n$. *Assume that* $f(\lambda_j) = \sigma_n(A_j)$, $j = 1, \dots, m$. *Then there exists an* $F \in \mathcal{O}(\mathbb{D}, \Omega_n)$ *such that* $F(\lambda_j) = A_j$, $j = 1, \dots, m$.

In particular, if $m = 1$, then (also using the density of $\mathcal{C}_n$ in Ω_n together with the continuity of the Kobayashi pseudodistance) we get the following:

Proposition A.9.3.

$$\begin{aligned} \boldsymbol{\ell}_{\Omega_n}(A_1, A_2) &= \boldsymbol{\ell}_{\mathbb{G}_n}(\sigma_n(A_1), \sigma_n(A_2)), \quad A_1, A_2 \in \mathcal{C}_n, \\ \boldsymbol{k}_{\Omega_n}(A_1, A_2) &= \boldsymbol{k}_{\mathbb{G}_n}(\sigma_n(A_1), \sigma_n(A_2)), \quad A_1, A_2 \in \Omega_n. \end{aligned}$$

Since $\boldsymbol{c}_{\mathbb{G}_2} \equiv \boldsymbol{\ell}_{\mathbb{G}_2}$ (cf. § 7.1), we get the following important, surprising example:

Example A.9.4 (cf. [379]).

$$\boldsymbol{c}_{\Omega_2} \equiv \boldsymbol{k}_{\Omega_2}, \quad \boldsymbol{c}_{\Omega_2} = \boldsymbol{\ell}_{\Omega_2} \text{ on } \mathcal{C}_n \times \mathcal{C}_n, \quad \boldsymbol{k}_{\Omega_2} \not\equiv \boldsymbol{\ell}_{\Omega_2}.$$

The problem of the discontinuity of $\boldsymbol{\ell}_{\Omega_n}$ has been studied, e.g., in [394, 500, 379].

Remark A.9.5. So far, the problem of the structure of the group $\mathrm{Aut}(\Omega_n)$ is not completely understood – cf. [321]

A.10 List of problems

A.1. Decide whether the estimate from Theorem A.4.7 is true for all pseudoconvex balanced Reinhardt domain in $\mathbb{C}^2$ 771

A.2. Decide whether the estimate from Theorem A.4.7 is satisfied in higher dimensions 771

A.3. Decide whether the domain G from Example A.5.4 is biholomorphic to a convex domain 773

A.4. Decide whether the estimate in Theorem A.8.1 remains true in higher dimensions 778

Appendix B

Addendum

B.1 Holomorphic functions

References: [105, 109, 203, 204, 265, 270, 327, 370, 445].

Since the whole book is based on the classical Schwarz lemma, this lemma should be given first.

B.1.1 (Classical Schwarz lemma). *Any* $f \in \mathcal{O}(\mathbb{D}, \mathbb{D})$ *with* $f(0) = 0$ *satisfies* $|f(\lambda)| \le |\lambda|$ *for all* $\lambda \in \mathbb{D}$ *and* $|f'(0)| \le 1$. *Moreover, if* $|f(\lambda_0)| = |\lambda_0|$ *for at least one* $\lambda_0 \in \mathbb{D}_*$ *or if* $|f'(0)| = 1$, *then* f *is just a rotation, i.e.,* $f(\lambda) = e^{i\theta}\lambda$, $\lambda \in \mathbb{D}$, *where* θ *is a suitable real number.*

B.1.2 (Classical Schwarz–Pick lemma). *If* $f \in \mathcal{O}(\mathbb{D}, \mathbb{D})$, *then*

$$\left|\frac{f(\lambda') - f(\lambda'')}{1 - \overline{f(\lambda'')}f(\lambda')}\right| \le \left|\frac{\lambda' - \lambda''}{1 - \overline{\lambda}''\lambda'}\right|, \qquad \lambda', \lambda'' \in \mathbb{D}, \tag{B.1.1}$$

$$\frac{|f'(\lambda'')|}{1 - |f(\lambda'')|^2} \le \frac{1}{1 - |\lambda''|^2}, \qquad \lambda'' \in \mathbb{D}. \tag{B.1.2}$$

Moreover, if there are $\lambda', \lambda'' \in \mathbb{D}$, $\lambda' \ne \lambda''$ (*resp.* $\lambda'' \in \mathbb{D}$) *such that equality holds in* (B.1.1) (*resp.* (B.1.2)), *then* f *is of the form*

$$f(\lambda) = e^{i\theta}\frac{\lambda - \lambda''}{1 - \overline{\lambda}''\lambda} \quad \text{with a suitable} \quad \theta \in \mathbb{R}.$$

Recall that the proof of B.1.1 is based on the maximum principle and the theorem on removable singularities. Therefore, the Schwarz lemma remains true in more general situations; for example, compare B.4.24.

B.1.3 (Fatou lemma). *Let* f *be a bounded holomorphic function on* $\mathbb{D}$. *Then for almost all* $\zeta \in \mathbb{T}$ *the function* f *has a non-tangential limit at* ζ, *i.e.,*

$$f^*(\zeta) = \lim_{\substack{\lambda \to \zeta \\ \sphericalangle}} f(\lambda) = \lim_{\Gamma_\alpha(\zeta) \ni \lambda \to \zeta} f(\lambda)$$

exists and is independent of $\alpha > 1$, *where* $\Gamma_\alpha(\zeta) := \{\lambda \in \mathbb{D} : |\lambda - \zeta| < \alpha(1 - |\lambda|)\}$.

B.1.4. *Any proper holomorphic map $f \in \mathcal{O}(P, P)$, where*

$$P := \{\lambda \in \mathbb{C} : 1/R < |\lambda| < R\}, \quad R > 1,$$

is biholomorphic.

For a proof, the reader is referred to theorem 14.22 in [461] and asked to make the necessary modifications to the proof there.

B.1.5 (Theorem of Bohr, cf. [72]). *Any continuous injective function on a domain $G \subset \mathbb{C}$ is holomorphic or antiholomorphic if for every $z_0 \in G$ the following limit exists:*

$$\lim_{z \nrightarrow z_0} \left| \frac{f(z) - f(z_0)}{z - z_0} \right| \in (0, \infty).$$

Observe that the assumption "injective" is important here, as the following example shows:

$$f(\lambda) := \begin{cases} \lambda & \text{if } \operatorname{Im} \lambda \geq 0 \\ \overline{\lambda} & \text{if } \operatorname{Im} \lambda \leq 0 \end{cases}.$$

The papers [200] and [502] may serve as a source of helpful information about B.1.5 and similar questions.

B.1.6. A *holomorphic covering* is a holomorphic map $\Pi : M_1 \longrightarrow M_2$ between two connected complex manifolds satisfying the following property:

any point $w_0 \in M_2$ has a neighborhood $U = U(w_0)$ such that $\Pi^{-1}(U)$ is the union of pairwise disjoint open subsets V_i of M_1 with $\Pi|_{V_i} : V_i \longrightarrow U$ being biholomorphic, $i \in I$; Π is called the *projection map*.

B.1.7 (Uniformization theorem). *Let G be an arbitrary domain in $\mathbb{C}$. Then there exists a holomorphic covering $\Pi : M \longrightarrow G$, where M is either $\mathbb{C}$ or $\mathbb{D}$, such that for any other holomorphic covering $\widetilde{\Pi} : \widetilde{M} \longrightarrow G$ and for any points $z' \in M$, $\widetilde{z}' \in \widetilde{M}$ with $\Pi(z') = \widetilde{\Pi}(\widetilde{z}')$, there is a unique holomorphic map $f : M \longrightarrow \widetilde{M}$ satisfying $\widetilde{\Pi} \circ f = \Pi$ and $f(z') = \widetilde{z}'$.*

Recall that $\Pi : M \longrightarrow G$ is uniquely defined up to a biholomorphic mapping commuting with its projections; $\Pi : M \longrightarrow G$ is called the *universal covering of G*. In short, the above result says that any domain G in the plane has $\mathbb{D}$ or $\mathbb{C}$ as its universal covering.

B.1.8 (Koebe distortion theorem). (a) (Cf. [437], Theorem 1.3) *If $f : \mathbb{D} \longrightarrow \mathbb{C}$ is holomorphic and injective, $f(0) = 0$, and $f'(0) = 1$, then,*

$$\frac{1-|\lambda|}{(1+|\lambda|)^3} \le |f'(\lambda)| \le \frac{1+|\lambda|}{(1-|\lambda|)^3},$$
$$\frac{|\lambda|}{(1+|\lambda|)^2} \le |f(\lambda)| \le \frac{|\lambda|}{(1-|\lambda|)^2}, \quad \lambda \in \mathbb{D}.$$

(b) (Cf. [437], Corollary 1.4) *If $f : \mathbb{D} \longrightarrow \mathbb{C}$ is holomorphic and injective, then*

$$\frac{1}{4}(1-|\lambda|^2)|f'(\lambda)| \le \operatorname{dist}(f(\lambda), \partial(f(\mathbb{D}))) \le (1-|\lambda|^2)|f'(\lambda)|, \quad \lambda \in \mathbb{D}.$$

So far, we have collected the main results from the classical complex analysis of the plane, which were applied in this book. Now, we discuss a few results from several complex variables.

B.1.9 (Hartogs' theorem). *A function $f : G \longrightarrow \mathbb{C}$, where G is a domain in $\mathbb{C}^n$, is holomorphic iff it is holomorphic in each variable separately, i.e., if $z^0 \in G$ and $1 \le j \le n$ are fixed, then the function $f(z_1^0, \dots, z_{j-1}^0, \cdot, z_{j+1}^0, \dots, z_n^0)$ is holomorphic on the open set $\{\lambda \in \mathbb{C} : (z_1^0, \dots, z_{j-1}^0, \lambda, z_{j+1}^0, \dots, z_n^0) \in G\}$.*

For more information on separately holomorphic functions, see [270].

B.1.10 (Montel). *Any locally bounded family $\mathcal{F}$ of holomorphic functions on a domain $G \subset \mathbb{C}^n$ is normal, i.e., any sequence $(f_\nu)_{\nu\in\mathbb{N}} \subset \mathcal{F}$ has a subsequence $(f_{\nu_j})_{j\in\mathbb{N}}$ that converges locally uniformly to an $f \in \mathcal{O}(G)$.*

B.1.11 (Holomorphic extension). *Let G be a domain in $\mathbb{C}^n$.*

(a) *If $n \ge 2$ and if $K \subset G$ is a compact subset of G such that $G \setminus K$ is connected, then any $f \in \mathcal{O}(G \setminus K)$ extends to an $\tilde{f} \in \mathcal{O}(G)$, i.e., $\tilde{f}|_{G\setminus K} = f|_{G\setminus K}$.*

(b) *If $F \subset G$ is a closed (with respect to G) subset of G with $H^{2n-2}(F) = 0$, then any $f \in \mathcal{O}(G \setminus F)$ has a holomorphic extension to G.*

(c) *If $F \subset G$ is relatively closed with $H^{2n-1}(F) = 0$, then every $f \in \mathcal{O}(G \setminus F) \cap L^\infty_{\mathrm{loc}}(G)$ is holomorphically continuable to the whole G.*

In particular, we recall that any proper analytic subset of G satisfies the condition in (c). Therefore, the classical Riemann removable singularity theorem is a particular case of (c).

B.1.12. *Let G and $\tilde{G}$ be domains in $\mathbb{C}^n$ such that any bounded holomorphic function on G extends to a holomorphic function $\tilde{f}$ on $\tilde{G}$ with $\tilde{f}|_{G_0} = f|_{G_0}$, where G_0 is a fixed non-empty open subset of $G \cap \tilde{G}$. Then, $\|\tilde{f}\|_{\tilde{G}} \le \|f\|_G$.*

B.1.13. *Any distribution T (in the sense of L. Schwartz) on $G \subset \mathbb{C}^n$ with $\bar{\partial}T = 0$ is given by a holomorphic function f, i.e., $\langle f, \varphi \rangle = \langle T, \varphi \rangle$ for all test functions $\varphi \in \mathcal{C}_0^\infty(G)$.*

We denote by $L_h^2(G)$ the set of all square integrable holomorphic functions on a domain G.

B.1.14 (Holomorphic extension for L^2-holomorphic functions). *Let A be a proper analytic subset of a domain $G \subset \mathbb{C}^n$. Then any function $f \in \mathcal{O}(G \setminus A) \cap L^2_{\mathrm{loc}}(G)$ extends to a holomorphic function on G.*

B.1.14 can be found in [445] as exercise E 3.2; a proof of it is contained in, for example, [50].

B.1.15. *Let $G_j \subset \mathbb{C}^{n_j}$ be arbitrary domains, $j = 1, 2$. Then the set*

$$\left\{ \sum_{j=1}^{N} f_j g_j : N \in \mathbb{N},\ f_j \in \mathcal{O}(G_1),\ g_j \in \mathcal{O}(G_2),\ 1 \le j \le N \right\}$$

is a dense subset in $\mathcal{O}(G_1 \times G_2)$ with respect to the topology of uniform convergence on compact subsets.

A proof of B.1.15 may be found in [369], Ch I, § 5.

B.1.1 Analytic sets

Cf. [105]. Let $\Omega \subset \mathbb{C}^n$ be open. A set $V \subset \Omega$ is said to be an *analytic subset of* Ω if for any point $a \in \Omega$ there exist a neighborhood $U_a \subset \Omega$ and a finite family $\mathcal{F}_a \subset \mathcal{O}(U_a)$ such that $V \cap U_a = \bigcap_{f \in \mathcal{F}_a} f^{-1}(0)$. Note that M is closed in Ω.

Suppose that $V \subset \Omega$ is analytic. A point $a \in V$ is *regular* ($a \in \mathrm{Reg}(V)$) if there exists a neighborhood $U_a \subset \Omega$ such that $V \cap U_a$ is a complex manifold. Points from $\mathrm{Sing}(V) := V \setminus \mathrm{Reg}(V)$ are called *singular*.

Obviously, $\mathrm{Reg}(V)$ is open in V and $\mathrm{Sing}(V)$ is closed in Ω. One can prove that $\mathrm{Reg}(V)$ is dense in V. We define the *dimension of V at a point $a \in V$*

$$\dim_a V := \limsup_{\mathrm{Reg}(V) \ni z \to a} \dim_z V$$

and the (global) *dimension* $\dim V := \max_{a \in V} \dim_a V$.

We say that V is a *proper analytic subset of* Ω if $\dim V < n$. One can prove that $\mathrm{Sing}(V)$ is an analytic subset of Ω and $\dim_z \mathrm{Sing}(V) < \dim_z V$, $z \in \mathrm{Sing}(V)$.

We say that a function $f : V \longrightarrow \mathbb{C}$ is *holomorphic* on V ($f \in \mathcal{O}(V)$) if for every point $a \in V$ there exist an open neighborhood $U_a \subset \mathbb{C}^n$ of a and a (standard) holomorphic function $f_a \in \mathcal{O}(U_a)$ such that $f_a = f$ on $V \cap U_a$.

B.2 Proper holomorphic mappings

B.2.1 (cf. [462]).

(a) *Let $G \subset \mathbb{C}^n$ be a domain and let $F : G \longrightarrow \mathbb{C}^n$ be a proper holomorphic mapping. Then F is open.*

(b) *Let $G, G' \subset \mathbb{C}^n$ be domains and let $F : G \longrightarrow G'$ be a proper holomorphic mapping. Let $\Delta := \{z \in G : \det F'(z) = 0\}$, $\Sigma := F(\Delta)$. Then,*

- *$F(G) = G'$.*
- *For every analytic subset $V \subset G$, the set $F(V)$ is an analytic subset of G'.*
- *Σ is a proper analytic subset of G'.*
- *There exists an $m \in \mathbb{N}$ (called the* multiplicity of F*) such that*
  - *for any* regular value $w_0 \in G' \setminus \Sigma$ *there exist an open neighborhood $U \subset G'$ and holomorphic mappings $\varphi_1, \dots, \varphi_m : U \longrightarrow G$ for which $F^{-1}(U) = \bigcup_{j=1}^m \varphi_j(U)$ and $\varphi_j(w) \neq \varphi_k(w)$, $j \neq k$, $w \in U$ ($\varphi_1, \dots, \varphi_m$ are branches of F^{-1} in U),*
  - *for any* critical value $w_0 \in \Sigma$ *we have $\#F^{-1}(w_0) < m$.*

B.2.2 (cf. [463]). *Let $F_j : \mathbb{C}^n \longrightarrow \mathbb{C}$ be a homogeneous polynomial,* $\deg F_j = d_j > 0$, $j = 1, \dots, n$, *$F := (F_1, \dots, F_n) : \mathbb{C}^n \longrightarrow \mathbb{C}^n$. Then,*

(a) *F is proper iff $F^{-1}(0) = 0$.*

(b) *If F is proper, then its multiplicity equals $d_1 \cdots d_n$.*

B.3 Automorphisms

References: [269, 370, 462]. Let G be a domain in $\mathbb{C}^n$, $z_0 \in G$. Set

$$\operatorname{Aut}(G) := \{F : G \longrightarrow G : F \text{ is biholomorphic}\},$$
$$\operatorname{Aut}_{z_0}(G) := \{F \in \operatorname{Aut}(G) : F(z_0) = z_0\}.$$

The group $\operatorname{Aut}(G)$ is called the *group of automorphisms of G*. We say that the group $\operatorname{Aut}(G)$ *acts transitively on* G if for any $z', z'' \in G$ there exists an automorphism $F \in \operatorname{Aut}(G)$ such that $F(z') = z''$.

B.3.1 Automorphisms of the unit disc

Let $\mathbb{D}$ be the unit disc in $\mathbb{C}$. For $a \in \mathbb{D}$ define

$$\boldsymbol{h}_a(\lambda) := \frac{\lambda - a}{1 - \overline{a}\lambda}.$$

B.3.1.

(a) $\operatorname{Aut}(\mathbb{D}) = \{e^{i\theta}\boldsymbol{h}_a : \theta \in \mathbb{R},\ a \in \mathbb{D}\}$.

(b) $\operatorname{Aut}_0(\mathbb{D}) = \{e^{i\theta}\,\mathrm{id} : \theta \in \mathbb{R}\}$ *is equal to the group of rotations.*

(c) *The group* $\operatorname{Aut}(\mathbb{D})$ *acts transitively on* $\mathbb{D}$; $(\boldsymbol{h}_a)^{-1} = \boldsymbol{h}_{-a}$, $\boldsymbol{h}_a(a) = 0$,

$$\boldsymbol{h}'_a(a) = \frac{1}{1-|a|^2}.$$

B.3.2 Automorphisms of the unit polydisc

B.3.2.

(a) $\operatorname{Aut}(\mathbb{D}^n) = \{\mathbb{D}^n \ni (z_1,\dots,z_n) \longrightarrow (e^{i\theta_1}\boldsymbol{h}_{a_1}(z_{\sigma(1)}),\dots,e^{i\theta_n}\boldsymbol{h}_{a_n}(z_{\sigma(n)})) :$
$\theta_1,\dots,\theta_n \in \mathbb{R},\ a_1,\dots,a_n \in \mathbb{D},\ \sigma$ *is an arbitrary permutation of* $(1,\dots,n)\}$.

(b) $\operatorname{Aut}_0(\mathbb{D}^n) = \{\mathbb{D}^n \ni (z_1,\dots,z_n) \longrightarrow (e^{i\theta_1}z_{\sigma(1)},\dots,e^{i\theta_n}z_{\sigma(n)}) :$
$\theta_1,\dots,\theta_n \in \mathbb{R},\ \sigma$ *is an arbitrary permutation of* $(1,\dots,n)\}$.

(c) *The group* $\operatorname{Aut}(\mathbb{D}^n)$ *acts transitively on* $\mathbb{D}^n$.

B.3.3 Automorphisms of the unit Euclidean ball

Let $\mathbb{B}_n$ be the unit Euclidean ball in $\mathbb{C}^n$. For $a \in \mathbb{B}_n$, define

$$\boldsymbol{h}_a(z) := \frac{1}{\|a\|^2}\,\frac{\sqrt{1-\|a\|^2}(\|a\|^2 z - \langle z,a\rangle a) - \|a\|^2 a + \langle z,a\rangle a}{1-\langle z,a\rangle} \qquad \text{if } a \neq 0,$$
$$\boldsymbol{h}_0(z) := \mathrm{id},$$

where $\|\ \|$ stands for the Euclidean norm and $\langle\cdot,\cdot\rangle$ for the standard complex scalar product in $\mathbb{C}^n$. We have

$$1 - \langle \boldsymbol{h}_a(z), \boldsymbol{h}_a(w)\rangle = \frac{(1-\langle a,a\rangle)(1-\langle z,w\rangle)}{(1-\langle z,a\rangle)(1-\langle a,w\rangle)}, \quad z,w \in \overline{\mathbb{B}}_n\ (a \in \mathbb{B}_n).$$

B.3.3.

(a) $\operatorname{Aut}(\mathbb{B}_n) = \{U \circ \boldsymbol{h}_a : U$ *is a unitary operator on* $\mathbb{C}^n$ *and* $a \in \mathbb{B}_n\}$.

(b) $\operatorname{Aut}_0(\mathbb{B}_n)$ *is the group of unitary operators on* $\mathbb{C}^n$.

(c) *The group* $\operatorname{Aut}(\mathbb{B}_n)$ *acts transitively on* $\mathbb{B}_n$; $(\boldsymbol{h}_a)^{-1} = \boldsymbol{h}_{-a}$, $\boldsymbol{h}_a(a) = 0$,

$$\boldsymbol{h}'_a(a)(X) = \frac{1}{\|a\|^2}\,\frac{\sqrt{1-\|a\|^2}(\|a\|^2 X - \langle X,a\rangle a) + \langle X,a\rangle a}{1-\|a\|^2},$$
$$X \in \mathbb{C}^n\ (a \neq 0).$$

B.4 Subharmonic and plurisubharmonic functions

References: [265, 305, 440].

For an open set $G \subset \mathbb{C}^1$, an upper semicontinuous function $u : G \longrightarrow [-\infty, +\infty)$ is said to be *subharmonic* (shortly *sh*) if for every relatively compact open subset G_0 of G and for every function h, continuous on $\overline{G_0}$ and harmonic on G_0, we have that if $u \le h$ on ∂G_0, then $u \le h$ on G_0. We will write $u \in \mathcal{SH}(G)$.

Now let $G \subset \mathbb{C}^n$ be open. An upper semicontinuous function $u : G \longrightarrow \mathbb{R}_{-\infty}$ is called *plurisubharmonic* (*psh*) on G if for each $a \in G$ and $X \in \mathbb{C}^n$ the function

$$\{\lambda \in \mathbb{C} : a + \lambda X \in G\} \ni \lambda \longmapsto u(a + \lambda X) \in [-\infty, +\infty)$$

is subharmonic (as a function of one complex variable). The set of all psh functions on G will be denoted by $\mathcal{PSH}(G)$. We say that a function $u : G \longrightarrow [0, +\infty)$ is *logarithmically-plurisubharmonic* (*log-psh*) on G if the function $\log u$ is psh on G.

B.4.1. *The set $\mathcal{PSH}(G)$ is a convex cone, i.e., for any $u_1, u_2 \in \mathcal{PSH}(G)$ and for any $t_1, t_2 > 0$ we have $t_1u_1 + t_2u_2 \in \mathcal{PSH}(G)$.*

B.4.2. *The plurisubharmonicity is a local property, i.e., a function $u : G \longrightarrow \mathbb{R}_{-\infty}$ is psh on G iff for each $a \in G$ there exists an open neighborhood U of a in G such that $u|_U \in \mathcal{PSH}(U)$.*

B.4.3. *Let $u : G \longrightarrow [-\infty, +\infty)$ be upper semicontinuous. Then $u \in \mathcal{PSH}(G)$ iff for any $a \in G$, $X \in \mathbb{C}^n$, and $r > 0$ with $a + r\overline{\mathbb{D}}X \subset G$ we have*

$$u(a) \le \frac{1}{2\pi} \int_0^{2\pi} u(a + re^{i\theta} X) d\theta.$$

B.4.4. *If $f \in \mathcal{O}(G)$, then $|f|$ is log-psh on G.*

B.4.5. *Let $D \subset \mathbb{C}^n$ be a domain and let $u \in \mathcal{PSH}(D), u \not\equiv -\infty$. Then, u belongs to $L^1_{\text{loc}}(D)$. In particular, the set $u^{-1}(-\infty)$ has Lebesgue measure zero.*

B.4.6 (Maximum principle). *Assume that $D \subset \mathbb{C}^n$ is a bounded domain and let $u \in \mathcal{PSH}(D)$, $u \not\equiv$ const. Then, $u(z) < \sup_{\zeta \in \partial D} \{\limsup_{D \ni z \to \zeta} u(z)\}$, $z \in D$.*

B.4.7. *Let $u \subset \mathcal{SH}(\mathbb{D}(R))$ be such that $u(z) = u(|z|)$, $z \in \mathbb{D}(R)$, i.e., u is a* radial function. *Then, $u|_{(0,R)}$ is monotonically increasing.*

B.4.8 (cf. [218]). *Let $H := \{\lambda \in \mathbb{C} : \operatorname{Re} \lambda < 0\}$, $b < 0$, and $M < 0$. Moreover, let $u \in \mathcal{SH}(H)$, $u < 0$, and $u(\lambda) \le M$ for all λ with $\operatorname{Re} \lambda = b$. Then, $u \le M$ on $\{\lambda \in \mathbb{C}\ \operatorname{Re} \lambda \le b\}$.*

To prove the above result, it suffices to use a biholomorphic mapping from H onto $\mathbb{D}$ and the extended maximum principle for subharmonic functions (see [446], Theorem 3.6.7).

B.4.9. *Let G_j be an open set in $\mathbb{C}^{n_j}$, $j = 1, 2$, and let $F : G_1 \longrightarrow G_2$ be a holomorphic mapping. Then for any $u \in \mathcal{PSH}(G_2)$ the function $u \circ F$ belongs to $\mathcal{PSH}(G_1)$.*

B.4.10. *Let $I \subset [-\infty, +\infty)$ be an interval and let $\varphi : I \longrightarrow \mathbb{R}$ be increasing and convex (if $-\infty \in I$, then we assume that φ is increasing, convex in the interval $I \cap \mathbb{R}$, and continuous at $-\infty$). Then for any psh function $u : G \longrightarrow I$ the function $\varphi \circ u$ belongs to $\mathcal{PSH}(G)$. In particular,*

- *if $u \in \mathcal{PSH}(G)$, then $e^u \in \mathcal{PSH}(G)$,*
- *if $u : G \longrightarrow \mathbb{R}_+$ is psh, then u^α is psh on G for every $\alpha \geq 1$,*
- *if u is log-psh on G, then u^α is psh on G for every $\alpha > 0$.*

B.4.11. *If u_1, u_2 are log-psh on G, then $u_1 + u_2$ is log-psh on G.*

B.4.12. *If $q : \mathbb{C}^n \longrightarrow \mathbb{R}_+$ is a $\mathbb{C}$-seminorm, then q is log-psh on $\mathbb{C}^n$.*

B.4.13. *Let $h : \mathbb{C}^n \longrightarrow \mathbb{R}_+$ be such that $h(\lambda z) = |\lambda| h(z)$, $\lambda \in \mathbb{C}$, $z \in \mathbb{C}^n$. Then, h is psh on $\mathbb{C}^n$ iff h is log-psh on $\mathbb{C}^n$.*

B.4.14. *If $(u_\nu)_{\nu=1}^\infty \subset \mathcal{PSH}(G)$ and $u_\nu \searrow u_0$ pointwise on G, then $u_0 \in \mathcal{PSH}(G)$.*

B.4.15. *If $(u_\nu)_{\nu=1}^\infty \subset \mathcal{PSH}(G)$ and $u_\nu \underset{\nu\to\infty}{\overset{K}{\Longrightarrow}} u_0$, then $u_0 \in \mathcal{PSH}(G)$.*

B.4.16. *If $(u_\alpha)_{\alpha\in A} \subset \mathcal{PSH}(G)$ is locally uniformly bounded from above, then the function $u_0 := (\sup\limits_{\alpha\in A} u_\alpha)^*$ is psh on G, where "$*$" denotes the upper semicontinuous regularization.*

B.4.17. *If $(u_\nu)_{\nu=1}^\infty \subset \mathcal{PSH}(G)$ is locally uniformly bounded from above, then the function $u_0 := (\limsup\limits_{\nu\to\infty} u_\nu)^*$ is psh on G.*

B.4.18. *Let $u \in \mathcal{PSH}(G), u_0 \in \mathcal{PSH}(G_0)$ with $G_0 \subset G$. Suppose that*

$$\limsup_{G_0 \ni z \to \zeta} u_0(z) \leq u(\zeta), \quad \zeta \in (\partial G_0) \cap G.$$

Then, the function

$$\tilde{u}(z) := \begin{cases} \max\{u(z), u_0(z)\}, & z \in G_0 \\ u(z), & z \in G \setminus G_0 \end{cases}$$

is psh on G.

Let $\Phi \in \mathcal{C}_0^\infty(\mathbb{C}^n, \mathbb{R}_+)$ be such that

$$\Phi(z) = \Phi(|z_1|, \dots, |z_n|), \quad z = (z_1, \dots, z_n) \in \mathbb{C}^n,$$
$$\operatorname{supp} \Phi = \overline{\mathbb{B}}_n, \quad \int_{\mathbb{C}_n} \Phi d\mathcal{L}^{2n} = 1.$$

Put

$$\Phi_\varepsilon(z) := \frac{1}{\varepsilon^{2n}} \Phi\left(\frac{z}{\varepsilon}\right), \quad \varepsilon > 0, \ z \in \mathbb{C}^n.$$

If G is an open set in $\mathbb{C}^n$, then we put $G_\varepsilon := \{z \in G : z + \varepsilon\mathbb{B}_n \subset G\}$.

B.4.19. *If $u \in \mathcal{PSH}(G)$, then $u * \Phi_\varepsilon \in \mathcal{PSH}(G_\varepsilon) \cap \mathcal{C}^\infty(G_\varepsilon)$ and $u * \Phi_\varepsilon \searrow u$ pointwise on G as $\varepsilon \searrow 0$.*

Here "$*$" denotes the convolution operator.

B.4.20. *Let $u_1, u_2 \in \mathcal{PSH}(G)$. If $u_1 = u_2$ almost everywhere in G (with respect to Lebesgue measure), then $u_1 \equiv u_2$.*

B.4.21 (Hartogs' lemma). *Let $(u_\nu)_{\nu=1}^\infty \subset \mathcal{PSH}(G)$ be locally uniformly bounded from above and suppose that $\limsup_{\nu\to\infty} u_\nu \le M$ on G. Then for each $\varepsilon > 0$ and for each compact set $K \subset G$ there exists a ν_0 such that*

$$\sup_K u_\nu \le M + \varepsilon, \quad \nu \ge \nu_0.$$

A set $P \subset \mathbb{C}^n$ is called *pluripolar* if for each $a \in P$ there exist a connected neighborhood U_a of a and a function $u_a \in \mathcal{PSH}(U_a)$, $u_a \not\equiv -\infty$, such that $P \cap U_a \subset u_a^{-1}(-\infty)$.

B.4.22. *Any proper analytic set is pluripolar. Pluripolar sets have Lebesgue measure zero.*

B.4.23 (Removable singularities theorem for psh functions). *Assume that $P \subset G$ is a closed (in G) pluripolar set.*

(a) *Let $u \in \mathcal{PSH}(G \setminus P)$ be locally bounded from above in G. Then the function*

$$\tilde{u}(z) := \begin{cases} u(z), & z \in G \setminus P \\ \limsup_{G\setminus P \ni z' \to z} u(z'), & z \in P \end{cases},$$

is psh on G.

(b) *If G is connected, then so is $G \setminus P$.*

(c) *For any* $u \in \mathcal{PSH}(G)$ *we have*

$$\limsup_{G\setminus P \ni z' \to z} u(z') = u(z), \quad z \in G.$$

(d) *Any* $f \in \mathcal{H}^\infty(G \setminus P)$ *extends holomorphically to* G.

B.4.24 (Schwarz lemma for log-sh functions). *Let u be a log-sh function on $\mathbb{D}$ such that*

- *the function $\lambda \longmapsto u(\lambda)/|\lambda|$ is bounded near zero and*
- $\limsup_{|\lambda|\to 1-} u(\lambda) \le 1$.

Then $u(\lambda) \le |\lambda|$, $\lambda \in \mathbb{D}$.

B.4.25 (Hadamard three circles theorem). *Let u be a log-sh function on the annulus* $P := \{\lambda \in \mathbb{C} : r_1 < |\lambda| < r_2\}$ *with* $0 < r_1 < r_2 < +\infty$. *Suppose that* $\limsup_{|\lambda|\to r_j} u(\lambda) \le M_j$, $j = 1, 2$. *Then,*

$$u(\lambda) \le M_1^{\frac{\log \frac{r_2}{|\lambda|}}{\log \frac{r_2}{r_1}}} M_2^{\frac{\log \frac{|\lambda|}{r_1}}{\log \frac{r_2}{r_1}}}, \quad \lambda \in P.$$

B.4.26 (Oka theorem). *Let $G \subset \mathbb{C}^1$ and let $u \in \mathcal{SH}(G)$. Suppose that $\alpha : [0,1] \longrightarrow G$ is a Jordan curve. Then* $\limsup_{t\to 0+} u(\alpha(t)) = u(\alpha(0))$.

B.4.27. *If $u \in \mathcal{PSH}(\mathbb{C}^n)$ is bounded from above, then* $u \equiv \text{const}$.

If $u \in \mathcal{C}^2(G)$, then $\mathcal{L}u : G \times \mathbb{C}^n \longrightarrow \mathbb{C}$ will denote the *Levi form* of u, that is,

$$(\mathcal{L}u)(a; X) := \sum_{j,k=1}^{n} \frac{\partial^2 u}{\partial z_j \partial \overline{z}_k}(a) X_j \overline{X}_k, \quad a \in G,\ X \in \mathbb{C}^n.$$

Observe that

$$(\mathcal{L}u)(a; X) = \frac{\partial^2 u_{a,X}}{\partial \lambda \partial \overline{\lambda}}(0),$$

where $u_{a,X}(\lambda) := u(a + \lambda X)$. In particular, in the case where $n = 1$ we get $(\mathcal{L}u)(a; X) = \frac{1}{4}\Delta u(a)|X|^2$, where Δ is the Laplace operator in $\mathbb{R}^2$.

B.4.28. *Let $u \in \mathcal{C}^2(G, \mathbb{R})$. Then $u \in \mathcal{PSH}(G)$ iff $(\mathcal{L}u)(a; X) \ge 0$ for any $a \in G$ and $X \in \mathbb{C}^n$.*

B.4.29. *If $u \in \mathcal{PSH}(G)$, then $\mathcal{L}u \ge 0$ in the sense of distributions, i.e.,*

$$\int_G u(z)(\mathcal{L}\varphi)(z; X) d\mathcal{L}^{2n}(z) \ge 0, \quad \varphi \in \mathcal{C}_0^\infty(G, \mathbb{R}_+),\ X \in \mathbb{C}^n.$$

B.4.30. *Let $u : G \longrightarrow [-\infty, +\infty]$ be a locally integrable function such that $\mathcal{L}u \geq 0$ in the sense of distributions. Then there exists a function $\widetilde{u} \in \mathcal{PSH}(G)$ such that $u = \widetilde{u}$ almost everywhere in G.*

(a) A function $u \in \mathcal{C}^2(G, \mathbb{R})$ is said to be *strictly plurisubharmonic* on G if

$$(\mathcal{L}u)(a; X) > 0 \text{ for any } a \in G \text{ and } X \in (\mathbb{C}^n)_*.$$

(b) A function $u \in \mathcal{C}(G, \mathbb{R})$ is called *strictly plurisubharmonic* on G if for any relatively compact open subset G_0 of G there exists an $\varepsilon > 0$ such that the function $G_0 \ni z \longmapsto u(z) - \varepsilon\|z\|^2$ is psh.

B.4.31. *A function $u \in \mathcal{C}^2(G, \mathbb{R})$ is strictly plurisubharmonic in the sense of (a) iff it is strictly plurisubharmonic in the sense of (b).*

B.4.32 (cf. [451]). *Let $u \in \mathcal{C}(G, \mathbb{R})$ be a strictly plurisubharmonic function and let $\varepsilon : G \longrightarrow (0, +\infty)$ be an arbitrary continuous function. Then there exists a strictly plurisubharmonic function $v \in \mathcal{C}^\infty(G, \mathbb{R})$ such that $u < v < u + \varepsilon$.*

A simple proof of B.4.32 can be found in [174].

B.4.33 (cf. [305], Proposition 2.9.26). *Let $F \in \mathcal{O}(G, D)$ be proper, $v \in \mathcal{PSH}(G)$. Put $u(w) := \max\{v(z) : z \in F^{-1}(w)\}$, $w \in D$. Then $u \in \mathcal{PSH}(D)$.*

The definition of a psh function may be extended to analytic sets. Let V be an analytic subset of an open set $\Omega \subset \mathbb{C}^n$. A function $u : V \longrightarrow [-\infty, +\infty)$ is said to be *plurisubharmonic* (*psh*) ($u \in \mathcal{PSH}(V)$) if for every $a \in V$ there exist an open neighborhood $U_a \subset \Omega$ and a function $u_a \in \mathcal{PSH}(U_a)$ (psh in the classical sense) such that $u = u_a$ on $V \cap U_a$.

B.4.34 (cf. [172]). *Let V be an analytic subset of an open set $\Omega \subset \mathbb{C}^n$ and let $u : V \longrightarrow [-\infty, +\infty)$. Then, the following conditions are equivalent:*

(i) *$u \in \mathcal{PSH}(V)$;*

(ii) *u is upper semicontinuous and $u \circ \varphi \in \mathcal{SH}(\mathbb{D})$ for an arbitrary holomorphic disc $\varphi : \mathbb{D} \longrightarrow V$.*

B.5 Green function and Dirichlet problem

References: [218, 221, 305, 333, 371, 446].

Let G be any bounded domain in $\mathbb{C}$. Then, the *Dirichlet problem* for G is the following question:

given a function $f \in \mathcal{C}(\partial G, \mathbb{R})$, find a function $h \in \mathcal{C}(\overline{G}, \mathbb{R})$ with $h|_{\partial G} = f$ such that h is harmonic in G.

Because of the maximum principle for harmonic functions, the solution of the Dirichlet problem, if it exists, is uniquely determined. A domain is called *regular with respect to the Dirichlet problem*, or simply, a *Dirichlet domain*, if for any continuous boundary function f there exists a solution of the Dirichlet problem. In the case $G = \mathbb{D}$ we have the following explicit solution. Let

$$P_r : \mathbb{D}(r) \times \partial\mathbb{D}(r) \longrightarrow \mathbb{R}_{>0}, \quad P_r(z,\zeta) = \frac{r^2 - |z|^2}{|\zeta - z|^2} = \operatorname{Re}\left(\frac{\zeta + z}{\zeta - z}\right), \qquad P := P_1,$$

be the *Poisson kernel*.

B.5.1 (Poisson integral formula). *For any* $f \in \mathcal{C}(\mathbb{T}, \mathbb{R})$ *the function*

$$h(z) := \begin{cases} \frac{1}{2\pi}\int_0^{2\pi} P(z, e^{i\theta}) f(e^{i\theta}) d\theta, & \text{if } z \in \mathbb{D} \\ f(z), & \text{if } z \in \mathbb{T} \end{cases}$$

solves the Dirichlet problem.

More generally, the following result holds:

B.5.2. *A bounded domain $G \subset \mathbb{C}$ is a Dirichlet domain iff every boundary point ζ of G is a local peak point for continuous subharmonic functions, i.e,. for any $\zeta \in \partial G$ there exist a neighborhood V of ζ and a function $u \in \mathcal{C}(G \cap V) \cap \mathcal{SH}(G \cap V)$ such that*

(i) $\displaystyle\lim_{G\cap V \ni z \to \zeta} u(z) = 0,$

(ii) $\displaystyle\limsup_{G\cap V \ni z \to \xi} u(z) < 0, \quad \xi \in \partial G \cap V, \ \xi \neq \zeta,$

(iii) $u(z) < 0, \quad z \in G \cap V.$

B.5.3 (Theorem of Bouligand). *Let G be bounded and let $a \in \partial G$. Suppose that there exist $V = V(a)$ and $u \in \mathcal{C}(G \cap V) \cap \mathcal{SH}(G \cap V)$ with*

$$\lim_{G\cap V \ni z \to a} u(z) = 0, \quad u(z) < 0, \ z \in G \cap V.$$

Then a is a local peak point for continuous subharmonic functions. Even more is true. Namely, there exists a harmonic function h on the whole G with

$$\lim_{G \ni z \to a} h(z) = 0, \quad \limsup_{G \ni z \to b} h(z) < 0, \ b \in (\partial G) \setminus \{a\}.$$

In particular, we have the following sufficient criterion:

B.5.4. *Any bounded domain G in $\mathbb{C}$ such that no connected component of $\mathbb{C} \setminus G$ reduces to a point is a Dirichlet domain.*

For a domain $G \subset \mathbb{C}$ fix $a \in G$. The (*classical*) *Green function of G with pole at a* is a function $\mathfrak{g}_G(a,\cdot) : G \setminus \{a\} \longrightarrow \mathbb{R}$ that satisfies the following three properties:

(1) $\mathfrak{g}_G(a,\cdot)$ is harmonic on $G \setminus \{a\}$;

(2) $\mathfrak{g}_G(a,\cdot) + \log|\cdot - a|$ extends to a harmonic function on G;

(3) there exists a polar set $F \subset \partial G$, such that:

- if $\zeta \in (\partial G) \setminus F$, then $\lim_{G \ni z \to \zeta} \mathfrak{g}_G(a,z) = 0$,
- if $\zeta \in F$ or $\zeta = \infty \in \partial G$, then $\mathfrak{g}_G(a,\cdot)$ is bounded near ζ.

B.5.5. *If $G \subset \mathbb{C}$ is a domain whose boundary is not a polar subset of $\mathbb{C}$, then for every $a \in G$ the Green function $\mathfrak{g}_G(a,\cdot)$ exists and is unique. Moreover, $\mathfrak{g}_G(a,z) > 0$ for $z \neq a$ and the function $\mathfrak{g}_G : G \times G \longrightarrow (0,+\infty]$ is symmetric and continuous.*

Moreover, by the maximum principle for subharmonic functions we obtain

B.5.6. *Let $G \subset \mathbb{C}$ be a domain and let $a \in G$ be such that the Green function $\mathfrak{g}_G(a,\cdot)$ exists. Then, for any subharmonic function $u : G \longrightarrow [-\infty, 0)$ satisfying $u(z) \le C + \log|z-a|$ near a, we have the inequality $u(z) \le -\mathfrak{g}_G(a,z)$, $z \in G$.*

Therefore,

B.5.7. *Let G be an arbitrary domain in $\mathbb{C}$, $a \in G$. If ∂G is polar, then $\mathfrak{g}_G(a,\cdot) \equiv 0$ on G. If ∂G is not polar, then $\boldsymbol{g}_G(a,\cdot) = \exp(-\mathfrak{g}_G(a,\cdot))$ on G, where $\boldsymbol{g}_G$ denotes the "complex Green function"; cf. Chapter 4. In particular, $\boldsymbol{g}_G^2(a,\cdot)$ is of class $\mathcal{C}^\infty$ in G.*

B.5.8. *For $G = \mathbb{D}(R)$ the Green function is given by*

$$\mathfrak{g}_G(a,z) = \log\left(\frac{1}{R}\left|\frac{R^2 - \bar{a}z}{z-a}\right|\right), \quad z \neq a.$$

B.5.9. *Let $G \subset \mathbb{C}$ be a domain such that $\mathfrak{g}_G(a,\cdot)$ exists for every $a \in G$. Then the Green function is symmetric, i.e., $\mathfrak{g}_G(z',z'') = \mathfrak{g}_G(z'',z')$ whenever $z',z'' \in G$, $z' \neq z''$.*

We conclude this part of the appendix with the Riesz representation theorem.

B.5.10 (Riesz theorem). *Let $u \in \mathcal{SH}(G)$, $u \not\equiv -\infty$. Then there exists a unique Borel measure μ on G such that for any compact subset $K \subset G$ with $\operatorname{int}(K) \neq \varnothing$ we have*

$$u(z) = \int_K \log|z-\zeta| d\mu(\zeta) + h(z), \quad z \in \operatorname{int}(K),$$

where h is a harmonic function on $\operatorname{int}(K)$.

Thus, many of the local properties of subharmonic functions can be deduced from those of logarithmic potentials.

B.6 Monge–Ampère operator

References: [89, 305].

Let $d = \partial + \bar\partial$ denote the operator of exterior differentiation in $\mathbb{C}^n$. Define $d^c := i(\bar\partial - \partial)$. Let G be an open set in $\mathbb{C}^n$. Then the *Monge–Ampère operator* is an operator acting on $\mathcal{PSH}(G) \cap L^\infty_{\text{loc}}(G)$, that assigns to each function $u \in \mathcal{PSH}(G) \cap L^\infty_{\text{loc}}(G)$ a non-negative Borel measure $(dd^c u)^n$ on G. In the case where $u \in \mathcal{PSH}(G) \cap \mathcal{C}^2(G)$, the definition of $(dd^c u)^n$ is elementary, namely,

$$(dd^c u)^n = \left(\det\left[\frac{\partial^2 u}{\partial z_j \partial \bar z_k}\right]_{j,k=1,\dots,n}\right) \cdot \mathcal{L}^{2n}.$$

In the general case, if $u_1, \dots, u_n \in \mathcal{PSH}(G) \cap L^\infty_{\text{loc}}(G)$, then (for $k = 1, \dots, n$) $dd^c u_1 \wedge \dots \wedge dd^c u_k$ is defined inductively as a positive (k,k)-current of order 0 by the formula

$$\int_G dd^c u_1 \wedge \dots \wedge dd^c u_k \wedge \chi = \int_G u_k dd^c u_1 \wedge \dots \wedge dd^c u_{k-1} \wedge dd^c \chi,$$

where χ is an arbitrary test form in G of bidegree $(n-k, n-k)$; see [305], § 3.4, for details. Then we set $(dd^c u)^n := \underbrace{dd^c u \wedge \dots \wedge dd^c u}_{n-\text{times}}$.

B.6.1 (Domination principle, cf. [45]). *Let G be a bounded open subset of $\mathbb{C}^n$ and let $u_+, u_- \in \mathcal{PSH}(G) \cap L^\infty(G)$ be such that*

$$(dd^c u_+)^n \le (dd^c u_-)^n \text{ in } G,$$
$$\liminf_{G \ni z \to \zeta} (u_+(z) - u_-(z)) \ge 0 \text{ for all } \zeta \in \partial G.$$

Then, $u_+ \ge u_-$ on G.

A plurisubharmonic function $u : G \longrightarrow \mathbb{R}$ is said to be *maximal* if for any relatively compact open subset G_0 of G and for every function v upper semicontinuous on $\overline{G_0}$ and plurisubharmonic in G_0, if $v \le u$ on ∂G_0, then $v \le u$ in G_0.

B.6.2. *Let $u \in \mathcal{PSH}(G) \cap L^\infty_{\mathrm{loc}}(G)$. Then u is maximal iff $(dd^c u)^n = 0$ in G.*

B.6.3 (cf. [553]). *Let Ω_j be an open subset of $\mathbb{C}^{n_j}$ and let $u_j \in \mathcal{PSH}(\Omega_j) \cap L^\infty_{\mathrm{loc}}(\Omega)$ be such that $(dd^c u_j)^{n_j} = 0$ on Ω_j, $j = 1, 2$. Define*

$$u(z_1, z_2) := \max\{u_1(z_1), u_2(z_2)\}, \quad (z_1, z_2) \in \Omega := \Omega_1 \times \Omega_2.$$

Then, $(dd^c u)^{n_1+n_2} = 0$ on Ω.

B.7 Domains of holomorphy and pseudoconvex domains

References: [204, 234, 265, 305, 327, 445].

A domain G in $\mathbb{C}^n$ is called a *domain of holomorphy* if there exists a holomorphic function f on G such that for every pair (U_1, U_2) of non-empty open sets $U_j \subset \mathbb{C}^n$ with $U_1 \subset U_2 \cap G \subsetneq U_2$, U_2 connected, the function $f|_{U_1}$ is never the restriction of an $\tilde{f} \in \mathcal{O}(U_2)$. Observe that any domain in the complex plane is a domain of holomorphy.

If $\mathcal{F} = \mathcal{F}(G)$ denotes a subfamily of $\mathcal{O}(G)$, we say that G is an *$\mathcal{F}(G)$-domain of holomorphy* (or, shortly, *$\mathcal{F}$-domain of holomorphy*) if the above definition holds with $f \in \mathcal{F}(G)$. For example, an *$\mathcal{H}^\infty(G)$-domain of holomorphy* is a domain which admits a bounded holomorphic function, which cannot be holomorphically extended through ∂G.

There is a long list of equivalent descriptions of domains of holomorphy, for example:

B.7.1 (Cartan–Thullen theorem). *A domain $G \subset \mathbb{C}^n$ is a domain of holomorphy iff G is holomorphically convex, i.e., whenever $K \subset G$ is compact, then its* holomorphically convex envelope $\hat{K} := \{z \in G : |f(z)| \le \|f\|_K,\ f \in \mathcal{O}(G)\}$ *is compact, too.*

Observe that such a characterization fails to hold for $\mathcal{H}^\infty$-domains of holomorphy, as the famous example of N. Sibony [471] has shown.

The most important characterization of domains of holomorphy is based on a more geometric condition. A domain $G \subset \mathbb{C}^n$ is said to be *pseudoconvex* if the function $-\log \operatorname{dist}(\cdot, \partial G)$ is psh on G.

B.7.2 (Solution of the Levi Problem). *A domain $G \subset \mathbb{C}^n$ is a domain of holomorphy iff G is pseudoconvex.*

B.7.3. *Let G be a domain in $\mathbb{C}^n$. Then, the following properties are equivalent:*

(i) *G is pseudoconvex,*

(ii) *there exists a psh $\mathcal{C}^\infty$-function u on G such that $\{z \in G : u(z) < k\} \subset\subset G$ for every $k \in \mathbb{R}_+$,*

(iii) *if $\varphi_\alpha \in \mathcal{C}(\overline{\mathbb{D}}, G) \cap \mathcal{O}(\mathbb{D}, G)$ with $\bigcup_{\alpha \in I} \varphi_\alpha(\mathbb{T}) \subset\subset G$, then $\bigcup_{\alpha \in I} \varphi_\alpha(\overline{\mathbb{D}}) \subset\subset G$.*

B.7.4.

(a) *If $G_j \subset \mathbb{C}^{n_j}$ is pseudoconvex, $j = 1, 2$, then $G_1 \times G_2$ is pseudoconvex.*

(b) *If $G = \bigcup_{\nu=1}^{\infty} G_\nu$, where $G_\nu \subset G_{\nu+1}$ is an increasing sequence of pseudoconvex domains, then G is pseudoconvex.*

For certain classes of domains in $\mathbb{C}^n$ the following characterizations of domains of holomorphy are known:

B.7.5 (cf. [269], Theorem 1.11.13). *A Reinhardt domain $G \subset \mathbb{C}^n$ is a domain of holomorphy if*

- *G is* logarithmically convex, *i.e., the* logarithmic image

$$\log G := \{x \in \mathbb{R}^n : (e^{x_1}, \dots, e^{x_n}) \in G\}$$

is convex in the usual sense, and

- *G is* relatively complete, *i.e., if $(a_1, \dots, a_{j-1}, 0, a_{j+1}, \dots, a_n) \in G$ (for some $j \in \{1, \dots, n\}$), then*

$$\{(z_1, \dots, z_{j-1}, \lambda z_j, z_{j+1}, \dots, z_n) : (z_1, \dots, z_n) \in G,\ \lambda \in \overline{\mathbb{D}}\} \subset G.$$

In particular,

- *if $0 \in G$, then G is a domain of holomorphy iff G is logarithmically convex and* complete, *i.e.,* $\{(\lambda_1 z_1, \dots, \lambda_n z_n) : (z_1, \dots, z_n) \in G,\ \lambda_1, \dots, \lambda_n \in \overline{\mathbb{D}}\} \subset G$;
- *if $G \subset \mathbb{C}_*^n$, then G is a domain of holomorphy iff G is logarithmically convex.*

B.7.6. *A balanced domain G in $\mathbb{C}^n$ given by its Minkowski function $\mathfrak{h}$ as $G = G_{\mathfrak{h}} = \{z \in \mathbb{C}^n : \mathfrak{h}(z) < 1\}$ is a domain of holomorphy iff $\mathfrak{h}$ is psh iff* $\log \mathfrak{h}$ *is psh.*

Let $D \subset \mathbb{C}^n$ be an arbitrary domain. A domain $G \subset D \times \mathbb{C}^m$ is called a *Hartogs domain over D (with m-dimensional balanced fibers)* if for any $z \in D$ the *fiber*

$$G_z := \{w \in \mathbb{C}^m : (z, w) \in G\}$$

is a non-empty balanced domain in $\mathbb{C}^m$.

B.7.7. *Let D and G be as above. Then,*

(a) *There exists exactly one upper semicontinuous function*

$$H = H_G : D \times \mathbb{C}^m \longrightarrow \mathbb{R}_+$$

with $H(z, \lambda w) = |\lambda| H(z, w)$ ($z \in D$, $w \in \mathbb{C}^m$, $\lambda \in \mathbb{C}$) such that

$$G = D_H = \{(z, w) \in D \times \mathbb{C}^m : H(z, w) < 1\}. \tag{†}$$

(b) *Conversely, any such a functions H defines via (†) a Hartogs domain over D with m-dimensional balanced fibers.*

(c) *$G = D_H$ is pseudoconvex iff D is pseudoconvex and $\log H \in \mathcal{PSH}(D \times \mathbb{C}^m)$.*

(d) *In particular, if $m = 1$, then the Hartogs domain*

$$G = \{(z, w) \in D \times \mathbb{C} : |w| < e^{-V(z)}\}$$

is pseudoconvex iff D is pseudoconvex and $V \in \mathcal{PSH}(D)$.

(e) *Assume that $G = D_H$ is bounded pseudoconvex. Then,*

(1) *G is taut if and only if D is taut and H is continuous;*

(2) *G is hyperconvex if and only if D is hyperconvex and H is continuous.*

Remark B.7.8. Hartogs domains may also be used to characterize pseudoconvex domains (see [393]). Let $D \subset \mathbb{C}^n$ be an arbitrary domain. Then, the following properties are equivalent:

(i) D is pseudoconvex;

(ii) the Hartogs domain $G := \{(z, w) \in D \times \mathbb{C}^n : z + \lambda w \in D,\ \lambda \in \overline{\mathbb{D}}\}$ over D with balanced fibers is pseudoconvex.

A bounded domain $G \subset \mathbb{C}^n$ with smooth $\mathcal{C}^2$-boundary is called *strongly pseudoconvex* (cf. [265], Definition 2.2.4) if

$$(\mathcal{L}r)(z; X) > 0 \text{ for all } z \in \partial G \text{ and } X \in (\mathbb{C}^n)_* \text{ with } \sum_{j=1}^{n} \frac{\partial r}{\partial z_j}(z) X_j = 0,$$

where r denotes an arbitrary function defining ∂G (i.e., r is a $\mathcal{C}^2$-function on an open neighborhood U of ∂G satisfying $U \cap G = \{z \in U : r(z) < 0\}$ and $\operatorname{grad} r(z) \neq 0$ for every $z \in \partial G$).

B.7.9. *Let G be a strongly pseudoconvex domain in $\mathbb{C}^n$. Then there exists a defining $\mathcal{C}^2$-function r on a neighborhood U of ∂G that is strictly psh on U.*

B.7.10. *Let ζ be a boundary point of a strongly pseudoconvex domain G. Then there exists a biholomorphic mapping $F : U \longrightarrow V$, U a neighborhood of ζ and V a neighborhood of 0, such that $F(U \cap G)$ is strictly convex.*

B.7.11. *Any pseudoconvex domain G can be exhausted by an increasing sequence of strongly pseudoconvex domains $G_\nu \subset\subset G$ with real analytic boundary.*

A bounded domain $G \subset \mathbb{C}^n$ is called *hyperconvex* if there exists a continuous negative psh function u on G such that whenever $\varepsilon < 0$, then $\{z \in G : u(z) < \varepsilon\}$ is relatively compact in G; u is called an *exhaustion function* of G.

B.7.12. *Any domain of holomorphy $G \subset \mathbb{C}^n$ is the union of an increasing sequence of hyperconvex subdomains.*

Let G be a bounded domain in $\mathbb{C}^n$. A boundary point $\zeta \in \partial G$ is a *peak point with respect to* $\mathcal{F} \subset \mathcal{C}(\overline{G})$ if there exists a function $f \in \mathcal{F}$ with $f(\zeta) = 1$ and $|f(z)| < 1$, $z \in \overline{G} \setminus \{\zeta\}$.

B.7.13. *Any boundary point of a strongly pseudoconvex domain $G \subset \mathbb{C}^n$ is a peak point with respect to $\mathcal{O}(\overline{G})$.*

For domains in $\mathbb{C}^2$, even more is known.

B.7.14 (cf. [39]). *If G is a bounded pseudoconvex domain in $\mathbb{C}^2$ with real analytic boundary, then any boundary point $\zeta \in \partial G$ is a peak point with respect to $\mathcal{A}(G) := \mathcal{C}(\overline{G}) \cap \mathcal{O}(G)$.*

? It seems to still be an open question whether B.7.14 remains true in higher dimensions.

Now, we repeat the main result of Hörmander's $\overline{\partial}$-theory. Let G be an arbitrary domain in $\mathbb{C}^n$ and let $\varphi \in \mathcal{PSH}(G)$. By $L^2(\Omega, \exp(-\varphi))$ we denote the space of functions in Ω that are square-integrable with respect to the measure $e^{-\varphi} d\mathcal{L}^{2n}$, where $\mathcal{L}^{2n}$ denotes, as usual, the Lebesgue measure. Let $L^2_{(0,q)}(G, \exp(-\varphi))$ be the space of $(0, q)$ forms $\alpha = \sum'_{|J|=q} \alpha_J d\overline{z}_J$ with coefficients $\alpha_J \in L^2(G, \exp(-\varphi))$; $\sum'$ means that summation is done only over strictly increasing multi-indices $J = (j_1, \dots, j_q)$ and $d\overline{z}_J = d\overline{z}_{j_1} \wedge \dots \wedge d\overline{z}_{j_q}$. Then, $L^2_{(0,q)}(G, \exp(-\varphi))$ is a Hilbert space with respect to the following scalar product:

$$\langle \alpha, \beta \rangle_{L^2_{(0,q)}(G, \exp(-\varphi))} := \sum_{|J|=q}{}' \int_G \alpha_J(z) \overline{\beta_J(z)} \exp(-\varphi(z)) d\mathcal{L}^{2n}(z).$$

For $\alpha = \sum_{j=1}^n \alpha_j d\overline{z}_j \in L^2_{(0,1)}(G, \exp(-\varphi))$, the formula

$$\overline{\partial}\alpha = \sum_{\nu=1}^n \sum_{j=1}^n \frac{\partial \alpha_\nu}{\partial \overline{z}_j} d\overline{z}_j \wedge d\overline{z}_\nu$$

defines a closed densely defined operator

$$\overline{\partial} : L^2_{(0,1)}(G, \exp(-\varphi)) \longrightarrow L^2_{(0,2)}(G, \exp(-\varphi)).$$

B.7.15. *Let G be a pseudoconvex domain and let $\varphi \in \mathcal{PSH}(G)$. Assume that a $(0,1)$-form $\alpha = \sum_{j=1}^n \alpha_j d\overline{z}_j \in L^2_{(0,1)}(G, \exp(-\varphi))$ satisfies*

(i) $\alpha_j \in \mathcal{C}^\infty(G)$, $1 \le j \le n$,

(ii) $\overline{\partial}\alpha = 0$.

Then there exists a $\mathcal{C}^\infty$-function $u \in \mathcal{C}^\infty(G)$ with $\overline{\partial} u = \sum_{j=1}^n \frac{\partial u}{\partial \overline{z}_j} d\overline{z}_j = \alpha$ and

$$\int_G |u(z)|^2 (1 + \|z\|^2)^{-2} \exp(-\varphi(z)) d\mathcal{L}^{2n}(z) \le \frac{1}{2} \langle \alpha, \alpha \rangle_{L^2_{(0,1)}(G, \exp(-\varphi))}.$$

B.7.1 Stein manifolds

For the convenience of the reader, we also quote the results from the theory of Stein manifolds that we used in Chapter 2.

Let M be a (connected) complex manifold. M is said to be a *Stein manifold* if

(i) $\mathcal{O}(M)$ separates the points of M,

(ii) M is $\mathcal{O}(M)$-convex, i.e., for every compact $K \subset\subset M$, the holomorphically convex envelope $\widehat{K} := \{z \in M : \forall_{f \in \mathcal{O}(M)} : |f(z)| \le \|f\|_K\}$ is compact,

(iii) for any point $p \in M$ there exists a holomorphic coordinate system near p that is given by global holomorphic functions on M.

Observe that a domain $G \subset \mathbb{C}^n$ is a Stein manifold iff G is a domain of holomorphy.

B.7.16. *Any open Riemann surface is a Stein manifold.*

The next fundamental theorem says that, in fact, the theory of Stein manifolds takes place in $\mathbb{C}^n$. To be precise, we have

B.7.17 (Remmert embedding theorem). *Let M be an n-dimensional Stein manifold. Then there exists a holomorphic embedding $F : M \longrightarrow \mathbb{C}^{2n+1}$, i.e., there is a closed submanifold $M' \subset \mathbb{C}^{2n+1}$ and a biholomorphism $F : M \longrightarrow M'$.*

B.7.18. *Let M be a closed submanifold of $\mathbb{C}^n$. Then there exist a neighborhood V of M and a holomorphic retraction $\varrho : V \longrightarrow M$.*

For a proof, see [204], Ch. VIII, C, Th. 8.

B.7.19 (cf. [480]). *Let M be a closed submanifold of $\mathbb{C}^n$. Then any open neighborhood V of M contains a domain of holomorphy G with $M \subset G$, i.e., M admits a neighborhood basis of domains of holomorphy.*

We mention that B.7.19 is a very special case of a general result in [480].

B.8 L^2-holomorphic functions

One of the most beautiful results in complex analysis is the following extension theorem due to Ohsawa–Takegoshi:

Theorem B.8.1 (cf. [405]). *Let D be a bounded pseudoconvex domain in $\mathbb{C}^n$ and H an affine subspace of $\mathbb{C}^n$. Then there is a positive constant C, which depends only on the diameter of D and n, such that for any $f \in L^2_h(D \cap H)$ there is an $F \in L^2_h(D)$ such that $F|_{D\cap H} = f$ and $\|F\|_{L^2_h(D)} \le C\|f\|_{L^2_h(D\cap H)}$.*

Recently, simple proofs of this result may be found in, for example, [8] or in [98]. The optimal constant C has been determined in [63] (see also [62]).

Moreover, we recall the following one-dimensional result (see [347, 94]):

Theorem B.8.2. *Let $D \subset \mathbb{C}$ be a bounded domain, $z_0 \in \partial D$, and $f \in L^2_h(D)$. Then for any $\varepsilon > 0$ there exist a neighborhood $U = U(z_0)$ and a function $g \in L^2_h(D \cup U)$ such that $\|f - g\|_{L^2_h(D)} \le \varepsilon$. In particular, the subspace of all functions in $L^2_h(D)$, bounded near z_0, is dense in $L^2_h(D)$.*

In [94], complete Kähler metrics were used to solve a corresponding $\bar\partial$-problem in order to find g. Here we give a proof that is based on Berndtsson's solution of a $\bar\partial$-problem (see [418]).

Proof. We may assume that $z_0 = 0 \in \partial D$ and that $\overline{D} \subset \mathbb{D}$. Fix $f \in L^2_h(D)$ and a sufficiently small $\varepsilon \in (0, 1/2)$. Put $\psi(z) := -\log\log(1/|z|)$, $z \neq 0$. Observe that $\psi \in \mathcal{C}^\infty(\mathbb{C}_*) \cap \mathcal{SH}(\mathbb{C}_*)$ and $|\frac{\partial\psi}{\partial z}|^2 = \frac{\partial^2\psi}{\partial z\partial\bar z} = (\log|z|^2)^{-2}|z|^{-2} > 0$.

Moreover, let $\chi \in \mathcal{C}^\infty(\mathbb{R}, [0, 1])$,

$$\chi(t) := \begin{cases} 1, & \text{if } t \le 1 - \log 2 \\ 0, & \text{if } t > 1 \end{cases},$$

be such that $|\chi'| \le 3$.

Finally, we define $\varrho_\varepsilon(z) := \chi(-\psi(z) - \log\log(1/\varepsilon) + 1)$, $z \in \mathbb{C}_*$. Observe that $\varrho_\varepsilon(z) = 0$ if $0 < |z| < \varepsilon$, and $\varrho_\varepsilon(z) = 1$ if $|z| > \sqrt{\varepsilon}$.

Then $\alpha := \bar\partial(\varrho_\varepsilon f)$ is a $\mathcal{C}^\infty$ $\bar\partial$-closed $(0,1)$-form on $D_\varepsilon := D \cup \mathbb{D}(\varepsilon)$. Now we like to apply the following theorem of Berndtsson:

Theorem (cf. [54]). *Let $\Omega \subset \mathbb{C}^n$ be a bounded pseudoconvex domain. Let $\varphi, \psi \in \mathcal{PSH}(\Omega)$, ψ strongly psh, be such that for any $X \in \mathbb{C}^n$ the inequality*

$$\sum_{j,k=1}^{n} \frac{\partial^2 \psi}{\partial z_j \partial \overline{z}_k}(z) X_j \overline{X}_k \geq \Big| \sum_{j=1}^{n} \frac{\partial \psi}{\partial z_j}(z) X_j \Big|^2$$

holds on Ω. Let $\delta \in (0,1)$ and $\alpha = \sum_{j=1}^{n} \alpha_j d\overline{z}_j$ a $\overline{\partial}$-closed $(0,1)$-form.

Then there exists a solution $u \in L^2_{loc}(\Omega)$ of $\overline{\partial} u = \alpha$ such that

$$\int_{\Omega} |u|^2 e^{-\varphi+\delta\psi} d\mathcal{L}^{2n}(z) \leq \frac{4}{\delta(1-\delta)^2} \int_{\Omega} \sum_{j,k=1}^{n} \psi^{j,k} \alpha_j \overline{\alpha}_k e^{-\varphi+\delta\psi} d\mathcal{L}^{2n}(z),$$

where $(\psi^{j,k})$ denotes the inverse matrix of $(\frac{\partial^2 \psi}{\partial z_j \partial \overline{z}_k})$.

Take $\varphi := (1/2)\psi$ and $\delta := 1/2$. Then there exists a function u_ε, $\overline{\partial} u_\varepsilon = \alpha$ on $D_\varepsilon \setminus \{0\}$ such that

$$\begin{aligned}\int_{D_\varepsilon \setminus \{0\}} |u|^2 e^{-\varphi+\delta\psi} d\mathcal{L}^2(z) &\leq \frac{4}{\delta(1-\delta)^2} \int_{D_\varepsilon \setminus \{0\}} \frac{|\alpha|^2}{|\frac{\partial^2 \psi}{\partial z \partial \overline{z}}|} e^{-\varphi+\delta\psi} d\mathcal{L}^2(z) \\ &= 16 \int_{z \in D,\ \varepsilon \leq |z| \leq \sqrt{\varepsilon}} |\chi'|^2 \cdot |f|^2 d\mathcal{L}^2(z).\end{aligned}$$

Then, the function $f_\varepsilon := u_\varepsilon - \varrho_\varepsilon f$ belongs to $L^2_h(D_\varepsilon \setminus \{0\})$ and

$$\begin{aligned}\|f - f_\varepsilon\|_{L^2_h(D)} &\leq \|(1-\varrho_\varepsilon) f\|_{L^2_h(D)} + 160 \|f\|_{L^2_h(D \cap \mathbb{D}(\sqrt{\varepsilon}))} \\ &\leq C \|f\|_{L^2_h(D \cap \mathbb{D}(\sqrt{\varepsilon}))} \underset{\varepsilon \to 0}{\longrightarrow} 0,\end{aligned}$$

where C is a general positive constant. It remains to note that $f_\varepsilon \in \mathcal{O}(D_\varepsilon)$ (use Laurent series), which finishes the proof. □

We note that under some proper assumptions, this result can be generalized to higher dimensions.

B.9 Hardy spaces

References: [141, 187, 195, 461].

Let $0 < p < +\infty$. We say that a function $h \in \mathcal{O}(\mathbb{D})$ is of class $\mathcal{H}^p(\mathbb{D})$ if

$$\|h\|_{\mathcal{H}^p} := \sup_{0<r<1} \left\{ \left(\int_0^{2\pi} |h(re^{i\theta})|^p d\theta \right)^{1/p} \right\} < +\infty.$$

B.9.1. *$\mathcal{H}^\infty(\mathbb{D}) \subset \mathcal{H}^p(\mathbb{D}) \subset \mathcal{H}^q(\mathbb{D})$ for any $0 < q < p < +\infty$.*

B.9.2. *If $h \in \mathcal{H}^p(\mathbb{D})$, then for almost all $\zeta \in \mathbb{T}$ (with respect to Lebesgue measure on $\mathbb{T}$) the function h has a non-tangential limit at ζ, i.e.,*

$$h^*(\zeta) = \lim_{\substack{\lambda \to \zeta \\ \sphericalangle}} h(\lambda) = \lim_{\Gamma_\alpha(\zeta) \ni \lambda \to \zeta} h(\lambda)$$

exists and is independent of $\alpha > 1$, where $\Gamma_\alpha(\zeta) := \{\lambda \in \mathbb{D} : |\lambda - \zeta| < \alpha(1 - |\lambda|)\}$. Moreover,

- $h^* \in L^p(\mathbb{T})$,
- $\|h^*\|_{L^p} = \lim_{r \to 1} \left(\int_0^{2\pi} |h(re^{i\theta})|^p d\theta \right)^{1/p} = \|h\|_{\mathcal{H}^p}$,
- $\lim_{r \to 1} \int_0^{2\pi} |h(re^{i\theta}) - h^*(e^{i\theta})|^p d\theta = 0$.

B.9.3 (Cauchy and Poisson integral formulas). *If $h \in \mathcal{H}^1(\mathbb{D})$, then*

$$h(\lambda) = \frac{1}{2\pi i} \int_{\mathbb{T}} \frac{h^*(\zeta)}{\zeta - \lambda} d\zeta, \quad \lambda \in \mathbb{D},$$
$$h(\lambda) = \frac{1}{2\pi} \int_0^{2\pi} \operatorname{Re} \left(\frac{e^{i\theta} + \lambda}{e^{i\theta} - \lambda} \right) h^*(e^{i\theta}) d\theta, \quad \lambda \in \mathbb{D}.$$

In particular,

– *if $|h^*| \le M$ almost everywhere on $\mathbb{T}$, then $|h| \le M$ on $\mathbb{D}$;*

– *if $h^* \in \mathbb{R}$ almost everywhere on $\mathbb{T}$, then $h \equiv \text{const} \in \mathbb{R}$.*

B.9.4 (Blaschke products). *Let $(a_\nu)_{\nu=1}^\infty \subset \mathbb{D}$ be such that*

$$\sum_{\nu=1}^\infty (1 - |a_\nu|) < +\infty. \tag{$*$}$$

Put $I := \{\nu : a_\nu \neq 0\}$ and let $k := \#(\mathbb{N} \setminus I)$. Then the product

$$B(\lambda) := z^k \prod_{\nu \in I} \frac{-\overline{a}_\nu}{|a_\nu|} \frac{\lambda - a_\nu}{1 - \lambda \overline{a}_\nu}, \quad \lambda \in \mathbb{D}, \tag{$**$}$$

is locally uniformly convergent, $B \in \mathcal{H}^\infty(\mathbb{D})$, $|B^*| = 1$ almost everywhere on $\mathbb{T}$, and the zeros of B (counted with multiplicities) coincide with $(a_\nu)_{\nu=1}^\infty$.

B.9.5. *If $h \in \mathcal{H}^p(\mathbb{D}), h \not\equiv 0$, then the zeros of h (counted with multiplicities) satisfy $(*)$. In particular, using $(**)$ one can define the Blaschke product B_h for h.*

B.9.6 (Decomposition theorem). *Let $h \in \mathcal{H}^p(\mathbb{D}), h \not\equiv 0$. Then,*

$$\log|h^*| \in L^1(\mathbb{T}),$$

$$\log|h(0)| \leq \frac{1}{2\pi}\int_0^{2\pi} \log|h^*(e^{i\theta})|d\theta, \qquad (***)$$

$$h = cB_h \cdot S_h \cdot Q_h \text{ in } \mathbb{D},$$

where $c \in \mathbb{T}$ is a constant,

$$S_h(\lambda) := \exp\left(-\int_0^{2\pi} \frac{e^{i\theta}+\lambda}{e^{i\theta}-\lambda} d\sigma(\theta)\right)$$

(σ is a non-negative Borel measure, singular w.r.t. Lebesgue measure), and

$$Q_h(\lambda) := \exp\left(\frac{1}{2\pi}\int_0^{2\pi} \frac{e^{i\theta}+\lambda}{e^{i\theta}-\lambda} \log|h^*(e^{i\theta})|d\theta\right), \quad \lambda \in \mathbb{D}.$$

Moreover,

- $S_h \in \mathcal{H}^\infty(\mathbb{D})$;
- $|S_h^*| = 1$ *almost everywhere on $\mathbb{T}$ with respect to Lebesgue measure;*
- $S_h^* = 0$ *almost everywhere on $\mathbb{T}$ with respect to the measure σ;*
- *the function $B_h S_h$ is constant iff we have equality in condition $(***)$.*

B.9.7 (Identity principle). *Let $h \in \mathcal{H}^p(\mathbb{D})$. Suppose that $h^* = 0$ on a set of positive Lebesgue measure. Then, $h \equiv 0$.*

B.9.8 (cf. [399], Chapter III). *Let $\varphi \in \mathcal{O}(\mathbb{D}, \mathbb{D})$ be an* inner function, *i.e., $|\varphi^*(\zeta)| = 1$ for a.a. $\zeta \in \mathbb{T}$. Assume that $\varphi \not\equiv$ const and φ is not a Blaschke product. Then there exists a $\zeta_0 \in \mathbb{T}$ such that $\varphi^*(\zeta_0) = 0$.*

B.9.9 (cf. [399], Chapter II). *Let $\varphi \in \mathcal{H}^\infty(\mathbb{D})$ and let $A \subset \mathbb{C}$ be a compact polar set. Assume that there exists a set $I \subset \mathbb{T}$ of positive measure such that $\varphi^*(\zeta) \in A$, $\zeta \in I$. Then, $\varphi \equiv$ const.*

B.9.10 (F.& M. Riesz theorem). *Let μ be a complex Borel measure on $[0, 2\pi)$. Then the following conditions are equivalent:*

(i) $\int_0^{2\pi} f(e^{i\theta})d\mu(\theta) = 0$ *for all* $f \in \mathcal{C}(\overline{\mathbb{D}}) \cap \mathcal{O}(\mathbb{D})$;

(ii) *there exists $h \in \mathcal{H}^1(\mathbb{D})$ such that $d\mu(\theta) = \frac{e^{i\theta}}{2\pi} h^*(e^{i\theta})d\theta$.*

B.9.11 (Hardy–Littlewood theorem). *Let $h \in \mathcal{O}(\mathbb{D})$ and let $0 < \alpha \le 1$. Then, h extends to an α-Hölder-continuous function on $\overline{\mathbb{D}}$ iff there exists a constant $M > 0$ such that*

$$|h'(\lambda)| \le \frac{M}{(1-|\lambda|)^{1-\alpha}}, \quad \lambda \in \mathbb{D}.$$

B.10 Kronecker theorem

References: [213]. There are the following two equivalent formulations of the Kronecker theorem (cf. [213], Theorems 442 and 444):

B.10.1. *Assume that $\alpha_1, \dots, \alpha_n, 1$ are linearly independent over $\mathbb{Q}$. Let $\mu_1, \dots, \mu_n \in \mathbb{R}$, $\varepsilon > 0$, and $C > 0$ be arbitrary. Then there exist $p_1, \dots, p_n, q \in \mathbb{Z}$ such that $q \ge C$ and $|q\alpha_j - p_j - \mu_j| \le \varepsilon$, $j = 1, \dots, n$.*

In particular, the set $\{(q\alpha_1 - \lfloor q\alpha_1 \rfloor, \dots, q\alpha_n - \lfloor q\alpha_n \rfloor) : q \in \mathbb{N}\}$ is dense in $[0,1]^n$.

For example, the set $\{e^{\alpha s 2\pi i} : s \in \mathbb{N}\}$ is dense in $\mathbb{T}$ when $\alpha \in \mathbb{R} \setminus \mathbb{Q}$.

B.10.2. *Assume that $\alpha_1, \dots, \alpha_n$ are linearly independent over $\mathbb{Q}$. Let $\mu_1, \dots, \mu_n \in \mathbb{R}$, $\varepsilon > 0$, and $C > 0$ be arbitrary. Then there exist $p_1, \dots, p_n \in \mathbb{Z}$, $q \in \mathbb{R}$, such that $q \ge C$ and $|q\alpha_j - p_j - \mu_j| \le \varepsilon$, $j = 1, \dots, n$.*

One can easily prove that Theorem B.10.1 $\iff$ Theorem B.10.2. In the case where $\mu_1 = \dots = \mu_n = 0$, as a direct consequence of Theorem B.10.1, we get the following approximation theorem:

B.10.3. *Let $\alpha_1, \dots, \alpha_n \in \mathbb{R}$, $\varepsilon > 0$, and $C > 0$ be arbitrary. Then there exist $p_1, \dots, p_n, q \in \mathbb{Z}$ such that $q \ge C$ and $|q\alpha_j - p_j| \le \varepsilon$, $\operatorname{sgn} p_j = \operatorname{sgn} \alpha_j$, $j = 1, \dots, n$.*

B.11 List of problems

Appendix C

List of problems

Chapter 1

Chapter 2

Chapter 3

Chapter 4

Chapter 5

Chapter 6

Chapter 7

Chapter 8

Chapter 14

Chapter 15

Chapter 16

Chapter 18

Chapter 19

Chapter 20

Appendix A

Appendix B

Bibliography

[1] M. Abate, Boundary behavior of invariant distances and complex geodesics, *Atti Acc. Lincei Rend. Fis., Ser. VIII* **80** (1986), 100–106.

[2] M. Abate, *The complex geodesics of the classical domains*, Manuscript, 1986.

[3] M. Abate, *Iteration Theory of Holomorphic Maps on Taut Manifolds*, Research and Lecture Notes in Mathematics. Complex Analysis and Geometry, Mediterranean Press, Rende, 1989.

[4] M. Abate, A characterization of hyperbolic manifolds, *Proc. Amer. Math.* **117** (1993), 789–79.

[5] M. Abate and G. Patrizio, *Holomorphic curvature of Finsler metrics*, Manuscript, 1992.

[6] M. Abate and G. Patrizio, Uniqueness of complex geodesics and characterization of circular domains, *Manuscripta Math.* **74** (1992), 277–297.

[7] A. A. Abouhajar, M. C. White, and N. J. Young, A Schwarz lemma for a domain related to μ-synthesis, *J. Geom. Anal.* **17** (2007), 717–749.

[8] K. Adachi, On the Ohsawa–Takegoshi extension theorem, *Bull. Fac. Educ. Nagasaki Univ., Nat. Sci.* **78** (2010), 1–16.

[9] J. Agler, Operator theory and the Carathéodory metric, *Invent. Math.* **101** (1990), 483–500.

[10] J. Agler and N. J. Young, The two-point spectral Nevanlinna–Pick problem, *Integr. Equ. Oper. Theory* **37** (2000), 375–385.

[11] J. Agler and N. J. Young, A Schwarz lemma for the symmetrized bidisc, *Bull. London Math. Soc.* **33** (2001), 175–186.

[12] J. Agler and N. J. Young, The hyperbolic geometry of the symmetrized bidisc, *J. Geom. Anal.* **14** (2004), 375–403.

[13] J. Agler and N. J. Young, The complex geodesics of the symmetrized bidisc, *Int. J. Math.* **17** (2006), 375–391.

[14] J. Agler and N. J. Young, The magic functions and automorphisms of a domain, *Compl. Anal. Oper. Theory* **2** (2008), 383–404.

[15] P. R. Ahern and R. Schneider, Isometries of H^∞, *Duke Math. J.* **42** (1975), 321–326.

[16] L. V. Ahlfors, An extension of Schwarz's Lemma, *Trans. Amer. Math. Soc.* **43** (1938), 359–364.

[17] L. V. Ahlfors and A. Beurling, Conformal invariants and function theoretic null sets, *Acta Math.* **83** (1950), 101–129.

[18] D. N. Akhiezer, Homogeneous complex manifolds, *Several complex variables, IV, Algebraic aspects of complex analysis, Encycl. Math. Sci.* **10** (1990), 195–244.

[19] G. J. Aladro, *Some consequences of the boundary behavior of the Carathéodory and Kobayashi metrics and applications to normal holomorphic functions*, Ph.D. thesis, Pennsylvania State University, 1985.

[20] M. Andersson, M. Passare, and R. Sigurdsson, *Complex Convexity and Analytic Functions*, Progress in Mathematics 225, Birkhäuser Verlag, Basel, 2004.

[21] N. U. Arakelian, Uniform and tangential approximations by analytic functions, *Amer. Math. Soc. Transl.* **122** (1984), 85–97.

[22] N. Arcozzi, R. Rochberg, E. Sawyer, and B. D. Wick, Distance functions for reproducing kernel Hilbert spaces, *Contemp. Math.* **547** (2011), 25–53.

[23] S. Axler, P. Bourdon, and W. Ramey, *Harmonic Function Theory*, Graduate Texts in Mathematics 137, Springer-Verlag, 1992.

[24] K. Azukawa, Hyperbolicity of circular domains, *Tôhuku Math. J.* **35** (1983), 403–413.

[25] K. Azukawa, Bergman metric on a domain of Thullen type, *Math. Rep. Toyama Univ.* **7** (1984), 41–65.

[26] K. Azukawa, Two intrinsic pseudo-metrics with pseudoconvex indicatrices and starlike domains, *J. Math. Soc. Japan* **38** (1986), 627–647.

[27] K. Azukawa, The invariant pseudo-metric related to negative plurisubharmonic functions, *Kodai Math. J.* **10** (1987), 83–92.

[28] K. Azukawa, A note on Carathéodory and Kobayashi pseudodistances, *Kodai Math. J.* **14** (1991), 1–12.

[29] K. Azukawa, The pluri-complex Green function and a covering mapping, *Michigan Math. J.* **42** (1995), 593–602.

[30] K. Azukawa, *The pluri-complex Green function and a covering mapping*, Geometric complex analysis (Hayama, 1995), World Sci. Publ., River Edge, NJ, 1996, pp. 43–50.

[31] K. Azukawa and M. Suzuki, The Bergman metric on a Thullen domain, *Nagoya Math. J.* **89** (1983), 1–11.

[32] T. J. Barth, Normality domains for families of holomorphic maps, *Math. Ann.* **190** (1971), 293–297.

[33] T. J. Barth, The Kobayashi distance induces the standard topology, *Proc. Amer. Math. Soc.* **35** (1972), 439–441.

[34] T. J. Barth, Some counterexamples concerning intrinsic distances, *Proc. Amer. Math. Soc.* **66** (1977), 49–59.

[35] T. J. Barth, Convex domains and Kobayashi hyperbolicity, *Proc. Amer. Math. Soc.* **79** (1980), 556–558.

[36] T. J. Barth, The Kobayashi indicatrix at the center of a circular domain, *Proc. Amer. Math. Soc.* **88** (1983), 527–530.

[37] T. J. Barth, Topologies defined by some invariant pseudodistances, *Banach Center Publ.* **31** (1995), 69–76.

[38] E. Bedford and J.-P. Demailly, Two counterexamples concerning the pluri-polar Green function, *Indiana Univ. Math. J.* **37** (1988), 865–867.

[39] E. Bedford and J. E. Fornæss, A construction of peak functions on weakly pseudoconvex domains, *Ann. of Math.* **107** (1978), 555–568.

[40] E. Bedford and J. E. Fornæss, Biholomorphic maps of weakly pseudoconvex domains, *Duke Math. J.* **45** (1979), 711–719.

[41] E. Bedford and S. Pinchuk, Domains in $\mathbb{C}^2$ with noncompact group of automorphisms, *Math. USSR Sbornik* **63** (1989), 141–151.

[42] E. Bedford and S. Pinchuk, Domains in $\mathbb{C}^{n+1}$ with noncompact automorphism group, *J. Geom. Anal.* **1** (1991), 165–191.

[43] E. Bedford and S. Pinchuk, *Domains in* $\mathbb{C}^{n+1}$ *with noncompact automorphism group II*, Manuscript, 1992.

[44] E. Bedford and S. I. Pinchuk, Convex domains with noncompact automorphism groups, *Russ. Acad. Sci., Sb., Math.* **82** (1995), 1–20, (translation from Mat. Sb. 185 (1994), 3–26).

[45] E. Bedford and B. A. Taylor, A new capacity for pluri-subharmonic functions, *Acta Math.* **149** (1982), 1–41.

[46] L. Belkhchicha, Caractérisation des isomorphes analytiques de certains domaines bornés, *C. R. Acad. Sci. Paris* **313** (1991), 281–284.

[47] L. Belkhchicha, Caractérisation des isomorphismes analytiques sur la boule-unité de $\mathbb{C}^n$ pour une norme, *Math. Z.* **215** (1994), 129–141.

[48] L. Belkhchicha and J.-P. Vigué, Sur les espaces complets pour la distance de Carathéodory, *Rend. Mat. Acc. Lincei* **5** (1994), 189–192.

[49] S. Bell, Differentiability of the Bergman kernel and pseudo-local estimates, *Math. Z.* **192** (1986), 467–472.

[50] S. R. Bell, The Bergman kernel function and proper holomorphic mappings, *Trans. Amer. Math. Soc.* **270** (1982), 685–691.

[51] S. R. Bell, Proper holomorphic mappings between circular domains, *Comment. Math. Helvetici* **57** (1982), 532–538.

[52] S. R. Bell and E. Ligocka, A simplification and extension of Fefferman's theorem on biholomorphic mappings, *Invent. Math.* **57** (1980), 283–289.

[53] S. Bergman, Zur Theorie von pseudokonformen Abbildungen, *Mat. Sbornik* **43** (1936), 79–96.

[54] B. Berndtsson, The extension theorem of Ohsawa–Takegoshi and the theorem of Donelly–Fefferman, *Ann. Inst. Fourier (Grenoble)* **46** (1996), 1083–1094.

[55] B. E. Blank, D. Fan, D. Klein, S. G. Krantz, D. Ma, and M.-Y. Peng, The Kobayashi metric of a complex ellipsoid in $\mathbb{C}^2$, *Experiment. Math.* (1992), 47–55.

[56] Z. Błocki, Estimates for the complex Monge–Ampère operator, *Bull. Pol. Acad. Sci. Math.* **41** (1993), 151–157.

[57] Z. Błocki, The complex Monge–Ampère operator in hyperconvex domains, *Ann. Sc. Norm. Super. Pisa Cl. Sci. (4)* **23** (1996), 721–747.

[58] Z. Błocki, Equilibrium measure of a product subset of $\mathbb{C}^n$, *Proc. Amer. Math. Soc.* **128** (2000), 3595–3599.

[59] Z. Błocki, Regularity of the pluricomplex Green function with several poles, *Indiana Univ. Math. J.* **50** (2001), 335–351.

[60] Z. Błocki, The Bergman metric and the pluricomplex Green function, *Trans. Amer. Math. Soc.* **357** (2005), 2613–2625.

[61] Z. Błocki, Some estimates for the Bergman kernel and metric in terms of logarithmic capacity, *Nagoya Math. J.* **185** (2007), 143–150.

[62] Z. Błocki, *On the Ohsawa–Takegoshi extension theorem*, http://gamma.im.uj.edu.pl/˜blocki, 2012.

[63] Z. Błocki, Suita conjecture and the Ohsawa–Takegoshi extension theorem, *Invent. Math.* **(to appear)** (2013).

[64] Z. Błocki and P. Pflug, Hyperconvexity and Bergman completeness, *Nagoya Math. J.* **151** (1998), 221–225.

[65] T. Bloom and I. Graham, A geometric characterization of points of type m on real submanifolds of $\mathbb{C}^n$, *J. Diff. Geom.* **12** (1977), 171–182.

[66] H. P. Boas, Counterexample to the Lu Qi-Keng conjecture, *Proc. Amer. Math. Soc.* **97** (1986), 374–375.

[67] H. P. Boas, Extension of Kerzman's theorem on differentiability of the Bergman kernel function, *Indiana Univ. Math. J.* **36** (1987), 495–499.

[68] H. P. Boas, The Lu Qi-Keng conjecture fails generically, *Proc. Amer. Math. Soc.* **124** (1996), 2021–2027.

[69] H. P. Boas, Lu Qi-Keng's problem, *J. Korean Math. Soc.* **37** (2000), 253–267.

[70] H. P. Boas, S. Fu, and E. J. Straube, The Bergman kernel function: Explicit formulas and zeroes, *Proc. Amer. Math. Soc.* **127** (1999), 805–811.

[71] S. Bochner and D. Montgomery, Locally compact groups of differentiable transformations, *Ann. of Math.* **47** (1946), 639–653.

[72] H. Bohr, Über streckentreue und konforme Abbildungen, *Math. Z.* **1** (1918), 403–420.

[73] L. Bos, N. Levenberg, and S. Waldron, Pseudometrics, distances and multivariate polynomial inequalities, *J. Approx. Theory* **153** (2008), 80–96.

[74] F. Bracci and A. Saracco, Hyperbolicity in unbounded convex domains, *Forum Math.* **21** (2009), 815–825.

[75] J. Bremermann, Holomorphic continuation of the kernel and the Bergman metric, in: *Lectures on functions of a complex variable*, pp. 349–383, University Michigan Press, 1955.

[76] A. Browder, Point derivations on function algebras, *J. Funct. Anal.* **1** (1967), 22–27.

[77] J. Burbea, The Carathéodory metric in plane domains, *Kodai Math. Sem. Rep.* **29** (1977), 157–166.

[78] J. Burbea, Inequalities between intrinsic metrics, *Proc. Amer. Math. Soc.* **67** (1977), 50–54.

[79] J. Burbea, On metric and distortion theorems, in: *Recent developments in several complex variables* (J. E. Fornæss, ed.), Ann. Math. Studies 100, pp. 65–92, 1981.

[80] H. Busemann and W. Mayer, On the foundations of calculus of variations, *Trans. Amer. Math. Soc.* **49** (1941), 173–193.

[81] L. Campbell and R. Ogawa, On preserving the Kobayashi pseudometric, *Nagoya Math. J.* **57** (1975), 37–47.

[82] C. Carathéodory, Über eine spezielle Metrik, die in der Theorie der analytischen Funktionen auftritt, *Atti Pontifica Acad. Sc., Nuovi Lincei* **80** (1927), 135–141.

[83] M. Carlehed, U. Cegrell, and F. Wikström, Jensen measures, hyperconvexity and boundary behaviour of the pluricomplex Green function, *Ann. Polon. Math.* **71** (1999), 87–103.

[84] M. Carlehed and J. Wiegerinck, Le cône des fonctions plurisousharmoniques négatives et une conjecture de Coman, *Ann. Polon. Math.* **80** (2003), 93–108.

[85] E. Cartan, Sur les domaines bornés homogènes de l'espace des n variables complexes, *Abh. Math. Sem. Hamburg* **11** (1935), 116–162.

[86] H. Cartan, *Sur les groupes des transformations analytiques*, Hermann, 1935.

[87] H. Cartan, Sur les fonctions de n variables complexes: les transformations du produit topologique de deux domaines borneés, *Bull. Soc. Math. France* **64** (1936), 37–48.

[88] D. Catlin, Estimates of invariant metrics on pseudoconvex domains of dimension two, *Math. Z.* **200** (1989), 429–466.

[89] U. Cegrell, *Capacities in Complex Analysis*, Aspects of Mathematics E14, Friedr. Vieweg & Sohn, Braunschweig, 1988.

[90] C. H. Chang, M. C. Hu, and H.-P. Lee, Extremal analytic discs with prescribed boundary data, *Trans. Amer. Math. Soc.* **310** (1988), 355–369.

[91] C. H. Chang and H.-P. Lee, Explicit solutions for some extremal analytic discs of the domain $D = \{z \in \mathbb{C}^n : \sum_1^n |z_j|^{2m_j} < 1,\ m_j \in \mathbb{N}\}$, *Several complex variables (Stockholm, 1987/1988), Math. Notes, 38* (1993), 194–204.

[92] B. Chen and J. Zhang, The Bergman metric on a Stein manifold with a bounded plurisubharmonic functions, *Trans. Amer. Math. Soc.* **354** (2002), 2997–3009.

[93] B.-Y. Chen, Completeness of the Bergman metric on non-smooth pseudoconvex domains, *Ann. Polon. Math.* **71** (1999), 241–251.

[94] B.-Y. Chen, A remark on the Bergman completeness, *Complex Var.* **42** (2000), 11–15.

[95] B.-Y. Chen, A note on the Bergman completeness, *Int. J. Math.* **12** (2001), 383–392.

[96] B.-Y. Chen, *On Bergman completeness of non-hyperconvex Hartogs domains*, Preprint, 2001.

[97] B.-Y. Chen, Weighted Bergman kernel: asymptotic behavior, applications and comparison results., *Studia Math.* **174** (2006), 111–130.

[98] B.-Y. Chen, *A simple proof of the Ohsawa–Takegoshi extension theorem*, arXiv:1105.2430, 2011.

[99] J.-H. Chen, *Estimates of the invariant metrics on convex domains in* $\mathbb{C}^n$, Ph.D. thesis, Purdue Univ., 1989.

[100] S.-C. Chen, A counterexample to the differentiability of the Bergman kernel function, *Proc. Amer. Math. Soc.* **124** (1996), 1807–1810.

[101] Z. H. Chen and Y. Liu, Schwarz–Pick estimates for bounded functions in the unit ball of $\mathbb{C}^n$, *Acta Math. Sinica, English Series* **26** (2010), 901–908.

[102] C. K. Cheung and K. T. Kim, Analysis of the Wu metric. I: The case of convex Thullen domains, *Trans. Amer. Math. Soc.* **348** (1996), 1421–1457.

[103] C. K. Cheung and K. T. Kim, Analysis of the Wu metric. II: The case of non-convex Thullen domains, *Proc. Amer. Math. Soc.* **125** (1997), 1131–1142.

[104] C. K. Cheung and K. T. Kim, The constant curvature property of the Wu invariant metric, *Pacific J. Math.* **211** (2003), 61–68.

[105] E. M. Chirka, *Complex Analytic Sets*, Mathematics and its Applications 46, Kluwer Acad. Publishers, 1989.

[106] S. Cho, Estimates of invariant metrics on some pseudoconvex domains in $\mathbb{C}^n$, *J. Korean Math. Soc.* **32** (1995), 661–678.

[107] S. Cohn-Vossen, Existenz kürzester Wege, *Comptes Rendus l'Acad. Sci. URSS, Vol. III (VIII)* **8 (68)** (1935), 239–242.

[108] D. Coman, The pluricomplex Green function with two poles of the unit ball of $\mathbb{C}^n$, *Pacific J. Math.* **194** (2000), 257–283.

[109] J. B. Conway, *Functions of One Complex Variable*, Graduate Texts in Mathematics 11, Springer Verlag, 1978.

[110] J. B. Conway, *Functions of One Complex Variable II*, Graduate Texts in Mathematics 159, Springer Verlag, 1995.

[111] C. Costara, *Le problème de Nevanlinna–Pick spectral*, Ph.D. thesis, Laval Univ., 2004.

[112] C. Costara, The symmetrized bidisc as a counterexample to the converse of Lempert's theorem, *Bull. London Math. Soc.* **36** (2004), 656–662.

[113] C. Costara, On the 2×2 spectral Nevanlinna–Pick problem, *J. London Math. Soc. (2)* **71** (2005), 684–702.

[114] C. Costara, On the spectral Nevanlinna–Pick problem, *Studia Math.* **170** (2005), 23–55.

[115] R. Courant and D. Hilbert, *Methoden der mathematischen Physik I*, Springer Verlag, 1931.

[116] E. Cygan, Factorization of polynomials, *Bull. Pol. Acad. Sci. Math.* **40** (1992), 45–52.

[117] S. Dai, H. Chen, and Y. Pan, The Schwarz–Pick lemma of higher order in several variables, *Michigan Math. J.* **59** (2010), 517–533.

[118] S. Dai and Y. Pan, Note on Schwarz–Pick estimates for bounded and positive real part analytic functions, *Proc. Amer. Math. Soc.* **136** (2008), 635–640.

[119] J. P. d'Angelo, A note on the Bergman kernel, *Duke Math. J.* **45** (1978), 259–265.

[120] J. P. d'Angelo, *Several Complex Variables and the Geometry of Real Hypersurfaces*, Studies in Advanced Mathematics, CRC Press, Inc., 1993.

[121] J.-P. Demailly, Measures de Monge–Ampeŕe et caractérisation géométrique des variétés algébriques affines, *Mém. Soc. Math. France (N. S.)* **19** (1985), 1–125.

[122] J.-P. Demailly, Mesures de Monge–Ampère et mesures plurisousharmoniques, *Math. Z.* **194** (1987), 519–564.

[123] F. Deng, Q. Guan, and L. Zhang, Some properties of squeezing functions on bounded domains, *Pacific J. Math.* **257** (2012), 319–341.

[124] K. Diederich, Das Randverhalten der Bergmanschen Kernfunktion und Metrik in streng pseudokonvexen Gebieten, *Math. Ann.* **187** (1970), 9–36.

[125] K. Diederich, Über die 1. und 2. Ableitungen der Bergmanschen Kernfunktion und ihr Randverhalten, *Math. Ann.* **203** (1973), 129–170.

[126] K. Diederich and J. E. Fornæss, Proper holomorphic maps onto pseudoconvex domains with real-analytic boundary, *Ann. of Math.* **110** (1979), 575–592.

[127] K. Diederich and J. E. Fornæss, Comparison of the Bergman and the Kobayashi metric, *Math. Ann.* **254** (1980), 257–262.

[128] K. Diederich, J. E. Fornæss, and G. Herbort, Boundary behavior of the Bergman metric, *Proc. Sympos. Pure Math.* **41** (1984), 59–67.

[129] K. Diederich and G. Herbort, On discontinuity of the Bergman kernel function, *Int. J. Math.* **10** (1999), 825–832.

[130] K. Diederich, G. Herbort, and T. Ohsawa, The Bergman kernel on uniformly extendable pseudoconvex domains, *Math. Ann.* **273** (1986), 471–478.

[131] K. Diederich and I. Lieb, *Konvexität in der komplexen Analysis*, Neue Ergebnisse und Methoden, DMV Seminar, Band 2, Birkhäuser Verlag, 1981.

[132] K. Diederich and T. Ohsawa, An estimate for the Bergman distance on pseudoconvex domains, *Ann. of Math.* **141** (1995), 181–190.

[133] K. Diederich and N. Sibony, Strange complex structures on Euclidean space, *J. Reine Angew. Math.* **311/312** (1979), 397–407.

[134] N. Q. Dieu, N. Nikolov, and P. J. Thomas, Estimates for invariant metrics near non–semipositive boundary points, *J. Geom. Anal.* **23** (2013), 598–610.

[135] N. Q. Dieu and N. V. Trao, Product property of certain extremal functions, *Complex Var.* **48** (2003), 681–694.

[136] N. Dieu and L. M. Hai, Some remarks about Reinhardt domains in $\mathbb{C}^n$, *Illinois J. Math.* **47** (2003), 699–708.

[137] S. Dineen, *The Schwarz Lemma*, Oxford Mathematical Monographs, Clarendon Press, 1989.

[138] S. Dineen and R. M. Timoney, Complex geodesics on convex domains, in: *Progress in Functional Analysis* (K. D. Bierstedt, J. Bonet, J. Horváth, and M. Maestre, eds.), Elsevier Science Publishers B. V., 1992.

[139] Z. Dinew, An example for the holomorphic sectional curvature of the Bergman metric, *Ann. Polon. Math.* **98** (2010), 147–167.

[140] F. Docquier and H. Grauert, Levisches Problem und Rungescher Satz für Teilgebiete Steinscher Mannigfaltigkeiten, *Math. Ann.* **140** (1960), 94–123.

[141] P. L. Duren, *Theory of H^p-spaces*, Pure and Applied Mathematics 38, Academic Press, 1970.

[142] C. J. Earle and R. S. Hamilton, A fixed point theorem for holomorphic mappings, *Proc. Symp. Pure Math.* **16** (1970), 61–65.

[143] C. J. Earle and L. A. Harris, Inequalities for the Carathéodory and Poincaré metrics in open unit balls, *Pure Appl. Math. Q., Special Issue: In honor of Frederick W. Gehring, Part 2* **7** (2011), 253–273.

[144] A. Eastwood, À propos des variétés hyperboliques completes, *C. R. Acad. Sci. Paris* **280** (1975), 1071–1075.

[145] A. Edigarian, On extremal mappings in complex ellipsoids, *Ann. Polon. Math.* **62** (1995), 83–96.

[146] A. Edigarian, A remark on the Lempert theorem, *Univ. Iagel. Acta Math.* **32** (1995), 83–88.

[147] A. Edigarian, Extremal mappings in convex domains, *Univ. Iagel. Acta Math.* **35** (1997), 151–156.

[148] A. Edigarian, On definitions of the pluricomplex Green function, *Ann. Polon. Math.* **67** (1997), 233–246.

[149] A. Edigarian, On the product property of the pluricomplex Green function, *Proc. Amer. Math. Soc.* **125** (1997), 2855–2858.

[150] A. Edigarian, On the product property of the pluricomplex Green function, II, *Bull. Pol. Acad. Sci. Math.* **49** (2001), 389–394.

[151] A. Edigarian, Analytic discs method in complex analysis, *Dissertationes Math.* **402** (2002), 1–56.

[152] A. Edigarian, A note on Rosay's paper, *Ann. Polon. Math.* **80** (2003), 125–132.

[153] A. Edigarian, A note on Costara's paper, *Ann. Polon. Math.* **83** (2004), 189–191.

[154] A. Edigarian, L. Kosiński, and W. Zwonek, The Lempert theorem and the tetrablock, *J. Geom. Anal.* **(to appear)** (2013).

[155] A. Edigarian and E. A. Poletsky, Product property of the relative extremal function, *Bull. Sci. Acad. Polon.* **45** (1997), 331–335.

[156] A. Edigarian and W. Zwonek, Invariance of the pluricomplex Green function under proper mappings with applications, *Complex Var.* **35** (1998), 367–380.

[157] A. Edigarian and W. Zwonek, Geometry of the symmetrized polydisc, *Arch. Math. (Basel)* **84** (2005), 364–374.

[158] A. Edigarian and W. Zwonek, Schwarz lemma for the tetrablock, *Bull. London Math. Soc.* **41** (2009), 506–514.

[159] D. A. Eisenman, *Intrinsic measures on complex manifolds*, Memoirs Amer. Math. Soc., 96, 1970.

[160] M. Engliš, Asymptotic behavior of reproducing kernels of weighted Bergman spaces, *Trans. Amer. Math. Soc.* **349** (1997), 3717–3735.

[161] M. Engliš, Zeroes of the Bergman kernel of Hartogs domains, *Comment. Math. Univ. Carolinae* **41** (2000), 199–202.

[162] A. A. Fadlalla, Quelques proprietés de la distance de Carathéodory, in: *7th Arab. Sc. Congr. Cairo II*, pp. 1–16, 1973.

[163] A. A. Fadlalla, On boundary value problem in pseudoconvex domains, in: *Analytic Functions, Błażejewko 1982* (J. Ławrynowicz, ed.), Lecture Notes in Math. 1039, pp. 168–176, Springer Verlag, 1983.

[164] H. Federer, *Geometric Measure Theory*, Die Grundlehren der mathematischen Wissenschaften 153, Springer Verlag, 1969.

[165] C. Fefferman, The Bergman kernel and biholomorphic mappings of pseudoconvex domains, *Invent. Math.* **26** (1974), 1–65.

[166] J. E. Fornæss, Strictly pseudoconvex domains in convex domains, *Amer. J. Math.* **98** (1976), 529–569.

[167] J. E. Fornæss, Short $\mathbb{C}^k$. Complex analysis in several variables – Memorial Conference of Kiyoshi Oka's Centennial Birthday, *Adv. Stud. Pure Math.* **42** (2004), 95–108.

[168] J. E. Fornæss, Comparison of the Kobayahi-Royden and Sibony metrics on ring domains, *Sci. China, Ser. A* **52** (2009), 2610–2616.

[169] J. E. Fornæss and L. Lee, Asymptotic behavior of the Kobayashi metric in the normal direction, *Math. Z.* **261** (2009), 399–408.

[170] J. E. Fornæss and L. Lee, Kobayahi, Carathéodory, and Sibony metrics, *Complex Var. Elliptic Equ.* **54** (2009), 293–301.

[171] J. E. Fornæss and J. D. McNeal, A construction of peak functions on some finite type domains, *Amer. J. Math.* **116** (1994), 737–755.

[172] J. E. Fornæss and R. Narasimhan, The Levi problem on complex spaces with singularities, *Math. Ann.* **248** (1980), 47–72.

[173] J. E. Fornæss and N. Sibony, Increasing sequences of complex manifolds, *Math. Ann.* **255** (1981), 351–360.

[174] J. E. Fornæss and B. Stensønes, *Lectures on Counterexamples in Several Complex Variables*, Mathematical Notes 33, Princeton University Press, 1987.

[175] F. Forstnerič, Proper holomorphic mappings: a survey, *Preprint Series University Ljubljana 27* (1989).

[176] F. Forstnerič and J.-P. Rosay, Localization of the Kobayashi metric and the boundary continuity of proper holomorphic mappings, *Math. Ann.* **279** (1987), 239–252.

[177] F. Forstnerič and J. Winkelmann, Holomorphic discs with dense images, *Math. Res. Lett.* **12** (2005), 265–268.

[178] T. Franzoni and E. Vesentini, *Holomorphic Maps and Invariant Distances*, North Holland Math. Studies 40, North-Holland Publishing Co., 1980.

[179] L. Frerick and G. Schmieder, Connectedness of the Carathéodory discs for doubly connected domains, *Ann. Polon. Math.* **85** (2005), 281–282.

[180] S. Fu, *Geometry of bounded domains and behavior of invariant metrics*, Ph.D. thesis, Washington Univ., 1994.

[181] S. Fu, On completeness of invariant metrics of Reinhardt domains, *Arch. Math. (Basel)* **63** (1994), 166–172.

[182] S. Fu, A sharp estimate on the Bergman kernel of a pseudoconvex domain, *Proc. Amer. Math. Soc.* **121** (1994), 979–980.

[183] S. Fu, The Kobayashi metric in the normal direction and the mapping problem, *Complex Var. Elliptic Equ.* **54** (2009), 303–316.

[184] B. A. Fuks, *Special Chapters in the Theory of Analytic Functions of Several Complex Variables*, Translations of Mathematical Monographs 14, AMS, Providence, R. I., 1965.

[185] T. W. Gamelin, *Uniform Algebras*, Chelsea Publishing Company, 1984.

[186] T. W. Gamelin and J. Garnett, Distinguished homomorphisms and fiber algebras, *Amer. J. Math.* **92** (1970), 455–474.

[187] J. Garnett, *Bounded Analytic Functions*, Pure and Applied Mathematics 96, Academic Press, 1981.

[188] J. B. Garnett and D. E. Marshall, *Harmonic Measure*, New Mathematical Monographs 2, Cambridge University Press, 2005.

[189] H. Gaussier, Tautness and complete hyperbolicity of domains in $\mathbb{C}^n$, *Proc. Amer. Math. Soc.* **127** (1999), 105–116.

[190] H. Gaussier and H. Seshadri, Totally geodesic discs in strongly convex domains, *Math. Z.* **(to appear)** (2013).

[191] G. Gentili, On non-uniqueness of complex geodesics in convex bounded domains, *Rend. Acad. Naz. Lincei* **79** (1985), 90–97.

[192] G. Gentili, On complex geodesics of balanced convex domains, *Ann. Mat. Pura Appl.* **144** (1986), 113–130.

[193] G. Gentili, Regular complex geodesics in the domain $D_n = \{(z_1, \dots, z_n) \in \mathbb{C}^n : |z_1| + |z_2| + \dots + |z_n| < 1\}$, in: *Complex Analysis III* (C. A. Berenstein, ed.), Lecture Notes in Math. 1277, pp. 235–252, 1987.

[194] J. Globevnik, On complex strict and uniform convexity, *Proc. Amer. Math. Soc.* **47** (1975), 175–178.

[195] G. M. Goluzin, *Geometric Theory of Functions of a Complex Variable*, Translations of Mathematical Monographs 26, Amer. Math. Soc., 1969.

[196] N. J. Gorenskiĭ, Some applications of the differentiable boundary distance in open sets of $\mathbb{C}^n$ (Russian), *Problems of holomorphic functions of several complex variables, Krasnojarsk* (1973), 203–214.

[197] I. Graham, Boundary behavior of the Carathéodory and Kobayashi metrics on strongly pseudoconvex domains in $\mathbb{C}^n$ with smooth boundary, *Trans. Amer. Math. Soc.* **207** (1975), 219–240.

[198] I. Graham, Holomorphic mappings into strictly convex domains which are Kobayashi isometries at one point, *Proc. Amer. Math. Soc.* **105** (1989), 917–921.

[199] I. Graham, Sharp constants for the Koebe theorem and for the estimates of intrinsic metrics on convex domains, *Proc. Symp. Pure Math.* **52 (Part 2)** (1991), 233–238.

[200] J. D. Gray and S. A. Morris, When is a function that satisfies the Cauchy–Riemann equations analytic?, *Amer. Math. Monthly* **85** (1978), 246–256.

[201] H. Grunsky, Eindeutige beschränkte Funktionen in mehrfach zusammenhängenden Gebieten I, *Jahresbericht der DMV* **50** (1940), 230–255.

[202] H. Grunsky, Eindeutige beschränkte Funktionen in mehrfach zusammenhängenden Gebieten II, *Jahresbericht der DMV* **52** (1942), 118–132.

[203] R. Gunning, *Introduction to Holomorphic Functions of Several Variables, vol. I* (*Function Theory*)*, vol. II* (*Local Theory*)*, vol. III* (*Homological Theory*), The Wadsworth & Brooks/Cole Mathematics Series, Wadsworth & Brooks/Cole, 1990.

[204] R. Gunning and H. Rossi, *Analytic Functions of Several Complex Variables*, Pentice-Hall, Englewood Cliffs, 1965.

[205] K. T. Hahn, On completeness of the Bergman metric and its subordinate metric, *Proc. Nat. Acad. Sci. USA* **73** (1976), 4294.

[206] K. T. Hahn, On completeness of the Bergman metric and its subordinate metrics II, *Pacific J. Math.* **68** (1977), 437–446.

[207] K. T. Hahn, Some remarks on a new pseudo-differential metric, *Ann. Polon. Math.* **39** (1981), 71–81.

[208] K. T. Hahn, Equivalence of the classical theorems of Schottky, Landau, Picard and hyperbolicity, *Proc. Amer. Math. Soc.* **89** (1983), 628–632.

[209] K. T. Hahn and P. Pflug, The Kobayashi and Bergman metrics on generalized Thullen domains, *Proc. Amer. Math. Soc.* **104** (1988), 207–214.

[210] K. T. Hahn and P. Pflug, On a minimal complex norm that extends the real Euclidean norm, *Monatsh. Math.* **105** (1988), 107–112.

[211] H. Hamada, F. Sakanashi, and T. Yasuoka, Note on a theorem of Wong–Rosay, *Memoirs Fac. Sci. Kyushu Univ., Ser. A* **39** (1985), 249–251.

[212] H. Hamada and H. Segawa, An elementary proof of a Schwarz lemma for the symmetrized bidisc, *Demonstratio Math.* **36** (2003), 329–334.

[213] G. H. Hardy and E. M. Wright, *An Introduction to the Theory of Numbers*, Clarendon, 1979.

[214] L. Harris, *Schwarz's lemma and the maximum principle in infinite dimensional spaces*, Ph.D. thesis, Cornell University, 1969.

[215] L. A. Harris, Schwarz–Pick systems of pseudometrics for domains in normed linear spaces, in: *Advances in Holomorphy* (J. A. Barosso, ed.), Math. Studies 34, pp. 345–406, North Holland, 1979.

[216] M. Hayashi, M. Nakai, and S. Segawa, Bounded analytic functions on two sheeted discs, *Trans. Amer. Math. Soc.* **333** (1992), 799–819.

[217] W. K. Hayman, *Multivalent Functions*, Cambridge Tracts in Mathematics 110, Cambridge University Press, 1994.

[218] W. K. Hayman and P. B. Kennedy, *Subharmonic Functions*, London Mathematical Society Monographs 9, London Math. Soc. Monographs 20, Academic Press, London, 1976.

[219] L. I. Hedberg, Bounded point evaluations and capacity, *J. Funct. Anal.* **10** (1972), 269–280.

[220] S. Helgason, *Differential Geometry, Lie Groups and Symmetric Spaces*, Pure and Applied Mathematics 80, Academic Press, 1978.

[221] L. L. Helms, *Introduction to Potential Theory*, R. E. Krieger Publishing Company, 1975.

[222] G. M. Henkin and E. M. Chirka, Boundary properties of holomorphic functions of several complex variables, *J. Sov. Math.* **5** (1976), 612–687.

[223] G. M. Henkin and J. Leiterer, *Theory of Functions on Complex Manifolds*, Monographs in Mathematics 79, Birkhäuser Verlag, 1984.

[224] G. Herbort, Logarithmic growth of the Bergman kernel for weakly pseudoconvex domains in $\mathbb{C}^3$ of finite type, *Manuscripta Math.* **45** (1983), 69–76.

[225] G. Herbort, *Über die Geodätischen der Bergmanmetrik*, Schriftenreihe d. Math. Inst. Univ. Münster, 2 Serie 26, Münster Univ., 1983.

[226] G. Herbort, *Wachstumsordnung des Bergmankerns auf pseudokonvexen Gebieten*, Schriftenreihe d. Math. Inst. Univ. Münster, 2 Serie 46, Münster Univ., 1987.

[227] G. Herbort, The growth of the Bergman kernel on pseudoconvex domains of homogeneous finite diagonal type, *Nagoya Math. J.* **126** (1992), 1–24.

[228] G. Herbort, Invariant metrics and peak functions on pseudoconvex domains of homogeneous finite diagonal type, *Math. Z.* **209** (1992), 223–243.

[229] G. Herbort, On the invariant differential metrics near pseudoconvex boundary points where the Levi form has corank one, *Nagoya Math. J.* **130** (1993), 25–54.

[230] G. Herbort, The Bergman metric on hyperconvex domains, *Math. Z.* **232** (1999), 183–196.

[231] G. Herbort, Localization lemmas for the Bergman metric at plurisubharmonic peak points, *Nagoya Math. J.* **171** (2003), 107–125.

[232] G. Herbort, Estimation on invariant distances on pseudoconvex domains of finite type in dimension two, *Math. Z.* **251** (2005), 673–703.

[233] M. Hervé, *Les fonctions analytiques*, Presses Universitaires de France, Paris, 1982.

[234] L. Hörmander, *An Introduction to Complex Analysis in Several Variables*, North-Holland Mathematical Library 7, North-Holland, 1990.

[235] L. Hörmander, *Notions of Convexity*, Progress in Mathematics 127, Birkhäuser, 1994.

[236] R. A. Horn and C. R. Johnson, *Topics in Matrix Analysis*, Cambridge University Press, 1991.

[237] L. Hua, *Harmonic Analysis of Functions of Several Complex Variables in the Classical Domains*, Amer. Math. Soc., 1963.

[238] X. Huang, A non-degeneracy property of extremal mappings and iterates of holomorphic self-mappings, *Ann. Sc. Norm. Super. Pisa Cl. Sci. (4)* **21** (1994), 399–419.

[239] A. Huckleberry and A. Isaev, On the Kobayashi hyperbolicity of certain tube domains, *Proc. Amer. Math. Soc.* **(to appear)** (2013).

[240] K. Ii, On the Bargmann-type transform and a Hilbert space of holomorphic functions, *Tohoku Math. J., II. Ser.* **38** (1986), 57–69.

[241] M. Irgens, Continuation of L^2-holomorphic functions, *Math. Z.* **247** (2004), 611–617.

[242] M. Ise, On the Thullen domains and Hirzebruch manifolds, *J. Math. Soc. Japan* **26** (1974), 508–522.

[243] D. Jacquet, $\mathbb{C}$-convex domains with $\mathcal{C}^2$ boundary, *Complex Var. Elliptic Equ.* **51** (2006), 303–312.

[244] P. Jakóbczak, The exceptional sets for functions from the Bergman space, *Portugaliae Math.* **50** (1993), 115–128.

[245] P. Jakóbczak, Description of exceptional sets in circles for functions from the Bergman space, *Czech. Math. J.* **47** (1997), 633–649.

[246] P. Jakóbczak, Complete pluripolar sets and exceptional sets for functions from the Bergman space, *Complex Variables, Theory Appl.* **42** (2000), 17–23.

[247] M. Jarnicki, W. Jarnicki, and P. Pflug, On extremal holomorphically contractible families, *Ann. Polon. Math.* **81** (2003), 183–199.

[248] M. Jarnicki and N. Nikolov, Behavior of the Carathéodory metric near strictly convex boundary points, *Univ. Iagel. Acta Math.* **40** (2002), 7–12.

[249] M. Jarnicki and P. Pflug, Existence domains of holomorphic functions of restricted growth, *Trans. Amer. Math. Soc.* **304** (1987), 385–404.

[250] M. Jarnicki and P. Pflug, Effective formulas for the Carathéodory distance, *Manuscripta Math.* **62** (1988), 1–22.

[251] M. Jarnicki and P. Pflug, Bergman completeness of complete circular domains, *Ann. Polon. Math.* **50** (1989), 219–222.

[252] M. Jarnicki and P. Pflug, The Carathéodory pseudodistance has the product property, *Math. Ann.* **285** (1989), 161–164.

[253] M. Jarnicki and P. Pflug, Three remarks about the Carathéodory distance, in: *Deformations of Mathematical Structures* (J. Ławrynowicz, ed.), pp. 161–170, Kluwer Academic Publisher, 1989.

[254] M. Jarnicki and P. Pflug, The simplest example for the non-innerness of the Carathéodory distance, *Results in Math.* **18** (1990), 57–59.

[255] M. Jarnicki and P. Pflug, A counterexample for the Kobayashi completeness of balanced domains, *Proc. Amer. Math. Soc.* **112** (1991), 973–978.

[256] M. Jarnicki and P. Pflug, The inner Carathéodory distance for the annulus, *Math. Ann.* **289** (1991), 335–339.

[257] M. Jarnicki and P. Pflug, Invariant pseudodistances and pseudometrics – completeness and product property, *Ann. Polon. Math.* **55** (1991), 169–189.

[258] M. Jarnicki and P. Pflug, Some remarks on the product property, *Proc. Symp. Pure Math.* **52 (Part 2)** (1991), 263–272.

[259] M. Jarnicki and P. Pflug, The inner Carathéodory distance for the annulus II, *Michigan Math. J.* **40** (1993), 393–398.

[260] M. Jarnicki and P. Pflug, *Invariant Distances and Metrics in Complex Analysis*, de Gruyter Expositions in Mathematics 9, Walter de Gruyter, 1993.

[261] M. Jarnicki and P. Pflug, Remarks on the pluricomplex Green function, *Indiana Univ. Math. J.* **44** (1995), 535–543.

[262] M. Jarnicki and P. Pflug, On balanced L^2 domains of holomorphy, *Ann. Polon. Math.* **63** (1996), 101–102.

[263] M. Jarnicki and P. Pflug, New examples of effective formulas for holomorphically contractible functions, *Studia Math.* **135** (1999), 219–230.

[264] M. Jarnicki and P. Pflug, A remark on the product property of the Carathéodory pseudodistance, *Ann. UMCS* **53** (1999), 89–94.

[265] M. Jarnicki and P. Pflug, *Extension of Holomorphic Functions*, de Gruyter Expositions in Mathematics 34, Walter de Gruyter, 2000.

[266] M. Jarnicki and P. Pflug, On automorphisms of the symmetrized bidisc, *Arch. Math. (Basel)* **83** (2004), 264–266.

[267] M. Jarnicki and P. Pflug, Invariant Distances and Metrics in Complex Analysis – revisited, *Dissertationes Math.* **430** (2005), 1–192.

[268] M. Jarnicki and P. Pflug, On the upper semicontinuity of the Wu metric, *Proc. Amer. Math. Soc.* **133** (2005), 239–244.

[269] M. Jarnicki and P. Pflug, *First Steps in Several Complex Variables: Reinhardt Domains*, EMS Textbooks in Mathematics, European Mathematical Society Publishing House, 2008.

[270] M. Jarnicki and P. Pflug, *Separately Analytic Functions*, Tracts in Mathematics 16, European Mathematical Society Publishing House, 2011.

[271] M. Jarnicki, P. Pflug, and J.-P. Vigué, The Carathéodory distance does not define the topology – the case of domains, *C. R. Acad. Sci. Paris* **312** (1991), 77–79.

[272] M. Jarnicki, P. Pflug, and J.-P. Vigué, A remark on Carathéodory balls, *Arch. Math. (Basel)* **56** (1992), 595–598.

[273] M. Jarnicki, P. Pflug, and J.-P. Vigué, An example of a Carathéodory complete but not finitely compact analytic space, *Proc. Amer. Math.* **118** (1993), 537–539.

[274] M. Jarnicki, P. Pflug, and R. Zeinstra, Geodesics for convex complex ellipsoids, *Ann. Sc. Norm. Super. Pisa Cl. Sci. (4)* **20** (1993), 535–543.

[275] M. Jarnicki, P. Pflug, and W. Zwonek, On Bergman completeness of non-hyperconvex domains, *Univ. Iagel. Acta Math.* **38** (2000), 169–184.

[276] W. Jarnicki, Kobayashi–Royden vs. Hahn pseudometric in $\mathbb{C}^2$, *Ann. Polon. Math.* **75** (2000), 289–294.

[277] W. Jarnicki, A note on the Lie ball, *Bull. Pol. Acad. Sci. Math.* **49** (2001), 177–179.

[278] W. Jarnicki, On Lempert functions in $\mathbb{C}^2$, *Univ. Iagel. Acta Math.* **39** (2001), 135–138.

[279] W. Jarnicki, Möbius function in complex ellipsoids, *Proc. Amer. Math. Soc.* **132** (2004), 3243–3250.

[280] W. Jarnicki and N. Nikolov, Concave domains with trivial biholomorphic invariants, *Ann. Polon. Math.* **79** (2002), 63–66.

[281] Z. Jinhao, Metric S on holomorphy domain, *Kexue Tongbao* **33** (1988), 353–356.

[282] F. John, Extremum problems with inequalities as subsidiary conditions, in: *Studies and Essays Presented to R. Courant on his 60th Birthday, January 8, 1948*, pp. 187–204, Interscience Publishers, Inc., New York, N. Y., 1948.

[283] P. Jucha, The Wu metric in elementary Reinhardt domains, *Univ. Iagel. Acta Math.* **40** (2002), 83–89.

[284] P. Jucha, Bergman completeness of Zalcman type domains, *Studia Math.* **163** (2004), 71–83.

[285] P. Jucha, On the Wu metric in unbounded domains, *Arch. Math. (Basel)* **90** (2008), 559–571.

[286] P. Jucha, The Wu metric is not upper semicontinuous, *Proc. Amer. Math. Soc.* **136** (2008), 1349–1358.

[287] P. Jucha, A remark on the dimension of the Bergman space of some Hartogs domains, *J. Geom. Anal.* **22** (2012), 23–37.

[288] H. Kang, L. Lee, and C. Zeager, Comparison of invariant metrics, *Rocky Mountain J. Math.* **(to appear)** (2013).

[289] L. Kaup and B. Kaup, *Holomorphic Functions of Several Variables*, de Gruyter Studies in Mathematics 3, Walter de Gruyter, 1983.

[290] W. Kaup, Über das Randverhalten von holomorphen Automorphismen beschränkter Gebiete, *Manuscripta Math.* **3** (1970), 257–270.

[291] W. Kaup and H. Upmeier, Banach spaces with biholomorphically equivalent unit balls are isomorphic, *Proc. Amer. Math. Soc.* **58** (1976), 129–133.

[292] W. Kaup and J.-P. Vigué, Symmetry and local conjugacy on complex manifolds, *Math. Ann.* **286** (1990), 329–340.

[293] L. D. Kay, On the Kobayashi–Royden metric for ellipsoids, *Math. Ann.* **289** (1991), 55–72.

[294] N. Kerzman, The Bergman kernel function. Differentiability at the boundary, *Math. Ann.* **195** (1972), 149–158.

[295] N. Kerzman and J. P. Rosay, Fonctions plurisousharmoniques d'exhaustion bornées et domaines taut, *Math. Ann.* **257** (1981), 171–184.

[296] J. J. Kim, I. G. Hwang, J. G. Kim, and J. S. Lee, On the higher order Kobayashi metrics, *Honam Math. J.* **26** (2004), 549–557.

[297] J. J. Kim, J. K. Kim, and J. S. Lee, On the higher order Kobayashi metrics, *Honam Math. J.* **28** (2006), 513–520.

[298] K.-T. Kim, Automorphism group of certain domains in $\mathbb{C}^n$ with a singular boundary, *Pacific J. Math.* **151** (1991), 57–64.

[299] K.-T. Kim, Biholomorphic mappings between quasicircular domains in $\mathbb{C}^n$, *Proc. Symp. Pure Math.* **52 (Part 2)** (1991), 283–290.

[300] K.-T. Kim, *The Wu metric and minimum ellipsoids*, Proc. 3rd Pacific Rim Geometry Conf. (J. Choe, ed.), Monogr. Geom. Topology 25, International Press, Cambridge, 1998, pp. 121–138.

[301] K.-T. Kim and S. G. Krantz, A Kobayashi metric version of Bun Wong's theorem, *Complex Var. Elliptic Equ.* **54** (2009), 355–369.

[302] P. Klembeck, Kähler metrics of negative curvature, the Bergman metric near the boundary and the Kobayashi metric on smooth bounded strictly pseudoconvex sets, *Indiana Univ. Math. J.* **27** (1978), 275–282.

[303] M. Klimek, Extremal plurisubharmonic functions and invariant pseudodistances, *Bull. Soc. Math. France* **113** (1985), 231–240.

[304] M. Klimek, Infinitesimal pseudometrics and the Schwarz Lemma, *Proc. Amer. Math. Soc.* **105** (1989), 134–140.

[305] M. Klimek, *Pluripotential Theory*, London Mathematical Society Monographs. New Series 6, Oxford University Press, 1991.

[306] P. Kliś, *Extremal mappings* (*Polish*), Ph.D. thesis, Jagiellonian Univ., 2013.

[307] G. Knese, Function theory on the Neil parabola, *Michigan Math. J.* **55** (2007), 139–154.

[308] M. Kobayashi, Convergence of invariant distances on decreasing domains, *Complex Var.* **47** (2002), 155–165.

[309] S. Kobayashi, Geometry of bounded domains, *Trans. Amer. Math. Soc.* **92** (1959), 267–290.

[310] S. Kobayashi, On complete Bergman metrics, *Proc. Amer. Math. Soc.* **13** (1962), 511–513.

[311] S. Kobayashi, Invariant distances on complex manifolds and holomorphic mappings, *J. Math. Soc. Japan* **19** (1967), 460–480.

[312] S. Kobayashi, *Hyperbolic Manifolds and Holomorphic Mappings*, Pure and Applied Mathematics 2, M. Dekker, 1970.

[313] S. Kobayashi, Some remarks and questions concerning the intrinsic distance, *Tôhoku Math. J.* **25** (1973), 481–486.

[314] S. Kobayashi, Intrinsic distances, measures and geometric function theory, *Bull. Amer. Math. Soc.* **82** (1976), 357–416.

[315] S. Kobayashi, A new invariant infinitesimal metric, *Int. J. Math.* **1** (1990), 357–416.

[316] S. Kobayashi, *Hyperbolic Complex Spaces*, Grundlehren der Mathematischen Wissenschaften 318, Springer Verlag, 1998.

[317] S. Kobayashi, *Hyperbolic Manifolds and Holomorphic Mappings. An introduction*, second ed, World Scientific Publishing Co. Pte. Ltd., Hackensack, NJ, 2005.

[318] A. Kodama, Boundedness of circular domains, *Proc. Japan Acad., Ser. A Math. Sci.* **58** (1982), 227–230.

[319] A. Kodama, S. G. Krantz, and D. Ma, A characterization of generalized complex ellipsoids in $\mathbb{C}^n$ and related results, *Indiana Univ. Math. J.* **41** (1992), 173–195.

[320] J. J. Kohn, Boundary behavior of $\overline{\partial}$ on weakly pseudoconvex manifolds of dimension two, *J. Diff. Geom.* **6** (1972), 523–542.

[321] Ł. Kosiński, The group of automorphisms of the spectral ball, *Proc. Amer. Math. Soc.* **140** (2012), 2029–2031.

[322] Ł. Kosiński and T. Warszawski, Lempert theorem for strictly linearly convex domains with real analytic boundaries, *Ann. Polon. Math.* **107** (2012), 167–216.

[323] Ł. Kosiński and W. Zwonek, Proper holomorphic mappings vs. peak points and Silov boundary, *Ann. Polon. Math.* **107** (2013), 97–108.

[324] P. Kot, A remark on the inner Carathéodory distance for the annulus, *Univ. Iagel. Acta Math.* **35** (1997), 211–212.

[325] S. G. Krantz, The boundary behavior of the Kobayashi metric, *Rocky Mountain J. Math.* **22** (1992), 227–233.

[326] S. G. Krantz, *Geometric Analysis and Function Spaces*, CBMS Regional Conference Series in Mathematics 81, American Mathematical Society, Providence, RI, 1993.

[327] S. G. Krantz, *Function Theory of Several Complex Variables*, reprint of the 1992 ed, AMS Chelsea Publishing, Providence, RI, 2001.

[328] S. G. Krantz, Pseudoconvexity, analytic discs and invariant metrics, *Bull. Allahabad Math. Soc.* **23** (2008), 245–262.

[329] S. G. Krantz and H. R. Parks, Distance to C^k hypersurfaces, *Journal of Diff. Equ.* **40** (1981), 116–120.

[330] N. Kritikos, Über analytische Abbildungen einer Klasse von vierdimensionalen Gebieten, *Math. Ann.* **99** (1927), 321–341.

[331] T. Kuczumow and W. O. Ray, Isometries in the Cartesian product of n unit open Hilbert balls with a hyperbolic metric, *Ann. Mat. Pura Appl.* **152** (1988), 359–374.

[332] M. H. Kwack, Generalization of the big Picard theorem, *Ann. of Math.* **19** (1969), 9–22.

[333] N. S. Landkof, *Foundations of Modern Potential Theory*, Grundlehren der mathematischen Wissenschaften 180, Springer Verlag, 1972.

[334] M. Landucci, On the proper holomorphic equivalence for a class of pseudoconvex domains, *Trans. Amer. Math. Soc.* **282** (1984), 807–811.

[335] S. Lang, *Introduction to Complex Hyperbolic Spaces*, Springer Verlag, 1987.

[336] F. Lárusson, P. Lassere, and R. Sigurdsson, Convexity of sublevel sets of plurisubharmonic extremal functions, *Ann. Polon. Math.* **68** (1998), 267–273.

[337] F. Lárusson and R. Sigurdsson, Plurisubharmonic functions and analytic discs on manifolds, *J. Reine Angew. Math.* **501** (1998), 1–39.

[338] L. Lee, Asymptotic behavior of the Kobayashi metric on convex domains, *Pacific J. Math.* **238** (2008), 105–118.

[339] P. Lelong, Fonction de Green pluricomplexe et lemme de Schwarz dans les espaces de Banach, *J. Math. Pures Appl.* **68** (1989), 319–347.

[340] L. Lempert, La métrique de Kobayashi et la représentation des domaines sur la boule, *Bull. Soc. Math. France* **109** (1981), 427–474.

[341] L. Lempert, Holomorphic retracts and intrinsic metrics in convex domains, *Analysis Mathematica* **8** (1982), 257–261.

[342] L. Lempert, Intrinsic distances and holomorphic retracts, *Complex Analysis and Applications '81, Sophia* (1984), 341–364.

[343] N. Levenberg and E. A. Poletsky, Pluripolar hulls, *Michigan Math. J.* **46** (1999), 151–162.

[344] J. Lewittes, A note on parts and hyperbolic geometry, *Proc. Amer. Math. Soc.* **17** (1966), 1087–1090.

[345] E. Ligocka, On the Forelli–Rudin construction and weighted Bergman projections, *Studia Math.* **94** (1989), 257–272.

[346] E. B. Lin and B. Wong, Curvature and proper holomorphic mappings between bounded domains in $\mathbb{C}^n$, *Rocky Mountain J. Math.* **20** (1990), 179–197.

[347] P. Lindberg, L^p approximation by analytic functions in an open region, *U. U.D. M. Report No.* **1977:7** (1977).

[348] J. J. Loeb, Applications harmoniques et hyperbolicité de domaines tube, *Enseign. Math.* **53** (2007), 347–367.

[349] S. Łojasiewicz, *Introduction to Complex Analytic Geometry*, Birkhäuser, 1991.

[350] K. H. Look, Schwarz Lemma and analytic invariants, *Sci. Sinica* **7** (1958), 435–504.

[351] D. Ma, Boundary behavior of invariant metrics and volume forms on strongly pseudoconvex domains, *Duke Math. J.* **63** (1991), 673–697.

[352] D. Ma, On iterates of holomorphic maps, *Math. Z.* **207** (1991), 417–428.

[353] D. Ma, Sharp estimates of the Kobayashi metric near strongly pseudoconvex points, *The Madison Symposium on Complex Analysis (Madison, WI, 1991), Contemp. Math., 137, Amer. Math. Soc., Providence, RI* (1992), 329–338.

[354] D. Ma, Smoothness of Kobayashi metric of ellipsoids, *Complex Var.* **26** (1995), 291–298.

[355] B. Maccluer, K. Stroethoff, and R. H. Zhao, Generalized Schwarz–Pick estimates, *Proc. Amer. Math. Soc.* **131** (2003), 593–599.

[356] M. Marden, *Geometry of Polynomials*, Mathematical Surveys 3, A. M.S., Providence, 1966.

[357] P. Mazet and J.-P. Vigué, Convexité de la distance de Carathéodory et points fixes d'applications holomorphes, *Bull. Sci. Math.* 2e *série* **116** (1992), 285–305.

[358] T. Mazur, P. Pflug, and M. Skwarczyński, Invariant distances related to the Bergman function, *Proc. Amer. Math. Soc.* **94** (1985), 72–76.

[359] J. D. McNeal, Local geometry of decoupled pseudoconvex domains, in: *Complex Analysis* (D. Diederich, ed.), pp. 223–230, Vieweg, 1990.

[360] G. Mengotti and E. H. Youssfi, The weighted Bergman projection and related theory on the minimal ball, *Bull. Sci. Math.* **123** (1999), 501–525.

[361] P. R. Mercer, Complex geodesics and iterates of holomorphic maps on convex domains in $\mathbb{C}^n$, *Trans. Amer. Math. Soc.* **338** (1993), 201–211.

[362] R. Meyer, The Carathéodory pseudodistance and positive linear operators, *Int. J. Math.* **8** (1997), 809–824.

[363] D. Minda, The Hahn metric on Riemann surfaces, *Kodai Math. J.* **6** (1983), 57–69.

[364] D. Minda, The strong form of Ahlfors' Lemma, *Rocky Mountain J. Math.* **17** (1987), 457–461.

[365] D. Minda and G. Schober, Another elementary approach to the theorems of Landau, Montel, Picard and Schottky, *Complex Analysis* **2** (1983), 157–164.

[366] G. Misra, S. S. Roy, and G. Zhang, Reproducing kernel for a class of weighted Bergman spaces on the symmetrized polydisc, *Proc. Amer. Math. Soc.* **(to appear)** (2013).

[367] D. Montgomery and L. Zippin, *Topological Transformation Groups*, Interscience Publishers, 1955.

[368] C. Müller, Sperical harmonics, *Lecture Notes in Math.* **17** (1966).

[369] R. Narasimhan, *Analysis on Real and Complex Manifolds*, Advanced Studies in Pure Mathematics 1, North Holland, 1968.

[370] R. Narasimhan, *Several Complex Variables*, Chicago Lectures in Mathematics, The University of Chicago Press, 1971.

[371] R. Narasimhan, *Complex Analysis in One Variable*, Birkhäuser, 1985.

[372] I. Naruki, The holomorphic equivalence problem for a class of Reinhardt domains, *Publ. RIMS Kyoto Univ.* **4** (1968), 527–543.

[373] V. A. Nguyên, The Lu Qi-Keng conjecture fails for strongly convex algebraic complete Reinhardt domains in $\mathbb{C}^n$ $(n \geq 3)$, *Proc. Amer. Math. Soc* **128** (2000), 1729–1732.

[374] N. Nikolov, Continuity and boundary behaviour of the Carathéodory metric, *Math. Notes* **67** (2000), 183–191.

[375] N. Nikolov, Behavior of invariant metrics near convexifiable boundary points, *Czech. Math. J.* **53** (2003), 1–7.

[376] N. Nikolov, The completeness of the Bergman distance of planar domains has a local character, *Complex Var.* **48** (2003), 705–709.

[377] N. Nikolov, The symmetrized polydisc cannot be exhausted by domains biholomorphic to convex domains, *Ann. Polon. Math.* **88** (2006), 279–283.

[378] N. Nikolov, Estimates of invariant distances on "convex" domains, *Ann. Mat. Pura Appl.* **(to appear)** (2013).

[379] N. Nikolov, Invariant functions and Metrics in Complex Analysis, *Dissertationes Math.* **486** (2012), 1–100.

[380] N. Nikolov and P. Pflug, Local vs. global hyperconvexity, tautness or k-completeness for unbounded open sets in $\mathbb{C}^n$, *Ann. Sc. Norm. Super. Pisa Cl. Sci. (5)* **IV** (2005), 601–618.

[381] N. Nikolov and P. Pflug, The multipole Lempert function is monotone under inclusion of pole sets, *Michigan Math. J.* **54** (2006), 111–116.

[382] N. Nikolov and P. Pflug, On the definition of the Kobayashi-Buseman pseudometric, *Int. J. Math.* **17** (2006), 1145–1149.

[383] N. Nikolov and P. Pflug, Invariant metrics and distances on generalized Neil parabolas, *Michigan Math. J.* **55** (2007), 255–268.

[384] N. Nikolov and P. Pflug, On the derivatives of the Lempert functions, *Ann. Mat. Pura Appl.* **187** (2008), 547–553.

[385] N. Nikolov and P. Pflug, Kobayashi-Royden pseudometric vs. Lempert function, *Ann. Mat. Pura Appl.* **190** (2011), 589–593.

[386] N. Nikolov, P. Pflug, and P. J. Thomas, Lipschitzness of the Lempert and the Green function, *Proc. Amer. Math. Soc.* **137** (2009), 2027–2036.

[387] N. Nikolov, P. Pflug, and P. J. Thomas, Upper bound for the Lempert function of smooth domains, *Math. Z.* **266** (2010), 425–430.

[388] N. Nikolov, P. Pflug, P. J. Thomas, and W. Zwonek, Estimates of the Carathéodory metric on the symmetrized polydisc, *J. Math. Anal. Appl.* **341** (2008), 140–148.

[389] N. Nikolov, P. Pflug, and W. Zwonek, The Lempert function of the symmetrized polydisc in higher dimesions is not a distance, *Proc. Amer. Math. Soc.* **135** (2007), 2921–2928.

[390] N. Nikolov, P. Pflug, and W. Zwonek, An example of a bounded $\mathbb{C}$-convex domain which is not biholomorphically to a convex domain, *Math. Scan.* **102** (2008), 149–155.

[391] N. Nikolov, P. Pflug, and W. Zwonek, Estimates for invariant metrics on $\mathbb{C}$-convex domains, *Trans. Amer. Math. Soc.* **363** (2012), 6245–6256.

[392] N. Nikolov and A. Saracco, Hyperbolicity of $\mathbb{C}$-convex domains, *C. R. Acad. Bulg. Sci.* **60** (2007), 935–938.

[393] N. Nikolov and P. J. Thomas, Rigid characterizations of pseudoconvex domains, *Indiana Univ. Math. J.* **(to appear)** (2013).

[394] N. Nikolov, P. J. Thomas, and W. Zwonek, Discontinuity of the Lempert function and the Kobayashi–Royden metric of the spectral ball, *Integr. Equ. Oper. Theory* **61** (2008), 401–412.

[395] N. Nikolov and W. Zwonek, Some remarks on the Green function and the Azukawa pseudometric, *Monatsh. Math.* **142** (2004), 341–350.

[396] N. Nikolov and W. Zwonek, On the product property for the Lempert function, *Complex Variables, Theory Appl.* **50** (2005), 939–952.

[397] N. Nikolov and W. Zwonek, The Bergman kernel of the symmetrized polydisc in higher dimensions has zeros, *Arch. Math. (Basel)* **87** (2006), 412–416.

[398] J. Noguchi and T. Ochiai, *Geometric Function Theory in Several Complex Variables*, Translations of Mathematical Monographs 80, Amer. Math. Soc., 1990.

[399] K. Noshiro, *Cluster Sets*, Ergebnisse der Mathematik und ihrer Grenzgebiete 28, Springer Verlag, 1960.

[400] K. Oeljeklaus, P. Pflug, and E. H. Youssfi, The Bergman kernel of the minimal ball and applications, *Ann. Inst. Fourier (Grenoble)* **74** (1997), 915–928.

[401] T. Ohsawa, Boundary behavior of the Bergman kernel function on pseudoconvex domains, *Publ. RIMS Kyoto Univ.* **20** (1984), 897–902.

[402] T. Ohsawa, On the Bergman kernel of hyperconvex domains, *Nagoya Math. J.* **129** (1993), 43–59.

[403] T. Ohsawa, Addendum to "On the Bergman kernel of hyperconvex domains", *Nagoya Math. J.* **137** (1995), 145–148.

[404] T. Ohsawa, On the extension of L^2 holomorphic function V – effects of generalization, *Nagoya Math. J.* **161** (2001), 1–21.

[405] T. Ohsawa and K. Takegoshi, On the extension of L^2 holomorphic functions, *Math. Z.* **195** (1987), 197–204.

[406] M. Overholt, Injective hyperbolicity of domains, *Ann. Polon. Math.* **62** (1995), 79–82.

[407] M.-Y. Pang, On infinitesimal behavior of the Kobayashi distance, *Pacific J. Math.* **162** (1994), 121–141.

[408] J. Park, New formulas of the Bergman kernels for complex ellipsoids in $\mathbb{C}^2$, *Proc. Amer. Math. Soc.* **136** (2008), 4211–4221.

[409] S.-H. Park, *Tautness and Kobayashi hyperbolicity*, Ph.D. thesis, Oldenburg Univ., 2003.

[410] S.-H. Park, On hyperbolicity of balanced domains, *Nagoya Math. J.* **176** (2007), 99–111.

[411] G. Patrizio, On holomorphic maps between domain in $\mathbb{C}^n$, *Ann. Sc. Norm. Super. Pisa Cl. Sci. (4)* **13** (1986), 267–279.

[412] K. Peters, Starrheitssätze für Produkte normierter Vektorräume endlicher Dimension und für Produkte hyperbolischer komplexer Räume, *Math. Ann.* **208** (1974), 343–354.

[413] P. Pflug, *Holomorphiegebiete, pseudokonvexe Gebiete und das Levi–Problem*, Lecture Notes in Math. 432, Springer Verlag, 1975.

[414] P. Pflug, Quadratintegrable holomorphe Funktionen und die Serre Vermutung, *Math. Ann.* **216** (1975), 285–288.

[415] P. Pflug, Various applications of the existence of well growing holomorphic functions, in: *Functional Analysis, Holomorphy and Approximation Theory* (J. A. Barroso, ed.), Math. Studies 71, pp. 391–412, North Holland, 1980.

[416] P. Pflug, About the Carathéodory completeness of all Reinhardt domains, in: *Functional Analysis, Holomorphy and Approximation Theory II* (G. I. Zapata, ed.), Math. Studies 86, pp. 331–337, North Holland, 1984.

[417] P. Pflug, Applications of the existence of well growing holomorphic functions, in: *Analytic Functions, Błażejewko 1983* (J. Ławrynowicz, ed.), Lecture Notes in Math. 1039, pp. 376–388, Springer Verlag, 1984.

[418] P. Pflug, Invariant metrics and completeness, *J. Korean Math. Soc.* **37** (2000), 269–284.

[419] P. Pflug and E. H. Youssfi, The Lu Qi-Keng conjecture fails for strongly convex algebraic domains, *Arch. Math. (Basel)* **71** (1998), 240–245.

[420] P. Pflug and E. H. Youssfi, Complex geodesics of the minimal ball in $\mathbb{C}^n$, *Ann. Sc. Norm. Super. Pisa Cl. Sci. (5)* **3** (2004), 53–66.

[421] P. Pflug and W. Zwonek, The Kobayashi metric for non-convex complex ellipsoids, *Complex Var.* **29** (1996), 59–71.

[422] P. Pflug and W. Zwonek, Effective formulas for invariant functions – case of elementary Reinhardt domains, *Ann. Polon. Math.* **69** (1998), 175–196 (erratum ibid., 69 (1998), 301).

[423] P. Pflug and W. Zwonek, L_h^2-domains of holomorphy and the Bergman kernel, *Studia Math.* **151** (2002), 99–108.

[424] P. Pflug and W. Zwonek, Logarithmic capacity and Bergman functions, *Arch. Math. (Basel)* **80** (2003), 536–552.

[425] P. Pflug and W. Zwonek, Bergman completeness of unbounded Hartogs domains, *Nagoya Math. J.* **180** (2005), 121–133.

[426] P. Pflug and W. Zwonek, Description of all complex geodesics in the symmetrized bidisc, *Bull. London Math. Soc.* **37** (2005), 575–584.

[427] P. Pflug and W. Zwonek, L_h^2-domains of holomorphy in the class of unbounded Hartogs domains, *Illinois J. Math.* **51** (2007), 617–624.

[428] P. Pflug and W. Zwonek, Boundary behavior of the Kobayashi–Royden metric in smooth pseudoconvex domains, *Michigan Math. J.* **60** (2011), 399–407.

[429] P. Pflug and W. Zwonek, Exhausting domains of the symmetrized bidisc, *Ark. Mat.* **50** (2012), 397–402.

[430] I. I. Piatetski-Shapiro, On a problem proposed by E. Cartan (Russian), *Dokl. Akad. Nauk. SSSR* **124** (1959), 272–273.

[431] S. Pinchuk, Proper holomorphic mappings of strictly pseudoconvex domains, *Soviet Math. Dokl* **19** (1978), 804–807.

[432] E. A. Poletskiĭ, The Euler–Lagrange equations for extremal holomorphic mappings of the unit disk, *Michigan Math. J.* **30** (1983), 317–333.

[433] E. A. Poletskiĭ, *Holomorphic currents*, Preprint, 1989.

[434] E. A. Poletskiĭ and B. V. Shabat, *Invariant metrics*, Several Complex Variables III (G. M. Khenkin, ed.), Springer Verlag, 1989, pp. 63–112.

[435] E. A. Poletsky, Plurisubharmonic functions as solutions of variational problems, *Proc. Sympos. Pure Math.* **52** (1991), 163–171.

[436] E. A. Poletsky, Holomorphic currents, *Indiana Univ. Math. J.* **42** (1993), 85–144.

[437] C. Pommerenke, *Boundary Behaviour of Conformal Maps*, Grundlehren der mathematischen Wissenschaften 299, Springer Verlag, 1992.

[438] L. Qi-Keng, On Kaehler manifolds with constant curvature, *Acta Math. Sinica* **16** (1966), 269–281.

[439] L. Qikeng, The conjugate points of $\mathbb{CP}^{\infty}$ and the zeroes of the Bergman kernel, *Acta Math. Sci., Ser. B, Engl. Ed.* **29** (2009), 480–492.

[440] T. Radó, *Subharmonic Functions*, Ergebnisse der Mathematik und ihre Grenzgebiete, Springer Verlag, 1937.

[441] Q. I. Rahman and G. Schmeisser, *Analytic Theory of Polynomials*, London Mathematical Society Monographs. New Series 26, Oxford University Press, 2002.

[442] I. Ramadanov, Sur une propriété de la fonction de Bergman, *Comt. Rend. Acad. Bulg. Sci.* **20** (1967), 759–762.

[443] I. Ramadanov, Some applications of the Bergman kernel to geometrical theory of functions, *Banach Center Publ.* **11** (1983), 275–286.

[444] R. M. Range, The Carathédory metric and holomorphic maps on a class of weakly pseudoconvex domains, *Pacific J. Math.* **78** (1978), 173–189.

[445] R. M. Range, *Holomorphic Functions and Integral Representations in Several Complex Variables*, Graduate Texts in Mathematics 108, Springer Verlag, 1986.

[446] T. Ransford, *Potential Theory in the Complex Plane*, London Mathematical Society Student Texts 28, Cambridge University Press, 1995.

[447] H.-J. Reiffen, *Die differentialgeometrischen Eigenschaften der invarianten Distanzfunktion von Carathéodory*, Schriftenreihe d. Math. Inst. Univ. Münster 26, Münster Univ., 1963.

[448] H.-J. Reiffen, Die Carathéodorysche Distanz und ihre zugehörige Differentialmetrik, *Math. Ann.* **161** (1965), 315–324.

[449] H.-J. Reiffen, Metrische Größen in C^q-Räumen, *Acad. Nazionale dei XL* (1977–1978), 1–29.

[450] K. Reinhardt, Über Abbildungen durch analytische Funktionen zweier Veränderlicher, *Math. Ann.* **83** (1921), 211–255.

[451] R. Richberg, Stetige streng pseudokonvexe Funktionen, *Math. Ann.* **175** (1968), 251–286.

[452] W. Rinow, *Die innere Geometrie der metrischen Räume*, Grundlehren der mathematischen Wissenschaften 105, Springer Verlag, 1961.

[453] R. M. Robinson, Analytic functions in circular rings, *Duke Math. J.* **10** (1943), 341–354.

[454] J.-P. Rosay, Sur une characterization de la boule parmi les domaines de $\mathbb{C}^n$ par son groupe d'automorphismes, *Ann. Inst. Fourier (Grenoble)* **29** (1979), 91–97.

[455] J.-P. Rosay, Un example d'ouvert borne de $\mathbb{C}^3$ "taut" mais non hyperbolique complet, *Pacific J. Math.* **98** (1982), 153–156.

[456] J.-P. Rosay, Poletsky theory of disks on holomorphic manifolds, *Indiana Univ. Math. J.* **52** (2003), 157–169.

[457] P. Rosenthal, On the zeros of the Bergman function in double-connected domains, *Proc. Amer. Math. Soc.* **21** (1969), 33–35.

[458] H. Royden, P.-M. Wong, and S. G. Krantz, The Carathéodory and Kobayashi/Royden metrics by way of dual extremal problems, *Complex Var. Elliptic Equ.* (2012), 1–16.

[459] H. L. Royden, Remarks on the Kobayashi metric, in: *Several complex variables, II*, Lecture Notes in Math. 189, pp. 125–137, Springer Verlag, 1971.

[460] H. L. Royden and P.-M. Wong, *Carathéodory and Kobayashi metric on convex domains*, Preprint, 1983.

[461] W. Rudin, *Real and Complex Analysis*, McGraw-Hill Series in Higher Mathematics, McGraw-Hill Book Company, 1974.

[462] W. Rudin, *Function Theory in the Unit Ball*, Grundlehren der Mathematischen Wissenschaften 241, Springer Verlag, 1980.

[463] W. Rudin, Proper holomorphic maps and finite reflection groups, *Indiana Univ. Math. J.* **31** (1982), 701–720.

[464] F. Rühs, *Funktionentheorie*, Hochschulbücher für Mathematik 56, Deutscher Verlag d. Wissenschaften, 1962.

[465] B. Schwarz, Bounds for the Carathéodory distance of the unit ball, *Linear and Multilinear Algebra* **32** (1992), 93–101.

[466] B. Schwarz, Carathéodory balls and norm balls of the domain $H = \{(z_1, z_2) \in \mathbb{C}^2 : |z_1| + |z_2| < 1\}$, *Isr. J. Math.* **84** (1993), 119–128.

[467] B. Schwarz and U. Srebro, Carathéodory balls and norm balls in $H_{p,n} = \{z \in \mathbb{C}^n : \|z\|_p < 1\}$, *Banach Center Publ.* **37** (1996), 75–83.

[468] M. A. Selby, On completeness with respect to the Carathéodory metric, *Canad. Math. Bull.* **17** (1974), 261–263.

[469] H. Seshadri and K. Verma, On isometries of the Carathéodory and Kobayashi metrics on strongly pseudoconvex domains, *Ann. Sc. Norm. Super. Pisa Cl. Sci.* (5) **5** (2006), 393–417.

[470] H. Seshadri and K. Verma, On the compactness of isometry groups in complex analysis, *Complex Var. Elliptic Equ.* **54** (2009), 387–399.

[471] N. Sibony, Prolongement de fonctions holomorphes bornées et metrique de Carathéodory, *Invent. Math.* **29** (1975), 205–230.

[472] N. Sibony, *Remarks on the Kobayashi metric*, Unpublished manuscript, 1979.

[473] N. Sibony, Personal communication, 1981.

[474] N. Sibony, A class of hyperbolic manifolds, in: *Recent developments in several complex variables* (J. E. Fornæss, ed.), Ann. Math. Studies 100, pp. 347–372, 1981.

[475] N. Sibony, Some aspects of weakly pseudoconvex domains, *Proc. Symp. Pure Math.* **52 (Part 1)** (1991), 199–231.

[476] J. Siciak, On some extremal functions and their applications in the theory of analytic functions of several variables, *Trans. Amer. Math. Soc.* **105** (1962), 332–357.

[477] J. Siciak, *Extremal plurisubharmonic functions and capacities in* $\mathbb{C}^n$, Sophia Kokyuroku in Mathematics 14, Sophia Univ., 1982.

[478] J. Siciak, Balanced domains of holomorphy of type H^∞, *Mat. Vesnik* **37** (1985), 134–144.

[479] R. R. Simha, The Carathéodory metric on the annulus, *Proc. Amer. Math. Soc.* **50** (1975), 162–166.

[480] Y. T. Siu, Every Stein subvariety admits a Stein neighborhood, *Invent. Math.* **38** (1976), 89–100.

[481] H. Skoda, Applications des techniques L^2 à la theorie de ideaux d'un algèbre de fonctions holomorphes avec poids, *Ann. Scient. Ec. Normale Sup.* **5** (1972), 548–580.

[482] M. Skwarczyński, The invariant distance in the theory of pseudoconformal transformations and the Lu Qi-keng conjecture, *Proc. Amer. Math.* **22** (1969), 305–310.

[483] M. Skwarczyński, Biholomorphic invariants related to the Bergman function, *Dissertationes Math.* **173** (1980), 1–59.

[484] M. Skwarczyński, Evaluation functionals on spaces of square integrable holomorphic functions, *Wyższa Szkoła Inżynierska w Radomiu, Prace Mat.-Fiz. Radom* (1982), 64–72.

[485] Z. Słodkowski, Polynomial hulls with convex fibers and complex geodesics, *J. Funct. Anal.* **94** (1990), 156–176.

[486] U. Srebro, Carathéodory balls and norm balls in $H_n = \{z \in \mathbb{C}^n : \|z\|_1 < 1\}$, *Isr. J. Math.* **89** (1995), 61–70.

[487] C. M. Stanton, A characterization of the polydisc, *Math. Ann.* **253** (1980), 129–135.

[488] C. M. Stanton, A characterization of the ball by its intrinsic metrics, *Math. Ann.* **264** (1983), 271–275.

[489] E. L. Stout, *The Theory of Uniform Algebras*, Bogden and Quigley Publishers, 1971.

[490] N. Suita, Capacities and kernels on Riemann surfaces, *Arch. Ration. Mech. Anal.* **46** (1972), 212–217.

[491] N. Suita and A. Yamada, On the Lu Qi-Keng conjecture, *Proc. Amer. Math. Soc.* **59** (1976), 222–224.

[492] T. Sunada, Holomorphic equivalence problem for bounded Reinhardt domains, *Math. Ann.* **235** (1978), 111–128.

[493] M. Suzuki, The holomorphic curvature of intrinsic metrics, *Math. Rep. Toyama Univ.* **4** (1981), 107–114.

[494] M. Suzuki, The intrinsic metrics on the domains in $\mathbb{C}^n$, *Math. Rep. Toyama Univ.* **6** (1983), 143–177.

[495] D. D. Thai and P. V. Duc, On the complete hyperbolicity and the tautness of the Hartogs domains, *Int. J. Math.* **11** (2000), 103–111.

[496] D. D. Thai and P. J. Thomas, $\mathbb{D}^*$-extension property without hyperbolicity, *Indiana Univ. Math. J.* **47** (1998), 1125–1130.

[497] D. D. Thai, P. J. Thomas, N. Trao, and A. Duc, On hyperbolicity and tautness modulo an analytic subset of Hartogs domains, *Proc. Amer. Math. Soc.* **(to appear)** (2013).

[498] P. J. Thomas, Green versus Lempert functions: a minimal example, *Pacific J. Math.* **257** (2012), 189–197.

[499] P. J. Thomas and N. V. Trao, Pluricomplex Green and Lempert functions for equally weighted poles, *Ark. Mat.* **41** (2003), 381–400.

[500] P. J. Thomas and N. V. Trao, Discontinuity of the Lempert function of the spectral ball, *Proc. Amer. Math. Soc.* **138** (2010), 2403–2412.

[501] D. K. Tishabaev, On regions biholomorphically equivalent to a ball (Russian), *Sib. Math. J.* **2** (1991), 193–196.

[502] J. J. Trokhimchuk, *Continuous Mappings and Conditions of Monogeneity*, Daniel Davey, 1964.

[503] M. Trybuła, *Bergman metric for the symmetrized polidisc*, Preprint, 2012.

[504] M. Tsuji, *Potential Theory in Modern Function Theory*, Chelsea Pub. Co., New York, 1975.

[505] N. T. Van and J. Siciak, Fonctions plurisousharmoniques extrémales et systèmes doublement orthogonaux de fonctions analytiques, *Bull. Sci. Math.* **115** (1991), 235–244.

[506] D. van Dantzig and B. L. van der Waerden, Über metrische homogene Räume, *Abhandlung Math. Sem. Univ. Hamburg* **6** (1928), 374–376.

[507] S. Venturini, Comparison between the Kobayashi and Carathéodory distances on strongly pseudoconvex bounded domains in $\mathbb{C}^n$, *Proc. Amer. Math. Soc.* **107** (1989), 725–730.

[508] S. Venturini, Pseudodistances and pseudometrics on real and complex manifold, *Ann. Mat. Pura Appl.* **154** (1989), 385–402.

[509] S. Venturini, Intrinsic metrics in complete circular domains, *Math. Ann.* **228** (1990), 473–481.

[510] E. Vesentini, Complex geodesics, *Compositio Math.* **44** (1981), 375–394.

[511] E. Vesentini, Complex geodesics and holomorphic mappings, *Sympos. Math.* **26** (1982), 211–230.

[512] E. Vesentini, Injective hyperbolicity, *Ricerche di Mat., Suppl.* **36** (1987), 99–109.

[513] J.-P. Vigué, La distance de Carathéodory n'est pas intérieure, *Resultate d. Math.* **6** (1983), 100–104.

[514] J.-P. Vigué, The Carathéodory distance does not define the topology, *Proc. Amer. Math. Soc.* **91** (1984), 223–224.

[515] J.-P. Vigué, Points fixes d'applications holomorphes dans un domaine borné, *Trans. Amer. Math. J.* **289** (1985), 345–355.

[516] J.-P. Vigué, Sur la caractérisation des automorphismes analytiques d'un domain borné, *Portugaliae Math.* **43** (1985), 439–453.

[517] J.-P. Vigué, Un lemme de Schwarz pour les domaines bornés symétriques irreductibles et certains domaines bornés strictement convexes, *Indiana Univ. Math. J.* **40** (1991), 293–304.

[518] J. P. Vigué, Les métriques invariantes et la caractérisation des isomorphismes analytiques, *Banach Center Publ.* **31** (1995), 373–382.

[519] J.-P. Vigué, Sur les domaines hyperboliques pour la distance intégrée de Carathéodory, *Ann. Inst. Fourier (Grenoble)* **46** (1996), 743–753.

[520] J. P. Vigué, Revêtements et isométries pour la métrique infinitésimale de Kobayashi, *Proc. Amer. Math. Soc.* **129** (2001), 3279–3284.

[521] B. Visintin, Carathéodory balls and norm balls in a class of convex bounded Reinhardt domains, *Isr. J. Math.* **110** (1999), 1–27.

[522] V. Vladimirov, *Methods of the Theory of Functions of Many Complex Variables*, M. I.T. Press, 1993.

[523] N. Vormoor, Topologische Fortsetung biholomorpher Funktionen auf dem Rande bei beschränkten streng-pseudokonvexen Gebieten im $\mathbb{C}^n$ mit C^∞-Rand, *Math. Ann.* **204** (1973), 239–261.

[524] R. Wada, On the Fourier–Borel transformations of analytic functionals on the complex sphere, *Tohoku Math. J., II. Ser.* **38** (1986), 417–432.

[525] X. Wang, Bergman completeness is not a quasi-conformal invariant, *Proc. Amer. Math. Soc.* **141** (2013), 543–548.

[526] T. Warszawski, Boundary behavior of the Kobayashi distance in pseudoconvex Reinhardt domains, *Michigan Math. J.* **61** (2012), 575–592.

[527] S. M. Webster, Biholomorphic mappings and the Bergman kernel off the diagonal, *Invent. Math.* **51** (1979), 155–169.

[528] J. Wiegerinck, Domains with finite dimensional Bergman space, *Math. Z.* **187** (1984), 559–562.

[529] F. Wikström, Non-linearity of the pluricomplex Green function, *Proc. Amer. Math. Soc.* **129** (2001), 1051–1056.

[530] F. Wikström, *Qualitative properties of biholomorphically invariant functions with multiple poles*, Manuscript, 2004.

[531] J. Winkelmann, Non-degenerate maps and sets, *Math. Z.* **249** (2005), 783–795.

[532] B. Wong, Characterization of the unit ball in $\mathbb{C}^n$ by its automorphism group, *Invent. Math.* **41** (1977), 253–257.

[533] B. Wong, On the holomorphic curvature of some intrinsic metrics, *Proc. Amer. Math. Soc.* **65** (1977), 57–61.

[534] H. Wu, Normal families of holomorphic mappings, *Acta Math.* **119** (1967), 194–233.

[535] H. Wu, A remark on holomorphic sectional curvature, *Indiana Univ. Math. J.* **22** (1973), 1103–1108.

[536] H. Wu, *Unpublished notes*, 1987.

[537] H. Wu, *Old and new invariant metrics*, Several complex variables: Proc. Mittag-Leffler Inst. 1987–88 (J. E. Fornæss, ed.), Math. Notes 38, Princeton Univ. Press, 1993, pp. 640–682.

[538] A. Yamamori, *Zeros of the Bergman kernel of the Fock-Bargmann-Hartogs domain and the interlacing property*, arXiv:1101.3135, 2011.

[539] W. Yin, A comparison theorem of the Kobayashi metric and the Bergman metric on a class of Reihardt domains, *Adv. in Math. (China)* **26** (1997), 323–334.

[540] N. J. Young, The automorphism group of the tetrablock, *J. London Math. Soc.* **77** (2008), 757–770.

[541] E. H. Youssfi, Proper holomorphic liftings and new formulas for the Bergman and Szegö kernels, *Studia Math.* **152** (2002), 161–186.

[542] J. Yu, Singular Kobayashi metrics and finite type conditions, *Proc. Amer. Math. Soc.* **123** (1995), 121–130.

[543] J. Yu, Weighted boundary limits of the generalized Kobayashi–Royden metrics on weakly pseudoconvex domains, *Trans. Amer. Math. Soc.* **347** (1995), 587–614.

[544] P. Zapałowski, Completeness of the inner k-th Reiffen pseudometric, *Ann. Polon. Math.* **79** (2002), 277–288.

[545] P. Zapałowski, Inner k-th Reiffen completeness of Reinhardt domains, *Atti Accad. Naz. Lincei Cl. Sci. Fis. Mat. Natur. Rend. Lincei (9) Mat. Appl.* **15** (2004), 87–92.

[546] P. Zapałowski, *Odległości Carathéodory'ego–Reiffena wyższych rzędów (Higher order Carathéodory–Reiffen distances)*, Ph.D. thesis, Jagielonian Univ., Cracow, 2004.

[547] P. Zapałowski, Invariant functions on Neil parabola in $\mathbb{C}^n$, *Serdica Math. J.* **33** (2007), 321–338.

[548] P. Zapałowski, Geometry of symmetrized ellipsoids, *Univ. Iagel. Acta Math.* **46** (2008), 105–116.

[549] P. Zapałowski, Proper holomorphic mappings between symmetrized ellipsoids, *Arch. Math. (Basel)* **97** (2011), 373–384.

[550] P. Zapałowski, Personal communication, 2012.

[551] P. Zapałowski, *Proper holomorphic mappings between complex ellipsoids and generalized Hartogs domains*, arXiv:1211.0786v1, 2012.

[552] E. Zeidler, *Nonlinear Functional Analysis and its Applications, vol. I: Fixed-Point Theorems*, Springer-Verlag, 1986.

[553] A. Zeriahi, *Potentiels capacitaires extremaux et inequalites polynomiales sur un sousensemble algebraic de* $\mathbb{C}^n$, Thése, Part B, Université Toulouse, 1986.

[554] Y. Zeytuncu, Weighted Bergman projections and kernels: L^p regularity and zeros, *Proc. Amer. Math.* **139** (2011), 2105–2112.

[555] W. Zwonek, A note on Carathéodory isometries, *Arch. Math. (Basel)* **60** (1993), 167–176.

[556] W. Zwonek, Carathéodory balls and norm balls of the domains $H_n = \{z \in \mathbb{C}^n\ |z_1| + \cdots + |z_n| < 1\}$, *Isr. J. Math.* **89** (1995), 71–76.

[557] W. Zwonek, The Carathéodory isometries between the products of balls, *Arch. Math. (Basel)* **65** (1995), 434–443.

[558] W. Zwonek, *Complex isometries (Polish)*, Ph.D. thesis, Jagiellonian Univ., 1995.

[559] W. Zwonek, Automorphism group of some special domain in $\mathbb{C}^n$, *Univ. Iagel. Acta Math.* **33** (1996), 185–189.

[560] W. Zwonek, Carathéodory balls in convex complex ellipsoids, *Ann. Polon. Math.* **64** (1996), 183–194.

[561] W. Zwonek, A note on the Kobayashi–Royden metric for real ellipsoids, *Proc. Amer. Math. Soc.* **125** (1997), 199–202.

[562] W. Zwonek, On an example concerning the Kobayashi pseudodistance, *Proc. Amer. Math. Soc.* **126** (1998), 2945–2948.

[563] W. Zwonek, On Bergman completeness of pseudoconvex Reinhardt domains, *Ann. Faculté Sci. Toulouse* **8** (1999), 537–552.

[564] W. Zwonek, On hyperbolicity of pseudoconvex Reinhardt domains, *Arch. Math. (Basel)* **72** (1999), 304–314.

[565] W. Zwonek, Completeness, Reinhardt domains and the method of complex geodesics in the theory of invariant functions, *Dissertationes Math.* **388** (2000), 1–103.

[566] W. Zwonek, On Carathéodory completeness of pseudoconvex Reinhardt domains, *Proc. Amer. Math. Soc.* **128** (2000), 857–864.

[567] W. Zwonek, Regularity properties of the Azukawa metric, *J. Math. Soc. Japan* **52** (2000), 899–914.

[568] W. Zwonek, An example concerning the Bergman completeness, *Nagoya Math. J.* **164** (2001), 89–101.

[569] W. Zwonek, Inner Carathéodory completeness of Reinhardt domains, *Rend. Mat. Acc. Lincei* **12** (2001), 153–157.

[570] W. Zwonek, Wiener's type criterion for Bergman exhaustiveness, *Bull. Pol. Acad. Sci. Math.* **50** (2002), 297–311.

[571] W. Zwonek, Asymptotic behavior of the sectional curvature of the Bergman metric for annuli, *Ann. Polon. Math.* **98** (2010), 291–299.

[572] W. Zwonek, Geometric properties of the tetrablock, *Arch. Math.* **100** (2013), 159–165.

List of symbols

General symbols

$\mathbb{N}$:= the set of natural numbers, $0 \notin \mathbb{N}$;
$\mathbb{N}_0 := \mathbb{N} \cup \{0\}$;
$\mathbb{N}_k := \{n \in \mathbb{N} : n \geq k\}$;
$\mathbb{Z}$:= the ring of integer numbers;
$\mathbb{Q}$:= the field of rational numbers;
$\mathbb{R}$:= the field of real numbers;
$\lfloor t \rfloor := \sup\{k \in \mathbb{Z} : k \leq t\}$ = the lower integer part of $t \in \mathbb{R}$;
$\lceil t \rceil := \inf\{k \in \mathbb{Z} : k \geq t\}$ = the upper integer part of $t \in \mathbb{R}$;
$\mathbb{R}_{\pm\infty} := \mathbb{R} \cup \{\pm\infty\}$;
$\overline{\mathbb{R}} := \mathbb{R} \cup \{-\infty, +\infty\} = \mathbb{R}_{-\infty} \cup \mathbb{R}_{+\infty}$;
$\mathbb{C}$:= the field of complex numbers;
$\overline{\mathbb{C}} := \mathbb{C} \cup \{\infty\}$ = the Riemann sphere;
$\operatorname{Re} z := x$ = the real part of $z = x + iy \in \mathbb{C}$;
$\operatorname{Im} z := y$ = the imaginary part of $z = x + iy \in \mathbb{C}$;
$\overline{z} := x - iy$ = the conjugate of $z = x + iy \in \mathbb{C}$;
$|z| := \sqrt{x^2 + y^2}$ = the modulus of $z = x + iy \in \mathbb{C}$;
$\arg z := \{\varphi \in \mathbb{R} : z = |z|e^{i\varphi}\}$ = the argument of $z \in \mathbb{C}$ ($\arg 0 = \mathbb{R}$);
$\operatorname{Arg} : \mathbb{C} \longrightarrow (-\pi, \pi]$, $\operatorname{Arg} 0 := 0$, if $z \neq 0$, then $\operatorname{Arg} z = \varphi \iff \varphi \in \arg z \cap (-\pi, \pi]$ (the main argument of $z \in \mathbb{C}$);
$\operatorname{Log} z := \log|z| + i \operatorname{Arg} z$ = the principal value of the logarithm of $z \in \mathbb{C} \setminus \{0\}$;
$\mathbb{C} \setminus (-\infty, 0] \ni z \overset{\operatorname{Log}}{\longmapsto} \operatorname{Log} z$ the principal branch of the logarithm;
A^n := the Cartesian product of n copies of the set A, e.g., $\mathbb{Z}^n$, $\mathbb{R}^n$, $\mathbb{C}^n$;
$\mathbb{M}(m \times n; A)$:= the set of all $m \times n$-matrices with entries in the set $A \subset \mathbb{C}$;
$\mathbb{I}_n$:= the unit matrix in $\mathbb{M}(n \times n; \mathbb{C})$;
$\mathbb{O}(n; \mathbb{K}) := \{A \in \mathbb{M}(n \times n; \mathbb{K}) : AA^t = \mathbb{I}_n\}$, $\mathbb{K} \in \{\mathbb{R}, \mathbb{C}\}$;
$\mathbb{O}(n) := \mathbb{O}(n; \mathbb{R})$;
$\mathbb{SO}(n; \mathbb{K}) := \{A \in \mathbb{O}(n; \mathbb{K}) : \det A = 1\}$;
$x \leq y :\iff x_j \leq y_j,\ j = 1, \dots, n$, where $x = (x_1, \dots, x_n),\ y = (y_1, \dots, y_n) \in \mathbb{R}^n$;
$A_* := A \setminus \{0\}$, e.g., $\mathbb{C}_*$, $(\mathbb{C}^n)_*$;
$A_*^n := (A_*)^n$, e.g., $\mathbb{C}_*^n$;
$A_+ := \{a \in A : a \geq 0\}$, e.g., $\mathbb{Z}_+$, $\mathbb{R}_+$;

$A_+^n := (A_+)^n$, e.g., $\mathbb{Z}_+^n$, $\mathbb{R}_+^n$;
$A_- := \{a \in A : a \le 0\}$;
$A_{>0} := \{a \in A : a > 0\}$, e.g., $\mathbb{R}_{>0}$;
$A_{>0}^n := (A_{>0})^n$, e.g., $\mathbb{R}_{>0}^n$;
$A + B := \{a + b : a \in A,\ b \in B\}$, where $A, B \subset X$, $(X, +)$ is a group;
$A \cdot B := \{a \cdot b : a \in A,\ b \in B\}$, where $A \subset \mathbb{C}$, $B \subset X$, $(X, +, \cdot)$ is a $\mathbb{C}$-vector space;
$\delta_{j,k} := \begin{Bmatrix} 0, & \text{if } j \ne k \\ 1, & \text{if } j = k \end{Bmatrix}$ = the Kronecker symbol;
$\boldsymbol{e} = (\boldsymbol{e}_1, \dots, \boldsymbol{e}_n) :=$ the canonical basis in $\mathbb{C}^n$, $\boldsymbol{e}_j := (\delta_{j,1}, \dots, \delta_{j,n})$, $j = 1, \dots, n$;
$\mathbf{1} = \mathbf{1}_n := (1, \dots, 1) \in \mathbb{N}^n$;
$\langle z, w \rangle := \sum_{j=1}^n z_j \overline{w}_j$ = the Hermitian scalar product in $\mathbb{C}^n$;
$\overline{w} := (\overline{w}_1, \dots, \overline{w}_n)$, $w = (w_1, \dots, w_n) \in \mathbb{C}^n$;
$z \bullet w := \langle z, \overline{w} \rangle = \sum_{j=1}^n z_j w_j$, $z = (z_1, \dots, z_n)$, $w = (w_1, \dots, w_n) \in \mathbb{C}^n$;
$z \cdot w := (z_1 w_1, \dots, z_n w_n)$, $z = (z_1, \dots, z_n)$, $w = (w_1, \dots, w_n) \in \mathbb{C}^n$;
$e^z := (e^{z_1}, \dots, e^{z_n})$, $z = (z_1, \dots, z_n) \in \mathbb{C}^n$;
$\|z\| := \langle z, z \rangle^{1/2} = \left(\sum_{j=1}^n |z_j|^2\right)^{1/2}$ = the Euclidean norm in $\mathbb{C}^n$;
$\|z\|_\infty := \max\{|z_1|, \dots, |z_n|\}$ = the maximum norm in $\mathbb{C}^n$;
$\|z\|_1 := |z_1| + \dots + |z_n|$ = the ℓ^1-norm in $\mathbb{C}^n$;
$\#A :=$ the number of elements of A;
$\operatorname{diam} A :=$ the diameter of the set $A \subset \mathbb{C}^n$ with respect to the Euclidean distance;
$\operatorname{dist}(A, B) := \inf\{\|a - b\| : a \in A,\ b \in B\}$, $A, B \subset \mathbb{C}^n$;
$\operatorname{dist}(a, B) := \operatorname{dist}(\{a\}, B)$;
$\chi_A :=$ the characteristic function of A;
$\operatorname{conv} A = \operatorname{conv}(A) :=$ the convex hull of A;
$A \subset\subset X :\Longleftrightarrow A$ is relatively compact in X;
$\operatorname{pr}_X : X \times Y \longrightarrow X$, $\operatorname{pr}_X(x, y) := x$
$B_d(a, r) := \{x \in X : d(x, a) < r\}$, $a \in X$, $r > 0$, where $d : X \times X \longrightarrow \mathbb{R}_+$ is a pseudodistance;
$\overline{B}_d(a, r) := \{x \in X : d(x, a) \le r\}$, $a \in X$, $r \ge 0$, where $d : X \times X \longrightarrow \mathbb{R}_+$ is a pseudodistance;
$B_q(a, r) := \{x \in X : q(x - a) < r\}$, $a \in X$, $r > 0$, where $q : X \longrightarrow \mathbb{R}_+$ is a seminorm;
$\overline{B}_q(a, r) := \{x \in X : q(x - a) \le r\}$, $a \in X$, $r \ge 0$, where $q : X \longrightarrow \mathbb{R}_+$ is a seminorm;
$\operatorname{top} G :=$ the Euclidean topology of $G \subset \mathbb{C}^n$;
$z^\alpha := z_1^{\alpha_1} \cdots z_n^{\alpha_n}$, $z = (z_1, \dots, z_n) \in \mathbb{C}^n$, $\alpha = (\alpha_1, \dots, \alpha_n) \in \mathbb{Z}^n$ $(0^0 := 1)$;
$\alpha! := \alpha_1! \cdots \alpha_n!$, $\alpha = (\alpha_1, \dots, \alpha_n) \in \mathbb{Z}_+^n$;
$|\alpha| := |\alpha_1| + \dots + |\alpha_n|$, $\alpha = (\alpha_1, \dots, \alpha_n) \in \mathbb{R}^n$;
$\binom{\alpha}{\beta} := \frac{\alpha(\alpha-1)\cdots(\alpha-\beta+1)}{\beta!}$, $\alpha \in \mathbb{C}$, $\beta \in \mathbb{Z}_+$;
$\binom{\alpha}{\beta} := \binom{\alpha_1}{\beta_1} \cdots \binom{\alpha_n}{\beta_n}$, $\alpha = (\alpha_1, \dots, \alpha_n) \in \mathbb{C}^n$, $\beta = (\beta_1, \dots, \beta_n) \in \mathbb{Z}_+^n$.

Euclidean balls:
$\mathbb{B}(a,r) = \mathbb{B}_n(a,r) := \{z \in \mathbb{C}^n : \|z-a\| < r\}$ = the open Euclidean ball in $\mathbb{C}^n$ with center $a \in \mathbb{C}^n$ and radius $r > 0$; $\mathbb{B}(a,+\infty) := \mathbb{C}^n$;
$\overline{\mathbb{B}}(a,r) = \overline{\mathbb{B}}_n(a,r) := \overline{\mathbb{B}_n(a,r)} = \{z \in \mathbb{C}^n : \|z-a\| \le r\}$ = the closed Euclidean ball in $\mathbb{C}^n$ with center $a \in \mathbb{C}^n$ and radius $r > 0$;
$\mathbb{B}(r) = \mathbb{B}_n(r) := \mathbb{B}_n(0,r)$; $\overline{\mathbb{B}}(r) = \overline{\mathbb{B}}_n(r) := \overline{\mathbb{B}}_n(0,r)$;
$\mathbb{B} = \mathbb{B}_n := \mathbb{B}_n(1)$ = the unit Euclidean ball in $\mathbb{C}^n$;
$\mathbb{D}(a,r) := \mathbb{B}_1(a,r)$; $\mathbb{D}(r) := \mathbb{D}(0,r)$;
$\overline{\mathbb{D}}(a,r) := \overline{\mathbb{B}}_1(a,r)$; $\overline{\mathbb{D}}(r) := \overline{\mathbb{D}}(0,r)$;
$\mathbb{D}_*(a,r) := \mathbb{D}(a,r) \setminus \{a\}$; $\mathbb{D}_*(r) := \mathbb{D}_*(0,r)$;
$\mathbb{D} := \mathbb{D}(1) = \{\lambda \in \mathbb{C} : |\lambda| < 1\}$ = the unit disc;
$\mathbb{T} := \partial\mathbb{D}$.

Polydiscs:
$\mathbb{P}(a,r) = \mathbb{P}_n(a,r) := \{z \in \mathbb{C}^n : \|z-a\|_\infty < r\}$ = the polydisc with center $a \in \mathbb{C}^n$ and radius $r > 0$; $\mathbb{P}_n(a,+\infty) := \mathbb{C}^n$;
$\overline{\mathbb{P}}(a,r) = \overline{\mathbb{P}}_n(a,r) := \overline{\mathbb{P}_n(a,r)}$; $\overline{\mathbb{P}}_n(a,0) := \{a\}$;
$\mathbb{P}(r) = \mathbb{P}_n(r) := \mathbb{P}_n(0,r)$;

$\mathbb{P}(a,\boldsymbol{r}) = \mathbb{P}_n(a,\boldsymbol{r}) := \mathbb{D}(a_1,r_1)\times\cdots\times\mathbb{D}(a_n,r_n)$ = the polydisc with center $a \in \mathbb{C}^n$ and multi-radius (polyradius) $\boldsymbol{r} = (r_1,\dots,r_n) \in \mathbb{R}^n_{>0}$; $\mathbb{P}(a,r) = \mathbb{P}(a,r\cdot\mathbf{1})$;
$\mathbb{P}(\boldsymbol{r}) = \mathbb{P}_n(\boldsymbol{r}) := \mathbb{P}_n(0,\boldsymbol{r})$;
$\partial_0\mathbb{P}(a,\boldsymbol{r}) := \partial\mathbb{D}(a_1,r_1)\times\cdots\times\partial\mathbb{D}(a_n,r_n)$ = the distinguished boundary of $\mathbb{P}(a,\boldsymbol{r})$.

Annuli:
$\mathbb{A}(a,r^-,r^+) := \{z \in \mathbb{C} : r^- < |z-a| < r^+\}$, $a \in \mathbb{C}$, $-\infty \le r^- < r^+ \le +\infty$, $r^+ > 0$; if $r^- < 0$, then $\mathbb{A}(a,r^-,r^+) = \mathbb{D}(a,r^+)$; $\mathbb{A}(a,0,r^+) = \mathbb{D}(a,r^+) \setminus \{a\}$;
$\mathbb{A}(r^-,r^+) := \mathbb{A}(0,r^-,r^+)$.

Functions:
$\|f\|_A := \sup\{|f(a)| : a \in A\}$, $f\colon A \longrightarrow \mathbb{C}$;
$f_k \overset{K}{\underset{k\to\infty}{\Longrightarrow}} f :\Longleftrightarrow f_k \longrightarrow f$ locally uniformly;
$\operatorname{supp} f := \overline{\{x : f(x) \neq 0\}}$ = the support of f;
$\mathcal{P}(\mathbb{K}^n)$:= the space of all polynomial mappings $F\colon \mathbb{K}^n \longrightarrow \mathbb{K}$, $\mathbb{K} \in \{\mathbb{R},\mathbb{C}\}$;
$\mathcal{P}_d(\mathbb{K}^n) := \{F \in \mathcal{P}(\mathbb{K}^n) : \deg F \le d\}$;
$\liminf_{x\nrightarrow a} f(x) := \liminf_{A\setminus\{a\}\ni x\to a} f(x)$, $\limsup_{x\nrightarrow a} f(x) := \limsup_{A\setminus\{a\}\ni x\to a} f(x)$, where $f : A \longrightarrow \overline{\mathbb{R}}$, A is a metric space;
$\lim_{x\nrightarrow a} f(x) := \lim_{A\setminus\{a\}\ni x\to a} f(x)$, where $f : A \longrightarrow Y$, A, Y are metric spaces;
$\mathcal{C}^\uparrow(X)$:= the set of all upper semicontinuous functions $u : X \longrightarrow \mathbb{R}_{-\infty}$;
$\frac{\partial f}{\partial z_j}(a) := \frac{1}{2}\Big(\frac{\partial f}{\partial x_j}(a) - i\frac{\partial f}{\partial y_j}(a)\Big)$, $\frac{\partial f}{\partial \bar z_j}(a) := \frac{1}{2}\Big(\frac{\partial f}{\partial x_j}(a) + i\frac{\partial f}{\partial y_j}(a)\Big)$ = the formal partial derivatives of f at a; sometimes, we write f_ξ instead of $\frac{\partial f}{\partial \xi}$;

$\operatorname{grad} u(a) := (\frac{\partial u}{\partial \overline{z}_1}(a), \dots, \frac{\partial u}{\partial \overline{z}_n}(a))$ = the gradient of u at a;
$D^{\alpha,\beta} := (\frac{\partial}{\partial z_1})^{\alpha_1} \circ \dots \circ (\frac{\partial}{\partial z_n})^{\alpha_n} \circ (\frac{\partial}{\partial \overline{z}_1})^{\beta_1} \circ \dots \circ (\frac{\partial}{\partial \overline{z}_n})^{\beta_n}$;
$\frac{\partial f}{\partial z_j}(a) := \lim_{\mathbb{C}_* \ni h \to 0} \frac{f(a+he_j)-f(a)}{h}$ = the j-th complex partial derivative of f at a;
$D^{\alpha} := (\frac{\partial}{\partial z_1})^{\alpha_1} \circ \dots \circ (\frac{\partial}{\partial z_n})^{\alpha_n}$ = α-th partial complex derivative;
$\overline{\partial}$ = the $\overline{\partial}$-operator;
$\mathcal{C}^k(X,Y)$:= the space of all $\mathcal{C}^k$-mappings $f : X \longrightarrow Y$, $k \in \mathbb{Z}_+ \cup \{\infty\} \cup \{\omega\}$ (ω stands for the real analytic case);
$\mathcal{C}^k(X) := \mathcal{C}^k(X,\mathbb{C})$;
$\mathcal{C}^k_0(X) := \{f \in \mathcal{C}^k(X) : \operatorname{supp} f \subset\subset X\}$;
$\mathcal{L}^N$:= the Lebesgue measure in $\mathbb{R}^N$;
$\mathfrak{m}_{\mathbb{T}} = \mathfrak{m}$:= the normalized Lebesgue measure on $\mathbb{T}$;
$L^p(X)$:= the space of all p-integrable functions on X, $1 \le p \le +\infty$;
$\| \ \|_{L^p(X)}$:= the norm in $L^p(X)$;
$L^p_{\mathrm{loc}}(X)$:= the space of all locally p-integrable functions on X;
$\mathcal{O}(X,Y)$:= the space of all holomorphic mappings $F : X \longrightarrow Y$;
$\mathcal{O}(X) := \mathcal{O}(X,\mathbb{C})$ = the space of all holomorphic functions $f : X \longrightarrow \mathbb{C}$;
$L^p_h(X) := \mathcal{O}(X) \cap L^p(X)$, $1 \le p \le +\infty$;
$\mathcal{H}^\infty(X) := L^\infty_h(X)$ = the space of all bounded holomorphic functions on X;
$\mathcal{A}(X) := \mathcal{C}(\overline{X}) \cap \mathcal{O}(X)$;
$\operatorname{Aut}(G)$:= the group of all automorphisms of the domain $G \subset \mathbb{C}^n$;
$\operatorname{Aut}_a(G) := \{h \in \operatorname{Aut}(G) : h(a) = a\}$;
$\operatorname{Aut}_{\mathrm{id}}(G)$:= the connected component of $\operatorname{Aut}(G)$ that contains the identity;
$\mathcal{SH}(\Omega)$:= the set of all subharmonic (sh) functions on the open set $\Omega \subset \mathbb{C}$;
$\mathcal{PSH}(X)$:= the set of all plurisubharmonic (psh) functions on X;
$\mathcal{L}u(a;\xi) := \sum_{j,k=1}^n \frac{\partial^2 u}{\partial z_j \partial \overline{z}_k}(a)\xi_j\overline{\xi}_k$ = the Levi form of u at a;
$\mathfrak{g}_G$:= the (classical) Green function of the domain $G \subset \overline{\mathbb{C}}$.

Symbols in individual chapters

Chapter 1

Chapter 2

Chapter 3

Chapter 4

Chapter 5

Chapter 6

Chapter 7

Chapter 8

Chapter 9

Chapter 11

Chapter 12

Chapter 13

Chapter 14

Chapter 15

Chapter 16

Chapter 17

Chapter 19

Appendix A

Appendix B

Index